Oxidation and Antioxidants in Organic Chemistry and Biology

Oxidation and Antioxidants in Organic Chemistry and Biology

Evgeny T. Denisov
Igor B. Afanas'ev

Taylor & Francis
Taylor & Francis Group

Boca Raton London New York Singapore

A CRC title, part of the Taylor & Francis imprint, a member of the
Taylor & Francis Group, the academic division of T&F Informa plc.

Published in 2005 by
CRC Press
Taylor & Francis Group
6000 Broken Sound Parkway NW
Boca Raton, FL 33487-2742

© 2005 by Taylor & Francis Group
CRC Press is an imprint of Taylor & Francis Group

No claim to original U.S. Government works
Printed in the United States of America on acid-free paper
10 9 8 7 6 5 4 3 2 1

International Standard Book Number-10: 0-8247-5356-9 (Hardcover)
International Standard Book Number-13: 978-0-8247-5356-6 (Hardcover)
Library of Congress Card Number 2004059305

This book contains information obtained from authentic and highly regarded sources. Reprinted material is quoted with permission, and sources are indicated. A wide variety of references are listed. Reasonable efforts have been made to publish reliable data and information, but the author and the publisher cannot assume responsibility for the validity of all materials or for the consequences of their use.

No part of this book may be reprinted, reproduced, transmitted, or utilized in any form by any electronic, mechanical, or other means, now known or hereafter invented, including photocopying, microfilming, and recording, or in any information storage or retrieval system, without written permission from the publishers.

For permission to photocopy or use material electronically from this work, please access www.copyright.com (http://www.copyright.com/) or contact the Copyright Clearance Center, Inc. (CCC) 222 Rosewood Drive, Danvers, MA 01923, 978-750-8400. CCC is a not-for-profit organization that provides licenses and registration for a variety of users. For organizations that have been granted a photocopy license by the CCC, a separate system of payment has been arranged.

Trademark Notice: Product or corporate names may be trademarks or registered trademarks, and are used only for identification and explanation without intent to infringe.

Library of Congress Cataloging-in-Publication Data

Denisov, E. T. (Evgeny Timofeevich)
  Oxidation and antioxidants in organic chemistry and biology / Evgeny T. Denisov, Igor B. Afanas'ev.
     p. cm.
  Includes bibliographical references and index.
     ISBN 0-8247-5356-9 (alk. paper)
  1. Hydrocarbons–Oxidation. 2. Antioxidants. 3. Chemistry, Organic. 4. Biology I. Afanas'ev, Igor B., 1935- II. Title.

QD305.H5D34 2004
547'.0104593–dc22   2004059305

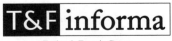

Taylor & Francis Group
is the Academic Division of T&F Informa plc.

Visit the Taylor & Francis Web site at
http://www.taylorandfrancis.com
and the CRC Press Web site at
http://www.crcpress.com

Dedication

We dedicate this book to the memory of pioneers of Oxidation Chemistry, Aleksey Nikolaevich Bach and Nikolay Aleksandrovich Shilov

Table of Contents

Preface xxiii
Authors xxvii
List of Chemical Symbols and Abbreviations xxix
List of Physicochemical Symbols xxxi
List of Biochemical Abbreviations xxxv

Part I
Chemistry and Kinetics of Organic Compounds Oxidation by Dioxygen 1

Chapter 1
Molecular Products and Thermochemistry of Hydrocarbon Oxidation 3
1.1 Earlier Concepts of Oxidation 3
1.2 Development of the Chain Theory of Oxidation of Organic Compounds 6
1.3 Hydroperoxides as Primary Molecular Products of Hydrocarbon Oxidation 9
    1.3.1 Hydroperoxides 9
    1.3.2 Dihydroperoxides 11
    1.3.3 Cyclic and Polymeric Peroxides 12
    1.3.4 Epoxides 13
1.4 Products of Hydroperoxide Decomposition 13
    1.4.1 Hydroperoxides as the Intermediates of Hydrocarbon Oxidation 13
    1.4.2 Alcohols 14
    1.4.3 Ketones 16
    1.4.4 Acids 16
References 18

Chapter 2
Chain Mechanism of Liquid-Phase Oxidation of Hydrocarbons 23
2.1 The Peculiarities of Chain Reactions 23
    2.1.1 Free-Valence Persistence in Reactions of Free Radicals with Molecules 23
    2.1.2 Condition of Cyclicity of Radical Conversions 23
    2.1.3 Priority of Chain Propagation Reaction 24
    2.1.4 Generation of Free Radicals 24
    2.1.5 Competition of Chain and Molecular Reactions 25
        2.1.5.1 High Chemical Reactivity of Free Radicals and Atoms 25
        2.1.5.2 Conservation of Orbital Symmetry in Chemical Reaction 26
        2.1.5.3 Configuration of Transition State 26
2.2 Chain Mechanism of Hydrocarbon Oxidation 27
2.3 Reaction of Alkyl Radicals with Dioxygen 34
2.4 Reactions of Peroxyl Radicals 39
    2.4.1 Structure and Thermochemistry of Peroxyl Radicals 39
    2.4.2 Reaction $RO_2^\bullet + RH \rightarrow ROOH + R^\bullet$ 43
    2.4.3 Intramolecular Hydrogen Atom Transfer in Peroxyl Radical 45
    2.4.4 Addition of Peroxyl Radical to the Double Bond 49

2.5   Chain Termination in Oxidized Hydrocarbons 55
   2.5.1   Tetroxides 55
   2.5.2   Disproportionation of Tertiary Peroxyl Radicals 57
   2.5.3   Disproportionation of Primary and Secondary Peroxyl Radicals 58
   2.5.4   Chemiluminescence 62
   2.5.5   Chain Termination via Alkyl Radicals 67
2.6   Reactions of Alkoxyl Radicals 69
2.7   Different Mechanisms of the Aliphatic Hydrocarbon Oxidation in Gas
      and Liquid Phases 72
References 76

Chapter 3
Initiation of Liquid-Phase Oxidation 85
3.1   Initiators 85
   3.1.1   Mechanisms of Decomposition of Initiators 85
      3.1.1.1   Unimolecular Decomposition of Initiator with One Bond Splitting 85
      3.1.1.2   Concerted Fragmentation of Initiators 86
      3.1.1.3   Anchimerically Assisted Decomposition of Peroxides 87
   3.1.2   Decay of Initiators to Molecular Products 88
   3.1.3   Chain Decomposition of Initiators 89
   3.1.4   Peroxides 90
   3.1.5   Azo-Compounds 90
3.2   Cage Effect 94
   3.2.1   Experimental Evidences for Cage Effect 94
      3.2.1.1   Quantum Yield 94
      3.2.1.2   Products of Radical Pair Combination 95
      3.2.1.3   Oxygen-18 Scrambling and Racemization 95
      3.2.1.4   Crossover Experiments 96
      3.2.1.5   Racemization 97
      3.2.1.6   Influence of Viscosity 98
      3.2.1.7   Influence of Pressure 99
      3.2.1.8   Spin Multiplicity Effects 99
   3.2.2   Mechanistic Schemes of the Cage Effect 99
3.3   Initiation of Oxidation by Chemically Active Gases 101
   3.3.1   Ozone 101
   3.3.2   Nitrogen Dioxide 110
   3.3.3   Halogens 113
3.4   Photoinitiation 118
   3.4.1   Intramolecular and Intermolecular Photophysical Processes 120
   3.4.2   Photosensitizers 123
   3.4.3   Photoinitiators 124
      3.4.3.1   Azo-compounds 124
      3.4.3.2   Peroxides 126
      3.4.3.3   Carbonyl Compounds 127
   3.4.4   Nonchain Photooxidation 127
3.5   Generation of Radicals by Ionizing Radiation 128
   3.5.1   Primary Radiation–Chemical Processes 128
   3.5.2   Radiolytic Initiation of Hydrocarbon Oxidation 129
References 132

Chapter 4
Oxidation as an Autoinitiated Chain Reaction 137
4.1 Chain Generation by Reaction of Hydrocarbon with Dioxygen 137
    4.1.1 Bimolecular Reaction of Dioxygen with the C—H Bond of the Hydrocarbon 137
    4.1.2 Trimolecular Reaction of Hydrocarbon with Dioxygen 141
    4.1.3 Bimolecular and Trimolecular Reactions of Dioxygen with the Double Bond of Olefin 141
4.2 Hydroperoxides 144
    4.2.1 Analysis 145
    4.2.2 Structure of Hydroperoxides 146
    4.2.3 Thermochemistry of Hydroperoxides 147
    4.2.4 Hydrogen Bonding between Hydroperoxides 147
4.3 Reactions of Free Radical Generation by Hydroperoxides 149
    4.3.1 Unimolecular Decomposition of Hydroperoxides 149
    4.3.2 Bimolecular Decomposition of Hydroperoxides 154
    4.3.3 Bimolecular Reactions of Hydroperoxides with π-Bonds of Olefins 154
    4.3.4 Bimolecular Reactions of Hydroperoxides with C—H, N—H, and O—H Bonds of Organic Compounds 157
4.4 Parabolic Model of Bimolecular Homolytic Reaction 158
    4.4.1 Main Equations of IPM 158
    4.4.2 Calculation of E and k of a Bimolecular Reaction 160
    4.4.3 Bimolecular Reactions of Radical Generation 163
        4.4.3.1 Bimolecular Splitting of Hydroperoxides 163
        4.4.3.2 Reaction of Hydroperoxides with Olefins 163
        4.4.3.3 Reaction of Hydroperoxides with Hydrocarbons 164
        4.4.3.4 Reaction of Hydroperoxides with Alcohols and Acids 164
4.5 Initiation by Reaction of Hydroperoxides with Ketones 167
4.6 Chain Decomposition of Hydroperoxides 168
4.7 Kinetics of Autoinitiated Hydrocarbon Oxidation 173
    4.7.1 Initial Stage of Autoxidation 173
    4.7.2 Autoxidation with Bimolecular Hydroperoxide Decay 176
    4.7.3 Initiation by Reactions of Hydroperoxide with Products of Oxidation 177
    4.7.4 Hydroperoxide as the Intermediate Product of Autoxidation 178
    4.7.5 Self-Inhibition of Hydrocarbon Oxidation 179
References 181

Chapter 5
Co-Oxidation of Hydrocarbons 185
5.1 Theory of Hydrocarbon Co-oxidation 185
5.2 Hydrocarbon Oxidation with Hydroperoxide 187
5.3 Co-oxidation of Hydrocarbons and Alcohols with Selective Inhibitor 191
5.4 Cross-disproportionation of Peroxyl Radicals 196
5.5 Cross-propagation Reactions of Peroxyl Radicals 199
5.6 High Reactivity of Haloidalkylperoxyl Radicals 203
5.7 Co-oxidation of Hydrocarbon and its Oxidation Intermediates 204
5.8 Catalysis by Nitroxyl Radicals in Hydrocarbon Oxidation 207
References 209

Chapter 6
Reactivity of the Hydrocarbons in Reactions with Peroxyl, Alkoxyl, and Alkyl Radicals 213
- 6.1 Semiempirical Model of Radical Reaction as an Intersection of Two Parabolic Potential Curves...213
  - 6.1.1 Introduction..213
  - 6.1.2 Intersecting Parabolas Model (IPM) 214
  - 6.1.3 Calculation of the Activation Energy 216
- 6.2 Factors Influencing the Activation Energy 219
  - 6.2.1 Reaction Enthalpy..219
  - 6.2.2 Force Constants of the Reacting Bonds 224
  - 6.2.3 Triplet Repulsion in the Transition State 225
  - 6.2.4 Electron Affinity of Atoms of Reaction Center 226
  - 6.2.5 Radii of Atoms of Reaction Center..228
  - 6.2.6 Influence of Adjacent π-Bonds...230
  - 6.2.7 Polar Effect 231
  - 6.2.8 Influence of Solvation...232
- 6.3 Geometry of the Transition State of Radical Abstraction Reaction 233
- 6.4 Intramolecular Hydrogen Transfer Reactions in Peroxyl, Alkoxyl, and Alkyl Radicals..236
  - 6.4.1 Peroxyl Radicals 236
  - 6.4.2 Alkoxyl Radicals...236
  - 6.4.3 Alkyl Radicals 238
- 6.5 Free Radical Addition Reactions..241
  - 6.5.1 Enthalpy of Free Radical Addition 241
  - 6.5.2 Parabolic Model of Radical Addition Reaction 241
  - 6.5.3 Triplet Repulsion in Radical Addition Reactions 242
  - 6.5.4 Influence of the Neighboring π-Bonds on the Activation Energy of Radical Addition 243
  - 6.5.5 Role of the Radius of the Atom Bearing a Free Valence 246
  - 6.5.6 Interaction of Polar Groups 247
  - 6.5.7 Steric Hindrance 250
- 6.6 Free Radical Substitution and Hydrogen Atom Transfer Reactions......251
  - 6.6.1 Free Radical Substitution Reactions 251
  - 6.6.2 Reaction of Peroxides with Ketyl Radicals 254
- References 257

Chapter 7
Oxidation of Alcohols and Ethers...261
- 7.1 Oxidation of Alcohols...261
  - 7.1.1 Introduction...261
  - 7.1.2 Chain Mechanism of Alcohol Oxidation 261
    - 7.1.2.1 Kinetics of Oxidation..263
    - 7.1.2.2 Reactions of Ketyl Radicals with Dioxygen 263
    - 7.1.2.3 Reactions of Alkylhydroxyperoxyl and Hydroperoxyl Radicals 263
    - 7.1.2.4 Chain Termination in Oxidized Alcohols 269
  - 7.1.3 Co-Oxidation of Alcohols and Hydrocarbons...........................270
  - 7.1.4 Reactivity of Alcohols in Reaction with Peroxyl Radicals 273

	7.1.5	Chain Generation in Autoxidation of Alcohols	278
		7.1.5.1 Chain Generation by Reaction with Dioxygen	278
		7.1.5.2 Decomposition of Hydrogen Peroxide into Free Radicals	278
		7.1.5.3 Chain Generation by Reaction of Hydrogen Peroxide with Carbonyl Compound	279
7.2	Oxidation of Ethers		281
	7.2.1	Introduction	281
	7.2.2	Chain Oxidation of Ethers	283
	7.2.3	Co-Oxidation of Ethers and Hydrocarbons	287
	7.2.4	Reactivity of Ethers in Reactions with Peroxyl Radicals	291
References			296

Chapter 8
Oxidation of Carbonyl Compounds and Decarboxylation of Acids ... 299

8.1	Oxidation of Aldehydes		299
	8.1.1	Introduction	299
	8.1.2	Chain Mechanism of Aldehyde Oxidation	300
	8.1.3	Co-Oxidation of Aldehydes with Hydrocarbons, Alcohols, and Aldehydes	303
	8.1.4	Reactivity of Aldehydes in Reactions with Peroxyl Radicals	306
8.2	Chemistry of Peracids		308
8.3	Oxidation of Ketones		311
	8.3.1	Chain Mechanism of Ketone Oxidation	311
	8.3.2	Reactivity of Ketones in Reactions with Peroxyl Radicals	314
	8.3.3	Interaction of Ketones with Hydroperoxides	317
	8.3.4	Chemistry of Ketone Oxidation	318
8.4	Oxidative Decarboxylation of Carboxylic Acids		320
	8.4.1	Attack of Peroxyl Radicals on C—H Bonds	320
	8.4.2	Oxidative Decarboxylation of Acids	321
References			324

Chapter 9
Oxidation of Amines, Amides, and Esters ... 329

9.1	Oxidation of Aliphatic Amines		329
	9.1.1	Introduction	329
	9.1.2	Chain Mechanism of Alkylamine Oxidation	331
	9.1.3	Reactivity of Amines in Reaction with Peroxyl Radicals	333
9.2	Oxidation of Amides		336
	9.2.1	Introduction	336
	9.2.2	Chain Mechanism of Amide Oxidation	338
	9.2.3	Decomposition of Hydroperoxides	338
9.3	Oxidation of Esters		341
	9.3.1	Introduction	341
	9.3.2	Chain Mechanism of Ester Oxidation	343
	9.3.3	Reactivity of Esters in Reactions with Peroxyl Radicals	349
	9.3.4	Effect of Multidipole Interaction in Reactions of Polyfunctional Esters	350
References			355

Chapter 10
Catalysis in Liquid-Phase Hydrocarbon Oxidation..................359
10.1 Catalysis by Transition Metal Ions and Complexes in Hydrocarbon Oxidation by Dioxygen..................359
    10.1.1 Introduction..................359
    10.1.2 Catalytic Decomposition of Hydrogen Peroxide by Ferrous Ions..................360
    10.1.3 Catalysis by Transition Metal Ions and Complexes in Liquid-Phase Oxidation of Hydrocarbons and Aldehydes by Dioxygen..................363
    10.1.4 Competition between Homolytic and Heterolytic Catalytic Decompositions of Hydroperoxides..................368
    10.1.5 Reactions of Transition Metals with Free Radicals..................370
    10.1.6 Reactions of Transition Metal Ions with Dioxygen..................377
    10.1.7 Catalytic Oxidation of Ketones..................382
10.2 Cobalt Bromide Catalysis..................383
10.3 Oscillating Oxidation Reactions..................386
10.4 Acid Catalysis in Liquid-Phase Oxidation of Hydrocarbons and Alcohols..................389
10.5 Catalytic Epoxidation of Olefins by Hydroperoxides..................390
10.6 Catalytic Oxidation of Olefins to Aldehydes..................394
    10.6.1 Catalysis by Palladium Salts..................394
    10.6.2 Kinetics and Mechanism of Reaction..................395
10.7 Heterogeneous Catalysis in Liquid-Phase Oxidation..................396
    10.7.1 Introduction..................396
    10.7.2 Decomposition of Hydroperoxides on Catalyst Surface..................397
    10.7.3 Activation of Dioxygen..................398
    10.7.4 Critical Phenomena in Heterogeneous Catalysis..................399
    10.7.5 Selectivity of Catalytic Oxidation..................400
10.8 Oxidation in Basic Solutions..................400
References..................403

Chapter 11
Oxidation of Hydrocarbons in Microheterogeneous Systems..................411
11.1 Emulsion Oxidation of Hydrocarbons..................411
11.2 Hydrocarbon Oxidation in Microheterogeneous Systems..................413
References..................415

Chapter 12
Sulfoxidation of Hydrocarbons..................417
12.1 Introduction..................417
12.2 Chain Mechanism of Sulfoxidation..................418
12.3 Elementary Steps of Sulfoxidation..................420
12.4 Decomposition of Alkylsulfonyl Peracids..................424
12.5 Oxidation by Alkylsulfonic Peracids..................425
    12.5.1 Oxidation of Aromatic Hydrocarbons..................425
    12.5.2 Oxidation of Olefins..................426
    12.5.3 Oxidation of Ketones..................428
References..................429

Chapter 13
Oxidation of Polymers 431
13.1   Initiated Oxidation of Polymers 431
       13.1.1   Cage Effect in Solid Polymers 431
       13.1.2   Migration of Free Valence in Solid Polymers 432
       13.1.3   Initiated Polymer Oxidation 433
       13.1.4   Diffusion of Dioxygen in Polymer 436
       13.1.5   Diffusion Regime of Polymer Oxidation 439
       13.1.6   Isomerization of Alkyl and Peroxyl Radicals of Polypropylene 441
13.2   Autoxidation of Polymers 443
       13.2.1   Chain Generation by Dioxygen 443
       13.2.2   Decomposition of Hydroperoxyl Groups 444
       13.2.3   Specificity of Formation of the Hydroperoxyl Group
                and its Decay in PP 447
       13.2.4   Chain Decay of Hydroperoxyl Groups of PP 449
13.3   Oxidative Degradation of Macromolecules 451
13.4   Heterogeneous Character of Polymer Oxidation 456
References 458

Part II
Chemistry of Antioxidants 463

Chapter 14
Theory of Inhibition of Chain Oxidation of Organic Compounds 465
14.1   Historical Introduction 465
14.2   Kinetic Classification of Antioxidants 466
       14.2.1   Classification 466
       14.2.2   Capacity, Strength, and Efficiency of Antioxidants 468
14.3   Mechanisms of Inhibition of Initiated Chain Oxidation 469
       14.3.1   Characteristics of Antioxidant Efficiency 475
14.4   Kinetics of Autoxidation of Organic Compounds Inhibited by Acceptors of
       Peroxyl Radicals 477
       14.4.1   Nonstationary Kinetics of Inhibited Autoxidation 477
       14.4.2   Quasistationary Kinetics of Inhibited Autoxidation 478
14.5   Topological Kinetics of Inhibited Oxidation of Hydrocarbons 480
       14.5.1   Initiated Oxidation 482
       14.5.2   Inhibited Autoxidation 486
References 488

Chapter 15
Antioxidants Reacting with Peroxyl Radicals 491
15.1   Reaction of Phenols with Peroxyl Radicals 491
       15.1.1   Reaction Enthalpy 491
       15.1.2   Triplet Repulsion in Transition State 494
       15.1.3   Steric Factor 496
       15.1.4   Effect of Hydrogen Bonding on Activity of Inhibitors 497
15.2   Reaction of Aromatic Amines with Peroxyl Radicals 501
       15.2.1   Dissociation Energies of N—H Bonds of Aromatic Amines 501
       15.2.2   Rate Constants of Reaction of $RO_2^\bullet$ with Aromatic Amines 501

	15.2.3	Electronegativity of Atoms in the Transition State 507
15.3	Reactions of Phenoxyl and Aminyl Radicals with $RO_2^\bullet$ 508	
	15.3.1	Recombination of Peroxyl with Phenoxyl Radicals 508
	15.3.2	Reaction of Peroxyl with Aminyl Radicals 511
15.4	Recombination and Disproportionation of Phenoxyl and Aminyl Radicals 512	
	15.4.1	Recombination and Disproportionation of Phenoxyl Radicals 512
	15.4.2	Reactions Controlled by Rotational Diffusion 515
	15.4.3	Disproportionation of Semiquinone Radicals 516
	15.4.4	Recombination and Disproportionation of Aminyl Radicals 516
15.5	Reactions of Phenoxyl and Aminyl Radicals with Hydrocarbons, Hydroperoxides, and Dioxygen 519	
	15.5.1	Triplet Repulsion 521
	15.5.2	Steric Factor 524
	15.5.3	Electron Affinity of Atoms in the TS 525
	15.5.4	Reactions of Phenoxyl Radicals with Dioxygen 527
15.6	Direct Oxidation of Phenols and Amines by Dioxygen and Hydroperoxides 528	
References 537		

Chapter 16
Cyclic Chain Termination in Oxidation of Organic Compounds 541
16.1 Cyclic Chain Termination by Aromatic Amines and Aminyl Radicals 541
16.2 Cyclic Chain Termination by Quinones 552
16.3 Cyclic Chain Termination by Nitroxyl Radicals 555
16.4 Acid Catalysis in Cyclic Chain Termination 561
16.5 Transition Metal Ions as Catalysts for Cyclic Chain Termination 564
References 569

Chapter 17
Hydroperoxide Decomposing Antioxidants 573
17.1 Introduction 573
17.2 Organophosphorus Antioxidants 573
    17.2.1 Reaction with Hydroperoxides 573
    17.2.2 Reaction with Peroxyl Radicals 577
17.3 Sulfur-Containing Antioxidants 580
    17.3.1 Reaction with Hydroperoxide 580
    17.3.2 Reaction with Peroxyl Radicals 583
17.4 Metal Thiophosphates and Thiocarbamates 588
17.5 Metal Complexes with Phosphites 591
References 595

Chapter 18
Synergism of Antioxidant Action 599
18.1 Introduction 599
18.2 Synergism of Chain Termination and Hydroperoxide Decomposing the Antioxidants 600
18.3 Synergism of Two Antioxidants Terminating Chains 603
    18.3.1 Combined Action of Phenol and Amine on Hydrocarbon Oxidation 603
    18.3.2 Combined Action of Two Phenols 607

	18.3.3	Synergism of Phenol and Nitroxyl Radical Action on Hydrocarbon Oxidation 611
	18.3.4	Synergistic Action of Quinone in Combination with Phenol or Amine 616
	18.3.5	The Combined Action of Fe and Cu Salts 620

References 625

Chapter 19
Peculiarities of Antioxidant Action in Polymers 627
19.1 Rigid Cage of Polymer Matrix 627
      19.1.1 Comparison of Bimolecular Reactions in Liquid and Solid Phases 627
      19.1.2 Conception of Rigid Cage of Polymer Matrix 631
      19.1.3 The Phenomena of Reactivity Leveling in Polymeric Matrix 641
      19.1.4 Reactions Limited by Rotational Diffusion in Polymer Matrix 643
19.2 Antioxidants Reacting with Peroxyl Radicals 644
19.3 Antioxidants Reacting with Alkyl Radicals 649
19.4 Cyclic Chain Termination in Oxidized Polymers 652
      19.4.1 Regeneration of Nitroxyl Radicals in Polymer Oxidation 652
      19.4.2 Cyclic Chain Termination in Oxidized Polypropylene 654
19.5 Inhibition of Synthetic Rubber Degradation 659
References 662

Chapter 20
Heterogeneous Inhibition of Oxidation 665
20.1 Retarding Action of Solids 665
20.2 Inhibitor Action in Catalyzed Hydrocarbon Oxidation 668
20.3 Dioxygen Acceptors as Stabilizers of Polymers 670
References 671

Part III
Biological Oxidation and Antioxidants 673

Chapter 21
Initiators of Free Radical-Mediated Processes 675
21.1 Superoxide 675
21.2 Hydroxyl Radical 676
21.3 Perhydroxyl Radical 677
21.4 Nitric Oxide 677
      21.4.1 Formation and Lifetime of Nitric Oxide 677
      21.4.2 Reaction with Dioxygen 678
      21.4.3 Reaction with Superoxide and Inorganic Nitrogen Compounds 678
      21.4.4 Reactions with Biomolecules 679
      21.4.5 Interaction with Enzymes 681
21.5 Other Reactive Nitrogen Oxide Species 681
21.6 Peroxynitrite 683
      21.6.1 Formation, Decomposition, and Reactions of Peroxynitrite 683
      21.6.2 Reaction of Peroxynitrite with Carbon Dioxide 687
      21.6.3 Examples of Biological Functions of Nitric Oxide and Peroxynitrite 688

21.7 Iron 689
    21.7.1 Structure and Origins of "Free" Iron 689
    21.7.2 Iron-Stimulated Free Radical-Mediated Damaging Processes....................690
    21.7.3 Iron-Stimulated Toxic Effects of Pathogenic Fibers and Particles 692
21.8 Possible Effects of Magnetic Field on Free Radical Formation 693
References 695

Chapter 22
Generation of Free Radicals by Prooxidant Enzymes 701
22.1 Xanthine Oxidase..701
    22.1.1 Mechanisms of Oxygen Radical Production 701
    22.1.2 Reactions of XO with Organic and Inorganic Substrates.......................703
    22.1.3 Free Radical-Mediated Biological Activity of Xanthine
           Oxidoreductase..704
22.2 NADPH Oxidases..704
    22.2.1 Leukocyte NADPH Oxidase...705
    22.2.2 NADPH Oxidase of Nonphagocytic Cells..707
22.3 Nitric Oxide Synthases 710
    22.3.1 Types and Structure of Nitric Oxide Synthases 710
    22.3.2 Free Radical Production by Nitric Oxide Synthases............................711
    22.3.3 Mitochondrial Nitric Oxide Synthase 714
22.4 Peroxidases..715
    22.4.1 Mechanism of Free Radical Production 715
    22.4.2 Oxidation of Phenolic Compounds 716
    22.4.3 Oxidation of Glutathione and Other Organic Substrates 718
    22.4.4 Reactions with Inorganic Compounds...718
    22.4.5 Mechanism of the Interaction of Peroxidases with Hydroperoxides.
           Role of Superoxide and Hydroxyl Radicals in Reactions Catalyzed
           by Peroxidases...719
    22.4.6 Peroxidase-Catalyzed Oxidative Processes 721
    22.4.7 Interaction with Nitrogen Oxides 721
References 723

Chapter 23
Production of Free Radicals by Mitochondria 731
23.1 Detection of Active Oxygen Species in Mitochondria 731
23.2 Rates and Mechanism of Superoxide Production in Mitochondria.......................732
23.3 Free Radical-Mediated Damage to Mitochondria 735
23.4 Mechanisms of Mitochondria Protection from Free Radical-Mediated
    Damage 736
23.5 Preconditioning as a Protection of Mitochondria from Oxygen
    Radical-Mediated Damage 736
23.6 Mitochondrial Oxygen Radical Production and Apoptosis..................................738
    23.6.1 Reactive Oxygen Species as Mediators of Apoptosis...........................738
    23.6.2 Mechanisms of the Activation of Apoptosis by Reactive
           Oxygen Species..739
    23.6.3 Protection Against Apoptosis Activated by Reactive Oxygen Species........740
23.7 Mitochondrial Nitrogen Oxide Production and Apoptosis...................................741
References 743

Chapter 24
Production of Free Radicals by Microsomes ... 747
24.1 Microsomal NADPH–Cytochrome P-450 Reductase and NADH cytochrome b_5 Reductase ... 747
24.2 Production of Free Radicals by Microsomes ... 749
24.3 Microsomal Free Radical-Mediated Oxidative Processes ... 750
24.4 Formation and Reactions of Nitrogen Oxygen Species in Microsomes ... 753
References ... 754

Chapter 25
Nonenzymatic Lipid Peroxidation ... 757
25.1 Initiation of Nonenzymatic Lipid Peroxidation ... 757
    25.1.1 $HO^\bullet$ and $O_2^{\bullet-}$ as Initiators of Lipid Peroxidation ... 757
    25.1.2 NO, NO Metabolites, and HOCL ... 760
    25.1.3 Xenobiotics as Initiators of Lipid Peroxidation ... 763
    25.1.4 Phagocytes as Initiators of Lipid Peroxidation ... 765
25.2 Lipid Peroxidation of Unsaturated Fatty Acids ... 765
25.3 Cholesterol Oxidation ... 766
25.4 The Formation of Prostanoids ... 769
25.5 Oxidation of Low-Density Lipoproteins ... 776
    25.5.1 Mechanism of LDL Oxidation ... 776
    25.5.2 Initiation of LDL Oxidation ... 777
    25.5.3 Role of Nitric Oxide and Peroxynitrite in the Initiation of LDL Oxidation ... 779
    25.5.4 Initiation of LDL Oxidation by MPO ... 779
    25.5.5 Major Products and Pathophysiological Effects of LDL Oxidation ... 782
References ... 783

Chapter 26
Enzymatic Lipid Peroxidation ... 789
26.1 Lipid Peroxidation and Free Radical Production Catalyzed by LOXs ... 789
    26.1.1 Mechanism of LOX Catalysis ... 789
    26.1.2 LOX-Catalyzed Oxidation of Unsaturated Acids ... 791
    26.1.3 LOX-Catalyzed Oxidation of Low-Density Lipoproteins ... 793
    26.1.4 LOX-Catalyzed Cooxidation of Substrates ... 795
    26.1.5 Oxygen Radical Formation by LOX-Catalyzed Reactions ... 796
    26.1.6 Effects of Nitrogen Oxides on LOX-Catalyzed Processes ... 796
    26.1.7 Comments on Pathophysiological Activities of LOXs ... 798
26.2 Lipid Peroxidation and Free Radical Production Catalyzed by Prostaglandin H Synthases ... 798
    26.2.1 Mechanism of Reactions Catalyzed by Prostaglandin H Synthases ... 798
    26.2.2 Production and Interaction with Oxygen Radicals ... 800
    26.2.3 Cyclooxygenase-Catalyzed Cooxidation of Substrates ... 801
    26.2.4 Effects of Reactive Nitrogen Species on Prostaglandin H-Catalyzed Processes ... 801
26.3 Oxidation by Linoleate Diol Synthase ... 803
References ... 805

Chapter 27
Oxidation of Proteins 809
27.1 Free Radical Mechanisms of Protein Oxidation 809
      27.1.1 Free Radical Initiation of Protein Oxidation 809
      27.1.2 Metal Ion-Catalyzed Oxidation of Proteins 811
      27.1.3 Protein Oxidation Initiated by Peroxynitrite, Nitric Oxide,
              and Hypoclorite 812
27.2 Some Examples of Oxidative Processes with the Participation of Proteins 814
27.3 Competition Between Protein and Lipid Oxidation 814
27.4 Inhibition of Protein Oxidation by Antioxidants and Free
     Radical Scavengers 815
27.5 Repairing and Proteolysis of Oxidized Proteins 815
References 816

Chapter 28
DNA Oxidative Damage 819
28.1 Hydroxyl Radical-Mediated DNA Damage 819
      28.1.1 Mechanism of Hydroxyl Radical and DNA Reactions 819
      28.1.2 Reactions of Hydroxyl and Hydroxyl-Like Radicals Produced
              by the Fenton Reaction with Nucleic Acids 820
28.2 Superoxide-Dependent DNA Damage 822
28.3 DNA Damage by Enzymatic Superoxide Production 823
      28.3.1 Phagocyte-Stimulated DNA Damage 823
28.4 Reactive Oxygen Species as Mediators of Drug- and Xenobiotic-Induced
     DNA Damage 825
      28.4.1 Effects of Prooxidants 825
      28.4.2 Damaging Effects of Antioxidants 826
      28.4.3 Damaging Effects of Mineral Dusts and Fibers 827
28.5 DNA Damage by Reactive Nitrogen Species 827
28.6 Repair of Free Radical-Mediated DNA Damage 828
References 830

Chapter 29
Antioxidants 835
29.1 Vitamin E 836
      29.1.1 Antioxidant and Prooxidant Activity of Vitamin E 836
      29.1.2 Biological Activity 838
      29.1.3 Effects of Vitamin E Supplementation in Aging and Heart Diseases 840
      29.1.4 Synthetic Analogs of Vitamin E 841
29.2 Vitamin C 841
      29.2.1 Antioxidant and Prooxidant Activity of Vitamin C 841
      29.2.2 Biological Activity 842
      29.2.3 Interaction Between Vitamins E and C 843
29.3 Flavonoids 844
      29.3.1 Free Radical Scavenging Activity 845
      29.3.2 Protection Against Free Radical-Mediated Damage 849
      29.3.3 Comparison of Free Radical Scavenging and Chelating Activities 850
      29.3.4 Inhibition of Free Radical-Mediated Damage in Cells 852
      29.3.5 Inhibition LDL Oxidation and Enzymatic Lipid Peroxidation 853

	29.3.6	Other Examples of Protective Activity of Flavonoids Against Free Radical-Mediated Damage in Biological Systems 853
	29.3.7	Antioxidant Effect of Metal–Flavonoid Complexes 854
	29.3.8	Comments on Prooxidant Activity of Flavonoids 856

29.4 Phenolic Compounds other than Flavonoids 857
29.5 Thiols 860
    29.5.1 Lipoic Acid 860
    29.5.2 Glutathione 862
    29.5.3 *N*-Acetylcysteine and Tetradecylthioacetic Acid 864
29.6 Ubiquinones 864
29.7 Quinones 866
29.8 Uric Acid 867
29.9 Steroids 867
29.10 Calcium Antagonists and β-Blockers 870
29.11 Pyrrolopyrimidines 873
29.12 NADPH 874
29.13 β-Carotene 874
29.14 Melatonin 876
29.15 Ebselen 877
29.16 Metallothioneins and Zinc 878
29.17 Metalloporphyrins 878
29.18 Lactate 879
29.19 Aspirin 879
29.20 Amino Acids 880
29.21 Vitamin B_6 880
29.22 Targeted Antioxidants 880
29.23 Natural Antioxidant Mixtures 880
29.24 Chelators 882
29.25 Synergistic Interaction of Antioxidants, Free Radical Scavengers, and Chelators 883
References 884

Chapter 30
Antioxidant Enzymes 895
30.1 Superoxide Dismutase: The Latest Developments 895
    30.1.1 Mechanism of Superoxide-Dismuting Activity of SODs 895
    30.1.2 SOD Mimics and SOD Modification 896
    30.1.3 Biological Effects of SOD 897
30.2 Superoxide Reductases 898
30.3 Catalase 899
30.4 Glutathione Redox Cycle 900
30.5 Thioredoxin Reductase 900
30.6 Phase 2 Antioxidant Enzymes and Alkenal/One Oxidoreductase 901
References 901

Chapter 31
Free Radicals and Oxidative Stress in Pathophysiological Processes 905
31.1 Cardiovascular Diseases 905
    31.1.1 Ischemia–Reperfusion 905

	31.1.2	Atherosclerosis 907
	31.1.3	Mitochondria and NADPH Oxidase as Initiators of Oxygen Radical Overproduction in Heart Diseases 908
	31.1.4	Antioxidants and Heart Disease 908
	31.1.5	Hypertension 910
		31.1.5.1 Preeclampsia 911
	31.1.6	Hyperglycemia and Diabetes Mellitus 911
		31.1.6.1 Hyperglycemia-Induced Oxidative Stress in Diabetes 911
		31.1.6.2 Free Radical-Mediated Processes in Diabetes Mellitus 913
		31.1.6.3 Antioxidants and Diabetes Mellitus 914
31.2	Cancer, Carcinogenesis, Free Radicals, and Antioxidants 915
	31.2.1	Mechanisms of Free Radical Reactions in Tumor Cells 915
	31.2.2	The Treatment of Cancer with Prooxidants 917
	31.2.3	The Treatment of Cancer with Antioxidants 918
31.3	Inflammation 920
	31.3.1	Mechanisms of Free Radical-Mediated Inflammatory Processes 920
	31.3.2	Rheumatoid Arthritis 921
	31.3.3	Lung Diseases 922
	31.3.4	Skin Inflammation 924
	31.3.5	Brain Inflammatory Diseases 925
		31.3.5.1 Multiple Sclerosis 925
		31.3.5.2 Amyotrophic Lateral Sclerosis 925
		31.3.5.3 Alzheimer's Disease 925
		31.3.5.4 Parkinson's Disease 926
	31.3.6	Kidney Diseases 927
	31.3.7	Liver Diseases 927
	31.3.8	Septic Shock, Pancreatitis, and Inflammatory Bowel Disease 928
31.4	Iron-Catalyzed Pathophysiological Disorders 928
	31.4.1	Hemochromatosis and Hemodialysis of Patients 929
	31.4.2	Thalassemia 929
	31.4.3	Sickle Cell Disease 931
	31.4.4	Fanconi Anemia 932
31.5	Bloom's Syndrome and Down Syndrome 933
31.6	Other Examples of Free Radical Formation in Pathophysiological Disorders 934
31.7	Free Radicals in Aging 934
	31.7.1	The Latest Developments 934
	31.7.2	The Mitochondrial Theory of Aging 935
References 936

Chapter 32
Comments on Contemporary Methods of Oxygen and Nitrogen Free Radical Detection 951
32.1	Detection of Superoxide in Biological Systems 951
	32.1.1	Cytochrome *c* Reduction 951
	32.1.2	Spin-Trapping 953
	32.1.3	Lucigenin-Amplified CL as a Sensitive and Specific Assay of Superoxide Detection 955
	32.1.4	Other Chemiluminescent Methods of Superoxide Detection 957

| | | 32.1.5 | Adrenochrome (Epinephrine) Oxidation | 959 |

  32.1.5 Adrenochrome (Epinephrine) Oxidation 959
  32.1.6 Nitroblue Tetrazolium Reduction 959
  32.1.7 Fluorescent Methods 960
  32.1.8 Interaction with Aconitase 960
32.2 Detection of Hydroxyl Radicals 960
32.3 Detection of Nitric Oxide 961
32.4 Detection of Peroxynitrite 961
References 962

Index 967

Preface

Oxidation of organic compounds by dioxygen is a phenomenon of exceptional importance in nature, technology, and life. The liquid-phase oxidation of hydrocarbons forms the basis of several efficient technological synthetic processes such as the production of phenol via cumene oxidation, cyclohexanone from cyclohexane, styrene oxide from ethylbenzene, etc. The intensive development of oxidative petrochemical processes was observed in 1950–1970. Free radicals participate in the oxidation of organic compounds. Oxidation occurs very often as a chain reaction. Hydroperoxides are formed as intermediates and accelerate oxidation. The chemistry of the liquid-phase oxidation of organic compounds is closely interwoven with free radical chemistry, chemistry of peroxides, kinetics of chain reactions, and polymer chemistry.

The science of oxidation processes developed intensively during the 20th century. In the very beginning of this century, A.N. Bach and C. Engler formulated the peroxide conception of oxidation by dioxygen. H. Backstrom proved the chain mechanism of slow liquid-phase oxidation (1927). N.N. Semenov formulated the theory of chain oxidation with degenerate branching of chains applied to hydrocarbon oxidation (1934). A line of brilliant kinetic experiments were performed in the period 1960–1980 concerning elementary steps of liquid-phase oxidation of hydrocarbons. The results of these studies were collected in two volumes edited by W.O. Lundberg, *Autoxidation and Antioxidants* (1961) and monograph *Liquid-Phase Oxidation of Hydrocarbons* written by N.M. Emanuel, E.T. Denisov, and Z.K. Maizus (1967). The results of the study of oxidation of oxygen-containing compounds were systematized in the monograph *Liquid-Phase Oxidation of Oxygen-Containing Compounds* (E.T. Denisov, N.I. Mitskevich, V.E. Agabekov, 1977). Several new excellent experimental methods were developed in the second half of the 20th century for the study of free radical reactions. A great body of kinetic data on reactions of alkyl, peroxyl, alkoxyl, and others radicals was obtained in many publications from 1960 to 1990 (see Database applied to the book).

It was in 1990 that one of the authors of this monograph (E.T. Denisov) formulated the semiempirical model of any bimolecular homolytic reaction as a result of the intersection of two parabolic curves for the potential energy of reacting bonds. This model uses empirical parameters and reaction enthalpy for the calculation of the activation energy, rate constants, and geometrical parameters of the transition state. The program is given in Database applied to the book as Electronic Application. This program helps to calculate automatically these parameters for more than 340,000 elementary free radical reactions. The parabolic model appeared to be an excellent tool for performing the analysis of the reactivity of reactants in the chain reactions of oxidation. The results of this analysis applied to reactions of peroxyl radicals with the C—H bond of different organic compounds are given in Chapters 7–9.

From the other side, oxidation processes in organic materials lead to negative consequences. Keeping and using various organic products in air often results in their rapid deterioration. The first product to be stabilized by addition of antioxidant was natural rubber. Antioxidants are widely used now to prevent the oxidation of fuel, lubricant oils, organic semiproducts, monomers, polymeric materials, etc. Several books devoted to the chemistry and kinetics of antioxidant action were written, including *Handbook of Antioxidants* (E.T. Denisov and T.G. Denisova, 2000). The reader finds in Part II of this book a main chemical mechanistic information about antioxidants and their action on the autoxidation of hydrocarbons, alcohols, and other organic compounds and polymers. The theory of antioxidant reactivity using the conception of intersection of parabolic potential curves is developed.

For a long time, oxidative processes in biology were considered only as damaging phenomena and "oxidative stress" as a cause of various pathological disorders. However, at present, free radicals and oxidative processes are no more regarded as only damaging ones. It has been proved that many physiological processes are mediated by oxygen and nitrogen free radicals and pathophysiological disorders may be the consequence not only of overproduction but also the insufficient formation of free radicals.

Numerous oxidative processes occur in living organisms, which are regulated or unregulated enzymatically. Many important biological oxidative processes catalyzed by various oxidases, oxygenases, and other enzymes do not proceed by free radical mechanism or, at least, there are no evidences of free radical participation. For a long time, the formation of free radicals in biological systems was considered to be an abnormal event, which originates from the undesirable intervention of various toxic agents such as pollutants, irradiation, drugs, toxic food components, etc. However, the discovery of enzymes catalyzing oxidative processes through free radical-mediated mechanisms (for example, cyclooxygenase and lipoxygenase) and the enzymes catalyzing the production of "physiological" free radicals superoxide and nitric oxide (xanthine oxidase, NADPH oxidase, etc.) has changed this point of view.

Despite numerous earlier hypotheses, the reliable experimental evidence of enzymatic free radical generation was obtained only in 1968–1969 by MacCord and Fridovich. These authors have shown that xanthine oxidase catalyzes the one-electron reduction of molecular oxygen (dioxygen) to dioxygen radical anion $O_2^{\bullet-}$ (superoxide) during the oxidation of xanthine and some other substrates. Later on, another "physiological" radical (i.e., formed under normal physiological conditions and not only as a result of a certain pathological disorder) nitric oxide (NO) has been identified. It was found that this free radical is formed during the oxidation of L-arginine by NO synthases. Interestingly, superoxide and nitric oxide have many similar properties. Both radicals are relatively long-living (from 0.001 to 100 msec) and not really very active species (the name "superoxide" is really a confusing because $O_2^{\bullet-}$ is a moderate reductant and not an oxidant at all). However superoxide and nitric oxide are the precursors of many highly reactive free radicals and diamagnetic molecules (hydroxyl radical, peroxyl radicals, peroxynitrite). Such "free radical cocktail" may transform the biological system into a new state now known as a system under oxidative stress. This vague term suggests that there is a certain deviation from the norm, which is characterized by the enhanced risk of free radical-mediated damage.

Now, we may differentiate the functions of free radicals in biological systems. The first one is a physiological function. Enzymatic production of superoxide and nitric oxide is the obligatory condition of normal biological activities including cell signaling, phagocytosis, bactericidal activity, etc. Furthermore, free peroxyl radicals mediate the enzymatic formation of prostaglandins, leukotrienes, and other biological molecules. The second one is a pathophysiological function, which depends on the damaging activity of reactive free radicals and their products resulting in the formation of oxidized lipids, proteins, DNA, and other biomolecules.

Both types of free radical activity include oxidative processes, which are considered in this book. It should be noted that the oxidation of biological molecules is not always a damaging process. There are some physiological functions, which depend on the formation of oxidized products, for example enzymatic lipid oxidation results in the formation biologically active prostaglandins and other biologically active compounds, etc. Nonetheless, "free radical cocktail" can be a cause of potent damaging events, and therefore, an organism had to develop powerful many-step antiradical protective systems. Traditionally, these systems are called the antioxidant systems. Antioxidant protective systems include numerous compounds of different structure, antioxidant enzymes (superoxide dismutase, catalase, peroxidases, etc.), endogenous low-molecular-weight antioxidants (glutathione, ubiquinone, uric acid, etc.) exogenous antioxidants, free radical scavengers and chelators supplied to an organism with food or intently administered as drugs or vitamins (for example, vitamins E and C).

The main goal of Part III of this book is to consider physiological and pathophysiological oxidative processes mediated by free radicals. In Chapter 21 the structures and important reactions of major initiators of oxidative processes in biological systems are considered. Major "prooxidant" enzymes are discussed in Chapter 22. Chapters 23 and 24 are dedicated to the role of mitochondria and microsomes in the production of free radicals. Nonenzymatic and enzymatic lipid oxidative processes are considered in Chapters 25 and 26. Oxidative destruction of proteins and DNA are described in Chapters 27 and 28. Antioxidants including free radical scavengers and chelators are considered in Chapter 29. Correspondingly, antioxidant enzymes are discussed in Chapter 30. Chapter 31 is dedicated to description of "free radical pathologies," i.e., pathologies associated with the formation of free radicals and to consideration of antioxidant and chelating therapy. In Chapter 32 major analytical methods of free radical determination under in vitro and in vivo conditions are described.

Parts I and II of this book are written by E.T. Denisov and Part III by I.B. Afanas'ev.

Symbols and units used in this book are in accordance with the IUPAC recommendations. We are very grateful to Elena Batova for attentive English editing of the manuscripts Parts I and II of this book.

All comments, criticisms, and suggestions will be welcomed by the authors. Address any comments to: E.T. Denisov, Institute of Problems of Chemical Physics, Chernogolovka, Moscow Region, 142432, Russia. E-mail: edenisov@icp.ac.ru and to I.B. Afanas'ev, Vitamin Research Institute, Moscow 117820, Russia. E-mail: iafan@aha.ru

Authors

Evgeny T. Denisov graduated from Moscow State University in 1953. He earned a Ph.D. in 1957 and Doctor of Science degree (chemistry) in 1966. Since 1956 he has been working at the Branch of Semenov Institute of Chemical Physics in Chernogolovka. From 1967 to 2000 he held the position of Head of the Laboratory of Kinetics of Free Radical Liquid-Phase Reactions and is currently a Principal Researcher of the Institute of Problems of Chemical Physics in Chernogolovka. Under his leadership 45 postgraduate students fulfilled their study requirements to obtain Ph.D. degrees. He was elected as an active member of the Academy of Creative Endeavors in 1991 and International Academy of Sciences in 1994. From 1979 to 1989 he was a member of the IUPAC Commission on Physicochemical Symbols, Terminology, and Units, and from 1989 to 1991 he was the Chairman of Commission on Chemical Kinetics. Dr. Denisov was Chairman of the Kinetic Section of the Scientific Council on Structure and Chemical Kinetics of the Academy of Sciences of USSR (1972–1991). He was awarded by the Academy of Creative Endeavors the Medal of Svante Arrhenius in 2000 and became Honored Scientist of Russia in 2001. His scientific interests lie in the following fields of chemical kinetics: free radical reactions in solution, reactivity of reactants in free radical reactions, bond dissociation energies, liquid-phase oxidation, degradation of polymers, mechanism of antioxidant action. His list of publications includes 8 monographs on oxidation and antioxidants, 4 handbooks, 5 manuals on chemical kinetics, and more than 500 publications in Russian and international scientific journals.

Igor B. Afanas'ev was born in Moscow, Russia (former USSR). He graduated from Moscow Chemico-Technological Institute in 1958 and received Ph.D. and D.Sc. degrees (Chemistry) in 1963 and 1972, respectively. He became Professor of Chemical Kinetics and Catalysis in 1982.

Professor Afanas'ev has been Chairman of the Russian Committee of Society for Free Radical Research (SFRR) (1989–1998), a member of the International Editorial Board of *Free Radical Biology & Medicine* (1994–1998), and the organizer of International Meetings on Free Radicals in Chemistry and Biology: Leningrad (1991), Moscow (1994), Kiev (1995), Moscow-Yaroslavl (together with Profs. A. Azzi and H. Sies) (1998).

He is the author of *Superoxide Ion: Chemistry and Biological Implications* (Volumes 1 and 2, CRC Press, 1989–1990) and the co-author of several jointed books on free radicals. He has published about 100 works.

At present, Professor Afanas'ev is head of the laboratory at the Vitamin Research Institute in Moscow.

List of Chemical Symbols and Abbreviations

AcacH	acetylacetone
AFR	acceptor of free radicals
AIBN	azodiizobutyronitrile
AmH	aromatic amine
Am$^\bullet$	aminyl radical
AmOH	hydroxylamine
AmO$^\bullet$	nitroxyl radical
AOT	sodium bis(2-ethylhexyl)sulfosuccinate
APP	atactic polyprpopylene
ArOH	phenol
ArO$^\bullet$	phenoxyl radical
Ar$_1$OH	sterically nonhindered phenol
Ar$_1$O$^\bullet$	sterically nonhindered phenoxyl radical
Ar$_2$OH	sterically hindered phenol
Ar$_2$O$^\bullet$	sterically hindered phenoxyl radical
BDE	bond dissociation energy
CBA	copolymer butadiene–acrylonitrile
CBDS	copolymer of butadiene with styrene
CEP	copolymer of ethylene with propylene
CL	chemiluminescence
CTAB	cetyltrimethyl ammonium bromide
CTC	molecular complex with charge transfere
DBP	peroxide, bis(1,1-dimethylethyl)-
DCHP	dicyclohexylperoxydicarbonate
DCP	peroxide, bis(1-methyl-1-phenylethyl)-
DFT	density functional theory
DIPP	deuterated isotactic polypropylene
DPE	deuterated polyethylene
EA	electron affinity
EPR	electron paramagnetic resonance
HDPE	polyethylene of high density
HQ$^\bullet$	semiquinone radical
I	initiator
InH	acceptor reacting with alkoxyl and peroxyl radicals
IPM	intersecting parabolas model of a homolytic reaction
IPP	isotactic polypropylene
IR	infrared spectroscopy
LDPE	low density polyethylene
MW	molecular weight
NMR	nuclear magnetic resonance spectroscopy
NR	natural rubber

P·	macroradical
PBD	polybutadiene
PDMB	polydimethylbutadiene
PE	polyethylene
PEA	polyethyl acrylate
PFE	polyfluoroethylene
PH	polymer
PIB	polyisobutylene
PIP	polyisopentene
PMMA	polymethyl methacrylate
PMP	polymethylpentene
$PO_2^·$	peroxyl macroradical
PP	polypropylene
PS	polystyrene
PVA	polyvinyl acetate
PVM	polymethylvinyl ether
Q	acceptor reacting with alkyl radicals, quinone
QH_2	hydroquinone
RH	organic substance reacting with its C—H bond
R^1H	aliphatic or alicyclic hydrocarbon
R^2H	olefin hydrocarbon
R^3H	alkylaromatic hydrocarbon
RN_2R	azo compound
ROOH	hydroperoxide
ROOR	peroxide
R·	alkyl radical
RO·	alkoxyl radical
$RO_2^·$	peroxyl radical
S	antioxidant decomposing hydroperoxide
SA	surfactant
SDS	sodium dodecylsulfate
SSR	stereoregular synthetic resin
TEMPO	2,2,6,6-tetramethylpiperidin-*N*-oxyl
TS	transition state
UV	ultraviolet light

List of Physicochemical Symbols

Symbol	Description	Unit
A	pre-exponential factor in Arrhenius equation of reaction rate constant $k = A \times \exp(-E/RT)$	s^{-1} (unimolecular reaction), $L\,mol^{-1}\,s^{-1}$ (bimolecular), $L^2\,mol^{-2}\,s^{-1}$ (trimolecular)
A_0	pre-exponential factor of reaction rate constant per attacked atom among bonds with equireactivity	same units as for A
b	$2b^2$ is the force constant of chemical bond	$kJ^{1/2}\,mol^{-1/2}\,m^{-1}$
D	diffusion coefficient	$m^2\,s^{-1}$
D_{Y-X}	dissociation energy of Y–X bond	$kJ\,mol^{-1}$
e	probability of formed free radical pair to escape the cage of solvent or polymer	
e	base of natural logarithms	$e = 2.718282$
E	activation energy of reaction in Arrhenius equation of reaction rate constant	$kJ\,mol^{-1}$
E_e	activation energy of reaction in parabolic model of bimolecular reaction; $E_e = E + 0.5hLv - 0.5RT$	$kJ\,mol^{-1}$
f	stoichiometric coefficient of free radical accepting by acceptor of free radicals	
G	radiochemical yield of products	Molecule/100 eV
ΔG^0	Gibbs energy of reaction under standard conditions (298 K, 1 atm)	$kJ\,mol^{-1}$
ΔH^0	enthalpy of reaction (298 K, 1 atm)	$kJ\,mol^{-1}$
$\Delta H^{\neq}$	enthalpy of transition state	$kJ\,mol^{-1}$
ΔH_e	enthalpy of reaction that includes difference of zero vibration energies of reacting bonds	$kJ\,mol^{-1}$
ΔH_f^0	enthalpy of molecule formation under standard conditions (298 K, 1 atm)	$kJ\,mol^{-1}$
ΔH_v^0	enthalpy of molecule evaporation under standard conditions (298 K, 1 atm)	$kJ\,mol^{-1}$
h	Planck constant, $h = 6.626075 \times 10^{-34}$	J s
I_{cl}	intencity of chemiluminescence	Quant s^{-1}
K	equilibrium constant, $RT \ln K = -\Delta G^0$	$(mol/L)^{\Delta n}$
k	reaction rate constant	s^{-1} (unimolecular), $L\,mol^{-1}\,s^{-1}$ (bimolecular), $L^2\,mol^{-2}\,s^{-1}$ (trimolecular)
k_a	rate constant of acceptor reaction with free radicals	$L\,mol^{-1}\,s^{-1}$
k_d	rate constant of decomposition of initiator to free radicals	s^{-1}
k_D	rate constant of diffusion controlled reaction	$L\,mol^{-1}\,s^{-1}$
k_i	rate constant of initiation (free radicals formation)	s^{-1} (unimolecular reaction), $L\,mol^{-1}\,s^{-1}$ (bimolecular)
k_{ind}	rate constant of induced decomposition of initiator	$L^{1/2}\,mol^{-1/2}\,s^{-1}$
k_{is}	rate constant of isomerization	s^{-1}

k_l	reaction rate constant in the liquid phase	s^{-1} or $L\,mol^{-1}\,s^{-1}$
k_m	rate constant of initiator decomposition to molecular products	s^{-1}
k_p	rate constant of chain propagation	$L\,mol^{-1}\,s^{-1}$
k_s	reaction rate constant in the solid phase	s^{-1} or $L\,mol^{-1}\,s^{-1}$
k_t	rate constant of chain termination	$L\,mol^{-1}\,s^{-1}$
L	Avogadro's number, $L = 6.02214 \times 10^{23}$	mol^{-1}
n_D	refractive index	
Δn	molecular change in a reaction	
p	pressure	Pa
R	gas constant, $R = 8.314510$	$J\,mol^{-1}\,K^{-1}$
r_1	ratio of rate konstants of chain propagation in co-oxidation, $r_1 = k_{p11}/k_{p12}$	
r_2	ratio of rate konstants of chain propagation in co-oxidation, $r_1 = k_{p22}/k_{p21}$	
r_R	radius of radical R	m
r_{A-B}	length of A–B bond	m
$S^0(RH)$	entropy of formation of RH in gas phase under standard conditions	$J\,mol^{-1}\,K^{-1}$
$\Delta S^{\neq}$	entropy of activation	$J\,mol^{-1}\,K^{-1}$
S	surface	$cm^2\,L^{-1}$
t	time	s
T	absolute temperature	K
$Q(RH)$	molecular partition function of RH	
$Q^{\neq}$	molecular partition function of transition state	
v	reaction rate	$mol\,L^{-1}\,s^{-1}$
v_i	rate of initiation reaction	$mol\,L^{-1}\,s^{-1}$
v_{i0}	rate of thermal initiation reaction	$mol\,L^{-1}\,s^{-1}$
v_{ind}	rate of induced decomposition of initiator	$mol\,L^{-1}\,s^{-1}$
$\Delta V^{\neq}$	change of molecular volume due to formation of the transition state, $\Delta V^{\neq} = V^{\neq}$ (transition state) $- V$ (reactants)	$cm^3\,mol^{-1}$
α	ratio b_i/b_f of the attacked (b_i) and forming (b_f) bonds	
α	percentage of polymer cristallinity	%
α_T	coefficient of linear temperature expansion	K^{-1}
γ	degree of stretching of polymer film	
ε	dielectric constant of solvent	
ε	molar absorption coefficient	$L\,mol^{-1}\,cm^{-1}$
η	viscosity	Pa s
η	quantum yield of chemiluminescence	
κ	ionic strength, $\kappa = 0.5 \Sigma c_i z_i^2$, c_i and z_i are concentration and charge of i-th ion in solution	$mol\,l^{-1}$
κ	effective rate of dioxygen solvation in the sample of oxidized substance	s^{-1}
λ	Henry's coefficient of dioxygen solving	$mol\,L^{-1}\,Pa^{-1}$
λ	Wavelength of light	m
μ	dipole moment of molecule	D
ν_i	frequency of valence vibration of the reacting bond	s^{-1}
ν_f	frequency of valence vibration of the forming bond	s^{-1}

ν_R	frequency of free radical R• rotation	2π s^{-1}
π	ratio of circumference to diameter of a circle	$\pi = 3.141592$
ρ	density	kg m^{-3}
τ	induction period of reaction	s
ϕ	ratio of rate constants of chain termination in co-oxidation, $\phi = k_{t12}/(k_{t11} k_{t22})^{1/2}$	
Φ	quantum yield	

List of Biochemical Abbreviations

AA, arachidonic acid
AE2, membrane-bound anion exchange protein 2
Ant1, adenine nucleotide translocator
AO, alkenal-one oxidoreductase
BDI, bleomycin-detectable iron
BHA, 2(3)-*tert*-butyl-4-hydroxyanisole
BHT, 5-di-*tert*-butyl-4-hydroxytoluene
BPAEC, bovine pulmonary artery endothelial cells
BPMVE, bovine pulmonary microvascular endothelial cells
BSA, bovine serum albumin
BZ-423, 1,4-benzodiazepine
CE, cholesteryl ester
CE–OOH, cholesteryl ester hydroperoxide
CGD, chronic granulomatous disease
cGP, classic glutathione peroxidase
Ch18:2-OOH, cholesteryl linoleate hydroperoxide
CIDNP, ^{15}N chemically induced dynamic nuclear polarization
CL, chemiluminescence
CLA, 2-methyl-6-phenyl-3,7-dihydroimidazo[1,2-α]pyrazin-3-one
CHF, chronic heart failure
CMEC, coronary microvascular endothelial cell
CNS, central nervous system
COX, cyclooxygenase
CPLA(2), cytosolic phospholipase A(2)
3-CP, 3-carbamoyl-2,2,5,5-tetramethylpyrrolidinoxyl
CP-3, 1-hydroxy-3-carboxypyrrolidine
DBPMPO, 5-(di-*n*-butoxyphosphoryl)-5-methyl-1-pyrroline-*N*-oxide
DCFH, dichlorodihydrofluorescin
DDC, diethyldithiocarbamate
DEMPO, 5-(diethoxyphosphoryl)-5-methyl-1-pyrroline-*N*-oxide
DFO, desferrioxamine
Dfx, desulfoferrodoxin
DHA, docosahexaenoic acid
DHE, dihydroethidium
DHLA, dihydrolipoic acid
DHR, dihydrorhodamine 123
5,15-DiHET, 5(*S*),15(*S*)-dihydroxy-6,13-*trans*-8,11-*cis*-eicosatetraenoic acid
DMPO, 5,5-dimethyl-l-pyrroline-*N*-oxide
DNPH, dinitrophenylhydrazine
DOPA, dihydroxyphenylalanine
DOPAC, 3,4-dihydroxyphenylacetic acid
Dox, doxorubicin
DPPMPO, 5-(di-*n*-propoxyphosphoryl)-5-methyl-l-pyrroline-*N*-oxide
DSBs, double-stranded breaks

EC-SOD, extracellular SOD
EDRF, the endothelium-derived relaxing factor
9EE-Ch18:2-OOH, cholesteryl 9-hydroperoxy-10E, 12E-octadecadienoate
13EE-Ch18:2-OOH, cholesteryl 13-hydroperoxy-9E,11E-octadecadienoate
EOP, eosinophil peroxidase
EPR, electron paramagnetic resonance
ERK, extracellular-signal-regulated kinase
ESR, electron spin resonance
9EZ-Ch18:2-OOH, cholesteryl 9-hydroperoxy-10E,12Z-octadecadienoate
FA, Fanconi anemia
FeDETC, iron-diethyldithiocarbamate
Fe-MGD, iron-methyl-D-glucamine dithiocarbamate
FMLP, N-Formyl-methionyl-leucyl-phenylalanine
FOXOS, protein kinase B-regulated Forkhead transcription factor
Fur, the repressor of iron uptake
GAPDH, glyceraldehyde-3-phosphate dehydrogenase
GBE, *Ginkgo biloba* extract
GP, glutathione peroxidase
GR, glutathione reductase
G-rutin, 4(G)-α-glucopyranosylrutin
GSH, glutathione
Hb, hemoglobin
H_4B, 6R-tetrahydrobiopterin
HDL, high-density lipoproteins
12-HETE, 12-hydroxyeicosatetraenoic acid
HL60, human promyelocytic leukemia cells
4-HNE, 4-hydroxynonenal
7,8-D-HODE, 7S,8S-dihydroxyoctadecadienoic acid
13-HODE, 13-hydroxyoctadecadienoic acid
5-HPETE, 5(S)-hydroperoxyeicosatetraenoic acid
12-HPETE, 12(S)-hydroperoxyeicosatetraenoic acid
15-HPETE, 15(S)-hydroperoxyeicosatetraenoic acid
4-HPNE, 4-hydroperoxy-2-nonenal
8-HPODE, 8R-hydroperoxyoctadecadienoic acid
9-HPODE, 9-hydroperoxy-10,12-octadecadienoic acid
13-HPODE, 13-hydroperoxy-9,11-octadecadienoic acid
HQ, 8-hydroxyquinoline
HRP, horseradish peroxidase
hsp90, heat shock protein 90
HUVEC, human umbilical-vein endothelial cells
ICDH, isocitrate dehydrogenase
IFN-γ, interferon-γ
IOL, iron-overload
IRP, iron regulatory protein
IsoLG, isolevuglandin
JNK, c-Jun N-terminal kinase
K_{ATP}, ATP-sensitive K^+ channel
L1, 1,2-dimethyl-3-hydroxypyrid-4-one
LA, lipoic acid
LacGer, glycosphingolipid lactosylceramide

LDL, low-density lipoprotein
LDS, linoleate diol oxidase
LLU-α, 2,7,7-trimethyl-2-(carboxyethyl)-6-hydroxychroman
LOX, lipoxygenase
LPC, lysophosphatidylcholine
LPS, lipopolysaccharide
LTA_4, leukotriene A_4
LTB_4, leukotriene B_4
LTC_4, leukotriene C_4
MARK, mitogen-activated protein kinase
Mb, myoglobin
MCLA, 2-methyl-6-(4-methoxyphenyl)-3,7-dihydroimidazo[1,2-α]pyrazin-3-one]
MDA, malondialdehyde
MF, magnetic field
MK-447, 2-aminomethyl-4-*tert*-butyl-6-iodophenol
MNNG, 1-methyl-3-nitro-1-nitrosoguanidine
MPO, myeloperoxide
MsrA, thioredoxin-dependent peptide methionine sulfoxide reductase
MT, metallothionein
MtDNA, mitochondrial DNA
MtNOS, mitochondrial nitric oxide synthase
NAC, *N*-acetylcysteine
NBT, 3,3′-(3,3′-dimethoxy-1,1′-biphenyl-4,4′-diyl)bis 2-(4-nitrophenyl)-5-phenyl-2H-tetrazolium dichloride, nitroblue tetrazolium
NDGA, nordihydroguaiaretic acid
NF-κB, redox-regulated transcription factor
NHA, N^w-hydroxy-L-arginine
8-nitro-G, 8-nitroguanine
NMDA, N-methyl-D-aspartate
NO, nitric oxide
NOS, nitric oxide synthase
NOS I (nNOS), neuronal nitric oxide synthase
NOS II (iNOS), inducible nitric oxide synthase
NOS III (eNOS), endothelial nitric oxide synthase
mtNOS, mitochondrial NO synthase
NOHA, N-hydroxyl-L-arginine
nNOSoxy, heme-containing oxygenase domain
NTBI, nontransferrin-bound iron
ODN, oligodeoxyribonucleotide
8-OHdG, 8-hydroxy-2′-deoxyguanosine
OXANO, 2-ethyl-l-hydroxy-2,5,5-trimethyl-3-oxazolidinoxyl
OXANOH, 2-ethyl-l-hydroxy-2,5,5-trimethyl-3-oxazolidine
OZ, opsonized zymosan
PAEC, pulmonary artery endothelial cell
PARS, poly(ADP–ribose) synthetase
PDTC, pyrrolidine dithiocarbamate
PBN, N-tert-butyl-phenylnitrone
PC, ischemic preconditioning
PEG-SOD, polyethylene glycol superoxide dismutase
PGE_2, prostaglandin

PGHS, prostaglandin endoperoxide H synthase
PGI$_2$, prostacycline synthase
PHGP, phospholipid hydroperoxide glutathione peroxidase
PIH, pyridoxal isonicotinoyl hydrazone
PKC, protein kinase C
PMA, phorbol
PMC, 2,2,5,7,8-pentamethylchroman-6-ol
PMN, polymorphonuclear leukocyte
PMS, phenazine methosulfate
PTIO, phenyl-4,4,5,5-tetramethylimidazoline-l-oxyl
p38 MAPK, p38 mitogen-activated protein kinase
p42/44 MAPK, p42/44 mitogen-activated protein kinase
RBCEC, rat brain capillary endothelial cells
RC, radical cation
SIN-1, 3-morpholinosydnonimine
SOD, superoxide dismutase
SOR, superoxide reductase
SOTs-1, di-(4-carboxybenzyl) hyponitrite
SSBs, single DNA strands
TBAP, 5,10,15,20-tetrakis-[4-carboxyphenyl] porphyrin
TBAR products, thiobarbituric acid reactive products
TBP, α-tocopherol-binding protein
TEMPO, 2,2,6,6-tetramethylpiperidinoxyl
TEMPOL, 4-hydroxy-2,2,6,6-tetramethylpiperidinoxyl
TEMPONE, 1-hydroxy-2,2,6,6-tetra-methyl-4-oxopiperidinoxyl
TEMPONEH, 1-hydroxy-2,2,6,6-tetra-methyl-4-oxopiperidine
TGF-β, transforming growth factor beta
Tiron, 1,2-dihydroxybenzene-3,5-sulfonate
TNF-α, tumor necrosis factor-α
TPA, tetradeconylphorbol acetate
TPO, thyroid peroxidase
TPO, 2,2,6,6-tetramethylpiperidinoxyl
Trx, thioredoxin
TTA, tetradecylthioacetic acid
TXB$_2$
U74006F, 21-[4-(2,6-di-l-pyrrolidinyl-4-pyrimidinyl)-1-piperazinyl]-16α-methylpregna-1,4,9(11)-tiene-3,20-dione monomethane sulfonate
U74500A, 21-[4-(3,6-bis(diethylamino)-2-pyridinyl)-1-piperazinyl]-16α-methylpregna-1,4,9(11)-triene-3,20-dione hydrochloride
VSMC, vascular smooth muscle cell
VVC, *v. vulnificus* cytolysin
WST-1, 4-[3-(4-iodophenyl)-2-(4-nitrophenyl)-2H-5-tetrazoliol]-1,3-benzene disulfonate sodium salt
XDH, xanthine dehydrogenase
XO, xanthine oxidase
XTT, sulfonated tetrazolium(2,3-bis(2-methoxy-4-nitro-5-sulfophenyl)-2-tetrazolium 5-carboxanilide
13ZE-Ch18:2-OOH, cholesteryl 13-hydroperoxy-9Z,11E-octadecadienoate

Part I

Chemistry and Kinetics of Organic Compounds Oxidation by Dioxygen

1 Molecular Products and Thermochemistry of Hydrocarbon Oxidation

1.1 EARLIER CONCEPS OF OXIDATION

Researchers encountered the phenomenon of the formation of an active intermediate during the oxidation of metals and organic compounds in the middle of the 19th century. Shonbein was the first to discover ozone formation in a slow oxidation of phosphine [1], diethyl ether, and ethyl alcohol [2]. He observed the oxidation of hydrogen iodide into diiodine and oxidation of indigo that accompanied the oxidation of Zn, SO_2, FeO, H_2S, AsH_3, SbH_3, and organic compounds [3–12]. Shonbein [13] put forward the hypothesis of participation of free charged oxygen atoms ("ozone" O^- and "antozone" O^+) in oxidation. He considered the oxygen of air as a monoatomic molecule, which can be in passive "O^o" and active (O^- and O^+) forms and proposed the following general scheme of oxidation:

$$A + 2O^o \longrightarrow A + O^- + O^+$$
$$A + O^- \longrightarrow AO \text{ (ozonide)}$$
$$H_2O + O^+ \longrightarrow H_2O_2 \text{ (antozonide)}$$

This scheme faced with a few contradictions. It was yet known at that time that an oxygen molecule consists of two oxygen atoms. The study of the ozone molecule showed that it includes more than two oxygen atoms. And, at last, Engler and Nasse [14] proved that gas "antozone" in Shonbein's experiment was nothing but vapors of the formed hydrogen peroxide. The proposed scheme with the participation of hypothetical polarized oxygen atoms ("ozone" and "antozone") in oxidation appeared false. However, its background, namely, the facts of active intermediate formation in oxidation were very important for the future development of the theory of oxidation.

Traube [15,16] performed the next important step in understanding of the oxidation mechanism. He studied the oxidation of metals in water and proposed hydrogen peroxide as the primary product of oxidation. Traube proposed the following scheme of metal oxidation:

$$Zn + 2H_2O + O_2 \longrightarrow Zn(OH)_2 + H_2O_2$$

The conception, proposed by Haber [17–19], was very close to Traube's hypothesis. Haber considered the formed hydrogen peroxide as an intermediate that can oxidize other substrate; for example, the induced oxidation of SO_2 during the oxidation of As_2O_3 was treated according to the following scheme:

$$As_2O_3 + 2O_2 + 2H_2 \longrightarrow As_2O_5 + 2H_2O_2$$
$$SO_2 + H_2O_2 \longrightarrow SO_3 + H_2O$$

Bach [20–22] and Engler [23–32] worked out the peroxide theory of oxidation. They performed many experiments on the oxidation of organic compounds (alcohols, aldehydes, fulvenes, etc.) by dioxygen and proved the formation of peroxide as the primary product of oxidation. Engler evidenced that peroxide was the product of oxidation in the absence of water (in contrast of Traube's scheme) and isolated peroxides among the products of oxidation of triethylphosphine, fulvenes, terpentine, and 3-hexene. The general scheme of substrate A oxidation in the peroxide theory had the following form:

$$A + O_2 \longrightarrow AO_2$$
$$AO_2 + A \longrightarrow 2AO$$

The phenomenon of chemical induction was intensively studied by Jorissen [33–37]. He discovered that indigo was not oxidized by dioxygen but was simultaneously oxidized in the presence of oxidized triethylphosphine or benzaldehyde. He measured the factor of chemical induction in these reactions as equal to unity. Later, he proved that the oxidation product of benzaldehyde, benzoic peracid, did not oxidize indigo under conditions of experiment. This shows that a very active intermediate was formed during the oxidation of benzaldehyde and that it was not perbenzoic acid. Engler assumed peroxide to be in two forms, namely, an active "moloxide" AO_2 and a more stable peroxide. A new correct interpretation of chemical induction in oxidation reactions was provided later by the chain theory of oxidation of organic compounds (see later).

Engler supposed that moloxide AO_2 was the primary product of the substrate A oxidation, which then reacts with acceptor B:

$$AO_2 + B \longrightarrow AO + BO$$

Water hydrolyzes the formed moloxide with the formation of hydrogen peroxide:

$$AO_2 + H_2O \longrightarrow AO + H_2O_2$$

Engler [38,39] proposed that peroxide produced by olefin oxidation has the structure of dioxetane:

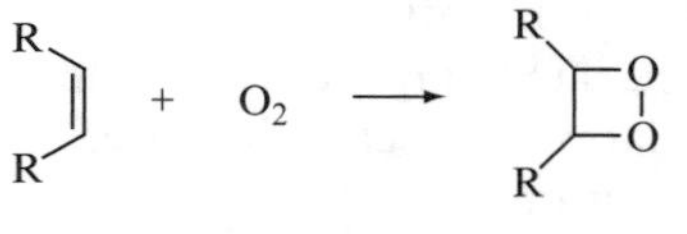

The same structure was proposed later by Hock and Schrader [40]. It became clear only in 1939 when Criegee et al. [41] proved that peroxide formed by cyclohexene oxidation has the structure of hydroperoxide. Later studies, performed by Farmer and Sutton [42], greatly extended the number of hydroperoxides as products of olefin oxidation. Beginning from the later part of the 20th century, the chain theory of organic compound oxidation became the theoretical ground for the experimental study in this field. The main events of the development of oxidation chemistry before the chain theory of oxidation are presented in Table 1.1.

TABLE 1.1
Main Discoveries and Concepts in the Field of Oxidation by Dioxygen, Which Appeared before the Chain Theory [43–46]

Year	Phenomena, Conception	Author
1768–1777	Discovery of dioxygen	K. W. Scheele
1774	Discovery of dioxygen	J. Priestley
1774–1777	Discovery of dioxygen	A. L. Lavoisier
1818	Discovery of hydrogen peroxide	L. J. Thenard
1835	Benzaldehyde was found to be oxidized by dioxygen to benzoic acid	J. Liebich
1858	Activation of oxygen in oxidation. Hypothetical scheme: $A + 2O^\circ \rightarrow A + O^+ + O^-$ $A + O^- \rightarrow AO, H_2O + O^+ \rightarrow H_2O_2$	C. F. Shonbein
1876	The first observation of hydrocarbon oxidation by dioxygen	C. Engler
1879	Paraffins were oxidized by dioxygen to carbon acids	C. Engler
1882	Hydrogen peroxide was isolated as the primary product of metal oxidation $Zn + 2H_2O + O_2 \rightarrow Zn(OH)_2 + H_2O_2$	W. Traube
1897	The peroxide theory of oxidation. General scheme: $A + O_2 \rightarrow AO_2, AO_2 + A \rightarrow 2AO$	A. N. Bach, C. Engler
1900	Hydrogen peroxide was supposed as an intermediate product of oxidation	F. Haber
1900	Perbenzoic acid was isolated as the product of benzaldehyde oxidation	A. Bayer, V. Villiger
1905	The theory of induced oxidation of organic and inorganic compounds was worked out.	N. A. Shilov
1912	First experiments on the autoxidation of toluene were performed; the benzoic acid was found as the product of oxidation	G. Ciamician, P. Silber
1925	Polymeric peroxide was synthesized by oxidation of 1,1-diphenylethylene $nO_2 + nPh_2CH=CH_2 \rightarrow (\sim Ph_2CCOO\sim)_n$	H. Staudinger.
1926	Methylbenzenes were oxidized, and substituted benzaldehydes were found among the products	H. N. Stephens
1928	Isolation of cyclohexene peroxide as the product of autoxidation	H. N. Stephens.
1932–1933	Tetralyl hydroperoxide was isolated as the product of tetralin autoxidation	M. Hartman, M. Seiberth, H. Hock, W. Susemihl
1936–1937	Peroxide formed in cyclohexene oxidation was proved to be the hydroperoxide	R. Criegee, H. Hock
1936	Oxidation of olefins was postulated as the reaction: $RCH=CHCH_2R^1 + O_2 \rightarrow RCH=CHCH(OOH)R^1$	A. Rieche
1942	Indanyl hydroperoxide was isolated as the product of indane oxidation	H. Hock and S. Lang
1943	Isolation of 1-phenylethyl hydroperoxide in the result of ethylbenzene oxidation	H. Hock and S. Lang
1944	Isolation of cumyl hydroperoxide	H. Hock and S. Lang
1950	Hydroperoxide $MeCH(OOH)CH_2Bu$ was isolated among the products of heptane autoxidation by dioxygen	K. I. Ivanov
1953	Decane was proved to be oxidized by all CH_2 groups with equal rates	J. L. Benton and M. M. Wirth

1.2 DEVELOPMENT OF THE CHAIN THEORY OF OXIDATION OF ORGANIC COMPOUNDS

The peroxide theory of Bach [20] and Engler [23] fixed the phenomenon of peroxide formation as the primary product of hydrocarbon oxidation by dioxygen. However, the problem of the mechanism of peroxide formation remained unsolved. The new stage of successful study of organic compound oxidation began after the discovery of free radicals as active intermediates of many chemical processes.

It was in the very beginning of the 20th century that Gomberg [47] synthesized the stable triphenylmethyl radical and observed its very fast reaction with dioxygen with production of peroxide. Bodenstein [48] was the first to discover the chain mechanism of reaction of dihydrogen with dichloride in 1913. From 1925 to 1928, Semenov [49] and Hinshelwood and Williamson [50] discovered the chain branching reactions. At the same time Backstrom [51] proved the chain mechanism of benzaldehyde photooxidation by dioxygen and later (in 1934) proposed the following mechanism for this reaction [52]:

$$PhC^{\bullet}(O) + O_2 \longrightarrow PhC(O)OO^{\bullet}$$

$$PhC(O)OO^{\bullet} + PhCH(O) \longrightarrow PhC(O)OOH + PhC^{\bullet}(O)$$

A similar mechanism of chain oxidation of olefinic hydrocarbons was observed experimentally by Bolland and Gee [53] in 1946 after a detailed study of the kinetics of the oxidation of nonsaturated compounds. Miller and Mayo [54] studied the oxidation of styrene and found that this reaction is *in essence* the chain copolymerization of styrene and dioxygen with production of polymeric peroxide. Rust [55] observed dihydroperoxide formation in his study of the oxidation of branched aliphatic hydrocarbons and treated this fact as the result of intramolecular isomerization of peroxyl radicals.

$$Me_2C^{\bullet}CH_2CHMe_2 + O_2 \longrightarrow Me_2C(OO^{\bullet})CH_2CHMe_2$$

$$Me_2C(OO^{\bullet})CH_2CHMe_2 \longrightarrow Me_2C(OOH)CH_2C^{\bullet}Me_2$$

$$Me_2C(OOH)CH_2C^{\bullet}Me_2 + O_2 \longrightarrow Me_2C(OOH)CH_2C(OO^{\bullet})Me_2$$

$$Me_2C(OOH)CH_2C(OO^{\bullet})Me_2 + RH \longrightarrow Me_2C(OOH)CH_2C(OOH)Me_2 + R^{\bullet}$$

The further study of the mechanism of hydrocarbon oxidation presented another important peculiarity of these reactions, namely, their self-accelerated character. The principal explanation of this phenomenon was suggested by Semenov [49]. He proposed that any active intermediate, formed as a result of hydrocarbon oxidation, slowly initiates new chains. The concentration of an intermediate (peroxide, aldehyde) increases during the initial stage of oxidation and this creates the increasing rate of chain initiation. This phenomenon is like chain branching in such chain branching reactions as $H_2 + O_2$ studied by Hinshelwood and Williamson [50] or combustion of phosphine vapors studied by Semenov [49]. The difference lies in the rate of chain branching. This rate is comparable with that of chain propagation for chain branching reactions ($H_2 + O_2$) and is much lower for chain reactions with degenerate chain branching (slow oxidation of hydrocarbons in the gas and the liquid phases). Later, during careful kinetic studies of a great variety of hydroperoxide reactions, free radical generation was discovered. In addition to the unimolecular decomposition of hydroperoxides with the splitting of the O—O bond, bimolecular decay of hydroperoxides exists: in reactions of hydroperoxide with the double bond of olefin, with hydrocarbons, alcohols, ketones, aldehydes, and acids (see Table 1.2).

TABLE 1.2
Chronological Table of the Main Concepts and Experimental Findings of Chain Theory of Hydrocarbon Oxidation

Year	Event	Author	Ref.
1900	Synthesis of the first stable free radical $Ph_3C^\bullet$	M. Gomberg	[47]
1913	Discovery of the chain mechanism of the reaction $H_2 + Cl_2$	M. Bodenstein	[48]
1925–1927	Semenov worked out the concept of chain branching reactions	N. N. Semenov	[49]
1927	Photochemically induced oxidation of benzaldehyde was proved to be the chain reaction ($\Phi \gg 1$)	H. Backstrom	[51]
1928–1930	Hinshelwood suggested the chain branching mechanism for reaction $H_2 + O_2$	C. Hinshelwood	[50]
1931	The formation of active hydroxyl radical in reaction Fe^{2+} with hydrogen peroxide was supposed: $Fe^{2+} + H_2O_2 \rightarrow Fe^{3+} + HO^\bullet + HO^-$	F. Haber and R. Willstatter	[61]
1932	The chain mechanism of hydrogen peroxide decay under catalytic action of transition metal ions was postulated	F. Haber and J. Weiss	[62]
1933	Radical $Ph_3C^\bullet$ initiates the oxidation of olefins and aldehydes	K. Ziegler	[63]
1934	Backstrom proposed the chain mechanism of benzaldehyde oxidation: $PhC^\bullet(O) + O_2 \rightarrow PhC(O)OO^\bullet$; $PhC(O)OO^\bullet + PhCH(O) \rightarrow PhC(O)OOH + PhC^\bullet(O)$	H. Backstrom	[52]
1934	Semenov put forward the concept of slow hydrocarbon oxidation as the chain reaction with degenerate branching	N. N. Semenov	[49]
1946	Bolland and Gee received the empirical equation for the rate of hydrocarbons oxidation: $v \sim [\text{Initiator}]^{1/2} \times [\text{RH}] \times F(pO_2)$ and proposed the kinetic scheme of chain mechanism of hydrocarbon oxidation	J. L. Bolland and G. Gee	[53]
1946	Experimental evidence of hydroxyl radicals generation in reaction $Fe^{2+} + H_2O_2$: $Fe^{2+} + H_2O_2 \rightarrow Fe^{3+} + HO^\bullet + HO^-$	J. H. Baxendale, M. G. Evans, and G. S. Park	[64]
1951	The reaction of chain generation: $RCHO + O_2 \rightarrow RC^\bullet(O) + HO_2^\bullet$ was kinetically evidenced	H. R. Cooper and H. W. Melville	[58]
1951	The cyclic mechanism of free radicals generation in oxidation of aldehydes catalyzed by transition metal ions was demonstrated	C. Bawn and J. Williamson	[65]
1951	Experimental evidence of alkoxyl radicals formation in the reaction of Fe^{2+} with hydroperoxide: $Fe^{2+} + ROOH \rightarrow Fe^{3+} + RO^\bullet + HO^-$	M. S. Kharash, F. S. Arimoto, and W. Nudenberg	[66]
1953	Bimolecular reaction of hydroperoxide decomposition to free radicals was discovered: $ROOH + ROOH \rightarrow RO^\bullet + H_2O + ROO^\bullet$	L. Bateman, H. Hughes, and A. Moris	[67]
1956	The reaction of chain generation $PhCH=CH_2 + O_2 \rightarrow$ free radicals was experimentally proved	A. A. Miller and F. R. Mayo	[54]
1956	Miller and Mayo studied the styrene oxidation and came to conclusion that this chain reaction occurs via addition of peroxyl radical to double bond of styrene with formation of polyperoxide as a product	A.A. Miller and F.R. Mayo	[54]
1957	Dihydroperoxides were found to be the primary products of branched alkanes oxidation. The intramolecular peroxyl radical reaction was proposed	F. F. Rust	[55]
1957	Russell proposed the mechanism of disproportionation of primary and secondary peroxyl radicals: $2R^1R^2CHOO^\bullet \rightarrow R^1R^2C(O) + O_2 + R^1R^2CHOH$	G. A. Russell	[68]

continued

TABLE 1.2
Chronological Table of the Main Concepts and Experimental Findings of Chain Theory of Hydrocarbon Oxidation—*continued*

Year	Event	Author	Ref.
1959	Chemiluminescence in the liquid-phase hydrocarbon oxidation was discovered. It was proved to be the result of secondary peroxyl radicals disproportionation	R. F. Vasi'ev, O. N. Karpukhin, and V. Ya Shlyapintokh	[69]
1960	Hydroperoxide reacts with hydrocarbon with free radical generation: $ROOH + RH \to RO^\bullet + H_2O + R^\bullet$	Z. K. Maizus, I. P. Skibida, and N. M. Emanuel	[70]
1960–1961	Trimolecular reaction of chain generation: $RH + O_2 + RH \to R^\bullet + H_2O_2 + R^\bullet$ was predicted and experimentally evidenced	E. T. Denisov	[59,60]
1961	Peroxyl radical of oxidized cumene was evidenced by EPR technique. The EPR spectrum was identified	Ya. S. Lebedev, V. F. Tsepalov, and V. Ya. Shlyapintokh	[71]
1962	Interaction of hydroperoxide with ketones with rapid chain generation was studied. The following mechanism: $ROOH + O{=}CR^1R^2 \Leftrightarrow$ Peroxide $\to$ Free radicals was proposed	E. T. Denisov	[72]
1964	Generation of free radicals by the reaction of hydroperoxide with alcohol was evidenced	E. T. Denisov	[73]
1964	Chain initiation by the following reaction of hydroperoxide with styrene was discovered: $PhCH{=}CH_2 + ROOH \to PhC^\bullet HCH_2OH + RO^\bullet$	E. T. Denisov and L. N. Denisova	[74]
1965	Chain generation by reaction ROOH with $R^1C(O)OH$ was observed	L. G. Privalova, Z. K. Maizus, and N. M. Emanuel	[75]
1967	Mineral acids were found to split hydroperoxides catalytically into free radicals: $2ROOH + H^+ \to RO^\bullet + ROO^\bullet + H_3O^+$	V. M. Solyanikov and E. T. Denisov	[76]
1967	The high rate of exchange reaction $RO_2^\bullet + R^1OOH \to ROOH + R^1O_2^\bullet$ was used to develop a special method for the measurement of rate constants of the reaction: $RO_2^\bullet + R_iH \to ROOH + R_i^\bullet$.	J. A. Howard, W. J. Schwalm, and K. U. Ingold	[77]
1968	The isotope effect was measured in the reaction of peroxyl radical with deuterated cumene ($PhMe_2CH$ and $PhMe_2CD$)	J. A. Howard, K. U. Ingold, and M. Symonds	[78]

One of the important problems of the chain oxidation of organic compounds was the problem of chain generation in the absence of hydroperoxide and other initiating agents. These reactions should be very slow due to their endothermicity. Two most probable reactions were predicted [56,57]:

$$RH + O_2 \longrightarrow R^\bullet + HO_2^\bullet$$
$$RCH{=}CH_2 + O_2 \longrightarrow RC^\bullet HCH_2OO^\bullet$$

Later, both reactions were proved experimentally by Cooper and Melville in 1951 [58] and Miller and Mayo in 1956 [54]. In addition, the trimolecular reaction $2\,RH + O_2$ was predicted in 1960 and experimentally proved in 1961 [59,60].

$$2RH + O_2 \longrightarrow 2R^\bullet + H_2O_2$$

The main scientific events in the science of chain oxidation of organic compounds are listed in Table 1.2.

1.3 HYDROPEROXIDES AS PRIMARY MOLECULAR PRODUCTS OF HYDROCARBON OXIDATION

1.3.1 Hydroperoxides

Hydroperoxides were proved to be the only primary molecular product of the oxidation of aliphatic and alkylaromatic hydrocarbons [79–84]. When the hydrocarbon is oxidized under mild conditions, in which the formed hydroperoxide is a stable product, the amount of produced ROOH was found to be nearly equal to the amount of consumed dioxygen [45,80,82].

RH	Ethyl linoleate	3-Heptene, 2,6-dimethyl-	Benzene, 1,4-diisopropyl-	Cumene
T (K)	328	333	363	363
$\Delta[O_2]$ (mol%)	1.2	0.97	3.5	2.6
[ROOH] (mol%)	98	99	99	98

The yield of the formed hydroperoxide depends on the structure of the oxidized hydrocarbon. The tertiary hydroperoxides appeared to be the most stable. Hence they can be received by hydrocarbon oxidation in high yield (see Table 1.3).

A very serious problem was to clear up the formation of hydroperoxides as the primary product of the oxidation of a linear aliphatic hydrocarbon. Paraffins can be oxidized by dioxygen at an elevated temperature (more than 400 K). In addition, the formed secondary hydroperoxides are easily decomposed. As a result, the products of hydroperoxide decomposition are formed at low conversion of hydrocarbon. The question of the role of hydroperoxide among the products of hydrocarbon oxidation has been specially studied on the basis of decane oxidation [82]. The kinetics of the formation of hydroperoxide and other products of oxidation in oxidized decane at 413 K was studied. In addition, the kinetics of hydroperoxide decomposition in the oxidized decane was also studied. The comparison of the rates of hydroperoxide decomposition and formation other products (alcohol, ketones, and acids) proved that practically all these products were formed due to hydroperoxide decomposition. Small amounts of alcohols and ketones were found to be formed in parallel with ROOH. Their formation was explained on the basis of the disproportionation of peroxide radicals in parallel with the reaction $RO_2^\bullet + RH$.

The oxidation of a hydrocarbon to hydroperoxide is an exothermic reaction. The values of the enthalpies of these reactions are collected in Table 1.4. The enthalpies of oxidation of different hydrocarbon groups by dioxygen to ROOH have the following values:

Group	—CH_2OOH	>CHOOH	≥COOH
ΔH (kJ mol^{-1})	-73.8 ± 3.4	-89.6 ± 2.4	-108.4
$\Delta\Delta H$ (kJ mol^{-1})	0	-15.8	-34.6

TABLE 1.3
Synthesis of ROOH by RH Oxidation in the Liquid Phase ($pO_2 = 10^5$ Pa)

Hydroperoxide	T (K)	Procedure of Oxidation	Yield, (mol %)	Ref.
Me_3COOH	398	Autoxidation	27.6	[84]
$Me_2C(OOH)(CH_2)_2CHMe_2$	390	Autoxidation	3.8	[55]
$Me_2C(OOH)(CH_2)_2C(OOH)Me_2$	390	Autoxidation	1.9	[55]
$Me_2C(OOH)(CH_2)_4CHMe_2$	352	Photooxidation		[85]
$Me_2C(OOH)CH(OOH)(CH_2)_3CHMe_2$	352	Photooxidation		[85]
cyclohexyl-O-O-H	428	Autoxidation	6.0	[86]
1-methylcyclohexyl-O-O-H	353	Photooxidation	5.0	[87]
cyclooctyl-OOH	423	Autoxidation	18.0	[88]
4-isopropylcyclohexyl-O-O-H	383	Autoxidation	3.0	[89]
decalinyl-O-O-H	343	Photooxidation	1.5	[90]
$CH_2=CHCH(OOH)Pr$	313	Autoxidation	2.0	[91]
$MeCH=CHCH(OOH)Pr$	338	Autoxidation	4.1	[92]
$EtCH=CHCH(OOH)Et$	333	Autoxidation	4.0	[92]
$CH_2=CHCH(OOH)CH_2Bu$	348	Autoxidation	1.35	[92]
$CH_2=CMeCH(OOH)CH_2Bu$	348	Autoxidation	0.55	[92]
cyclopentenyl-O-O-H	296	Photooxidation		[93]
cyclohexenyl-O-O-H	328	Autoxidation	3.5	[92]
cyclohexenyl-O-O-H	303	Autoxidation	20.0	[41,94]
methylcyclohexenyl-O-O-H	308	Autoxidation	21.0	[95,96]
cyclopentenyl-O-O-H derivative	333	Autoxidation	8.0	[97]

TABLE 1.3
Synthesis of ROOH by RH Oxidation in the Liquid Phase ($pO_2 = 10^5$ Pa)—continued

Hydroperoxide	T (K)	Procedure of Oxidation	Yield, (mol %)	Ref.
(methylcyclohexenyl hydroperoxide)	323	Photooxidation		[98]
PhCH$_2$OOH	363	Autoxidation	0.03	[99]
(4-methylbenzyl hydroperoxide)	333	Photooxidation		[100]
PhCH(OOH)CH$_3$	363	Autoxidation	0.59	[99]
PhMe$_2$COOH	363	Autoxidation	28.5	[101]
PhMe$_2$COOH	363	Autoxidation	2.5	[99]
PhCH(OOH)Pr	353	Photooxidation	1.5	[45]
Ph$_2$CHOOH	338	Photooxidation		[102]
Ph$_2$MeCOOH	338	Autoxidation	18.5	[103]
1-phenylcyclopentyl hydroperoxide	363	Autoxidation	1.9	[99]
1-phenylcyclohexyl hydroperoxide	388	Autoxidation	8.3	[104]
1,4-bis(2-hydroperoxypropan-2-yl)benzene derivative	363	Autoxidation	24.0	[99]
indanyl hydroperoxide	363	Autoxidation	3.1	[99]
tetralinyl hydroperoxide	363	Autoxidation	7.8	[99]

1.3.2 Dihydroperoxides

Rust [55] studied the oxidation of branched alkanes and was the first to observe the formation of dihydroperoxides as primary products of the hydrocarbon oxidation [55]. Dihydroperoxide was found to be the main product of 2,4-dimethylpentane oxidation by dioxygen at 388 K:

$$Me_2CHCH_2CH_2CHMe_2 + 2O_2 \longrightarrow Me_2C(OOH)CH_2CH_2C(OOH)Me_2$$

The yield of dihydroperoxide depends on the common position of two tertiary C—H bonds [55].

TABLE 1.4
Comparison of the Formation Enthalpies ΔH_f^0 of Hydroperoxides and Hydrocarbons [105,106]

Hydroperoxide	$-\Delta H_f^0(RH)$ (kJ mol^{-1})	$-\Delta H_f^0(RO_2H)$ (kJ mol^{-1})	$\Delta H_f^0(RH)-\Delta H_f^0(RO_2H)$ (kJ mol^{-1})	$S^0(ROOH)$ (J mol^{-1} K^{-1})
CH$_3$OOH	74.5	130.5	56.0	282.4
MeCH$_2$OOH	84.1	169.4	85.3	
Me$_2$CHOOH	104.6	198.3	93.7	
PrCH$_2$OOH	126.4	205.0	78.6	393.8
Me$_3$COOH	134.3	242.7	108.4	360.6
BuCH$_2$OOH	146.4	217.6	71.2	433.2
PrMeCHOOH	146.4	234.3	87.9	430.7
Bu(CH$_2$)$_2$OOH	166.9	238.5	71.6	472.6
BuMeCHOOH	167.2	255.2	88.0	470.1
BuCH$_2$CHMeOOH	187.4	276.1	88.7	509.5
cyclohexyl-OOH	123.1	229.9	106.8	
1-methylcyclohexyl-OOH	154.8	263.3	108.5	
decalinyl-OOH	182.3	277.8	95.5	
PhMe$_2$COOH	−3.9	85.4	89.3	

Hydrocarbon	Pentane, 2,3-dimethyl-	Pentane, 2,4-dimethyl-	Hexane, 2,5-dimethyl-
[ROOH]/[O$_2$] (mol%)	52	89	83
[di-ROOH]/[all ROOH] (%)	16	95	60

The formation of dihydroperoxides as the primary products of hydrocarbon oxidation is the result of peroxyl radical isomerization (see Chapter 2).

1.3.3 CYCLIC AND POLYMERIC PEROXIDES

Hydrocarbons with conjugated double bonds are oxidized with the formation of cyclic peroxides [46,80,82], for example:

In parallel with cyclic peroxides polymeric peroxides are formed:

$$\text{cyclopentadiene} + O_2 \longrightarrow (-\text{cyclopentene-O-O-})_n$$

Monomers, such as styrene or methyl methacrylate, are oxidized to oligomeric peroxides also [107]:

$$n\,PhCH{=}CH_2 + nO_2 \longrightarrow (-CHPhCH_2O_2-)_n$$

Olefins with α-C—H bond, which is easily attacked by the peroxyl radical, are oxidized in parallel to hydroperoxide and oligomeric peroxide [108,109]. The competition between these two routes of oxidation was studied by Hargrave and Morris [92] (333–348 K, conversion 0.1–0.2 wt.%):

Hydrocarbon	Hexadecene, 3, 7, 11, 15-tetramethyl-	2-Octene, 2,6-dimethyl-	2-Heptene	1-Nonene, 2-methyl-
ROOH (mol%)	50	68	79	42

The mechanism of olefin oxidation is discussed in Chapter 2.

1.3.4 Epoxides

Epoxide is formed side by side with oligomeric peroxide during monomer oxidation. Miller and Mayo [54] assumed the following mechanism of decomposition of formed radicals:

$$\sim CH_2C^\bullet HPh + O_2 \longrightarrow \sim CH_2CH(OO^\bullet)Ph$$
$$\sim CH_2CHPhOO^\bullet + CH_2{=}CHPh \longrightarrow \sim CH_2CHPhOOCH_2C^\bullet HPh$$
$$\sim CH_2CHPhOOCH_2C^\bullet HPh \longrightarrow \sim CH_2CHPhO^\bullet + \underset{}{\triangle}Ph$$

The yield of epoxide strongly depends on the substituent in benzene ring of styrene. The values of the ratio of the formation rates of epoxide and polyperoxide during styrene oxidation at 368 K are given below [110]:

p-Substituent	MeO	Me$_3$C	H	Br	Cl	CF$_3$
v(oxide)/v(ROOR)	0.11	0.56	0.68	0.48	1.19	0.38

1.4 PRODUCTS OF HYDROPEROXIDE DECOMPOSITION

1.4.1 Hydroperoxides as the Intermediates of Hydrocarbon Oxidation

Under the action of heat and free radicals, hydroperoxides are decomposed into alcohols and carbonyl compounds. The primary hydroperoxide RCH_2OOH is an unstable molecule and is decomposed into aldehyde, acid, and dihydrogen through the interaction with formed aldehyde [111].

$$RCH_2OOH + RCH(O) \rightleftharpoons RCH_2OOCH(OH)R$$
$$RCH_2OOCH(OH)R \longrightarrow RCH(O) + H_2 + RC(O)OH$$

The formed aldehyde plays the role of an active intermediate and, therefore, the decomposition of the hydroperoxide occurs autocatalytically. Ester is formed in parallel, apparently, by the ionic reaction.

$$RCH_2OOCH(OH)R + AH \rightleftharpoons RCH_2OOCH(OH_2^+)R + A^-$$
$$RCH_2OOCH(OH_2^+)R \longrightarrow RCH_2OOC^+HR + H_2O$$
$$RCH_2OOC^+HR + A^- \longrightarrow RCH_2OC(O)R + HA$$

The hydroperoxide group weakens the α-C—H bonds, and the peroxyl radical of the oxidized hydrocarbon attacks this group with aldehyde formation.

$$RO_2^{\bullet} + RCH_2OOH \longrightarrow ROOH + RCH(O) + HO^{\bullet}$$

As a result, the primary hydroperoxide is decomposed into aldehyde, carbonic acid, ester, and dihydrogen.

Secondary hydroperoxides are decomposed into alcohols and ketones (scheme of Langenbeck and Pritzkow [112–114]):

$$R^1R^2CHOOH = R^1R^2CHOH + 0.5O_2$$
$$R^1R^2CHOOH = R^1R^2C(O) + H_2O$$

1.4.2 ALCOHOLS

Alcohols are formed as a result of reactions with the homolytic splitting of the O—O bond of hydroperoxides, for example (see Chapter 4):

$$ROOH \longrightarrow RO^{\bullet} + HO^{\bullet}$$
$$ROOH + HY \longrightarrow RO^{\bullet} + H_2O + Y^{\bullet}$$
$$ROOH + CH_2{=}CHR \longrightarrow RO^{\bullet} + HOCH_2C^{\bullet}HR$$
$$RO^{\bullet} + RH \longrightarrow ROH + R^{\bullet}$$

In parallel, they are formed by the disproportionation of peroxyl radicals [68].

$$R^1R^2CHO_2^{\bullet} + R^1R^2CHO_2^{\bullet} \longrightarrow R^1R^2CHOH + O_2 + R^1R^2C(O)$$

In addition, hydroperoxides are hydrolyzed under the catalytic action of acid formed in the oxidized hydrocarbon [46,83].

$$ROOH + HA \rightleftharpoons ROOH_2^+ + A^-$$
$$ROOH_2^+ + H_2O \longrightarrow ROH + H_2O_2 + H^+$$

The hydrocarbon with a tertiary C—H bond is oxidized to stable tertiary hydroperoxide. This hydroperoxide is decomposed homolytically with the formation of alcohol [82]:

$$R^1R^2R^3COOH \longrightarrow R^1R^2R^3CO^{\bullet} + HO^{\bullet}$$
$$R^1R^2R^3CO^{\bullet} + RH \longrightarrow R^1R^2R^3COH + R^{\bullet}$$

TABLE 1.5
Comparison of the Formation Enthalpies ΔH_f^0 of Hydrocarbons and Alcohols [105]

Alcohol (ROH)	$-\Delta H_f^0$(RH) (kJ mol^{-1})	$-\Delta H_f^0$(ROH) (kJ mol^{-1})	ΔH_f^0(RH)$-\Delta H_f^0$(ROH) (kJ mol^{-1})
CH$_3$OH	74.5	201.7	127.2
MeCH$_2$OH	84.1	234.7	150.6
EtCH$_2$OH	104.6	254.8	150.2
Me$_2$CHOH	104.6	272.4	167.8
PrCH$_2$OH	126.4	274.9	148.5
Me$_3$COH	134.3	312.5	178.2
BuCH$_2$OH	146.4	296.4	150.2
PrMeCHOH	146.4	313.8	167.4
Et$_2$CHOH	146.4	315.5	169.7
cyclohexyl-OH (secondary)	123.4	290.0	166.6
cyclohexyl-OH (tertiary, methylated)	154.8	359.8	205.0
CH$_2$=CHCH$_2$OH	−20.1	125.5	145.6
PhCH$_2$OH	−50.2	100.4	150.6
PhMeCHOH	−29.3	138.2	167.5
PhMe$_2$COH	−3.9	191.4	195.3

Along with this reaction, the alkoxyl radicals are formed by the recombination of the tertiary peroxyl radical (see Chapter 2).

$$R^1R^2R^3COO^\bullet + R^1R^2R^3COO^\bullet \longrightarrow 2R^1R^2R^3CO^\bullet + O_2$$

Under the catalytic action of acid, tertiary hydroperoxide is hydrolyzed to alcohol and hydrogen peroxide [46,83].

The oxidation of the hydrocarbon to alcohol is an exothermic reaction (see Table 1.5). The heat of oxidation depends on the structure of the oxidized group.

Group	—CH$_2$OH	>CHOH	>COH
ΔH (kJ mol^{-1})	−149.9±0.8	−167.6±0.2	−178.2
$\Delta\Delta H$ (kJ mol^{-1})	0	−17.7	−28.3

1.4.3 KETONES

Ketones are formed in oxidized hydrocarbons by the following ways.

1. Secondary hydroperoxides are attacked by peroxyl radicals followed by the splitting of the O—O bond [82].

$$RO_2^{\bullet} + R^1R^2CHOOH \longrightarrow ROOH + R^1R^2C^{\bullet}OOH$$
$$R^1R^2C^{\bullet}OOH \longrightarrow R_1R_2C(O) + HO^{\bullet}$$

2. Acids catalyze the decomposition of secondary hydroperoxide with the formation of carbonyl compounds [46,83].

$$R^1R^2CHOOH + HA \rightleftharpoons R^1R^2CHOOH_2^+ + A^-$$
$$R^1R^2CHOOH_2^+ \longrightarrow R^1R^2C(O) + H_3O^+$$

3. Acid catalyzes the decomposition of tertiary α-arylhydroperoxide and gives phenol and ketone [46,84].

$$PhR^1R^2COOH + HA \rightleftharpoons PhR^1R^2COOH_2^+ + A^-$$
$$PhR^1R^2COOH_2^+ + H_2O \longrightarrow PhOH + R^1R^2C(O) + H_3O^+$$

4. Tertiary hydroperoxide is decomposed to alkoxyl and peroxyl radicals, for example [67]:

$$2R^1R^2R^3COOH \longrightarrow R^1R^2R^3CO^{\bullet} + H_2O + R^1R^2R^3COO^{\bullet}$$

In turn, the formed alkoxyl radical splits into ketone and the alkyl radical.

$$R^1R^2R^3CO^{\bullet} \longrightarrow R^1R^2C(O) + R_3^{\bullet}$$

Tertiary alkoxyl radicals are formed in the oxidized hydrocarbon by peroxyl radical recombination also (see earlier).

The values of the formation enthalpies of aldehydes, ketones, and parent hydrocarbons are presented in Table 1.6. The last column contains the values of enthalpies of the reactions

$$RCH_3 + O_2 = RCH(O) + H_2O$$

and

$$R^1CH_2R^2 + O_2 = R^1C(O)R^2 + H_2O$$

The mean ΔH value for the first reaction is $\Delta H = -325.3 \pm 2.4 \text{ kJ mol}^{-1}$ and for the second one $\Delta H = -354.8 \pm 1.5 \text{ kJ mol}^{-1}$.

1.4.4 ACIDS

Carboxylic acids are the products of the oxidation of aldehydes (see Chapter 6).

$$RC(O)H + O_2 = RC(O)OOH$$
$$RC(O)H + RC(O)OOH = 2RC(O)OH$$

Ketone is oxidized to α-ketohydroperoxide; and the latter is decomposed into acid and aldehyde according to the stoichiometric equations [112,114]

$$R^1C(O)CH_2R^2 + O_2 = R^1C(O)CH(OOH)R^2$$

$$R^1C(O)CH(OOH)R^2 = R^1C(O)OH + R^2CH(O)$$

Other mechanisms of ketone oxidation are also known and will be discussed in Chapter 8. Peracid, which is formed from aldehyde, oxidizes ketones with lactone formation (Bayer–Villiger reaction).

$$RC(O)OOH + R^1C(O)CH_2R^2 \longrightarrow RC(O)OH + R^1C(O)OCH_2R^2$$

$$R^1C(O)OCH_2R^2 + H_2O \longrightarrow R^1C(O)OH + R^2CH_2OH$$

The formation enthalpies of a few acids and parent hydrocarbons are given in Table 1.6. The oxidation of the methyl group of hydrocarbon to carboxyl group is a highly exothermic reaction. The enthalpy of the reaction

TABLE 1.6
Comparison of Formation Enthalpies ΔH_f^0 of Aldehydes, Ketones, Acids, and Parent Hydrocarbons [105]

RCH(O)	$-\Delta H_f^0$ (RMe) (kJ mol^{-1})	$-\Delta H_f^0$(RCHO) (kJ mol^{-1})	ΔH_f^0(RMe) $-\Delta H_f^0$(RCHO) (kJ mol^{-1})	$-\Delta H$(RMe + O$_2$) (kJ mol^{-1})
H$_2$C(O)	74.5	108.8	34.3	276.1
MeCH(O)	84.1	165.7	81.6	323.4
EtCH(O)	104.6	187.4	82.8	324.6
PrCH(O)	126.4	207.5	81.1	322.9
BuCH(O)	146.4	230.5	84.1	325.9
CH$_2$=CHCHO	−20.1	75.3	95.4	337.2
PhCH(O)	−50.2	37.7	87.9	329.7

R^{1}C(O)R^2	$-\Delta H_f^0$(RH) (kJ mol^{-1})	$-\Delta H_f^0$(R^{1}COR^2) (kJ mol^{-1})	ΔH_f^0(RH) $-\Delta H_f^0$(R^{1}COR^2) (kJ mol^{-1})	$-\Delta H$(RH + O$_2$) (kJ mol^{-1})
MeC(O)Me	104.6	217.1	112.5	354.3
MeC(O)Et	126.4	238.4	112.0	353.8
MeC(O)Pr	146.4	259.0	112.6	354.4
EtC(O)Et	146.4	258.2	111.8	353.6
PhCH$_2$Me	−29.3	86.6	115.9	357.7

RC(O)OH	$-\Delta H_f^0$ (RMe) (kJ mol^{-1})	$-\Delta H_f^0$(RCOOH) (kJ mol^{-1})	ΔH_f^0(RMe) $-\Delta H_f^0$(RCOOH) (kJ mol^{-1})	$-\Delta H$(RMe + 1.5O$_2$) (kJ mol^{-1})
HC(O)OH	74.5	378.7	304.2	546.0
CH$_3$C(O)OH	84.1	432.2	348.1	589.9
EtC(O)OH	104.6	447.7	343.1	582.9
PrC(O)OH	126.4	472.8	346.4	588.2
CH$_2$=CHCO$_2$H	−20.1	336.2	356.3	598.1
PhC(O)OH	−50.2	294.1	344.3	586.1

TABLE 1.7
The Thermochemical Scale of the Oxidation of Aliphatic Hydrocarbons

Group	ΔH (kJ mol^{-1})	Group	ΔH (kJ mol^{-1})	Group	ΔH (kJ mol^{-1})
—CH$_3$	0.0	>CH$_2$	0.0	>CH	0.0
—CH$_2$OOH	−74				
		>CHOOH	−90		
				>COOH	−108
—CH$_2$OH	−150				
		>CHOH	−168		
				>COH	−178
—CH(O) + H$_2$O	−325				
		>C=O + H$_2$O	−355		
—COOH + H$_2$O	−587				

$$RCH_3 + 1.5O_2 = RC(O)OH + H_2O$$

is $\Delta H = -586.8 \pm 2.6$ kJ mol^{-1}. The general stoichiometric scheme of the oxidation of aliphatic and alkylaromatic hydrocarbons includes the following stages:

$$\text{Hydrocarbon} + O_2 \longrightarrow \text{Hydroperoxide}$$
$$sec\text{-Hydroperoxide} \longrightarrow \text{Alcohol} + \text{ketone}$$
$$prim\text{-Hydroperoxide} \longrightarrow \text{Aldehyde} + \text{acid}$$
$$\text{Alcohol} + 0.5O_2 \longrightarrow \text{Ketone}$$
$$\text{Ketone} \longrightarrow \text{Acid} + \text{aldehyde}$$
$$\text{Aldehyde} \longrightarrow \text{Peracid}$$
$$\text{Peracid} + \text{ketone} \longrightarrow \text{Acid} + \text{lactone}$$
$$\text{Peracid} + \text{aldehyde} \longrightarrow 2\text{Acid}$$

Detailed information about molecular products of hydrocarbon oxidation is given in monographs [45,46,80,82]. The kinetic schemes of the oxidation of alcohols, ketones, aldehydes, and acids are discussed in Chapters 7,8. The thermochemical scale of hydrocarbon oxidation is given in Table 1.7.

REFERENCES

1. CF Shonbein. *Ber Naturf Ges Basel* 6:16, 1844.
2. CF Shonbein. *Ber Naturf Ges Basel* 7:4, 1845.
3. CF Shonbein. *J Prakt Chem* 56:354, 1852.
4. CF Shonbein. *J Prakt Chem* 93:25, 1864.
5. CF Shonbein. *J Prakt Chem* 53:65, 1851.
6. CF Shonbein. *J Prakt Chem* 53:72, 1851.
7. CF Shonbein. *J Prakt Chem* 53:321, 1851.
8. CF Shonbein. *J Prakt Chem* 74:328, 1858.
9. CF Shonbein. *J Prakt Chem* 54:75, 1851.
10. CF Shonbein. *J Prakt Chem* 55:11, 1852.
11. CF Shonbein. *J Prakt Chem* 84:406, 1861.

12. CF Shonbein. *J Prakt Chem* 105:226, 1868.
13. CF Shonbein. *Liebich Ann* 108:157, 1858.
14. C Engler, H Nasse. *Liebich Ann* 154:215, 1870.
15. W Traube. *Berichte* 15:2421, 1882.
16. W Traube. *Berichte* 16:123, 1883.
17. F Haber. *Z Phys Chem* 34:513, 1900.
18. F Haber. *Z Phys Chem* 35:81, 1900.
19. F Haber. *Z Phys Chem* 35:608, 1900.
20. AN Bach. *Compt Rend* 124:2, 1897.
21. AN Bach. *Compt Rend* 124:951, 1897.
22. AN Bach. *Zh Russian Fiz–Khim Obschestva* 44: applic. 1:79, 1897.
23. C Engler, E Wild. *Berichte* 30:1669, 1897.
24. C Engler, J Weissberg. *Berichte* 31:3046, 1898.
25. C Engler, J Weissberg. *Berichte* 31:3055, 1898.
26. C Engler, J Weissberg. *Berichte* 33:1090, 1900.
27. C Engler, J Weissberg. *Berichte* 33:1097, 1900.
28. C Engler, J Weissberg. *Berichte* 33:1109, 1900.
29. C Engler, C Frankenstein. *Berichte* 34:293, 1901.
30. C Engler, C Frankenstein. *Berichte* 36:2642, 1903.
31. C Engler, C Frankenstein. *Berichte* 36:2642, 1903.
32. C Engler, C Frankenstein. *Berichte* 37:49, 1904.
33. WP Jorissen. *Z Phys Chem* 21:1707, 1896.
34. WP Jorissen. *Berichte* 29:1708, 1896.
35. WP Jorissen. *Z Phys Chem* 22:34, 1897.
36. WP Jorissen. *Z Phys Chem* 22:44, 1897.
37. WP Jorissen. *Induced Oxidation*. New York: Elsevier, 1959.
38. C Engler. *Berichte* 33:1090, 1900.
39. C Engler. *Berichte* 34:2933, 1901.
40. H Hock, O Schrader. *Naturwiss* 24:159, 1936.
41. R Criegee, H Pilz, H Flygare. *Berichte* 72:1799, 1939.
42. EH Farmer, DA Sutton. *J Chem Soc* 10, 1966.
43. JR Partington. *A History of Chemistry*. London: Macmillan, 1961.
44. NA Shilov. *About Reactions of Cooxidation*. Moscow: Tipografiya Mamontova, 1905 [in Russian].
45. KI Ivanov. *Intermediate Products and Reactions of Hydrocarbon Autoxidation*. Moscow: GNTI NGTL, 1949 [in Russian].
46. EGE Hawkins. *Organic Peroxides*. London: Spon, 1961.
47. M Gomberg. *Berichte* 33:3150, 1900.
48. M Bodenstein. *Z Phys Chem* 85:329, 1913.
49. NN Semenov. *Chemical Kinetics and Chain Reactions*. London: Oxford University Press, 1935.
50. C Hinshelwood, A Williamson. *The Reaction between Hydrogen and Oxygen*. Oxford: Clarendon Press, 1934.
51. H Backstrom. *J Am Chem Soc* 49:1460, 1927.
52. H Backstrom. *Z Phys Chem* 25B:99, 1934.
53. JL Bolland, G Gee. *Trans Faraday Soc* 42:236, 1946.
54. AA Miller, FR Mayo. *J Am Chem Soc* 78:1017, 1956.
55. FF Rust. *J Am Chem Soc* 79:4000, 1957.
56. EH Farmer. *Trans Faraday Soc* 42:228, 1946.
57. JL Bolland, G Gee. *Trans Faraday Soc* 42:236, 1946.
58. HR Cooper, HW Melville. *J Chem Soc* 1984, 1951.
59. ET Denisov. *Dokl AN SSSR* 130:1055, 1960.
60. ET Denisov. *Dokl AN SSSR* 141:131, 1961.
61. F Haber, R Willstatter. *Berichte* 64B:2844, 1931.
62. F Haber, J Weiss. *Naturwiss* 20:948, 1932.
63. K Ziegler. *Annalen* 504:131, 1933.

64. JH Baxendale, MG Evans, GS Park. *Trans Faraday Soc* 42:155, 1946.
65. C Bawn, JW Williamson. *Trans Faraday Soc* 47:735, 1951.
66. MS Kharash, FS Arimoto, W Nudenberg. *J Org Chem* 16:1556, 1951.
67. L Bateman, H Hughes, A Moris. *Disc Faraday Soc* 14:190, 1953.
68. GA Russell. *J Am Chem Soc* 79:3871, 1957.
69. RF Vasil'ev, ON Karpukhin, V Ya Shlyapintokh. *Dokl AN SSSR* 125:106, 1959.
70. ZK Maizus, IP Skibida, NM Emanuel. *Dokl AN SSSR* 131:880, 1960.
71. YaS Lebedev, VF Tsepalov, VYa Shyapintokh. *Dokl AN SSSR* 139:1409, 1961.
72. ET Denisov. *Dokl AN SSSR* 146:394, 1962.
73. ET Denisov. *Zh Fiz Khim* 38:2085, 1964.
74. ET Denisov, LN Denisova. *Dokl AN SSSR* 157:907, 1964.
75. LG Privalova, ZK Maizus, NM Emanuel. *Dokl AN SSSR* 161:1135, 1965.
76. VM Solyanikov, ET Denisov. *Dokl AN SSSR* 173:1106, 1967.
77. JA Howard, WJ Schwalm, KU Ingold. *Adv Chem Ser* 75:6, 1968.
78. JA Howard, KU Ingold, M Symonds. *Can J Chem* 46:1017, 1968.
79. A Rieche. Die Bedeutung der organische Peroxiden fur die Chemische Wissenschaft und Technik. Stuttgart, 1936.
80. D Swern. In: WO Lundberg (ed.). *Autoxidation and Antioxidants*, vol. 1. New York: Interscience, 1961, pp. 1–54.
81. IV Berezin, ET Denisov, NM Emanuel. *The Oxidation of Cyclohexane*. Oxford: Pergamon Press, 1966.
82. NM Emanuel, ET Denisov, ZK Maizus. *Liquid-Phase Oxidation of Hydrocarbons*. New York: Plenum Press, 1967.
83. S Patai. *The Chemistry Peroxides*. New York: Wiley, 1983.
84. VL Antonovskii. *Organic Peroxide Initiators*. Moscow: Khimiya, 1972 [in Russian].
85. KI Ivanov, VK Savinova, VP Zhakhovskaya. *Dokl AN SSSR* 59:703,1948.
86. IV Berezin, BG Dzantiev, NF Kazanskaya, LN Sinochkina, NM Emanuel. *Zh Fiz Khim* 31:554, 1957.
87. KI Ivanov, VK Savinova. *Dokl AN SSSR* 59:493, 1948.
88. VG Bykovchenko. Kinetics and Mechanism of Cyclododecane Oxidation Directed to Hydroperoxide and Cyclododecanone. Ph.D. thesis, Moscow State University, Moscow, 1962, pp. 3–11 [in Russian].
89. GS Fisher, JS Stinson. *Ind Eng Chem* 47:1368, 1955.
90. KI Ivanov, VK Savinova. *Dokl AN SSSR* 48:32, 1945.
91. H Hock, A Neuwitz. *Berichte* 72:1562, 1939.
92. KR Hargrave, AL Morris. *J Chem Soc* 89, 1956.
93. H Hock, O Schrader. *Brenst Chem* 18:6, 1937.
94. H Hock, O Schrader. *Naturwiss* 24:159, 1936.
95. EH Farmer, A Sundralingam. *J Chem Soc* 121, 1942.
96. R Criegee, H Pilz, H Flygare. *Berichte* 72:1799, 1939.
97. AI Chirko. *Zh Org Khim* 1:1984, 1965.
98. H Hock, S Lang. *Berichte* 75:300, 1942.
99. GA Russel. *J Am Chem Soc* 78:1047, 1956.
100. H Hock, S Lang. *Berichte* 76:169, 1943.
101. BA Redoshkin, VA Shushunov. Trudy Khimii i khimicheskoy Tekhnologii, Gorkii, N1:157, 1961.
102. H Hock, S Lang. *Berichte* 77:257, 1944.
103. TI Yurzhenko, DK Tolopko, VA Puchin. In: SP Sergienko (ed.). *Problems of Hydrocarbon Oxidation*. Moscow: Izdatelsvo Akad Nauk SSSR 1954, p. 111 [in Russian].
104. G Caprara, G Lemetre. *Chim E Ind* (Milan) 42:974, 1960.
105. SG Lias, JF Liebman, RD Levin, SA Kafafi. NIST Standard Reference Database, 19A, NIST Positive Ion Energetics, Version 2.0. Gaithersburg: NIST, 1993.
106. SW Benson, HE O'Neal. *Kinetic Data on Gas Phase Unimolecular Reactions*. Washington: NSRDS, 1970.

107. MM Mogilevich, EM Pliss. *Oxidation and Oxidative Polymerization of Nonsaturated Compounds.* Moscow: Khimiya, 1990 [in Russian].
108. T Mill, DG Hendry. In: CH Bamford and CFH Tipper (eds.). *Comprehensive Chemical Kinetics,* vol. 16. Amsterdam: Elsevier, 1980, pp. 1–87.
109. W Sun, K Shiomori, Y. Kawano, Y. Hatate. Kagaku Kogaku Ronbunshu, 26:869, 2000.
110. W Suprun. *J Prakt Chem* 338:231, 1996.
111. CF. Wurster, LJ Durham, HS Mosher. *J Am Chem Soc* 80:327, 1958.
112. W Langenbeck, W Pritzkow. *Chem Tech* 2:116, 1950.
113. W Pritzkow. *Berichte* 87:1668, 1954.
114. W Pritzkow. *Berichte* 88:575, 1955.

2 Chain Mechanism of Liquid-Phase Oxidation of Hydrocarbons

2.1 THE PECULARITIES OF CHAIN REACTIONS

In addition to oxidation, many other reactions occur as free radical chain reactions: polymerization, decomposition, fluorination, chlorination, etc. All chain reactions have a few important general peculiarities [1–3].

2.1.1 Free-Valence Persistence in Reactions of Free Radicals with Molecules

A free atom or a radical possesses an odd number of electrons, except atoms of noble gases. Typically, a valence-saturated molecule has an even number of electrons. Therefore, the reaction of a radical or an atom with a molecule will inevitably give rise to another atom or radical [1–3]:

$$RO_2^\bullet + HR \longrightarrow ROOH + R^\bullet$$
$$RO_2^\bullet + CH_2{=}CHR \longrightarrow ROOCH_2C^\bullet HR$$

Free valence also persists in unimolecular reactions of radicals, such as decomposition and isomerization.

$$Me_2C(OO^\bullet)CH_2CHMe_2 \longrightarrow Me_2C(OOH)CH_2C^\bullet Me_2$$
$$R^1R^2C(OO^\bullet)OH \longrightarrow R^1R^2C{=}O + HO_2^\bullet$$

Thus, free valence persists whenever an atom or a radical undergoes a unimolecular reaction or interacts with valence-saturated molecules (possessing an even number of electrons). This is a natural consequence of conservation of the number of electrons in chemical reactions. Therefore, free valence cannot persist when a radical reacts with a radical. Both reactants have an odd numbers of electrons, and the product formed has an even number of electrons, for example,

$$RO_2^\bullet + R^\bullet \longrightarrow ROOR$$
$$RO_2^\bullet + RO_2^\bullet \longrightarrow ROOR + O_2$$

2.1.2 Condition of Cyclicity of Radical Conversions

The comparison of various radical reactions suggests that the generation of free radicals is not sufficient for the chain process to occur. For example, 1,1-dimethylethylperoxide in a hydrocarbon (RH) solution undergoes the following reactions (given in p. 24):

$$(CH_3)_3COOC(CH_3)_3 \longrightarrow 2(CH_3)_3CO^\bullet$$
$$(CH_3)_3CO^\bullet + RH \longrightarrow (CH_3)_3COH + R^\bullet$$
$$(CH_3)_3CO^\bullet \longrightarrow CH_3^\bullet + CH_3COCH_3$$
$$CH_3^\bullet + R^\bullet \longrightarrow CH_3R$$
$$R^\bullet + R^\bullet \longrightarrow RR$$
$$R^\bullet + R^\bullet \longrightarrow RH + \text{alkene}$$

It can be seen that the decomposing peroxide gives rise to a number of radicals, viz, $(CH_3)_3CO^\bullet$, $C^\bullet H_3$, and $R^\bullet$, but the chain reaction fails to be initiated in this system. At the same time, the addition of oxygen to the peroxide–RH system will initiate chain oxidation with chain propagation through the cycle of reactions:

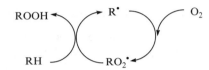

What is the difference between the systems initiator–RH and initiator–RH–O_2 that is responsible for the initiation of the chain reaction? It is apparent that, in contrast to the former system, the radical $R^\bullet$ in the latter system is regenerated via a cyclic sequence of chemical conversions. The analysis of this system and other similar systems suggests that a chain reaction can be sustained only in a system where radicals are involved in a cyclic sequence of conversions with the conservation of free valence. It is the multiple repetition of conversion cycles that leads to the chain reaction [2,3].

A sequence of elementary steps of radical reaction leading to the regeneration of the original radical is called the chain cycle, whereas the particular reaction steps are the events of chain propagation.

2.1.3 Priority of Chain Propagation Reaction

When free radicals appear in a system, two basically different types of reactions are possible: reactions with conservation of free valence and reactions in which radicals (or atoms) interact with each other without conservation of free valence. For example, the peroxyl radical propagates the chain by the reaction

$$RO_2^\bullet + HR \longrightarrow ROOH + R^\bullet$$

and the termination of chains occur via reaction

$$RO_2^\bullet + RO_2^\bullet \longrightarrow ROOR + O_2$$

A chain reaction can proceed if the rate of propagation is higher than the rate of chain termination. Hence, the necessary condition of a chain process is that the radicals generated in the system preferentially undergo reactions with conservation of free valence [2].

2.1.4 Generation of Free Radicals

The first three conditions of chain reaction assume that a chemical system contains free radicals. Therefore, a mechanism providing a continuous generation of radicals must exist. For instance, vinyl monomers $CH_2=CHX$ are oxidized by dioxygen only in the presence of an initiator, such as a thermally unstable compound, or physical methods of radical generation (light, ionizing radiation, etc.). An increase in the reaction temperature can also initiate the generation of radicals from reactants.

To conclude, the propagation of chain radicals is possible for systems in which

(i) The principle of free-valence persistence holds
(ii) Radicals are continuously generated
(iii) Particular reactions are cyclic and
(iv) Cyclic radical reactions with conservation of free valence occur more rapidly than the reactions in which free-valence carriers are eliminated.

Typically, a chain reaction involves a number of steps which, depending on their role in the overall chain process, are classified as chain initiation, chain propagation, and chain termination reactions.

2.1.5 Competition of Chain and Molecular Reactions

The problem of competition of the molecular reaction (direct route) and chain reaction (complicated, multistage route) was firstly considered in the monograph by Semenov [1]. The new aspect of this problem appeared recently because the quantum chemistry formulated the rule of conservation of orbital symmetry in chemical and photochemical reactions (Woodward–Hofmann rule [4]). Very often the structure of initial reactants suggests their direct interaction to form the same final products, which are also obtained in the chain reaction, and the thermodynamics does not forbid the reaction with $\Delta G < 0$. However, the experiment often shows that many reactions of this type occur in a complicated manner through several intermediate stages. For example, the reaction

$$Me_3CH + O_2 \longrightarrow Me_3COOH$$

is exothermic ($\Delta H = -108.4\,\text{kJ}\,\text{mol}^{-1}$) but occurs only by the chain route through the intermediate stages involving alkyl and peroxyl radicals. Now, analyzing numerous cases where a molecular transformation is possible but the reaction occurs via the chain route, one can distinguish several reasons for which the chain route of transformation has a doubtless advantage over the molecular route.

2.1.5.1 High Chemical Reactivity of Free Radicals and Atoms

The high reactivity of radicals and atoms is clearly seen from the comparison of rate constants of reactions of the same type involving molecules and radicals with a closely related structure. High reactivity is manifested by free atoms and radicals in abstraction reactions. For example, the peroxyl radical abstracts the H atom from cyclohexane with the rate constant $k = 2.4\,\text{L}\,\text{mol}^{-1}\,\text{s}^{-1}$ (400 K, see later), and the oxygen molecule does it with $k = 1.2 \times 10^{-9}\,\text{L}\,\text{mol}^{-1}\,\text{s}^{-1}$ (400 K, Chapter 4). Such a great difference is related to the fact that the first reaction proceeds with enthalpy $\Delta H = 43.3\,\text{kJ}\,\text{mol}^{-1}$, and the second reaction is very endothermic ($\Delta H = 188.8\,\text{kJ}\,\text{mol}^{-1}$). This distinction again follows from the structure of the species and strength of the formed O—H bonds: in the secondary alkyl peroxyl radical $D(\text{ROO—H}) = 365.5\,\text{kJ}\,\text{mol}^{-1}$, and in radical H—$O_2$ it is only $220\,\text{kJ}\,\text{mol}^{-1}$.

Radicals also exhibit high activity in addition reactions. For example, the peroxyl radical of oxidizing styrene adds to the double bond of styrene with the rate constant $k = 68\,\text{L}\,\text{mol}^{-1}\,\text{s}^{-1}$, and dioxygen adds with $k = 5.6 \times 10^{-10}\,\text{L}\,\text{mol}^{-1}\,\text{s}^{-1}$ (298 K). As in the case of abstraction reactions, the distinction results from the fact that the first reaction is exothermic ($\Delta H = -100$ kJ mol^{-1}) and the second reaction is endothermic ($\Delta H = 125$ kJ mol^{-1}). In this case, the differences are due to the fact that the chemical energy is stored in the free radical in the form of free valence. To illustrate this, below we present the ΔH values for molecules (RH) and radicals (R$^\bullet$) [5,6].

RH	MeCH$_3$	PhMe$_2$CH	MeCH$_2$OH	PhMe$_2$COOH
ΔH(RH) (kJ mol^{-1})	−84.7	3.9	−234.8	−85.3
R$^\bullet$	MeCH$_2^\bullet$	PhMe$_2$C$^\bullet$	MeCH$_2$O$^\bullet$	PhMe$_2$COO$^\bullet$
ΔH(R$^\bullet$) (kJ mol^{-1})	119.0	140.6	−53.0	55.3
ΔH(R$^\bullet$) − ΔH(RH) (kJ mol^{-1})	203.7	136.7	181.8	140.6

It is seen that this difference ranges from 137 to 204 kJ mol^{-1}, which is very significant. The high chemical reactivity of free atoms and radicals in various chemical reactions is one of the reasons for which the radical chain reactions occur much more rapidly than the direct molecular transformation of reactants into products. Although the radical formation is an endothermic reaction but, when appearing in the system, radicals rapidly enter into the reaction, and each radical induces the chain of transformations.

2.1.5.2 Conservation of Orbital Symmetry in Chemical Reaction

One more reason for which chain reactions have an advantage over molecular reactions is the restrictions that are imposed on the elementary act by the quantum-chemical rule of conservation of symmetry of orbits of bonds, which undergo rearrangement in the reaction [4]. If this rule is applied, the reaction, even if it is exothermic, requires very high activation energy to occur. For example, the reaction

$$Me_3CH + O_2 \longrightarrow Me_3COOH$$

is exothermic ($\Delta H = -108.4$ kJ mol^{-1}) but does not proceed due to the huge activation energy. At the same time, the reaction

$$Me_2PhCOO^\bullet + Me_2PhCH \longrightarrow Me_2PhCOOH + Me_2PhC^\bullet$$

has $\Delta H = -3.9$ kJ mol^{-1} and occurs with an activation energy of only 43 kJ mol^{-1} because in this act, the products conserve the symmetry of orbits of the reactants. According to the rule of conservation, the reaction is allowed and occurs with low activation energy (if it is exothermic) if the symmetries of orbits of the dissociated and formed bonds coincide. If this symmetry is disturbed, unoccupied high-energy orbits of reactants should participate in the formation of the transition state (TS) and this results in a very high activation energy of transformation. For example, the reaction RH + O$_2$ → ROOH mentioned earlier is forbidden by the rule of conservation because oxygen exists in the triplet state and the hydroperoxide formed is in the singlet state, that is, the spin of the system is not retained. The reaction of this type can involve the exited singlet dioxygen.

2.1.5.3 Configuration of Transition State

In abstraction reactions, atoms and some radicals have one more advantage over molecules. When two reacting species form the TS, the fragments of these species arranged near the reaction center are repulsed. The repulsion energy depends on the configuration of the TS. In the reaction of an X atom abstraction from a RX molecule, the minimum repulsion is provided by the configuration of the TS close to linear. This configuration is possible during the attack at the R—X bond of the atom or alkyl radical (see Chapter 6). As a rule, in reactions between molecules more compact nonlinear configurations appear, which increases the activation energy due to a higher repulsion energy of fragments of molecules in the TS.

2.2 CHAIN MECHANISM OF HYDROCARBON OXIDATION

The experimental proofs of the chain mechanism of hydrocarbon oxidation are the following [2,3,7–15].

1. Initiation by light accelerates oxidation due to the photochemical generation of free radicals, which was noticed by Backstrom [16] and repeated by many others [9,11–13]. The quantum yield (Φ) of photooxidation products is sufficiently higher than unity. Here are several examples [12].

Hydrocarbon	Cyclohexene	Cyclohexene, 1-methyl-	Dihydromyrcene	Ethyl linoleate
Φ (298 K)	15	23	10	90

For detailed information and bibliography on photooxidation of hydrocarbons, see Chapter 3.

2. The photochemical *after-effect*, considering that after light is switched off, the light oxidation continues for some time (time of chain growth), also indicates the chain nature of the process. This after-effect was observed in the photochemical oxidation of unsaturated hydrocarbons [9,12,15].
3. *Initiators* (peroxides, azo-compounds, polyphenylbutanes) accelerate the oxidation of hydrocarbons. The rate of initiated oxidation is much higher than the rate of initiator decomposition [9,10,12,13].
4. Salts and complexes of transition metals accelerate hydrocarbon oxidation due to the catalytic decomposition of hydroperoxides to free radicals (see Chapter 10).
5. Acceptors of peroxyl radicals (phenols, hydroquinones, aromatic amines) retard hydrocarbon oxidation, terminating the chains (see Part II).
6. Electron paramagnetic resonance (EPR) spectroscopy proves the formation of peroxyl radicals in oxidized hydrocarbons [12–15].
7. Oxidation of many organic compounds induces *chemiluminescence* (CL), initiated by the disproportionation of peroxyl radicals [17].

Chain oxidation of hydrocarbons occurs by the following elementary steps [2,3,10–15]:

$$I \longrightarrow r^\bullet \quad (v_i)$$
$$R^\bullet + O_2 \longrightarrow RO_2^\bullet \quad (k_{p1})$$
$$RO_2^\bullet + RH \longrightarrow ROOH + R^\bullet \quad (k_p)$$
$$R^\bullet + R^\bullet \longrightarrow RR \quad (2k_{t1})$$
$$R^\bullet + RO_2^\bullet \longrightarrow ROOR \quad (2k_{t2})$$
$$RO_2^\bullet + RO_2^\bullet \longrightarrow \text{Molecular products} \quad (2k_t)$$

The reaction of the alkyl radical with dioxygen proceeds with a high rate constant of 10^8 to 10^9 L mol^{-1} s^{-1}, and alkyl radicals are rapidly converted into peroxyl radicals, providing the concentration of dissolved dioxygen in hydrocarbon at $pO_2 \approx 0.1$–1 atm higher than

10^{-4} mol L^{-1}. Because of this, the concentration of alkyl radicals is much lower than that of peroxyl radicals, and chains are terminated only by the reaction between two peroxyl radicals, while chain propagation is limited by the reaction $RO_2^{\bullet} + RH$. In the presence of the initiator I, initiation by the formed hydroperoxide is insignificant, so that the rate of initiation $v_i \approx k_i[I]$. When conditions are quasistationary, $v_i = 2k_t[RO_2^{\bullet}]^2$, chains are long ($k_p[RH][RO_2^{\bullet}] \gg v_i$), and the rate of chain oxidation is the following:

$$v = k_p(2k_t)^{-1/2}[RH]v_i^{1/2} \qquad (2.1)$$

The values of the $k_p(2k_t)^{-1/2}$ parameters for the oxidation of hydrocarbons with different structures are collected in Table 2.1. This formula was verified by a number of experiments [9,12,13,15]. The deviation from linearity between the oxidation rate v and [RH], which takes place only when RH or solvent has polar groups, affects transient solvation and parameters k_p and $2k_t$ [18]. The methods of kinetic study of hydrocarbon oxidation and elementary steps of chain mechanism are described in detail in monographs [12–15,19].

If chains are short, initiation can contribute considerably to the overall oxidation rate, so that the oxidation rate looks like

$$v = v_i + k_p(2k_t)^{-1/2}[RH]v_i^{1/2}. \qquad (2.2)$$

The chain length is the following:

$$\nu = k_p[RH][RO_2^{\bullet}]/v_i = k_p(2k_t)^{-1/2}[RH]v_i^{-1/2}. \qquad (2.3)$$

Therefore, ν diminishes with the increasing initiation rate. At $v_i \geq k_p^2(2k_t)^{-1}[RH]^2$, hydrocarbon oxidation mainly proceeds as a nonchain radical process, with the predominance of chain termination products. Parameter $k_p(2k_t)^{-1/2}$ increases with temperature, so that for each hydrocarbon the temperature T_{min} exists, below which oxidation occurs as a nonchain process at v_i = const. The values of T_{min} calculated for several hydrocarbons oxidized with $v_i = 10^{-7}$ mol L^{-1} s^{-1} are given below [2].

Hydrocarbon	Cyclohexane	Ethylbenzene	Tetralin	Cumene
T_{min} (K)	363	283	247	226

The border conditions between chain and nonchain mechanisms of oxidation depends not only on temperature but also on the hydrocarbon concentration and the rate of chain initiation. The following equation describes this dependence:

$$RT_{min} = (E_p - 0.5E_t)/\{\ln[RH] - 0.5\ln v_i + \ln A_p - 0.5\ln A_t\} \qquad (2.4)$$

The quasistationary pattern of chain oxidation of RH is established in a definite time $\tau = 0.74(v_i/2k_t)^{-1/2}$ [12]. For v_i varying from 10^{-8} to 10^{-6} mol L^{-1} s^{-1} and $2k_t$ varying from 10^4 to 10^6 L mol^{-1} s^{-1}, τ ranges from 0.1 to 100 s.

The kinetics of an increase in the peroxyl radical concentration during the initial period of oxidation with a constant rate of initiation obeys the following equation [12]:

$$[RO_2^{\bullet}] = \sqrt{\frac{v_i}{2k_t}} \times \frac{e^{2\tau} - 1}{e^{2\tau} + 1}, \qquad (2.5)$$

where $\tau = t\sqrt{2k_t v_i}$.

TABLE 2.1
Values of Ratio $k_p/\sqrt{2k_t}$ for Oxidation of Hydrocarbons

Hydrocarbon	Solvent	T (K)	$k_p/\sqrt{2k_t}$ (L mol^{-1} s^{-1})$^{1/2}$	Ref.
		Aliphatic Hydrocarbons		
Butane	Butane	373–398	$7.02 \times 10^5 \exp(-64.0/RT)$	[20]
Pentane	Pentane	253–303	$1.34 \times 10^5 \exp(-59.0/RT)$	[21]
Decane	Decane	283–355	$2.97 \times 10^5 \exp(-58.3/RT)$	[22]
Decane	Decane	323	6.50×10^{-5}	[23]
Tetradecane	Tetradecane	323	1.37×10^{-4}	[23]
Hexadecane	Hexadecane	323	1.53×10^{-4}	[23]
Butane, 2-methyl-	Butane, 2-methyl-	387–423	$5.9 \times 10^3 \exp(-50.6/RT)$	[24]
Butane, 2,3-dimethyl-	Butane, 2,3-dimethyl	333	6.10×10^{-4}	[25]
Pentane, 2,4-dimethyl-	Pentane, 2,4-dimethyl-	323–398	$1.0 \times 10^4 \exp(-44.8/RT)$	[26]
Methylcyclopentane	Methylcyclopentane	333	1.60×10^{-4}	[25]
Cyclohexane	Cyclohexane	403–433	$1.9 \times 10^4 \exp(-54.0/RT)$	[27]
Cyclohexane	Cyclohexane	333	2.50×10^{-5}	[25]
Methylcyclohexane	Methylcyclohexane	333	9.50×10^{-5}	[25]
Pinane	Pinane	333	1.72×10^{-3}	[25]
Decalin	Decalin	333	9.0×10^{-5}	[25]
		Olefins		
1-Butene	1-Butene	333	$1.68 \times 10^4 \exp(-45.8/RT)$	[25]
1-Butene	1-Butene	333	1.10×10^{-3}	[25]
2-Butene	2-Butene	333	1.24×10^{-3}	[25]
2-Butene 2,3-dimethyl-	2-Butene 2,3-dimethyl-	333	$2.73 \times 10^5 \exp(-47.4/RT)$	[25]
2-Butene 2,3-dimethyl-	2-Butene 2,3-dimethyl-	333	1.00×10^{-2}	[25]
1-Hexene	1-Hexene	333	3.70×10^{-4}	[25]
3-Heptene	3-Heptene	303	5.40×10^{-4}	[28]
1-Octene	1-Octene	303	6.20×10^{-5}	[28]
1-Octene	1-Octene	373–393	$1.69 \times 10^3 \exp(-41.1/RT)$	[29]
1,4-Pentadiene	1,4-Pentadiene	303	4.20×10^{-4}	[28]
Methyl oleate	Methyl oleate	303	8.90×10^{-4}	[28]
Methyl linoleate	Methyl linoleate	303	2.10×10^{-2}	[28]
Ethyl linoleate	Ethyl linoleate	284	8.10×10^{-2}	[30]
Methyl linolenate	Methyl linolenate	303	3.90×10^{-2}	[28]
Cyclopentene	Cyclopentene	303	2.80×10^{-3}	[28]
Cyclohexene	Cyclohexene	323	$1.0 \times 10^2 \exp(-26.5/RT)$	[31]
Cyclohexene	Cyclohexene	303	2.30×10^{-3}	[28]
Cyclohexene	Cyclohexene	288	6.70×10^{-4}	[29]
Cyclohexene	Cyclohexene	333	5.10×10^{-3}	[25]
1-Methylcyclohexene	1-Methylcyclohexene	288	1.55×10^{-3}	[30]
1,3-Cyclohexadiene	1,3-Cyclohexadiene	303	0.10	[28]
1,4-Cyclohexadiene	Decane	303	$3.61 \times 10^4 \exp(-38.3/RT)$	[32]
1,4-Cyclohexadiene	1,4-Cyclohexadiene	303	3.90×10^{-2}	[28]
Limonene	Limonene	313–353	$1.10 \times 10^4 \exp(-38.1/RT)$	[33]
Dihydromyrcene	Dihydromyrcene	288	5.00×10^{-4}	[30]
5-Decyne	5-Decyne	303	7.40×10^{-4}	[28]

continued

TABLE 2.1
Values of Ratio $k_p/\sqrt{2k_t}$ for Oxidation of Hydrocarbons—continued

Hydrocarbon	Solvent	T (K)	$k_p/\sqrt{2k_t}$ ((L mol^{-1} s^{-1})$^{1/2}$)	Ref.
\multicolumn{4}{c}{Alkylaromatic Hydrocarbons}	
Toluene	Toluene	323–353	$8.12 \times 10^4 \exp(-58.0/RT)$	[34]
Toluene	Toluene	303	1.40×10^{-5}	[28]
Toluene	Toluene	333	5.30×10^{-5}	[25]
Toluene	Toluene	343–363	$9.1 \times 10^3 \exp(-53.8/RT)$	[35]
m-Xylene	m-Xylene	348	2.50×10^{-4}	[34]
m-Xylene	m-Xylene	303	2.80×10^{-5}	[28]
m-Xylene	m-Xylene	333	1.18×10^{-4}	[25]
o-Xylene	o-Xylene	348	3.30×10^{-4}	[34]
o-Xylene	o-Xylene	303	3.30×10^{-5}	[28]
o-Xylene	o-Xylene	333	1.63×10^{-4}	[25]
p-Xylene	p-Xylene	348	3.20×10^{-4}	[34]
p-Xylene	p-Xylene	303	4.90×10^{-5}	[28]
p-Xylene	p-Xylene	333	1.61×10^{-4}	[25]
Benzene, 1,3,5-trimethyl-	Benzene, 1,3,5-trimethyl-	348	3.30×10^{-4}	[34]
Benzene, 1,3,5-trimethyl-	Benzene, 1,3,5-trimethyl-	333	1.39×10^{-4}	[25]
Benzene, 1,2,4-trimethyl-	Benzene, 1,2,4-trimethyl-	348	6.00×10^{-4}	[34]
Benzene, 1,2,4-trimethyl-	Benzene, 1,2,4-trimethyl-	333	2.40×10^{-4}	[25]
Benzene, 1,2,4,5-tetramethyl-	Benzene, 1,2,4,5-tetramethyl-	323–353	$4.26 \times 10^3 \exp(-44/RT)$	[34]
Pentamethyl benzene	Pentamethyl benzene	323–353	$8.80 \times 10^2 \exp(-37.0/RT)$	[34]
Hexamethyl benzene	Hexamethyl benzene	323–353	$8.04 \times 10^2 \exp(-34.0/RT)$	[34]
Ethylbenzene	Ethylbenzene/Chlorobenzene	338–358	$4.02 \times 10^3 \exp(-44/RT)$	[36]
Ethylbenzene	Ethylbenzene	303–377	$2.51 \times 10^2 \exp(-36.0/RT)$	[37]
Ethylbenzene	Ethylbenzene	303	2.10×10^{-4}	[28]
Ethylbenzene	Ethylbenzene	323–373	$2.19 \times 10^2 \exp(-35.6/RT)$	[38]
Ethylbenzene	Ethylbenzene	323–353	$5.71 \times 10^2 \exp(-38.0/RT)$	[34]
Ethylbenzene	Ethylbenzene	333	5.30×10^{-4}	[25]
Propylbenzene	Propylbenzene	323–353	$9.90 \times 10^3 \exp(-55.0/RT)$	[34]
Propylbenzene	Propylbenzene	333	2.22×10^{-4}	[25]
Butylbenzene	Butylbenzene	333	3.92×10^{-4}	[25]
Hexylbenzene	Hexylbenzene	323–353	$2.36 \times 10^3 \exp(-43/RT)$	[34]
Cumene	Cumene	308–338	$1.70 \times 10^2 \exp(-29.9/RT)$	[39]
Cumene	Cumene	313–368	$1.17 \times 10^2 \exp(-28.4/RT)$	[38]
Cumene	Cumene	323–353	$2.46 \times 10^2 \exp(-31.0/RT)$	[34]
Cumene	Cumene	303	1.50×10^{-3}	[28]
Cumene	Cumene	303	2.60×10^{-3}	[40]
Cumene	Cumene	333	3.56×10^{-3}	[25]
Cumene	Cumene	323–353	$2.83 \times 10^2 \exp(-62.0/RT)$	[34]
Izoamylbenzene	Izoamylbenzene	323–353	$2.17 \times 10^4 \exp(-50.0/RT)$	[34]
Neopentylbenzene	Neopentylbenzene	323–353	$2.97 \times 10^9 \exp(-89.0/RT)$	[34]
p-Cymene	p-Cymene	333	1.10×10^{-3}	[25]
Diphenylmethane	Diphenylmethane	303	4.00×10^{-4}	[41]
Diphenylmethane	Diphenylmethane	323–353	$1.12 \times 10^3 \exp(-38.9/RT)$	[38]
Diphenylmethane	Diphenylmethane	323–353	$4.17 \times 10^1 \exp(-36.0/RT)$	[34]
1,1-Diphenylethane	1,1-Diphenylethane	313–353	$1.07 \times 10^3 \exp(-36.0/RT)$	[42]
1,1-Diphenylethane	1,1-Diphenylethane	303	1.10×10^{-3}	[41]
1,1-Diphenylethane	1,1-Diphenylethane	323–353	4.60×10^{-3}	[42]

TABLE 2.1
Values of Ratio $k_p/\sqrt{2k_t}$ for Oxidation of Hydrocarbons—*continued*

Hydrocarbon	Solvent	T (K)	$k_p/\sqrt{2k_t}$ ((L mol^{-1} s^{-1})$^{1/2}$)	Ref.
1-Methylpropylbenzene	1-Methylpropylbenzene	323–353	$4.87 \times 10^4 \exp(-53.0/RT)$	[42]
1-Methylpropylbenzene	1-Methylpropylbenzene	303	1.80×10^{-4}	[41]
1-Methylpropylbenzene	1-Methylpropylbenzene	333	4.48×10^{-4}	[25]
1-Methylbutylbenzene	1-Methylbutylbenzene	303	1.10×10^{-4}	[28]
Dicumylmethane	Dicumylmethane	348	5.75×10^{-3}	[34]
Phenylcyclopentane	Phenylcyclopentane	333	2.90×10^{-3}	[25]
Phenylcyclohexane	Phenylcyclohexane	303	1.50×10^{-4}	[28]
Phenylcyclohexane	Phenylcyclohexane	333	6.00×10^{-4}	[25]
Phenylcyclohexane	Phenylcyclohexane	323–353	$5.05 \times 10^2 \exp(-38.0/RT)$	[34]
Allylbenzene	Allylbenzene	303	4.90×10^{-4}	[28]
Crotylbenzene	Crotylbenzene	303	4.20×10^{-3}	[28]
Indan	Indan	303	1.70×10^{-3}	[28]
Indene	Indene	303	2.84×10^{-2}	[28]
Tetralin	Tetralin	286–323	$4.4 \times 10^1 \exp(-25.1/RT)$	[43]
Tetralin	Tetralin	303	2.30×10^{-3}	[28]
Tetralin	Tetralin	303	2.35×10^{-3}	[44]
Tetralin	Tetralin	303	2.85×10^{-3}	[45]
Tetralin	Tetralin	333	6.16×10^{-3}	[25]
Tetralin	Tetralin	348	1.09×10^{-2}	[36]
1,2-Dihydronaphtalene	1,2-Dihydronaphtalene	303	2.75×10^{-2}	[44]
1,4-Dihydronaphtalene	1,4-Dihydronaphtalene	303	3.50×10^{-2}	[44]
9,10-Dihydroanthracene	9,10-Dihydroanthracene	303	7.90×10^{-2}	[28]
9,10-Dihydroanthracene	9,10-Dihydroanthracene	303	6.93×10^{-2}	[44]
1,2,3,4,5,6,7,8-Octahydroanthracene	1,2,3,4,5,6,7,8-Octahydroanthracene	348	3.12×10^{-2}	[36]
m-Isopropylpyridine	m-Isopropylpyridine	323–353	$9.90 \times 10^2 \exp(-36.0/RT)$	[34]
o-Isopropylpyridine	o-Isopropylpyridine	323–353	$3.73 \times 10^2 \exp(-38.0/RT)$	[34]
p-Isopropylpyridine	p-Isopropyl pyridine	323–353	$1.79 \times 10^2 \exp(-32.0/RT)$	[34]
m-Methylpyridine	m-Methylpyridine	348	1.90×10^{-4}	[34]
o-Methylpyridine	o-Methylpyridine	348	7.00×10^{-5}	[34]
p-Methylpyridine	p-Methylpyridine	348	1.10×10^{-4}	[34]
m-Ethylpyridine	m-Ethylpyridine	323–353	$6.93 \times 10^2 \exp(-39.0/RT)$	[34]
o-Ethylpyridine	o-Ethylpyridine	323–353	$2.70 \times 10^2 \exp(-38.0/RT)$	[34]
p-Ethylpyridine	p-Ethylpyridine	323–353	$8.70 \times 10^2 \exp(-41.0/RT)$	[34]

If the concentration of an initiator does not virtually change throughout the experiment, then $v_i = $ const. and oxidation must occur at a constant rate. Actually, the initiation rate decreases. If the initiator is a sole source of radicals in the system, the kinetics of oxidation is described by the equation [2]:

$$\Delta[O_2] = 2a[RH]k_d^{-1}\sqrt{k_i[I]_0}\{1 - \exp(-0.5k_d t)\}, \qquad (2.6)$$

where $a = k_p(2k_t)^{-1/2}$, $k_i = 2ek_d$.

If the partial pressure of dioxygen varies greatly, reactions (t1) and (t2) contribute considerably to chain termination at low concentration of dioxygen. So, in the general case, the oxidation rate becomes dependent on the partial pressure of dioxygen. At a very low concentration of the dissolved dioxygen, chains propagate according to reaction (p1) and are terminated according to reaction (t1), so that the chain reaction rate becomes equal to [2,12]

$$v = k_{p1}[O_2]\sqrt{\frac{v_i}{2k_{t1}}} \tag{2.7}$$

In a wide range of $[O_2]$, the oxidation rate is related to $[O_2]$ and $[RH]$ as [3,12,13]

$$\frac{[O_2]}{[RH]}\left(\frac{v_\infty^2}{v^2} - 1\right) = \frac{k_p k_{t2}}{2k_{p1} k_t} + \frac{k_p^2 k_{t1}}{k_{p1}^2 k_t}\frac{[RH]}{[O_2]}, \tag{2.8}$$

where $v_\infty = k_p(2k_t)^{-1/2}[RH]v_i^{1/2}$.

This sophisticated dependence can be approximated by a simpler formula [14]

$$v = v_\infty(1 + \beta[O_2]^{-1})^{-1}, \tag{2.9}$$

where β is equal to $[O_2]$ such that $v = 0.5v_\infty$. The ratio of chains terminated by each of reactions: $R^\bullet + R^\bullet$, $R^\bullet + RO_2^\bullet$ and $RO_2^\bullet + RO_2^\bullet$ depends on the concentration of dissolved dioxygen. As an example, the percentage of chain termination by each of these reactions in oxidized cyclohexane at 350 K is presented.

$[O_2]$ (mol L^{-1})	10^{-8}	10^{-7}	10^{-6}	10^{-5}	10^{-4}
$v(R^\bullet + R^\bullet)/v_i$	10.0	1.1	0.1	2×10^{-3}	3×10^{-5}
$v(R^\bullet + RO_2^\bullet)/v_i$	89.7	95.4	72.8	21.0	2.7
$v(RO_2^\bullet + RO_2^\bullet)/v_i$	0.3	3.5	27.1	79.0	97.3

Hydrocarbon oxidized by dioxygen in a reaction vessel is a two-phase system. Dioxygen should be dissolved in hydrocarbon as a first step. When oxidation occurs slowly and diffusion of dioxygen in hydrocarbon proceeds rapidly, we deal with the kinetic regime of oxidation. The process of dioxygen diffusion in hydrocarbon does not influence the oxidation rate. The concentration of dioxygen in hydrocarbon is close to equilibrium: $[O_2] = \gamma \times pO_2$ where γ is Henry's coefficient for dioxygen in hydrocarbon (see Table 2.2). The rate of dioxygen diffusion into hydrocarbon depends, first, on the mode and rate of mixing the gas and liquid, and second, on the partial pressure of dioxygen. The lower the pO_2, the slower the process of dioxygen diffusion from the gas phase into the liquid. Therefore, at low pO_2 the process of dioxygen diffusion in hydrocarbon can manifest the limiting stage of oxidation, and is the diffusion pattern of oxidation in this case. Taking into account that dioxygen dissolution in hydrocarbon is the rate-determining step of the whole process, we can express the oxygen consumption as a two-stage process [7]:

$$(O_2)_{gas} \rightleftharpoons [O_2]_{liquid} \tag{κ}$$
$$R^\bullet + O_2 \longrightarrow RO_2^\bullet \tag{k_{p1}}$$

In the absence of free radical initiation when $[R^\bullet] = 0$, the dynamics of dioxygen dissolution in the hydrocarbon obeys the simple equation

$$\frac{d[O_2]}{dt} = \kappa(\gamma pO_2 - [O_2]) \tag{2.10}$$

$$[O_2] = \gamma pO_2(1 - e^{-\kappa t}), \tag{2.11}$$

TABLE 2.2
Henry's Coefficients γ for Solubility of Dioxygen in Organic Solvents (ρ is Density of Solvent at 298 K in kg m^{-3}) [46]

Solvent	T (K)	$\gamma \times 10^3$ / (mol L^{-1} atm^{-1})
Acetic acid	293	8.11
Acetone	195–313	9.11 exp(−141/T)
Acetonitrile	297	9.10
Benzene	283–333	16.9 exp(−393/T)
m-Xylene	298	9.69
o-Xylene	298	9.22
p-Xylene	298	10.0
Benzene, chloro-	273–353	10.0 exp(−285/T)
Benzene, nitro-	291	1.52
1-Butanol	298	8.65
2-Butanone	298	11.2
Carbon tetrachloride	298	12.4
Chloroform	293	11.6
Cumene	298	9.90
Cyclohexane	298	11.5
Cyclohexane, methyl-	298	12.5
Cyclohexanol	299	8.27
Cyclohexanone	298	6.11
Decane	298	11.2
Diethyl ether	293	14.7
1,4-Dioxane	298	6.28
Dodecane	298	8.14
Ethanol	298	9.92
Ethyl acetate	293	8.89
Ethylbenzene	298	9.91
Heptane	298	13.2
Hexane	298	14.7
Isooctane	289	15.3
Kerosene ($\rho = 810$)	291	3.44
Methanol	298	10.2
Methyl acetate	298	11.2
Nonane	298	11.2
Octane	293	13.3
Paraffinic oil ($\rho = 880$)	298	2.47
Pentane	298	17.7
Petrol ($\rho = 709$)	291	6.32
Piperidine	298	7.39
1-Propanol	298	6.69
2-Propanol	298	10.2
Pyridine	298	5.66
Toluene	298	9.88
2,2,4-Trimethylpentane	298	15.5
Water	298	1.27

where κ is an effective coefficient of dioxygen dissolution in the hydrocarbon at the chosen experimental conditions. The mean time of liquid saturation by dioxygen is $t_s \approx \kappa^{-1}$. When free radicals are initiated in the hydrocarbon, the hydrocarbon is oxidized at the rate v. The dioxygen concentration in the oxidized hydrocarbon depends on the rates of two processes (given by Equations [2.10] and [2.11]): the rate of dioxygen solvation and the rate of dioxygen consumption due to oxidation. The rates of these two processes are equal under the quasistate conditions and as a result, the dioxygen concentration is the following:

$$[O_2] = \gamma pO_2 - v\kappa^{-1} \qquad (2.12)$$

Oxidation occurs in the kinetic regime when $v \ll \kappa\gamma pO_2$, and in diffusion regime when $v \approx \kappa\gamma pO_2$. The dependence of the oxidation rate v on dioxygen concentration is nonlinear (see Equation [2.9]). Taking this into account, we obtain the following equation for the rate of initiated hydrocarbon oxidation, including the diffusion regime of oxidation:

$$\gamma pO_2 = \frac{\beta v}{av_i^{1/2} - v} + \frac{v}{\kappa} \qquad (2.13)$$

Equation (2.13) can be used for the estimation of the β and κ coefficients from experimental measurements. The data on the solubility of dioxygen in organic compounds are presented in Table 2.2.

2.3 REACTION OF ALKYL RADICALS WITH DIOXYGEN

Radicals produced from the initiator either directly attack the organic compound RH (for instance, this is the case during the decomposition of peroxides) or first react with dioxygen, and then, already as peroxyl radicals, attack RH (for instance, this is the case of decomposition of azo-compounds). RH gives rise to alkyl radicals when attacked by these radicals.

The reaction of dioxygen addition to an alkyl radical,

$$R^\bullet + O_2 \longrightarrow RO_2^\bullet,$$

is exothermic. The change in the enthalpy of this reaction depends on structure of alkyl radical (see Table 2.3). These values are close to those calculated from enthalpies of peroxyl radical formation (see Table 2.7).

In solution the reaction of alkyl radical with dioxygen occurs extremely rapidly with a diffusion rate constant (see Table 2.4). The data on solubility of dioxygen in different organic solvents are collected in Table 2.2.

The recent quantum-chemical analysis of the reaction of dioxygen with ethyl radical in the gas phase provided evidence for two pathways of interaction [83]:

$$CH_3C^\bullet H_2 + O_2 \longrightarrow CH_3CH_2OO^\bullet$$
$$k = 2.02 \times 10^{10} T^{0.98} \times \exp(31.8/T) \text{L mol}^{-1}\text{s}^{-1}$$

$$CH_3C^\bullet H_2 + O_2 \longrightarrow CH_2{=}CH_2 + HOO^\bullet$$
$$k = 1.41 \times 10^4 T^{1.09} \times \exp(990/RT) \text{L mol}^{-1}\text{s}^{-1}$$

$$CH_3CH_2OO^\bullet \longrightarrow CH_2{=}CH_2 + HOO^\bullet$$
$$k = 7.14 \times 10^4 T^{2.32} \times \exp(-14100/T)\text{s}^{-1}$$

TABLE 2.3
Enthalpies of Dioxygen Addition to Alkyl Radicals [13,47,48]

R•	$-\Delta H_1$ (kJ mol^{-1})	R•	$-\Delta H_1$ (kJ mol^{-1})
C•H$_3$	137.0	cyclohexyl•	142.6
MeC•H$_2$	148.4	cyclohex-2-enyl•	63.0
Me$_2$C•H	155.4	cyclohexa-2,4-dienyl•	31.0
Me$_3$C•	152.8	CH$_2$=CHCH$_2$•	110.0
PhMe$_2$C•	47.3	1,2-dihydronaphthalenyl•	45.0
C•H$_2$Cl	122.4	C•Cl$_3$	92.0
C•HCl$_2$	108.2	MeC•HCl	131.2

At temperatures below 500 K, the alkyl radical adds to dioxygen forming the stable peroxyl radical, which reacts with hydrocarbon, resulting in hydroperoxide formation. Between 500 and 800 K, the predominant reaction is the formation of olefin and HO$_2$•. Hydroperoxyl radicals disproportionate very rapidly. These reactions produce the "negative temperature coefficient regime" of gas-phase oxidation. At temperatures higher than 800 K, hydrogen peroxide formed by the reaction of HO$_2$• with RH dissociates rapidly into free radicals and becomes the degenerate branching agent. As a result, hydrocarbon oxidation occurs as the reaction with positive activation energy.

Under conditions of liquid-phase oxidation, alkylperoxyl radicals are stable and react rapidly with RH. They do not decompose into olefin and hydroperoxyl radical. However, some peroxyl radicals have a weak C—OO bond and decompose back to R• and dioxygen:

$$R^\bullet + O_2 \rightleftharpoons RO_2^\bullet \qquad K$$
$$RO_2^\bullet \longrightarrow R^\bullet + O_2 \qquad k_d$$

For example, cyclohexadienyl peroxyl radicals decompose back sufficiently rapidly, so that the rate constant of decomposition $k_d > k_p[RH]$ [13]:

Radical	T (K)	K (L mol^{-1})	k_d (s^{-1})	E_d (kJ mol^{-1})
cyclohexa-2,4-dienyl peroxyl	303	0.16	1.3×10^2	31.0
cyclohexa-2,5-dienyl peroxyl	323	0.10	1.0×10^4	
1,2-dihydronaphthalenyl peroxyl	303	0.16	4.6	45.0

TABLE 2.4
Rate Constants of Dioxygen Addition to Alkyl Radicals (Experimental Data)

R•	Solvent	T (K)	k (L mol^{-1} s^{-1})	Ref.
C•H$_3$	Water	296	4.7×10^9	[49]
C•H$_3$	Water, pH = 1.0	298	4.1×10^9	[50]
C•H$_3$	Water, pH = 3.0	298	3.7×10^9	[51]
CH$_3$C•H$_2$	Water, pH = 0.0	298	2.1×10^9	[50]
CH$_3$C•H$_2$	Water	298	2.9×10^9	[52]
EtC•H$_2$	Water, pH = 8.5	298	1.9×10^9	[53]
Me$_2$C•H	Water, pH = 1.0	298	3.8×10^9	[50]
PrC•H$_2$	Water, pH = 0.0	298	1.8×10^9	[50]
PrC•H$_2$	Water, pH = 8.5	298	1.3×10^9	[53]
EtMeC•H	Water, pH = 1.0	298	3.2×10^9	[50]
Me$_2$CHC•H$_2$	Water, pH = 1.0	298	3.2×10^9	[50]
Me$_3$C•	Cyclohexane	300	4.9×10^9	[54]
Me(CH$_2$)$_3$C•H$_2$	Water, pH = 0.0	298	3.8×10^9	[50]
Me(CH$_2$)$_4$C•H$_2$	Water, pH = 0.0	298	3.9×10^9	[50]
Me$_3$CC•H$_2$	Water, pH = 1.0	298	2.7×10^9	[50]
Me(CH$_2$)$_6$C•H$_2$	Water, pH = 1.0	298	2.4×10^9	[50]
Me(CH$_2$)$_5$C•HMe	Water, pH = 1.0	298	2.4×10^9	[50]
Me(CH$_2$)$_7$C•HMe	Decane	298	4.8×10^9	[55]
Me(CH$_2$)$_{13}$C•HMe	Hexadecane	298	1.5×10^9	[56]
Me(CH$_2$)$_{14}$C•HMe	Heptadecane	298	1.5×10^9	[56]
Me(CH$_2$)$_4$CH=CHC•H CH=CH(CH$_2$)$_7$COOH	Linoleic acid	295	3.0×10^8	[57]
Et(CH=CHCH$_2$)$_2$CH=CH C•H(CH$_2$)$_6$COOH	Linolenic acid	295	3.0×10^8	[57]
cyclopentyl•	Water, pH = 1.0	298	3.5×10^9	[50]
cyclohexyl•	Cyclohexane	298	2.0×10^9	[58]
cyclohexenyl•	Benzene	300	1.6×10^9	[54]
2-hydroxycyclohexadienyl•	Water	298	1.5×10^8	[59]
2,6-di-tert-butyl-4-methylphenoxy• type	Toluene, 3,5-bis-1,1-dimethylethyl-	298	9.0×10^7	[58]

TABLE 2.4
Rate Constants of Dioxygen Addition to Alkyl Radicals (Experimental Data)—continued

R·	Solvent	T (K)	k (L mol^{-1} s^{-1})	Ref.
C·H$_2$Cl	Water, pH = 1.0	298	1.9×10^9	[50]
C·H$_2$Br	Water, pH = 0.0	298	2.0×10^9	[50]
CH$_3$C·HCl	Water	298	9.0×10^8	[60]
CH$_3$C·Cl$_2$	Water	298	1.5×10^9	[60]
CF$_3$C·HCl	Water/tert-Butanol	298	1.3×10^9	[61]
CH$_2$ClC·HCl	Water	298	9.7×10^8	[60]
N≡CC·H$_2$	Water/Acetonitrile	298	1.3×10^9	[62]
Me$_2$NC·H$_2$	Trimethylamine	298	3.5×10^9	[63]
C·H$_2$OH	Water, pH = 10.7	298	4.2×10^9	[64]
C·H$_2$CH$_2$OH	Water, pH = 1.0	298	6.6×10^9	[65]
CH$_3$C·HOH	Water, pH = 7.0	298	4.6×10^9	[66]
CH$_3$CH$_2$C·HOH	Water, pH = 7.0	298	4.7×10^9	[66]
Me$_2$C·OH	2-Propanol	300	3.9×10^9	[54]
Me$_2$C·OH	Acetonitrile	297	6.6×10^9	[67]
Me$_2$CHC·HOH	Water, pH = 7.0	298	3.4×10^9	[66]
EtMeC·OH	Water, pH = 7.0	298	4.0×10^9	[66]
HOCH=C·H	Ethanol	295	1.0×10^9	[68]
C·H(OH)CH$_2$OH	Water, pH = 7.0	298	3.2×10^9	[66]
cyclohexyl-OH (1-hydroxycyclohexyl radical)	Acetonitrile	297	5.4×10^9	[67]
C·(O)NH$_2$	Water	298	2.7×10^9	[69]
HC·(COOH)$_2$	Water, pH = 0	298	1.6×10^9	[70]
HC·(COO$^-$)$_2$	Water, pH = 0	298	1.3×10^9	[70]
C·H$_2$CO$_2^-$	Water, pH = 8.0	298	1.7×10^9	[71]
(CO$_2^-$)$_2$C·OH	Water	293	1.6×10^9	[72]
CH$_3$C(O)C·H$_2$	Water	298	3.1×10^9	[73]
CH$_3$C(O)OC·H$_2$	Water	298	1.4×10^{10}	[74]
C·H$_2$C(O)OCH$_3$	Water	RT	1.8×10^9	[74]
C$_6$H$_5$·	Water	293	3.3×10^9	[75]
CN–C$_6$H$_4$·	Water	288	2.7×10^9	[75]
H$_3$CO–C$_6$H$_4$·	Water	293	3.2×10^9	[75]
C$_6$H$_5$C·H$_2$	Water, pH = 1.0	298	2.8×10^9	[50]
C$_6$H$_5$C·H$_2$	Benzene	300	2.9×10^9	[54]
C$_6$H$_5$C·H$_2$	Hexane	300	2.8×10^9	[54]
C$_6$H$_5$C·H$_2$	Cyclohexane	300	2.4×10^9	[54]
C$_6$H$_5$C·H$_2$	Hexadecane	300	1.0×10^9	[54]

continued

TABLE 2.4
Rate Constants of Dioxygen Addition to Alkyl Radicals (Experimental Data)—*continued*

R•	Solvent	T (K)	k (L mol^{-1} s^{-1})	Ref.
$C_6H_5C^•H_2$	Acetonitrile	300	3.4×10^9	[54]
$C_6H_5C^•H_2$	2-Propanol	300	2.5×10^9	[54]
$C_6H_5C^•H_2$	Toluene	294	2.0×10^9	[76]
$C_6H_5C^•H_2$	Hexane	298	2.6×10^9	[77]
4-Me-C$_6$H$_4$-C$^•$H$_2$	Hexane	298	3.4×10^9	[77]
4-Ph-C$_6$H$_4$-C$^•$H$_2$	Hexane	298	1.7×10^9	[77]
4-F-C$_6$H$_4$-C$^•$H$_2$	Hexane	298	2.9×10^9	[77]
4-Br-C$_6$H$_4$-C$^•$H$_2$	Hexane	298	1.3×10^9	[77]
4-NC-C$_6$H$_4$-C$^•$H$_2$	Hexane	298	5.8×10^8	[77]
4-O$_2$N-C$_6$H$_4$-C$^•$H$_2$	4-Nitrotoluene	294	9.0×10^8	[76]
PhC$^•$HCH$_3$	Ethylbenzene	323	8.8×10^8	[78]
PhC$^•$HOCH$_3$	Water	293	3.2×10^9	[79]
PhMe$_2$C$^•$	Cumene	323	9.0×10^8	[54]
Ph$_2$C$^•$H	Diphenylmethane	294	7.5×10^8	[76]
Ph$_3$C$^•$	Triphenylmethane	293	1.2×10^9	[80]
EtCH$_2$C$^•$HC$_6$H$_5$	1-Phenylpropane	323	8.0×10^7	[78]
Ph$_2$NC$^•$HCH$_3$	Ethyldiphenylamine	300	4.9×10^9	[54]
glycine anhydride radical	Water	298	2.0×10^9	[81]
pyridinyl radical	Water/Methanol	293	2.2×10^9	[82]
morpholine N-isopropyl radical	Acetonitrile	297	6.3×10^9	[67]
morpholine N-allyl radical	Acetonitrile	297	4.5×10^9	[67]
morpholine N-allyl radical	Acetonitrile	297	3.9×10^9	[67]
morpholine N-(Ph-allyl) radical	Acetonitrile	297	2.3×10^9	[67]
morpholine N-dimethylaminoethyl radical	Acetonitrile	297	4.3×10^9	[67]

The oxidation of such nonsaturated compounds proceeds through hydroperoxyl radical formation by the reaction [13]:

$$\text{dihydronaphthalene}^\bullet + O_2 \longrightarrow \text{naphthalene} + H\dot{O}_2$$

The situation with polyarylmethanes is very similar. Due to the stabilization of free valence in arylmethyl radicals, the bond dissociation energy (BDE) of the bond C—O_2, for example, in triphenylmethyl radical is sufficiently lower than in alkylperoxyl radicals. This radical is decomposed under oxidation conditions (room temperature), and the reaction of $Ph_3C^\bullet$ with dioxygen is reversible:

$$Ph_3C^\bullet + O_2 \rightleftharpoons Ph_3COO^\bullet$$

Howard and Ingold studied this equilibrium reaction in experiments on the oxidation of tetralin and 9,10-dihydroanthracene in the presence of specially added triphenylmethyl hydroperoxide[41]. They estimated the equilibrium constant K to be equal to 60 atm^{-1} (8 × 10^3 L mol^{-1}, 303 K). This value is close to $K = 25$ atm^{-1} at 300 K ($\Delta H = 38$ kJ mol^{-1}), which was found in the solid crystal lattice permeable to dioxygen [84]. The reversible addition of dioxygen to the diphenylmethyl radical absorbed on MFI zeolite was evidenced and studied recently by the EPR technique [85].

$$\text{cyclohexadienyl}^\bullet + O_2 \rightleftharpoons \text{cyclohexadienyl-OO}^\bullet$$

The addition of dioxygen to the cyclohexadienyl radical is also reversible [86]. The following thermodynamic parameters were estimated for such equilibrium: $\Delta H = 21$ kJ mol^{-1} and $\Delta S = -84$ J mol^{-1} K^{-1}.

2.4 REACTIONS OF PEROXYL RADICALS

2.4.1 Structure and Thermochemistry of Peroxyl Radicals

The interatomic distances in peroxyl radicals were calculated by quantum-chemical methods. The experimental measurements were performed only for the hydroperoxyl radical and the calculated values were close to the experimental measurements (see Table 2.5). The length of the O—O bond in the peroxyl radical lies between that in the dioxygen molecule ($r_{O-O} = 1.20 \times 10^{-10}$ m) and in hydrogen peroxide ($r_{O-O} = 1.45 \times 10^{-10}$ m).

The internal rotation around the C—O bond in the peroxyl radical occurs with the energetic barrier E_{rot}. The height of the barrier depends on the substituent: the greater the volume of substituent, the higher the barrier of internal rotation [97]:

$RO_2^\bullet$	$CH_3O_2^\bullet$	$MeCH_2O_2^\bullet$	$Me_3CO_2^\bullet$
E_{rot} (kJ mol^{-1})	2.5	5.7	9.2
$RO_2^\bullet$	$PhMeCHCO_2^\bullet$	$Ph_2CHO_2^\bullet$	$Ph_2MeCO_2^\bullet$
E_{rot} (kJ mol^{-1})	7.9	81.1	77.2

TABLE 2.5
The Structural Parameters of Peroxyl Radicals

HOO•	r(O—O)[$\times 10^{10}$ (m)]	r(H—O)[$\times 10^{10}$ (m)]	θ(OOH) (deg)	Ref.
		Experimental		
HOO•	1.34	0.968	106	[87]
HOO•	1.30	0.96	108	[88]
HOO•	1.30	0.968	106	[89]
HOO•	1.335	0.977	104	[90]
		Quantum-Chemical Calculation		
HOO•	1.19	1.05	110.7	[91]
HOO•	1.19	0.96	111	[92]
HOO•	1.384	0.968	106.8	[93]
HOO•	1.23	0.96	113.8	[94]
HOO•	1.39	0.968	106.8	[95]
HOO•	1.349	0.972	117.6	[96]
HOO•	1.198	1.050	110.4	[97,98]
HOO•	1.183	1.011	110	[34]
HOO•	1.326	0.975	104.4	[99]

ROO•	r(O—O)[$\times 10^{10}$ (m)]	r(C—O)[$\times 10^{10}$ (m)]	θ(COO) (deg)	Ref.
CH_3O_2•	1.335	1.335	122.1	[96]
CH_3O_2•	1.20	1.38	112.5	[92]
CH_3O_2•	1.195	1.44	112.5	[99]
CH_3O_2•	1.19	1.44	111	[100]
CH_3O_2•	1.19	1.44	110	[34]
$MeCH_2O_2$•	1.197		114	[101]
$MeCH_2O_2$•	1.333	1.363	127.1	[96]

The peroxyl radical is polar and possesses a high dipole moment μ. For the values of μ calculated by the quantum-chemical method, see below [98, 102]:

Radical	HO_2•	CH_3O_2•	$MeCH_2O_2$•	Me_2CHO_2•
μ (Debye)	1.94	2.07	2.38	2.45
Radical	Me_3CO_2•	$PhCH_2O_2$•	Ph_2CHO_2•	$PhMe_2CO_2$•
μ (Debye)	2.40	2.41	2.26	2.55

The spin density is concentrated on the last oxygen atom in the peroxyl radical $R(^2O)(^1O^•)$ (about 80%). However, about 20% of the spin density is concentrated on the (^2O) atom also (see Table 2.6). This proves the existence of rather strong interaction between electronic clouds of two oxygen atoms and agrees with the length of O—O bond in the peroxyl radical (see Table 2.5).

TABLE 2.6
Spin and Electron Densities in Peroxyl Radicals [97]

ROO·	Spin Density			Electron Density		
	H	^{2}O	^{1}O	H	^{2}O	^{1}O
HO$_2$·	−0.021	0.2165	0.8065	0.1399	−0.0400	−0.0999
CH$_3$O$_2$·	0.0061	0.1952	0.8084	−0.0223	−0.0396	−0.1198
MeCH$_2$O$_2$·	0.0059	0.1944	0.8105	−0.0345	−0.0061	−0.1258
Me$_2$CHO$_2$·	0.0037	0.2064	0.7999	−0.0387	−0.0633	−0.1321
Me$_3$CO$_2$·	0.0	0.2099	0.7964		−0.0723	−0.1399
PhCH$_2$O$_2$·	0.0052	0.1770	0.8113	−0.0611	−0.0911	−0.1340
PhMeCHO$_2$·	0.0061	0.2027	0.8038	−0.056	−0.0710	−0.1350
PhMe$_2$CO$_2$·	0.0	0.1878	0.8029		−0.0995	−0.1402
Ph$_2$CHO$_2$·	0.0052	0.1770	0.8113	−0.0611	−0.0911	−0.1340
Ph$_2$MeCO$_2$·	0.0	0.1878	0.8029		−0.0955	−0.1402

The electronic spectrum of the cyclohexylperoxyl radical has a maximum at $\lambda = 275$ nm with molar absorption coefficient $\varepsilon = 2.0 \times 10^3$ L mol^{-1} cm^{-1} [103]. The dissociation energy of the O—H bond in a hydroperoxide ROOH depends on the R structure [104–106]:

R	H	RCH$_2$	R$_1$R$_2$CH	R$_1$R$_2$R$_3$C
D(O—H) (kJ mol^{-1})	369.0	365.5	365.5	358.6

The values of enthalpies of peroxyl radical formation (ΔH_f^0) calculated from the enthalpies of hydroperoxide formation according to the thermochemical equation:

$$\Delta H_f^0(\text{ROO}^\bullet) = \Delta H_f^0(\text{ROOH}) + D(\text{O—H}) - \Delta H_f^0(\text{H}^\bullet) \quad (2.14)$$

are presented in Table 2.7.

The enthalpy of peroxyl radical formation is related to ΔH_f^0(ROOH) by the following relationship:

$$\Delta H_f^0(\text{ROO}^\bullet) = \Delta H_f^0(\text{ROOH}) + \Delta\Delta H, \quad (2.15)$$

where the increment $\Delta\Delta H$ is 147.5 kJ mol^{-1} for primary and secondary alkyl hydroperoxides, and 140.6 kJ mol^{-1} for tertiary alkyl hydroperoxides. The increments for various groups in the additive scheme of enthalpy ΔH_f calculation for hydroperoxides have the following values: ΔH(O—(C)(O)) = −20.75 ± 0.81 kJ mol^{-1}, ΔH[(O)—OH] = −72.26 ± 1.2 kJ mol^{-1} [116]. The $\Delta\Delta H$ increments for various groups in the calculation of enthalpy ΔH_v of peroxides evaporation are ΔH[(O)—O—(C)] = 6.26 kJ mol^{-1}, ΔH_v[(O)—OH] = 27.4 kJ mol^{-1} [117].

TABLE 2.7
Enthalpies of Formation ΔH_f^0 (gas, 298 K) of Hydroperoxides and Peroxyl Radicals and C—$O_2^{\bullet}$ Bond Dissociation Energies in Peroxyl Radicals

Hydroperoxide	$-\Delta H_f^0$(ROOH) (kJ mol^{-1})	ΔH_f^0(ROO$^{\bullet}$) (kJ mol^{-1})	D(R–$O_2^{\bullet}$) (kJ mol^{-1})	Ref.
Methyl-, MeOOH	132.2	15.3	131.7	[107,108]
ethyl-, EtOOH	169.4	−21.9	140.9	[108]
1-Methylethyl-, Me$_2$CHOOH	198.3	−50.8	140.8	[108]
Butyl-, PrCH$_2$OOH	205.0	−57.5	135.5	[107]
1,1-Dimethylethyl-, Me$_3$COOH	242.0	−101.4	149.4	[107–109]
Pentyl-, Me(CH$_2$)$_3$CH$_2$OOH	217.6	−70.1	127.7	[107]
1-Methylbutyl-, MePrCHOOH	234.3	−86.8	135.4	[107]
1-Hexyl-, Me(CH$_2$)$_4$CH$_2$OOH	237.8	−90.3	127.3	[107,110]
1-Methylpentyl, BuCH(OOH)Me	252.6	−105.1	133.1	[107,110]
1-Ethylbutyl-, PrCH(OOH)Et	245.1	−97.6		[110]
1-Heptyl-, Bu(CH$_2$)$_3$OOH	282.2	−134.7	151.3	[107,110]
1-Methylhexyl-, BuCH$_2$CH(OOH)Me	278.8	−131.3	139.7	[107,110]
1-Ethylpentyl-, BuCH(OOH)Et	282.1	−134.6	142.8	[110]
1-Propylbutyl-, PrCH(OOH)Pr	269.1	−121.6	131.5	[110]
1,1-Dimethyl-2-pentyn-4-enyl-, CH$_2$=CHC≡CCMe$_2$OOH	−127.1	267.7		[111]
1,1,4,4-tetramethyl-1,4-dihydro peroxybutane, [HOOCMe$_2$CH$_2$]$_2$	428.6	−288.0		[112]
cyclohexyl-,	214.9	−67.4	135.1	[110]

TABLE 2.7
Enthalpies of Formation ΔH_f^0 (gas, 298 K) of Hydroperoxides and Peroxyl Radicals and C—$O_2^\bullet$ Bond Dissociation Energies in Peroxyl Radicals—*continued*

Hydroperoxide	$-\Delta H_f^0$(ROOH) (kJ mol^{-1})	ΔH_f^0(ROO$^\bullet$) (kJ mol^{-1})	D(R–$O_2^\bullet$) (kJ mol^{-1})	Ref.
1-Methylcyclohexyl-, 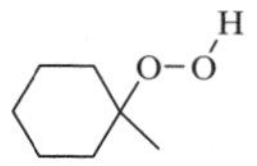	270.3	−129.7	152.4	[110]
E-9-Decalyl-, 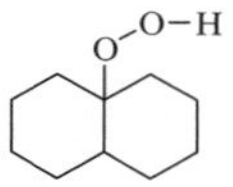	277.8	−137.2	136.9	[113]
1-Methyl-1-phenylethyl-,	85.3	55.3	85.3	[109,114]
PhMe$_2$COOH				[115]
1-Tetralyl-, 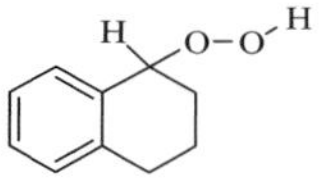	104.5	43.0	91.2	[110]

2.4.2 Reaction $RO_2^\bullet + RH \rightarrow ROOH + R^\bullet$

Peroxyl radicals can undergo various reactions, e.g., hydrogen abstraction, isomerization, decay, and addition to a double bond. Chain propagation in oxidized aliphatic, alkylaromatic, alicyclic hydrocarbons, and olefins with weak C—H bonds near the double bond proceeds according to the following reaction as a limiting step of the chain process [2–15]:

$$RO_2^\bullet + RH \longrightarrow ROOH + R^\bullet.$$

The isotope effect in peroxyl radical reactions with C—H/C—D bonds of attacked hydrocarbon shows the direct hydrogen atom abstraction as the limiting step of this reaction [15]. For example, the cumylperoxyl radical reacts with the C—D bond of α-deuterated cumene (PhMe$_2$CD) ninefold slower than with the C—H bond (cumene, 303 K [118]). The second isotope effect (ratio k_p(PhMe$_2$CH)/k_p(Ph(CD$_3$)$_2$CH) is close to unity, i.e., 1.06 per C—D bond [118].

	RD	PhMe$_2$CD	PhMe$_2$CD	Tetralin C$_{10}$D$_{12}$
RO$_2^\bullet$		Me$_3$COO$^\bullet$	PhMe$_2$COO$^\bullet$	Me$_3$COO$^\bullet$
k_p(H)/k_p(D)		10	9	15.5
E_H-E_D (kJ mol^{-1})		5.8	5.5	6.9

The difference of activation energies in $RO_2^\bullet$ reactions with C—D and C—H bonds (6.1 ± 0.6 kJ mol^{-1}) is close to that of dissociation energies of these bonds ($D_{C-D} - D_{C-H} = 6.3$ kJ mol^{-1}).

The enthalpy of the $RO_2^\bullet + RH$ reaction is determined by the strengths of disrupted and newly formed bonds: $\Delta H = D_{R-H} - D_{ROO-H}$. For the values of O—H BDEs in hydroperoxides, see the earlier discussion on page 41. The dissociation energies of the C—H bonds of hydrocarbons depend on their structure and vary in the range 300 – 440 kJ mol^{-1} (see Chapter 7). The approximate linear dependence (Polany–Semenov relationship) between activation energy E and enthalpy of reaction ΔH was observed with different E_0 values for hydrogen atom abstraction from aliphatic (R^1H), olefinic (R^2H), and alkylaromatic (R^3H) hydrocarbons [119]:

$$E = E_0 + 0.45\Delta H. \tag{2.16}$$

The pre-exponential factor A_{C-H} for the reaction $RO_2^\bullet + RH$ per attacked C—H bond differs for aliphatic hydrocarbons and for hydrocarbons, where the attacked C—H bond is in the α-position to π-C—C bond. This difference is the result of additional loss of the activation entropy due to retardation of group rotation, resulting from the interaction of π-electrons with electrons of reaction center. When the peroxyl radical attacks the C—H bond in neighborhood with the π-C—C bond, the retardation of free rotation around the C—C bond in the transition state additionally lowers the entropy of the transition state. The values of E_0 and A_{C-H} are given here [119]:

	R^iH	R^1H	R^2H	R^3H
E_0 (kJ mol^{-1})		40.4	54.3	46.3
A_{C-H} (L mol^{-1} s^{-1})		10^9	10^8	10^8

Opeida proposed the following empirical equation for the rate constant of the peroxyl radical reaction with C—H bonds of alkylaromatic hydrocarbons [34]:

$$\log(k(348\ K)) = 23.03 - 0.0807 \times D_{R-H} + 31.48(IP - EA)^{-1} - 0.96 V_R, \tag{2.17}$$

where V_R (cm^3 mol^{-1}) is the volume of the substituent in the α-position to the reacting C—H bond, the values of D_{R-H}, IP (ionization potential), and EA (electron affinity) are expressed in kJ mol^{-1} and k in L mol^{-1} s^{-1}, respectively. The influence of different structural and physical factors on the activation energy of the radical reaction of peroxyl radicals with C—H bonds of organic compounds will be discussed in Chapter 6.

According the latest quantum-chemical calculations, the TS of the peroxyl radical reaction with the C—H bond (reaction $EtOO^\bullet + HEt$) has the following characteristics (Figure 2.1):

Reaction	r(C—H)[× 10^{-10} (m)]	r(O—H)[× 10^{-10} (m)]	θ(CHO) (°)	Method	Ref.
$HO_2^\bullet + CH_4$	1.295	1.176	180	MP2/6-16**	[99]
$EtO_2^\bullet + EtH$	1.470	1.115	176.2	B3LYP	[120]

The peroxyl radical is a polar reagent and is solvated in polar solvents. As a result, polar solvents influence the reaction of peroxyl radicals with C—H bonds of a polar molecule

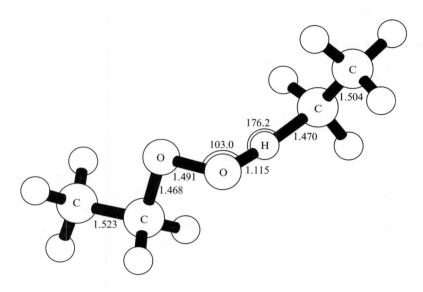

FIGURE 2.1 Geometry of transition state of reaction $EtO_2^\bullet$ + EtH calculated by DFT method [120].

(see Chapter 7) and with O—H and N—H bonds of antioxidants (see Part II). The solvent influence on the reaction of any peroxyl radical with the C—H bond of hydrocarbon is moderate. The experimental results on the reaction of $RO_2^\bullet$ + RH, where RH is 2-methylpentene-2, $T = 333$ K [121], are presented below:

Solvent (mol%)	RH (100%)	C_6H_6 (50%)	C_6H_6 (90%)	$PhNO_2$ (10%)	$PhNO_2$ (20%)
k_p (L mol^{-1} s^{-1})	4.2	8.1	11.5	1.8	0.9

The experimental values of rate constants and activation energies for the reaction of $RO_2^\bullet$ with hydrocarbons are given in Table 2.8. For the experimental methods of k_p estimation, see elsewhere [7,9,12,13,15,17,19].

2.4.3 Intramolecular Hydrogen Atom Transfer in Peroxyl Radical

The peroxyl radical of a hydrocarbon can attack the C—H bond of another hydrocarbon. In addition to this bimolecular abstraction, the reaction of intramolecular hydrogen atom abstraction is known when peroxyl radical attacks its own C—H bond to form as final product dihydroperoxide. This effect of intramolecular chain propagation was first observed by Rust in the 2,4-dimethylpentane oxidation experiments [130]:

TABLE 2.8
Rate Constants of the Hydrogen Atom Abstraction by Peroxyl Radicals from the Hydrocarbons (RO$_2^\bullet$ + RH → ROOH + R$^\bullet$)

RH	T (K)	E (kJ mol^{-1})	log A, A (L mol^{-1} s^{-1})	k (350 K) (L mol^{-1} s^{-1})	Ref.
Me(CH$_2$)$_8$Me	323			0.10	[23]
Me(CH$_2$)$_{12}$Me	323			0.22	[23]
Me(CH$_2$)$_{14}$Me	323			0.30	[23]
MeEt$_2$CH	303			8.0 × 10^{-3}	[122]
Me$_2$CHCH$_2$CHMe$_2$	373			0.23	[26]
(Me$_2$CHCH$_2$)$_2$CHMe	313–353	38.1	5.28	0.39	[123]
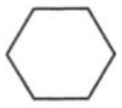 cyclohexane	333			0.71	[25]
 cyclohexane	403–433	54.0	7.43	0.35	[27]
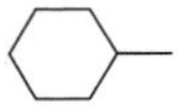 methylcyclohexane	333			0.43	[25]
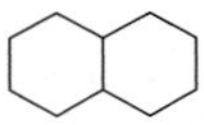 decalin	333			0.40	[25]
 bicyclic	323–358	71.4	10.70	1.11	[124]
Me$_2$C=CMe$_2$	303			2.60	[28]
Me$_2$C=CHCH$_2$Me	313–333	36.8	6.40	8.09	[125]
MeCH$_2$CH=CHCH$_2$Et	303			1.40	[28]
Me$_2$CHCH=CHCHMe$_2$	303			2.30	[28]
CH$_2$=CHCH$_2$CH$_2$Bu	303			1.00	[28]
CH$_2$=CHCH$_2$CH=CH$_2$	303			14.0	[28]
PrCH$_2$C≡CCH$_2$Pr	303			2.80	[28]
 cyclopentene	303			7.00	[28]
 cyclohexene	303–323	29.3	5.46	12.2	[126]
 cyclohexene	333			10.0	[25]
 cycloheptene	303			5.40	[28]
 cycloheptene	288			0.65	[30]
 methylcyclohexene	298–318	25.1	4.93	15.30	[126]
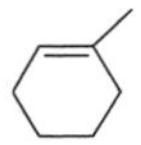 methylcyclohexene	288			1.10	[25]
 methylcyclohexene	303–323	39.7	7.61	48.44	[126]

TABLE 2.8
Rate Constants of the Hydrogen Atom Abstraction by Peroxyl Radicals from the Hydrocarbons (RO$_2^\bullet$ + RH → ROOH + R$^\bullet$)—*continued*

RH	T (K)	E (kJ mol^{-1})	log A, A (L mol^{-1} s^{-1})	k (350 K) (L mol^{-1} s^{-1})	Ref.
(cyclohexane-like)	288–308	34.7	7.05	7.44	[126]
(benzene ring)	303			8.1 × 10^2	[28]
(benzene ring)	303			1.5 × 10^3	[28]
PhCH$_3$	303			0.24	[28]
PhCH$_3$	348			2.30	[34]
(o-xylene)	303			0.42	[28]
(o-xylene)	333			2.80	[25]
(m-xylene)	303			0.48	[28]
(m-xylene)	333			2.05	[25]
(p-xylene)	303			0.84	[28]
(p-xylene)	333			2.80	[25]
PhCH$_2$Me	323–363	35.6	5.98	4.65	[38]
PhCH$_2$Me	303			1.30	[28]
PhCHMe$_2$	308–338	42.4	6.51	1.52	[39]
PhCHMe$_2$	313–363	41.0	6.60	3.03	[127]
PhCHMe$_2$	303			0.18	[28,41]
PhCHMe$_2$	303			0.41	[40]
PhCHMe$_2$	348			1.44	[34]
PhCH$_2$Pr	303			0.56	[28]
PhCHMeEt	303			7.6 × 10^{-2}	[41]
Ph$_2$CH$_2$	323–353	48.9	8.30	10.05	[38]
Ph$_2$CH$_2$	348			23.4	[34]
Ph$_2$CH$_2$	303			2.10	[28]
PhCH$_2$CH$_2$Ph	303			0.28	[28]
PhMePrCH	303			0.07	[128]
MePh$_2$CH	313–353	50.6	8.57	10.43	[42]
Ph$_2$CHMe	303			0.34	[28,41]

continued

TABLE 2.8
Rate Constants of the Hydrogen Atom Abstraction by Peroxyl Radicals from the Hydrocarbons (RO$_2^\bullet$ + RH → ROOH + R$^\bullet$)—*continued*

RH	T (K)	E (kJ mol^{-1})	log A, A (L mol^{-1} s^{-1})	k (350 K) (L mol^{-1} s^{-1})	Ref.
cyclopentyl-phenyl	348			5.10	[128]
cyclohexyl-phenyl	348			1.00	[128]
cyclohexyl-phenyl	303			0.06	[28]
cycloheptyl-phenyl	348			2.20	[128]
cyclooctyl-phenyl	348			24.0	[128]
indane	303			4.8	[28]
tetralin	291–333	23.0	4.74	20.30	[129]
tetralin	273–298	18.8	4.40	39.29	[45]
tetralin	303			6.35	[28,41]
tetralin	348			55.6	[36]
dihydroanthracene	303			3.30 × 10^2	[28,41]
octahydroanthracene	348			1.56 × 10^2	[36]
PhCH$_2$CH=CH$_2$	303			10.0	[28]
PhCH$_2$CH=CHMe	303			8.2	[28]
indene	303			1.42 × 10^2	[28]
dihydronaphthalene	303			9.0 × 10^2	[28]

The rate of this intramolecular isomerization depends on the chain length, with the maximum in the case of a six-atomic transition state, i.e., when the tertiary C—H bond is in the β-position with respect to the peroxyl group [13]. For the values of rate constants of intramolecular attack on the tertiary and secondary C—H bond, see Table 2.9. The parameters of peroxyl radical reactivity in reactions of intra- and intermolecular hydrogen atom abstraction are compared and discussed in Chapter 6.

TABLE 2.9
Rate Constants of Isomerization of Peroxyl Radicals (Experimental Data)

Peroxyl Radical	Product	T, (K)	k, (s^{-1})	Ref.
Me$_2$C(O$_2^•$)CH$_2$CHMe$_2$	Me$_2$C(OOH)CH$_2$C$^•$Me$_2$	373	18	[26]
MeCH(O$_2^•$)CH$_2$CH$_2$Me	MeCH(OOH)CH$_2$C$^•$HMe	373	0.87	[131]
Me$_2$C(O$_2^•$)(CH$_2$)$_2$CHMe$_2$	Me$_2$C(OOH)(CH$_2$)$_2$C$^•$Me$_2$	373	8.0	[13]
Me$_2$CHCH$_2$C(O$_2^•$)Me CH$_2$CHMe$_2$	Me$_2$C$^•$CH$_2$C(OOH)MeCH$_2$CHMe$_2$	373	46	[132]
MeCH(O$_2^•$)(CH$_2$)$_{11}$Me	MeCH(OOH)CH$_2$C$^•$H(CH$_2$)$_9$Me	413	1.7×10^2	[133]
MeCH(O$_2^•$)(CH$_2$)$_{13}$Me	MeCH(OOH)CH$_2$C$^•$H(CH$_2$)$_{11}$Me	433	1.4×10^2	[134]
PhCH(O$_2^•$)OCH$_2$Ph	PhCH(OOH)OC$^•$HPh	303	86	[135]
PhCH(O$_2^•$)OCH$_2$Ph	PhCH(OOH)OC$^•$HPh	323	1.1×10^2	[136]
PhCH(OOH)OCH(O$_2^•$)Ph	PhC$^•$(OOH)OCH(OOH)Ph	323	9.5	[136]

2.4.4 Addition of Peroxyl Radical to the Double Bond

Olefin possesses two reaction centers to be attacked by the peroxyl radical. The peroxyl radicals abstract the hydrogen atom from the weakest C–H bonds in the α-position to the double bond of these compounds with the formation of hydroperoxides. In addition to this reaction, they attack the double bond of the olefin with the formation of oligomeric polyperoxides [12,13,15,137]:

$$RO_2^• + CH_2{=}CHX \longrightarrow ROOCH_2C^•HX$$
$$ROOCH_2C^•HX + O_2 \longrightarrow ROOCH_2CHXOO^•$$
$$ROOCH_2CHXOO^• + CH_2{=}CHX \longrightarrow ROOCH_2CHXOOCH_2C^•HX$$

Oligomeric polyperoxide is formed as a result of such copolymerization of the monomer and dioxygen. During oxidation of many unsaturated hydrocarbons, both reactions (abstraction and addition) occur in parallel to produce a mixture of hydroperoxides and oligomeric polyperoxides. The relative amounts of the products of RO$_2^•$ addition to the π-bond of olefins are given below [13]:

Olefin	Ethylene	Propylene	Hexene-1	Butene-1	Cyclohexene
T (K)	383	383	363	343	333
[ROOR]/[ROOH] (%)	100	50	33	26	4.4

Olefin	Isobutene	Butene-1, Trimethyl-	Trimethyl ethylene	Cyclopentene	Cyclo Octene
T (K)	353	343	333	323	343
[ROOR]/[ROOH] (%)	80	6	52	11	71

The rate constant of RO$_2^•$ addition to the π-bond of the unsaturated hydrocarbon depends on its structure and, hence, varies widely (see Table 2.10). The problems of reactivity of reactants in such reactions will be discussed in Chapter 6.

Alkylperoxyl radicals produced by addition reactions can be destructively isomerized with the formation of epoxides [13,139,154,155]:

TABLE 2.10
Rate Constants of the Addition of Peroxyl Radicals to the Double Bond of Olefins (Experimental Data)

Oxidizing Compound	Peroxyl Radical	T (K)	k (L mol^{-1} s^{-1})	Ref.
$CH_2=CHEt$	$CH_2=CHCHMeO_2^{\bullet}$	343	6.3	[13]
$E\text{-}MeCH=CHMe$	$E\text{-}MeCH=CHCH_2O_2^{\bullet}$	343	1.9	[13]
$CH_2=CMe_2$	$CH_2=CMeCH_2O_2^{\bullet}$	353	3.8	[13]
$CH_2=CHBu$	$(CH_3)_3CO_2^{\bullet}$	393	0.83	[138]
$Me_2C=CMe_2$	$Me_3CO_2^{\bullet}$	393	22	[138]
$Me_2C=CMe_2$	$\sim CH_2CH(O_2^{\bullet})Ph$	323	1.3×10^2	[139]
$Me_2C=CMe_2$	$CF_3CCl_2O_2^{\bullet}$	294	3.1×10^7	[140]
$Me_2C=CMe_2$	$CF_2ClO_2^{\bullet}$	294	8.3×10^7	[140]
$Me_2C=CMe_2$	$(CF_3)_2CFO_2^{\bullet}$	294	5.0×10^7	[140]
$CH_2=CHCMe_3$	$Me_3CO_2^{\bullet}$	393	0.56	[138]
$CH_2=CHCMe_3$	$PhMe_2CO_2^{\bullet}$	393	0.66	[138]
$PhCH=CH_2$	$HO_2^{\bullet}$	323	78.8	[141]
$PhCH=CH_2$	$Me_3CO_2^{\bullet}$	323	13.8	[141]
$PhCH=CH_2$	$Me_3CO_2^{\bullet}$	303	1.3	[142]
$PhCH=CH_2$	$Me_3CO_2^{\bullet}$	393	40.0	[138]
$PhCH=CH_2$	$PrCH_2CO_2^{\bullet}$	303	14.0	[44]
$PhCH=CH_2$	$PhMe_2CO_2^{\bullet}$	323	21.1	[141]
$PhCH=CH_2$	$PhMe_2CO_2^{\bullet}$	368	76.6	[143]
$PhCH=CH_2$	$Ph_2CHO_2^{\bullet}$	303	16.0	[44]
$PhCH=CH_2$	$CF_3CCl_2O_2^{\bullet}$	294	57×10^5	[140]
$PhCH=CH_2$	tetralinyl-OO$^{\bullet}$	303	8.65	[144]
$PhCH=CH_2$	cyclo-$C_8H_{15}O_2^{\bullet}$	323	1.10×10^2	[141]
$PhCH=CH_2$	$\sim CH_2CH(O_2^{\bullet})Ph$	286–333	$4.70 \times 10^7 \exp(-35.1/RT)$	[145]
$PhCH=CH_2$	$\sim CH_2CH(O_2^{\bullet})Ph$	323	1.1×10^2	[146]
$PhCD=CH_2$	$\sim CH_2CH(O_2^{\bullet})Ph$	286–333	$2.29 \times 10^8 \exp(-39.7/RT)$	[147]
$PhCD=CH_2$	$\sim CH_2CD(O_2^{\bullet})Ph$	286–313	$2.516 \times 10^8 \exp(-39.7/RT)$	[147]
$PhCH=CD_2$	$\sim CD_2CMe(O_2^{\bullet})Ph$	313	78	[147]
$PhCD=CD_2$	$\sim CD_2CD(O_2^{\bullet})Ph$	313	55	[147]
$C_6D_5CH=CH_2$	$\sim CH_2CH(O_2^{\bullet})\ C_6D_5$	313	69	[147]
$Br\text{-}C_6H_4\text{-}CH=CH_2$	$\sim CH_2CH(O_2^{\bullet})Ph$	323	56	[139]
$Br\text{-}C_6H_4\text{-}CH=CH_2$	$PhMe_2CO_2^{\bullet}$	368	57.9	[143]

TABLE 2.10
Rate Constants of the Addition of Peroxyl Radicals to the Double Bond of Olefins (Experimental Data)—*continued*

Oxidizing Compound	Peroxyl Radical	T (K)	k (L mol^{-1} s^{-1})	Ref.
4-tert-butyl styrene	PhMe$_2$CO$_2^\bullet$	368	84.4	[143]
4-CN styrene	~CH$_2$CH(O$_2^\bullet$)Ph	323	45	[139]
4-CN styrene	4-CN-C$_6$H$_4$-CH(OO$^\bullet$)CH$_2$R	313	91	[147]
3-Cl styrene	3-Cl-C$_6$H$_4$-CH(OO$^\bullet$)CH$_2$R	303	101	[147]
3-Cl styrene	3-Cl-C$_6$H$_4$-CH(OO$^\bullet$)CH$_2$R	313	1.0×10^2	[147]
4-Cl styrene	PhMe$_2$CO$_2^\bullet$	368	55.5	[143]
4-Cl styrene	~CH$_2$CH(O$_2^\bullet$)Ph	323	55	[139]
4-Cl styrene	4-Cl-C$_6$H$_4$-CH(OO$^\bullet$)CH$_2$R	313	1.2×10^2	[147]
4-Cl styrene	4-Cl-C$_6$H$_4$-CH(OO$^\bullet$)CH$_2$R	313	123	[147]
4-F styrene	~CH$_2$CH(O$_2^\bullet$)Ph	323	56	[139]
4-I styrene	~CH$_2$CH(O$_2^\bullet$)Ph	323	70	[139]
4-MeO styrene	PhMe$_2$CO$_2^\bullet$	368	1.11×10^2	[143]
4-MeO styrene	4-MeO-C$_6$H$_4$-CH(OO$^\bullet$)CH$_2$R	286–333	$1.41 \times 10^7 \exp(-30.1/RT)$	[145]

continued

TABLE 2.10
Rate Constants of the Addition of Peroxyl Radicals to the Double Bond of Olefins (Experimental Data)—*continued*

Oxidizing Compound	Peroxyl Radical	T (K)	k (L mol^{-1} s^{-1})	Ref.
4-MeO-C$_6$H$_4$-CH=CH$_2$	4-MeO-C$_6$H$_4$-CH(OO·)CH$_2$R	313	123	[147]
4-Me-C$_6$H$_4$-CH=CH$_2$	~CH$_2$CH(O$_2$·)Ph	323	69	[139]
4-Me-C$_6$H$_4$-CH=CH$_2$	4-Me-C$_6$H$_4$-CH(OO·)CH$_2$R	313	83	[147]
3-NO$_2$-C$_6$H$_4$-CH=CH$_2$	~CH$_2$CH(O$_2$·)Ph	323	40	[139]
4-NO$_2$-C$_6$H$_4$-CH=CH$_2$	~CH$_2$CH(O$_2$·)Ph	323	41	[139]
4-CF$_3$-C$_6$H$_4$-CH=CH$_2$	PhMe$_2$CO$_2$·	368	35.6	[143]
PhMeC=CH$_2$	PrCH$_2$O$_2$·	303	25.5	[44]
PhMeC=CH$_2$	Me$_3$CO$_2$·	303	2.9	[142]
PhMeC=CH$_2$	Me$_3$CO$_2$·	323	17	[141]
PhMeC=CH$_2$	Me$_3$CO$_2$·	393	47.0	[138]
PhMeC=CH$_2$	PhMe$_2$CO$_2$·	323	34	[141]
PhMeC=CH$_2$	cyclo-C$_8$H$_{15}$O$_2$·	323	2.19×10^2	[141]
PhMeC=CH$_2$	tetralinyl-O-O·	303	16.0	[44]
PhMeC=CH$_2$	Ph$_2$CHO$_2$·	303	25.0	[44]
PhMeC=CH$_2$	~CH$_2$CMe(COOMe)O$_2$·	323	13.0	[141]
PhMeC=CH$_2$	~CH$_2$CMe(O$_2$·)Ph	286–323	6.6×10^6 exp(−33.7/RT)	[148]
PhMeC=CH$_2$	~CH$_2$CH(O$_2$·)Ph	323	98.0	[141]
PhClC=CH$_2$	PhMe$_2$CO$_2$·	303	3.9	[44]
MeCH=CHPh	Me$_3$CO$_2$·	303	2.5	[149]
MeCH=CHPh	PhMe$_2$CO$_2$·	303	4.2	[118]
MeCH=CHPh	~CHMeCHPhO$_2$·	303	51	[148]
E-PhCH=CHPh	HO$_2$·	313–393	1.20×10^8 exp(−37.7/RT)	[141]
E-PhCH=CHPh	Me$_3$CO$_2$·	303	0.44	[149]
Z-PhCH=CHPh	Me$_3$CO$_2$·	303	0.12	[149]
CH$_2$=CPh$_2$	Me$_3$CO$_2$·	303	9.2	[149]

TABLE 2.10
Rate Constants of the Addition of Peroxyl Radicals to the Double Bond of Olefins (Experimental Data)—continued

Oxidizing Compound	Peroxyl Radical	T (K)	k (L mol^{-1} s^{-1})	Ref.
cyclohexene	$CF_3CCl_2O_2^\bullet$	294	2.1×10^5	[118]
1-methylcyclohexene	$CF_3CCl_2O_2^\bullet$	294	3.2×10^6	[118]
indene	indanyl-O-O$^\bullet$ (R)	303	1.2×10^2	[148]
1,2-dihydronaphthalene	tetralinyl-O-O$^\bullet$ (R)	303	2.9×10^2	[28]
phenylcyclohexene	R,Ph-cyclohexyl-O-O$^\bullet$	323–343	$1.4 \times 10^8 \exp(-35.1/RT)$	[150]
anthracene	$PhCH(O_2^\bullet)Me$	323–358	$1.1 \times 10^6 \exp(-18.6/RT)$	[151]
anthracene	$PhCH_2O_2^\bullet$	323–353	$1.2 \times 10^{13} \exp(-60.6/RT)$	[151]
$MeCH=CHCH_2OH$	$CF_3CCl_2O_2^\bullet$	294	4.9×10^4	[140]
$CH_2=CMe\ CH_2CH_2OH$	$CF_3CCl_2O_2^\bullet$	294	5.3×10^4	[140]
$CH_2=CHOEt$	$Me_3CO_2^\bullet$	303	4.0×10^{-2}	[149]
$CH_2=CHC(O)Et$	$Me_3CO_2^\bullet$	303	1.3×10^{-2}	[149]
$CH_2=CHC(O)Et$	$\sim CH_2CH(C(O)Et)O_2^\bullet$	303	2.7	[149]
$Me_2C=CHC(O)Me$	$HO_2^\bullet$	313–393	$2.0 \times 10^2 \exp(-18.8/RT)$	[141]
$CH_2=CHOC(O)Me$	$HO_2^\bullet$	323	6.8	[141]
$CH_2=CHOC(O)Me$	$Me_3CO_2^\bullet$	323	0.10	[141]
$CH_2=CHOC(O)Me$	$PhMe_2CO_2^\bullet$	323	0.20	[141]
$CH_2=CHOC(O)Me$	$cyclo\text{-}C_8H_{15}O_2^\bullet$	323	1.50	[141]
$CH_2=CHOC(O)Me$	$\sim CH_2CH(O_2^\bullet)OC(O)Me$	323	2.80	[141]
$CH_2=CHC(O)OMe$	$\sim CH_2CH(O_2^\bullet)Ph$	323	9.8	[141]
$CH_2=CMeC(O)OMe$	$PhMe_2CO_2^\bullet$	323	1.8	[141]
$CH_2=CMeC(O)OMe$	$Me_3CO_2^\bullet$	323	1.1	[141]
$CH_2=CMeC(O)OMe$	$HO_2^\bullet$	323	40	[141]
$CH_2=CMeC(O)OMe$	$cyclo\text{-}C_8H_{15}O_2^\bullet$	323	10.4	[141]
$CH_2=CMeC(O)OMe$	$\sim CH_2CH(O_2^\bullet)Ph$	323	12	[141]
$CH_2=CMeC(O)OMe$	$\sim CH_2CMe(O_2^\bullet)C(O)Me$	303–323	$8.3 \times 10^8 \exp(-53.5/RT)$	[152]

continued

TABLE 2.10
Rate Constants of the Addition of Peroxyl Radicals to the Double Bond of Olefins (Experimental Data)—*continued*

Oxidizing Compound	Peroxyl Radical	T (K)	k (L mol^{-1} s^{-1})	Ref.
Me$_2$C=CHOC(O)Me	Me$_3$CO$_2^\bullet$	303	1.0 10^{-2}	[149]
Me$_2$C=CHOC(O)Me	~CMe$_2$CH(COOMe)O$_2^\bullet$	303	0.20	[149]
E-MeCH=CHC(O)OMe	HO$_2^\bullet$	313–393	3.72 10^5 exp(–25.1/RT)	[141]
E-MeCH=CH C(O)OEt	HO$_2^\bullet$	313–393	1.82 10^5 exp(–25.1/RT)	[141]
CH$_2$=CMeC(O)OBu	~CH$_2$CMe(COOBu)O$_2^\bullet$	303–323	6.3 10^7 exp(–45.6/RT)	[152]
CH$_2$=CMeC(O)OCH$_2$CHMe$_2$	~CH$_2$CMe(C(O)OCH$_2$CHMe$_2$)O$_2^\bullet$	323	2.4	[152]
Z-EtOC(O)CH=CHC(O)OEt	HO$_2^\bullet$	313–393	7.08 10^6 exp(41.4/RT)	[141]
E-EtOC(O)CH=CHC(O)OEt	HO$_2^\bullet$	313–393	5.37 10^9 exp(61.5/RT)	[141]
[(CH$_2$=CHC(O)OCH$_2$]$_2$CMe$_2$	PhMe$_2$CO$_2^\bullet$	323	0.76	[153]
[CH$_2$=CHC(O)OCH$_2$]$_4$C	PhMe$_2$CO$_2^\bullet$	323	1.3	[153]
[CH$_2$=C(CH$_3$)C(O)OCH$_2$]$_4$C	PhMe$_2$CO$_2^\bullet$	323	3.1	[154]
CH$_2$=CHC(O)NH$_2$	PhMe$_2$CO$_2^\bullet$	323	0.40	[141]
CH$_2$=CHC(O)NH$_2$	Me$_3$CO$_2^\bullet$	323	0.20	[141]
CH$_2$=CHC(O)NH$_2$	cyclo-C$_8$H$_{15}$O$_2^\bullet$	323	3.30	[141]
CH$_2$=CMeC(O)NH$_2$	Me$_3$CO$_2^\bullet$	323	0.30	[141]
CH$_2$=CMeC(O)NH$_2$	cyclo-C$_8$H$_{15}$O$_2^\bullet$	323	3.90	[141]
CH$_2$=CHCN	HO$_2^\bullet$	323	35.0	[141]
CH$_2$=CHCN	Me$_3$CO$_2^\bullet$	323	0.70	[141]
CH$_2$=CHCN	PhMe$_2$CO$_2^\bullet$	323	1.5	[141]
CH$_2$=CHCN	cyclo-C$_8$H$_{15}$O$_2^\bullet$	323	5.1	[141]
CH$_2$=CHCN	~CH$_2$CH(O$_2^\bullet$)Ph	323	25	[141]
CH$_2$=CHCN	~CH$_2$CH(O$_2^\bullet$)CN	323	11	[141]
CH$_2$=CHCN	~CH$_2$CH(O$_2^\bullet$)CN	303	3.2	[149]
MeCH=CHCN	~CHMeCH(CN)O$_2^\bullet$	303	4.5	[149]

TABLE 2.11
Rate Constants and Activation Energies of the Decomposition of Peroxyalkyl Radicals with Epoxide Formation ([13,139])

Monomer	T (K)	k (s^{-1})	E (kJ mol^{-1})	log A, A (s^{-1})
E-MeCH=CHMe	363	9.1×10^6	33.9	12.1
CH$_2$=CMePr	343	9.1×10^6	33.9	12.1
Me$_2$C=CHMe	333	6.7×10^6	33.8	12.1
CH$_2$=CHPh	323	7.6×10^3	53.6	12.5
CH$_2$=CMePh	323	6.8×10^4	47.7	12.5
MeO-C$_6$H$_4$-CH=CH$_2$	368	3.8×10^4		
tBu-C$_6$H$_4$-CH=CH$_2$	368	6.6×10^4		
Br-C$_6$H$_4$-CH=CH$_2$	368	4.7×10^4		
Cl-C$_6$H$_4$-CH=CH$_2$	368	7.6×10^4		
CF$_3$-C$_6$H$_4$-CH=CH$_2$	368	2.4×10^4		
cyclopentene	323	4.5×10^6	38.2	12.8
cycloheptene	333	2.9×10^7	34.2	12.8
cyclooctene	343	4.5×10^8	27.4	12.8

$$R-O-C-C(X)-O-O^{\bullet} \longrightarrow \triangle(O)X + RO^{\bullet}$$

Such isomerization occurs rapidly (see Table 2.11).

Chain propagation in oxidized 1,2-substituted ethylenes proceeds via addition of dioxygen followed by the elimination of the hydroperoxyl radical [156]:

$$HOO^{\bullet} + RCH=CHR \longrightarrow RC^{\bullet}HCH(OOH)R$$
$$RC^{\bullet}HCH(OOH)R + O_2 \longrightarrow RC(OO^{\bullet})HCH(OOH)R$$
$$RC(OO^{\bullet})HCH(OOH)R \longrightarrow RCH=C(OOH)R + HOO^{\bullet}$$

2.5 CHAIN TERMINATION IN OXIDIZED HYDROCARBONS

2.5.1 TETROXIDES

Chain termination during the oxidation of hydrocarbons usually is a result of the interaction of two peroxyl radicals by a multistep mechanism. The mechanism of dispropotionation is

TABLE 2.12
Thermodynamic Characteristics of Equilibrium ROOOOR ⇔ 2 RO$_2^\bullet$

R—	Solvent	ΔH^0 (kJ mol^{-1})	ΔS^0 (J mol^{-1} K^{-1})	K (200 K) (mol L^{-1})	Ref.
Me$_3$C—	Dichloro methane	46.6	222	0.27	[157]
Me$_3$C—	Dichlorodifluoromethane	35.1	136	8.6 × 10^{-3}	[158]
Me$_3$C—	2-Methylbutane	35.1	125	2.3 × 10^{-3}	[158]
EtMe$_2$C—	2-Methylbutane	37.2	151	1.5 × 10^{-2}	[159]
EtMe$_2$C—	Dichlorodifluoromethane	31.4	121	1.3 × 10^{-2}	[158]
MeEt$_2$C—	2-Methylbutane	40.6	184	0.10	[159]
Me$_2$CH—	Dichlorodifluoromethane	33.5	105	5.4 × 10^{-4}	[160]
Me$_2$CH—	Cyclopropane	33.0	63	4.7 × 10^{-6}	[161]
Me$_2$CD—	Dichlorodifluoromethane	33.5	138	2.9 × 10^{-2}	[160]
PrMe$_2$C—	2-Methylpentane	37.2	163	6.3 × 10^{-2}	[159]
Me$_2$CHMe$_2$C—	2,3-Dimethylbutane	34.3	142	3.9 × 10^{-2}	[159]
Me$_2$CHMe$_2$C—	2-Methylbutane	36.0	138	6.4 × 10^{-3}	[159]
Me$_3$CMe$_2$C—	2,2,3-Trimethylbutane	36.4	159	6.3 × 10^{-2}	[159]
Me$_3$CCH$_2$Me$_2$C—	Dichlorodifluoromethane	32.6	130	1.9 × 10^{-2}	[159]
cyclopentyl—	Dichlorodifluoromethane	31.4	59	7.6 × 10^{-6}	[161]
cyclopentyl—	Cyclopropane	31.0	84	1.9 × 10^{-4}	[161]
1-methylcyclopentyl—	Dichlorodifluoromethane	33.5	125	6.0 × 10^{-3}	[158]
1-methylcyclohexyl—	Dichlorodifluoromethane	29.3	146	0.94	[158]
PhMe$_2$C—	Cumene	46.9	201	1.8 × 10^{-2}	[159]
PhMe$_2$C—	2-Methylbutane	44.3			[162]
PhMe$_2$C—	Dichlorodifluoromethane	38.5	134	8.8 × 10^{-4}	[158]

different for tertiary and secondary alkylperoxyl radicals. The first step involves the reversible formation of unstable tetroxide [12,13,15]:

$$2RO_2^\bullet \rightleftharpoons ROOOOR$$

The equilibrium between 1,1-dimethylethylperoxyl radicals and 1,1-dimethylethyl tetroxide was first evidenced by Bartlett and Guaraldi [157] for peroxyl radicals generated by irradiation of bis(1,1-dimethylethyl) peroxycarbonate in CH$_2$Cl$_2$ at 77 K and oxidation of 1,1-dimethylethyl hydroperoxide with lead tetraacetate at 183 K in CH$_2$Cl$_2$. A series of studies of this equilibrium were performed later using the EPR technique (see Table 2.12). It is seen that the enthalpy of tetroxide decomposition ranges from 29 to 47 kJ mol^{-1}.

The irreversible decay of tetroxide occurs as the result of the dissociation of the O—OR bond to produce RO$_3^\bullet$ and RO$^\bullet$ radicals:

$$ROOOOR \longrightarrow RO_3^\bullet + RO^\bullet$$

The results of estimation of the decomposition rate constants k_d are collected in Table 2.13.

Chain Mechanism of Liquid-Phase Oxidation of Hydrocarbons

TABLE 2.13
The Rate Constants and Activation Energies of Tetroxides Decomposition ROOOOR → RO$_3^{\bullet}$ + RO$^{\bullet}$

R—	Solvent	E (kJ mol^{-1})	log A, A (s^{-1})	k_d(200 K) (s^{-1})	Ref.
Me$_2$CH—	Dichlorodifluoromethane	40.2	10.3	0.63	[160]
Me$_3$C—	Dichlorodifluoromethane	71.5	16.7	1.06×10^{-2}	[157]
Me$_3$C—	2-Methylbutane	73.2	16.6	3.04×10^{-3}	[157]
Me$_2$PrC—	2-Methylpentane	76.1	19.6	0.53	[162]
BuCH$_2$CHMe—	Heptane	44.8	11.6	0.79	[158]
Me$_3$CMe$_2$C—	2,2,3-Trimethylbutane	67.8	17.5	0.21	[158]
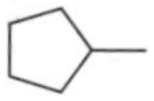 cyclopentyl	Cyclopentane	41.8	12.3	24.2	[156]
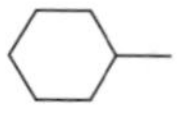 cyclohexyl	Cyclohexane	40.6	11.6	9.9	[156]
PhMe$_2$C—	Cumene	69.0	17.1	0.12	[157]

According to the quantum-chemical calculations and experimental findings, the BDEs were found to be the following [21]:

$$D(\text{ROO}-\text{OOR}) = 35.0 \pm 2.3 \text{ kJ mol}^{-1}$$
$$D(\text{ROOO}-\text{OH}) = 105.0 \pm 2.0 \text{ kJ mol}^{-1}$$
$$D(\text{ROOO}-\text{OR}) = 87.0 \pm 5.8 \text{ kJ mol}^{-1}$$
$$D(\text{ROOOO}-\text{R}) = 266.1 \pm 5.0 \text{ kJ mol}^{-1}$$

For the enthalpies of formation of tetroxides calculated by the method of increments [163,164], see Table 2.14.

2.5.2 Disproportionation of Tertiary Peroxyl Radicals

When R$^{\bullet}$ is a tertiary alkyl radical, the formed tetroxide decomposes with the formation of two RO$^{\bullet}$ and O$_2$. The chain termination includes the following stages in hydrocarbon oxidation by tertiary the C—H bond [12,13,15,165,166]:

TABLE 2.14
The Enthalpies of Formation of Tetroxides [21]

Tetroxide	$-\Delta H^0$ (kJ mol^{-1})	Tetroxide	$-\Delta H^0$ (kJ mol^{-1})
HOOOOH	136.0	EtOOOOMe	156.5
MeOOOOH	131.0	Me$_3$COOOOMe	231.4
EtOOOOH	163.2	EtOOOOEt	192.9
Me$_3$COOOOH	246.0	Me$_3$COOOOEt	265.7
MeOOOOMe	125.5	Me$_3$COOOOCMe$_3$	349.0

$$RO_2^{\bullet} + RO_2^{\bullet} \rightleftharpoons ROOOOR$$
$$ROOOOR \longrightarrow [RO_3^{\bullet} + RO^{\bullet}]$$
$$[RO_3^{\bullet} + RO^{\bullet}] \longrightarrow [RO^{\bullet} + RO^{\bullet}] + O_2$$
$$[RO^{\bullet} + RO^{\bullet}] \longrightarrow RO^{\bullet} + RO^{\bullet}$$
$$[RO^{\bullet} + RO^{\bullet}] \longrightarrow ROOR$$

The chain termination is a result of tertiary alkylperoxyl radical recombination in the solvent cage. The values of the rate constants for chain termination through the disproportionation of tertiary peroxyl radicals are collected in Table 2.15. They vary in the range 10^3 to 10^5 L mol^{-1} s^{-1} at room temperature. The probability of a pair of alkoxyl radicals to escape cage recombination is sufficiently higher than that of cage recombination. The values of rate constants of the reaction $2\,RO_2^{\bullet} \rightarrow 2\,RO^{\bullet} + O_2$ measured by the EPR technique are presented in Table 2.16.

The chain termination by reactions of tertiary alkylperoxyl radicals is complicated by their decomposition with production of ketone and the alkyl radical, for example:

$$Me_2PhCO^{\bullet} \longrightarrow PhC(O)Me + CH_3^{\bullet}$$

The formed methyl radical adds dioxygen, and the methylperoxyl radical participates in chain termination:

$$CH_3^{\bullet} + O_2 \longrightarrow CH_3O_2^{\bullet}$$
$$Me_2PhCOO^{\bullet} + CH_3O_2^{\bullet} \longrightarrow Me_2PhCOH + CH_2(O) + O_2$$

The last reaction occurs much rapidly than the disproportionation of two cumylperoxyl radicals and accelerates chain termination in oxidized cumene [15]. The addition of cumene hydroperoxide helps to avoid the influence of the cross termination reaction $Me_2PhCOO^{\bullet} + CH_3O_2^{\bullet}$ on the oxidation of cumene and to measure the "pure disproportionation" of cumylperoxyl radicals [15].

2.5.3 Disproportionation of Primary and Secondary Peroxyl Radicals

Russell [179] proposed the following mechanism of chain termination by primary and secondary peroxyl radicals with coordinated decomposition of formed tetroxide to alcohol, ketone, and O_2:

$$2R^1R^2HCOO^{\bullet} \longrightarrow R^1R^2HCOOOOCHR_1R_2$$
$$R^1R^2HCOOOOCHR^1R^2 \longrightarrow R^1R_2C{=}O + O_2 + HOCHR^1R^2$$

This reaction is very exothermic (e.g., $\Delta H = -405\,kJ\,mol^{-1}$ for cyclohexyltetroxide) and supposed to occur rapidly. The values of rate constants for primary and secondary $RO_2^{\bullet}$ cover the range at 300 K, $2k_t = 10^6$ to 10^8 L mol^{-1} s^{-1} (see Table 2.15). According to Russell's mechanism, tetroxide decomposition proceeds via the cyclic transition state and includes the abstraction of the C—H bond:

TABLE 2.15
Rate Constants of Chain Termination by the Disproportionation of Tertiary Peroxyl Radicals in Hydrocarbon Solutions (Experimental Data)

Peroxyl Radical	Solvent	T (K)	k (L mol^{-1} s^{-1})	Ref.
Me$_3$CO$_2^\bullet$	Me$_2$CHPr	225–249	$1.30 \times 10^{12} \exp(-35.1/RT)$	[167]
Me$_3$CO$_2^\bullet$	Me(CH$_2$)$_4$Me	303	$8.51 \times 10^{10} \exp(-30.1/RT)$	[167]
Me$_3$CO$_2^\bullet$	Et$_2$MeCH	303	$1.23 \times 10^{12} \exp(-35.1/RT)$	[167]
Me$_3$CO$_2^\bullet$	cyclopentane	303	$1.10 \times 10^{12} \exp(-35.6/RT)$	[167]
Me$_3$CO$_2^\bullet$	methylcyclohexane	303	$1.48 \times 10^{12} \exp(-35.6/RT)$	[167]
Me$_3$CO$_2^\bullet$	C$_6$H$_6$	309–338	$1.41 \times 10^{10} \exp(-42.7/RT)$	[168]
Me$_3$CO$_2^\bullet$	C$_6$H$_6$	295–333	$2.00 \times 10^{7} \exp(-22.6/RT)$	[169]
Me$_3$CO$_2^\bullet$	C$_6$H$_6$	295–333	$2.51 \times 10^{6} \exp(-18.8/RT)$	[169]
Me$_3$CO$_2^\bullet$	CF$_2$Cl$_2$	253–303	$3.16 \times 10^{5} \exp(-21.3/RT)$	[158]
Me$_3$CO$_2^\bullet$	CCl$_4$	303	1.0×10^{3}	[170]
Me$_3$CO$_2^\bullet$	CH$_3$OH	295–333	$1.58 \times 10^{8} \exp(-26.4/RT)$	[169]
Me$_3$CO$_2^\bullet$	Me$_3$COOH	338	2.80×10^{3}	[15]
Me$_3$CO$_2^\bullet$	Me$_3$COOH	295	4.2×10^{3}	[168]
Me$_3$CO$_2^\bullet$	Me$_3$COOCMe$_3$	303	$3.63 \times 10^{11} \exp(-34.7/RT)$	[167]
EtMe$_2$CO$_2^\bullet$	CF$_2$Cl$_2$	253–303	$5.01 \times 10^{5} \exp(-19.7/RT)$	[158]
EtMe$_2$CO$_2^\bullet$	CCl$_4$	303	4.50×10^{2}	[158]
Me$_2$(Me$_2$CH)CO$_2^\bullet$	CCl$_4$	303	1.00×10^{3}	[170]
Me$_2$(Me$_2$CH)CO$_2^\bullet$	CF$_2$Cl$_2$	253–303	$5.01 \times 10^{5} \exp(-16.7/RT)$	[158]
Me$_2$(Me$_2$CH)CO$_2^\bullet$	CH$_3$OH	253–303	$1.58 \times 10^{5} \exp(-29.3/RT)$	[158]
Me(CH$_2$)$_2$C(O$_2^\bullet$)Me$_2$	PrCHMe$_2$	213–273	$1.30 \times 10^{11} \exp(-39.0/RT)$	[171]
Me$_3$CC(O$_2^\bullet$)Me$_2$	Me$_3$CCHMe$_2$	243–293	$1.60 \times 10^{9} \exp(-31.4/RT)$	[171]
Me$_2$C(O$_2^\bullet$)CH$_2$CHMe CH$_2$CHMeCHMe$_2$	Me$_2$CHCH$_2$CHMe CH$_2$CHMeCHMe$_2$	243–293	$2.60 \times 10^{13} \exp(-46.0/RT)$	[172]
Me$_3$CCH$_2$Me$_2$CO$_2^\bullet$	CF$_2$Cl$_2$	253–303	$5.00 \times 10^{5} \exp(-16.7/RT)$	[158]
Me$_3$CCH$_2$Me$_2$CO$_2^\bullet$	CCl$_4$	303	1.00×10^{3}	[158]
Me$_3$CCH$_2$Me$_2$CO$_2^\bullet$	CH$_3$OH	303	$1.58 \times 10^{9} \exp(-29.3/RT)$	[170]
cyclopentyl-O-O$^\bullet$	CF$_2$Cl$_2$	253–303	$3.16 \times 10^{7} \exp(-23.8/RT)$	[158]
cyclohexyl-O-O$^\bullet$	CF$_2$Cl$_2$	253–303	$5.01 \times 10^{7} \exp(-25.5/RT)$	[158]
methylcyclohexyl-O-O$^\bullet$	methylcyclohexane	187–263	$7.40 \times 10^{8} \exp(-25.2/RT)$	[173]
Me$_2$C=CHC(O$_2^\bullet$)Me CH$_2$CHMeCHMe$_2$	Me$_2$C=CHCHMe CH$_2$CHMeCHMe$_2$	234–293	$1.60 \times 10^{10} \exp(-21.0/RT)$	[172]
PhMe$_2$CO$_2^\bullet$	PhMe$_2$CH	164–243	$1.60 \times 10^{9} \exp(-25.1/RT)$	[127]
PhMe$_2$CO$_2^\bullet$	C$_6$H$_6$	253–303	$3.98 \times 10^{9} \exp(-28.7/RT)$	[158]
PhMe$_2$CO$_2^\bullet$	CCl$_4$	303	8.00×10^{3}	[170]
PhMe$_2$CO$_2^\bullet$	PhMe$_2$COOH	398	4.40×10^{4}	[174]
PhMe$_2$CO$_2^\bullet$	PhMe$_2$COOH	338	8.00×10^{4}	[15]
PhMe$_2$CO$_2^\bullet$	H$_2$O	303	4.40×10^{4}	[158]
Me$_2$(PhCH$_2$)CO$_2^\bullet$	CCl$_4$	303	1.70×10^{4}	[170]
Ph$_2$MeCO$_2^\bullet$	CCl$_4$	303	6.40×10^{5}	[170]
Ph$_2$MeCO$_2^\bullet$	C$_6$H$_6$	175–200	$1.26 \times 10^{8} \exp(-10.5/RT)$	[158]
MeEtPhCO$_2^\bullet$	MeEtPhCH	303–329	$2.00 \times 10^{9} \exp(-37.7/RT)$	[41]

continued

TABLE 2.15
Rate Constants of Chain Termination by the Disproportionation of Tertiary Peroxyl Radicals in Hydrocarbon Solutions (Experimental Data)—*continued*

Peroxyl Radical	Solvent	T (K)	k (L mol^{-1} s^{-1})	Ref.
(4-methylcumyl peroxyl)	4-methylcumene	308–338	2.00×10^5	[175]
(3-methoxycumyl peroxyl)	3-methoxycumene	303	6.00×10^4	[118]
(4-methoxycumyl peroxyl)	4-methoxycumene	303	4.00×10^4	[118]
(4-methoxycarbonylcumyl peroxyl)	4-methoxycarbonylcumene	303	3.00×10^4	[118]
(4-ethyl-α,α-dimethylbenzyl peroxyl)	4-ethylcumene	348	6.60×10^6	[176]
(4-isopropylcumyl peroxyl)	4-isopropylcumene	348	1.10×10^5	[177]
(3-isopropylcumyl peroxyl)	3-isopropylcumene	348	1.00×10^5	[177]
(3,5-diisopropylcumyl peroxyl)	3,5-diisopropylcumene	348	1.00×10^5	[177]
(2,4-diisopropylcumyl peroxyl)	2,4-diisopropylcumene	348	3.20×10^5	[177]
(2,4,6-triisopropylcumyl peroxyl)	2,4,6-triisopropylcumene	348	8.00×10^6	[178]
Et$_2$PhCO$_2^\bullet$	PhCHEt$_2$	348	2.70×10^6	[176]
(1-phenylcyclopropyl peroxyl)	cyclopropylbenzene	348	1.90×10^6	[176]
(1-phenylcyclobutyl peroxyl)	cyclobutylbenzene	348	8.00×10^3	[176]
(1-phenylcyclopentyl peroxyl)	cyclopentylbenzene	348	8.1×10^5	[176]
(1-phenylcyclohexyl peroxyl)	cyclohexylbenzene	348	8.9×10^5	[176]

TABLE 2.16
Rate Constants of the Reaction: $2RO_2^\bullet \to 2RO^\bullet + O_2$ (Experimental Data)

Peroxyl Radical	Solvent	T (K)	k (L mol^{-1} s^{-1})	Ref.
$Me_3CO_2^\bullet$	CF_2Cl_2	253–303	$5.00 \times 10^9 \exp(-36.4/RT)$	[158]
$Me_3CO_2^\bullet$	CH_3OH	295–333	$2.51 \times 10^{10} \exp(-39.7/RT)$	[169]
$Me_3CO_2^\bullet$	CH_3OH	295	8.20×10^3	[168]
$Me_3CO_2^\bullet$	H_2O	295	2.50×10^4	[168]
$Me_3CCH_2Me_2CO_2^\bullet$	CH_3OH	295–303	$1.00 \times 10^{11} \exp(-41.0/RT)$	[169]
$PhMe_2CO_2^\bullet$	CH_3OH	295–303	$5.00 \times 10^{10} \exp(-32.6/RT)$	[169]
$PhMe_2CO_2^\bullet$	CF_2Cl_2	295–303	$5.00 \times 10^{10} \exp(-30.5/RT)$	[158]
$MeEtPhCO_2^\bullet$	CH_3OH	295–303	$5.00 \times 10^{10} \exp(-23.0/RT)$	[169]

It can be seen that primary and secondary $RO_2^\bullet$ radicals disproportionate with the participation of the α-C—H bond. This explains why the substitution of D in the α-position for H retards the recombination of $RO_2^\bullet$ [$k_H/k_D = 1.9$ for ethylbenzene, $k_H/k_D = 2.1$ for styrene, and $k_H/k_D = 1.37$ for diphenylmethane [179]). Because of this, $RO_2^\bullet$ radicals of unsaturated compounds with a double bond in the α-position to the peroxyl free valence disproportionate more rapidly than structurally analogous aliphatic peroxyl radicals (at 300 K, $2k_t = 2 \times 10^7$ and 3.8×10^6 L mol^{-1} s^{-1} for $RO_2^\bullet$ radicals of cyclohexene and cyclohexane, respectively [180]). Among the products of secondary peroxyl radicals disproportionation, carbonyl compound and alcohol were found in a ratio of 1:1 at room temperature (in experiments with ethylbenzene [181], tetralin [103], and cyclohexane [182–184].

Later, doubts were cast if the decay of tetroxide proceeds as an elementary decay with synchronous scission of two bonds:

1. Nangia and Benson [185] noted that such one decomposition act should have high activation energy (more than 18 kJ mol^{-1}) and does not occur rapidly. The preexponential factor A of concerted decomposition should be less than value 10^{13} s^{-1} [186]. Consequently such one-step decay of tetroxide should occur slowly.
2. According to the quantum-chemical law of conservation of orbital symmetry in products and reactants of the elementary reaction, all products of tetroxide decomposition should have the singlet orbits including dioxygen [4]. The singlet dioxygen is formed as a result of $RO_2^\bullet$ disproportionation, however, in a yield sufficiently less than unity [15].
3. Hydrogen peroxide was identified as the product of secondary peroxyl radical disproportionation [187–192]. It cannot be explained by the concerted mechanism of tetroxide decomposition.

The mechanism of successive homolytic fragmentation of tetroxide and formed radicals seems to be the very probable [167,193,194]:

$$R^1R^2HCOOOOCHR^1R^2 \longrightarrow [R^1R^2HCOOO^\bullet + R^1R^2CHO^\bullet]$$
$$[R^1R^2HCOOO^\bullet + R^1R^2CHO^\bullet] \longrightarrow R^1R^2CHOH + R^1R^2C^\bullet OOO^\bullet$$
$$[R^1R^2HCOOO^\bullet + R^1R^2CHO^\bullet] \longrightarrow R^1R^2C(O) + R^1R^2HCOOOH$$
$$R^1R^2C^\bullet OOO^\bullet \longrightarrow R^1R^2C(O)^* + {}^1O_2$$
$$[R^1R^2HCOOO^\bullet + R^1R^2CHO^\bullet] \longrightarrow R^1R^2C(O) + R^1R^2HCOH + O_2$$

$$[R^1R^2HCOOO^\bullet + R^1R^2CHO^\bullet] \longrightarrow R^1R^2HCOOOH + R^1R^2C(O)$$
$$R^1R^2HCOOOH \longrightarrow [R^1R^2HCO^\bullet + HOO^\bullet]$$
$$[R^1R^2HCO^\bullet + HOO^\bullet] \longrightarrow R^1R^2C(O) + H_2O_2$$
$$[R^1R^2HCO^\bullet + HOO^\bullet] \longrightarrow R^1R^2CHOH + O_2$$

This mechanism of homolytic decomposition of formed tetroxide explains all known facts about secondary peroxyl radicals disproportionation:

1. The CL that accompanied the disproportionation of secondary and primary peroxyl radicals (see later 2.5.4). The $R^1R^2C^\bullet OOO^\bullet$ biradical formed in the cage is a predecessor of exited carbonyl compound and singlet dioxygen.
2. The formation of hydrogen peroxide as the product of peroxyl radical reactions in the cage [187–192].
3. All reactions of this scheme are in accordance with the quantum-chemical law of conservation of orbital symmetry [4].
4. This scheme explains the high values of rate constants and the low activation energy of such reactions (see Table 2.15).
5. It can be seen that primary and secondary $RO_2^\bullet$ radicals disproportionate in the cage involving the α-C—H bond, which explains why the substitution of D in the α-position for H retards the disproportionation of $RO_2^\bullet$. Because of this, $RO_2^\bullet$ radicals of unsaturated compounds with a double bond in the α-position to the peroxyl free valence disproportionate more rapidly than structurally analogous aliphatic peroxyl radicals [195].
6. The limiting step of the reaction between two peroxyl radicals is the homolytic splitting of tetroxide. Such reactions occur with high pre-exponential factors (10^{14} to 10^{16} s^{-1}).
7. According to the concerted mechanism of tetroxide decomposition, the ratio of formed amounts [alcohol]/[ketone] = 1 should be temperature-independent. However, the experimental study provided evidence for the dependence of this ratio on temperature.

$RO_2^\bullet$	BuOO$^\bullet$	BuMeCHOO$^\bullet$	cyclo-C$_5$H$_9$OO$^\bullet$	cyclo-C$_6$H$_{11}$OO$^\bullet$
T (K)	148	148	148	243
[ROH]/[R^1R^2CO]	0.10	0.10	0.10	0.43
T (K)	373	373	373	343
[ROH]/[R^1R^2CO]	2.50	2.50	2.50	1.90
Ref.	[196]	[197]	[197]	[198]

The increase in the yield of alcohol among the products of peroxyl radicals disproportionation with increasing temperature is the result of acceleration of hydrotrioxide decomposition to the alcoxyl radical and $HO_2^\bullet$. The proposed scheme is valid for the disproportionation of tertiary peroxyl radicals as well (see earlier). The rate constants of disproportionation of primary and secondary peroxyl radicals are presented in Table 2.17.

2.5.4 Chemiluminescence

Luminescence in chemical reactions occurs at the expense of the energy released by exothermic elementary steps of the reaction. The number of excited molecules formed in the exothermic reaction and CL intensity are proportional to the reaction rate. Only a small

TABLE 2.17
Rate Constants of Primary and Secondary Peroxyl Radicals Disproportionation in Hydrocarbon Solutions (Experimental Data)

Peroxyl Radical	T (K)	k (L mol^{-1} s^{-1})	Ref.
$HO_2^{\bullet}$	323	1.0×10^8	[157]
$HO_2^{\bullet}$	303	1.34×10^9	[197]
$CH_3O_2^{\bullet}$	295	7.7×10^8	[198]
$Me_2CHO_2^{\bullet}$	210–300	$7.24 \times 10^7 \exp(-10.3/RT)$	[199]
$Me_2CHO_2^{\bullet}$	293–396	$5.0 \times 10^9 \exp(-20.0/RT)$	[200]
$PrCH_2O_2^{\bullet}$	303	4.0×10^7	[201]
$BuCH_2O_2^{\bullet}$	253–303	$5.0 \times 10^9 \exp(-9.1/RT)$	[21]
$EtMeCHO_2^{\bullet}$	193–257	$1.0 \times 10^9 \exp(-11.3/RT)$	[202]
$MePrCHO_2^{\bullet}$	253–303	$2.0 \times 10^7 \exp(-6.9/RT)$	[21]
$BuMeCHO_2^{\bullet}$	283–320	$2.88 \times 10^7 \exp(-8.4/RT)$	[203]
$Me(CH_2)_4MeCHO_2^{\bullet}$	294–324	$2.88 \times 10^7 \exp(-8.4/RT)$	[203]
$Me(CH_2)_5MeCHO_2^{\bullet}$	283–356	$2.88 \times 10^7 \exp(-8.4/RT)$	[203]
$Me(CH_2)_6MeCHO_2^{\bullet}$	283–324	$2.88 \times 10^7 \exp(-8.4/RT)$	[203]
$Me(CH_2)_7MeCHO_2^{\bullet}$	283–355	$2.88 \times 10^7 \exp(-8.4/RT)$	[203]
$Me(CH_2)_7MeCHO_2^{\bullet}$	323	2.2×10^6	[23]
$Me(CH_2)_9MeCHO_2^{\bullet}$	284–355	$2.88 \times 10^7 \exp(-8.4/RT)$	[203]
$Me(CH_2)_9MeCHO_2^{\bullet}$	323	2.9×10^6	[23]
$CH_3CH(O_2^{\bullet})(CH_2)_{10}CH_3$	293–358	$2.88 \times 10^7 \exp(-8.4/RT)$	[203]
$Me(CH_2)_{13}MeCHO_2^{\bullet}$	323	3.1×10^6	[23]
cyclopentyl-OO$^{\bullet}$	175–200	$4.0 \times 10^9 \exp(-10.9/RT)$	[204]
cyclohexyl-OO$^{\bullet}$	285–333	$1.95 \times 10^7 \exp(-5.4/RT)$	[205]
cycloheptyl-OO$^{\bullet}$	298	8.6×10^6	[206]
cyclooctyl-OO$^{\bullet}$	298	1.40×10^6	[206]
cyclononyl-OO$^{\bullet}$	345–417	$1.30 \times 10^8 \exp(-7.8/RT)$	[207]
$CH_2=CHCH(O_2^{\bullet})CH=CH_2$	303	1.10×10^9	[201]
$Me_2C=CHCH(O_2^{\bullet})Me$	313–333	$2.50 \times 10^7 \exp(-5.0/RT)$	[125]
$EtCH=CHCH(O_2^{\bullet})Et$	303	6.40×10^6	[201]
cis-$PrCH=CHCH(O_2^{\bullet})Et$	323	1.50×10^7	[206]
$PhCH=CHCH_2O_2^{\bullet}$	303	4.4×10^7	[208]
$CH_2=CHCH(O_2^{\bullet})(CH_2)_4Me$	298	1.00×10^7	[209]
cyclopentenyl-OO$^{\bullet}$	193–257	$6.30 \times 10^7 \exp(-4.2/RT)$	[202]
$PhCH_2O_2^{\bullet}$	303	3.0×10^8	[210]
cyclohexenyl-OO$^{\bullet}$	282–319	$5.62 \times 10^7 \exp(-8.3/RT)$	[211]
p-methylbenzyl-OO$^{\bullet}$	348	2.6×10^8	[212]

continued

TABLE 2.17
Rate Constants of Primary and Secondary Peroxyl Radicals Disproportionation in Hydrocarbon Solutions (Experimental Data)—continued

Peroxyl Radical	T (K)	k (L mol^{-1} s^{-1})	Ref.
(methylcyclohexenyl peroxyl)	313	8.6×10^5	[213]
(tetralinyl peroxyl)	286–323	$8.70 \times 10^9 \exp(-18.0/RT)$	[208]
(tetralinyl peroxyl)	348	2.60×10^7	[214]
(octahydroanthracenyl peroxyl)	348	2.50×10^7	[214]
PhMeCHO$_2^\bullet$	343–363	1.9×10^6	[38]
PhMeCHO$_2^\bullet$	303	4.0×10^6	[210]
PhMeCHO$_2^\bullet$	323–353	$1.30 \times 10^9 \exp(-9.4/RT)$	[215]
(4-ethylcumyl peroxyl)	348	1.0×10^7	[216]
(phenylcyclohexenyl peroxyl)	313–343	2.5×10^8	[150]
(3-ethylcumyl peroxyl)	348	1.10×10^7	[217]
(3,4-diethylcumyl peroxyl)	348	1.0×10^7	[218]
(3,5-diethylcumyl peroxyl)	348	1.0×10^7	[218]
(triethyl cumyl peroxyl)	348	1.9×10^6	[218]
(triethyl cumyl peroxyl)	348	3.3×10^6	[218]
(tetraethyl cumyl peroxyl)	348	6.0×10^5	[218]

TABLE 2.17
Rate Constants of Primary and Secondary Peroxyl Radicals Disproportionation in Hydrocarbon Solutions (Experimental Data)—*continued*

Peroxyl Radical	T (K)	k (L mol^{-1} s^{-1})	Ref.
(pentaethyl-substituted phenyl CH(CH$_3$)O–O•)	348	4.5×10^5	[218]
(m-methoxy phenyl CH(CH$_3$)O–O•)	348	1.3×10^7	[217]
(p-methoxy phenyl CH(CH$_3$)O–O•)	348	9.0×10^6	[217]
(p-chloro phenyl CH(CH$_3$)O–O•)	348	1.6×10^7	[217]
(p-acetyl phenyl CH(CH$_3$)O–O•)	348	1.5×10^7	[217]
(m-nitro phenyl CH(CH$_3$)O–O•)	348	2.3×10^7	[217]
(p-nitro phenyl CH(CH$_3$)O–O•)	348	2.4×10^7	[217]
(3-pyridyl CH$_2$O–O•)	348	8.3×10^7	[34]
(2-pyridyl CH(CH$_3$)O–O•)	348	2.3×10^7	[34]
(3-pyridyl CH(CH$_3$)O–O•)	348	1.0×10^7	[34]
(4-pyridyl CH(CH$_3$)O–O•)	348	3.2×10^7	[34]
EtPhCHO$_2$•	348	3.0×10^7	[216]
Ph$_2$CHO$_2$•	303	1.6×10^8	[208]

fraction of molecular products gives up the reaction energy as light. In the initial chemical stage, the formation of excited molecules compete with processes of formation of nonexcited molecules. In the following physical stage, the emission of light by the excited molecule competes with routes of deactivation, such as quenching, energy transfer, re-absorption of light, etc. Gas-phase CL is well-known and was intensively studied during the 20th century. CL is very weak and became the object of study only in the second half of the 20th century after the invention of effective photoelectric multipliers.

CL accompanies many reactions of the liquid-phase oxidation of hydrocarbons, ketones, and other compounds. It was discovered in 1959 for liquid-phase ethylbenzene oxidation [219,220]. This phenomenon was intensively studied in the 1960s and 1970s, providing foundation for several methods of study of oxidation, decay of initiators, and kinetics of antioxidant action [12,17,221]. Later this technique was effectively used to study the mechanism of solid polymer oxidation (see Chapter 13).

CL arises in the very exothermic reaction of secondary and primary peroxyl radicals disproportionation [221,222]:

$$R^1R^2HCOOOOCHR^1R^2 \longrightarrow [R^1R^2HCOOO^\bullet + R^1R^2CHO^\bullet]$$
$$[R^1R^2HCOOO^\bullet + R^1R^2CHO^\bullet] \longrightarrow R^1R^2CHOH + R^1R^2C^\bullet OOO^\bullet$$
$$R^1R^2C^\bullet OOO^\bullet \longrightarrow R^1R^2C(O)^* + O_2$$
$$R^1R^2C(O)^* \longrightarrow R^1R^2C(O) + h\nu$$

The decay of the biradical produces ketone molecule in the triplet state, which is an emitter of light [222]. The CL intensity was proved to be propotional to the rate of chain initiation, which is equal to the rate of chain termination. The observed luminescence spectra were found to be identical with the spectra of the subsequent ketone in the triplet state. The intensity of CL (I_{chl}) produced by oxidized hydrocarbon is the following:

$$I_{chl} = 2k_t\eta[RO_2^\bullet]^2, \qquad (2.18)$$

where η is the overall CL yield.

Dioxygen has a dual effect on CL in the oxidized hydrocarbon [17,221,223]. On one hand, the higher the [O_2], the higher the proportion of chain termination by peroxyl radical disproportionation and the yield of the excited ketone. On the other hand, dioxygen quenches CL. The combined effects of dioxygen give the following result when the hydrocarbon oxidation is performed in a sealed reaction vessel. The gradual consumption of dissolved dioxygen increases the intensity of CL. At the end of experiment a sharp drop in luminescence intensity is observed due to the disappearance of the peroxyl radical. The kinetics of quenching is described by the Stern–Volmer equation [223]:

$$\frac{1}{I} = \frac{1}{I_0} + \frac{k\tau}{I_0}[O_2], \qquad (2.19)$$

where τ is the lifetime of the ketone in exited state (P*) and k is the rate constant of quenching:

$$P^* + O_2 \rightarrow P + O_2$$

If quenching is a diffusion-controlled process ($k \approx 3 \times 10^9$ L mol^{-1} s^{-1}), the lifetime $\tau \approx 3 \times 10^{-7}$ s coincides with the lifetime of triplet acetophenone (product of peroxyl radical disproportionation in oxidized ethylbenzene).

Some initiators, for example, AIBN (azobisisobutyronitrile) and dibenzoyl peroxide exhibit a strong quenching effect. That is why, the dependence of the CL intensity on the initiator concentration ($I = \eta v_i$) is nonlinear in these cases.

CL is weak in liquid-phase hydrocarbon oxidation and can be intensified. To increase the CL, activators are added to the oxidized hydrocarbon [17,220,223]. The activator takes an excess of energy from the excited ketone molecule and emits light in high yield:

$$P^* + A \longrightarrow P + A^*$$
$$A^* \longrightarrow A + h\nu$$

In the presence of an activator, naturally, the spectrum, yield, and lifetime are characteristics of the activator molecule. Indeed, the luminescence spectrum of the oxidized ethylbenzene was found to be identical to that of activator fluorescence [221]. 9,10-Dibromanthracene, 9,7-dipropylanthracene, and derivatives of oxazole were used as activators [221,223].

The kinetic scheme of CL in the oxidized hydrocarbon ($R^1CH_2R^2$) in the presence of quencher Q and activator A includes the following stages [221–225]:

$$2R^1R^2CHOO^\bullet \rightleftharpoons R^1R^2CHOOOOCHR^1R^2$$
$$R^1R^2CHOOOOCHR^1R^2 \longrightarrow [R^1R^2CHOOO^\bullet + {}^\bullet OCHR^1R^2]$$
$$[R^1R^2CHOOO^\bullet + {}^\bullet OCH R^1R^2] \longrightarrow R^1R^2C^\bullet OOO^\bullet + R^1R^2CHOH$$
$$R^1R^2C^\bullet OOO^\bullet \longrightarrow R^1R^2C(O)^* + O_2$$
$$R^1R^2C(O)^* \longrightarrow R^1R^2C(O) + h\nu_P \quad (f_P)$$
$$R^1R^2C(O)^* \longrightarrow R^1R^2C(O) \quad (d_P)$$
$$R^1R^2C(O)^* + Q \longrightarrow R^1R^2C(O) + Q \quad (k_{PQ})$$
$$R^1R^2C(O)^* + A \longrightarrow R^1R^2C(O) + A^* \quad (k_A)$$
$$A^* \longrightarrow h\nu_A + A \quad (f_A)$$
$$A^* \longrightarrow A \quad (d_A)$$
$$A^* + Q \longrightarrow A + Q \quad (k_{AQ})$$

According to this scheme, the intensity of CL is proportional to the rate of initiation (ν_i) and depends on the concentrations of quenchers and activators. Some initiators, for example AIBN and dibenzoyl peroxide, act as strong quenchers [221]:

$$I = f_P[P^*] + f_A[A^*] = \{\eta_P(1 - \eta_{PA}) + \eta_A \eta_{PA}\} \eta_P^{exc} \nu_i \quad (2.20)$$

where η_P^{exc} is the quantum yield of ketone excitation, $\eta_P = f_P(f_P + d_P + k_{PQ}[Q])^{-1}$ is the radiation quantum yield for P*, $\eta_A = f_A(f_A + d_A + k_{AQ}[Q])^{-1}$ is the radiation quantum yield for A*, $\eta_{PA} = k_{PA}[A](f_P + f_d + k_{PQ}[Q] + k_{PA}[A])^{-1}$ is the probability of energy transfer. The values of the overall CL quantum yield η are very low [221].

Substance	Cyclohexane	Decane	Ethylbenzene	Butanone-2
η	10^{-10}	10^{-10}	10^{-9}	10^{-10}

The reason for the low values of η lies in the distribution of the released energy between the three molecules formed (ketone, dioxygen, and alcohol) in the reaction of two peroxyl radicals.

2.5.5 Chain Termination via Alkyl Radicals

Due to the very fast reaction of alkyl radicals with dioxygen, $[R^\bullet] \ll [RO_2^\bullet]$ in oxidized hydrocarbons at $[O_2] > 10^{-4}$ mol L^{-1} (see 2.3). Therefore, alkyl radicals do not participate in chain termination when dioxygen concentrations in solution are higher than $10^{-4}\,\text{mol}\,L^{-1}$. This corresponds to the dioxygen partial pressure higher than 0.01 atm. The steady-state concentration of the alkyl radicals becomes close to that of peroxyl radicals in three cases:

(i) When the hydrocarbon is oxidized at a very low dioxygen partial pressure (less than 0.01 atm) in the kinetic regime of oxidation
(ii) When oxidation is limited by dioxygen diffusion into the bulk of oxidized hydrocarbon (diffusion regime of oxidation)
(iii) When the formed C—O bond in the peroxyl radical is weak and we deal with the equilibrium reaction, which increases the steady-state concentration of alkyl radicals in oxidized hydrocarbon:

$$R^\bullet + O_2 \rightleftharpoons RO_2^\bullet$$

The rate constants of the recombination of alkyl radicals with peroxyl radicals

$$R^\bullet + RO_2^\bullet \longrightarrow ROOR \qquad (k_{t1})$$

were measured by Nikolaev using flash photolysis technique [226]. These reactions appeared to proceed very rapidly with rate constants close to self-diffusion of radicals in the hydrocarbon (see Table 2.18).

The reaction of two alkyl radicals leads either to one molecule by combination, for example,

$$CH_3C^\bullet H_2 + CH_3C^\bullet H_2 \longrightarrow C_4H_{10}$$

or to two molecules by the transfer of a hydrogen atom from one radical to the second one (disproportionation reaction), for example [227–229]:

$$CH_3C^\bullet H_2 + CH_3C^\bullet H_2 \longrightarrow C_2H_6 + CH_2{=}CH_2.$$

The recombination and the disproportionation of alkyl radicals play an important role in many other chain reactions, for example, pyrolysis, photolysis, and radiolysis of organic

TABLE 2.18
Rate Constants of Chain Termination in Oxidized Hydrocarbons by the Reaction of Alkyl Radicals with Peroxyl Radicals at 293 K in RH Solution

$R^\bullet$	$RO_2^\bullet$	$2k_{t1}$ (L mol^{-1} s^{-1})	$2k_{t1}/2k_D$	Ref.
cyclohexyl	$Me_3CO_2^\bullet$	4.5×10^9	0.36	[226]
BuC$^\bullet$HMe	$Me_3CO_2^\bullet$	6.3×10^9	0.19	[226]
Me(CH$_2$)$_7$C$^\bullet$HMe	$Me_3CO_2^\bullet$	1.9×10^9	0.15	[226]
Me(CH$_2$)$_7$C$^\bullet$HMe	$Me_3CO_2^\bullet$	1.8×10^9	0.28	[226]
cyclohexyl	cyclohexyl-OO$^\bullet$	1.6×10^9	0.13	[226]
Ph$_3$C$^\bullet$	Ph$_3$CO$_2^\bullet$	1.5×10^8		[41]

compounds, free radical polymerization, chlorination of hydrocarbons, various organic free radical syntheses, polymer degradation, etc.

The reactions of free radical recombination are very exothermic. Entropies of these reactions are negative due to the association of two species. For example, the recombination of two ethyl radicals occurs (in the gas phase) with $\Delta H^0 = -366.5$ kJ mol^{-1} and $\Delta S^0 = -192.4$ J mol^{-1} K^{-1}. The disproportionation of two ethyl radicals in the gas phase occurs with enthalpy $\Delta H^0 = -270$ kJ mol^{-1} and entropy $\Delta S^0 = -36.3$ J mol^{-1} K^{-1}.

The study of the recombination (k_c) and the disproportionation (k_{dis}) of ethyl radicals proved the dependence of the ratio k_{dis}/k_c on the solvent [230–232]. Stefani [231] found that $\log(k_{dis}/k_c)$ is a linear function of the solubility δ_s, which is equal to the square root of the cohesive energy density of the solvents (D_{ce}):

$$\log(k_{dis}/k_c) = a\delta_s + b \tag{2.21}$$

The latter depends on the energy of evaporation (U_{ev}):

$$D_{ce} = \frac{U_{ev}}{V} = \frac{\Delta H_{ev} - RT}{M/\rho} \tag{2.22}$$

where V is the molecular volume, ΔH_{ev} is the enthalpy of solvent evaporation, M and ρ are the molecular weight and density of solvent, respectively. The dependence of the ratio k_{dis}/k_c on the cohesive energy density of the solvents proves that the TS of radical recombination has a loose structure and that of the disproportionation reaction has a compact structure. The more the pressure of molecules of the cage on the pair of reacting radicals, the faster the reaction of disproportionation. Stefani [231] found evidence for the ratio k_{dis}/k_c increasing with decreasing temperature. The increase of this ratio can be easily explained on the basis of an increase in the internal pressure of the liquid with decreasing temperature.

Alkyl radicals react in solution very rapidly. The rate of their disappearance is limited by the frequency of their encounters. This situation is known as microscopic diffusion control or encounter control, when the measured rate is almost exactly equal to the rate of diffusion [230]. The rate of diffusion-controlled reaction of free radical disappearance is the following (the stoichiometric coefficient of reaction is two [233]):

$$v = 2k_D[R^\bullet]^2 = 32\pi D_R r_R [R^\bullet]^2 \tag{2.23}$$

where D_R and r_R are the diffusion coefficient and radius of the radical, respectively. According to Stokes's law

$$D_R = kT/6\pi r_R \eta, \tag{2.24}$$

and the rate constant of diffusion-controlled reaction is inversely proportional to viscosity:

$$k_D = 16kT/3\eta = \text{const.} \times T/\eta \tag{2.25}$$

The values of rate constants of alkyl radical recombination in solution are collected in Table 2.19.

2.6 REACTIONS OF ALKOXYL RADICALS

Alkoxyl radicals are formed in oxidized hydrocarbons and play an important role in the chain mechanism of oxidation. They are formed by the following reactions [13,15]:

TABLE 2.19
Rate Constants ($2k_t$) of Alkyl Radical Combination in Solution ($k_t = k_c + k_{dis}$)

R·	Solvent	T (K)	E (kJ mol^{-1})	log (2A), A (L mol^{-1} s^{-1})	$2k_t$ (298 K) (L mol^{-1} s^{-1})	Ref.
C·H$_3$	Cyclohexane	298			8.9×10^9	[234]
C·H$_3$	Cyclohexane	218			1.4×10^9	[235]
C·H$_3$	Water, pH = 5.5	298			2.48×10^9	[236]
C·H$_3$	Water	279–340	16.0	12.30	3.13×10^9	[237]
MeC·H$_2$	Ethane	98–196	3.5	10.11	3.14×10^9	[238]
MeC·H$_2$	Water, pH = 5.5	298			1.92×10^9	[236]
MeC·H$_2$	Water, pH = 4.4	278–341	16.0	12.19	2.43×10^9	[237]
EtC·H$_2$	Benzene	298			2.1×10^8	[239]
EtC·H$_2$	Cyclohexane	298			1.7×10^9	[239]
EtC·H$_2$	Water	298			1.26×10^9	[240]
BuC·H$_2$	Benzene	298			2.0×10^9	[239]
BuC·H$_2$	Cyclohexane	298			4.2×10^8	[241]
BuC·H$_2$	Toluene	192–292	10.5	12.30	2.88×10^{10}	[242]
BuCH$_2$C·H$_2$	Cyclohexane	298			2.2×10^9	[235]
BuCH$_2$C·H$_2$	Cyclopropane	190–232	8.8	11.6	1.14×10^{10}	[243]
BuCH$_2$C·H$_2$	Hexane		7.1	11.26	1.04×10^{10}	[244]
BuCH$_2$C·H$_2$	Hexane	298			6.2×10^9	[245]
BuCH$_2$C·H$_2$	Water	298			2.4×10^9	[245]
Me(CH$_2$)$_5$C·H$_2$	Heptane	273–333	8.1	11.27	7.08×10^9	[246]
Me(CH$_2$)$_6$C·H$_2$	Octane	279–349	6.8	10.94	5.60×10^9	[246]
Me(CH$_2$)$_7$C·H$_2$	Nonane	284–322	9.4	11.34	4.92×10^9	[246]
Me(CH$_2$)$_8$C·H$_2$	Decane	283–333	10.9	11.62	5.12×10^9	[246]
Me(CH$_2$)$_{10}$C·H$_2$	Dodecane	278–347	13.5	11.80	2.71×10^9	[246]
Me(CH$_2$)$_{10}$C·H$_2$	Cyclohexane	298			2.4×10^9	[243]
Me(CH$_2$)$_{11}$C·H$_2$	Tridecane	291–352	11.3	11.44	2.88×10^9	[246]
Me(CH$_2$)$_{14}$C·H$_2$	Hexadecane	293–358	13.0	11.74	2.89×10^9	[246]
Me(CH$_2$)$_{16}$C·H$_2$	Cyclohexane	298			6.2×10^9	[244]
Me$_2$C·H	Heptane	313–353	5.6	10.87	7.73×10^9	[247]
Me$_2$C·H	Hexadecane	312			3.5×10^9	[247]
Me$_2$C·H	3-Methyl-3-pentanol	333			5.2×10^9	[247]
Me$_2$C·H	Tetraethyl siloxane	333			8.0×10^9	[248]
Me$_3$CC·HCHMe$_2$	Isooctane	298			3.0×10^9	[249]
Me$_3$C·	Acetonitrile	266–349	8.3	11.30	7.00×10^9	[250]
Me$_3$C·	Benzene	298			7.3×10^9	[251]
Me$_3$C·	Benzene	292			5.7×10^9	[252]
Me$_3$C·	Benzene	281–351	10.2	11.57	6.05×10^9	[250]
Me$_3$C·	Benzene	298			8.0×10^8	[247]
Me$_3$C·	Benzene	298			7.2×10^8	[240]
Me$_3$C·	Cyclohexane	298			2.5×10^9	[253]
Me$_3$C·	Cyclopentane	170–330	4.3	10.78	1.06×10^{10}	[254]
Me$_3$C·	Decane	291–351	10.7	11.65	5.95×10^9	[255]
Me$_3$C·	1,1-Dimethyl ethanol	287–356	3.3	11.16	3.82×10^{10}	[250]
Me$_3$C·	Dodecane	294–366	11.3	11.62	4.36×10^9	[250]

TABLE 2.19
Rate Constants ($2k_t$) of Alkyl Radical Combination in Solution ($k_t = k_c + k_{dis}$)—continued

R•	Solvent	T (K)	E (kJ mol^{-1})	log (2A), A (L mol^{-1} s^{-1})	$2k_t$(298 K) (L mol^{-1} s^{-1})	Ref.
Me$_3$C•	Heptane	291–362	9.6	11.63	8.86×10^9	[250]
Me$_3$C•	Hexadecane	296–363	14.4	12.03	3.21×10^9	[250]
Me$_3$C•	Isobutane	170–330	4.3	10.78	1.1×10^9	[254]
Me$_3$C•	3-Methyl- 3-pentanol	248–293	3.6	11.32	4.89×10^{10}	[250]
Me$_3$C•	2-Methylpropane	188–262	4.2	10.2	2.91×10^9	[256]
Me$_3$C•	Octane	294–364	9.9	11.58	6.99×10^9	[250]
Me$_3$C•	Pentane	298			5.4×10^9	[236]
Me$_3$C•	Tetradecane	294–365	13.5	11.93	3.66×10^9	[250]
Me$_3$C•	Tridecane	298			1.8×10^9	[236]
Me$_3$C•	Toluene	218			9.8×10^9	[254]
cyclo-C$_6$H$_{11}$•	Cyclohexane	295			2.0×10^9	[228]
cyclo-C$_6$H$_{11}$•	Benzene	298			3.6×10^8	[239]
cyclo-C$_6$H$_{11}$•	Cyclohexane	298			6.0×10^8	[239]
PhC•H$_2$	Benzene	298			1.8×10^9	[239]
PhC•H$_2$	Cyclohexane	283–332	8.1	10.70	1.91×10^9	[249]
PhC•H$_2$	Cyclohexane	298			2.0×10^9	[239]
PhC•H$_2$	Cyclohexane	298			4.0×10^9	[257]
PhC•H$_2$	Methanol	298			1.36×10^9	[258]
PhC•H$_2$	Toluene	298			2.4×10^9	[259]
PhC•H$_2$	Toluene	222–331	12.5	11.90	5.12×10^9	[247]

1. Due to the unimolecular and bimolecular homolytic decomposition of hydroperoxide, such as

$$ROOH \longrightarrow RO^\bullet + HO^\bullet$$
$$ROOH + RH \longrightarrow RO^\bullet + H_2O + R^\bullet$$
$$ROOH + CH_2=CHR \longrightarrow RO^\bullet + HOCH_2C^\bullet HR$$

2. In reactions of recombination of the tertiary peroxyl radical (see 2.1.3)

$$RO_2^\bullet + RO_2^\bullet \longrightarrow RO^\bullet + O_2 + RO^\bullet$$

The last reaction occurs more rapidly than the reaction of chain termination and as a result two simultaneous chain reactions occur, one with the formation of hydroperoxide and other with alcohol production:

$$R^\bullet + O_2 \longrightarrow RO_2^\bullet$$
$$RO_2^\bullet + RH \longrightarrow ROOH + R^\bullet$$
$$2RO_2^\bullet \longrightarrow 2RO^\bullet + O_2$$
$$RO^\bullet + RH \longrightarrow ROH + R^\bullet$$

3. In reactions of decay of alkyl radicals with the peroxide group (see 2.4.4):

$$RO_2^\bullet + CH_2=CHR^1 \longrightarrow ROOCH_2C^\bullet HR^1$$
$$ROOCH_2C^\bullet HR^1 \longrightarrow RO^\bullet + cyclo\text{-}[OCH_2CHR^1]$$

Alkoxyl radicals are very active in reactions of hydrogen atom abstraction (see Table 2.20). The problems of their reactivity will be discussed in Chapter 6.

Tertiary alkoxyl radicals are unstable and decomposed into the carbonyl compound and the alkyl radical:

$$R^1R^2RCO^\bullet \longrightarrow R^1C(O)R^2 + R^\bullet$$

The formed alkyl radical reacts with dioxygen in the oxidized hydrocarbon. Participating in chain termination, the newly formed peroxyl radical accelerates it:

$$R^\bullet + O_2 \longrightarrow RO_2^\bullet$$
$$RO_2^\bullet + R^1R^2RCOO^\bullet \longrightarrow R^1R^2RCOH + O_2 + \text{Ketone}$$

The values of rate constants of decay of tertiary alkoxyl radicals are collected in Table 2.21.

Like peroxyl radicals (see Section 2.4.4.) alkoxyl radicals with a long hydrocarbon substituent can isomerize to the alkyl radical

$$R^1R^2C(O^\bullet)CH_2CH_2CH_2R \longrightarrow R_1R_2C(OH)(CH_2)_2C^\bullet HR$$

The latter reacts with dioxygen, giving the bifunctional product of oxidation. The rate constants of these reactions are given in Table 2.22.

2.7 DIFFERENT MECHANISMS OF THE ALIPHATIC HYDROCARBON OXIDATION IN GAS AND LIQUID PHASES

The temperature increases the rate of the one-stage simple reaction. The situation is different in the case of chain reaction. The change in temperature and other conditions not only change the reaction rate but can change the mechanism of the reaction and composition of the formed products. This can be illustrated by analysis of the mechanism of the hydrocarbon oxidation at different temperatures, concentration of the reactants, and the rates of initiation [288]. Varying the conditions of oxidation, the mechanism and products of the reaction can be changed.

The traditional chain oxidation with chain propagation via the reaction $RO_2^\bullet + RH$ occurs at a sufficiently elevated temperature when chain propagation is more rapid than chain termination (see earlier discussion). The main molecular product of this reaction is hydroperoxide. When tertiary peroxyl radicals react more rapidly in the reaction $RO_2^\bullet + RO_2^\bullet$ with formation of alkoxyl radicals than in the reaction $RO_2^\bullet + RH$, the mechanism of oxidation changes. Alkoxyl radicals are very reactive. They react with parent hydrocarbon and alcohols formed as primary products of hydrocarbon chain oxidation. As we see, alkoxyl radicals decompose with production of carbonyl compounds. The activation energy of their decomposition is higher than the reaction with hydrocarbons (see earlier discussion). As a result, heating of the system leads to conditions when the alkoxyl radical decomposition occurs more rapidly than the abstraction of the hydrogen atom from the hydrocarbon. The new chain mechanism of the hydrocarbon oxidation occurs under such conditions, with chain

TABLE 2.20
Rate Constants of Reactions $R^1O^{\bullet} + RH \rightarrow R^1OH + R^{\bullet}$ (Selected Experimental Data)

RH	Solvent	T (K)	k (L mol^{-1} s^{-1})	Ref.
\multicolumn{5}{c}{$CH_3O^{\bullet} + RH$}				
CH_4	Gas phase	298	51.3	[260]
Me_2CH_2	Gas phase	298	6.4×10^5	[261]
Me_3CH	Gas phase	298	1.8×10^5	[262]
$Me_2CHCHMe_2$	Gas phase	373	4.0×10^5	[263]
$Z—MeCH=CHMe$	C_6H_6	393	3.65×10^7	[264]
$PhCH_2Me$	C_6H_6	393	1.22×10^7	[264]
$PhCH_2Me$	C_6H_6	291	4.50×10^6	[264]
o-xylene	C_6H_6	393	2.09×10^7	[264]
p-xylene	C_6H_6	393	1.51×10^7	[264]
$PhCHMe_2$	C_6H_6	393	1.33×10^7	[264]
\multicolumn{5}{c}{$Me_3CO^{\bullet} + RH$}				
$MeCH_2CH_2Me$	Gas phase	298	2.14×10^5	[265]
Me_4C	Gas phase	298	4.48×10^4	[266]
$Me(CH_2)_3Me$	C_6H_6	318	7.1×10^5	[267]
$Me(CH_2)_4Me$	C_6H_6	318	9.1×10^5	[267]
$Me(CH_2)_5Me$	C_6H_6	318	1.1×10^6	[267]
$Me_2CHCHMe_2$	C_6H_6	318	7.8×10^5	[267]
$Me_2CHCH_2CHMe_2$	C_6H_6	318	4.6×10^5	[267]
$Me_3CCH_2CHMe_2$	C_6H_6	318	2.3×10^5	[267]
cyclopentane	cyclopentane	295	8.8×10^5	[267]
cyclohexane	Gas phase	298	6.45×10^5	[266]
cyclohexane	cyclohexane	408	3.2×10^7	[268]
methylcyclohexane	C_6H_6	408	2.4×10^7	[268]
$CH_2=CHCH_3$	Gas phase	298	1.15×10^5	[266]
$Me_2C=CMe_2$	C_6H_6	300	1.5×10^7	[269]
$Me_2C=CMe_2$	Gas phase	298	3.48×10^6	[266]
$CH_2=CH(CH_2)_5Me$	C_6H_6	300	1.5×10^6	[269]
cyclopentadiene	C_6H_5Cl	273	2.7×10^6	[270]
cyclohexene	C_6H_6	300	5.8×10^6	[269]
1,3-cyclohexadiene	C_6H_6	295	4.2×10^7	[271]
1,4-cyclohexadiene	C_6H_6	295	5.3×10^7	[271]
cyclooctene	C_6H_6	300	2.6×10^6	[269]
$PhCH_3$	$PhCH_3/C_6H_6$	295	2.3×10^5	[272]
$PhCH_2Me$	$PhCH_2Me$	408	9.3×10^6	[273]

continued

TABLE 2.20
Rate Constants of Reactions $R^1O^\bullet + RH \to R^1OH + R^\bullet$ (Selected Experimental Data)—continued

RH	Solvent	T (K)	k (L mol^{-1} s^{-1})	Ref.
PhCH$_2$Me	C$_6$H$_5$Cl	273	3.6×10^5	[270]
PhCH$_2$Me	PhCH$_2$Me/C$_6$H$_6$	295	1.05×10^6	[272]
PhCHMe$_2$	PhCHMe$_2$/C$_6$H$_6$	295	8.7×10^5	[274]
PhCHMe$_2$	C$_6$H$_6$	318	1.2×10^6	[267]
PhCHMe$_2$	PhCHMe$_2$	313	9.3×10^5	[275]
PhCH$_2$CH=CH$_2$	PhCl	273	4.5×10^5	[270]
	EtMe$_2$CO$^\bullet$ + RH			
⬡	CCl$_4$	273	3.89×10^4	[276]
	PhMe$_2$CO$^\bullet$ + RH			
Me$_3$CH$_2$CHMe	Me$_3$CH$_2$CHMe$_2$	313	1.02×10^6	[275]
Me(CH$_2$)$_{10}$Me	Me(CH$_2$)$_{10}$Me	426	2.75×10^7	[277]
⬡	C$_6$H$_6$	303	1.22×10^6	[278]
Me-C$_6$H$_4$	CCl$_4$	313	1.04×10^6	[279]
PhMe$_2$CH	PhMe$_2$CH	313	4.22×10^6	[275]

TABLE 2.21
Rate Constants of Decomposition of Alcoxyl Radicals: $R^1R^2RCO^\bullet \to R^1R^2C(O) + R^\bullet$

RO$^\bullet$	Solvent	T (K)	E (kJ mol^{-1})	log A, A (s^{-1})	k (300 K) (s^{-1})	Ref.
MeCH$_2$O$^\bullet$	Gas phase	468–488	73.2	13.40	4.52	[108]
Me$_3$CO$^\bullet$	Gas phase	393–453	69.0	13.50	30.6	[108]
Me$_3$CO$^\bullet$	H$_2$O, pH = 8.5	298			1.36×10^6	[280]
Me$_3$CO$^\bullet$	Cl$_2$C=CHCl	303–343	72.8	15.5	6.68×10^2	[13]
MeEtCHO$^\bullet$	Gas phase	298	59.8	14.30	7.73×10^3	[281]
EtMe$_2$CO$^\bullet$	Cl$_2$C=CHCl	303–343	38.5	11.80	1.25×10^3	[13]
cyclo-C$_5$H$_9$O$^\bullet$	Gas phase	298			2.90×10^6	[281]
cyclo-C$_6$H$_{11}$O$^\bullet$	Gas phase	298			6.39×10^4	[279]
cyclo-C$_7$H$_{13}$O$^\bullet$	Gas phase	298			1.40×10^6	[281]
Me$_2$(CN)CO$^\bullet$	C$_6$H$_5$Cl	353			1.60×10^5	[282]
PhCH$_2$CH$_2$O$^\bullet$	C$_6$H$_6$	238–300	20.2	10.89	2.36×10^7	[283]
PhMe$_2$CO$^\bullet$	C$_6$H$_6$	303			3.75×10^5	[279]
PhMe$_2$CO$^\bullet$	C$_6$H$_5$Cl	303			5.54×10^5	[279]
PhMe$_2$CO$^\bullet$	Me$_3$COH	303			5.84×10^5	[279]
PhMe$_2$CO$^\bullet$	CH$_3$CN	303			6.33×10^5	[279]
PhMe$_2$CO$^\bullet$	MeCOOH	303			1.96×10^6	[279]
PhMe$_2$CO$^\bullet$	CCl$_4$	296			2.27×10^5	[284]
PhMe$_2$CO$^\bullet$	CCl$_4$	303			2.63×10^5	[279]
PhMe$_2$CO$^\bullet$	C$_6$H$_6$	234–300	36.0	12.36	1.24×10^6	[285]

TABLE 2.22
Enthalpies, Activation Energies and Rate Constants of Hydrogen Atom Intramolecular Abstraction in Alcoxyl Radicals (Experimental and Calculated)

RO$^\bullet$	ΔH (kJ mol^{-1})	E (kJ mol^{-1})	k (350 K) (s^{-1})	Ref.
(C—H)H$_2$(CH$_2$)$_2$Me$_2$CO$^\bullet$	−15.0	30.1	6.44 × 10^4	[286]
(C—H)H$_2$MeCHCH$_2$Me$_2$CO$^\bullet$	−15.0	29.8	7.14 × 10^4	[286]
Me(C—H)H(CH$_2$)$_2$Me$_2$CO$^\bullet$	−24.2	23.4	6.44 × 10^5	[287]
Me$_2$C(O$^\bullet$)(CH$_2$)$_2$(C—H)H$_2$	−15.2	29.3	2.5 × 10^5	[287]
Me$_2$C(O$^\bullet$)(CH$_2$)$_2$(C—H)HMe	−25.2	25.4	6.4 × 10^5	[287]
Me$_2$C(O$^\bullet$)(CH$_2$)$_2$(C—H)Me$_2$	−37.2	21.1	1.4 × 10^6	[287]
Me$_2$C(O$^\bullet$)CH$_2$NH(C—H)HMe	−58.6	14.0	3.3 × 10^7	[287]
Me$_2$C(O$^\bullet$)CH$_2$NMe(C—H)HMe	−70.1	10.5	1.1 × 10^8	[287]
Me$_2$C(O$^\bullet$)CH$_2$NH(C—H)Me$_2$	−80.2	7.7	1.4 × 10^8	[287]
Me$_2$C(O$^\bullet$)CH$_2$NMe(C—H)Me$_2$	−85.3	4.7	4.0 × 10^8	[287]
Me$_2$C(O$^\bullet$)(CH$_2$)$_2$(C—H)HOH	−39.8	20.1	3.9 × 10^6	[287]
Me$_2$C(O$^\bullet$)(CH$_2$)$_2$(C—H)MeOH	−45.7	18.1	3.9 × 10^6	[287]
Me$_2$C(O$^\bullet$)CH$_2$C(O)(C—H)H$_2$	−27.2	24.7	1.2 × 10^6	[287]
Me$_2$C(O$^\bullet$)CH$_2$C(O)(C—H)HMe	−39.4	20.3	3.8 × 10^6	[287]
Me$_2$C(O$^\bullet$)CH$_2$C(O)(C—H)Me$_2$	−44.5	18.5	3.4 × 10^6	[287]
Me$_2$C(O$^\bullet$)(CH$_2$)$_2$(C—H)(O)	−51.6	16.2	7.6 × 10^6	[287]
Me$_2$C(O$^\bullet$)(CH$_2$)$_2$(C—H)HPh	−68.5	14.0	3.3 × 10^6	[287]
Me$_2$C(O$^\bullet$)(CH$_2$)$_2$(C—H)MePh	−82.5	9.9	6.6 × 10^6	[287]
Me$_2$C(O$^\bullet$)(CH$_2$)$_2$(C—H)HCH=CH$_2$	−87.4	13.4	4.0 × 10^6	[287]
Me$_2$C(O$^\bullet$)(CH$_2$)$_2$(C—H)MeCH=CH$_2$	−97.6	10.6	5.2 × 10^6	[287]

propagation in the reaction of alkoxyl radical decomposition. Carbonyl compounds are the main products of such oxidation, which occurs in the gas phase. Another mechanism of the hydrocarbon oxidation appears at high temperatures as a result of thermal instability of the peroxyl radical. The bond R—O$_2$ is weak (see earlier) and at $T > 450$ K, the alkyl peroxyl radical promptly decomposes into alkyl radical and dioxygen. The reaction between the alkyl radical and dioxygen occurs more rapidly with the formation of olefin and hydroperoxyl radical. So, olefin appears to be the main product of oxidation. The following four mechanisms can be formulated for the chain hydrocarbon oxidation [288]:

Mechanism I. Hydrocarbon oxidizes by consecutive reactions R$^\bullet$ + O$_2$ and RO$_2^\bullet$ + RH, with the formation of hydroperoxide as the primary product of oxidation.

Mechanism II. Reaction RO$_2^\bullet$ + RH occurs slowly and tertiary peroxyl radicals react more rapidly with the formation of alkoxyl radicals. Chain propagation includes the following steps:

$$RO_2^\bullet + RO_2^\bullet \longrightarrow RO^\bullet + RO^\bullet + O_2$$
$$RO^\bullet + RH \longrightarrow ROH + R^\bullet$$
$$R^\bullet + O_2 \longrightarrow RO_2^\bullet$$

The main product of the hydrocarbon oxidation is alcohol.

Mechanism III. When the temperature is sufficiently high for the prompt decomposition of the alkoxyl radical, the chain oxidation of the hydrocarbon in the gas phase includes the following steps:

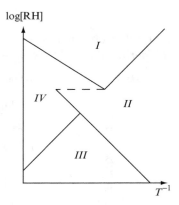

FIGURE 2.2 Topology of different mechanisms of hydrocarbon oxidation in gas and liquid phases in coordinates: hydrocarbon oxidation (log[RH]) versus temperature (1/T) [290].

$$RO_2^{\bullet} + RO_2^{\bullet} \longrightarrow RO^{\bullet} + RO^{\bullet} + O_2$$
$$RR^1R^2CO^{\bullet} \longrightarrow R^1R^2C(O) + RO^{\bullet}$$
$$RO^{\bullet} + RH \longrightarrow ROH + R^{\bullet}$$
$$R^{\bullet} + O_2 \longrightarrow RO_2^{\bullet}$$

The main products of oxidation are ketones and low-molecular weight alcohols.

Mechanism IV. When the peroxyl radical becomes unstable at high temperatures, it reacts with dioxygen to form an olefin. The latter becomes the main product of oxidation:

$$RCH_2C^{\bullet}HR^1 + O_2 \longrightarrow HO_2^{\bullet} + RCH{=}CHR^1$$
$$HO_2^{\bullet} + RCH_2CH_2R^1 \longrightarrow H_2O_2 + RCH_2C^{\bullet}HR^1$$
$$H_2O_2 \longrightarrow 2HO^{\bullet}$$
$$HO^{\bullet} + RCH_2CH_2R^1 \longrightarrow H_2O + RCH_2C^{\bullet}HR^1$$

The topology of all the four mechanisms in the coordinates log[RH] versus 1/T is shown in Figure 2.2.

Among the different factors determing the mechanism of oxidation, the BDE of the weakest C—H bond of the oxidizing hydrocarbon is very important [288].

REFERENCES

1. NN Semenov. *Some Problems of Chemical Kinetics and Reactivity*, vols 1 and 2. London: Pergamon Press, 1958–1959.
2. ET Denisov, VV Azatyan. *Inhibition of Chain Reactions*. London: Gordon and Breach, 2000.
3. ET Denisov, OM Sarkisov, GI Likhtenstein. *Chemical Kinetics*. Amsterdam: Elsevier, 2003.
4. RG Pearson. *Symmetry Rules for Chemical Reactions*. New York: Wiley Interscience, 1976.
5. JB Pedley, RD Naylor, SP Kirby. *Thermochemical Data of Organic Compounds*. London: Chapman & Hall, 1986.
6. W Tsang. In: A. Greenberg, J. Liebman, eds. *Energetics of Free Radicals*. New York: Blackie Academic & Professional, 1996, p. 22.
7. IV Berezin, ET Denisov, NM Emanuel. *The Oxidation of Cyclohexane*. Oxford: Pergamon Press, 1966.
8. C Walling. *Free Radicals in Solution*. New York: Chapman & Hall, 1957.

9. RB Mesrobian, AV Tobolsky. In: WO Lundberg, ed. *Autoxidation and Antioxidants*, vol 1. New York: Interscience, Wiley, 1961, pp 107–132.
10. N Uri. In: WO Lundberg, ed. *Autoxidation and Antioxidants*, vol 1. New York: Interscience, Wiley, 1961, pp 55–106.
11. R Livingston, In: WO Lundberg, ed. *Autoxidation and Antioxidants*, vol 1. New York: Interscience, Wiley, 1961, pp 249–298.
12. NM Emanuel, ET Denisov, ZK Maizus. *Liquid-Phase Oxidation of Hydrocarbons*. New York: Plenum Press, 1967.
13. T Mill, DG Hendry. In: CH Bamford and CFH Tipper, eds. *Comprehensive Chemical Kinetics*, vol 16. Amsterdam: Elsevier, 1980, pp 1–87.
14. ET Denisov, NI Mitskevich, VE Agabekov. *Liquid-Phase Oxidation of Oxygen-Containing Compounds*. New York: Consultants Bureau, 1977.
15. JA Howard. *Adv Free Radical Chem* 4:49–173, 1972.
16. H Backstrom. *J Am Chem Soc* 49:1460–1472, 1927.
17. VYa Shlyapintokh, ON Karpukhin, LM Postnikov, VF Tsepalov, AA Vichutinskiy, IV Zakharov. *Chemiluminescence Techniques in Chemical Reactions*. New York: Consultants Bureau, 1968.
18. NM Emanuel, GE Zaikov, ZK Maizus. *Oxidation of Organic Compounds. Medium Effect in Radical Reaction*. Oxford: Pergamon Press, 1984.
19. GG Hammes, ed. *Investigation of Rates and Mechanisms of Reactions*. New York: Wiley, 1974.
20. JK Thomas. *J Phys Chem* 71:1919–1925, 1967.
21. SL Khursan, RL Safiullin, SYu Serenko. *Khim Fiz* 9:375–379, 1990.
22. SI Maslennikov, AI Nikolaev, VD Komissarov. *Kinet Katal* 20:326–329, 1979.
23. VA Ickovic, VM Potekhin, W Pritzkow, VA Proskurjakov, D Schnurpfeil. Autoxidation von Kohlenwasserstoffen. Leipzig: VEB Deutscher Verlag fur Grundstoffindustrie, 1981.
24. TG Degtyareva. Mechanism of Oxidation of 2-Methylbutane. PhD Thesis Dissertation, Institute of Chemical Physics, Chernogolovka, 1972, pp 3–17 [in Russian].
25. L Sajus. *Adv Chem Ser* 75:59–77, 1968.
26. T Mill, G. Montorsi. *Int J Chem Kinet* 5:119–136, 1973.
27. AE Semenchenko, VM Solyanikov, ET Denisov. *Neftekhimiya* 11:555–561, 1971.
28. JA Howard, KU Ingold. *Can J Chem* 45:793–802, 1967.
29. BI Chernyak, MV Nikipanchuk, MG Kotur, SI Kozak. *Kinet Katal* 37:818–820, 1996.
30. L Bateman, G Gee. *Proc Roy Soc* 195A:391–402, 1948.
31. AV Nikitin, LI Murza, IF Rusina, VL Rubailo. *Neftekhimiya* 30:639–647, 1990.
32. JA Howard, KU Ingold. *Can J Chem* 45:785–792, 1967.
33. GM Kuznetsova, TV Lobanova, IF Rusina, OT Kasaikina. *Russ Chem Bull* 45:1586–1591, 1996.
34. IA Opeida. *Cooxidation of Alkylaromatic Hydrocarbons in the Liquid Phase*. Doctoral Dissertation, Institute of Chemical Physics, Chernogolovka, 1982, pp 1–336 [in Russian].
35. G Vasvari, D Gal. *Ber Bunsenges Phys Chem* 97:22, 1993.
36. IA Opeida, AG Matvienko, IV Yefimova, IO Kachurin. *Zh Org Khim* 24:572–576, 1988.
37. NM Emanuel, D Gal. *Oxidation of Ethylbenzene*. Moscow: Nauka, 1984 [in Russian].
38. VF Tsepalov. The Study of Elementary Reactions of Liquid-Phase Oxidation of Alkylaromatic Hydrocarbons. Doctoral Thesis Dissertation, Institute of Chemical Physics, Chernogolovka, 1975, pp 3–40 [in Russian].
39. DG Hendry. *J Am Chem Soc* 89:5433–5438, 1967.
40. HW Melville, S Richards. *J Chem Soc* 1954:944–952.
41. JA Howard, KU Ingold. *Can J Chem* 46:2655–2660, 1968.
42. AF Guk, VF Tsepalov, VF Shuvalov, VYa Shlyapintokh. *Izv AN SSSR Ser Khim* 2250–2253, 1968.
43. JA Howard, KU Ingold. *Can J Chem* 44:1119–1130, 1966.
44. JA Howard, KU Ingold. *Can J Chem* 46:2661–2666, 1968.
45. CH Bamford, MJS Dewar. *Proc Roy Soc* A198:252–267, 1949.
46. VV Kafarov, ed. *Handbook of Solubility*, vol 1, sub/vol 1. Moscow: Academia Nauk SSSR, 1961, pp 570–575 [in Russian].

47. VD Knyazev, A Bensura, IR Stagle. Fourth International Conference on Chemical Kinetics. Gaithersburg: NIST, 1997, pp 6–7.
48. EI Finkelshtein, GN Gerasimov. *Zh Fiz Khim* 58:942–946, 1984.
49. JK Thomas. *J Phys Chem* 71:1919–1925, 1967.
50. A Marchaj, DG Kelly, A Bakach, JH Esperson. *J Phys Chem* 95:4440–4441, 1991.
51. A Sauer, H Cohen, D Meyerstein. *Inorg Chem* 27:4578–4579, 1988.
52. B Hickel. *J Phys Chem* 79:1054–1059, 1975.
53. N Getoff. *Radiat Phys Chem* 37:673–680, 1991.
54. B Maillard, KU Ingold, JC Scaiano. *J Am Chem Soc* 105:5095–5099, 1983.
55. RW Fessenden, PM Carton, H Shimamori, JC Scaiano. *J Phys Chem* 86:3803–3811, 1982.
56. O Brede, R Herman, R Mehnert. *J Chem Soc Faraday Trans 1*, 83:2365–2368, 1987.
57. K Hasegawa, LK Patterson. *Photochem Photobiol* 28:817–822, 1978.
58. O Brede, L Wojnarovits. *Radiat Phys Chem* 37:537–548, 1991.
59. X-M Pan, MN Schuchmann, C von Sonntag. *J Chem Soc Perkin Trans 2*, 289–297, 1993.
60. M Lal, C Schoeneich, J Moenig, K-D Asmus. *Int J Radiat Biol* 54:773–785 1988.
61. J Moenig, K-D Asmus, M Schaeffer, TF Slater, RL Willson. *J Chem Soc Perkin Trans 2*, 1133–1137, 1983.
62. S Mossery, P Neta, D Meisel. *Radiat Phys Chem* 36:683–692, 1990.
63. S Das, MN Schuchmann, C von Sonntag. *Chem Ber* 120:319–323, 1987.
64. J Rabani, D Klug-Roth, A Henglein. *J Phys Chem* 78:2089–2093, 1974.
65. CF Cullis, JM Francis, Y Raef, AJ Swallow. *Proc Roy Soc (London) Ser A* 300:443–454, 1967.
66. GE Adams, RL Willson. *Trans Faraday Soc* 65:2981–2987, 1969.
67. S Jockusch, NJ Turro. *J Am Chem Soc* 121:3921–3925, 1999.
68. D Schulte-Frohlinde, R Anker, E Bothe. In: M. A. J. Rogers, E. L. Powers, eds. *Oxygen and Oxy-Radicals in Chemistry and Biology*. New York: Academic Press, 1981, p 61.
69. F Munoz, MN Schuchman, G Olbrich, C von Sonntag. *J Chem Soc Perkin Trans 2*, 655–659, 2000.
70. MN Schuchmann, R Rao, M Hauser, SC Muller, C von Sonntag. *J Chem Soc Perkin Trans 2*, 491–493, 2000.
71. MN Schuchmann, H Zegota, C von Sonntag. *Z Naturforsch B Anorg Chem Org Chem B*, 40: 215–221, 1985.
72. MN Schuchmann, HP Schuchmann, C von Sonntag. *J Phys Chem* 99:9122–9129 1995.
73. H Zegota, MN Schuchmann, D Schulz, C von Sonntag. Z Naturforsch B, *Chem Sci* 41B: 1015–1022, 1986.
74. MT Nenadovic, OI Micic. *Radiat Phys Chem* 12:85–89, 1978.
75. X Fang, R Mertens, C von Sonntag. *J Chem Soc Perkin Trans 2*, 1033–1036, 1995.
76. P Neta, RE Huie, S Mosseri, LV Shastry, JP Mittal, P Maruthamuthu, S Steenken. *J Phys Chem* 93:4099–4104, 1989.
77. K Tokumara, H Nosaka, T Ozaki. *Chem Phys Lett* 169:321–328, 1990.
78. AL Aleksandrov, EM Pliss, VF Shuvalov. *Izv AN SSSR Ser Khim* 2446–2450, 1979.
79. JL Faria, S Steenken. *J Chem Soc Perkin Trans 2*, 1153–1159, 1997.
80. E Zador, J Warman, A Hummel. *J Chem Soc Faraday Trans 1*, 75:914–921, 1979.
81. O Mieden, MN Schuchmann, C von Sonntag, *J Phys Chem* 97:3783–3790, 1993.
82. ZB Alfassi, GI Khaikin, P Neta. *J Phys Chem* 99:4544–4545, 1995.
83. JA Miller, SJ Klippenstein. *Int J Chem Kinet* 33:654–668, 2001.
84. EG Janzen, FJ Johnston, CL Ayers. *J Am Chem Soc* 89:1176–1183, 1967.
85. T Hirano, W Li, L Abrams, P Krusie, MF Ottaviani. *J Am Chem Soc* 121:7170–7171, 1999.
86. DG Hendry, D Schuetzle. *J Am Chem Soc* 97:7123–7126, 1975.
87. HE Hunziker, HR Wendt. *J Chem Phys* 60:4622–4623, 1974.
88. TT Paukert, HS Johnston. *J Chem Phys* 56:2824–4637, 1972.
89. DW Smith, L Andrews. *J Chem Phys* 60:81–85, 1974.
90. Y Beers, CJ Howard. *J Chem Phys* 64:1541–1543, 1976.
91. MS Gordon, JA Pople. *J Chem Phys* 49:4643–4650, 1968.
92. K Ohkubo, F Kitagava. *Bull Chem Soc Jpn* 48:703–704, 1975.

93. DH Liskow, HF Schaefer, CF. Bender. *J Am Chem Soc* 93:6734–6737, 1971.
94. JL Gole, EF Hayes. *J Chem Phys* 57:360–363, 1972.
95. RJ Blint, MD Newton. *J Chem Phys* 59:6220–6223, 1973.
96. K Ohkubo, T Fujita, H Sato. *J Mol Struct* 36:101–110, 1977.
97. RV Kucher, IA Opeida, AF Dmitruk, LT Kholoimova. *Oxid Commun* 5:75–87, 1983.
98. RV Kucher, IA Opeida, AF Dmitruk, LT Kholoimova, VV Lobanov, VV Shimanovskiy. *Theor Experim Khim* 19:22–30, 1983.
99. A Delabie, S Creve, B Coussens, MT Ngueyen, *J Chem Soc Perkin Trans 2*, 977–981, 2000.
100. S Biscupic, L Valco. *J Mol Struct* 27:97–103, 1975.
101. K Ohkubo, F Kitagawa. *Bull Chem Soc Jpn* 46:2942–2949, 1973.
102. RV Kucher, IA Opeida, AF Dmitruk. *Neftekhimiya* 18:519–524, 1978.
103. RL McCarthy, A MacLachlan. *J Phys Chem* 35:1625–1627, 1961.
104. DR Lide, ed. *Handbook of Chemistry and Physics*, 72nd ed. Boca Raton: CRC Press, 1991–1992.
105. ET Denisov, TG Denisova. *Kinet Catal* 34:173–179, 1993.
106. JL Holmes, FP Lossing, PM Mayer. *J Am Chem Soc* 113:9723–9728, 1991.
107. SG Lias, JF Liebman, RD Levin, SA Kafafi. NIST Standard Reference Database, 19A, NIST Positive Ion Energetics, Version 2.0. Gaithersburg: NIST, 1993.
108. SW Benson, HE O'Neal. *Kinetic Data on Gas Phase Unimolecular Reactions*. Washington: NSRDS, 1970.
109. NA Kozlov, IB Rabinovich. *Trudy po Khimii i Khimicheskoy Tekhnologii*, vol 2 (10) Gorkiy: Gorkiy State University, 1964, p 189 [in Russian].
110. W Pritzkow, KA Muller. *Chem Ber* 89:2318–2325, 1956.
111. YY Van-Chin-Syan, TN Dolbneva, MA Dikiy, SK Chuchmaev. *Zh Fiz Khim* 58:2937–2940, 1984.
112. YY Van-Chin-Syan, NS Kachurina. Vestnik Nizhegorodskogo Gosudarstvennogo Universiteta. Organic and Elementoorganic Peroxides, Nizhniy Novgorod: Izd-vo Nizhegorodskogo Universiteta, 1996, pp 29–35 [in Russian].
113. JW Breitenbach, J Derkosch. *Monatsh Chem* 81:689–697, 1950.
114. JP Fortuin, HI Waterman, *Chem Eng Sci* 2:182–189, 1953.
115. W Pritzkow, R Hoffman. *J Prakt Chem N* 14, 13–21, 1961.
116. ES Dombrovski, ED Hearing. *J Phys Chem Ref Data* 22:805–1174, 1993.
117. YaA Lebedev, EA Miroshnichenko. *Thermochemistry of Evaporation of Organic Compounds*. Moscow: Nauka, 1981 [in Russian].
118. JA Howard, KU Ingold, HI Symonds. *Can J Chem* 46:1017–1022, 1968.
119. ET Denisov. *Russ Chem Rev* 66:859–876, 1997.
120. TG Denisova, ES Emel'yanova. *Kin Catal* 44:441–449, 2003
121. GE Zaikov, AA Vichutinskiy, ZK Maizus, NM Emanuel. *Izv AN SSSR Ser Khim* 1743–1748, 1968.
122. JA Howard, JHB Chenier, DA Holden, *Can J Chem* 56:170–175, 1978.
123. AL Buchachenko, KYa Kaganskaya, MB Neuman, AA Petrov, *Kinet Katal* 2:44–49, 1961.
124. VA Belyakov, NM Zalevskaya, IO Kachurin, IA Opeida, *Neftekhimiya* 24:670–675, 1986.
125. GE Zaikov, ZK Mauzus, NM Emanuel. *Izv AN SSSR* 2265–2271, 1968.
126. JC Robb, M Shahin. *Trans Faraday Soc* 55:1753–1759, 1959.
127. IS Gaponova, TV Fedotova, VF Tsepalov, VF Shuvalov, YaS Lebedev. *Kinet Katal* 12:1137–1143, 1971.
128. VA Belyakov, G. Lautenbach, W Pritzkow, V Voerckel. *J Prakt Chem* 334:373–382, 1992.
129. S Fukuzumi, Y. Ono. *J Phys Chem* 80:2973–2978, 1976.
130. FF Rust. *J Am Chem Soc* 79:4000–4003, 1957.
131. DE Van Sickle, T Mill, FR Mayo, H Richardson, CW Gould. *J Org Chem* 38:4435–4440, 1973.
132. DE Van Sickle. *J Org Chem* 37:755–760, 1972.
133. IN Demidov, VM Solyanikov. *Neftekhimiya* 26:406–408, 1986.
134. RK Jensen, S Korcek, LR Mahoney, M Zinbo. *J Am Chem Soc* 103:1742–1749, 1981.
135. JA Howard, KU Ingold. *Can J Chem* 48:873–880, 1970.
136. SG Voronina, LV Krutskaya, AL Perkel, BG Freydin. *Zh Prikl Khim* 63:1376–1383, 1990.

137. MM Mogilevich, EM Pliss. *Oxidation and Oxidative Polymerization of Nonsaturated Compounds.* Moscow: Khimiya, 1990 [in Russian].
138. P Koelewijn. *Rec Trav Chim Pay–Bas* 91:759–779, 1972.
139. FR Mayo, MG Syz, T Mill, JK Castleman. *Adv Chem Ser* 75:38–58, 1968.
140. LCT Shoute, ZB Alfassi, P Neta, RE Huie. *J Phys Chem* 98:5701–5704, 1994.
141. VA Machtin. Reactions of Peroxyl Radicals in Oxidizing Vinyl Monomers and Reactivity of Double Bonds. Ph.D. thesis Dissertation, Institute of Chemical Physics, Chernogolovka, 1984, pp 1–18 [in Russian].
142. JA Howard. *Can J Chem* 50:2798–2304, 1972.
143. W Suprun. *J Prakt Chem* 338:231-237, 1996.
144. C Chevriau, P Naffa, JC Balaceanu. *Bull Soc Chim Fr* 1, 3002–3010, 1964.
145. JA Howard, KU Ingold. *Can J Chem* 43:2729–2736, 1965.
146. NN Pozdeeva, IK Yakushchenko, AL Aleksandrov, ET Denisov. *Kinet Catal* 32:1162–1169, 1991.
147. JA Howard, KU Ingold. *Can J Chem* 43:2737–2743, 1965.
148. JA Howard, KU Ingold. *Can J Chem* 44:1113–1118, 1966.
149. JA Howard. *Can J Chem* 50:2298–2304, 1972.
150. VL Rubailo, AB Gagarina, NM Emanuel. *Dokl AN SSSR* 224:883–886, 1975.
151. IA Opeida, LG Nechitaylo. *Kinet Katal* 19:1581–1585, 1978.
152. EM Pliss, AL Aleksandrov, MM Mogilevich. *Izv AN SSSR Ser Khim* 1971–1974, 1975.
153. VA Machtin, EM Pliss, ET Denisov. *Izv AN SSSR Ser Khim* 746–750, 1981.
154. GA Twigg. *Chem Eng Sci Suppl* 3:5–13, 1954.
155. A Fish. *Adv Chem Ser* 76:69–78, 1968.
156. RE Pliss, VA Machtin, EM Pliss. In: Abstracts of Conference *Regulation of Biological Processes by Free Radicals: Role of Antioxidants, Free Radical Scavengers, and Chelators*. Moscow, Yaroslavl, Yaroslavl State Technical University, 1998, p 17.
157. PD Bartlett, G Guaraldi. *J Am Chem Soc* 89:4799–4801, 1967.
158. K Adamic, JA Howard, KU Ingold. *Can J Chem* 47:3803–3808, 1969.
159. JE Bennett, DM Brown, B Mile. *Trans Faraday Soc* 66:386–396, 1970.
160. E Furimsky, JA Howard, J Selwyn. *Can J Chem* 58:677–680, 1980.
161. JE Bennett, G Brunton, JRL Smith, TMF Salmon, DJ Waddington. *J Chem Soc Faraday Trans* 1, 83:2421–2432, 1987.
162. PD Bartlett, P Gunther. *J Am Chem Soc* 88:3288–3294, 1966.
163. S Patai, ed. *The Chemistry of Peroxides*. Chichester: Wiley, 1983.
164. SW Benson. *Thermochemical Kinetics*, 2nd ed. New York: Wiley, 1972.
165. ET Denisov. *Usp Khim* 39:62–93, 1970.
166. SL Khursan, VS Martemyanov, ET Denisov. *Kinet Katal* 31:1031–1040, 1990.
167. WJ Maguire, RC Pink. *Trans Faraday Soc* 64:1097–1105, 1967.
168. JR Thomas. *J Am Chem Soc* 87:3935–3940, 1965.
169. JR Thomas, KU Ingold. *Adv Free Radical Chem* 4:258, 1972.
170. JA Howard, K Adamic, KU Ingold. *Can J Chem* 47:3793–3802, 1969.
171. JE Bennett, DM Brown, B Mile. *Trans Faraday Soc* 66:386–396, 1970.
172. A Faucitano, A Buttafava, F Martinotti, V Comincoli, P Bortolus. *Polymer Photochem* 7:491–502, 1986.
173. LA Tavadyan, MV Musaelyan, VA Mardoyan. *Khim Fiz* 10:511–515, 1991.
174. JJ Zwolenik. *J Phys Chem* 71:2464–2469, 1967.
175. JA Howard, JC Robb. *Trans Faraday Soc* 59:1590–1599, 1963.
176. VA Belyakov, G Lauterbach, W Pritzkow, V Voerckel. *J Prakt Chem*, 340:475–483, 1992.
177. SA Gerasimova, IO Kachurin, AG Matvienko, IA Opeida. *Neftekhimiya* 30:476–481, 1990.
178. IA Opeida, IV Yefimova, AG Matvienko, AF Dmitruk, OM Zarechnaya. *Kinet Katal* 31: 1342–1348, 1990.
179. GA Russel. *J Am Chem Soc* 79:3871–3877, 1957.
180. RL Vardanyan, RL Safiullin, VD Komissarov. *Kinet Katal* 26:1327–1331, 1985.
181. A Baignee, JHB Chenier, JA Howard. *Can J Chem* 61:2037–2043, 1983.
182. SI Maslennikov, AI Nikolaev, VD Komissarov. *Kinet Katal* 20:326–329, 1979.

183. SI Maslennikov, LG Galimova, VD Komissarov. *Izv AN SSSR Ser Khim* 631–634, 1979.
184. LA Tavadyan, TK Nubaryan, AK Tonikyan, AB Nalbandyan. *Arm Khim Zh* 40:343–347, 1987.
185. PS Nangia, SW Benson. *Int J Chem Kinet* 12:43–54, 1980.
186. IV Aleksandrov. *Theor Experim Khimiya* 12:299–306, 1976.
187. E Bothe, MN Suchman, D Schulz, Frohlinde, C von Sonntag. *Z Naturforsch* 38b:212–220, 1983.
188. LA Tavadyan. *Khim Fiz* 5:63–71, 1986.
189. MN Suchman, H Zegota, C von Sonntag. *Z Naturforsch* 40b:215–221, 1985.
190. H Zegota, MN Suchman, D Schulz, C von Sonntag. *Z Naturforsch* B41:1015–1022, 1986.
191. MN Suchman, H Zegota, C von Sonntag. *J Phys Chem* 86:1995–2000, 1982.
192. JE Bennett, R Summers. *Can J Chem* 52:1377–1379, 1974.
193. ET Denisov. *Neftekhimiya* 26:723–735, 1986.
194. ET Denisov, TG Denisova. *Handbook of Antioxidants.* Boca Raton: CRC Press. 2000.
195. RL Vardanyan. Oxidation and Stabilization of Cholesterol Esters: Kinetics, Mechanism, Physical Properties. Doctor Thesis Dissertation, Institute of Chemical Physics, Chernogolovka, 1986, pp 1–42 [in Russian].
196. JE Bennett, G Brunton, JRL Smith, DJ Waddington. *J Chem Soc Faraday Trans 1*, 83:2433–2441, 1987.
197. JA Howard, KU Ingold. *Can J Chem* 45:785–795, 1967.
198. AI Nikolaev, RL Safiullin, LR Enikeeva, VD Komissarov. *Khim Fizika* 11:69–72, 1992.
199. JE Bennett, G Brunton, JRL Smith, TMF Salmon, DJ Waddington. *J Chem Soc Faraday Trans 1*, 83:2421–2432, 1987.
200. JE Bennett. *J Chem Soc Faraday Trans 1*, 83:1805–1813, 1987.
201. JA Howard, KU Ingold. *J Am Chem Soc* 90:1058–1059, 1968.
202. JA Howard, JE Bennett. *Can J Chem* 50:2374–2377, 1972.
203. SI Maslennikov, AI Nikolaev, VD Komissarov. *Kinet Katal* 20:326–329, 1979.
204. E Furimsky, JA Howard, J Selwyn. *Can J Chem* 58:677–680, 1980.
205. SI Maslennikov, LG Galimova, VD Komissarov. *Izv Akad Nauk SSSR Ser Khim* 631–634, 1979.
206. B Smaller, JR Retko, EC Avery. *J Chem Phys* 48:5174–5181, 1968.
207. BYa Ladygin, GM Zimina, AV Vannikov. *Khim Vysok Energii* 18:301–309, 1984.
208. JA Howard, KU Ingold. *Can J Chem* 44:1119–1130, 1966.
209. RL McCarthy, A MacLachlan. *Trans Faraday Soc* 57:1107–1116, 1961.
210. JA Howard, KU Ingold. *Can J Chem* 45:793–802, 1967.
211. RL Vardanyan, RL Safiullin, VD Komissarov. *Kinet Katal* 26:1327–1331, 1985.
212. TP Kenisberg, NG Ariko, NI Mitskevich. *Dokl Akad Nauk SSR* 24:817–218, 1980.
213. JC Robb, M Shahin. *Trans Faraday Soc* 55:1753–1759, 1959.
214. SA Gerasimova, AG Matvienko, IA Opeida, IO Kachurin. *Zh Org Khim* 34:781–781, 1998.
215. AN Nikolayevskii, VG Koloyerova, RV Kucher. *Neftekhimiya* 16:752–757, 1976.
216. VA Belyakov, G Lauterbach, W Pritzkow, V Voerckel. *J Prakt Chem* 340:475, 1992.
217. IA Opeida, AG Matvienko, IV Yefimova, IO Kachurin. *Zh Org Khim* 24:572–576, 1988.
218. IA Opeida, AG Matvienko, IV Yefimova. *Kinet Katal* 28:1341–1346, 1987.
219. RF Vasil'ev, ON Karpukhin, VYa Shlyapintokh. *Dokl Akad Nauk SSR* 125:106–109, 1959.
220. VYa Shlaypintokh, RF Vasil'ev, ON Karpukhin, LM Postnikov, LA Kibalko. *J Chim Phys* 57:1113–1119, 1960.
221. RF Vasil'ev. *Prog React Kinet* 4:305–352, 1967.
222. GF Fedorova. The Study of Chemiluminescence in Liquid-Phase Oxidation of Hydrocarbons. Thesis Dissertation, Institute of Chemical Physics, Moscow, 1979, pp 1–30 [in Russian].
223. VA Belyakov, RF Vasil'ev, GF Fedorova. *Izv AN SSSR Ser Fiz* 37:743–752, 1973.
224. VA Belyakov, RF Vasil'ev, GF Fedorova. *Khim Vysok Energii* 12:247–252, 1978.
225. VA Belyakov, RF Vasil'ev, GF Fedorova. *Izv AN SSSR Ser Fiz* 32:1325–1331, 1968.
226. AI Nikolaev. Kinetics of Self-Reactions of Alkyl, Peroxyl, and Sulfonyl Radicals. PhD Thesis Dissertation, Institute of Chemical Physics, Chernogolovka, 1984, pp 1–18 [in Russian].
227. VN Kondratiev, EE Nikitin. *Gas-Phase Reactions.* Berlin: Springer-Valley, 1981.
228. JA Kerr. In: JK Kochi ed. *Free Radicals*, vol 1. New York: Wiley, 1973, pp 1–36.
229. KU Ingold. In: JK Kochi ed. *Free Radicals*, vol 1. New York: Wiley, 1973, pp 37–112.

230. ZB Alfassi. In: ZB Alfassi ed. *General Aspects of the Chemistry of Radicals*. Chichester: Wiley, 1999, pp 139–173.
231. AP Stefani. *J Am Chem Soc* 90:1694–1708, 1968.
232. A Bazkac, JH Espenson. *J Am Chem Soc* 90:325–331, 1986.
233. KJ Laidler. *Chemical Kinetics*, New York: Harper and Row, 1987.
234. DJ Carlsson, KU Ingold. *J Am Chem Soc* 90:7047–7055, 1968.
235. DJ Carlsson, KU Ingold, LC Bray. *Int J Chem Kinet* 1:315–323, 1969.
236. GC Stevens, RM Clarke, EJ Hart. *J Phys Chem* 76:3863–3867, 1972.
237. B. Hickol. *J Phys Chem* 79:1054–1059, 1975.
238. RW Fessenden. *J Phys Chem* 68:1508–1515, 1964.
239. RD Burkhart. *J Phys Chem* 73:2703–2706, 1969.
240. N Getoff. *Radiat Phys Chem* 37:673–679, 1991.
241. RD Burkhart, RF Boynton, JC Merril. *J Am Chem Soc* 93:5013–5017, 1971.
242. KU Ingold, B Maillard, JC Walten. *J Chem Soc Perkin Trans 2*, 970–974, 1981.
243. P Schmid, D Griller, KU Ingold. *Int J Chem Kinet* 11:333–338, 1979.
244. AI Nikolaev, RL Safiullin, VD Komissarov. *Khim Fizika* 3:257–261, 1984.
245. MC Sauer, I Mani. *J Phys Chem* 72:3856–3862, 1968.
246. AI Nikolaev, RL Safiullin, LR Enikeeva. *Khim Fizika* 3:711–714, 1984.
247. Landolt-Bornstein. In: H. Fischer ed. *Numerical Data and Functional Relationships in Science and Technology, New Series, Group II: Atomic and Molecular Physics*, vol 13, sub/vol a, *Radical Reaction Rates in Liquids*, Berlin: Springer-Verlag, 1984, p 16.
248. C Huggenberger, H Fischer. *Helv Chim Acta* 64:338–353, 1981.
249. GM Zimina, R Cech. *Radiochem Radioanal Lett* 38:119–124, 1979.
250. H-H Schuh, H Fischer. *Helv Chim Acta* 61:2130–2164, 1978.
251. SA Weiner, GS Hammond. *J Am Chem Soc* 91:986–991, 1969.
252. EJ Hamilton, H Fischer. *J Phys Chem* 77:722–724, 1973.
253. M Ebert, JP Keene, EJ Land, AJ Swallow. *Proc Roy Soc (London) Ser A* 287:1–14, 1965.
254. JE Bennett, R Summers. *J Chem Soc Perkin Trans 2*, 1504–1508, 1977.
255. H Schuh, H Fischer. *Helv Chim Acta* 61:2130–2164, 1978.
256. GB Watts, KU Ingold. *J Am Chem Soc* 94:491–494, 1972.
257. RJ Hagemann, HA Schwarz. *J Phys Chem* 71:2694–2699, 1967.
258. TO Meiggs, LI Grossweiner, SI Miller. *J Am Chem Soc* 94:7986–7991, 1972.
259. HC Christensen, K Schested, EJ Hart. *J Phys Chem* 77:983–987, 1973.
260. W Tsang, RF Hampson. *J Phys Chem Ref Data* 15:1087, 1986.
261. W Tsang. *J Phys Chem Ref Data* 17:887, 1988.
262. W Tsang. *J Phys Chem Ref Data* 19:1–68, 1990.
263. WG Alcock, B Mile. *Combust Flame* 24:125–131, 1975.
264. Landolt-Bornstein. In: H. Fischer ed. Numerical Data and Functional Relationships in Science and Technology, New Series, Group II: Atomic and Molecular Physics, vol 13, sub/vol d, *Radical Reaction Rates in Liquids*, Berlin: Springer-Verlag, 1984, p 431.
265. JL Brokenshire, A Nechvatal, JM Tedder. *Trans Faraday Soc* 66:2029–2037, 1970.
266. MI Sway, DJ Waddington. *J Chem Soc Perkin Trans 2* 63–69, 1984.
267. E Niki, Y Kamiya. *J Org Chem* 38:1404–1406, 1973.
268. E Patmore, P Gritter. *J Org Chem* 27:4196–4200, 1962.
269. MV Encina, JC Scaino. *J Am Chem Soc* 103:6393–6397, 1981.
270. P Wagner, C Walling. *J Am Chem Soc* 87:5179–5185, 1965.
271. A Effio, D Griller, KU Ingold, JC Scaiano, S Sheng. *J Am Chem Soc* 102:6063–6068, 1980.
272. H Paul, R Small, JC Scaiano. *J Am Chem Soc* 100:4520–4527, 1978.
273. A Williams, E Oberright, J Brooks. *J Am Chem Soc* 78:1190–1193, 1956.
274. R Small, JC Scaiano. *J Am Chem Soc* 100:296–298, 1978.
275. L Dulog, P Klein. *Chem Ber* 104:895–901, 1971.
276. C Walling, P Wagner. *J Am Chem Soc* 86:3368–3375, 1964.
277. L Loan. *J Polymer Sci* 2:3053–3066, 1964.
278. D Avila, C Brown, KU Ingold, J Lusztyk. *J Am Chem Soc* 115:466–470, 1993.

279. R Kennedy, KU Ingold. *Can J Chem* 44:2381–2385, 1966.
280. M Erben-Russ, C Michel, W Bors, M Saran. *J Phys Chem* 91:2362–2365, 1987.
281. R Atkinson. *Int J Chem Kinet* 29:99–111, 1997.
282. G Vasvari, E Kuramshin, S Holly, T Vidiczy, D Gal. *J Phys Chem* 92:3810–3818, 1988.
283. G Mendenhall, L Stewart, JC Scaiano. *J Am Chem Soc* 104:5109–5114, 1982.
284. A Neville, C Brown, D Rayner, J Lusztyk, KU Ingold. *J Am Chem Soc* 111:3269–3270, 1989.
285. A Baignee, JA Howard, JC Scaiano, L Stewart. *J Am Chem Soc* 105:6120–6123, 1983.
286. JK Kochi. In: JK Kochi ed. *Free Radicals*, vol 2. New York: Wiley, 1973, pp 665–710.
287. TG Denisova, ET Denisov. *Kinet Katal* 42:684–695, 2001.
288. ET Denisov. *Khim Fiz* 105–112, 1982.

3 Initiation of Liquid-Phase Oxidation

3.1 INITIATORS

3.1.1 MECHANISMS OF DECOMPOSITION OF INITIATORS

Radical *initiators* are molecules bearing one or several weak bonds with a BDE of about 100–200 kJ mol^{-1}. When the temperature of the reaction is sufficiently high, the initiator decomposes with homolysis of the weakest bond and produces free radicals, which initiate a chain or nonchain free radical reaction.

The following bonds have sufficiently low values of BDE [1]:

Compound	MeOOMe	MeONO	MeNO	PrN$_2$CH$_2$CH=CH$_2$
Bond	O—O	O—N	C—N	C—N
D (kJ mol^{-1})	157	175	167	141

Different mechanisms of free radical formation as a result of the decomposition of initiators are known.

3.1.1.1 Unimolecular Decomposition of Initiator with One Bond Splitting

Most initiators decompose with dissociation of the weakest bond, for example [2–4]:

$$R^1O\text{—}OR^2 \longrightarrow R^1O^\bullet + R^2O^\bullet$$

According to the transition state theory, the rate constant of unimolecular reaction (at high pressure in the gas phase) is the following [5]:

$$k_d = k_\infty = e\left(\frac{RT}{Lh}\right)e^{\Delta S^\#/R}e^{-E/RT} \qquad (3.1)$$

When a polyatomic molecule, for example, peroxide ROOR, decomposes into two free radicals RO$^\bullet$, the following changes in the energy distribution are observed [6]:

1. One stretching vibration along the O—O bond disappears
2. One inner rotation of the O—O bond disappears
3. Two C—O—O angles vibrations disappear.

As a result, the activation entropy of unimolecular decomposition $\Delta S^{\#} > 0$ and the pre-exponential factor $(A = eRT(Lh)^{-1} \exp(\Delta S^{\#}/R))$ is sufficiently higher than $eRT(Lh)^{-1} \approx 10^{13} \, s^{-1}$. For many unimolecular reactions, $\Delta S^{\#} \approx 20$–$70 \, J \, mol^{-1} \, K^{-1}$ [6].

Due to the elongation of the dissociating bond (for example, O—O in peroxide), the volume of the transition state $V^{\#}$ is greater than the volume of reactant V. As a result, the difference in the volumes $\Delta V^{\#} = V^{\#} - V$ is positive. The study of the decomposition of initiators with one bond dissociation under high pressure gives evidence that $\Delta V^{\#}$ is positive [2,7].

3.1.1.2 Concerted Fragmentation of Initiators

Initiators that decompose with the simultaneous dissociation of two or more bonds are known, for example [2–4]:

$$RC(O)O\text{—}OR^1 \longrightarrow R^{\bullet} + CO_2 + R^{1\bullet}$$

Such decay is known as *concerted fragmentation*. Peroxides have the weak O—O bond and usually decompose with dissociation of this bond. The rate constants of such decomposition of ROOR into RO$^{\bullet}$ radicals demonstrate a low sensitivity of the BDE of the O—O bond to the structure of the R fragment [4]. Bartlett and Hiat [8] studied the decay of many peresters and found that the rate constants of their decomposition covered a range over $10^5 \, s^{-1}$. The following mechanism of decomposition was proposed in parallel with a simple dissociation of one O—O bond [3,4]:

$$RC(O)OOR^1 \longrightarrow R^{\bullet} + CO_2 + R^{1\bullet}$$

This decay of the molecule into fragments was named concerted fragmentation. The energy needed for the activation of the molecule is concentrated simultaneously on the two cleaving bonds: R—C(O) and O—O. Among other products, carbon dioxide is formed as a result of perester fragmentation. The formation of the π-bond of formed carbon dioxide compensates partially for the energy of dissociation of the C—C and O—O bonds of perester. The decay of perester to four fragments is known [9]:

$$Me_3COOCMe_2C(O)OOCMe_3 \longrightarrow Me_3CO^{\bullet} + MeCOMe + CO_2 + Me_3CO^{\bullet}$$

The following peculiarities characterize the concerted fragmentation in comparison with the decay of a molecule with dissociation of one bond [3,4]:

Decay with Dissociation of One Bond	Concerted Fragmentation
1. The activation energy is equal to the dissociation energy of the weakest bond: ($E \approx D \approx 140$–$160 \, kJ \, mol^{-1}$ for peresters).	The activation energy of decay is sufficiently lower than the BDE of the weakest bond: ($E \approx 90$–$125 \, kJ \, mol^{-1}$ for peresters).
2. The entropy of activation $\Delta S^{\#} > 0$: $\Delta S^{\#} \approx 40$–$100 \, J \, mol^{-1} \, K^{-1}$ for peroxide decomposition.	The entropy of activation is low; for perester decomposition $\Delta S^{\#} \approx -10$ to $+10 \, J \, mol^{-1} \, K^{-1}$.
3. The dissociation of one bond leads to an increase in the volume of the molecule; $\Delta V^{\#} \approx 10 \pm 3 \, cm^3 \, mol^{-1}$.	Concerted decomposition occurs through the compact transition state, and $\Delta V^{\#}$ is close to zero; $\Delta V^{\#} \approx -5$ to $10 \, cm^3 \, mol^{-1}$.

4. The rate constant of decay of symmetric RN=NR and nonsymmetrical RN=NR^1 molecules are close due to a low difference in the BDE of the R—N bonds in these two molecules.

The rate constants of decay of RN=NR and RN=NR^1 azo-compounds are very different due to their simultaneous dissociation.

The model of interacting oscillators was developed to describe the concerted decomposition of a molecule [10]. The decomposing molecule is treated as a collection of oscillators. The reaction of concerted decomposition is described as the transition of the system of oscillators from the thermal vibration with amplitude d to the vibration with critical amplitude d^*. If n bonds participate in the concerted decomposition and the activated energy is equal to E_n the rate constant of concerted decomposition k_n depends on n and E_n according to the equation:

$$k_n = A_0 \frac{n}{2n-1} \left(\frac{nRT}{\pi E_n}\right)^{(n-1)/2} \exp(-E_n/RT), \qquad (3.2)$$

where $A_0 = A$ at $n=1$. When $n=2$,

$$k_2 = A_0 \sqrt{\frac{2RT}{\pi E_2}} \exp(-E_2/RT) \qquad (3.3)$$

3.1.1.3 Anchimerically Assisted Decomposition of Peroxides

There are experimental evidences that some *ortho*-substituents in 1,1-dimethylethylbenzoyl peresters strongly accelerate the decomposition of peresters. The *ortho*-substituents in 1,1-dimethylethylbenzoyl peresters and values illustrating the ratio of rate constants k_d(*o*-substituted)/k_d(H) at 333 K are given below [3,4]:

o-CH=CHPh$_2$	*o*-SCH$_3$	*o*-SPh
67	140.000	250.000

We see a very strong influence of the *ortho*-substituent containing sulfur. The following mechanism was proposed [3]:

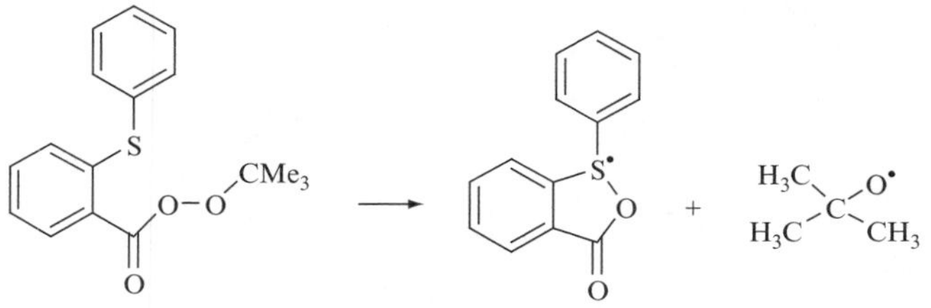

Disulfide was found to be the main product (yield 52.5%) of this perester decomposition. Accelerating action of *ortho*-substituents with *p*- or π-electrons is due to the formation of an intermediate bond of the O$^\bullet \cdots$ S or O$^\bullet \cdots$ C=C type in the transition state:

$$2 \quad \text{[structure]} \longrightarrow \text{[structure]}$$

This bond formation compensates (partially) the activation energy for dissociation of the O—O bond in perester. The empirical peculiarities of *anchimeric assistance decomposition* are the following [3,4]:

(a) The activation energy of this decomposition is lower than the BDE of the O—O bond, and the rate constant is much higher than typical k_d for compounds of this class.
(b) Solvent changes dramatically the rate constants of this decomposition.
(c) The products of recombination of radical pairs in the cage differ from the parent compounds.

3.1.2 Decay of Initiators to Molecular Products

Having a weak O—O bond, peroxides split easily into free radicals. In addition to homolytic reactions, peroxides can participate in heterolytic reactions also, for example, they can undergo hydrolysis under the catalytic action of acids. Both homolytic and heterolytic reactions can occur simultaneously. For example, perbenzoates decompose into free radicals and simultaneously isomerize to ester [11]. The *para*-substituent slightly influences the rate constants of homolytic splitting of perester. The rate constant of heterolytic isomerization, by contrast, strongly depends on the nature of the *para*-substituent. Polar solvent accelerates the heterolytic isomerization. Isomerization reaction was proposed to proceed through the cyclic transition state [11].

Parallel reactions of homolytic splitting and heterolytic isomerization were observed for triphenylcumylperoxysilane (anisole, $T = 433$–463 K) [11]:

$$Ph_3SiOOCMe_2Ph \longrightarrow Ph_3SiO^\bullet + PhMe_2CO^\bullet \quad (k_d)$$
$$Ph_3SiOOCMe_2Ph \longrightarrow PhOSiPh_2OCMe_2Ph \quad (k_{is})$$

The rate constants k_d and k_{is} were estimated and found to be: $\log(k_d/s^{-1}) = 15.5 - 167/RT$ and $\log(k_{is}/s^{-1}) = 8.2 - 104/RT$, $\Delta S_d^\# = 39.7$ J mol^{-1} K^{-1} and $\Delta S_{is}^\# = -39.7$ J mol^{-1} K^{-1}. It is seen that $E_d > E_{is}$, $\Delta S_d^\# > 0$ and $\Delta S_{is}^\# < 0$. Isomerization of this kind was not observed in the case of aliphatic silane peroxides. Polyarylperoxysilanes take part in isomerization of this type [11]:

$$Ph_2Si(OOCMe_3)_2 \longrightarrow Ph(PhO)Si(OOCMe_3)(OCMe_3)$$

Alkyl peroxyphosphates also undergo heterolytic isomerization simultaneously with homolytic splitting [12]:

$$Me_3COOP(O)Et_2 \longrightarrow MeOMe_2PCOP(O)Et_2 \quad (k_{is})$$
$$Me_3COOP(O)Et_2 \longrightarrow Me_3CO^\bullet + Et_2P(O)O^\bullet \quad (k_d)$$

The rate constants of these reactions were found to be very close: $k_d = 2.0 \times 10^{-5}$ s^{-1} and $k_{is} = 2.2 \times 10^{-5}$ s^{-1} (*n*-nonane, 403 K). The competition between homolytic and heterolytic reactions influences the effectiveness of initiation. When the heterolytic isomerization of initiator occurs, the effectiveness on initiation is $e = k_i(2k_d + 2k_{is})^{-1}$. It can be increased by changing the solvent and temperature.

3.1.3 Chain Decomposition of Initiators

Initiators are introduced into the reactant, as a rule, in very small amounts. The initiator produces free radicals, most of which react with the reactant or solvent or recombine with other free radicals. Radicals formed from the initiator or reactant react with the initiator very negligibly. However, systems (initiator–reactant) are known where free radicals induce the chain decomposition of initiators [4,13–15]. Nozaki and Bartlett [16,17] were the first to provide evidence for the induced decomposition of benzoyl peroxide in different solvents. They found that the empirical rate constant of benzoyl peroxide decomposition increases with an increase in the peroxide concentration in a solution. The dependence of the rate of peroxide decomposition on its concentration was found to be

$$v = k_d[\text{PhC(O)OOC(O)Ph}] + k_{ind}[\text{PhC(O)OOC(O)Ph}]^{3/2} \tag{3.4}$$

This dependence is the result of general occurrence of the homolytic decay of peroxide with the rate constant k_d and chain decomposition of peroxide due to reactions with the radical formed from the solvent RH according to the following kinetic scheme:

$$\text{PhC(O)OOC(O)Ph} \longrightarrow [\text{PhCO}_2^\bullet + \text{PhCO}_2^\bullet] \quad (k_i)$$
$$[\text{PhCO}_2^\bullet + \text{PhCO}_2^\bullet] \longrightarrow 2\text{PhCO}_2^\bullet$$
$$[\text{PhCO}_2^\bullet + \text{PhCO}_2^\bullet] \longrightarrow [\text{PhCO}_2^\bullet + \text{Ph}^\bullet] + \text{CO}_2$$
$$[\text{PhCO}_2^\bullet + \text{Ph}^\bullet] \longrightarrow \text{PhCO}_2^\bullet + \text{Ph}^\bullet$$
$$[\text{PhCO}_2^\bullet + \text{Ph}^\bullet] \longrightarrow \text{PhC(O)OPh}$$
$$\text{PhCO}_2^\bullet + \text{RH} \longrightarrow \text{PhCOOH} + \text{R}^\bullet$$
$$\text{Ph}^\bullet + \text{RH} \longrightarrow \text{PhH} + \text{R}^\bullet$$
$$\text{R}^\bullet + \text{PhC(O)OOC(O)Ph} \longrightarrow \text{PhC(O)OR} + \text{PhCO}_2^\bullet \quad (k_p)$$
$$\text{R}^\bullet + \text{R}^\bullet \longrightarrow \text{RR or RH} + \text{R}'\text{H} \quad (2k_t)$$

Empirical $k_{ind} = k_i^{1/2}k_p(2k_t)^{-1/2}$ and $k_i = 2ek_d$. The chain length depends on the ratio $k_p(2k_t)^{1/2}$: the faster the reaction of the radical R$^\bullet$ with peroxide, the longer the chain. Intensive chain decay of the peroxide was observed at a sufficiently high peroxide concentration. For example, $k_d = 6.36 \times 10^{-1}\,\text{s}^{-5}$ and $k_{ind} = 3.35 \times 10^{-4}\,\text{L mol}^{-1/2}\,\text{s}^{-1}$ for dibenzoyl peroxide decomposition in cyclohexane at 353 K [16]; and $k_d[\text{I}] = k_{ind}[\text{I}]^{3/2}$ at $[\text{PhC(O)OOC(O)Ph}] = 3.6 \times 10^{-2}\,\text{mol L}^{-1}$.

The very intensive chain decomposition of benzoyl peroxide was found in alcoholic solutions [16,18,19]. This is the result of the very high reductive activity of ketyl radicals formed from alcohol. They cause the chain decomposition of peroxide by the following mechanism:

$$\text{PhC(O)OOC(O)Ph} \longrightarrow 2e\,\text{PhCO}_2^\bullet \quad (k_d)$$
$$\text{PhCO}_2^\bullet + \text{R}^1\text{R}^2\text{CHOH} \longrightarrow \text{PhCOOH} + \text{R}_1\text{R}_2\text{C}^\bullet\text{OH}$$
$$\text{R}^1\text{R}^2\text{C}^\bullet\text{OH} + \text{PhC(O)OOC(O)Ph} \longrightarrow \text{R}^1\text{R}^2\text{C(O)} + \text{PhCO}_2\text{H} + \text{PhCO}_2^\bullet \quad (k_p)$$
$$2\text{R}^1\text{R}^2\text{C}^\bullet\text{OH} \longrightarrow \text{R}^1\text{R}^2\text{C(O)} + \text{R}^1\text{R}^2\text{CHOH} \quad (2k_t)$$

The kinetics of peroxide decomposition are described by Equation (3.4). However, the values of $k_{ind} = k_p k_i^{1/2}(2k_t)^{-1/2}$ are sufficiently higher.

3.1.4 PEROXIDES

Dialkyl peroxides decompose with splitting of the weakest O—O bond [3,4]. The pair of forming alkoxyl radicals recombine or disproportionate in the cage or go out the cage:

$$ROOR \longrightarrow [RO^\bullet + RO^\bullet]$$
$$[RO^\bullet + RO^\bullet] \longrightarrow ROOR$$
$$[RO^\bullet + RO^\bullet] \longrightarrow ROH + \text{carbonyl compound}$$
$$[RO^\bullet + RO^\bullet] \longrightarrow RO^\bullet + RO^\bullet$$

The rate constants of dialkyl peroxide decomposition (k_d and k_i, $k_i = 2ek_d$) are presented in Table 3.1–Table 3.3.

The homolytic decomposition of diacyl peroxides proceeds via splitting of the weakest O—O bond. The acyloxy radicals formed are very unstable and a cascade of cage reactions follows this decomposition [4,42–46]:

$$CH_3C(O)OOC(O)CH_3 \longrightarrow [CH_3CO_2^\bullet + CH_3CO_2^\bullet]$$
$$[CH_3CO_2^\bullet + CH_3CO_2^\bullet] \longrightarrow CH_3C(O)OOC(O)CH_3$$
$$[CH_3CO_2^\bullet + CH_3CO_2^\bullet] \longrightarrow CH_3CO_2^\bullet + CH_3CO_2^\bullet$$
$$[CH_3CO_2^\bullet + CH_3CO_2^\bullet] \longrightarrow [CH_3^\bullet + CH_3CO_2^\bullet] + CO_2$$
$$[C^\bullet H_3 + CH_3CO_2^\bullet] \longrightarrow C^\bullet H_3 + CH_3CO_2^\bullet$$
$$[C^\bullet H_3 + CH_3CO_2^\bullet] \longrightarrow CH_3C(O)OCH_3$$
$$[C^\bullet H_3 + CH_3CO_2^\bullet] \longrightarrow [C^\bullet H_3 + C^\bullet H_3] + CO_2$$
$$[C^\bullet H_3 + C^\bullet H_3] \longrightarrow C^\bullet H_3 + C^\bullet H_3$$
$$[C^\bullet H_3 + C^\bullet H_3] \longrightarrow C_2H_6$$

The yield of cage reaction products increases with increasing viscosity of the solvent. The decomposition of diacyl peroxides was the object of intensive study. The values of rate constants of diacyl peroxides (diacetyl and dibenzoyl) decomposition (k_d) and initiation ($k_i = 2ek_d$) are collected in Tables 3.4 and Table 3.5. The values of e are collected in the *Handbook of Radical Initiators* [4].

Three different mechanisms of perester homolytic decay are known [3,4]: splitting of the weakest O—O bond with the formation of alkoxyl and acyloxyl radicals, concerted fragmentation with simultaneous splitting of O—O and C—C(O) bonds [3,4], and some *ortho*-substituted benzoyl peresters are decomposed by the mechanism of decomposition with anchimeric assistance [3,4]. The rate constants of perester decomposition and values of $e = k_i/2k_d$ are collected in the *Handbook of Radical Initiators* [4]. The yield of cage reaction products increases with increasing viscosity of the solvent.

3.1.5 AZO-COMPOUNDS

Along with peroxides, azo-compounds are widely used as initiators of liquid-phase oxidation at mild temperatures [2–4,66,67]:

TABLE 3.1
Rate Constants of Thermal Decay of Bis(1,1-Dimethylethyl) Peroxide in the Gas Phase and Various Solvents

Phase, Solvent	T (K)	E (kJ mol^{-1})	log A, A (s^{-1})	k_d, k_i (400 K) (s^{-1})	Ref.
		k_d			
Gas phase	374–418	161.9	16.40	1.81×10^{-5}	[20]
Gas phase	433–551	160.2	16.08	1.45×10^{-5}	[21]
Gas phase	403–433	156.5	15.60	1.46×10^{-5}	[22]
Gas phase	363–403	158.2	15.80	1.39×10^{-5}	[23]
Gas phase	363–623	158.6	15.81	1.26×10^{-5}	[23]
Gas phase	413–433	163.6	16.50	1.37×10^{-5}	[24]
Gas phase	500–600	156.5	15.60	1.46×10^{-5}	[25]
Acetic acid	388–403	137.1	13.54	4.33×10^{-5}	[26]
Acetonitrile	388–403	133.0	13.00	4.29×10^{-5}	[26]
Aniline, N,N-dimethyl-	393–408	160.1	16.27	2.31×10^{-5}	[26]
Benzene	393–408	151.0	15.10	2.41×10^{-5}	[26]
Benzoic acid, ethyl ester	393–408	151.7	15.19	2.40×10^{-5}	[26]
Cyclohexane	393–408	174.1	17.98	1.76×10^{-5}	[26]
Cyclohexene	393–408	159.7	16.11	1.80×10^{-5}	[26]
Ethanol, 1,1-dimethyl-	393–408	146.3	14.60	3.13×10^{-5}	[26]
Neopentyl butanoate	408–438	122.0	11.65	5.23×10^{-5}	[27]
Neopentyl 2,2-dimethylpropanoate	408–438	144.8	14.32	2.58×10^{-5}	[27]
Neopentyl 3,3-dimethylbutanoate	408–438	141.5	14.05	3.73×10^{-5}	[27]
Neopentyl 2,2-dimethylbutanoate	408–438	148.2	14.74	2.44×10^{-5}	[27]
Neopentyl 2-mehylbutanoate	408–438	115.2	10.73	4.86×10^{-5}	[27]
Neopentyl 2-methylpropanoate	408–438	135.7	13.18	2.88×10^{-5}	[27]
Nitrobenzene	393–408	152.3	15.41	3.33×10^{-5}	[26]
Nitroethane	393–408	151.9	15.31	2.98×10^{-5}	[26]
Propanol, 1,1-dimethyl-	393–408	159.7	16.11	1.80×10^{-5}	[26]
Tetrahydrofuran	393–408	158.6	16.13	2.63×10^{-5}	[26]
Triethylamine	393–408	172.6	17.87	2.14×10^{-5}	[26]
		$k_i = 2ek_d$			
Benzene	373–398	150.6	15.10	2.72×10^{-5}	[28]
Ethanol, 1-methyl-	359–411	146.4	14.64	3.33×10^{-5}	[29]
Styrene	313–371	146.4	14.45	2.15×10^{-5}	[30]
Styrene	353–373	161.5	16.21	1.32×10^{-5}	[31]
Styrene	363–373	157.3	15.86	2.08×10^{-5}	[32]

$$RN{=}NR \longrightarrow R^\bullet + N_2 + R^\bullet$$

Two mechanisms of azo-compound decomposition were discussed intensively in the literature: concerted decomposition with simultaneous dissociation of two C—N bonds

$$RN{=}NR \longrightarrow R \cdots N{=}N \cdots R \longrightarrow R^\bullet + N_2 + R^\bullet$$

and nonconcerted decomposition with dissociation of one C—N bond followed by the fast decomposition of the formed unstable $RN{=}N^\bullet$ radical [3,4,66,67]:

TABLE 3.2
Rate Constants of the Thermal Decay of Bis(1-Methyl-1-Phenylethyl) Peroxide in Various Solvents

Solvent	T (K)	E (kJ mol^{-1})	log A, A (s^{-1})	k_d, k_i (400 K) (s^{-1})	Ref.
		k_d			
Cumene	383–423	144.3	14.63	6.12×10^{-5}	[33]
Decane	373–413	143.5	14.57	6.78×10^{-5}	[26]
Dodecane	401–431	140.2	14.17	7.28×10^{-5}	[34]
Styrene	368–378	166.9	17.76	9.24×10^{-5}	[33]
		$k_i = 2ek_d$			
Benzene	373–398	146.0	14.70	4.31×10^{-5}	[28]
Chlorobenzene	368–398	133.8	13.18	5.10×10^{-5}	[35]
Paraffin oil	393–423	135.6	13.51	6.35×10^{-5}	[36]
Octadiene-1,7 2,6-dimethyl-	393–423	133.9	13.26	5.95×10^{-5}	[37]
Z-1,4-Polyisoprene	393–423	135.9	13.15	2.53×10^{-5}	[38]
Mineral oil	393–423	141.8	14.43	8.19×10^{-5}	[39]

TABLE 3.3
Rate Constants of the Thermal Decay of Dialkyl Peroxides (k_i) in the Gas Phase and Various Solvents

Peroxide	Solvent	T (K)	E (kJ mol^{-1})	log A, A(s^{-1})	k_i (400 K) (s^{-1})	Ref.
Diethyl-, EtOOEt	Benzene/Styrene*	333–353	142.7	14.16	3.35×10^{-5}	[40]
Bis(1-methylethyl)-, Me$_2$CHOOCHMe$_2$	Benzene/Styrene*	333–353	156.0	15.27	7.92×10^{-6}	[40]
Dibutyl-, BuOOBu	Benzene/Styrene*	333–353	143.5	13.98	1.74×10^{-5}	[41]
Bis(1-methylpropyl)-, EtMeCHOOCHEtMe	Benzene/Styrene*	333–353	142.3	13.56	9.50×10^{-6}	[40]
Bis(2,2-dimethylpropyl)-, Me$_3$CCH$_2$OOCH$_2$CMe$_3$	Styrene	353–373	145.6	15.48	2.93×10^{-4}	[31]
Bis(1,1-dimethylpropyl)-, EtMe$_2$COOCCEtMe$_2$	Styrene	353–383	157.8	16.19	3.83×10^{-5}	[32]
Bis(1-methyl-1-ethylpropyl)-, Et$_2$MeCOOCEt$_2$Me	Styrene	353–383	159.0	16.55	6.12×10^{-5}	[32]

*-Benzene/Styrene (1:1 V/V).

$$RN{=}NR \longrightarrow R^{\bullet} + RN{=}N^{\bullet}$$
$$RN{=}N^{\bullet} \longrightarrow R^{\bullet} + N_2$$

The nonconcerted mechanism of decomposition was observed by Szwarc et al. [67]. They studied the photolytic decomposition of cyclopropyltrifluoromethyl diazene in the gas phase

TABLE 3.4
The Rate Constants of the Thermal Decay of Diacetyl Peroxide in Different Solvents

Solvent	T (K)	E (kJ mol^{-1})	log A, A(s^{-1})	k (353 K) (s^{-1})	Ref.
		k_d			
Gas phase	363–463	123.4	14.25	9.78×10^{-5}	[47]
Acetic acid	328–358	126.4	14.51	6.40×10^{-5}	[48]
Benzene	328–358	135.1	15.93	8.69×10^{-5}	[48]
Carbon tetrachloride	333–373	139.7	16.43	5.73×10^{-5}	[49]
Cyclohexane	328–358	131.4	15.27	6.71×10^{-5}	[48]
Cyclohexene	333–373	133.5	15.59	6.85×10^{-5}	[50]
1-Hexene	343–373	132.6	15.56	8.69×10^{-5}	[50]
Isooctane	328–358	134.7	15.82	7.73×10^{-5}	[48]
2-Methy-1-pentene	343–373	126.8	14.75	9.71×10^{-5}	[50]
1-Pentene	343–363	133.9	15.71	7.88×10^{-5}	[50]
Propionic acid	337.9–358.2	122.6	14.1	9.09×10^{-5}	[48]
Toluene	323–353	138.1	16.31	7.50×10^{-5}	[51]
Toluene	323–353	129.7	15.05	7.21×10^{-5}	[52]
		$k_i = 2ek_d$			
Styrene	343			2.02×10^{-5}	[53]

and in 2,3-dimethylbutane solution and found 2-pyrazoline with a yield of 15–20% among the products. The apparent mechanism is the following:

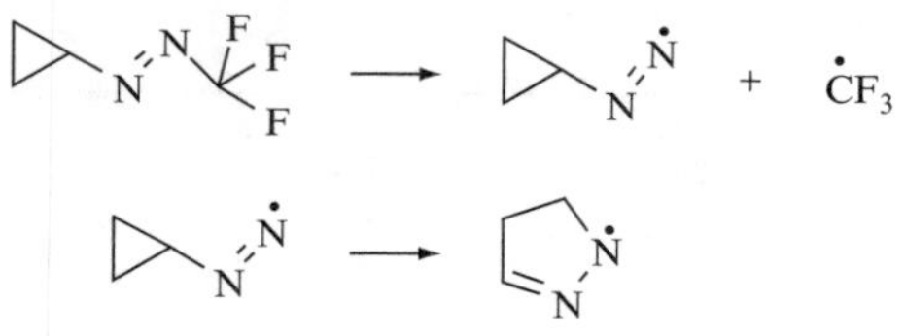

It is considered that the decomposition of symmetrical azo-compounds proceeds via the concerted mechanism, and some unsymmetrical azo-compounds are decomposed by the concerted and nonconcerted mechanisms simultaneously. Phenyl-substituted azo-compounds are decomposed by the nonconcerted mechanism [3].

Recently the two-step decomposition of azomethane was proved in the study of the femtosecond dynamics of this reaction [68]. The intermediate $CH_3N_2^{\cdot}$ radical was detected and isolated in time. The reaction was found to occur via the occurrence of the first and the second C—N bond breakages. The lifetime of $CH_3N_2^{\cdot}$ radical is very short, i.e., 70 fsec. The quantum-chemical calculations of cis- and trans-azomethane dissociation was performed [69].

The linear dependence between the activation energy of decomposition of the azo-compounds RN$_2$R and the BDE of the R—H bond (D(R—H)) was established [3]. The rate constants of the decomposition of azo-compounds in the gas phase and hydrocarbon solvents have close values. The mean value of the rate constant of AIBN decomposition in hydrocarbon and aromatic solutions was recommended to be $k_d = 10^{15} \times \exp(-127.5/RT)$ s^{-1} [2]. The values of the activation energies and the rate constants of the decomposition of azo-compounds in the gas and liquid phases can be found in the *Handbook of Radical Initiators* [4].

TABLE 3.5
The Rate Constants of the Thermal Decay of Dibenzoyl Peroxide in Different Solvents

Solvent	T (K)	E (kJ mol^{-1})	log A, A(s^{-1})	k_d, k_i (353 K)(s^{-1})	Ref.
		k_d			
Acetic anhydride	333–353	128.5	14.90	7.69×10^{-5}	[15]
Benzene	353–363	128.5	14.04	1.06×10^{-5}	[54]
Benzene	333–353	123.8	13.64	2.10×10^{-5}	[55]
tert-Butylbenzene	353–363	127.2	14.48	4.55×10^{-5}	[54]
Cumene	318–353	120.5	13.08	1.78×10^{-5}	[56]
Cyclohexane	353–363	118.0	13.48	1.05×10^{-4}	[54]
Dioxane	333–353	125.5	14.18	4.07×10^{-5}	[57]
Ethylbenzene	346.5–358	130.3	14.90	4.16×10^{-5}	[58]
Methylcyclohexane	353–363	128.5	14.79	5.97×10^{-5}	[54]
Phenyl acetate	343–367.5	126.4	14.33	4.23×10^{-5}	[59]
Styrene	343–363	137.0	15.99	5.22×10^{-5}	[60]
Styrene	346–358	129.7	14.76	3.70×10^{-5}	[58]
Vinyl acetate	318–353	118.4	13.42	7.94×10^{-5}	[17]
		$k_i = 2ek_d$			
Acetone	373–393	130.1	15.31	1.15×10^{-4}	[61]
Benzene	324–349	123.8	13.48	1.45×10^{-5}	[62]
Carbon tetrachloride	332–350	136.8	15.68	2.74×10^{-5}	[63]
Chlorobenzene	378–388	110.2	12.12	6.51×10^{-5}	[64]
Methyl acetate	322–327	123.8	13.82	3.17×10^{-5}	[62]
Nitrobenzene	322–343	123.8	13.66	2.19×10^{-5}	[62]
Styrene	333–353	128.0	14.26	2.09×10^{-5}	[65]
Styrene	313–343	140.6	16.30	3.13×10^{-5}	[53]
Toluene	322–343	123.8	13.86	3.48×10^{-5}	[62]

3.2 CAGE EFFECT

The decomposition of an initiator in the liquid phase leads to the formation of two radicals that exist side by side for a certain time, surrounded by solvent molecules. The solvent molecules create a "cage" around a pair of formed radicals due to intermolecular forces of the solvent molecules [3,4,15,70,71]. The *cage effect* was discovered by Frank and Rabinowitch [72]. The cage effect is very important for the understanding of the chemistry and the kinetics of initiators decomposition in various solvents and solid polymers. The rate constant for the interaction of two radicals is very high. As a result, a few of such radicals react with each other before the pair of radicals formed from initiator is separated by diffusion. In the 1960s and 1970s, widespread experimental evidence was obtained concerning the cage effect in liquids.

3.2.1 Experimental Evidences for Cage Effect

3.2.1.1 Quantum Yield

Quantum yield (Φ) of molecular photodissociation in the gas phase is equal to unity according to the Einstein law. Frank and Rabinowitch [72] predicted the reduction of the quantum yield in a solution due to the cage effect. The quantum yield $\Phi < 1$ was observed in the photodissociation of I_2, Br_2, RN_2R:

Molecule	Solvent	λ (nm)	T (K)	Φ	Ref.
I_2	CCl_4	440	298	0.14	[73]
Br_2	CCl_4	440	298	0.22	[74]
cyclohexyl-N=N-cyclohexyl	CCl_4		298	0.24	[75]

3.2.1.2 Products of Radical Pair Combination

Free radicals formed from an initiator in the gas phase take part in other reactions and recombine with a very low probability (0.1–2%). The decomposition of the initiator in the liquid phase leads to the formation of radical pairs, and the probability of recombination of formed radicals in the liquid phase is high. For example, the photolysis of azomethane in the gas phase in the presence of propane (RH) gives the ratio $[C_2H_6]/[N_2] = 0.015$ [76]. This ratio is low due to the fast reactions of the formed methyl radicals with propane:

$$CH_3N{=}NCH_3 + h\nu \longrightarrow C^{\bullet}H_3 + N_2 + C^{\bullet}H_3$$
$$C^{\bullet}H_3 + Me_2CH_2 \longrightarrow CH_4 + Me_2C^{\bullet}H[R^{\bullet}]$$
$$R^{\bullet} + R^{\bullet} \longrightarrow R{-}R \text{ and } RH + CH_3CH{=}CH_2$$

In the liquid phase, this ratio was 0.65 due to the recombination of the methyl radical pair in the cage:

$$CH_3N{=}NCH_3 + h\nu \longrightarrow [C^{\bullet}H_3 + C^{\bullet}H_3] + N_2$$
$$[C^{\bullet}H_3 + C^{\bullet}H_3] \longrightarrow C_2H_6$$
$$[C^{\bullet}H_3 + C^{\bullet}H_3] \longrightarrow 2C^{\bullet}H_3$$

When molecules of a solvent are stable toward free radicals formed from the initiator, a scavenger of radicals is added to consume all the free radicals that escape from recombination in the cage. The yield of the product of primary radicals formed illustrates the probability of cage radical pair recombination. For example, the thermolysis of azobisisobutyronitrile in CCl_4 gives tetramethylsuccinonitrile in 80% yield; this yield falls to a value of 19% in the presence of butanethiol as a free radical acceptor [77]. The value of 19% characterizes the recombination of radicals in the cage. Formed from the initiator, free radicals start chain oxidation in the presence of dioxygen:

$$RN{=}NR \longrightarrow [R^{\bullet} + R^{\bullet}] + N_2$$
$$[R^{\bullet} + R^{\bullet}] \longrightarrow R{-}R$$
$$[R^{\bullet} + R^{\bullet}] \longrightarrow 2R^{\bullet}$$
$$R^{\bullet} + O_2 \longrightarrow RO_2^{\bullet}$$

3.2.1.3 Oxygen-18 Scrambling and Racemization

The back recombination of the pair of acetoxyl radicals with the formation of parent diacetyl peroxide was observed in special experiments on the decomposition of acetyl peroxide labelled by the ^{18}O isotope on the carbonyl group [78,79]. The reaction of acetyl peroxide with $NaOCH_3$ produces methyl acetate and all ^{18}O isotopes are contained in the carbonyl group. The transition of ^{18}O atoms from the carbonyl to peroxide group of acetyl peroxide was detected by hydrolysis of peroxide to hydrogen peroxide and conversion of the resulting hydrogen peroxide to dioxygen by permanganate oxidation. The amount of ^{18}O in the formed O_2 was analyzed by mass spectometry. It was found that the scrambling of ^{18}O occurs during the decomposition of peroxide. After some time of decomposition, the residue of peroxide was found to contain peroxide with the ^{18}O isotope in the carbonyl as well as peroxide groups. This scrambling of ^{18}O isotopes proves the cage recombination of acetoxyl radicals:

$$CH_3C(^{18}O)OOC(^{18}O)CH_3 \longrightarrow [CH_3C^{18}OO^\bullet + CH_3C^{18}OO^\bullet]$$
$$[CH_3C^{18}OO^\bullet + CH_3C^{18}OO^\bullet] \longrightarrow CH_3C(O)^{18}O^{18}OC(O)CH_3 \quad (k_{sc})$$
$$[CH_3C^{18}OO^\bullet + CH_3C^{18}OO^\bullet] \longrightarrow C^\bullet H_3 + CO_2 + CH_3C^{18}OO^\bullet$$

The amount of ^{18}O in the carbonyl group of peroxide was found to decrease with time of decomposition, and amount of ^{18}O in the peroxide group was found to increase (isooctane, 353 K, $[MeCOOOCOMe]_0 = 0.05$ mol L^{-1} [79]):

t (s)	0	870	17400
$^{18}O_2/^{16}O_2$	0.00428	0.01451	0.02228
$C(^{18}O)/C(^{16}O)$	1.000	0.712	0.499

The rate constant of scrambling k_{sc} was found to increase with increasing viscosity of the solvent [79]:

Solvent	Isooctane	Dodecane	Octadecane	Mineral Oil
k_{sc} (s^{-1})	4.00×10^5	4.68×10^5	5.25×10^5	6.37×10^5

The higher the viscosity of the solvent, the longer the period of time of radical pair existence in the cage and the higher the observed value of scrambling rate constant [3,80–82]. The same phenomenon was observed during the photolysis of benzoyl peroxide and 1,1-dimethylethyl perbenzoate [3].

3.2.1.4 Crossover Experiments

When a mixture of perdeuterio and protioazomethane in a ratio of 1:1 was photolyzed in the gas phase, the isolated ethane was found to be a statistical mixture of three possible dimers: C_2H_6, $C_2H_3D_3$, and C_2D_6 in the ratio 1:2:1 [83,84]. This is the result of the reactions

$$CH_3N=NCH_3 + h\nu \longrightarrow 2C^\bullet H_3 + N_2$$
$$CD_3N=NCD_3 + h\nu \longrightarrow 2C^\bullet D_3 + N_2$$
$$C^\bullet H_3 + C^\bullet H_3 \longrightarrow C_2H_6$$
$$C^\bullet H_3 + C^\bullet D_3 \longrightarrow CH_3CD_3$$
$$C^\bullet D_3 + C^\bullet D_3 \longrightarrow C_2D_6$$

In an isooctane solution, 75% of C_2H_6 and C_2D_6 were formed, which was due to the recombination of the pairs $C^•H_3$ and $C^•D_3$ in the cage [83]:

$$CH_3N=NCH_3 + h\nu \longrightarrow [C^•H_3 + C^•H_3] + N_2$$
$$[C^•H_3 + C^•H_3] \longrightarrow C_2H_6$$
$$CD_3N=NCD_3 + h\nu \longrightarrow [C^•D_3 + C^•D_3] + N_2$$
$$[C^•D_3 + C^•D_3] \longrightarrow C_2D_6$$

This difference in the yields of the symmetrical dimers in the liquid [75%] and the gas (25%) phases (75−50 = 25%) is because of the geminate recombination in the cage of the liquid. Similar results were reported for ethane and methyl acetate isolated by thermolysis of a mixture of protio- and perdeuterioacetyl peroxide [85,86].

3.2.1.5 Racemization

The existence of cage effect was proved in the experiments on photolysis of the optically active azo-compounds. The photodecomposition of these compounds is accompanied by *racemization* [3]. For example, the partial (40%) photolysis of optically active 2-phenylazo-(2-phenyl)-butane in a hexadecane solution provides racemization to 26% [87]. The fraction of geminate recombination was found to be 52% (hexadecane, room temperature):

$$PhMeEtC^*N=NPh + h\nu \longrightarrow [PhMeEtC^• + {}^•NNPh] \longrightarrow Products$$
$$[PhMeEtC^• + {}^•NNPh] \longrightarrow PhMeEtCN=NPh$$

Trans-isomer was found to be transformed virtually into the *Z*-isomer due to geminate recombination:

$$E\text{-}PhMeEtCN=NPh + h\nu \longrightarrow [PhMeEtC^• + {}^•NNPh] \longrightarrow Products$$
$$[PhMeEtC^• + {}^•NNPh] \longrightarrow Z\text{-}PhMeEtCN=NPh$$

The problem of retention of asymmetry of the formed free radical in the fast geminate recombination of radicals was studied by photolysis of the optically active azo-compound $PhMeCH-N=NCH_2Ph$ [88,89]. The radical pair of two alkyl radicals was initiated by the photolysis of the azo-compound in benzene in the presence of 2-nitroso-2-methylpropane as a free radical acceptor. The yield of the radical pair combination product was found to be 28%. This product $PhMeEtCCH_2Ph$ was found to be composed of 31% S,S-(−)(double retention), 48% *meso* (one inversion), and 21% R,R(+) double inversion. These results were interpreted in terms of the competition between recombination (k_c), diffusion (k_D), and rotation (k_{rot}) of one of the optically active radicals with respect to another. The analysis of these data gave: $k_{rot}/k_c = 15$, $k_D/k_c = 2.5$ [89]:

$$(-)DD - AZO \longrightarrow [D^• + D^•] \longrightarrow D-D(-)$$
$$[D^• + D^•] \rightleftharpoons [L^• + D^•] \longrightarrow L-D(meso)$$
$$[L^• + D^•] \rightleftharpoons [L^• + L^•] \longrightarrow L-L(+)$$

The study of the decomposition of optically active 1-methyl-2,2-diphenylcyclopropanoyl peroxide proved the retention (37%) of the product of the geminate radical pair recombination [90]. The radical center in the formed cyclopropyl radical is so strained that the racemization rate is unusually slow.

3.2.1.6 Influence of Viscosity

When an initiator splits into a pair of radicals, this pair can recombine with the formation of the parent molecule, for example:

$$RC(O)OOCMe_3 \rightleftharpoons [RCO_2^\bullet + Me_3CO^\bullet] \longrightarrow R^\bullet + CO_2 + Me_3CO^\bullet$$

The higher the viscosity of the solvent, the higher the amount of the parent molecules formed due to the geminate recombination of radicals. The observed rate constant of decomposition of the initiator decreases with an increase in viscosity [3,90]. This was observed in the decomposition of peresters and diacetyl peroxide in various solutions. Subsequently, the fraction f_r of the radical pairs recombining to the parent molecule increases with an increase in the viscosity:

	Diacetyl Peroxide, 353 K [80]		
Solvent	Isooctane	Dodecane	Octadecane
f_r (%)	28	39	49

	1,1-Dimethylethyl Peracetate, 403 K [82]		
Solvent	Hexane	Nonane	Paraffin
f_r (%)	6	12	42

The introduction of a free radical acceptor (scavenger) helps to measure the probability (e) of radical pairs to escape from geminate combination and diffuse out of the cage. The value of e for the fixed initiator or photoinitiator depends on the viscosity η of the solvent. The following empirical dependence for the photodecomposition of initiators was found [91,92]:

Photoinitiator	Equation	Equation
$CH_3N=NCH_3$	$(1-e)^{-1} = 1.1 + 6.5 \times 10^{-6}\, T^{1/2}\, \eta^{-1}$	(3.5)
$CF_3N=NCF_3$	$(1-e)^{-1} = 1.1 + 4.0 \times 10^{-5}\, T^{1/2}\eta^{-1}$	(3.6)
$PhMe_2CC(O)OOCMe_3$	$(1-e)^{-1} = 1.24 + 5.75 \times 10^{-5}\, T^{1/2}\eta^{-1}$	(3.7)

There are initiators that split into two radicals only and initiators that dissociate with the formation of two radicals and one or two molecules. The formation of molecules simultaneously with radicals influences the efficiency of initiation ($T = 318$ K [93,94]).

Radical Pair	Solvent	$e/(1-e)$
$Me_3CO^\bullet, Me_3CO^\bullet$	Isooctane	7.32
$Me_3CO^\bullet, N_2, Me_3CO^\bullet$	Isooctane	9.01
$Me_3CO^\bullet, CO_2, CO_2, Me_3CO^\bullet$	Isooctane	10.8
$Me_3CO^\bullet, Me_3CO^\bullet$	White oil	0.32
$Me_3CO^\bullet, N_2, Me_3CO^\bullet$	White oil	0.47
$Me_3CO^\bullet, CO_2, CO_2, Me_3CO^\bullet$	White oil	0.85

3.2.1.7 Influence of Pressure

The influence of pressure on the cage effect was studied by Neuman and colleagues [95–98]. They measured the influence of pressure on the cage effect for competition between recombination and diffusion for the 1,1-dimethylethoxy radical pairs generated from bis(1,1-dimethylethyl)hyponitrite. The empirical activation volume difference ($\Delta V_d^\# - \Delta V_c^\#$) for the competition of initiator decomposition and radical pair recombination was found to depend on pressure [99,100]. A fairly good correlation between the yield of the recombination product and the fluidity of the solvent (octane) at various pressures was obtained.

3.2.1.8 Spin Multiplicity Effects

Thermal and direct photolytic decomposition of the initiator produce a radical pair in the singlet state when back geminate recombination occurs rapidly. The radical pair is formed in the triplet state when a triplet sensitizer is used. Before geminate recombination, this radical pair should reach the singlet state [101,102]. This is the reason why the fraction of geminate recombination of the triplet pair is less than that of the singlet pair. For example, direct photolysis of N-(1-cyanocyclohexyl)-pentamethyleneketenimine gives 24% of succinonitryl in CCl_4, whereas triplet-sensitized photolysis gives only 8% [3]. The photolysis of phenyl benzoate in benzene gives 8% phenyl benzoate due to geminate recombination, and triplet-sensitized photolysis gives only 3% of phenyl benzoate. ^{18}O randomization is also less important in the triplet-sensitized decomposition of diacyl peroxides.

The chemically induced dynamic nuclear polarization (DNP) opened perspective to study products formed from free radicals [102]. The basis of this study is the difference in NMR spectra of normal molecules and those formed from free radicals and radical pairs. The molecules formed from radicals have an abnormal NMR spectrum with lines of emission and abnormal absorption [102]. DNP spectra help to obtain the following mechanistic information:

1. The proof of radical ancestry for particular products
2. The identity and characteristics of radical precursors
3. The identity of pairs in which polarization occurs
4. Some information on radical pair separation
5. The direct observation of unstable diamagnetic reaction intermediates.

3.2.2 Mechanistic Schemes of the Cage Effect

The cage effect can be interpreted within the scope of a simple kinetic scheme [15]. For example, azo-compound decomposes according to the kinetic scheme given below:

$$RN=NR \longrightarrow [R^\bullet + N_2 + R^\bullet] \qquad (k_d)$$
$$[R^\bullet + N_2 + R^\bullet] \longrightarrow RR + N_2 \qquad (k_C)$$
$$[R^\bullet + N_2 + R^\bullet] \longrightarrow R^\bullet + N_2 + R^\bullet \qquad (k_D)$$

The probability of the radical pair to escape from geminate recombination (e) depends on the rate constants according to the following equation:

$$e^{-1} = 1 + \frac{k_c}{k_D} = 1 + \frac{6\pi L r_R k_c \eta}{10^6 RT} = 1 + \text{const.} \times \eta \qquad (3.8)$$

where r_R is the radius of the diffusing $R^\bullet$ radical.

The model of cage effect in the photodissociation of the initiator was analyzed by Noyes [103]. Free radicals were assumed to be spherical particles, which move in a viscous continuum. The separation r_0 is the distance achieved during the initiating event. This separation is calculated by assuming that the initial kinetic energy has a magnitude greater than the average thermal energy. The second encounter of formed radicals is taken into account. The reciprocal of the total probability of geminate recombination has the following form:

$$\frac{1}{1-e} = \frac{r_0}{2r_R}\left\{1 + \frac{\alpha A_E + A_T}{\alpha\eta} + \frac{A_T A_E}{\alpha\eta^2}\right\}, \qquad (3.9)$$

where $A_E = m_R^{1/2}(h\nu - E_I)^{1/2}/6\pi r_R^2$, $A_T = (1.5\, m_R kT)^{1/2}/6\pi r_R^2$, α is the probability of the reaction between two R$^\bullet$ per collision, m_R is the weight of radical R$^\bullet$, E_I is the dissociation energy of the initiator, r_0 is the initial displacement of the formed radicals (depends on the formation of the third particle due to initiator decomposition).

When free radicals formed from the initiator are unstable, they decompose in the cage, for example:

$$\text{RC(O)OOCMe}_3 \longrightarrow [\text{RCO}_2^\bullet + \text{Me}_3\text{CO}^\bullet] \qquad (k_d)$$
$$[\text{RCO}_2^\bullet + \text{Me}_3\text{CO}^\bullet] \longrightarrow [\text{R}^\bullet + \text{CO}_2 + \text{Me}_3\text{CO}^\bullet] \qquad (k_\beta)$$
$$[\text{R}^\bullet + \text{CO}_2 + \text{Me}_3\text{CO}^\bullet] \longrightarrow \text{ROCMe}_3 + \text{CO}_2 \qquad (k_c)$$
$$[\text{R}^\bullet + \text{CO}_2 + \text{Me}_3\text{CO}^\bullet] \longrightarrow \text{R}^\bullet + \text{CO}_2 + \text{Me}_3\text{CO}^\bullet \qquad (k_D)$$

The yield (y) of the product of geminate pair recombination (ROCMe$_3$) as a function of the subsequent rate constants [104,105] is given by

$$\frac{1}{y} - 1 = \frac{k_D}{k_c} + \frac{k_\beta}{k_c} \qquad (3.10)$$

If one takes into account the time dependence of k_D in the form ($r^2 = 2Dt$)

$$k_D = \frac{r}{2r_R - r_0} \times \frac{1}{t} = \frac{\sqrt{2D}}{2r_R - r_0} \times \frac{1}{\sqrt{t}} \qquad (3.11)$$

where r_R is the radius of the formed radical R$^\bullet$, r_0 is the distance between two radicals at $t=0$, D is the diffusion coefficient of the radical. If $D \sim \eta^{-1}$, we obtain the equation:

$$\frac{1}{y} - 1 = \frac{\text{const}}{\sqrt{\eta}} + \frac{k_\beta}{k_c} \qquad (3.12)$$

This equation helps to distinguish between the concerted and the nonconcerted decomposition of the initiator (for concerted decomposition, $k_\beta = 0$).

The cage effect was also analyzed for the model of diffusion of two particles (radical pair) in viscous continuum using the diffusion equation [106]. Due to initiator decomposition, two radicals R$^\bullet$ formed are separated by the distance r_0 at $t=0$. The acceptor of free radicals Q is introduced into the solvent; it reacts with radicals with the rate constant k_a. Two radicals recombine with the rate constant k_c when they come into contact at a distance $2r_R$, where r_R is the radius of the radical R$^\bullet$. Solvent is treated as continuum with viscosity η. The distribution of radical pairs (n) as a function of the distance x between them obeys the equation of diffusion:

$$\frac{D}{x}\frac{d^2(nx)}{dx^2} - 2k_a n[Q] = 0 \qquad (3.13)$$

The solution of this equation at $4k_Q[Q]r_R/D \ll 1$ has the following simple form:

$$e = 1 - \frac{2r_R}{r_0}(1 + 8\pi r_R D/k_c)^{-1}, \qquad (3.14)$$

Substituting the diffusion coefficient D into its expression in the Stokes–Einstein equation, we have the equation:

$$\frac{1}{1-e} = \frac{r_0}{2r_R}\left(1 + \frac{4 \times 10^6 RT}{3Lk_c\eta}\right) \qquad (3.15)$$

3.3 INITIATION OF OXIDATION BY CHEMICALLY ACTIVE GASES

The initiation of hydrocarbon liquid-phase oxidation by chemically active gases was proposed by Emanuel [107,108]. This method consists in passing a small quantity of gas initiator into oxidized hydrocarbon dioxygen or air. The introduced gas reacts with hydrocarbon and its intermediates with the generation of free radicals. Then the free radicals initiate the chain reaction of oxidation. Some time after the beginning of the chain oxidation, the supply of initiating gas is stopped and formed hydroperoxide initiates oxidation. (see Chapter 4). The following gases were studied as initiators [109]: ozone, nitrogen dioxide, halogens, and hydrogen bromide.

3.3.1 OZONE

The initiating action of ozone on hydrocarbon oxidation was demonstrated in the case of oxidation of paraffin wax [110] and isodecane [111]. The results of these experiments were described in a monograph [109]. The detailed kinetic study of cyclohexane and cumene oxidation by a mixture of dioxygen and ozone was performed by Komissarov [112]. Ozone is known to be a very active oxidizing agent [113–116]. Ozone reacts with C—H bonds of hydrocarbons and other organic compounds with free radical formation, which was proved by different experimental methods.

1. Ozone initiates the chain oxidation of hydrocarbons in the gas [117] and the liquid phases [118].
2. Peroxyl radicals were identified as products of hydrocarbon and polymer oxidation by an O_3–O_2 mixture and were proved by EPR spectroscopy [118,119].
3. Ozone induces CL in the oxidized hydrocarbons (RH) by disproportionation of the formed peroxyl radicals [113,120,121]. This reaction produces a carbonyl compound in the triplet state, which is the source of luminescence.
4. Oxidation of hydrocarbons in a tetrachloride solution produces alkyl chloride [122–125]:

$$R^\bullet + CCl_4 \longrightarrow RCl + C^\bullet Cl_3$$

Hydrochloric acid is formed in the oxidation of alcohols by ozone in carbon tetrachloride as solvent [126]:

TABLE 3.6
Kinetic Isotope Effect in Oxidation of Aliphatic Compounds by Ozone

Gas	RD	Solvent	T (K)	k_H/k_D	Ref.
O_3–O_2	C_6D_{12}	Carbon tetrachloride	295	4.5	[122]
O_3–O_2	C_6D_{12}	Carbon tetrachloride	300	5.4	[127]
O_3–O_2	$PhCD_2OCMe_3$	Freon-11	273	4.1	[128]
O_3–O_2	$PhCD_2OCMe_3$	Freon-11	351	6.3	[128]
O_3–N_2	$PhCD_2OCMe_3$	Freon-11	273	3.8	[128]
O_3–N_2	$PhCD_2OCMe_3$	Freon-11	195	6.7	[128]
O_3–O_2	$PhCD_2OCMe_3$	Freon-11/Pyridine	195	4.2	[128]
O_3–N_2	$PhCD_2OCMe_3$	Freon-11/Pyridine	195	3.7	[128]
O_3–N_2	$PhCD_2OCMe_3$	Dichloromethane/Acetone	195	3.2	[128]
O_3–O_2	$PhCD_2OCMe_3$	Acetone	273	4.5	[128]
O_3–He	$PhCD_2OCMe_3$	Acetone	273	2.4	[128]
O_3–O_2	$PhCD_2OCMe_3$	Pyridine	273	2.6	[128]
O_3–N_2	C_6H_5COD	Butanone-2	273	1.4	[129]
O_3–N_2	C_6H_5COD	Carbon tetrachloride	273	2.2	[129]
O_3–N_2	C_6H_5COD	Pyridine	273	2.0	[129]
O_3–O_2	CH_3CD_2OH	Chloroform	293	5.9	[130]
O_3–O_2	CD_3CH_2OH	Carbon tetrachloride	293	1.2	[130]
O_3–O_2	CH_3CH_2OD	Carbon tetrachloride	293	0.9	[130]
O_3–O_2	CD_3CD_2OD	Carbon tetrachloride	293	7.1	[130]
O_3–O_2	CH_3CH_2OD	D_2O	293	1.2	[130]
O_3–O_2	CD_3CD_2OD	D_2O	293	6.3	[130]

$$R^1R^2C^\bullet OH + CCl_4 \longrightarrow R^1R^2C{=}O + HCl + C^\bullet Cl_3$$

The kinetic isotope effect proves the attack of ozone on the C—H bond and consequently the C—D bond of the oxidized compound. The values of the kinetic isotope effect (k_H/k_D) are collected in Table 3.6.

The initiation of free radical reactions by ozone in the gas phase at elevated temperatures occurs due to ozone monomolecular decomposition [131,132]:

$$M + O_3 \longrightarrow O_2 + O + M$$

accompanied by the reaction

$$O + RH \longrightarrow R^\bullet + HO^\bullet$$

The enthalpy of ozone decomposition $\Delta H = D_{O_2-O} = 107\,kJ\,mol^{-1}$. The most probable reaction of initiation by ozone in solution is the abstraction reaction [133]:

$$RH + O_3 \longrightarrow [R^\bullet + HO_3^\bullet]$$
$$[R^\bullet + HO_3^\bullet] \longrightarrow R^\bullet + HO_3^\bullet$$
$$[R^\bullet + HO_3^\bullet] \longrightarrow ROOOH$$

accompanied by the fast decomposition of the unstable hydrotrioxide and the hydrotrioxyl radical:

$$HO_3^{\bullet} \longrightarrow HO^{\bullet} + O_2$$
$$ROOOH \longrightarrow RO_2^{\bullet} + HO^{\bullet}$$

The rate constant of $HO_3^{\bullet}$ decay is $k \approx 10^{10}\,s^{-1}$ (300 K, quantum-chemical calculation [134]):

$$RH + O_3 \rightleftharpoons [R^{\bullet} + HO_3^{\bullet}]$$
$$[R^{\bullet} + HO_3^{\bullet}] \longrightarrow R^{\bullet} + HO_3^{\bullet}$$
$$[R^{\bullet} + HO_3^{\bullet}] \longrightarrow ROOOH$$
$$HO_3^{\bullet} \longrightarrow HO^{\bullet} + O_2$$
$$HO^{\bullet} + RH \longrightarrow H_2O + R^{\bullet}$$
$$ROOOH \longrightarrow RO_2^{\bullet} + HO^{\bullet}$$

The study of the detailed mechanism of free radical initiation (rate constant k_i) and ozone decay (rate constant k_d) by the reaction with cyclohexane, cumene, and aldehydes gave the following results ($T = 298$ K):

RH	cyclo-C_6H_{12}	Me_2PhCH	$Me_2PhCOOH$	PhCHO	MeCHO
k_i (L $mol^{-1}\,s^{-1}$)	1.1×10^{-3}	1.6×10^{-2}	0.15	2.0	1.6
k_d (L $mol^{-1}\,s^{-1}$)	1.8×10^{-3}	0.32	0.11	2.0	2.7
e	0.30	2.5×10^{-2}	0.68	0.50	0.30
Ref.	[112]	[135]	[136]	[137]	[137]

The very low yield of radicals by the reaction of ozone with cumene was found to be the result of the intensive ozone reaction with the benzene ring of cumene with molozonide formation. The values of the parameter e in other reactions are typical of the cage effect of radical pairs in solutions. The rate constants of ozone reactions with various compounds are presented in Table 3.7 and Table 3.8.

Ozone, as a very strong oxidizing agent, reacts very rapidly with free radicals. The reactions of ozone with atoms and small radicals ($HO^{\bullet}$, $N^{\bullet}O_2$, $HO_2^{\bullet}$) were the object of intensive study due to their important role in the chemistry of stratosphere [145].

Due to high activity in reactions with free radicals, ozone undergoes the chain decomposition in solutions also. The chain reaction of ozone decomposition was evidenced in 1973 in the kinetic study of cyclohexane and butanone-2 oxidation by a mixture of O_2 and O_3 [146–151]. It was observed that the rate of ozone consumption obeys the equation [112]:

$$v_{O_3} = k_1[RH][O_3] + k_2[RH][O_3]^2[O_2]^{-1}. \qquad (3.16)$$

The first term characterizes the rate of the ozone reaction with the substrate and the second term characterizes the reaction with chain propagation

$$R^{\bullet} + O_3 \longrightarrow RO^{\bullet} + O_2$$

and with chain termination in the reactions

TABLE 3.7
Rate Constants of Ozone Reaction with Organic Compounds (Experimental Data)

RH	Solvent	T (K)	E (kJ mol^{-1})	log A, A (L mol^{-1} s^{-1})	k (L mol^{-1} s^{-1})	Ref.
Me$_3$CH	Carbon tetrachloride	298			8.7×10^{-2}	[138]
Me(CH$_2$)$_3$Me	Carbon tetrachloride	293			1.5×10^{-2}	[139]
Me(CH$_2$)$_3$Me	Carbon tetrachloride	298			6.5×10^{-3}	[138]
Me$_3$CCH$_3$	Carbon tetrachloride	298			3.8×10^{-5}	[138]
Me(CH$_2$)$_4$Me	Carbon tetrachloride	293	51.3	7.42	1.9×10^{-2}	[139]
Me$_2$CHCHMe$_2$	Carbon tetrachloride	298			0.20	[138]
Et$_2$CHMe	Carbon tetrachloride	298			0.10	[138]
Et$_2$CHMe	Carbon tetrachloride	293			0.15	[138]
Me(CH$_2$)$_5$Me	Carbon tetrachloride	293	52.0	7.67	2.1×10^{-2}	[139]
Me$_2$CHCHMeEt	Carbon tetrachloride	293			0.29	[139]
Me$_2$CHCH$_2$CHMe$_2$	Carbon tetrachloride	293			8.0×10^{-2}	[139]
Me$_2$CH(CH$_2$)$_3$Me	Carbon tetrachloride	293			0.13	[139]
EtMeCH(CH$_2$)$_2$Me	Carbon tetrachloride	293			0.20	[139]
Me(CH$_2$)$_6$Me	Carbon tetrachloride	298			1.4×10^{-2}	[138]
Me(CH$_2$)$_6$Me	Carbon tetrachloride	293	52.0	7.63	2.3×10^{-2}	[139]
Me(CH$_2$)$_6$Me	Carbon tetrachloride	298			1.4×10^{-2}	[138]
Me$_3$CCHMeEt	Carbon tetrachloride	293			0.59	[139]
Me$_3$C(CH$_2$)$_3$Me	Carbon tetrachloride	293			1.5×10^{-2}	[139]
Me$_3$CCMe$_3$	Octane	293			2.0×10^{-4}	[139]
Me(CH$_2$)$_7$Me	Carbon tetrachloride	293			2.6×10^{-2}	[139]
Me$_3$CCH$_2$CHMeEt	Carbon tetrachloride	293			0.13	[139]
MeEtCH(CH$_2$)$_5$Me	Carbon tetrachloride	293			0.20	[139]
Me$_3$C(CH$_2$)$_2$CHMe$_2$	Carbon tetrachloride	293			0.19	[139]
Me(CH$_2$)$_8$Me	Carbon tetrachloride	293	50.7	7.51	2.91×10^{-2}	[139]
Me(CH$_2$)$_8$Me	Carbon tetrachloride	293	56.5	8.60	3.33×10^{-2}	[140]
Me(CH$_2$)$_{12}$Me	Carbon tetrachloride	293			3.61×10^{-2}	[139]
Me(CH$_2$)$_{16}$Me	Carbon tetrachloride	293			4.62×10^{-2}	[139]
cyclopentane (H,H)	Carbon tetrachloride	293			2.61×10^{-2}	[138]
cyclopentane (H,H)	Carbon tetrachloride	293	51.2	7.52	2.50×10^{-2}	[140]
cyclohexane (H,H)	Carbon tetrachloride	298			1.05×10^{-2}	[138]
cyclohexane (H,H)	Carbon tetrachloride	298	56.9	7.83	4.81×10^{-3}	[126]
cyclohexane (H,H)	Carbon tetrachloride	293	54.9	8.03	1.72×10^{-2}	[140]
cyclohexane (H,H)	Carbon tetrachloride	293	57.7	8.30	1.04×10^{-2}	[127]
cycloheptane (H,H)	Carbon tetrachloride	293	45.4	7.45	0.23	[127]

TABLE 3.7
Rate Constants of Ozone Reaction with Organic Compounds (Experimental Data)—continued

RH	Solvent	T (K)	E (kJ mol^{-1})	log A, A (L mol^{-1} s^{-1})	k (L mol^{-1} s^{-1})	Ref.
cyclooctane (CH)	Carbon tetrachloride	293	43.0	7.34	0.48	[127]
cyclononane (CH)	Carbon tetrachloride	293	54.7	9.36	0.41	[127]
cyclodecane (CH)	Carbon tetrachloride	293	59.0	9.32	6.22×10^{-2}	[139]
isopropylcyclopropane	Carbon tetrachloride	293			5.01×10^{-2}	[139]
methylcyclohexane	Carbon tetrachloride	293	56.6	8.25	1.42×10^{-2}	[139]
norbornane	Carbon tetrachloride	293			1.41×10^{-2}	[139]
adamantane	Carbon tetrachloride	293			0.22	[139]
PhCH$_2$Me	Carbon tetrachloride	293	33.4	5.78	0.67	[140]
PhCHMe$_2$	Carbon tetrachloride	293	35.0	6.02	0.60	[140]
Ph$_2$CH$_2$	Carbon tetrachloride	293	39.2	6.60	0.41	[140]
MeCH$_2$OH	Carbon tetrachloride	298	40.2	6.72	0.35	[141]
Me$_2$CHOH	Carbon tetrachloride	298	41.2	7.30	0.89	[141]
MeCH(OH)CH$_2$Me	Carbon tetrachloride	298	38.8	7.12	1.61	[141]
PrCH$_2$OH	Carbon tetrachloride	298			0.39	[141]
PrCH$_2$OH	Carbon tetrachloride	298	42.4	7.30	0.54	[141]
PhMe$_2$COH	Carbon tetrachloride	298	27.6	4.82	0.78	[135]
Me$_3$COH	Carbon tetrachloride	298			5.0×10^{-2}	[138]
Me$_3$COH	Carbon tetrachloride	298	37.1	4.63	1.02×10^{-2}	[141]
1-methylcyclopentanol	Carbon tetrachloride	298			1.35	[138]
cyclohexanol	Carbon tetrachloride	298	34.5	6.62	2.92	[126]

continued

TABLE 3.7
Rate Constants of Ozone Reaction with Organic Compounds (Experimental Data)—*continued*

RH	Solvent	T (K)	E (kJ mol^{-1})	log A, A (L mol^{-1} s^{-1})	k (L mol^{-1} s^{-1})	Ref.
MeC(O)CH$_2$Me	Carbon tetrachloride	293	69.8	9.56	1.30×10^{-3}	[142]
MeC(O)CH$_2$Me	Water	313			8.5×10^{-3}	[143]
(cyclohexanone)	Carbon tetrachloride	295			5.9×10^{-3}	[126]
PhCHO	Carbon tetrachloride	282	38.2	7.42	2.21	[137]
PhCHO	Carbon tetrachloride	300			5.92	[137]
MeCHO	Carbon tetrachloride	300			4.30	[137]
(1,3-dioxolane, H,H)	Carbon tetrachloride	298			10.1	[144]
(1,3-dioxane, H,H)	Carbon tetrachloride	298			0.90	[144]
(1,3-dioxepane, H,H)	Carbon tetrachloride	298	32.1	6.30	4.70	[144]
(1,3-dioxolane, Me,H)	Carbon tetrachloride	298	31.7	6.92	23.0	[144]
(1,3-dioxane, Me,H)	Carbon tetrachloride	298	30.1	6.03	5.7	[144]
(1,3-dioxepane, Me,H)	Carbon tetrachloride	298			11.5	[144]
(1,3-dioxolane, Ph,H)	Carbon tetrachloride	298			56.3	[144]
(cyclohexyl Ph,H)	Carbon tetrachloride	298			19.5	[144]

$$R^\bullet + O_2 \longrightarrow RO_2^\bullet$$
$$RO_2^\bullet + RO_2^\bullet \longrightarrow ROH + O_2 + \text{ketone}$$

Chain decomposition of ozone was also observed in the oxidation of cumene by an O_3–O_2 mixture [151]. The rate of ozone consumption was found to be

$$v_{O_3} = k_1[\text{RH}][O_3] + k_2[\text{RH}]^{1/2}[O_3]^{3/2} \quad (3.17)$$

with $k_1 = 1.0$ L mol^{-1} s^{-1}, $k_2 = 18.0$ L mol^{-1} s^{-1} (cumene, 395 K [151]).

TABLE 3.8
Rate Constants and Activation Energies of Reaction $RH + O_3 \rightarrow R^{\cdot} + HO_3^{\cdot}$ in the Hydrocarbon Solution Calculated by the IPM Method (see Chapter 4, [133])

RH	D (kJ mol^{-1})	n	ΔH_e (kJ mol^{-1})	E (kJ mol^{-1})	log A, A (L mol^{-1} s^{-1})	k (300 K) (L mol^{-1} s^{-1})
EtMeCH—H	413.0	4	66.2	74.9	9.60	3.62×10^{-4}
Me$_3$C—H	400.0	1	53.2	67.2	9.00	1.99×10^{-3}
cyclobutane-H,H	418.5	8	71.7	78.3	9.90	1.85×10^{-4}
cyclopentane-H,H	408.4	10	61.6	72.1	10.00	2.80×10^{-3}
cycloheptane-H,H	403.9	14	57.1	69.5	10.15	1.12×10^{-2}
cyclohexane-H	395.5	1	48.7	64.7	9.00	5.43×10^{-3}
Z-Decalin	387.6	2	40.8	60.3	9.30	6.32×10^{-2}
CH$_2$=CHCH$_2$—H	368.0	3	21.2	63.9	8.48	2.26×10^{-3}
CH$_2$=CHMeCH—H	349.8	2	3.0	55.4	8.30	4.51×10^{-2}
CH$_2$=CHMe$_2$C—H	339.6	1	−7.2	50.8	8.00	0.14
Z-MeCH=CHMeCH—H	344.0	2	−2.8	52.8	9.30	1.28
Me$_2$C=CHMeCH—H	332.0	2	−14.8	47.6	8.30	1.03
Me$_2$C=CMeMe$_2$C—H	322.8	1	−24.0	43.7	8.00	2.46
(CH$_2$=CH)$_2$C—HMe	307.2	1	−39.6	37.6	8.00	28.4
cyclohexene-H,H	341.5	4	−5.3	51.7	9.60	3.97
cyclohexadiene-H,H	330.9	4	−15.9	47.1	9.60	25.1
cyclohexadiene-H,H	312.6	4	−34.2	39.6	9.60	5.07×10^2
cycloheptatriene-H,H	301.0	2	−45.8	35.2	9.30	1.47×10^3
MeC≡CMe$_2$C—H	329.4	1	−17.4	46.5	8.00	0.80
PhMeCH—H	364.1	2	17.3	54.2	8.30	7.31×10^{-2}
PhMe$_2$C—H	354.7	1	7.9	49.7	8.00	0.22
tetralin-H,H	345.6	4	−1.2	45.6	9.60	45.7

continued

TABLE 3.8
Rate Constants and Activation Energies of Reaction RH + O$_3$ → R$^\bullet$ + HO$_3^\bullet$ in the Hydrocarbon Solution Calculated by the IPM Method (see Chapter 4, [133])—*continued*

RH	D (kJ mol^{-1})	n	ΔH_e (kJ mol^{-1})	E (kJ mol^{-1})	log A, A (L mol^{-1} s^{-1})	k(300 K) (L mol^{-1} s^{-1})
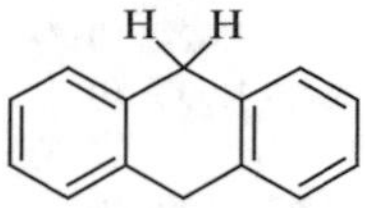 anthracene-H,H	322.0	4	−24.8	35.6	9.60	2.52 × 10^3
Me$_2$NH$_2$C—H	379.5	1	32.7	56.0	9.00	0.18
(CH$_2$=CHCH—H)$_3$N	345.6	6	−1.2	53.5	8.78	0.29
Me$_2$(HO)C—H	390.5	1	4.4	61.9	9.00	1.67 × 10^{-2}
 cyclohexenyl-H,OH	329.7	1	−17.1	32.7	9.00	2.02 × 10^3
MeCH=CMeC—HMeOH	325.2	1	−21.6	44.7	8.00	1.65
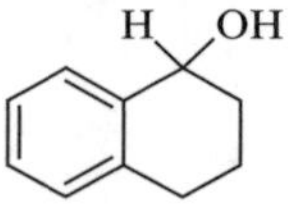 tetralin-H,OH	337.5	1	−9.3	42.0	9.00	48.7
PhC(O)—H	348.0	1	1.2	54.5	9.00	0.32
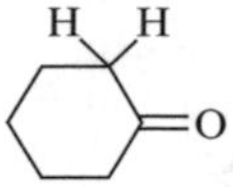 cyclohexanone-H,H	394.1	4	47.3	63.9	9.60	2.98 × 10^{-2}
Me$_2$CHOC—HMe$_2$	390.8	2	44.0	62.0	9.30	3.21 × 10^{-2}
(CH$_2$=CHCH—H)$_2$O	360.0	4	13.2	60.1	8.60	1.37 × 10^{-2}
Ph$_2$C—HOMe	354.2	1	7.4	49.5	8.00	0.24
PhCOOHCH—H	367.0	2	20.2	55.6	8.30	4.17 × 10^{-2}
Me$_3$COO—H	358.6	1	15.6	31.2	9.00	3.69 × 10^3
(Me$_3$C)$_3$Si—H	351.0	1	−1.0	46.5	9.00	8.01
Ph$_3$Ge—H	322.5	1	−29.1	43.3	9.00	28.9
Ph$_3$Sn—H	296.9	1	−55.2	31.2	9.00	3.69 × 10^3

The second term characterizes the chain decomposition of ozone. The mechanism of chain reaction was found to include the following reactions:

$$RH + O_3 \longrightarrow R^\bullet + HO_3^\bullet$$
$$HO_3^\bullet \longrightarrow HO^\bullet + O_2$$
$$HO^\bullet + RH \longrightarrow H_2O + R^\bullet$$
$$R^\bullet + O_2 \longrightarrow RO_2^\bullet$$
$$RO_2^\bullet + O_3 \longrightarrow RO^\bullet + 2O_2$$
$$RO^\bullet \longrightarrow PhC(O)Me + C^\bullet H_3$$
$$CH_3^\bullet + O_2 \longrightarrow CH_3O_2^\bullet$$
$$CH_3O_2^\bullet + O_3 \longrightarrow CH_3O^\bullet + 2O_2$$
$$CH_3O^\bullet + RH \longrightarrow CH_3OH + R^\bullet$$
$$CH_3O_2^\bullet + RO_2^\bullet \longrightarrow CH_2O + O_2 + ROH$$

The similar kinetic scheme was proposed for the ozone reaction with acetaldehyde [150]. The reaction rate obeys the equation

$$v_{O_3} = k_1[\text{MeCHO}][O_3] + k_2[\text{MeCHO}]^{1/2}[O_3]^{3/2} \tag{3.18}$$

with $k_1 = 4.3\,\text{L mol}^{-1}\,\text{s}^{-1}$, $k_2 = 15.0\,\text{L mol}^{-1}\,\text{s}^{-1}$ (CCl$_4$, 300 K [150]).

The more complicated kinetic equation was found for ozone consumption in oxidized cyclohexane [295 K, CCl$_4$ [146]]:

$$v_{O_3} = 3.2 \times 10^{-3}[\text{RH}][O_3] + 0.3[\text{RH}]^{1/2}[O_3]^{3/2} + 0.75[\text{RH}][O_3]^2[O_2]^{-1} \tag{3.19}$$

The first term characterizes the initiation of free radicals by the reaction of ozone with cyclohexane and the second chain reaction with chain propagation

$$\text{RO}_2^\bullet + O_3 \longrightarrow \text{RO}^\bullet + 2O_2^\bullet$$

and the third one with chain propagation by the reaction

$$\text{R}^\bullet + O_3 \longrightarrow \text{RO}^\bullet + O_2^\bullet$$

Ozone chain decomposition occurs in the reaction of ozone with cumyl hydroperoxide [146]. The rate of this reaction is

$$v_{O_3} = k_1[\text{ROOH}][O_3] + k_2[\text{ROOH}]^{1/2}[O_3]^{3/2} \tag{3.20}$$

$k_1 = 1.0\,\text{L mol}^{-1}\,\text{s}^{-1}$ and $k_2 = 18.0\,\text{L mol}^{-1}\,\text{s}^{-1}$ (CCl$_4$, 298 K [146]). The chain reaction includes the following steps:

$$\text{ROOH} + O_3 \longrightarrow \text{RO}_2^\bullet + \text{HO}_3^\bullet$$
$$\text{RO}_2^\bullet + O_3 \longrightarrow \text{RO}^\bullet + 2O_2^\bullet$$
$$\text{HO}^\bullet + \text{ROOH} \longrightarrow \text{H}_2\text{O} + \text{RO}_2^\bullet$$
$$\text{RO}^\bullet + \text{ROOH} \longrightarrow \text{ROH} + \text{RO}_2^\bullet$$
$$\text{RO}^\bullet + O_3 \longrightarrow \text{RO}_2^\bullet + O_2$$
$$\text{RO}_2^\bullet + \text{RO}_2^\bullet \longrightarrow 2\text{RO}^\bullet + O_2$$
$$\text{RO}_2 + \text{RO}_2^\bullet \longrightarrow \text{ROOR} + O_2$$

For the values of rate constants of the free radical reactions in the gas and the liquid phases, see Table 3.9.

So, three different chain reactions of ozone decomposition were observed in solutions:

1. Reaction with chain propagation

$$\text{R}^\bullet + O_3 \longrightarrow \text{RO}^\bullet + O_2$$
$$\text{RO}^\bullet + \text{RH} \longrightarrow \text{ROH} + \text{R}^\bullet$$

2. Reaction with chain propagation

$$\text{RO}^\bullet + O_3 \longrightarrow \text{RO}_2^\bullet + O_2$$
$$\text{RO}_2^\bullet + \text{RO}_2^\bullet \longrightarrow 2\text{RO}^\bullet + O_2^\bullet$$

TABLE 3.9
Rate Constants of Free Radical Reactions with Ozone (Experimental Data)

Radical	Phase, solvent	T (K)	E (kJ mol^{-1})	log A, A (L mol^{-1} s^{-1})	k(298 K) (L mol^{-1} s^{-1})	Ref.
C•H$_3$	Gas phase	298			1.57×10^9	[152]
CH$_3$C•H$_2$	Gas phase	298			2.0×10^7	[153]
Me$_3$C•	Carbon tetrachloride	298			3.28×10^{10}	[154]
cyclo-C$_6$H$_{11}^•$	Carbon tetrachloride	295			5.0×10^8	[146]
MeC(O)C•HMe	Carbon tetrachloride	413			4.0×10^9	[146]
HO$_2$•	Gas phase		8.4	7.30	6.74×10^5	[155]
HO$_2$•	Gas phase	300			1.8×10^6	[156]
CH$_3$O$_2$•	Gas phase	296			6.03×10^3	[157]
CH$_3$O$_2$•	Carbon tetrachloride	300			6.0×10^4	[151]
CH$_3$O$_2$•	Carbon tetrachloride	313–338	25	8.90	3.30×10^4	[151]
cyclo-C$_6$H$_{11}$O$_2$•	Carbon tetrachloride	295			5.0×10^3	[148]
Me$_2$PhCO$_2$•	Carbon tetrachloride	295			1.7×10^3	[149]

3. Reaction with chain propagation

$$RO_2^• + O_3 \longrightarrow RO^• + 2O_2$$
$$RO^• + RH \longrightarrow ROH + R^•$$
$$R^• + O_2 \longrightarrow RO_2^•$$

3.3.2 Nitrogen Dioxide

Nitrogen dioxide accelerates the liquid-phase oxidation of hydrocarbons. This was observed in the oxidation of butane [109], hexadecane [110], cyclohexane [158], and paraffin wax [159]. For the experimental data on the oxidation with nitrogen dioxide, see the monograph on *Liquid-Phase Oxidation of Hydrocarbons* [109]. Nitrogen dioxide influences the oxidation of engine lubricants oxidizing antioxidants and stimulating lubricant degradation, and deposit formation [160–164]. Several reactions of free radical generation by nitrogen dioxide in the oxidized hydrocarbons are known. Nitrogen dioxide initiates the chain liquid-phase oxidation of hydrocarbons as a result of the reactions [109,165,166]:

$$RH + NO_2^• \longrightarrow R^• + HONO$$
$$ONOH \longrightarrow NO^• + HO^•$$
$$HO^• + RH \longrightarrow H_2O + R^•$$
$$R^• + O_2 \longrightarrow RO_2^•$$
$$R^• + NO_2^• \longrightarrow RONO$$
$$RO_2^• + NO_2^• \longrightarrow ROONO_2$$
$$RONO \longrightarrow RO^• + NO^•$$
$$RO_2^• + NO^• \longrightarrow RO^• + NO_2^•$$
$$ROONO_2 \longrightarrow RO^• + NO_3^•$$
$$RO^• + RH \longrightarrow ROH + R^•$$
$$NO_3^• + RH \longrightarrow HNO_3 + R^•$$
$$ROOH + ONOH \longrightarrow RO^• + H_2O + NO_2^•$$

Nitrogen dioxide is a stable radical with an unpaired electron and forms the dimer N_2O_4 in the equilibrium reaction:

$$2N^\bullet O_2 \rightleftharpoons N_2O_4$$

This equilibrium reaction occurs with enthalpy $\Delta H = -58.2$ kJ mol^{-1} and entropy $\Delta S = -177$ J mol^{-1} K^{-1} [167]. Nitrogen dioxide decays at elevated temperatures in the gas phase by the reaction

$$2N^\bullet O_2 \longrightarrow 2N^\bullet O + O_2$$

with rate constant $k = 2.0 \times 10^9 \exp(-111.0/RT)$ L mol^{-1} s^{-1} [166].

Three mechanisms of radical initiation by nitrogen dioxide are known [5]:

1. Hydrogen abstraction by $NO_2^\bullet$ from RH

$$NO_2^\bullet + HR \longrightarrow ONOH + R^\bullet$$

This reaction is endothermic, its enthalpy is $\Delta H = D_{R-H} - 327.6$ kJ mol^{-1}. The experimental rate constants of these reactions are collected in Table 3.10 and those calculated by the IPM method [168] in Table 3.11.

The reactions of nitrogen dioxide addition to the double bond of olefins occur much more rapidly. However, this reaction is reversible and, hence, the formed radical is stabilized due to the addition of the dioxygen molecule:

$$NO_2^\bullet + CH_2{=}CHR \rightleftharpoons NO_2CH_2C^\bullet HR$$
$$O_2 + NO_2CH_2C^\bullet HR \longrightarrow NO_2CH_2CH(O_2^\bullet)R$$

The enthalpy of $NO_2^\bullet$ addition to ethylene is $\Delta H = 14.6$ kJ mol^{-1}. Due to the reverse reaction, the pre-exponential factor A measured experimentally is very low (see Table 3.12).

TABLE 3.10
Rate Constants of Reactions $N^\bullet O_2 + HR \longrightarrow ONOH + R^\bullet$ in the Gas Phase (Experimental Data)

RH	T (K)	E (kJ mol^{-1})	log A, A (L mol^{-1} s^{-1})	k(298 K) (L mol^{-1} s^{-1})	Ref.
CH_4	1300–1900	125.5	8.84	6.93×10^{-14}	[169]
$MeCH_2Me$	423–498	94.6	8.38	6.27×10^{-9}	[170]
$(CH_2{=}CH)_2CH_2$	296			1.22	[171]
CH_3OH	640–713	89.5	8.56	7.45×10^{-8}	[172]
CH_3OH	900–1100	94.5	8.30	3.64×10^{-9}	[173]
EtCHO	295–390	51.9	7.40	2.02×10^{-2}	[174]
PrCHO	295–390	51.9	7.40	2.02×10^{-2}	[174]
CF_3CHO	533–584	98.7	8.94	4.35×10^{-9}	[175]
CH_3COCH_3	298–373	29.8	2.58	2.24×10^{-3}	[176]
Et_2NOH	298			3.31×10^3	[177]
Me_2NNH_2	298			1.39×10^4	[178]

TABLE 3.11
Enthalpies, Activation Energies, and Rate Constants of Reaction RH + N·O$_2$ → R· + HONO Calculated by the IPM Method [168]

RH	D (kJ mol^{-1})	n	ΔH$_e$ (kJ mol^{-1})	E (kJ mol^{-1})	log A, A (L mol^{-1} s^{-1})	k (300 K) (L mol^{-1} s^{-1})
EtMeCH—H	413.0	4	81.3	86.6	10.67	3.77 × 10^{-5}
Me$_3$C—H	400.0	1	68.3	73.6	9.77	8.80 × 10^{-4}
cyclobutane-H$_2$	418.5	8	86.8	82.0	11.06	4.84 × 10^{-5}
methylcyclopentane-H	408.4	1	76.7	82.0	10.98	4.84 × 10^{-4}
cycloheptane-H$_2$	403.9	1.4	72.2	77.5	11.02	3.26 × 10^{-3}
methylcyclohexane-H	395.5	1	63.8	69.1	9.63	3.88 × 10^{-3}
Z-Decalin	387.6	2	55.9	61.2	9.60	8.51 × 10^{-2}
CH$_2$=CHCH$_2$—H	368.0	3	36.3	59.0	8.48	1.57 × 10^{-2}
CH$_2$=CHCH—HMe	349.8	2	18.1	49.8	8.30	0.42
CH$_2$=CHC—HMe$_2$	339.6	1	7.9	45.0	8.00	1.45
Z-MeCH=CHCH—HMe	344.0	2	12.3	47.1	8.30	1.27
Me$_2$C=CHCH—HMe	332	2	0.3	41.6	8.30	11.58
Me$_2$C=CMeC—HMe$_2$	322.8	1	−8.9	37.5	8.00	28.92
(CH$_2$=CH)$_2$C—HMe	307.2	1	−24.5	31.1	8.00	3.76 × 10^2
cyclohexene-H$_2$	341.5	4	9.8	45.9	9.60	40.8
cyclohexadiene-H$_2$	330.9	4	−0.8	41.1	9.60	2.82 × 10^2
cyclohexadiene-H$_2$	312.6	4	−19.1	33.3	9.60	6.33 × 10^3
cycloheptatriene-H$_2$	301.0	2	−30.7	28.7	9.30	1.97 × 10^4
MeC≡CC—HMe$_2$	329.4	1	−2.3	40.4	8.00	9.19
MePhCH—H	364.1	2	32.4	50.6	8.30	0.31
Me$_2$PhC—H	354.7	1	23.0	45.8	8.00	1.07
tetralin-H$_2$	345.6	4	13.9	41.3	9.60	2.56 × 10^2
dihydroanthracene-H$_2$	322.0	4	−9.7	30.7	9.60	1.82 × 10^4

TABLE 3.11
Enthalpies, Activation Energies, and Rate Constants of Reaction RH + N·O$_2$ → R· + HONO Calculated by the IPM Method [168]—continued

RH	D (kJ mol^{-1})	n	ΔH_e (kJ mol^{-1})	E (kJ mol^{-1})	log A, A (L mol^{-1} s^{-1})	k(300 K) (L mol^{-1} s^{-1})
Me$_2$NH$_2$C—H	379.5	1	47.8	53.1	8.66	0.25
(CH$_2$=CHCH—H)$_3$N	345.6	6	13.9	47.8	8.78	2.82
Me$_2$C—H(OH)	390.5	1	58.8	64.1	9.44	1.84 × 10^{-2}
cyclohexenyl-H(OH)	329.7	1	−2.0	40.5	9.00	87.2
MeCH=CMeC—HMeOH	325.2	1	−6.5	38.6	8.00	19.1
tetralin-H(OH)	337.5	1	5.8	37.5	9.00	2.94 × 10^2
PhC(O)—H	348.0	1	16.3	42.5	8.00	4.03
cyclohexanone-H	394.1	4	62.4	67.7	10.19	2.43 × 10^{-2}
Me$_2$CHOC—HMe$_2$	390.8	2	59.1	64.4	9.75	3.37 × 10^{-2}
(CH$_2$=CHCH—H)$_2$O	360.0	4	28.3	54.9	8.60	0.11
Ph$_2$C—HOMe	354.2	1	22.5	45.5	8.00	1.18
Ph(COOH)CH—H	367.0	2	35.3	52.2	8.30	0.16
Me$_3$COO—H	358.6	1	30.7	45.9	9.00	10.34
(Me$_3$C)$_3$Si—H	351.0	1	15.0	40.8	9.00	77.6
Ph$_3$Ge—H	322.5	1	−14.0	29.8	9.00	6.38 × 10^3
Ph$_3$Sn—H	296.9	1	−40.1	20.4	9.00	2.76 × 10^5

3.3.3 Halogens

Dichlorine shortens the induction period of autoxidation of paraffin wax [187] and accelerates the oxidation of hydrocarbons [109]. Difluorine is known as very active initiator of gas-phase chain reactions, for example, chlorination [188,189].

Difluorine is an extremely active reagent and reacts with organic molecules at low (200 K and lower) temperatures. Such a high activity of difluorine is due to the very high BDE of the formed hydrogen fluoride molecule (D_{F-H} = 570 kJ mol^{-1}) and the relatively low BDE in the difluorine molecule (D_{F-F} = 158.7 kJ mol^{-1}). Due to this great difference in the BDE of the reactants and the products, bimolecular reactions of hydrogen atom abstraction

$$RH + F_2 \longrightarrow R^\cdot + HF + F^\cdot$$

are exothermic for most organic molecules. The rate constant of the difluorine reaction with methane, which contains C—H bonds of high strength (D_{C-H} = 440 kJ mol^{-1}), is $k = 2.0 \times 10^9 \exp(-47/RT)$ L mol^{-1} s^{-1} (298 K, [190]. The values of the activation energies and the rate constants of the difluorine bimolecular reaction with several organic molecules calculated by the IPM model [191] are presented in Table 3.13. The enthalpy of these

TABLE 3.12
Rate Constants of Addition Reactions $N^•O_2 + CH_2=CHR \rightarrow RC^•HCH_2NO_2$ in the Gas Phase (Experimental Data)

$CH_2=CHR$	T (K)	E (kJ mol^{-1})	log A, (A (L mol^{-1} s^{-1}))	k(298 K) (L mol^{-1} s^{-1})	Ref.
$CH_2=CH_2$	298–382	58.6	9.97	0.50	[179]
$CH_2=CHMe$	323			4.73×10^{-2}	[180]
$CH_2=CHMe$	293–373	23.4	2.38	1.90×10^{-2}	[181]
$CH_2=CHMe$	298–373	33.0	3.50	5.19×10^{-3}	[176]
$CH_2=CHEt$	298–382	30.5	3.40	1.13×10^{-2}	[176]
$CH_2=CHPr$	298–373	30.1	0.20	8.39×10^{-6}	[176]
$CH_2=CHBu$	293–333	29.3	4.89	0.57	[181]
$CH_2=CMe_2$	298–373	16.7	1.60	4.71×10^{-2}	[176]
Z-2-MeCH=CHMe	298–382	47.7	8.64	1.90	[179]
E-2-MeCH=CHMe	298–382	48.5	8.87	2.34	[179]
Z-CH_2=CHCH=CHMe	298			1.0×10^2	[182]
E-CH_2=CHCH=CHMe	298			1.26×10^2	[182]
$CH_2=CMeCH=CH_2$	298			1.09×10^2	[183]
$CH_2=CMeCH=CH_2$	298			1.08×10^2	[182]
$CH_2=CMeCH=CH_2$	295			6.21×10^1	[184]
$Me_2C=CMe_2$	298			6.44	[185]
$Me_2C=CMe_2$	295			9.27	[184]
E-CH_2=CHCH=CH_2	298			18.7	[184]
E-CH_2=CHCH=CH_2	298			17.1	[171]
$CH_2=CMeCMe=CH_2$	298			1.81×10^2	[183]
$CH_2=CMeCMe=CH_2$	298			1.50×10^2	[171]
$CH_2=CHCMe=CHMe$	298			3.31×10^2	[183]
$Me_2C=CHCH=CH_2$	298			2.59×10^2	[183]
$Me_2C=CHCH=CMe_2$	298			1.32×10^3	[183]
E-CH_2=CHCH=CHEt	298			1.21×10^2	[183]
E,Z-CH_2=MeCH=CHCH=CHMe	298			3.25×10^2	[183]
E,E-CH_2=MeCH=CHCH=CHMe	298			3.56×10^2	[183]
Z-CH_2=CHCH=CHCH=CH_2	298			4.04×10^2	[183]
E-CH_2=CHCH=CHCH=CH_2	298			5.78×10^2	[183]
$CH_2=CMe(CH_2)_3C(CH=CH_2)=CH_2$	294			1.57×10^2	[183]
1,3-cyclohexadiene	298			1.10×10^3	[183]
1,3-cyclohexadiene	295			1.07×10^3	[184]
α-terpinene (isopropyl-methyl-cyclohexadiene)	295			3.92×10^3	[184]
1,3,5-cycloheptatriene	295			1.46×10^{-2}	[184]

TABLE 3.12
Rate Constants of Addition Reactions $N^{\cdot}O_2 + CH_2{=}CHR \rightarrow RC^{\cdot}HCH_2NO_2$ in the Gas Phase (Experimental Data)—continued

$CH_2{=}CHR$	T (K)	E (kJ mol^{-1})	log A, (A (L mol^{-1} s^{-1}))	k(298 K) (L mol^{-1} s^{-1})	Ref.
cyclooctene	298			30.1	[183]
cycloheptadiene (H H)	298			54.2	[183]
ClCH=CCl$_2$	303–343	36.8	2.61	1.44×10^{-4}	[186]
HC≡CH	298–382	60.2	9.77	0.16	[179]

reactions varies from −6 to −118 kJ mol^{-1} and activation energy ranges from 12 to 36 kJ mol^{-1} for the chosen molecules.

Another fast reaction of radical initiation is the difluorine addition to the double bond of the unsaturated compounds [192–194]:

$$F_2 + CH_2{=}CHR \longrightarrow F^{\cdot} + FCH_2C^{\cdot}HR$$

The C—F bond is strong (($D_{C-F} = 464$ kJ mol^{-1} in CH$_3$F) and its formation compensates the energy of difluorine dissociation and π-C—C bond splitting. The values of the rate constants of the difluorine reaction with ethene in the gas phase were found to be the following: $k = 7.76 \times 10^7 \exp(-19.2/RT) = 2.04 \times 10^4$ (298 K, [192]), $k = 1.81 \times 10^4$ L mol^{-1} s^{-1} (298 K, [193]), $k = 4.16 \times 10^4$ L mol^{-1} s^{-1} (315 K, [194]).

For the values of the rate constants and the activation energies of difluorine with olefins calculated by the IPM method [303], see Table 3.14.

The chlorination of hydrocarbons proceeds via the chain mechanism [195]. Chlorine atoms are generated photochemically or by the introduction of the initiator. However, liquid-phase chlorination occurs slowly in the dark in the absence of an initiator. The most probable reaction of thermal initiation in RH chlorination is the bimolecular reaction

$$Cl_2 + HR \longrightarrow Cl^{\cdot} + HCl + R^{\cdot}$$

The rate constant of the reaction

$$Cl_2 + HCF_3 \longrightarrow Cl^{\cdot} + HCl + C^{\cdot}F_3$$

in the gas phase is $k = 5.0 \times 10^9 \exp(-142.2/RT) = 2.1 \times 10^{-3}$ (600 K) L mol^{-1} s^{-1} [196]. The values of the activation energies of the reactions $Cl_2 + RH$ are close to the reaction enthalpy. The rate constants of the reactions $Cl_2 + RH$ calculated by the IPM method are collected in Table 3.15.

TABLE 3.13
Enthalpies, Activation Energies, and Rate Constants of Reaction: $F_2 + HR \rightarrow F^{\bullet} + HF + R^{\bullet}$ Calculated by the IPM Method [191]

RH	D (kJ mol^{-1})	n	ΔH_e (kJ mol^{-1})	E (kJ mol^{-1})	log A (A (L mol^{-1} s^{-1}))	k(300 K) (L mol^{-1} s^{-1})
EtMeCH—H	413.0	4	−6	36.5	10.94	8.02×10^3
Me$_3$C—H	400.0	1	−19	31.5	10.34	1.60×10^4
cyclobutane-H	418.5	8	−0.5	38.7	11.25	6.60×10^3
cyclopentane-H	408.4	10	−10.6	33.0	11.34	8.60×10^4
cycloheptane-H	403.9	14	−15.1	31.3	11.49	2.46×10^5
methylcyclohexane-H	395.5	1	−23.5	28.2	10.34	6.38×10^4
Z-Decalin	387.6	2	−31.4	25.4	10.64	4.12×10^5
CH$_2$=CHCH$_2$—H	368.0	3	−51	26.7	10.82	3.62×10^5
CH$_2$=CHCH—HMe	349.8	2	−69.2	21.1	10.64	2.53×10^6
CH$_2$=CHC—HMe$_2$	339.6	1	−79.4	18.1	10.34	4.57×10^6
Z-MeCH=CHCH—HMe	344.0	2	−75	23.5	10.64	9.19×10^5
Me$_2$C=CHCH—HMe	332.0	2	−87	20.1	10.64	3.88×10^6
Me$_2$C=CMeC—HMe$_2$	322.8	1	−96.2	17.6	10.34	5.66×10^6
(CH$_2$=CH)$_2$C—HMe	307.2	1	−111.8	13.7	10.34	3.06×10^7
cyclohexene-CH$_2$	341.5	4	−77.5	22.8	10.94	2.46×10^6
cyclohexadiene-CH$_2$	330.9	4	−88.1	19.8	10.94	8.80×10^6
cyclohexadiene-CMe	312.6	4	−106.4	15.0	10.94	6.92×10^7
cycloheptatriene-CH$_2$	301.0	2	−118	12.2	10.64	1.18×10^8
MeC≡CC—HMe$_2$	329.4	1	−89.6	19.4	10.34	2.62×10^5
MePhCH—H	364.1	2	−54.9	24.8	10.64	5.31×10^5
Me$_2$PhC—H	354.7	1	−64.3	21.9	10.34	9.05×10^5
tetralin-CH$_2$	345.6	4	−73.4	19.3	10.94	1.09×10^7

TABLE 3.13
Enthalpies, Activation Energies, and Rate Constants of Reaction: $F_2 + HR \rightarrow F^\bullet + HF + R^\bullet$ Calculated by the IPM Method [191]—continued

RH	D (kJ mol^{-1})	n	ΔH_e (kJ mol^{-1})	E (kJ mol^{-1})	log A (A (L mol^{-1} s^{-1}))	k(300 K) (L mol^{-1} s^{-1})
(anthracene H,H)	322.0	4	−97	13.0	10.94	1.66×10^8
Me$_2$NH$_2$C—H	379.5	1	−39.5	26.7	10.34	1.20×10^5
(CH$_2$=CHCH—H)$_3$N	345.6	6	−73.4	23.9	11.12	2.35×10^6
Me$_2$C—H(OH)	390.5	1	−28.5	30.5	10.34	2.44×10^4
(cyclohexenol H,OH)	329.7	1	−89.3	19.5	10.34	2.51×10^6
MeCH=CMeC—HMeOH	325.2	1	−93.8	18.3	10.34	4.20×10^6
(tetralinol H,OH)	337.5	1	−81.5	17.1	10.34	7.03×10^6
PhC(O)—H	348.0	1	−71	20.0	10.34	2.03×10^6
(cyclohexanone H,H)	394.1	4	−24.9	31.8	10.94	5.64×10^4
Me$_2$CHOC—HMe$_2$	390.8	2	−28.2	30.6	10.64	4.68×10^4
(CH$_2$=CHCH—H)$_2$O	360.0	4	−59	28.3	10.94	2.44×10^5
Ph$_2$C—HOMe	354.2	1	−64.8	21.8	10.34	9.45×10^5
Ph(COOH)CH—H	367.0	2	−52	25.7	10.64	3.64×10^5
Me$_3$COO—H	358.6	1	−56	18.4	10.34	4.02×10^6
(Me$_3$C)$_3$Si—H	351.0	1	−72.8	16.8	10.34	7.98×10^6

Thermal initiation in the reaction of dichlorine with olefins proceeds via the bimolecular reaction of addition [4]:

$$Cl_2 + CH_2=CHR \longrightarrow Cl^\bullet + ClCH_2C^\bullet HR$$

These reactions of diclorine with olefins are endothermic, ΔH varies from 63 kJ mol^{-1} (Ph$_2$C=CH$_2$) to 191 kJ mol^{-1} (MeC≡CMe). The rate constants of chlorine reactions with olefins calculated by the IPM method are presented in Table 3.16.

Hydrogen bromide was found to initiate the autoxidation of decane [197]. The reactions of free radicals generation are the following:

1. The reaction of HBr with dioxygen with the formation of the HO$_2^\bullet$ radical and bromine atom:

$$HBr + O_2 \longrightarrow Br^\bullet + HO_2^\bullet$$

The reaction is endothermic, its $\Delta H = 146.3$ kJ mol^{-1}.

2. The reaction of peroxyl radical with HBr is followed by the reactions [109]:

TABLE 3.14
Enthalpies (ΔH_e), Activation Energies, and Rate Constants of the Addition Reactions
$F_2 + CH_2{=}CHR \rightarrow F^\bullet + FCH_2C^\bullet HR$ Calculated by the IPM Method [191]

Olefin	ΔH_e (kJ mol^{-1})	E (kJ mol^{-1})	log A, A (L mol^{-1} s^{-1})	k(300 K) (L mol^{-1} s^{-1})
$CH_2{=}CH_2$	−29.1	38.6	11.00	1.90×10^4
$Me_2C{=}CH_2$	−32.3	36.9	10.70	1.88×10^4
$EtCH{=}CH_2$	−27.7	39.3	10.70	7.20×10^3
cyclohexene (H,H)	−33.7	36.2	11.00	4.98×10^4
$CH_2{=}CHCH{=}CH_2$	−74.5	13.4	11.00	4.64×10^7
$CH_2{=}CHCH{=}CMe_2$	−80.1	11.0	10.70	6.09×10^8
$Me_2C{=}CMeCMe{=}CMe_2$	−109.5	0.03	11.00	9.88×10^{10}
cyclohexadiene (H,H)	−91.3	6.4	11.00	7.68×10^9
$PhCH{=}CH_2$	−70.0	26.7	10.70	1.12×10^6
$PhCH{=}CHMe$	−66.2	28.4	10.70	5.69×10^5
$Ph_2C{=}CH_2$	−137.2	1.9	10.70	2.34×10^{10}
dihydronaphthalene	−66.2	28.4	10.70	5.69×10^5

$$RO_2^\bullet + HBr \longrightarrow ROOH + Br^\bullet$$
$$Br^\bullet + RO_2^\bullet \longrightarrow ROOBr$$
$$ROOBr \longrightarrow RO^\bullet + BrO^\bullet$$
$$RO^\bullet + RH \longrightarrow ROH + R^\bullet$$
$$BrO^\bullet + RH \longrightarrow BrOH + R^\bullet$$
$$R^\bullet + O_2 \longrightarrow RO_2^\bullet$$

3. Free radicals are produced by the reactions of HBr and the bromide anion with hydroperoxide [198]:

$$ROOH + HBr \longrightarrow RO^\bullet + H_2O + Br^\bullet$$
$$HBr \rightleftharpoons Br^- + H^+$$
$$Br^- + ROOH \longrightarrow Br^\bullet + RO^\bullet + OH^-$$
$$RO^\bullet + RH \longrightarrow ROH + R^\bullet$$
$$Br^\bullet + RH \longrightarrow BrH + R^\bullet$$

3.4 PHOTOINITIATION

Photooxidation plays a crucial role in the discovery and study of chain reactions of oxidation (see Chapter 1). Photooxidation has two important peculiarities to study the mechanism of oxidation as a chain process.

1. The rate of photooxidation does not virtually depend on temperature. Therefore, photooxidation gives a possibility to oxidize the hydrocarbon at room or lower

TABLE 3.15
Enthalpies, Activation Energies, and Rate Constants of Reaction RH + Cl$_2$ → R$^•$ + HCl + Cl$^•$ Calculated by the IPM Method [191]

RH	D (kJ mol^{-1})	n	ΔH (kJ mol^{-1})	log A, A (L mol^{-1} s^{-1})	k(400 K) (L mol^{-1} s^{-1})
EtMeCH—H	413.0	4	224.0	11.95	2.94 × 10^{-19}
Me$_3$C—H	400.0	1	211.0	11.33	3.62 × 10^{-19}
cyclobutane (H,H)	418.5	8	229.5	12.26	1.12 × 10^{-19}
cyclopentane (H,H)	408.4	10	219.4	12.34	2.87 × 10^{-18}
cycloheptane (H,H)	403.9	14	214.9	12.48	1.55 × 10^{-17}
methylcyclohexane (H)	395.5	1	206.5	11.32	1.37 × 10^{-17}
Z-Decalin	387.6	2	198.6	11.61	4.88 × 10^{-16}
CH$_2$=CHCH$_2$—H	368.0	3	179.0	11.59	9.85 × 10^{-13}
CH$_2$=CHCH—HMe	349.8	2	160.8	11.37	1.61 × 10^{-11}
CH$_2$=CHC—HMe$_2$	339.6	1	150.6	11.03	1.64 × 10^{-10}
Z-MeCH=CHCH—HMe	344.0	2	155.0	11.35	8.95 × 10^{-11}
Me$_2$C=CHCH—HMe	332.0	2	143.0	11.31	3.14 × 10^{-9}
Me$_2$C=CMeC—HMe$_2$	322.8	1	133.8	10.97	2.36 × 10^{-8}
(CH$_2$=CH)$_2$C—HMe	307.2	1	118.2	10.90	2.33 × 10^{-6}
cyclohexene (H,H)	341.5	4	152.5	11.60	3.40 × 10^{-10}
cyclohexadiene (H,H)	330.9	4	141.9	11.60	8.53 × 10^{-9}
cyclohexadiene (H,H)	312.6	4	123.6	11.53	1.91 × 10^{-6}
cycloheptatriene (H,H)	301.0	2	112.0	11.17	2.86 × 10^{-5}
MeC≡CC—HMe$_2$	329.4	1	140.4	11.00	3.39 × 10^{-9}
MePhCH—H	364.1	2	175.1	11.52	2.95 × 10^{-13}
Me$_2$PhC—H	354.7	1	165.7	11.19	2.40 × 10^{-12}
tetralin (H,H)	345.6	4	156.6	11.77	1.45 × 10^{-10}

continued

TABLE 3.15
Enthalpies, Activation Energies, and Rate Constants of Reaction $RH + Cl_2 \rightarrow R^\bullet + HCl + Cl^\bullet$ Calculated by the IPM Method [191]—*continued*

RH	D (kJ mol^{-1})	n	ΔH (kJ mol^{-1})	log A, A (L mol^{-1} s^{-1})	k(400 K) (L mol^{-1} s^{-1})
9,10-dihydroanthracene (H H)	322.0	4	133.0	11.7	1.61×10^{-7}
Me$_2$NH$_2$C—H	379.5	1	190.5	11.29	1.40×10^{-15}
(CH$_2$=CHCH—H)$_3$N	345.6	6	156.6	11.83	1.67×10^{-10}
Me$_2$C—H(OH)	390.5	1	201.5	11.31	6.08×10^{-17}
cyclohexenol (H, OH)	329.7	1	140.7	11.00	3.10×10^{-9}
MeCH=CMeC—HMeOH	325.2	1	136.2	10.98	1.16×10^{-8}
tetralinol (H OH)	337.5	1	148.5	11.15	4.10×10^{-10}
PhC(O)—H	348.0	1	159.0	11.18	1.80×10^{-11}
cyclohexanone (H H, =O)	394.1	4	205.1	11.92	8.31×10^{-17}
Me$_2$CHOC—HMe$_2$	390.8	2	201.8	11.61	1.11×10^{-16}
(CH$_2$=CHCH—H)$_2$O	360.0	4	171.0	11.70	1.56×10^{-12}
Ph$_2$C—HOMe	354.2	1	165.2	11.19	2.08×10^{-12}
Ph(COOH)CH—H	367.0	2	178.0	11.52	1.22×10^{-13}
Me$_3$COO—H	358.6	1	169.6	10.63	2.02×10^{-13}
(Me$_3$C)$_3$Si—H	351.0	1	162.0	10.92	3.98×10^{-12}

temperatures and synthesize unstable intermediates (hydroperoxides, hydrotrioxides, tetroxides, etc.). *Photoinitiation* of chain oxidation opens up the possibility to measure the activation energy of the chain reaction $E_v = E_p - 0.5E_t$ directly as the slope $R\, d \ln v/d(T^{-1})$ from the experimental values of the oxidation rate $v(T)$ (see Chapter 2).

2. Photoinitiation can be switched on and off extremely rapidly. For example, the time of laser flash can be as short as 1 psec (10^{-12} s) and shorter. The practical absence of time inertia of photoinitiation lies in the timescales of the experimental techniques for studying fast free radical reactions (flash photolysis, rotating sector technique, photo after-effect [109]).

3.4.1 Intramolecular and Intermolecular Photophysical Processes

Two laws form the basis of interaction of light and substance:

1. Grotthus–Draper law: radiation should be absorbed by the substance to perform a chemical change [199,200].
2. Stark–Einstein law of photochemical equivalence: one photon of radiation can be absorbed only by one molecule [201,202].

TABLE 3.16
Enthalpies, Activation Energies, and Rate Constants of the Reactions $Cl_2 + CH_2=CHR \rightarrow Cl^\bullet + ClCH_2C^\bullet HR$, Calculated by the IPM Method [191]

RCH=CH$_2$	ΔH (kJ mol^{-1})	E (kJ mol^{-1})	k(300 K) (L mol^{-1} s^{-1})
\multicolumn{4}{|c|}{RCH=CH$_2$ + Cl$_2$ $\rightarrow$ RC$^\bullet$HCH$_2$Cl + Cl$^\bullet$}			
CH$_2$=CH$_2$	168.2	169.3	5.96×10^{-20}
Me$_2$C=CH$_2$	165.0	166.5	1.83×10^{-19}
EtCH=CH$_2$	169.6	170.5	3.65×10^{-20}
cyclohexene (CH)	155.6	158.3	4.80×10^{-18}
HC≡CH	183.7	205.5	6.03×10^{-26}
MeHC≡CMe	199.4	218.8	2.91×10^{-28}
CH$_2$=CHCH=CH$_2$	122.8	131.3	2.42×10^{-13}
CH$_2$=CHCH=CMe$_2$	118.2	127.5	1.14×10^{-12}
Me$_2$C=CMeCMe=CMe$_2$	88.8	103.4	1.81×10^{-8}
cyclohexadiene	98.0	110.8	9.13×10^{-10}
PhCH=CH$_2$	126.9	136.7	2.87×10^{-14}
PhCH=CHMe	121.7	132.5	1.55×10^{-13}
Ph$_2$C=CH$_2$	66.5	89.8	4.18×10^{-6}
dihydronaphthalene	123.2	133.7	9.54×10^{-14}

A molecule exhibits a great difference in the speeds of electronic transitions and vibrational atomic motions. The absorbtion of photon and a change in the electronic state of a molecule occurs in 10^{-15}–10^{-18} s. The vibrational motion of atoms in a molecule takes place in 10^{-13} s. Therefore, an electronically excited molecule has the interatomic configuration of the nonexited state during some period of time. Different situations for the exited molecule can exist. Each situation is governed by the Franck–Condon principle [203,204].

The general scheme of photophysical processes followed by the photon absorption by the molecule induces the below-mentioned elementary stages [205–209]:

Process	Scheme	
Excitation	$A(S_0) + h\nu \longrightarrow A^*(S_1^\bullet)$	$k = 10^{15}$–10^{18} s^{-1}
Internal conversion	$A^*(S_1^\bullet) \longrightarrow A(S_0)$	$k_{ic} = 10^4$–10^{12} s^{-1}
Fluorescence	$A^*(S_1^\bullet) \longrightarrow A(S_0) + h\nu$	$k \approx 10^6$–10^9 s^{-1}
Dissociation	$A^*(S_1) \longrightarrow B + C$	$k_d \approx 10^{13}$–10^{15} s^{-1}
Transition to triplet state	$A^*(S_1^\bullet) \longrightarrow A^*(T_1)$	$k_{isc} \approx 10^{-4}$–10^{-12} s^{-1}
Phosphorescence	$A^*(T_1) \longrightarrow A(S_0) + h\nu$	$k_{ph} \approx 10^{-2}$–10^4 s^{-1}
Intersystem crossing	$A^*(T_1) \longrightarrow A(S_0)$	$k'_{isc} = 10^4$–10^{12} s^{-1}

When a substance is illuminated with a constant intensity, a steady state is reached. If we express the absorbed intensity of light I_a in moles of photons per unit volume per second, the steady-state concentration of $A^*(S_1^\bullet)$ is

$$[A^*(S_1^\bullet)] = I_a(k_{ic} + k_f + k_{isc})^{-1} \tag{3.21}$$

and the steady-state concentration of the triplet state $A^*(T_1)$ is

$$A^*(T_1) = k_{isc} I_a (k'_{isc} + k_{ph})^{-1}(k_{ic} + k_f + k_{isc})^{-1} \tag{3.22}$$

The observed singlet lifetime is equal to

$$\tau_s^{-1} = k_{ic} + k_f + k_{isc} \tag{3.23}$$

The quantum yield for *fluorescence* is equal to

$$\Phi_{fl} = k_f(k_{ic} + k_f + k_{isc})^{-1} \tag{3.24}$$

and that for *phosphorescence* is

$$\Phi_{ph} = k_{ph} k_{isc} (k'_{isc} + k_{ph})^{-1}(k_{ic} + k_f + k_{isc})^{-1} \tag{3.25}$$

Excited molecules A* may be rapidly deactivated by other molecules (quenchers, Q):

$$A^*(T_1) + Q(S_0) \longrightarrow A(S_0) + Q^*(T_1) \qquad k_Q$$

Quencher Q lowers the intensity of phosphorescence. The dependence of the light intensity of phosphorescence I_{ph} on the quencher concentration obeys the Stern–Volmer equation [210]:

$$I_{ph}^0 / I_{ph} = 1 + k_Q \tau_T [Q] \tag{3.26}$$

where τ_T is the phosphorescence lifetime ($\tau_T^{-1} = k_{ph} + k'_{isc}$).

Quenching an excited molecule (A*) with another molecule (Q) may result in the electronic excitation of Q with the concomitant deactivation of A [205]. This electronic energy transfer proceeds in accordance with the Wigner spin conservation rule. The overall spin angular momentum of the interacting pair of molecules must be unchanged in an electronic energy transfer. This rule is not absolute, but it is an important guide. According to this rule, the following processes of energy transfer are possible:

$$A(S_1) + Q(S_0) \longrightarrow A(S_0) + Q(S_1)$$
$$A(T_1) + Q(S_0) \longrightarrow A(S_0) + Q(T_1)$$

A dioxygen molecule is a very active quencher ($k_Q \approx 10^9$–10^{10} L mol^{-1} s^{-1}).

Energy transfer occurs in a long-lived collision complex. An exited molecule is often very polarizable and may form a collision complex with the Q molecule in the ground state. The collision complex A*Q has a longer lifetime than the corresponding AQ collision complex. The formation of an exciplex provides the energy transfer by a collision mechanism.

Initiation of Liquid-Phase Oxidation

3.4.2 Photosensitizers

To perform the dissociation of the hydrocarbon to alkyl radicals with C—C bond scission, a hydrocarbon molecule should absorb light with the wavelength $\lambda \approx 270$–370 nm. However, alkanes do not absorb light with such wavelength. Therefore, *photosensitizers* are used for free radical initiation in hydrocarbons. Mercury vapor has been used as a sensitizer for the generation of free radicals in the oxidized hydrocarbon [206–212]. Nalbandyan [212–214] was the first to study the photooxidation of methane, ethane, and propane using Hg vapor as photosensitizer. Hydroperoxide was isolated as the product of propane oxidation at room temperature. The quantum yield of hydroperoxide was found to be ≥ 2, that is, oxidation occurs with short chains. The following scheme of propane photoxidation was proposed [117]:

$$Hg(^1S_0) + h\nu \longrightarrow Hg(^3P_1)$$
$$Hg(^3P_1) + RH \longrightarrow Hg(^1S_0) + RH^*$$
$$RH^* \longrightarrow R^\bullet + H^\bullet$$
$$R^\bullet + O_2 \longrightarrow RO_2^\bullet$$
$$RO_2^\bullet + RH \longrightarrow ROOH + R^\bullet$$
$$RO_2^\bullet + Wall \longrightarrow Molecular\ products$$

Gray [215] studied the photooxidation of methane and ethane at room temperature in the gas phase using Hg as a photosensitizer. Hydroperoxide was found as the product of oxidation. Along with hydroperoxide, ozone was also found as the product of photooxidation. It was supposed to be due to the reaction of excited mercury atoms with dioxygen.

$$Hg(^1S_0) + h\nu \longrightarrow Hg(^3P_1)$$
$$Hg(^3P_1) + O_2 \longrightarrow HgO + O$$
$$O_2 + O + M \longrightarrow O_3 + M$$

The free radical mechanism of hydroperoxide formation is close to that proposed by Nalbandyan.

In addition to mercury atoms, cadmium and zinc are used as sensitizers of gas-phase free radical reactions. Their photophysical characteristics are given here [5]:

Metal	Transition	Wavelength (nm)	Energy of Excited Atoms (kJ mol^{-1})
Mercury	$6^3P_1 \to 6^1S_0$	253.7	469.4
Mercury	$6^1P_1 \to 6^1S_0$	184.9	643.9
Cadmium	$5^3P_1 \to 5^1S_0$	326.1	365.3
Cadmium	$5^1P_1 \to 5^1S_0$	228.8	520.5
Zinc	$4^3P_1 \to 4^1S_0$	307.6	387.0
Zinc	$4^1P_1 \to 4^1S_0$	213.9	558.1

The reaction of excited cadmium with dihydrogen was proved to proceed with cadmium hydride formation [216]:

$$Cd(^3P_1) + H_2 \longrightarrow CdH + H^\bullet$$

Various organic molecules are used as photosensitizers in liquid-phase reactions, for example, anthraquinones, aryl ketones, polycyclic aromatic hydrocarbons, dyes, etc. The following mechanism, as the most probable, was suggested for the initiation by the organic photosensitizer Q with the aromatic ring [204–208]:

$$Q + h\nu \longrightarrow Q^*$$
$$Q^* + RH \longrightarrow {}^\bullet QH + R^\bullet$$
$${}^\bullet QH + O_2 \longrightarrow Q + HO_2^\bullet$$
$$R^\bullet + O_2 \longrightarrow RO_2^\bullet$$
$$RO_2^\bullet + RH \longrightarrow ROOH + R^\bullet$$

or

$$Q^* + O_2 \longrightarrow {}^\bullet QOO^\bullet$$
$${}^\bullet QOO^\bullet + RH \longrightarrow {}^\bullet QOOH + R^\bullet$$
$${}^\bullet QOOH \longrightarrow Q + HO_2^\bullet$$
$$HO_2^\bullet + RH \longrightarrow H_2O_2 + R^\bullet$$

The study of the mechanism of photoinitiation is complicated by the quenching action of dioxygen (see page 122). The values of the triplet state of selected compounds used as photosensitizers are given in Table 3.17.

3.4.3 PHOTOINITIATORS

Photoinitiators are compounds which decompose to free radicals under the action of light. They are widely used for the study of peroxyl radical fast reactions, for example, disproportionation (see Chapter 2).

3.4.3.1 Azo-compounds

The most popular is AIBN, which is used as an initiator at elevated temperatures (330–380 K). Photoinitiation by azo-compound has the following stages in the liquid [205]:

$$RN=NR + h\nu \longrightarrow RN=NR^*$$
$$RN=NR^* \longrightarrow [R^\bullet + R^\bullet] + N_2$$
$$[R^\bullet + R^\bullet] \longrightarrow RR$$
$$[R^\bullet + R^\bullet] \longrightarrow R^\bullet + R^\bullet$$
$$R^\bullet + O_2 \longrightarrow RO_2^\bullet$$
$$RO_2^\bullet + R_iH \longrightarrow ROOH + R_i^\bullet$$

Photodissociation is accompanied by the *cis–trans* isomerization of azoalkanes. Azoalkanes have the *trans*-configuration. During photodecomposition, they are transformed into the *cis*-configuration [66]. The excited molecule of *trans*-asopropane is transformed into *cis*-asopropane with $\Phi = 0.31$, into the *trans*-configuration with $\Phi = 0.51$, and into free radicals with $\Phi = 0.18$ (gas phase, 600 tor CO_2, room temperature). The following scheme of photophysical stages was proposed [205]:

TABLE 3.17
The Values of Triplet Energies of Selected Organic Compounds [209]

Compound	E_T (kJ mol^{-1})	Compound	E_T (kJ mol^{-1})
$CH_2=CH_2$	330	C_6H_6	353
$CH_2=CHCOOMe$	372	PhCl	342
$CH_2=CHCH=CH_2$	250	PhCOOH	324
$PhCH=CH_2$	258	$PhNH_2$	297
$PhMeC=CH_2$	260	$PhC\equiv N$	320
$Ph_2C=CH_2$	247	$PhNO_2$	243
$PhNH_2$	297	MeC(O)Me	332
PhCH(O)	301	PhC(O)Me	310
PhC(O)Ph	287	PhC(O)Et	312
cyclohexene	219	p-benzoquinone	224
indene	264	tetrachloro-p-benzoquinone	206
naphthalene	253	1,4-naphthoquinone	241
anthracene	178	anthrone	301

$$E\text{-}RN_2R + h\nu \longrightarrow E\text{-}RN_2R(^1n, \pi^*)$$

$$E\text{-}RN_2R(^1n, \pi^*) \longrightarrow RN_2R(^3\pi, \pi^*)$$

$$RN_2R(^3\pi, \pi^*) \longrightarrow E\text{-}RN_2R(^3n, \pi^*)$$

$$RN_2R(^3\pi, \pi^*) \longrightarrow Z\text{-}RN_2R(^3n, \pi^*)$$

$$RN_2R(^3\pi, \pi^*) \longrightarrow E\text{-}RN_2R + h\nu$$

$$RN_2R(^3\pi, \pi^*) \longrightarrow Z\text{-}RN_2R + h\nu$$

$$E\text{-}RN_2R(^3n, \pi^*) \longrightarrow E\text{-}RN_2R + h\nu$$

$$Z\text{-}RN_2R(^3n, \pi) \longrightarrow Z\text{-}RN_2R + h\nu$$

The quantum yield of the selected azoalkanes photodecomposition are given in Table 3.18. The extinction coefficient ε depends on the wavelength of light, and for AIBN in benzene solution (room temperature) has the values as given below [205]:

λ (nm)	400	360	347	330	310
ε (L mol^{-1} cm^{-1})	0.85	11.9	14.7	11.6	4.3

TABLE 3.18
Quantum Yields of Azoalkanes Photolysis in Solution at Room Temperature: $RN_2R + h\nu \rightarrow N_2$ + Products ([205])

Azoalkane	Solvent	Φ
E-Me$_2$CHN=NCHMe$_2$	Isooctane	0.021
E-Me$_2$(CN)CN=NC(CN)Me$_2$	Benzene	0.44
E-Me$_3$CN=NCMe$_3$	Benzene	0.46
E-MeOMe$_2$CN=NCMe$_2$OMe	Benzene	0.21
E-MeSMe$_2$CN=NCMe$_2$SMe	Benzene	0.38
E-HC≡CMe$_2$CN=NCMe$_2$C≡CH	Benzene	0.47
E-H$_2$C=CHMe$_2$CN=NCMe$_2$CH=CH$_2$	Benzene	0.57
E-AcOMe$_2$CN=NCMe$_2$OAc	Benzene	0.005
E-EtO(O)CMe$_2$CN=NCMe$_2$C(O)OEt	Benzene	0.42
E-PhCH$_2$N=NCH$_2$Ph	Benzene	0.02
E-PhCH$_2$N=NCH$_2$Ph	Isooctane	0.049
E-PhMeHCN=NCHMePh	Isooctane	0.035
E-PhMeHCN=NCHMePh	Benzene	0.043
E-PhMe$_2$CN=NCMe$_2$Ph	Benzene	0.36
E-Me$_3$CN=NCMe$_2$CH=CH$_2$	Benzene	0.49
E-Me$_3$CN=NCMe$_2$CH$_2$CMe$_3$	Benzene	0.42

3.4.3.2 Peroxides

The absorption of photon causes homolysis of peroxide with the generation of free radicals [205–208]:

$$ROOR^1 + h\nu \longrightarrow RO^\bullet + R^1O^\bullet$$

The formed radicals are excited: at $\lambda = 313$ nm and an excess of energy is equal to 320 kJ mol^{-1}, at $\lambda = 254$ nm, it is equal to 230 kJ mol^{-1}. This is the reason for the gas-phase photolysis leading to intensive decomposition of the formed alkoxyl radicals. The values of λ_{max} and extinction coefficients for several peroxides are presented [205]:

1,1-Dimethylethylperoxide: ε (L mol^{-1} cm^{-1}) = 1.6 ($\lambda = 301$ nm), 2.5($\lambda = 289$ nm), 4.0($\lambda = 275$ nm), 4.5($\lambda = 270$ nm), 5.4($\lambda = 262$ nm), 7.1($\lambda = 245$ nm).
1-Methyl-1-phenylethyl peroxide: ε (L mol^{-1} cm^{-1}) = 415($\lambda = 257$ nm)
1,1-Dimethylethyl hydroperoxide: ε (L mol^{-1} cm^{-1}) = 6($\lambda = 260$ nm), 17($\lambda = 240$ nm), 70($\lambda = 200$ nm).
1-Methyl-1-phenyl hydroperoxide: ε (L mol^{-1} cm^{-1}) = 200 ($\lambda = 258$ nm)
Acetyl peroxide: ε (L mol^{-1} cm^{-1}) = 20($\lambda = 265$ nm), 56($\lambda = 250$ nm), 89($\lambda = 240$ nm).
Benzoyl peroxide: $\varepsilon = 2.3 \times 10^3 (\lambda = 275$ nm), 2.7×10^4 ($\lambda = 230$ nm).

3.4.3.3 Carbonyl Compounds

Two parallel photolytic reactions of aldehydes decomposition are known [205]:

$$RCHO + h\nu \longrightarrow R^\bullet + HC^\bullet(O)$$

and

$$RCHO + h\nu \longrightarrow RH + CO$$

The ratio of the rate constants of these reactions depends on the energy of photon absorbed by the aldehyde molecule. The values of the quantum yield for acetaldehyde are the following [205]:

λ (nm)	313	280	265	254	238
$\Phi(C^\bullet H_3 + HC^\bullet O)$	0.20	0.39	0.36	0.38	0.31
$\Phi(CH_4 + CO)$	0.001	0.15	0.28	0.66	0.37

Four parallel photolytic reactions are known for ketones [205]:

$$RC(O)R^1 + h\nu \longrightarrow R^\bullet + R^1 C^\bullet(O)$$
$$RC(O)R^1 + h\nu \longrightarrow RC^\bullet(O) + R^{1\bullet}$$
$$Me_2CHCMe_2CMe_2C(O)R^1 + h\nu \longrightarrow Me_2C=CMe_2 + Me_2C=C(OH)R^1$$

$$Me_2CHCMe_2CMe_2C(O)R^1 + h\nu \longrightarrow \text{[cyclobutane with } R_1, OH\text{ substituents]}$$

The homolytic photochemical reaction occurs preferentially with the splitting of the weakest C—C bond. The ratio of the quantum yields depends on the energy of absorbed photon. For example, for butanone-2 photolysis, the ratio $\Phi(Et^\bullet)/\Phi(Me^\bullet)$ is equal [205]:

λ (nm)	313	265	254
$\Phi(Et^\bullet)/\Phi(Me^\bullet)$	40	5.5	2.4

The acyl radicals formed in ketone photolysis are excited and, therefore, rapidly splits into CO and alkyl radical (in the gas phase). Since aldehydes and ketones are products of oxidation, continuous hydrocarbon photooxidation is an autoaccelerated process.

3.4.4 NONCHAIN PHOTOOXIDATION

Polynuclear aromatic hydrocarbons can be oxidized photolytically with the formation of cyclic peroxide. For example, anthracene is photooxidized to peroxide with the quantum yield $\Phi = 1.0$ [205]. The introduction of quenchers lowers the peroxide yield.

Dienic hydrocarbons are photooxidized to cyclic peroxides. For example, terpinene is photooxidized to ascaridole peroxide [217]:

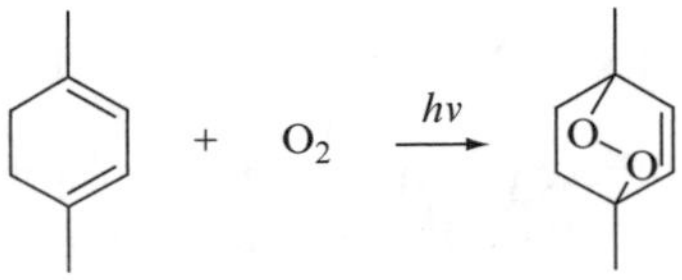

Such compounds with conugated π-bonds are excited by light to the triplet state, and then such a triplet molecule reacts with dioxygen with the formation of peroxide.

3.5 GENERATION OF RADICALS BY IONIZING RADIATION

3.5.1 Primary Radiation–Chemical Processes

The discovery of atomic energy and the progress in atomic physics and technology initiated the origin and developing of radiation chemistry [218–225]. The very active agents in the radiation chemistry are electrons (β-rays), γ-rays, x-rays, α-particles, neutrons, and protons. The formed particles and radiation are high-energy agents. Their interaction with a substrate (gas, liquid, solid) leads to the ionization of molecules and formation of electrons and unstable ions. The recombination of ions forms electronically excited molecules. These molecules decompose into free radicals. The processes induced by electrons (e^-) or γ-rays ($h\nu$) in deaerated water are the following [222–224]:

$$H_2O + e^- \longrightarrow H_2O^+ + e^- + e^-$$
$$H_2O + H_2O^+ \longrightarrow HO^\bullet + H_3O^+$$
$$e^- \longrightarrow e^-_{aq} \text{(solvated electron)}$$
$$e^-_{eq} + H_3O^+ \longrightarrow H_2O + H^\bullet$$
$$e^-_{aq} + H_2O \longrightarrow H^\bullet + HO^-$$
$$H^\bullet + H^\bullet \longrightarrow H_2$$
$$H^\bullet + HO^\bullet \longrightarrow H_2O$$
$$HO^\bullet + HO^\bullet \longrightarrow H_2O_2$$
$$H_2O^+ + e^- \longrightarrow H_2O^*$$
$$H_2O^* \longrightarrow H_2 + O$$

When dioxygen is dissolved in irradiated water, the following reactions occur in addition:

$$e^-_{eq} + O_2 \longrightarrow O_2^{-\bullet}$$
$$O_2^{-\bullet} + H_3O^+ \longrightarrow HO_2^\bullet + H_2O$$
$$HO_2^\bullet + H^\bullet \longrightarrow H_2O_2$$
$$HO_2^\bullet + HO_2^\bullet \longrightarrow H_2O_2 + O_2$$

Initiation of Liquid-Phase Oxidation

Hydrogen peroxide formed reacts with atoms and radicals:

$$H_2O_2 + H^\bullet \longrightarrow HO_2^\bullet + H_2$$
$$H_2O_2 + HO^\bullet \longrightarrow H_2O + HO_2^\bullet$$
$$H_2O_2 + H^\bullet \longrightarrow H_2O + HO^\bullet$$
$$H_2O_2 + HO_2^\bullet \longrightarrow H_2O + HO^\bullet + O_2$$

Due to these reactions, hydrogen peroxide is an intermediate product of radiolysis of aerated water. Rate constants of free radical reactions with dioxygen and hydrogen peroxide are collected in Table 3.19. For the characteristics of solvated electron and information about its reactions, see monographs [219–223].

3.5.2 RADIOLYTIC INITIATION OF HYDROCARBON OXIDATION

The hydrocarbon molecule RH is ionized under the action of fast electron (or γ-photon). The formed electrons are retarded due to collisions with molecules and become solvated electrons e_s^-. Excited molecules of the hydrocarbon are produced due to recombination of solvated electrons e_s^- and positive ions RH^+ [223,224].

$$e^- + RH \longrightarrow e^- + e^- + RH^+$$
$$e^- \longrightarrow e_s^-$$
$$RH^+ + e_s^- \longrightarrow RH^*$$
$$RH^* \longrightarrow R_1^\bullet + R_2^\bullet$$
$$RH^+ + e_s^- \longrightarrow R^\bullet + H^\bullet$$

TABLE 3.19
Rate Constants of Solvated Electron, H$^\bullet$, and HO$^\bullet$ Reactions with Dioxygen and Hydrogen Peroxide in Water at Room Temperature [223–225]

Radical	Reagent	k (L mol^{-1} s^{-1})
e^-_{aq}	H_2O	16
e^-_{aq}	D_2O	1.2
e^-_{aq}	H_3O^+	2.4×10^{10}
e^-_{aq}	H_2O_2	1.3×10^{10}
e^-_{aq}	HO_2^-	3.5×10^{10}
e^-_{aq}	O_2	1.9×10^{10}
e^-_{aq}	e^-_{aq}	1.2×10^{10}
e^-_{aq}	$H^\bullet$	2.5×10^{10}
e^-_{aq}	$HO^\bullet$	3.0×10^{10}
e^-_{aq}	$O_2^{-\bullet}$	1.3×10^{10}
$H^\bullet$	H_2O_2	5.8×10^7
$H^\bullet$	O_2	2.0×10^{10}
$H^\bullet$	HO^-	1.8×10^7
$H^\bullet$	$H^\bullet$	2.0×10^{10}
$H^\bullet$	$HO^\bullet$	2.2×10^{10}
$H^\bullet$	$HO_2^\bullet$	2.0×10^{10}
$HO^\bullet$	H_2O_2	4.5×10^7
$HO^\bullet$	HO_2^-	8.3×10^{10}
$HO^\bullet$	$HO^\bullet$	1.1×10^{10}
$HO^\bullet$	$HO_2^\bullet$	1.0×10^{10}
$HO^\bullet$	$O_2^{-\bullet}$	9.0×10^9

The radiation yield of ionized molecules of cyclohaxane was found to be equal to five molecules per 100 eV [223].

The formed solvated electrons were studied by the method of pulse radiolysis [225]. The following table presents the characteristics of e_s^- in different hydrocarbons:

Hydrocarbon	λ_{max}(nm)	ε_{max} (L mol^{-1}cm^{-1})	Ref.
Hexane	≥1600	$\varepsilon_{1000} = 6.7 \times 10^3$	225
Methylcycloxane-d$_{14}$	3400	1.9×10^4	226
Octane, 3-ethyl–	2100	2.1×10^4	227

Solvated electron reacts with dioxygen with a high rate:

$$e_s^- + O_2 \longrightarrow O_2^{-\bullet}$$
$$O_2^{-\bullet} + RH^+ \longrightarrow HO_2^\bullet + R^\bullet$$
$$R^\bullet + O_2 \longrightarrow RO_2^\bullet$$

Hydrocarbon	Hexane	Cyclohexane	Isooctane
$k(e_s^- + O_2)$ (L mol^{-1} s^{-1})	1.5×10^{11}	1.7×10^{11}	1.4×10^{11}
Ref.	[228]	[229]	[228]

Positive charged ions RH^+ react with other R_iH molecules in a hydrocarbon solution. This charge transfer is very fast:

$$RH^+ + R_iH \longrightarrow RH + R_iH^+$$

The rate constants of charge transfer from RH^+ (RH is heptane) to R_iH at room temperature have the following values [227]:

R_iH	cyclo-C_6H_{10}	$CH_2=CHBu$	C_6H_6	$PhCH_3$	$Me_2C=CMe_2$
k (L mol^{-1}s^{-1})	3×10^9	4×10^{10}	6×10^{10}	6×10^{10}	7.5×10^{10}

The excited molecules are formed in the singlet (S) and triplet (T) states. The following table presents the values of *radiation yield G* (molecule/100 eV) of these states for some hydrocarbons [223]:

RH	Toluene	o-Xylene	m-Xylene	p-Xylene	Mesitylene
$G(S)$	2.1	2.5	2.7	1.6	1.6
$G(T)$	2.4	1.7	1.8	1.8	1.8

Free radicals are formed in the hydrocarbons are the result of decay of excited molecules (see earlier). The value of $G(R^\bullet)$ from cyclohexane (RH) is 5.7 [222]. Various alkyl radicals are

Initiation of Liquid-Phase Oxidation

produced from linear alkane. The $G(R_i^\bullet)$ values for different radicals formed from heptane is given in the following table [223]:

$R_i^\bullet$	Me$^\bullet$	Et$^\bullet$	Pr	Bu$^\bullet$	$C_6H_{13}^\bullet$
$G(R_i^\bullet)$	0.7	0.3	0.3	0.27	4.1

Oxygen-containing molecules are formed in the irradiated hydrocarbon due to the reactions of peroxyl radicals [228–232]:

$$R^\bullet + O_2 \longrightarrow RO_2^\bullet$$
$$RO_2^\bullet + RH \longrightarrow ROOH + R^\bullet$$
$$RO_2^\bullet + R^\bullet \longrightarrow ROOR$$
$$RO_2^\bullet + RO_2^\bullet \longrightarrow ROH + O_2 + \text{Ketone}$$
$$H^\bullet + O_2 \longrightarrow HO_2^\bullet$$
$$HO_2^\bullet + HO_2^\bullet \longrightarrow H_2O_2 + O_2$$

Due to the high initiation rate and low (room) temperature, chains for oxidation of alkanes are short and many products are formed by disproportionation of peroxyl and hydroperoxyl radicals. The G values of the products of radiolytic oxidation of four alkanes are given in the following table [233]:

Hydrocarbon	ROOH	ROOR	H_2O_2	ROH	$R^1C(O)R^2$
Hexane	1.22	1.0	0.44	1.21	1.55
Heptane	1.2	2.2	0.3		2.0
Isooctane	0.7	1.3	0.3		1.2
Cyclohexane	1.0	0.2	0		0.6

The radiation yield depends on the temperature of oxidation and the initiation rate, i.e., the intensity of radiation I_r [233]. Radoxidation occurs as an initiated chain reaction at an elevated temperature when peroxyl radicals react more rapidly with hydrocarbon RH than disproportionate, $k_p^2(2k_t)^{-1}[RH]^2 > v_I$ (see Chapter 2)]. Radoxidation proceeds as a nonchain reaction at low temperatures when peroxyl radicals disproportionate more rapidly than react with hydrocarbon. The temperature boundary T_v between these two regimes of oxidation depends on the value of radiation intensity I_r. The values of T_v for irradiated heptane oxidation is as follows [233]:

I_r (eV L^{-1} s^{-1})	7.5×10^{18}	4.1×10^{17}	1.2×10^{17}	7.8×10^{16}
T_v (K)	276	252	228	218

Pulse radiolysis is widely used for the study of fast free radical reactions and reactions of solvated electrons [223].

REFERENCES

1. DR Lide, ed. *Handbook of Chemistry and Physics*, 72nd ed. Boca Raton, FL: CRC Press, 1991–1992.
2. ET Denisov. *Liquid-Phase Reaction Rate Constants*. New York: IFI/Plenum, 1974.
3. T Koenig. In: JK Kochi ed. *Free Radicals*, vol 1. New York: Wiley, 1973, pp 113–155.
4. ET Denisov, TG Denisova, TS Pokidova. *Handbook of Radical Initiators*. New York: Wiley, 2003.
5. KJ Laidler. *Chemical Kinetics*. New York: Harper and Row, 1987.
6. SW Benson. *Thermochemical Kinetics*, 2nd ed. New York: Wiley, 1972.
7. MG Gonikberg. *Chemical Equilibrium and Reaction Rate under High Pressures*. Moscow: Khimiya, 1969 [in Russian].
8. PD Bartlett, RR Hiatt. *J Am Chem Soc* 80:1398–1405, 1958.
9. WH Richardson, WC Koskinen. *J Org Chem* 41:3182–3187, 1976.
10. IV Aleksandrov. *Theor Experim Khimiya* 12:299–306, 1976.
11. VA Yablokov. Isomerisation of Organic and Elementoorganic Peroxides of IV Group Elements. Dissertation, LTI, Leningrad, 1976.
12. VP Maslennikov. Organic Peroxides of Elements of II–V Groups of Periodic System. Dissertation, MGU, Moscow, 1976.
13. KU Ingold, BP Roberts. *Free-Radical Substitution Reaction*. New York: Wiley-Interscience, 1971.
14. ML Poutsma. In: JK Kochi ed. *Free Radicals*, vol 2. New York: Wiley, 1973, pp 113–158.
15. ET Denisov. *Mechanisms of Homolytic Decomposition of Molecules in the Liquid Phase*, Itogi Nauki i Tekhniki, Kinetika i Kataliz, vol 9. Moscow: VINITI, 1981, pp 1–158 [in Russian].
16. K Nozaki, PD Bartlett. *J Am Chem Soc* 68:1686–1692, 1946.
17. K Nozaki, PD Bartlett. *J Am Chem Soc* 68:2377–2380, 1946.
18. PD Bartlett, K Nozaki. *J Am Chem Soc* 69:2299–2306, 1947.
19. PD Bartlett, WE Vaughan. *J Phys Chem* 51:942–948, 1947.
20. AR Blake, KO Kutschke. *Can J Chem* 37:1462–1470, 1959.
21. MFR Mulcahy, DJ Williams. *Aust J Chem* 14:534–544, 1961.
22. L Batt, SW Benson. *J Chem Phys* 36:895–901, 1962.
23. DH Shaw, HO Pritchard. *Can J Chem* 46:2721–2724, 1968.
24. JH Ralley, FF Rust, WE Vaughan. *J Am Chem Soc* 70:1336–1338, 1948.
25. MJ Perona, DM Golden. *Int J Chem Kinet* 5:55–65, 1973.
26. ES Huyser, RM Van Scoy. *J Org Chem* 33:3524–3525, 1968.
27. JRL Smith, E Nagatomi, A Stead, DJ Waddington, SD Beviere. *J Chem Soc Perkin Trans 2*, 1193–1198, 2000.
28. SM Kavun, AL Buchachenko. *Izv AN SSSR Ser Khim* 1483–1485, 1966.
29. ET Denisov, VM Solyanikov. *Neftekhimiya* 3:360–366, 1963.
30. JA Offenbach, AV Tobolsky. *J Am Chem Soc* 79:278–281, 1957.
31. VV Zaitseva, AI Yurzhenko, VD Enaliev. *Zh Org Khim* 4:1402–1406, 1968.
32. OM Mashnenko, AE Batog, NJ Mironenko, MK Romantsevich. *Vysokomol Soedin* B10:444–446, 1968.
33. HC Bailey, GW Godin. *Trans Faraday Soc* 52:68–73, 1956.
34. MS Kharasch, A Fono, W Nudenberg. *J Org Chem* 16:105–112, 1951.
35. PA Ivanchenko, VV Kharitonov, ET Denisov. *Vysokomol Soedin* A11:1622–1630, 1969.
36. EM Dannenberg, ME Jordan, HM Cole. *J Polym Sci* 31:127–134, 1958.
37. CR Parks, O Lorenz. *J Polym Sci* 50:287–298, 1961.
38. DK Thomas. *J Appl Polym Sci* 6:613–618, 1963.
39. BME Van der Hoff. *Ind Eng Chem Prod Res Dev* 2:273–282, 1963.
40. WA Pryor, DM Huston, TR Fiske, TL Rickering, E Ciuffarin. *J Am Chem Soc* 86:4237–4243, 1964.
41. WA Pryor, GL Kaplan. *J Am Chem Soc* 86:4234–4236, 1964.
42. AG Davis, *Organic Peroxides*. London: Butterworths, 1961.
43. D Swern, ed. *Organic Peroxides*. New York: Wiley, 1970.
44. VL Antonovskii. *Organic Peroxides as Initiators*. Moscow: Khimiya, 1972 [in Russian].

45. EGE Hawkins. *Organic Peroxides*. London: Spon, 1961.
46. S Patai, ed. *The Chemistry Peroxides*. New York: Wiley, 1983.
47. A Rembaum, M Szwarc. *J Am Chem Soc* 76:5975–5978, 1954.
48. M Levy, M Steinberg, M Szwarc. *J Am Chem Soc* 76:5978–5981, 1954.
49. FG Edwards, FR Mayo. *J Am Chem Soc* 72:1265–1269, 1950.
50. MJ Shine, JA Waters, DM Hoffman. *J Am Chem Soc* 85:3613–3621, 1963.
51. SD Ross, MA Fineman. *J Am Chem Soc* 73:2176–2181, 1951.
52. OJ Walker, GLE Wold. *J Chem Soc* 1132–1139, 1937.
53. W Cooper. *J Chem Soc* 3106–3113, 1951.
54. PF Hartman, HG Sellers, D Turnbull. *J Am Chem Soc* 69:2416–2417, 1947.
55. KhS Bagdasaryan, RI Milutinskaya. *Zh Fiz Khim* 27:420–432, 1953.
56. HC Bailey, GW Godin. *Trans Faraday Soc* 52:68–73, 1956.
57. CG Swain, WH Stockmayer, JT Clarke. *J Am Chem Soc* 72:5426–5434, 1950.
58. VI Galibey. Study of Acyl Peroxides as Initiators of Radical Polymerisation. Ph.D. thesis Dissertation, OGU, Odessa, 1965 [in Russian].
59. AT Blomquist, AJ Buselli. *J Am Chem Soc* 73:3883–3888, 1951.
60. RV Pankevich, VS Dutka. *Kinet Katal* 33:1087–1092, 1992.
61. ET Denisov. *Izv AN SSSR Ser Khim* 812–815, 1960.
62. CEH Bawn, SF Mellish. *Trans Faraday Soc* 47:1216–1227, 1951.
63. JC McGowan, T Powell. *J Chem Soc* 238–241, 1960.
64. IS Voloshanovskii, YuN Anisimov, SS Ivanchev. *Zh Obshch Khim* 43:354–359, 1973.
65. YuA Oldekop, GS Bylina. *Vysokomol Soedin* A6:1617–1623, 1964.
66. PS Engel. *Chem Rev* 80:99–150, 1980.
67. K Chakravorty, JM Pearson, M Szwarc. *J Phys Chem* 73:746–748, 1969.
68. E W-G Diau, OK Abou-Zied, AA Scala, AH Zewail. *J Am Chem Soc* 120:3245–3246, 1998.
69. NWC Hon, Z-D Chen, Z-F Liu. *J Am Chem Soc* 124:6792–6801, 2002.
70. ET Denisov, OM Sarkisov, GI Likhtenstein. *Chemical Kinetics*. Amsterdam: Elsevier, 2003.
71. AM North. *The Collision Theory of Chemical Reactions in Liquids*. London: Methuen, 1964.
72. J Frank, E Rabinowitch. *Trans Faraday Soc* 30:120, 1934.
73. RM Noyes. *Z Electrochem* 69:153, 1960.
74. RL Strong. *J Am Chem Soc* 87:3563–3567, 1965.
75. JP Fisher, G Mucke, GV Schulz. *Ber Bunsenges Phys Chem* 73:154–163, 1969.
76. L Herk, M Feld, M Szwarc. *J Am Chem Soc* 83:2998–3005, 1961.
77. PD Bartlett, GN Fickes, FC Haupt, R Helgeson. *Acc Chem Res* 3:177–185, 1970.
78. JC Martin, JV Taylor, EH Dew. *J Am Chem Soc* 89:129–135, 1967.
79. JC Martin, SA Dombchik. *Adv Chem Ser* 75(1):269-281, 1968.
80. WA Pryor, K Smith. *J Am Chem Soc* 92:5403–5412, 1970.
81. LF Meadows, RM Noyes. *J Am Chem Soc* 82:1872–1876, 1960.
82. T Koenig, J Huntington, R Cruthoff. *J Am Chem Soc* 92:5413–5418, 1970.
83. RK Lyon, DH Levy. *J Am Chem Soc* 83:4290–4293, 1961.
84. RE Rebert, P Ausloos. *J Phys Chem* 66:2253–2259, 1962.
85. JC Martin, JW Taylor. *J Am Chem Soc* 89:6904–6911, 1969.
86. T Koenig, R Cruthoff. *J Am Chem Soc* 91:2562–2569, 1969.
87. D Gegiou, K Muszkat, E Fischer. *J Am Chem Soc* 90:12–18, 1968.
88. KR Kopecky, T Gillan. *Can J Chem* 47:2371–2386, 1969.
89. FD Green, MA Berwick, TC Stowell. *J Am Chem Soc* 92:867–875, 1970.
90. HM Walborsky, J Chen. *J Am Chem Soc* 93:671–680, 1971.
91. O Dobis, I Nemes, M Szwarc. Lecture on International Conference on Mechanism of Reactions in Solution. England: Kent, 1970.
92. FE Herkes, J Friedman, PD Bartlett. *Int J Chem Kinet* 1:193–207, 1969.
93. H Kiefer, TG Traylor. *J Am Chem Soc* 89:6667–6671, 1967.
94. CS Wu, GS Hammond, JM Wright. *J Am Chem Soc* 82:5386–5394, 1960.
95. RC Neuman, JV Behar. *J Am Chem Soc* 91:6024–6031, 1969.
96. RC Neuman, ES Alhadeff. *J Org Chem* 35:3401–3405, 1970.

97. RC Neuman, RJ Bussey. *J Am Chem Soc* 92:2440–2445, 1970.
98. RC Neuman, RC Pankratz. *J Am Chem Soc* 95:8372–8374, 1973.
99. RC Neuman, JV Behar. *J Org Chem* 36:657–661, 1971.
100. MYa Botnikov, VM Zhulin, LG Bubnova, GA Stashina. *Izv AN SSSR Ser Khim* 229–231, 1977.
101. AL Buchachenko. *Chemical Polarisation of Electrons and Nucleus*. Moscow: Nauka, 1974 [in Russian].
102. HR Ward. In: JK Kochi ed. *Free Radicals*, vol 1. New York: Wiley, 1973, pp 239–273.
103. RM Noyes. *Prog React Kinet* 1:129, 1961.
104. T Koenig, M Deinzer. *J Am Chem Soc* 90:7014–7019, 1968.
105. T Koenig. *J Am Chem Soc* 91:2558–2562, 1969.
106. KhS Bagdasaryan. *Zh Fiz Khim* 41:1679–1682, 1967.
107. NM Emanuel. *Zh Fiz Khim* 30:847–855, 1956.
108. NM Emanuel. *Izv Akad Nauk SSSR Otd Khim Nauk* 1298–1302, 1957.
109. NM Emanuel, ET Denisov, ZK Maizus. *Liquid-Phase Oxidation of Hydrocarbons*. New York: Plenum Press, 1967.
110. EA Blyumberg, VG Voronkov, NM Emanuel. *Izv AN SSSR Otd Khim Nauk* 25–31, 1959.
111. RF Vasil'ev, ON Karpukhin, VYa Shlyapintokh, NM Emanuel. *Dokl AN SSSR* 124:1258–1260, 1959.
112. VD Komissarov. Mechanism of Oxidation of Saturated Compounds by Ozone, Doctoral Thesis Dissertation. Institute of Khimii, Ufa, 1989 [in Russian].
113. SD Razumovsky, GE Zaikov. *Ozone Reactions with Organic Compounds*. Moscow: Nauka, 1974 [in Russian].
114. RS Bailey. *Ozonation in Organic Chemistry*, vol 1. New York: Academic Press, 1978.
115. MC Whiting, AJ Bolt, JH Parish. *Adv Chem Ser* 77:4–18, 1968.
116. GA Hamilton, BS Ribner, TM Hellman. *Adv Chem Ser* 77:15–22, 1968.
117. VYa Shtern. *Mechanism of Hydrocarbons Oxidation in Gas Phase*. Moscow: Izdatelstvo AN SSSR, 1960 [in Russian].
118. VV Shereshovets, EE Zaev, VD Komissarov. *Izv AN SSSR Ser Khim* 983, 1975.
119. SK Rakovskii, SD Razumovskii, GE Zaikov. *Izv AN SSSR Ser Khim* 701–703, 1976.
120. G Wagner. *J Pract Chem* 27:297–303, 1965.
121. RF Vasil'ev, S. M. Petukhov, P. I. Zhuchkova. Optiko-mekhanich prom N 37, 2, 1963.
122. AV Ruban, SK Rakovskii, AA Popov. *Izv AN SSSR, Ser Khim* 1950–1956, 1976.
123. D Barnard, GP McSweeney, JF Smith. *Tetrahedron Lett* N 1:1, 1960.
124. HM White, PS Bailey. *J Org Chem* 30:3037–3044, 1965.
125. TM Hellman, GA Hamilton. *J Am Chem Soc* 96:1530–1535, 1974.
126. LG Galimova. Mechanism of Cyclohexane Oxidation by Ozone, Thesis Dissertation, BSU, Ufa, 1975 pp 3–25 [in Russian].
127. AA Popov, SK Rakovskii, DM Shopov, AV Ruban. *Izv AN SSSR Ser Khim* 982–490, 1976.
128. RE Erickson, RT Hansen, J Harkins. *J Am Chem Soc* 90:6777–6783, 1968.
129. RE Erickson, D Bakalik, C Richards., M Scanlon, G Huddleston. *J Org Chem* 31:461–468, 1966.
130. VV Shereshovets, NN Shafikov, VD Komissarov. *Kinet Katal* 21:1596–1598, 1980.
131. NA Kleimenov, IN Antonova, AM Markevich, AB Nalbandyan. *Zh Fiz Khim* 30:794–797, 1956.
132. NA Kleimenov, AB Nalbandyan. *Dokl AN SSSR* 122:103–105, 1958.
133. ET Denisov, TG Denisova. *Kinet Catal* 37:46–50, 1996.
134. SL Khursan. Organic Polyoxides. Thesis Dissertation, Institute of Organic Chemistry, Ufa, 1999.
135. VV Shereshovets. Mechanism of Oxidation of Cumene by Mixture of Ozone–Oxygen. Ph.D. thesis Dissertation, Institute of Chemical Physics, Chernogolovka, 1978, pp 1–21 [in Russian].
136. VV Shereshovets, VV Komissarov, ET Denisov. *Izv AN SSSR Ser Khim* 2482–2487, 1978.
137. IN Komissarova. Mechanism of Oxidation of Aldehydes by Mixture of Ozone–Oxygen. PhD Thesis Dissertation, Institute of Chemical Physics, Chernogolovka, 1978, pp 3–25 [in Russian].
138. DG Williamson, RJ Cvetanovic. *J Am Chem Soc* 92:2949–2952, 1970.
139. SD Razumovskii, SK Rakovskii, GE Zaikov. *Izv AN SSSR Ser Khim* 1963–1967, 1975.
140. AA Kefeli. Reactions of Ozone with Saturated Polymers and Model Compounds. Thesis Dissertation, Institute of Chemical Physics, Moscow, 1973 [in Russian].

141. AYa Gerchikov, EP Kuznetsova, ET Denisov. *Kinet Katal* 15:509–511, 1974.
142. AYa Gerchikov, VD Komissarov, ET Denisov, GB Kochemasova. *Kinet Katal* 13:1126–1130, 1972.
143. AYa Gerchikov, EM Kuramshin, VD Komissarov, ET Denisov. *Kinet Katal* 15:230–232, 1974.
144. BM Brudnik, SS Zlotskii, UB Imashev, DL Rakhmankulov. *Dokl AN SSSR* 241:129–130, 1978.
145. R Atkinson, DL Baulch, RA Cox, RF Hampson, JA Kerr, MJ Rossi, J Troe. *J Phys Chem Ref Data* 26:1329, 1997.
146. VD Komissarov, AYa Gerchikov, LG Galimova, ET Denisov. *Dokl AN SSSR* 213:881–883, 1973.
147. LG Galimova, VD Komissarov, ET Denisov. *Izv AN SSSR Ser Khim* 307–311, 1973.
148. VD Komissarov, LG Galimova, IN Komissarova, VV Shereshovets, ET Denisov. *Dokl AN SSSR* 235:1350–1352, 1977.
149. VV Shereshovets, VD Komissarov, ET Denisov. *Izv AN SSSR Ser Khim* 2482–2487, 1978.
150. VD Komissarov, IN Komissarova, GK Farrakhova, ET Denisov. *Izv AN SSSR Ser Khim* 1205–1212, 1979.
151. VV Shereshovets, VD Komissarov, ET Denisov. *Izv AN SSSR Ser Khim* 1212–1219, 1979.
152. SG Cheskis, AA Iogansen, IYu Razuvaev, OM Sarkisov, AA Titov. *Chem Phys Lett* 155:37–41, 1989.
153. R Simonaitis, J Heicklen. *J Phys Chem* 79:298–302, 1975.
154. AA Turniseed, SB Barone, RA Ravinshankara. *J Phys Chem* 97:5926–5934, 1993.
155. R Paltenghi, EA Ogryzlo, KD Bayes. *J Phys Chem* 88:2595–2599, 1984.
156. R Simonaitis, J Heicklen. *J Phys Chem* 77:1932–1935, 1973.
157. WB De More. *Science* 180:735–740, 1973.
158. ET Denisov, NM Emanuel. *Zh Fiz Khim* 31:1266–1275, 1957.
159. EA Blyumberg, NM Emanuel. *Izv AN SSSR Otd Khim Nauk* 274–280, 1957.
160. E Dimitroff, JV Moffitt, RD Quillian. *Trans ASME* 91:406–410, 1969.
161. JB Hanson, SW Harris, CT West. SAE Technical Paper 1988:881581.
162. RR Kuhn. *ACS Preprints Div Petrol Chem* 18:697–698, 1973.
163. MD Johnson, S Korcek. *Lubrication Sci* N 3:95–101, 1991.
164. MD Johnson, S Korcek, MJ Rokosz. *Lubrication Sci* N 6:247–251, 1994.
165. IV Berezin, ET Denisov, NM Emanuel. *The Oxidation of Cyclohexane*. Oxford: Pergamon Press, 1966.
166. RE Huie. *Toxicology* 89:193–216, 1994.
167. SW Benson, HE O'Neal. *Kinetic Data on Gas Phase Unimolecular Reactions*. Washington: NSRDS, 1970.
168. TG Denisova, ET Denisov. *Petrol Chem* 37:187–194, 1997.
169. MW Slack, AR Grillo. *Proc Int Symp Shock Tubes Waves* 11:408, 1978.
170. TA Titarchuk, AP Ballod, VYa Shtern. *Kinet Katal* 17:1070–1070, 1976.
171. WA Glasson, CS Tuesday. *Environ Sci Technol* 4:752–761, 1970.
172. C Anastasi, DU Hancock. *J Chem Soc Faraday Trans 2*, 84:1697–1706, 1988.
173. S Koda, M Tanaka. *Combust Sci Technol* 47:165–176, 1986.
174. S Jaffe, E Wan. *Environ Sci Technol* 8:1024–1031, 1974.
175. PC Jackson, R Silverwood, JH Thomas. *Trans Faraday Soc* 67:3250–3258, 1971.
176. S Jaffe. *Chem React Urban Atmos Proc Symp* 1969, 1971.
177. JG Gleim, J Heicklen. *Int J Chem Kinet* 14:699–710, 1982.
178. EC Tuazon, WPL Carter, RV Brown, AM Winer, JN Pitts. *J Phys Chem* 87:1600–1605, 1983.
179. JL Sprung, H Akimoto, JN Pitts. *J Am Chem Soc* 96:6549–6554, 1974.
180. V Sotiropoulou, N Katsanos, H Metuxa, F Roubani-Kalantzopoulou. *Chromatographia* 42:441, 1996.
181. VA Gryaznov, AI Rozowskii. *Dokl AN SSSR* 230:1129–1132, 1976.
182. T Ohta, H Nagura, S Suzuki. *Int J Chem Kinet* 18:1–6, 1986.
183. SE Paulson, RC Flagan, JH Seinfeld. *Int J Chem Kinet* 24:79–88, 1992.
184. R Atkinson, SM Aschmonn, AM Winer, JN Pitts. *Int J Chem Kinet* 16:697–706, 1984.
185. H Niki, PD Maker, CM Savage, LP Breitenbach, MD Hurley. *Int J Chem Kinet* 18:1235–1242, 1986.

186. J Cazarnowski. *Int J Chem Kinet* 24:679–688, 1992.
187. IV Berezin, G Vagner, NM Emanuel. *Zh Prikl Khim* 32:173–179, 1959.
188. WT Miller, A L Dittman. *J Am Chem Soc* 78:2793–2797, 1956.
189. WT Miller, SD Koch, FW McLafferty. *J Am Chem Soc* 78:4992–4995, 1956.
190. C Seeger, G Rotzoll, A Lubbert, K Schugerl. *Int J Chem Kinet* 13:39–58, 1981.
191. ET Denisov, *Chem. Phys. Rep.* 17:705–711, 1998.
192. GA Kapralova, AM Chaikin, AE Shilov. *Kinet Katal* 8:485–492, 1967.
193. ZhKh Gyulbekyan, OM Sarkisov, VI Vedeneev. *Kinet Katal* 15:1115–1118, 1974.
194. VL Orkin, AM Chaikin. *Kinet Katal* 23:529–535, 1982.
195. ML Poutsma, In: JK Kochi ed. *Free Radicals*, vol 2. New York: Wiley, 1973, pp 159–229.
196. GF Zhdanow. *Kinet Katal* 34:973–979, 1993.
197. NM Emanuel. *Dokl AN SSSR* 102:559–562, 1955.
198. ET Denisov. *Izv AN SSSR Ser Khim* 1608–1610, 1967.
199. CDJ von Grotthuss. *Ann Phys* 61:50, 1819.
200. JW Draper. *Philos Mag* 19:195, 1841.
201. J Stark. *Phys Z* 9:889, 894, 1908.
202. A Einstein. *Ann Phys* 37:832, 1912.
203. J Frank. *Trans Faraday Soc* 21:536, 1925.
204. EV Condon. *Phys Rev* 32:858, 1928.
205. JC Calvert, JN Pitts. *Photochemistry*. New York: Wiley, 1966.
206. RB Cundal, A Gilbert. *Photochemistry*. New York: Appleton Century Crofts, 1970.
207. NJ Turro. *Modern Molecular Photochemistry*. Menlo Park: Benjamin/Cummings, 1978.
208. HJ Wells. *Introduction in Molecular Photochemistry*. London: Chapman & Hall, 1972.
209. SL Marlov, I Carmochael, GL Hug. *Handbook of Photochemistry*. New York: Dekker, 1993.
210. O Stern, M Volmer. *Phys Z* 20:183, 1919.
211. R Livingston. In: WO Landberg ed. *Autoxidation and Antioxidants*, vol 1. New York: Interscience, 1961, p 249.
212. AB Nalbandyan. *Zh Fiz Khim* 22:1443–1449, 1948.
213. AB Nalbandyan. *Dokl AN SSSR* 66:413–416, 1949.
214. NV Fok, AB Nalbandyan. *Dokl AN SSSR* 85:1093–1095, 1952.
215. JA Gray. *J Chem Soc* 1952:3150–3158.
216. P Bender. *Phys Rev* 36:1535, 1930.
217. W Bergman, MJ McLean. *Chem Rev* 28:397, 1941.
218. G Hughes. *Radiation Chemistry*. Oxford: Clarendon, 1973.
219. JWT Spinks, RJ Wood. *An Introduction to Radiation Chemistry*. New York: Wiley, 1976.
220. AJ Swallow. *Radiation Chemistry: An Introduction*. London: Longman, 1973.
221. G Draganic, ZD Draganic. *The Radiation Chemistry of Water*. New York: Academic, 1971.
222. MS Matheson, LM Dorfman. *Pulse Radiolysis*. Cambridge MA: MIT Press, 1966.
223. AK Pikaev. *The Modern Radiation Chemistry*. Moscow: Nauka, 1986 [in Russian].
224. G Cserep, I Gyorgy, M Roder, L Wojnarovits. *Radiation Chemistry of Hydrocarbons*. Budapest: Akad. Kiado, 1981.
225. AK Pikaev, SA Kabakchi, IE Makarov, BG Ershov. *Pulse Radiolysis and its Application*. Moscow: Atomizdat, 1980, pp 3–290 [in Russian].
226. MS Ahmad, SJ Atherton, JH Baxendale. In: *Proc Congr Radiat Res*. Tokyo: JARR, 1979, p 220.
227. NV Klassen, GG Teather. *J Phys Chem* 89:2048–2051, 1985.
228. RA Holroyd, TE Gangwer. *Rad Phys Chem* 15:283–290, 1980.
229. JH Baxendale, JP Keene, EJ Rusburn. *J Chem Soc Faraday Trans 1*. 70:718–728, 1974.
230. MF Romantsev, VA Lavin. *Oxidation of Organic Compounds Induced by Radiation*. Moscow: Atomizdat, 1972 [in Russian].
231. JF Mead. In: WO Landberg ed. *Autoxidation and Antioxidants*, vol 1. New York: Interscience, 1961, p 299.
232. JT Kunjappa, KN Rao. *Rad Phys Chem* 13:97–105, 1979.
233. VV Saraeva. Radiation Induced Oxidation of Organic Compounds. Doctoral Thesis Dissertation, Institute of Electrochemistry, Moscow, 1970, pp 1–36 [in Russian].

4 Oxidation as an Autoinitiated Chain Reaction

4.1 CHAIN GENERATION BY REACTION OF HYDROCARBON WITH DIOXYGEN

Liquid-phase oxidation of organic compounds is performed in laboratories and technological installations at 300–500 K. Under these conditions, organic compounds are quite stable, and their decomposition with dissociation at the C—C bond and in the reaction of retrodisproportionation

$$RH + CH_2{=}CHR^1 \longrightarrow R^\bullet + CH_3C^\bullet HR^1$$

does not virtually occur. Chain generation in the absence of initiating additives and those formed by hydrocarbon oxidation occur preferentially via the reactions involving dioxygen. The mechanism of chain initiation in an oxidized RH in the absence of ROOH was intensively discussed in the 1950s and 1960s [1–4].

4.1.1 BIMOLECULAR REACTION OF DIOXYGEN WITH THE C—H BOND OF THE HYDROCARBON

Bolland and Gee [5] proposed the following reaction of free radical initiation in oxidized hydrocarbon as the most probable:

$$RH + O_2 \longrightarrow R^\bullet + HO_2^\bullet$$

The first experimental study of this reaction was performed by Cooper and Melville [6]. The main experimental evidences of this mechanism are the following [7–9]:

1. The rate of free radical generation in the oxidized RH in the absence of ROOH or other initiators was found to be proportional to the product of the concentrations of the hydrocarbon and dioxygen [10–13]:

$$v = k_i[RH][O_2]. \tag{4.1}$$

2. The rate constant k_i of this reaction increases with a decrease in the BDE of the weakest C—H bond in the oxidized hydrocarbon. This is in agreement with the high endothermicity of these reactions.

RH	cyclo-C_6H_{12}	Me_2CHCH_2Me	$PhCH_3$	$PhMe_2CH$	cyclo-C_6H_{10}
D (kJ mol^{-1})	408.8	400.0	375.0	354.7	341.5
ΔH (kJ mol^{-1})	188.8	180.0	155.0	134.7	121.5
k_i ($T = 403$ K) (L mol^{-1} s^{-1})	1.5×10^{-10}	3.7×10^{-9}	1.6×10^{-8}	2.0×10^{-7}	1.3×10^{-7}
Ref.	[10]	[11]	[12]	[13]	[12]

3. Compounds with C—D bonds react with dioxygen more slowly than those with C—H bonds. For example, the *kinetic isotope effect* $k_H/k_D = 5.2$ (diethyl sebacinate, 413 K) [14].
4. The reaction of RH with dioxygen occurs more rapidly in polar solvents [15] due to the polar structure of the transition state $R(\delta+) \cdots H \cdots O_2(\delta-)$.
5. The *effect of multidipole interaction* was observed in the reactions of dioxygen with C—H bonds of polyatomic esters. The strong influence of some polar groups on the rate constant of this reaction was observed ($\Delta G_{n\mu}^{\neq} = -RT \ln(k_{in}/k_{il})$) [16,17].

$Me_{4-n}C(CH_2OC(O)CH_2Me)_n + O_2 \longrightarrow$ Free Radicals				
n	1	2	3	4
k (423 K) (L mol^{-1} s^{-1})	2.8×10^{-8}	1.1×10^{-7}	1.7×10^{-7}	8.5×10^{-7}
$\Delta G_{n\mu}^{\neq}$ (kJ mol^{-1})	0.0	−4.6	−6.1	−11.8

The values of the rate constants and the activation energies of radical generation by dioxygen measured experimentally are collected in Table 4.1. The values of the rate constants of these reactions can be calculated by the intersecting parabolas model (IPM) method [18]. According to this method, the activation energy of the bimolecular homolytic reaction with high enthalpy ($\Delta H_e > \Delta H_{e\ max}$, $D(R-H) > D_{max}$) is almost equal to the enthalpy of the reaction ($E = \Delta H + 0.5RT$). The values of $\Delta H_{e\ max}$ are the following for aliphatic (R^1H), nonsaturated (R^2H), and alkylaromatic (R^3H) hydrocarbons:

RH	R^1H	R^2H	R^3H
$\Delta H_{e\ max}$ (kJ mol^{-1})	76.3	110.2	90.6
D_{max} (kJ mol^{-1})	300.1	334.0	314.4

Practically, all the hydrocarbons have BDE of C—H bonds higher than 300 kJ mol^{-1}, and this method of calculation can be used for them. The enthalpy of reaction was calculated as: $\Delta H = D(R-H) - 220$ (kJ mol^{-1}). The weakest bonds participate in this reaction. The pre-exponential factor depends on the reaction enthalpy value for the reactions with high enthalpy [18].

$$A = A_0 \left\{ 1 + \beta \left[\sqrt{\Delta H_e} - \sqrt{\Delta H_{e\ max}} \right] \right\}^2, \tag{4.2}$$

where the coefficient $\beta = 1.3$ and the factor $A_0 = 10^{10}$ (L mol^{-1} s^{-1}) per reacting C—H bond. The values of the enthalpies and the rate constants calculated by the IPM method [18] are given in Table 4.2.

TABLE 4.1
Rate Constants of the Reaction: RH + O_2 → Free Radicals (Experimental Data)

RH	Solvent	T (K)	k (L mol^{-1} s^{-1})	Ref.
CH_3—H	Gas phase	300–2500	$3.98 \times 10^{10} \exp(-238/RT)$	[19]
CH_3—H	Gas phase	300–2500	$3.98 \times 10^{10} \exp(-238.1/RT)$	[20]
CH_3—H	Gas phase	300–2500	$7.59 \times 10^{8} \exp(-217.4/RT)$	[21]
CH_3—H	Gas phase	1000–2500	$7.94 \times 10^{10} \exp(-234.3/RT)$	[22]
CH_3—H	Gas phase	300–1000	$4.27 \times 10^{10} \exp(-246.1/RT)$	[23]
CH_3CH_2—H	Gas phase	500–2000	$6.03 \times 10^{10} \exp(-217.0/RT)$	[19]
CH_3CH_2—H	Gas phase	300–2000	$3.98 \times 10^{10} \exp(-212.8/RT)$	[20]
CH_3CH_2—H	Gas phase	865–905	$1.00 \times 10^{10} \exp(-213.4/RT)$	[24]
Me_2CH—H	Gas phase	300–2500	$3.98 \times 10^{10} \exp(-238/RT)$	[25]
$Me_2CH(C$—$H)Me_2$	Gas phase	773–813	$2.04 \times 10^{10} \exp(-173.0/RT)$	[26]
Me_3C—H	Gas phase	300–2500	$3.98 \times 10^{10} \exp(-184.1/RT)$	[27]
Me_2C—HCH_2Me	Benzene	410–439	$1.50 \times 10^{12} \exp(-159.0/RT)$	[11]
$BuCH_2(CH$—$H)Me$	Benzene	378–433	$3.16 \times 10^{14} \exp(-181.0/RT)$	[12]
Me_3CCH_2C—HMe_2	Benzene	400–466	$1.00 \times 10^{12} \exp(-159.0/RT)$	[11]
cyclohexane (CH$_2$)	Chlorobenzene	383–413	$7.94 \times 10^{12} \exp(-167.4/RT)$	[10]
CH_2=$CHCH_2$—H	Gas phase	600–1000	$1.90 \times 10^{9} \exp(-163.8/RT)$	[28]
CH_2=$CMeCH_2$—H	Gas phase	673–793	$4.79 \times 10^{9} \exp(-161.2/RT)$	[29]
cyclohexene (CH$_2$)	Benzene	373–413	$1.58 \times 10^{7} \exp(-104.6/RT)$	[12]
cyclohexadiene (CH$_2$)	Gas phase	573–623	$8.13 \times 10^{8} \exp(-104.0/RT)$	[30]
cyclopentadiene (H)	Benzene, 1,2-dichloro-	353	3.70×10^{-7}	[31]
vinylcyclopentadiene	Benzene, 1,2-dichloro-	348–363	$2.00 \times 10^{8} \exp(-96.7/RT)$	[31]
$PhCH_2$—H	Benzene	378–433	$7.08 \times 10^{9} \exp(-133.9/RT)$	[12]
o-MeC$_6$H$_4$CH$_2$—H	Chlorobenzene	383–418	$3.16 \times 10^{9} \exp(-130.0/RT)$	[32]
m-MeC$_6$H$_4$CH$_2$—H	Chlorobenzene	383–418	$3.16 \times 10^{10} \exp(-135.0/RT)$	[32]
p-MeC$_6$H$_4$CH$_2$—H	Chlorobenzene	383–418	$3.16 \times 10^{9} \exp(-127.0/RT)$	[32]
$PhMe_2C$—H	Chlorobenzene	373–423	$3.50 \times 10^{9} \exp(-114.6/RT)$	[10]
$PhMe_2C$—H	Benzene	378–433	$9.55 \times 10^{6} \exp(-113.0/RT)$	[12]
$PhMe_2C$—H	Benzene	353–413	$2.57 \times 10^{3} \exp(-77.5/RT)$	[13]

TABLE 4.2
Enthalpies and Rate Constants of the Reaction RH + O_2 → $R^\bullet$ + $HO_2^\bullet$ in Hydrocarbon Solution Calculated by the IPM Method [18]

RH	D (kJ mol^{-1})	n_{C-H}	ΔH (kJ mol^{-1})	log A_iA (L mol^{-1} s^{-1})	k (450 K) (L mol^{-1} s^{-1})
Me$_3$CCH$_2$—H	422.0	12	202.0	12.58	8.17 × 10^{-12}
EtMeCH—H	413.0	4	193.0	12.06	2.68 × 10^{-11}
Me$_3$C—H	400.0	1	180.0	11.38	1.86 × 10^{-10}
cyclobutane-H	418.5	8	198.5	12.39	1.32 × 10^{-11}
cyclopentane-H	408.4	10	188.4	12.43	2.16 × 10^{-10}
cycloheptane-H	403.9	14	183.9	12.55	9.49 × 10^{-10}
methylcyclohexane-H	395.5	1	175.5	11.35	5.68 × 10^{-10}
Z-Decalin	387.6	2	167.6	11.60	8.31 × 10^{-9}
CH$_2$=CHCH$_2$—H	368.0	3	148.0	11.12	5.25 × 10^{-7}
CH$_2$=CH(CH—H)Me	349.8	2	129.8	10.58	1.96 × 10^{-5}
CH$_2$=CH(C—H)Me$_2$	339.6	1	119.6	9.96	4.47 × 10^{-4}
Z-MeCH=CH(CH—H)Me	344.0	2	124.0	10.41	6.27 × 10^{-5}
Me$_2$C=CH(CH—H)Me	332.0	2	112.0	9.89	4.70 × 10^{-4}
Me$_2$C=CMe(C—H)Me$_2$	322.8	1	102.8	8.63	3.02 × 10^{-4}
CH$_2$=C(MeC—H)CH=CH$_2$	307.2	1	87.2	9.47	1.34 × 10^{-1}
cyclohexene-H	341.5	4	121.5	10.63	2.02 × 10^{-4}
cyclohexadiene-H	330.9	4	110.9	10.12	1.06 × 10^{-3}
cyclohexadiene-H	312.6	4	92.6	9.48	3.27 × 10^{-2}
cycloheptatriene-H	301.0	2	81.0	10.17	3.55
MeC≡CC—HMe$_2$	329.4	1	109.4	9.41	3.09 × 10^{-4}
PhCH$_2$—H	375.0	3	155.0	11.02	6.47 × 10^{-8}
PhMeCH—H	364.1	2	144.1	11.21	1.81 × 10^{-6}
PhMe$_2$C—H	354.7	1	134.7	10.79	8.59 × 10^{-6}
tetralin-H	345.6	4	125.6	11.25	2.82 × 10^{-4}
dihydroanthracene-H	322.0	4	102.0	10.66	3.94 × 10^{-2}

4.1.2 Trimolecular Reaction of Hydrocarbon with Dioxygen

In addition to the bimolecular reactions of organic compounds with dioxygen, free radicals are generated in an oxidized substrate in the liquid phase by the trimolecular reaction [3,8,9]

$$RH + O_2 + HR \longrightarrow R^\bullet + H_2O_2 + R^\bullet$$

This reaction was proposed in 1960 for compounds with a weak C—H bond [33] and was experimentally proved in the reactions of oxidation of cyclohexanol and tetralin [34,35]. The rate of this reaction was found to obey the following equation:

$$v = k_i[RH]^2[O_2]. \qquad (4.3)$$

The enthalpy of this reaction is $\Delta H = 2D_{R-H} - 572\,\text{kJ mol}^{-1}$. The enthalpy of the bimolecular reaction $(RH + O_2)\Delta H = D_{R-H} - 220\,\text{kJ mol}^{-1}$. So, $\Delta H(2\,RH + O_2) < \Delta H(RH + O_2)$ at $D_{R-H} < 352\,\text{kJ mol}^{-1}$. The frequency factor of the trimolecular collisions (z_{03}) in liquids is close to that for the bimolecular collisions (z_{02}). The pre-exponential factor A_{03} is less than A_{02} due to the concerted mechanism of simultaneous energy concentration on the two reacting C—H bonds in the trimolecular reaction [36]. The ratio $A_{03}/A_{02} = (z_{03}/z_{02}) \times (2RT/\pi E_3)^{1/2}$. As a result, the ratio $k_3[RH]^2[O_2]/k_2[RH][O_2] = (z_{03}[RH]/z_{02}) \times (2RT/\pi E_3)^{1/2} \exp[-(E_3 - E_2)/RT] \approx (2RT/\pi E_3)^{1/2} \exp[-(D_{R-H} - 352)/RT]$, and the trimolecular reaction should be faster than the bimolecular reaction for substrates with $D_{R-H} < 340\,\text{kJ mol}^{-1}$. Polar solvents accelerate the trimolecular reaction due to the polar structure of the transition state $C(\delta+) \cdots H \cdots (\delta-)O—O(\delta-) \cdots H \cdots (\delta+)C$. The values of the rate constants of the trimolecular reactions $2RH + O_2$ are collected in Table 4.3.

The trimolecular reaction of two dioxygen molecules with two C—H bonds of one hydrocarbon was observed in ethylbenzene oxidation [43].

$$PhCH_2CH_3 + 2O_2 \longrightarrow HO_2^\bullet + PhCH=CH_2 + HO_2^\bullet$$

The rate of this reaction was found to be $v_i = k_i\,[RH][O_2]^2$ with the rate constant $k_i = 6.0 \times 10^8 \exp(-108/RT) = 4.73 \times 10^{-6}$ (400 K) $L^2\,\text{mol}^{-2}\,s^{-1}$ [43].

4.1.3 Bimolecular and Trimolecular Reactions of Dioxygen with the Double Bond of Olefin

In addition to the reaction with the C—H bond, dioxygen attacks the double bond of olefin with free radical formation [9].

$$RCH=CH_2 + O_2 \longrightarrow RC^\bullet HCH_2OO^\bullet$$
$$RC^\bullet HCH_2OO^\bullet + O_2 \longrightarrow RC(OO^\bullet)HCH_2OO^\bullet$$

This reaction was proposed by Farmer [44]. Miller and Mayo observed this reaction experimentally in oxidized styrene [45]. The rate constant of this reaction was measured by the free radical acceptor method by Denisova and Denisov [46]. This reaction is endothermic. The activation energies of these reactions are sufficiently higher than their enthalpy values.

Olefin	$CH_2=CH_2$	$CH_2=CHEt$	$CH_2=CHPh$	$CH_2=CHCOOMe$
ΔH (kJ mol^{-1})	80.8	80.4	20.4	84.8
E (kJ mol^{-1})			118	117

TABLE 4.3
Rate Constants and Activation Energies of the Trimolecular Reaction $2RH + O_2 \rightarrow$ Free Radicals (Experimental Data)

RH	Solvent	T (K)	k (L^2 mol^{-2} s^{-1})	Ref.
$CH_2=C(CH-H)BuMe$	1-Octene	373–393	$5.38 \times 10^8 \exp(-142/RT)$	[37]
$C_8H_{17}CH=CH(CH-H)$ $(CH_2)_6COOMe$	Chlorobenzene	313–333	$2.0 \times 10^{12} \exp(-130/RT)$	[38]
$Bu(CH_2CH=CH)_2$ $(CH-H)(CH_2)_7COOMe$	Chlorobenzene	313–333	$2.0 \times 10^6 \exp(-93.0/RT)$	[38]
$Et(CH=CHCH-H)_3$ $(CH_2)_6COOMe$	Chlorobenzene	313–333	$6.30 \times 10^3 \exp(-76.5/RT)$	[38]
Cholesteryl pelargonate	Chlorobenzene	364–388	$2.69 \times 10^5 \exp(-73.0/RT)$	[39]
(tetralin)	Decane	403–423	$3.47 \times 10^3 \exp(-86.5/RT)$	[35]
(tetralin)	Decane	378–397	$2.40 \times 10^5 \exp(-99.0/RT)$	[40]
(indene)	Decane	345–365	$3.89 \times 10^3 \exp(-78.5/RT)$	[40]
(cyclohexadiene)	Chlorobenzene	313–348	$7.08 \times 10^6 \exp(-74.5/RT)$	[41]
(cyclopentadiene)	Benzene, 1,2-dichloro-	353	1.13×10^{-7}	[31]
(methyl cyclohexenyl ketone)	Chlorobenzene	353–393	$1.70 \times 10^9 \exp(-114.0/RT)$	[42]

The transition state of this reaction has a polar structure and therefore this reaction occurs more rapidly in polar solvents (compare rate constants in chlorobenzene and N,N-dimethylformamide for reactions of styrene and butyl methacrylate in Table 4.4). The effect of multidipole interaction was observed for reactions of polyatomic esters [47–49].

Monomer	$CH_2=CHCO_2Bu$	$CH_2=CHC(O)OCH_2$ $C(CH_2OC(O)Et)_3$	$(CH_2=CHC(O)$ $OCH_2)_4C$
k (363 K) (L mol^{-1} s^{-1})	1.72×10^{-7}	1.10×10^{-8}	1.4×10^{-8}
$\Delta G^{\neq}_{n\mu}$ (kJ mol^{-1})	0.0	8.2	7.8

The charge transfer complex (CTC) of dioxygen with the monomer is very probable as a precursor of this reaction [48].

$$RCH=CH_2 + O_2 \rightleftharpoons RCH=CH_2(\delta +) \cdots O_2(\delta -)$$
$$RCH=CH_2(\delta +) \cdots O_2(\delta -) \longrightarrow RC^{\bullet}HCH_2OO^{\bullet}$$

For the values of the rate constants of dioxygen reaction with monomers, see Table 4.4.

Unsaturated compounds react with dioxygen by trimolecular reaction also [48]. It is very probable that this reaction proceeds via preliminary formation of a CTC. The formed complex reacts with another olefin molecule.

TABLE 4.4
Rate Constants of the Bimolecular Reaction $RCH=CH_2 + O_2 \rightarrow$ Free Radicals (Experimental Data)

Monomer	Solvent	T (K)	k (L mol^{-1} s^{-1})	Ref.
$PhCH=CH_2$	Chlorobenzene	343–363	$1.78 \times 10^{10} \exp(-118.1/RT)$	[45]
$PhCH=CH_2$	Chlorobenzene	378–398	$3.63 \times 10^{11} \exp(-125.5/RT)$	[46]
$PhCH=CH_2$	N,N-Dimethyl-formamide	343	4.0×10^{-8}	[47]
$PhMeC=CH_2$	Chlorobenzene	343	2.5×10^{-8}	[47]
$PhMeC=CH_2$	Benzonitrile	343	5.5×10^{-8}	[47]
$MeOCO(Me)C=CH_2$	Chlorobenzene	353	1.3×10^{-7}	[48]
$BuOCOCH=CH_2$	Chlorobenzene	343–363	$1.10 \times 10^{6} \exp(-88.6/RT)$	[48]
$BuOCO(Me)C=CH_2$	Chlorobenzene	342–363	$5.30 \times 10^{6} \exp(-91.6/RT)$	[49]
$BuOCO(Me)C=CH_2$	Chlorobenzene	363	3.4×10^{-7}	[48]
$BuOCO(Me)C=CH_2$	N,N-Dimethyl-formamide	363	8.1×10^{-7}	[48]
$Me_3COCOCH=CH_2$	Chlorobenzene	353	7.1×10^{-8}	[48]
$MeOCO(Et)C=CH_2$	Chlorobenzene	353	1.2×10^{-7}	[48]
$MeCH=CHCOOEt$	Chlorobenzene	363	5.1×10^{-8}	[48]
$NH_2COCH=CH_2$	Chlorobenzene	353	1.5×10^{-6}	[48]
$CH_2=CHCOOCH_2C(CH_2OCOEt)_3$	Chlorobenzene	363	1.40×10^{-8}	[50]
$(CH_2=CHC(O)OCH_2)_4C$	Chlorobenzene	363	1.10×10^{-8}	[50]
$(CH_2=C(Me)C(O)OCH_2)_4C$	Chlorobenzene	343–393	$1.60 \times 10^{18} \exp(-180.6/RT)$	[49]
2-vinylpyridine	N,N-Dimethyl-formamide	343	4.00×10^{-8}	[48]
6-methyl-2-vinylpyridine	N,N-Dimethyl-formamide	343	8.00×10^{-8}	[48]
3-vinylpyridine	N,N-Dimethyl-formamide	343	4.40×10^{-8}	[48]
Retinal acetate	Chlorobenzene	303–333	$6.30 \times 10^{3} \exp(-53.2/RT)$	[51]
E-β-Carotene	Benzene, 1,3-dimethyl-	323–343	$4.00 \times 10^{12} \exp(-107.5/RT)$	[52]

TABLE 4.5
Rate Constants of the Trimolecular Reaction RCH=CH$_2$ + O$_2$ + CH$_2$=CHR → R˙CHCH$_2$OOCH$_2$CHR˙

RH	Solvent	T (K)	k (L^2 mol^{-2} s^{-1})	Ref.
PhCH=CH$_2$	N,N-Dimethyl-formamide	343	2.4 × 10^{-8}	[48]
PhCH=CH$_2$	Chlorobenzene	333–353	3.40 × 10^8 exp(−113.2/RT)	[47]
PhMeC=CH$_2$	Benzonitrile	343	3.7 × 10^{-8}	[48]
BuOCOMeC=CH$_2$	N,N-Dimethyl-formamide	363	6.0 × 10^{-8}	[48]
MeCH=CHC(O)OEt	Chlorobenzene	363	4.2 × 10^{-7}	[48]
2-vinylpyridine	Chlorobenzene	343–363	3.02 × 10^2 exp(−55.6/RT)	[48]
2-vinylpyridine	N,N-Dimethyl-formamide	343	1.5 × 10^{-6}	[48]
4-vinylpyridine	Chlorobenzene	343–453	9.12 × 10^3 exp(−64.4/RT)	[48]
4-vinylpyridine	N,N-Dimethyl-formamide	343	1.3 × 10^{-6}	[48]
4-vinylpyridine	Benzene, 1,2-dichloro-	348–363	2.00 × 10^8 exp(−96.7/RT)	[48]
methyl-2-vinylpyridine	Chlorobenzene	343	6.75 × 10^{-5}	[48]
methyl-2-vinylpyridine	N,N-Dimethyl-formamide	343	2.8 × 10^{-6}	[48]
furyl methyl ester	Chlorobenzene	303–363	50 exp(−54.0/RT)	[53]
furyl propyl ester	Chlorobenzene	313–363	1.40 × 10^2 exp(−62.0/RT)	[53]
furylvinyl ester	Chlorobenzene	313–363	1.82 exp(−52.0/RT)	[53]

$$RCH=CH_2 + O_2 \rightleftharpoons RCH=CH_2(\delta +) \cdots O_2(\delta -)$$
$$RCH=CH_2 \cdots O_2 + CH_2=CHR \longrightarrow RC\dot{}HCH_2OOCH_2\dot{C}HR$$

The values of the rate constants of the trimolecular reactions 2RCH=CH$_2$ + O$_2$ are collected in Table 4.5.

4.2 HYDROPEROXIDES

Hydroperoxide is the first product of hydrocarbon oxidation and plays a key role in the chain mechanism of autoxidation. Hydroperoxide possesses a weak O—O bond and decomposes with free radical production. Besides this, hydroperoxide as active oxidizing agent reacts with the oxidized hydrocarbon and the products of oxidation with free radical generation [1–3,8,9,54–59]. So, it acts as the autoinitiator accelerating the chain reaction of oxidation [1]. All aspects of the chemistry of hydroperoxides will be discussed in this chapter: their structure, thermochemistry, self-association, association with other oxidation products, unimolecular decomposition, generation of free radicals via various bimolecular reactions, and mechanisms of chain decay.

4.2.1 ANALYSIS

Many different analytical techniques were developed for the estimation of hydroperoxides. Among them, the iodometric technique has been used for a long period of time [60]. According to this method, peroxide is reduced by HI, and diiodine is formed in stoichiometric quantity. The amount of the formed I_2 is measured by reduction with thiosulfate using any titrometric technique or photometrically.

$$ROOH + 2HI = ROH + H_2O + I_2$$
$$2Na_2S_2O_3 + I_2 = 2NaI + Na_2S_4O_6$$

The following analytical procedure for the accurate estimation of hydroperoxides was proposed [61]. The reduction of peroxide occurs in a solution of isopropanol saturated with NaI in the presence of acetic acid and CO_2 atmosphere at 373 K (in water bath). The reaction ceases after 15 min. The relative standard deviation equals 0.2%.

The fast reduction of hydroperoxide by Sn(II) is used in stannometric technique of hydroperoxide estimation.

$$ROOH + SnCl_2 + 2HCl = ROH + H_2O + SnCl_4$$

The analytical procedure is the following [62]. Acetic acid is added to the sample of hydroperoxide, then dissolved air is removed by vacuum; the vessel is filled with dinitrogen, and an aqueous solution of $SnCl_2$ is added. The reaction occurs for 1 h at room temperature; after this an excess of $NH_4Fe(SO_4)_2$ solution is introduced, and in 30 min the amount of formed Fe(II) is estimated by titration with potassium dichromate.

The spectrophotometric technique of hydroperoxide estimation was developed with triphenyl phosphite [63].

$$ROOH + Ph_3P = ROH + Ph_3PO$$

Triphenyl phosphite is added in an excess and the rest of it is estimated colorimetrically after the introduction of formaldehyde and HCl. The latter forms colored triphenyl phosphonyl chloride $[Ph_3PCH_2OH]^+Cl^-$.

Polarography is successfully used for the estimation of hydroperoxides and peracids [60]. The reduction of hydroperoxide proceeds according to the electrochemical equation:

$$ROOH + 2e^- \longrightarrow RO^- + HO^-$$

The individual polarographic characteristic of the analyzed compound is the potential $E_{1/2}$ at which the current strength equals 50% of its maximum value. The binary solvent benzene–methanol or benzene–ethanol is used. The amount of peroxide is proportional to the maximum strength of the electric current at peroxide concentration in solution lower than

$0.01\,\text{mol}\,L^{-1}$. The values of $E_{1/2}$ for several hydroperoxides (ROOH) in benzene–ethanol solution (1:1 v/v) at $[\text{LiCl}] = 0.3\,\text{mol}\,L^{-1}$ are the following [60]:

R—	Me₃C—	EtMe₂C—	Me₂PhC—	Ph₂CH—	Tetralyl—
$E_{1/2}$ (V)	−1.22	−1.19	−0.97	−0.82	−0.88

Liquid and paper chromatographies as well as mass spectrometry (MS) are used for the identification and analysis of hydroperoxides [60]. Nuclear magnetic resonance (NMR) spectroscopy is used for identification of diacyl peroxides.

4.2.2 Structure of Hydroperoxides

The structure of hydroperoxides was studied by the x-ray structural analysis method. The results of the experimental measurements are collected in Table 4.6.

The analysis of the IR spectrum of hydrogen peroxide and cumyl hydroperoxide gave the following values of frequencies (cm^{-1}) of valence and bond angle vibrations [60].

Bond Angle	O—O	O—H	C—O	θ(C—O—O)	θ(O—O—H)
H₂O₂	880	3598 (symm.)			1390 (symm.)
		3610 (nonsym.)			1266 (nonsym.)
Me₂PhCOOH	880	3350	1270	585	

The hydrogen bond formation decreases the frequency of the O—H bond valence vibration (see Section 4.2.3). Two configurations of tertiary hydroperoxides are known: E- and Z-configurations. The activation barrier for transition from Z- to E-configuration is found to be equal to $195\,\text{kJ}\,\text{mol}^{-1}$ (quantum-chemical calculation [64]).

The dipole moment of hydrogen peroxide is $\mu = 1.573\,\text{D}$ [64]. The values of the dipole moments and the polarization of hydroperoxides and peracids are given in Table 4.7.

TABLE 4.6
Bond Lengths and Angles of the Hydroperoxide Groups [59,64]

Compound	Length (10^{-10} m)		Angle (deg)		
	O—O	O—C	O—O—H	O—O—C	C—OO—H
Hydrogen peroxide, H₂O₂	1.453	0.97 (O—H)	94.8		90.2
Hydrogen peroxide, H₂O₂	1.464	0.965 (O—H)	99.4		
Hydroperoxide, methyl-, MeOOH	1.443	1.437	99.6	105.7	114
Trifluoromethyl-, CF₃OOH	1.447	1.376	100.0	107.6	95
1,1-Dimethylethyl-, Me₃COOH	1.472	1.463	100	109.6	100
1-Methyl-1-phenylethyl-, MePh₂COOH	1.477	1.461	100.9	108.7 (Me), 109.7 (Ph)	109
Triphenylmethyl-, Ph₃COOH	1.455	1.454	102.9	109.7 (Ph)	101

TABLE 4.7
Dipole Moments and Polarization of Hydroperoxides in Benzene [65]

Hydroperoxide	T (K)	μ (D)	P (cm^3 mol^{-1})
1,1-Dimethylethyl-, Me$_3$COOH	303	1.82	91.1
1,1-Dimethylethyl-, Me$_3$COOH	323	1.81	86.5
1-Methyl-1-phenylethyl-, Me$_2$PhCOOH	303	1.76	106.1

4.2.3 THERMOCHEMISTRY OF HYDROPEROXIDES

The values of the enthalpies of hydroperoxides formation are given in Table 4.8. The increments of groups in the additive scheme of enthalpy ΔH_f calculation for hydroperoxides have the following values [66]: $\Delta H[\text{O}-(\text{C})(\text{O})] = -19.2 \pm 0.81$ kJ mol^{-1}, $\Delta H[(\text{O})-\text{OH}] = -54.7$ kJ mol^{-1}, and additional increment of tertiary atom C $\Delta H[(\text{C})_3\text{C}-\text{OO}-\text{C}(\text{C})_3] = -5.72$ kJ mol^{-1}. The increments of groups for calculation of enthalpy ΔH_v of peroxides evaporation are: $\Delta H[(\text{O})-\text{O}-(\text{C})] = 6.26$ kJ mol^{-1}, $\Delta H_v[(\text{O})-\text{OC(O)}-(\text{C}_B)] = 25.4$ kJ mol^{-1}, $\Delta H_v[(\text{O})-\text{OH}] = 27.4$ kJ mol^{-1}, and $\Delta H_v[(\text{O})-\text{OC(O)}-(\text{C})] = 18.3$ kJ mol^{-1}.

A linear correlation between $\Delta H_f(\text{ROOH})$ and $\Delta H_f(\text{ROH})$ was found [76]:

$$\Delta H_f^0(\text{ROOH}) = \Delta H_f^0(\text{ROH}) + 74.0 \text{ kJ mol}^{-1} \tag{4.4}$$

The dissociation energies of the O—O bond in hydroperoxides ROOH poorly depend on the structure of R and are the following:

R	CH$_3$	CH$_3$CH$_2$	Me$_3$C	Me$_2$PhC
D (kJ mol^{-1})	191.0	193.0	182.5	175.0

The dissociation energy of the O—H bond depends on the R structure [77,78].

R	sec-R	tert-R	CF$_3$	CCl$_3$	CCl$_3$CCl$_2$
D (kJ mol^{-1})	365.5	358.6	418.0	407.2	413.1

4.2.4 HYDROGEN BONDING BETWEEN HYDROPEROXIDES

Hydroperoxides have the OOH group and, therefore, form dimers and trimers connected by hydrogen bonds. The following scheme of self-association was proposed [59]:

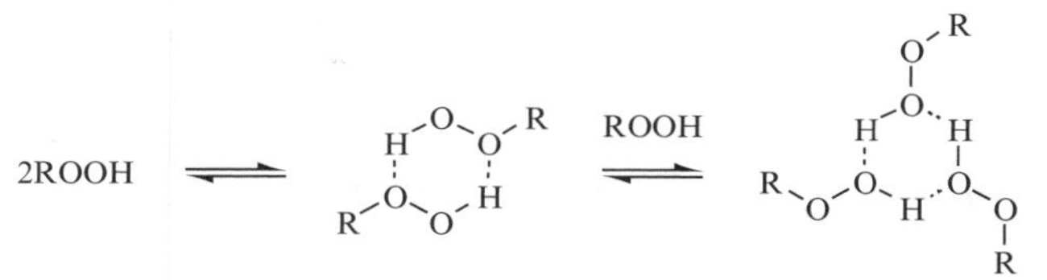

$$2\text{ROOH} \rightleftharpoons \text{[dimer]} \rightleftharpoons \text{[trimer]}$$

The equilibrium constants K_H and the thermodynamic parameters of self-association of several hydroperoxides are given in Table 4.9.

TABLE 4.8
Enthalpies of Formation ΔH_f^0 (Gas, 298 K) and Evaporation ΔH_v of Hydroperoxides

Hydroperoxide	$-\Delta H_f^0$ (kJ mol^{-1})	ΔH_v (kJ mol^{-1})	ΔS^0 (J mol^{-1} K^{-1})	Ref.
Hydrogen peroxide, H$_2$O$_2$	136.4	51.0	232.8	[67]
Methyl-, MeOOH	130.5		282.4	[67]
Methyl-, MeOOH	133.9			[68]
Ethyl-, EtOOH	169.4	43.7	321.8	[68]
1-Methylethyl-, Me$_2$CHOOH	198.3		358.7	[68]
Butyl-, BuOOH	205.0		393.8	[67]
1,1-Dimethylethyl-, Me$_3$COOH	242.7		360.6	[67]
1,1-Dimethylethyl-, Me$_3$COOH	236.4	48.7		[68]
1,1-Dimethylethyl-, Me$_3$COOH	243.3			[69]
1,1-Dimethylethyl-, Me$_3$COOH	245.8			[59]
1,1-Dimethylethyl-, Me$_3$COOH	218.3			[70]
1,1-Dimethylethyl-, Me$_3$COOH	225.7			[66]
Pentyl-, Me(CH$_2$)$_3$CH$_2$OOH	217.6		433.2	[67]
1-Methylbutyl-, MePrCHOOH	234.3		430.7	[67]
1-Hexyl-, Me(CH$_2$)$_4$CH$_2$OOH	238.5		472.6	[67]
1-Hexyl-, Me(CH$_2$)$_4$CH$_2$OOH	237.1	62.7		[71]
1-Hexyl-, Me(CH$_2$)$_4$CH$_2$OOH	259.2			[59]
1-Methylpentyl, Me(CH$_2$)$_3$CH(OOH)Me	255.2		470.1	[67]
1-Methylpentyl, Me(CH$_2$)$_3$CH(OOH)Me	250.1	60.1		[71]
1-Methylpentyl, Me(CH$_2$)$_3$CH(OOH)Me	267.5			[59]
1-Ethylbutyl-, Me(CH$_2$)$_2$CH(OOH)Et	245.1	60.1		[71]
1-Ethylbutyl-, Me(CH$_2$)$_2$CH(OOH)Et	263.3			[59]
1-Heptyl-, Me(CH$_2$)$_5$CH$_2$OOH	288.7		512.0	[67]
1-Heptyl-, Me(CH$_2$)$_5$CH$_2$OOH	275.7	67.5		[71]
1-Methylhexyl-, Me(CH$_2$)$_4$CH(OOH)Me	276.1		509.5	[67]
1-Methylhexyl-, Me(CH$_2$)$_4$CH(OOH)Me	281.6	64.9		[71]
1-Ethylpentyl-, Me(CH$_2$)$_3$CH(OOH)Et	282.1	64.9		[71]
1-Propylbutyl-, Me(CH$_2$)$_2$CH(OOH)Pr	269.1	64.9		[71]
1,1-Dimethyl-2-pentyn-4-enyl-, CH$_2$=CHC≡CCMe$_2$OOH	−127.1	63.5		[72]
Cyclohexyl-,	214.9	58.5		[71]
Cyclohexyl-,	229.9			[59]
1-Methylcyclohexyl-,	270.3	60.2		[71]

TABLE 4.8
Enthalpies of Formation ΔH_f^0 (Gas, 298 K) and Evaporation ΔH_v of Hydroperoxides—continued

Hydroperoxide	$-\Delta H_f^0$ (kJ mol^{-1})	ΔH_v (kJ mol^{-1})	ΔS^0 (J mol^{-1} K^{-1})	Ref.
1-Methylcyclohexyl-,	263.3			[71]
E-9-Decalyl-,	277.8	67.2		[73]
1-Methyl-1-phenylethyl-, PhMe$_2$COOH	82.5	66.0		[69]
1-Methyl-1-phenylethyl-, PhMe$_2$COOH	87.9	66.0		[74]
1-Methyl-1-phenylethyl-, PhMe$_2$COOH	85.4	66.0		[75]
1-Tetralyl-,	104.5	77.2		[71]
1,1,4,4-Tetramethyl-1,4-dihydroperoxy butane, HOOCMe$_2$(CH$_2$)$_2$CMe$_2$OOH	428.6	138.6		[66]
1,1,4,4-Tetramethyl-1,4-dihydroperoxy butyne, HOOCMe$_2$C≡CCMe$_2$OOH	130.2	127.4		[66]

Polar molecules and molecules with π-bonds form complexes with hydroperoxides via hydrogen bond also, for example:

$$R^1OOH + NR_3 \rightleftharpoons R^1OOH \cdots NR_3$$

The values of the equilibrium constants K_H are listed in Table 4.10.

In addition to hydroperoxides having the OH group, they also possesses acidic properties. The pK_a values of hydroperoxides fall over a rather narrow range of 11.5–13.2 and for peracids 7.2–8.2 in water. The values of pK_a for hydroperoxides ROOH in water at 298 K are presented below [59].

R	Me	Et	Me$_3$C	Me$_2$PhC	PhCH$_2$CMe$_2$	MePh$_2$C
pK_a	11.50	11.80	13.27	13.08	13.25	12.94

4.3 REACTIONS OF FREE RADICAL GENERATION BY HYDROPEROXIDES

4.3.1 Unimolecular Decomposition of Hydroperoxides

Hydroperoxides have a weak O—O bond and split under heating with the dissociation of this bond forming two active free radicals [54–59].

TABLE 4.9
Equilibrium Constants (K), Enthalpies (ΔH), and Entropies (ΔS) of Self-Hydrogen Bonding of Hydroperoxides

Hydroperoxide	Solvent	T (K)	$-\Delta H$ (kJ mol^{-1})	$-\Delta S$ (J mol^{-1} K^{-1})	K (298 K) (L mol^{-1})	Ref.
ROOH + ROOH $\Leftrightarrow$ ROOH $\cdots$ O(H)OR						
1-Cyclohexyl-1-ethinyl- (C≡CH, OOH)	CCl$_4$	258–312	17.6	48.9	0.29	[60]
1,1-Dimethylethyl-, Me$_3$COOH	n-C$_7$H$_{16}$	363			1.90	[79]
1,1-Dimethylethyl-, Me$_3$COOH	CCl$_4$	258–313	23.8	82.0	0.77	[80]
1,1-Dimethylethyl-, Me$_3$COOH	CCl$_4$	303	24.9	77.0	2.20	[81]
1,1-Dimethylethyl-, Me$_3$COOH	C$_6$H$_6$	343	6.6	37.7	0.15	[81]
1,1-Dimethylethyl-, Me$_3$COOD	CCl$_4$	303	20.5	62.8	2.06	[59]
1,1-Dimethylethyl-, Me$_3$COOD	C$_6$H$_6$	343			0.64	[82]
1,1-Dimethylethyl-, Me$_3$COOH	CCl$_4$	313	26.4	90.4	0.80	[59]
ROOH + ROOH + ROOH $\Leftrightarrow$ 3(ROOH)						
1,1-Dimethylethyl-, Me$_3$COOH	cyclo-C$_6$H$_{12}$	288–343	50.2	108.7	1.32×10^3	[83]
1,1-Dimethylethyl-, Me$_3$COOH	C$_7$H$_{14}$	258–343	50.2	111.3	9.67×10^2	[83]
1,1-Dimethylethyl-, Me$_3$COOH	cyclo-C$_6$H$_{11}$Me	288–343	50.2	112.1	8.78×10^2	[83]
1,1-Dimethylethyl-, Me$_3$COOH	CCl$_4$	258–343	46.9	106.3	4.66×10^2	[83]

$$ROOH \longrightarrow RO^\bullet + HO^\bullet$$

In addition to this reaction, many other reactions of hydroperoxide decay occur in solution and they will be discussed later. This is the reason why the unimolecular decomposition of hydroperoxides was studied preferentially in the gas phase. The rate constants of the unimolecular decomposition of some hydroperoxides in the gas phase and in solution are presented in Table 4.11. The decay of 1,1-dimethylethyl hydroperoxide in solution occurs more rapidly. This demonstrates the interaction of ROOH with the solvent.

TABLE 4.10
Equilibrium Constants (K), Enthalpies (ΔH), and Entropies (ΔS) of Hydrogen Bonding of the Hydroperoxides: ROOH + Y $\Leftrightarrow$ ROOH...Y

Hydroperoxide	Y	Solvent	T (K)	$-\Delta H$ (kJ mol^{-1})	$-\Delta S$ (J mol^{-1} K^{-1})	K (298 K) (L mol^{-1})	Ref.
AcCH(OOH)Me	Me$_3$COH	MeC(O)Et	343			0.65	[84]
AcCH(OOH)Me	MeC(O)OH	MeC(O)Et	343			3.11	[84]
Me$_3$COOH	C$_6$H$_6$	C$_6$H$_6$	303–343	6.6	37.7	0.15	[81]
Me$_3$COOH	C$_6$H$_6$	C$_6$H$_6$	288–343	8.2	42.3	0.17	[83]
Me$_3$COOH	PhCH$_3$	PhCH$_3$	288–343	5.7	29.9	0.27	[83]
Me$_3$COOH	PhCHMe$_2$	PhCHMe$_2$	288–343	11.7	52.1	0.21	[83]
Me$_3$COOH	C$_6$H$_5$Cl	C$_6$H$_5$Cl	288–343	7.6	42.5	0.13	[83]
Me$_3$COOH	C$_6$H$_5$Cl	C$_6$H$_5$Cl	343			0.10	[82]
Me$_3$COOH	o-C$_6$H$_4$Cl$_2$	o-C$_6$H$_4$Cl$_2$	343			0.13	[82]
Me$_3$COOH	PhCH=CH$_2$	PhCH=CH$_2$	303–343	10.0	43.9	0.29	[81]
Me$_3$COOH	PhCH=CH$_2$	PhCH=CH$_2$	288–343	7.5	35.4	0.29	[83]
Me$_3$COOD	PhCH=CH$_2$	PhCH=CH$_2$	343			0.50	[82]
Me$_3$COOH	MeCOMe	CCl$_4$	249–293	12.5	31.0	3.73	[85]
Me$_3$COOH	MeCOEt	MeCOEt	288–343	15.0	57.9	0.40	[83]
Me$_3$COOH	MeOH	CCl$_4$	249–293	12.5	31.4	3.55	[85]
Me$_3$COOH	cyclohexanone	CCl$_4$	249–293	9.2	18.8	4.27	[85]
Me$_3$COOH	EtOEt	EtOEt	233–289	13.0	25.5	8.84	[86]
Me$_3$COOH	EtOEt	EtOEt	288–343	19.7	68.9	0.71	[83]
Me$_3$COOH	Me$_2$CHOCHMe$_2$	Me$_2$COCMe$_2$	288–343	20.5	67.9	1.11	[83]
Me$_3$COOH	BuOBu	BuOBu	288–343	19.5	68.2	0.72	[83]
Me$_3$COOH	Me$_2$EtCHO CHMe$_2$Et	Me$_2$EtCHO CHMe$_2$Et	288–343	19.7	66.1	1.00	[83]
Me$_3$COOH	tetrahydrofuran	tetrahydrofuran	288–343	21.8	81.5	0.37	[83]
Me$_3$COOH	1,4-dioxane	1,4-dioxane	288–343	17.8	66.9	0.42	[83]
Me$_3$COOH	MeOCH$_2$CH$_2$OMe	MeOCH$_2$CH$_2$OMe	288–343	18.0	70.4	0.30	[83]
Me$_3$COOH	PhOMe	PhOMe	288–343	10.6	30.0	1.95	[83]
Me$_3$COOH	PhOEt	PhOEt	288–343	10.5	29.6	1.97	[83]
Me$_3$COOH	PhCH$_2$OMe	PhCH$_2$OMe	288–343	20.1	67.7	0.97	[83]
Me$_3$COOH	Me$_3$CO OCMe$_3$	Me$_3$CO OCMe$_3$	288–343	13.5	43.7	1.21	[83]
Me$_3$COOH	propylene oxide	propylene oxide	288–343	17.3	61.6	0.65	[83]

continued

TABLE 4.10
Equilibrium Constants (K), Enthalpies (ΔH), and Entropies (ΔS) of Hydrogen Bonding of the Hydroperoxides: ROOH + Y $\Leftrightarrow$ ROOH...Y—*continued*

Hydroperoxide	Y	Solvent	T (K)	$-\Delta H$ (kJ mol^{-1})	$-\Delta S$ (J mol^{-1} K^{-1})	K (298 K) (L mol^{-1})	Ref.
Me$_3$COOH	PhNH$_2$	PhNH$_2$	298	3.3	−7.28	9.09	[87]
Me$_3$COOH	PhNHEt	PhNHEt	298	4.8	−0.83	7.67	[87]
Me$_3$COOH	AcNMe$_2$	CCl$_4$	294–341	22.6	51.1	19.6	[88]
Me$_3$COOH	AcNHBu	CCl$_4$	292–341	23.9	55.7	19.0	[88]
Me$_3$COOH	AcNHCHMe$_2$	CCl$_4$	294–332	14.7	43.5	2.02	[88]
Me$_3$COOH	Me$_2$NC(O)H	Me$_2$NC(O)H	288–343	22.0	93.1	9.84 × 10^{-2}	[83]
Me$_3$COOH	CH$_2$=CHCOOMe	CH$_2$=CHCOOMe	293–343	13.8	35.6	3.62	[48]
Me$_3$COOH	CH$_2$=CMeCOOMe	CH$_2$=CMeCOOMe	293–343	18.8	52.3	3.66	[48]
Me$_3$COOH	CH$_3$CN	CH$_3$CN	288–343	12.0	44.3	0.62	[83]
Me$_3$COOH	CH$_2$=CHCN	CH$_2$=CHN	293–343	12.5	33.5	2.76	[48]
Me$_3$COOH	piperidine (NH)	piperidine (NH)	298	28.8	85.8	3.68	[87]
Me$_3$COOH	pyridine	pyridine	248–289	28.0	72.0	14.03	[86]
Me$_3$COOH	pyridine	pyridine	298	11.0	9.7	26.39	[87]
Me$_3$COOH	quinoline	Me$_3$COOH	298	10.7	12.0	17.73	[87]
Me$_3$COOH	morpholine (O,N-H)	morpholine (O,N-H)	298	21.4	52.5	10.20	[87]
Me$_3$COOH	Me$_2$SO	Me$_2$SO	288–343	23.7	107.6	3.42 × 10^{-2}	[83]
EtMe$_2$COOH	EtMe$_2$COOH	CCl$_4$	258–313	22.6	75.3	1.07	[80]
EtMe$_2$COOH	MeCOMe	MeCOMe	270–313	15.9	37.2	6.98	[86]
EtMe$_2$COOH	EtOEt	EtOEt	248–288	10.5	10.9	18.66	[86]
EtMe$_2$COOH	pyridine	pyridine	253–288	28.0	74.0	11.03	[86]
cyclohexyl-OOH (dimer)	cyclohexyl-OOH	n-C$_{10}$H$_{14}$	293	23.4	79.5	0.89	[59]
cyclohexyl-OOH	cyclohexyl-OOH	n-C$_{10}$H$_{14}$	293	11.7	26.8	4.48	[59]
Me$_2$PhCOOH	Me$_2$PhCOOH	Me$_2$PhCH	273–313	25.1	66.6	8.33	[88]
Me$_2$PhCOOH	Me$_2$PhCOOH	CCl$_4$	265–323	28.5	96.2	0.93	[80]
Me$_2$PhCOOH	PhOCH$_3$	CCl$_4$	293			0.46	[59]
Me$_2$PhCOOH	MeCOMe	CCl$_4$	249–293	15.1	53.5	0.43	[85]
Me$_2$PhCOOH	MeCOMe	MeCOMe	270–313	15.9	35.1	9.00	[89]

TABLE 4.10
Equilibrium Constants (K), Enthalpies (ΔH), and Entropies (ΔS) of Hydrogen Bonding of the Hydroperoxides: ROOH + Y $\Leftrightarrow$ ROOH...Y—continued

Hydroperoxide	Y	Solvent	T (K)	$-\Delta H$ (kJ mol^{-1})	$-\Delta S$ (J mol^{-1} K^{-1})	K (298 K) (L mol^{-1})	Ref.
Me$_2$PhCOOH	MeOH	CCl$_4$	249–293	15.9	50.2	1.46	[85]
Me$_2$PhCOOH	cyclohexanone	CCl$_4$	249–293	10.9	30.1	0.46	[85]
Me$_2$PhCOOH	EtOEt	EtOEt	248–288	13.8	25.5	12.2	[86]
Me$_2$PhCOOH	PhNH$_2$	PhNH$_2$	298	8.0	13.7	4.86	[87]
Me$_2$PhCOOH	AcNMe$_2$	CCl$_4$	294–350	20.9	49.8	11.54	[88]
Me$_2$PhCOOH	AcNHBu	CCl$_4$	294–337	20.5	49.0	10.81	[88]
Me$_2$PhCOOH	AcNHCHMe$_2$	CCl$_4$	294–332	21.3	55.6	6.75	[88]
Me$_2$PhCOOH	pyridine	pyridine	298	17.7	49.4	3.33	[87]
Me$_2$PhCOOH	CH$_2$=CHCOOMe	CH$_2$=CHCOOMe	293–343	14.6	37.2	4.13	[48]
Me$_2$PhCOOH	CH$_2$=CMeCOOMe	CH$_2$=CMeCOOMe	293–343	13.8	42.7	1.54	[48]
Me$_2$PhCOOH	CH$_2$=CHCN	CH$_2$=CHCN	293–343	12.5	33.5	2.76	[48]
Me$_2$PhCOOH	pyridine		253–289	29.3	75.3	15.9	[86]
cyclohexyl-Ph-OOH	cyclohexyl-Ph-OOH	CCl$_4$	269–312	21.8	83.6	0.28	[80]
Me$_2$C(OOH)CH=CHMe	Me$_2$C(OOH)CH=CHMe	CCl$_4$	269–304	18.8	73.7	0.28	[89]
Ph$_3$COOH	Ph$_3$COOH	CCl$_4$	269–313	23.8	82.0	0.77	[85]
Ph$_3$COOH	MeCOMe	CCl$_4$	269–313	14.6	50.2	0.86	[85]
Ph$_3$COOH	MeOH	CCl$_4$	269–313	16.3	54.4	1.04	[85]
Ph$_3$COOH	cyclohexanone	CCl$_4$	269–313	10.9	31.4	1.86	[85]

TABLE 4.11
Rate Constants of the Unimolecular Decomposition of Hydroperoxides and Peracids

Hydroperoxide	Solvent	T (K)	E (kJ mol^{-1})	log A, A (s^{-1})	k (400 K) (s^{-1})	Ref.
Methyl-, MeOOH	Gas	565–651	180.0	14.91	2.54×10^{-9}	[68]
Ethyl-, EtOOH	Gas	553–653	180.0	15.35	6.99×10^{-9}	[68]
1-Methylethyl-, Me$_2$CHOOH	Gas	553–653	180.0	15.50	9.88×10^{-9}	[68]
1,1-Dimethylethyl-, Me$_3$COOH	Gas	553–653	180.0	15.60	1.24×10^{-8}	[68]
1,1-Dimethylethyl-, Me$_3$COOH	Toluene	423	180.0	16.10	3.94×10^{-8}	[90]

4.3.2 BIMOLECULAR DECOMPOSITION OF HYDROPEROXIDES

The bimolecular reaction of hydroperoxide decomposition to free radicals

$$ROOH + HOOR \longrightarrow RO^\bullet + H_2O + RO_2^\bullet$$

was first observed by Bateman and coworkers [91,92] in the kinetic study of the autoxidation of nonsaturated acids and their esters. He found that the rate of chain initiation in an oxidized substrate was proportional to the square of hydroperoxide concentration ($v_i \sim [ROOH]^2$). This result was proved later by the free radical acceptor method [79]. The rate of initiation by Me_3COOH was found to be

$$v_{03} = k_{i1}[ROOH] + k_{i2}[ROOH]^2 \times (1 + K[ROOH])^{-1} \qquad (4.5)$$

where k_{i1} characterizes the unimolecular decay of hydroperoxide (or decay by the reaction of hydroperoxide with the solvent) and k_{i2} characterizes the decomposition of a complex between two molecules of ROOH, forming the hydrogen bond $ROOH \cdots O(H)OR$. The coefficient K is the equilibrium constant of formation of this complex. The values of K are given in Table 4.12. So, the "bimolecular" decomposition of hydroperoxide occurs in two steps. Two molecules of hydroperoxide form a complex through the hydrogen bond. Then this complex is decomposed into free radicals and water molecule.

$$ROOH + ROOH \rightleftharpoons ROOH \cdots O(H)OR \qquad (K)$$
$$ROOH \cdots O(H)OR \longrightarrow RO_2^\bullet + H_2O + RO^\bullet \qquad (k_{i2})$$

This bimolecular decomposition of Me_3COOH occurs with endothermicity $\Delta H = D_{O-H} + D_{O-O} - D_{HO-H} = 358.6 + 175.0 - 498 = 35.6 \text{ kJ mol}^{-1}$ in comparison with $\Delta H = 180 \text{ kJ mol}^{-1}$ for unimolecular decay. Hence the bimolecular decay occurs much more rapidly than the unimolecular decay. The equilibrium constant K and k_{i2} were found to be $K = 1.9 \text{ L mol}^{-1}$ (363 K) and $k_{i2} = 5.8 \times 10^7 \exp(-96.1/RT) = 8.61 \times 10^{-7} \text{ L mol}^{-1} \text{ s}^{-1}$ (363 K) for 1,1-dimethylethyl hydroperoxide [79] and $K = 3.3 \times 10^{-4} \exp(25.0/RT) = 1.3 \text{ L mol}^{-1}$ (363 K), $k_{i2} = 2.2 \times 10^8 \exp(-106.0/RT) = 1.23 \times 10^{-7} \text{ L mol}^{-1} \text{ s}^{-1}$ (363 K) for 1-methyl-1-phenylethyl hydroperoxide [93]. The experimental data on the bimolecular homolytic decomposition of hydroperoxides are collected in Table 4.12.

Let us compare the rate constants of decay of one molecule (k_{i1}) and a complex of two molecules (k_{i2}) of 1,1-dimethylethyl hydroperoxide.

T (K)	350	375	400	450
k_{i1} (s^{-1})	5.48×10^{-12}	3.39×10^{-10}	1.25×10^{-8}	5.11×10^{-6}
k_{i2} (s^{-1})	3.33×10^{-8}	3.78×10^{-7}	3.16×10^{-6}	1.09×10^{-4}

It is seen that the dimer is decomposed into free radicals by two to four orders of magnitude more rapidly than one molecule of hydroperoxide. Due to the difference in their activation energies, this difference increases with lowering of temperature.

4.3.3 BIMOLECULAR REACTIONS OF HYDROPEROXIDES WITH π-BONDS OF OLEFINS

The reaction of hydroperoxide (Me_3COOH) with the π-bond of styrene with free radical generation

TABLE 4.12
Rate Constants and Activation Energies of the Bimolecular Reactions of Hydroperoxides $2ROOH \rightarrow RO_2^{\bullet} + H_2O + RO^{\bullet}$ (Experimental Data)

ROOH	Solvent	T (K)	E (kJ mol^{-1})	log A, A (L mol^{-1} s^{-1})	k (350 K) (L mol^{-1} s^{-1})	Ref.
HOOH	Cyclohexanol	393–413	121.7	9.84	4.76 × 10^{-9}	[94]
Me$_3$COOH	Heptane	333–363	96.1	8.04	4.99 × 10^{-7}	[78]
Me$_2$C(OOH)Et	2-Methylbutane	333–363	100.0	7.80	7.52 × 10^{-8}	[95]
CH$_2$=CHCHMeOOH	1-Butene	338–353	108.8	11.70	2.90 × 10^{-5}	[96]
MeCH=CHCH$_2$OOH	2-Butene	338–353	117.1	12.70	1.67 × 10^{-5}	[96]
Cyclohexyl-OOH	Benzene	333–353	120.0	11.40	3.10 × 10^{-7}	[97]
Me$_2$PhCOOH	Cumene	333–368	80.7	4.86	6.55 × 10^{-8}	[93]
Me$_2$PhCOOH	Chlorobenzene	373–393	68.6	2.38	1.38 × 10^{-8}	[98]
Me$_2$PhCOOH	Cumene	393			1.90 × 10^{-6}	[99]
Me$_2$PhCOOH	Chlorobenzene	393			6.02 × 10^{-7}	[99]
Me$_2$PhCOOH	Nitrobenzene	393			4.47 × 10^{-5}	[99]
PhCH(OOH)CH=CH$_2$	Benzene, 2-propenyl-	388			2.82 × 10^{-7}	[91]
CH$_2$=CHCH(OOH)CH$_2$Bu	1-Octene	388			2.88 × 10^{-7}	[91]
Cyclohexenyl-OOH	Cyclohexene	388			5.37 × 10^{-7}	[91]
3-methyl-cyclohexenyl-OOH	Cyclohexene, 3-methyl-	388			1.15 × 10^{-6}	[91]
3,3,6-trimethyl-cyclohexenyl-OOH	Cyclohexene, 3,3,6-trimethyl-	388			3.24 × 10^{-6}	[91]
EtCH=CMeCH(OOH)Et	Heptene-3, -4-methyl	388			1.7 × 10^{-6}	[91]
Me$_2$C=CHCH(OOH)Bu	Octene-2, -2-methyl	388			5.0 × 10^{-7}	[91]
Methyloleate	Methyloleate	388			4.5 × 10^{-6}	[91]
Ethyllinoleate	Ethyllinoleate	388			4.7 × 10^{-6}	[91]

$$ROOH + CH_2=CHPh \longrightarrow RO^{\bullet} + HOCH_2C^{\bullet}HPh$$

was observed by the free radical acceptor method in 1964 by Denisov and Denisova [100]. The rate of free radical initiation was found to obey the following equation:

$$v_i = k_i[ROOH][PhCH=CH_2], \qquad (4.6)$$

with the rate constant $k_i = 1.2 \times 10^4 \exp(-72.0/RT)$ L mol^{-1} s^{-1}. The enthalpy of this reaction is $\Delta H = D_{O-O} - D_{R-OH} = 175 - 165 = 10$ kJ mol^{-1}. The rate constants of free radical generation by this reaction measured by different methods are in good agreement [101].

It is very probable that the preliminary formation of the CTC between hydroperoxide and olefin precedes the reaction.

$$\text{ROOH} + \text{CH}_2=\text{CHR}^1 \rightleftharpoons \text{ROOH}(\delta^-) \cdots \text{CH}_2=\text{CHR}^1(\delta^+)$$

$$\text{ROOH}(\delta^-) \cdots \text{CH}_2=\text{CHR}^1(\delta^+) \longrightarrow [\text{RO}^\bullet + \text{HOCH}_2\text{C}^\bullet\text{HR}^1]$$

$$[\text{RO}^\bullet + \text{HOCH}_2\text{C}^\bullet\text{HR}^1] \longrightarrow \text{ROH} + \text{HOCH}=\text{CHR}^1$$

$$[\text{RO}^\bullet + \text{HOCH}_2\text{C}^\bullet\text{HR}^1] \longrightarrow \text{HOCH}_2\text{C}(\text{OR})\text{HR}^1$$

$$[\text{RO}^\bullet + \text{HOCH}_2\text{C}^\bullet\text{HR}^1] \longrightarrow \text{RO}^\bullet + \text{HOCH}_2\text{C}^\bullet\text{HR}^1$$

The formation of the hydrogen bond between hydroperoxide and polar monomer, for example, methyl acrylate or acrylonitrile, does not influence the rate constant of the reaction of hydroperoxide with the double bond of monomer [101]. The values of the rate constants of the reaction of hydroperoxide with olefins are given in Table 4.13. The effect of multidipole interaction was observed for reactions of hydroperoxide with polyfunctional monomers (see Table 4.14, $\Delta G_\mu^{\neq}$ is the Gibbs energy of multidipole interaction in the transition state).

TABLE 4.13
Rate Constants and Activation Energies of the Reactions of Hydroperoxides with Monomers: ROOH + CH$_2$=CHR → RO$^\bullet$ + HOCH$_2$C$^\bullet$HR (Experimental Data)

ROOH	CH$_2$=CHR	Solvent	T (K)	E (kJ mol^{-1})	log A, A (L mol^{-1} s^{-1})	k (350 K) (L mol^{-1} s^{-1})	Ref.
Me$_3$COOH	CH$_2$=CHPh	Chlorobenzene	328–363	72.0	4.08	2.16×10^{-7}	[100]
Me$_3$COOH	CH$_2$=CHPh	Styrene	323–353	79.5	5.21	2.21×10^{-7}	[48]
Me$_2$PhCOOH	CH$_2$=CHPh	Styrene	323–353	84.2	6.35	6.08×10^{-7}	[48]
Me$_2$PhCOOH	CH$_2$=CHCOOMe	Methyl acrylate	323–353	82.6	5.88	3.12×10^{-7}	[101]
Me$_2$PhCOOH	CH$_2$=CHCN	Acrylonitrile	323			5.4×10^{-8}	[101]
Me$_2$PhCOOH	CH$_2$=CMeCOOMe	Methyl methacrylate	323–363	78.2	5.01	2.18×10^{-7}	[101]
Me$_2$PhCOOH	CH$_2$=CMeCOOBu	Butyl methacrylate	323			1.9×10^{-8}	[101]
Me$_2$PhCOOH	CH$_2$=CEtCOOMe	Methylethacrylate	323			3.2×10^{-8}	[101]
Me$_2$PhCOOH	CH$_2$=CMePh	Benzene	363			7.1×10^{-7}	[101]
Me$_2$PhCOOH	(CH$_2$=CHC(O)OCH$_2$)$_4$C	Chlorobenzene	343			9.40×10^{-8}	[102]
Me$_2$PhCOOH	(CH$_2$=CHC(O)OCH$_2$)$_3$CH	Chlorobenzene	343			1.83×10^{-7}	[102]
Me$_2$PhCOOH	CH$_2$=CHC(O)OCH$_2$C(CH$_2$OCOEt)$_3$	Chlorobenzene	343			1.9×10^{-8}	[103]

TABLE 4.14
The Reactivity of Polyfunctional Esters in Bimolecular Reaction with 1-Methyl-1-Phenyl Hydroperoxide

Monomer	k_i (343 K) (L mol^{-1} s^{-1})	$\Delta G_\mu^{\neq}$ (kJ mol^{-1})	Refs
CH$_2$=CHCOOMe	17.5×10^{-8}	0	[101]
(CH$_2$=CHCOOCH$_2$)$_3$CEt	18.3×10^{-8}	3.0	[102]
(CH$_2$=CHCOOCH$_2$)$_4$C	11.2×10^{-8}	5.6	[102]
CH$_2$=CHCOOCH$_2$(CH$_2$OCOEt)$_3$	1.9×10^{-8}	6.3	[102]

4.3.4 BIMOLECULAR REACTIONS OF HYDROPEROXIDES WITH C—H, N—H, AND O—H BONDS OF ORGANIC COMPOUNDS

The bimolecular reaction of the type

$$\text{ROOH} + \text{HY} \longrightarrow \text{RO}^\bullet + \text{H}_2\text{O} + \text{Y}^\bullet$$

has a sufficiently lower enthalpy than the unimolecular decomposition of hydroperoxide ($D_{O-O} = 180$ kJ mol^{-1}) because in most cases, $D_{Y-H} < D_{HO-H} = 498$ kJ mol^{-1}. These reactions can occur with an activation energy lower than the D_{O-O} of hydroperoxide. A few reactions of this type with the initiation rate $v_i = k_i$ [ROOH] [YH] were observed where YH was either a hydrocarbon [93,98], or an alcohol [95,104], or a carboxylic acid [105,106]. The experimental data are presented in Table 4.15. These data will be interpreted later in comparison with the results of the rate constants calculated by the IPM method.

TABLE 4.15
Rate Constants and Activation Energies of the Bimolecular Reactions of Hydroperoxides with Substrates ROOH + YH → Free Radicals (Experimental Data)

ROOH	Y—H	Solvent	T (K)	E (kJ mol^{-1})	log A, A (L mol^{-1} s^{-1})	k (400 K) (L mol^{-1} s^{-1})	Ref.
EtMe$_2$COOH	Me$_3$CCH$_2$CHMe$_2$	Isooctane	333–363	115.1	8.42	2.43×10^{-7}	[95]
Me$_2$PhCOOH	Me$_2$PhC—H	C$_6$H$_5$Cl	373–393	109.0	7.70	2.92×10^{-7}	[98]
Me$_2$PhCOOH	Me$_2$PhC—H	Me$_2$PhCH	333–368	110.0	8.00	4.32×10^{-7}	[93]
⌬—OOH	cyclo-C$_6$H$_{12}$	cyclo-C$_6$H$_{12}$	353–393	133.9	10.83	2.23×10^{-7}	[104]
EtMe$_2$COOH	CH$_3$CH$_2$O—H	Isooctane	348			2.0×10^{-7}	[95]
EtMe$_2$COOH	EtMe$_2$CO—H	Isooctane	333–396	93.3	8.54	2.27×10^{-4}	[95]
⌬—OOH	⌬—OH	C$_6$H$_5$Cl	353–393	92.0	8.64	4.23×10^{-4}	[104]
Me$_2$PhCOOH	PhC(O)O—H	Me$_2$PhCH	363–383	86.0	9.85	4.17×10^{-2}	[105]
MeCH(OOH)(CH$_2$)$_7$Me	BuCH$_2$COO—H	Me(CH$_2$)$_8$Me	293–433	67.4	6.30	3.15×10^{-3}	[106]
Me$_3$COOH	C$_5$H$_5$N	C$_6$H$_5$Cl	333–368	53.0	3.54	4.16×10^{-4}	[107]

4.4 PARABOLIC MODEL OF BIMOLECULAR HOMOLYTIC REACTION

4.4.1 Main Equations of IPM

The IPM as a semiempirical model of an elementary bimolecular reaction appeared to be very useful and efficient in the analysis and calculation of the activation energies for a wide variety of radical abstraction and addition reactions [108–113]. As a result, it became possible to classify diverse radical abstraction reactions and to differentiate in each class the groups of isotypical reactions. Later this conception was applied to the calculations of activation energies and rate constants of bimolecular reactions of chain generation [114]. In the IPM, the radical abstraction reaction, for example,

$$ROOH + CH_2=CHY \longrightarrow RO^{\bullet} + HOCH_2C^{\bullet}HY$$

in which the hydroxyl radical is transferred from the initial molecule (ROOH) to the final radical (HOCH$_2$C$^{\bullet}$HY), is regarded as the result of an intersection of two potential curves, one of which describes the potential energy $U_i(r)$ of the vibration of the O atom along the dissociating bond (O—O bond) in the initial molecule, and another describes the potential energy $U_f(r)$ of the vibration of the same atom along the forming bond (C—O bond) in the reaction product (U is the potential energy and r is the amplitude of the atomic vibrations along the valence bond). The stretching vibrations of the H atom in RH and XH are regarded as harmonic and described by the parabolic law:

$$\sqrt{U(r)} = br \qquad (4.7)$$

The following parameters are used to characterize the elementary step [109,110]:

1. The enthalpy of reaction ΔH_e includes the difference between the zero-point energies of the broken and generated bonds

$$\Delta H_e = D_i - D_f + 0.5hL(\nu_i - \nu_f), \qquad (4.8)$$

   where D_i and D_f are the dissociation energies of the broken (i) and generated (f) bonds, respectively, and ν_i and ν_f are the stretching vibrational frequencies of the formed and generated bonds, respectively.

2. The activation energy E_e is related to the experimentally determined Arrhenius energy E by the equation

$$E_e = E + 0.5(hL\nu_i - RT). \qquad (4.9)$$

3. The coefficients b_i and b_f describe the dependence of the potential energy on the atomic vibrational amplitude along the valence bonds. There is a parabolic relationship between the potential energy and the vibration amplitude

$$U_i = b_i^2 r^2 \quad \text{and} \quad U_f = b_f^2 (r_e - r)^2 \qquad (4.8)$$

   The quantity $2b^2$ is the force constant of the corresponding bond with $b = \pi\nu(2\mu)^{1/2}$, where μ is the reduced mass of the atoms forming the bond.

4. The parameter r_e characterizes the distance between the two minimum points of the intersecting parabolas. This parameter is equal to the sum of the elongation of the two transformed bonds in the TS of the reaction.

In the IPM, these parameters are related by the equation:

$$br_e = \alpha\sqrt{E_e - \Delta H_e} + \sqrt{E_e} \qquad (4.11)$$

where $b = b_i$, i.e., refers to the attacked bond in the molecule, while $\alpha = b_i/b_f$. In the case of structurally isotypical reactions ($br_e = const.$), a thermally neutral reaction ($\Delta H_e = 0$) occurs with the activation energy E_{e0}, which is determined by two parameters, namely, α and br_e (or r_e, b_i, and b_f):

$$\sqrt{E_{e0}} = \frac{br_e}{1+\alpha} \qquad (4.12)$$

On substituting the parameter br_e in Equation (4.12) by its value from Equation (4.11), we get

$$(1+\alpha)\sqrt{E_{e0}} = \alpha\sqrt{E_e - \Delta H_e} + \sqrt{E_e} \qquad (4.13)$$

The transition state in the IPM is characterized by the distance $r^{\neq}$, which is related to other parameters by the equation

$$r^{\neq}/r_e = E_e^{1/2}/br_e = \left[1 + \alpha(1 - \Delta H_e E_e^{-1})\right]^{-1} \qquad (4.14)$$

In the case of a thermally neutral reaction, $r^{\neq}$ is equal to the ratio

$$(r^{\neq}/r_e)_0 = (1+\alpha)^{-1} \qquad (4.15)$$

The parabolic model is, in essence, empirical because the parameter α is calculated from spectroscopic (ν_i and ν_f) and atomic (μ_i and μ_f) data, while the parameter br_e (or E_{e0}) is found from the experimental activation energies $E(E = RT\ln(A/k))$, where A is the pre-exponential factor typical of the chosen group of reactions, and k is the rate constant. The enthalpy of reaction is calculated by Equation (4.6). The calculations showed that $br_e = const.$ for structurally similar reactions. The values of α and br_e for reactions of different types are given in Table 4.16.

The activation energies for highly endothermic reactions are known to be virtually equal to the enthalpy of the reaction. According to IPM, each group of reactions is characterized by the critical value of the enthalpy of the reaction $\Delta H_{e\,max}$. When the reaction enthalpy $\Delta H_e > \Delta H_{e\,max}$, the activation energy $E = \Delta H + 0.5RT$, whereas $\Delta H_{e\,max}$ depends on parameters α and br_e [115].

$$\Delta H_{e\,max} = (br_e)^2 - 2\alpha br_e\sqrt{0.5hL\nu_f} + 0.5hL\nu_f(\alpha^2 - 1) \qquad (4.16)$$

Special attention should be paid to the pre-exponential factors of the reactions [115]. For a separate group of reactions, the quantity A calculated for a single attacked bond is constant $A = A_0$ when $\Delta H_e < \Delta H_{e\,max}$. For reactions with $\Delta H_e > \Delta H_{e\,max}$, the parameter r_e increases with an increase in ΔH_e and, hence, the pre-exponential factor A also increases. The dependence of A on ΔH is described by the formulas [115]:

$$A = A_0\left\{1 + \beta\left[\sqrt{\Delta H_e} - \sqrt{\Delta H_{e\,max}}\right]\right\}^2 s \qquad (4.17)$$

The mean value of the coefficient β is equal to 1.3. Table 4.17 presents the values of $\Delta H_{e\,max}$ calculated by Equation (4.16). Evidently, the range of ΔH_e in which $A = A_0 = const.$ depends on the parameter br_e and varies within wide limits.

TABLE 4.16
Kinetic Parameters of the Bimolecular Reactions of the Free Radicals and Molecules with Dioxygen and Hydroperoxide Calculated by the IPM Method [109]

Reaction	α	b $((kJ\ mol^{-1})^{1/2}\ m^{-1})$	$0.5hL\nu_i$ (kJ mol^{-1})	$0.5hL(\nu_i-\nu_f)$, (kJ mol^{-1})	$(r^{\neq}/r_e)_0$
$O_2\ (O_3) + HY \rightarrow HO_2^{\bullet}\ (HO_3^{\bullet}) + Y^{\bullet}$ [116–119]					
$O_2\ (O_3) + HR$	0.814	3.743×10^{11}	17.4	−3.8	0.551
$O_2\ (O_3) + HNR_2$	0.936	4.306×10^{11}	20.0	−1.2	0.516
$O_2\ (O_3) + HOR$	1.022	4.701×10^{11}	21.7	0.5	0.495
$O_2\ (O_3) + HOOR$	1.000	4.600×10^{11}	21.2	0.0	0.500
$O_2\ (O_3) + HSiR_3$	0.624	2.871×10^{11}	13.1	−8.1	0.616
$O_2\ (O_3) + HSR$	0.658	3.026×10^{11}	13.8	−7.4	0.603
$O_2\ (O_3) + HGeR_3$	0.608	2.796×10^{11}	12.6	−8.6	0.622
$O_2\ (O_3) + HSnR_3$	0.586	2.695×10^{11}	12.1	−9.1	0.630
$O_2 + CH_2{=}CHR \rightarrow {}^{\bullet}OOCH_2C^{\bullet}HR$ [120]					
$O_2 + CH_2{=}CHR$	1.737	5.389×10^{11}	9.9	4.6	0.366
$ROOH + HY \rightarrow RO^{\bullet} + H_2O + Y^{\bullet}$ [121–124]					
$ROOH + HR^1$	0.788	3.743×10^{11}	17.4	−4.8	0.559
$ROOH + HNR_2$	0.907	4.306×10^{11}	20.0	−2.2	0.524
$ROOH + HOR^1$	0.990	4.701×10^{11}	21.7	−0.5	0.502
$ROOH + HOOR^1$	0.969	4.600×10^{11}	21.2	−1.0	0.508
$ROOH + HSiR_3$	0.604	2.871×10^{11}	13.1	−9.1	0.623
$ROOH + HSR^1$	0.637	3.026×10^{11}	13.8	−8.4	0.611
$ROOH + HGeR_3$	0.589	2.796×10^{11}	12.6	−9.6	0.629
$ROOH + HSnR_3$	0.567	2.695×10^{11}	12.1	−10.1	0.638
$RC^{\bullet}HYH + R^1OOH \rightarrow RCH{=}Y + H_2O + R^1O^{\bullet}$ [121,125]					
$R^1OOH + RC^{\bullet}HCH_2{-}H$	0.788	3.743×10^{11}	17.4	−4.8	0.559
$R^1OOH + RC^{\bullet}HNH{-}H$	0.907	4.306×10^{11}	20.0	−2.2	0.524
$R^1OOH + RC^{\bullet}HO{-}H$	0.990	4.701×10^{11}	21.7	−0.5	0.502
$R^{\bullet} + HOOR^1 \rightarrow ROH + R^1O^{\bullet}$ [126]					
$R^{\bullet} + HOOR^1$	0.889	3.238×10^{11}	5.1	−1.1	0.529
$ROOH + CH_2{=}CHR \rightarrow RO^{\bullet} + HOCH_2C^{\bullet}HR$ [126]					
$ROOH + CH_2{=}CHR$	0.889	3.238×10^{11}	5.1	−1.1	0.541

4.4.2 Calculation of E and k of a Bimolecular Reaction

Within the framework of the IPM, the values of E and k are calculated by means of the following formulas [110–112]:

(a) Reactions with $\Delta H_e < \Delta H_{e\ max}$.

(1) for $\alpha = 1$
$$\sqrt{E_e} = \frac{br_e}{2} + \frac{\Delta H_e}{2br_e} \tag{4.18}$$

(2) for $\alpha \neq 1$
$$\sqrt{E_e} = \frac{br_e}{1-\alpha^2}\left[1 - \alpha\sqrt{1 - \frac{1-\alpha^2}{(br_e)^2}\Delta H_e}\right] \tag{4.19}$$

(3) for $\Delta H_e(1-\alpha^2) \ll (br_e)^2$
$$\sqrt{E_e} = \frac{br_e}{1+\alpha} + \frac{\alpha \Delta H_e}{2br_e} \tag{4.20}$$

TABLE 4.17
Kinetic Parameters of the Bimolecular Reactions of Free Radical Generation Calculated by the IPM Method [116]

Reaction	br_e ($[\text{kJ mol}^{-1}]^{1/2}$)	E_{e0} (kJ mol^{-1})	A_0 ($\text{L mol}^{-1}\text{s}^{-1}$)	$\Delta H_{e\,max}$ (kJ mol^{-1})
	$O_2 + HY \rightarrow HO_2^\bullet + Y^\bullet$ [116–119]			
$O_2 + HR_1$	13.61	56.3	5.0×10^9	76.1
$O_2 + HR_2$	15.20	70.2	5.0×10^9	110.1
$O_2 + HR_3$	14.32	62.3	5.0×10^9	90.6
$O_2 + NH_3$	9.36	23.4	4.0×10^9	4.3
$O_2 + H_2NR$	10.80	31.1	4.0×10^9	20.9
$O_2 + HNR_2$	11.73	36.7	4.0×10^9	33.9
$O_2 + ROH$	14.45	51.1	5.0×10^9	73.7
$O_2 + ROOH$	13.13	43.1	5.0×10^9	51.5
$O_2 + HSiR_3$	12.41	58.4	5.0×10^9	69.7
$O_2 + RSH$	10.39	39.3	5.0×10^9	33.0
$O_2 + HGeR_3$	12.74	62.8	5.0×10^9	77.6
$O_2 + HSnR_3$	12.35	60.6	5.0×10^9	72.0
	$O_2 + CH_2{=}CHR \rightarrow {}^\bullet OOCH_2C^\bullet HR$ [120]			
$CH_2{=}CHR$	26.04	90.5	2.0×10^9	408.5
$CH_2{=}CHPh$	27.23	99.0	2.0×10^8	540.3
	$O_3 + HY \rightarrow HO_3^\bullet + Y^\bullet$ [127]			
$O_3 + HR_1$	13.61	56.3	2.0×10^9	76.1
$O_3 + HR_2$	15.20	70.2	2.0×10^8	110.1
$O_3 + HR_3$	14.32	62.3	2.0×10^8	90.6
$O_3 + NH_3$	9.36	23.4	2.0×10^9	4.3
$O_3 + H_2NR$	10.80	31.1	2.0×10^9	20.9
$O_3 + HNR_2$	11.73	36.7	2.0×10^9	33.9
$O_3 + HOR$	14.45	51.1	2.0×10^9	73.7
$O_3 + HOOR$	13.13	43.1	2.0×10^9	51.5
$O_3 + HSiR_3$	12.41	58.4	2.0×10^9	69.7
$O_3 + RSH$	10.39	39.3	6.4×10^8	33.0
$O_3 + HGeR_3$	12.74	62.8	1.0×10^9	77.6
$O_3 + HSnR_3$	12.35	60.6	1.0×10^9	72.0
	$ROOH + HY \rightarrow RO^\bullet + H_2O + Y^\bullet$ [122–124]			
$ROOH + HR_1$	19.85	123.2	1.0×10^9	239.7
$ROOH + HR_2$	21.15	139.9	1.0×10^8	283.5
$ROOH + HR_3$	20.35	129.5	1.0×10^8	256.2
$ROOH + H_2NR^1$	17.33	82.6	1.0×10^9	149.7
$ROOH + HNR_2$	18.48	93.9	1.0×10^9	181.2
$ROOH + HOR_1$	19.94	101.4	1.0×10^9	213.0
$ROOH + HOOR$	20.45	107.9	1.0×10^9	228.6
$ROOH + HSiR_3$	15.80	97.0	1.0×10^9	145.6
$ROOH + HSR^1$	18.40	126.3	1.0×10^9	214.9
$ROOH + HGeR_3$	16.71	110.6	1.0×10^9	172.0
$ROOH + HSnR_3$	17.15	119.8	1.0×10^9	187.4

continued

TABLE 4.17
Kinetic Parameters of the Bimolecular Reactions of Free Radical Generation Calculated by the IPM Method [116]—continued

Reaction	br_e ($[\text{kJ mol}^{-1}]^{1/2}$)	E_{e0} (kJ mol^{-1})	A_0 (L mol^{-1} s^{-1})	$\Delta H_{e\,max}$ (kJ mol^{-1})
RC·HYH + R'OOH → RCH=Y + H₂O + R'O· [121]				
R₁C·HCH₃ + R'OOH	19.85	123.2	1.0×10^9	239.7
R₂C·HCH₃ + R'OOH	21.15	139.9	1.0×10^8	283.5
R₃C·HCH₃ + R'OOH	20.35	129.5	1.0×10^8	256.2
RC·HNH₂ + R'OOH	17.33	82.6	1.0×10^{10}	149.7
RC·HNRH + R'OOH	18.48	93.9	1.0×10^{10}	181.2
RC·HOH + R'OOH	20.45	107.9	1.0×10^{10}	228.6
R· + R'OOH → ROH + R'O· [125]				
R₁· + R'OOH	18.93	100.4	1.0×10^9	273.2
R₂· + R'OOH	19.92	111.2	1.0×10^9	307.3
R₃· + R'OOH	19.32	104.6	1.0×10^9	286.4
RCH=CH₂ + HOOR' → RC·HCH₂OH + R'O· [126]				
ROOH + CH₂=CHR'	18.93	100.4	1.0×10^9	273.2
ROOH + CH₂=CHCH=CHR'	19.92	111.2	1.0×10^9	307.3
ROOH + CH₂=CHPh	19.32	104.6	1.0×10^9	286.4

The activation energy E and the rate constant k are calculated by the formulas:

$$E = E_e - 0.5hL\nu_i + 0.5RT \qquad (4.21)$$

$$k = n_{C-H} A_0 \exp(-E/RT) = A \exp(-E/RT), \qquad (4.22)$$

where n_{C-H} is the number of attacked C—H bonds with the same reactivity, and A_0 is the standard pre-exponential factor per attacked C—H bond for the chosen group of reactions.

(b) Reactions with $\Delta H_e > \Delta H_{e\,max}$.

The activation energy for these reactions is close to the enthalpy of the reaction:

$$E = \Delta H + 0.5RT \qquad (4.23)$$

The pre-exponential factor depends on the reaction enthalpy, and the rate constant is equal to the following:

$$k = n_{C-H} \times A_0 e^{-1/2}\left\{1 + \beta\left[\sqrt{\Delta H_e} - \sqrt{\Delta H_{e\,max}}\right]\right\}^2 \exp(-\Delta H/RT) \qquad (4.24)$$

(c) Reactions with concerted dissociation of several bonds.

In the bimolecular reactions of the type

$$ROOH + HY \longrightarrow RO^{\bullet} + H_2O + Y^{\bullet}$$

more than one bond, i.e., two bonds O—O and Y—H, dissociate simultaneously. The concerted concentration of the activation energy on the two breaking bonds is necessary to perform this reaction. The probability of this event depends on the activation energy and the number of broken bonds. According to the oscillation theory of decay of a polyatomic molecule, this probability $P(E)$ for the reaction with the simultaneous dissociation of two bonds is described by the equation [36]

$$P(E) = \sqrt{\frac{2RT}{\pi E}} \tag{4.25}$$

and the rate constant of this reaction has the following form:

$$k = n_{Y-H} A_0 \sqrt{\frac{2RT}{\pi E}} \exp(-E/RT) = A \sqrt{\frac{2RT}{\pi E}} \exp(-E/RT) \tag{4.26}$$

where A_0 is the pre-exponential factor typical of similar reactions with dissociation of one Y—H bond. The activation energy is calculated as described above.

The parameters used in the IPM are presented in Table 4.16 and Table 4.17. In these tables, the additional parameters for the reactions of hydroperoxides with molecules and free radicals are given. The reactions of hydroperoxide with free radicals are important for the chain processes of the decomposition of hydroperoxides (see later). The results of the calculation of rate constants of various hydroperoxide reactions are collected in Tables 4.18–4.21. The comparison of the calculated values with the experimental values helped to introduce a few corrections in the traditional view on the bimolecular reactions of hydroperoxides.

4.4.3 BIMOLECULAR REACTIONS OF RADICAL GENERATION

4.4.3.1 Bimolecular Splitting of Hydroperoxides

The calculated values of the rate constants are close to those from experimental measurements (Table 4.12 and Table 4.18).

Hydroperoxide	H_2O_2	Me_3COOH	$Me_2PhCOOH$
k_i (L mol^{-1} s^{-1}) (calculated)	7.97×10^{-9}	2.90×10^{-8}	2.90×10^{-8}
k_i (L mol^{-1} s^{-1}) (experimental)	4.76×10^{-9}	0.50×10^{-8}	6.55×10^{-8}

4.4.3.2 Reaction of Hydroperoxides with Olefins

We observe close values for the reaction of 1,1-dimethylethyl hydroperoxide at $T = 350$ K with styrene: $k_i = 2.2 \times 10^{-7}$ L mol^{-1} s^{-1} (Table 4.13, experiment) and $k_i = 0.85 \times 10^{-7}$ L mol^{-1} s^{-1} (Table 4.19, calculation).

TABLE 4.18
Enthalpies, Activation Energies, and Rate Constants of Bimolecular Reactions of Hydroperoxides 2ROOH → ROO• + H$_2$O + RO• Calculated by the IPM Method [122–124]

ROOH	n	ΔH_e (kJ mol^{-1})	E (kJ mol^{-1})	Log A, A (L mol^{-1} s^{-1})	k (350 K) (L mol^{-1} s^{-1})
HOOH	2	50.0	114.1	8.93	7.97 × 10^{-9}
Me$_2$CHOOH	1	46.5	112.2	8.63	7.68 × 10^{-9}
C$_6$H$_{11}$—OOH	1	46.5	112.2	8.63	7.68 × 10^{-9}
Me$_3$COOH	1	39.6	108.4	8.64	2.90 × 10^{-8}
PhMe$_2$COOH	1	39.6	108.4	8.64	2.90 × 10^{-8}
Me$_2$C(OH)OOH	1	52.6	115.5	8.63	2.47 × 10^{-9}
CCl$_3$OOH	1	88.2	135.8	8.60	2.15 × 10^{-12}
CBr$_3$OOH	1	88.0	135.7	8.60	2.23 × 10^{-12}
CCl$_3$CCl$_2$OOH	1	88.2	135.8	8.60	2.15 × 10^{-12}

4.4.3.3 Reaction of Hydroperoxides with Hydrocarbons

The enthalpy of the reaction

$$ROOH + HR^1 \longrightarrow RO^• + H_2O + R^{1•}$$

is $\Delta H = D_{C-H} + 175 - 498 = D_{C-H} - 323$ kJ mol^{-1} and decreases with the decreasing value of the BDE of the C—H bond. The activation energy of this reaction is sufficiently higher than its enthalpy (see Table 4.15) due to the influence of the triplet repulsion and nonlinear configuration of the reaction center of the transition state on the activation energy. The experimental values of the activation energy of the hydroperoxide reaction with cumene (Table 4.15) are sufficiently lower than the calculated values (see Table 4.20). Probably, this is the result of another parallel reaction of initiation, namely, the reaction of hydroperoxide with the aromatic ring.

$$ROOH + C_6H_5R^1 \longrightarrow RO^• + HOC_6^•H_5R^1$$
$$HOC_6^•H_5R^1 + O_2 \longrightarrow HO_2^• + HOC_6H_4R^1$$

As in the case of the reaction of hydroperoxide with the π-bond of the olefin, the reaction of ROOH with the π-bond of the aromatic ring occurs more rapidly than the attack of ROOH on the C—H bond of alkylaromatic hydrocarbon.

4.4.3.4 Reaction of Hydroperoxides with Alcohols and Acids

The reaction of alcohol with ROOH

$$ROOH + HOR^1 \longrightarrow RO^• + H_2O + R^1O^•$$

TABLE 4.19
Enthalpies, Activation Energies, and Rate Constants of the Reactions of 1,1-Dimethylethyl Hydroperoxide with Monomers: $Me_3COOH + CH_2=CHR \rightarrow Me_3CO^\bullet + HOCH_2C^\bullet HR$ Calculated by the IPM Method [124,126]

$CH_2=CHR$	n_C	ΔH_e (kJ mol^{-1})	E (kJ mol^{-1})	log A, A (L mol^{-1} s^{-1})	k (400 K) (L mol^{-1} s^{-1})
$CH_2=CH_2$	2	53.5	124.0	9.30	1.28×10^{-7}
$CH_2=CHMe$	1	55.2	124.9	9.00	4.90×10^{-8}
$CH_2=CMe_2$	1	50.3	122.3	9.00	1.07×10^{-7}
$CH_2=CHEt$	1	54.9	124.8	9.00	5.05×10^{-8}
cyclohexene	2	55.2	124.9	9.30	9.77×10^{-8}
cycloheptene	2	52.2	123.3	9.30	1.58×10^{-7}
$HC\equiv CH$	2	33.2	113.3	9.60	6.38×10^{-6}
$HC\equiv CMe$	1	36.9	115.2	9.30	1.81×10^{-6}
$MeC\equiv CMe$	2	48.9	121.5	9.60	5.42×10^{-7}
$CH_2=CHCH=CH_2$	2	63.0	110.8	9.30	6.78×10^{-6}
$CH_2=CHCMe=CH_2$	1	54.0	110.4	9.00	3.83×10^{-6}
$CH_2=CHCH=CMe_2$	1	−0.3	107.7	9.00	8.63×10^{-6}
1,3-cyclohexadiene	2	−13.4	101.1	9.30	1.26×10^{-4}
$CH_2=CHPh$	1	13.6	107.7	9.00	8.63×10^{-6}
$CH_2=CMePh$	1	13.1	107.5	9.00	9.16×10^{-6}
$MeCH=CMePh$	1	3.0	102.6	9.00	4.00×10^{-5}
$Me_2C=CMePh$	1	3.0	102.6	9.00	4.00×10^{-5}
$CH_2=CPh_2$	1	−45.7	80.5	9.00	3.07×10^{-2}
dihydronaphthalene	1	12.9	107.4	9.00	9.44×10^{-6}

occurs with a rate constant (calculated by the IPM method) of few orders of magnitude lower than the rate constant measured experimentally. Moreover, the experimental values of the activation energy are lower than ΔH of hydrogen abstraction reaction. This implies that a quite different reaction was observed experimentally instead of the reaction written above. This reaction cannot be the reaction of hydroperoxide with the C—H bond of the alcohol because the rate constants of free radical generation in the reaction of ROOH with secondary and tertiary alcohols are close (see Table 4.15), whereas secondary and tertiary alcohols have C—H bonds of different strengths and the rate constants of ROOH reactions with C—H bonds should be quite different. Experiment demonstrates the close rate constants (see Table 4.15). This implies that a quite different mechanism of free radical generation predominates in the system $ROOH + R^1OH$. The same situation is observed in the reaction of hydroperoxides with acids (see Table 4.15 and Table 4.21). The most probable explanation lies in the acid

TABLE 4.20
Enthalpies, Activation Energies, and Rate Constants of the Bimolecular Reactions of Hydroperoxides with Hydrocarbons ROOH + HR → RO• + H$_2$O + R• Calculated by the IPM Method [122–124]

RH	n_{RH}	ΔH_e (kJ mol^{-1})	E (kJ mol^{-1})	log A, A (L mol^{-1} s^{-1})	k (400 K) (L mol^{-1} s^{-1})
EtMeCH—H	4	90.2	153.2	9.60	4.01 × 10^{-11}
Me$_3$C—H	1	77.2	146.3	9.00	7.96 × 10^{-11}
cyclobutane-H	8	95.7	156.1	9.90	3.28 × 10^{-11}
cyclopentane-H	10	85.6	150.7	10.00	2.10 × 10^{-10}
cycloheptane-H	14	81.1	148.3	10.15	6.02 × 10^{-10}
cyclohexane-H	1	72.7	143.9	9.00	1.61 × 10^{-10}
Z-Decalin	2	64.8	139.9	9.30	1.09 × 10^{-9}
CH$_2$=CHCH$_2$—H	3	45.2	146.7	9.48	2.12 × 10^{-10}
CH$_2$=CH(CH—H)Me	2	27.0	138.0	9.30	1.88 × 10^{-9}
CH$_2$=CH(C—H)Me$_2$	1	16.8	133.4	9.00	3.87 × 10^{-9}
Z-MeCH=CH(CH—H)Me	2	21.2	135.4	9.30	4.22 × 10^{-9}
Me$_2$C=CH(CH—H)Me	2	9.2	129.9	9.30	2.17 × 10^{-8}
Me$_2$C=CMe(C—H)Me$_2$	1	0.0	125.8	9.00	3.72 × 10^{-8}
CH$_2$=C(MeC—H)CH=CH$_2$	1	−15.6	119.1	9.00	2.85 × 10^{-7}
cyclohexene-H	4	18.7	134.2	9.60	1.19 × 10^{-8}
cyclohexadiene-H	4	8.1	129.4	9.60	7.57 × 10^{-8}
cyclohexadiene-H	4	−10.2	121.4	9.60	5.67 × 10^{-7}
cycloheptatriene-H	2	−21.8	116.4	9.30	1.26 × 10^{-6}
MeC≡CC—HMe$_2$	1	6.6	128.7	9.00	1.54 × 10^{-8}
PhMeCH—H	2	41.3	134.5	9.30	5.53 × 10^{-9}
PhMe$_2$C—H	1	31.9	130.0	9.00	1.06 × 10^{-8}
tetralin-H	4	22.8	125.7	9.60	1.53 × 10^{-7}
dihydroanthracene-H	4	−0.8	115.1	9.60	3.75 × 10^{-6}
Me$_2$NH$_2$C—H	1	56.7	135.8	9.00	1.86 × 10^{-9}
(CH$_2$=CHCH—H)$_3$N	1	22.8	136.1	9.78	1.01 × 10^{-8}
Me$_2$(HO)C—H	1	67.7	141.4	9.00	3.49 × 10^{-10}

TABLE 4.20
Enthalpies, Activation Energies, and Rate Constants of the Bimolecular Reactions of Hydroperoxides with Hydrocarbons ROOH + HR → RO$^\bullet$ + H$_2$O + R$^\bullet$ Calculated by the IPM Method [122–124]—continued

RH	n_{RH}	ΔH_e (kJ mol^{-1})	E (kJ mol^{-1})	log A, A (L mol^{-1} s^{-1})	k (400 K) (L mol^{-1} s^{-1})
 cyclohexyl-CH(H)OH	1	6.9	112.2	9.00	2.27 × 10^{-6}
MeCH=CMe(C—H)MeOH	1	2.4	126.9	9.00	2.71 × 10^{-6}
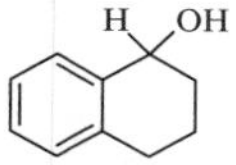 tetralin-H,OH	1	14.7	122.0	9.00	1.17 × 10^{-7}
PhC(O)—H	1	25.2	126.8	9.00	2.73 × 10^{-8}
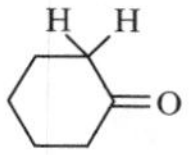 cyclohexanone-H,H	4	71.3	143.2	9.60	8.00 × 10^{-10}
(Me$_2$C—H)$_2$O	2	68.0	141.5	9.30	6.66 × 10^{-10}
(CH$_2$=CHCH—H)$_2$O	4	37.2	142.8	9.60	8.94 × 10^{-10}
Ph$_2$C—HOMe	1	31.4	129.8	9.00	1.14 × 10^{-8}
Ph(COOH)CH—H	2	44.2	135.9	9.30	3.63 × 10^{-9}
(Me$_3$C)$_3$Si—H	1	23.9	94.0	9.00	5.24 × 10^{-4}
Ph$_3$Ge—H	1	−5.1	96.4	9.00	2.60 × 10^{-4}
Ph$_3$Sn—H	1	−31.2	96.6	9.00	2.42 × 10^{-4}

catalysis of the reactions of hydroperoxide with acids and alcohols with formation of free radicals (see Chapter 10).

4.5 INITIATION BY REACTION OF HYDROPEROXIDES WITH KETONES

Hydroperoxides undergo reversible addition across the carbonyl group of a ketone with the formation of a new peroxide.

$$ROOH + R^1C(O)R^2 \rightleftharpoons R^1C(OH)(OOR)R^2$$

The formed hydroxyperoxide decomposes into free radicals much more rapidly than alkyl hydroperoxide [128]. So, the equilibrium addition of the hydroperoxide to the ketone changes the rate of formation of the radicals. This effect was first observed for cyclohexanone and 1,1-dimethylethyl hydroperoxide [128]. In this system, the rate of radical formation increases with an increase in the ketone concentration. The mechanism of radical formation is described by the following scheme:

$$ROOH \longrightarrow RO^\bullet + HO^\bullet \qquad (k_{i1})$$
$$ROOH + R^1C(O)R^2 \rightleftharpoons R^1C(OH)(OOR)R^2 \qquad (K)$$
$$R^1C(OH)(OOR)R^2 \longrightarrow R^1C(OH)(O^\bullet)R^2 + RO^\bullet \qquad (k_{i2})$$

TABLE 4.21
Enthalpies, Activation Energies, and Rate Constants of the Bimolecular Rreactions of Hydroperoxides with Molecules ROOH + HY → RO$^\bullet$ + H$_2$O + Y$^\bullet$ Calculated by the IPM Method [122–124]

Reaction	D (kJ mol^{-1})	n	ΔH$_e$ (kJ mol^{-1})	E (kJ mol^{-1})	log A, A (L mol^{-1} s^{-1})	k (400 K) (L mol^{-1} s^{-1})
ROOH + HNR^{1}R^2 → RO$^\bullet$ + H$_2$O + R^{1}R^{2}N$^\bullet$						
H$_2$NNH—H	366.1	4	45.9	139.3	8.59	2.51 × 10^{-10}
EtNH—H	418.4	2	98.2	173.8	8.24	3.50 × 10^{-15}
Et$_2$N—H	382.8	1	62.6	160.9	7.95	8.69 × 10^{-14}
ROOH + HOR^1 → RO$^\bullet$ + H$_2$O + R^{1}O$^\bullet$						
EtO—H	440.0	1	121.5	205.7	7.90	1.09 × 10^{-19}
Me$_2$CHO—H	443.0	1	124.5	207.8	7.90	5.81 × 10^{-20}
Me$_3$CO—H	446.0	1	127.5	209.9	7.89	3.02 × 10^{-20}
C$_6$H$_{11}$—OH	437.1	1	118.6	203.6	7.90	2.05 × 10^{-19}
ONO—H	327.6	1	9.1	134.2	7.99	2.92 × 10^{-10}
MeC(O)O—H	442.7	1	124.2	207.6	7.90	6.17 × 10^{-20}
PhCH$_2$C(O)O—H	443.5	1	125.0	208.1	7.90	5.31 × 10^{-20}

and the rate of radical generation is given by the following equation:

$$v_i = k_{i1}[\text{ROOH}] + k_{i2}[\text{R}^1\text{C(OH)(OOR)R}^2]$$
$$= (k_{i1} + k_{i2}K[\text{R}^1\text{C(O)R}^2])[\text{ROOH}](1 + K[\text{R}^1\text{COR}^2])^{-1} \quad (4.27)$$

The values of k_{i1}, k_{i2}, and K are collected in Table 4.22.

4.6 CHAIN DECOMPOSITION OF HYDROPEROXIDES

Different chain mechanisms of hydroperoxide decomposition are known with the participation of alkyl, alkoxyl, and peroxyl radicals [9].

Due to the ability of tertiary peroxyl radicals to disproportionate with the formation of alkoxyl radicals, the chain decomposition of tertiary hydroperoxides proceeds via the action of intermediate alkoxyl radicals [9,135].

$$\text{Me}_3\text{COOH} \longrightarrow \text{Me}_3\text{CO}^\bullet + \text{HO}^\bullet \qquad k_1$$
$$2\text{Me}_3\text{COOH} \longrightarrow \text{Me}_3\text{COO}^\bullet + \text{H}_2\text{O} + \text{Me}_3\text{CO}^\bullet \qquad k_2$$
$$\text{HO}^\bullet + \text{HOOCMe}_3 \longrightarrow \text{H}_2\text{O} + \text{Me}_3\text{CO}_2^\bullet \qquad \text{Fast}$$
$$\text{Me}_3\text{CO}^\bullet + \text{HOOCMe}_3 \longrightarrow \text{Me}_3\text{COH} + \text{Me}_3\text{CO}_2^\bullet \qquad \text{Fast}$$
$$\text{Me}_3\text{CO}_2^\bullet + \text{Me}_3\text{CO}_2^\bullet \longrightarrow 2\text{Me}_3\text{CO}^\bullet + \text{O}_2 \qquad k_3$$
$$\text{Me}_3\text{CO}_2^\bullet + \text{Me}_3\text{CO}_2^\bullet \longrightarrow \text{Me}_3\text{COOCMe}_3 + \text{O}_2 \qquad k_4$$

TABLE 4.22
Rate Constants of the Decomposition of ROOH (k_1) and $R^1C(OH)(OOR)R^2$ (k_2) and Equilibrium Constants K of Hydroxyalkyl Peroxide Formation from ROOH and Ketone

Hydroperoxide	Ketone	Solvent	T (K)	k_{i1} (s^{-1}) or k_{i2} (s^{-1}) or K (L mol^{-1})	Ref.
1,1-Dimethylethyl-, Me$_3$COOH	cyclohexanone	C$_6$H$_5$Cl	383–400	$k_{i1} = 3.6 \times 10^{12} \exp(-138.1/RT)$ $k_{i2} = 3.6 \times 10^9 \exp(-108.8/RT)$ $K = 6.9 \times 10^{-7} \exp(46.0/RT)$	[128]
1,1-Dimethylethyl-, Me$_3$COOH	cyclopentanone	CCl$_4$	295–313	$K = 7.61 \times 10^{-3} \exp(10.5/RT)$	[129]
EtMe$_2$COOH	EtC(O)Me	Isooctane	343	$k_{i2} K = 1.3 \times 10^{-7}$	[130]
1-Methyl-1-phenylethyl-, Me$_2$PhCOOH	cyclohexanone	C$_6$H$_5$Cl	393	$k_{i1} = 2.1 \times 10^{-6}$ $k_{i2} = 5.7 \times 10^{-6}$, $K = 1.0$	[131]
1-Methyl-1-phenylethyl-, Me$_2$PhCOOH	cyclohexanone	CCl$_4$	295–313	$K = 1.56 \times 10^{-3} \exp(12.5/RT)$	[129]
Cyclohexyl-,	cyclohexanone	C$_6$H$_5$Cl	353–383	$k_1 = 6.3 \times 10^{11} \exp(-133.9/RT)$ $k_2 = 4.0 \times 10^4 \exp(-63.6/RT)$ $K = 1.4 \times 10^{-5} \exp(33.0/RT)$	[132]
2-Oxocyclohexyl-,	cyclohexanone	C$_6$H$_5$Cl	393	$k_{i1} = 1.2 \times 10^{-5}$ $k_{i2} = 3.0 \times 10^7 \exp(-85.3/RT)$	[132]
1-Acetylethyl-, MeC(O)CH(OOH)Me	MeC(O)Et	C$_6$H$_6$	293–343	$k_{i1} = 3.4 \times 10^{12} \exp(-114.6/RT)$ $k_{i2} = 1.15 \times 10^{10} \exp(96.2/RT)$ $K = 0.8(293K)$	[133]

continued

TABLE 4.22
Constants of the Decomposition of ROOH (k_1) and R^{1}C(OH)(OOR)R^2 (k_2) and Equilibrium Constants K of Hydroxyalkyl Peroxide Formation from ROOH and Ketone—continued

Hydroperoxide	Ketone	Solvent	T (K)	k_{i1} (s^{-1}) or k_{i2} (s^{-1}) or K (L mol^{-1})	Ref.
1-Methyl-2-oxocyclohexyl-	2-methylcyclohexanone	2-methylcyclohexanone	373–403	$k_{i2} = 1.9 \times 10^6 \exp(-76.1/RT)$	[134]
2-Methyl-5-oxocyclohexyl-	3-methylcyclohexanone	3-methylcyclohexanone	373–403	$k_{i2} = 6.9 \times 10^6 \exp(-79.5/RT)$	[134]
3-Methyl-5-oxocyclohexyl-	4-methylcyclohexanone	4-methylcyclohexanone	373–403	$k_{i2} = 8.9 \times 10^5 \exp(-72.4/RT)$	[134]
1,1-Dimethylethyl-, Me$_3$COOH	4-methylcyclohexanone	CCl$_4$	295–313	$K = 7.60 \times 10^{-3} \exp(10.5/RT)$	[129]
1,1-Dimethylethyl-, Me$_3$COOH	3-methylcyclohexanone	CCl$_4$	295–313	$K = 1.17 \times 10^{-2} \exp(8.4/RT)$	[129]
1-Tetralyl-	MeC(O)Me	CCl$_4$/Tetralin (1:1)	293	$K = 0.53$	[129]
1-Tetralyl-	MeC(O)Et	CCl$_4$/Tetralin (1:1)	293	$K = 0.38$	[129]

1-Tetralyl-, cyclohexanone	CCl$_4$/Tetralin (1:1)	293–313	$K = 4.07 \times 10^{-2} \exp(10.5/RT)$	[129]
1-Tetralyl-, 2-methylcyclohexanone	CCl$_4$/Tetralin (1:1)	293	$K = 0.29$	[129]
1-Tetralyl-, 3-methylcyclohexanone	CCl$_4$/Tetralin (1:1)	293–313	$K = 0.23 \exp(8.4/RT)$	[129]
1-Tetralyl-, 4-methylcyclohexanone	CCl$_4$/Tetralin (1:1)	293–313	$K = 4.74 \times 10^{-2} \exp(9.6/RT)$	[129]

The rate of tertiary hydroperoxide decomposition is equal to

$$v = (k_1[\text{ROOH}] + 2k_2[\text{ROOH}]^2)(1 + k_3/k_4) \qquad (4.28)$$

The ratio $k_3/k_4 > 1$, for example, $k_3/k_4 = 7.2 \pm 1.1$ for the decay of 1,1-dimethylethyl hydroperoxide in benzene at 318 K [136], and $k_3/k_4 = 10$ for MeEtCHOOH in chlorobenzene (318 K) [137].

Secondary hydroperoxides are decomposed in oxidizing hydrocarbons in the chain reaction with peroxyl radicals [138].

$$R^1R^2\text{CHOOH} + RO_2^\bullet \longrightarrow \text{ROOH} + R^1R^2\text{C}=\text{O} + \text{HO}^\bullet$$
$$\text{HO}^\bullet + \text{RH} \longrightarrow H_2O + R^\bullet$$
$$R^\bullet + O_2 \longrightarrow RO_2^\bullet$$

In the absence of dioxygen when hydroperoxide initiates the formation of alkyl radicals, the following chain reaction of ROOH decomposition occurs [139].

$$\text{ROOH} \longrightarrow RO^\bullet + \text{HO}^\bullet$$
$$2\text{ROOH} \longrightarrow RO^\bullet + H_2O + RO_2^\bullet$$
$$RO^\bullet + R^1H \longrightarrow \text{ROH} + R^{1\bullet}$$
$$RO_2^\bullet + R^1H \longrightarrow \text{ROOH} + R^{1\bullet}$$
$$R^{1\bullet} + \text{ROOH} \longrightarrow R^1\text{OH} + RO^\bullet$$
$$R^{1\bullet} + R^{1\bullet} \longrightarrow R^1R^1$$

The chain decomposition of hydroperoxides was proved and studied for hydroperoxides produced by the oxidation of polyesters such as dicaprilate of diethylene glycol and tetravalerate of erythritol [140]. The retarding action of phenolic antioxidant on the decay of hydroperoxides was observed. The initial rate of hydroperoxide decomposition was found to depend on the hydroperoxide concentration in accordance with the kinetic equation typical for the induced chain decomposition.

$$v = k_d[\text{ROOH}] + k_{\text{ind}}[\text{ROOH}]^{3/2} \qquad (4.29)$$

The kinetics of hydroperoxide decomposition obeys the following equation:

$$\ln\frac{[\text{ROOH}]_0}{[\text{ROOH}]} + \ln\frac{[\text{ROOH}]^{1/2} + k_d/(k_{p1}+k_{p2})}{[\text{ROOH}]_0^{1/2} + k_d/(k_{p1}+k_{p2})} = k_d t, \qquad (4.30)$$

where k_{p1} and k_{p2} are the rate constants of the following chain propagation reactions:

$$R^\bullet + R^1\text{OOH} \longrightarrow \text{ROH} + R^1O^\bullet \qquad (k_{p1})$$
$$R^\bullet + R^1R^2\text{CHOOH} \longrightarrow \text{RH} + R^1R^2\text{C(O)} + \text{HO}^\bullet \qquad (k_{p2})$$

RH	$[\text{BuCH}_2\text{C(O)OCH}_2]_2$	$[\text{BuCOOCH}_2]_4\text{C}$
k_d (s^{-1})	$2.4 \times 10^8 \exp(-98.0/RT)$	$6.0 \times 10^8 \exp(-100.0/RT)$
$2ek_d$ (s^{-1})	$4.9 \times 10^{11} \exp(-131.0/RT)$	$2.1 \times 10^{11} \exp(-132.0/RT)$
$k_{p1}+k_{p2}$ (L mol^{-1} s^{-1})	$6.9 \times 10^7 \exp(-33.0/RT)$	$2.6 \times 10^7 \exp(-25.5/RT)$

The chain decomposition of hydroperoxide was proved for the hydroperoxide of dimethylacetamide [141]. The rate of chain decay $v_v = k_{ind}[ROOH]_0^{3/2}$. The values of the rate constant k_{ind} in dimethylacetamide were found to be the following [141,142]:

ROOH	Me$_3$COOH	Me$_2$PhCOOH	AcN(CH$_3$)CH$_2$OOH
T (K)	405	405	300–405
k_{ind} (L$^{1/2}$ mol$^{-1/2}$ s^{-1})	2.0×10^{-4}	9.0×10^{-4}	$8.5 \times 10^{-7} \exp(-87.0/RT)$

The primary and secondary alcohols induce the chain decomposition of hydroperoxides with the participation of ketyl radicals that possess high reducing activity [143–145].

$$ROOH \longrightarrow RO^\bullet + HO^\bullet$$
$$RO^\bullet + R^1CH(OH)R^2 \longrightarrow ROH + R^1C^\bullet(OH)R^2$$
$$R^1C^\bullet(OH)R^2 + ROOH \longrightarrow R^1R^2C(O) + H_2O + RO^\bullet \quad (k_p)$$
$$2R^1C^\bullet(OH)R^2 \longrightarrow R^1C(O)R^2 + R^1CH(OH)R^2 \quad (k_t)$$

The rate of chain decomposition of hydroperoxide $v = (k_p/\sqrt{2k_t})[ROOH]v_i^{1/2}$. Dioxygen reacts with ketyl radicals with the formation of hydroxyperoxyl radicals. The latter is decomposed into ketone and HO$_2^\bullet$. Hydroperoxyl radicals also possesses reducing activity and induce the chain decomposition of hydroperoxide [121,143].

$$RO^\bullet + R^1CH(OH)R^2 \longrightarrow ROH + R^1C^\bullet(OH)R^2$$
$$R^1C^\bullet(OH)R^2 + O_2 \longrightarrow R^1C(OH)(OO^\bullet)R^2$$
$$R^1C(OH)(OO^\bullet)R^2 \longrightarrow R^1C(O)R^2 + HO_2^\bullet$$
$$R^1C(OH)(OO^\bullet)R^2 + ROOH \longrightarrow R^1C(O)R^2 + H_2O + RO^\bullet + O_2$$
$$HO_2^\bullet + ROOH \longrightarrow H_2O + RO^\bullet + O_2$$
$$2R^1C(OH)(OO^\bullet)R^2 \longrightarrow R'C(O)R^2 + O_2 + R'C(OH)(OOH)R^2$$
$$HO_2^\bullet + HO_2^\bullet \longrightarrow H_2O_2 + O_2$$

The values of the rate constants of the chain propagation in chain reactions of ROOH decomposition are collected in Table 4.23.

4.7 KINETICS OF AUTOINITIATED HYDROCARBON OXIDATION

4.7.1 Initial Stage of Autoxidation

A characteristic feature of hydrocarbon autoxidation is the auto-accelerated kinetics of the chain-chemical process. In the absence of an initiator or an initiating agent (light, radiation), the initial rate of the chain initiation, v_{i0}, in an oxidized organic compound, RH, is very low [3,9,56]. During oxidation, hydroperoxide is formed and accumulates. With the increasing concentration of ROOH, the rate of initiation increases due to the decomposition of hydroperoxide into radicals. Therefore the rate of chain oxidation increases in time. We observe the positive feedback between the proceeding reaction and the reaction rate. The typical kinetic curve of hydroperoxide formation in an oxidized hydrocarbon is presented in Figure 4.1.

The kinetic curve has a sigmoidal shape and can be intuitively divided into four stages: (a) induction period, (b) stage of accelerated hydroperoxide formation, (c) stage of retarded ROOH formation, and (d) stage with prevalence of ROOH decomposition after the point

TABLE 4.23
Rate Constants of Free Radical Reactions with Hydroperoxides Calculated by the IPM Method

Radical	Hydroperoxide	E (kJ mol^{-1})	log A, A (L mol^{-1} s^{-1})	k(400 K) (L mol^{-1} s^{-1})
\multicolumn{5}{c}{$R^\bullet + R_1OOH \rightarrow ROH + R_1O^\bullet$ [126]}
$C^\bullet H_3$	H_2O_2	36.7	9.30	3.22×10^4
$CH_3C^\bullet H_2$	H_2O_2	35.4	9.30	4.77×10^4
$Me_2C^\bullet H$	H_2O_2	32.8	9.30	1.04×10^5
$Me_3C^\bullet$	H_2O_2	33.6	9.30	8.19×10^4
$PhC^\bullet H_2$	H_2O_2	48.6	9.30	9.01×10^2
$C^\bullet H_3$	Me_3COOH	27.5	9.00	2.56×10^5
$CH_3C^\bullet H_2$	Me_3COOH	26.4	9.00	3.57×10^5
$Me_2C^\bullet H$	Me_3COOH	24.1	9.00	7.13×10^5
$Me_3C^\bullet$	Me_3COOH	24.8	9.00	5.77×10^5
$PhC^\bullet H_2$	Me_3COOH	37.9	9.00	1.12×10^4
\multicolumn{5}{c}{$RCH_2C^\bullet H_2 + ROOH \rightarrow RCH=CH_2 + H_2O + RO^\bullet$ [121]}
$CH_3C^\bullet H_2$	Me_3COOH	47.0	8.54	2.55×10^2
$Me_2C^\bullet H$	Me_3COOH	46.2	8.85	6.58×10^2
$EtMeC^\bullet H$	Me_3COOH	43.4	8.40	5.38×10^2
$Me_3C^\bullet$	Me_3COOH	47.4	9.00	6.46×10^2
$MePhC^\bullet H$	Me_3COOH	64.7	8.48	1.06
$EtPhC^\bullet H$	Me_3COOH	62.0	8.30	1.60
\multicolumn{5}{c}{$RC^\bullet HOH + ROOH \rightarrow RCHO + H_2O + RO^\bullet$ [121]}
$C^\bullet H_2OH$	H_2O_2	13.7	8.82	1.07×10^7
$MeC^\bullet HOH$	H_2O_2	12.8	8.88	1.62×10^7
$Me_2C^\bullet OH$	H_2O_2	11.8	8.90	2.30×10^7
$PhC^\bullet HOH$	H_2O_2	20.5	8.78	1.26×10^6
$MePhC^\bullet OH$	H_2O_2	19.4	8.79	1.82×10^6
$MeC^\bullet HOH$	Me_3COOH	3.3	8.89	2.89×10^8
$Me_2C^\bullet OH$	Me_3COOH	2.4	8.95	4.37×10^8
$PhC^\bullet HOH$	Me_3COOH	9.8	8.65	2.36×10^7
$MePhC^\bullet OH$	Me_3COOH	8.8	8.67	3.33×10^7
\multicolumn{5}{c}{$HO_2^\bullet + ROOH \rightarrow RO^\bullet + H_2O + O_2$ [121]}
$HO_2^\bullet$	H_2O_2	47.1	8.60	2.83×10^2
$HO_2^\bullet$	$ROOH$	33.2	8.30	9.24×10^3

with the maximum hydroperoxide concentration. Let us compare the rates of chain initiation by the reaction of dioxygen v_{i0} and hydroperoxide with the same hydrocarbon (k_i(ROOH + RH), experimental values).

RH	Me_2CHEt	cyclo-C_6H_{12}	Me_2PhCH	Tetralin
T (K)	410	403	403	403
v_{i0} (mol L^{-1} s^{-1})	2.20×10^{-9}	1.53×10^{-9}	4.20×10^{-9}	1.00×10^{-8}
k_i[RH] (s^{-1})	3.48×10^{-6}	5.28×10^{-6}	2.09×10^{-6}	2.44×10^{-5}
[ROOH] = v_{i0}/k_i[RH]	6.32×10^{-4}	2.90×10^{-4}	2.01×10^{-3}	4.10×10^{-4}
Ref.	[11,95]	[10,146]	[13,105]	[147]

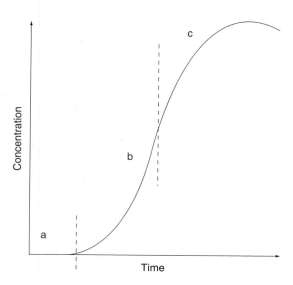

FIGURE 4.1 Typical kinetic curve of hydroperoxide formation during hydrocarbon oxidation.

It is seen that the initiation by the reaction of hydrocarbon with dioxygen is very slow and it is enough to introduce the very low hydroperoxide concentration (2×10^{-3}–6×10^{-4} mol L^{-1}) to succeed this value.

If radicals are produced in the reactions of unimolecular hydroperoxide decomposition and the reaction of ROOH with hydrocarbon whose concentration at the initial stages of oxidation is virtually constant, the production of radicals from ROOH can be regarded as a pseudo-monomolecular process occurring at the rate $v_i = k_i[\text{ROOH}] = (k_{id} + k_{iRH}[\text{RH}])$. The unimolecular splitting of hydroperoxides occurs in solutions more slowly than the bimolecular reaction with the C—H or C—C bond of the hydrocarbon (see earlier) and consequently $k_{id} \ll k_{iRH}[\text{RH}]$. Since the rate of chain oxidation, $v = a[\text{RH}]\, v_i^{1/2}$ (see Equation (2.1) in Chapter 2), and $v_i = v_{i0} + k_3[\text{ROOH}]$, autoxidation occurs with the following rate [3,9,146].

$$v = a(v_{i0} + k_{iRH}[\text{RH}][\text{ROOH}])^{1/2}[\text{RH}] \qquad (4.31)$$

When all the dioxygen consumed appears in hydroperoxide groups and the number of decomposed hydroperoxide molecules is low, the kinetics of oxygen uptake can be described by the following equation [3].

$$\frac{\Delta[\text{O}_2]}{t} = a[\text{RH}]\sqrt{v_{i0}} + b^2 t, \qquad (4.32)$$

where $a = k_p(2k_t)^{-1/2}$ and $b = 0.5a[\text{RH}](k_i\text{RH})^{1/2}$. At a very low value of $a[\text{RH}]\, v_{i0}^{1/2}$ (this is the case of a very low v_{i0}), the equation takes a simple form [3].

$$\sqrt{\Delta[\text{O}_2]} = bt \qquad (4.33)$$

The validity of this equation is confirmed by a large amount of experimental data on the oxidation of various individual hydrocarbons and hydrocarbon mixtures (see Chapter 5).

When the rate of initiation is very low the important moment of chain reaction becomes the kinetics of the establishment of the stationary concentration of free radicals. This time is comparable with the lifetime of the radical that reaction limits the chain propagation. The reaction of peroxyl radical limits the chain propagation in oxidized hydrocarbon. The mean lifetime of peroxyl radical $\tau = (2k_t v_i)^{-1/2}$ and the less initiation rate the longer the period of time for the process: $[RO_2^{\bullet}] \rightarrow [RO_2^{\bullet}]_{st} = \sqrt{v_i/2k_t}$. This process is described by the following equation [3]:

$$[RO_2^{\bullet}] = [RO_2^{\bullet}]_{st}\frac{e^{ct}-1}{e^{ct}+1} \approx \sqrt{\frac{v_{i0}}{2k_t}}\frac{e^{ct}-1}{e^{ct}+1} \quad (4.34)$$

where coefficient $c = 2(2k_t v_{i0})^{1/2} = \tau^{-1}$. The values of τ and $[RO_2^{\bullet}]_{st}$ calculated for different rates of initiation at $2k_t = 10^6$ L mol^{-1} s^{-1} are presented.

v_{i0} (mol L^{-1} s^{-1})	10^{-12}	10^{-11}	10^{-10}	10^{-9}	10^{-8}	10^{-6}
τ (s)	1.0×10^3	3.2×10^2	10^2	31.6	10.0	1.0
$[RO_2^{\bullet}]_{st}$ (mol L^{-1})	10^{-9}	3.2×10^{-9}	10^{-8}	3.2×10^{-8}	10^{-7}	10^{-6}

It is seen that τ can be 100 s and higher at $v_{i0} = 10^{-10}$ mol L^{-1} s^{-1} and less. When the period τ is longer than the time of reactor heating, one can take it into account and transform Equation (4.33) into the following form:

$$\Delta[O_2]^{1/2} = b(t - \tau) \quad (4.35)$$

If the taken hydrocarbons possess inhibiting impurities, period τ includes additionally the time of annihilation of this inhibitor.

4.7.2 Autoxidation with Bimolecular Hydroperoxide Decay

At relatively high concentrations hydroperoxide breaks down into free radicals by a bimolecular reaction (see earlier). The rate of initiation in oxidized hydrocarbon is equal to:

$$v_i = k_{i1}[RH][ROOH] + k_{i2}[ROOH]^2 \quad (4.36)$$

The oxidation rate v and [ROOH] are related in the following way [3].

$$\frac{v^2}{a^2[RH]^2[ROOH]} = k_{i1}[RH] + k_{i2}[ROOH] \quad (4.37)$$

This equation was used for the estimation of the rate constants k_{i1} and k_{i2} using experimental measurements [3]. The rate of chain oxidation is the following:

$$\frac{d[ROOH]}{dt} = a[RH]\left(v_{i0} + k_{i1}[RH][ROOH] + k_{i2}[ROOH]^2\right)^{1/2} \quad (4.38)$$

The kinetics of dioxygen consumption obeys the following equation when the amount of decomposed hydroperoxide is negligible in comparison with hydroperoxide formed:

$$\ln\left\{1 + 2\beta\Delta[O_2]\left(1 + \sqrt{1+(\Delta[O_2]\beta)^{-1}}\right)\right\} = \gamma t, \quad (4.39)$$

where $\beta = k_{i2}/k_{i1}[RH]$ (L mol^{-1}) and $\gamma = k_p[RH](k_{i2}/2k_t)^{1/2}$ (s^{-1}). When the hydroperoxide concentration is low and the initiation by reaction ROOH with RH predominate, i.e., $k_{i1}[RH]$

$\gg k_{i2}$[ROOH], Equation (4.39) is transformed into Equation (4.33) (see earlier). When the hydroperoxide concentration increases and bimolecular reaction of initiation becomes the main route of free radical generation, the kinetics of oxidation obeys the exponential law [3].

$$\Delta[\text{ROOH}] = [\text{ROOH}]_0 \exp \varphi t, \qquad (4.40)$$

where $\varphi = ak_{i2}^{1/2}$[RH]. The concentration of hydroperoxide [ROOH]$_{1/2}$ at this initiation rate by both reactions can be calculated from the values of the rate constants k_{i2} and k_{i1}: [ROOH]$_{1/2}$ = k_{i1}[RH]/k_{i2}. Here are a few examples.

RH	cyclohexene	Me$_2$PhCH	Ethyloleate
T (K)	378	350	328
k_{i1} (L mol^{-1} s^{-1})	4.63×10^{-8}	2.71×10^{-9}	6.10×10^{-9}
k_{i2} (L mol^{-1} s^{-1})	1.56×10^{-6}	1.38×10^{-8}	4.70×10^{-6}
[ROOH]$_{1/2}$ (mol l^{-1})	3.0×10^{-2}	0.19	1.47×10^{-3}
Ref.	[3]	[93]	[91]

4.7.3 Initiation by Reactions of Hydroperoxide with Products of Oxidation

The study of the interaction of hydroperoxide with other products of hydrocarbon oxidation showed the intensive initiation by reactions of hydroperoxide with formed alcohols, ketones, and acids [6,134]. Consequently, with the developing of the oxidation process the variety of reactions of initiation increases. In addition to reactions of hydroperoxide with the hydrocarbon and the bimolecular reaction of ROOH, reactions of hydroperoxide with alcohol and ketone formed from hydroperoxide appear. The values of rate constants (in L mol^{-1} s^{-1}) of these reactions for three oxidized hydrocarbons are given below.

RH	EtMe$_2$CH [96]	cyclo-C$_6$H$_{12}$ [104,138]	Me$_2$PhCH [105]
T (K)	350	350	350
ROOH + RH	1.24×10^{-9}	7.09×10^{-10}	2.71×10^{-9}
ROOH + ROOH	7.52×10^{-8}		1.38×10^{-8}
ROOH + ROH	4.17×10^{-6}	8.13×10^{-6}	
ROOH + ketone	1.3×10^{-7} (343 K)	5.42×10^{-6}	
ROOH + RC(O)OH	3.69×10^{-6}		1.04×10^{-3}

The concentrations of the products change during oxidation and, consequently, the absolute value as well as the ratio of partial initiation rates increases. The following table provides the example of dynamics of free radical generation in the autoxidation of 2-methylbutane at $T = 412$ K [130].

t (s)	10.5×10^3	16.5×10^3	19.5×10^3	25.5×10^3	31.5×10^3
[ROOH] (mol L^{-1})	0.04	0.14	0.23	0.52	0.86
v_i (ROOH + RH) (mol L^{-1} s^{-1})	1.40×10^{-7} (87%)	4.91×10^{-7} (58%)	8.02×10^{-7} (19%)	1.91×10^{-6} (5%)	3.00×10^{-6} (2%)
v_i (ROOH + ROOH) (mol L^{-1} s^{-1})	2.11×10^{-8} (13%)	2.59×10^{-7} (30%)	6.98×10^{-7} (16%)	3.57×10^{-6} (9%)	9.76×10^{-6} (5%)

continued

t (s)	10.5×10^3	16.5×10^3	19.5×10^3	25.5×10^3	31.5×10^3
[ROH] (mol L^{-1})	0.0	0.0	0.02	0.11	0.28
v_i(ROOH + ROH) (mol L^{-1} s^{-1})	0.0	0.0	2.35×10^{-6} (55%)	2.93×10^{-5} (76%)	1.23×10^{-4} (70%)
[MeCO$_2$H] (mol l^{-1})	0.0	0.003	0.008	0.03	0.19
v_i(ROOH + RCO$_2$H) (mol L^{-1} s^{-1})	0.0	1.03×10^{-7} (12%)	4.51×10^{-7} (10%)	3.82×10^{-6} (10%)	4.00×10^{-5} (23%)
$\Sigma\, v_i$ (mol L^{-1} s^{-1})	1.61×10^{-7}	8.53×10^{-7}	4.30×10^{-6}	3.86×10^{-5}	1.76×10^{-4}

We can see that the rate of initiation increases during autoxidation from 10^{-7} to 10^{-4} mol L^{-1} s^{-1} (about 1000 times). Due to increasing concentrations of 2-methylbutyl alcohol and acetic acid, the latter becomes a very important reactant in the reactions of free radical generation.

4.7.4 Hydroperoxide as the Intermediate Product of Autoxidation

In the oxidized hydrocarbon, hydroperoxides break down via three routes. First, they undergo homolytic reactions with the hydrocarbon and the products of its oxidation to form free radicals. When the oxidation of RH is chain-like, these reactions do not decrease [ROOH]. Second, the hydroperoxides interact with the radicals R$^\bullet$, RO$^\bullet$, and RO$_2^\bullet$. In this case, ROOH is consumed by a chain mechanism. Third, hydroperoxides can heterolytically react with the products of hydrocarbon oxidation. Let us consider two of the most typical kinetic schemes of the hydroperoxide behavior in the oxidized hydrocarbon. The description of 17 different schemes of chain oxidation with different mechanisms of chain termination and intermediate product decomposition can be found in a monograph by Emanuel et al. [3].

Scheme A. This scheme is typical of the hydrocarbons, which are oxidized with the production of secondary hydroperoxides (nonbranched paraffins, cycloparaffins, alkylaromatic hydrocarbons of the PhCH$_2$R type) [3,146]. Hydroperoxide initiates free radicals by the reaction with RH and is decomposed by reactions with peroxyl and alkoxyl radicals. The rate of initiation by the reaction of hydrocarbon with dioxygen is negligible. Chains are terminated by the reaction of two peroxyl radicals. The rates of chain initiation by the reactions of hydroperoxide with other products are very low (for simplicity). The rate of hydroperoxide accumulation during hydrocarbon oxidation should be equal to:

$$\frac{d[\text{ROOH}]}{dt} = (k_p[\text{RH}] - k_{p\text{ROOH}}[\text{ROOH}])\sqrt{\frac{k_{i1}[\text{ROOH}][\text{RH}]}{2k_t}}, \qquad (4.41)$$

and for the kinetics of hydroperoxide accumulation we have the following formula [3]:

$$[\text{ROOH}] = \frac{k_p[\text{RH}]}{k_{p\text{ROOH}}} \times \left\{\frac{e^{t/\tau} - 1}{e^{t/\tau} + 1}\right\}^2, \qquad (4.42)$$

where the characteristic time is $\tau^{-1} = k_{p\text{ROOH}}[\text{RH}]\,(k_{i1}k_p/2k_t k_{p\text{ROOH}})^{1/2}$. The maximum concentration of hydroperoxide is $[\text{ROOH}]_{\max} = k_p[\text{RH}]/k_{p\text{ROOH}}$. It is proportional to hydrocarbon oxidation and depends on the relative reactivity of the hydrocarbon and hydroperoxide toward peroxyl and alkoxyl radicals. The temperature dependence of $[\text{ROOH}]_{\max}$ is determined by the difference of activation energies between the free radical reactions with RH and ROOH. The BDE of the α-C—H bond of the hydroperoxide is lower than that of the C—H bond of the hydrocarbon and, consequently, $E_{p\text{ROOH}} < E_p$. Hence $[\text{ROOH}]_{\max}$ increases with temperature. For example, the values of [ROOH]$_{max}$ in oxidized cyclohexane were found to be the following [146]:

T (K)	428	418	408	398
t (s) × 10^{-3}	7.2	10.8	23.4	36.0
[ROOH]$_{max}$ (mol L^{-1})	0.115	0.076	0.093	0.034

When oxidation occurs for a long period of time, the hydrocarbon is consumed and this influences on the concentration of hydroperoxide and the rate of hydrocarbon oxidation. The exact solution for the description of hydrocarbon consumption during oxidation can be found by the common integration of two differential equations: one for hydrocarbon consumption and another for hydroperoxide accumulation. The approximation for the time of oxidation $t > t_{max}$, where t_{max} is the moment when [ROOH] = [ROOH]$_{max}$ gives the following equation [3,56]:

$$\frac{[RH]_0}{[RH]} = \left\{ \frac{k_p}{k_{p\,ROOH}} \left[\ln(e^{t/\tau} + 1) - \ln 2 - 0.5t/\tau \right] + 1 \right\}^2 \quad (4.43)$$

Scheme B. Oxidation occurs as a chain reaction in scheme A. However, hydroperoxide formed is decomposed not by the reaction with free radicals but by a first-order molecular reaction with the rate constant k_m [3,56]. This scheme is valid for the oxidation of hydrocarbons where tertiary C—H bonds are attacked. For $k_m \gg k_{i1}[RH]$ the maximum [ROOH] is attained at the hydroperoxide concentration when the rate of the formation of ROOH becomes equal to the rate of ROOH decay: $a[RH](k_{i1}[ROOH][RH])^{1/2} = k_m[ROOH]$; therefore, [ROOH]$_{max} = a^2 k_{i1} k_m^{-2} [RH]^3$. The kinetics of ROOH formation and RH consumption are described by the following equations [3].

$$[ROOH] = \frac{k_p^2 k_{i1} [RH]^3}{2k_t k_m^2} \left\{ 1 - e^{-0.5 k_m t} \right\} \quad (4.44)$$

$$[RH]^{-3/2} - [RH]_0^{-3/2} = \frac{3 k_p^2 k_{i1}}{4 k_t k_m} \left\{ t + 2 k_m^{-1} \left(e^{-0.5 k_m t} - 1 \right) \right\} \quad (4.45)$$

The rate constant is an effective characteristic of the heterolytic decomposition of hydroperoxide. Its value can depend on the concentration of acid, for example, and increase during oxidation.

4.7.5 Self-Inhibition of Hydrocarbon Oxidation

In the later stages, the oxidation of hydrocarbons is often *self-inhibited* due to the accumulation of such oxidation products that retard chain oxidation. Each hydrocarbon has the individual peculiarities of oxidation including the mechanism of self-inhibition. A few of such main peculiarities will be mentioned here [3,56].

1. Acids are the final products of all hydrocarbon oxidations. They catalyze the heterolytic decomposition of hydroperoxides. The sharp decrease in the hydroperoxide concentration in oxidizing the hydrocarbon is observed as soon as acids are formed in the oxidized hydrocarbon. Consequently, the rate of initiation and the rate of

oxidation decrease with an increase in the concentration of acids in oxidizing the hydrocarbon. The most probable mechanism of acid-catalyzed decay of hydroperoxide includes the following stages for the secondary and tertiary hydroperoxides:

$$ROOH + HA \rightleftharpoons ROOH_2^+ + A^-$$
$$R^1R^2CHOOH_2^+ + H_2O \longrightarrow R^1R^2C(O) + H_2O + H_3O^+$$
$$R^1R^2R^3COOH_2^+ + H_2O \longrightarrow R^2R^3C(O) + H_3O^+R^1OH$$

Any oxygen-containing product of the hydrocarbon oxidation can act as an acceptor of a proton.

2. When alkylaromatic hydrocarbon is oxidized, acids catalyze the decomposition of hydroperoxide with production of phenolic compounds [57,58].

$$PhR_2COOH + HA \rightleftharpoons PhR_2COOH_2^+ + A^-$$
$$PhR_2COOH_2^+ + H_2O \longrightarrow PhOH + R_2C(O) + H_3O^+$$

Phenols are antioxidants (see Chapter 12) and terminate chains retarding oxidation.

3. In addition, phenols are formed by the reaction of hydroxyl radical addition to the aromatic ring of oxidized alkylaromatic hydrocarbon [56].

The generation of hydroxyl radicals occurs by the reactions:

$$ROOH \longrightarrow RO^\bullet + HO^\bullet$$
$$RO_2^\bullet + R^1R^2CHOOH \longrightarrow ROOH + R^1R^2C(O) + HO^\bullet$$
$$HO_2^\bullet + HOOH \longrightarrow H_2O + HO^\bullet + O_2$$

4. Acids catalyze the hydrolysis of hydroperoxides with hydrogen peroxide formation.

$$H_2O + ROOH \longrightarrow ROH + H_2O_2$$

The latter generates hydroperoxyl radicals possessing the reducing activity. Hydroperoxyl radicals reduce hydroperoxide and accelerate chain termination by the reactions:

$$HO_2^\bullet + ROOH \longrightarrow RO^\bullet + H_2O + O_2$$
$$RO_2^\bullet + HO_2^\bullet \longrightarrow ROOH + O_2$$
$$HO_2^\bullet + HO_2^\bullet \longrightarrow HOOH + O_2$$

In addition, hydroperoxyl radicals become the active chain termination agents in the presence of ions and salts of transition metals (see Chapter 17).

5. Hydrocarbon oxidation in the presence of intermediate products proceeds as a chain reaction of co-oxidation of hydrocarbon with intermediate products of oxidation. The

accumulation of various intermediates in the oxidized RH leads to the appearance of alternative radicals (peroxyls and others). The accumulation of radicals that are less reactive than $RO_2^\bullet$ diminishes the total activity of the radicals and the effective value of parameter $a = k_p(2k_t)^{-1/2}$. The important peculiarity of a few types of radicals with different activities in chain propagation is that more active radicals cannot more than twice accelerate the reaction and less active can decrease the rate principally to zero [148,149].

The accumulation of hydroxyl-containing products, such as hydroperoxides, alcohols, acids, and water, also reduce the total activity of peroxyl radicals due to the hydrogen bonding with $RO_2^\bullet$ [150]. When acting together, these factors cause self-inhibition of autoxidation at conversion levels of 40–50% [3].

REFERENCES

1. NN Semenov. *Some Problems of Chemical Kinetics and Reactivity*, vols 1 and 2. London: Pergamon Press, 1958–1959.
2. WO Landberg (ed.). *Autoxidation and Antioxidants*, vols 1 and 2. New York: Wiley Interscience, 1962.
3. NM Emanuel, ET Denisov, ZK Maizus. *Liquid-Phase Oxidation of Hydrocarbons*. New York: Plenum Press, 1967.
4. C Walling. *Free Radicals in Solution*. New York: Wiley, 1962.
5. JL Bolland, G Gee. *Trans Faraday Soc* 42:236, 244, 1946.
6. HR Cooper, HW Melville. *J Chem Soc* 1984–1992, 1951.
7. ET Denisov. *Zh Fiz Khim* 42:1585–1597, 1978.
8. ET Denisov. In: CH Bamford and CFH Tipper (eds.), *Comprehensive Chemical Kinetics*, vol. 16. Amsterdam: Elsevier, 1980, pp. 125–203.
9. ET Denisov. In: EA Blyumberg (ed.), *Itogi Nauki i Tekhniki, Ser. Kinetika i Kataliz*, vol. 9. Moscow: VINITI, 1981, pp. 1–158 [in Russian].
10. LN Denisova, ET Denisov. *Kinet Katal* 10:1244–1248, 1969.
11. TG Degtyareva, LN Denisova, ET Denisov. *Kinet Katal* 13:1400–1404, 1972.
12. ET Denisov, LN Denisova. *Int J Chem Kinet* 8:123–130, 1975.
13. VL Antonovskii, ET Denisov, IA Kuznetsov, Yu Ya Mekhryushev, LV Solntseva. *Kinet Katal* 6:607–610, 1965.
14. GV Butovskaya, VE Agabekov, NI Mitskevich. *Dokl AN BSSR* 25:722–723, 1981.
15. VE Agabekov, GV Butovskaya, NI Mitskevich. *Neftekhimiya* 22:272–277, 1982.
16. ET Denisov. *Usp Khim* 44:1466–1486, 1985.
17. NN Pozdeeva, ET Denisov. *Izv AN SSSR Ser Khim* 2681–2686, 1987.
18. ET Denisov. *Pet Chem* 37:99–104, 1997.
19. DL Baulch, CJ Cobos, RA Cox, C Esser, P Frank, Th Just, JA Kerr, MJ Pilling, J Troe, RW Walker, J Warnatz. *J Phys Chem Ref Data* 21:411–429, 1992.
20. W Tsang, RF Hampson. *J Phys Chem Ref Data* 15:1087, 1986.
21. R Shaw. *J Phys Chem Ref Data* 7:1179, 1978.
22. GB Skinner, A Lipshitz, K Scheller, A Burcat. *J Chem Phys* 56:3853–3861, 1972.
23. SW Mayer, L Schieler. *J Phys Chem* 72:2628–2634, 1968.
24. JE Taylor, DM Kulich. *Int J Chem Kinet* 5:455–462, 1973.
25. W Tsang. *J Phys Chem Ref Data* 17:887, 1988.
26. RR Baldwin, GR Drewery, RW Walker. *J Chem Soc Faraday Trans* 1, 80:3195–3204, 1984.
27. W Tsang. *J Phys Chem Ref Data* 19:1, 1990.
28. DL Baulch, CJ Cobos, RA Cox, C Esser, P Frank, Th Just, JA Kerr, MJ Pilling, J Troe, RW Walker, J Warnatz. *J Phys Chem Ref Data* 23:847, 1994.
29. T Ingham, RW Walker, RE Woolforel. *Symp Int Combust Proc* 25:767, 1994.

30. P Mulder, R Louw. *J Chem Soc Perkin Trans* 2:1135–1142, 1985.
31. VA Machtin, MI Mishustina, EM Pliss, LA. Kashina, VI. Mishustin. *Vestn VVOATN RF Ser Khim and Khim Techn* 63, 1996.
32. AA Kutuev, NN Terpilovskii. *Kinet Katal* 16:372–376, 1975.
33. ET Denisov. *Dokl AN SSSR* 130:1055–1058, 1960.
34. ET Denisov. *Dokl. AN SSSR* 141:131–134, 1961.
35. ET Denisov. *Kinet Katal* 4:53–59, 1963.
36. IV Aleksandrov. *Teor Experim Khim* 12:299–306, 1976.
37. BI Chernyak, MV Nikipanchuk, MG Kotur, SI Kozak. *Kinet Katal* 37:818–820, 1996.
38. N Yanishilieva, IP Skibida, ZK Maizus. *Izv Otd Khim Nauk Bulg AN* 4:1, 1971.
39. GE Dingchan, NS Khanukova, RL Vardanyan. *Arm Khim Zh* 30:644–649, 1977.
40. DJ Carlsson, JC Robb, *Trans Faraday Soc* 62:3403–3412, 1966.
41. RL Vardanyan, IG Verner, ET Denisov. *Kinet Katal* 14:575–578, 1973.
42. VE Agabekov. The Reactions and Reactivity of Oxygen-containing Compounds in Free Radical Reactions of Oxidation. Doctoral Thesis, Institute of Chemical Physics, Chernogolovka, 1980 pp. 3–43 [in Russian].
43. LN Denisova, NN Shafikov, ET Denisov. *Dokl AN SSSR* 213:376–378, 1973.
44. EH Farmer. *Trans Faraday Soc* 42:228–236, 1946.
45. AA Miller, FR Mayo. *J Am Chem Soc* 78:1017–1023, 1956.
46. LN Denisova, ET Denisov. *Izv AN SSSR Ser Khim* 1702–1704, 1965.
47. VA Machtin, MI Mishustina, EM Pliss. *Vestn. VVO ATN RF, Ser Khim and Khim Techn* 59, 1996.
48. MM Mogilevich, EM Pliss. *Oxidation and Oxidative Polymerization of Nonsaturated Compounds.* Moscow: Khimiya, 1990 [in Russian].
49. EM Pliss, VM. Troshin, ET Denisov. *Dokl AN SSSR* 264:368–370, 1982.
50. AV Sokolov, EM Pliss, ET Denisov. *Bull Acad Sci USSR* 219–223, 1988.
51. EI Finkel'shteyn, NA Mednikova, IS Panasenko, EI Kozlov. *Zh Org Khim* 17:933–936, 1981.
52. AB Gagarina, OT Kasaikina, NM Emanuel. *Dokl AN SSSR* 212:399–402, 1973.
53. VD Sukhov, MM Mogilevich. *Zh Fiz Khim* 53:1477–1480, 1979.
54. G. Scott. *Atmospheric Oxidation and Antioxidants.* Amsterdam: Elsevier, 1965.
55. ET Denisov, TG Denisova. *Handbook of Antioxidants.* Boca Raton, FL: CRC Press, 2000.
56. ET Denisov, VV Azatyan. *Inhibition of Chain Reactions.* London: Gordon and Breach, 2000.
57. EGE Hawkins. *Organic Peroxides.* London: Spon, 1961.
58. VL Antonovskii. *Organic Peroxides as Initiators.* Moscow: Khimiya, 1972 [in Russian].
59. S Patai (ed.), *The Chemistry of Peroxides.* New York: Wiley, 1983.
60. VL Antonovskii, MM Buzlanova. *Analytical Chemistry of Organic Peroxides.* Moscow: Khimiya, 1978 [in Russian].
61. C Wagner, R Smith, D Peters. *Anal Chem* 19:976–981, 1947.
62. D Barnard, N Hargrave. *Anal Chim Acta* 10:476–484, 1951.
63. L Dulog, K-H Burg. *Z Anal Chem* 203:184–195, 1964.
64. W Gase, I Boggs. *J Molec Struct.* 116:207–214, 1984.
65. W Lobuncze, JR Rittenhouse, JG Miller. *J Am Chem Soc* 80:3505–3509, 1958.
66. YuYa Van-Chin-Syan, NS Kachurina. In: Vestnik Nizhegorodskogo Gosudarstvennogo Universiteta. Organic and Elementoorganic Peroxides, Nizhniy Novgorod: Izd-vo Nizhegorodskogo Universiteta, 1996, p. 29 [in Russian].
67. SG Lias, JF Liebman, RD Levin, SA Kafafi. *NIST Standard Reference Database, 19A, NIST Positive Ion Energetics, Version 2.0.* Gaithersburg: NIST, 1993.
68. SW Benson, HE O'Neal. *Kinetic Data on Gas Phase Unimolecular Reactions.* Washington: NSRDS, 1970.
69. NA Kozlov, IB Rabinovich. In: Trudy po Khimii i Khimicheskoy Tekhnologii, vol. 2 (10) Gorkiy: Gorkiy State University, 1964, p. 189 [in Russian].
70. ER Bell, FH Dickey, JH Raley. *Ind Eng Chem* 41:2597–2604, 1949.
71. W Pritzkow, K A Muller. *Chem Ber* 89:2318–2324, 1956.
72. YuYa Van-Chin-Syan, TN Dolbneva, MA Dikiy, SK Chuchmarev. *Zh Fiz Khim* 58:2937–2940, 1984.

73. JW Breitenbach, J Derkosch. *Monatsh Chem* 81:689–695, 1950.
74. JP Fortuin, HI Waterman. *Chem Eng Sci* 2:182–189, 1953.
75. W Pritzkow, R Hoffman. *J Prakt Chem N* 14:13–22, 1961.
76. ET Denisov, TG Denisova. *Zh Fiz Khim* 57:304–309, 1988.
77. JL Holmes, FP Lossing, PM Mayer. *J Am Chem Soc* 113:9723–9728, 1991.
78. ET Denisov, TG Denisova. *Kinet Catal* 34:173–179, 1993.
79. ET Denisov. *Zh Fiz Khim* 38:2085–2087, 1964.
80. OP Yablonsky, NS Lastochkina, VA Belyaeva. *Neftekhimiya* 13:851–855, 1973.
81. C Walling, LD Heaton. *J Am Chem Soc* 87:48–51, 1965.
82. C Walling, LD Heaton. *J Am Chem Soc* 87:38–47, 1965.
83. GL Bitman. Intermolecular Association of *tert*-Butyl Hydroperoxide. Ph.D. thesis, ITKhT, Moscow, 1973.
84. GE Zaikov, ZK Maizus, NM Emanuel. *Izv AN SSSR Ser Khim* 256–260, 1968.
85. IF Franchuk, LI Kalinina. *Teor Experim Khim* 8:553–557, 1972.
86. OP Yablonsky, AK Kobyakov, VA Belyaev. *Neftekhimiya* 14:446–449, 1974.
87. EI Solov'eva. Physico-Chemical Study of Assotiative Interaction of Organic Peroxides with Bases. PhD Thesis, GGU, Gor'kiy, 1979.
88. TI Drozdova, VT Varlamov, VP Lodygina, AL Aleksandrov. *Zh Fiz Khim* 52:735–736, 1978.
89. OP Yablonskii, LF Lapuka, NM Rodionova, ZA Pokrovskaya, VA Belyaeva. *Neftekhimiya* 14:101–105, 1974.
90. RR Hiatt, WMJ Stachan. *J Org Chem* 28:1893–1902, 1963.
91. L Bateman. *Quart Rev* 8:147–162, 1954.
92. L Bateman, H Hughes, A Moris. *Disc Faraday Soc* 14:190–202, 1953.
93. NI Solomko, VF Tsepalov, AI Yurzhenko. *Kinet Katal* 9:766–772, 1968.
94. ET Denisov, VV Kharitonov. *Kinet Katal* 5:781–786, 1964.
95. TG Degtyareva, VM Solyanikov, ET Denisov. *Neftekhimiya* 12:854–861, 1972.
96. A Chauvel, G Clement, JC Balaceanu. *Bull Soc Chim Fr* 2025–2032, 1963.
97. AV Tobolsky, RE Mesrobian, *Organic Peroxides: their Chemistry, Decomposition and Role in Polymerization.* New York: Interscience, 1954.
98. VL Antonovskii, ET Denisov, LV Solntseva. *Kinet Katal* 6:815–819, 1965.
99. IP Shevchuk. The Study of Solvent Influence on Liquid-Phase Oxidation of Cumene, 1,1-Diphenylethane and Nonene-1. PhD thesis, L'vov Polytech. Inst.:L'vov, 1968.
100. ET Denisov, LN Denisova. *Dokl AN SSSR* 157:907–909, 1964.
101. EM Pliss, VM Troshin. *Neftekhimiya* 22:539–542, 1982.
102. VM Troshin, EM Pliss, ET Denisov. *Izv AN SSSR Ser Khim* 2191–2194, 1984.
103. AV Sokolov, EM Pliss, ET Denisov. *Bull Acad Sci SSSR Ser Khim* 37:219–223, 1988.
104. AE Semenchenko, VM Solyanikov, ET Denisov. *Neftekhimiya* 10:864–869, 1970.
105. VL Antonovskii, ET Denisov, LV Solntseva. *Kinet Katal* 7:409–413, 1966.
106. LG Privalova, ZK Maizus, NM Emanuel. *Dokl AN SSSR* 161:1135–1137, 1965.
107. NV Zolotova, ET Denisov. *Izv AN SSSR Ser Khim* 767–768, 1966.
108. ET Denisov. *Mendeleev Commun* 2:1, 1992.
109. ET Denisov. *Kinet Catal* 33:50–57, 1992.
110. ET Denisov. *Kinet Catal* 35:614–635, 1994.
111. ET Denisov. *Russian Chem Rev* 66:859–876, 1997.
112. ET Denisov. *Russian Chem Rev* 69:153–164, 2000.
113. ET Denisov. In: ZB Alfassi (ed.), *General Aspects of the Chemistry of Radicals.* Chichester: Wiley, 1999, pp. 79–137.
114. ET Denisov, TG Denisova. *Russ Chem Rev* 71:417–438, 2002.
115. ET Denisov. *Kinet Catal* 37:519–523, 1996.
116. ET Denisov. *Pet Chem* 37:99–104, 1997.
117. ET Denisov. *Kinet Catal* 39:17–23, 1998.
118. ET Denisov, TI Drozdova. *Kinet Catal* 43:14–22, 2002.
119. ET Denisov, TG Denisova. *Kinet Catal* 37:46–50 1996.
120. TG Denisova, ET Denisov. *Pet Chem* 38:12–18, 1998.

121. ET Denisov. *Neftekhimiya* 39:434–444, 2000.
122. TG Denisova, ET Denisov. *Kinet Catal* 40:223–232, 1999.
123. ET Denisov. *Kinet Catal* 40:217–222, 1999.
124. TG Denisova, ET Denisov. *Pet Chem* 40:65–72, 2000.
125. ET Denisov. *Russ Chem Bull* 2110–2116, 1998.
126. TS Pokidova, ET Denisov. *Neftekhimiya* 38:269–276, 1998.
127. ET Denisov, TG Denisova. *Kinet Catal* 37:46–50, 1996.
128. ET Denisov. *Dokl AN SSSR* 146:394–397, 1962.
129. VL Antonovskii, VA Terentiev. *Zh Fiz Khim* 43:2727–2729, 1969.
130. TG Degtyareva. Mechanism of Liquid-phase Oxidation of 2-Methlbutane. Ph.D. thesis, Institute of Chemical Physics, Chernogolovka, 1972.
131. ET Denisov. *Zh Fiz Khim* 37:1896–1899, 1963.
132. ET Denisov, LN Denisova. *Izv AN SSSR Ser Khim* 1731–1737, 1963.
133. GE Zaikov, ZK Maizus, NM Emanuel. *Izv AN SSSR Ser Khim* 53–58, 1968.
134. ET Denisov, NI Mitskevich, VE Agabekov. *Liquid-Phase Oxidation of Oxygen-Containing Compounds*. New York: Consultants Bureau, 1977.
135. T Mill, DG Hendry. In: CH Bamford and CFH Tipper (eds.), *Comprehensive Chemical Kinetics*, vol. 16. Amsterdam: Elsevier, 1980, pp. 1–88.
136. R Hiatt, J Clipsham, T Visser. *Can J Chem* 42:2754–2757, 1964.
137. TG Degtyareva, VM Solyanikov, ET Denisov, VV Komkin. *Neftekhimiya* 13:229–234, 1973.
138. AE Semenchenko, VM Solyanikov, ET Denisov. *Neftekhimiya* 11:555–561, 1971.
139. ML Poutsma, In: JK Kochi (Ed.), *Free Radicals*, vol. 2. New York: Wiley, 1973, pp. 113–158.
140. GG Agliullina, VS Martem'yanov, IA Ivanova, ET Denisov. *Izv AN SSSR Ser Khim* 2221–2224, 1977.
141. TI Drozdova, AL Aleksandrov, ET Denisov. *Izv AN SSSR Ser Khim* 965–967, 1978.
142. TI Drozdova, AL Aleksandrov, ET Denisov. *Izv AN SSSR Ser Khim* 1213–1216, 1978.
143. PD Bartlett, K Nozaki. *J Am Chem Soc* 69:2299–2306, 1947.
144. IP Hajdu, I Nemes, D Gal, VL Rubailo, NM Emanuel. *Can J Chem* 55:2677–2731, 1977.
145. JSB Park, PM Wood, BC Gilbert, AC Whitwood. *J Chem Soc Perkin Trans 2*, 923–931, 1999.
146. IV Berezin, ET Denisov, NM Emanuel. *The Oxidation of Cyclohexane*. Oxford: Pergamon Press, 1966.
147. JR Thomas. *J Am Chem Soc* 77:246–248, 1955.
148. ET Denisov. *Izv AN SSSR Otd Khim Nauk N* 2:195–203, 1960.
149. ET Denisov. *Izv AN SSSR Otd Khim Nauk N* 5:796–803, 1961.
150. ET Denisov. *Izv AN SSSR Otd Khim Nauk N* 1:53–58, 1960.

5 Co-Oxidation of Hydrocarbons

5.1 THEORY OF HYDROCARBON CO-OXIDATION

The chain mechanism is complicated when two hydrocarbons are oxidized simultaneously. Russell and Williamson [1,2] performed the first experiments on the co-oxidation of hydrocarbons with ethers. The theory of these reactions is close to that for the reaction of free radical copolymerization [3] and was developed by several researchers [4–9]. When one hydrocarbon R^1H is oxidized in the liquid phase at a sufficiently high dioxygen pressure chain propagation is limited only by one reaction, namely, $R^1OO^\bullet + R^1H$. For the co-oxidation of two hydrocarbons R^1H and R^2H, four propagation reactions are important, viz,

$$R_1OO^\bullet + R^1H \longrightarrow R_1OOH + R_1^\bullet \qquad (k_{p11})$$
$$R_1OO^\bullet + R^2H \longrightarrow R_1OOH + R_2^\bullet \qquad (k_{p12})$$
$$R_2OO^\bullet + R^2H \longrightarrow R_2OOH + R_2^\bullet \qquad (k_{p22})$$
$$R_2OO^\bullet + R^1H \longrightarrow R_2OOH + R_1^\bullet \qquad (k_{p21})$$

In addition to cross-propagation reactions, one cross-termination reaction is introduced in addition to two self-termination reactions, viz,

$$R_1OO^\bullet + R_1OO^\bullet \longrightarrow \text{Termination} \qquad (k_{t11})$$
$$R_1OO^\bullet + R_2OO^\bullet \longrightarrow \text{Termination} \qquad (k_{t12})$$
$$R_2OO^\bullet + R_2OO^\bullet \longrightarrow \text{Termination} \qquad (k_{t22})$$

When the kinetic chain length is high, the rate of total oxygen consumption is

$$\frac{d\Delta[O_2]}{dt} = \frac{(r_1[R^1H]^2 + 2[R^1H][R^2H] + r_2[R^2H]^2)\sqrt{v_i}}{(r_1^2\delta_1^2[R^1H]^2 + \phi r_1 r_2 \delta_1 \delta_2[R^1H][R^2H] + r_2^2\delta_2^2[R^2H]^2)^{1/2}} \qquad (5.1)$$

where

$$r_1 = \frac{k_{p11}}{k_{p12}}; \; r_2 = \frac{k_{p22}}{k_{p21}}; \; \delta_1 = \frac{\sqrt{2k_{t11}}}{k_{p11}}; \; \delta_2 = \frac{\sqrt{2k_{t22}}}{k_{p22}} \text{ and } \phi = \frac{k_{t12}}{\sqrt{k_{t11}k_{t22}}}.$$

Figure 5.1 illustrates the effect of hexamethylbenzene that produces a secondary peroxyl radical on the oxidation of cumene [9].

The linear dependence of dioxygen consumption on the composition of a hydrocarbon mixture is observed when

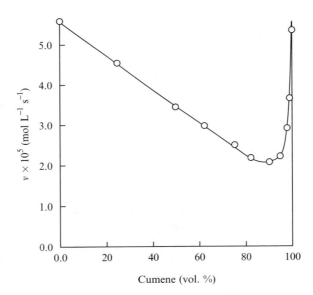

FIGURE 5.1 Co-oxidation of cumene with hexamethylbenzene: dependence of oxidation rate on cumene concentration (vol. %) ($T = 348$ K, [AIBN] $= 0.04$ mol L^{-1} $pO_2 = 1$ atm. [46]).

$$r_1 \frac{\delta_1}{\delta_2} + \phi r_2 \frac{\delta_2}{\delta_1} = 2 \quad \text{or} \quad \phi r_1 \frac{\delta_1}{\delta_2} + r_2 \frac{\delta_2}{\delta_1} = 2. \tag{5.2}$$

The linear co-oxidation dependence was observed for the following pairs of hydrocarbons (333 K, initiator AIBN): tetralin–ethylbenzene, phenylcyclopentane–ethylbenzene, phenylcyclohexane–ethylbenzene, tetralin–phenylcyclohexane, cyclohexene–2-butene, 2,3-dimethyl, and cyclohexene–pinane [8].

Carrying out a number of co-oxidation experiments with various hydrocarbon mixtures, one compares the results of experiment with that of computer simulations and step by step estimates of the parameters r_1, r_2, δ_1, δ_2, and ϕ. This technique is an effective qualitative method for estimating these coefficients. This technique is described in monographs and papers [5–9].

Along with dioxygen consumption, the rate of hydrocarbon can be measured during co-oxidation. At long chains and high dioxygen pressure this rate is equal to [6]:

$$-\frac{d[R^1H]}{dt} = \frac{(r_1[R^1H] + [R^2H])[R^1H]\sqrt{v_i}}{(r_1^2\delta_1^2[R^1H]^2 + \phi r_1 r_2 \delta_1 \delta_2 [R^1H][R^2H] + r_2^2\delta_2^2[R^2H]^2)^{1/2}} \tag{5.3}$$

This rate is measured experimentally and can be used for the estimation of the parameters of co-oxidation. When the concentration of R^1H is low and approaches zero, the ratio is

$$\frac{v_{R^1H}}{[R^1H]} = \left\{ \frac{1}{r_2\delta_2} + \frac{r_1[R^1H]}{r_2\delta_2[R^2H]} - \frac{\phi r_1 \delta_1 [R^1H]}{2r_2^2\delta_2^2[R^2H]} \right\} \sqrt{v_i} \tag{5.4}$$

from the measured consumption of two hydrocarbons, the values of r_1 and r_2, the ratios of the propagation coefficients as defined earlier, may be determined. At low conversions and long chains, the ratio of concentrations of two peroxyl radicals remains virtually constant.

$$k_{p21}[R^2O_2^\bullet][R^1H] = k_{p12}[R^1O_2^\bullet][R^2H] \qquad (5.5)$$

As a result we obtained the following equation:

$$\frac{\Delta[R^1H]}{\Delta[R^2H]} = \frac{(r_1[R^1H]/[R^2H]) + 1}{(r_2[R^2H]/[R^1H]) + 1} \qquad (5.6)$$

Equation (5.6) may be simplified as follows (Fineman and Ross [10]):

$$\frac{\rho - 1}{R} = r_1 - \frac{\rho}{R^2}r_2, \qquad (5.7)$$

where $R = [R^1H]/[R^2H]$ and $\rho = \Delta[R^1H]/\Delta[R^2H]$. Thus, the plot of the left-hand side of Equation (5.4) versus ρ/R^2 gives r_2 as the negative slope and r_1 as the intercept.

The values of r_1 and r_2 are measures of the relative reactivity of two hydrocarbons toward each peroxyl radical. Thus, if both the self-propagation rate constants (k_{p11} and k_{p22}) are known, the corresponding cross-propagation rate constants (k_{p12} and k_{p21}) can be determined. The differences in r_1 and $1/r_2$ are the results of the differences in the structure of two peroxyl radicals and the quantity $r_1r_2 = (k_{p11} \times k_{p22})/(k_{p12} \times k_{p21})$ is a measure of the differences in selectivity. Studying the co-oxidation for a series of hydrocarbons with one standard hydrocarbon, it is possible to determine the reactivity of the series R^iH toward the peroxyl radical of this standard. Once r_1, r_2, δ_1, and δ_2 are determined, then it is possible to determine coefficient ϕ by substituting into Equation (5.1). Table 5.1 summarizes the values obtained in the co-oxidation of various hydrocarbons and Table 5.2 summarizes the values obtained in the co-oxidation of monomers.

5.2 HYDROCARBON OXIDATION WITH HYDROPEROXIDE

Peroxyl radicals react very rapidly with hydroperoxide. This peculiarity of the system: $R^1H + ROOH + O_2$ was used by Howard et al. [21] to determine the following method for the measurement of the rate constants of the reaction of one peroxyl radical with several hydrocarbons. Hydroperoxide (ROOH) is introduced into the oxidized hydrocarbon R^iH in such a concentration (0.2–1.0 mol L^{-1}) that it is sufficient for the rapid exchange of all peroxyl and alkoxyl radicals by reactions with ROOH into $RO_2^\bullet$ radicals.

$$R^iOO^\bullet + R^1OOH \longrightarrow R^1O_2^\bullet + R^iOOH$$

$$R^iO^\bullet + R^1OOH \longrightarrow R^1O_2^\bullet + R^iOH$$

$$R^1O^\bullet + R^1OOH \longrightarrow R^1O_2^\bullet + R^1OH$$

The rate of the initiated chain oxidation of R^iH under such conditions is determined only by two reactions:

$$R^1O_2^\bullet + R^iH \longrightarrow R^1OOH + R^{i\bullet} \qquad (k_{p1i})$$

$$R^1O_2^\bullet + R^1O_2^\bullet \longrightarrow \text{Molecular products} \qquad (2k_{t11})$$

TABLE 5.1
The Parameters r_1, r_2, and ϕ of Hydrocarbon Co-Oxidation

R^1H	R^2H	T (K)	r_1	r_2	ϕ	Ref.
PhMe$_2$CH	PhCH$_3$	333	3.0	0.2	0.8	[11]
PhMe$_2$CH	PhCH$_3$	348	2.60	0.21	1.0	[12]
PhMe$_2$CH	o-xylene	348	1.12	0.65	0.8	[12]
PhMe$_2$CH	m-xylene	348	1.5	0.52	0.93	[12]
PhMe$_2$CH	p-xylene	333	2.10	1.0	2.1	[11]
PhMe$_2$CH	p-xylene	348	1.24	0.72	1.0	[12]
PhMe$_2$CH	mesitylene	333	2.2	0.8	1.1	[11]
PhMe$_2$CH	1,2,4-trimethylbenzene	348	0.63	1.24	0.9	[12]
PhMe$_2$CH	1,2,4,5-tetramethylbenzene	348	0.32	2.12	0.9	[12]
PhMe$_2$CH	pentamethylbenzene	348	0.112	3.42	0.8	[12]
PhMe$_2$CH	hexamethylbenzene	348	0.087	8.83	1.0	[12]
PhMe$_2$CH	PhCH$_2$Me	333	0.5	0.8	1.3	[13]
PhMe$_2$CH	PhCH$_2$Me	333	0.44	0.97	0.73	[14]
PhMe$_2$CH	PhCH$_2$Me	333	1.3	1.0	1.2	[11]
PhMe$_2$CH	PhCH$_2$Me	341	0.56	1.0	1.3	[15]
PhMe$_2$CH	PhCH$_2$Me	348	0.59	1.0	1.2	[15]
PhMe$_2$CH	PhCH$_2$Me	353	0.49	0.98	0.79	[14]
PhMe$_2$CH	PhCH$_2$Et	333	2.8	0.4	1.2	[11]
PhMe$_2$CH	PhCH$_2$Et	348	1.4	0.72	1.0	[16]
PhMe$_2$CH	PhCH$_2$Pr	333	1.9	0.9	1.2	[11]
PhMe$_2$CH	PhCH$_2$CHMe$_2$	348	3.5	0.1	1.0	[16]
PhMe$_2$CH	PhCHEtMe	348	2.0	0.30	0.8	[16]
PhMe$_2$CH	PhCHEtMe	333	1.15	0.03	–	[2]
PhMe$_2$CH	PhCH$_2$CHEtMe	348	2.3	0.50	4.6	[16]
PhMe$_2$CH	PhCH$_2$CMe$_3$	348	6.8	0.12	2.1	[16]
PhMe$_2$CH	Ph$_2$CH$_2$	348	0.42	2.2	1.4	[15]

TABLE 5.1
The Parameters r_1, r_2, and ϕ of Hydrocarbon Co-Oxidation—*continued*

R^{1}H	R^{2}H	T (K)	r_1	r_2	ϕ	Ref.
PhMe$_2$CH	Ph$_2$CHMe	348	0.18	2.8	8.3	[17]
PhMe$_2$CH	phenylcyclopentane	333	0.1	4.0	–	[13]
PhMe$_2$CH	phenylcyclohexane	333	0.55	0.75	0.75	[18]
PhMe$_2$CH	tetralin	333	0.35	0.75	0.75	[18]
PhMe$_2$CH	tetralin	363	0.04	16.0	6	[19]
PhMe$_2$CH	tetralin		0.13	4.6	–	3	[7]
PhMe$_2$CH	tetralin	353	0.21	5.0	4.3	[20]
PhMe$_2$CH	tetralin	329	0.18	4.5	4.8	[21]
PhMe$_2$CH	tetralin	303	0.11	8.3	5.9	[21]
PhMe$_2$CH	2-methylpyridine	348	3.2	0.053	1.6	[22]
PhMe$_2$CH	3-methylpyridine	348	3.0	0.17	1.6	[22]
PhMe$_2$CH	4-methylpyridine	348	4.0	0.09	1.3	[22]
PhMe$_2$CH	2-ethylpyridine	348	0.63	0.46	1.2	[22]
PhMe$_2$CH	3-ethylpyridine	348	0.17	0.69	0.64	[22]
PhMe$_2$CH	4-ethylpyridine	348	0.30	0.45	0.8	[22]
o-xylene	m-xylene	377	3.0	0.1	0.9	[11]
o-xylene	p-xylene	377	1.6	1.0	0.6	[11]
o-xylene	PhCH$_2$Pr	377	0.5	1.8	1.1	[11]
m-xylene	p-xylene	377	1.0	1.6	0.6	[11]

continued

TABLE 5.1
The Parameters r_1, r_2, and ϕ of Hydrocarbon Co-Oxidation—*continued*

R^1H	R^2H	T (K)	r_1	r_2	ϕ	Ref.
PhCH$_2$Me	PhCH$_3$	348	2.2	0.2	3.0	[23]
PhCH$_2$Me	(mesitylene)	333	0.5	1.2	1.0	[11]
PhCH$_2$Me	PhCH$_2$Et	333	1.0	0.8	1.0	[11]
PhCH$_2$Me	PhCH$_2$Pr	377	0.8	1.2	1.2	[11]
PhCH$_2$Me	PhCHEtMe	377	1.6	0.6	1.0	[11]
PhCH$_2$Me	PhCHEtMe	348	0.58	1.14	1.0	[21]
PhCH$_2$Me	PhCH$_2$CHMe$_2$	348	2.5	0.35	1.0	[24]
PhCH$_2$Me	PhCH$_2$CMe$_3$	348	4.9	0.54	1.8	[24]
PhCH$_2$Me	Ph(CH$_2$)$_2$Bu	348	1.8	0.62	1.4	[24]
PhCH$_2$Me	(p-cymene)	333	0.8	1.25	0.9	[11]
PhCH$_2$Me	Ph$_2$CH$_2$	348	0.31	1.2	2.5	[23]
PhCH$_2$Me	PhCHMePr	348	0.7	0.7	2.0	[24]
PhCH$_2$Me	(cyclohexylbenzene)	348	0.38	0.9	1.8	[24]
Ph$_2$CH$_2$	PhCH$_3$	348	6.3	0.10	3.0	[17]
Ph$_2$CHCH$_3$	PhCH$_3$	348	5.2	0.07	3.6	[17]
Ph$_2$CHCH$_3$	Ph$_2$CH$_2$	348	1.3	0.51	3.2	[23]
(p-cymene)	PhCH$_3$	333	2.2	0.4	1.0	[11]
PhCHEtMe	PhCH$_2$Pr	377	0.8	2.0	1.1	[11]
(tetralin)	(mesitylene)	333	6.5	0.15	1.0	[11]
(tetralin)	PhCH$_2$Et	333	6.7	0.1	1.0	[11]
(tetralin)	(p-xylene)	333	4.6	0.15	0.6	[11]
(tetralin)	(xylene)	333	4.3	0.15	1.4	[11]
(tetralin)	PhCHEtMe	333	4.0	0.25	0.9	[11]
(tetralin)	PhCH$_2$Pr	333	3.5	0.15	1.4	[11]
(tetralin)	(o-xylene)	333	3.1	0.15	0.6	[11]
(tetralin)	PhCH$_2$Me	333	3.0	0.2	0.8	[11]

TABLE 5.1
The Parameters r_1, r_2, and ϕ of Hydrocarbon Co-Oxidation—*continued*

R^{1}H	R^{2}H	T (K)	r_1	r_2	ϕ	Ref.
(tetralin)	(xylene)	333	2.6	0.1	0.6	[11]
(tetralin)	(cumene deriv.)	333	1.8	0.4	1.1	[11]

This rate is equal (at $v \gg v_i$) to

$$v = k_{p1i}(2k_{t11})^{-1/2}[R^iH]v_i^{1/2}. \tag{5.8}$$

If the values of $2k_{t11}$ and v_i are known, one can easily calculate the rate constant k_{p1i}. Such measurements were performed for several hydrocarbons [5]. The comparison of the values of $k_{p1i}(2k_{t11})^{-1/2}$ measured by the co-oxidation and oxidation of hydrocarbons with hydroperoxides are given in Refs. [5,9]. The dependence of the initiated oxidation rate on the added concentration of hydroperoxide [R^{1}OOH] is described by the equation [5]:

$$v = (k_{p1i}[R^iH] + k_{exc}[ROOH]) \times v_i^{1/2} \times F^{-1/2}, \tag{5.9}$$

where F is the function of the hydroperoxide and hydrocarbon concentrations:

$$F = 2k_{t1i} + 4k_{t1i}\frac{k_{exc}[R^1OOH]}{k_{p1i}[R^1H]} + 2k_{t11}\left\{\frac{k_{exc}[R^1OOH]}{k_{p1i}[R^iH]}\right\}^2. \tag{5.10}$$

Equation (5.10) can be used for the additional estimation of the values of the rate constants $2k_{t1i}$ (cross-termination) and k_{exc} (exchange between R^{i}O$_2^{\bullet}$ and R^{1}OOH) from experimental measurements. The values of k_{exc} are given in Table 5.3.

5.3 CO-OXIDATION OF HYDROCARBONS AND ALCOHOLS WITH SELECTIVE INHIBITOR

Chain propagation in oxidizing alcohols is performed by hydroperoxyl and alkylhydroxyperoxyl radicals. These radicals, unlike alkyl peroxyl radicals, exhibit a dual character: they can bring about oxidation and reduction. Thus hydroperoxyl radicals react not only with radicals but also with such oxidizing agents as quinones and transition metal ions (see Chapter 16). The kinetics of the oxidation of isopropyl alcohol in the presence of benzoquinone was studied [37]. In an alcohol undergoing oxidation, quinone brings about chain termination by the reaction with hydroxyperoxyl and hydroxyalkyl radicals, as can be seen from the dependence of the rate of oxidation on pO_2 in the presence of quinone. At $pO_2 = 760$ Torr and $T = 344$ K, 86% of chain termination takes place by the reaction HOO$^{\bullet}$ + quinone and 14% by the reaction R$^{\bullet}$ + quinone. The rate constant of this reaction is $k_Q = 3.2 \times 10^3$ L mol^{-1} s^{-1},

TABLE 5.2
The Parameters r_1 and r_2 of the Monomer and Hydrocarbon Co-Oxidation

Monomer 1 (R^1H)	Monomer 2 (R^2H)	T (K)	r_1	r_2	$r_1 \times r_2$	Ref.
$PhMe_2CH$	$PhCH=CH_2$	303	0.1	13	1.3	[25]
$PhMe_2CH$	$PhCH=CH_2$	303	0.05	20	1.0	[7]
$PhMe_2CH$	$PhMeC=CH_2$	303	0.1	9.3	0.93	[26]
$PhMe_2CH$	$CH_2=CHCH=CH_2$	303	0.03	23	0.69	[7]
tetralin	$CH_2=CHCH=CH_2$	303	0.22	3.5	0.77	[7]
$PhCH=CH_2$	$PhMe_2CH$	333	20.5	0.053	1.09	[7]
$PhCH=CH_2$	$PhMe_2CH$	333	13	0.1	1.3	[25]
$PhCH=CH_2$	$PhMe_2CH$	333	6.5	0.2	1.3	[2]
$PhCH=CH_2$	tetralin	333	16	0.044	0.70	[19]
$PhCH=CH_2$	tetralin	333	4.6	0.13	0.60	[7]
$PhCH=CH_2$	tetralin	303	2.3	0.43	0.99	[7]
$PhCH=CH_2$	tetralin	303	2.7	0.49	1.32	[20]
$PhCH=CH_2$	tetralin	333	4.2	0.85	3.6	[25]
$PhCH=CH_2$	tetralin	333	2.8	1.2	3.4	[2]
$PhCH=CH_2$	$CH_2=CHCH=CH_2$	303	0.45	2.2	0.99	[7]
$PhCH=CH_2$	4-MeO-styrene	323	0.38	2.10	0.8	[27]
$PhCH=CH_2$	4-MeO-styrene	323	0.55	1.2	0.66	[1]
$PhCH=CH_2$	4-Me-styrene	323	0.71	1.51	1.1	[27]
$PhCH=CH_2$	4-Br-styrene	323	0.88	0.97	0.85	[27]
$PhCH=CH_2$	4-Cl-styrene	323	0.89	0.98	0.87	[27]
$PhCH=CH_2$	4-F-styrene	323	0.88	0.92	0.81	[27]
$PhCH=CH_2$	4-I-styrene	323	0.705	0.995	0.70	[27]
$PhCH=CH_2$	4-NC-styrene	323	1.10	0.50	0.55	[27]

TABLE 5.2
The Parameters r_1 and r_2 of the Monomer and Hydrocarbon Co-Oxidation—continued

Monomer 1 (R^1H)	Monomer 2 (R^2H)	T (K)	r_1	r_2	$r_1 \times r_2$	Ref.
PhCH=CH$_2$	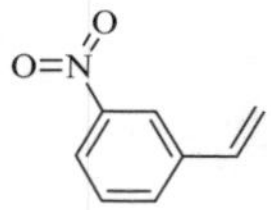 3-nitrostyrene (O=N(O)-C$_6$H$_4$-CH=CH$_2$)	323	1.24	0.32	0.40	[27]
PhCH=CH$_2$	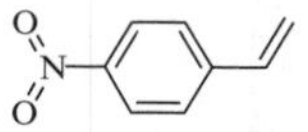 4-nitrostyrene	323	1.20	0.42	0.50	[27]
PhCH=CH$_2$	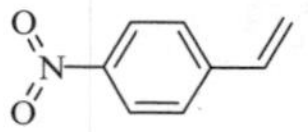 4-nitrostyrene	323	1.5	0.22	0.33	[1]
PhCH=CH$_2$	PhMeC=CH$_2$	323	0.48	2.1	1.0	[28]
PhCH=CH$_2$	PhMeC=CH$_2$	323	0.9	1.2	1.1	[1]
PhCH=CH$_2$	MeOC(O)MeC=CH$_2$	323	9.2	0.09	0.83	[27]
PhCH=CH$_2$	CH$_2$=CMeCMe=CH$_2$	323	0.37	2.75	1.02	[27]
PhMeC=CH$_2$	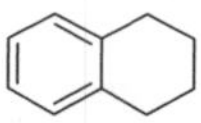 tetralin	303	1.3	0.85	1.1	[2]
PhMeC=CH$_2$	MeOC(O)MeC=CH$_2$	288	15	0.041	0.62	[28]
PhMeC=CH$_2$	PhCH=CH$_2$	303	1.2	0.9	1.1	[2]
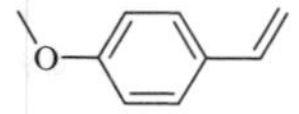 4-methoxystyrene	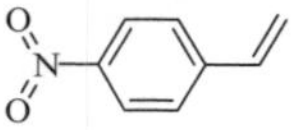 4-nitrostyrene	323	2.3	0.17	0.39	[1]

and the ratio $k_Q(RO_2^{\bullet} + Q)/k_p = 1.0 \times 10^4$. As in the case of aromatic amines, for quinone $f = 23$, that is, quinone is regenerated by the following reactions:

$$Q + HOO^{\bullet} \longrightarrow {}^{\bullet}QH + O_2$$
$${}^{\bullet}QH + HOO^{\bullet} \longrightarrow Q + HOOH$$
$$Q + R_2C(OH)OO^{\bullet} \longrightarrow {}^{\bullet}QH + O_2 + R_2C(O)$$
$${}^{\bullet}QH + R_2C(OH)OO^{\bullet} \longrightarrow Q + R_2C(OH)OOH$$

Thus, quinone is a *selective inhibitor*, reacting selectively with the hydroxyperoxyl and hydroxyalkyl radicals of alcohol, and this has become the basis of a new method for measuring the rate constants of the reactions of peroxyl radicals with alcohols — the method of co-oxidation of hydrocarbon and alcohol in the presence of a selective inhibitor [38]. Alcohol and hydrocarbon (or any organic compound), the oxidation of which does not involve the production of hydroxyalkylperoxyl or hydroperoxyl radicals, are oxidized together in the presence of an initiator and quinone. Quinone is added in concentrations at which almost all chains are terminated by the reaction of quinone with the hydroxyalkylperoxyl radicals, that is, practically independent of the quinone concentration [38]. The partial pressure of dioxygen is sufficiently high for the reaction of the alkyl radicals with quinone to occur very slowly. Under these conditions, the oxidation of alcohol (HR_iOH) with hydrocarbon (RH) involves the following reactions:

TABLE 5.3
The Rate Constants of Peroxyl Radical Exchange Reaction with Hydroperoxides

$RO_2^\bullet$	R^1OOH	T (K)	k_{exc} (L mol^{-1}s^{-1})	Ref.
Me$_3$COO$^\bullet$	EtMeCOOH	294	4.85×10^2	[29]
Me$_3$COO$^\bullet$	cyclohexyl-O-OH	353	2.0	[30]
Me$_3$COO$^\bullet$	cyclohexyl-O-OH	353	9.8	[31]
Me$_3$COO$^\bullet$	Ph$_3$COOH	294	7.00×10^2	[29]
Me$_3$COO$^\bullet$	tetralinyl-OOH	190	11.2	[29]
Me$_3$COO$^\bullet$	anthracenyl-OOH	294	7.40×10^2	[29]
EtMe$_2$COO$^\bullet$	tetralinyl-OOH	200	12	[29]
PhMe$_2$COO$^\bullet$	EtOCH(OOH)Me	303	66	[32]
PhMe$_2$COO$^\bullet$	tetralinyl-OOH	303	6.00×10^2	[21]
PhMe$_2$COO$^\bullet$	tetralinyl-OOH	329	1.10×10^3	[21]
PhMe$_2$COO$^\bullet$	tetralinyl-OOH	329	3.00×10^2	[33]
PhMe$_2$COO$^\bullet$	tetrahydrofuranyl-O-OH	303	38	[34]
PhMe$_2$COO$^\bullet$	tetrahydropyranyl-O-OH	303	47	[34]
PhMe$_2$COO$^\bullet$	EtOCH$_2$Me	303	66	[34]
PhMe$_2$COO$^\bullet$	tetralinyl-OOD	303	12	[21]

TABLE 5.3
The Rate Constants of Peroxyl Radical Exchange Reaction with Hydroperoxides—*continued*

$RO_2^\bullet$	R^1OOH	T (K)	k_{exc} (L mol^{-1}s^{-1})	Ref.
tetralyl-OO•	Me_3COOH	333	5.30×10^2	[34]
tetralyl-OO•	Me_3COOH	293	4.80×10^2	[35]
tetralyl-OO•	$PhMe_2COOH$	303	2.50×10^3	[21]
tetralyl-OO•	$PhMe_2COOH$	329	2.80×10^3	[21]
~$CH_2CHPh(OO^\bullet)$	Me_3COOH	333	6.45×10^2	[36]
~$CH_2CHPh(OO^\bullet)$	$PhMe_2COOD$	303	1.40×10^2	[21]

$$I \longrightarrow r^\bullet \qquad k_i$$
$$r^\bullet + RH \longrightarrow rH + R^\bullet \qquad \text{Fast}$$
$$R^\bullet + O_2 \longrightarrow RO_2^\bullet \qquad \text{Fast}$$
$$RO_2^\bullet + RH \longrightarrow ROOH + R^\bullet \qquad k_p$$
$$RO_2^\bullet + HR_iOH \longrightarrow ROOH + R_i^\bullet OH \qquad k_{pi}$$
$$R_i^\bullet OH + O_2 \longrightarrow R_i(OH)OO^\bullet \qquad \text{Fast}$$
$$R_i(OH)OO^\bullet + Q \longrightarrow R_i{=}O + O_2 + {}^\bullet QH \qquad k_Q$$
$$RO_2^\bullet + {}^\bullet QH \longrightarrow ROOH + Q \qquad \text{Fast}$$

Under stationary conditions for long chains and at sufficiently high quinone concentrations when $2k_Q[Q] \gg (2k_t v_i)^{1/2}$

$$k_{pi}[HR_iOH][RO_2^\bullet] = 2k_Q[Q][R_i(OH)OO^\bullet]$$
$$v_i = 2k_Q[Q][R_i(OH)OO^\bullet] \qquad (5.11)$$
$$v = \frac{k_p[RH]v_i}{2k_{pi}[HR_iOH]}$$

By measuring the oxidation rate v for different ratios $[RH]/[HR_iOH]$, it is possible to find the ratio k_p/k_{pi} and calculate k_{pi} when k_p is known. The results of co-oxidation of different alcohols with cyclohexene are given below ($T = 333$ K, $pO_2 = 1$ atm, initiator AIBN [38]).

HR_iOH	MeOH	EtOH	BuOH	Me_2CHOH	cyclo-$C_6H_{11}OH$	$PhCH_2OH$
k_{pi} (L mol^{-1} s^{-1})	0.3 (313 K)	1.9	1.2	2.0	2.5	5.6

This method was used for the estimation of k_p in oxidized polyatomic alcohols [39].

5.4 CROSS-DISPROPORTIONATION OF PEROXYL RADICALS

The mechanisms of chain termination by disproportionation of secondary and tertiary peroxyl radicals are sufficiently different (see Chapter 2). Secondary $RO_2^{\bullet}$ disproportionate by reaction [4–6]

$$2R^1R^2CHOO^{\bullet} \longrightarrow R^1R^2C(O) + O_2 + R^1R^2CHOH$$

Tertiary peroxyl radicals recombine by consecutive reactions [4–6].

$$2RO_2^{\bullet} \rightleftharpoons ROOOOR$$
$$ROOOOR \longrightarrow [RO^{\bullet} + O_2 + RO^{\bullet}]$$
$$[RO^{\bullet} + O_2 + RO^{\bullet}] \longrightarrow 2RO^{\bullet} + O_2$$
$$[RO^{\bullet} + O_2 + RO^{\bullet}] \longrightarrow ROOR + O_2$$

The values of $2k_t$ for secondary peroxyl radicals lies in the range 10^6–10^8 L mol^{-1} s^{-1} and that of tertiary peroxyl radicals are sufficiently less, i.e., in the range of 10^3–10^5 L mol^{-1} s^{-1}. Cross-termination by the reaction of *tert*-$RO_2^{\bullet}$ + *sec*-$RO_2^{\bullet}$ predominantly occurs according to the mechanism of disproportionation of secondary peroxyl radicals.

$$R^1R^2CHOO^{\bullet} + R^1R^2R^3COO^{\bullet} \longrightarrow R^1R^2C(O) + O_2 + R^1R^2R^3COH$$

The importance of the cross-termination reaction on the rate of oxygen consumption was observed by Russell [19,40]. The different propagation rate constants (k_{p11}/k_{p21} and k_{p22}/k_{p12}) do not vary significantly for alkylperoxyl radicals. However, the rate of two self-termination reactions can vary by as much as 10^4. Thus, as the composition of the mixture is varied from 100% of one hydrocarbon to 100% of the other, the importance of various termination reactions changes accordingly. The effect of cross-terminations is most dramatic when coefficient $\phi \gg 2$. Small amounts of hydrocarbon, which has the high self-termination rate coefficient, can drastically reduce the rate of oxidation of hydrocarbon, which has a very low termination rate constant. Figure 5.1 illustrates the effect of hexamethylbenzene that produces secondary peroxyl radicals on the oxidation of cumene. The latter can terminate more rapidly than cumyl peroxyl radical (see Chapter 2). In all the cases, small amounts of the hydrocarbon with oxidized CH_2 group reduce the rate of oxidation of cumene. In the case of hydrocarbons that are oxidized more rapidly than cumene, a distinct minimum rate is observed upon addition of a few percent of these compounds. The rate constants of such disproportionation are close to those of the secondary peroxyl radicals. Hence coefficients $\phi = k_{t12}(k_{t11}k_{t22})^{-1/2}$ are much higher than unity when $R^1O_2^{\bullet}$ is the secondary and $R^2O_2^{\bullet}$ is the tertiary peroxyl radical (see Table 5.4). When both peroxyl radicals ($R^1O_2^{\bullet}$ and $R^2O_2^{\bullet}$) are secondary or tertiary, coefficients ϕ are close to unity. The values of the rate constants of peroxyl radicals of cross-termination are given in Table 5.4.

TABLE 5.4
Rate Constants of the Cross-Disproportionation of Two Different Peroxyl Radicals (Experimental Data)

$R_1O_2^\bullet$	$R_2O_2^\bullet$	Solvent	T (K)	k (L mol^{-1} s^{-1})	Ref.
$CH_3OO^\bullet$	$Me_3COO^\bullet$	H_2O	293	1.8×10^6	[41]
$CH_3OO^\bullet$	$PhMe_2COO^\bullet$	C_6H_5Cl	330	1.0×10^7	[42]
$CH_3OO^\bullet$	$Me_3COOCMe_2CH_2OO^\bullet$	C_6H_6	295	2.4×10^8	[43]
$BuCH_2OO^\bullet$	$MePrCHOO^\bullet$	RH	303	4.9×10^7	[44]
$PhMe_2COO^\bullet$	$PrCH(OO^\bullet)OBu$	R_1H/R_2H	333	2.9×10^6	[45]
$PhMe_2COO^\bullet$	$PhCH_2OO^\bullet$	R_1H/R_2H	348	5.7×10^6	[46]
$Me_2CH(CH_2)_4CMe_2OO^\bullet$	$MePhCHOO^\bullet$	R_1H/R_2H	350	1.2×10^6	[47]
$PhMe_2COO^\bullet$	$Ph_2CHOO^\bullet$	R_1H/R_2H	348	2.8×10^6	[46]
$PhMe_2COO^\bullet$	$PhMeCHOO^\bullet$	R_1H/R_2H	348	1.5×10^6	[46]
$PhMe_2COO^\bullet$	(pentaethylphenyl-CHMe-O-O$^\bullet$)	R_1H/R_2H	348	9.0×10^4	[48]
$PhMe_2COO^\bullet$	(4-methylbenzyl-O-O$^\bullet$)	R_1H/R_2H	348	2.7×10^6	[46]
$PhMe_2COO^\bullet$	(3,5-dimethylbenzyl-O-O$^\bullet$)	R_1H/R_2H	333	3.3×10^6	[11]
$PhMe_2COO^\bullet$	$EtMePhCOO^\bullet$	R_1H/R_2H	348	8.8×10^4	[46]
$PhMe_2COO^\bullet$	$Ph_2MeCOO^\bullet$	R_1H/R_2H	348	5.8×10^6	[46]
$Ph_2MeCOO^\bullet$	$Ph_2CHOO^\bullet$	R_1H/R_2H	348	4.1×10^7	[46]
$Ph_2MeCOO^\bullet$	$PhCH_2OO^\bullet$	R_1H/R_2H	348	1.3×10^8	[46]
$PhMe_2COO^\bullet$	(1-tetralyl-O-O$^\bullet$)	R_1H/R_2H	363	8.4×10^6	[19]
$PhMe_2COO^\bullet$	(1-tetralyl-O-O$^\bullet$)	R_1H/R_2H	353	6.2×10^6	[20]
$PhMe_2COO^\bullet$	(4-isopropyl-α,α-dimethylbenzyl-O-O$^\bullet$)	R_1H/R_2H	348	9.0×10^4	[49]

continued

TABLE 5.4
Rate Constants of the Cross-Disproportionation of Two Different Peroxyl Radicals (Experimental Data)—*continued*

$R_1O_2^\bullet$	$R_2O_2^\bullet$	Solvent	T (K)	k (L mol^{-1} s^{-1})	Ref.
PhMeC(OH)OO$^\bullet$	1-tetralinylperoxyl	R_1H/R_2H	333	1.4×10^7	[50]
PhMe$_2$COO$^\bullet$	EtPhCHOO$^\bullet$	R_1H/R_2H	348	1.8×10^6	[46]
PhMe$_2$COO$^\bullet$	Ph(CMe$_3$)CHOO$^\bullet$	R_1H/R_2H	348	3.9×10^6	[46]
PhMe$_2$COO$^\bullet$	1-phenylcyclohexylperoxyl	R_1H/R_2H	348	8.1×10^5	[46]
PhMe$_2$COO$^\bullet$	PhMe(CH$_2$)$_4$CHOO$^\bullet$	R_1H/R_2H	348	6.9×10^6	[46]
PhMe$_2$COO$^\bullet$	PhC(O)OCH(OO$^\bullet$)Pr	R_1H/R_2H	348	5.0×10^6	[51]
PhMe$_2$COO$^\bullet$	MeC(O)OCH(OO$^\bullet$)Ph	R_1H/R_2H	348	1.5×10^7	[51]
PhMe$_2$COO$^\bullet$	PhCH(OO$^\bullet$)OBu	R_1H/R_2H	348	8.2×10^6	[51]
PhMe$_2$COO$^\bullet$	(4-pyridyl)CH$_2$OO$^\bullet$	R_1H/R_2H	348	3.9×10^6	[51]
PhMeCHOO$^\bullet$	EtMePhCOO$^\bullet$	R_1H/R_2H	348	1.9×10^6	[24]
PhMeCHOO$^\bullet$	PhCH$_2$OO$^\bullet$	R_1H/R_2H	348	2.9×10^8	[46]
Ph$_2$CHOO$^\bullet$	PhMeC(OH)OO$^\bullet$	R_1H/R_2H	348	2.9×10^7	[46]
PhMeCHOO$^\bullet$	$\sim$CH$_2$PhCHOO$^\bullet$	R_1H/R_2H	338	3.6×10^8	[52]
Ph$_2$CHOO$^\bullet$	PhCH$_2$OO$^\bullet$	R_1H/R_2H	348	3.1×10^8	[46]
PhMeCHOO$^\bullet$	PhCH(OH)OO$^\bullet$	R_1H/R_2H	348	2.8×10^7	[46]
PhMe$_2$COO$^\bullet$	Me$_2$CHOMe$_2$COO$^\bullet$	R_1H/R_2H	333	1.8×10^5	[45]
Me$_2$CHOMe$_2$COO$^\bullet$	3-cyclohexenylperoxyl	R_1H/R_2H	333	2.0×10^5	[45]
2-tetrahydrofuranylperoxyl	3-cyclohexenylperoxyl	R_1H/R_2H	333	1.0×10^7	[45]
2-tetrahydrofuranylperoxyl	1-tetralinylperoxyl	R_1H/R_2H	333	1.2×10^7	[45]
3-cyclohexenylperoxyl	2-tetrahydrofuranylperoxyl	R_1H/R_2H	333	1.1×10^7	[8]
3-cyclohexenylperoxyl	PrCH(OO$^\bullet$)OBu	R_1H/R_2H	333	8.2×10^6	[8]

TABLE 5.4
Rate Constants of the Cross-Disproportionation of Two Different Peroxyl Radicals (Experimental Data)—continued

$R_1O_2^\bullet$	$R_2O_2^\bullet$	Solvent	T (K)	k (L mol^{-1}s^{-1})	Ref.
~CH$_2$PhCHOO$^\bullet$	~CH$_2$CH(CN)OO$^\bullet$	R$_1$H/R$_2$H	323	4.8×10^7	[53]
~CH$_2$PhCHOO$^\bullet$	~CH$_2$CMe(OO$^\bullet$)CO$_2$Me	R$_1$H/R$_2$H	323	2.5×10^6	[53]
~CH$_2$PhCHOO$^\bullet$	~CH$_2$CH(OO$^\bullet$)CO$_2$Me	R$_1$H/R$_2$H	323	1.5×10^6	[53]
~CH$_2$PhMeCOO$^\bullet$	~CH$_2$CMe(OO$^\bullet$)CO$_2$Me	R$_1$H/R$_2$H	323	2.7×10^5	[53]
~CH$_2$PhMeCOO$^\bullet$	~CH$_2$CH(OO$^\bullet$)CO$_2$Me	R$_1$H/R$_2$H	323	6.9×10^4	[53]

5.5 CROSS-PROPAGATION REACTIONS OF PEROXYL RADICALS

The methods of co-oxidation and oxidation of hydrocarbon (R$_i$H) in the presence of hydroperoxide (ROOH) opened the way to measure the rate constants of the same peroxyl radical with different hydrocarbons. Both the methods give close results [5,9]. The activity of different secondary peroxyl radicals is very close. It is seen from comparison of rate constants of *prim*-RO$_2^\bullet$ and *sec*-RO$_2^\bullet$ reactions with cumene at 348 K [9].

RO$_2^\bullet$	PhCH$_2$O$_2^\bullet$	3-pyridyl-CH$_2$OO$^\bullet$	2-pyridyl-CHMeOO$^\bullet$	3-pyridyl-CHMeOO$^\bullet$	Ph$_2$CHO$_2^\bullet$
k (L mol^{-1} s^{-1})	11	10	5.7	4.1	10.5

Howard and Ingold [54] observed the same tendency in reactions of primary and secondary peroxyl radicals with several hydrocarbons. The following table gives the values of rate constants (L mol^{-1} s^{-1}) at 303 K.

Hydrocarbon	PrCH$_2$O$_2^\bullet$	EtMeCHO$_2^\bullet$	cyclohexyl-O$_2^\bullet$	tetralyl-O$_2^\bullet$	Ph$_2$CHO$_2^\bullet$
PhMeCH$_2$	0.55	0.5	0.5	0.5	0.6
tetralin	6.4	4.2	4.5	6.3	6.5
Me$_2$PhCH	0.45	0.40	–	0.5	–
dihydroanthracene	310	140	160	240	330

The reactivity of tertiary peroxyl radicals of different structures is very close, as it was evidenced by measuremets of Howard and Ingold [55,56]. The rate constants (L mol^{-1} s^{-1}) of four tertiary peroxyl radicals with a few alkylaromatic hydrocarbons ($T = 303$ K) are presented below.

Hydrocarbon	Me$_3$COO$^\bullet$	Me$_3$CCMe$_2$OO$^\bullet$	EtMePhCOO$^\bullet$	Me$_2$PhCOO$^\bullet$
PhCH$_3$	0.05	0.04	0.05	0.034
PhMeCH$_2$	0.20	0.32	0.22	0.21
(tetralin)	2.0	2.0	1.3	1.65
Me$_2$PhCH	0.22	0.14	0.15	0.18

The activity of secondary and tertiary peroxyl radicals is different due to different BDEs of the forming O—H bond: $D(\text{O—H}) = 365.5$ kJ mol^{-1} for secondary hydroperoxide and $D(\text{O—H}) = 358.6$ kJ mol^{-1} for tertiary hydroperoxide [57]. The comparison of the rate constants of secondary and tertiary RO$_2^\bullet$ reactions with different hydrocarbons is given below (rate constants are given in L mol^{-1} s^{-1} at 348 K) [9].

Hydrocarbon	PhCH$_3$	PhCH$_2$Me	Ph$_2$CH$_2$	PhCH$_2$Et	PhMe$_2$CH	PhEtMeCH
PhMeCHO$_2^\bullet$	2.2	4.9	15.9	4.5	5.1	4.3
PhMe$_2$CO$_2^\bullet$	0.55	2.4	3.4	1.03	1.44	0.72

The mean value of the ratio $k_p(\text{sec-RO}_2^\bullet)/k_p(\text{tert-RO}_2^\bullet) = 4.5 \pm 1.0$ [57].

The reactivity of the hydrocarbons with the C—H bond in reactions with peroxyl radicals depends, first of all, on the strength of the attacked bond. The problem of reactivity of the substrates in free radical reactions will be discussed in Chapter 7. In this chapter, the empirical data are given. The rate constants of the reactions R$_i$O$_2^\bullet$ + R$_j$H are collected in Table 5.5. Pritzkow and Suprun [58] recently performed the detailed study of reactivity of alkanes and cycloalkanes toward the cumylperoxyl radical. According to these data, the relative reactivity (RR) of CH$_3$ and CH$_2$ groups in alkane is the following (RR = ratio of rate constants of the cumylperoxyl radical with one C—H bond of the chosen group and tertiary C—H bond of cumene at 373 K).

Position	1	2	3	4
Group	CH$_3$	CH$_2$	CH$_2$	CH$_2$
Nonane	0.0016	0.015	0.012	0.010
Decane	0.0014	0.014	0.011	0.009
Dodecane	0.0015	0.013	0.010	0.008
Tetradecane	0.0009	0.011	0.009	0.007

The reactivity of the C—H bond of cycloalkane depends on the ring size and the alkyl substituent.

TABLE 5.5
Rate Constants of the Cross-Propagation in Co-Oxidation of Hydrocarbons

RH	RO$_2^\bullet$	T (K)	E (kJ mol^{-1})	log A, A (L mol^{-1} s^{-1})	k (350 K) (L mol^{-1} s^{-1})	Ref.
EtMe$_2$CH	Me$_3$COO$^\bullet$	303			7.3 × 10^{-3}	[59]
Me(CH$_2$)$_4$Me	PhMe$_2$COO$^\bullet$	404			41.7	[60]
Me(CH$_2$)$_4$Me	Me$_3$COO$^\bullet$	303			3.6 × 10^{-3}	[59]
Me$_2$PrCH	Me$_3$COO$^\bullet$	303			7.86 × 10^{-3}	[59]
Me$_2$CHCHMe$_2$	Me$_3$COO$^\bullet$	303			1.86 × 10^{-2}	[59]
Me$_3$CCHMe$_2$	Me$_3$COO$^\bullet$	303–353	55.2	7.50	0.18	[59]
Me(CH$_2$)$_7$Me	PhMe$_2$COO$^\bullet$	383			8.72 × 10^{-2}	[61]
Me(CH$_2$)$_{11}$Me	Me$_3$COO$^\bullet$	313			8.6 × 10^{-4}	[62]
Me(CH$_2$)$_{14}$Me	Me$_3$COO$^\bullet$	303–343	74.5	10.50	0.24	[62]
cyclo-C$_5$H$_{10}$	Me$_3$COO$^\bullet$	303–353	70.1	9.80	0.22	[59]
cyclo-C$_6$H$_{12}$	Me$_3$COO$^\bullet$	313			6.5 × 10^{-4}	[62]
cyclo-C$_6$H$_{12}$	Me$_3$COO$^\bullet$	353			0.17	[31]
cyclo-C$_7$H$_{14}$	Me$_3$COO$^\bullet$	303	81.6	11.48	0.20	[59]
Me-cyclo-C$_5$H$_9$	Me$_3$COO$^\bullet$	303			2.58 × 10^{-2}	[59]
Me-cyclo-C$_6$H$_{11}$	Me$_3$COO$^\bullet$	303			1.32 × 10^{-2}	[59]
CH$_2$=CHCH$_3$	Me$_3$COO$^\bullet$	393			0.90	[63]
CH$_2$=CHCH$_2$Pr	Me$_3$COO$^\bullet$	393			4.8	[63]
Me$_2$C=CMe$_2$	Me$_3$COO$^\bullet$	393			35.0	[63]
CH$_2$=CH(CH$_2$)$_5$Me	Me$_3$COO$^\bullet$	283–333	47.7	7.40	1.91	[62]
CH$_2$=CPhCH$_3$	Me$_3$COO$^\bullet$	393			4.0	[63]
cyclohexene	Me$_3$COO$^\bullet$	273–303	49.0	9.30	97.1	[62]
cyclohexene	HOO$^\bullet$	303			3.4 × 10^2	[64]
PhCH$_3$	BuOO$^\bullet$	303			0.10	[54]
PhCH$_3$	Me$_3$COO$^\bullet$	303			5.0 × 10^{-2}	[55]
PhCH$_3$	Me$_3$CCH$_2$C(OO$^\bullet$)Me$_2$	303			0.40	[55]
PhCH$_3$	MeEtPhCOO$^\bullet$	303			5.0 × 10^{-2}	[55]
p-xylene	Me$_3$COO$^\bullet$	283–343	46.0	6.90	1.08	[62]
p-xylene	PhMe$_2$COO$^\bullet$	393–408	45.6	6.34	0.34	[65]

continued

TABLE 5.5
Rate Constants of the Cross-Propagation in Co-Oxidation of Hydrocarbons—*continued*

RH	RO$_2^\bullet$	T (K)	E (kJ mol^{-1})	log A, A (L mol^{-1} s^{-1})	k (350 K) (L mol^{-1} s^{-1})	Ref.
methyl 4-methylbenzoate	PhMe$_2$COO$^\bullet$	303–393	53.1	6.95	0.11	[65]
PhCH$_2$Me	BuOO$^\bullet$	303			0.55	[20]
PhCH$_2$Me	Me$_3$COO$^\bullet$	303			0.20	[55]
PhCH$_2$Me	Me$_3$CCH$_2$C(O$_2^\bullet$)Me$_2$	303			0.32	[55]
PhCH$_2$Me	PhMeEtCOO$^\bullet$	303			0.22	[55]
PhCHMe$_2$	Me$_3$COO$^\bullet$	303			0.22	[55]
PhCHMe$_2$	Me$_3$CCH$_2$C(OO$^\bullet$)Me$_2$	303			0.14	[55]
PhCHMe$_2$	PhMeEtCOO$^\bullet$	303			0.15	[55]
PhCHMe$_2$	Ph$_2$MeCOO$^\bullet$	303			0.18	[55]
PhCHMe$_2$	Me$_3$COO$^\bullet$	303–333	55.2	8.70	2.90	[66]
PhCHMe$_2$	BuOO$^\bullet$	303			0.45	[54]
tetralin	Me$_3$COO$^\bullet$	293–333	51.5	9.20	32.7	[62]
tetralin	Me$_3$COO$^\bullet$	303			2.0	[55]
tetralin	Me$_3$CCH$_2$C(OO$^\bullet$)Me$_2$	303			2.0	[55]
tetralin	PhMeEtCOO$^\bullet$	303			1.3	[55]
tetralin	HOO$^\bullet$	303			9.0 × 10^2	[64]
tetralin	Me$_3$COO$^\bullet$	333			11.0	[67]
tetralin	BuOO$^\bullet$	303			0.64	[54]
tetralin	PhMe$_2$COO$^\bullet$	348			14.6	[68]
octahydroanthracene	PhMe$_2$COO$^\bullet$	348			73.5	[68]
dodecahydrotriphenylene	PhMe$_2$COO$^\bullet$	348			1.17 × 10^2	[68]
9,10-dihydroanthracene	BuOO$^\bullet$	303			3.10 × 10^2	[54]

Ring Size	6	7	8	9	10	11	12
RR of CH_2		0.032	0.045	0.038	0.029	0.028	0.011
RR of CHMe	0.102	0.261	0.258	0.115			
RR of CHEt	0.072	0.122	0.148	0.128			

The reactivity of equatorial and axial C—H bonds in dimethylcyclohaxanes appeared to be sufficiently different [58]. Tertiary C—H bonds in the equatorial position react more readily with peroxyl radicals than similar bonds in the axial position. For example, in cis-1,2-dimethylcyclohexane RR (equatorial C—H) = 0.39 and RR (axial C—H) = 0.064; in trans-1,3-dimethylcyclohexane RR (axial C—H) = 0.150 and RR (equatorial C—H) = 0.328, in cis-decalin RR (axial C—H) = 0.056 and RR (equatorial C—H) = 0.468.

The reactivity of CH_3, CH_2, and CH groups in alkylaromatic compounds increases from the CH_3 to CH group. However, the difference in reactivity is not so high as in aliphatic compounds. A few examples are given below [58].

Hydrocarbon					
$RR(CH_3)$	0.13	0.08	0.13	0.13	—
$RR(CH_2)$	0.75	0.23	—	—	—
RR(CH)	—	—	0.93	1.50	1.00

Opeida [46] compared the values of the rate constants of peroxyl radical reactions with hydrocarbons with the BDE of the oxidized hydrocarbon, electron affinity of peroxyl radical, $EA(RO_2^\bullet)$ ionization potential of hydrocarbon (I_{RH}), and steric hindrance of α-substituent $R(V_R)$. They had drawn out the following empirical equation:

$$\log k_p(348K) = a + b \times D(R—H) + c \times (I_{RH} - EA(RO_2^\bullet)) + d \times V_R, \quad (5.12)$$

where $a = 23.03$, $b = -0.0807$, $c = 31.48$, and $d = -0.962$. The results of the calculation are close to experimental values of rate constants. The values of the rate constants of peroxyl radicals $R_iO_2^\bullet$ with hydrocarbons R_jH are presented in Table 5.5.

5.6 HIGH REACTIVITY OF HALOIDALKYLPEROXYL RADICALS

The study of co-oxidation of hydrocarbons and haloidalkanes proved the extremely high activity of peroxyl radicals possessing a few chlorine substituents in the α-position ($CCl_3OO^\bullet$, $CCl_3CCl_2OO^\bullet$ etc.). This is the result of exothermicity of the reactions

$$RCCl_2OO^\bullet + RH \longrightarrow RCCl_2OOH + R^\bullet$$

due to the formation of the strong O—H bond in the forming haloidalkyl hydroperoxide [69].

ROOH	MeOOH	CF$_3$OOH	CCl$_3$OOH	CCl$_3$CCl$_2$OOH	CF$_2$ClOOH
D (kJ mol^{-1})	365.5	418.0	407.2	413.1	415.4

The results of the study of co-oxidation of different hydrocarbons with CCl$_3$CCl$_2$H and CHCl$_2$CHCl$_2$ are presented in Table 5.6.

5.7 CO-OXIDATION OF HYDROCARBON AND ITS OXIDATION INTERMEDIATES

In the initial period the oxidation of hydrocarbon RH proceeds as a chain reaction with one limiting step of chain propagation, namely reaction RO$_2^{\bullet}$ + RH. The rate of the reaction is determined only by the activity and the concentration of peroxyl radicals. As soon as the oxidation products (hydroperoxide, alcohol, ketone, etc.) accumulate, the peroxyl radicals react with these products. As a result, the peroxyl radicals formed from RH (RO$_2^{\bullet}$) are replaced by other free radicals. Thus, the oxidation of hydrocarbon in the presence of produced and oxidized intermediates is performed in co-oxidation with complex composition of free radicals propagating the chain [4]. A few examples are given below.

1. Reaction of peroxyl radical with secondary hydroperoxide produces very active hydroxyl radical. The latter attacks immediately the hydrocarbon molecule.

$$RO_2^{\bullet} + R^1R^2CHOOH \longrightarrow ROOH + R^1R^2C(O) + HO^{\bullet}$$
$$HO^{\bullet} + RH \longrightarrow H_2O + R^{\bullet}$$

2. The alkylhydroxyperoxyl and hydroperoxyl radicals formed from alcohol possess a reducing activity and attack hydroperoxide with the formation of the alkoxyl radical. This radical is very active and propagates the chain reacting with hydrocarbon.

$$RO_2^{\bullet} + R^1R^2CHOH \longrightarrow ROOH + R^1R^2C^{\bullet}(OH)$$
$$R^1R^2C^{\bullet}(OH) + O_2 \longrightarrow R^1R^2C(OH)OO^{\bullet}$$
$$R^1R^2C(OH)OO^{\bullet} + ROOH \longrightarrow R^1R^2C(O) + H_2O + O_2 + RO^{\bullet}$$
$$R^1R^2C(OH)OO^{\bullet} \longrightarrow R^1R^2C(O) + HOO^{\bullet}$$
$$HOO^{\bullet} + ROOH \longrightarrow H_2O + O_2 + RO^{\bullet}$$
$$RO^{\bullet} + RH \longrightarrow ROH + R^{\bullet}$$

3. Aldehyde produces the acylperoxyl radical as given by the following reactions:

$$RO_2^{\bullet} + R^1CH(O) \longrightarrow ROOH + R^1C^{\bullet}(O)$$
$$R^1C^{\bullet}(O) + O_2 \longrightarrow R^1C(O)OO^{\bullet}$$
$$R^1C(O)OO^{\bullet} + RH \longrightarrow R^1C(O)OOH + R^{\bullet}$$

The acylperoxyl radical is extremely active due to the high dissociation energy of O—H bond ($D_{O-H} = 418$ kJ mol^{-1} in benzaldehyde [73]) and accelerates the chain propagation.

4. The addition of hydroxyl radicals to benzene ring of alkylaromatic hydrocarbon gives phenolic compounds. Phenols retard oxidation, terminating the chains (see Part II).

TABLE 5.6
Rate Constants of Haloidalkylperoxyl Radicals Reactions with Hydrocarbons (348 K, Solvent is Oxidized Hydrocarbon)

Hydrocarbon	k_p ($C_2Cl_5OO^\bullet$) (L mol^{-1} s^{-1})	k_p ($Cl_2CHCCl_2OO^\bullet$) (L mol^{-1} s^{-1})	Ref.
Me(CH$_2$)$_6$Me	2.30×10^3	1.78×10^3	[70]
CH$_3$(CH$_2$)$_7$CH$_3$	2.48×10^3	2.22×10^3	[70]
CH$_3$(CH$_2$)$_8$CH$_3$	3.11×10^3	2.54×10^3	[70]
CH$_3$(CH$_2$)$_9$CH$_3$	3.23×10^3	2.95×10^3	[70]
CH$_3$(CH$_2$)$_{10}$CH$_3$	3.81×10^3	2.83×10^3	[70]
CH$_3$(CH$_2$)$_{12}$CH$_3$	3.39×10^3	3.33×10^3	[70]
CH$_3$(CH$_2$)$_{14}$CH$_3$	4.06×10^3	5.18×10^3	[70]
CH$_3$(CH$_2$)$_{18}$CH$_3$	5.55×10^3	5.83×10^3	[70]
cyclohexane	2.20×10^3	1.40×10^3	[71]
cyclohexane-d_{12}		6.30×10^2	[71]
PhCH$_3$	2.60×10^2	2.10×10^2	[72]
PhCD$_3$		1.20×10^2	[72]
o-xylene	7.10×10^2	5.20×10^2	[72]
1,3,5-trimethylbenzene	2.10×10^3	1.20×10^3	[72]
1,2,4,5-tetramethylbenzene	4.10×10^3	2.60×10^3	[72]
pentamethylbenzene	8.80×10^3	6.30×10^3	[72]
hexamethylbenzene	1.20×10^4	1.10×10^4	[72]
PhCH$_2$Me	8.20×10^2	7.30×10^2	[72]
1,4-diethylbenzene		1.66×10^3	[72]
4-ethylacetophenone		6.2×10^2	[72]
4-ethylnitrobenzene		2.3×10^2	[72]

continued

TABLE 5.6
Rate Constants of Haloidalkylperoxyl Radicals Reactions with Hydrocarbons (348 K, Solvent is Oxidized Hydrocarbon)—continued

Hydrocarbon	k_p ($C_2Cl_5OO^\bullet$) (L mol^{-1} s^{-1})	k_p ($Cl_2CHCCl_2OO^\bullet$) (L mol^{-1} s^{-1})	Ref.
PhCHMe$_2$	1.3×10^3	9.7×10^2	[72]
PhCMe$_3$	2.7×10^2	2.4×10^2	[72]
(fluorene)	5.9×10^3	7.2×10^3	[72]
(acenaphthene)	4.5×10^3	4.3×10^3	[72]

$$HO^\bullet + C_6H_6 \longrightarrow HO\text{-}C_6H_6^\bullet$$

$$HO\text{-}C_6H_6^\bullet + ROO^\bullet \longrightarrow C_6H_5\text{-}OH + ROOH$$

The composition of the intermediates changes during the oxidation and with it the composition of radicals participating in chain propagation. This affects the rate of hydrocarbon oxidation. The theory of this phenomenon was developed by Denisov [74–78] and described in a monograph [4]. The experimental evidences of change in free radical activity were provided by the studies of the oxidation kinetics of decane [79], cyclohexane (see Figure 5.2), and cyclododecanone [80]. For example, the relative change of activity of radicals propagating the chain in oxidized cyclohexane was as much as 2.5 times (383 K, $pO_2 = 6$ atm atm [81]).

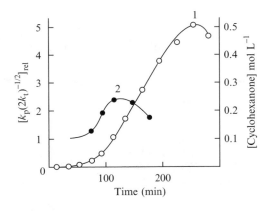

FIGURE 5.2 Cyclohexane oxidation: the kinetic curve of cyclohexanone formation (1) and relative change in the ratio $k_p(2k_t)^{-1/2}$ (2) during the oxidation ($T = 428$ K, $pO_2 = 5.4$ atm. [81]).

Co-Oxidation of Hydrocarbons

The kinetic analysis proves that formation of very active radical from intermediate product can increase the reaction rate not more than twice. However, the formation of inactive radical can principally stop the chain reaction [77]. Besides the rate, the change of composition of chain propagating radicals can influence the rate of formation and decay of intermediates in the oxidized hydrocarbon. In its turn, the concentrations of intermediates (alcohols, ketones, aldehydes, etc.) influence autoinitiation and the rate of autoxidation of the hydrocarbon (see Chapter 4).

5.8 CATALYSIS BY NITROXYL RADICALS IN HYDROCARBON OXIDATION

Catalysis by nitroxyl radicals in hydrocarbon oxidation was discovered and studied recently [82–89]. The introduction of N-hydroxyphthalimide into oxidized alkylaromatic hydrocarbon was found to accelerate the oxidation. The formation of the stable phthalimide-N-oxyl (PINO) radical was evidenced by the EPR method [90]. The following kinetic scheme was put forward to explain the accelerating effect of PINO on the chain oxidation of hydrocarbons [82–84].

The introduction of hydroxylamine into oxidizing hydrocarbon adds the new cycle of chain propagation reactions to the traditional $R^\bullet \to RO_2^\bullet \to R^\bullet$ cycle. This scheme is similar to that of hydrocarbon oxidation with the addition of another hydroperoxide (see earlier).

The free radical equilibrium between PINO and hydroxylamines and phenols with known BDE of the O—H bonds was used for to estimate the BDE of the O—H bond of PINO and other hydroxylamines [90]. The values of BDE (kJ mol^{-1}) are given below.

| 368.6 | 295.4 | 297.5 | 291.6 |
| 304.6 | 328.4 | 331.4 | 335.6 |

TABLE 5.7
Rate Constants and Kinetic Isotope Effects of PINO Reaction with Hydrocarbons ($T = 298$ K)

Hydrocarbon	Solvent	k (L mol^{-1} s^{-1})	k_H/k_D	Ref.
cyclohexane	AcOH	2.32×10^{-2}	24.0	[91]
cyclohexane	C$_6$H$_6$ + 10%MeCN	0.56		[90]
PhCH$_3$	AcOH	0.62	27.1	[91]
PhCH$_3$	C$_6$H$_6$ + 10%MeCN	1.14		[90]
PhCH$_2$Me	AcOH	5.36	21.6	[91]
PhCH$_2$Me	C$_6$H$_6$ + 10%MeCN	4.48	8.7	[90]
PhMe$_2$CH	AcOH	26.6		[91]
PhMe$_2$CH	C$_6$H$_6$ + 10%MeCN	3.25		[90]
Ph$_2$CH$_2$	AcOH	13.3		[91]
Ph$_3$CH	AcOH	1.17×10^2		[91]
1-methylnaphthalene	AcOH	4.29		[91]
fluorene	AcOH	40.6	13.3	[91]
9,10-dihydroanthracene	AcOH	5.02×10^3		[91]
adamantane	C$_6$H$_6$ + 10%MeCN	9.4×10^{-2}		[90]
PhCH$_2$OH	AcOH	11.3		[91]
PhCH$_2$OH	C$_6$H$_6$ + 10%MeCN	28.3		[90]
PhCHO	AcOH	21.2		[91]
Ph$_2$CHOH	AcOH	57.5		[91]
2-methyl-1,4-benzoquinone	AcOH	2.66×10^{-2}		[91]
o-xylene	AcOH	2.99		[92]
m-xylene	AcOH	3.06		[92]
p-xylene	AcOH	5.95	24.8	[92]
p-tolualdehyde	AcOH	12.4		[92]
o-toluic acid	AcOH	0.113		[92]
m-toluic acid	AcOH	0.20		[92]

The kinetic scheme of a hydrocarbon RH oxidation catalyzed by PINO in the presence of initiator I includes the following elementary steps [90]:

$$I \longrightarrow r^{\bullet} \xrightarrow{RH} R^{\bullet}$$
$$R^{\bullet} + O_2 \longrightarrow RO_2^{\bullet} \quad \text{fast}$$
$$RO_2^{\bullet} + RH \longrightarrow ROOH + R^{\bullet} \quad k_p$$
$$RO_2^{\bullet} + RO_2^{\bullet} \longrightarrow \text{Products} \quad 2k_t$$
$$RO_2^{\bullet} + PINOH \longrightarrow ROOH + PINO^{\bullet} \quad k_{PINO1}$$
$$PINO^{\bullet} + RH \longrightarrow R^{\bullet} + PINOH \quad k_{PINO2}$$

In accordance with this scheme, the oxidation rate is equal to [90]:

$$v = \{k_p[RH] + k_{PINO1}[PINOH]\}v_i^{1/2}(2k_t)^{-1/2} \qquad (5.12)$$

The accelerating effect of PINO appears to be the result of the following two peculiarities:

1. PINO does not participate in chain termination. So, the introduction of PINO in oxidized hydrocarbon decreases the steady-state concentration of peroxyl radicals and consequently the rate of chain termination.
2. PINO possesses a high reactivity in the reaction with the C—H bond of the hydrocarbon. Hence, the substitution of peroxyl radicals to nitroxyl radicals accelerates the chain reaction of oxidation. The accumulation of hydroperoxide in the oxidized hydrocarbon should decrease the oxidation rate because of the equilibrium reaction.

$$PINO^{\bullet} + ROOH \rightleftharpoons PINOH + RO_2^{\bullet}$$

The rate constants of the reaction of PINO$^{\bullet}$ with hydrocarbons and alcohols, and the values of k_H/k_D are given in Table 5.7. A very high kinetic isotope effect is seen in this reaction.

REFERENCES

1. GA Russell, RC Williamson. *J Am Chem Soc* 86:2357–2363, 1964.
2. GA Russell, RC Williamson. *J Am Chem Soc* 86:2364–2367, 1964.
3. GC Eastmond. In: CH Bamford and CFH Tipper (eds.), *Comprehensive Chemical Kinetics*, vol. 14A. Amsterdam: Elsevier, 1980, pp. 1–103.
4. NM Emanuel, ET Denisov, ZK Maizus. *Liquid-Phase Oxidation of Hydrocarbons*. New York: Plenum Press, 1967.
5. JA Howard. *Adv Free Radical Chem* 4:49–173, 1972.
6. T Mill, DG Hendry. In: CH Bamford and CFH Tipper (Eds.), *Comprehensive Chemical Kinetics*, vol. 16. Amsterdam: Elsevier, 1980, pp. 1–88.
7. FR Mayo, MG Syz, T Mill, JK Castleman. *Adv Chem Ser* 75:38–58, 1968.
8. L Sajus. *Adv Chem Ser* 75:59–77, 1968.
9. RV Kucher, IA Opeida. *Co-oxidation of Organic Compounds in Liquid Phase*. Kiev: Naukova Dumka, 1989 [in Russian].
10. MA Fineman, SD Ross. *J Polymer Sci* 5:259–262, 1950.
11. C Gadell, G Clement. *Bull Soc Chim Fr N* 1:44–54, 1968.
12. RV Kucher, IA Opeida, AG Matvienko. *Oxid Commun* 3:115–124, 1983.
13. J Alagy, G Clement, JC Balaceanu. *Bull Soc Chim Fr* 10, 1792–1799, 1961.
14. VF Tsepalov, VYa Shlyapintokh, CP Khuan. *Zh Fiz Khim* 38:52–58, 1964.

15. RV Kucher, IA Opeida. *Neftekhimiya* 10:54–58, 1970.
16. IA Opeida, VI Timokhin. *Ukr Khim Zh* 44:187–190, 1978.
17. IA Opeida, RV Kucher. *Ukr Khim Zh* 36:1041–1043, 1970.
18. J Alagy, G Clement, JC Balaceanu. *Bull Soc Chim Fr* 8/9, 1495–1499, 1960.
19. GA Russell. *J Am Chem Soc* 77:4583–4590, 1955.
20. E Niki, J Kamia, N Ohta. *Bull Chem Soc Jpn* 42:512–520, 1969.
21. JA Howard, WJ Schwalm, KU Ingold. *Adv Chem Ser* 75:6–23, 1968.
22. AG Matvienko, IA Opeida, RV Kucher. *Neftekhimiya* 23:115–117, 1970.
23. RV Kucher, IA Opeida, AN Nikolaevskiy. *Kinet Katal* 11:1568–1570, 1970.
24. VI Timokhin, IA Opeida, RV Kucher. *Neftekhimiya* 17:555–558, 1977.
25. C Chevriau, P Naffa, JC Balaceanu. *Bull Soc Chim Fr* 3002–3010, 1964.
26. E Niki, Y Kamiya, N Ohta. *Kogyo Kaguku Zasshi* 71:1187–1192, 1968.
27. L Dulog, J Szita, W Kern. *Fette Seifen Anstrichmittel* 65:108–113, 1963.
28. FR Mayo, AA Miller, GA Russell. *J Am Chem Soc* 80:2500–2507, 1958.
29. JHB Chenier, JA Howard. *Can J Chem* 53:623–627, 1975.
30. JA Howard, JHB Chenier. *Can J Chem* 58:2808–2812, 1980.
31. DG Hendry, CW Gould, D Schuetzle, D Syz, FR Mayo. *J Org Chem* 41:1–10, 1976.
32. JA Howard, KU Ingold. *Can J Chem* 47:3809–3815, 1969.
33. JR Thomas, CA Tolman. *J Am Chem Soc* 84:2079–2080, 1962.
34. E Niki, K Okayasu, Y. Kamiya. *Int J Chem Kinet* 6:279–290, 1974.
35. S Fukusumi, Y Ono. *J Phys Chem* 81:1895–1900, 1977.
36. R. Hiatt, CW Gould, FR Mayo. *J Org Chem* 29:3461–3472, 1964.
37. ET Denisov. *Izv AN SSSR Ser Khim* 328–331, 1969.
38. RL Vardanyan, ET Denisov, VI Zozulya. *Izv AN SSSR Ser Khim* 611–613, 1972.
39. TG Degtyareva, ET Denisov, VS Martem'yanov, LA Badretdinova. *Izv AN SSSR Ser Khim* 1219–1225, 1979.
40. GA Russell. *J Am Chem Soc* 78:1047–1054, 1956.
41. JE Bennett. *J Chem Soc Faraday Trans 2*, 86:3247–3252, 1990.
42. JR Thomas. *J Am Chem Soc* 89:4872–4875, 1967.
43. SL Khursan, RL Safiullin, VS Martemianov, AI Nikolaev, IA Urozhai. *React Kinet Catal Lett* 39:261–266, 1989.
44. SL Khursan, RL Safiullin, SYu Serenko. *Khim Fiz* 9:375–379, 1990.
45. P Grosborne, I Seree de Roch. *Bull Soc Chim Fr* 2260–2267, 1967.
46. IA Opeida. Cooxidation of Alkylaromatic Hydrocarbons in the Liquid Phase. Doctoral Dissertation, Institute of Chemical Physics, Chernogolovka, 1982, pp. 1–336 [in Russian].
47. VS Rafikova, EF Brin, IP Skibida. *Kinet Katal* 12:1374–1379, 1971.
48. IA Opeida, AG Matvienko, IV Efimova. *Kinel Katal* 28:1341–1346, 1987.
49. IA Opeida, IV Efimova, AG Matvienko, AF Dmitruk, OM Zarechnaya. *Kinet Katal* 31:1342–1348, 1990.
50. C Parlant. *Rev Inst Fr Petrol* 19:1–40, 1964.
51. IA Opeida, VI Timokhin, OV Nosyreva, VG Kaloerova, AG Matvienko. *Neftekhimiya* 21:110–113, 1981.
52. KE Kharlampidi, FI Nigmatullina, FI Batyrshin, NM Lebedeva. *Neftekhimiya* 24:676–681, 1986.
53. VA Machtin. Reactions of Peroxyl Radicals in Oxidizing Vinyl Monomers and Reactivity of Double Bonds. Ph.D. dissertation, Institute of Chemical Physics, Chernogolovka, 1984, pp. 1–130 [in Russian].
54. JA Howard, KU Ingold. *Can J Chem* 46:2661–2666, 1968.
55. JA Howard, KU Ingold. *Can J Chem* 46:2655–2660, 1968.
56. JA Howard, KU Ingold, M Symonds. *Can J Chem* 46:1017–1022, 1968.
57. ET Denisov, TG Denisova. *Kinet Catal* 34:173–179, 1993.
58. WW Pritzkow, VYa Suprun. *Russ Chem Rev* 65:503–546, 1996.
59. JHB Chenier, SB Tong, JA Howard. *Can J Chem* 56:3047–3053, 1978.
60. NN Pozdeeva, ET Denisov, VS Martem'yanov. *Kinet Katal* 22:912–919, 1981.
61. G Lautenbach, F Karabet, M Makhoul, W Pritzkow. *J Prakt Chem* 336:712–713, 1994.

62. S Korcek, JHB Chenier, JA Howard, KU Ingold. *Can J Chem* 50:2285–2297, 1972.
63. P Koelewijn. *Rev Trav Chim Pays-Bas* 91:759–770, 1972.
64. JA Howard, KU Ingold. *Can J Chem* 45:785–792, 1967.
65. NI Mitskevich, NG Ariko, VE Agabekov, NN Kornilova. *React Kinet Catal Lett* 1:467, 1974.
66. JA Howard, JHB Chenier, DA Holden. *Can J Chem* 56:170–175, 1978.
67. JA Howard, JHB Chenier. *Int J Chem Kinet* 6:527–530, 1974.
68. SA Gerasimova, AG Matvienko, IA Opeida, IO Kachurin. *Zh Org Khim* 34:781–781, 1998.
69. TG Denisova, ET Denisov, *Petr Chem* 44:250–255, 2004.
70. RV Kucher, VI Timokhin, RI Flunt. *Dokl AN SSSR* 305:134–137, 1989.
71. RV Kucher, VI Timokhin, AP Pokutsa, RI Flunt. *Neftekhimiya* 28:701–706, 1988.
72. RV Kucher, RI Flunt, VI Timokhin, OI Makogon, AG Matvienko. *Dokl AN SSSR* 310:137–140, 1990.
73. ET Denisov, TG Denisova. *Kinet Catal* 34:883–889, 1993.
74. ET Denisov. *Izv AN SSSR Ser Khim* 195–203, 1960.
75. ET Denisov. *Izv AN SSSR Ser Khim* 2100–2111, 1959.
76. ET Denisov. *Zh Fiz Khim* 36:2352–2361, 1962.
77. ET Denisov. *Izv AN SSSR Ser Khim* 796–803, 1961.
78. ET Denisov. *Izv AN SSSR Ser Khim* 980–991, 1963.
79. LG Privalova, ZK Maizus, ET Denisov. *Zh Fiz Khim* 39:1965–1969, 1965.
80. VG Bykovchenko, IV Berezin. *Neftekhimiya* 3:565–671, 1963.
81. ET Denisov. *Zh Fiz Khim* 33:1198–1208, 1959.
82. Y Yoshino, Y Hayashi, T Iwahama, S Sakaguchi, Y Ishii. *J Org Chem* 62:6810–6813, 1997.
83. T Iwahama, Y Yoshima, T Keitoku, S Sakaguchi, Y Ishii. *J Org Chem* 65:6502, 2000.
84. S Sakaguchi, Y Nishiwaki, T Kitamura, T Yshii. *Angew Chem Int Ed.* 40, 222, 2001.
85. Y Ishii, S Sakaguchi, T Iwahama. *Adv Synth Catal* 343:393–427, 2001.
86. A Cecchetto, F Fontana, F Minisci, F Recupero. *Tetrahedron Lett* 42:6651, 2001.
87. F Minisci, C Punta, F Recupero, F Fontana, GF Pedulli. *J Org Chem* 67:2671–2676, 2002.
88. A Cecchetto, F Minisci, F Recupero, F Fontana, GF Pedulli. *Tetraedron Lett* 43:3605, 2002.
89. F Minisci, C Punta, F Recupero, F Fontana, GF Pedulli. *Chem Commun* 7:688–689, 2002.
90. R Amorati, M Lucarini, V Mugnaini, GF Pedulli, F Minisci, F Recupero, F Fontana, P Astolfi, L Greci. *J Org Chem* 68:5198–5204, 2003.
91. N Koshino, Y Cai, JH Espenson. *J Phys Chem A* 107:4262–4267, 2003.
92. N Kohino, B Saha, JH Espenson. *J Org Chem* 68:9364–9370, 2003.

6 Reactivity of the Hydrocarbons in Reactions with Peroxyl, Alkoxyl, and Alkyl Radicals

6.1 SEMIEMPIRICAL MODEL OF RADICAL REACTION AS AN INTERSECTION OF TWO PARABOLIC POTENTIAL CURVES

6.1.1 INTRODUCTION

Several empirical correlations are known for rate constants and activation energies of bimolecular radical reactions [1–4]. Evans and Polyany [5] were the first to derive the linear correlation between the activation energy and the enthalpy of reaction of R_iX with Na. Later Semenov [1] generalized this empirical equation for different free radical reactions in the following form:

$$E_i = A' + \alpha' \Delta H_i, \tag{6.1}$$

Linear correlations proposed by Hammett [6] and Taft [7] in the form

$$\log(k_i/k_0) = \rho\sigma \tag{6.2}$$

are widely used in physical organic chemistry (σ is the Hammett or Taft constant and ρ is the coefficient of the chosen reaction series). Since the pre-exponential factor A is constant for one free radical reaction series (for example, $ROO^\bullet + MeCH_2C_6H_4Y$), Equation (6.2) can be transformed into the form close to Polyany–Semenov equation [1]:

$$E_i = E_0 + 2.303RT\rho\sigma \tag{6.3}$$

Nonlinear hyperbolic dependence between the activation energy and the enthalpy of reaction was proposed by Rudakov [8]:

$$E(E - \Delta H) = E_0^2 \tag{6.4}$$

$$E_0 = p_X p_Y \tag{6.5}$$

The parameters p_X and p_Y are empirical coefficients of the reaction

$$X^\bullet + HY \longrightarrow XH + Y^\bullet$$

A very effective instrument for the analysis of the reactivity of the reactants is the intersecting parabolas model (IPM). This model was used in analysis of the activation energies for a wide variety of radical abstraction reactions [4,9–12]. By using IPM, it became possible to create the empirical hierarchical system for diverse radical abstraction reactions (individual reaction–group of reactions–class of reactions–type of reactions). The parameters for groups and classes paves way for the separation of physical factors that influence the height of the activation barrier of the reaction. The creation of a hierarchical system of reactions made it possible to identify, at an empirical level, the physical characteristics of the reactants that determine the height of the activation barrier.

6.1.2 Intersecting Parabolas Model (IPM)

In the IPM model the radical abstraction reaction, for example,

$$RO_2^{\bullet} + R_iH \longrightarrow ROOH + R_i^{\bullet}$$

in which a hydrogen atom is transferred from the initial (R_iH) to the final (ROOH) molecule, is assumed to be resulting from the intersection of two potential curves [9–12]. One of these curves describes the potential energy $U_i(r)$ of the vibration of the H atom along the dissociating bond in the initial molecule (R_iH) as the function of vibration amplitude r. The other curve $U_f(r)$ describes the potential energy of the vibration of the same atom along the forming bond in the reaction product ROOH. The stretching vibrations of the H atom in the C—H bond of the attacked RH, and O—H bond of the formed ROOH, are regarded as harmonic and described by the parabolic law

$$U(r)^{1/2} = br \qquad (6.6)$$

The following parameters are used to characterize the elementary step in the parabolic model (see Figure 6.1).

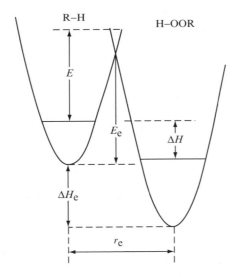

FIGURE 6.1 Parabolic model of the reaction $RO_2^{\bullet} + RH$ in coordinates: potential energy versus amplitude of vibration of the reacting bonds (for symbols, see text).

1. The enthalpy of reaction ΔH_e, which includes the difference between the zero-point energies of the broken and generated bonds,

$$\Delta H_e = D_i - D_f + 0.5hL(\nu_i - \nu_f), \quad (6.7)$$

where D_i and D_f are the dissociation energies of the cleaved (i) and the generated (f) bonds, ν_i and ν_f are the stretching vibrational frequencies of these bonds, h is the Planck constant, and L is the Avogadro's number.

2. The activation energy E_e is related to the experimentally determined Arrhenius activation energy E by the relation:

$$E_e = E + 0.5(hL\nu_i - RT), \quad (6.8)$$

3. The coefficients b_i and b_f describe the dependence of the potential energy on the atomic vibration amplitude along the initial (i) and final (f) valence bonds. There is a parabolic relationship between the potential energy and the vibrational amplitude:

$$U_i = b_i^2 r^2 \quad \text{and} \quad U_f = b_f^2 (r_e - r)^2 \quad (6.9)$$

The quantity $2b^2$ is the force constant of the corresponding bond with $b = \pi \nu (2\mu)^{1/2}$, where μ is the reduced mass of atoms forming the bond.

4. The distance r_e characterizes the displacement of the abstracted atom in the elementary step. The main equations of IPM are given in Chapter 4.

All known free radical bimolecular reactions can be divided into the following five types:

1. Atom abstraction by a radical or an atom from a molecule

$$R^1O^\bullet + RH \longrightarrow R^1OH + R^\bullet$$

2. Atom transfer from a radical to a molecule

$$R_2C^\bullet(OH) + O_2 \longrightarrow R_2C(O) + HOO^\bullet$$

3. Free radical substitution

$$R^\bullet + R^1OOH \longrightarrow ROH + R^1O^\bullet$$

4. Free radical addition and

$$ROO^\bullet + CH_2{=}CHX \longrightarrow ROOCH_2CHX$$

5. Direct atom substitution reaction.

$$D^\bullet + HOH \longrightarrow DOH + H^\bullet$$

The force constants of the stretching vibrations of the bonds ($2\pi^2 \nu^2 \mu$) of each type (C—H, O—H, etc.) and the energies of the zero vibrations of these bonds ($0.5hL\nu_i$ and $0.5hL\nu_f$) are nearly identical for the entire class of isotypical compounds. These important factors make it possible to classify radical abstraction reactions in terms of the type of reacting bonds [11]. A pair of coefficients, b_i and b_f, or the coefficient α and b corresponds to each class of such reactions. In the calculation of the activation energy E_e from E and conversely, the zero-point vibrational energy of the ruptured bond also becomes important (see Equation (6.8)). The values of α, b, $0.5hL\nu_i$, and $0.5hL\nu_f$ for the radical abstraction reactions of compounds in different classes are listed in Table 6.1.

Knowing these parameters, it is easy to find the analogous characteristics of the reverse reactions. For a reverse reaction to which the index f corresponds, we have $\alpha_f = \alpha^{-1}$, $b_f = b/\alpha$, and $0.5hL\nu_f = 0.5hL\nu_i/\alpha$.

TABLE 6.1
Kinetic Parameters of Different Classes of Free Radical Abstraction Reactions (R· + RH, RO· + RH, ROO· + RH [4,11,12])

Reaction	α	b (kJ mol^{-1})$^{1/2}$	$0.5hL\nu_i$ (kJ mol^{-1})	$0.5\,hL(\nu_i - \nu_f)$ (kJ mol^{-1})	$(r^{\#}/r_e)_0$
R· + HR	1.00	3.743×10^{11}	17.4	0.0	0.500
R· + HOR	1.256	4.701×10^{11}	21.7	4.3	0.443
R· + HOOR	1.229	4.600×10^{11}	21.2	3.8	0.448
RO· + RH	0.796	3.743×10^{11}	17.4	−4.3	0.443
RO· + HOR	1.000	4.701×10^{11}	21.7	0.0	0.500
RO· + HOOR	0.978	4.600×10^{11}	21.2	−0.5	0.505
RO$_2$· + HR	0.814	3.743×10^{11}	17.4	−3.8	0.449
RO$_2$· + HOR	1.022	4.701×10^{11}	21.7	0.5	0.494
RO$_2$· + HOOR	1.00	4.600×10^{11}	21.2	0.0	0.500

The class of radical abstraction reactions may include a single reaction (for example, H· + HCl), one group of reactions (for example, H· + RH), or several such groups (for example, the class of reactions $R_iOO· + R_jH$). All the reactions belonging to one group inside one class are characterized by a single parameter r_e or br_e. The quantities br_e, calculated on the basis of parabolic model for reactions ROO· + HR^1 involving only aliphatic hydrocarbons having different structures, are very close; the average value of br_e for reaction ROO· + HR^1 is equal to 13.62 ± 0.39 (kJ mol^{-1})$^{1/2}$ [13]. The quantity A calculated for each of the reacting C—H bond is 10^8 L mol^{-1} s^{-1} for the liquid phase and 2×10^7 L mol^{-1} s^{-1} for the gas phase. Knowing the parameter br_e, it is possible to calculate an important characteristic of the reactivity of each group of reactions, such as the activation energy for a thermally neutral reaction E_{e0}, which is written as:

$$\sqrt{E_{e0}} = \frac{br_e}{1 + \alpha} \qquad (6.10)$$

The activation energies E_{e0} may differ significantly for reactions of the same class. The values of the IPM parameters for different groups of reactions of R·, RO·, and ROO· are collected in Table 6.2.

The classification of radical reactions carried out in this way makes it possible to observe empirically (on the basis of the parameter br_e) the structural differences between the reactants within one class, to compare the classes and groups of reactions in terms of reactivity (in terms of the parameter br_e or E_{e0}), and to identify the physical and structural factors determining the reactivities of groups and classes of reactants.

6.1.3 Calculation of the Activation Energy

The activation energy for any individual reaction within the limits of the given group of reactions may be calculated correctly from the parameter br_e. Within the framework of the parabolic model, the quantity E_e was calculated by one of the following equations [4,11]:

(1) for $\alpha = 1$,

$$\sqrt{E_e} = \frac{br_e}{2} + \frac{\Delta H_e}{2br_e} \qquad (6.11)$$

TABLE 6.2
Parameters of the IPM Method for the Reactions of H-Atom Abstraction. R^1H is Aliphatic Hydrocarbon, R^2H is Olefin, and R^3H is Alkylaromatic Hydrocarbon [4,11,13,14]

Reaction	br_e (kJ mol^{-1})$^{1/2}$	E_{e0} (kJ mol^{-1})	$-\Delta H_{e\ min}$ (kJ mol^{-1})	$\Delta H_{e\ max}$ (kJ mol^{-1})	A (L mol^{-1} s^{-1})
$R^{1\bullet} + R^1H$	17.30	74.8	155.0	155.0	1.0×10^9
$R^{1\bullet} + R^2H$	18.60	86.5	190.8	190.8	1.0×10^8
$R^{1\bullet} + R^3H$	17.80	79.2	168.3	168.3	1.0×10^8
$R^{2\bullet} + R^2H$	19.25	92.6	190.8	190.8	1.0×10^8
$R^{2\bullet} + R^3H$	18.45	85.1	210.0	210.0	1.0×10^8
$R^{3\bullet} + R^3H$	18.05	81.4	186.5	186.5	1.0×10^8
$RO^\bullet + R^1H$	13.10	53.2	108.5	66.6	1.0×10^9
$RO^\bullet + R^2H$	14.14	62.0	139.4	87.1	1.0×10^8
$RO^\bullet + R^3H$	13.50	56.5	120.0	74.2	1.0×10^8
$RO_2^\bullet + R^1H$	13.61	56.3	117.2	76.1	1.0×10^8
$RO_2^\bullet + R^2H$	15.20	70.2	166.4	110.1	1.0×10^7
$RO_2^\bullet + R^3H$	14.32	62.3	138.0	90.6	1.0×10^7
$R^{1\bullet} + H-OR$	16.46	53.2	66.6	108.5	1.0×10^8
$R^{2\bullet} + H-OR$	17.76	62.0	87.1	139.4	1.0×10^8
$R^{3\bullet} + H-OR$	16.96	56.5	74.2	120.0	1.0×10^8
$R^{1\bullet} + H-OOR$	16.73	56.3	76.1	117.2	1.0×10^7
$R^{2\bullet} + H-OOR$	18.67	70.2	110.1	166.4	1.0×10^7
$R^{3\bullet} + H-OOR$	17.59	62.3	90.6	138.0	1.0×10^7
$RO^\bullet + H-OR$	13.62	46.4	61.3	61.3	1.0×10^9
$RO^\bullet + H-OOR$	14.13	51.0	73.7	70.1	1.0×10^9
$RO_2^\bullet + H-OR$	14.45	51.1	70.1	73.7	1.0×10^9
$RO_2^\bullet + H-OOR$	13.13	43.1	51.5	51.5	1.0×10^8

(2) for $\alpha \neq 1$,

$$\sqrt{E_e} = \frac{br_e}{1-\alpha^2}\left\{1 - \alpha\sqrt{1 - \frac{1-\alpha^2}{(br_e)^2}\Delta H_e}\right\} \quad (6.12)$$

(3) for $\Delta H_e(1-\alpha^2) \ll (br_e)^2$,

$$\sqrt{E_e} = \frac{br_e}{1+\alpha} + \frac{\alpha \Delta H_e}{2br_e} \quad (6.13)$$

Equations (6.11)–(6.13) are valid for the reactions of radical abstraction with enthalpy varying inside the limits: $\Delta H_{e\ min} < \Delta H_e < \Delta H_{e\ max}$ [11,15]. For example, for reactions of the class $ROO^\bullet + R^1H$, $\Delta H_{e\ min} = 117.3$ kJ mol^{-1} and $\Delta H_{e\ max} = 76.3$ kJ mol^{-1} [11]. For highly endothermic reactions with $\Delta H_e > \Delta H_{e\ max}$, the activation energy is $E = \Delta H + 0.5RT$, while for exothermic reactions with $\Delta H_e < \Delta H_{e\ min}$, it is close to $0.5RT$. The limiting values of enthalpies of reaction $\Delta H_{e\ max}$ and $\Delta H_{e\ max}$ are related to the parameters br_e and α in the following way:

$$\Delta H_{e\ max} = (br_e)^2 - 2\alpha br_e(0.5hL\nu_f)^{1/2} + 0.5(\alpha^2 - 1)hL\nu_f \quad (6.14)$$

$$\Delta H_{e\ min} = -(br_e/\alpha)^2 + 2\alpha^{-2}br_e(0.5hL\nu_i)^{1/2} + 0.5(1 - \alpha^{-2})hL\nu_i \quad (6.15)$$

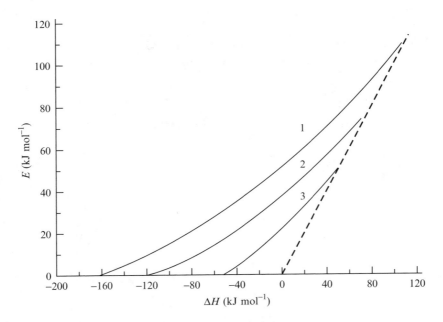

FIGURE 6.2 Dependence of the activation energy on the reaction enthalpy for reactions: $RO_2^\bullet + R^2H$ (1), $RO_2^\bullet + R^1H$ (2), $RO_2^\bullet + ROOH$ (3) calculated by the IPM method (see Equations [6.11] and [6.12]).

As can be seen from Equations (6.14) and (6.15), for the wider range of variation of ΔH_e in which $br_e = const$, the greater the parameter br_e. This illustrated by Figure 6.2.

Another important peculiarity of reactions with high values of $|\Delta H_e|$ is the dependence of the pre-exponential factor A on the reaction enthalpy [11,15]. This factor $A = const$ for all reactions of one group with ΔH_e in the limits: $\Delta H_{e\,min} < \Delta H_e < \Delta H_{e\,max}$. Outside these limits, the factor A varies with varying ΔH_e. For exothermic reactions with high values of $|\Delta H_e|$,

$$\sqrt{\frac{A}{A_0}} = 1 + 1.3 \times \left(\sqrt{|\Delta H_e|} - \sqrt{|\Delta H_{e\,min}|}\right), \qquad (6.16)$$

and for endothermic reactions with high values of ΔH_e,

$$\sqrt{\frac{A}{A_0}} = 1 + 1.3 \times \left(\sqrt{\Delta H_e} - \sqrt{\Delta H_{e\,max}}\right). \qquad (6.17)$$

where $A_0 = A$ for reactions inside the limits: $\Delta H_{e\,min} < \Delta H_e < \Delta H_{e\,max}$, $\Delta H_{e\,max/min} = \Delta H_{e\,max}$ for endothermic reactions, and $|\Delta H_{e\,min}|$ for exothermic reactions (see Figure 6.3).

The algorithm for rate constant calculation has the following form:

ΔH	E	A	Rate Constant
$\Delta H_e < \Delta H_{e\,min}$	$E = 0.5RT$	Equation (6.16)	$k = AF(\Delta H_e)\,e^{-1/2}$
$\Delta H_{e\,min} < \Delta H_e < \Delta H_{e\,max}$	Equations (6.11)–(6.13)	$A = const$	$k = A\exp(-E/RT)$
$\Delta H_e > \Delta H_{e\,max}$	$E = \Delta H + 0.5RT$	Eqn 6.17	$k = A\,e^{-1/2}F(\Delta H_e)\exp(-\Delta H/RT)$

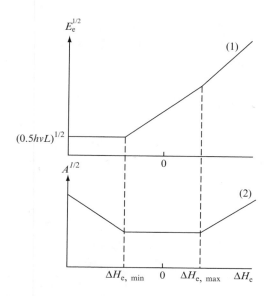

FIGURE 6.3 Dependence of the activation energy $E_e^{1/2}$ (1) and the pre-exponencial factor $A^{1/2}$ (2) on the reaction enthalpy ΔH_e according to the IPM.

The results of the calculation of the activation energies and the rate constants of peroxyl, alkoxyl, and alkyl radicals with alkanes and cycloalkanes are presented in Table 6.3–Table 6.5.

6.2 FACTORS INFLUENCING THE ACTIVATION ENERGY

6.2.1 REACTION ENTHALPY

A clear-cut dependence of the activation energy on the heat (enthalpy) of the reaction, which is equal, in turn, to the difference between the dissociation energies of the ruptured (D_i) and the formed (D_f) bonds, was established for a great variety of radical abstraction reactions [1,2,16]. In parabolic model, the values of D_{ei} and D_{ef}, incorporating the zero-point energy of the bond vibrations, are examined. The enthalpy of reaction ΔH_e, therefore, also includes the difference between these energies (see Equation [6.7]).

As noted above, all radical abstraction reactions can be divided into groups and the activation energy E_{e0} for a thermally neutral reaction can be calculated for each group (see Equation [6.11]). This opens up the possibility of calculating of the enthalpy contribution (ΔE_H) to the activation energy for the given (ith) reaction and a thermally neutral reaction characterized by the quantity E_{e0} [4,11]:

$$\Delta E_H = E_e - E_{e0} \qquad (6.18)$$

As an example, Table 6.6 presents the values of ΔE_H for the reactions of peroxyl radicals with aliphatic hydrocarbons. It follows from the table that the slope of this dependence dE/dH for these groups of reactions changes from zero to unity with the increasing reaction enthalpy (see Figure 6.2).

In terms of the parabolic model, it is possible to obtain simple and physically clear equations for the estimation of ΔE_H as a function of α, br_e, and ΔH_e. The following simple

TABLE 6.3
Rate Constants and Activation Energies of Ethylmethylperoxyl Radical Reactions with C—H Bonds of Organic Compounds (See Equations (6.7), (6.8), (6.12), and Parameters in Table 6.1 and Table 6.2)

RH	n	ΔH_e (kJ mol^{-1})	E (kJ mol^{-1})	A (L mol^{-1} s^{-1})	k (300 K) (L mol^{-1} s^{-1})
EtMeCH—H	4	43.7	61.9	4.0×10^8	6.68×10^{-3}
Me$_3$C—H	1	30.7	54.9	1.0×10^8	2.73×10^{-2}
cyclobutane H	8	49.2	65.0	8.0×10^8	3.92×10^{-3}
cyclopentane H	1	39.1	59.4	1.0×10^9	4.58×10^{-2}
cyclohexane H	1	26.2	52.6	1.0×10^8	6.89×10^{-2}
decalin	2	18.3	48.7	2.0×10^8	0.67
CH$_2$=CHCH$_2$—H	3	−1.3	53.4	3.0×10^7	1.49×10^{-2}
CH$_2$=CHCH—HMe	2	−19.5	45.6	2.0×10^7	0.23
CH$_2$=CHC—HMe$_2$	1	−29.7	41.5	1.0×10^7	0.61
Z-MeCH=CHCH—HMe	2	−25.3	43.2	2.0×10^7	0.60
Me$_2$C=CHCH—HMe	2	−37.3	38.5	2.0×10^7	4.01
Me$_2$C=CMeC—HMe$_2$	1	−46.5	35.0	1.0×10^7	8.09
CH$_2$=CHMeC—HCH=CH$_2$	1	−62.1	29.4	1.0×10^7	76.1
cyclohexene H,H	4	−27.8	42.2	4.0×10^8	17.9
cyclohexadiene H,H	4	−38.4	38.0	4.0×10^8	95.1
cyclohexadiene H,H	4	−56.7	31.3	4.0×10^8	1.43×10^3
MeC≡CC—HMe$_2$	1	−39.9	37.5	1.0×10^7	3.0
PhCH$_3$	3	5.7	44.9	3.0×10^7	0.46
PhMeCH—H	2	−5.2	40.0	2.0×10^7	2.17
PhMe$_2$C—H	1	−14.6	35.9	1.0×10^7	5.61
tetralin H H	4	−23.7	32.1	4.0×10^8	1.03×10^3
dihydroanthracene H H	4	−47.3	22.9	4.0×10^8	4.12×10^4
PhMeCHOO—H	1	0.0	23.1	1.0×10^8	9.51×10^3

expressions follow from the combination of Equations (6.11)–(6.13), and (6.18) for the reactions of one group with ΔH_e inside the interval $\Delta H_{e\ min} < \Delta H_e < \Delta H_{e\ max}$:

(1) for $\alpha = 1$,

$$\Delta E_H = 0.5\Delta H_e + 0.25(br_e)^{-2}\Delta H_e^2, \quad (6.19)$$

TABLE 6.4
Rate Constants and Activation Energies of 1-Methyl-1-Phenylethoxyl Radical (Me$_2$PhCO$^\bullet$) Reactions with C—H Bonds of Organic Compounds (See Equations (6.7), (6.8), (6.12), and Parameters in Tables 6.1 and 6.2)

RH	n	ΔH_e (kJ mol^{-1})	E (kJ mol^{-1})	A (L mol^{-1} s^{-1})	k (300 K) (L mol^{-1} s^{-1})
EtMeCH—H	4	−31.6	24.1	4.0×10^9	2.53×10^5
Me$_3$C—H	1	−44.6	19.4	1.0×10^9	4.13×10^5
cyclobutane (H, H)	8	−26.1	26.2	8.0×10^9	2.19×10^5
cyclopentane (H, H)	1	−36.2	22.4	10×10^{10}	1.25×10^6
cyclohexane (H)	1	−49.1	17.9	1.0×10^9	7.65×10^5
decalin	2	−57.0	15.3	2.0×10^9	4.35×10^6
CH$_2$=CHCH$_2$—H	3	−76.6	17.3	3.0×10^8	2.92×10^5
CH$_2$=CHCH—HMe	2	−94.8	12.0	2.0×10^8	1.62×10^6
CH$_2$=CHC—HMe$_2$	1	−105.0	9.3	1.0×10^8	2.42×10^6
Z-MeCH=CHCH—HMe	2	−100.6	10.4	2.0×10^8	3.05×10^6
Me$_2$C=CHCH—HMe	2	−112.6	7.3	2.0×10^8	1.05×10^7
Me$_2$C=CMeC—HMe$_2$	1	−121.8	5.1	1.0×10^8	1.28×10^7
CH$_2$=CHCMe—HCH=CH$_2$	1	−137.4	1.6	1.0×10^8	5.20×10^7
cyclohexene (H, H)	4	−103.1	9.8	4.0×10^8	7.95×10^7
cyclohexadiene (H, H)	4	−113.7	7.1	4.0×10^9	2.35×10^8
cyclohexadiene (H, H)	4	−132.0	2.8	4.0×10^9	1.30×10^9
MeC≡CC—HMe$_2$	1	−115.2	6.7	1.0×10^8	6.81×10^6
PhMeCH—H	2	−80.5	11.1	2.0×10^8	2.26×10^6
PhMe$_2$C—H	1	−89.9	8.6	1.0×10^8	3.23×10^6
tetralin (H H)	4	−99.0	6.2	4.0×10^9	3.36×10^8
dihydroanthracene (H H)	4	−122.6	1.0	4.0×10^9	2.68×10^9
PhMeCHOO—H	1	−78.0	1.2	1.7×10^9	1.05×10^9

for $\alpha \neq 1$,

$$(1 - \alpha^2)\Delta E_H = 2\alpha(br_e)^2 - \alpha^2 \Delta H_e - 2\alpha(br_e)^2[1 - (1 - \alpha^2)(br_e)^{-2}\Delta H_e]^{1/2} \quad (6.20)$$

for $\Delta H_e (1-\alpha^2) \ll (br_e)^2$,

$$\Delta E_H = \alpha(1+\alpha)^{-1}\Delta H_e + 0.25\alpha(br_e)^{-2}\Delta H_e^2 \quad (6.21)$$

TABLE 6.5
Activation Energies of Alkyl Radical Reactions with the C—H Bonds of Organic Compunds $R^{1\cdot} + R_iH \rightarrow R^1H + R_i^{\cdot}$ Calculated by the IPM Method (See Equations (6.7), (6.8), (6.11), and Parameters in Tables 6.1 and 6.2)

	E (kJ mol^{-1})					
R_iH	$C^{\cdot}H_3$	$MeC^{\cdot}H_2$	$Me_2C^{\cdot}H$	$Me_3C^{\cdot}$	cyclopropyl$^{\cdot}$	cyclohexyl$^{\cdot}$
EtMeCH—H	45.8	54.3	59.2	65.4	50.9	67.7
Me$_3$C—H	40.1	48.1	52.8	58.7	44.9	70.0
cyclobutane (H,H)	48.4	57.0	62.0	68.3	53.5	70.7
cyclopentane (H,H)	43.8	52.1	56.9	63.0	48.8	65.3
cyclohexane (H)	38.1	46.1	50.7	56.5	42.9	58.7
decalin (H)	34.8	42.5	47.0	52.6	39.5	54.8
CH$_2$=CHCH$_2$—H	38.1	46.0	49.8	55.1	42.6	57.2
CH$_2$=CHCH—HMe	31.2	38.1	42.1	47.1	35.3	49.0
CH$_2$=CHC—HMe$_2$	27.5	34.1	38.0	42.8	31.5	44.7
Z-MeCH=CHCH—HMe	29.0	35.8	39.7	44.7	33.1	46.6
Me$_2$C=CHCH—HMe	24.8	31.2	35.0	39.7	28.7	41.5
Me$_2$C=CMeC—HMe$_2$	21.7	27.9	31.5	36.1	25.4	37.9
CH$_2$=CHCMe—HCH=CH$_2$	16.7	22.5	25.9	30.2	20.2	31.9
cyclohexene (H,H)	28.2	34.8	38.7	43.6	32.2	45.5
cyclohexadiene (H,H)	24.4	30.8	34.6	39.3	28.3	41.1
cyclohexadiene-2 (H,H)	18.4	24.3	27.8	32.2	22.0	33.9
MeC≡CC—HMe$_2$	23.9	30.3	34.0	38.7	27.8	40.5
PhMeCH—H	29.7	36.9	41.0	46.2	34.0	48.2
PhMe$_2$C—H	26.2	33.0	37.0	42.1	30.3	44.0
tetralin (H,H)	22.9	29.5	33.4	38.3	26.9	40.1

It is seen from Equations (6.19)–(6.21) that for low absolute values of $|\Delta H_e|$, the increment ΔE_H is directly proportional to ΔH_e, which agrees with the Polany–Semenov equation (Equation [6.1]). The slope of the linear plot of ΔE_H as a function of ΔH_e depends on α, i.e., on the force constants of the reacting bonds. The ratio $\Delta E_H/\Delta H_e = \alpha\,(1+\alpha)^{-1}$ at low $|\Delta H_e|$ may serve as a parameter of the sensitivity of the activation energy to ΔH_e for low

TABLE 6.6
Contribution of Enthalpy to the Activation Energy of Secondary Peroxyl Radical Reactions with the C—H Bonds of Hydrocarbons Calculated by the IPM Method (See Equations (6.7), (6.8), (6.12), (6.20), and Parameters in Tables 6.1 and 6.2)

RH	ΔH (kJ mol^{-1})	E (kJ mol^{-1})	ΔE_H (kJ mol^{-1})	$dE/d\Delta H$
EtMeCH—H	47.5	61.9	21.8	0.53
Me$_3$C—H	34.5	54.9	14.8	0.51
cyclobutane-H	53.0	65.0	24.9	0.54
cyclopentane-H	42.9	59.4	19.3	0.52
cycloheptane-H	38.4	57.0	16.9	0.51
methylcyclohexane-H	30.0	52.6	12.5	0.50
decalin	22.1	48.7	8.6	0.48
CH$_2$=CHCH—HMe	−15.7	45.6	−8.4	0.42
CH$_2$=CHC—HMe$_2$	−25.9	41.5	−12.5	0.41
Z-MeCH=CHCH—HMe	−21.5	43.2	−10.8	0.41
Me$_2$C=CHCH—HMe	−33.5	38.5	−15.5	0.39
Me$_2$C=CMeC—HMe$_2$	−42.7	35.0	−19.0	0.38
CH$_2$=CHMeC—HCH=CH$_2$	−58.3	29.4	−24.6	0.36
cyclohexene-H	−24.0	42.2	−11.8	0.41
cyclohexadiene-H	−34.6	38.0	−16.0	0.39
cyclohexadiene-H	−52.9	31.3	−22.7	0.37
MeC≡CC—HMe$_2$	−36.1	37.5	−16.5	0.39
PhCH$_3$	9.5	47.5	5.2	0.46
PhMeCH—H	−1.4	41.6	−0.7	0.44
PhMe$_2$C—H	−10.8	36.4	−5.9	0.43
tetralin-H	−19.9	36.0	−6.3	0.41
dihydroanthracene-H	−43.5	27.0	−15.3	0.37

values of ΔH_e. For reactions of different classes, the coefficient α varies in the range from 0.6 to 1.7 (see Table 6.1) and the coefficient $\alpha(1+\alpha)^{-1}$, therefore, varies from 0.38 to 0.63. For reactions of radicals R$^•$, RO$^•$, and RO$_2^•$ with C—H bonds, the values of $\alpha(1+\alpha)^{-1}$ are given here:

Reaction	$R^\bullet + RH$	$RO^\bullet + RH$	$RO_2^\bullet + RH$	$RO_2^\bullet + ROOH$
$\alpha(1+\alpha)^{-1}$	0.50	0.47	0.45	0.50

The activation energy increases with an increase in ΔH_e and diminishes with an increase in $|\Delta H_e|$ for $\Delta H_e < 0$. On the other hand, according to the law of conservation of energy, $E \geq 0$ for exothermic reactions and $E \geq \Delta H$ for endothermic reactions. It follows from the law of conservation of energy and Equations (6.11)–(6.13) that the parameter br_e for one group of reactions is constant, while ΔH_e varies in the range $\Delta H_{e\,min} < \Delta H_e < \Delta H_{e\,max}$. For very exothermic reactions with $\Delta H_e < \Delta H_{e\,min}$, the slope dE/dH is close to zero, and for very endothermic reactions with $\Delta H_e > \Delta H_{e\,max}$, the slope dE/dH is close to unity. Inside the limits $\Delta H_{e\,min} < \Delta H_e < \Delta H_{e\,max}$, on the slope

$$\frac{dE_H}{dH_e} \approx \frac{\alpha}{1+\alpha} + \frac{\alpha \Delta H_e}{(br_e)^2} \tag{6.22}$$

dE/dH depends on the enthalpy of the reaction.

Since the coefficient α' in Polany–Semenov equation (1) represents the ratio $\Delta E/(\Delta H_2 - \Delta H_1)$ for the chosen group of reactions, it follows from Equation (6.26) that coefficient α' is given by

$$\alpha' \approx \frac{\alpha}{1+\alpha} + \frac{\alpha^2(\Delta H_{e2} - \Delta H_{e1})}{4(br_e)^2} \tag{6.23}$$

For not very high values of the reaction enthalpies $\alpha' \approx \alpha(1+\alpha)^{-1}$ (see Equation [6.23]). In the general case inside interval $\Delta H_{e\,min} < \Delta H_e < \Delta H_{e\,max}$, the coefficient α' is the following function of the reaction enthalpy values [11] (see Figure 6.2):

$$\alpha' = \frac{(br_e)^2}{\Delta H_{e2} - \Delta H_{e1}} \left\{ \sqrt{1 - \frac{(1-\alpha^2)\Delta H_{e1}}{(br_e)^2}} - \sqrt{1 - \frac{(1-\alpha^2)\Delta H_{e2}}{(br_e)^2}} \right\} \tag{6.24}$$

It is follows from Equations (6.23) and (6.24) that the coefficient α' in the Polany–Semenov equation depends not only on parameters br_e and α but also on the incidental factor such as the range of variation of enthalpies of the selected series of reactions. The same situation is observed for coefficient ρ in the Hammett and Taft equations (see Equation [6.2]) [11].

6.2.2 Force Constants of the Reacting Bonds

Another important characteristic of radical abstraction reactions is the force constants of the ruptured and the generated bonds. The dependence of the activation energy for the reactions of the type $R^\bullet + R^1X \rightarrow RX + R^{1\bullet}$, where $X = H, Cl, Br$, or I, on the coefficients b_i and b_f was demonstrated experimentally [17]. It was found that parameter $r_e = const$ in these reactions, while the square root of the activation energy for a thermally neutral reaction is directly proportional to the force constant of the ruptured bond. The smaller the force constant of the C—X bond, the lower the E_{e0}, and the relationship $E_{e0}^{1/2}$ to $b(1+\alpha)^{-1}$ is linear (see Figure 6.4). The same result was also obtained for the reactions of hydrogen atoms with RCl, RBr, and RI [17].

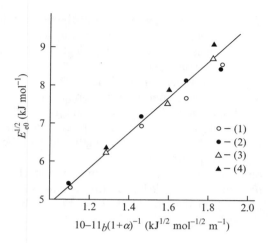

FIGURE 6.4 Correlation of the activation energy of the thermoneutral reaction with the coefficient $b(1+\alpha)^{-1}$ for the reaction $R^\bullet + RX$ (X = H, Cl, Br, I) on the basis of the IPM (1) and the model of intersecting Morse curves (2), and for the reaction $H^\bullet + RX$ (R = Cl, Br, I) on the basis of the IPM (3) and the model of intersecting Morse curves (4) [11].

These results confirm the important role of the force constant of the reacting bonds in the formation of the activation barrier. The activation energies E_{e0} for the $R^\bullet + RX$ reactions can easily be estimated from the empirical formulas [17] (units are given in brackets):

$$E_{e0}^{1/2}\ [\text{kJ mol}^{-1}]^{1/2} = 4.80 \times 10^{-11}\ (b[\text{kJ}^{1/2}\text{mol}^{-1/2}\text{m}^{-1}])(1+\alpha)^{-1}[r(\text{C—X})[\text{m}]] \quad (6.25)$$

6.2.3 Triplet Repulsion in the Transition State

The hydrogen atom migrates from Y to X in the $X^\bullet + HY$ reaction. The transition state of this reaction can be regarded conventionally as a labile formation containing the $X \cdots H \cdots Y$ pseudobond. The characteristics of this bond may influence the activation energy. This influence was evidenced with the aid of the parabolic model [4,11,18] since the model makes possible the conversion of the activation energy for an individual reaction into the activation energy for a thermally neutral reaction (see Equation [6.11]) and also to take into account the influence of the force constants of the X—H and Y—H bonds on E_{e0}. When the hydrogen abstraction reactions are compared in relation to different classes of compounds, it is useful to employ the parameter r_e, in which the influence of ΔH_e, b_i, and b_f on the activation energy for the reaction has already been taken into account.

In the $X \cdots H \cdots Y$ transition state, the $X \cdots Y$ pseudobond is formed by three electrons. According to the Pauli principle, one molecular orbital can be occupied by only two electrons with opposite spins. Two molecular orbitals therefore participate in the transition state: the bonding orbital of the X—Y bond, in which two electrons are accommodated, and its nonbonding orbital containing the third electron. The higher the energy of the nonbonding orbital D_T the stronger the X—Y bond; its value is described by the Sato formula [19]:

$$D_T = D_e\{\exp(-brD_e^{-1/2}) + 0.5\exp(-2brD_e^{-1/2})\}, \quad (6.26)$$

where D_e and D_T are the energies of the bonding and nonbonding X—Y orbitals, respectively, $2b^2$ is the force constant, and r is the amplitude of the vibration of this bond. In the given case, $r = r_{XH} + r_{YH} - r_{XY} + r_e$. The involvement of the nonbonding orbital of the X—Y bond in the formation of the activation barrier has come to be referred to as *triplet repulsion* [20]. This effect constitutes the basis of three semiempirical methods for the calculation of the activation energies for radical substitution reactions: the "bond energy–bond order" (BEBO) method [20], the "bond energy–bond length" (BEBL) method [21], and the Zavitsas method [22]. In all these methods the activation energy corresponds to the maximum on the "potential energy–bond order" (BEBO) curves or the "potential energy–bond length" (BEBL, Zavitsas method) curves while the potential energy curve represents a superposition of two Morse curves corresponding to the Y—H and X—H reacting bonds and the curve for the nonbonding orbital of the X—Y bond (Equation [6.26]).

The role of triplet repulsion in radical abstraction can clearly be traced on comparing reactions in which the energies of the X—Y bonds differ significantly. The values of E_{e0} and r_e for a series of radical abstraction reactions found by the parabolic method as well as the energies D_e of the X—Y bond are presented below.

X ··· H ··· Y	E_{e0} (kJ mol^{-1})	r_e (m)	D_e (kJ mol^{-1})
R ··· H ··· R	74.8	4.622×10^{11}	381
H$_2$N ··· H ··· NH$_2$	81.3	4.188×10^{11}	285
RO$_2$ ··· H ··· OAr	45.6	2.895×10^{11}	~0
RO$_2$ ··· H ··· OAr	45.3	2.885×10^{11}	~0
HOO ··· H ··· OOH	44.3	2.894×10^{11}	80
RO$_2$ ··· H ··· O$_2$R	43.1	2.854×10^{11}	88
ArO ··· H ··· OAr	41.8	2.772×10^{11}	~0

The difference between the reactions with a high energy of the X—Y (C—C and C—N) bond for which $E_{e0} \sim 75$–80 kJ mol^{-1} and the reactions with a very low energy of this bond (O—O) for which $E_{e0} \sim 42$–46 kJ mol^{-1} can be clearly traced. Within the limits of the error of measurement, the parameter r_e for the last four reactions is constant: $r_e = (2.86 \pm 0.05) \times 10^{-11}$ m. This value is characteristic of reactions with zero triplet repulsion in the transition state. On substituting this quantity in Equation (6.10), we obtain the following equation for the estimation of the contribution of triplet repulsion ΔE_T to the activation energy E_{e0}[11]:

$$\Delta E_T \text{ (kJ mol}^{-1}) = (bb_f)^2(b+b_f)^{-2}[(r_e[m])^2 - 8.18 \times 10^{-22}] \quad (6.27)$$

For reactions of the R$^{\bullet}$ + RH class, this contribution is 39.6 kJ mol^{-1}, i.e., a considerable proportion of the activation energy $E_{e0} = 76$ kJ mol^{-1} is due precisely to the triplet repulsion in the transition state. The following empirical dependence of the r_e parameter on the D_e(X—Y) has been established [17,18]:

$$10^{22}\Delta(r_e[m])^2 = 13.7 \times [D_e(X-Y)/D_e(H-H)] \quad (6.28)$$

6.2.4 Electron Affinity of Atoms of Reaction Center

Another important characteristic of the X—Y bond is its polarity induced by the different electron affinities of the atoms or the radicals X and Y. The greater the difference, the greater the polarity of the bond, its strength, and its dipole moment. According to Pauling [23], the polarity of the X—Y bond is be characterized by the extent to which its strength differs from half the sum of the bond dissociation energies of the XX and YY molecules by virtue of the different electronegativities of the atoms X and Y [23]:

$$\Delta EA(XY) = D_{XY} - 0.5(D_{XX} + D_{YY}) \qquad (6.29)$$

The question arises whether the polarity of the X—Y bond affects the $X^\bullet + YH$ abstraction reactions. We shall compare two reactions: $H^\bullet + H_2$ and $Cl^\bullet + H_2$ [24]:

Reaction	E_{e0} (kJ mol^{-1})	r_e (m)	D_{XY} (kJ mol^{-1})
$H^\bullet + H_2$	58.2	3.69×10^{11}	436.0
$Cl^\bullet + H_2$	36.7	3.04×10^{11}	431.6

Evidently, the dissociation energies of the H—H and Cl—H bonds are very close and the triplet repulsion in the transition states of these reactions is, therefore, almost identical. Nevertheless, the quantities E_{e0} and r_e in these two reactions differ very considerably. The reason for this is that the H—H bond is nonpolar, while the Cl—H bond is polarized its $\Delta EA = 92.3$ kJ mol^{-1} (Equation [6.29]). As in the HCl molecule, in the transition state there is evidently a strong attraction between Cl and H, which in fact induces a decrease in r_e and E_{e0}. If the $Cl^\bullet + H_2$ reaction was characterized by the same parameter $r_e = 3.69 \times 10^{-11}$ m as the $H^\bullet + H_2$ reaction, an activation energy of $E_{e0} = 56.5$ kJ mol^{-1} would be obtained for that reaction. The difference between the observed and expected activation energies ($\Delta E_{EA} = 36.7 - 56.5 = -19.8$ kJ mol^{-1}) must be attributed to the influence of the unequal electronegativities of the hydrogen and the chlorine atoms on E_{e0} in the $Cl^\bullet + H_2$ reaction.

The $R^\bullet + RH$ and $RO^\bullet + RH$ reactions may serve as another example [18]:

Reaction	E_{e0} (kJ mol^{-1})	r_e (m)	$D_e(X—Y)$ (kJ mol^{-1})
$R^\bullet + RH$	74.8	4.622×10^{11}	381
$RO^\bullet + RH$	53.2	2.787×10^{11}	372
$ROO^\bullet + RH$	56.3	2.959×10^{11}	295

In this case, the similarity of the R—R and RO—R bond dissociation energies also leads to the similarity of the triplet repulsion energies in the R $\cdots$ H $\cdots$ R and RO $\cdots$ H $\cdots$ R transition states. However, in these reactions also, the quantities E_{e0} and r_e differ appreciably. The polarity of the O—C bond in the $RO^\bullet + HR$ reaction is manifested here. On substituting the dissociation energies $D(CH_3—CH_3)$, $D(CH_3O—OCH_3)$, and $D(CH_3—OCH_3)$ in Equation (6.29), we obtain $\Delta E_A = 80$ kJ mol^{-1}. The calculation of the contribution of the electronegativities of the O and C atoms to the activation energy for the $RO^\bullet + RH$ reaction yields $\Delta E_{EA} = -26.1$ kJ mol^{-1}, which should be regarded as very considerable bearing in mind that $E_{e0} = 53.2$ kJ mol^{-1}. The empirical correlation between the r_e parameter and difference in electron affinities of X and Y atoms has the following form [17,18]:

$$10^{22}(\Delta r_e[m])^2 = -22.4 \times \{\Delta E_A(X—Y)/D_e(H—H)\} \qquad (6.30)$$

6.2.5 Radii of Atoms of Reaction Center

When the fragments X and Y approach one another in the X ... H ... Y transition state, their outer electron shells begin to repel one another. It is to be expected that the repulsion will be stronger when the radii of the atoms X and Y is larger. The examples presented in Table 6.7 confirm this conclusion.

An evident parallel variation of the increment in r_e and in the bond length r_{XY} is observed. On the other hand, the influence of the strengths of the X—Y bonds on the activation energies in these reactions were taken into account. The electronegativities of C and Si, Ge, Sn atoms are close. The empirical dependence of the parameter r_e (in m) on D_{XY} and r_{XY} in the interaction of radicals carrying a free valence on the C and O atoms with the C—H, Si—H, Sn—H, Ge—H, and P—H bonds is presented on Figure 6.5.

The common dependence of the parameter r_e on BDE, electron affinity, and length of X—Y bond has the following form [11,26]:

$$10^{22}(r_e[m])^2 = 13.7(D_{XY}/D_{HH}) - 22.4(\Delta E_A/D_{HH}) + 9.4(r_{XY}/r_{HH}) - 12.4 \qquad (6.31)$$

This formula makes possible the estimation of the contribution of each factor, i.e., triplet repulsion ΔE_T, electronegativity, ΔE_{EA}, and repulsion of the electron shells of the X and Y atoms ΔE_R in the TS, to the activation barrier E_{e0} for each class of radical abstraction reactions. Since $E_{e0}^{1/2} = br_e(1+\alpha)^{-1}$, it follows that, by employing the corresponding increments from Equation (6.31), it is possible to calculate the contribution of a particular factor. Table 6.8 presents the results of such calculation for 17 classes of radical abstraction reactions.

It is seen from the data presented in Table 6.8 that the triplet repulsion ΔE_T makes an important contribution to the activation energy E_{e0}. The difference between the electronegativities of the fragments X and Y lowers the activation energy. When the discrepancy between the electron affinities is large, this decrease may be very considerable. For example,

TABLE 6.7
Comparison of r_e Parameters for the Reactions $R_3EH + X \rightarrow R_3E^\bullet + XH$ with BDE and Length of the E—H Bonds of the Reactants Calculated by the IPM Method [25,26]

Reactant	r_e (m)	r(E—X) (m)	D_e(E—X) (kJ mol^{-1})	$r_e^2 \times 10^{22} - 13.7 \frac{D_e(E-X)}{D_e(H-H)}$ (m^2)
\multicolumn{5}{c}{$X^\bullet = H^\bullet$}
CH$_4$	3.87×10^{-11}	1.091×10^{-10}	457.4	1.42
SiH$_4$	4.09×10^{-11}	1.480×10^{-10}	408.0	4.64
PH$_3$	4.09×10^{-11}	1.420×10^{-10}	364.9	5.91
SeH$_2$	4.06×10^{-11}	1.470×10^{-10}	348.0	6.14
GeH$_4$	4.28×10^{-11}	1.525×10^{-10}	380.4	7.04
\multicolumn{5}{c}{$X^\bullet = R^\bullet$}
R$_3$CH	4.62×10^{-11}	1.536×10^{-10}	384.2	9.96
R$_2$PH	4.16×10^{-11}	1.858×10^{-10}	306.6	8.22
R$_3$SiH	4.77×10^{-11}	1.870×10^{-10}	381.3	11.45
R$_3$GeH	4.93×10^{-11}	1.945×10^{-10}	350.4	10.92
R$_3$SnH	5.07×10^{-11}	2.144×10^{-10}	300.2	16.81

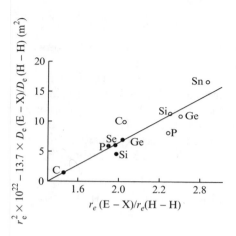

FIGURE 6.5 Correlation of the parameter r_e^2 on the length of the bond E—X for the abstraction reactions $X^\bullet + HER_n$: black points for reactions with $X^\bullet = H^\bullet$ and white points for reactions with $X^\bullet = R^\bullet$ and E = C, P, Si, Se, Ge, and Sn [26].

$\Delta E_{EA} = -45$ kJ mol^{-1} for the HO$^\bullet$ + SiH$_4$ reaction and $\Delta E_{EA} = -43$ kJ mol^{-1} in the reaction of hydrogen atom with water. The repulsion of the electron orbitals of the atoms forming the reaction center ΔE_R plays an important role in all the radical abstraction reactions. In the interaction of radicals with molecules the contribution of this repulsion ranges from 25 to 46 kJ mol^{-1}. In reactions of molecules with hydrogen atoms the contribution is naturally smaller, varying from 8 to 16 kJ mol^{-1}.

TABLE 6.8
The Contributions of Triplet Repulsion (ΔE_T), the Electron Affinity of the Fragments X and Y (ΔE_{EA}), and the Repulsion of the Atoms Forming the X—Y Bond (ΔE_R) to the Activation Energy E_{e0} for the $X^\bullet$ + HY Reaction [11,25]

Reaction	$b(1+\alpha)^{-1}$ (kJ mol^{-1})$^{1/2}$	r_e (m)	ΔE_T (kJ mol^{-1})	ΔE_{EA} (kJ mol^{-1})	ΔE_R (kJ mol^{-1})
R$^\bullet$ + HR	1.87×10^{11}	4.62×10^{-11}	41	0	34
R$^\bullet$ + H$_2$NR	2.05×10^{11}	3.84×10^{-11}	45	-8	32
R$^\bullet$ + HOOR	2.06×10^{11}	3.80×10^{-11}	37	-7	29
R$^\bullet$ + HSiR'$_3$	1.60×10^{11}	4.78×10^{-11}	30	-3	27
R$^\bullet$ + HSR	2.13×10^{11}	4.09×10^{-11}	44	0	46
HO$^\bullet$ + HR	2.09×10^{11}	3.67×10^{-11}	53	-22	30
HO$^\bullet$ + SiH$_4$	1.79×10^{11}	3.30×10^{-11}	55	-45	28
RO$^\bullet$ + HR	2.08×10^{11}	3.55×10^{-11}	49	-20	30
RO$^\bullet$ + HSiR'$_3$	1.74×10^{11}	3.68×10^{-11}	45	-34	26
RO$_2^\bullet$ + HR	2.06×10^{11}	3.80×10^{-11}	37	-7	29
RO$_2^\bullet$ + HOOR	2.30×10^{11}	2.85×10^{-11}	15	0	28
R$_3$Si$^\bullet$ + HSiR'$_3$	1.38×10^{11}	4.97×10^{-11}	18	0	27
H$^\bullet$ + HR	1.96×10^{11}	3.76×10^{-11}	53	-8	10
H$^\bullet$ + NH$_3$	2.11×10^{11}	3.45×10^{-11}	63	-21	9
H$^\bullet$ + H$_2$O	2.20×10^{11}	3.91×10^{-11}	76	-43	8
H$^\bullet$ + SiH$_4$	1.69×10^{11}	4.09×10^{-11}	34	-2	15
H$^\bullet$ + H$_2$S	1.88×10^{11}	3.60×10^{-11}	42	-5	16

6.2.6 Influence of Adjacent π-Bonds

An aromatic ring and a double or triple bond in the α-position relative to the C—H bond weaken this bond by virtue of the delocalization of the unpaired electron in its interaction with the π-bond. The weakening of the C—H bond is very considerable: for example, $D(\text{C—H})$ is 422 kJ mol^{-1} in ethane [27], 368 kJ mol^{-1} in the methyl group of propene [27] ($\Delta D = 54$ kJ mol^{-1}), and 375 kJ mol^{-1} in the methyl group of toluene [27] ($\Delta D = 47$ kJ mol^{-1}). Such decrease in the strength of the C—H bond diminishes the enthalpy of the radical abstraction reaction and, hence, its activation energy. This effect is illustrated below for the reactions of the ethylperoxyl radical with hydrocarbons:

Hydrocarbon	C_2H_6	$CH_3CH=CH_2$	$C_6H_5CH_3$
ΔH (kJ mol^{-1})	56.5	2.5	9.5
E (kJ mol^{-1})	61.9	53.4	44.9

A comparative analysis of the kinetics of the reactions of atoms and radicals with paraffinic (R^1H), olefinic (R^2H), and aromatic alkyl-substituted (R^3H) hydrocarbons within the framework of the parabolic model permitted a new important conclusion. It was found that the π-C—C bond occupying the α-position relative to the attacked C—H bond increases the activation energy for thermally neutral reaction [11]. The corresponding results are presented in Table 6.9.

Evidently, the activation energy for a thermally neutral reaction with participation of a hydrogen atom or a radical (alkyl, alkoxyl, etc.) is higher in these cases where there is a π-bond or an aromatic ring adjacent to the attacked C—H bond. This effect is a property of the structures themselves, and the π-bond exerts a dual effect on the reaction center. On the one hand, by weakening the C—H bond the π-bond in the α-position lowers the enthalpy

TABLE 6.9
Influence of the Adjacent π-Bond on the Activation Energy of Hydrogen Atom Abstraction Reaction [13,18,24,28,29]

R•	RH	br_e (kJ mol^{-1})$^{1/2}$	E_{e0} (kJ mol^{-1})	ΔE_π (kJ mol^{-1})
H•	R^1H	14.49	57.9	0.0
H•	R^2H	15.58	66.9	9.0
H•	R^3H	15.34	64.9	7.0
R•	R^1H	17.23	74.2	0.0
R•	R^2H	18.86	88.9	14.7
R•	R^3H	18.11	82.0	7.8
RO•	R^1H	13.37	55.2	0.0
RO•	R^2H	14.41	64.1	8.9
RO•	R^3H	14.16	61.9	6.7
ROO•	R^1H	14.23	61.5	0.0
ROO•	R^2H	15.68	74.7	13.2
ROO•	R^3H	14.74	66.0	4.5
AmO•	R^1H	13.72	58.0	0.0
AmO•	R^2H	15.66	75.5	17.5
AmO•	R^3H	14.42	64.0	6.0

and, hence, the activation energy of the reaction. On the other hand, by interacting with electrons of the reacting bonds, the π-orbital increases the strength of the C—Y bond in the Y• + HR reaction, which increases the energy of the nonbonding C—Y orbital and intensifies triplet repulsion.

6.2.7 Polar Effect

The extensive literature was devoted to polar effects in chemical reactions. The IPM permits a fresh approach to this important problem. The introduction of a functional group into a hydrocarbon molecule alters the dissociation energy of the neighboring C—H bonds, which is indicated by the examples presented below:

Compound	Me$_2$CH—H	MeCH—H(OH)	CH$_3$C—H(O)	EtOCH—HMe
D (kJ mol^{-1})	412	400	338	399

The change in the C—H bond strength affects the enthalpies and through them the activation energies for radical abstraction reactions. If a molecule is attacked by a nonpolar radical (hydrogen atom, alkyl radical), then the influence of the polar group on the activation energy is confined to a change in ΔH. This was confirmed by the data on the reactions of the methyl radical with the C—H bonds of hydrocarbons and their derivatives (alcohols, ethers, etc.), which are characterized by the virtually identical parameter $br_e = 17.30$ (kJ mol^{-1})$^{1/2}$ [11]. Thus, a polar group (OH, OR, etc.) influences only the enthalpy of the reaction and not the energy of the transition state of the type C $\cdots$ H $\cdots$ C and H $\cdots$ H $\cdots$ C.

A different picture is observed when a polar radical reacts with a C—H bond of a polar molecule. For example, the reaction of an oxygen atom with the methane C—H bond is characterized by the activation energy of thermoneutral reaction $E_{e0} = 54.6$ kJ mol^{-1} and parameter $br_e = 13.11$ (kJ mol^{-1})$^{1/2}$ while the reaction with the methanol C—H bond is characterized by $E_{e0} = 50$ kJ mol^{-1} and parameter $br_e = 12.55$ (kJ mol^{-1})$^{1/2}$ [30]. For these values of br_e, the difference between the activation energies is 4.6 kJ mol^{-1}. The decrease in the activation energy can be explained by the fact that the polar O—H group in the O $\cdots$ H $\cdots$ C—OH transition state interacts with the O $\cdots$ H $\cdots$ C polar reaction center.

The IPM model makes possible the isolation from the overall effect of the substituent its contribution to E due to polar interaction. In the calculation, the values of ΔH and the activation energy for a thermally neutral reaction of the given group (reaction series) are used, because the activation energy can be represented by the sum $E_e = E_{e0} + \Delta E_H$ (see above), where the second term takes into account only the influence of enthalpy. The polar interaction in the transition state may be inferred by comparing the values of E_{e0} in various reactions of the same radical involving a nonpolar hydrocarbon and the corresponding polar molecule. The contribution of the polar interaction to the activation energy (ΔE_μ) can be estimated from the formula [4,11,30]:

$$\Delta E_\mu = [(br_e)^2_\mu - (br_e)^2_{RH}](1 + \alpha)^{-2}, \qquad (6.32)$$

where the parameters $(br_e)_\mu$ and $(br_e)_{RH}$ refer to reactions involving polar compound YH and reference hydrocarbon RH, respectively. The results of the calculation of the energy ΔE_μ (in kJ mol^{-1}) for the X• + HY gas-phase reactions are presented below:

X•/HY	CH$_3$OH	CH$_2$(O)	RCH(O)	CH$_3$C(O)CH$_3$	(CH$_3$)$_2$O
O	4.6	−4.6	−7.9	−9.4	−5.1
HO•	2.4	1.8	1.5	−11.4	2.8
CH$_3$O•	−6.0	0	−12.2	—	—
(CH$_3$)$_3$CO•	—	−4.2	−3.1	−17.8	—

Extensive data concerning the influence of the polar groups on the activation energies for abstraction reactions were obtained by analyzing the experimental data on reactions involving alkoxyl and peroxyl radicals in the liquid phase. These data are discussed in Chapters 7–9.

Another effect called *multidipole interaction* is manifested in the reactions of the polar radicals with polyfunctional compounds [31]. The unequal reactivities of the same group, for example, R$_2$CH(OH), in compounds with one or several functional groups forms the basis for this effect. The corresponding kinetic data were analyzed within the framework of the parabolic model. The multidipole interaction is manifested, in particular, in the reactions of the peroxyl radicals RO$_2$•, containing functional groups, with oxygen-containing compounds YH (see Chapter 9). The magnitude of this effect in the transition state may be inferred from the change in the activation energy for a thermally neutral reaction on passing from a monofunctional compound YH to a polyfunctional compound ZH.

$$\Delta\Delta E_\mu = [(br_e)^2_{ZH} - (br_e)^2_{YH}](1 + \alpha)^{-2} \tag{6.33}$$

For the steric effect in reactions of peroxyl radicals, see Chapter 12.

6.2.8 Influence of Solvation

Another important effect observed when reactions take place in the liquid phase is associated with the solvation of the reactants. Theoretical comparison showed that the collision frequencies of the species in the gas and liquid phases are different, which is due to the difference between the free volumes. In the gas phase, the free volume is virtually equal to the volume occupied by the gas species ($V_f \approx V$), while in the liquid phase, it is much smaller than the volume of the liquid species ($V_f \ll V$). Since the motion and collision of the species occur in the free volume, the collision frequency in the liquid is higher than in the gas by the amount $(V/V_f)^{1/3}$ [32,33]. The activation energies for the reactions of radicals and atoms with hydrocarbon C—H bonds in the gas and the liquid phases are virtually identical, and that in the liquid is independent of the solvent polarity. This also applies to the parameter br_e, which can be seen from the following examples referring to the interaction of the hydroxyl radical with hydrocarbons [30]:

RH	C$_2$H$_6$	C$_3$H$_8$
br_e (kJ mol^{-1})$^{1/2}$ (gas)	13.74	13.52
br_e (kJ mol^{-1})$^{1/2}$ (H$_2$O)	13.48	13.40

A different picture is observed when a polar molecule is attacked by a polar radical (HO•, RO•, RO$_2$•). The reaction in a polar solvent is slower than in a nonpolar hydrocarbon solution

TABLE 6.10
Contributions of the Solvation Effect ΔE_{sol} to the Activation Energies for Radical Abstraction Reactions [30]

RH	$(br_e)_{gas}$ (kJ mol^{-1})$^{1/2}$	$(br_e)_{liquid}$ (kJ mol^{-1})$^{1/2}$	ΔE_{sol} (kJ mol^{-1})
Radical: HO$^{\bullet}$, Solvent: H$_2$O			
CH$_3$OH	13.91	14.26	3.1
CH$_3$CH$_2$OH	13.48	14.01	4.6
(CH$_3$)$_2$CHOH	13.31	14.12	7.1
(CH$_3$)$_3$COH	13.39	14.13	6.5
CH$_3$OCH$_3$	13.95	14.70	6.8
(C$_2$H$_5$)$_2$O	13.47	14.33	7.6
CH$_2$O	13.84	14.98	10.4
CH$_3$CHO	13.81	14.99	10.8
CH$_3$COCH$_3$	13.81	14.50	6.3
CH$_3$COOCH$_3$	14.38	14.97	5.5
Radical: (CH$_3$)$_3$CO$^{\bullet}$, Solvent: CH$_3$COCH$_3$			
CH$_3$COCH$_3$	12.67	13.17	7.2

or in the gas phase. From the change in the parameter br_e, it is possible to estimate the extent to which the activation energy increases as a result of the solvation of the polar reactants [30]:

$$\Delta E_{sol} = [(br_e)_l^2 - (br_e)_g^2] \times (1 + \alpha)^{-2}, \quad (6.34)$$

The subscripts l and g in Equation (6.38) refer to the liquid and gas phases, respectively. The results of the comparison are presented in Table 6.10. If the HO$^{\bullet}$ + YH reaction takes place in an aqueous solution and not in the gas phase, the parameter br_e and hence the activation energy increase. This is associated with the solvation of the reactants and the need to overcome the solvation shell by the reacting component in order to effect the elementary step. The contribution of ΔE_{sol} is particularly large in the reaction of the hydroxyl radical with aldehydes.

When a radical or atom attacks a polar O—H or N—H bond, the reactant Y forms a hydrogen bond of the type O—H $\cdots$ Y or N—H $\cdots$ Y in polar solvents. The hydrogen bond shields the reactant and slows down the reaction regardless of of the type of radical (polar or nonpolar) attacking it (see Chapters 12 and 13).

6.3 GEOMETRY OF THE TRANSITION STATE OF RADICAL ABSTRACTION REACTION

The IPM model of the bimolecular reaction introduces geometric parameters r_e and $r^{\neq}$ connected with such empirical characteristics of reaction as ΔH and E (see Equations [6.10] and [6.13]). The quantum-chemical calculation by the functional density method gave the following configuration of the transition state for the reaction EtOO$^{\bullet}$ + EtH: $r(\text{C} \cdots \text{H}) = 1.470 \times 10^{-10}$ m, $r(\text{O} \cdots \text{H}) = 1.115 \times 10^{-10}$ m, and $\varphi(\text{CHO}) = 176°$, which is very close to 180° [34]. These quantum-chemical parameters are compared with C $\cdots$ H, O $\cdots$ H, and C $\cdots$ O distances, which were calculated by the IPM method [34].

Distance (m)	In Molecule	In TS	Δr	Difference (%)
C—H	1.092×10^{-10}	1.336×10^{-10} (IPM)	0.244×10^{-10}	22
C—H	1.092×10^{-10}	1.470×10^{-10} (DFT*)	0.378×10^{-10}	35
O—H	0.970×10^{-10}	1.090×10^{-10} (IPM)	0.120×10^{-10}	12
O—H	0.970×10^{-10}	1.115×10^{-10} (DFT*)	0.145×10^{-10}	15
C—H—O	2.062×10^{-10}	2.426×10^{-10} (IPM)	0.364×10^{-10}	18
C—H—O	2.062×10^{-10}	2.585×10^{-10} (DFT*)	0.523×10^{-10}	25

*Density functional theory.

It is seen from this comparison that the geometrical parameters obtained by the IPM and by the density functional theory (DFT) methods are close. It is enough to multiply r_e by 1.44 to obtain the same C ··· H ··· O distance for the TS calculated by the DFT method. The following parametric equations were proposed for the estimation of the C—H, O—H, and C—O distances in the reaction center of TS for reactions ROO$^\bullet$ + RH in the IPM method [34,35].

$$r(\text{C—O}) \times 10^{10}[\text{m}] = 2.06 + 3.85 \times 10^{-2} \left(\sqrt{E_e} + 0.814\sqrt{E_e - \Delta H_e} \right) \quad (6.35)$$

$$r(\text{C—H}) \times 10^{10}[\text{m}] = 1.09 + 3.85 \times 10^{-2}\sqrt{E_e} \quad (6.36)$$

$$r(\text{O—H}) \times 10^{10}[\text{m}] = 0.97 + 3.13 \times 10^{-2}\sqrt{E_e - \Delta H_e} \quad (6.37)$$

The results of calculations of the geometric parameters by the IPM method for $RO_2^\bullet$ reactions with several hydrocarbons are presented in Table 6.11.

When the peroxyl radical reacts with the C—H bond of a polar molecule, for example alcohol, the polar interaction of the hydroxyl group influences the activation energy (see earlier) and the geometry of the transition state [34]. The parameters of the transition state of reaction EtOO$^\bullet$ + MeCH$_2$OH calculated by the DFT method are the following [35]:

Parameter	r(C—H) (m)	r(O—H) (m)	φ(C—H—O) (°)	E (kJ mol^{-1})
Value	1.373×10^{-10}	1.20×10^{-10}	160	52.7

The straight line distance calculated from these data is $r(\text{C—O}) = 2.547 \times 10^{-10}$ m and sum of $r(\text{C—H}) + r(\text{O—H}) = 2.579 \times 10^{-10}$ m. Thus, the difference in interatomic distances $r(\text{C—O})$ in reactions $RO_2^\bullet$ + EtH and $RO_2^\bullet$ + EtOH is equal to $\Delta r = 2.579 - 2.547 = 0.032 \times 10^{-10}$ m. The calculation of parameter r_e for reactions ROO$^\bullet$ + RH and ROO$^\bullet$ + EtOH shows that the decrease in the activation energy ($\Delta E_\mu = E_{e0}(RO_2^\bullet + RH) - E_{e0}(RO_2^\bullet + EtOH) = 4$ kJ mol^{-1} [35]). As a result, the interaction between the polar hydroxyl group and the reaction center of the transition state of the reaction EtOO$^\bullet$ + EtOH transforms the linear configuration of atoms O ··· H ··· C of the reaction $RO_2^\bullet$ + RH into angle configuration with φ(C—H—O) < 180°. This leads to a decrease in the activation energy. Such a change in geometry was associated with a change in the activation energy and provided a pathway to estimate the geometric parameters from experimental data (ΔH, E) of the reaction in the scope of IPM. The equation for the estimation of the O—H—C angle in TS of reaction $RO_2^\bullet$ + $R_\mu H$ ($R_\mu H$ is polar molecule with the attacked C—H bond) is the following [35]:

TABLE 6.11
Geometric Parameters of Transition State Calculated for Radical Abstraction Reactions sec-$RO_2^\bullet$ + RH by Equations (6.11), (6.35)–(6.37) [35]

RH	ΔH_e (kJ/mol^{-1})	E_e (kJ/mol^{-1})	r(C—H) (m)	r(O—H) (m)	r(C—O) (m)
MeCH$_2$—H	52.7	83.4	1.442 × 10^{-10}	1.143 × 10^{-10}	2.585 × 10^{-10}
EtMeCH—H	43.7	78.1	1.432 × 10^{-10}	1.154 × 10^{-10}	2.586 × 10^{-10}
Me$_3$C—H	30.7	71.1	1.416 × 10^{-10}	1.169 × 10^{-10}	2.585 × 10^{-10}
cyclobutane-H,H	49.2	81.2	1.439 × 10^{-10}	1.147 × 10^{-10}	2.586 × 10^{-10}
cyclopentane-H,H	39.1	75.6	1.426 × 10^{-10}	1.159 × 10^{-10}	2.585 × 10^{-10}
cycloheptane H,H	34.6	73.2	1.421 × 10^{-10}	1.165 × 10^{-10}	2.586 × 10^{-10}
cyclohexane-H	26.2	68.8	1.411 × 10^{-10}	1.174 × 10^{-10}	2.585 × 10^{-10}
decalin	18.3	64.9	1.402 × 10^{-10}	1.184 × 10^{-10}	2.586 × 10^{-10}
CH$_2$=CHCH$_2$—H	−1.3	69.6	1.413 × 10^{-10}	1.234 × 10^{-10}	2.647 × 10^{-10}
CH$_2$=CHCH—HMe	−19.5	61.8	1.394 × 10^{-10}	1.252 × 10^{-10}	2.646 × 10^{-10}
CH$_2$=CHC—HMe$_2$	−29.7	57.7	1.384 × 10^{-10}	1.263 × 10^{-10}	2.647 × 10^{-10}
Z-MeCH=CHCH—HMe	−25.3	59.4	1.388 × 10^{-10}	1.258 × 10^{-10}	2.646 × 10^{-10}
Me$_2$C=CHCH—HMe	−37.3	54.7	1.377 × 10^{-10}	1.270 × 10^{-10}	2.647 × 10^{-10}
Me$_2$C=CMeC—HMe$_2$	−46.5	51.2	1.367 × 10^{-10}	1.279 × 10^{-10}	2.646 × 10^{-10}
CH$_2$=CHCMe—HCH=CH$_2$	−62.1	45.6	1.352 × 10^{-10}	1.294 × 10^{-10}	2.647 × 10^{-10}
cyclohexene-H,H	−27.8	58.4	1.386 × 10^{-10}	1.261 × 10^{-10}	2.647 × 10^{-10}
cyclohexene-H,H	−38.4	54.2	1.375 × 10^{-10}	1.271 × 10^{-10}	2.646 × 10^{-10}
cyclohexadiene-H,H	−56.7	47.5	1.357 × 10^{-10}	1.290 × 10^{-10}	2.647 × 10^{-10}
MeC≡CC—HMe$_2$	−39.9	53.7	1.374 × 10^{-10}	1.273 × 10^{-10}	2.647 × 10^{-10}
PhMeCH—H	−5.2	60.0	1.390 × 10^{-10}	1.223 × 10^{-10}	2.613 × 10^{-10}
PhMe$_2$C—H	−14.6	56.0	1.380 × 10^{-10}	1.233 × 10^{-10}	2.613 × 10^{-10}
tetralin-H,H	−23.7	52.2	1.370 × 10^{-10}	1.243 × 10^{-10}	2.613 × 10^{-10}
anthracene-H,H	−47.3	43.2	1.345 × 10^{-10}	1.268 × 10^{-10}	2.613 × 10^{-10}

$$\cos(\pi - \varphi) = \frac{r_\mu(\text{C—O})^2 - r(\text{C—H})^2 - r(\text{O—H})^2}{2r(\text{C—H}) \times r(\text{O—H})}, \quad (6.38)$$

where r_μ(C—O) is the C—O distance in the reaction center of reaction $RO_2^\bullet + R_\mu H$, and r(C—H) and r(O—H) are subsequent distances in the reaction center of the reaction

$RO_2^{\bullet} + RH$. For examples of the calculation of the geometric characteristics of the reactions $RO_2^{\bullet} + R_\mu H$, see Chapter 7.

6.4 INTRAMOLECULAR HYDROGEN TRANSFER REACTIONS IN PEROXYL, ALKOXYL, AND ALKYL RADICALS

6.4.1 Peroxyl Radicals

In addition to hydrogen abstraction by peroxyl radical from another molecule, the reaction of intramolecular hydrogen transfer occurs in peroxyl radicals in oxidized hydrocarbons (see Chapter 2).

These reactions are very important in the oxidation of carbon-chain polymers (see Chapter 19). The available experimental data on the rates of such reactions are summarized in Table 2.9. The parameters for intramolecular hydrogen transfer in peroxyl radicals calculated by the IPM are presented in Table 6.12.

They were used for the calculation of the activation energies for isomerization of several peroxyl radicals. Peroxyl radical isomerization involving the formation of a six-membered activated complex is energetically more favorable: the activation energy of a thermally neutral reaction E_e is 53.2 kJ mol^{-1}. For the seven-membered transition state, the E_{e0} value (54.8 kJ mol^{-1}) is slightly higher. The calculated br_e parameter for the six-membered transition state (13.23 (kJ mol^{-1})$^{1/2}$) is close to the br_e value (13.62 (kJ mol^{-1})$^{1/2}$) for the transition state of the bimolecular H atom abstraction from the aliphatic C—H bond by the peroxyl radical. Therefore, the kinetic parameters for isomerization are close to those for bimolecular H-atom abstraction by the peroxyl radical. This allows the estimation of the kinetic parameters for peroxyl radical isomerization. Relevant results of calculation via Eqns. (6.7, 6.8, 6.12) are presented in Table 6.13. The data in this table show that the activation energy for peroxyl radical isomerization ranges from 18 to 64 kJ mol^{-1}, whereas the rate constant k(350 K) varies from 1.5 to 3.8 × 10^6 s^{-1}.

6.4.2 Alkoxyl Radicals

Alkoxyl radicals are very active and rapidly enter into bimolecular reaction (see Chapter 2). Moreover, alkoxyl radicals with sufficiently long alkyl substituents react with intramolecular hydrogen atom transfer, for example [37]:

TABLE 6.12
IPM Characteristics of Intramolecular Hydrogen Abstraction Reactions [36]

Radical	br_e (kJ mol^{-1})$^{1/2}$	E_{e0} (kJ mol^{-1})	r_e (m)	A_{C-H} (s^{-1})	E_{e0} (bimol.) (kJ mol^{-1})
R$^\bullet$ → R$^{1\bullet}$ (6-Membered Transition State)					
RC$^\bullet$H(CH$_2$)$_3$(C—H)HY	14.63	53.5	3.91 × 10^{-11}	1.0 × 10^{10}	74.8
RC$^\bullet$H(CH$_2$)$_3$(C—H)HPh	15.13	57.2	4.04 × 10^{-11}	1.0 × 10^9	79.2
RC$^\bullet$H(CH$_2$)$_3$(C—H)HCH=CH$_2$	15.93	63.4	4.26 × 10^{-11}	1.0 × 10^9	86.5
R$^\bullet$ → R$^{1\bullet}$ (5-Membered Transition State)					
RC$^\bullet$H(CH$_2$)$_2$(C—H)HY	16.17	65.4	4.32 × 10^{-11}	3.5 × 10^{10}	74.8
RC$^\bullet$H(CH$_2$)$_2$(C—H)HPh	16.67	69.5	4.45 × 10^{-11}	3.5 × 10^9	79.2
RC$^\bullet$H(CH$_2$)$_2$(C—H)HCH=CH$_2$	17.47	76.3	4.67 × 10^{-11}	3.5 × 10^9	86.5
Ph$^\bullet$CMe$_2$CH$_2$—H	14.98	59.3	4.00 × 10^{-11}	3.5 × 10^9	75.3
R$^\bullet$ → R$^{1\bullet}$ (7-Membered Transition State)					
RC$^\bullet$H(CH$_2$)$_4$(C—H)HY	15.90	63.2	4.25 × 10^{-11}	2.8 × 10^9	74.8
RC$^\bullet$H(CH$_2$)$_4$(C—H)HPh	16.40	67.2	4.38 × 10^{-11}	2.8 × 10^8	79.2
RC$^\bullet$H(CH$_2$)$_4$(C—H)HCH=CH$_2$	17.20	74.0	4.59 × 10^{-11}	2.8 × 10^8	86.5
RO$^\bullet$ → R$^{1\bullet}$ (6-Membered Transition State)					
Me$_2$C(O$^\bullet$)(CH$_2$)$_2$(C—H)HR	13.13	53.4	3.51 × 10^{-11}	2.0 × 10^9	53.2
Me$_2$C(O$^\bullet$)(CH$_2$)$_2$(C—H)HPh	13.53	56.8	3.61 × 10^{-11}	2.0 × 10^8	56.5
Me$_2$C(O$^\bullet$)(CH$_2$)$_2$(C—H)HCH=CH$_2$	14.17	62.2	3.79 × 10^{-11}	2.0 × 10^8	62.0
RO$_2^\bullet$ → R$^{1\bullet}$ (6-Membered Transition State)					
RCH(OO$^\bullet$)Y(C—H)HR	13.23	53.2	3.53 × 10^{-11}	2.0 × 10^9	56.3
RCH(OO$^\bullet$)Y(C—H)HPh	14.38	62.8	3.84 × 10^{-11}	2.0 × 10^8	62.3
RCH(OO$^\bullet$)Y(C—H)HCH=CH$_2$	14.82	66.7	3.96 × 10^{-11}	2.0 × 10^8	70.2
RO$_2^\bullet$ → R$^{1\bullet}$ (7-Membered Transition State)					
RCH(OO$^\bullet$)YCH$_2$(C—H)HR	13.43	54.8	3.59 × 10^{-11}	5.6 × 10^8	56.3
RCH(OO$^\bullet$)YCH$_2$(C—H)HPh	14.58	64.6	3.90 × 10^{-11}	5.6 × 10^7	62.3
RCH(OO$^\bullet$)YCH$_2$(C—H)HCH=CH$_2$	15.02	68.6	4.01 × 10^{-11}	5.6 × 10^7	70.2
R$^{1\bullet}$ → RO$_2^\bullet$ (6-Membered Transition State)					
RCH(OO—H)YC$^\bullet$HR	16.25	53.2	3.53 × 10^{-11}	2.0 × 10^8	56.3
RCH(OO—H)YC$^\bullet$HPh	17.67	62.9	3.84 × 10^{-11}	2.0 × 10^8	62.3
RCH(OO—H)YC$^\bullet$HCH=CH$_2$	18.21	66.8	3.96 × 10^{-11}	2.0 × 10^8	70.2
R$^{1\bullet}$ → RO$_2^\bullet$ (7-Membered Transition State)					
RCH(OO—H)YCH$_2$C$^\bullet$HR	16.50	54.8	3.59 × 10^{-11}	5.6 × 10^7	56.3
RCH(OO—H)YCH$_2$C$^\bullet$HPh	17.91	64.6	3.89 × 10^{-11}	5.6 × 10^7	62.3
RCH(OO—H)YCH$_2$C$^\bullet$HCH=CH$_2$	18.45	68.6	4.01 × 10^{-11}	5.6 × 10^7	70.2

Therefore, the chain chlorination of hydrocarbons with alkyl hypochlorite results in the formation of chlorine-containing alcohols because of the faster bimolecular reaction compared to isomerization [37].

$$RC^\bullet H(CH_2)_2 Me_2 COH + R_1 OCl \longrightarrow RCHCl(CH_2)_2 Me_2 COH + R_1 O^\bullet$$

TABLE 6.13
Enthalpies, Activation Energies, and Rate Constants of Intramolecular Hydrogen Atom Transfer in Peroxyl Radicals Calculated by the IPM Method [36], (ΔE_μ is Increment of Polar Interaction in the Transition State)

Radical RO$_2^\bullet$	ΔH (kJ mol^{-1})	ΔE_μ (kJ mol^{-1})	E (kJ mol^{-1})	k (350 K) (s^{-1})	E (bimol.)* (kJ mol^{-1})
MeCH$_2$(OO$^\bullet$)CMe$_2$CH$_2$—H	56.5	0	64.3	1.5	67.2
MeCH(OO$^\bullet$)CH$_2$(C—H)HMe	46.5	0	58.7	7.0	61.6
Me$_2$C(OO$^\bullet$)CH$_2$(C—H)Me$_2$	41.4	0	55.9	9.2	58.9
MeCH(OO$^\bullet$)NH(C—H)HMe	13.1	0	41.6	2.5×10^3	44.7
MeCH(OO$^\bullet$)NMe(C—H)HMe	1.6	0	36.3	1.5×10^4	39.4
Me$_2$C(OO$^\bullet$)NH(C—H)Me$_2$	−1.6	0	34.9	1.2×10^4	38.0
Me$_2$C(OO$^\bullet$)NMe(C—H)Me$_2$	−13.1	0	30.0	6.6×10^4	33.1
CH(OH)(OO$^\bullet$)CH$_2$(C—H)HOH	25.8	−2.2	45.5	6.4×10^2	48.6
MeC(OH)(OO$^\bullet$)CH$_2$(C—H)MeOH	19.9	−2.2	42.6	8.7×10^2	45.7
CH$_2$(OO$^\bullet$)C(O)(C—H)H$_2$	44.5	−15.4	42.2	3.1×10^3	45.1
MeCH(OO$^\bullet$)C(O)(C—H)HMe	32.3	−15.4	35.7	1.9×10^4	38.7
Me$_2$C(OO$^\bullet$)C(O)(C—H)Me$_2$	34.1	−15.7	36.3	7.7×10^3	39.3
C(O)(OO$^\bullet$)CH$_2$(C—H)(O)	−20.5	−8.8	18.3	3.8×10^6	21.3
PhCH(OO$^\bullet$)CH$_2$(C—H)HPh	−1.4	0	44.6	8.7×10^1	44.1
PhC(OO$^\bullet$)CH$_2$(C—H)MePh	−3.9	0	43.5	6.3×10^1	43.0
CH$_2$=CHCH(OO$^\bullet$)CH$_2$(C—H)HCH=CH$_2$	−15.7	0	42.4	1.9×10^2	45.9
CH$_2$=CHC(OO$^\bullet$)MeCH$_2$(C—H)MeCH=CH$_2$	−19.0	0	41.2	1.5×10^2	44.5

*calculated via Eqns. (6.7, 6.8, 6.12)

The IPM parameters for hydrogen transfer atom in alkoxyl radicals are presented in Table 6.12. Isomerization proceeds via the formation of a six-membered activated complex, and the activation energy for the thermally neutral isomerization of alkoxyl radicals is equal to 53.4 kJ mol^{-1}. These parameters were used for the calculation of the activation energies for isomerization of several alkoxyl radicals via Eqns. (6.7, 6.8, 6.12) (see Table 6.14). The activation energies for the bimolecular reaction of hydrogen atom (H-atom) abstraction by the alkoxyl radical and intramolecular isomerization are virtually the same.

6.4.3 ALKYL RADICALS

Alkyl radical isomerization is accompanied by a free-valence transfer from one carbon atom to another

It occurs during the chain cracking and radiolysis of hydrocarbons [38], radical polymerization and oligomerization of monomers [39], thermal and thermooxidative destruction of polymers (see Chapter 19) and hydrocarbon oxidation at low dioxygen pressure.

TABLE 6.14
Enthalpies, Activation Energies, and Rate Constants of Intramolecular H-Atom Abstraction in Alkoxyl Radicals Calculated by the IPM Method [36]

Radical RO•	D_i (kJ mol^{-1})	ΔH_e (kJ mol^{-1})	E (kJ mol^{-1})	k (350 K) (s^{-1})	E (bimol.)* (kJ mol^{-1})
Me$_2$C(O•)(CH$_2$)$_2$(C—H)H$_2$	422.0	−19.5	29.3	2.5×10^5	29.1
Me$_2$C(O•)(CH$_2$)$_2$(C—H)HMe	412.0	−29.5	25.4	6.4×10^5	25.2
Me$_2$C(O•)(CH$_2$)$_2$(C—H)Me$_2$	400.0	−41.5	21.1	1.4×10^6	20.8
Me$_2$C(O•)CH$_2$NH(C—H)HMe	378.6	−62.9	14.0	3.3×10^7	13.7
Me$_2$C(O•)CH$_2$NMe(C—H)HMe	367.1	−74.4	10.5	1.1×10^8	10.3
Me$_2$C(O•)CH$_2$NH(C—H)Me$_2$	357.0	−84.5	7.7	1.4×10^8	7.5
Me$_2$C(O•)CH$_2$NMe(C—H)Me$_2$	345.5	−96.0	4.7	4.0×10^8	4.5
Me$_2$C(O•)(CH$_2$)$_2$(C—H)HOH	397.4	−44.1	20.1	3.9×10^6	19.9
Me$_2$C(O•)(CH$_2$)$_2$(C—H)MeOH	391.5	−50.0	18.1	3.9×10^6	17.9
Me$_2$C(O•)CH$_2$C(O)(C—H)H$_2$	410.0	−31.5	24.7	1.2×10^6	24.5
Me$_2$C(O•)CH$_2$C(O)(C—H)HMe	397.8	−43.7	20.3	3.8×10^6	20.1
Me$_2$C(O•)CH$_2$C(O)(C—H)Me$_2$	392.7	−48.8	18.5	3.4×10^6	18.3
Me$_2$C(O•)(CH$_2$)$_2$(C—H)(O)	385.6	−55.9	16.2	7.7×10^6	15.9
Me$_2$C(O•)(CH$_2$)$_2$(C—H)HPh	368.7	−72.8	14.0	3.3×10^6	13.7
Me$_2$C(O•)(CH$_2$)$_2$(C—H)MePh	354.7	−86.8	9.9	6.6×10^6	9.7
Me$_2$C(O•)(CH$_2$)$_2$(C—H)HCH=CH$_2$	349.8	−91.7	13.4	4.0×10^6	13.2
Me$_2$C(O•)(CH$_2$)$_2$(C—H)MeCH=CH$_2$	339.6	−101.9	10.6	5.2×10^6	10.4

*calculated via Eqns. (6.7, 6.8, 6.12)

The reaction of butyl radical isomerization

was analyzed by DFT method and was compared with reaction of H-atom abstraction from ethane by ethyl radical [36]. The activation energies of both reactions had very close values. The geometric parameters of these two reactions are given here.

Reaction	r(C—H) (m)	θ(C—H$^\#$—C) (deg)	θ(C—C—H$^\#$) (deg)	E(kJ mol^{-1})
Et• + EtH	1.353×10^{-10}	180	107.8	62.5
Bu• → Bu•	1.377×10^{-10}	132.5	94.9	65.8

It appeared very surprising that the geometries of the compared reactions are different but their activation energies are practically the same.

The experimental data on free radical isomerization reactions with H-atom abstraction were analyzed by the IPM method [36]. The kinetic parameters of isomerization were compared with those of the intermolecular H-atom abstraction (see Chapter 2). The results

TABLE 6.15
Enthalpies, Activation Energies, and Rate Constants of Intramolecular H-Atom Abstraction in Alkyl Radicals Calculated by IPM Method [36]

Radical R˙	ΔH (kJ mol^{-1})	E (kJ mol^{-1})	k (350 K) (s^{-1})	E (bimol.)* (kJ mol^{-1})
C˙H$_2$(CH$_2$)$_3$(C—H)HMe	−10.0	32.6	2.72×10^5	54.0
MeC˙H(CH$_2$)$_3$(C—H)HMe	0	37.5	5.06×10^4	58.9
MeC˙H(CH$_2$)$_3$(C—H)Me$_2$	−12.0	31.7	1.86×10^4	53.0
MeC˙H(CH$_2$)$_2$NH(C—H)HMe	−33.4	22.1	1.01×10^7	43.2
MeC˙H(CH$_2$)$_2$NMe(C—H)HMe	−44.9	17.4	5.06×10^7	38.2
Me$_2$C˙(CH$_2$)$_2$NH(C—H)Me$_2$	−43.0	18.2	1.92×10^7	39.0
Me$_2$C˙(CH$_2$)$_2$NMe(C—H)Me$_2$	−54.5	13.7	9.02×10^7	34.2
MeC˙H(CH$_2$)$_3$(C—H)HOH	−14.6	30.5	5.61×10^5	51.8
MeC˙H(CH$_2$)$_3$(C—H)MeOH	−20.5	27.7	7.35×10^5	49.0
MeC˙H(CH$_2$)$_2$C(O)(C—H)H$_2$	−2.0	36.5	1.07×10^5	57.9
MeC˙H(CH$_2$)$_2$C(O)(C—H)HMe	−14.2	30.6	5.42×10^5	52.0
MeC˙H(CH$_2$)$_2$C(O)(C—H)Me$_2$	−19.3	28.3	5.98×10^5	49.6
MeC˙H(CH$_2$)$_3$(C—H)(O)	−26.4	25.1	1.79×10^6	46.3
MeC˙H(CH$_2$)$_3$(C—H)HPh	−43.3	21.6	1.19×10^6	43.1
Me$_2$C˙(CH$_2$)$_3$(C—H)MePh	−45.3	20.8	7.87×10^5	42.3
Me$_2$C˙(CH$_2$)$_3$(C—H)HCH=CH$_2$	−50.2	24.8	4.98×10^5	47.3
Me$_2$C˙(CH$_2$)$_3$(C—H)MeCH=CH$_2$	−60.4	20.8	8.87×10^5	43.0

*calculated via Eqns. (6.7, 6.8, 6.12)

of this analysis and their comparison are given in Table 6.15. It is seen from the activation energy of the thermoneutral reactions that the six-membered activated complex is the most favored intermediate for the isomerization of alkyl radicals. Obviously, this configuration is the best for the intramolecular radical isomerization involving H-atom abstraction. The effect of the cycle size on the activation energy was estimated for intramolecular H-atom abstraction in alkyl radicals. A comparison of the ΔE_e values with the cycle strain energy in cyclo-C$_n$H$_{2n}$ suggests their similarity.

Transition State	(5-membered)	(6-membered)	(7-membered)
E_e (kJ mol^{-1})	59.4	46.6	57.1
ΔE_e (kJ mol^{-1})	11.9	0.0	9.7
E (ring strength) (kJ mol^{-1})	26.4	0.0	26.8

The intramolecular hydrogen atom transfer occurs with lower activation energies in comparison with the intermolecular transfer (see the values of E_e for both types of reactions in Table 6.11). The values of the activation energies of intramolecular radical H-atom abstraction calculated by the IPM method are given in Table 6.15.

6.5 FREE RADICAL ADDITION REACTIONS

6.5.1 ENTHALPY OF FREE RADICAL ADDITION

The radical addition reactions, for example

$$ROO^\bullet + CH_2{=}CHY \longrightarrow ROOCH_2C^\bullet HY$$

involve the rupture of the C—C π-bond and the formation of the C—O σ-bond. A π-bond is normally stronger than a σ-bond; hence, radical addition is an exothermic reaction. The values of enthalpies of the addition of ethyl peroxyl radical to different olefins are given below [40]:

$CH_2{=}CXY$; X, Y	H, H	H, Me	H, Et	Me, Me	H, Ph	Me, Ph	Ph, Ph
$-\Delta H$ (kJ mol^{-1})	42	39	40	42	86	82	90

The stabilization energy of the formed radical is a very important factor (compare the ΔH values for the addition to olefins and styrenes).

All the addition reactions are accompanied by a decrease in entropy (two species are combined to give one species). Therefore, for addition reactions, $|\Delta G| < |\Delta H|$ and, when the temperature is sufficiently high, exothermic addition is reversible because $\Delta G = \Delta H - T\Delta S$ [41]. For example, the addition reaction of the HOO$^\bullet$ radical to ethylene proceeds with the following values: $\Delta H = -63$ kJ mol^{-1}, $\Delta S = -134$ J mol^{-1} K^{-1}, and $\Delta G = -23$ kJ mol^{-1} at 298 K. Free radical addition can be considered reversible if the equilibrium concentration of the X$^\bullet$ radicals is commensurable with the concentration of the XCH$_2$C$^\bullet$HY radicals formed, i.e., if the [X$^\bullet$]/[XCH$_2$C$^\bullet$HY] ratio in the equilibrium state is not less than 0.05. When [CH$_2{=}$CHY] = 10 mol L^{-1} and $T = 300$ K, this condition results in the inequality $K \geq 200$ (K is the equilibrium constant). In turn, $-RT \ln K = -\Delta G$; therefore, $\Delta G \geq -RT \ln 200 = -13$ kJ mol^{-1} > -23. For the addition reactions of peroxyl radicals, $\Delta G \approx -23$ kJ mol^{-1}, and the addition proceeds as an irreversible reaction. Naturally, as the temperature rises, the boundary value of K increases in absolute magnitude.

6.5.2 PARABOLIC MODEL OF RADICAL ADDITION REACTION

Within the framework of the parabolic model, the radical addition

$$ROO^\bullet + CH_2{=}CHY \longrightarrow ROOCH_2C^\bullet HY$$

is represented as a result of an intersection of two potential curves, each of them describing the potential energy of the stretching vibrations of atoms as a parabolic function of the amplitudes of vibrations of atoms in the outgoing (C=C) and incoming (C—O) bonds [40–50]. In terms of the IPM, the radical addition can be characterized by the same parameters as the reaction of free radical abstraction (see earlier): reaction enthalpy ΔH_e, activation energy E_e, coefficient b of the attacked π-bond, coefficient α, zero-point vibrational energy of the attacked π-bond, and parameter r_e. Equations (6.7)–(6.15) are valid for the addition reactions as well. The physical, thermodynamic, and kinetic parameters for 15 classes of reactions involving the addition of atoms and radicals to carbon–carbon and carbon–oxygen multiple bonds are listed in Table 6.16.

TABLE 6.16
Parameters of Various Classes of the Addition of Atoms and Radicals to Multiple Bonds Used in the Parabolic Model [40–45]

Reaction	α	b ((kJ mol^{-1})$^{1/2}$ m^{-1})	$0.5hL\nu$ (kJ mol^{-1})	$0.5hL(\nu_i-\nu_f)$ (kJ mol^{-1})	A (L mol^{-1} s^{-1})
H$^{\bullet}$ + CH$_2$=CHR	1.440	5.389×10^{11}	9.9	−7.5	10×10^{10}
H$^{\bullet}$ + CH≡CR	1.847	6.912×10^{11}	12.7	−4.7	40×10^{10}
H$^{\bullet}$ + O=CR^{1}R^2	1.600	5.991×10^{11}	10.3	−11.4	10×10^{10}
D$^{\bullet}$ + CH$_2$=CHR	1.461	5.389×10^{11}	9.9	−2.7	10×10^{10}
Cl$^{\bullet}$ + CH$_2$=CHR	1.591	5.389×10^{11}	9.9	4.8	9×10^{10}
Br$^{\bullet}$ + CH$_2$=CHR	1.844	5.389×10^{11}	9.9	5.8	5×10^{10}
R$^{\bullet}$ + CH$_2$=CHR	1.202	5.389×10^{11}	9.9	1.7	0.1×10^{10}
R$^{\bullet}$ + CH≡CR	1.542	6.912×10^{11}	12.7	3.8	0.1×10^{10}
R$^{\bullet}$ + O=CR^{1}R^2	1.570	5.991×10^{11}	10.3	3.7	0.05×10^{10}
N$^{\bullet}$H$_2$ + CH$_2$=CHR	1.410	5.389×10^{11}	9.9	3.1	0.008×10^{10}
RO$^{\bullet}$ + CH$_2$=CHR	1.413	5.389×10^{11}	9.9	3.3	0.05×10^{10}
RO$_2^{\bullet}$ + CH$_2$=CHR	1.737	5.389×10^{11}	9.9	4.6	0.1×10^{10}
R$_3$Si$^{\bullet}$ + CH$_2$=CHR	2.012	5.389×10^{11}	9.9	4.1	0.1×10^{10}
R$_3$Si$^{\bullet}$ + O=CR^{1}R^2	2.518	5.991×10^{11}	10.3	7.0	0.08×10^{10}
PhS$^{\bullet}$ + CH$_2$=CHR	2.282	5.389×10^{11}	9.9	6.3	0.07×10^{10}

The parabolic model makes possible the development of an empirical hierarchy of addition reactions. All known addition reactions are divided a priori into classes in accordance with the atomic structure of the reaction center in the transition state. Each class is characterized by a pair of force constants of the outgoing and incoming bonds or by the parameters $b = b_i$ and $\alpha = b_i/b_f$ (see above). Groups of reactions are distinguished in each class. Each group is characterized by the parameter $r_e = const$ or $br_e = const$, which is confirmed by the analysis of a large array of experimental results. Each group of reactions can be described additionally by the energy of the thermally neutral reaction E_{e0} (see Equation [6.10]) and by the threshold value $\Delta H_{e\,min}$, for which $E = 0.5RT$ provided that $\Delta H_e < \Delta H_{e\,min}$ (see Equation (6.15)). The kinetic parameters for various groups of addition reactions are listed in Table 6.17.

The values of ΔH_e, activation energies E, and the rate constants of the addition of peroxyl and alkoxyl radicals to olefins calculated by Equations (6.7), (6.8), and (6.12) are presented in Table 6.18 and Table 6.19. The contribution of enthalpy to the activation energy ΔE_H can be estimated as the difference $\Delta E_{e0} = E_e - E_{e0}$. As an example, Table 6.20 presents the results of this comparison for three groups of addition reactions. When reactions are highly exothermic, the contribution of enthalpy to the activation energy is fairly high; it varies from −44 to −99 kJ mol^{-1}, i.e., the activation energy amounts to 40–50% of ΔH.

6.5.3 Triplet Repulsion in Radical Addition Reactions

The activation energy of radical abstraction is influenced by the so-called triplet repulsion in the transition state. This influence is manifested by the fact that the stronger the X—R bond towards which the hydrogen atom moves in the thermally neutral reaction X$^{\bullet}$ + RH, the higher the activation energy of this reaction. The triplet repulsion is due to the fact that three electrons cannot be accommodated in the bonding orbital of X—C; therefore, one electron

TABLE 6.17
Kinetic Parameters E_{e0}, $b(1+\alpha)^{-1}$, r_e, $(r^{\#}/r_e)_0$, and $-\Delta H_{e\,min}$ Calculated from Equations (9.10) and (9.12) for the Addition Reactions of Various Subclasses [40–45]

Y$^{\bullet}$	E_{e0} (kJ mol^{-1})	$b(1+\alpha)^{-1}$ (kJ$^{1/2}$ mol$^{-1/2}$ m^{-1})	r_e (m)	$(r^{\#}/r_e)_0$	$-\Delta H_{e\,min}$ (kJ mol^{-1})
		Y$^{\bullet}$ + CH$_2$=CHX → YCH$_2$C$^{\bullet}$HX			
H$^{\bullet}$	101.6	2.21 × 10^{11}	4.56 × 10^{-11}	0.41	211.9
D$^{\bullet}$	99.6	2.19 × 10^{11}	4.56 × 10^{-11}	0.41	204.9
Cl$^{\bullet}$	50.5	1.86 × 10^{11}	3.82 × 10^{-11}	0.39	82.9
Br$^{\bullet}$	31.2	1.69 × 10^{11}	3.30 × 10^{-11}	0.35	37.9
C$^{\bullet}$H$_3$	82.6	2.45 × 10^{11}	3.71 × 10^{-11}	0.45	194.4
C$_6$H$_5^{\bullet}$	105.3	2.45 × 10^{11}	4.19 × 10^{-11}	0.45	252.0
N$^{\bullet}$H$_2$	61.0	2.24 × 10^{11}	3.49 × 10^{-11}	0.41	226.0
RO$^{\bullet}$	65.2	2.23 × 10^{11}	3.62 × 10^{-11}	0.41	119.0
RO$_2^{\bullet}$	90.5	1.97 × 10^{11}	4.83 × 10^{-11}	0.36	150.4
R$_3$Si$^{\bullet}$	76.6	1.78 × 10^{11}	4.92 × 10^{-11}	0.33	117.5
PhS$^{\bullet}$	31.4	1.64 × 10^{11}	3.42 × 10^{-11}	0.30	34.7
		Y$^{\bullet}$ + HC≡CX → YCH=C$^{\bullet}$X			
H$^{\bullet}$	125.1	2.43 × 10^{11}	4.60 × 10^{-11}	0.35	229.3
C$^{\bullet}$H$_3$	97.7	2.72 × 10^{11}	3.63 × 10^{-11}	0.39	184.7
		Y$^{\bullet}$ + O=CR^{1}R^2			
H$^{\bullet}$	102.9	2.30 × 10^{11}	4.41 × 10^{-11}	0.38	149.7
C$^{\bullet}$H$_3$	72.9	2.33 × 10^{11}	3.66 × 10^{-11}	0.39	112.2
R$_3$Si$^{\bullet}$	114.5	1.70 × 10^{11}	6.29 × 10^{-11}	0.28	176.8

occupies the nonbonding X—C orbital. Meanwhile, the stronger the X—C bond, the higher the energy of the nonbonding orbital and the higher the activation energy of abstraction.

How do matters stand with radical addition? A comparison of E_{e0} and r_e with the energy of dissociation of the resulting bond D_e showed that this influence certainly does exist [41–46]. The parameters E_{e0} and r_e are juxtaposed with the dissociation energy of the bond formed D_e(X—C) in Table 6.21. For reactions of one class, namely, X$^{\bullet}$ + CH$_2$=CHY, the linear correlation $E_{e0} = const \times [D_e(X—C)]^2$ holds; $const = 5.95 \times 10^{-4}$ mol kJ^{-1}. The following linear correlation was found to be fulfilled for the reactions of all the 13 classes considered (see Figure 6.6):

$$r_e(m) = (0.98 \times 10^{-13} [m\,mol\,kJ^{-1}]) \times D_e(X—C) \quad (6.39)$$

This correlation should be regarded as an empirical proof of the fact that the nonbonding orbital of the bond formed actually does participate in the generation of the activation energy of addition. The stronger this bond, the higher the energy of the nonbonding orbital and the higher the activation energy. Empirical correlation (6.39) can be used to estimate roughly the activation energies of diverse addition reactions.

6.5.4 Influence of the Neighboring π-Bonds on the Activation Energy of Radical Addition

Radical abstraction reactions that involve molecules with π-bonds in the vicinity of the reaction center are characterized by higher E_{e0} values than the corresponding reactions

TABLE 6.18
Enthalpies (ΔH_e), Activation Energies (E), Pre-Exponential Factors (A), and Rate Constants (k) of Methylethylperoxyl Radical Addition to Olefins Calculated by the IPM Method (See Equations (6.7), (6.8), (6.12), and Tables 6.15 and 6.16)

CH_2=CHY	$-\Delta H_e$ (kJ mol^{-1})	E (kJ mol^{-1})	A (L mol^{-1} s^{-1})	k(300 K) (L mol^{-1} s^{-1})
CH_2=CH_2	37.1	59.2	2.0×10^9	9.83×10^{-2}
CH_2=CHMe	34.7	60.6	1.0×10^9	2.80×10^{-2}
CH_2=CHEt	35.2	60.3	1.0×10^9	3.15×10^{-2}
CH_2=CMe_2	36.9	59.3	1.0×10^9	4.70×10^{-2}
E-MeCH=CHMe	34.2	60.9	2.0×10^9	5.00×10^{-2}
Z-MeCH=CHMe	37.6	58.9	2.0×10^9	0.11
CH_2=CHCl	42.9	55.8	1.0×10^9	0.19
CH_2=CCl_2	55.9	48.5	1.0×10^9	3.55
CH_2=CHF	38.9	58.2	1.0×10^9	7.47×10^{-2}
CH_2=CHOAc	43.4	55.6	1.0×10^9	0.21
CH_2=CHOEt	35.1	60.4	1.0×10^9	3.08×10^{-2}
CH_2=CMeOMe	47.6	53.2	1.0×10^9	0.55
CH_2=CMeOAc	55.6	48.7	1.0×10^9	3.32
CH_2=CHC(O)OMe	40.6	57.2	1.0×10^9	0.11
CH_2=CMeC(O)OMe	66.9	42.6	1.0×10^9	39.0
E-MeCH=CHC(O)OH	66.2	42.9	1.0×10^9	33.6
Z-MeCH=CHC(O)OH	70.4	40.7	1.0×10^9	82.1
CH_2=CHCH$_2$OAc	40.9	57.0	1.0×10^9	0.12
CH_2=CHC(O)OCH$_2$Ph	43.4	55.6	1.0×10^9	0.21
CH_2=CHCN	68.3	41.8	1.0×10^9	52.6
CH_2=CMeCN	65.6	43.3	1.0×10^9	29.5
CH_2=CHCH=CH_2	83.8	44.1	2.0×10^9	42.9
MeCH=CHCH=CHMe	75.3	48.5	2.0×10^9	7.13
CH_2=CMeCMe=CH_2	87.8	42.0	2.0×10^9	98.0
CH_2=CHPh	81.6	42.9	1.0×10^9	34.0
CH_2=CMePh	77.6	45.0	1.0×10^9	14.7
HC≡CH	43.0	78.0	4.0×10^9	1.04×10^{-4}
MeC≡CH	43.9	77.5	2.0×10^9	6.54×10^{-5}
MeC≡CMe	32.6	84.8	4.0×10^9	6.82×10^{-6}

involving similar hydrocarbons without these bonds. This is due to triplet repulsion. In these reactions, the three electrons of the reaction center interact with the neighboring π-electrons; as a consequence, the energy of the nonbonding orbital in the transition state increases and hence the activation energy also increases. Do neighboring π bonds influence the E_{e0} values of radical addition? In order to answer this question, let us compare the parameters br_e and E_{e0} for the addition of the hydrogen atom to the monomers CH_2=CHY and monomers with double bonds in the α-position relative to the attacked double bond. If the monomer contains an aromatic ring or a double bond in the α-position, the activation energy of a thermally neutral reaction increases; if there are two aromatic rings in the α-position, E_{e0} increases to even a greater extent (Table 6.22). The difference between $E_{e0}(\pi)$ and $E_{e0}(CH_2$=CHY$)$ can be used to evaluate the contribution of this factor to the activation barrier [41]:

$$\Delta E(\pi) = E_{e0}(\pi) - E_{e0}(CH_2\text{=CHY}) \tag{6.40}$$

TABLE 6.19
Enthalpies (ΔH_e), Activation Energies (E), Pre-Exponential Factors (A), and Rate Constants (k) of Dimethylethoxyl Radical Addition to Olefins Calculated by the IPM Method (See Equations (6.7), (6.8), (6.12), and Tables 6.15 and 6.16)

$CH_2=CHY$	$-\Delta H_e$ (kJ mol^{-1})	E (kJ mol^{-1})	A (L mol^{-1} s^{-1})	k(300 K) (L mol^{-1} s^{-1})
$CH_2=CH_2$	85.8	13.7	1.0×10^9	4.17×10^6
$CH_2=CHMe$	83.4	14.6	5.0×10^8	1.44×10^6
$CH_2=CHEt$	83.9	14.4	5.0×10^8	1.55×10^6
$CH_2=CMe_2$	85.6	13.8	5.0×10^8	2.02×10^6
E-MeCH=CHMe	82.9	14.8	1.0×10^9	2.65×10^6
Z-MeCH=CHMe	86.3	13.5	1.0×10^9	4.50×10^6
$CH_2=CHCl$	91.6	11.5	5.0×10^8	5.00×10^6
$CH_2=CCl_2$	104.6	6.9	5.0×10^8	3.10×10^7
$CH_2=CHF$	87.6	13.0	5.0×10^8	2.75×10^6
$CH_2=CHOAc$	92.1	11.3	5.0×10^8	5.39×10^6
$CH_2=CHOEt$	83.8	14.4	5.0×10^8	1.53×10^6
$CH_2=CMeOMe$	96.3	9.8	5.0×10^8	9.90×10^6
$CH_2=CMeOAc$	104.3	7.0	5.0×10^8	2.98×10^7
$CH_2=CHC(O)OMe$	89.3	12.3	5.0×10^8	3.55×10^6
$CH_2=CMeC(O)OMe$	115.6	3.5	5.0×10^8	1.24×10^8
E-MeCH=CHC(O)OH	114.9	3.7	5.0×10^8	1.14×10^8
Z-MeCH=CHC(O)OH	119.1	2.4	5.0×10^8	1.87×10^8
$CH_2=CHCH_2OAc$	89.6	12.2	5.0×10^8	3.71×10^6
$CH_2=CHC(O)OCH_2Ph$	92.1	11.3	5.0×10^8	5.39×10^6
$CH_2=CHCN$	117.0	3.1	5.0×10^8	1.47×10^8
$CH_2=CMeCN$	114.3	3.9	5.0×10^8	1.06×10^8
$CH_2=CHCH=CH_2$	132.5	8.5	1.0×10^9	3.38×10^7
$MeCH=CHCH=CHMe$	124.0	11.2	1.0×10^9	1.10×10^7
$CH_2=CMeCMe=CH_2$	136.5	7.2	1.0×10^9	5.57×10^7
$CH_2=CHPh$	130.3	7.3	5.0×10^8	2.67×10^7
$CH_2=CMePh$	126.3	8.6	5.0×10^8	1.61×10^7
$HC\equiv CH$	91.7	26.9	2.0×10^9	4.09×10^4
$MeC\equiv CH$	92.6	26.5	1.0×10^9	2.44×10^4
$MeC\equiv CMe$	81.3	32.1	2.0×10^9	5.08×10^3

Triplet repulsion also manifests itself in the reactions in question; for example, in the reaction

$$R^{\bullet} + CH_2=CHCH=CH_2 \longrightarrow RCH_2C^{\bullet}HCH=CH_2$$

a multicenter multielectron $R \cdots C \cdots C \cdots C \cdots C$ bond is formed in the transition state; this increases the strength of the bond, and, hence, enhances the triplet repulsion and its contribution to the activation energy. It can be seen from the data in Table 6.22 that this contribution can reach 20 kJ mol^{-1}. The increase in the dissociation energy $D_e(X-C)$ caused by this electron delocalization can be estimated using relation (6.40).

The observed difference in the parameters is due most likely to the electron delocalization in the aromatic ring and, hence, to the increase in the triplet repulsion. The difference between the E_{e0} values for the reactions considered amounts to 13 kJ mol^{-1}. With allowance for the examples considered above, Equation (6.39) can be extended and written in the following form:

$$r_e(m) = (0.98 \times 10^{-13} [\text{m mol kJ}^{-1}]) \times [D_e(X-C) + \Delta E(\pi)]. \quad (6.41)$$

TABLE 6.20
Contribution ΔE_H of the Reaction Enthalpy ΔH to the Activation Energy E of Radical Addition Reaction (See Equation [6.20])

R	Y	$-\Delta H$ (kJ mol^{-1})	E_e (kJ mol^{-1})	$-\Delta E_H$ (kJ mol^{-1})
		H$^\bullet$ + CH$_2$=CRY $\longrightarrow$ CH$_3$C$^\bullet$RY		
H	H	162.1	22.5	79.1
H	Me	165.3	21.0	80.6
Me	Me	173.3	18.0	83.6
H	Cl	179.5	19.8	81.8
H	Ph	204.1	19.2	93.0
Ph	Ph	213.3	19.0	98.7
		C$^\bullet$H$_3$ + CH$_2$=CRY $\longrightarrow$ CH$_3$CH$_2$C$^\bullet$RY		
H	H	98.5	39.0	43.6
H	Me	96.1	36.9	45.7
Me	Me	98.3	35.7	46.9
H	Cl	104.3	33.6	49.0
H	Ph	143.3	26.6	63.4
Ph	Ph	154.0	24.5	69.3
		ROO$^\bullet$ + CH$_2$=CRY $\longrightarrow$ ROOCH$_2$C$^\bullet$RY		
H	H	32.5	67.9	22.6
H	Me	30.1	69.3	21.2
Me	Me	32.3	68.0	22.5
H	Cl	38.3	64.5	26.0
H	Ph	77.0	51.6	47.4

6.5.5 ROLE OF THE RADIUS OF THE ATOM BEARING A FREE VALENCE

The greater the radius of the atom carrying the free valence, the higher the E_{e0} for radical abstraction. Does the radius of the atom that attacks the double bond influence the activation energy of addition reactions? This question was answered by the analysis of the experimental data on the addition of triethylsilyl and phenylthiyl radicals to alkenes [49]. Since the strength of the C—X (D_e) bond formed influences the parameter r_e (see Equation (6.39)), the characteristic to be compared for the addition of various radicals to alkenes is the r_e/D_e ratio (Table 6.23). The lengths r(C—X) of the bonds formed upon the addition of radicals, such as C$^\bullet$H$_3$, N$^\bullet$H$_2$, and RO$^\bullet$ to the C=C bond and upon the addition of C$^\bullet$H$_3$ to the C=O bond are close; therefore, the ratio $r_e/D_e = const$. However, in the case of addition of PhS$^\bullet$ and R$_3$Si$^\bullet$ to the C=C bond, the r_e/D_e ratio is much greater and a linear correlation is observed between r_e/D_e and r(C—X). The correlation has the following form (see Figure 6.7):

$$r_e = (8.81 \times 10^{-4} [\text{mol kJ}^{-1}]) \times [r(X-C) - 0.42 \,(\times 10^{-10} \text{m})](D_e[\text{kJ mol}^{-1}]) \qquad (6.42)$$

Thus, the radius of the atom carrying the free valence has a substantial influence on the activation barrier to the addition reaction: the greater the radius of this atom, the higher the activation energy. Apparently, this effect is due to the repulsion in the transition state, which is due to the interaction between the electron shells of the attacked double bond and the atom that attacks this bond.

TABLE 6.21
Parameters of Various Classes of Radical Addition (E_{e0}, r_e) and Strengths of the Formed Bonds [42–46]

Reaction	E_{e0} (kJ mol^{-1})	r_e (m)	D_e(X—C) (kJ mol^{-1})	(r_e/D_e(X—C)) (m mol kJ^{-1})
H$^\bullet$ + CH$_2$=CHR	101.6	4.563 × 10^{-11}	439	1.04 × 10^{-13}
D$^\bullet$ + CH$_2$=CHR	99.6	4.557 × 10^{-11}	439	1.04 × 10^{-13}
Cl$^\bullet$ + CH$_2$=CHR	50.5	3.427 × 10^{-11}	357	0.96 × 10^{-13}
Br$^\bullet$ + CH$_2$=CHR	31.2	2.949 × 10^{-11}	299	0.99 × 10^{-13}
H$^\bullet$ + HC≡CR	125.2	4.608 × 10^{-11}	462	1.00 × 10^{-13}
H$^\bullet$ + O=CR^{1}R^2	102.9	4.402 × 10^{-11}	459	0.96 × 10^{-13}
C$^\bullet$H$_3$ + CH$_2$=CHR	82.6	3.713 × 10^{-11}	378	0.98 × 10^{-13}
Me$_2$C$^\bullet$H + CH$_2$=CHR	78.3	3.617 × 10^{-11}	372	0.97 × 10^{-13}
Me$_3$C$^\bullet$ + CH$_2$=CHR	68.3	3.377 × 10^{-11}	360	0.94 × 10^{-13}
C$_6$H$_5^\bullet$ + CH$_2$=CHR	105.3	4.194 × 10^{-11}	436	0.96 × 10^{-13}
N$^\bullet$H$_2$ + CH$_2$=CHR	62.1	3.530 × 10^{-11}	360	0.98 × 10^{-13}
RO$^\bullet$ + CH$_2$=CHR	65.2	3.617 × 10^{-11}	359	1.01 × 10^{-13}
R$^\bullet$ + O=CR^{1}R^2	68.1	3.540 × 10^{-11}	365	0.97 × 10^{-13}

6.5.6 Interaction of Polar Groups

The polar effect involved in radical addition has been repeatedly discussed in the scientific literature. The parabolic model opens up new prospects for the correct estimation of the polar effect (see Section 6.2.7). It permits one to determine the contribution of this effect to the activation energy using experimental data. This contribution (ΔE_μ) is estimated by choosing a reference reaction that involves the same reaction center but in which one or both reactants

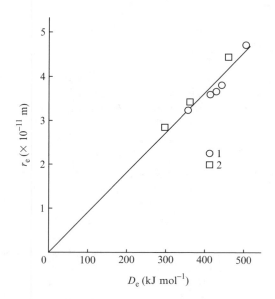

FIGURE 6.6 Correlation of the parameter r_e with the BDE of the formed bond upon atom and radical addition to C=C and C=O multiple bonds: (1) atom + C=C, (2) radical + C=C or C≡C [47].

TABLE 6.22
Influence of a π-Bond Adjacent to the Reaction Center on the Activation Energy of the Radical Addition to Alkenes [40–48]

Reaction	br_e (kJ mol^{-1})$^{1/2}$	E_{e0} (kJ mol^{-1})	ΔE_π (kJ mol^{-1})	$\Delta D_e(\pi)$ (kJ mol^{-1})
H$^\bullet$ + CH$_2$=CHR	24.59	101.6	0	0
H$^\bullet$ + CH$_2$=CHPh	25.85	112.2	10.6	24.1
H$^\bullet$ + CH$_2$=CPh$_2$	26.47	117.7	16.1	36.0
H$^\bullet$ + 1,3-cyclo-C$_6$H$_8$	26.47	117.7	16.1	36.0
H$^\bullet$ + CH$_2$=CHCN	25.45	108.8	7.2	16.4
C$^\bullet$H$_3$ + CH$_2$=CHR	20.01	82.6	0	0
C$^\bullet$H$_3$ + CH$_2$=CHPh	20.82	89.4	6.8	15.5
C$^\bullet$H$_3$ + CH$_2$=CMePh	20.67	88.7	6.1	12.6
C$^\bullet$H$_3$ + CH$_2$=CPh$_2$	21.33	94.4	11.8	25.2
C$^\bullet$H$_3$ + CH$_2$=CHCH=CH$_2$	20.88	89.9	7.3	16.6
C$^\bullet$H$_3$ + MeCH=CHCH=CHMe	21.15	92.2	9.6	21.8
C$^\bullet$H$_3$ + CH$_2$=CMeCMe=CH$_2$	21.03	91.2	8.6	19.5
C$^\bullet$H$_3$ + 1,3-cyclo-C$_6$H$_8$	22.27	102.3	19.7	43.2
C$^\bullet$H$_3$ + 1,4-cyclo-C$_6$H$_8$	22.23	101.9	19.3	42.5
N$^\bullet$H$_2$ + CH$_2$=CHR	18.82	61.0	0	0
N$^\bullet$H$_2$ + (E)-CH$_2$=CHCH=CH$_2$	20.52	72.5	10.4	23.7
RO$^\bullet$ + CH$_2$=CHR	19.49	61.8	0	0
RO$^\bullet$ + CH$_2$=CHPh	21.09	76.4	14.6	30.6
RO$_2^\bullet$ + CH$_2$=CHR	26.04	90.5	0	0
RO$_2^\bullet$ + CH$_2$=CHPh	27.23	99.0	8.5	22.8
RO$_2^\bullet$ + CH$_2$=CPh$_2$	28.04	105.0	14.5	38.3
R$_3$Si$^\bullet$ + CH$_2$=CHR	26.46	76.6	0	0
R$_3$Si$^\bullet$ + CH$_2$=CHPh	28.52	89.0	12.4	27.9
R$_3$Si$^\bullet$ + O=CR^{1}R^2	37.62	114.5	0	0
R$_3$Si$^\bullet$ + O=CMePh	39.74	127.6	13.1	22.5
R$_3$Si$^\bullet$ + O=CPh$_2$	40.02	129.3	14.8	25.5

are nonpolar. The reference reaction is characterized by the parameter br_e and the reaction with two polar reactants is characterized by the parameter $(br_e)_\mu$. The component of the activation energy caused by polar interaction (ΔE_μ) is calculated from the equation [11]:

$$\Delta E_\mu = [(br_e)_\mu^2 - (br_e)^2](1 + \alpha)^{-2} \qquad (6.43)$$

Table 6.24 presents the results of the calculation of ΔE_μ for the reactions of six polar radicals with a number of polar monomers. It can be seen that the polar interaction in the transition states for the addition can either decrease ($\Delta E_\mu < 0$) or, in other cases, increase ($\Delta E_\mu > 0$) the activation energy. The ΔE_μ values vary from $+19.5$ to -23 kJ mol^{-1}, i.e., they can be rather large.

The excellent semiempirical description of the polar effect in radical addition reactions was proposed recently by Fischer and Radom [54]. According to the quantum-chemical treatment of addition reaction, the four lowest doublet configurations of the three-center–three-electron system influence the energy of the transition state. These are the reactant ground-state configuration with a singlet electron pair on the alkene (R$^\bullet$ + C=C), the excited reactant configuration with the triplet electron pair (R$^\bullet$ + C=C^3), and two polar charge

TABLE 6.23
Strengths (D_e) and Lengths ($r(C-X)$ or $r(O-X)$) of the Bonds Formed Upon the Addition of X$^\bullet$ Radicals to C=C or C=O Double Bonds [41–43,49]

X	D_e (kJ mol^{-1})	$r(C-X)$ (or $r(O-X)$)(m)	r_e (m)	$(r_e/D_e) \times 10^{11}$ (m mol kJ^{-1})
		X$^\bullet$ + CH$_2$=CHY $\longrightarrow$ XCH$_2$C$^\bullet$HY		
C$^\bullet$H$_3$	378	1.52 × 10^{-10}	3.71 × 10^{-11}	0.98
N$^\bullet$H$_2$	360	1.47 × 10^{-10}	3.53 × 10^{-11}	0.98
RO$^\bullet$	359	1.43 × 10^{-10}	3.62 × 10^{-11}	1.01
RS$^\bullet$	284	1.79 × 10^{-10}	3.41 × 10^{-11}	1.20
R$_3$Si$^\bullet$	378	1.89 × 10^{-10}	4.91 × 10^{-11}	1.30
		X$^\bullet$ + O=CR^{1}R^2 $\longrightarrow$ XOC$^\bullet$R^{1}R^2		
C$^\bullet$H$_3$	365	1.43 × 10^{-10}	3.54 × 10^{-11}	0.97
R$_3$Si$^\bullet$	487	1.64 × 10^{-10}	6.28 × 10^{-11}	1.20

transfer configurations (R$^+$ + C=C and R$^-$ + C=C). The polar effects are expected to increase with decreasing energy of either of the charge transfer configurations, namely $E_I(R) - E_{EA}(C=C)$ or $E_I(C=C) - E_{EA}(R)$ of the reactants, where E_I is the potential of ionization and E_{EA} is the electron affinity. The following equation was proposed for the calculation of the activation energy of radical addition reactions:

$$E [\text{kJ mol}^{-1}] = 50 [\text{kJ mol}^{-1}] + 0.22 F_n F_e \qquad (6.44)$$

where factor F_n depends on $E_I(R) - E_{EA}(C=C)$ and factor F_e depends on $E_I(C=C) - E_{EA}(R)$. The results of this calculation are in good agreement with the experimental measurements [54].

The multidipole interaction in a bimolecular reaction arises if one or both reactants contain several polar groups [31]. The multidipole effect shows itself as a deviation of the rate constant for the addition of a polar radical to a polyfunctional compound (calculated in

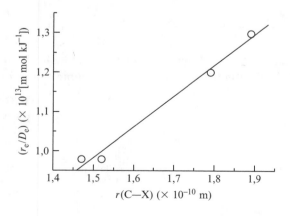

FIGURE 6.7 Correlation between the ratio of parameters r_e/D_e and the length $r(X-C)$ of the incoming bond for the addition to C=C bond of the radicals: R$^\bullet$, NH$_2^\bullet$, RO$^\bullet$, RS$^\bullet$, and R$_3$Si$^\bullet$ [49].

TABLE 6.24
Contribution of the Polar Effect ΔE_μ to the Activation Energy of the Addition of Polar Radicals to Polar Monomers $CH_2=CRY$ (Calculated from the Data of Several Studies [40,51–53])

	ΔE_μ (kJ mol^{-1})					
X, Y	Me$_2$(HO)C$^\bullet$	Me$_2$(CN)C$^\bullet$	Me$_3$COC(O)C$^\bullet$H$_2$	sec-RO$_2^\bullet$	tert-RO$_2^\bullet$	HO$_2^\bullet$
H, EtO	−1.9	−4.9	13.2			
Me, MeO	−6.5	−9.4	10.1			
H, AcO	−15.1	−11.2	17.5	9.5	7.3	11.5
Me, AcO	−16.4	−4.3	19.5			
Me, Cl	−8.2	−4.4	10.8			
Cl, Cl	−10.2	−3.8	13.4			
H, C(O)OMe	−20.1	−1.4	13.1	−0.3	−2.0	
Me, C(O)OMe		−3.4		−5.2	−6.4	−4.7
H, CN	−23.2	−0.9	13.8	−9.0	−10.8	−7.9

relation to one reaction center) from the rate constant for the addition to a monofunctional compound. The multidipole effect was discovered for the radical abstraction of a hydrogen atom from polyfunctional esters induced by peroxyl radicals. Later, this effect has also been found in the addition of peroxyl radicals to the double bonds of polyatomic unsaturated esters. Table 6.25 summarizes the results of calculations of the contribution of the multidipole effect to the activation energy performed using the relation (6.38). A monofunctional ester of the corresponding structure served as the reference compound. It can be seen that the role of the multidipole interaction in these reactions is fairly low: the contribution of the interaction of several polar groups $\Delta\Delta E_\mu$ varies from 0.8 to 2.3 kJ mol^{-1}.

6.5.7 STERIC HINDRANCE

The addition of trialkylsilyl radicals to 1,2-disubstituted ethylene derivatives is subject to a steric effect [49]. This shows itself in the greater E_{e0} value for Et$_3$Si$^\bullet$ addition to RCH=CHR compared with that for the addition of the same radical to CH$_2$=CHR. The contribution of

TABLE 6.25
Multidipole Effect $\Delta\Delta E_\mu$ in the Addition of Cumylperoxyl Radicals to the Unsaturated Polyatomic Esters [55] (n is the Number of Reacting Double Bonds in the Ester Molecule)

Ester	k (323 K) (L mol^{-1} s^{-1})	k (323 K)/n (L mol^{-1} s^{-1})	$\Delta\Delta E_\mu$ (kJ mol^{-1})
CH$_2$=CHC(O)OMe	0.50	0.50	0
(CH$_2$=CHC(O)OCH$_2$)$_2$CMe$_2$	0.76	0.38	0.8
(CH$_2$=CHC(O)OCH$_2$)$_2$CHOC(O)CH=CH$_2$	1.05	0.35	1.0
(CH$_2$=CHC(O)OCH$_2$)$_4$C	1.28	0.32	1.2
CH$_2$=CMeC(O)OMe	1.79	1.79	0
(CH$_2$=CMeC(O)OCH$_2$)$_2$CHOC(O)MeC=CH$_2$	3.93	1.31	0.9
(CH$_2$=CMeC(O)OCH$_2$)$_4$C	3.08	0.77	2.3

steric repulsion to the activation energy can be characterized by the increment ΔE_S ($\Delta E_S = E_{e0} - E_{e01}$):

Alkene	br_e (kJ mol^{-1})$^{1/2}$	E_{e0} (kJ mol^{-1})	ΔE_S (kJ mol^{-1})
CH$_2$=CHR	26.46	76.6	0.0
RCH=CHR	27.54	83.0	6.4

No effect of this type is manifested for the addition of alkyl radicals to the same alkenes. Evidently, the steric effect involved in the addition of trialkylsilyl radicals to 1,2-disubstituted ethylene derivatives is due to the repulsion between the carbon and silicon atom, caused by the large size of the silicon atom in the reaction center of the transition state.

The addition of tris-(1,1-dimethylethyl)phenoxyl radical to styrene can serve as an example of reaction subject to the steric effect. This reaction is characterized by $br_e = 23.89$ (kJ mol^{-1})$^{1/2}$, whereas the addition of the 1,1-dimethylethoxyl radical to styrene (a reaction with a similar structure of the reaction center) has the parameter $br_e = 21.09$ (kJ mol^{-1})$^{1/2}$. The difference between these values is matched by the difference between the activation energies of the corresponding thermally neutral reactions, $E_{e0} = 21.6$ kJ mol^{-1}; this can be regarded as a rough estimate of the energy of additional repulsion in the transition state of the reacting system. This repulsion in the transition state is brought about by two 1,1-dimethylethyl groups of the phenoxyl radical.

6.6 FREE RADICAL SUBSTITUTION AND HYDROGEN ATOM TRANSFER REACTIONS

6.6.1 FREE RADICAL SUBSTITUTION REACTIONS

Along with free radical atom abstraction reactions, reactions of radical substitution are known where a free radical attacks the weak Y—Y bond and abstracts radical Y [56]:

$$R^\bullet + YY \longrightarrow RY + Y^\bullet$$

These substitution reactions were discovered by Nozaki and Bartlett [57,58] in their study of benzoyl peroxide decomposition in different solvents. When benzoyl peroxide is decomposed, the formed benzoyloxyl radical attacks the solvent (RH), and the formed alkyl radical (R$^\bullet$) induces the chain decomposition of the peroxide (see Chapter 3).

$$PhC(O)OOC(O)Ph \longrightarrow PhCO_2^\bullet + PhCO_2^\bullet$$
$$PhCO_2^\bullet + RH \longrightarrow PhC(O)OH + R^\bullet$$
$$R^\bullet + PhC(O)OOC(O)Ph \longrightarrow ROC(O)Ph + PhCO_2^\bullet$$

Hydrogen atoms also react with the O—O bond of any peroxide, for example with hydrogen peroxide [59–61]:

$$H^\bullet + H_2O_2 \longrightarrow H_2O + HO^\bullet$$

This reaction is very exothermic and, therefore, very fast.

TABLE 6.26
Rate Constants of Free Radical Substitution Reactions of Hydrogen Atom, Alkyl, and Stannyl Radicals with Peroxides: $R^\bullet + R^1OOR^1 \longrightarrow ROR^1 + R^1O^\bullet$

Reaction	Solvent	T (K)	k (L mol^{-1} s^{-1})	Ref.
$H^\bullet + H_2O_2$	Water	298	2.35×10^7	[59]
$H^\bullet + H_2O_2$	Water	298	2.44×10^7	[60]
$H^\bullet + H_2O_2$	Water	298	2.93×10^7	[61]
$C^\bullet H_3 + [PhC(O)O]_2$	Acetic acid	353	1.93×10^3	[57]
$MeC^\bullet H_2 + [PhC(O)O]_2$	Ethyliodide	353	6.81×10^2	[57]
$cyclo\text{-}C_6H_{11}^\bullet$	Cyclohexane	353	3.32×10^3	[57]
$PhC^\bullet H_2 + [PhC(O)O]_2$	Toluene	353	4.50×10^2	[57]
$PhMeC^\bullet H + [PhC(O)O]_2$	Ethylbenzene	353	1.27×10^3	[57]
$PhMe_2C^\bullet + [PhC(O)O]_2$	Cumene	353	1.17×10^3	[57]
$PhCMe_2C^\bullet H_2 + [PhC(O)O]_2$	2-Methyl-2-phenylpropane	353	1.36×10^3	[57]
$C^\bullet H_3 + [PhC(O)O]_2$	p-Xylene	353	9.81×10^2	[57]
$Ph_3C^\bullet + [PhC(O)O]_2$	Benzene	298	1.2	[62]
$(4\text{-}MeOC_6H_4)_2PhC^\bullet + [PhC(O)O]_2$	Benzene	298	0.2	[62]
$(4\text{-}MeC_6H_4)_2PhC^\bullet + [PhC(O)O]_2$	Benzene	298	0.4	[62]
$(3\text{-}MeC_6H_4)_2PhC^\bullet + [PhC(O)O]_2$	Benzene	298	0.7	[62]
$(3\text{-}MeOC_6H_4)_2PhC^\bullet + [PhC(O)O]_2$	Benzene	298	1.5	[62]
$(4\text{-}FC_6H_4)_2PhC^\bullet + [PhC(O)O]_2$	Benzene	298	2.3	[62]
$(4\text{-}ClC_6H_4)_2PhC^\bullet + [PhC(O)O]_2$	Benzene	298	5.9	[62]
$(3\text{-}ClC_6H_4)_2PhC^\bullet + [PhC(O)O]_2$	Benzene	298	15.0	[62]
$Bu_3Sn^\bullet + EtOOEt$	Benzene	283	2.5×10^4	[63]
$Bu_3Sn^\bullet + AcOOAc$	Benzene	283	7.0×10^4	[63]
$Bu_3Sn^\bullet + AcOOCMe_3$	Benzene	283	4.0×10^4	[63]
$Bu_3Sn^\bullet + PhC(O)OOCMe_3$	Benzene	283	1.4×10^5	[63]

The decomposition of peroxides in the presence of stannum hydride is accompanied by the chain decomposition of peroxide [56]. Chain propagation occurs by the substitution reaction.

$$ROOR \longrightarrow RO^\bullet + RO^\bullet$$
$$RO^\bullet + Bu_3SnH \longrightarrow ROH + Bu_3Sn^\bullet$$
$$Bu_3Sn^\bullet + ROOR \longrightarrow Bu_3SnOR + RO^\bullet$$

The rate constants of these reactions are collected in Table 6.26.

Experimental data on the substitution reactions of free radicals with peroxides were analyzed by the IPM method [64]. The calculated parameters are collected in Table 6.27. The activation energies and the rate constants of radical substitution reactions calculated by the IPM method are presented in Table 6.28.

The studied free radicals have the following activation energy values in a thermoneutral reaction [64]: $R_3Sn^\bullet$ ($E_{e0} = 75$ kJ mol^{-1}), $H^\bullet$ ($E_{e0} = 94$ kJ mol^{-1}), and $R^\bullet$ ($E_{e0} = 120$ kJ mol^{-1}). The comparison of E_{e0} with electronegativity (EA) of the atoms in the reaction center of the reaction proved the following empirical dependence [64]:

$$br_e[\text{kJ mol}^{-1}]^{1/2} = 25.63 - 6.89 \times 10^{-2}(\Delta EA\,[\text{kJ mol}^{-1}]) \qquad (6.45)$$

that is, the more the difference in electron affinity of R (H, R^1, $R_3Sn^\bullet$), the lower the activation energy of the thermoneutral reaction.

TABLE 6.27
Kinetic Parameters of Free Radical Substitution Reactions: $R^\bullet + YY \rightarrow RY + Y^\bullet$ [64]

Radical	α	br_e (kJ mol^{-1})$^{1/2}$	E_{e0} (kJ mol^{-1})	$0.5hL(\nu_i - \nu_f)$ (kJ mol^{-1})	$0.5hL\nu$ (kJ mol^{-1})	A (L mol^{-1} s^{-1})
			YY = ROOH			
H$^\bullet$	0.682	16.06	91.2	−16.7	5.1	1.0×10^{-10}
R$^\bullet$	0.889	18.85	99.6	−1.1	5.1	1.0×10^{-9}
R$_3$Sn$^\bullet$	0.651	12.90	61.0	−0.9	5.1	1.0×10^{-9}
			YY = ROOR			
H$^\bullet$	0.689	17.84	111.6	−16.6	5.1	2.0×10^{-10}
R$^\bullet$	0.826	20.23	122.7	−1.6	5.1	2.0×10^{-9}
R$_3$Sn$^\bullet$	0.651	14.27	74.7	−0.9	5.1	2.0×10^{-9}

TABLE 6.28
Enthalpies, Activation Energies, and Rate Constants of the Substitution Reactions $R^\bullet + R^1OOH \rightarrow ROH + R^1O^\bullet$ in Hydrocarbon Solution Calculated by the IPM Method (See Table 6.27 and Equations (6.7), (6.8), and (6.12))

Reaction	$-\Delta H$ (kJ mol^{-1})	E (kJ mol^{-1})	A (L mol^{-1} s^{-1})	k (350 K) (L mol^{-1} s^{-1})
C$^\bullet$H$_3$ + H$_2$O$_2$	173.1	31.8	1.0×10^9	1.79×10^4
C$^\bullet$H$_3$ + Me$_3$COOH	205.5	23.6	5.0×10^8	1.50×10^5
C$^\bullet$H$_3$ + Me$_3$COOCMe$_3$	183.5	51.4	1.0×10^9	21.3
C$^\bullet$H$_3$ + MeC(O)OOC(O)Me	217.9	42.1	2.0×10^9	1.04×10^3
C$^\bullet$H$_3$ + PhC(O)OOC(O)Ph	242.7	36.0	2.0×10^9	8.48×10^3
MeC$^\bullet$H$_2$ + H$_2$O$_2$	178.1	30.4	1.0×10^9	2.90×10^4
MeC$^\bullet$H$_2$ + Me$_3$COOH	210.5	22.5	5.0×10^8	2.19×10^5
MeC$^\bullet$H$_2$ + Me$_3$COOCMe$_3$	190.3	49.5	1.0×10^9	41.0
MeC$^\bullet$H$_2$ + MeC(O)OOC(O)Me	234.3	38.0	2.0×10^9	4.26×10^3
MeC$^\bullet$H$_2$ + PhC(O)OOC(O)Ph	249.5	34.4	2.0×10^9	1.47×10^4
Me$_2$C$^\bullet$H + H$_2$O$_2$	188.1	27.8	1.0×10^9	7.10×10^4
Me$_2$C$^\bullet$H + Me$_3$COOH	220.5	20.2	5.0×10^8	4.83×10^5
Me$_2$C$^\bullet$H + Me$_3$COOCMe$_3$	198.6	47.2	1.0×10^9	90.3
Me$_2$C$^\bullet$H + MeC(O)OOC(O)Me	232.9	38.4	2.0×10^9	3.72×10^3
Me$_2$C$^\bullet$H + PhC(O)OOC(O)Ph	257.8	32.5	2.0×10^9	2.82×10^4
Me$_3$C$^\bullet$ + H$_2$O$_2$	185.1	28.6	1.0×10^9	5.39×10^4
Me$_3$C$^\bullet$ + Me$_3$COOH	207.5	23.2	5.0×10^8	1.72×10^5
Me$_3$C$^\bullet$ + Me$_3$COOCMe$_3$	165.1	58.8	1.0×10^9	1.68
Me$_3$C$^\bullet$ + MeC(O)OOC(O)Me	208.6	44.5	2.0×10^9	4.57×10^2
Me$_3$C$^\bullet$ + PhC(O)OOC(O)Ph	233.8	38.1	2.0×10^9	4.13×10^3
PhC$^\bullet$H$_2$ + H$_2$O$_2$	132.1	43.7	1.0×10^9	3.01×10^2
PhC$^\bullet$H$_2$ + Me$_3$COOH	164.5	34.1	5.0×10^8	4.07×10^3
PhC$^\bullet$H$_2$ + MeC(O)OOC(O)Me	177.8	53.0	2.0×10^9	24.6
PhC$^\bullet$H$_2$ + PhC(O)OOC(O)Ph	202.6	46.1	2.0×10^9	2.64×10^2

6.6.2 Reaction of Peroxides with Ketyl Radicals

The ketyl radicals ($R^1R^2C^\bullet OH$) have a weak O—H bond and are very active reducing agents. They react as donors of hydrogen atoms with carbonyl compounds, dioxygen, and free radicals.

$$R^1R^2C^\bullet OH + O_2 \longrightarrow R^1R^2C{=}O + HO_2^\bullet$$

$$R^1R^2C^\bullet OH + ROO^\bullet \longrightarrow R^1R^2C{=}O + ROOH$$

Ketyl radicals reduce peroxides with the splitting of the O—O bond [57,65–67].

$$R^1R^2C^\bullet OH + ROOH \longrightarrow R^1R^2C{=}O + H_2O + RO^\bullet$$

Ketyl radicals react with hydrogen peroxide to form water and the hydroxyl radical.

$$R^1R^2C^\bullet OH + H_2O_2 \longrightarrow R^1R^2C{=}O + H_2O + HO^\bullet$$

Bartlett and Nozaki [65] discovered the chain reaction of benzoyl peroxide decomposition in the presence of alcohols where the chain propagating step is the reaction of the ketyl radical with peroxide.

$$R^1R^2C^\bullet OH + PhC(O)OOC(O)Ph \longrightarrow R^1R^2C{=}O + PhCOOH + PhCO_2^\bullet$$

The rate constants of these reactions are given in Table 6.29. These reactions are very exothermic and occur with high rate constants ($7\text{–}14 \times 10^5$ L mol^{-1} s^{-1}).

The experimental data on the reactions of ketyl radicals with hydrogen and benzoyl peroxides were analyzed within the framework of IPM [68]. The elementary step was treated as a reaction with the dissociation of the O—H bond of the ketyl radical and formation of the same bond in acid (from acyl peroxide), alcohol (from alkyl peroxide), and water (from hydrogen peroxide). The hydroperoxyl radical also possesses the reducing activity and reacts with hydrogen peroxide by the reaction

$$HO_2^\bullet + H_2O_2 \longrightarrow O_2 + H_2O + HO^\bullet$$

The parameters calculated for these reactions are collected in Table 6.30.

TABLE 6.29
Rate Constants of Reducing Reactions of Ketyl Radicals with Peroxides

Reaction	Solvent	T (K)	k (L mol^{-1} s^{-1})	Ref.
$H_2C^\bullet OH + H_2O_2 \to H_2C(O) + H_2O + HO^\bullet$	Water	298	1.75×10^5	[67]
$MeC^\bullet HOH + H_2O_2 \to MeCH(O) + H_2O + HO^\bullet$	Water	298	2.80×10^5	[67]
$Me_2C^\bullet OH + H_2O_2 \to Me_2C(O) + H_2O + HO^\bullet$	Water	298	2.30×10^5	[67]
$Me_2CHC^\bullet HOH + H_2O_2 \to Me_2CHCH(O) + H_2O + HO^\bullet$	Water	298	2.20×10^5	[67]
$H_2C^\bullet OH + [PhC(O)O]_2 \to H_2C(O) + PhCO_2H + PhC(O)O^\bullet$	Methanol	353	3.84×10^5	[65]
$MeC^\bullet HOH + [PhC(O)O]_2 \to MeCH(O) + PhCO_2H + PhC(O)O^\bullet$	Ethanol	353	1.38×10^6	[65]
$Me_2C^\bullet OH + [PhC(O)O]_2 \to Me_2C(O) + PhCO_2H + PhC(O)O^\bullet$	2-Propanol	353	8.82×10^6	[65]
$PrC^\bullet HOH + [PhC(O)O]_2 \to PrCH(O) + PhCO_2H + PhC(O)O^\bullet$	Butanol	353	7.12×10^5	[65]

TABLE 6.30
Kinetic Parameters of Reduction of Peroxides by Ketyl and Alkyl Radicals [68]

Reaction	α	br_e (kJ mol^{-1})$^{1/2}$	E_{e0} (kJ mol^{-1})	A (L mol^{-1} s^{-1})
$R^1R^2C^\bullet OH + H_2O_2$	1.000	21.24	112.8	2.0×10^9
$R^1R^2C^\bullet OH + ROOR$	1.000	21.51	115.7	2.0×10^9
$RC^\bullet HCH_3 + H_2O_2$	0.778	20.70	135.7	2.0×10^9
$RC^\bullet HCH_3 + ROOR$	0.796	20.90	135.4	2.0×10^9
$RC^\bullet HCH_3 + ROOH$	0.778	20.70	135.7	1.0×10^9
$RCH=CHC^\bullet HCH_3 + H_2O_2$	0.778	22.16	155.3	2.0×10^9
$RCH=CHC^\bullet HCH_3 + ROOR$	0.796	22.36	155.0	2.0×10^9
$RCH=CHC^\bullet HCH_3 + ROOH$	0.778	22.16	155.3	1.0×10^9
$PhC^\bullet HCH_3 + H_2O_2$	0.778	21.18	141.9	2.0×10^9
$PhC^\bullet HCH_3 + ROOR$	0.796	21.40	142.0	2.0×10^9
$PhC^\bullet HCH_3 + ROOH$	0.778	21.18	141.9	1.0×10^9
$HOC_6H_4O^\bullet + H_2O_2$	0.982	21.24	112.3	1.0×10^8
$HOC_6H_4O^\bullet + ROOR$	0.992	21.51	126.4	1.0×10^8
$HOC_6H_4O^\bullet + ROOH$	0.982	21.00	112.3	5.0×10^7
$HO_2^\bullet + H_2O_2$	0.969	21.24	116.4	1.0×10^8
$HO_2^\bullet + ROOR$	0.978	21.51	118.3	1.0×10^8
$HO_2^\bullet + ROOH$	0.969	21.24	116.4	5.0×10^7

The comparison of the values of E_{e0} for reactions of ketyl radicals with peroxides with those for reactions $R^1R^2C^\bullet OH + R_2C(O)$ show that the reaction of the ketyl radical with peroxide occurs with sufficiently higher E_{e0} (see Table 6.30). Obviously, this is the result of strong additional repulsion in the nonlinear polyatomic transition state of the ketyl radical reaction with peroxide

$$\begin{array}{c} \diagdown \\ O\cdots H\cdots O \diagup \\ \diagdown R \end{array} \begin{array}{c} O-R \end{array}$$

compared to the linear transition state: $-O \cdots H \cdots O \cdots$ of the ketyl radical reaction with a carbonyl compound. The increase in the activation energy due to the additional repulsion in the transition state of the ketyl radical reaction with hydrogen peroxide is $104.6 - 41.5 = 63.1$ kJ mol^{-1} and 78.8 kJ mol^{-1} for the reaction of the ketyl radical with peroxide (ROOR). The activation energies and the rate constants of the reactions of different ketyl radicals and $HO_2^\bullet$ with peroxides calculated by the IPM method are presented in Table 6.31.

The alkyl radicals possess reducing activity as well. In the disproportionation reaction, one alkyl radical reacts as an acceptor and another as a donor of the hydrogen atom. This is the reason for the reducing action of alkyl radicals in reactions with peroxides.

$$RCH_2C^\bullet H_2 + R^1OOR^1 \longrightarrow RCH=CH_2 + R^1OH + R^1O^\bullet$$

The activation energies and the rate constants of such reactions were calculated by the IPM method [68] (see Table 6.32). The parameters used for calculations are given in Table 6.29.

As we have seen earlier, alkyl radicals react with peroxides in two different ways: substitution and reduction. The comparison of both these reactions is presented in Table 6.33. It is seen

TABLE 6.31
Enthalpies, Activation Energies, and Rate Constants of the Reduction of Peroxides by Ketyl, Semiquinone, and Hydroperoxyl Radicals Calculated by the IPM Model [68]

Reaction	$-\Delta H$ (kJ mol^{-1})	E (kJ mol^{-1})	log A, A (L mol^{-1} s^{-1})	k (350 K) (L mol^{-1} s^{-1})
MeC$^\bullet$HOH + H$_2$O$_2$	178.5	20.7	8.78	4.91 × 10^5
MeC$^\bullet$HOH + Me$_3$COOH	210.9	11.5	8.62	8.01 × 10^6
MeC$^\bullet$HOH + EtOOEt	173.9	24.6	8.72	1.12 × 10^5
MeC$^\bullet$HOH + Me$_3$COOCMe$_3$	178.7	23.1	8.75	2.01 × 10^5
MeC$^\bullet$HOH + PhC(O)OOC(O)Ph	180.1	22.6	8.76	7.44 × 10^5
Me$_2$C$^\bullet$OH + H$_2$O$_2$	182.6	19.5	8.79	7.58 × 10^5
Me$_2$C$^\bullet$OH + Me$_3$COOH	215.0	10.4	8.63	1.20 × 10^7
Me$_2$C$^\bullet$OH + EtOOEt	178.0	23.3	8.75	1.87 × 10^5
Me$_2$C$^\bullet$OH + Me$_3$COOCMe$_3$	182.8	21.8	8.66	2.55 × 10^5
Me$_2$C$^\bullet$OH + PhC(O)OOC(O)Ph	184.2	21.4	8.76	3.68 × 10^5
PhC$^\bullet$HOH + H$_2$O$_2$	150.4	29.6	8.70	1.92 × 10^4
PhC$^\bullet$HOH + Me$_3$COOH	182.8	19.4	8.50	4.02 × 10^5
PhC$^\bullet$HOH + EtOOEt	145.8	33.8	8.66	4.13 × 10^3
PhC$^\bullet$HOH + Me$_3$COOCMe$_3$	150.6	32.1	8.67	7.57 × 10^3
PhC$^\bullet$HOH + PhC(O)OOC(O)Ph	152.0	31.6	8.68	9.20 × 10^3
PhMeC$^\bullet$OH + H$_2$O$_2$	152.9	28.8	8.70	2.52 × 10^4
PhMeC$^\bullet$OH + Me$_3$COOH	185.3	18.7	8.51	5.24 × 10^5
PhMeC$^\bullet$OH + EtOOEt	148.3	32.9	8.67	5.75 × 10^3
PhMeC$^\bullet$OH + Me$_3$COOCMe$_3$	153.1	31.3	8.62	8.89 × 10^3
PhMeC$^\bullet$OH + [PhC(O)O]$_2$	154.5	30.8	8.69	1.24 × 10^4
4-HOC$_6$H$_4$O$^\bullet$ + H$_2$O$_2$	56.8	68.1	7.21	1.13 × 10^{-3}
4-HOC$_6$H$_4$O$^\bullet$ + Me$_3$COOH	89.2	54.7	6.96	6.26 × 10^{-2}
4-HOC$_6$H$_4$O$^\bullet$ + EtOOEt	52.2	71.8	7.21	3.13 × 10^{-4}
4-HOC$_6$H$_4$O$^\bullet$ + Me$_3$COOCMe$_3$	57.0	69.7	7.21	6.56 × 10^{-4}
4-HOC$_6$H$_4$O$^\bullet$ + [PhC(O)O]$_2$	58.4	69.0	7.21	8.13 × 10^{-4}
HO$_2^\bullet$ + H$_2$O$_2$	63.1	67.4	7.22	1.43 × 10^{-3}
HO$_2^\bullet$ + Me$_3$COOH	95.5	54.2	6.97	7.60 × 10^{-2}
HO$_2^\bullet$ + EtOOEt	58.5	71.1	7.21	3.90 × 10^{-4}
HO$_2^\bullet$ + PhC(O)OOC(O)Ph	64.7	68.5	7.22	9.93 × 10^{-4}

that the alkyl radicals (primary and secondary) react with hydrogen peroxide, hydroperoxide, and dibenzoyl peroxide more rapidly (100 to 1000 times) as substituting rather than reducing agents.

Ketyl radicals are more active reducing agents than alkyl radicals. It is seen from the comparison of the rate constants and the activation energies for reactions of ketyl and alkyl radicals with hydrogen peroxide [68].

Radical	k(350 K) (L mol^{-1} s^{-1})	E (kJ mol^{-1})	$-\Delta H$ (kJ mol^{-1})	E_{e0} (kJ mol^{-1})
MeC$^\bullet$HOH	1.4 × 10^7	11.8	185.6	104.6
EtMeC$^\bullet$H	3.7	54.1	147.0	125.5

TABLE 6.32
Enthalpies, Activation Energies, and Rate Constants of the Reduction of Hydroperoxides by Alkyl Radicals Calculated by IPM Model [68]

Reaction	$-\Delta H$ (kJ mol^{-1})	E (kJ mol^{-1})	log A, A (L mol^{-1} s^{-1})	k(350 K) (L mol^{-1} s^{-1})
Et$^{\bullet}$ + H$_2$O$_2$	132.8	68.8	9.00	5.40×10^{-2}
M$_2$C$^{\bullet}$H + H$_2$O$_2$	144.4	65.0	9.31	0.40
MeEtC$^{\bullet}$H + H$_2$O$_2$	145.8	64.6	8.83	0.15
Me$_3$C$^{\bullet}$ + H$_2$O$_2$	129.5	69.9	9.47	0.11
PhC$^{\bullet}$HMe + H$_2$O$_2$	92.8	88.7	8.94	5.04×10^{-5}
PhC$^{\bullet}$HEt + H$_2$O$_2$	104.6	84.5	8.77	1.47×10^{-4}
Et$^{\bullet}$ + Me$_3$COOH	165.1	55.3	8.74	6.12
Me$_2$C$^{\bullet}$H + Me$_3$COOH	176.8	51.8	9.05	42.3
MeEtC$^{\bullet}$H + Me$_3$COOH	178.2	51.4	8.58	16.3
Me$_3$C$^{\bullet}$ + Me$_3$COOH	161.9	56.3	9.21	13.0
PhMeC$^{\bullet}$H + Me$_3$COOH	125.2	74.0	8.67	8.58×10^{-3}
PhEtC$^{\bullet}$H + Me$_3$COOH	137.0	70.0	8.51	2.29×10^{-2}

TABLE 6.33
Comparison of the Thermodynamic and the Kinetic Characteristics of Substitution and Reduction of Peroxides by Alkyl Radicals

Reactants	Reaction	$-\Delta H$ (kJ mol^{-1})	E (kJ mol^{-1})	k(350 K) (L mol^{-1} s^{-1})
Et$^{\bullet}$, H$_2$O$_2$	Substitution	178.1	30.4	2.9×10^4
Et$^{\bullet}$, H$_2$O$_2$	Reduction	132.8	68.8	5.3×10^{-2}
Me$_2$C$^{\bullet}$H, H$_2$O$_2$	Substitution	188.1	27.8	4.8×10^5
Me$_2$C$^{\bullet}$H, H$_2$O$_2$	Reduction	144.4	65.0	0.40
Me$_3$C$^{\bullet}$, H$_2$O$_2$	Substitution	185.1	28.6	5.4×10^4
Me$_3$C$^{\bullet}$, H$_2$O$_2$	Reduction	129.5	69.9	0.11
Et$^{\bullet}$, Me$_3$COOH	Substitution	210.5	22.5	2.2×10^5
Et$^{\bullet}$, Me$_3$COOH	Reduction	165.1	55.3	6.1
Me$_2$C$^{\bullet}$H, Me$_3$COOH	Substitution	220.5	20.2	4.8×10^5
Me$_2$C$^{\bullet}$H, Me$_3$COOH	Reduction	176.8	51.8	42
Me$_3$C$^{\bullet}$, Me$_3$COOH	Substitution	207.5	23.2	1.7×10^5
Me$_3$C$^{\bullet}$, Me$_3$COOH	Reduction	161.9	56.3	13

There are two reasons for such a great difference:

1. The reaction of the ketyl radical is more exothermic (-186 versus -147 kJ mol^{-1}).
2. The triplet repulsion is greater for the reaction of the alkyl radical ($E_{e0} = 125$ kJ mol^{-1}) compared with that of the ketyl radical ($E_{e0} = 105$ kJ mol^{-1}).

REFERENCES

1. NN Semenov. *Some Problems of Chemical Kinetics and Reactivity*, vols 1 and 2. London: Pergamon Press, 1958–1959.
2. KU Ingold. In: JK Kochi, ed. *Free Radicals*, vol 1. New York: Wiley, 1973, pp 37–112.

3. IB Afanasiev. *Usp Khim* 40:385–416, 1971.
4. ET Denisov. In: ZB Alfassi, ed. *General Aspects of the Chemistry of Radicals*. New York: Wiley, 1999, pp 79–138.
5. MG Evans, M Polyany. *Trans Faraday Soc* 34:11–23, 1938.
6. LP Hammett. *Physical Organic Chemistry*. New York: McGraw-Hill, 1940.
7. RW Taft. *J Am Chem Soc* 74:2729–2735, 1952.
8. ES Rudakov, LK Volkova. *Dokl AN Ukr SSR*, Ser B, N 10:912–914, 1978.
9. ET Denisov. *Kinet Catal* 32:406–440, 1991.
10. ET Denisov. *Mendeleev Commun* 2, 1–2, 1992.
11. ET Denisov. *Russ Chem Rev* 66:859–876, 1997.
12. ET Denisov, TG Denisova. *Handbook of Antioxidants*. Boca Raton: CRC Press, 2000.
13. ET Denisov, TG Denisova. *Kinet Catal* 34:173–179, 1993.
14. TG Denisova, ET Denisov. *Zh Fiz Khim* 65:1208–1213, 1991.
15. ET Denisov. *Kinet Catal* 37:519–523, 1996.
16. JA Kerr. In: CH Bamford and CFH Tipper, eds. *Comprehensive Chemical Kinetics*, vol 18, Amsterdam: Elsevier, 1976, pp 39–166.
17. ET Denisov. *Kinet Catal* 35:293–298, 1994.
18. ET Denisov. *Kinet Catal* 35:617–635, 1994.
19. S Sato. *J Chem Phys* 23:592–600, 1955.
20. HS Johnston. *Gas Phase Reaction Rate Theory*. New York: Ronald, 1966.
21. T Berces, J Dombi. *Int J Chem Kinet* 12:123–140, 1980.
22. AA Zavitsas. *J Am Chem Soc* 94:2779–2789, 1972.
23. L Pauling. *General Chemistry*. San Francisco: Freeman and Co, 1970.
24. ET Denisov. *Russ J Phys Chem* 68:1089–1093, 1994.
25. ET Denisov. *Russ Chem Bull* 47:1274–1279, 1998.
26. TI Drozdova, ET Denisov. *Kinet Katal* 43:14–22, 2002.
27. W Tsang. In: A. Greenberg, J. Liebman, eds. *Energetics of Free Radicals*. New York: Blackie Academic and Professional, 1996, p. 22.
28. ET Denisov. *Khim Fiz* 11:1328–1337, 1992.
29. ET Denisov. *Kinet Catal* 36:351–356, 1995.
30. ET Denisov, TG Denisova. *Russ Chem Bull* 43:29–34, 1994.
31. ET Denisov. *Izv AN SSSR Ser Khim* 1746–1750, 1978.
32. AM North. *The Collision Theory of Chemical Reactions in Liquids*. London: Methuen, 1964.
33. ET Denisov, OM Sarkisov, GI Likhtenshtein. *Chemical Kinetics*. Amsterdam: Elsevier, 2003.
34. AF Shestakov, ET Denisov. *Russ Chem Bull* 52:320–329, 2003.
35. TG Denisova, NS Emel'yanova. *Kinet Katal* 44:441–449, 2003.
36. TG Denisova, ET Denisov. *Kinet Catal* 42:620–630, 2001.
37. JK Kochi. In: JK Kochi, ed. *Free Radicals*, vol 2. New York: Wiley, 1973, pp 665–710.
38. KJ Laidler, LF Loucks. In: CH Bamford, CFH Tipper, eds. *Comprehensive Chemical Kinetics*, vol 5. Amsterdam: Elsevier, 1972, pp 1–148.
39. GC Eastmond. In: CH Bamford, CFH Tipper, eds. *Comprehensive Chemical Kinetics*, vol 14. Amsterdam: Elsevier, 1976, pp 105–152.
40. TG Denisova, ET Denisov. *Petrol Chem* 38, 12–18, 1998.
41. ET Denisov. *Russ Chem Rev* 69:153–164, 2000.
42. ET Denisov. In: Book of Abstracts of the 4th International Conference on Chemical Kinetics. Gaithersburg: NIST, 1997, p 9.
43. ET Denisov. In: Proceedings of 5th Arab International Conference on Polymer Science and Technology, Part 1. Luxor–Aswan, Egypt, 1999, pp 241–251.
44. ET Denisov. *Kinet Catal* 33, 50–57, 1992.
45. ET Denisov, TG Denisova. *Chem Phys Rep* 17:2105–2117, 1998.
46. ET Denisov. *Russ Chem Bull* 48:442–447, 1999.
47. ET Denisov. *Kinet Catal* 40:756–763, 1999.
48. ET Denisov. *Kinet Catal* 41:293–297, 2000.
49. ET Denisov. *Kinet Catal* 42:23–29, 2001.

50. IM Borisov, ET Denisov. *Petrol Chem* 41:20–25, 2001.
51. K Heberger, H Fischer. *Int J Chem Kinet* 25:249–256, 1993.
52. K Heberger, H Fischer. *Int J Chem Kinet* 25:913–920, 1993.
53. JQ Wu, I Beranek, H Fischer. *Helv Chim Acta* 78:194–202, 1995.
54. H Fisher, L Radom. *Angew Chem Intern Ed* 40:1340, 2001.
55. VA Machtin, EM Pliss, ET Denisov. *Izv AN SSSR Ser Khim* 746–750, 1981.
56. KU Ingold, BP Roberts. *Free-Radical Substitution Reactions*. New York: Wiley, 1971.
57. K Nozaki, PD Bartlett. *J Am Chem Soc* 68:1686–1692, 1946.
58. K Nozaki, PD Bartlett. *J Am Chem Soc* 68:2377–2382, 1946.
59. J Warnats. In: WC Gardiner, ed. *Combustion Chemistry*. New York: Springer-Verlag, 1984, p 197.
60. DL Baulen, CJ Cobos, RA Cox. *J Phys Chem Ref Data* 21:411, 1992.
61. W Tsang, RF Hampson. *J Phys Chem Ref Data* 15:1087, 1986.
62. T Suehiro, A Konoya, H Hara, T Nakahama, M Omori, T Komori. *Bull Chem Soc Jpn* 40:668, 1967.
63. JL Brokenshire, KU Ingold. *Int J Chem Kinet* 3:343–357, 1971.
64. TS Pokidova, ET Denisov. *Petrol Chem* 38:244–250, 1998.
65. PD Bartlett, K Nozaki. *J Am Chem Soc* 69:2299–2306, 1947.
66. B Barnett, WE Vaughan. *J Phys Chem* 51:942–955, 1947.
67. K Kishore, PN Moorthy, KN Rao. *Radiat Phys Chem* 39:309–326, 1987.
68. ET Denisov. *Petrol Chem* 39:395–405, 1999.

7 Oxidation of Alcohols and Ethers

7.1 OXIDATION OF ALCOHOLS

7.1.1 INTRODUCTION

The hydroxyl group of alcohol weakens the α-C—H bond. Therefore, free radicals attack preferentially the α-C—H bonds of the secondary and primary alcohols. The values of bond dissociation energy (BDE) of C—H bonds in alcohols are presented in Table 7.1. The BDE values of C—H bonds of the parent hydrocarbons are also presented. It is seen from comparison that the hydroxyl group weakens BDE of the C—H bond by 23.4 kJ mol^{-1} for aliphatic alcohols and by 8.0 kJ mol^{-1} for allyl and benzyl alcohols.

Alcohols are polar compounds. They have dipole moment and this influences their reactivity in reactions with polar peroxyl radicals (see later). The values of the dipole moments μ for selected alcohols are given below [6].

Alcohol	MeOH	EtOH	Me$_2$CHOH	PhCH$_2$OH	CH$_2$=CHCH$_2$OH
μ (Debye)	1.70	1.69	1.58	1.71	1.60

Alcohols having a hydroxyl group form hydrogen bonds with polar compounds such as hydroperoxides, ketones, etc. This causes some peculiarities of the oxidation kinetics of alcohols. On the other hand, alcohols are weak acids and dissociate as acids in polar alcoholic media. Protonated alcohol molecules induce several heterolytic and homolytic reactions complicating the mechanism of oxidation and composition of the products. The hydroxyl group of the alkyl radical formed from alcohol complicates the mechanism of alcohol oxidation also. One of the peculiarities of alcohol oxidation is the production of hydrogen peroxide as a primary intermediate, and another is a high reducing activity of peroxyl radicals formed from oxidized alcohols. The chemistry and kinetics of alcohol oxidation are discussed in detail in monographs [7–10].

7.1.2 CHAIN MECHANISM OF ALCOHOL OXIDATION

Alcohols, like hydrocarbons, are oxidized by the chain mechanism. The composition of the molecular products of oxidation indicates that oxidation involves first the alcohol group and the neighboring C—H bond. This bond is broken more readily than the C—H bond of the corresponding hydrocarbon, since the unpaired electron of the formed hydroxyalkyl radical interacts with the p electrons of the oxygen atom.

TABLE 7.1
Bond Dissociation Energies of α-C—H Bonds in Alcohols [1–5]

Alcohol	Bond	D (kJ mol^{-1})	D (R–H) (kJ mol^{-1})
Methanol	HOCH$_2$—H	411.0	440.0
Ethanol	Me(HO)CH—H	399.8	422.0
Propanol	Et(HO)CH—H	399.5	422.0
Butanol	Pr(HO)CH—H	397.0	422.0
Pentanol	Bu(HO)CH—H	397.0	422.0
Ethanol, 1-methyl-	Me$_2$(HO)C—H	390.5	412.0
Propanol, 2,2-dimethyl-	Me$_3$C(HO)CH—H	395.2	422.0
Ethanol, 1,1-dimethyl-	Me$_2$(HO)CCH$_2$—H	417.4	422.0
Cyclohexanol	(cyclohexane with H, OH)	388.4	408.8
Cyclopentenol	(cyclopentene with H, OH)	330.7	342.5
Cyclohexenol	(cyclohexene with H, OH)	329.7	341.5
Ethylenglicol	HOCH$_2$(HO)CH—H	401.5	422.0
1,3-Butanediol	HOCH$_2$CH$_2$(HO)MeC—H	387.7	412.0
1,4-Butanediol	HOCH$_2$(CH$_2$)$_2$(HO)CH—H	404.2	422.0
2,3-Butanediol	MeCH(OH)C—H(OH)Me	383.1	412.0
Allyl alcohol	CH$_2$=CH(HO)CH—H	360.0	368.0
2-Propen-1-ol, 2-methyl-	CH$_2$=CMeCH—H(OH)	349.9	357.9
2-Buten-2-ol, 2-methyl-	CH$_2$=CMeC—HMe(OH)	339.2	347.2
Z-2-Buten-1-ol	Z-MeCH=CHCH—H(OH)	347.8	355.8
E-2-Buten-1-ol	E-MeCH=CHCH—H(OH)	348.8	356.8
2-Buten-1-ol, 2-methyl-	MeCH=CMeCH—H(OH)	344.4	354.4
2-Buten-1-ol, 2,3-dimethyl-	Me$_2$C=CMeCH—H(OH)	342.2	350.2
Z-3-Penten-2-ol	Z-MeCH=CHC—HMe(OH)	336.0	344.0
E-3-Penten-2-ol	E-MeCH=CHC—HMe(OH)	338.2	346.2
3-Penten-2-ol, 4-methyl-	Me$_2$C=CHC—HMe(OH)	324.0	332.0
3-Penten-2-ol, 3-methyl-	MeCH=CMeC—HMe(OH)	325.2	333.2
2-Propin-1-ol	CH≡CCH—H(OH)	351.8	359.8
3-Butin-2-ol	CH≡CC—HMe(OH)	338.8	346.8
2-Butin-1-ol	MeC≡CCH—H(OH)	346.7	354.7
3-Butin-2-ol	MeC≡CC—HMe(OH)	333.8	341.8
Benzyl alcohol	Ph(HO)CH—H	366.9	375.0
α-Phenylethanol	PhMeC(OH)—H	356.0	364.1
Butanol, 1-phenyl-	PhPrC(OH)—H	360.6	368.7
Diphenylmethanol	Ph$_2$C—H(OH)	348.7	356.8

TABLE 7.1
Bond Dissociation Energies of α-C—H Bonds in Alcohols [1–5]—*continued*

Alcohol	Bond	D (kJ mol^{-1})	D(R—H) (kJ mol^{-1})
1-Indanol	H OH (indanol structure)	351.3	359.4
1-Tetralol	H OH (tetralol structure)	337.5	345.6

7.1.2.1 Kinetics of Oxidation

The regular features of the initiated oxidation of alcohols are similar to those of the oxidation of hydrocarbons (see Chapter 2). The rate of oxidation of the alcohol (RH(OH)) in the presence of an initiator (I) is given by the expression [8,9]:

$$v = k_i^{1/2} k_p (2k_t)^{-1/2} [\text{RH(OH)}][\text{O}_2]^0 [\text{I}]^{1/2}, \qquad (7.1)$$

where k_p is the rate constant of chain propagation via the reaction of the peroxyl radical with the C—H bond of alcohol and k_t is the rate constant of chain termination by the reaction of disproportionation of two peroxyl radicals. This equation is applicable under the following conditions: (i) Dioxygen pressure pO_2 must be sufficiently high, so that [R˙(OH)] ≪ [R(OH)O$_2$˙], and the chain termination reactions HOR˙ + HOR˙ and HOR˙ + R(OH)O$_2$˙ do not play a significant role compared with the reaction R(OH)O$_2$˙ + R(OH)O$_2$˙. (ii) The rate of radical formation from hydrogen peroxide produced in the course of the reaction must be very low compared with k_i[I]. (iii) The chains of oxidation should be long. The values of the ratio $k_p(2k_t)^{-1/2}$ for the oxidation of different alcohols are listed in Table 7.2.

7.1.2.2 Reactions of Ketyl Radicals with Dioxygen

Practically one reaction of alkyl radicals with dioxygen is known, namely addition reaction (see Chapter 2). Ketyl radicals having a free valence on the carbon atom add dioxygen also. However, they possess high reducing activity and easily react with dioxygen by the abstraction reaction.

$$R^1 R^2 \text{C˙OH} + O_2 \longrightarrow R^1 R^2 \text{C(OO˙)OH}$$
$$R^1 R^2 \text{C˙OH} + O_2 \longrightarrow R^1 R^2 \text{C(O)} + \text{HO}_2\text{˙}$$

The addition reaction prevails in solutions at moderate temperatures and abstraction reaction occurs in the gas phase at elevated temperatures. The rate constants of both reactions are collected in Table 7.3.

7.1.2.3 Reactions of Alkylhydroxyperoxyl and Hydroperoxyl Radicals

The reaction of hydrogen atom abstraction by the alkylhydroxyperoxyl radical from alcohol limits chain propagation in oxidized alcohol [8,9].

$$R^1 R^2 \text{C(OO˙)OH} + R^1 R^2 \text{CH(OH)} \longrightarrow R^1 R^2 \text{C(OOH)OH} + R^1 R^2 \text{C˙(OH)}$$

TABLE 7.2
Values of $k_p (2k_t)^{-1/2}$ for Oxidation of Alcohols

Alcohol	T (K)	pO_2 (Pa)	$k_p(2k_t)^{-1/2}$ (L mol^{-1} s^{-1})$^{1/2}$	Ref.
CH$_3$OH	354–418	(4.0–8.0) × 10^5	6.0 × 10^4 exp(−57.3/RT)	[11]
CH$_3$OH	348	1.01 × 10^5	2.3 × 10^{-5}	[12]
CH$_3$CH$_2$OH	348	1.01 × 10^5	3.3 × 10^{-5}	[8]
CH$_3$CH$_2$OH	418	1.01 × 10^6	3.8 × 10^{-3}	[11]
Me$_2$CHOH	353	(0.13–1.01) × 10^5	1.2 × 10^{-3}	[8]
Me$_2$CHOH	359–411	(2.0–2.9) × 10^5	3.0 × 10^4 exp(−50.2/RT)	[13]
Me$_2$CHOH	293	1.01 × 10^5	7.9 × 10^{-5}	[14]
Me$_2$CHOH	418	1.01 × 10^6	1.8 × 10^{-2}	[11]
EtMeCHOH	338–348	1.01 × 10^5	1.0 × 10^5 exp(−54.4/RT)	[15]
Me$_2$CHCH$_2$MeCHOH	333	1.01 × 10^5	2.46 × 10^{-4}	[16]
MeCH(OH)CMe$_3$	357–377	1.01 × 10^5	5.0 × 10^4 exp(−52.7/RT)	[15]
Me(CH$_2$)$_8$CH$_2$OH	353–403	1.01 × 10^5	2.10 × 10^4 exp(−51.2/RT)	[17]
Me(CH$_2$)$_5$CH(OH)Me	339–401	1.01 × 10^5	2.14 × 10^4 exp(−49.9/RT)	[17]
HOCH$_2$CH$_2$CH$_2$CH$_2$OH	353–403	1.01 × 10^5	1.58 × 10^4 exp(−38.2/RT)	[17]
MeCH(OH)CH(OH)Me	403	1.01 × 10^5	3.41 × 10^{-3}	[17]
HOCH$_2$CH$_2$CH(OH)Me	403	1.01 × 10^5	3.23 × 10^{-3}	[17]
cyclohexanol	353	(0.13–1.01) × 10^5	1.6 × 10^{-3}	[8]
cyclohexanol	323–348	(0.27–1.07) × 10^5	7.6 × 10^3 exp(−45.2/RT)	[18]
cyclohexanol	338–384	(0.40–1.27) × 10^5	7.5 × 10^3 exp(−46.0/RT)	[19]
cyclohexanol	363–393	1.01 × 10^5	5.1 × 10^3 exp(−43.1/RT)	[20]
cyclohexanol	323–348	1.01 × 10^5	4.0 × 10^3 exp(−43.9/RT)	[21]
cyclohexanol	353–373	1.01 × 10^5	2.1 × 10^4 exp(−50.2/RT)	[22]
2-methylcyclohexanol	323–373	1.01 × 10^5	2.1 × 10^8 exp(−77.8/RT)	[21]
2-methylcyclohexanol	353–373	1.01 × 10^5	1.8 × 10^4 exp(−50.2/RT)	[22]
3-methylcyclohexanol	333–348	1.01 × 10^5	2.3 × 10^5 exp(−54.3/RT)	[21]
3-methylcyclohexanol	353–373	1.01 × 10^5	9.3 × 10^4 exp(−53.5/RT)	[22]
4-methylcyclohexanol	323–348	1.01 × 10^5	1.3 × 10^4 exp(−46.0/RT)	[21]

TABLE 7.2
Values of $k_p (2k_t)^{-1/2}$ for Oxidation of Alcohols—continued

Alcohol	T (K)	pO$_2$ (Pa)	$k_p (2k_t)^{-1/2}$ (L mol^{-1} s^{-1})$^{1/2}$	Ref.
4-methylcyclohexanol (H, OH)	353–373	1.01 × 10^5	2.7 × 10^4 exp(−51.0/RT)	[22]
cyclohexanol (H, OH)	328–348	1.01 × 10^5	7.6 × 10^3 exp(−37.7/RT)	[15]
cyclohexanol (H, OH)	333	1.01 × 10^5	9.70 × 10^{-2}	[16]
cyclohexanol (OH)	353–373	1.01 × 10^5	3.7 × 10^4 exp(−56.1/RT)	[22]
PhCH$_2$OH	323–343	1.01 × 10^5	3.47 × 10^2 exp(−38.6/RT)	[23]
PhCH$_2$OH	303	1.01 × 10^5	8.50 × 10^{-4}	[24]
PhCH$_2$OH	348	1.01 × 10^5	4.70 × 10^{-3}	[25]
o-Cl-C$_6$H$_4$CH$_2$OH	348	1.01 × 10^5	2.0 × 10^{-3}	[26]
MePhCHOH	338–368	1.01 × 10^5	1.15 × 10^4 exp(−43.9/RT)	[15]
MePhCHOH	338–368	1.01 × 10^5	1.46 × 10^2 exp(−37.4/RT)	[23]
MePhCHOH	333	1.01 × 10^5	1.40 × 10^{-4}	[16]

The analysis of alcohol reactivities in reactions with the peroxyl radicals will be discussed later. The values of the rate constants of chain propagation in oxidized alcohols are collected in Table 7.4.

In addition to the abstraction reaction, alkylhydroxyperoxyl radicals are decomposed into carbonyl compound and hydroperoxyl radical.

$$R^1R^2C(OO^\bullet)OH \longrightarrow R^1R^2C(O) + HO_2^\bullet$$

The rate constant of their decomposition depends on the peroxyl radical structure and temperature (see Table 7.5).

As a result, two different peroxyl radicals take part in chain propagation in oxidized alcohol, namely, alkylhydroxyperoxyl and hydroperoxyl radicals.

$$\begin{array}{c}\text{CHOH} \\ \text{COH} \end{array} \rightleftarrows \begin{array}{c} \text{C(OO}^\bullet\text{)OH} \\ \text{C(OOH)OH} \end{array} \longrightarrow \begin{array}{c} \text{HO}_2^\bullet \\ \text{H}_2\text{O}_2 \end{array} \rightleftarrows \begin{array}{c} \text{CHOH} \\ \text{COH} \end{array}$$

The reactivities of both the radicals are apparently very close due to very the close BDE of the OH bonds in the formed hydroperoxides.

TABLE 7.3
Rate Constants of Ketyl Radical Reactions with Dioxygen

Radical	Solvent	T (K)	k (L mol^{-1} s^{-1})	Ref.
Addition Reaction				
C·H$_2$OH	H$_2$O, pH = 7	298	4.9×10^9	[27]
C·H$_2$OH	H$_2$O, pH = 10.7	298	4.2×10^9	[28]
MeC·HOH	H$_2$O, pH = 7	298	4.6×10^9	[27]
EtC·HOH	H$_2$O, pH = 7	298	4.7×10^9	[27]
Me$_2$C·OH	H$_2$O, pH = 7	298	4.2×10^9	[27]
Me$_2$C·OH	H$_2$O, pH = 5.6	298	3.5×10^9	[29]
Me$_2$C·OH	H$_2$O, pH = 0.3	298	4.5×10^9	[30]
Me$_2$C·OH	Me$_2$CHOH	298	3.9×10^9	[31]
Me$_2$CHC·HOH	H$_2$O, pH = 7	298	3.4×10^9	[27]
C·H(OH)$_2$	H$_2$O, pH = 5.7	298	7.7×10^8	[32]
C·H(OH)$_2$	H$_2$O, pH = 3.5	294	4.5×10^9	[33]
C·H(OH)$_2$	H$_2$O	293	3.5×10^9	[34]
HOCH$_2$C·HOH	H$_2$O, pH = 7	298	3.2×10^9	[27]
HOCH$_2$CHOHC·HOH	H$_2$O, pH = 7	298	3.3×10^9	[27]
HOC·HCO$_2^-$	H$_2$O, pH = 7	298	1.8×10^9	[27]
HOC·HCH(NH$_2$)CO$_2^-$	H$_2$O, pH = 7	298	2.4×10^9	[27]
Reaction of Hydrogen Atom Abstraction				
C·H$_2$OH	Gas phase	200–300	5.48×10^9	[35]
MeC·HOH	Gas phase	298	1.15×10^{10}	[36]
EtC·HOH	Gas phase	296	1.57×10^{10}	[37]
Me$_2$C·OH	Gas phase	296	2.23×10^{10}	[37]

Hydroperoxide	sec-ROOH	HOOH	R^{1}R^{2}C(OOH)OH	cyclohexyl-OO-H / OH
D_{OO-H} (kJ mol^{-1})	365.5	369.0	370.7	362.1
Ref.	[45]	[6]	[46]	[47]

Let us compare the ratio of radicals in oxidized 2-propanol and cyclohexanol at different temperatures when oxidation occurs with long chains and chain initiation and termination do not influence the stationary state concentration of radicals. The values of the rate constants of the reactions of peroxyl radicals (k_p) with alcohol and decomposition of the alkylhydroxyperoxyl radical (k_d) are taken from Table 7.4 and Table 7.5.

TABLE 7.4
Rate Constants of Chain Termination and Propagation in Oxidized Alcohols

Alcohol	T (K)	$2k_t$ (L mol^{-1} s^{-1})	k_p (L mol^{-1} s^{-1})	Ref.
CH_3OH	354–418	3.7×10^9	$3.65 \times 10^9 \exp(-57.3/RT)$	[11,38]
$MeCH_2OH$	348	3.3×10^9	$2.00 \times 10^9 \exp(-51.1/RT)$	[8,38]
Me_2CHOH	359–411	1.1×10^9	$1.00 \times 10^9 \exp(-50.2/RT)$	[13,39]
$PhCH_2OH$	323–343	$9.91 \times 10^{12} \exp(-29.2/RT)$	$1.04 \times 10^9 \exp(-54.3/RT)$	[23]
$PhCH_2OH$	303	3.2×10^7	4.8	[24]
$MePhCHOH$	323–343	$8.98 \times 10^{12} \exp(-31.3/RT)$	$4.61 \times 10^8 \exp(-52.0/RT)$	[23]
$MePhCHOH$	333	5.0×10^9	10.0	[16]
cyclohexanol (H, OH)	323–348	$5.60 \times 10^8 \exp(-15.1/RT)$	$1.80 \times 10^8 \exp(-52.7/RT)$	[18]
methylcyclohexanol (H, OH)	328–387		$3.16 \times 10^8 \exp(-54.0/RT)$	[39]
methylcyclohexanol (H, OH)	403	1.8×10^7		[40]
cyclooctanol (H, OH)	328–387		$7.90 \times 10^8 \exp(-74.0/RT)$	[41]

T (K)	300	350	400	450
2-Propanol k_d (s^{-1})	956	1.87×10^4	1.74×10^5	9.83×10^5
k_p[HROH] (s^{-1})	23.8	4.22×10^2	3.65×10^3	1.95×10^4
[HO$_2^\bullet$]/[HR(OH)O$_2^\bullet$]	40.2	44.3	47.7	50.4
Cyclohexanol				
k_d (s^{-1})	0.88	14.6	119	616
k_p[HROH] (s^{-1})	1.15	23.6	227	1.32×10^3
[HO$_2^\bullet$]/[HR(OH)O$_2^\bullet$]	0.76	0.62	0.52	0.47

A different situation in the oxidation of these two alcohols is seen. The hydroperoxyl radical is the main chain propagating species in oxidized 2-propanol; the portion of alkylhydroxyperoxyl radicals in this reaction is less than 2.5%. In oxidized cyclohexanol, on the contrary, the stationary state concentrations of both radicals are close and both of them take important part in chain propagation.

In the presence of specially added hydrogen peroxide, the stationary concentration of hydroperoxyl radicals increases due to the exchange reaction.

$$R^1R^2C(OO^\bullet)OH + H_2O_2 \longrightarrow R^1R^2C(OOH)OH + HO_2^\bullet$$

This reaction can be used as a kinetic test to determine which radicals predominate in an alcohol oxidized under given conditions. If hydroxyperoxide radicals predominate, the

TABLE 7.5
Rate Constants of Alkylhydroxy Peroxyl Radical Decomposition

Peroxyl Radical	Products	T (K)	k (s^{-1})	Ref.
$HOCH_2OO^{\bullet}$	$CH_2O + HOO^{\bullet}$	295	10	[42]
$MeCH(OO^{\bullet})OH$	$MeCHO + HOO^{\bullet}$	295	52	[42]
$CH(OO^{\bullet})(OH)CH_2OH$	$CH(O)CH_2OH + HOO^{\bullet}$	295	1.9×10^2	[43]
$Me_2C(OO^{\bullet})OH$	$Me_2CO + HOO^{\bullet}$	295	6.7×10^2	[42]
1-hydroxycyclohexyl-1-peroxyl	$cyclo\text{-}C_6H_{10}O + HOO^{\bullet}$	328	4.5	[41]
1-hydroxycyclohexyl-1-peroxyl	$cyclo\text{-}C_6H_{10}O + HOO^{\bullet}$	357	16.0	[41]
1-hydroxycyclohexyl-1-peroxyl	$cyclo\text{-}C_6H_{10}O + HOO^{\bullet}$	387	100	[41]
1-hydroxycyclododecyl-1-peroxyl	$cyclo\text{-}C_{12}H_{22}O + HOO^{\bullet}$	387	5.15×10^4	[41]
$CH(OO^{\bullet})(OH)(CHOH)_2CH_2OH$	$CH(O)(CHOH)_2CH_2OH + HOO^{\bullet}$	295	1.9×10^2	[43]
$CH(OO^{\bullet})(OH)(CHOH)_3CH_2OH$	$CH(O)(CHOH)_3CH_2OH + HOO^{\bullet}$	295	2.2×10^2	[43]
$CH(OO^{\bullet})(OH)(CHOH)_4CH_2OH$	$CH(O)(CHOH)_4CH_2OH + HOO^{\bullet}$	295	2.1×10^2	[43]
$(CO_2^-)_2C(OH)O_2^{\bullet}$	$(CO_2^-)_2C(O) + HOO^{\bullet}$	293	1.1×10^4	[44]

addition of hydrogen peroxide should influence the kinetics of the reaction by displacing the equilibrium. In fact, the addition of H_2O_2 is reflected in the kinetics of the inhibited oxidation of cyclohexanol [48]. In the presence of bases (Na_2HPO_4), however, no difference is observed in the initiated oxidation of cyclohexanol with or without the addition of H_2O_2 [49]. Consequently, the decomposition of alkylhydroxyperoxyl radicals can be accelerated by acids and bases and may be changed during oxidation. Thus, depending on the conditions of the oxidation of alcohol, the chain develops through either hydroxyperoxide or hydroperoxide radicals.

Another factor complicating the situation in composition of peroxyl radicals propagating chain oxidation of alcohol is the production of carbonyl compounds due to alcohol oxidation. As a result of alcohol oxidation, ketones are formed from the secondary alcohol oxidation and aldehydes from the primary alcohols [8,9]. Hydroperoxide radicals are added to carbonyl compounds with the formation of alkylhydroxyperoxyl radical. This addition is reversible.

$$HO_2^{\bullet} + R^1R^2C(O) \rightleftharpoons R^1R^2C(OO^{\bullet})OH$$

Oxidation of Alcohols and Ethers

According to the Benson estimations [46], this equilibrium in the case of acetone has the following thermodynamic functions: $\Delta H = -51.9 \text{ kJ mol}^{-1}$, $\Delta S = -161.1 \text{ J mol}^{-1} \text{ K}^{-1}$, and $\Delta G = -3.8 \text{ kJ mol}^{-1}$ (350 K). The calculated ratios of the equilibrium concentrations of two kinds of peroxyl radicals at $T = 350$ K and different acetone concentrations are presented below.

[Me$_2$CO] (mol L^{-1})	0.0	0.2	0.5	1.0	2.0
[HR(OH)O$_2^\bullet$]/[HO$_2^\bullet$]	0.023	0.042	0.105	0.21	0.42

Beginning from an acetone concentration of 1 mol L^{-1}, the influence of this equilibrium on the peroxyl radicals composition becomes important.

7.1.2.4 Chain Termination in Oxidized Alcohols

In addition to two peroxyl radicals, $HO_2^\bullet$ and $R^1R^2C(OH)OO^\bullet$, participating in chain propagation in the oxidized alcohols, there are three reactions that are guilty of chain termination in the oxidized alcohols. The most probable reaction between them is disproportionation.

$$HO_2^\bullet + HO_2^\bullet \longrightarrow HOOH + O_2$$
$$HO_2^\bullet + R^1R^2C(OO^\bullet)OH \longrightarrow HOOH + R^1R^2C(O) + O_2$$
$$R^1R^2C(OO^\bullet)OH + R^1R^2C(OO^\bullet)OH \longrightarrow R^1R^2C(OOH)OH + O_2 + R^1R^2C(O)$$

The rate constant k_t measured experimentally at high dioxygen pressures characterizes the rates of of all the three reactions. The rate constants $2k_t$ are close to diffusion-controlled reactions (see Table 7.4).

In addition to disproportionation, another mechanism of the alkylhyroxyperoxyl radical chain termination is recombination [38,39].

$$HOO^\bullet + HOO^\bullet \longrightarrow HOOOOH$$
$$R^1R^2C(OO^\bullet)OH + HOO^\bullet \longrightarrow R^1R^2C(OOOOH)OH$$

The reaction between two peroxyl radicals gives a labile compound, which Stockhausen and coworkers [38,39] regard as a hydrotetroxide. The formed tetroxides are very unstable and their decomposition results in the formation of active free radicals.

$$HOOOOH \longrightarrow HO^\bullet + HO_3^\bullet$$
$$R^1R^2C(OOOOH)OH \longrightarrow HO^\bullet + R^1R^2C(OOO^\bullet)OH$$
$$HO_3^\bullet \longrightarrow HO^\bullet + O_2$$
$$R^1R^2C(OOO^\bullet)OH \longrightarrow R^1R^2C(O^\bullet)OH + O_2$$

So, these reactions cannot lead to effective chain termination in oxidized alcohol. The decomposition of tetroxides depends on pH and apparently proceeds homolytically as well as heterolytically in an aqueous solution. The values of the rate constants (s^{-1}) of tetroxide decomposition at room temperature in water at different pH values are given below [38,39].

Alcohol	CH_3OH	C_2H_5OH	Me_2CHOH
pH = 4	13	22	140
pH = 10	4.6	5.3	5.3

The decomposition of hydroxyhydrotetroxide is slower in an alkaline medium than in an acidic medium.

7.1.3 CO-OXIDATION OF ALCOHOLS AND HYDROCARBONS

The theory of chain co-oxidation of binary mixtures of organic compounds was described in Chapter 5. The experimental study of co-oxidation of alcohols (HR_iOH) and hydrocarbon R^1H opens the way to measure the rate constants of one chosen peroxyl radical $R^1OO^\bullet$ with several alcohols HR_iOH and on the reverse, the chosen alcohol HR^1OH with several peroxyl radicals $R_iOO^\bullet$. The parameters of co-oxidation of alcohols and hydrocarbons are collected in Table 7.6. The absolute values of peroxyl radical reactions with alcohols were calculated from these data using the values of k_p from Table 2.8 (see Table 7.7).

What is the main factor that influences the reactivity of alcohols in reactions with one peroxyl radical? Let us compare the rate constants of reaction of $R^1O_2^\bullet + HR_iOH$ (R^1H = cyclohexene, T = 333 K) with the strength of the α-C—H bond.

Alcohol	CH_3OH	$MeCH_2OH$	Me_2CHOH	$PhCH_2OH$
D_{C-H} (kJ mol^{-1})	411	400	390	367
k_H (L mol^{-1} s^{-1})	0.35	0.95	1.00	2.80

It is seen that the weaker the attacked α-C—H bond the higher the partial rate constant. Therefore, the strength of the bond of alcohol attacked by the peroxyl radical is a very important factor of alcohol reactivity. It is seen from the data in Table 7.7 that peroxyl radicals of different structures react with the same alcohol with different rate constants. The comparison of the rate constants of reactions $R_iOO^\bullet + HR^1OH$ with the strength of the formed O—H bond in hydroperoxide shows the higher the BDE of the O—H bond in R_iOOH the more rapidly peroxyl radical $R_iOO^\bullet$ reacts with cyclohexanol (T = 348 K [52]).

Radical	$Me_2PhCOO^\bullet$	cyclohexyl-OO$^\bullet$	$CHCl_2CCl_2OO^\bullet$	$CCl_3CCl_2OO^\bullet$
D_{O-H} (kJ mol^{-1})	358.6	365.5	411.6	413.1
k_{p12} (L mol^{-1} s^{-1})	0.68	5.25	1000	2600

Chain propagation in oxidized alcohols proceeds with the participation of a mixture of $HO_2^\bullet$ and $R(OH)O_2^\bullet$ radicals. Using the data of Table 7.4 and Table 7.7 we can compare the activity of the $HO_2^\bullet/HORO_2^\bullet$ mixture and alkylperoxyl radicals toward the same alcohol ($R^1OO^\bullet$ is the peroxyl radical of cyclohexene, T = 333 K).

TABLE 7.6
The Parameters r_1, r_2 and ϕ of Hydrocarbon Co-Oxidation with Alcohols

R¹H	R²H	T (K)	r_1	r_2	ϕ	Ref.
cyclohexene	CH_3OH	313	10.0			[50]
cyclohexene	CH_3CH_2OH	333	4.4			[50]
cyclohexene	$PrCH_2OH$	333	6.8			[50]
cyclohexene	Me_2CHOH	333	4.0			[50]
$Me(CH_2)_{14}Me$	$Me(CH_2)_8CH_2OH$	403	2.8			[17]
$Me(CH_2)_{14}Me$	$Me(CH_2)_5CH(OH)Me$	403	4.5			[17]
cyclohexene	cyclohexanol	333	3.2			[50]
Me_2PhCH	cyclohexanol	333	0.2	0.4	–	[51]
$Me(CH_2)_{14}Me$	$MeCH(OH)CH_2CH_2OH$	403	2.2			[17]
$Me(CH_2)_{14}Me$	$EtC(CH_2OH)_3$	403	12.0			[17]
cyclohexene	$PhCH_2OH$	333	1.4			[50]
$MePhCH_2$	$PhCH_2OH$	348	0.8	2.3	0.7	[26]
Ph_2CH_2	$PhCH_2OH$	348	0.26	8.0	0.7	[26]
$PhCH_2CH_2Ph$	$PhCH_2OH$	348	0.27	5.7	1.6	[26]
Me_2PhCH	$PhCH_2OH$	348	0.50	12	13	[26]
Me_2PhCH	2-Cl-C$_6$H$_4$CH$_2$OH	348	0.97	2.2	6.5	[26]
tetralin	$MePhCHOH$	333	0.25	0.59	1.45	[51]
$MePhCHOH$	$PrCH_2OH$	333	1.1	0.34	1.44	[51]
$MePhCHOH$	cyclohexanol	333	1.8	0.43	0.7	[51]

Alcohol	CH_3OH	$MeCH_2OH$	Me_2CHOH	$PhCH_2OH$
$k(R^1OO^\bullet)$ (L mol^{-1} s^{-1})	0.35	0.95	1.00	2.80
$k(HOO^\bullet/HOROO^\bullet)$ (L mol^{-1} s^{-1})	3.75	19.3	13.36	3.16

TABLE 7.7
The Rate Constants of Peroxyl Radical Reaction with Alcohols

Alcohol	Peroxyl Radical	T (K)	k (L mol^{-1} s^{-1})	Ref.
CH_3OH	cyclohexenyl-OO•	313	$3.01 \times 10^8 \exp(-53.9/RT)$	[50]
$MeCH_2OH$	cyclohexenyl-OO•	333	$2.00 \times 10^8 \exp(-51.1/RT)$	[50]
Me_2CHOH	cyclohexenyl-OO•	333	$1.00 \times 10^8 \exp(-49.1/RT)$	[50]
$Me(CH_2)_2CH_2OH$	cyclohexenyl-OO•	333	$2.00 \times 10^8 \exp(-52.4/RT)$	[50]
$MeCH(OH)CH_2CH_2OH$	$Me(CH_2)_{13}CH(OO•)Me$	403	$1.00 \times 10^8 \exp(-47.8/RT)$	[17]
$EtC(CH_2OH)_3$	$Me(CH_2)_{13}CH(OO•)Me$	403	$6.00 \times 10^8 \exp(-48.2/RT)$	[17]
$PhCH_2OH$	cyclohexenyl-OO•	333	$2.00 \times 10^7 \exp(-41.8/RT)$	[50]
$PhCH_2OH$	MePhCHOO•	348	$2.00 \times 10^7 \exp(-44.4/RT)$	[26]
$PhCH_2OH$	$Ph_2CHOO•$	348	$2.00 \times 10^7 \exp(-46.1/RT)$	[26]
$PhCH_2OH$	$Me_2PhCOO•$	348	$2.00 \times 10^7 \exp(-47.7/RT)$	[26]
MePhCHOH	tetralinyl-OO•	333	$1.00 \times 10^7 \exp(-41.2/RT)$	[51]
cyclohexanol (H,OH)	cyclohexenyl-OO•	333	$1.00 \times 10^8 \exp(-48.5/RT)$	[50]
cyclohexanol (H,OH)	$Me_2PhCOO•$	333	$1.00 \times 10^8 \exp(-54.4/RT)$	[51]
cyclohexanol (H,OH)	$CCl_3CCl_2OO•$	348	$1.00 \times 10^8 \exp(-30.7/RT)$	[52]
cyclohexanol (H,OH)	$CHCl_2CCl_2OO•$	348	$1.00 \times 10^8 \exp(-33.3/RT)$	[52]
4-Me-C$_6$H$_4$-CH$_2$OH	$Me_2PhCOO•$	348	30.8	[26]
4-MeO-C$_6$H$_4$-CH$_2$OH	$Me_2PhCOO•$	348	25.9	[26]
4-Br-C$_6$H$_4$-CH$_2$OH	$Me_2PhCOO•$	348	21.6	[26]
2-Cl-C$_6$H$_4$-CH$_2$OH	$Me_2PhCOO•$	348	8.90	[26]
4-O_2N-C$_6$H$_4$-CH$_2$OH	$Me_2PhCOO•$	348	18.8	[26]

We observe the sufficiently higher activity of $HO_2•/HORO_2•$ radicals in their reactions with alcohols than that of $R^1OO•$. The next section will be devoted to detailed analysis of alcohol reactivity in reactions with peroxyl radicals.

7.1.4 Reactivity of Alcohols in Reaction with Peroxyl Radicals

As described earlier, the reaction enthalpy is a very important factor that influences the reactivity of alcohol in free radical abstraction reactions. The IPM model helps to estimate the increment of activation energy ΔE_H which characterizes the influence of enthalpy on the activation energy (see Equation [6.20] in Chapter 6). The parameters br_e and values ΔE_H for reactions of different peroxyl radicals with alcohols are given in Table 7.8. The mean value

TABLE 7.8
Kinetic Parameters of Peroxyl Radicals Reactions with Alcohols Calculated by the IPM

Reaction	T (K)	k (333 K) (L mol^{-1} s^{-1})	E (kJ mol^{-1})	br_e (kJ mol^{-1})$^{1/2}$	ΔE_μ (kJ mol^{-1})	Ref.
MeOH + HO$_2^\bullet$	333	3.75	50.35	12.47	−9.15	[20,21]
MeOH + cyclohexenyl-OO$^\bullet$	333	1.05	53.87	12.68	−7.50	[22]
EtOH + HO$_2^\bullet$	333	19.3	44.69	12.52	−8.76	[3,21]
EtOH + cyclohexenyl-OO$^\bullet$	333	1.93	51.07	13.11	−4.11	[22]
Me$_2$CHOH + HO$_2^\bullet$	333	13.36	43.80	13.02	−4.89	[23,24]
Me$_2$CHOH + cyclohexenyl-OO$^\bullet$	333	1.99	49.06	13.46	−1.32	[22]
cyclohexyl-O-H + HO$_2^\bullet$	333	0.97	51.05	14.03	3.44	[25]
cyclohexyl-O-H + HO$_2^\bullet$	333	1.07	50.78	14.00	3.16	[24]
cyclohexyl-O-H + cyclohexenyl-OO$^\bullet$	333	2.47	48.47	13.51	−0.88	[22]
cyclohexyl-O-H + CCl$_3$CCl$_2$OO$^\bullet$	348	2500	30.64	13.60	−0.19	[26]
cyclohexyl-O-H + CHCl$_2$CCl$_2$OO$^\bullet$	348	1000	33.29	13.85	1.91	[26]
cyclohexyl-O-H + MePhCHOO$^\bullet$	333	0.29	54.39	13.82	1.62	[26]
CH$_2$OHCH$_2$OH + MeCH(OO$^\bullet$)(CH$_2$)$_{13}$Me	403	28.1	55.15	13.48	−1.14	[17]
CH$_2$OH(CH$_2$)$_2$CH$_2$OH + MeCH(OO$^\bullet$)(CH$_2$)$_{13}$Me	403	30.4	54.89	13.27	−2.88	[17]
CH$_2$OHCH$_2$CHOHMe + MeCH(OO$^\bullet$)(CH$_2$)$_{13}$Me	403	36.0	49.68	13.67	0.40	[17]
Me$_2$C(CH$_2$OH)$_2$ + MeCH(OO$^\bullet$)(CH$_2$)$_{13}$Me	403	964	43.31	11.68	−14.89	[17]
EtC(CH$_2$OH)$_3$ + MeCH(OO$^\bullet$)(CH$_2$)$_{13}$Me	403	296	48.63	12.44	−9.35	[17]
Me(CHOH)$_2$OPr + MeCH(OO$^\bullet$)(CH$_2$)$_{13}$Me	403	56	50.52	14.03	3.48	[17]

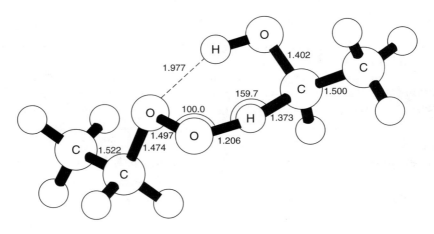

FIGURE 7.1 Geometry of transition state of abstraction reaction $EtO_2^\bullet + MeCH_2OH$ calculated by DFT method [54].

of the $\Delta E_H/\Delta H$ ratio is 0.52 ± 0.13. Along with reaction enthalpy, the activation energy of the reaction of peroxyl radical with the C—H bond of alcohol depends on the force constants of the reacting bonds (parameter $b_{C-H} = 3.473 \times 10^{11}\,kJ^{1/2}\,mol^{-1/2}\,m^{-1}$ and $b_{O-H} = 4.600 \times 10^{11}\,kJ^{1/2}\,mol^{-1/2}\,m^{-1}$), triplet repulsion in the TS, difference in the electron affinity of C and O atoms in the TS, radii of C and O atoms, and π-C—C bonds in the vicinity of the reaction center (see Chapter 6).

In addition to these factors, a new one, namely the interaction of the polar O—H group with the polar C $\cdots$ H $\cdots$ O—O reaction center of the TS, appears to be important [53]. The contribution of polar interaction ΔE_μ to the activation energy can be estimated by IPM in Equation (6.32) (Chapter 6) or as difference: $\Delta E_\mu = E_{e0}(R^1O_2^\bullet + HROH) - E_{e0}(R^1O_2^\bullet + R^1H)$. Polar interaction in the reaction of peroxyl radicals with alcohols change the activation energy value of thermoneutral reaction (see Table 7.8). The values of ΔE_μ vary from -15 to $+3\,kJ\,mol^{-1}$ and most of them are negative. This means that the polar interaction, as a rule, decreases the energy of the TS.

The quantum-chemical calculation of the activation energies of reactions: $EtOO^\bullet + EtH$ and $EtOO^\bullet + EtOH$ by the DFT method supports this result [54]. For the reaction of $EtOO^\bullet$ with ethane, the calculated E value is $82\,kJ\,mol^{-1}$, $\Delta H = 76\,kJ\,mol^{-1}$, and $E_{e0} = 56\,kJ\,mol^{-1}$. For the reaction of $EtOO^\bullet$ with ethanol $E = 52.7$. The structure of TS is presented in Figure 7.1.

The polar interaction of the hydroxyl and the peroxyl groups in the TS changes the geometry of the reaction center. This center has virtually a linear configuration in the reaction $EtOO^\bullet + EtH$ [54]: atoms C $\cdots$ H $\cdots$ O are situated practically on the straight line ($\varphi(CHO) = 176°$). The polar interaction of the O—H and O—O bonds in the TS of the reaction $EtOO^\bullet + EtH$ transforms the geometry of the reaction center decreasing $\varphi(CHO)$ from $176°$ to $160°$ [54]. The IPM method gives simple equations for the calculation of angle $\varphi(OCHO)$ for such reactions using experimental values of the activation energy and the enthalpy of reaction (Equations [6.35]–[6.38] in Chapter 6). The results of calculation of the $r(C-H)$ and $r(O-H)$ distances and $\varphi(CHO)$ angle in the reaction center for reactions $R^1OO^\bullet + HROH$ and $HOO^\bullet + HROH$ are presented in Table 7.9.

TABLE 7.9
Kinetic and Geometric Parameters of TS of Peroxyl Radicals Reactions with Alcohols Calculated by the IPM Method (Equations [6.35–6.38]) [54]

Reaction	ΔH_e (kJ mol^{-1})	E_e (kJ mol^{-1})	$r(C-H) \times 10^{10}$ (m)	$r(O-H) \times 10^{10}$ (m)	$r(C-O) \times 10^{10}$ (m)	$\varphi°$
		$A_{C-H} = 10^8$ (L mol^{-1} s^{-1})				
MeOH + HO$_2^\bullet$	38.2	66.4	1.404	1.180	2.540	158
MeOH + cyclohexyl-OO$^\bullet$	41.7	69.9	1.412	1.172	2.548	161
EtOH + HO$_2^\bullet$	27.0	60.7	1.390	1.194	2.542	159
EtOH + cyclohexyl-OO$^\bullet$	30.5	67.1	1.405	1.179	2.565	166
Me$_2$CHOH + HO$_2^\bullet$	17.7	59.8	1.388	1.196	2.561	165
Me$_2$CHOH + cyclohexyl-OO$^\bullet$	21.2	65.1	1.401	1.183	2.578	172
cyclohexyl-O-H + HO$_2^\bullet$	15.6	67.1	1.405	1.195	2.600	180.0
cyclohexyl-O-H + HO$_2^\bullet$	15.6	66.8	1.405	1.194	2.599	180.0
cyclohexyl-O-H + cyclohexyl-OO$^\bullet$	19.1	64.5	1.399	1.185	2.580	174
cyclohexyl-O-H + CCl$_3$CCl$_2$OO$^\bullet$	−22.6	46.6	1.353	1.231	2.583	177
cyclohexyl-O-H + CHCl$_2$CCl$_2$OO$^\bullet$	−21.2	49.2	1.360	1.233	2.593	180
cyclohexyl-O-H + Me$_2$PhCHOO$^\bullet$	26.0	70.4	1.413	1.179	2.592	180
CH$_2$OHCH$_2$OH + MeCH(OO$^\bullet$)(CH$_2$)$_{13}$Me	32.2	70.9	1.414	1.170	2.579	173
CH$_2$OH(CH$_2$)$_2$CH$_2$OH + MeCH(OO$^\bullet$)(CH$_2$)$_{13}$Me	34.9	70.6	1.413	1.171	2.570	168
CH$_2$OHCH$_2$CHOHMe + MeCH(OO$^\bullet$)(CH$_2$)$_{13}$Me	18.4	65.4	1.401	1.185	2.586	180
Me$_2$C(CH$_2$OH)$_2$ + MeCH(OO$^\bullet$)(CH$_2$)$_{13}$Me	34.9	59.0	1.386	1.198	2.509	152
EtC(CH$_2$OH)$_3$ + MeCH(OO$^\bullet$)(CH$_2$)$_{13}$Me	34.9	64.3	1.399	1.185	2.538	158
Me(CHOH)$_2$OPr + MeCH(OO$^\bullet$)(CH$_2$)$_{13}$Me	13.8	66.2	1.403	1.197	2.600	180

continued

TABLE 7.9
Kinetic and Geometric Parameters of TS of Peroxyl Radicals Reactions with Alcohols Calculated by the IPM Method (Equation [6.35–6.38]) [54]—continued

Reaction	ΔH_e (kJ mol^{-1})	E_e (kJ mol^{-1})	r(C—H) × 10^{10}(m)	r(O—H) × 10^{10}(m)	r(C—O) × 10^{10}(m)	φ°
$A_{C-H} = 10^8$ (L mol^{-1} s^{-1})						
[furanose diol structure] + MeCH(OO$^\bullet$)(CH$_2$)$_{13}$Me	13.8	57.9	1.383	1.178	2.561	164
$A_{C-H} = 10^7$ (L mol^{-1} s^{-1})						
PhCH$_2$OH + HO$_2^\bullet$	−5.9	59.3	1.386	1.225	2.609	174
PhCH$_2$OH + cyclohexenylperoxyl	−2.4	57.7	1.382	1.229	2.595	167
PhCH$_2$OH + Me$_2$PhCOO$^\bullet$	4.5	63.7	1.397	1.214	2.608	173
PhMeCHOH + HO$_2^\bullet$	−16.8	57.4	1.382	1.239	2.621	180
PhMeCHOH + tetralylperoxyl	−13.3	57.2	1.381	1.233	2.614	180
PhCH$_2$OH + MePhCHOO$^\bullet$	−2.4	60.4	1.389	1.222	2.607	172
PhCH$_2$OH + Ph$_2$CHOO$^\bullet$	−2.4	62.1	1.393	1.221	2.615	180

A good linear dependence between $\cos[180° − \varphi(CHO)]$ and ΔE_μ is observed (Figure 7.2). It has the following analytical expression:

$$\cos[180° − \varphi(CHO)] = 1 + 8.14 \times 10^{-3} \times \Delta E_\mu \quad (7.2)$$

An additional polar interaction called *multidipole interaction* is observed in reactions of peroxyl radicals with polyatomic alcohols [55]. A few polar O—H groups interact with the polar reaction center C···H···O in such systems. A few examples of such interaction are given here [17]. Multidipole interaction sufficiently changes the thermoneutral activation energy of the reaction HOO$^\bullet$ + alcohol [54] and can be characterized by increment $\Delta\Delta E_\mu$.

Alcohol	E_{e0} (kJ mol^{-1})	Alcohol	E_{e0} (kJ mol^{-1})	$\Delta\Delta E_\mu$ (kJ mol^{-1})
HOCH$_2$CH$_2$OH	55.2	MeCH$_2$OH	47.6	7.6
HOCH$_2$(CH$_2$)$_2$CH$_2$OH	53.5	MeCH$_2$OH	47.6	5.9
MeCHOHCH$_2$CH$_2$OH	56.8	Me$_2$CHOH	51.4	5.4
Me$_2$C(CH$_2$OH)$_2$	41.4	MeCH$_2$OH	47.6	−6.2
EtC(CH$_2$OH)$_3$	47.0	MeCH$_2$OH	47.6	−0.6
MeC(CH$_2$OH)$_2$OPr	59.8	MeCH$_2$OH	47.6	12.2

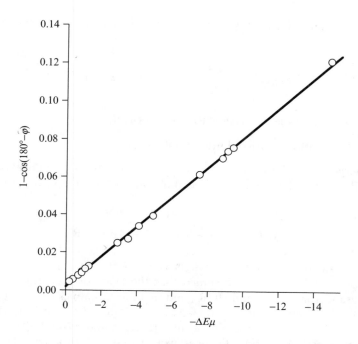

FIGURE 7.2 The dependence of $1-\cos(180-\varphi)$ on ΔE_μ for reactions of peroxyl radicals with alcohols.

Polar solvents have no effect on the rate constant of the reaction $RO_2^\bullet + RH$ [56]. This means that the solvation energies of the peroxyl radical $RO_2^\bullet$ and TS $RO_2 \cdots HR$ are very close. A different situation was observed for the reaction of cumylperoxyl radical with benzyl alcohol (see Table 7.10). The rate constant of this reaction is twice in polar dimethylsulfoxide ($\varepsilon = 33.6$) than that in cumene ($\varepsilon = 2.25$). It was observed that the very important property of the solvent is basicity (B), that is, affinity to proton. A linear correlation

TABLE 7.10
Comparison of Solvent in Fluence on Rate Constants and Activation Energy of Cumylperoxyl Radical with Cumene and Benzyl Alcohol [56]

Solvent	ε	k (348 K) (L mol^{-1} s^{-1})	E (kJ mol^{-1})	k (348 K) (L mol^{-1} s^{-1})	E (kJ mol^{-1})
		Me$_2$PhCH		PhCH$_2$OH	
Cumene	2.25	1.1	36.5	11.4	13.5
Chlorobenzene	3.48	1.1	27.8	10.4	16.1
Bromobenzene	3.44	1.2	32.3	10.3	10.9
Acetophenone	10.93	1.2	41.3	10.4	9.3
Benzonitrile	6.71	1.4	36.6	11.2	19.4
Dimethylsulfoxide	33.6	1.4	33.8	21.6	13.5
Pyridine	5.8	1.0	34.7	25.8	19.1

between the rate constant of the cumylperoxyl reaction with benzyl alcohol and basicity B was found [56].

$$\ln k = 6.54 - \frac{12450}{RT} + 7.71 \times 10^{-3} \exp(449/T)B \qquad (7.3)$$

7.1.5 CHAIN GENERATION IN AUTOXIDATION OF ALCOHOLS

7.1.5.1 Chain Generation by Reaction with Dioxygen

In alcohol undergoing oxidation in the absence of initiators, free radicals are formed by bimolecular and trimolecular reactions of alcohol with dioxygen [8,9]:

$$\text{HROH} + \text{O}_2 \longrightarrow \text{HOR}^\bullet + \text{HOO}^\bullet$$
$$2\text{HROH} + \text{O}_2 \longrightarrow \text{HOR}^\bullet + \text{H}_2\text{O}_2 + \text{R}^\bullet$$

The enthalpy of the bimolecular reaction is $\Delta H = D(\text{H}-\text{ROH}) - 220 - \Delta\Delta H$ (solvation reactants and products). It may be assumed that the heats of dissolution in the alcoholic medium are the same for HROH and HOR$^\bullet$, and that the enthalpies of dissolution are -34 kJ mol^{-1} for HO$_2^\bullet$, -16 kJ mol^{-1} for O$_2$, and -54 kJ mol^{-1} for H$_2$O$_2$ (as in H$_2$O). Thus $\Delta H = D(\text{H}-\text{ROH}) - 238$ kJ mol^{-1} for the bimolecular reaction and $\Delta H = 2D(\text{H}-\text{ROH}) - 608$ kJ mol^{-1} for the trimolecular reaction. One can expect that TS of the trimolecular reaction is more polar and more solvated than that of the bimolecular reaction. So, the trimolecular reaction can predominate in the oxidation of many alcohols.

The mechanism of chain initiation in cyclohexanol by the reaction with dioxygen was studied by the inhibitor method [57]. It was established that

$$v_{i0} = k_i [\text{HROH}]^2 [\text{O}_2] \qquad (7.4)$$

that is, chain initiation takes place by the trimolecular reaction with the rate constant $k = 8.3 \exp(-66.9/RT)$ L^2 mol^{-2} s^{-1}. The pre-exponential factor seems to be very low (10 compared with 10^3–10^5 for tetralin, see Chapter 2). If we suppose $E = \Delta H = 170$ kJ mol^{-1} so we obtain the following expression for the trimolecular rate constant: $k_i = 2.4 \times 10^{14} \exp(-170/RT)$ L^2 mol^{-2} s^{-1}. It is very probable that the trimolecular reaction proceeds through the preliminary association of two alcohol molecules via hydrogen bond.

7.1.5.2 Decomposition of Hydrogen Peroxide into Free Radicals

In the absence of an initiator, alcohols are oxidized with self-acceleration [7–9]. As in the oxidation of hydrocarbons, the increase in the reaction rate is due to the formation of peroxides initiating the chains. The kinetics of radical formation from peroxides was studied for the oxidation of isopropyl alcohol [58] and cyclohexanol [59,60].

In the oxidation of isopropyl alcohol, the kinetics of formation of hydrogen peroxide for [H$_2$O$_2$] = 0.7 mol L^{-1} is described by the equation [58]

$$\frac{\sqrt{[\text{H}_2\text{O}_2]}}{[\text{HROH}]} = \frac{\sqrt{[\text{H}_2\text{O}_2]_0}}{[\text{HROH}]_0} + \frac{k_p \sqrt{k_i}}{\sqrt{2k_t}} \times t \qquad (7.5)$$

This equation agrees with the chain mechanism for the oxidation of the 2-propanol, under the condition that H$_2$O$_2$ is the only source of radicals and the rate of initiation is $v_i = k_i[\text{H}_2\text{O}_2]$.

Oxidation of Alcohols and Ethers

The experimental results were used to estimate the value of effective unimolecular rate constant k_i [58]. The latter was found to be equal to $k_i = 1.1 \times 10^7 \exp(-96.2/RT)$ s^{-1}.

The formation of radicals from hydrogen peroxide in cyclohexanol was measured by the free radical acceptor method [60]; the effective rate constant of initiation was found to be equal to $k_i = 9.0 \times 10^6 \exp(-90.3/RT)$ s^{-1}. For the first-order decomposition of H_2O_2 in an alcohol medium, the following reactions were discussed.

Reaction	ΔH (kJ mol^{-1})
HOOH $\longrightarrow$ HO$^\bullet$ + $^\bullet$OH	215
HOOH + H—OCHMe$_2$ $\longrightarrow$ HO$^\bullet$ + H$_2$O + Me$_2$CHO$^\bullet$	150
HOOH + H—C(OH)Me$_2$ $\longrightarrow$ HO$^\bullet$ + H$_2$O + Me$_2$C$^\bullet$OH	107

Since for an endothermic reaction the activation energy $E \geq \Delta H$, all such reactions cannot explain the experimental value of the activation energy (see Chapter 4). The following mechanism seems to be the most probable now. Hydrogen peroxide is protonized in a polar alcohol solution. Protonization of H_2O_2 intensifies its oxidizing reactivity. Protonized hydrogen peroxide reacts with alcohol with free radical formation.

$$ROH + ROH \rightleftharpoons RO^- + ROH_2^+$$
$$HOOH + ROH_2^+ \rightleftharpoons HOOH_2^+ + ROH$$
$$HOOH_2^+ + ROH \longrightarrow HO^\bullet + H_3O^+ + RO^\bullet$$

The phenomenon of acid catalysis of peroxide homolytic splitting will be discussed in Chapter 10.

In addition to hydroperoxide decomposition by the reaction of the first-order bimolecular decomposition was observed in cyclohexanol at $[H_2O_2] > 1\,M$ [60]. The bimolecular radical generation occurs with the rate constant $k_i = 6.8 \times 10^8 \exp(-121.7/RT)$ L mol^{-1} s^{-1}. The following mechanism was suggested as the most probable.

$$HOOH + HOOH \rightleftharpoons HOO(H) \cdots HOOH$$
$$HOO(H) \cdots HOOH \longrightarrow HO^\bullet + H_2O + HOO^\bullet$$

The enthalpy of this reaction ($\Delta H = 86$ kJ mol^{-1}) is in good agreement with the experimental activation energy ($E = 122$ kJ mol^{-1}).

7.1.5.3 Chain Generation by Reaction of Hydrogen Peroxide with Carbonyl Compound

Oxidation of alcohol produces hydrogen peroxide and carbonyl compound simultaneously (aldehyde and ketone from primary and secondary alcohols, respectively). Carbonyl compound is formed as a result of alkylhydroxyperoxyl radical decomposition

$$R^1R^2C(OO^\bullet)OH \longrightarrow R^1R^2C(O) + HO_2^\bullet$$

and from hydroxyhydroperoxide via reversible decomposition

$$R^1R^2C(OH)OOH \rightleftharpoons R^1R^2C(O) + HOOH$$

During the early period of oxidation, this equilibrium is shifted to the right due to the very low concentration of carbonyl compound. The concentration of carbonyl compound is increasing during oxidation and in parallel the concentration of hydroxyperoxide increases. The thermodynamic parameters of this equilibrium are the following for formaldehyde and acetone in the gas phase [46].

ΔH (kJ mol^{-1})	ΔS (J mol^{-1} K)	ΔG (350 K) (kJ mol^{-1})	K (300 K) (L mol^{-1})	K (350 K) (L mol^{-1})
		$H_2C(O) + HOOH \rightleftharpoons H_2C(OH)OOH$		
-67.9	-133.4	-21.2	7.15×10^4	1.46×10^3
		$Me_2C(O) + HOOH \rightleftharpoons Me_2C(OH)OOH$		
-53.6	-157.8	1.6	12.3	0.57

One can see that the equilibrium is shifted to the right in the case of formaldehyde and it is nearly 50/50 for acetone (350 K).

Ketones play an important role in the decomposition of peroxides to form radicals in alcohols undergoing oxidation. The formed hydroxyhydroperoxide decomposes to form radicals more rapidly than hydrogen peroxide. With an increase in the ketone concentration, there is an increase in the proportion of peroxide in the form of hydroxyhydroperoxide, with the corresponding increase in the rate of formation of radicals. This was proved by the acceptor radical method in the cyclohexanol–cyclohexanone–hydrogen peroxide system [59]. The equilibrium constant was found to be $K = 0.10$ L mol^{-1} (373 K), 0.11 L mol^{-1} (383 K), and 0.12 L mol^{-1} (393 K). The rate constant of free radical generation results in the formation of cyclohexylhydroxy hydroperoxide decomposition and was found to be $k_i = 2.2 \times 10^4 \exp(-67.8/RT)$ s^{-1} [59].

In isopropyl alcohol hydrogen peroxide and acetone evidently form the hydroxyhydroperoxide, which decomposes rapidly to form radicals. The rate of initiation is

$$v_{i0} = k_i K [H_2O_2] [Me_2C(O)], \tag{7.6}$$

and $k_i K = 1.2 \times 10^{-6}$ L mol^{-1} s^{-1} at 391 K [58]. The decomposition of peroxides ArCH(OH)OOH occurs with the following rate constants [61].

ArCH(O)	T (K)	E (kJ mol^{-1})	log A, A (s^{-1})	k_1 (s^{-1}) (333K)
PhCH(O)	333–348	122.5	14.06	7.00×10^{-6}
4-F-C$_6$H$_4$-CH(O)	323–343	129.2	15.43	1.46×10^{-5}
2,3,5,6-F$_4$-C$_6$H-CH(O)	353–373	160.0	18.26	1.54×10^{-7}

7.2 OXIDATION OF ETHERS

7.2.1 INTRODUCTION

The oxidation of ethers by dioxygen is very similar to hydrocarbon oxidation. The following three peculiarities of ethers can be mentioned since they influence the mechanism of their oxidation by dioxygen.

1. The BDE of the α-C—H bonds of ethers $R^1R^2CHOCHR^1R^2$ are weaker in comparison with the C—H bonds of the parent hydrocarbons [2]. When α-C—H of ether is cleaved, the formed α-alkoxyalkyl radical is stabilized by the interaction of an unpaired electron with p-electrons of the oxygen atom. Therefore, the attack of the peroxyl radical on the ether molecule occurs more rapidly than on hydrocarbon. Most of the ethers are easily oxidized by dioxygen at moderate temperatures. Table 7.11 contains a list of BDEs of ethers, as well as BDEs of hydrocarbons, and values of enthalpies of peroxyl radical abstraction with ether.
2. A molecule of linear alkyl ether possesses a very convenient geometry for intramolecular hydrogen atom abstraction by the peroxyl radical. Therefore, chain propagation is performed by two ways in oxidized ethers: intermolecular and intramolecular. As a result, two peroxides as primary intermediates are formed from ether due to oxidation, namely, hydroperoxide and dihydroperoxide [62].

3. Ethers are polar compounds. A molecule of ether possesess a dipole moment. μ (Debye) values of a few ethers are given below [6].

MeOMe	Z—MeOEt	EtOEt	PrOPr	[Me$_2$CH]$_2$O	(tetrahydrofuran)	(1,3-dioxane)
1.30	1.17	1.15	1.21	1.13	1.75	2.06

The polar media influences the reaction of peroxyl radicals with ether as on the reaction of two polar reagents [10] and on chain generation by the reaction of ether with dioxygen as reaction with the polar TS.

The mechanism and kinetics of ether oxidation are discussed in monographs [7–10]. The valuable information about the chemistry of ether oxidation is given in Ref. [8]. The photo- and radiation-induced oxidation of ethers are described.

TABLE 7.11
Energies of α-C—H Bonds Dissociation in Ethers and Enthalpies of Secondary Alkylperoxyl Radical Reactions with Ethers [2]

Ether	Bond	D (kJ mol^{-1})	ΔH (kJ mol^{-1})
Methane, oxybis-	MeOCH$_2$—H	411.9	46.4
Ethane, oxybis-	EtOMeCH—H	399.5	34.0
Propane, 2,2'-oxybis-	Me$_2$CHOMe$_2$C—H	390.8	25.3
Butane, 1,1'-oxybis-	BuOPrCH—H	392.2	26.7
Cyclohexane, methoxy-		399.6	34.1
1-Propene, 3,3'-oxybis-	[CH$_2$=CH(CH—H)]$_2$O	360.0	−5.5
1-Propene-2-methyl, 3,3'-oxybis-	[CH$_2$=CMeCH—H]$_2$O	349.9	−15.6
1-Butene-2-methyl, 3,3'-oxybis-	[CH$_2$=CMe(C—H)Me]$_2$O	339.2	−26.3
1-Propene-2-ethyl, 3,3'-oxybis-	[CH$_2$=CEtCH—H]$_2$O	359.7	−5.8
Z-2-Butene, 1,1'-oxybis-	(Z-MeCH=CHCH—H)$_2$O	347.8	−17.7
E-2-Butene, 1,1'-oxybis-	(E-MeCH=CHCH—H)$_2$O	348.8	−16.7
2-Butene-2-methyl, 1,1'-oxybis-	[MeCH=CMeCH—H]$_2$O	344.4	−21.1
1-Propine, 3,3'-oxybis-	[CH≡CCH—H]$_2$O	351.8	−13.7
1-Butine, 3,3'-oxybis-	[CH≡CC—HMe]$_2$O	338.8	−26.7
2-Butine, 4,4'-oxybis-	[MeC≡CCH—H]$_2$O	346.7	−18.8
2-Pentine, 4,4'-oxybis-	[MeC≡CC—HMe]$_2$O	333.8	−31.7
Anisole	PhOCH$_2$—H	385.0	19.5
Methane, phenylmethoxy-	PhCH—HOMe	359.0	−6.5
Benzene, bis-1,1'-[oxybis(methylene)]	[Ph(C—H)H]$_2$O	359.0	−6.5
Methane, diphenylmethoxy-	Ph$_2$(C—H)OMe	354.1	−11.4
1,1-Dimethoxyethane	(MeO)$_2$MeC—H	387.2	21.7
Diethoxymethane	(EtO)$_2$CH—H	390.1	24.6
Dipropoxymethane	(PrO)$_2$CH—H	391.8	26.3
Dibutoxymethane	(BuO)$_2$MeC—H	390.2	24.7
Dimethoxyphenylmethane	PhC—H(OMe)$_2$	353.9	−11.6
Oxirane, 2-methyl-		376.0	10.5
Furane		408.0	42.5
Tetrahydrofurane		391.6	26.1
Tetrahydrofurane, 2-methyl-		384.1	18.6
Tetrahydrofurane, 2,5-dimethyl-		387.5	22.0
1,3-Dioxalane, 2-methyl-		377.1	11.6
1,3-Dioxalane, 4,4-dimethyl-		394.8	29.3
1,3-Dioxalane, 2-methyl-2-propyl-		381.7	16.2

TABLE 7.11
Energies of α-C—H Bonds Dissociation in Ethers and Enthalpies of Secondary Alkylperoxyl Radical Reactions with Ethers [2]—*continued*

Ether	Bond	D (kJ mol^{-1})	ΔH (kJ mol^{-1})
γ-Pyrane, tetrahydro-		401.7	36.2
γ-Pyrane, tetrahydro-2,2-dimethyl-4-oxo-		389.6	24.1
Dioxane		405.4	39.9
1,3-Dioxane		390.8	25.3
1,3-Dioxane, 2,4-dimethyl-		373.0	7.5
1,3-Dioxane, 2-propyl-		377.7	12.2
1,3-Dioxane, 2-propyl-5,5-dimethyl-		365.7	0.2
1,3-Dioxane, 2-methylethyl-4, 4-dimethyl-		382.0	16.5
Spiro[5.5]undecane, 2,5-dioxa-		396.2	30.7
1,3-Dioxepane		393.5	28.0
1,3-Dioxepane, 2-methylethyl-		378.2	12.7

7.2.2 CHAIN OXIDATION OF ETHERS

Like hydrocarbons, ethers are oxidized by dioxygen via the chain mechanism. The mechanism of ether oxidation with initiator I includes the following elementary steps [8,9]:

$$I \longrightarrow r^\bullet$$
$$r^\bullet + R^1OCH_2R \longrightarrow rH + R^1OC^\bullet HR$$
$$R^1OC^\bullet HR + O_2 \longrightarrow R^1OCH(OO^\bullet)R$$
$$R^1OCH(OO^\bullet)R + R^1OCH_2R \longrightarrow R^1OCH(OOH)R + R^1OC^\bullet HR \qquad k_p$$

$$\text{RCH(OO}^{\bullet}\text{)OCH}_2\text{R} \longrightarrow \text{RCH(OOH)OC}^{\bullet}\text{HR} \qquad k_p'$$

$$\text{RCH(OOH)OC}^{\bullet}\text{HR} + \text{O}_2 \longrightarrow \text{RCH(OOH)OCH(OO}^{\bullet}\text{)R}$$

$$\text{RCH(OOH)OCH(OO}^{\bullet}\text{)R} + \text{R}_i\text{H} \longrightarrow \text{RCH(OOH)OCH(OOH)R} + \text{R}_i^{\bullet}$$

$$2\text{RCH(OO}^{\bullet}\text{)OCH}_2\text{R} \longrightarrow \text{RCH(OH)OCH}_2\text{R} + \text{RC(O)OCH}_2\text{R} + \text{O}_2 \qquad k_t$$

The rate of initiated ether oxidation at high dioxygen pressures obeys the equation [8,9]:

$$v = (k_p[\text{RCH}_2\text{OCH}_2\text{R}] + k_p')(2k_t)^{-1/2}[\text{O}_2]^0 v_i^{1/2} \qquad (7.7)$$

The values of rate constant ratio $k_p(2k_t)^{-1/2}$ are collected in Table 7.12, and absolute rate constants k_p and $2k_t$ are given in Table 7.13.

Intramolecular hydrogen atom abstraction occurs rapidly in oxidized ethers. The rate constants of intramolecular hydrogen atom abstraction have the following values.

Reaction	T (K)	$k_p'(\text{s}^{-1})$	Ref.
PhCH(OO$^{\bullet}$)OCH$_2$Ph $\longrightarrow$ Ph(CH(OOH)OC$^{\bullet}$HPh	303	86	[62]
PhCH(OO$^{\bullet}$)OCH$_2$Ph $\longrightarrow$ Ph(CH(OOH)OC$^{\bullet}$HPh	323	110	[69]
PhCH(OOH)OCH(OO$^{\bullet}$)Ph $\longrightarrow$ PhCH(O)OCH(OOH)Ph + HO$^{\bullet}$	323	9.5	[69]

Let us compare the competition of intermolecular and intramolecular chain propagations in oxidized dibenzyl ether. The rate constant $k_p(\text{RO}_2^{\bullet} + \text{RH}) = 95$ L mol^{-1}s^{-1} ($T = 303$ K, Table 7.13), ether concentration [PhCH$_2$OCH$_2$Ph] = 5.26 mol L^{-1}, $k_p'(\text{RO}_2^{\bullet} \rightarrow \text{R}^{\bullet}) = 86$ (s^{-1}), and ratio $k_p[\text{PhCH}_2\text{OCH}_2\text{Ph}]/k_p' = 5.8$. The ratio of the formed dihydroperoxide to monohydroperoxide was estimated as 0.35 for the oxidation of diisopropyl ether ($T = 303$ K) and 0.60 for oxidation of the dibenzyl ether ($T = 303$ K) [8].

Chain generation by the reaction of diacetals of different structures with dioxygen was studied by the method of free radical acceptors. Free radical generation was found to occur with the rate constant [68]

$$v_i = k_{i0}[(\text{RO})_2\text{CHR}^1]^2[\text{O}_2] \qquad (7.8)$$

This shows that the free radical generation proceeds via the trimolecular reaction (see Chapter 2).

$$(\text{RO})_2\text{CHR}^1 + \text{O}_2 + (\text{RO})_2\text{CHR}^1 \longrightarrow (\text{RO})_2\text{C}^{\bullet}\text{R}^1 + \text{H}_2\text{O}_2 + (\text{RO})_2\text{C}^{\bullet}\text{R}^1$$

The preference of the trimolecular reaction over bimolecular RH + O$_2$ is the result of weak C—H bond in diacetals and a polar media (see Chapter 4). The last factor is important for the energy of formation of very polar TS of the trimolecular reaction. The rate constants of trimolecular reactions are presented in Table 7.14.

The autoxidation of ethers occurs with self-acceleration as autoxidation of hydrocarbons. The kinetics of such reactions was discussed earlier (see Chapter 2). The autoacceleration of ether oxidation occurs by the initiating activity of the formed hydroperoxide. The rate constants of initiation formed by hydroperoxides were estimated from the parabolic kinetic

TABLE 7.12
The Values of $k_p(2k_t)^{-1/2}$ for Oxidation of Ethers

Ether	T (K)	$k_p(2k_t)^{-1/2}$ (L mol^{-1} s^{-1})$^{1/2}$	Ref.
Me$_2$CHOCHMe$_2$	303	3.7×10^{-3}	[63]
Me$_2$CHOCHMe$_2$	333	1.4×10^{-3}	[64]
Me$_2$CHOCHMe$_2$	333	1.41×10^{-3}	[16]
PrCH$_2$OCH$_2$Pr	303	1.0×10^{-4}	[63]
PrCH$_2$OCH$_2$Pr	343	1.0×10^{-3}	[64]
PrCH$_2$OCH$_2$Pr	333	6.7×10^{-4}	[16]
Me$_2$CHOCMe$_3$	303	1.0×10^{-3}	[63]
cyclohexyl-H, OCH$_3$	323–354	$1.6 \times 10^3 \exp(-41.0/RT)$	[18]
PhCH$_2$OCH$_2$Ph	273–303	$8.5 \times 10^2 \exp(-28.4/RT)$	[65]
PhCH$_2$OCH$_2$Ph	308–338	$1.1 \times 10^4 \exp(-37.7/RT)$	[65]
PhCH$_2$OCH$_2$Ph		$2.3 \times 10^3 \exp(-33.5/RT)$	[66]
PhCH$_2$OCH$_2$Ph	303	7.1×10^{-3}	[63]
PhCH$_2$OCH$_2$Ph	333	1.18×10^{-2}	[65]
PhCH$_2$OCH$_2$Ph	333	1.33×10^{-2}	[16]
PhMeCHOCHMePh	303	3.0×10^{-2}	[63]
PhMeCHOCHMePh	333	2.0×10^{-3}	[16]
PhCH$_2$OPh	303	3×10^{-4}	[63]
PhCH$_2$OPh	303	1.48×10^{-4}	[67]
PhCH$_2$OCMe$_3$	303	2.75×10^{-3}	[63]
dicyclohexyl ether	303–343	$3.98 \times 10^4 \exp(-47.8/RT)$	[68]
CH$_2$(OMe)$_2$	303	1.0×10^{-4}	[68]
CH$_2$(OEt)$_2$	303–343	$8.91 \times 10^3 \exp(-45.6/RT)$	[68]
CH$_2$(OCH$_2$Et)$_2$	303–343	$7.41 \times 10^3 \exp(-46.0/RT)$	[68]
CH$_2$(OCH$_2$Pr)$_2$	303–343	$1.02 \times 10^4 \exp(-46.0/RT)$	[68]
CH$_2$(OCH$_2$CHMe$_2$)$_2$	303–343	$1.23 \times 10^4 \exp(-47.0/RT)$	[68]
MeCH(OMe)$_2$	303–343	3.30×10^{-4}	[68]
MeCH(OCH$_2$Me)$_2$	303–343	$5.75 \times 10^2 \exp(-35.0/RT)$	[68]
MeCH(OCH$_2$Et)$_2$	303–343	$2.09 \times 10^4 \exp(-45.6/RT)$	[68]
MeCH(OCH$_2$Pr)$_2$	303–343	$2.95 \times 10^4 \exp(-46.0/RT)$	[68]
MeCH(OCH$_2$CHMe$_2$)$_2$	303–343	$2.63 \times 10^3 \exp(-41.0/RT)$	[68]
tetrahydrofuran-2,2-H,H	303	7.8×10^{-4}	[63]
2-methyltetrahydrofuran	303	2.25×10^{-2}	[63]
1,3-dioxolane	303–343	$2.01 \times 10^3 \exp(-32.7/RT)$	[68]

continued

TABLE 7.12
The Values of $k_p(2k_t)^{-1/2}$ for Oxidation of Ethers—*continued*

Ether	T (K)	$k_p(2k_t)^{-1/2}$ (L mol^{-1} s^{-1})$^{1/2}$	Ref.
	303	1.02×10^{-2}	[68]
	303–343	$4.00 \times 10^3 \exp(-30.6/RT)$	[68]
	303–343	$3.16 \times 10^2 \exp(-35.2/RT)$	[68]
	303	7.8×10^{-4}	[63]
	303	7.2×10^{-5}	[63]
	303–343	$1.58 \times 10^2 \exp(-32.7/RT)$	[68]
	323–368	$3.1 \times 10^3 \exp(-40.6/RT)$	[68]
	303–343	$1.26 \times 10^3 \exp(-35.6/RT)$	[68]
	323–368	$2.5 \times 10^2 \exp(-31.4/RT)$	[68]
	323–368	$1.3 \times 10^4 \exp(-43.3/RT)$	[68]
	303–343	$1.58 \times 10^2 \exp(-33.9/RT)$	[68]
	303–343	$3.98 \times 10^2 \exp(-34.8/RT)$	[68]
	303–343	$2.51 \times 10^2 \exp(-35.6/RT)$	[68]
	323–368	$3.48 \times 10^2 \exp(-31.8/RT)$	[68]
	303–343	$1.26 \times 10^3 \exp(-36.9/RT)$	[68]
	303–343	$5.01 \times 10^3 \exp(-41.5/RT)$	[68]
	303–343	$3.16 \times 10^3 \exp(-27.2/RT)$	[68]
	303–343	$6.31 \times 10^3 \exp(-45.3/RT)$	[68]

TABLE 7.12
The values of $k_p(2k_t)^{-1/2}$ for Oxidation of Ethers—continued

Ether	T (K)	$k_p(2k_t)^{-1/2}$ (L mol^{-1} s^{-1})$^{1/2}$	Ref.
(structure)	303–343	$1.26 \times 10^3 \exp(-38.1/RT)$	[68]
(structure)	303–343	$1.26 \times 10^3 \exp(-38.1/RT)$	[68]
(structure)	303–343	$3.16 \times 10^3 \exp(-37.3/RT)$	[68]
(structure)	323–368	$3.0 \times 10^2 \exp(-30.5/RT)$	[68]
(structure)	323–368	$5.1 \times 10^3 \exp(-42.7/RT)$	[68]
(structure)	303–343	$6.31 \times 10^2 \exp(-34.3/RT)$	[68]
(structure)	303	4.25×10^{-2}	[63]
(structure)	323–368	$4.8 \times 10^3 \exp(-42.3/RT)$	[68]
(structure)	303–343	$6.31 \times 10^2 \exp(-35.2/RT)$	[68]

curves of dioxygen consumption of autoxidized ethers [68]. The measured rate constants are given in Table 7.15.

7.2.3 Co-Oxidation of Ethers and Hydrocarbons

The theory of chain co-oxidation of two reactants R^{1}H and R^{2}H was described earlier in Chapter 5. The results of the study of ether co-oxidation with hydrocarbons and ether (1) + ether (2) are collected in Table 7.16.

These data appeared to be very useful for the estimation of the relative O—H bond dissociation energies in hydroperoxides formed from peroxyl radicals of oxidized ethers. All reactions of the type RO$_2^{\bullet}$ + RH (RH is hydrocarbon) are reactions of the same class (see Chapter 6). All these reactions are divided into three groups: RO$_2^{\bullet}$ + R^{1}H (alkane, parameter $br_e = 13.62$ (kJ mol^{-1})$^{1/2}$, RO$_2^{\bullet}$ + R^{2}H (olefin, $br_e = 15.21$ (kJ mol^{-1})$^{1/2}$, and RO$_2^{\bullet}$ + R^{3}H (akylaromatic hydrocarbon), $br_e = 14.32$ (kJ mol^{-1})$^{1/2}$ [71]. Only one factor, namely reaction enthalpy, determines the activation energy of the reaction inside one group of reactions. Also,

TABLE 7.13
Rate Constants of Chain Propagation and Termination in Oxidized Ethers

Ether	T (K)	k_p (L mol^{-1} s^{-1})	$2k_t$ (L mol^{-1} s^{-1})	Ref.
MeCH$_2$OCH$_2$Me	303		$5.0 \times 10^8 \exp(-3.2/RT)$	[63,64]
Me$_2$CHOCHMe	303–343	$2.0 \times 10^8 \exp(-50.3/RT)$	$2.0 \times 10^{10} \exp(-24.1RT)$	[63]
PrCH$_2$OCH$_2$Pr	303–343	$4.4 \times 10^8 \exp(-49.7/RT)$	$5.0 \times 10^8 \exp(-3.2/RT)$	[63]
Me$_2$CHOCMe$_3$	303–333	$1.0 \times 10^8 \exp(-50.4/RT)$	$2.0 \times 10^{10} \exp(-32.9/RT)$	[63]
PrCH$_2$OCH$_2$Ph	303–343	$1.9 \times 10^7 \exp(-29.7/RT)$	$5.0 \times 10^8 \exp(-2.6/RT)$	[63,65]
PhCH$_2$OPh	303–343	$2.0 \times 10^7 \exp(-41.4/RT)$	$5.0 \times 10^8 \exp(-7.6/RT)$	[63]
PhCH$_2$OCMe	303–343	$2.0 \times 10^7 \exp(-35.6/RT)$	$5.0 \times 10^8 \exp(-7.3/RT)$	[63]
EtOCH$_2$OEt	303	0.9	5.0×10^7	[68]
(tetrahydrofuran)	303–343	$4 \times 10^8 \exp(-46.2/RT)$	$5.0 \times 10^8 \exp(-7.0/RT)$	[63]
(2-methyltetrahydrofuran)	303–343	$2 \times 10^8 \exp(-44.2/RT)$	$2.0 \times 10^{10} \exp(-32.7/RT)$	[63]
(tetrahydropyran)	303–343	$4.0 \times 10^8 \exp(-46.7/RT)$	$5.0 \times 10^8 \exp(-8.0/RT)$	[63]
(1,3-dioxane)	303–343	$8.0 \times 10^8 \exp(-53.3/RT)$	$5.0 \times 10^8 \exp(-5.8/RT)$	[63]
(2-methyl-1,3-dioxane)	303–343	$5.0 \times 10^7 \exp(-41.5/RT)$	$1.7 \times 10^9 \exp(-20.1/RT)$	[68]
	303–343	$5.0 \times 10^7 \exp(-42.3/RT)$	$5.7 \times 10^8 \exp(-9.2/RT)$	[68]
	303–343	$4.0 \times 10^7 \exp(-41.5/RT)$	$8.0 \times 10^8 \exp(-9.6/RT)$	[68]
	303	1.4	2.4×10^6	[68]
	303–343	$1.7 \times 10^7 \exp(-44.4/RT)$	$4.0 \times 10^9 \exp(-19.3/RT)$	[68]
	303–343	$1.6 \times 10^7 \exp(-40.2/RT)$	$1.7 \times 10^9 \exp(-17.2/RT)$	[68]
	303	34.3	3.5×10^7	[68]
	343	2.3	1.1×10^7	[68]
	303–343	$7.0 \times 10^5 \exp(-36.9/RT)$	$1.0 \times 10^8 \exp(-17.2/RT)$	[68]
	303	0.59	2.0×10^7	[68]

TABLE 7.14
Rate Constants of Radical Generation by Trimolecular Reaction of Dioxygen with Ethers [68]

Ether	T (K)	k_i (L^2 mol^{-2} s^{-1})	E (kJ mol^{-1})	log A, A (L^2 mol^{-2} s^{-1})
(BuO)$_2$CH$_2$	393	1.4×10^{-8}	67.0	1.00
(BuO)$_2$CHMe$_2$	373	1.7×10^{-8}	88.0	4.00
(tetrahydropyran derivative)	393	1.6×10^{-8}	74.6	2.6
(dioxane derivative)	393	1.2×10^{-8}	59.9	0.50
(Ph-dioxolane derivative)	373	7.9×10^{-8}	72.9	2.80
(vinyl tetrahydropyran derivative)	393	5.8×10^{-7}	84.6	5.1

TABLE 7.15
Effective Rate Constants of Hydroperoxide Decomposition in Solution of Subsequent Either [68]

Hydroperoxide	T (K)	k_d ((s^{-1}) s^{-1})	E (kJ mol^{-1})	log A, A ((s^{-1}) s^{-1})
(EtO)$_2$CHOOH	343	3.60×10^{-7}	85.0	6.5
(BuO)$_2$CHOOH	343	2.51×10^{-7}	83.4	6.1
[Me$_2$CHCH$_2$O]$_2$CHOOH	343	2.34×10^{7}	79.0	5.4
(EtO)$_2$CMeOOH	343	7.52×10^{-6}	117.7	12.8
(BuO)$_2$CMeOOH	343	2.17×10^{-6}	97.6	9.2
(dioxolane-OOH)	373	1.75×10^{-6}	81.1	5.6
(dioxane-OOH)	343	1.57×10^{-7}	115.6	10.8
(methylated dioxane-OOH)	343	2.38×10^{-8}	123.6	11.2
(spiro dioxane-OOH)	373	1.37×10^{-6}	114.0	10.1
(oxepane-OOH)	373	1.96×10^{-6}		

TABLE 7.16
Parameters of Co-Oxidation of Ethers and Hydrocarbons

Ether	Hydrocarbon	T (K)	r_1	r_2	ϕ	Ref.
PrCH$_2$OCH$_2$Pr	cyclohexene	333	0.41	3.2	1.3	[69]
PrCH$_2$OCH$_2$Pr	Me$_2$PhCH	333	2.3	0.17	0.21	[69]
Me$_2$CHOCHMe$_2$	cyclohexene	333	0.72	4.3	0.14	[69]
Me$_2$CHOCHMe$_2$	tetralin	333	0.67	2.2	0.40	[69]
Me$_2$CHOCHMe$_2$	Me$_2$PhCH	333	2.20	0.11	0.55	[69]
Me$_2$CHOCHMe$_2$	PrCH$_2$OCH$_2$Pr	333	1.60	0.65	1.70	[69]
Me$_2$CHOCHMe$_2$	Me$_2$PhCOCPhMe$_2$	333	0.61	0.84	2.10	[69]
PrCH$_2$OCH$_2$Me	PhCH=CH$_2$	333	0.20	3.8		[67]
PrCH$_2$OCH$_2$Me	Me$_2$PhCH	333	2.10	0.10		[67]
PrCH$_2$OCH$_2$Me	tetralin	333	0.48	0.88		[67]
PhCH$_2$OCH$_2$Ph	Me$_2$PhCH	348	54.0	0.16	3.0	[70]
PhCH$_2$OCH$_2$Ph	MePhCH$_2$	348	20.0	0.15	2.8	[70]
PhCH$_2$OCH$_2$Ph	PhCH$_3$	348	20.0	0.03	0.9	[70]
PhCH$_2$OCH$_2$Ph	o-Me-C$_6$H$_4$-CH$_2$–H	348	68.0	0.04	3.9	[70]
PhCH$_2$OCH$_2$Ph	p-Me-C$_6$H$_4$-CH$_2$–H	348	52.0	0.07	0.9	[70]
Me$_2$PhCOCPhMe$_2$	cyclohexene	333	0.10	5.0	1.10	[69]
Me$_2$PhCOCPhMe$_2$	Me$_2$PhCH	333	9.5	0.5	0.87	[69]
2,2-dimethyl-tetrahydrofuran	cyclohexene	333	0.32	2.7	1.30	[69]
2,2-dimethyl-tetrahydrofuran	tetralin	333	1.20	5.8	0.5	[69]
2,2-dimethyl-tetrahydrofuran	Me$_2$PhCH	333	5.80	0.11	0.07	[69]
2,2-dimethyl-tetrahydrofuran	tetrahydrofuran	333	1.60	0.56	0.80	[69]
2-phenyl-tetrahydrofuran	PhCH=CH$_2$	333	1.50	0.20		[69]
PhCH$_2$OCH$_2$Ph	Me$_2$PhCOCPhMe$_2$	333	4.30	0.45	0.30	[69]

the pre-exponential factor A is the same for all reactions of the same group. Therefore, one can estimate the difference in $\Delta D = D_i(R_i—H) - D_1(R_1—H)$ using the experimental ratio of rate constants of one peroxyl radical with different hydrocarbons R_iH of the same group. If the rate constants of reactions of different peroxyl radicals with the same hydrocarbon are

Oxidation of Alcohols and Ethers

measured, one can calculate the difference $\Delta E_i = E_i(R_iO_2^{\bullet} + R_1H) - E_1(R_1O_2^{\bullet} + R_1H)$ and $\Delta D = D_i(R_iOO{-}H) - D_1(R_1OO{-}H)$ using ratio of rate constants. When hydrocarbon R_1H is chosen with the known BDE of the O—H bond of the formed hydroperoxide R_1OOH, one can estimate the BDE of the O—H bond of studied R_iOOH. This method was used to estimate the BDE of a few hydroperoxides [47,72]. When ether is the first component of co-oxidation (R^1H) and hydrocarbon is the second (R^2H), the parameter $r_1 = k_{p11}/k_{p12}$ and, therefore, one can calculate the absolute value k_{p12} as ratio: $k_{p12} = k_{p11}/r_1$. The difference of activation energies between the reactions $R^2O_2^{\bullet} + R^2H$ and $R^1O_2^{\bullet} + R^2H$ was calculated according to the equation:

$$\Delta E = E_{p12} - E_{p22} = RT \ln(k_{p22}/k_{p12}) \quad (7.9)$$

The difference of reaction enthalpies is $\Delta\Delta H = \Delta H_{p22} - \Delta H_{p12} = D(R_2{-}H) - D(R_2OO{-}H) - D(R_2{-}H) + D(R_1OO{-}H) = D(R_2OO{-}H) - D(R_1OO{-}H) = \Delta D$. This difference ΔD is connected with ΔE by the following equation within the scope of IPM [71,72]:

$$\Delta D = 2br_e\alpha^{-1}\left\{\sqrt{E_{e1} + \Delta E} - \sqrt{E_{e1}}\right\} + (\alpha^{-2} - 1)\Delta E \quad (7.10)$$

where parameters α and br_e are characteristics of the chosen group of reactions and $E_{e1} = E_{p22} + 17.4 - 0.5\, RT$ kJ mol^{-1} (see Chapter 6). The results of calculation of ΔD for the studied reactions (see Table 7.16 and Table 2.8) are presented in Table 7.17.

The following BDEs of the O—H bonds can be recommended for hydroperoxides formed from ethers.

ROOH	RCH(OOH)OR	R$_2$C(OOH)OR	PhCH(OOH)OR	tetrahydrofuranyl-OOH
D(O—H) (kJ mol^{-1})	367.3	358.4 ± 1.9	374.8 ± 0.4	367.6

Along with tertiary hydroperoxide of ether, the BDE of the O—H bonds of alkoxy hydroperoxides are higher than that of similar hydrocarbons. Very valuable data were obtained in experiments on ether oxidation (R_iH) in the presence of hydroperoxide (R_1OOH). Peroxyl radicals of oxidized ether exchange very rapidly to peroxyl radicals of added hydroperoxide ROOH and only $RO_2^{\bullet}$ reacts with ether (see Chapter 5). The rate constants of alkylperoxyl radicals with several ethers are presented in Table 7.18. The reactivity of ethers in reactions with peroxyl radicals will be analyzed in next section.

7.2.4 Reactivity of Ethers in Reactions with Peroxyl Radicals

The reaction enthalpy is known as a very important factor that determines the reactivity of reactants in free radical abstraction reactions [71]. The IPM method helps to calculate the increment of ΔE_H that enthalpy determines in the activation energy of the individual reaction. This increment can be estimated within the scope of IPM through the comparison of activation energy E_e of the chosen reaction and activation energy of the thermoneutral reaction E_{e0} (see Equation [6.18] in Chapter 6). This increment was calculated for several reactions of different peroxyl radicals with ethers (Table 7.19).

TABLE 7.17
Calculation of O—H Bond Dissociation Energies in Ether Hydroperoxides from Kinetic Data (See Table 2.8 and Table 7.16)

R^1H	R^2H	k_{p12} (L mol^{-1} s^{-1})	k_{p22} (L mol^{-1} s^{-1})	ΔE (kJ mol^{-1})	ΔD (kJ mol^{-1})	$D(R^1OO-H)$ kJ mol^{-1})
PrCH$_2$OCH$_2$Pr	cyclohexene	17.1 (333 K)	7.30	2.35	5.8	371.9
PrCH$_2$OCH$_2$Pr	Me$_2$PhCH	2.87 (333 K)	1.47	1.85	4.6	363.2
PrCH$_2$OCH$_2$Me	Me$_2$PhCH	4.90 (333 K)	1.47	3.33	8.2	366.8
Me$_2$CHOCHMe$_2$	cyclohexene	3.58 (333 K)	7.30	−1.97	−45.5	360.0
Me$_2$CHOCHMe$_2$	tetralin (H,H)	3.85 (333 K)	13.60	−3.50	−8.7	356.8
tetrahydrofuran (O,H,H)	tetralin (H,H)	18.83 (333 K)	13.60	0.89	2.1	367.6
PhCH$_2$OCH$_2$Ph	Me$_2$PhCH	21.1 (348 K)	2.79	5.85	14.2	372.8
PhCH$_2$OCH$_2$Ph	MePhCH$_2$	22.5 (348 K)	4.33	4.76	11.2	376.7

TABLE 7.18
Rate Constants of Peroxyl Radical Reactions with Ethers

Ether	RO$_2^\bullet$	T (K)	k (L mol^{-1} s^{-1})	Ref.
PrCH$_2$OCH$_2$Pr	Me$_3$COO$^\bullet$	303	6.4×10^{-2}	[62]
Me$_2$CHOCMe$_3$	Me$_3$COO$^\bullet$	303	2.0×10^{-2}	[62]
Me$_2$CHOCHMe$_2$	Me$_3$COO$^\bullet$	303	0.11	[62]
tetrahydropyran (H,H,O)	Me$_3$COO$^\bullet$	303	2.4×10^{-2}	[62]
PhCH$_2$OCH$_2$Ph	Me$_3$COO$^\bullet$	303	1.3	[62]
Me$_2$PhCOCPhMe$_2$	Me$_3$COO$^\bullet$	303	8.4×10^{-2}	[62]
PhCH$_2$OPh	Me$_3$COO$^\bullet$	303	0.20	[62]
tetrahydrofuran (O,H,H)	Me$_3$COO$^\bullet$	303	0.34	[62]

TABLE 7.18
Rate Constants of Peroxyl Radical Reactions with Ethers—*continued*

Ether	RO$_2^\bullet$	T (K)	k (L mol^{-1} s^{-1})	Ref.
PhCH$_2$OCH$_2$Ph	EtOCH(OO$^\bullet$)Me	303	36.0	[62]
(tetrahydrofuran)	EtOCH(OO$^\bullet$)Me	303	4.50	[62]
(tetrahydrofuran)	(tetrahydropyranyl-OO$^\bullet$)	303	4.40	[62]
PhCH$_2$OCH$_2$Ph	(tetrahydropyranyl-OO$^\bullet$)	303	25.7	[62]
(tetrahydrofuran)	(tetralinyl-OO$^\bullet$)	303	3.0	[62]
PhCH$_2$OCH$_2$Ph	(tetralinyl-OO$^\bullet$)	303	19.0	[62]
PhCH$_2$OCMe$_3$	Me$_3$COO$^\bullet$	303	1.10	[62]
(2-methyltetrahydrofuran)	Me$_3$COO$^\bullet$	303	0.80	[62]
(EtO)$_2$CH$_2$	Me$_2$PhCOO$^\bullet$	343	1.2	[68]
(BuO)$_2$CH$_2$	Me$_2$PhCOO$^\bullet$	343	1.5	[68]
(EtO)$_2$CHMe	Me$_2$PhCOO$^\bullet$	343	3.4	[68]
(BuO)$_2$CHMe	Me$_2$PhCOO$^\bullet$	343	3.7	[68]
(1,3-dioxolane)	Me$_2$PhCOO$^\bullet$	303–343	$1.3 \times 10^7 \exp(-37.0/RT)$	[68]
(2,2-dimethyl-1,3-dioxolane)	Me$_2$PhCOO$^\bullet$	303–343	$6.3 \times 10^6 \exp(-41.0/RT)$	[68]
(1,3-dioxane)	Me$_2$PhCOO$^\bullet$	303–343	$1.6 \times 10^6 \exp(-37.0/RT)$	[68]
(2,2-dimethyl-1,3-dioxane)	Me$_2$PhCOO$^\bullet$	303–343	$2.5 \times 10^5 \exp(-35.0/RT)$	[68]
(5,5-dimethyl-1,3-dioxane)	Me$_2$PhCOO$^\bullet$	303–343	$2.5 \times 10^6 \exp(-38.0/RT)$	[68]
(2,2,5,5-tetramethyl-1,3-dioxane)	Me$_2$PhCOO$^\bullet$	303–343	$3.2 \times 10^6 \exp(-40.0/RT)$	[68]
(1,3-dioxepane)	Me$_2$PhCOO$^\bullet$	303–343	$1.6 \times 10^6 \exp(-38.0/RT)$	[68]
(2,2-dimethyl-1,3-dioxepane)	Me$_2$PhCOO$^\bullet$	303–343	$7.4 \times 10^6 \exp(-44.0/RT)$	[68]

TABLE 7.19
Reaction Enthalpy and Polar Interaction as Factors that Influence on Activation Energy of Reaction $RO_2^\bullet$ + Ether (Equation [6.32])

Ether	ROO$^\bullet$	ΔH_e (kJ mol^{-1})	E_e (kJ mol^{-1})	E_{e0} (kJ mol^{-1})	ΔE_H (kJ mol^{-1})	ΔE_μ (kJ mol^{-1})
(PrCH$_2$)$_2$O	Me$_3$COO$^\bullet$	29.8	72.8	58.5	14.3	2.1
(PrCH$_2$)$_2$O	PrCH(OO$^\bullet$)OCHPr	17.1	65.5	57.5	8.0	1.1
(Me$_2$CH)$_2$	Me$_3$COO$^\bullet$	28.4	74.0	60.4	13.6	4.0
(Me$_2$CH)$_2$	Me$_2$C(OO$^\bullet$)OCHMe$_2$	15.7	66.3	59.0	7.3	2.6
(PhCH$_2$)$_2$O	Me$_3$COO$^\bullet$	−3.4	59.4	60.9	−1.5	1.4
(PhCH$_2$)$_2$O	EtOCH(OO$^\bullet$)Me	−16.1	51.1	58.0	−6.9	−4.3
tetrahydrofuran (α-H)	Me$_3$COO$^\bullet$	28.5	68.6	54.9	13.7	−1.5
tetrahydrofuran (α-H)	tetrahydrofuran-2-yl-OO$^\bullet$	20.2	62.2	52.6	9.6	−3.8
tetrahydrofuran (α-H)	EtOCH(OO$^\bullet$)Me	16.5	62.1	54.4	7.7	−2.0
tetrahydropyran (α-H)	Me$_3$COO$^\bullet$	39.3	75.3	55.9	19.4	−0.5
tetrahydropyran (α-H)	tetrahydropyran-2-yl-OO$^\bullet$	30.3	62.7	47.9	14.8	−8.5
(BuO)$_2$CH$_2$	Me$_2$PhCOO$^\bullet$	29.4	69.3	55.1	14.2	−1.3
(EtO)$_2$CHMe	Me$_2$PhCOO$^\bullet$	24.8	65.8	54.0	11.8	−2.4
(BuO)$_2$CHMe	Me$_2$PhCOO$^\bullet$	24.8	64.8	52.9	11.9	−3.5
1,3-dioxolane (α-H)	Me$_2$PhCOO$^\bullet$	29.2	66.8	52.7	14.1	−3.7
1,3-dioxolane (α-H)	1,3-dioxolan-2-yl-OO$^\bullet$	20.2	61.4	51.8	9.6	−4.6
1,3-dioxane (α-H)	Me$_2$PhCOO$^\bullet$	28.8	66.8	52.9	13.9	−3.5
2-methyl-1,3-dioxane	Me$_2$PhCOO$^\bullet$	10.6	68.1	63.2	4.9	6.8
2,2-dimethyl-1,3-dioxane	Me$_2$PhCOO$^\bullet$	28.8	66.5	52.66	13.9	−3.8
1,3-dioxepane (α-H)	Me$_2$PhCOO$^\bullet$	31.1	68.0	52.9	15.1	−3.5

TABLE 7.20
Geometrical Parameters of Transition State of Peroxyl Radical Reactions with Ethers (Equations [6.35]–[6.38])

Ether	ROO•	ΔH_e (kJ mol^{-1})	E_e (kJ mol^{-1})	ΔE_μ (kJ mol^{-1})	r(C—H) × 10^{10} (m)	r(O—H) × 10^{10} (m)	$\varphi°$
(PrCH$_2$)$_2$O	Me$_3$COO•	37.1	72.8	−1.8	1.414	1.170	171
(PrCH$_2$)$_2$O	PrCH(OO•)OCHPr	28.4	65.5	−4.6	1.400	1.183	165
(Me$_2$CH)$_2$	Me$_3$COO•	28.4	74.0	4.0	1.412	1.172	180
(Me$_2$CH)$_2$	Me$_2$C(OO•)OCHMe$_2$	28.6	66.3	−3.9	1.412	1.172	166
(PhCH$_2$)$_2$O	Me$_3$COO•	−3.4	59.4	−1.4	1.390	1.221	172
(PhCH$_2$)$_2$O	EtOCH(OO•)Me	−12.1	51.1	0.0	1.377	1.234	180
tetrahydrofuran (CH$_2$)	Me$_3$COO•	29.2	68.6	−1.8	1.412	1.172	171
tetrahydrofuran (CH$_2$)	tetrahydrofuran-OO•	20.5	62.2	−3.9	1.402	1.182	166
tetrahydrofuran (CH$_2$)	EtOCH(OO•)Me	20.5	62.1	−4.2	1.403	1.180	166
tetrahydropyran (CH$_2$)	Me$_3$COO•	39.3	75.3	−0.5	1.425	1.159	175
tetrahydropyran (CH$_2$)	tetrahydropyran-OO•	34.4	62.7	−10.8	1.419	1.165	157
(BuO)$_2$CH$_2$	Me$_2$PhCOO•	27.7	69.3	−0.4	1.424	1.182	176
(EtO)$_2$CHMe	Me$_2$PhCOO•	24.8	65.8	−2.4	1.408	1.176	169
(BuO)$_2$CHMe	Me$_2$PhCOO•	24.8	64.8	−3.4	1.408	1.178	167
1,3-dioxolane	Me$_2$PhCOO•	28.8	66.8	−3.5	1.424	1.182	167
1,3-dioxolane	1,3-dioxolan-OO•	20.1	61.4	−4.5	1.414	1.176	165
1,3-dioxane	Me$_2$PhCOO•	28.8	66.8	−3.0	1.412	1.170	168
1,3-dioxane (OCHO)	Me$_2$PhCOO•	25.1	68.1	−0.2	1.391	1.193	177
Me-1,3-dioxolane	Me$_2$PhCOO•	28.8	66.5	−3.3	1.411	1.170	167
1,3-dioxepane	Me$_2$PhCOO•	31.1	68.0	−3.4	1.415	1.169	167

Another important factor, which was noticed in reactions of peroxyl radicals with alcohols, is the polar factor. The polar reaction center O — O $\cdots$ H $\cdots$ C of the TS interacts with the adjacent polar alkoxy group of ether (see dipole moments of ether in Section 7.2.1). This interaction influences the activation energy of the reaction. The contribution of polar interaction to the activation energy was calculated by the IPM method (Chapter 6, Equation [6.43]). The values of this contribution ΔE_μ are presented in Table 7.19. There are positive and negative values of ΔE_μ. They vary from 4 to -8 kJ mol^{-1}. The increment of ΔE_μ is negative for reactions of benzyl ether, acetals, and cyclic ethers with the attacked $>$CH$_2$ group. Vice versa, reactions of cyclic ethers with attacked the tertiary C—H bond have in most cases positive ΔE_μ. The values ΔH_e and E_e calculated from experimental data can be used for the evaluation of the geometric parameters of the TS, namely distances O $\cdots$ H, C $\cdots$ H, and angle φ(CHO). These values were calculated by the IPM method (Chapter 6, Equations [6.35]–[6.38]) and are presented in Table 7.20. The angle φ(CHO) in TS varies from 180° to 157°.

In addition to the enthalpy and polar interaction, the following factors influence the activation energy of these reactions: triplet repulsion, electron affinity of atoms C and O in the TS, radii of C and O atoms, and π-bonds in the vicinity of the reaction center. These factors were discussed in Chapter 6 in application to the reaction of peroxyl radicals with hydrocarbons.

REFERENCES

1. ET Denisov. *Russ J Phys Chem* 68:24–28, 1994.
2. VE Tumanov, EA Kromkin, ET Denisov. *Russ Chem Bull* 51:1641–1650, 2002.
3. EA Kromkin, VE Tumanov, ET Denisov. *Neftekhimiya* 42:3–13, 2002.
4. BE Tumanov, ET Denisov. *Pet Chem* 41:93–102, 2001.
5. W Tsang. In: A. Greenberg, J. Liebman, (eds.). *Energetics of Free Radicals*. New York: Blackie Academic and Professional, 1996, p. 22.
6. DR Lide, (ed.) *Handbook of Chemistry and Physics*, 72nd ed. Boca Raton: CRC Press, 1991–1992.
7. NM Emanuel, ET Denisov, ZK Maizus. *Liquid-Phase Oxidation of Hydrocarbons*. New York: Plenum Press, 1967.
8. ET Denisov, NI Mitskevich, VE Agabekov. *Liquid-Phase Oxidation of Oxygen-Containing Compounds*. New York: Consultants Bureau, 1977.
9. ET Denisov. In: CH Bamford, CFH Tipper (eds.). *Comprehensive Chemical Kinetics*, vol 16. Amsterdam: Elsevier, 1980, pp. 125–204.
10. NM Emanuel, GE Zaikov, ZK Maizus. *Oxidation of Organic Compounds. Medium Effect in Radical Reaction*. Oxford: Pergamon Press, 1984.
11. LV Shibaeva, DI Metelitsa, ET Denisov. *Zh Fiz Khim* 44:2793–2798, 1970.
12. VYa Ladygin, VV Saraeva. *Kinet Katal* 7:967–976, 1966.
13. ET Denisov, VM Solyanikov. *Neftekhimiya* 3:360–366, 1963.
14. G Hughes, HA Makada. *Adv Chem Ser* 75:102–111, 1968.
15. C Parlant. *Rev Inst France Petrol* 19:1–40, 1964.
16. L Sajus. *Adv Chem Ser* 75:59–77, 1968.
17. TG Degtyareva, ET Denisov, VS Martemyanov, IA Kaftan, LR Enikeeva. *Izv AN SSSR Ser Khim* N 4:735–741, 1981.
18. GA Kovtun, AV Kazantsev, AL Aleksandrov. *Izv AN SSSR Ser Khim* 2635–2637, 1974.
19. C Parlant, I Sere de Roch, J-C Balaceanu. *Bull Soc Chim France* 2452–2661, 1963.
20. KA Zhavnerko. Liquid-Phase Oxidation of Cyclohexanol Initiated by Hydrogen Peroxide. Ph.D. thesis, Institute of Physical Organic Chemistry, Minsk, 1969, pp. 3–19 [in Russian].
21. VYa Ladygin, MS Furman, VI Mogilev. *Khim Vys Energ* 6:447–450, 1972.

22. VM Potekhin. To the problem of mechanism of oxidation of paraffin and alkylcycloparaffin hydrocarbons. Doctor diss. Thesis, LTI, Leningrad, 1972 [in Russian].
23. A Keszler, G Irinyi, K Heberger, D Gal. *Ber Bunsenges Phys Chem* 96:175–179, 1992.
24. JA Howard, S Korcek. *Can J Chem* 48:2165–2176, 1970.
25. AN Shendrik, NP Mytsyk, IA Opeida. *Kinet Katal* 18:1077–1078, 1977.
26. IA Opeida. Co-oxidation of alkylaromatic hydrocarbons in the liquid phase. Doctoral dissertation, Institute of Chemical Physics, Chernogolovka, 1982, pp. 1–336 [in Russian].
27. GE Adams, RL Willson. *Trans Faraday Soc* 65:2981–2987, 1969.
28. J Rabany, D Kug-Roth, A Heglein. *J Phys Chem* 78:2089–2093, 1974.
29. RL Willson. *Trans Faraday Soc* 67:3008–3019, 1971.
30. J Butler, GG Jayson, AJ Swallow. *J Chem Soc Faraday Trans 1*, 70:1394–1401, 1974.
31. B Mailard, KU Ingold, JC Scaiano. *J Am Chem Soc* 105:5095–5099, 1983.
32. K Stockhausen, A Henglein. *Ber Bunsenges Phys Chem* 75:833–840, 1971.
33. E Bothe, D Schulte-Frohlinde. *Z Naturforsch B* 35:1035–1039, 1980.
34. WJ McElroy, SJ Waygood. *J Chem Soc Faraday Trans 1*, 87:1513–1521, 1991.
35. WG Mallard, F Westley, JT Herron, RF Hempson. NIST Chemical Kinetics Database, Version 6.0. Gaithersburg: NIST, 1994.
36. R Atkinson, DL Baulch, RA Cox, Mailard, RF Hempson, JA Kerr, MJ Rossi, J Troe. *J Phys Chem Ref Data* 26:3029, 1997.
37. A Miyoshi, H Matsui, N Washida. *J Phys Chem* 94:3016–3022, 1990.
38. K Stockhausen, A Fojtic, A Henglein. *Ber Bunsenges Phys Chem* 74:34–40, 1970.
39. K Stockhausen, A Henglein, G Beck. *Ber Bunsenges Phys Chem* 73:567–571, 1969.
40. RL McCarthy, A MacLachlan. *Trans Faraday Soc* 1107–1116, 1961.
41. B Ya Ladygin, AA Revin. *Izv AN SSSR Ser Khim* 282–290, 1985.
42. E Bothe, G Behrens, D Schulte-Frohlide. *Z Naturforsch B* 32:886–889, 1977.
43. E Bothe, D Schulte-Frohlide, C von Sonntag. *J Chem Soc Perkin Trans II*, 416–420, 1978.
44. X Zhang, N Zhang, HP Schuchmann, C von Sonntag. *J Phys Chem* 98:6541–6547, 1994.
45. ET Denisov, TG Denisova. *Kinet Catal* 34:173–179, 1993.
46. SW Benson. *Int J Chem Kinet* 33:509–512, 2001.
47. TG Denisova, ET Denisov. *Petrol Chem* 44:250–255, 2004.
48. RL Vardanyan, VV Kharitonov, ET Denisov. *Izv AN SSSR Ser Khim* 1536–1542, 1970.
49. AL Aleksandrov, ET Denisov. *Izv AN SSSR Ser Khim* 2322–2324, 1969.
50. RL Vardanyan, ET Denisov, VI Zozulya. *Izv AN SSSR Ser Khim* 611–613, 1972.
51. C Parlant, I Sere de Roch, J-C Balaceanu. *Bull Soc Chim France* 3161–3169, 1974.
52. AN Shendrik, IA Opeida, RV Kucher, VI Dubina. *Neftekhimiya* 22:760–763, 1982.
53. ET Denisov, TG Denisova. *Kinet Catal* 34:738–744, 1993.
54. TG Denisova, NS Emel'yanova. *Kinet Katal* 44:441–449, 2003
55. ET Denisov. *Usp Khim* 54:1466–1486, 1985.
56. AN Shendrik, VN Dubina, IA Opeida. *Kinet Katal* 25:745–748, 1984.
57. ET Denisov. *Dokl AN SSSR* 141:131–134, 1961.
58. ET Denisov, VM Solyanikov. *Neftekhimiya* 4:458–465, 1964.
59. ET Denisov, VV Kharitonov, EN Raspopova. *Kinet Katal* 5:981–988, 1964.
60. ET Denisov, VV Kharitonov. *Kinet Katal* 5:781–786, 1964.
61. AI Rakhimov. In: Vestnik Nizhegorodskogo Gosudarstvennogo Universiteta. Organic and Elementoorganic Peroxides, Nizhniy Novgorod: Izd-vo Nizhegorodskogo Universiteta, 1996, p. 115 [in Russian].
62. JA Howard, KU Ingold. *Can J Chem* 48:873–880, 1970.
63. JA Howard. *Adv Free Radical Chem* 4:49–173, 1972.
64. N Kulevsky, CT Wang, VJ Steinberg. *J Org Chem* 34:1345–1352, 1969.
65. L Debiais, P Horstmann, M. Niclause, M Letort. *J Chim Phys* 56:69–77, 1959.
66. F Refmers. *Q J Pharm Pharmacol* 19:27–32, 1946.
67. GA Russel, RC Williamson. *J Am Chem Soc* 86:2357–2367, 1964.

68. DL Rakhmankulov, VV Zorin, EM Kuramshin, SS Zlotskii. Rate Constants of Homolytic Liquid-Phase Reactions of 1,3-Diheterocycloalkanes and Analogs. Ufa: Reaktiv, 1999 [in Russian].
69. P Gosborne, I Seree de Roch. *Bull Soc Chim France* 2260–2267, 1967.
70. RV Kucher, NA Kravchuk, VI Timokhin, AA Berlin, VA Romanov. *Neftekhimiya* 25:678–684, 1986.
71. ET Denisov. *Russ Chem Rev* 66:859–876, 1997.
72. ET Denisov, TG Denisova. *Kinet Catal* 34:173–179, 1993.

8 Oxidation of Carbonyl Compounds and Decarboxylation of Acids

8.1 OXIDATION OF ALDEHYDES

8.1.1 INTRODUCTION

Aldehydes are oxidized very rapidly at moderate temperatures including room temperature. This is the result of the common action of the following three factors:

1. Chain propagation in an oxidized aldehyde is limited by the reaction of the acylperoxyl radical with the aldehyde. The dissociation energy of the O—H bond of the formed peracid is sufficiently higher than that of the alkyl hydroperoxide. For example, in hydroperoxide PhMeCHOOH, $D_{O-H} = 365.5$ kJ mol^{-1} and in benzoic peracid PhC(O)OOH, $D_{O-H} = 403.9$ kJ mol^{-1} [1]. Therefore, acylperoxyl radicals are more active in chain propagation reactions compared to alkylperoxyl radicals.
2. Carbonyl group of the aldehyde decreases the BDE of the adjacent C—H bond. This is due to the stabilization of the formed acyl radical, resulting from the interaction of the formed free valence with π-electrons of the carbonyl group. For example, $D_{C-H} = 422$ kJ mol^{-1} in ethane and $D_{C-H} = 373.8$ kJ mol^{-1} in acetaldehyde. The values of D_{C-H} in aldehydes of different structures are presented in Table 8.1. In addition, the values of the enthalpies of acylperoxyl radical reactions with aldehydes were calculated ($D_{O-H} = 387.1$ kJ mol^{-1} in RC(O)OO—H).
3. Carbonyl groups are very polar due to the high difference in the electron affinities of C and O atoms connected by the easily polarized π-bond. Aldehydes have high values of dipole moment (μ) [3].

Aldehyde	CH$_2$(O)	MeCH(O)	CH$_2$=CHCH(O)	EtCH(O)	PhCH(O)
μ (Debye)	2.33	2.75	3.12	2.52	3.00

The polar carbonyl group interacts with the polar transition state of the reaction between the peroxyl radical and the C—H bond of the aldehyde. This interaction lowers the activation energy of this reaction (see Section 8.1.4). As a result, all the three factors, viz., the strong RC(O)OO—H bond formed, the weak C—H bond of the oxidized aldehyde, and the polar interaction in the transition state, contribute to lowering the activation energy of the reaction RC(O)OO$^\bullet$ + RCH(O) and increasing the rate constant of the chain propagation reaction (see Section 8.1.4).

TABLE 8.1
The Values of the C—H Bond Dissociation Energies in Aldehydes D_{C-H} and Enthalpies ΔH of the Reaction of Acylperoxyl Radical (RC(O)OO•) with Aldehydes [2]

Aldehyde	Bond	D_{C-H} (kJ mol^{-1})	ΔH (kJ mol^{-1})
Formaldehyde	H(O)C—H	377.8	−9.2
Acetaldehyde	Me(O)C—H	373.8	−13.2
Acetaldehyde	H—CH$_2$CH(O)	394.3	7.3
Propanal	Et(O)C—H	372.0	−15.0
Butanal	Pr(O)C—H	371.2	−15.8
Pentanal	Bu(O)C—H	372.0	−15.0
Propanal, 2-methyl-	Me$_2$CH(C—H)(O)	364.5	−22.5
Butanal, 3-methyl-	Me$_2$CHCH$_2$(C—H)(O)	362.6	−24.4
Butanal, 2-methyl-	MeCH$_2$CH(CH$_3$)(C—H)(O)	360.8	−26.2
Propanal, 2,2-dimethyl-	Me$_3$C(C—H)(O)	375.1	−11.9
Furfurol	furfural structure	387.5	0.5
Propenal	CH$_2$=CHC—H(O)	339.4	−47.6
Benzaldehyde	PhC—H(O)	348.0	−39.0
Propanal, 2-methyl-2-phenyl-	PhMe$_2$CC—H(O)	362.9	−24.1
Arabinose	CH$_2$OH(CHOH)$_3$C—H(O)	369.8	−17.2
Glucose	CH$_2$OH(CHOH)$_4$C—H(O)	371.4	−15.6
D-Ribose	CH$_2$OH(CHOH)$_3$C—H(O)	370.0	−17.0
Glyoxylic acid	(C—H)(O)C(O)OH	375.3	−11.7

Another peculiarity of the aldehyde oxidation is connected with the chemistry of the primary molecular product of oxidation, acyl peracid. The formed peracid interacts with the aldehydes to form peroxide with the structure RC(O)OOCH(OH)R [4,5]. This peroxide is unstable and decomposes into acids and free radicals.

$$RC(O)OOCH(OH)R \longrightarrow 2RC(O)OH$$
$$RC(O)OOCH(OH)R \longrightarrow RCO_2^\bullet + RCH(OH)O^\bullet$$

Therefore, aldehyde autoxidation produces an efficient autoinitiator.

Peracids react heterolytically with olefins with the formation of epoxides by the Prilezhaev reaction. So, the co-oxidation of aldehydes with olefins has technological importance. Peracids react with ketones with formation of lactones. These reactions will be discussed in Section 8.2. The oxidation of aldehydes are discussed in monographs [4–8].

8.1.2 Chain Mechanism of Aldehyde Oxidation

Aldehydes are oxidized by dioxygen by the chain mechanism in reactions brought about in different ways: initiated, thermal, photochemical, and induced by radiation as well as in the presence of transition metal compounds [4–8]. Oxidation chains are usually very long: from 200 to 50,000 units [4]. Acyl radicals add dioxygen very rapidly with a rate constant of 10^8–10^9 L mol^{-1} s^{-1} [4]. Therefore, the initiated chain oxidation of aldehyde includes the following elementary steps at high dioxygen pressures [4–7]:

$$I \longrightarrow r^{\bullet} \qquad 2ek_d$$
$$r^{\bullet} + RC(O)H \longrightarrow rH + RC^{\bullet}(O) \qquad \text{Fast}$$

$$RC^{\bullet}(O) + O_2 \longrightarrow RC(O)OO^{\bullet} \qquad \text{Fast}$$
$$RC(O)OO^{\bullet} + RC(O)H \longrightarrow RC(O)OOH + RC^{\bullet}(O) \qquad k_p$$
$$2\, RC(O)OO^{\bullet} \longrightarrow \text{Products} \qquad 2k_t$$

According to this scheme, the rate of absorption of dioxygen is given by

$$v = k_p(2k_t)^{-1/2}[\text{RCHO}][\text{O}_2]^0 v_i^{1/2} \qquad (8.1)$$

The values of $k_p(2k_t)^{-1/2}$ for various aldehydes obtained from experiments on the initiated oxidation of aldehydes are given in Table 8.2.

The chain unit in the thermal and photochemical oxidation of aldehydes by molecular dioxygen consists of two consecutive reactions: addition of dioxygen to the acyl radical and abstraction reaction of the acylperoxyl radical with aldehyde. Experiments confirmed that the primary product of the oxidation of aldehyde is the corresponding peroxyacid. Thus, in the oxidation of n-heptaldehyde [10,16,17], acetaldehyde [4,18], benzaldehyde [13,14,18], p-tolualdehyde [19], and other aldehydes, up to 90–95% of the corresponding peroxyacid were detected in the initial stages. In the oxidation of acetaldehyde in acetic acid [20], chain propagation includes not only the reactions of $RC^{\bullet}(O)$ with O_2 and $RC(O)OO^{\bullet}$ with $RC(O)H$, but also the exchange of radicals with solvent molecules ($R = CH_3$).

TABLE 8.2
The Values $k_p (2k_t)^{-1/2}$ for Chain Oxidation of Aldehydes

Aldehyde	Solvent	T (K)	$k_p(2k_t)^{-1/2}$ (L mol^{-1} s^{-1})$^{1/2}$	Ref.
MeC(O)H	Chlorobenzene	273	0.26	[9]
Me(CH$_2$)$_5$C(O)H	Chlorobenzene	273	0.39	[9]
Me(CH$_2$)$_5$C(O)H	Decane	276	0.10	[10]
Me(CH$_2$)$_6$C(O)H	Chlorobenzene	273	0.47	[9]
Me(CH$_2$)$_8$C(O)H	Decane	275–318	$3.44 \times 10^2 \exp(-17.6/RT)$	[11]
Me$_2$CHC(O)H	Chlorobenzene	283–303	$9.7 \exp(-15.9/RT)$	[12]
Me$_3$CC(O)H	Chlorobenzene	273	0.96	[9]
cyclohexyl-C(O)H	Chlorobenzene	273	0.44	[9]
PhC(O)H	Benzaldehyde	278–293	$3.46 \times 10^5 \exp(-7.5/RT)$	[13]
PhC(O)H	Chlorobenzene	273	0.29	[9]
PhC(O)H	Decane	278	0.13	[13]
PhC(O)H	o-Dichlorobenzene	316	0.29	[14]
4-substituted-PhC(O)H	Chlorobenzene	333–363	$2.4 \times 10^4 \exp(-15.9/RT)$	[15]

$$RC(O)OO^\bullet + RC(O)OH \longrightarrow RC(O)OOH + RC(O)O^\bullet$$
$$RC(O)O^\bullet \longrightarrow R^\bullet + CO_2$$
$$R^\bullet + O_2 \longrightarrow ROO^\bullet$$
$$ROO^\bullet + RC(O)OO^\bullet \longrightarrow \text{Products}$$

The rate constants of chain propagation are presented in Table 8.3.

The recombination of acyl peroxide radicals leads to the formation of the corresponding peroxide and the liberation of a dioxygen molecule. In the case of the photooxidation of acetaldehyde and propionaldehyde in the gaseous phase [22,23], it was established using ^{18}O that dioxygen was liberated during the recombination of the peroxide radicals and it was formed from oxygen atoms of both radicals. The mechanism of recombination of these radicals was studied in more detail by Ingold, who proposed the following scheme [24]:

$$RC(O)OO^\bullet + RC(O)OO^\bullet \longrightarrow RC(O)OOOOC(O)R$$
$$RC(O)OOOOC(O)R \longrightarrow [RC(O)O^\bullet + RC(O)O^\bullet] + O_2$$
$$[RC(O)O^\bullet + RC(O)O^\bullet] \longrightarrow RC(O)OOC(O)R$$
$$[RC(O)O^\bullet + RC(O)O^\bullet] \longrightarrow 2RC(O)O^\bullet$$
$$[RC(O)O^\bullet + RC(O)O^\bullet] \longrightarrow [R^\bullet + R^\bullet] + 2CO_2$$
$$[R^\bullet + R^\bullet] \longrightarrow RR$$
$$[R^\bullet + R^\bullet] \longrightarrow R^\bullet + R^\bullet$$

The rate constants of chain termination by disproportionation of two acylperoxyl radicals are collected in Table 8.4.

In thermal oxidation, chain initiation takes place by the reaction of the aldehyde with dioxygen. Two reactions of chain generation in autoxidized aldehydes, namely, bimolecular and trimolecular, were proved [25].

TABLE 8.3
Rate Constants for Chain Propagation in the Oxidation of Aldehydes

Aldehyde	Solvent	T (K)	k_p (L mol^{-1} s^{-1})	Ref.
MeC(O)H	Chlorobenzene	273	2.7×10^2	[9]
Me(CH$_2$)$_5$C(O)H	Chlorobenzene	273	3.1×10^2	[9]
Me(CH$_2$)$_6$C(O)H	Chlorobenzene	273	3.9×10^2	[9]
Me(CH$_2$)$_8$C(O)H	Decane	278	2.7×10^2	[11,13]
Me(CH$_2$)$_8$C(O)H	Decane	283–299	$1.4 \times 10^6 \exp(-17.6/RT)$	[11,13]
Me$_3$CC(O)H	Chlorobenzene	273	2.5×10^2	[9]
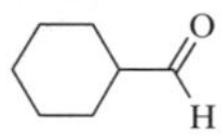 cyclohexyl-C(O)H	Chlorobenzene	273	1.1×10^2	[9]
PhC(O)H	o-Dichlorobenzene	316	$1.78 \times 10^2 \exp(-4.2/RT)$	[14]
PhC(O)H	Decane	278	1.91×10^2	[11,13]
PhC(O)H	Decane	283–299	$4.8 \times 10^4 \exp(-7.5/RT)$	[11,13]
PhC(O)H	Chlorobenzene	303	33.02	[1]

TABLE 8.4
Rate Constants for Chain Termination in the Oxidation of Aldehydes by the Reaction $2RC(O)OO^{\bullet} \rightarrow$ Products

Aldehyde	Solvent	T (K)	$2k_t$ (L mol^{-1} s^{-1})	Ref.
MeC(O)H	Chlorobenzene	273	1.04×10^8	[9]
Me(CH$_2$)$_5$C(O)H	Chlorobenzene	273	5.4×10^7	[9]
Me(CH$_2$)$_8$C(O)H	Chlorobenzene	273	6.9×10^7	[9]
Me(CH$_2$)$_8$C(O)H	Decane	278	3.4×10^7	[13]
Me(CH$_2$)$_8$C(O)H	Decane	278	7.5×10^6	[11]
Me$_2$CHC(O)H	Chlorobenzene	283–303	$9.5 \times 10^7 \exp(-17.1/RT)$	[12]
Me$_3$CC(O)H	Chlorobenzene	273	6.6×10^6	[9]
cyclohexyl-C(O)H	Chlorobenzene	273	6.8×10^6	[9]
PhC(O)H	o-Dichlorobenzene	313	4.0×10^8	[14]
PhC(O)H	Decane	278	2.1×10^8	[11,13]
PhC(O)H	Chlorobenzene	303	1.25×10^9	[21]

$$RC(O)H + O_2 \longrightarrow RC^{\bullet}(O) + HO_2^{\bullet}$$
$$RC(O)H + O_2 + RC(O)H \longrightarrow RC^{\bullet}(O) + H_2O_2 + RC^{\bullet}(O)$$

The trimolecular reaction predominates in oxidized aldehydes with the lowest BDE of the C—H bond: benzaldehyde (348 kJ mol^{-1}) and nonsaturated aldehydes. This is in agreement with the estimation of the reaction enthalpy (see Chapter 4). For benzaldehyde, $\Delta H(\text{PhC(O)H} + O_2) = 348 - 220 = 148$ kJ mol^{-1} and $\Delta H(\text{2PhC(O)H} + O_2) = 2 \times 348 - 570 = 126$ kJ mol^{-1}. For acrolein, $\Delta H(\text{RC(O)H} + O_2) = 339 - 220 = 119$ kJ mol^{-1} and ΔH (2 RC(O)H + O$_2$) = $2 \times 339 - 570 = 108$ kJ mol^{-1}. Thus, the experimental data are in good agreement with the enthalpies of these two endothermic reactions. The values of the rate constants of these reactions calculated from the initial rates of the oxidation of aldehydes for the known values of $k_p(2k_t)^{-1/2}$ are given in Table 8.5. The details of photochemical and radiation-induced oxidation of aldehydes are available elsewhere [4–8].

8.1.3 Co-Oxidation of Aldehydes with Hydrocarbons, Alcohols, and Aldehydes

Aldehydes do not co-oxidize alkanes due to a huge difference in the reactivity of these two classes of organic compounds. Alkanes are almost inert to oxidation at room temperature and can be treated as inert solvents toward oxidized aldehydes [35]. Olefins and alkylaromatic hydrocarbons are co-oxidized with aldehydes. The addition of alkylaromatic hydrocarbon (R^{2}H) to benzaldehyde (R^{1}H) retards the rate of the initiated oxidation [36–39]. The rate of co-oxidation obeys the equation [37]:

$$\frac{k_{p12}[R^2H]}{\sqrt{2k_{t11}v_i}} = \frac{v_0}{v} - \frac{v}{v_0} \tag{8.2}$$

(R^{1}H is aldehyde, R^{2}H is alkylaromatic hydrocarbon.)

TABLE 8.5
Rate Constants and Activation Energies of Reactions RC(O)H + O_2 → Free Radicals and 2RC(O)H + O_2 → Free Radicals

RH	Solvent	T (K)	k (L mol^{-1} s^{-1})	Ref.
\multicolumn{5}{c}{RC(O)H + O_2 → RC$^\bullet$(O) + HOO$^\bullet$}				
PrC(O)H	Chlorobenzene	283–303	4.0×10^4 exp(−117/RT)	[26]
Me_2CHC(O)H	Chlorobenzene	283–303	1.3×10^8 exp(−92/RT)	[27]
Me$(CH_2)_8$C(O)H	Decane	278–299	3.2×10^5 exp(−65/RT)	[11]
CH_2=CHC(O)H	Benzene	413	6.3×10^{-2}	[28]
MeCH=CHC(O)H	Chlorobenzene	288–308	1.3×10^{15} exp(−124/RT)	[29]
CH_2=CEtC(O)H	Chlorobenzene	288–308	2.5×10^{13} exp(−96.0/RT)	[30]
CH_2=CBuC(O)H	Chlorobenzene	288–308	1.0×10^{10} exp(−94/RT)	[31]
MeCH=CHC(O)H	Chlorobenzene	273–298	7.9×10^{13} exp(−119/RT)	[31]
\multicolumn{5}{c}{2RC(O)H + O_2 → RC$^\bullet$(O) + HOOH + RC$^\bullet$(O), (k (L^2 mol^{-2} s^{-1}))}				
PhCHO	Acetic acid	303	1.1×10^{-4}	[32]
PhCHO	Carbon tetrachloride	300	5.4×10^{-5}	[33]
MeCH=CHC(O)H	Decane	278–298	4.4×10^{13} exp(−105/RT)	[34]
CH_2=C(CH_3)C(O)H	Chlorobenzene	288–308	1.3×10^8 exp(−88/RT)	[29]

The ratio of the rate constants $k_{p12}(2k_{t11})^{-1/2}$ calculated from experimental data was found to be the following (at 313 K):

R^2H	$PhCH_3$	Me_2PhCH	(3,4-diMe)	(2,4,5-triMe)	(pentaMe)	(hexaMe)
$k_{p12}(2k_{t11})^{-1/2}$ (L mol^{-1} $s^{-1})^{1/2}$	0.24	0.27	4.3	5.6	58	130

Alcohols retard the oxidation of aldehydes. The parameters of aldehyde co-oxidation with cycloolefins, alcohols, and aldehydes are collected in Table 8.6.

The values of r_1 can be used for the estimation of the BDE of O—H bonds in peracids formed from acylperoxyl radicals. The method of D_{O-H} calculation was described previously (see Chapter 7, Equation [7.9] and Equation [7.10]). The values of k_{p12} and the calculated values of ΔE and ΔD are given here (R^2H = 1,4-cyclohexadiene, T = 273 K, $R^2OO^\bullet = HO_2^\bullet$, $k_{p11} = 376$ L mol^{-1} s^{-1}) [9].

$R^1OO^\bullet$	$PhCO_3^\bullet$	$Bu(CH_2)_3CO_3^\bullet$	$Bu(CH_2)_4CO_3^\bullet$	$Me_3CCO_3^\bullet$	cyclohexyl-C(O)$O_2^\bullet$
k_{p12} (L $mol^{-1}s^{-1}$)	7.1×10^4	4.8×10^3	6.1×10^3	1.2×10^3	1.2×10^3
ΔE (kJ mol^{-1})	12.0	5.8	6.3	2.6	2.6
ΔD (kJ mol^{-1})	34.9	17.4	18.8	7.9	7.9

TABLE 8.6
Parameters of Aldehydes Co-oxidation with Hydrocarbons, Alcohols, and Aldehydes

R^{1}H	R^{2}H	T (K)	r$_1$	r$_2$	Ref.
BuC(O)H	cyclohexane	243	2.94	0.14	[40]
C$_6$H$_{13}$C(O)H	cyclohexene	273	0.64	1.65	[9]
C$_7$H$_{15}$C(O)H	cyclohexene	273	0.64	1.65	[9]
C$_6$H$_{11}$C(O)H (cyclohexanecarbaldehyde)	cyclohexene	273	0.95	0.44	[9]
Me$_3$CC(O)H	cyclohexene	273	2.15	0.36	[9]
PhC(O)H	cyclohexene	273	0.17	4.75	[9]
PhC(O)H	Me$_2$CHOH	273	5.00	0.067	[41]
PhC(O)H	PhCH$_2$OH	273	0.80	0.43	[41]
PhC(O)H	PhCH$_2$OH	303	0.20	–	[42]
C$_6$H$_{13}$C(O)H	PhC(O)H	273	10.0	0.10	[9]
C$_9$H$_{19}$C(O)H	PhC(O)H	273	8.60	0.20	[13]
4-Cl-C$_6$H$_4$C(O)H	PhC(O)H	303	0.28	10.70	[43]
4-MeO-C$_6$H$_4$C(O)H	PhC(O)H	303	2.22	1.00	[43]
3-Cl-C$_6$H$_4$C(O)H	PhC(O)H	303	0.38	1.95	[43]
4-Cl-C$_6$H$_4$C(O)H	Me$_2$CHOH	303	0.28	1.87	[43]
4-Cl-C$_6$H$_4$C(O)H	PhC(O)H	303	0.54	1.62	[43]
4-Cl-C$_6$H$_4$C(O)H	3-Me-C$_6$H$_4$C(O)H	303	0.33	2.20	[43]
4-Cl-C$_6$H$_4$C(O)H	4-Me-C$_6$H$_4$C(O)H	303	0.24	2.15	[43]
4-Cl-C$_6$H$_4$C(O)H	4-MeO-C$_6$H$_4$C(O)H	303	0.15	2.36	[43]
4-Cl-C$_6$H$_4$C(O)H	4-NC-C$_6$H$_4$C(O)H	303	1.15	0.46	[43]

The difference of BDE, $\Delta D = D(R^1OO-H) - D(HOO-H) = D(R^1OO-H) - 369.0$ kJ mol^{-1}. The values of $\Delta D = 17.4$ and 18.8 kJ mol^{-1} are very close and the mean value of ΔD for RC(O)OO—H bond is 18.1 kJ mol^{-1}. The following D_{O-H} can be recommended for peracids:

Bond	PhC(O)OO—H	RC(O)OO—H	Me₃CC(O)OOH—H	C₆H₁₁C(O)—H
D(O—H) (kJ mol^{-1})	403.9	387.0	377.0	377.0

These BDEs are higher than that for alkyl hydroperoxides (see Chapter 2) and this is the main reason for the extremely high reactivity of peroxyl radicals formed from aldehydes. The absolute rate constants of the reactions of different peroxyl radicals with aldehydes are collected in Table 8.7.

There are two channels of the reaction of acylperoxyl radicals with olefins: hydrogen atom abstraction and addition to the double bond with epoxide formation [5,35]:

PhC(O)OO$^\bullet$ + RCH$_2$CH=CH$_2$ ⟶ PhC(O)OOH + RC$^\bullet$HCH=CH$_2$

PhC(O)OO$^\bullet$ + RCH$_2$CH=CH$_2$ ⟶ PhC(O)OOCH$_2$C$^\bullet$HCH$_2$R

PhC(O)OOCH$_2$C$^\bullet$HCH$_2$R ⟶ PhC(O)O$^\bullet$ + (epoxide)R

Both these reactions occur rapidly. The kinetics and the products of co-oxidation of aldehydes and olefins were studied by Emanuel and coworkers [47–49]. The values of the rate constants of the addition of acylperoxyl radical to olefins are presented in Table 8.8. The experimental data on aldehyde co-oxidation are discussed in monographs [4–6].

8.1.4 Reactivity of Aldehydes in Reactions with Peroxyl Radicals

The important role of reaction enthalpy in the free radical abstraction reactions is well known and was discussed in Chapters 6 and 7. The BDE of the O—H bonds of alkyl hydroperoxides depends slightly on the structure of the alkyl radical: $D_{O-H} = 365.5$ kJ mol^{-1} for all primary and secondary hydroperoxides and $D_{O-H} = 358.6$ kJ mol^{-1} for tertiary hydroperoxides (see Chapter 2). Therefore, the enthalpy of the reaction R$_i$OO$^\bullet$ + R$_j$H depends on the BDE of the attacked C—H bond of the hydrocarbon. But a different situation is encountered during oxidation and co-oxidation of aldehydes. As proved earlier, the BDE of peracids formed from acylperoxyl radicals is much higher than the BDE of the O—H bond of alkyl hydroperoxides and depends on the structure of the acyl substituent. Therefore, the BDEs of both the attacked C—H and O—H of the formed peracid are important factors that influence the chain propagation reaction. This is demonstrated in Table 8.9 where the calculated values of the enthalpy of the reaction R$_i$O$_2^\bullet$ + R$_j$H for different R$_j$Hs including aldehydes and different peroxyl radicals are presented. One can see that the value ΔH(RO$_2^\bullet$ + RH) is much lower in the reactions of the same compound with acylperoxyl radicals.

Another factor that influences the reactivity of two polar reactants, acylperoxyl radical with aldehyde, is the polar interaction of carbonyl group with reaction center in the transition state. Aldehydes are polar compounds, their dipole moments are higher than 2.5 Debye (see Section 8.1.1). The dipole moment of the acylperoxyl radical is about 4 Debye ($\mu = 3.87$ Debye for PhC(O)OO$^\bullet$ according to the quantum-chemical calculation [54]). Due to this, one can expect a strong polar effect in the reaction of peroxyl radicals with aldehydes. The IPM helps to evaluate the increment ΔE_μ in the activation energy E_e of the chosen reaction using experimental data [1]. The results of ΔE_μ calculation are presented in Table 8.10.

TABLE 8.7
Rate Constants for Reaction of Different Peroxyl Radicals with Aldehydes

Aldehyde	Peroxyl Radical	T (K)	k (L mol^{-1} s^{-1})	Ref.
$CH_3C(O)H$	$Me_3CO_2^\bullet$	213–263	$1.7 \times 10^4 \exp(-18.2/RT)$	[44]
$EtC(O)H$	$Me_3CO_2^\bullet$	213–263	$3.0 \times 10^3 \exp(-15.1/RT)$	[44]
$PrC(O)H$	$Me_3CO_2^\bullet$	213–263	$1.1 \times 10^3 \exp(-12.5/RT)$	[44]
$Me(CH_2)_5C(O)H$	$HO_2^\bullet$	273	50	[45]
$Me_2CHC(O)H$	$Me_3CO_2^\bullet$	213–263	$1.5 \times 10^3 \exp(-12.7/RT)$	[44]
$Me_3CC(O)H$	$HO_2^\bullet$	273	2.3×10^3	[45]
$PhC(O)H$	$Me_3CO_2^\bullet$	213–263	$1.6 \times 10^5 \exp(-26.0/RT)$	[44]
$PhC(O)H$	$Me_2PhCO_2^\bullet$	348	68	[46]
4-MeO-C$_6$H$_4$-CHO	$Me_3CO_2^\bullet$	213–263	$1.5 \times 10^3 \exp(-15.4/RT)$	[44]
4-MeO-C$_6$H$_4$-CHO	$Me_2PhCO_2^\bullet$	348	1.1×10^3	[46]
2,5-(MeO)$_2$-C$_6$H$_3$-CHO	$Me_2PhCO_2^\bullet$	348	32	[46]
4-Br-C$_6$H$_4$-CHO	$Me_2PhCO_2^\bullet$	348	55	[46]
4-Cl-C$_6$H$_4$-CHO	$Me_3CO_2^\bullet$	213–263	$1.0 \times 10^3 \exp(-15.9/RT)$	[44]
4-F-C$_6$H$_4$-CHO	$Me_2PhCO_2^\bullet$	348	66.4	[46]
4-O$_2$N-C$_6$H$_4$-CHO	$Me_2PhCO_2^\bullet$	348	51.3	[46]
3-O$_2$N-C$_6$H$_4$-CHO	$Me_2PhCO_2^\bullet$	348	41.9	[46]
2,4-Cl$_2$-C$_6$H$_3$-CHO	$Me_2PhCO_2^\bullet$	348	10.4	[46]
$MeCH=CHC(O)H$	$Me_3CO_2^\bullet$	213–263	$6.0 \times 10^3 \exp(-16.7/RT)$	[44]
$CH_2=CHC(O)H$	$Me_3CO_2^\bullet$	213–263	$1.9 \times 10^3 \exp(-15.9/RT)$	[44]

The values of ΔE_μ are negative and varies from -1.6 to -11 kJ mol^{-1}. The reaction center C $\cdots$ H $\cdots$ O for the reaction of the peroxyl radical with the C—H bond of the hydrocarbon has a nearly linear geometry (see Chapter 2). The polar interaction changes the geometry of the TS (see Chapter 7). The geometric parameters for the reactions of peroxyl radicals with aldehydes

TABLE 8.8
Rate Constants of Addition of Acylperoxyl Radical to Olefins with Epoxide Formation

Olefin	RC(O)OO$^\bullet$	T (K)	k (L mol^{-1} s^{-1})	Ref.
$CH_2=CH_2$	MeC(O)OO$^\bullet$	393	5.0×10^3	[50]
$CH_2=CHMe$	MeC(O)OO$^\bullet$	393	6.4×10^3	[50]
$CH_2=CHEt$	MeC(O)OO$^\bullet$	393	1.1×10^4	[46]
Z-MeCH=CHMe	MeC(O)OO$^\bullet$	393	7.5×10^4	[51]
E-MeCH=CHMe	MeC(O)OO$^\bullet$	393	1.2×10^5	[51]
Z-MeCH=CHEt	MeC(O)OO$^\bullet$	393	1.4×10^5	[50]
E-MeCH=CHEt	MeC(O)OO$^\bullet$	393	1.4×10^5	[50]
$CH_2=CMe_2$	MeC(O)OO$^\bullet$	393	9.2×10^4	[46]
$CH_2=CMeEt$	MeC(O)OO$^\bullet$	393	1.5×10^5	[50]
$Me_2C=CHMe$	MeC(O)OO$^\bullet$	393	8.4×10^5	[50]
$CH_2=CHCHMe_2$	MeC(O)OO$^\bullet$	393	1.3×10^4	[50]
$CH_2=CHBu$	MeC(O)OO$^\bullet$	393	2.2×10^4	[50]
$CH_2=CHBu$	PhC(O)OO$^\bullet$	343	3.2×10^3	[52]
$CH_2=CHBu$	PhC(O)OO$^\bullet$	293	1.2×10^3	[53]
MeCH=CHPr	PhC(O)OO$^\bullet$	293	1.3×10^3	[53]
MeCH=CH(CH$_2$)$_4$Me	PhC(O)OO$^\bullet$	293	1.3×10^3	[53]
$CH_2=CH(CH_2)_5Me$	PhC(O)OO$^\bullet$	293	2.1×10^3	[53]
$CH_2=CH(CH_2)_5Me$	PhC(O)OO$^\bullet$	343	4.5×10^2	[52]
cyclopentene	PhC(O)OO$^\bullet$	293	2.8×10^3	[53]
cyclohexene	PhC(O)OO$^\bullet$	293	1.8×10^5	[45]

were calculated using Equations (6.35)–(6.37) of the IPM. The results of the calculation of interatomic distances $r(C-H)$, $r(O-H)$, and angle $\varphi(C-H-O)$ for the reactions $R_iO_2^\bullet + RCH(O)$ are presented in Table 8.10. The angle $\varphi(C-H-O)$ of the TS is less than 180° and changes from 172° to 160°. So, polar interaction sufficiently changes the activation energy as well as geometry of the TS in the reaction of the peroxyl radical with the aldehyde. It is seen that polar interaction has strong influence and changes in angle φ varies from 180° to 160°.

8.2 CHEMISTRY OF PERACIDS

Peracids were isolated as the primary products of the oxidation of aldehydes [5]. Peracid reacts with aldehyde with the formation of an intermediate hydroxyoxoperoxide [4,5].

$$RC(O)OOH + RC(O)H \longrightarrow RCH(OH)OOC(O)R$$

The structure of the formed peroxide as hydroxyoxoperoxide was proved by NMR spectroscopy for peroxyacetic and peroxybutyric acids [55]. On the basis of IR spectra, it was sugested

TABLE 8.9
Enthalpies of Reactions of Different Peroxyl Radicals with Hydrocarbons and Aldehydes (See Tables 6.3 and 8.1 for BDE Values)

R—H	Me$_3$COO$^\bullet$	Me$_2$CHOO$^\bullet$	Me$_3$C(O)OO$^\bullet$	RC(O)OO$^\bullet$	PhC(O)OO$^\bullet$
EtMeCH—H	54.4	47.5	36.0	26.0	9.0
Me$_3$C—H	41.4	34.5	23.0	13.0	−4.0
CH$_2$=CH(CH—H)Me	−9.2	−16.1	−27.3	−37.6	−54.3
cyclohexane	−17.1	−24.0	−35.5	−45.5	−62.5
MePhCH—H	5.5	−1.4	−12.9	−22.9	−39.9
Me$_2$PhC—H	−3.9	−10.8	−22.3	−32.3	−49.3
tetralin	−9.0	−15.9	−27.4	−37.4	−54.4
dihydroanthracene	−36.6	−43.5	−55.0	−65.0	−82.0
MeC(O)—H	15.2	8.3	−3.2	−13.2	−30.2
PrC(O)—H	12.6	5.7	−5.8	−15.7	−32.8
PhC(O)—H	−10.6	−17.5	−29.0	−39.0	−56.0
PhCH$_2$C(O)—H	3.4	−3.5	−15.0	−25.0	−42.0
4-Me-benzaldehyde	−12.8	−19.7	−31.2	−41.2	−58.2
4-Cl-benzaldehyde	−6.3	−13.2	−24.7	−34.7	−51.7
4-O$_2$N-benzaldehyde	−18.2	−25.1	−36.6	−46.6	−63.4

that the peroxide formed from heptanoic peracid and heptaldehyde exists in three isomeric forms: free, with intramolecular and intermolecular hydrogen bonds [56].

TABLE 8.10
Geometric Parameters of the TS and Increments ΔE_μ of Polar Ineractions for Peroxyl Radical Reactions with Aldehydes (Equations [6.32], [6.35]–[6.38])

Aldehyde	Peroxyl Radical	ΔH_e (kJ mol^{-1})	ΔE_μ (kJ mol^{-1})	r(C—H) (m)	r(O—H) (m)	φ(CHO)°
$CH_3C(O)H$	$Me_3CO_2^\bullet$	11.4	−7.5	1.373×10^{-10}	1.211×10^{-10}	161
$CH_3C(O)H$	$MeC(O)O_2^\bullet$	−17.1	−3.7	1.349×10^{-10}	1.235×10^{-10}	167
$EtC(O)H$	$Me_3CO_2^\bullet$	2.6	−5.8	1.375×10^{-10}	1.209×10^{-10}	163
$PrC(O)H$	$Me_3CO_2^\bullet$	8.8	−5.7	1.375×10^{-10}	1.209×10^{-10}	163
$Me(CH_2)_5C(O)H$	$HO_2^\bullet$	−1.6	−6.5	1.360×10^{-10}	1.224×10^{-10}	162
$Me(CH_2)_5C(O)H$	$Me(CH_2)_5C(O)O_2^\bullet$	−19.7	−2.9	1.348×10^{-10}	1.236×10^{-10}	168
$Me_3CC(O)H$	$Me_3CC(O)O_2^\bullet$	−5.7	−8.3	1.350×10^{-10}	1.234×10^{-10}	160
cyclohexyl-C(O)H	cyclohexyl-C(O)O-O$^\bullet$	−8.8	−5.1	1.341×10^{-10}	1.243×10^{-10}	164
$PhC(O)H$	$Me_3CO_2^\bullet$	−14.4	−4.4	1.378×10^{-10}	1.233×10^{-10}	166
$PhC(O)H$	$Me_2PhCO_2^\bullet$	−14.4	−5.7	1.378×10^{-10}	1.233×10^{-10}	165
$PhC(O)H$	$PhC(O)O_2^\bullet$	−59.7	−1.6	1.330×10^{-10}	1.281×10^{-10}	172
$MeCH=CHC(O)H$	$Me_3CO_2^\bullet$	−23.0	−10.7	1.389×10^{-10}	1.257×10^{-10}	160
$CH_2=CHC(O)H$	$Me_3CO_2^\bullet$	−23.0	−8.9	1.389×10^{-10}	1.257×10^{-10}	161

The interaction of peracid and aldehyde was found to be reversible [4].

$$RC(O)OOH + RC(O)H \rightleftharpoons RC(O)OOCH(OH)R$$

The equilibrium constants for the reaction of acetaldehyde with peracetic acid are: $K = 6.7$ mol L^{-1} (acetic acid, 298 K [20]) and $K = 6.25 \times 10^{-4} \exp(23.0/RT)$ mol^{-1} L (toluene, 253–303 K [57]). Hydroxyoxoperoxide decomposes with acetic acid formation [4,5].

$$RC(O)OOCH(OH)R \longrightarrow 2RC(O)OH$$

The thermal decomposition of peracetic acid in an aqueous solution produces acetic acid, CO_2, and dioxygen [4]. The detailed data on the chemistry and the kinetics of peracids decay are presented elsewhere [4–7].

Perbenzoic acid reacts with benzaldehyde by the bimolecular reaction to form benzoic acid as the main product.

$$PhC(O)OOH + PhC(O)H \longrightarrow 2PhC(O)OH$$

The rate of the reaction was found to be equal:

$$v = k[PhC(O)OOH][PhC(O)H]. \tag{8.3}$$

The rate constant of the reaction depends on the solvent: $k = 2.0 \times 10^{-4}$ L mol^{-1} s^{-1} (benzene, 303 K [58]), $k = 7.18 \times 10^4 \exp(-46.0/RT) = 6.2 \times 10^{-4}$ L mol^{-1} s^{-1} (acetic acid, 298 K [59]), $k = 2.2 \times 10^3 \exp(-41.8/RT) = 3.0 \times 10^{-4}$ L mol^{-1} s^{-1} (o-dichlorobenzene, 313 K [14]). In addition to benzoic acid, phenol was found to be produced by the reaction of benzoic acid with benzaldehyde [60,61]. Acids accelerate the reaction of perbenzoic acid with benzaldehyde.

The study of benzaldehyde and cyclohaxanone co-oxidation showed the formation of ε-caprolactone as the main product of cyclohexanone oxidation [5]. Cyclohexanone was found not to react practically with peroxyl radicals under mild conditions. The oxidation of benzaldehyde produces perbenzoic acid. The latter oxidizes the benzaldehyde to benzoic acid and cyclohexanone to ε-caprolactone.

$$\text{PhC(O)OOH} + \text{PhC(O)H} \longrightarrow 2\text{PhC(O)OH} \quad (k1)$$

$$\text{PhC(O)OOH} + \text{cyclohexanone} \longrightarrow \text{PhC(O)OH} + \text{ε-caprolactone} \quad (k2)$$

The last reaction occurs much rapidly: $k_1 = 2.17 \times 10^3 \exp(-41.8/RT)$ L mol^{-1} s^{-1} and $k_2 = 1.78 \times 10^4 \exp(-41.8/RT)$ L mol^{-1} s^{-1} [5].

8.3 OXIDATION OF KETONES

8.3.1 CHAIN MECHANISM OF KETONE OXIDATION

Ketones, like hydrocarbons and other organic compounds, are oxidized by dioxygen via the chain mechanism [4,62]. The carbonyl group weakens the adjacent C—H bond. Therefore, a peroxyl radical attacks the α-C—H bond as this bond is the most reactive in a ketone. The peculiarities of ketone oxidation are the same as aldehyde oxidation.

1. Ketones are more reactive than aliphatic hydrocarbons due to the influence of the carbonyl group on the BDE of the adjacent C—H bond (see Table 8.11).
2. The formed hydroperoxide reacts with the carbonyl group. This interaction influences the kinetics of the oxidation of ketones due to the fast splitting of the formed peroxide to free radicals.
3. Ketone and the formed α-ketoperoxyl radical are polar molecules. Hence the polar effect influences the reactivity of the ketones and the peroxyl radicals. Polar solvents also influence the reactions of peroxyl radicals with ketones as well as other free radical reactions.

The kinetic scheme of the oxidation of ketones RCH$_2$C(O)R^1 is similar to that for hydrocarbons, aldehydes, etc. It includes the presence of initiator I and a high concentration of the dioxygen ($>10^{-5}$ mol L^{-1}). Oxidation procceds by the following elementary steps [4,62]:

$$\text{I} \longrightarrow \text{r}^\bullet$$
$$\text{r}^\bullet + \text{RCH}_2\text{C(O)R}^1 \longrightarrow \text{RH} + \text{RC}^\bullet\text{HC(O)R}^1$$
$$\text{O}_2 + \text{RC}^\bullet\text{HC(O)R}^1 \longrightarrow \text{RCH(OO}^\bullet\text{)C(O)R}^1$$
$$\text{RCH(OO}^\bullet\text{)C(O)R}^1 + \text{RCH}_2\text{C(O)R}^1 \longrightarrow \text{RCH(OOH)C(O)R}^1 + \text{RC}^\bullet\text{HC(O)R}^1$$
$$2\text{RCH(OO}^\bullet\text{)C(O)R}^1 \longrightarrow \text{RCH(OH)C(O)R}^1 + \text{O}_2 + \text{RC(O)C(O)R}^1$$

TABLE 8.11
The Values of C—H bond Dissociation Energies in Ketones [1,2] and Enthalpies ΔH of Reactions: $sec\text{-}RO_2^\bullet + R^1CH_2C(O)R^2 \longrightarrow sec\text{-}ROOH + R^1C^\bullet HC(O)R^2$

Ketone	Bond	D_{C-H} (kJ mol^{-1})	ΔH (kJ mol^{-1})
Acetone	MeC(O)CH$_2$—H	411.8	46.3
2-Butanone	MeC(O)(Me)CH—H	397.2	31.7
2-Butanone, 3-methyl-	MeC(O)C—HMe$_2$	387.6	22.1
2-Pentanone	MeC(O)CH—HEt	394.6	29.1
3-Pentanone	EtC(O)CH—HMe	396.5	31.0
4-Heptanone	PrC(O)CH—HEt	394.6	29.1
3-Pentanone, 2,4-dimethyl-	Me$_2$CHC(O)Me$_2$C—H	387.5	22.0
5-Nonanone	BuC(O)CH—HPr	395.8	30.3
Diacetyl	CH$_3$COCOCH$_2$—H	411.8	46.3
2,4-Pentanedione	MeC(O)CH—HC(O)Me	399.1	33.6
Ethylacetoacetate	MeCOCH—HCOEt	396.5	31.0
Cyclopentanone	(cyclopentanone α-CH)	401.2	35.7
Cyclohexanone	(cyclohexanone α-CH)	394.1	28.6
Acetophenone	PhC(O)CH$_2$—H	402.8	37.3
Ethylphenyl ketone	PhC(O)(Me)CH—H	394.4	28.9
1-Methylethylphenylketone	PhC(O)Me$_2$C—H	385.3	19.8
Dibenzyl ketone	PhCH$_2$C(O)(Ph)CH—H	353.6	−11.9
1-Penten-4-one	CH$_2$=CHCH—HC(O)CH$_3$	346.6	−18.9

The values of the $k_p(2k_t)^{-1/2}$ ratio are given in the Table 8.12. The absolute values of k_p and $2k_t$ are collected in Table 8.13.

Chain generation in autoxidized ketones proceeds via the bimolecular reaction [4]. The BDE of the α-C—H bonds of the alkyl and benzyl ketones are higher than 330 kJ mol^{-1} and, therefore, the bimolecular reaction should prevail as the main reaction of radical generation (see Chapter 4).

$$RCH_2C(O)R^1 + O_2 \longrightarrow RC^\bullet HC(O)R^1$$

Chain generation by the reaction with dioxygen was studied for cyclohexanone by the acceptor radical method and was proved to proceed through the bimolecular reaction with the rate

$$v_i = k_i[RCH_2C(O)R^1][O_2] \tag{8.4}$$

TABLE 8.12
The Values of Ratio $k_p(2k_t)^{-1/2}$ for Ketone Oxidation

Ketone	T (K)	$k_p(2k_t)^{-1/2}$ ($L^{1/2}$ mol$^{-1/2}$ s$^{-1/2}$)	$k_p(2k_t)^{-1/2}$, (343 K) ($L^{1/2}$ mol$^{-1/2}$ s$^{-1/2}$)	Ref.
Propanone-2	368–493	$7.5 \times 10^6 \exp(-64.8/RT)$	1.02×10^{-3}	[63]
Butanone-2	308–348	$27.5 \exp(-31.8/RT)$	3.95×10^{-4}	[64]
Pentanone-2	323–363	$1.5 \times 10^2 \exp(-37.7/RT)$	2.72×10^{-4}	[65]
Pentanone-3	343	5.45×10^{-4}	5.45×10^{-4}	[66]
Butanone-2, 3-methyl-	313–353	$2.8 \times 10^2 \exp(-31.8/RT)$	4.02×10^{-3}	[65]
Heptanone-4	343	4.52×10^{-4}	4.52×10^{-4}	[66]
Nonanone-5	343	6.63×10^{-4}	6.63×10^{-4}	[66]
Decanone-2	343	4.40×10^{-4}	4.40×10^{-4}	[66]
Decanone-3	333–353	$7.6 \times 10^2 \exp(-41.0/RT)$	4.34×10^{-4}	[66]
Decanone-5	343	4.16×10^{-4}	4.16×10^{-4}	[66]
Undecanone-6	343	4.56×10^{-4}	4.56×10^{-4}	[66]
Tetradecanone-8	343	4.00×10^{-4}	4.00×10^{-4}	[66]
Cyclohexanone	328–353	$4.6 \times 10^3 \exp(-46.9/RT)$	3.32×10^{-4}	[67]
2-Methylcyclohexanone	333–353	$6.7 \times 10^5 \exp(-57.7/RT)$	1.09×10^{-3}	[68]
3-Methylcyclohexanone	338–358	$6.5 \times 10^4 \exp(-54.0/RT)$	3.88×10^{-4}	[68]
4-Methylcyclohexanone	338–358	$7.2 \times 10^3 \exp(-46.9/RT)$	5.19×10^{-4}	[68]

TABLE 8.13
The Absolute Values of k_p and $2k_t$ for Ketone Oxidation

Ketone	T (K)	k_p (L mol^{-1} s^{-1})	$2k_t$ (L mol^{-1} s^{-1})	Ref.
$CH_3C(O)CH_3$	368–393	$4.0 \times 10^{-9} \exp(-68.2/RT)$		[63]
$MeC(O)CH_2Me$	308–348	$1.3 \times 10^{-5} \exp(-35.2/RT)$	$2.27 \times 10^7 \exp(-3.4/RT)$	[64]
$MeC(O)CH_2Et$	323–363	$4.0 \times 10^{-5} \exp(-41.0/RT)$	$7.11 \times 10^6 \exp(-4.6/RT)$	[65]
$EtCH_2C(O)(CH_2)_6Me$	333–353	$3.4 \times 10^{-6} \exp(-44.4/RT)$	$2.00 \times 10^7 \exp(-6.8/RT)$	[66]
$MeC(O)CHMe_2$	313–353	$5.7 \times 10^{-4} \exp(-30.1/RT)$	$3.00 \times 10^7 \exp(-6.7/RT)$	[65]
⬡=O	328–353	$2.4 \times 10^{-7} \exp(-43.5/RT)$	$2.70 \times 10^7 \exp(-6.7/RT)$	[67]
$PhC(O)CH_2Ph$	333–363	$1.6 \times 10^{-6} \exp(-30.5/RT)$		[69]
$PhC(O)CH_2Ph$	348	12		[70]
$PhCH_2C(O)CH_2Ph$	348	10		[70]
$PhCH_2C(O)CH_2Ph$	303	3.28	3.6×10^7	[21]

and rate constant $k_i = 1.05 \times 10^9 \exp(-109.6[\text{kJ mol}^{-1}]/RT) = 5.2 \times 10^{-6}$ L mol^{-1} s^{-1} at 400 K [71].

Co-oxidation was studied for cumene–2-heptylcyclohexanone pair [72] in different solvents (see Table 8.14). The parameters of co-oxidation (r_1, r_2, Φ) vary slightly from one solvent to another.

As mentioned above, the parameters of co-oxidation can be used for the estimation of the BDE of the O—H bond formed in the chain propagation reaction. The ratio

TABLE 8.14
Parameters of Cumene Co-Oxidation with 2-Heptylcyclohexanone (348 K) [72]

Solvent	r_1	r_2	ϕ
Cumene	0.7	2.1	0.5
Benzene	1.3	2.8	0.7
Acetophenone	0.8	1.6	1.1
Acetonitrile	1.4	2.0	0.8
1,1-Dimethylethanol	0.5	1.2	1.0
Nitrobenzene	0.7	1.7	0.5
Pyridine	0.6	1.6	0.7
Acetic acid	0.3	2.0	0.4

$k_{p22}/k_{p21} = 0.4$ (333 K) for the co-oxidation of cumene with cyclohexanone and, hence, $k_{p21} = 9.0$ L mol^{-1} s^{-1} (for the value of k_{p22}, see Table 8.13). The rate constant $k_{p11} = 1.72$ L mol^{-1} s^{-1} (see Table 2.8) the difference $\Delta E = E_{11} - E_{21} = 4.6$ kJ mol^{-1}, $\Delta D = 11.2$ kJ mol^{-1}, and $D(O—H) = 369.8$ kJ mol^{-1} for α-ketohydroperoxide of cyclohexanone.

This means that the carbonyl group increases the strength of O—H bond of the adjacent hydroperoxyl group. Therefore, the following BDEs for the secondary and tertiary O—H groups of α-ketohydro peroxides can be recommended.

Hydroperoxide	RCH(OO—H)C(O)R^1	R^{1}R^{2}C(OO—H)C(O)R^3
D (kJ mol^{-1})	369.8	363.0

8.3.2 Reactivity of Ketones in Reactions with Peroxyl Radicals

Like for aldehydes, two factors are important for the reactivity of ketones in reactions with peroxyl radicals: reaction enthalpy and polar interaction. The enthalpy of the reaction of the peroxyl radical with ketone is $\Delta H = D_{C—H} - D_{O—H}$. The BDE of the α-C—H bonds of ketones are lower than those of the C—H bonds of the hydrocarbons (see Table 8.11) and the BDEs of the O—H bonds in α-ketohydroperoxides are marginally higher than those of alkylhydroperoxides. Therefore, the enthalpies of RO$_2^\bullet$ + RH reactions are lower than those of parent hydrocarbons (Table 8.15).

The polar effect plays an important role in reactions involving polar molecules and radicals [1,73–76]. In this case, the energy of the transition state is governed not only by the enthalpy of reaction, but also by the interaction of two polar fragments, the carbonyl and peroxyl groups. In addition to polar effect, the carbonyl group having π-orbitals increases the triplet repulsion due to the interaction of π-electrons with electrons of the transition state. This effect is observed in free radical reactions with olefins and alkylaromatic hydrocarbons (see Chapter 6). The IPM model allows one to estimate the common influence of both these factors, i.e., polar interaction and triplet repulsion on activation energy. This estimation is performed by the comparison of the activation energy of the two thermoneutral reactions (see Chapter 6). One of the thermoneutral reactions is the reaction of the peroxyl radical with polar ketone YH, and another is the reaction of RO$_2^\bullet$ with hydrocarbon RH, which has no polar effect. The comparison of the two E_{e0} or parameters $(br_e)_{RH}$ and $(br_e)_{YH}$ helps to estimate the increment $\Delta E_\mu = (E_{e0})_{YH} - (E_{e0})_{RH}$ (Chapter 6, Equation [6.32]). The values of

TABLE 8.15
Geometric Parameters of the TS and Increments ΔE_μ for Peroxyl Radical Reactions with Ketones (Equations [6.32], [6.35]–[6.38])

Ketone	ΔH_e (kJ mol^{-1})	E_{e0} (kJ mol^{-1})	ΔE_μ (kJ mol^{-1})	r (C—H) (m)	r (O—H) (m)	$\varphi°$
MeC(O)CH$_2$Me	19.7	65.8	−13.8	1.426 × 10^{-10}	1.219 × 10^{-10}	155
MeC(O)CH$_2$Et	19.7	68.4	−11.1	1.431 × 10^{-10}	1.214 × 10^{-10}	158
MeC(O)CHMe$_2$	17.9	61.1	−17.5	1.417 × 10^{-10}	1.228 × 10^{-10}	152
C$_6$H$_{10}$=O	20.3	61.1	−18.9	1.427 × 10^{-10}	1.218 × 10^{-10}	151
PhC(O)CH$_2$Ph	−9.1	53.8	−12.5	1.386 × 10^{-10}	1.259 × 10^{-10}	157
PhCH$_2$C(O)CH$_2$Ph	−9.1	60.2	−14.3	1.386 × 10^{-10}	1.259 × 10^{-10}	155
PhCH$_2$C(O)CH$_2$Ph	−9.1	57.1	−9.2	1.386 × 10^{-10}	1.259 × 10^{-10}	160

ΔE_μ are collected in Table 8.15. It is seen that ΔE_μ is negative and varies from −9 to −19 kJ mol^{-1}.

Peroxyl radicals are known to react with olefins by an addition reaction (see Chapter 2). The same reaction was proposed for compounds with carbonyl groups (ketones, esters [77]).

$$RO_2^\bullet + R^1C(O)R^2 \longrightarrow R^1C(OOR)(O^\bullet)R^2$$

Such reactions were analyzed recently by Borisov and Denisov [78]. These reactions appeared to be very endothermic. The enthalpies of these reactions (kJ mol^{-1}) for a series of additions of HO$_2^\bullet$ and MeO$_2^\bullet$ to carbonyl compounds are given below:

RC(O)R^1	MeC(O)H	Me$_2$C(O)	PhC(O)H	PhC(O)Me
HO$_2^\bullet$	40.3	41.4	47.0	48.0
CH$_3$O$_2^\bullet$	44.1	48.8	51.0	52.0

All these reactions are endothermic and, in addition, occur with a loss of entropy. Back unimolecular reactions are exothermic and occur with an increase in entropy. So, the role of such reactions should be negligible due to high activation energy and very fast back reaction. The values of the rate constants for addition reactions CH$_3$O$_2^\bullet$ + carbonyl compound, calculated by the IPM method are presented in the following table:

RC(O)R^1	MeC(O)H	Me$_2$C(O)	PhC(O)H	PhC(O)Me
k (298 K)(L mol^{-1} s^{-1})	2 × 10^{-13}	5 × 10^{-13}	2 × 10^{-14}	2 × 10^{-14}

The rates of these reactions are a few orders magnitude less (10^{10} – 10^{12}) than the rates of chain termination. So, they are negligible for the fate of free radical transformation in the oxidized ketones.

TABLE 8.16
Solvent Influence on the Values of k_p and k_t of Butanone-2 Oxidation [87]

Solvent	k_p (313 K) (L mol^{-1} s^{-1})	E_p (kJ mol^{-1})	log A, A (L mol^{-1} s^{-1})	$2k_t$ (313 K) (L mol^{-1} s^{-1})	E_t (kJ mol^{-1})	log A, A (L mol^{-1} s^{-1})
Decane	9.0			5.00×10^4		
Benzene	5.7	57.3	8.38	7.50×10^4	11.3	6.78
Carbon, tetrachloride	24.0 (333K)			1.30×10^5 (333 K)		
Chlorobenzene	11.5	43.1	6.25	1.16×10^5	12.5	7.14
p-Dichloro benzene	27.6 (333K)			9.10×10^4 (333 K)		
Acetic acid	12.3	40.2	5.80	7.20×10^4	12.1	7.85
Butanone-2	17.0	35.1	5.10	1.58×10^6	6.7	7.30

The polar interaction changes the geometry of the transition state of the reaction $RO_2^\bullet + RH$. Atoms C, H, O of the reaction center $O \cdots H \cdots C$ of this reaction are in a straight line for the reaction of the peroxyl radical with a hydrocarbon. The reaction center $O \cdots H \cdots C$ has an angular geometry in the reaction of the polar peroxyl radical with a polar molecule of the ketone. The interatomic distances r_{C-H} and r_{O-H} and angles $\varphi(CHO)$ for peroxyl radical reactions with ketones calculated by the IPM method [79,80] are given in Table 8.15.

Since the reactants ($RO_2^\bullet$, ketone) and the transition state have a polar character, they are solvated in a polar solvent. Hence polar solvents influence the rate constants of the chain propagation and termination reactions. This problem was studied for reactions of oxidized butanone-2 by Zaikov [81–86]. It was observed that k_p slightly varies from one solvent to another. On the contrary, k_t changes more than ten times from one solvent to another. The solvent influences the activation energy and pre-exponential factor of these two reactions (see Table 8.16).

The dependence of the rate constants (k_p and k_t) on the dielectric constant ε is described by the Kirkwood equation [87]:

$$\ln k = \ln k_o - \frac{L}{RT} \frac{\varepsilon - 1}{2\varepsilon + 1} \left(\frac{\mu_{RH}^2}{r_{RH}^3} + \frac{\mu_{RO_2}^2}{r_{RO_2}^3} - \frac{\mu_{\neq}^2}{r_{\neq}^3} \right) \quad (8.5)$$

This relationship proves that the influence of solvents (decane, benzene, acetic acid, p-dichlorobenzene) on the rate constant is due to the nonspecific solvation of the reactants and the transition state. The values of the dipole moments of the TS calculated from experimenntal data within the scope of Equation (18.2) have the following values [87]:

Solvent	PhH	MeCOOH	$C_{10}H_{22}$	CCl_4	p-ClC$_6$H$_4$Cl
$\mu_p^{\neq}$ (Debye)	8.1	8.4	8.0	8.0	8.1
$\mu_t^{\neq}$ (Debye)	11.1	11.4	11.5	11.3	12.5

The dependencies of k_p and k_t on ε were found to obey the following equations [87]:

$$\log[k_p(323\,K)\,(L\,mol^{-1}\,s^{-1})] = 1.55 + 2.1 \frac{\varepsilon - 1}{2\varepsilon + 1} \quad (8.6)$$

Oxidation of Carbonyl Compounds and Decarboxylation of Acids

$$\log[(2k_t)(323\,\text{K})\,(\text{L mol}^{-1}\text{s}^{-1})] = 3.18 + 6.2\frac{\varepsilon - 1}{2\varepsilon + 1} \qquad (8.7)$$

In addition to nonspecific solvation, a specific solvation of the polar reactants through the formation of intermolecular complexes exists. Peroxyl radicals produce complexes through hydrogen bonding with hydroxyl-containing molecules, such as water and alcohols. The decrease in the peroxyl radical activity due to hydrogen bonding was observed in the experiments of the oxidation of acetone with addition of water [88]. A volume of 6% (vol) of water added lowers the ratio $k_p(2k_t)^{-1/2}$ of acetone oxidation fivefold (393 K [88]). The dependence of the ratio $v/v_i^{1/2}$ on water concentration is described by the following equation [88]:

$$\frac{v}{\sqrt{v_i}} = \frac{1 + aK[\text{H}_2\text{O}]}{1 + K[\text{H}_2\text{O}]} \frac{k_p[RH]}{\sqrt{2k_t}} \qquad (8.8)$$

where $a = k_p'/k_p$ and K is the equilibrium constant. This effect was described by the following kinetic scheme.

$$\text{RO}_2^\bullet + \text{H}_2\text{O} \rightleftharpoons \text{ROO}^\bullet \cdots \text{H}_2\text{O} \qquad (K)$$
$$\text{RO}_2^\bullet + \text{RH} \longrightarrow \text{ROOH} + \text{R}^\bullet \qquad (k_p)$$
$$\text{RO}_2^\bullet \cdots \text{H}_2\text{O} + \text{RH} \longrightarrow \text{ROOH} + \text{H}_2\text{O} + \text{R}^\bullet \qquad (k_p')$$
$$\text{RO}_2^\bullet + \text{RO}_2^\bullet \longrightarrow \text{ROH} + \text{R}'(\text{O}) + \text{O}_2 \qquad (2k_t)$$

The parameters a and K were found to be $a = 0.8$, $K = 4\,\text{L mol}^{-1}$.

Similar results on the influence of hydrogen bonding on chain propagation and chain termination were obtained in the study of butanone-2 oxidation [83,89,90]. In addition to reactions discussed above, chain termination by the following reactions were added.

$$\text{RO}_2^\bullet \cdots \text{H}_2\text{O} + \text{RO}_2^\bullet \longrightarrow \text{Mol. products} \qquad (2k_t')$$
$$\text{RO}_2^\bullet \cdots \text{H}_2\text{O} + \text{RO}_2^\bullet \cdots \text{H}_2\text{O} \longrightarrow \text{Mol. products} \qquad (2k_t'')$$

The results of the interpretation of experimental data within the scope of this scheme are collected in Table 8.17. Hydrogen bonding decreases the value of k_p by 20 times and $2k_t$ by 18 times. For additional data on this problem, see the monograph by Emanuel et al. [87].

8.3.3 Interaction of Ketones with Hydroperoxides

Ketones react with hydroperoxides with formation a series of labile peroxides. This addition is performed reversibly across the carbonyl group of the ketone with the formation of a new peroxide.

$$\text{ROOH} + \text{R}^1\text{C(O)R}^2 \rightleftharpoons \text{R}^1\text{C(OH)(OOR)R}^2$$

The formed hydroxyperoxide decomposes into free radicals much more rapidly than alkyl hydroperoxide [91]. So, the equilibrium addition of the hydroperoxide to the ketone increases the rate of formation of radicals. This effect was first observed for cyclohexanone and 1,1-dimethylethyl hydroperoxide [91]. In this system, the rate of radical formation increases with the ketone concentration. The mechanism of radical formation is described by the following kinetic scheme:

TABLE 8.17
The Influence of Hydrogen Bonding on Reactions of Peroxyl Radicals with Butanone-2 [87]

Reaction	k (333 K) (L mol^{-1} s^{-1})	E (kJ mol^{-1})	log A, A (L mol^{-1} s^{-1})
MeCH(OO$^{\bullet}$)C(O)Me + MeCH$_2$C(O)Me	0.39	35.1	5.10
MeCH(OO$^{\bullet}$ $\cdots$ H$_2$O)C(O)Me + MeCH$_2$C(O)Me	2.02×10^{-2}	69.0	9.11
MeCH(OO$^{\bullet}$ $\cdots$ HOCMe$_3$)C(O)Me + MeCH$_2$C(O)Me	0.10	59.8	8.32
MeCH(OO$^{\bullet}$ $\cdots$ HOMe)C(O)Me + MeCH$_2$C(O)Me	6.7×10^{-2}		
MeCH(OO$^{\bullet}$)C(O)Me + H$_2$O $\rightleftharpoons$ MeCH(OO$^{\bullet}$ $\cdots$ H$_2$O)C(O)Me	1.07 (L mol^{-1})	-20	-3.12
MeCH(OO$^{\bullet}$)C(O)Me + HOCMe$_3$ $\rightleftharpoons$ MeCH(OO$^{\bullet}$HOCMe$_3$)C(O)Me	0.64	-20.5	-3.58
2 MeCH(OO$^{\bullet}$)C(O)Me $\rightarrow$ Mol. products	1.80×10^{6}	6.7	7.30
2 MeCH(OO$^{\bullet}$ $\cdots$ H$_2$O)C(O)Me $\rightarrow$ Mol. products	7.0×10^{4}	19.2	7.86
2 MeCH(OO$^{\bullet}$ $\cdots$ HOCMe$_3$)C(O)Me $\rightarrow$ Mol. products	7.0×10^{4}	13.0	7.06
2 MeCH(OO$^{\bullet}$ $\cdots$ HOMe)C(O)Me $\rightarrow$ Mol. products	7.0×10^{4}		

$$\text{ROOH} \longrightarrow \text{RO}^{\bullet} + \text{HO}^{\bullet} \qquad (k_{i1})$$

$$\text{ROOH} + \text{R}^1\text{C(O)R}^2 \rightleftharpoons \text{R}^1\text{C(OH)(OOR)R}^2 \qquad (K)$$

$$\text{R}^1\text{C(OH)(OOR)R}^2 \longrightarrow \text{R}^1\text{C(OH)(O}^{\bullet}\text{)R}^2 + \text{RO}^{\bullet} \qquad (k_{i2})$$

and the rate of radical generation is given by the equation

$$\begin{aligned} \nu_i &= 2ek_1[\text{ROOH}] + 2ek_2[\text{R}^1\text{C(OH)(OOR)R}^2] \\ &= (2ek_1 + 2ek_2K[\text{R}^1\text{C(O)R}^2])[\text{ROOH}](1 + K[\text{R}^1\text{COR}^2])^{-1} \end{aligned} \qquad (8.9)$$

The values of k_1, k_2 and K are collected in Table 8.18.

8.3.4 Chemistry of Ketone Oxidation

The first intermediate product of ketone oxidation is α-ketohydroperoxide. All other molecular products are formed by decay and reactions of this hydroperoxide and its adduct with ketone. Among these products, aldehydes, diketones, α-hydroxyketones, acids, esters, and CO$_2$ were observed. The information about the products of the oxidation of ketones by dioxygen are available in monographs [4,7].

The important problem for the oxidation of ketones and hydrocarbons is the mechanism of acid formation [4,7]. Langenbeck and Pritzkow proposed the following α-mechanism of acid formation [96]:

$$\text{RC(O)CH}_2\text{R}^1 \longrightarrow \text{RC(O)CH(OOH)R}^1 \longrightarrow \text{RC(O)OH} + \text{R}^1\text{C(O)H}$$

$$\text{R}^1\text{C(O)H} \longrightarrow \text{R}^1\text{C(O)OH}$$

According to this mechanism as the single mechanism of C—C bond splitting, the ratio [C$_{n-x}$ acid]/[C$_x$acid] should be equal to unity. But the experimental data on the composition of acids

TABLE 8.18
Rate Constants of Decomposition of ROOH(k_1) and $R^1C(OH)(OOR)R^2$ (k_2) and Equilibrium Constant K of Hydroxyalkyl Peroxide Formation from ROOH and Ketone

ROOH	Ketone	Solvent	T (K)	k_1 (s^{-1}) or k_2 (s^{-1}) or K (L mol^{-1})	Ref.
Me$_3$COOH	cyclohexanone	PhCl	383–400	$k_1 = 3.6 \times 10^{12} \exp(-138.1/RT)$ $k_2 = 3.6 \times 10^9 \exp(-108.8/RT)$ $K = 6.9 \times 10^{-7} \exp(46.0/RT)$	[91]
Me$_3$COOH	cyclohexanone	CCl$_4$	295–313	$K = 7.61 \times 10^{-3} \exp(10.5/RT)$	[92]
Me$_2$PhCOOH	cyclohexanone	PhCl	393	$k_1 = 2.1 \times 10^{-6}$ $k_2 = 5.7 \times 10^{-6}$, $K = 1.0$	[93]
Me$_2$PhCOOH	cyclohexanone	CCl$_4$	295–313	$K = 1.56 \times 10^{-3} \exp(12.5/RT)$	[92]
cyclohexyl-OOH	cyclohexanone	PhCl	353–383	$k_1 = 6.3 \times 10^{-11} \exp(-133.9/RT)$ $k_2 = 4.0 \times 10^{-4} \exp(-63.6/RT)$ $K = 1.4 \times 10^{-5} \exp(33.0/RT)$	[94]
2-hydroperoxycyclohexanone	cyclohexanone	PhCl	393	$k_1 = 1.2 \times 10^{-5}$ $k_2 = 3.0 \times 10^7 \exp(-85.3/RT)$	[94]
MeC(O)CH(OOH)Me	MeC(O)Et	C$_6$H$_6$	293–343	$k_1 = 3.4 \times 10^{-12} \exp(-114.6/RT)$ $k_2 = 1.15 \times 10^{-10} \exp(96.2/RT)$ $K = 0.8$ (293 K)	[95]
2-hydroperoxy-2-methylcyclohexanone	2-methylcyclohexanone	2-methylcyclohexanone	373–403	$k_2 = 1.9 \times 10^6 \exp(-76.1/RT)$	[4]
2-hydroperoxy-3-methylcyclohexanone	3-methylcyclohexanone	3-methylcyclohexanone	373–403	$k_2 = 6.9 \times 10^6 \exp(-79.5/RT)$	[4]
2-hydroperoxy-4-methylcyclohexanone	4-methylcyclohexanone	4-methylcyclohexanone	373–403	$k_2 = 8.9 \times 10^5 \exp(-72.4/RT)$	[4]
Me$_3$COOH	4-methylcyclohexanone	CCl$_4$	295–313	$K = 7.60 \times 10^{-3} \exp(10.5/RT)$	[92]
Me$_3$COOH	3-methylcyclohexanone	CCl$_4$	295–313	$K = 1.17 \times 10^{-2} \exp(8.4/RT)$	[92]
1-hydroperoxy-1,2,3,4-tetrahydronaphthalene	MeC(O)Me	CCl$_4$/Tetralin (1:1)	293	$K = 0.53$	[92]
1-hydroperoxy-1,2,3,4-tetrahydronaphthalene	Me(CO)Et	CCl$_4$/Tetralin (1:1)	293	$K = 0.38$	[92]
1-hydroperoxy-1,2,3,4-tetrahydronaphthalene	cyclohexanone	CCl$_4$/Tetralin (1:1)	293–313	$K = 4.07 \times 10^{-2} \exp(10.5/RT)$	[92]
1-hydroperoxy-1,2,3,4-tetrahydronaphthalene	3-methylcyclohexanone	CCl$_4$/Tetralin (1:1)	293	$K = 0.29$	[92]
1-hydroperoxy-1,2,3,4-tetrahydronaphthalene	3-methylcyclohexanone	CCl$_4$/Tetralin (1:1)	293–313	$K = 0.23 \exp(8.4/RT)$	[92]
1-hydroperoxy-1,2,3,4-tetrahydronaphthalene	4-methylcyclohexanone	CCl$_4$/Tetralin (1:1)	293–313	$K = 4.74 \times 10^{-2} \exp(9.6/RT)$	[92]

formed from oxidized hydrocarbons and ketones gave different ratios. It was firmly found that the ratio $[C_{n-x}$ acid$]/[C_x$acid$] > 1$ at $x < 0.5n$ [97–100]. For example, $C_2:C_4 = 2.52$ for hexane oxidation [98], $C_2:C_4 = 2.61$ for hexanone-2 oxidation [100], $C_2:C_5 = 2.35$ for nonanone-2 oxidation [100], $C_5:C_6 = 1.30$ for 5-decanone oxidation [101]. This discrepancy means that other mechanisms of acid formation with parallel formation of CO_2 exist in addition to the α-mechanism. The most probable seems to be the following mechanism. The α-ketoperoxyl radical formed from dialkyl ketone attacks the weak α-C—H bond of the same molecule.

$$RCH(OO^\bullet)C(O)CH_2R \longrightarrow RCH(OOH)C(O)C^\bullet HR$$

The peroxyl radical formed from this alkyl radical attacks the α-C—H bond, which is weak due to the dual influence of the adjacent oxo and hydroperoxy groups.

$$RCH(OOH)C(O)C^\bullet HR + O_2 \longrightarrow RCH(OOH)C(O)CH(OO^\bullet)R$$

$$RCH(OOH)C(O)CH(OO^\bullet)R \rightleftharpoons \text{[cyclic intermediate]}$$

$$\text{[cyclic intermediate]} \longrightarrow \text{[cyclic intermediate]}$$

$$\text{[cyclic intermediate]} \longrightarrow RC^\bullet(O) + RCH(OOH)C(O)OH$$

$$RCH(OOH)COOH \longrightarrow RC(O)H + H_2O + CO_2$$

Or

$$RCH(OOH)C(O)CH(OO^\bullet)R \longrightarrow RC^\bullet(OOH)C(O)CH(OOH)R$$

$$RC^\bullet(OOH)C(O)CH(OOH)R \longrightarrow RC(O)C(O)CH(OOH)R + HO^\bullet$$

$$RC(O)C(O)CH(OOH)R \longrightarrow \text{[cyclic intermediate]}$$

$$\text{[cyclic intermediate]} \longrightarrow RC(O)OH + RCH(O) + CO$$

$$RO_2^\bullet + CO \longrightarrow RO^\bullet + CO_2$$

The last reaction occurs rapidly. Cumylperoxyl radicals react with carbon oxide in the liquid phase with the rate constant $k = 6.8 \times 10^5 \exp(-41.8/RT) = 0.19$ L mol^{-1} s^{-1} ($T = 333$ K) [102].

8.4 OXIDATIVE DECARBOXYLATION OF CARBOXYLIC ACIDS

8.4.1 ATTACK OF PEROXYL RADICALS ON C—H BONDS

The carboxyl group has influence on the BDE of the α-C—H bond. These bonds are weaker than that of hydrocarbons. The values of the BDE of α-C—H bonds of carboxylic acids are

TABLE 8.19
The Values of α-C—H bond Dissociation Energy in Acids (D_{C-H}) and Enthalpies (ΔH) of Secondary Alkylperoxyl Radical ($R^1R^2CHOO^\bullet$) Reaction with Acids [2]

Acid	Bond	ΔH (kJ mol^{-1})	D_{C-H} (kJ mol^{-1})
Formic acid	C—H(O)OH	27.2	392.7
Acetic acid	HOOCCH$_2$—H	48.6	414.1
Propanoic acid	HOOC(Me)CH—H	33.4	398.9
Propanoic acid, 1-methyl-	HOOC(Me)$_2$C—H	22.8	388.3
Butanoic acid	EtCH—HCOOH	33.2	398.7
Propanoic acid, 2,2-dimethyl-	Me$_2$CH$_2$—HCCOOH	49.7	415.2
Pentanoic acid	PrCH—HCOOH	33.9	399.4
Hexanoic acid	BuCH—HCOOH	31.6	397.1
Cyclobutanecarboxylic acid	H, COOH (cyclobutyl)	33.5	399.0
Cyclohexanecarboxylic acid	H, COOH (cyclohexyl)	22.7	388.2
Benzenacetic acid	Ph(COOH)CH—H	1.5	367.0
Glycolic acid	HOOC(HO)CH—H	32.3	397.8
Propionic acid, 1-hydroxy-	MeC—H(OH)COOH	20.5	386.0
Aceticacid, cloro-	HOOCClCH—H	33.4	398.9
Propanoic acid, 2,2-dimethyl-	Me$_2$(COOH)CCH$_2$—H	40.6	415.1
1,4-Butanedioic acid	HOOCCH—HCH$_2$COOH	32.3	397.8
1,4-Butanedioic acid, 2,3-dihydroxy-	HOOCC—H(OH)CH(OH)COOH	27.2	392.7
Cyclohexene,1-carboxylic acid	cyclohexenyl-COOH	4.5	370.0

given in Table 8.19. It is seen from comparison of the BDE in acids and hydrocarbons (see ΔD values) that the BDE of the carboxylic group (D_{C-H}) decreases to 8–12 kJ mol^{-1}.

Peroxyl radicals attack all groups of acid Me(CH$_2$)$_n$COOH and preferencially the α-CH$_2$ group. The more the number of CH$_2$ groups in carboxylic acid, the higher is the rate constant (see Table 8.20).

The attack of peroxyl radicals on β-CH$_2$ groups produces the same functional groups (hydroperoxyl, hydroxy, oxo) as in the case of subsequent hydrocarbon oxidation. The oxidation of unsaturated acids proceeds similarly to the oxidation of olefins [4,7].

8.4.2 Oxidative Decarboxylation of Acids

Carboxylic acids are thermally stable. Decarboxylation of carboxylic acids is observed at 600 K and higher in the absence of dioxygen [4]. At the same time, the decarboxylation of fatty acids in oxidized cumene was observed at 350 K [71]. The study of CO_2 production from oxidized acetic, butanoic, isobutanoic, pentanoic, and stearic acids labeled with ^{14}C in the

TABLE 8.20
Rate Constants of Cumylperoxyl Radical Reaction with Mono- and Dicarboxylic Acids [104,105]

Acid	E (kJ mol^{-1})	log A, A (L mol^{-1} s^{-1})	k (398 K) (L mol^{-1} s^{-1})	k (CH$_2$)(398 K) (L mol^{-1} s^{-1})
Acetic	67.8	8.35	0.28	
Propanoic	63.2	7.85	0.36	0.12
Butanoic	61.9	7.83	0.49	0.13
Pentanoic	63.2	8.10	0.62	0.13
Heptanoic	63.2	8.24	0.87	0.12
Decanoic	63.6	8.51	1.45	0.16
Isobutiric	56.1	7.28	0.81	
Pentanedioic	47.7	6.44	1.51	
Heptanedioic	46.0	6.34	2.00	0.24
Octanedioic	47.7	6.67	2.57	0.35
Nonanedioic	47.7	6.73	2.95	0.36
Decanedioic	47.7	6.78	3.31	0.36

carboxyl group showed that a portion of CO_2 was formed from the labeled carboxyl group and another part from carbons of the acid "tail" [4]. The decarboxylation of deuterated acid RC(O)OD due to oxidation occurs more slowly than that of nondeuterated acid. The kinetic isotope effect was found to vary from 2 to 4 (co-oxidation of acids with cumene, 408 K, [acid] = 0.1 mol L^{-1})[103].

Acid	MeCOOH	PrCOOH	BuCOOH	Me(CH$_2$)$_8$COOH	Me(CH$_2$)$_{16}$CO$_2$H
k_H/k_D	4.2(383 K)	2.1	2.1	2.2	2.3

The kinetic study of the decarboxylation of aliphatic acids in co-oxidation with cumene showed the following two chemical channels of CO_2 production [104].

$$RO_2^\bullet + RCH_2C(O)OH \longrightarrow \begin{cases} RCH_2C(O)O^\bullet \longrightarrow R\dot{C}H_2 + CO_2 \\ R\dot{C}HC(O)OH \longrightarrow RCH(OOH)C(O)OH \end{cases}$$

$$RO_2^\bullet \longrightarrow RCH_2C(O)O^\bullet \longrightarrow R\dot{C}H_2 + CO_2$$
$$RCH_2C(O)OH$$
$$RO_2^\bullet \longrightarrow R\dot{C}HC(O)OH \xrightarrow{O_2, RH} RCH(OOH)C(O)OH \longrightarrow CO_2 + H_2O + RCHO$$

1. The attack of the cumylperoxyl radical (CuOO$^\bullet$) on the O—H group of the monomeric acid.

$$2\ RC(O)OH \rightleftharpoons R-C\underset{O\cdots H-O}{\overset{O-H\cdots O}{\diagup\diagdown}}C-R$$

$$RC(O)OH + CuOO^\bullet \longrightarrow CuOOH + RCO_2^\bullet \qquad (k_{pCOOH})$$

$$RCO_2^\bullet \longrightarrow R^\bullet + CO_2 \qquad \text{Fast}$$

The decomposition of carboxyl radical occurs very rapidly, and CO_2 is formed with a constant rate in the initiated co-oxidation of cumene and acid [104].

2. Cumylperoxyl radical attacks the α-CH_2 group of the carboxylic acid with the formation of a labile hydroperoxide. The concentration of this hydroperoxide increases during oxidation till it reaches a stationary concentration $[RCH(OOH)\text{-}COOH]_{st} = k_{p12}[RCH_2COOH][CuOO^\bullet]/k_d$. This reaction produces CO_2 with acceleration during some period of time equal to the time of increasing the α-carboxyhydroperoxide concentration.

$$CuOO^\bullet + RCH_2COOH \longrightarrow CuOOH + RC^\bullet HCOOH \qquad (k_{p12})$$

$$RC^\bullet HCOOH + O_2 \longrightarrow RCH(OO^\bullet)COOH \qquad F$$

$$RCH(OO^\bullet)COOH + CuH \longrightarrow RCH(OOH)COOH + Cu^\bullet \qquad (k_{p21})$$

$$RCH(OOH)COOH \longrightarrow RC(O)H + H_2O + CO_2 \qquad (k_d)$$

The values of the rate constants of all the three key reactions, $k_{pCHOH}(CuOO^\bullet + RC(O)OH)$, $k_{p12}(CuOO^\bullet + RCH_2COOH)$, and the rate of decomposition of the formed hydroperoxide (k_d) are presented in Table 8.21.

The decarboxylation via the peroxyl radical reaction with the carboxylic group was the main channel of CO_2 production (78–82%). The attack of the peroxyl radical on the CH_2 group adds 18–22% of CO_2. However, in the case of isobutyric acid with a weak tertiary C—H bond, the attack on the C—H bond appeared to be the main reaction of decarboxylation.

Dicarboxylic acids form very strong intramolecular hydrogen bonds.

$$n(H_2C)\diagup\diagdown_{\text{COOH}}^{\text{COOH}} \rightleftharpoons \left[\overset{O-H\cdots O}{\underset{O\cdots H-O}{C\diagup\diagdown C}}\right]_{(CH_2)_n}$$

As a result, the carboxyl groups are blocked, and the peroxyl radicals react only with the C—H bonds of the dicarboxylic acid. The results of the experimental study of acid decarboxylation in reactions with cumylperoxyl radicals are presented below (393 K, [acid] = 0.8 mol L^{-1}, cumene [CuOOH] = 0.1 mol L^{-1} [106]).

TABLE 8.21
Rate Constants of Cumylperoxyl Radical (CuOO·) Reactions with C—H Bond (k_{p12}) and Carboxyl Group (k_{COOH}) of Fatty Acids and Rate Constants of Decomposition (k_d) of Formed Hydroperoxide [104]

Acid	Rate Constant	k (408 K) (L mol^{-1} s^{-1})	E (kJ mol^{-1})	log A, A (L mol^{-1} s^{-1})
Butanoic	k_{COOH}	0.55	95.8	12.00
Butanoic	k_{p12}	2.5	66.9	8.96
Butanoic	k_d*	6.3×10^{-4}		
Pentanoic	k_{COOH}	0.59	94.6	11.88
Pentanoic	k_{p12}	2.6	69.9	11.88
Pentanoic	k_d*	6.4×10^{-4}		
Decanoic	k_{COOH}	0.78	92.0	11.67
Decanoic	k_{p12}	2.9	73.2	9.83
Decanoic	k_d*	6.9×10^{-4}		
Isobutanoic	k_{p12}	4.1	64.8	9.01
Isobutanoic	k_d ($T = 403$ K)*	1.04×10^{-3}		
Octadecanoic	k_{COOH}	0.88	87.9	11.20
Octadecanoic	k_{p12}	3.2	74.9	10.09
Octadecanoic	k_d*	7.4×10^{-4}		

* Values are given in s^{-1}.

	Heptanedioic Acid	Nonanedioic Acid	Decanedioic Acid
k_{p12} (L mol^{-1} s^{-1})	0.25	0.44	0.58
k_d (s^{-1})	6.7×10^{-4}	6.4×10^{-4}	6.7×10^{-4}

The decarboxylation of decanedioic acid was studied in co-oxidation with cyclohexanol [106]. This reaction was seen to proceed through the attack of radicals (HO$_2^·$ and cyclohexyl–hydroxyperoxyl radical), see Chapter 7) on the C—H bonds of the acid and decomposition of the formed hydroperoxide. It was found that $k_{p12} = 3.8 \times 10^4 \exp(-50.2/RT)$ L mol^{-1} s^{-1} and $k_d = 1.4 \times 10^{15} \exp(-108.8/RT)$ s^{-1}. The quasistationary concentration of the intermediate hydroperoxide decomposing with the formation of CO$_2$ was estimated to be as small as 3.4×10^{-3} mol L^{-1} (393 K, cyclohexanol, $v_i = 4 \times 10^{-6}$ mol L^{-1} s^{-1}).

Oxalic acid does not form intramolecular hydrogen bonds. The decarboxylation of oxalic acid occurs by the direct reaction of the peroxyl radical with the carboxylic group. The rate constants of the peroxyl radicals of the oxidized cyclohexanol in co-oxidation with oxalic acid was found to be $k_{p12} = 7.4 \times 10^7 \exp(-60.2/RT)$ L mol^{-1} s^{-1} in cyclohexanol at 348–368 K [106].

REFERENCES

1. TG Denisova, ET Denisov. *Kinet Catal* 44:278–283, 2004.
2. VE Tumanov, EA Kromkin, ET Denisov. *Russ Chem Bull* 51:1641–1650, 2002.
3. DR Lide, ed. *Handbook of Chemistry and Physics*, 72nd ed. Boca Raton: CRC Press, 1991–1992.

4. ET Denisov, NI Mitskevich, VE Agabekov. *Liquid-Phase Oxidation of Oxygen-Containing Compounds*. New York: Consultants Bureau, 1977.
5. L Sajus, I Seree de Roch. In: CH Bamford and CFH Tipper, eds. *Comprehensive Chemical Kinetics*, vol 16. Amsterdam: Elsevier, 1980, pp 89–124.
6. JA Howard. *Adv Free Radical Chem* 4:49–173, 1972.
7. NM Emanuel, ET Denisov, ZK Maizus. *Liquid-Phase Oxidation of Hydrocarbons*, New York: Plenum Press, 1967.
8. L Horner. In: ed. WO Landberg. *Autoxidation and Antioxidants*, vol 1, New York: Wiley, 1961, pp 171–232.
9. GE Zaikov, JA Howard, KU Ingold. *Can J Chem* 47:3017–3029, 1969.
10. J Lemaire, M Niclause, M Dzierzynski. *Bull Soc Chim Belg* 71:780–791, 1962.
11. HR Cooper, HW Melville. *J Chem Soc* 1984–2002, 1951.
12. LA Andrianova, BI Chernyak. *Ukr Khim Zh* 39:199–201, 1973.
13. TA Ingles, HW Melville. *Proc Roy Soc Lond*, A218:163–189, 1953.
14. JP Franck, I Seree de Roch, L Sajus. *Bull Soc Chim Fr* 6:1947–1957, 1969.
15. NG Ariko, NI Mitskevich, VA Lashitskii, MK Buslova, MD Koval'kov. *Neftekhimiya* 10:48–53, 1970.
16. JR McNesby, TW Davies. *J Am Chem Soc* 76:2148–2152, 1954.
17. J Lemaire, M.Niclause, E Paraut. *Comput Rend* 260:2203–2209, 1965.
18. Z Csuzos, J Morgos, B Losonczi. *Acta Chim Acad Sci Hung* 43:271–283, 1965.
19. K Nakao, T Matsumoto, T Otake. *Kogyo Kagaku Zasshi* 72:1880–1892, 1969.
20. CEH Bawn, JB Williamson. *Trans Faraday Soc* 47:721, 1961.
21. JA Howard, S Korcek. *Can J Chem* 48:2165–2176, 1970.
22. CA McDowell, IK Sharples. *Can J Chem* 36:251–259, 1958.
23. CA McDowell, S Sifniades. *Can J Chem* 41:300–307, 1963.
24. KU Ingold. *Acc Chem Res* 2:1–9, 1969.
25. ET Denisov. *Zh Phys Khim* 52:1585–1597, 1978.
26. BI Chernyak, LA Andrianova. *Zh Org Khim* 11:1800–1802, 1975.
27. BI Chernyak, LA Andrianova. *Neftekhimiya* 14:97–100, 1974.
28. T Hava, Y Ohkatsu, T Osa. *Chem Lett* 1953–1955, 1973.
29. MA Maaruan. Synthesis of Alkylacroleins and Study of its Liquid-Phase Oxidation. PhD Thesis Dissertation, L'vov Politechnical Institute, L'vov, 1978 [in Russian].
30. MA Maaruan, DK Tolopko, BI Chernyak. *Kinet Katal* 18:224–229, 1977.
31. MA Maaruan, BI Chernyak, MV Nikipanchuk. *Zh Org Khim* 16:1573–1577, 1978.
32. E Boga, F Marta. *Acta Chim Acad Sci Hung* 78:105–112, 1973.
33. N Komissarova. Mechanism of Oxidation of Aldehydes by Ozone–Oxygen Mixture. PhD Thesis Dissertation, Institute of Chemical Physics, Chernogolovka, 1978, pp 3–25 [in Russian].
34. MA Maaruan, MV Nikipanchuk, BI Chernyak. *Kinet Katal* 19:499–501, 1978.
35. RV Kucher, IA Opeida. *Cooxidation of Organic Compounds in Liquid-Phase*. Kiev: Naukova Dumka, 1989 [in Russian].
36. AM Novak. Liquid-Phase Oxidation of Benzaldehyds and Tetrahydrobenzaldehydes and Cooxidation of Benzaldehydes with Organic Compounds. Thesis Dissertation, Institute Physical Organic Chemistry, Donetsk, 1979 [in Russian].
37. AM Novak, IA Opeida, RV Kucher. *Neftekhimiya* 19:446–451, 1979.
38. IA Opeida, VG Koloerova, AM Novak. *Kinet Katal* 23:489–491, 1982.
39. AM Novak, RV Kucher, IA Opeida. *Dokl AN USSR Ser B* 995–998, 1977.
40. KE Simmons, DE Van Sickle. *J Am Chem Soc* 95:7759–7763, 1973.
41. HLJ Backstrom, CF Aquist. *Acta Chem Scand* 24:1431–1444, 1970.
42. AM Ivanov. Role of Aldehydes in Selective Liquid-Phase Oxidation of Alkylaromatic Compounds and Oils. Thesis Dissertation, LPI, Leningrad, 1980 [in Russian].
43. C Walling, EA McElhill. *J Am Chem Soc* 73:2927–2931, 1951.
44. VA Mardoyan, LA Tavadyan, NB Nalbandyan. *Khim Fizika* 4:1107–1111, 1985.
45. T Ikava, T Fukushima, M Muto, T. Yanagihara. *Can J Chem* 44:1817–1825, 1966.

46. K Selby, DJ Waddington. *J Chem Soc Perkin Trans II*, 1715–1718, 1975.
47. PI Valov, EA Blumberg, NM Emanuel. *Izv AN SSSR Ser Khim* 1334–1339, 1966.
48. PI Valov, EA Blumberg, NM Emanuel. *Izv AN SSSR Ser Khim* 791–796, 1969.
49. TV Filippova, EA Blumberg. *Usp Khim* 51:1017–1033, 1982.
50. RR Dias, K Selby, DJ Waddington. *J Chem Soc Perkin Trans II*, 360, 1977.
51. RR Dias, K Selby, DJ Waddington. *J Chem Soc Perkin Trans II*, 758, 1975.
52. AD Vreugdenhil, H Reit. *Rec Trav Chim* 91:237–244, 1972.
53. T Ikawa, H Tomizawa, T Yamagihara. *Can J Chem* 45:1900–1902, 1967.
54. RV Kucher, IA Opeida, AF Dmitruk. *Neftekhimiya* 18:519–524, 1978.
55. OP Yablonskii, MG Vinogradov, RV Kereselidze, GI Nikishin. *Izv AN SSSR Ser Khim* 318–321, 1969.
56. J Lemaire, M Niclause, E Parlaut. *Compt Rend* 260:2203–2205, 1965.
57. RF Vasiliev, AI Terenin, NM Emanuel. *Izv AN SSSR Ser Khim* 403–410, 1956.
58. Y Ogata, I Tabushi. *J Am Chem Soc* 83:3444–3448, 1961.
59. CEH Bawn, JE Jolley. *Proc Roy Soc Lond* A236:297–305, 1956.
60. WE Parker, LP Witnauer, D Swern. *J Am Chem Soc* 80:323–327, 1958.
61. D Lefort, J Sorba, D Rouillard. *Bull Soc Chim Fr* 2219–2225, 1961.
62. ET Denisov. In: CH Bamford and CFH Tipper, eds. *Comprehensive Chemical Kinetics*, vol 16. Amsterdam: Elsevier, 1980, pp 125–204.
63. ET Denisov. *Izv AN SSSR Ser Khim* 812–815, 1960.
64. GE Zaikov, AA Vichutinskii, ZK Maizus. *Kinet Katal* 8:675–676, 1967.
65. GE Zaikov. *Kinet Katal* 9:1166–1169, 1968.
66. LK Obukhova. *Neftekhimiya* 5:97–100, 1965.
67. AL Aleksandrov, ET Denisov. *Kinet Katal* 10:904–906, 1969.
68. VA Itskovich. The Liquid-Phase Oxidation of Isomeric Cyclohexanones by Dioxygen. PhD Thesis Dissertation, Institute of Chemical Technology, Leningrad, 1970 [in Russian].
69. AA Fokin. Oxidation of Aromatic Ketones in Liquid-Phase by Dioxygen. PhD Thesis Dissertation, Kemerovo State University, Kemerovo, 1985, pp 1–22 [in Russian].
70. AM Romantsevitch, MA Simonov, IA Opeida, VV Petrenko, VN Matvienko. *Zh Org Khim* 25:805–810, 1989.
71. GV Butovskaya, VE Agabekov, OP Dmitrieva, NI Mitskevich. *Dokl AN BSSR*, 20:1103–1105, 1976.
72. IA Opeida, LG Nechitailo, AG Matvienko. *Ukr Khim Zh* 48:542–551, 1982.
73. ET Denisov, TG Denisova. *Kinet Catal* 34:883–889, 1993.
74. ET Denisov, TG Denisova. *Russ Chem Bull* 43:29–34, 1994.
75. ET Denisov, TG Denisova. *Kinet Catal* 35:305–311, 1994.
76. ET Denisov. *Kinet Catal* 35:617–635, 1994.
77. VS Martemyanov. Kinetics and Mechanism of High-Temperature Oxidation of Esters of Polyatomic Alcohols. Doctoral Thesis Dissertation, Institute of Chemical Physics, Chernogolovka, 1987, pp 1–46 [in Russian].
78. IM Borisov, ET Denisov. *Neftekhimiya* 41:24–29, 2001.
79. AF Shestakov, ET Denisov. *Russ Chem Bull* 320–329, 2003.
80. TG Denisova, NS Emel'yanova. *Kinet Catal* 44:441–449, 2003.
81. GE Zaikov. *Izv AN SSSR Ser Khim* 1692–1697, 1967.
82. GE Zaikov, ZK Maizus. *Adv Chem Ser*, vol 1. Washington: American Chemical Society Publications, 1968, pp 150–165.
83. GE Zaikov, ZK Maizus, NM Emanuel. *Zh Teoret Eksper Khim* 3:612–619, 1967.
84. GE Zaikov, ZK Maizus. *Izv AN SSSR Ser Khim* 47–51, 1968.
85. GE Zaikov, ZK Maizus. *Zh Fiz Khim* 43:115–119, 1969.
86. GE Zaikov, ZK Maizus. *Kinet Katal* 9:511–515, 1968.
87. NM Emanuel, GE Zaikov, ZK Maizus. *Oxidation of Organic Compounds. Medium Effect in Radical Reaction.* Oxford: Pergamon Press, 1984.
88. ET Denisov. *Izv AN SSSR Ser Khim* 53–58, 1960.
89. LM Andronov, GE Zaikov, ZK Maizus. *Zh Fiz Khim* 41:1122–1127, 1967.

90. GE Zaikov, ZK Maizus, NM Emanuel. *Neftekhimiya* 8:217–223, 1967.
91. E. T. Denisov. *Dokl AN SSSR* 146:394–397, 1962.
92. VL Antonovskii, VA Terentiev. *Zh Fiz Khim* 43:2727–2729, 1969.
93. ET Denisov. *Zh Fiz Khim* 37:1896–1899, 1963.
94. ET Denisov, LN Denisova. *Izv AN SSSR Ser Khim* 1731–1737, 1963.
95. GE. Zaikov, ZK Maizus, NM Emanuel. *Izv AN SSSR Ser Khim* 53–58, 1968.
96. W Langenbeck, W Pritzkow. *Chem Tech* 2:116, 1950.
97. LK Obukhova, NM. Emanuel. *Izv AN SSSR Ser Khim* 1544–1548, 1960.
98. AE Robson, D Yang. British Patent No 771992, 1957.
99. A Elhill, AE Robson, D Yang. British Patent No 771991, 1957.
100. II Rif, VM Potekhin, VA Proskuryakov. *Zh Prikl Khim* 45:2043–2047, 1972.
101. AM Syroezhko, VM Potekhin, VA Proskuryakov, VV Serov. *Zh Prikl Khim* 44:597–602, 1971.
102. VE Agabekov, II Korsak, ET Denisov, NI Mitskevich. *Dokl AN SSSR* 217:116–118, 1974.
103. VE Agabekov, NI Mitskevich, VA Azarko, NL Budeyko. *Dokl AN BSSR* 17:826–825, 1973.
104. VE Agabekov, NI Mitskevich, ET Denisov, VA Azarko, NL Budeyko. *Dokl AN BSSR* 18:38–41, 1974.
105. VE Agabekov, ET Denisov. *Kinet Katal* 10:731–734, 1969.
106. VE Agabekov, ET Denisov, NI Mitskevich. *Izv AN SSSR Ser Khim* 2254–2261, 1968.

9 Oxidation of Amines, Amides, and Esters

9.1 OXIDATION OF ALIPHATIC AMINES

9.1.1 INTRODUCTION

Aromatic amines are known as to be efficient inhibitors of hydrocarbon and polymer oxidation (see Chapters 15 and 19). Aliphatic amines are oxidized by dioxygen via the chain mechanism under mild conditions [1,2]. Peroxyl and hydroperoxyl radicals participate as chain propagating species in the chain oxidation of amines. The weakest C—H bonds in aliphatic amines are adjacent to the amine group. The bond dissociation energy (BDE) of C—H and N—H bonds of amines are collected in Table 9.1. One can see that the BDE of the N—H bond of the NH_2 group is higher than the BDE of the α-C—H bond in the amine molecule. For example, $D_{N-H} = 418.4\,kJ\,mol^{-1}$ and $D_{C-H} = 400\,kJ\,mol^{-1}$ in methaneamine. However, the BDE of N—H bond of dialkylamine is lower than that of the C—H bond of the α-CH_2 group. For example, $D_{N-H} = 382.8\,kJ\,mol^{-1}$ and $D_{C-H} = 389.0\,kJ\,mol^{-1}$ in N-ethylethaneamine.

The BDE of the α-C—H bond in alkylamines depends on alkyl substituents at the N and C atoms of the amine. The more the number of alkyl substituents, lower the BDE of the α-C—H and N—H bonds. The values of $\Delta D = D_{C-H}(CH_3NH_2) - D_{C-H}$ (amine) (kJ mol^{-1}) are given below.

Group	—NH_2	—NHR	—NR_2
CH_3—	0.0	11.0	23.4
RCH_2—	10.4	15.4	20.5
R_2CH—	20.5	25.0	30.0

The comparison of the BDE of hydrocarbons and amines proves the strong effect of the amino group on the BDE of the α-C—H bond. This is the result of the strong interaction of the α-C atom free valence of the aminoalkyl substituent with the p-electron pair of nitrogen atom. The energy of such interaction demonstrates the difference in the BDE of the α-C—H bond in hydrocarbons and amines ($\Delta D = D_{C-H}(RCH_3) - D_{C-H}(RNH_2)$).

R	CH_3—	$MeCH_2$—	Me_2CH—
ΔD (kJ mol^{-1})	22.0	22.4	20.5

TABLE 9.1
Bond Dissociation Energies of C—H and N—H Bonds in Amines [3,4]

Amine	Bond	D (kJ mol^{-1})
	C—H Bonds	
Methanamine	H_2NCH_2—H	400.0
Methanamine, N-methyl-	MeNHCH$_2$—H	389.0
Methanamine, N,N-dimethyl-	Me$_2$NCH$_2$—H	376.6
Ethanamine	Me(NH$_2$)CH—H	389.6
Ethanamine, N-ethyl-	(EtNH)MeCH—H	384.6
Ethanamine, N-diethyl-	(Et$_2$N)MeCH—H	379.5
2-Propanamine	Me$_2$(NH$_2$)C—H	379.5
2-Propanamine, N-2-propyl-	(Me$_2$CHNH)Me$_2$C—H	375.0
2-Propanamine, N,N-di-2-propyl-	(Me$_2$CH)$_2$NMe$_2$C—H	370.0
Cyclohexanamine	cyclohexyl-NH$_2$ (α-CH)	395.9
Cyclohexanamine, N-methyl-	cyclohexyl-NHMe (α-CH)	388.0
Cyclohexanamine, N,N-dimethyl-	cyclohexyl-NMe$_2$ (α-CH)	381.0
Cyclohexanamine, N-cyclohexyl-	(cyclohexyl)$_2$NH (α-CH)	387.3
Piperidine	piperidine (α-CH)	385.6
2-Propen-1-amin, N,N-di-2-propenyl-	CH_2=CH(CH—H)N(CH$_2$CH=CH$_2$)$_2$	345.6
Benzenamin, N,N-di-2-propenyl-	CH_2=CH(CH—H)NPh(CH$_2$CH=CH$_2$)	339.3
Benzenmethanamin	Ph(NH$_2$)CH—H	368.0
Benzenmethanamin, N,N-dimethyl-	Me$_2$NPhCH—H	370.8
Benzenmethanamin, N,N-dibenzyl-	(PhCH$_2$)$_2$NPhCH—H	372.8
Benzenamin, N,N-dimethyl-	PhNMeCH$_2$—H	383.7
Benzenamin, N,N-diethyl-	PhEtNCH—HMe	383.2
Benzenamin, N-methyl-N-phenyl-	Ph$_2$NCH$_2$—H	379.5
Morpholin, N-phenyl-	N-phenylmorpholine (α-CH)	361.0
	N—H Bonds	
Ammonia	H_2N—H	449.4
Hydrazine	(H$_2$N)NH—H	366.1
Methylamine	CH$_3$NH—H	418.4
Ethylamine	CH$_3$CH$_2$NH—H	418.4
Methylamine, N-methyl-	(CH$_3$)$_2$N—H	382.8
Ethylamine, N-ethyl-	(CH$_3$CH$_2$)$_2$N—H	382.8

Oxidation of Amines, Amides, and Esters

The substitution of the methyl group by the amino group decreases the BDE of the α-C—H bond to $21.6 \pm 0.8 \, \text{kJ mol}^{-1}$. As a result, peroxyl radicals attack preferentially the α-C—H bonds of the oxidized amines.

Amines possess a pair of p-electrons on the nitrogen atom. The nitrogen atom has a low electron affinity in comparison with oxygen. Therefore, amine can be the electron donor reactant in a charge-transfer complex (CTC) in association with oxygen-containing molecules and radicals. It will be shown that the formation of CTC complexes of amines with peroxyl radicals is important in the low-temperature oxidation of amines.

The α-aminoalkylperoxyl radicals RCH(OO$^\bullet$)NHR possess a dual reactivity: oxidative (due to the peroxyl group) and reducing (due to the amino group) [5]. As a result, many antioxidants terminate the chains of oxidized amines by the mechanisms of cyclic chain termination (see Chapter 16).

9.1.2 Chain Mechanism of Alkylamine Oxidation

Amines and hydrocarbons are oxidized by dioxygen by the chain mechanism [1,2]. Their oxidation is initiated by radical initiators. In the presence of a sufficient concentration of dissolved dioxygen ($[O_2] > 10^{-4} \, \text{mol L}^{-1}$), the rate of initiated oxidation does not depend on the dioxygen pressure [2]. The lengths of chains (ν) are high. For example, at $T = 323 \, \text{K}$ and initiation rate $v_i = 10^{-8} \, \text{mol L}^{-1} \text{s}^{-1}$ in the liquid phase at $pO_2 = 1 \, \text{atm}$ the chain length is given below [2]:

Amine	Bu_2NH	$BuNMe_2$	Et_3N	$(PhCH_2)_3N$	$PhNMe_2$
ν	53	1090	3000	1250	1930

Similarly alcohols and amines are oxidized with chain propagation via two kinds of peroxyl radicals, namely aminoalkyl peroxyl and hydroperoxyl according the following scheme:

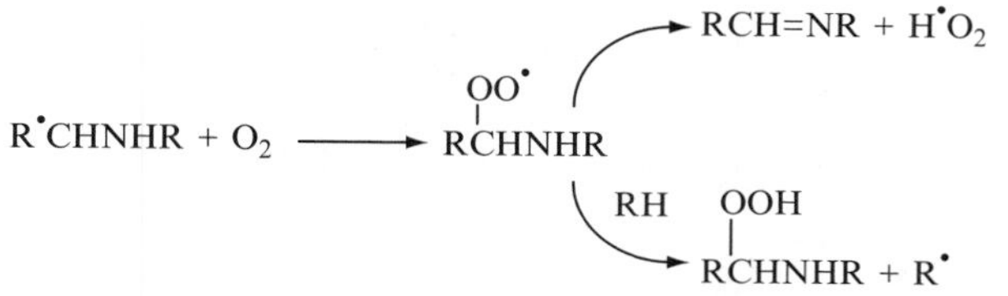

$$R^\bullet CHNHR + O_2 \longrightarrow RCH(OO^\bullet)NHR \begin{cases} \longrightarrow RCH=NR + H^\bullet O_2 \\ \xrightarrow{RH} RCH(OOH)NHR + R^\bullet \end{cases}$$

The kinetic scheme of amine oxidation includes the following steps [1,2]:

$$I \longrightarrow r^\bullet$$
$$r^\bullet + RCH_2NHR^1 \longrightarrow rH + RC^\bullet HNHR^1$$
$$RC^\bullet HNHR^1 + O_2 \longrightarrow RCH(OO^\bullet)NHR^1$$
$$RCH(OO^\bullet)NHR^1 + RCH_2NHR^1 \longrightarrow RCH(OOH)NHR^1 + RC^\bullet HNHR^1$$
$$RCH(OO^\bullet)NHR^1 \longrightarrow RCH=NR + HO_2^\bullet$$
$$HO_2^\bullet + RCH_2NHR^1 \longrightarrow H_2O_2 + RC^\bullet HNHR^1$$
$$2RCH(OO^\bullet)NHR^1 \longrightarrow RCH(OH)NHR^1 + O_2 + RC(O)NHR^1$$
$$RCH(OO^\bullet)NHR^1 + HO_2^\bullet \longrightarrow RCH(OOH)NHR^1 + O_2$$
$$2HO_2^\bullet \longrightarrow HOOH + O_2$$

In accordance with this scheme, the rate of initiated amine oxidation (v) obeys the following kinetic equation [1,2]:

$$v = k_p(2k_t)^{-1/2}[RCH_2NHR^1][O_2]^0 v_i^{1/2} \qquad (9.1)$$

The values of $k_p(2k_t)^{-1/2}$, k_p, and $2k_t$ are collected in Table 9.2.

Since two different peroxyl radicals, namely, $HO_2^{\bullet}$ and α-aminoalkyl peroxyl, take part in chain propagation and termination, the values of k_p and k_t rate constants are effective. They depend on the ratio of the concentrations of these two peroxyl radicals. The decay of

TABLE 9.2
The Values of Ratio and Rate Constants of Chain Propagation and Termination for Oxidation of Amines $k_p(2k_t)^{-1/2}$

Amine	$k_p/(2k_t)^{-1/2}$ (L mol^{-1} s^{-1})$^{1/2}$	k_p/(328 K) (L mol^{-1} s^{-1})	$2k_t$/(323 K) (L mol^{-1} s^{-1})	Ref.
(PrCH$_2$)$_2$NH	$1.58 \times 10^5 \exp(-51.0/RT)$	4.5	2.9×10^7	[6,7]
PrCH$_2$NMe$_2$	$1.10 \times 10^2 \exp(-23.8/RT)$	46.0	9.0×10^6	[8]
Me$_2$NCH$_2$CH$_2$NMe$_2$	$3.55 \times 10^2 \exp(-25.1/RT)$	2.45×10^2	6.3×10^7	[9]
(MeCH$_2$)$_3$N	$31.6 \times \exp(-17.8/RT)$	2.80×10^2	4.4×10^7	[10]
cyclohexyl-NH$_2$	$2.57 \times 10^4 \exp(-49.4/RT)$	1.6	1.8×10^7	[6,7]
dicyclohexylamine	$4.27 \times 10^4 \exp(-45.2/RT)$	8.7	1.8×10^7	[2]
N-methylcyclohexylamine	$2.29 \times 10^3 \exp(-30.8/RT)$	54.0	5.0×10^6	[11]
N,N-dimethylcyclohexylamine	$1.05 \times 10^3 \exp(-27.5/RT)$	80.0	4.5×10^6	[10]
N-phenylmorpholine	$3.63 \times 10^3 \exp(-31.0/RT)$	7.69	5.0×10^8	[2]
EtC(O)OCH$_2$CH$_2$NMe$_2$	$6.17 \times 10^2 \exp(-28.8/RT)$	40.0	9.0×10^6	[8]
PhNMe$_2$	$25.1 \times \exp(-16.7/RT)$	7.76×10^2	2.4×10^8	[12]
PhNMe$_2$	$65.2 \times \exp(-21.0/RT)$			[12]
4,4'-methylenebis(N-methylaniline)	$40.0 \times \exp(-15.4/RT)$			[13]
PhCH$_2$NH$_2$	$5.25 \times 10^4 \exp(-46.5/RT)$	23	2.6×10^8	[14]
PhCH$_2$NHMe	$3.09 \times 10^4 \exp(-45.3/RT)$	22	2.0×10^8	[2]
PhCCH$_2$NMe$_2$	$1.86 \times 10^2 \exp(-24.0/RT)$	3.48×10^2	2.0×10^8	[2]
(PhCH$_2$)$_3$N	$20.4 \times \exp(-17.0/RT)$	5.15×10^2	2.0×10^8	[2]

α-aminoperoxyl radicals was not studied. The decay of aminocarboxy (anion) methylperoxyl radical

$$NH_2CH(OO^{\bullet})CO_2^- \longrightarrow HN{=}CHCO_2^- + HO_2^{\bullet}$$

occurs very rapidly in water. The rate constant of this reaction is $k_d = 1.5 \times 10^5\,s^{-1}$ at $T = 293$ K [15]. If the decay of the aminoalkylperoxyl radical would occur so rapidly in an amine solution, the main chain propagating intermediate would be the hydroperoxyl radical. However, one can expect in this case the close values of rate constants of chain termination in different oxidized amines. It is not so as we see from Table 9.2. The values of $2k_t$ vary from 5×10^6 to $5 \times 10^8\,s^{-1}$. Since $HO_2^{\bullet}$ radicals disproportionate very rapidly with the diffusion-limited rate constant [16], one can suppose that $HO_2^{\bullet}$ is the main chain propagating species in amines with values of $2k_t \approx 10^8\,s^{-1}$ (see Table 9.2).

The mechanism of $HO_2^{\bullet}$ formation from peroxyl radicals of primary and secondary amines is clear (see the kinetic scheme). The problem of $HO_2^{\bullet}$ formation in oxidized tertiary amines is not yet solved. The analysis of peroxides formed during amine oxidation using catalase, Ti(IV) and by water extraction gave controversial results [17]. The formed hydroperoxide appeared to be labile and is hydrolyzed with H_2O_2 formation. The analysis of hydroperoxides formed in co-oxidation of cumene and 2-propaneamine, N-bis(ethyl methyl) showed the formation of two peroxides, namely H_2O_2 and $(Me_2CH)_2NC(OOH)Me_2$ [16]. There is no doubt that the two peroxyl radicals are acting: $HO_2^{\bullet}$ and α-aminoalkylperoxyl. The difficulty is to find experimentally the real proportion between them in oxidized amine and to clarify the way of hydroperoxyl radical formation.

The study of co-oxidation of both hydrocarbons and amines proved the strong retarding effect of aliphatic amines on hydrocarbon oxidation [18]. The rate of hydrocarbon oxidation was found to drop with an increase in the amine concentration. The rate of oxidation v depends on the amine concentration in accordance with equation [19]:

$$\frac{v_i}{v} - \frac{v_i}{v_0} = \frac{2k_{p12}[R_2H]}{k_{p11}[R_1H]} \qquad (9.2)$$

where v_i is the initiation rate, v_0 is the rate of hydrocarbon oxidation without amine, k_{p11} and k_{p12} are rate constants of the $R_1OO^{\bullet}$ reaction with hydrocarbon (R_1H) and amine (R_2H). It was found that the cumylperoxyl radical reacts with ethaneamine, N-diethyl with the rate constant $k_{p12} = 135\,L\,mol^{-1}\,s^{-1}$ at $T = 333$ K [20].

This rate constant helps to estimate the BDE of the O—H bond formed in aminoalkylperoxyl radical in the reaction with the hydrocarbon C—H bond (see Chapter 7). The rate constant $k_{p22} = 336\,L\,mol^{-1}\,s^{-1}$ (see Table 9.2). The ratio of two rate constants k_{p12}/k_{p22} helps to calculate the difference in activation energies of these two reactions ($Me_2PhCOO^{\bullet}$ and $Et_2NCH(OO^{\bullet})Me$ with Et_3N): $\Delta E = -RT\ln(k_{p12}/k_{p22}) = 2.5\,kJ\,mol^{-1}$. Using Equation (7.10) we estimate $\Delta D_{O-H} = 5.5\,kJ\,mol^{-1}$. Since $D_{O-H}(Me_2PhCOO{-}H) = 358.6\,kJ\,mol^{-1}$ (see Chapter 2), one can calculate $D_{O-H}(Et_2NCH(OO{-}H)Me) = 358.6 + 5.5 = 364.1\,kJ\,mol^{-1}$. For the tertiary aminoperoxyl radical we can suppose the same difference $\Delta D = 6.9\,kJ\,mol^{-1}$ as for alkylperoxyl radicals and propose $D_{O-H} = 364.1 - 6.9 = 357.2\,kJ\,mol^{-1}$. Both these values should be treated as approximate due to complex peroxyl radical composition in oxidized amines. The values of the rate constants of different peroxyl radicals with amines are collected in Table 9.3.

9.1.3 Reactivity of Amines in Reaction with Peroxyl Radicals

The reactions of peroxyl radicals with the C—H bond of amine belong to one class of abstraction reactions, namely, $RO_2^{\bullet} + RH$ (see the IPM model in Chapters 4 and 6). The

TABLE 9.3
Rate Constants of Peroxyl Radical Reactions with Amines

Amine	Peroxyl Radical	T (K)	Solvent	k (L mol^{-1}s^{-1})	Ref.
Et$_2$NH	Me$_3$COO$^\bullet$	220–263	C$_7$H$_{16}$	$3.39 \times 10^2 \exp(-11.1/RT)$	[21]
Et$_3$N	Me$_3$COO$^\bullet$	220–263	C$_7$H$_{16}$	$1.74 \times 10^3 \exp(-15.0/RT)$	[21]
Et$_3$N	Me$_3$COO$^\bullet$	303	Et$_2$MeCH	23.0	[22]
EtMe$_2$N	Me$_2$NCH(OO$^\bullet$)Me	303–323	EtMe$_2$N	$1.86 \times 10^6 \exp(-28.8/RT)$	[8]
Pr$_2$NH	Me$_3$COO$^\bullet$	220–263	C$_7$H$_{16}$	$5.75 \times 10^2 \exp(-11.8/RT)$	[21]
EtMeCHNH$_2$	Me$_3$COO$^\bullet$	220–263	C$_7$H$_{16}$	$1.20 \times 10^3 \exp(-15.9/RT)$	[21]
BuMe$_2$N	Me$_2$NCH(OO$^\bullet$)Pr	303–323	BuMe$_2$N	$3.31 \times 10^5 \exp(-23.3/RT)$	[8]
Bu$_2$NH	Me$_3$COO$^\bullet$	220–263	C$_7$H$_{16}$	$2.19 \times 10^3 \exp(-14.1/RT)$	[21]
Bu$_2$NH	Me$_3$COMe$_2$COO$^\bullet$	187–263	C$_7$H$_{16}$/MeCO$_2$Et	$1.17 \times 10^7 \exp(-7.2/RT)$	[23]
Bu$_2$NH	Me$_2$PhCOO$^\bullet$	187–263	C$_7$H$_{16}$/MeCO$_2$Et	$1.26 \times 10^4 \exp(-4.9/RT)$	[23]
Bu$_2$NH	Me$_3$COO$^\bullet$	187–263	C$_7$H$_{16}$/MeCO$_2$Et	$3.89 \times 10^3 \exp(-4.4/RT)$	[23]
Bu$_2$NH	cyclohexyl-OO$^\bullet$	187–263	C$_7$H$_{16}$/MeCO$_2$Et	$3.80 \times 10^3 \exp(-4.3/RT)$	[23]
Bu$_2$NH	Me$_3$COO$^\bullet$	241	PhCH$_3$	1.68	[24]
Bu$_2$NH	Me$_3$COO$^\bullet$	241	PhCl	1.20	[24]
Bu$_2$NH	Me$_3$COO$^\bullet$	241	ClCH$_2$CH$_2$Cl	1.20	[24]
Bu$_2$NH	Me$_3$COO$^\bullet$	241	CCl$_3$H	0.64	[24]
Bu$_2$NH	Me$_3$COO$^\bullet$	241	BuOBu	3.24	[24]
Bu$_2$NH	Me$_3$COO$^\bullet$	241	MeCOOEt	1.12	[24]
Bu$_2$NH	Me$_3$COO$^\bullet$	241	MeCOMe	2.64	[24]
Bu$_2$NH	Me$_3$COO$^\bullet$	241	BuOH	1.24	[24]
Bu$_2$NH	Me$_3$COO$^\bullet$	241	BuCOOH	0.80	[24]
Bu$_2$NH	Me$_3$COO$^\bullet$	241	CH$_3$NO$_2$	0.56	[24]
Bu$_2$NH	Et$_2$N(CHOO$^\bullet$)Me	313	Bu$_2$NH	4.7	[11]
Bu$_2$NH	piperidinyl-OO$^\bullet$	318	Bu$_2$NH	3.0	[11]
Bu$_2$NH	Ph(NMe$_2$)CHOO$^\bullet$	318	Bu$_2$NH	21.0	[11]
Bu$_3$N	Me$_3$COO$^\bullet$	220–263	C$_7$H$_{16}$	$1.26 \times 10^3 \exp(-15.0/RT)$	[21]
BuCH$_2$NH$_2$	Me$_3$COO$^\bullet$	220–263	C$_7$H$_{16}$	$81.3 \exp(-11.5/RT)$	[21]
Me(CH$_2$)$_6$NH$_2$	Me$_3$COO$^\bullet$	220–263	C$_7$H$_{16}$	$5.37 \times 10^2 \exp(-15.0/RT)$	[21]
HO(CH$_2$)$_2$NEt$_2$	Me$_2$PhCOO$^\bullet$	333	Me$_2$PhCH	1.00×10^2	[11]
cyclohexyl-NH$_2$	Me$_3$COO$^\bullet$	209–263	C$_7$H$_{16}$	$2.0 \times 10^3 \exp(-16.3/RT)$	[24]

TABLE 9.3
Rate Constants of Peroxyl Radical Reactions with Amines—continued

Amine	Peroxyl Radical	T (K)	Solvent	k (L mol^{-1} s^{-1})	Ref.
piperidine (H,H / N-H)	piperidyl-OO•	323	piperidine	7.5	[11]
piperidine	Me$_2$PhCOO•	348	Me$_2$PhCH	2.90×10^2	[25]
piperidine	Me$_3$COO•	220–263	C$_7$H$_{16}$	$2.24 \times 10^3 \exp(-14.9/RT)$	[21]
morpholine	Me$_3$COO•	220–263	C$_7$H$_{16}$	$5.89 \times 10^2 \exp(-14.3/RT)$	[21]
morpholine	Me$_2$PhCOO•	348	Me$_2$PhCH	66.0	[25]
MePhNH	MePhNCH$_2$OO•	313	MePhNH	1.2×10^2	[11]
PhCH$_2$NH$_2$	Me$_3$COO•	220–263	C$_7$H$_{16}$	$3.63 \times 10^3 \exp(-15.4/RT)$	[21]
PhCH$_2$NHMe	Ph(NMe$_2$)CHOO•	318	PhCH$_2$NHMe	29.0	[11]

parameters of this class are the following [26]: $\alpha = 0.814$, $b = 3.743 \times 10^{11}$ (kJ mol^{-1})$^{1/2}$, $0.5hL\nu_i = 17.4$ kJ mol^{-1}, and $0.5hL(\nu_i - \nu_f) = -3.8$ kJ mol^{-1}. This class is divided into groups. Each group has its individual parameter br_e. The analysis of peroxyl radical reactions with aliphatic amines [4] gave the value of $br_e = 13.95$ (kJ mol^{-1})$^{1/2}$. This value differs from that for the reaction of the peroxyl radical with the C—H bond of the aliphatic hydrocarbon (R^{1}H) and lies between br_e for reactions RO$_2^\bullet$ + R^{1}H ($br_e = 13.62$ (kJ mol^{-1})$^{1/2}$) and RO$_2^\bullet$ + R^{3}H (R^{3}H is alkylaromatic hydrocarbon RCH$_2$Ph, $br_e = 14.32$ (kJ mol^{-1})$^{1/2}$). This consequence of br_e, br_e(RO$_2^\bullet$ + R^{1}H) < br_e(RO$_2^\bullet$ + amine) < br_e(RO$_2^\bullet$ + R^{3}H), can be interpreted as the result of different triplet repulsions in TS of these groups of reactions (see Chapter 4). In the TS of the reaction RO$_2^\bullet$ + R^{3}H π-electrons of aromatic ring interact with three electrons of TS. This interaction via triplet repulsion increases the energy of the TS O $\cdots$ H $\cdots$ C and increases the activation energy of the thermoneutral reaction E_{e0}. This is the reason why the activation energy of the thermoneutral reaction E_{e0}(RO$_2^\bullet$ + amine) > E_{e0}(RO$_2^\bullet$ + R^{1}H). The values of br_e (kJ mol^{-1})$^{1/2}$ and E_{e0} (kJ mol^{-1}) for four groups of reactions RO$_2^\bullet$ + R^{i}H [4,26] are as follows:

R^{i}H	R^{1}H	RCH$_2$NHR^1	RCH$_2$Ph	RCH$_2$CH=CH$_2$
br_e (kJ mol^{-1})$^{1/2}$	13.62	13.95	14.32	15.21
E_{e0} (kJ mol^{-1})	56.4	59.1	62.3	70.3

We see that the triplet repulsion in the case of aliphatic amines increases E_{e0} to 2.7 kJ mol^{-1}. It is less than that of alkylaromatic hydrocarbons where $\Delta E_{e0} = 5.9$ kJ mol^{-1}. So, one can conclude that the pair of p-electrons of nitrogen interacts with TS not so strongly as π-electrons of the aromatic ring.

There are not so many data on peroxyl radical reactions with amines. Therefore, the values of activation energy and rate constants of reactions of $HO_2^\bullet$ radical with series of amines were calculated using the IPM model (see Chapter 4). The following parameters were used for calculation [4,26].

Amine	Aliphatic	Benzylic	Allylic
br_e (kJ mol^{-1})$^{1/2}$	13.95	14.65	15.54
E_{e0} (kJ mol^{-1})	59.1	65.2	73.4
A_{C-H} (L mol^{-1} s^{-1})	1×10^8	1×10^7	1×10^7

The results of calculation as well as the values of ΔE_H, which illustrate the contribution of the enthalpy into the activation energy, are presented in Table 9.4.

The sufficiently higher values of br_e ($br_e = 14.82$ (kJ mol^{-1})$^{1/2}$ and E_{e0} ($E_{e0} = 66.7$ kJ mol^{-1}) were estimated from the experimental rate constants of the reaction of the 1,1-dimethylethyl peroxyl radical with N-diethylethaneamine. This reaction was studied by Tavadyan et al. [21] using EPR method at low temperatures (210–260 K). This discrepancy in E_{e0} values for reactions of one group is the result of a great difference in the temperature of the experiment (210–260 K [21] and 330–380 K for other experiments).

It is most likely that this difference of the E_{e0} values is caused by the CTC formation due to the donor–acceptor interactions between peroxyl oxygen with a free valence and amine nitrogen of the type

$$RO_2^\bullet + NR_3 \rightleftharpoons RO_2^\bullet(\delta-) \cdots NR_3(\delta+)$$

At low temperature, this equilibrium shifts to the right, and only a small fraction of free peroxyl radicals participate in the reaction, whereas at a high temperature the equilibrium is shifted to the left, and the fraction of bound radicals is small. If we assume that at 210–260 K almost all $RO_2^\bullet$ radicals are bound in complexes with amines, then it can be shown that the difference of low- and high-temperature E_{e0} values is close to the enthalpy of this donor–acceptor interaction in the CTC complex: $\Delta E_{e0} \cong \Delta H_{DA} = 66.7 - 59.1 = 7.6$ kJ mol^{-1}.

9.2 OXIDATION OF AMIDES

9.2.1 INTRODUCTION

Aliphatic amides possess more strong α-C—H bonds in comparison with amines. This is the result of the carbonyl group influence on the stabilization of the formed α-amidoalkyl radical formed from amide in the reaction with the peroxyl radical. This influence is not so strong as that of the amine group. The values of the α-C—H bond in a few amides were estimated recently by the IPM method [4] and are given here.

Amide	AcNHCH—HMe	AcNHC—HMe$_2$	AcMeNCH$_2$—H
D_{C-H} (kJ mol^{-1})	399.2	395.5	386.9
Amine	EtNHCH—HMe	Me$_2$CHNHC—HMe$_2$	Me$_2$NCH$_2$—H
D_{C-H} (kJ mol^{-1})	384.6	375.0	376.6
ΔD (kJ mol^{-1})	14.6	20.5	10.3

TABLE 9.4
Activation Energies, Rate Constants, and Contribution Enthalpy in Activation Energy for Peroxyl Radical Reactions with Amines, Calculated by IPM Method (See Chapter 6)

Amine	ΔH_e (kJ mol^{-1})	E (kJ mol^{-1})	log A, A (L mol^{-1} s^{-1})	k (350 K) (L mol^{-1} s^{-1})	ΔE_H (kJ mol^{-1})
H$_2$NCH$_2$—H	27.2	56.0	8.48	1.33	12.9
MeNHCH$_2$—H	16.2	50.6	8.78	16.9	7.5
Me$_2$NCH$_2$—H	3.8	44.9	8.95	1.77×10^2	1.8
Me(NH$_2$)CH—H	16.8	50.9	8.30	5.07	7.8
EtNHMeCH—H	11.8	48.5	8.60	23.1	5.4
Et$_2$NMeCH—H	6.7	46.2	8.78	76.8	3.1
(Me$_2$CH)$_2$NMe$_2$C—H	−2.8	41.9	8.48	1.69×10^2	−1.2
cyclohexyl-N(Me)H (structure)	−4.2	41.3	8.00	68.6	−1.8
piperidine α-CH (structure)	12.8	49.0	8.60	19.4	5.9
MePhNCH$_2$—H	−14.4	36.8	8.78	1.94×10^3	−6.3
morpholine N-Ph α-CH (structure)	−24.2	32.8	8.60	5.07×10^3	−10.3
Ph(NH$_2$)CH—H	−4.8	47.1	7.30	1.87	−2.1
Me$_2$NPhCH—H	−2.0	48.3	7.30	1.24	−0.9
(PhCH$_2$)$_2$NPhCH—H	0.0	49.2	7.78	2.74	0.0
PhNMeCH$_2$—H	10.9	48.1	8.78	40.0	5.0
PhEtNMeCH—H	10.4	47.9	8.30	14.2	4.8
Ph$_2$NCH$_2$—H	6.7	46.2	8.48	38.2	3.1
CH$_2$=CH(CH—H)N(CH$_2$CH=CH$_2$)$_2$	−27.2	45.7	7.78	9.1	−11.7
CH$_2$=CH(CH—H)NPh(CH$_2$CH=CH$_2$)	−33.5	43.1	7.48	11.2	−14.3
Me$_2$C=NMeCH—H	−18.8	49.2	7.30	0.91	−8.2
MeN=CMeCH—HMe	−22.2	47.8	7.30	1.47	−9.6

The values of BDE of the weakest C—H bonds in amides, we considered, are by 10–20 kJ mol^{-1} higher than those in the corresponding amines. Naturally, this difference influences the oxidation of amides.

Amides have carbonyl group and nitrogen atom. Both these groups can form hydrogen bonds with O—H and N—H-containing compounds. Therefore, hydroperoxides form associates with amides via hydrogen bonding. This association influences the kinetics of the oxidation of amides and decay of hydroperoxides.

Amides possess high dipole moments μ due to the presence of the amide group. Some values are given below.

Amide	$AcNH_2$	$HC(O)NHMe$	$HC(O)NMe_2$
μ (Debye)	3.76	3.83	3.82

As a result, polar reactants and polar TS are solvated in the amide medium.

9.2.2 Chain Mechanism of Amide Oxidation

Amides, as amines and hydrocarbons, are oxidized by dioxygen according to the chain mechanism [1]. The initiated oxidation of amides proceeds according to the classical scheme of chain oxidation of organic compounds [2].

$$I \longrightarrow R^\bullet$$
$$r^\bullet + RC(O)NHCH_2R^1 \longrightarrow rH + RC(O)NHC^\bullet HR^1$$
$$RC(O)NHC^\bullet HR^1 + O_2 \longrightarrow RC(O)NHCH(OO^\bullet)R^1$$
$$RC(O)NHCH(OO^\bullet)R^1 + RC(O)NHCH_2R^1 \longrightarrow RC(O)NHCH(OOH)R^1 + RC(O)NHC^\bullet HR$$
$$2RC(O)NHCH(OO^\bullet)R^1 \longrightarrow RC(O)NHCH(OH)R^1 + O_2$$
$$+ RC(O)NHC(O)R^1$$

In contrast to aminoperoxyl radicals, amidoperoxyl radicals do not decompose with the formation of $HO_2^\bullet$ radicals [2]. According to this scheme the rate of amides oxidation obeys the following equation (at $[O_2] > 10^{-4}$ mol L^{-1}):

$$v = k_p (2k_t)^{-1/2} [RC(O)NHCH_2R^1][O_2]^0 v_i^{-1/2} \tag{9.3}$$

The value of ratio $k_p(2k_t)^{-1/2}$ are collected in Table 9.5.

The absolute values of k_p and $2k_t$ were estimated for a few amides. They are collected in Table 9.6.

9.2.3 Decomposition of Hydroperoxides

Hydroperoxides formed due to the oxidation of amides are decomposed into free radicals and accelerate oxidation. Hydroperoxides form hydrogen bonds with amides. The enthalpies, entropies, and equilibrium constants of hydrogen bonding are presented in Table 9.7.

Depending on amide and hydroperoxide, the equilibrium constants at room temperature vary from 0.1 to 20 L mol^{-1}. The enthalpy of hydrogen bond hydroperoxide and amide is around 20 to 24 kJ mol^{-1}. Three types of hydrogen bonds are formed: ROOH $\cdots$ O=C<, ROOH $\cdots$ N<, and NH $\cdots$ O(H)OR.

Apparently, hydrogen bonding is a preliminary stage of free radical generation through hydroperoxide decomposition. This is in agreement with the kinetic equation for hydroperoxides decomposition to free radicals [33]:

$$v_i = k_i [ROOH][AcNHR] \tag{9.4}$$

TABLE 9.5
The Values of $k_p(2k_t)^{-1/2}$ for Initiated Oxidation of Amides

Amide	T (K)	$k_p(2k_t)^{-1/2}$ (357 K) (L mol^{-1} s^{-1})	E (kJ mol^{-1})	log A, A (L mol^{-1} s^{-1})	Ref.
AcNMe$_2$	323–363	2.0×10^{-3}	46.0	4.04	[27]
AcNEt$_2$	328–363	3.3×10^{-3}	31.0	2.05	[2]
AcNHCHMe$_2$	333–393	4.0×10^{-4}	54.4	4.56	[27]
AcNHBu	323–396	9.0×10^{-4}	50.2	4.30	[27]
EtC(O)NHPr	323–393	1.7×10^{-3}	54.4	(5.20)	[2]
AcNHEt	357	2.1×10^{-3}			[2]
PrC(O)NH(CH$_2$)$_7$Me	323–393	1.7×10^{-3}	55.0	(5.27)	[28]
2-pyrrolidinone	364	1.6×10^{-3}			[29]
2-piperidinone	323–393	1.1×10^{-3}	55.5	(5.18)	[2]

TABLE 9.6
Rate Constants k_p and $2k_t$ for Chain Oxidation of Amides

Amide	T (K)	k (353 K) (L mol^{-1} s^{-1})	E (kJ mol^{-1})	log A, A (L mol^{-1} s^{-1})	Ref.
		k_p			
AcNHCHMe$_2$	333–393	0.98	54.4	8.04	[27]
AcNHBu	323–396	3.81	50.0	7.98	[27]
AcNMe$_2$	308–328	20.1	46.0	8.11	[27]
AcNEt$_2$	323–396	20.5	31.0	5.90	[2]
2-pyrrolidinone	364	11.8			[29]
2-azepanone	364	9.8			[29]
		$2k_t$			
AcNHCHMe$_2$	353	8.7×10^6			[27]
AcNMe$_2$	353	1.4×10^8			[27]
AcNHBu	353	2.3×10^7			[27]
AcNEt$_2$	353	5.0×10^7			[2]
2-pyrrolidinone	364	5.6×10^7			[29]
2-azepanone	364	6.6×10^7			[29]

TABLE 9.7
Equilibrium Constants, Enthalpies, and Entropies of Hydrogen Bonding of Hydroperoxides with Amides [30–32]

Amide	Hydroperoxide	Solvent	T (K)	$-\Delta H$ (kJ mol^{-1})	$-\Delta S$ (J mol^{-1} K)	K (298 K) (L mol^{-1})
AcNMe$_2$	Me$_3$COOH	CCl$_4$	294–341	22.6	51.1	19.6
AcNMe$_2$	Me$_2$PhCOOH	CCl$_4$	294–350	20.9	49.8	11.5
AcNBu	Me$_3$COOH	CCl$_4$	292–341	23.9	55.7	19.0
AcNBu	Me$_2$PhCOOH	CCl$_4$	294–337	20.5	49.0	10.8
AcNHCHMe$_2$	Me$_3$COOH	CCl$_4$	294–332	14.7	43.5	2.02
AcNHCHMe$_2$	Me$_2$PhCOOH	CCl$_4$	294–332	21.3	55.6	6.75
Me$_2$NC(O)H	Me$_3$COOH	Me$_2$NCOH	288–343	22.0	93.1	9.84×10^{-2}

The rate constants of homolytic decomposition of the two hydroperoxides in butylacetamide media are given in Table 9.8.

Quite another equation was established for the initiation rate for hydroperoxide decomposition in dimethylacetamide [34]:

$$v_i = k_i[\text{ROOH}][\text{AcNMe}_2]^{-1} \quad (9.5)$$

The formation of two forms of bonding between hydroperoxide and amide was proposed [34].

$$\text{ROOH} + \text{AcNMe}_2 \rightleftharpoons \text{X} \qquad K_1$$
$$\text{X} \longrightarrow \text{Free radicals} \qquad k_i$$
$$\text{X} + \text{AcNMe}_2 \rightleftharpoons \text{Y} \qquad K_2$$

The rate of initiation is

$$v_i = k_i[\text{X}] \approx k_i K_2^{-1}[\text{ROOH}][\text{AcNMe}_2]^{-1} \quad (9.6)$$

TABLE 9.8
Rate Constants of Free Radical Generation by Hydroperoxides in Amides

Amide	Hydroperoxide	T (K)	E (kJ mol^{-1})	log A, A (L mol^{-1} s^{-1})	k_i (350 K) (L mol^{-1} s^{-1})	Ref.
\multicolumn{7}{c}{$v_i = k_i[\text{ROOH}][\text{AcNHR}]$, A and k_i (L mol^{-1} s^{-1})}
AcNHBu	Me$_3$COOH	354–396	70.5	3.34	6.58 × 10^{-8}	[2]
AcNHBu	Me$_2$PhCOOH	353–397	126.0	10.27	2.92 × 10^{-9}	[2]
AcNHCHMe$_2$	Me$_3$COOH	363–398	39.0	2.30	3.02 × 10^{-4}	[34]
AcNHCHMe$_2$	Me$_2$PhCOOH	375–402	105.8	7.69	7.68 × 10^{-9}	[34]
\multicolumn{7}{c}{$v_i = k_i[\text{ROOH}][\text{AcNMe}_2]^{-1}$, A and k_i (mol L^{-1} s^{-1})}
AcNMe$_2$	Me$_3$COOH	325–363	60.7	4.22	1.45 × 10^{-5}	[34]
AcNMe$_2$	AcNMeCH$_2$OOH	357–388	158.0	16.70	1.32 × 10^{-7}	[35]

TABLE 9.9
Chain Lenth ν, Probability of Radical Pair to Escape the Cage e and Rate Constant of Induced Chain Decomposition k_{ind} at 405 K [33–36]

Amide	ROOH	k_{ind} (L$^{1/2}$ mol$^{-1/2}$ s$^{-1/2}$)	ν	e (%)
AcNHBu	Me$_3$COOH	1.66×10^{-5}		
AcNHBu	Me$_2$PhCOOH	2.4×10^{-5}	6	100
AcNHCHMe$_2$	Me$_3$COOH	2.8×10^{-5}		
AcNHCHMe$_2$	Me$_2$PhCOOH	3.5×10^{-5}		
AcNMe$_2$	Me$_3$COOH	2.0×10^{-4}	5	36
AcNMe$_2$	Me$_2$PhCOOH	9.0×10^{-4}	50	34
AcNMe$_2$	AcNMeCH$_2$OOH	5.0×10^{-4}	6	18

at $K_2 > K_1$ and $K_1 > 1$, $K_2 > 1$. The mechanism of initiation remains unclear. Probably, it includes the electron transfer from nitrogen of the amide group to hydroperoxide.

$$ROOH + AcNMe_2 \longrightarrow RO^\bullet + HO^- + AcNMe_2^{+\bullet}$$

Along with homolytic decomposition, hydroperoxides are decomposed in an acetamide solution by the chain mechanism under the action of formed free radicals [33,35].

$$RO^\bullet + AcNHCH_2R \longrightarrow ROH + AcNHC^\bullet HR$$
$$ROOH + AcNHC^\bullet HR \longrightarrow RO^\bullet + AcNHCH(OH)R$$
$$2\,AcNHC^\bullet HR \longrightarrow \text{Molecular products}$$

The chain length varies from 5 to 50 (see Table 9.9). The kinetic equation for the rate of chain decomposition of hydroperoxide has the following forms:

$$v = k_{ind}[ROOH]^{3/2}[AcNMe_2]^{-1/2} \tag{9.7}$$

$$v = k_{ind}[ROOH]^{1/2} \tag{9.8}$$

9.3 OXIDATION OF ESTERS

9.3.1 INTRODUCTION

A molecule of aliphatic ester possesses two substituents around the ester group —C(O)O—, namely, alcohol and acid residues. Ester group decreases the BDE of α-C—H bonds in both substituents: alcoholic and acidic. Therefore, an ester molecule has two different types of weak C—H bonds that are attacked by peroxyl radicals: α-C—H bonds of the alcohol substituent —CH$_2$OC(O)R and the α-C—H bonds of the acid substituent —CH$_2$C(O)OR. The values of BDE of these types of C—H bonds are close but not the same. The values of BDE of the C—H bonds are collected in Table 9.10.

The BDEs of α-C—H bonds of alcoholic and acidic substituents are compared with those of ethers and acids. Both types of BDE of C—H bonds are lower than those of the hydrocarbons.

TABLE 9.10
Dissociation Energies of C—H Bonds in Esters [37]

Ester	Bond	D (kJ mol^{-1})
	Alcoholic Group	
Acetic acid, methyl ester	MeC(O)OCH$_2$—H	404.0
Acetic acid, ethyl ester	MeC(O)OCH—HMe	399.5
Pentanoic acid, methylethyl ester	BuCOOC—HMe$_2$	387.9
Acetic acid, methylethyl ester	MeC(O)OC—HMe$_2$	392.3
Acetic acid, 3-propenyl ester	MeC(O)O(CH—H)CH=CH$_2$	348.3
Acetic acid, benzyl ester	MeC(O)O(CH—H)Ph	359.0
Oxalic acid, dimethyl ester	MeOC(O)C(O)OCH$_2$—H	403.9
Oxalic acid, diethyl ester	EtOC(O)C(O)CH—HMe	396.2
Oxalic acid, bis(1-methylethyl) ester	Me$_2$CHOC(O)C(O)OC—HMe$_2$	396.2
	Acidic Group	
Butanoic acid, methyl ester	EtCH—HCOOMe	394.9
Malonic acid, dimethyl ester	[MeOC(O)]$_2$CH—H	390.4
1,1-Dimethylpropanoic acid, 1,1-dimethylester	Me$_3$CC(O)OCMe$_2$CH$_2$—H	415.0
1,3-Dipropanoic acid, dimethyl ester	MeOC(O)CH—HC(O)OMe	391.7
1,4-Dibutanoic acid, dimethyl ester	MeOC(O)CH—HCH$_2$C(O)OMe	404.8
Acetic acid, anhydride	MeC(O)OC(O)CH$_2$—H	406.0

R	CH$_3$	OR'	OC(O)R'	C(O)OH	C(O)OR'
RCH$_2$—H	422.0	411.9	404.0	414.1	403.9
RCH—HMe	412.0	399.5	399.5	398.9	394.9
RC—HMe$_2$	400.0	390.8	392.3	388.3	—
CH$_2$=CHCH—HR	349.8	339.2	348.3	—	—

Ester group is polar and possesses a high dipole moment μ [38].

Ester	HC(O)OMe	HC(O)OEt	MeC(O)OMe	MeC(O)OEt	PhC(O)OEt
μ (Debye)	1.77	1.98	1.72	1.78	2.00

Therefore, the polar group influences the reactivity of ester in reactions with peroxyl radicals (see later). Due to the polar groups, the effect of multidipole interaction was observed in reactions of polyesters with RO$_2^\bullet$, O$_2$, and ROOH (see Section 9.3.4). Ester as a polar media solvates the polar TS and influences the reactivity of polar reagents.

The ability of esters to form hydrogen bonds with polar reactants is especially important for the reactions of peroxyl radicals with antioxidants such as phenols and amines. Amines form hydrogen bonds with ester groups. The hydrogen bonding lowers the activity of antioxidants as acceptors of peroxyl radicals (see Chapters 14 and 15).

9.3.2 CHAIN MECHANISM OF ESTER OXIDATION

Aliphatic esters are oxidized by dioxygen through the chain mechanism similar to the mechanism of hydrocarbon oxidation. In the presence of initiator I at such dioxygen pressure when $[O_2] > 10^{-4}$ mol L^{-1} in an ester solution, ester $AcOCH_2R$ oxidation includes the following elementary steps [39,40].

$$I \longrightarrow r^\bullet \qquad k_i$$
$$r^\bullet + AcOCH_2R \longrightarrow rH + AcOC^\bullet HR \qquad \text{Rapidly}$$
$$AcOC^\bullet HR + O_2 \longrightarrow AcOCH(OO^\bullet)R \qquad \text{Rapidly}$$
$$AcOCH(OO^\bullet)R + AcOCH_2R \longrightarrow AcOCH(OOH)R + AcOC^\bullet HR \qquad k_p$$
$$2AcOCH(OO^\bullet)R \longrightarrow AcOCH(OH)R + O_2 + AcOC(O)R \qquad 2k_t$$

In accordance with this scheme, the rate of ester oxidation was found to obey the equation:

$$v = k_p(2k_t)^{-1/2}[AcOCH_2R][O_2]^0 v_i^{-1/2} \qquad (9.9)$$

The values of rate constants ratio $k_p(2k_t)^{-1/2}$ are given in Table 9.11.

The nonsaturated esters with π-C=C bonds and without activated α-C—H bonds (esters of acrylic acid (CH_2=CHCOOR) and esters of vinyl alcohols (RC(O)OCH=CH_2)) are oxidized by the chain mechanism with chain propagation via the addition of peroxyl radicals to the double bond. Oligomeric peroxides are formed as primary products of this chain reaction. The kinetic scheme includes the following steps in the presence of initiator I and at pO_2 sufficient to support $[O_2] > 10^{-4}$ mol L^{-1} in the liquid phase [49].

$$I \longrightarrow r^\bullet$$
$$r^\bullet + CH_2=CHCOOR \longrightarrow rCH_2C^\bullet HCOOR \qquad \text{Rapidly}$$
$$RCH_2C^\bullet HCOOR + O_2 \longrightarrow rCH_2CH(OO^\bullet)COOR \qquad \text{Rapidly}$$
$$RCH_2CH(OO^\bullet)COOR + CH_2=CHCOOR \longrightarrow RCH_2CH(COOR)OOCH_2C^\bullet HCOOR \qquad k_p$$
$$2RCH_2CH(OO^\bullet)COOR \longrightarrow RCH_2CH(OH)COOR + O_2 + RCH_2C(O)COOR \qquad 2k_t$$

As in the case of the oxidation of saturated esters, the rate of chain copolymerization monomer and dioxygen obeys the equation similar to that for aliphatic ester oxidation.

The values of k_p for abstraction and addition reactions of peroxyl radicals were measured by different methods and are presented in Table 9.12.

The oxidation of nonsaturated esters with double bonds far away from the ester group occurs like the oxidation of olefins (see Chapter 2). Esters like methyl oleate have weak bonds near the double bond. The peroxyl radical attacks these bonds, and the oxidation reaction occurs far from the ester group. The ester group influences the oxidation rate through its solvent properties.

Disproportionation of ester peroxyl radicals occurs very rapidly. Apparently, the ester group influences the rate constant of this reaction by increasing it. The rate constants of chain termination in oxidized esters are collected in Table 9.13.

TABLE 9.11
The Values of $k_p(2k_t)^{-1/2}$ for Ester Oxidation

Ester	T (K)	E (kJ mol^{-1})	log A, A (L mol^{-1} s^{-1})	$k_p (2k_t)^{-1/2}$ (348 K) (L mol^{-1} s^{-1})$^{1/2}$	Ref.
Acetic acid, butyl ester	353–368	36.0	1.94	3.44×10^{-4}	[41]
Acetic acid, pentyl ester	353–368	38.5	2.35	3.73×10^{-4}	[41]
Acetic acid, decyl ester	353–368	40.2	2.89	7.18×10^{-4}	[41]
Acetic acid, benzyl ester	348			7.8×10^{-4}	[42]
Acetic acid, benzyl ester	393–413	47.0	4.52	2.92×10^{-3}	[43]
Acetic acid, 1-cyclohexenyl ester	333–363	30.5	2.63	1.13×10^{-2}	[44]
Acetic acid, 3-cyclohexenyl ester	333–363	19.6	3.95	10.2	[44]
Butyric acid, benzyl ester	348			6.0×10^{-4}	[42]
Phenylacetic acid, propyl ester	348			4.1×10^{-4}	[42]
Oleic acid, methyl ester	333–373	45.2	4.64	7.17×10^{-3}	[45]
Linoleic acid, methyl ester	333–373	23.8	2.20	4.24×10^{-2}	[45]
Linolenic acid, methyl ester	333–373	33.5	4.11	0.12	[45]
Adipic acid, methyl ester	418–443	54.4	4.22	1.13×10^{-4}	[46]
Adipic acid, ethyl ester	418–443	51.5	4.10	2.34×10^{-4}	[46]
Adipic acid, diethyl ester	418–443	92.5	9.38	3.13×10^{-5}	[46]
Adipic acid, dipropyl ester	418–443	97.1	9.91	2.16×10^{-5}	[46]
Adipic acid, (methylethyl)ester	418–443	43.9	5.36	5.90×10^{-2}	[46]
Adipic acid, bis(methylethyl)ester	418–443	66.1	9.45	0.34	[46]
Adipic acid, bis(2-methylpropyl)ester	418–443	70.7	9.94	0.21	[46]
Diethyleneglicol, dihexanoic ester	383–413	54.0	4.27	1.46×10^{-4}	[47]
Pentaeritritol, tetrapentanoic ester	383–423	50.2	3.90	2.32×10^{-4}	[47]
Pentanoic acid, ethyl ester	385–409	49.6	3.63	1.54×10^{-4}	[43]
Ethyleneglicol, dihexanoic ester	413–470	65.9	6.20	2.03×10^{-4}	[43]
1,10-Didecanoic acid, bis(2-ethylhexyl) ester	413–470	59.3	5.29	2.45×10^{-4}	[43]
1,6-Dihexanoic acid, bis(2-ethylhexyl) ester	413–470	39.6	2.63	4.85×10^{-4}	[43]
Phtalic acid, bis(2-ethylhexyl) ester	413–470	28.5	0.85	3.82×10^{-4}	[44]
Formic acid, cholesterol ester	348			6.35×10^{-3}	[48]
Acetic acid, cholesterol ester	348			4.45×10^{-3}	[48]
Propanoic acid, cholesterol ester	348			3.40×10^{-3}	[48]
Butanoic acid, cholesterol ester	338–371	29.8	2.41	7.60×10^{-3}	[48]
Hexanoic acid, cholesterol ester	338–371	34.8	3.12	7.90×10^{-3}	[48]
Nonanoic acid, cholesterol ester	338–371	62.6	7.24	5.20×10^{-3}	[48]
Dodecanoic acid, cholesterol ester	338–371	31.6	2.60	7.00×10^{-3}	[48]
Oleic acid, cholesterol ester	308–348	31.8	3.16	2.32×10^{-2}	[48]
Cinnamic acid, cholesterol ester	308–348	34.0	3.28	1.48×10^{-2}	[48]
p-Methoxybenzoic acid, cholesterol ester	348			3.57×10^{-3}	[48]
Pentanoic acid, cholesterol ester	348			6.09×10^{-3}	[48]]
Octadecanoic acid, cholesterol ester	348			7.80×10^{-3}	[48]
Benzoic acid, cholesterol ester	348			7.20×10^{-3}	[48]

TABLE 9.12
Rate Constants of Chain Propagation k_p in Oxidized Esters

Ester	T (K)	E (kJ mol^{-1})	log A_p, A_p (L mol^{-1} s^{-1})	k_p(348 K) (L mol^{-1} s^{-1})	Ref.
		Hydrogen Atom Abstraction			
Acetic acid, benzyl ester	393–413	59.3	10.30	25.1	[43]
Pentanoic acid, ethyl ester	385–409	55.7	9.40	10.9	[43]
Benzoic acid, ethyl ester	373–403	96.2	13.65	0.16	[50]
Benzoic acid, benzyl ester	373–403	57.8	9.75	11.9	[50]
2-Methylpropionic acid, methyl ester	328–353	87.9	13.11	0.82	[51]
Phenylacetic acid, 2-dimethylethyl ester	373–403	41.9	7.15	7.26	[49]
Dodecanoic acid, cholesterol ester	338–371	39.4	7.12	16.1	[52]
Ethyleneglicol, dihexanoic ester	413–470	73.0	12.00	11.0	[43]
Diethyleneglicol, dihexanoic ester	383–413	60.9	10.20	11.5	[43]
Pentaeritritol, tetrapentanoic ester	383–413	67.7	10.60	2.74	[43]
Didecanoic acid, *bis*(2-ethylhexyl) ester	413–470	73.0	11.50	3.49	[43]
Dihexanoic acid, *bis*(2-ethylhexyl) ester	413–470	53.3	8.80	6.31	[43]
Phtalic acid, *bis*(2-ethylhexyl) ester	413–470	42.2	7.10	5.83	[43]
		Addition			
Acetic acid, vinyl ester	323			2.80	[53]
Acrylic acid, methyl ester	323			1.70	[53]
Acrylic acid, butyl ester	323			1.40	[53]
Methacrylic acid, methyl ester	303–323	53.5	8.92	7.76	[54]
Methacrylic acid, butyl ester	303–323	45.6	7.80	9.03	[54]
Methacrylic acid, 2-methylpropyl ester	323			2.40	[54]
3-Methylbutenoic acid, methyl ester	303			0.20	[55]

The chain generation by the reaction with dioxygen was studied for esters of different structures. Four mechanisms of free radical formation were evidenced.

1. The bimolecular reaction with hydrogen atom abstraction by dioxygen from the weakest C—H bond. Ester as a polar medium accelerates this reaction due to solvation of the polar TS.

$$RC(O)OCH_2R^1 + O_2 \longrightarrow RC(O)OC^{\bullet}HR^1 + HO_2^{\bullet}$$

2. The bimolecular addition of dioxygen to the double bond of nonsaturated ester. This reaction seems to be preceded by CTC formation.

$$O_2 + CH_2=CHCOOR \rightleftharpoons O_2\delta(-) \cdots \delta(+)CH_2=CHCOOR$$
$$O_2 \cdots CH_2=CHCOOR \longrightarrow {}^{\bullet}OOCH_2C^{\bullet}HCOOR$$

3. The trimolecular reaction of dioxygen with the weakest C—H bonds of two nonsaturated esters.

$$2RCH=CHCH_2CH=CHR^1 + O_2 \longrightarrow 2RCH=CHC^{\bullet}HCH=CHR^1 + H_2O_2$$

TABLE 9.13
Rate Constants of Disproportionation of Ester Peroxyl Radical

Ester	T (K)	E_t (kJ mol^{-1})	log ($2A_t$), A_t (L mol^{-1} s^{-1})	$2k_t$ (348 K) (L mol^{-1} s^{-1})	Ref.
Acetic acid, methyl ester	293			2.8×10^8	[56]
Acetic acid, ethyl ester	293			3.4×10^8	[56]
Acetic acid, propyl ester	293			6.7×10^7	[56]
Acetic acid, methylethyl ester	293			1.6×10^8	[56]
Acetic acid, butyl ester	293			7.6×10^6	[56]
Acetic acid, 1,1-dimethylethyl ester	293			4.1×10^8	[56]
Acetic acid, pentyl ester	293			1.3×10^7	[56]
Acetic acid, benzyl ester	393–413	24.6	11.50	6.42×10^7	[43]
Pentanoic acid, ethyl ester	385–409	18.2	11.50	5.86×10^8	[43]
Dodecanoic acid, cholesterol ester	281–308	15.70	9.04	4.82×10^6	[52]
Dodecanoic acid, cholestanol ester	282–319	14.20	8.27	1.38×10^6	[52]
Ethyleneglicol, dihexanoic ester	383–413	14.2	11.50	2.34×10^9	[43]
Diethyleneglicol, dihexanoic ester	383–413	14.2	11.00	7.39×10^8	[43]
Pentaeritritol, tetrapentanoic ester	383–413	27.4	12.40	1.94×10^8	[43]
Didecanoic acid, bis(2-ethylhexyl) ester	383–413	27.4	12.40	1.94×10^8	[43]
Dihexanoic acid, bis(2-ethylhexyl) ester	383–413	27.4	12.40	1.94×10^8	[43]
Phtalic acid, bis(2-ethylhexyl) ester	383–413	27.4	12.40	1.94×10^8	[43]
Vinyl acetate	303			4.9×10^7	[55]
Acrylic acid, methyl ester	303			1.7×10^7	[55]
Methyl acrylate	303–323	19.3	9.32	1.6×10^7	[54]
Isopropenyl acetate	303			1.6×10^6	[55]
Butyl acrylate	323			9.0×10^6	[54]
Isobutyl methacrylate	303–323	18.4	9.41	4.45×10^6	[54]
Butyl methacrylate	303–323	10.9	8.34	5.06×10^6	[56]

4. The trimolecular reaction of dioxygen with double bonds of two molecules of nonsaturated esters. As in the case of a similar bimolecular reaction, this reaction seems to be preceded by CTC formation.

$$O_2 + CH_2=CHCO_2R \rightleftharpoons O_2\delta(-) \cdots \delta(+)CH_2=CHCO_2R$$

$$CH_2=CHCO_2R + O_2CH_2=CHCO_2R \rightleftharpoons CH_2=CHCO_2R \cdots O_2 \cdots CH_2=CHCO_2R$$

$$CH_2=CHCO_2R \cdots O_2 \cdots CH_2=CHCO_2R \rightarrow RO_2CC^\bullet HCH_2OOCH_2C^\bullet HCO_2R$$

The rate constants of these reactions are collected in Table 9.14. Detailed information about these reactions are given in Chapter 4.

The accumulation of hydroperoxide accelerates the ester oxidation. As in hydrocarbon oxidation, this acceleration is the result of hydroperoxide decomposition into free radicals. The most probable is the bimolecular reaction of hydroperoxide with the weakest C—H bond of saturated ester (see Chapter 4).

$$R^1OOH + RC(O)OCH_2R \rightarrow R^1O^\bullet + H_2O + RC(O)OC^\bullet HR$$

TABLE 9.14
Rate Constants of Free Radical Generation by Reaction of Esters with Dioxygen

Ester	Solvent	T (K)	E (kJ mol^{-1})	log A, (L mol^{-1} s^{-1})	k (450 K) (L mol^{-1} s^{-1})	Ref.
		RC(O)OCH$_2$R + O$_2$ → RC(O)OC$^\bullet$HR + HO$_2^\bullet$				
EtC(O)O(CH—H)Me	Propionic acid, methyl ester	343–363	133.8	13.81	1.91×10^{-2}	[57]
Ph(CH—H)OC(O)Ph	Benzene, 1,2-dichloro-	373–403	96.5	5.90	5.00×10^{-6}	[58]
Ph(CH—H)OCH$_2$Ph	Chlorobenzene	398			7.80×10^{-9}	[59]
Ph(CH—H)OCH$_2$Ph	Dibenzyl ether	398			3.7×10^{-9}	[59]
Ph(CH—H)OCH$_2$Ph	Benzene, 1,2-dichloro-	398			9.50×10^{-9}	[59]
Ph(CH—H)OCH$_2$Ph	Nitrobenzene	398			4.1×10^{-8}	[59]
Ph(CH—H)OCH$_2$Ph	Benzene, 1,2-dichloro-	403			2.86×10^{-8}	[60]
Ph(CH—H)OCH$_2$Ph	Nitrobenzene	413			1.15×10^{-7}	[60]
MeOC(O)(CH$_2$)$_8$(CH—H)C(O)OMe	Chlorobenzene	413–443	164.3	13.90	6.74×10^{-6}	[61]
EtOC(O)(CH$_2$)$_8$C(O)O(CH—H)Me	Chlorobenzene	413–443	141.7	11.72	1.87×10^{-5}	[61]
EtOC(O)(CH$_2$)$_8$C(O)O(CH—H)Me	Chlorobenzene	413			4.90×10^{-7}	[62]
C$_2$D$_5$OC(O)(CH$_2$)$_8$C(O)O(CD—D)CD$_3$	Chlorobenzene	413			9.00×10^{-8}	[62]
EtOC(O)(CH$_2$)$_8$C(O)O(CH—H)Me	Chlorobenzene	433			2.30×10^{-6}	[62]
C$_2$D$_5$OC(O)(CH$_2$)$_8$C(O)O(CD—D)CD$_3$	Chlorobenzene	433			3.00×10^{-7}	[62]
Me$_2$CHOC(O)(CH$_2$)$_8$C(O)OC—HMe$_2$	Chlorobenzene	413–443	145.1	12.00	1.44×10^{-5}	[62]
EtC(O)O(CH—H)Me	Chlorobenzene	343–363	134.0	13.80	1.76×10^{-2}	[61]
(Me(CH—H)C(O)OCH$_2$)$_4$C	Pentaerythritol tetrapropionate	423			3.4×10^{-6}	[63]
(CH$_3$C(O)O(CH—H))$_4$C	Pentaerythritol tetraacetate	423			9.4×10^{-7}	[63]
(Me(CH—H)C(O)OCH$_2$)$_3$CEt	1,2,3-Propane-triol-2-ethyl tripropionate	423			5.2×10^{-7}	[63]
(Me(CH—H)C(O)OCH$_2$)$_2$CMe$_2$	1,3-Propanediol-2,2-dimethyl tripropionate	423			2.2×10^{-7}	[63]
		CH$_2$=CHC(O)OR + O$_2$ → $^\bullet$OOCH$_2$C$^\bullet$HC(O)OR				
BuOC(O)CH=CH$_2$	Chlorobenzene	343–363	88.6	6.04	6.57×10^{-8}	[49]
BuOC(O)(Me)C=CH$_2$	Chlorobenzene	342–363	91.5	6.72	1.16×10^{-7}	[49]
BuOC(O)(Me)C=CH$_2$	Chlorobenzene	363			3.4×10^{-7}	[49]
BuOC(O)(Me)C=CH$_2$	N,N-Dimethylformamide	363			8.1×10^{-7}	[49]
Me$_3$COCOCH=CH$_2$	Chlorobenzene	353			7.1×10^{-8}	[49]
MeOCO(Me)C=CH$_2$	Chlorobenzene	353			1.3×10^{-7}	[49]
MeOCO(Et)C=CH$_2$	Chlorobenzene	353			1.2×10^{-7}	[49]
MeCH=CHC(O)OEt	Chlorobenzene	363			5.1×10^{-8}	[49]
NH$_2$COCH=CH$_2$	Chlorobenzene	353			1.5×10^{-6}	[49]

continued

TABLE 9.14
Rate Constants of Free Radical Generation by Reaction of Esters with Dioxygen—continued

Ester	Solvent	T (K)	E (kJ mol^{-1})	log A, (L mol^{-1} s^{-1})	k (450 k) (L mol^{-1} s^{-1})	Ref.
CH$_2$=CHCOOCH$_2$C(OCOEt)$_3$	Chlorobenzene	363			1.40 × 10^{-8}	[64]
(CH$_2$=CHC(O)OCH$_2$)$_4$C	Chlorobenzene	363			1.10 × 10^{-8}	[64]
(CH$_2$=C(Me)C(O)OCH$_2$)$_4$C	Chlorobenzene	343–393	180.6	18.20	1.77 × 10^{-9}	[65]
			2 RH + O$_2$ → R$^\bullet$ + H^{2}O$_2$ + R$^\bullet$			
C$_8$H$_{17}$CH=CH(CH—H)(CH$_2$)$_6$C(O)OMe	Chlorobenzene	313–333	142.0	12.30	5.71 × 10^{-7} *	[45]
Bu(CH$_2$CH=CH)$_2$(CH—H)(CH$_2$)$_7$C(O)OMe	Chlorobenzene	313–333	93.0	6.30	1.43 × 10^{-6} *	[45]
Et(CH=CHCH—H)$_3$(CH$_2$)$_6$C(O)OMe	Chlorobenzene	313–333	76.5	3.80	6.46 × 10^{-7} *	[45]
Cholesteryl pelargonate	Chlorobenzene	364–388	73.0	5.43	7.89 × 10^{-5}	[66]
			2 CH$_2$=CHC(O)OR + O$_2$ → RO(O)CC$^\bullet$HCH$_2$OOCH$_2$C$^\bullet$HC(O)OR			
BuOC(O)MeC=CH$_2$	N,N-Dimethylformamide	363			6.0 × 10^{-8} *	[49]
MeCH=CHC(O)OEt	Chlorobenzene	363			4.2 × 10^{-7} *	[49]

*k is expressed in L^2 mol^{-2} s^{-1}.

Another probable reaction of homolytic decomposition of ester hydroperoxide is the intramolecular interaction of the hydroperoxide group with the carbonyl group of ester with the formation of labile hydroxyperoxide succeeded the splitting of the weak O—O bond (see decomposition of hydroperoxides in oxidized ketones in Chapter 8).

$$RC(O)OCH(OOH)R' \rightleftharpoons \begin{array}{c} R \quad O \quad H \\ \diagdown \diagup \diagdown \diagup \\ HO \quad O-O \quad R' \end{array}$$

$$\begin{array}{c} R \quad O \quad H \\ \diagdown \diagup \diagdown \diagup \\ HO \quad O-O \quad R' \end{array} \longrightarrow RC(O)O^\bullet + RC(O)H + HO^\bullet$$

In oxidized nonsaturated esters the chain generation proceeds via the reaction of formed hydroperoxides with the double bond of ester (see Chapter 4).

$$R^1OOH + CH_2 = CHC(O)OR \longrightarrow R^1O^\bullet + HOCH_2C^\bullet HC(O)OR$$

The rate constants of these reactions are collected in Table 9.15.

The decay of ester hydroperoxides was found to occur more rapidly than free radical generation. The probability e for the radical pair formed from hydroperoxide to escape from

TABLE 9.15
Rate Constants of Ester Hydroperoxides Decomposition into Free Radicals

Ester	T (K)	E, (kJ mol^{-1})	log A, (L mol^{-1} s^{-1})	k_i(400 K)	Ref.
		RC(O)OCH(OOH)R → Free radicals, A, k_i (s^{-1})			
Acetic acid, benzyl ester	393–413	65.4	3.94	2.51×10^{-5}	[43]
Pentanoic acid, ethyl ester	385–409	95.3	6.86	2.60×10^{-6}	[43]
Pentaeritritol, tetrapentanoic ester	413–492	122.1	10.08	1.37×10^{-6}	[43]
Diethyleneglicol, dihexanoic ester	413–470	129.5	11.52	4.07×10^{-6}	[43]
Dihexanoic acid, *bis*(2-ethylhexyl) ester	413–470	94.4	7.12	6.20×10^{-6}	[43]
Phtalic acid, *bis*(2-ethylhexyl) ester	413–470	132.4	11.84	3.55×10^{-6}	[43]
Didecanoic acid, *bis*(2-ethylhexyl) ester	413–470	70.2	4.10	8.57×10^{-6}	[43]
		Me$_2$PhCOOH + CH$_2$=CHC(O)OR → HOCH$_2$C$^•$HC(O)OR + Me$_2$PhCO$^•$ (A, k_i (L mol^{-1} s^{-1}))			
Acrylic acid, methyl ester	313–343	82.6	5.88	1.24×10^{-5}	[67]
Acrylic acid, butyl ester	323			2.70×10^{-8}	[67]
Methacrylic acid, methyl ester	313–343	78.2	5.01	6.28×10^{-6}	[67]
Methacrylic acid, butyl ester	323			1.90×10^{-8}	[67]
2-Methylacrylic acid, methyl ester	323			3.20×10^{-8}	[67]

the cage was measured for two hydroperoxides [43]. The results are presented in the following table.

Ester	T (K)		
	423	443	463
(Me(CH$_2$)$_4$COOCH$_2$CH$_2$)$_2$O	0.12	0.15	0.23
C(CH$_2$OC(O)Bu)$_4$	0.018	0.042	0.046

9.3.3 Reactivity of Esters in Reactions with Peroxyl Radicals

Chain propagation proceeds in ester oxidation via the reaction of the ester peroxyl radical with weakest C—H bond of the ester. Is the activity of the ester peroxyl radical same as alkyl peroxyl radical? Let us compare the rate constants of two different peroxyl radicals reactions with the same ester group, namely, RCH$_2$C(O)OCH$_2$R.

Ester	Peroxyl Radical	k (348 K) (L mol^{-1} s^{-1})
PrCH$_2$C(O)OCH$_2$Me	PrCH$_2$C(O)OCH(OO$^•$)Me	10.9
MeCH$_2$C(O)OCH$_2$Me	Me$_2$PhCOO$^•$	2.2

We see that the ester peroxyl radical is nearly five times more active than the cumyl peroxyl radical. Two reasons are possible for such a difference:difference in the BDE of the formed O—H bond and polar influence of the ester group on the polar transition state. One can suppose that the BDE of the formed O—H bond in ester hydroperoxide is more than that of the cumyl hydroperoxide as it is in peroxyl radicals of carbonyl compounds (see Chapter 8). We can estimate BDE of this O—H bond using the ratio of rate constants of two peroxyl radicals with the same hydrocarbon. The hydrocarbon molecule has no polar group. Therefore, the difference in reaction enthalpies can be the reason for this difference in activation energies and rate constants of these two reactions. Data on cumene (R_1H) co-oxidation with benzylic esters ((R_2H), $T = 348$ K, [68]) helps to perform such estimation (for ΔD calculation, refer to Equation [7.10]).

Ester (R_2H)	r_2	k_{p21}	k_{p21}/k_{p11}	ΔE	ΔD
AcOCH$_2$Ph	0.55	45.6	16.3	8.1	19.4
EtC(O)OCH$_2$Ph	0.90	27.9	10.0	6.7	16.1

The values of ΔD are close and the mean value is $\Delta D = 17.8 \pm 1.6$ kJ mol^{-1}. As a result, we obtain the value for D_{O-H} of secondary ester hydroperoxide $D = 358.6 + 17.8 = 376.4$ kJ mol^{-1}. The difference in D_{O-H} of the secondary and tertiary alkyl hydroperoxides is equal to 6.9 kJ mol^{-1} [79]. Supposing the same difference for the secondary and tertiary ester hydroperoxides we estimate for the last $D_{O-H} = 376.4 - 6.9 = 369.5$ kJ mol^{-1}. The experimental data on reactions of 1,1-dimethylethylperoxyl and cumyl peroxyl radicals with esters are collected in Table 9.16.

The role and influence of polar interaction in reactions of alkyl peroxyl radicals with esters can be studied within the scope of the IPM (see Chapter 7). The increment of polar interaction in activation energy of peroxyl radical with polar compound can be calculated from parameters br_e (Equation (6.33)). The results of the calculation of ΔE_μ are presented in Table 9.17.

We see that the polar effect is strong in reactions of peroxyl radicals with monoesters. It lowers the activation energy of the reaction of the peroxyl radical in most cases. This means that the geometry of atoms C $\cdots$ H $\cdots$ O of the TS is nonlinear. The geometric parameters of the TS of peroxyl radical reactions with C—H bonds of esters are presented in Table 9.18.

The study of the attack of the 1,1-dimethylethyl radical on different C—H bonds of acetylcyclohexyl ester gave the following interesting picture of reactivity (333 K [73]).

Position	α	β	γ	δ
n_{C-H}	1	4	4	2
k_{C-H} (L mol^{-1} s^{-1})	1.89×10^{-2}	3.3×10^{-4}	8.0×10^{-4}	3.7×10^{-3}

The most remote δ-C—H bonds appeared to have the same reactivity as that of cyclohexane ($k_{C-H} = 3.7 \times 10^{-3}$ (333 K)). The C—H bonds of α- and β-positions lose their reactivity under the polar influence of the ester group. So, one can conclude that the electric field effect is a very important factor for the reactivity in reactions of RO$_2^\bullet$ with such polar compounds as esters.

9.3.4 Effect of Multidipole Interaction in Reactions of Polyfunctional Esters

When a molecule consists of a few similar fragments n, the rate constant of the reactant reaction with this molecule can be expressed as the product of the partial rate constants k_j: $k = n \times k_j$. This was proved many times for free radical reactions for groups of reactants where both reactants or one of them are nonpolar. For example, the rate constants of peroxyl radical reactions with nonbranched aliphatic hydrocarbons Me(CH$_2$)$_n$Me can be presented in

TABLE 9.16
Rate Constants for Peroxyl Radical Reactions with Esters

Ester	Peroxyl Radical	T (K)	E (kJ mol^{-1})	log A, A (L mol^{-1} s^{-1})	k (333 K) (L mol^{-1} s^{-1})	Ref.
		Hydrogen Abstraction				
BuC(O)OCH$_2$Me	Me$_2$PhCO$_2^\bullet$	404			1.04	[43]
BuC(O)OCH$_2$Me	Me$_2$PhCO$_2^\bullet$	398–404	60.0	8.30	0.077	[69]
Me$_2$CHC(O)OCMe$_3$	Me$_2$PhCO$_2^\bullet$	358–388	67.8	9.06	0.027	[70]
AcOCH$_2$Ph	Me$_3$CO$_2^\bullet$	303			0.08	[71]
AcOCH$_2$Ph	Me$_2$PhCO$_2^\bullet$	348			0.37	[68]
BuC(O)OCH$_2$Ph	Me$_2$PhCO$_2^\bullet$	348			0.20	[68]
Me$_3$CC(O)OCH$_2$Ph	Me$_3$CO$_2^\bullet$	333			0.29	[50]
PhC(O)OCH$_3$	Me$_2$PhCO$_2^\bullet$	403–423	68.8	8.50	5.1×10^{-3}	[72]
PhC(O)OCH$_2$Me	Me$_2$PhCO$_2^\bullet$	413–433	63.9	8.30	0.019	[72]
PhC(O)OCD$_2$Me	Me$_2$PhCO$_2^\bullet$	413–433	66.3	8.30	8.0×10^{-3}	[72]
cyclohexyl acetate	Me$_3$CO$_2^\bullet$	333			0.082	[73]
PhC(O)OCH$_2$Ph	Me$_3$CO$_2^\bullet$	303–333	50.4	7.30	0.25	[50]
4-Cl-C$_6$H$_4$C(O)OCH$_2$Ph	Me$_3$CO$_2^\bullet$	333			0.24	[50]
4-Me-C$_6$H$_4$C(O)OCH$_2$Ph	Me$_3$CO$_2^\bullet$	333			0.26	[50]
PhCH$_2$C(O)OCH$_2$Ph	Me$_3$CO$_2^\bullet$	333			0.46	[50]
CH$_3$OC(O)C(O)OCH$_3$	Me$_2$PhCO$_2^\bullet$	348–373	23.0	3.93	2.12	[74]
MeOC(O)CH$_2$C(O)OMe	Me$_2$PhCO$_2^\bullet$	383–418	66.9	9.53	0.11	[74]
MeOC(O)(CH$_2$)$_2$C(O)OMe	Me$_2$PhCO$_2^\bullet$	403–418	78.2	10.62	0.022	[74]
MeOC(O)(CH$_2$)$_3$C(O)OMe	Me$_2$PhCO$_2^\bullet$	403–418	69.0	9.59	0.059	[74]
EtOC(O)(CH$_2$)$_3$C(O)OEt	Me$_2$PhCO$_2^\bullet$	398–418	69.0	9.57	0.056	[75]
PrOC(O)(CH$_2$)$_3$C(O)OPr	Me$_2$PhCO$_2^\bullet$	398–418	57.3	8.15	0.14	[75]
BuOC(O)(CH$_2$)$_3$C(O)OBu	Me$_2$PhCO$_2^\bullet$	398–418	55.6	8.00	0.18	[75]
Me$_3$COC(O)(CH$_2$)$_2$C(O)OCMe$_3$	Me$_2$PhCO$_2^\bullet$	408–418	72.8	9.65	0.017	[76]
MeOC(O)(CH$_2$)$_4$C(O)OMe	Me$_2$PhCO$_2^\bullet$	413			7.57	[74]
MeOC(O)(CH$_2$)$_4$C(O)OCMe$_3$	Me$_2$PhCO$_2^\bullet$	413–433	74.9	8.60	7.13×10^{-4}	[75]
EtOC(O)(CH$_2$)$_4$C(O)OCMe$_3$	Me$_2$PhCO$_2^\bullet$	413–433	72.4	8.28	8.36×10^{-4}	[75]
Me$_2$CHOC(O)(CH$_2$)$_4$COOCMe$_3$	Me$_2$PhCO$_2^\bullet$	413–433	69.9	7.97	1.01×10^{-3}	[75]
Me$_3$COC(O)(CH$_2$)$_4$C(O)OCMe$_3$	Me$_2$PhCO$_2^\bullet$	413–433	69.9	8.15	1.52×10^{-3}	[70]
MeOC(O)(CH$_2$)$_5$C(O)OMe	Me$_2$PhCO$_2^\bullet$	403–418	66.9	9.40	0.028	[74]
MeOC(O)(CH$_2$)$_7$C(O)OMe	Me$_2$PhCO$_2^\bullet$	403–418	59.4	8.50	0.15	[74]
MeOC(O)(CH$_2$)$_8$C(O)OH	Me$_2$PhCO$_2^\bullet$	403–418	66.9	9.15	0.045	[74]
MeOC(O)(CH$_2$)$_8$C(O)OMe	Me$_2$PhCO$_2^\bullet$	403–418	57.3	8.28	0.19	[74]

continued

TABLE 9.16
Rate Constants for Peroxyl Radical Reactions with Esters—continued

Ester	Peroxyl Radical	T (K)	E (kJ mol^{-1})	log A, A (L mol^{-1} s^{-1})	k (333 K) (L mol^{-1} s^{-1})	Ref.
[Me(CH$_2$)$_4$C(O)OCH$_2$CH$_2$]$_2$O	Me$_2$PhCO$_2^\bullet$	383–413	97.8	13.23	7.76 × 10^{-3}	[77]
C(CH$_2$OC(O)Bu)$_4$	Me$_2$PhCO$_2^\bullet$	393–413	100.5	13.65	7.74 × 10^{-3}	[77]
C(CH$_2$OAc)$_4$	Me$_2$PhCO$_2^\bullet$	404			1.92	[43]
C(CH$_2$OC(O)Et)$_4$	Me$_2$PhCO$_2^\bullet$	404			3.24	[43]
C(CH$_2$OC(O)(CH$_2$)$_8$Me)$_4$	Me$_2$PhCO$_2^\bullet$	404			20.9	[43]
EtC(CH$_2$OC(O)Et)$_3$	Me$_2$PhCO$_2^\bullet$	404			1.52	[43]
Me$_2$C(CH$_2$OC(O)Et)$_2$	Me$_2$PhCO$_2^\bullet$	404			0.70	[43]
[Me(CH$_2$)$_6$C(O)OCH$_2$]$_2$	Me$_2$PhCO$_2^\bullet$	404			6.89	[43]
Addition						
CH$_2$=CHOC(O)Me	Me$_2$PhCO$_2^\bullet$	323			0.20	[53]
CH$_2$=CHOC(O)Me	Me$_3$CO$_2^\bullet$	323			0.10	[53]
CH$_2$=CHOC(O)Me	HO$_2^\bullet$	323			6.8	[53]
CH$_2$=CHC(O)Me	Me$_2$PhCO$_2^\bullet$	323			0.50	[53]
CH$_2$=CHC(O)Me	Me$_3$CO$_2^\bullet$	323			0.40	[53]
CH$_2$=CHC(O)Me	~CH$_2$CH(O$_2^\bullet$)Ph	323			9.8	[53]
CH$_2$=CHC(O)Me	~CH$_2$CH(O$_2^\bullet$)C(O)Me	323			3.4	[53]
CH$_2$=CMeC(O)OMe	Me$_2$PhCO$_2^\bullet$	323			1.8	[53]
CH$_2$=CMeC(O)OMe	Me$_3$CO$_2^\bullet$	323			1.1	[53]
CH$_2$=CMeC(O)OMe	HO$_2^\bullet$	323			40	[53]
CH$_2$=CMeC(O)OMe	~CH$_2$CH(O$_2^\bullet$)Ph	323			12	[53]
CH$_2$=CMeC(O)OMe	~CH$_2$CMe(O$_2^\bullet$)C(O)OMe	303–323	53.5	8.92	3.4	[54]
Me$_2$C=CHOC(O)Me	Me$_3$CO$_2^\bullet$	303			1.0 × 10^2	[55]
E- MeCH=CHC(O)OMe	HO$_2^\bullet$	323			13	[53]
E- MeCH=CHC(O)OEt	HO$_2^\bullet$	323			14	[53]
[(CH$_2$=CHCO$_2$CH$_2$]$_2$CMe$_2$	Me$_2$PhCO$_2^\bullet$	323			0.76	[78]
[CH$_2$=CHC(O)OCH$_2$]$_4$C	Me$_2$PhCO$_2^\bullet$	323			1.3	[78]
[CH$_2$=CMeC(O)OCH$_2$]$_4$C	Me$_2$PhCO$_2^\bullet$	323			3.1	[78]

the form: $k = n \times k_{CH_2} + 2 k_{CH_3}$ [80]. The problem appears when both reactants are polar. Is the principle of partial rate constants valid in these reactions? Can the molecular rate constant be presented as the sum or the product of the partial rate constants? This problem was checked at first in experiments on rate constant measurements of peroxyl radical reactions with polyatomic esters and dicarboxylic acids [39,43,44,78]. These results gave impetus to

TABLE 9.17
Reactions of Peroxyl Radicals with Esters: Contribution of Enthalpy and Polar Interaction into Activation Energy (Equations [6.21] and [6.32])

Ester	$RO_2^\bullet$	ΔH_e (kJ mol^{-1})	E_e (kJ mol^{-1})	E_{e0} (kJ mol^{-1})	ΔE_H (kJ mol^{-1})	ΔE_μ (kJ mol^{-1})
BuCOOCH$_2$Me	Me$_2$PhCO$_2^\bullet$	37.1	78.7	60.6	18.1	4.2
Me$_2$CHCOOMe	Me$_2$C(OO$^\bullet$)COOMe	2.7	69.9	68.7	1.2	12.3
Me$_2$CHCOOCMe$_3$	Me$_2$PhCO$_2^\bullet$	25.1	77.0	65.1	11.9	9.5
AcOCH$_2$Ph	Me$_3$CO$_2^\bullet$	−3.4	70.5	72.0	−1.5	9.7
AcOCH$_2$Ph	Me$_2$PhCO$_2^\bullet$	−3.4	74.2	75.7	−1.5	12.9
AcOCH$_2$Ph	AcOCH(OO$^\bullet$)Ph	−21.2	62.0	71.1	−9.1	8.8
PhCOOCH$_2$Me	PhCOOCH(OO$^\bullet$)Me	16.7	76.6	68.8	7.8	12.4
PhCOOCH$_2$Me	Me$_2$PhCO$_2^\bullet$	37.1	79.9	61.8	18.1	5.4
PhCH$_2$COOCMe$_3$	PhCH(OO$^\bullet$)COOCMe$_3$	−13.2	65.6	71.4	−5.8	9.1
PhCOOCH$_2$Ph	PhCOOCH(OO$^\bullet$)Ph	−21.2	64.1	73.2	−9.1	10.9
PhCOOCH$_2$Ph	Me$_3$CO$_2^\bullet$	−3.4	72.9	74.4	−1.5	10.6

TABLE 9.18
Geometric Parameters of Peroxyl Radical Reaction with Esters (Equations [6.35]–[6.38])

Ester	ROO$^\bullet$	ΔH_e (kJ mol^{-1})	E_e (kJ mol^{-1})	ΔE_μ (kJ mol^{-1})	r(C—H) × 10^{10} m	r(O—H) × 10^{10} m
BuCOOCH$_2$Me	Me$_2$PhCO$_2^\bullet$	37.1	78.7	4.2	1.431	1.172
Me$_2$CHCOOCMe$_3$	Me$_2$PhCO$_2^\bullet$	25.1	77.0	9.5	1.428	1.195
AcOCH$_2$Ph	Me$_3$CO$_2^\bullet$	−3.4	70.5	9.5	1.413	1.239
AcOCH$_2$Ph	Me$_2$PhCO$_2^\bullet$	−3.4	74.2	12.9	1.422	1.246
AcOCH$_2$Ph	AcOCH(OO$^\bullet$)Ph	−21.2	62.0	8.8	1.393	1.255
PhCOOCH$_2$Me	Me$_2$PhCO$_2^\bullet$	37.1	79.9	5.4	1.434	1.162
PhCOOCH$_2$Ph	PhCOOCH(OO$^\bullet$)Ph	−21.2	64.1	10.9	1.398	1.259
PhCOOCH$_2$Ph	Me$_3$CO$_2^\bullet$	−3.4	72.9	10.6	1.419	1.243

formulate the conception of multidipole interaction in TS of reactions of the polar reactant with the polyfunctional molecule [81,82]. A series of reactions was studied later where the effect of multidipole interaction was identified.

Let us consider a polyfunctional molecule as a community of interacting dipoles μ_i. Due to the rotation of groups, each dipole can change orientation but this possibility is restricted due to a fixed length of the bonds and constant valence angles in the molecule. The distance between two chosen dipole groups i and j is supposed to be longer than the length of each group. The potential of interaction of these two dipoles U_{ij} in this simple electrostatic model of the polyfunctional molecule is given by

$$U_{ij} = \frac{\mu_i \mu_j}{\varepsilon r_{ij}^3} \left\{ -2\cos\vartheta_i \cos\vartheta_j + \sin\vartheta_i \sin\vartheta_j \cos(\varphi_i - \varphi_j) \right\} \quad (9.10)$$

TABLE 9.19
Effect of Multidipole Interaction in Reactions of Polyesters and Polyatomic Alcohols

Ester, Alcohol	T (K)	$k_{n\mu}$ (L mol^{-1} s^{-1})	$k_{n\mu}/nk_{\mu}$	$\Delta G_{n\mu}^{\#}$ (kJ mol^{-1})	Ref.
\multicolumn{6}{c}{$Me_2PhCO_2^{\bullet}$ + Ester (Abstraction)}
$Me_3CCH_2OC(O)Bu$	404	1.0	1.0	0.0	[83]
$Me_2C[CH_2OC(O)Et]_2$	404	3.0	1.5	−1.4	[83]
$EtC[CH_2OC(O)Et]_3$	404	6.3	2.1	−2.5	[83]
$C(CH_2OC(O)Et)_4$	404	13.6	3.4	−4.1	[83]
\multicolumn{6}{c}{$Me(CH_2)_{13}CH(OO^{\bullet})Me$ + Alcohol (Abstraction)}
RCH_2OH	403	32.3	1.0	0.0	[84]
$RCH(OH)R$	403	41.9	1.0	0.0	[84]
$HOCH_2CH_2OH$	403	15.0	0.23	4.9	[84]
$HOCH_2CH_2CH(OH)Me$	403	16.8	0.23	4.9	[84]
$HOCH_2CH_2CH_2CH_2OH$	403	22.4	0.30	4.0	[84]
$Me_2C(CH_2OH)_2$	403	662.0	10.25	−7.8	[84]
$Et_2C(CH_2OH)_3$	403	181.0	1.86	−2.1	[84]
\multicolumn{6}{c}{$R(O)COCH_2CH_2OCH_2CH(OO^{\bullet})OC(O)R$ + Alcohol (Abstraction)}
RCH_2OH	403	20.4	1.0	0.0	[84]
$RCH(OH)R$	403	21.0	1.0	0.0	[84]
$HOCH_2CH_2OH$	403	14.3	0.34	3.6	[84]
$HOCH_2CH_2CH(OH)Me$	403	15.6	0.37	3.3	[84]
$HOCH_2CH_2CH_2CH_2OH$	403	26.7	0.63	1.5	[84]
$Me_2C(CH_2OH)_2$	403	121	2.97	−3.7	[84]
$Et_2C(CH_2OH)_3$	403	118	1.92	−2.2	[84]
\multicolumn{6}{c}{$Me_2PhCO_2^{\bullet}$ + Ester (Addition)}
$CH_2=CHCOOMe$	323	0.50	1.0	0.0	[85]
$Me_2C(OC(O)CH=CHMe)_2$	323	0.76	0.76	0.7	[85]
$EtC(OC(O)CH=CHMe)_3$	323	1.05	0.70	1.0	[85]
$C(CH_2OCOCH=CHMe)_4$	323	1.28	0.64	1.2	[85]
$MeCH=CMeCOOMe$	323	1.79	1.0	0.0	[85]
$EtC(OC(O)CMe=CHMe)_3$	323	3.93	0.77	0.8	[85]
$C(CH_2OCOCH=CHMe)_4$	323	3.08	0.45	2.3	[85]
\multicolumn{6}{c}{$RO_2^{\bullet}$ + 2,4,6–tri-tret.butylphenol (Abstraction)}
$\sim CH_2CH(OO^{\bullet})C(O)OMe$	343	8.1	1.0	0.0	[86]
$\sim CH_2CH(OO^{\bullet})C(O)OC(OCOCH=CH_2)_3$	343	1.9	0.23	4.1	[86]
$\sim CH_2CMe(OO^{\bullet})C(O)OMe$	343	3.8	1.0	0.0	[86]
$\sim CH_2CMe(OO^{\bullet})C(O)OC(OCOCMe=CH_2)_3$	343	1.1	2.9	−2.5	[86]
\multicolumn{6}{c}{O_2 + Ester (Addition)}
$CH_2=CMeCOOBu$	473	7.6×10^{-7}	1.0	0.0	[87]
$C(CH_2OCOCH=CH_2)_4$	473	2.9×10^{-7}	0.09	7.1	[87]
\multicolumn{6}{c}{$Me_2PhCOOH$ + Ester (Addition)}
$CH_2=CHCOOMe$	343	1.75×10^{-7}	1.0	0.0	[88]
$EtC(OC(O)CMe=CHMe)_3$	343	1.83×10^{-7}	0.35	3.0	[88]
$C(CH_2OCOCH=CHMe)_4$	343	9.6×10^{-8}	0.14	5.6	[88]

where ϑ_i is the angle between the dipole axis and the line connected the centers of two dipoles and φ_i is angle between two axes of interacting dipoles i and j. The potential energy of all the interacting dipoles can be presented as the sum of partial U_{ij} energies:

$$U_{ij} = \sum_i^n \sum_j^n U_{ij} \qquad (9.11)$$

The energy of this electrostatic interaction in molecules with four polar groups with $\mu_i = 2$ Debye can vary from -25 to $+25\,\text{kJ}\,\text{mol}^{-1}$ [81]. When the polar free radical attacks this molecule the formed transition state can be regarded as a multidipole system of $n+1$ interacting dipoles with the potential energy U_{n+1}. So, the multidipole interaction in TS and reactant introduces the increment $\Delta E_\mu = U_{n+1} - U_n$ into the activation energy. This increment can be negative or positive. In addition to the energy of a molecule and TS, the multidipole interaction changes the entropy of each system (reactant and TS). So, it would be correct to compare the difference in the partial rate constants with the change in the activation Gibbs potential:

$$\Delta\Delta G_{n\mu}^{\#} = \Delta G_{n\mu}^{\#} - \Delta G_\mu^{\#} = -RT \ln(k_{n\mu}/k_\mu), \qquad (9.12)$$

where symbols μ and $n\mu$ designate the monofunctional and polyfunctional reactants. The quantum-chemical interpretation of multidipole interaction in TS of peroxyl radical in addition reactions are discussed by Machtin [53]. The compensation effect follows from the conception of multidipole interaction [81]. This effect was observed experimentally and can be illustrated by the results on the polyesters reaction with cumyl peroxyl radicals [77].

Ester	PrC(O)OEt	[EtOC(O)CH$_2$]$_2$	[BuCH$_2$COO(CH$_2$)$_2$]$_2$O	(BuCOOCH$_2$)$_4$C
$\Delta G_{n\mu}^{\#}$ (kJ mol^{-1})	0.0	4.8	6.6	10.0
E_μ (kJ mol^{-1})	37.0	69.0	98.0	100.0
$\Delta S_{n\mu}^{\#}$ (J mol^{-1} K^{-1})	0.0	71.0	127.0	135.0
$\Delta E_\mu/\Delta S_{n\mu}^{\#}$ (K)	0.0	451	480	467

One can see that the higher the activation energy, the higher the entropy of activation. A good linear dependence is observed between the relative increase in the activation energy and entropy of activation related to multidipole interaction (as demonstrate values of last stroke).

The effect of multidipole interaction was observed and studied for a series of different reactions of polyatomic esters and alcohols with peroxyl radicals, hydroperoxides, and dioxygen. The results of rate constants measurements and estimation of $\Delta G_{n\mu}^{\#}$ are collected in Table 9.19.

We see that the effect of multidipole interaction plays an important role in all reactions of abstraction and addition of polar reactants. This interaction can increase or decrease the activation energy of the reaction. However, the multidipole interaction does not influence the reactions of nonpolar trichloromethyl radicals with mono- and polyatomic esters due to the nonpolar character of the attacking radical [89].

REFERENCES

1. DL Trimm. In: CH Bamford and CFH Tipper, (eds.) *Comprehensive Chemical Kinetics*, vol 16. Amsterdam: Elsevier, 1980, pp. 205–248.

2. AL Aleksandrov. Negative Catalysis in Chain Reactions Oxidation of N and O-containing Compounds, Doctoral Thesis, Institute of Chemical Physics, Chernogolovka, 1987, pp. 1–40 [in Russian].
3. SG Lias, JF Liebman, RD Levin, SA Kafafi. NIST Standard Reference Database, 19A, NIST Positive Ion Energetics, Version 2.0. Gaithersburg: NIST, 1993.
4. ET Denisov, SL Khursan. *Russ J Phys Chem* 74(35):5491–5497, 2000.
5. ET Denisov. *Russ Chem Rev* 65:505–520, 1996.
6. GA Kovtun, AL Aleksandrov. *Izv AN SSSR Ser Khim* 2208–2211, 1973.
7. GA Kovtun, AV Kazantsev, LL Aleksandrov. *Izv AN SSSR Ser Khim* 2635–2637, 1974.
8. EM Pliss, AL Aleksandrov, VS Mikhlin, MM Mogilevich. *Izv AN SSSR Ser Khim* 2259–2262, 1978.
9. UYa Samatov, AL Aleksandrov, IR Akhunov. *Izv AN SSSR Ser Khim* 2254–2258, 1978.
10. AL Aleksandrov. *Izv AN SSSR Ser Khim* 2720–2725, 1980.
11. AL Aleksandrov. *Izv AN SSSR Ser Khim* 2469–2474, 1980.
12. GS Bakhturidze, IL Edilashvili, AL Aleksandrov. *Izv AN SSSR Ser Khim* 515–519, 1979.
13. GS Bakhturidze, IL Edilashvili, AL Aleksandrov. *Izv AN SSSR Ser Khim* 69–73, 1980.
14. GA Kovtun, AL Aleksandrov. *Izv AN SSSR Ser Khim* 38–44, 1977.
15. S Abramovich, J Rabani. *J Phys Chem* 80:1562–1565, 1976.
16. JA Howard. *Adv Free Radical Chem* 4:49–173, 1972.
17. AL Aleksandrov. *Izv AN SSSR Ser Khim* 1736–1741, 1986.
18. JR Thomas. *J Am Chem Soc* 85:593–594, 1963.
19. RV Kucher, IA Opeida, *Cooxidation of Organic Compounds in the Liquid Phase*. Kiev: Naukova Dumka, 1989 [in Russian].
20. JA Howard, T Yamada. *J Am Chem Soc* 103:7102–7106, 1981.
21. LA Tavadyan, MV Musaelyan, VA Mardoyan. *Khim Fiz* 9:806–811, 1990.
22. D Griller, JA Howard, PR Marriott, JC Scaiano. *J Am Chem Soc* 103:619–623, 1981.
23. LA Tavadyan, MV Musaelyan, VA Mardoyan. *Khim Fiz* 10:511–517, 1991.
24. LA Tavadyan, MV Musaelyan, VA Mardoyan. *Int J Chem Kinet* 22:555–563, 1996.
25. RV Kucher, VI Timokhin, NA Kravchuk. *Dokl AN SSSR* 294:1411–1412, 1987.
26. ET Denisov. *Russ Chem Rev* 64:859–876, 1997.
27. NN Pozdeeva, TI Sapacheva, AL Aleksandrov. *Izv AN SSSR Ser Khim* 1738–1741, 1976.
28. B Lanska, LM Postnikov, AL Aleksandrov, J Sebenda. *Coll Czech Chem Commun* 46:2650–2658, 1981.
29. AL Aleksandrov, B Lanska. *Kinet Catal* 35:608–610, 1994.
30. TI Drozdova, VT Varlamov, VP Lodygina, AL Aleksandrov. *Zh Fiz Khim* 52:735–736, 1978.
31. GI Bitman. *Intermolecular Association of tert-Butyl Hydroperoxide*. PhD Thesis, ITKhT, Moscow, 1973 [in Russian].
32. S Patai. *The Chemistry of Peroxides*. New York: Wiley, 1983.
33. AL Aleksandrov. *Izv AN SSSR Ser Khim* 2223–2230, 1982.
34. TI Drozdova, AL Aleksandrov, ET Denisov. *Izv AN SSSR Ser Khim* 1213–1216, 1978.
35. TI Drozdova, AL Aleksandrov, ET Denisov. *Izv AN SSSR Ser Khim* 967–970, 1978.
36. TI Drozdova, AL Aleksandrov, ET Denisov. *Izv AN SSSR Ser Khim* 965–967, 1978.
37. VE Tumanov, EA Kromkin, ET Denisov. *Russ Chem Bull* 1641–1650, 2002.
38. DR Lide (ed.). *Handbook of Chemistry and Physics*, 72nd ed. Boca Raton, FL: CRC Press, 1991–1992.
39. ET Denisov, NI Mitskevich, VE Agabekov. *Liquid-Phase Oxidation of Oxygen-Containing Compounds*. New York: Consultants Bureau, 1977.
40. ET Denisov. In: CH Bamford and CFH Tipper (eds.) *Comprehensive Chemical Kinetics*, vol 16. Amsterdam: Elsevier, 1980, pp. 125–204.
41. VM Potekhin. To the Problem of Mechanism of Oxidation of Paraffin and Alkylcycloparaffin Hydrocarbons. Doctoral Thesis, LTI, Leningrad, 1972 [in Russian].
42. IA Opeida, VI Timokhin, OV Nosyreva, VG Koloerova, AG Matvienko. *Neftekhimiya* 21:110–113, 1981.
43. VS Martemyanov. Kinetics and Mechanism of High-Temperature Oxidation of Esters of Polyatomic Alcohols. Doctoral Thesis, Institute of Chemical Physics, Chernogolovka, 1987, pp. 1–46 [in Russian].
44. VE Agabekov. Reactivity and Reactions of Oxygen-Containing Compounds in Free Radical Reactions of Oxidation. Doctoral Thesis, Institute of Chemical Physics, Chernogolovka, 1980, pp. 3–43 [in Russian].

45. N Yanishilieva, IP Skibida, ZK Maizus. *Izv Otd Khim Nauk Bulg AN* 4:1, 1971.
46. TG Kosmacheva, VE Agabekov, NI Mitskevich, MN Fedorischeva. *Vestsi AN BSSR Ser Khim Navuk* 2:102–109, 1976.
47. GG Agliullina, VS Martemyanov, ET Denisov. *Neftekhimiya* 16:262–268, 1976.
48. GE Dingchyan, NS Khanukaev, RL Vardanyan. *Arm Khim Zh* 30:644–651, 1977.
49. MM Mogilevich, EM Pliss. *Oxidation and Oxidative Polymerisation of Nonsaturated Compounds.* Moscow: Khimiya, 1990 [in Russian].
50. SG Voronina. The Kinetics, Reactivity and Peculiarities of Mechanism of Oxidative Destruction of Esters. Ph.D. Thesis, Kemerovo State University, Kemerovo, 1991, pp. 3–18 [in Russian].
51. BI Chernyak, IM Krip, GI Donets. *Ukr Khim Zh* 47:1103–1112, 1981.
52. RL Vardanyan, RL Safiullin, VD Komissarov. *Kinet Katal* 29:1327–1331, 1985.
53. VA Machtin. Reactions of Peroxyl Radicals in Oxidizing Vinyl Monomers and Reactivity of Double Bonds. Ph.D. Thesis, Institute of Chemical Physics, Chernogolovka, 1984, pp. 1–18 [in Russian].
54. EM Pliss, AL Aleksandrov, MM Mogilevich. *Izv AN SSSR Ser Khim* 1971–1974, 1975.
55. JA Howard. *Can J Chem* 50:2298–2304, 1972.
56. AI Nikolaev, VS Martemyanov, RL Safiullin. *React Kinet Catal Lett* 24:19–23, 1984.
57. GV Butovskaya, VE Agabekov, NI Mitskevich. *Izv AN BSSR Ser Khim* 36–42, 1981.
58. SG Voronina, AL Perkel, BG Freidin. *Kinet Katal* 33:266–274, 1992.
59. GV Butovskaya. Chain Generation in Oxidized Oxygen-Containing Compounds. Ph.D. Thesis, Institute General and Inorganic Chemistry, Minsk, 1982 [in Russian].
60. VE Agabekov, GV Butovskaya, NI Mitskevich. *Neftekhimiya* 22:272–277, 1982.
61. GV Butovskaya, VE Agabekov, NI Mitskevich. *React Kinetic Catal Lett* 4:105–111, 1976.
62. GV Butovskaya, VE Agabekov, NI Mitskevich. *Dokl AN BSSR* 25:722–727, 1981.
63. NN Pozdeeva, ET Denisov. *Bull Acad Sci USSR* 36:2485–2489, 1987.
64. AV Sokolov, EM Pliss, ET Denisov. *Bull Acad Sci USSR* 37:219–223, 1988.
65. EM Pliss, VM Troshin, ET Denisov. *Dokl AN SSSR* 264:368–370, 1982.
66. GE Dingchan, NS Khanukova, RL Vardanyan. *Arm Khim Zh* 30:644–649, 1977.
67. EM Pliss, VM Troshin. *Neftekhimiya* 22:539–542, 1982.
68. IA Opeida. Co-oxidation of Alkylaromatic Hydrocarbons in the Liquid Phase. Doctoral Dissertation, Institute of Chemical Physics, Chernogolovka, 1982, pp. 1–336 [in Russian].
69. NN Pozdeeva, ET Denisov, VS Martemyanov. *Kinet Katal* 22:912–919, 1981.
70. VE Agabekov, NI Mitskevich, GV Butovskaya, TG Kosmacheva. *React Kinet Catal Lett* 2:123–128, 1975.
71. S Korcek, JHB Chenier, JA Howard, KU Ingold. *Can J Chem* 50:2285–2297, 1972.
72. NN Kornilova, NG Arico, VE Agabekov, NI Mitskevich. *React Kinet Catal Lett* 7:241–248, 1977.
73. SV Puchkov, AL Perkel, EI Buneeva. *Kinet Katal* 42:828–835, 2001.
74. VE Agabekov, ET Denisov, NI Mitskevich, TG Kosmacheva, GV Butovskaya. *Kinet Katal* 15:883–887, 1974.
75. VE Agabekov, NI Mitskevich, MN Fedorischeva. *Kinet Katal* 15:1149–1151, 1974.
76. VE Agabekov, MN Fedorischeva, NI Mitskevich. *Vestsi AN BSSR Ser Khim* 3:31–38, 1972.
77. GG Agliullina, VS Martemyanov, ET Denisov, TI Eliseeva. *Izv AN SSSR Ser Khim* 50–57, 1977.
78. VA Machtin, EM Pliss, ET Denisov. *Izv AN SSSR Ser Khim* 746–750, 1981.
79. ET Denisov, TG Denisova. *Kinet Catal* 34:173–179, 1993.
80. WW Pritzkow, VYa Suprun. *Russ Chem Rev* 64:497–519, 1996.
81. ET Denisov. *Izv AN SSSR Ser Khim* 1746–1750, 1978.
82. ET Denisov. *Usp Khim* 54:1466–1486, 1985.
83. NN Pozdeeva, ET Denisov, VS Martemyanov. *Izv AN SSSR Ser Khim* 912–919, 1981.
84. TG Degtyareva, ET Denisov, VS Martemyanov, IA Kaftan, LR Enikeeva. *Izv AN SSSR Ser Khim* 735–741, 1981.
85. VA Machtin, AV Sokolov, EM Pliss, ET Denisov. *Izv AN SSSR Ser Khim* 543–546, 1985.
86. AV Sokolov, AA Nikanorov, EM Pliss, ET Denisov. *Izv AN SSSR Ser Khim* 778–781, 1985.
87. EM Pliss, ET Denisov, VM Troshin. *Dokl AN SSSR* 264:368–370, 1982.
88. VM Troshin, EM Pliss, ET Denisov. *Izv AN SSSR Ser Khim* 2191–2194, 1984.
89. NN Pozdeeva, ET Denisov. *Izv AN SSSR Ser Khim* 2029–2035, 1983.

10 Catalysis in Liquid-Phase Hydrocarbon Oxidation

10.1 CATALYSIS BY TRANSITION METAL IONS AND COMPLEXES IN HYDROCARBON OXIDATION BY DIOXYGEN

10.1.1 INTRODUCTION

Transition metal ions react with other ions, radicals, and molecules in electron transfer reactions. Such reactions often occur very rapidly. Redox catalysis is based on this capability of transition metal ions to serve as efficient mediators in electron transfer reactions. This type of catalysis found a wide application for accelerating the liquid-phase oxidation of hydrocarbons and aldehydes. Since a free radical is always formed in the reaction of an ion with a molecule during electron transfer, these redox systems became efficient generators of free radicals. These reactions are used for the initiation of reactions of radical polymerization and oxidation. The oscillating regime of the process is often observed in these systems. Redox reactions play a very important role in living organisms where they form the basis for enzymatic processes of respiration, nitrogen fixation, and removal of products harmful for the organism.

The primary product of the oxidation of organic compounds is hydroperoxide, which is known as an effective electron acceptor. Hydroperoxides are decomposed catalytically by transition metal salts and complexes with the generation of free radicals via the following cycle of reactions [1–6]:

$$H^+ + RO_2^\bullet \leftarrow M^{n+} \leftarrow ROOH$$
$$ROOH \rightarrow M^{n+1} \rightarrow RO^\bullet + HO^-$$

During this process, the metal catalyst transforms the slow self-accelerated oxidation into the fast accelerated chain process. One can lower the temperature of oxidation and decrease undesirable side reactions and products.

The transition metal salts and complexes in the lower valence state react rapidly with peroxyl radicals [5].

$$M^{n+} + ROO^\bullet \longrightarrow M^{n+1} + ROO^-$$

This leads to chain termination in the absence of hydroperoxide. There are experimental examples when the introduction of transition metal salt does not accelerate oxidation but generates an induction period [4,5]. When hydrocarbon oxidation with the metal catalyst is developed, the transition metal ions react with hydroperoxide, peroxyl radicals, other products of oxidation, and participate in reactions of chain generation and propagation [6,7]. The catalyst influences the rate and the composition of products of oxidation [6–9]. Since the activity of the catalyst depends on the composition of ligands of the catalyst, a feedback exists between the accumulation of oxidation products and activity of the feedback metal catalyst [4,8,9].

10.1.2 Catalytic Decomposition of Hydrogen Peroxide by Ferrous Ions

The catalysis of hydrogen peroxide decomposition by iron ions occupies a special place in redox catalysis. This was precisely the reaction for which the concept of redox cyclic reactions as the basis for this type of catalysis was formulated [10–13]. The detailed study of the steps of this process provided a series of valuable data on the mechanism of redox catalysis [14–17]. The catalytic decomposition of H_2O_2 is an important reaction in the system of processes that occur in the organism [18–22].

The thermodynamic functions (ΔH, ΔS, ΔG(298 K)) of hydrogen peroxide reactions with transition metal ions in aqueous solutions are presented in Table 10.1. We see that ΔG(298 K) has negative values for reactions of hydroxyl radical generation with Cu^{1+}, Cr^{2+}, and Fe^{2+} ions and for reactions of hydroperoxyl radical generation with Ce^{4+}, Co^{3+}, and Mn^{3+}.

The catalytic decomposition of hydrogen peroxide under the action of the Fe^{2+}/Fe^{3+} ions includes the following steps (H_2O, $T = 298$ K, acidic medium) [14–17]:

$$Fe^{2+} + H_2O_2 \longrightarrow Fe^{3+} + HO^\bullet \quad k = 50 \text{ L mol}^{-1}\text{ s}^{-1}$$
$$HO^\bullet + Fe^{2+} \longrightarrow HO^- + Fe^{3+} \quad k = 3.0 \times 10^8 \text{ L mol}^{-1}\text{ s}^{-1}$$
$$HO^\bullet + H_2O_2 \longrightarrow H_2O + HO_2^\bullet \quad k = 4.5 \times 10^7 \text{ L mol}^{-1}\text{ s}^{-1}$$
$$HO_2^\bullet + Fe^{3+} \longrightarrow Fe^{2+} + O_2 + H^+ \quad k = 3.3 \times 10^5 \text{ L mol}^{-1}\text{ s}^{-1}$$
$$HO_2^\bullet + Fe^{2+} \longrightarrow HO_2^- + Fe^{3+} \quad k = 7.2 \times 10^5 \text{ L mol}^{-1}\text{ s}^{-1}$$

TABLE 10.1
Enthalpies, Entropies, and Gibb's Energies of Transition Metal Ion Oxidation–Reduction Reactions with Hydrogen Peroxide in Aqueous Solution ($T = 298$ K) [23]

Reaction	ΔG (kJ mol^{-1})	ΔH (kJ mol^{-1})	ΔS (J mol^{-1} K^{-1})
$Ce^{3+} + H_2O_2 + H^+ = Ce^{4+} + HO^\bullet + H_2O$	88.8	32.4	−189
$Co^{2+} + H_2O_2 + H^+ = Co^{3+} + HO^\bullet + H_2O$	108.0	48.0	−201
$Cr^{2+} + H_2O_2 + H^+ = Cr^{3+} + HO^\bullet + H_2O$	−37.0	−88.5	−173
$Cu^+ + H_2O_2 + H^+ = Cu^{2+} + HO^\bullet + H_2O$	−62.4	−171.4	−366
$Fe^{2+} + H_2O_2 + H^+ = Fe^{3+} + HO^\bullet + H_2O$	−2.8	−41.2	−129
$Mn^{2+} + H_2O_2 + H^+ = Mn^{3+} + HO^\bullet + H_2O$	71.5	49.1	−75
$Ce^{4+} + H_2O_2 = Ce^{3+} + HO_2^\bullet + H^+$	−23.2	59.7	278
$Co^{3+} + H_2O_2 = Co^{2+} + HO_2^\bullet + H^+$	−42.4	44.1	290
$Cu^{2+} + H_2O_2 = Cu^+ + HO_2^\bullet + H^+$	128.0	180.6	176.5
$Fe^{3+} + H_2O_2 = Fe^{2+} + HO_2^\bullet + H^+$	68.4	133.3	218
$Mn^{3+} + H_2O_2 = Mn^{2+} + HO_2^\bullet + H^+$	−5.9	43.0	164

The first step, which is the slowest, limits the process and, hence, the rate of H_2O_2 decomposition is

$$v = k[H_2O_2][Fe^{2+}] \qquad (10.1)$$

and the rate constant k depends on the pH of the medium. The stoichiometric ratio of the consumed H_2O_2 and Fe^{2+} depends on the ratio of the reactant concentrations [16]:

$$\Delta[H_2O_2]/\Delta[Fe^{2+}] = 0.5 + const \times [H_2O_2]/[Fe^{2+}] \qquad (10.2)$$

The transfer of an electron from Fe^{2+} to hydrogen peroxide is intraspheric, preceded by the incorporation of H_2O_2 into the internal coordination sphere of the Fe_{aq}^{2+} ion.

$$H_2O_2 + Fe^{2+}(H_2O)_6 \rightleftharpoons Fe^{2+}(H_2O)_5(H_2O_2) + H_2O \qquad K$$
$$Fe^{2+}(H_2O)_5(H_2O_2) \longrightarrow Fe^{3+}(H_2O)_5(OH^-) + HO^\bullet \qquad k$$

$K = 1.8 \times 10^{-2} \, L \, mol^{-1}$, $k = 2.8 \times 10^3 \, s^{-1}$, and $kK = 50 \, L \, mol^{-1} \, s^{-1}$ [17].

The reaction is accelerated with an increase in the pH of the solution. This is caused first by the hydrolysis of the $Fe(H_2O_2)^{2+}$

$$Fe^{2+}(H_2O_2)(H_2O)_5 \rightleftharpoons Fe^{2+}(H_2O_2)(H_2O)_4(OH^-) + H^+$$

and the faster electron transfer in this complex. Two variants are accepted in the literature, namely, one-electron transfer

$$Fe^{2+}(H_2O_2)(OH^-) \longrightarrow Fe^{3+}(OH^-)_2 + HO^\bullet$$

and two-electron transfer to form the ferryl ion

$$Fe^{2+}(H_2O_2)(OH^-) \longrightarrow FeO^{2+}(OH^-) + H_2O$$

Then the ferryl ion either reacts with water to form the hydroxyl radical, or oxidizes another Fe^{2+} ion [14,16].

$$FeO^{2+}(OH^-) + H_2O \longrightarrow FeO^{3+}(OH^-)_2 + HO^\bullet$$
$$FeO^{2+}(OH^-) + Fe^{2+} + H^+ \longrightarrow 2Fe^{3+}(OH^-) + H_2O$$

Thus, two routes of transformation are possible for the $Fe^{2+}(H_2O_2)$ complex: one-electron transfer to form the hydroxyl radical and two-electron transfer to form the ferryl ion. It is difficult to prove experimentally the formation of the ferryl ions because they are very reactive, so that this route of interaction of H_2O_2 with Fe^{2+} remains hypothetical to a great extent. Another change in the mechanism of H_2O_2 decomposition with increasing pH is related to the acidic dissociation of $HO_2^\bullet$ ($pK_a = 4.4$)

$$HO_2^\bullet \rightleftharpoons O_2^{-\bullet} + H^+ \qquad K$$

The $O_2^{-\bullet}$ superoxide ion possesses a low electron affinity (42 kJ/mol in vacuum) and is an active reducing agent. Therefore, it rapidly reacts, in particular, with Fe^{3+}.

$$O_2^{-\bullet} + Fe^{3+} \longrightarrow O_2 + Fe^{2+}$$

The values of the rate constants for the reactions of transition metal ions with hydrogen peroxide in an aqueous solution are presented in Table 10.2.

The oxidation of organic compounds by Fenton reagent is associated, as it is clear now, with the generation of hydroxyl radicals. For example, the oxidation of tartaric acid includes the following reactions as key steps [16]:

TABLE 10.2
Rate Constants of Transition Metal Ion Reactions with Hydrogen Peroxide in Aqueous Solutions

Ion	Conditions	T (K)	E (kJ mol^{-1})	log A, A (L mol^{-1} s^{-1})	k(298 K) (L mol^{-1} s^{-1})	Ref.
Ce^{4+}		298			8.61×10^5	[24]
Ce^{4+}	0.8 N H_2SO_4	298			1.00×10^6	[25]
Cu^+	pH = 2.3	298	39.2	10.0	1.35×10^3	[26]
Cu(I)	$\kappa = 1$; Cl$^-$	298			49.0	[27]
Cu(Dipy)$_2$	pH = 5.0	298			8.51×10^2	[28]
Fe^{2+}		285–308	42.3	9.25	68.5	[13]
Fe^{2+}	1.35 M $HClO_4$	273–298	39.3	8.65	57.7	[29]
Fe^{2+}		298			50.0	[30]
Fe^{2+}	0.8 N H_2SO_4	291			41.7	[31]
Fe^{2+}	0.8 N H_2SO_4	298			64.6	[31]
Fe^{2+}	1.0 N H_2SO_4	293			61.7	[32]
Fe^{2+}	0.5 N H_2SO_4	298			60.3	[33]
Fe^{2+}	0.5–0.8 N H_2SO_4	288–314	35.4	8.02	65.3	[34]
Fe^{2+}	0.01–1.0 N $HClO_4$	273–313	39.5	8.72	62.6	[35]
Fe^{2+}	0.8N H_2SO_4	273–313	40.8	8.98	67.4	[35]
Fe^{2+}	0.001–1.0 N $HClO_4$	273–318	30.5	7.14	62.2	[36]
Fe^{2+}		298			53.0	[37]
Fe^{2+}		298			58.0	[38]
Fe^{2+}		298			66.0	[39]
Fe^{2+}		298			49.5	[40]
Fe^{2+}		298			73.0	[41]
Fe^{2+}		298			76.0	[42]
FeF$_2$		273–318	58.6	12.4	1.34×10^2	[36]
FeCl$^+$		283–318	35.1	8.08	84.6	[36]
FeBr$^+$		273–298	40.2	8.95	80.1	[36]
FePF$_6^+$		273–298	42.7	11.0	3.28×10^3	[36]
Fe(Dipy)$^{2+}$		290			2.29×10^2	[43]
Fe(II)(DTPA)		298			1.37×10^3	[22]
MnO$_4^-$	0.1 N H_2SO_4	291			3.02×10^3	[24]
Mn^{3+}		298			7.24×10^4	[44]
MnOH^{2+}		298			3.16×10^4	[44]
Ti^{3+}		298			5.00×10^2	[24]
U(IV)	2 M $HClO_4$	275–307	52.3	11.0	67.6	[45]

$$Fe^{2+} + H_2O_2 \longrightarrow Fe(OH)^{2+} + HO^{\bullet}$$
$$HO^{\bullet} + HOOCCHOHCHOHCOOH \longrightarrow HOOCC^{\bullet}(OH)CHOHCOOH + H_2O$$
$$HOOCC^{\bullet}(OH)CHOHCOOH + Fe^{3+} \longrightarrow HOOCC(O)CHOHCOOH + Fe^{2+} + H^+$$

which in combination form a cycle. Other organic compounds are oxidized similarly.

10.1.3 Catalysis by Transition Metal Ions and Complexes in Liquid-Phase Oxidation of Hydrocarbons and Aldehydes by Dioxygen

Salts of transition metals are widely used in technological processes for the preparation of various oxygen-containing compounds from hydrocarbon raw materials. The principal mechanism of acceleration of RH oxidation by dioxygen in the presence of salts of heavy metals was discovered by Bawn [46–49] for benzaldehyde oxidation (see Chapter 1). Benzaldehyde was oxidized with dioxygen in a solution of acetic acid, with cobalt acetate as the catalyst. The oxidation rate was found to be [50]:

$$v = k[PhCHO]^{3/2}[Co^{3+}]^{1/2}[O_2]^0 \qquad (10.3)$$

The rate of radical generation was measured by the inhibitor method and turned out to be [46–50]

$$v_i = k_i[PhCHO][Co^{3+}],$$
$$k_i = 3.0 \times 10^9 \exp(-61.9/RT) \text{ L mol}^{-1} \text{ s}^{-1} \qquad (10.4)$$

The reaction rate of Co^{3+} with benzaldehyde was measured in independent experiments from the consumption of Co^{3+} in the absence of oxygen. The rate constant of this bimolecular reaction was found to coincide with k_i. Thus, in this process the limiting step of initiation is the reduction of Co^{3+} by aldehydes, and the complete cycle of initiation reactions includes the reactions [50,51]:

and the formation of perbenzoic acid proceeds via the chain reaction including the steps

$$PhC^{\bullet}(O) + O_2 \longrightarrow PhC(O)OO^{\bullet}$$
$$PhC(O)OO^{\bullet} + PhCH(O) \longrightarrow PhC(O)OOH + PhC^{\bullet}(O)$$
$$2PhC(O)OO^{\bullet} \longrightarrow PhC(O)OOC(O)Ph.$$

Catalysis is demonstrated by the process that the radicals are generated by the oxidized form of the catalyst in the reaction with aldehyde, and the reduced form of the catalyst is rapidly oxidized by perbenzoic acid formed in the chain reaction. Data on the catalytic oxidation of aldehydes of different structures are found in Refs. [50,51].

Alkylaromatic hydrocarbons, such as tetralin, ethylbenzene, and cumene, are oxidized in a solution of acetic acid in the presence of cobalt acetate by a different mechanism. In these systems, after rather a short acceleration period, a constant oxidation rate $v \sim [RH]^2$ is established: it is independent of either the catalyst concentration or the partial pressure of oxygen [4–9]. The stationary concentration of hydroperoxide $[ROOH]_{st} \sim ([RH]/[Co^{2+}])^2$ corresponds to the constant oxidation rate. In a nitrogen atmosphere the hydroperoxide decomposes with the rate

$$v = k_d[ROOH][Co(OAc)_2], \tag{10.5}$$

These results agree with the following scheme of RH catalytic chain oxidation [5]:

$$\begin{aligned}
R^\bullet + O_2 &\longrightarrow ROO^\bullet & &\text{Rapidly} \\
ROO^\bullet + RH &\longrightarrow ROOH + R^\bullet & &k_p \\
2ROO^\bullet &\longrightarrow ROOR + O_2 & &2k_t \\
ROOH + Co^{2+} &\rightleftharpoons ROOH.Co^{2+} & &K \\
ROOH.Co^{2+} + Co^{2+} &\longrightarrow RO^\bullet + Co^{3+} + HO^- + Co^{2+} & &k_I \\
ROOH + Co^{3+} &\longrightarrow ROO^\bullet + Co^{2+} + H^+ & &\text{Rapidly} \\
RO^\bullet + RH &\longrightarrow ROH + R^\bullet & &\text{Rapidly}
\end{aligned}$$

The reaction of Co^{2+} with ROOH limits initiation. In the quasistationary regime, the rate constants of formation and decomposition of hydroperoxide given by the following equations:

$$k_p[RH][RO_2^\bullet] = k_i K[ROOH][Co^{2+}]^2, \tag{10.6}$$

$$[RO_2^\bullet] = \{k_i K[ROOH][Co^{2+}]^2/2k_t\}^{1/2} \tag{10.7}$$

$$[ROOH]_{st} = k_p^2[RH]^2/2k_t k_i K[Co^{2+}]^2, \tag{10.8}$$

$$v_{st} = (k_p^2/2k_t)[RH]^2, \tag{10.9}$$

where $[ROOH]_{st}$ is the stationary hydroperoxide concentration, and v_{is} is the stationary rate of catalytic oxidation of hydrocarbon. The dependence of the oxidation rate v on the catalyst concentration looks somewhat paradoxical: the reaction is catalyzed by the cobalt ions but the reaction rate is independent of their concentration. This is due to the fact that initiation occurs by the reaction of hydroperoxide with the catalyst but the process is limited in the quasistationary regime by the rate of chain oxidation of hydrocarbon. In essence, the catalyst transforms the chain process into a nonchain process due to the high initiation rate.

As in the case with catalytic decomposition of hydrogen peroxide, radical generation by the reaction of metal ions with hydroperoxide consists of several steps. In an aqueous solution, first ROOH is substituted in the internal coordination sphere of the ion followed by the transfer of an electron from the ion to ROOH accompanied by the subsequent cleavage of hydroperoxide to $RO^\bullet$ and OH^-, for example,

$$\begin{aligned}
ROOH + Fe^{2+}(H_2O)_6 &\rightleftharpoons Fe^{2+}(H_2O)_5(ROOH) + H_2O & &K \\
Fe^{2+}(H_2O)_5(ROOH) &\longrightarrow Fe^{3+}(H_2O)_5(OH^-) + RO^\bullet & &k \\
Fe^{3+}(H_2O)_5(OH^-) + H_2O &\rightleftharpoons HO^- + Fe^{3+}(H_2O)_6 &&
\end{aligned}$$

The total rate constant $k_{exp} = kK$ weakly depends on the hydrocarbon residue, which is seen from the data collected in Table 10.3. However, the ligand environment, as well as the solvent, has a substantial effect on the k value [4–9].

TABLE 10.3
Rate Constants of the Reaction of Transition Metal Ions and Complexes with Hydroperoxides

ROOH	Solvent	T (K)	E (kJ mol^{-1})	log A, A (L mol^{-1} s^{-1})	k (298 K) (L mol^{-1} s^{-1})	Ref.
\multicolumn{7}{c}{Co^{2+}}					
Me$_3$COOH	CH$_3$C(O)OH/H$_2$O	308	82.8	9.91	2.49×10^{-5}	[52]
\multicolumn{7}{c}{Co[O(O)CCHEtBu]$_2$}					
Me$_3$COOH	CH$_3$C(O)OH	313			6.10×10^{-3}	[53]
Me$_3$COOH	PhCl	293	50.2	8.38	0.38	[54]
\multicolumn{7}{c}{Co(EDTA)$^{2-}$}					
Me$_3$COOH	CH$_3$C(O)OH	333			2.19×10^{-2}	[55]
Me$_3$COOH	PhCl	293			10.5	[55]
Me$_2$PhCOOH	PhCl	318	50.2	7.58	6.04×10^{-2}	[5]
\multicolumn{7}{c}{Co[O(O)C(CH$_2$)$_{14}$Me]$_2$}					
Me$_2$PhCOOH	n-C$_9$H$_{20}$	364–389	39.3	3.86	9.36×10^{-4}	[56]
\multicolumn{7}{c}{Co(II)Phthalocyanine}					
Me$_2$PhCOOH	n-C$_9$H$_{20}$	323–393	39.7	7.21	1.78	[57]
\multicolumn{7}{c}{Co[O(O)C(CH$_2$)$_{16}$CH$_3$]$_2$}					
cyclo-C$_6$H$_{11}$OOH	cyclo-C$_6$H$_{12}$	303–342	42.0	7.93	3.70	[58]
tetralin-OOH	PhCl	323			6.17	[59]
\multicolumn{7}{c}{Co[O(O)C(CH$_2$)$_{16}$Me]$_3$}					
cyclo-C$_6$H$_{11}$OOH	cyclo-C$_6$H$_{12}$	298			0.01	[58]
\multicolumn{7}{c}{Co(O(O)CMe)$_2$}					
MeC(O)OOH	cyclo-C$_6$H$_{12}$	303–342	79.5	12.38	2.79×10^{-2}	[58]
tetralin-OOH	Tetralin	323			2.88	[59]
\multicolumn{7}{c}{Cu[O(O)C(CH$_2$)$_{16}$CH$_3$]$_2$}					
MePhCHOOH	MePhCH$_2$	343–363	66.1	9.92	2.16×10^{-2}	[60]
\multicolumn{7}{c}{Cu[O(O)C(CH$_2$)$_7$CH=CH(CH$_2$)$_7$Me]$_2$}					
tetralin-OOH	Tetralin	330			1.0×10^{-2}	[61]
\multicolumn{7}{c}{Fe^{2+}}					
Me$_3$COOH	H$_2$O	273–298	41.0	8.31	13.3	[62]
Me$_3$COOH	H$_2$O, pH=1.0	273–298	38.6	7.97	16.0	[63]

continued

TABLE 10.3
Rate Constants of the Reaction of Transition Metal Ions and Complexes with Hydroperoxides—*continued*

ROOH	Solvent	T (K)	E (kJ mol^{-1})	log A, A (L mol^{-1} s^{-1})	k (298 K) (L mol^{-1} s^{-1})	Ref.
Me$_3$COOH	H$_2$O, pH=2.0	298			14.0	[63]
Me$_2$PhCOOH	H$_2$O	273–298	37.9	8.55	80.7	[64]
Me$_2$PhCOOH	D$_2$O	273–298	45.3	8.97	10.7	[64]
Me$_2$PhCOOH	H$_2$O	273–298	50.2	10.03	17.0	[65]
Me$_2$PhCOOH	H$_2$O	273–298	46.4	9.59	28.6	[66]
Me$_2$PhCOOH	H$_2$O	273–298	41.6	8.49	15.8	[63]
(CH$_3$)$_2$(C$_6$H$_4$)C–OOH	H$_2$O; pH = 1.2	273–298	32.2	7.00	22.5	[63]
(CH)(CH$_3$)(C$_6$H$_4$)C–OOH	H$_2$O; pH = 4.2	284–299	45.2	9.60	47.6	[67]
C–(C$_6$H$_4$)–C–OOH	H$_2$O; pH = 4.2	273–288	41.4	9.255	99.6	[67]
O$_2$N–(C$_6$H$_4$)–C–OOH	H$_2$O; pH = 4.2	273–298	54.8	10.90	19.7	[68]
cyclohexyl–C–OOH	H$_2$O; pH = 4.2	273–289	46.0	9.80	54.6	[68]
1-phenylcyclohexyl-OOH	H$_2$O; pH = 4.2	273–298	45.2	9.38	28.7	[68]
n-C$_8$H$_{17}$CH(OOH)Me	n-C$_{10}$H$_{22}$ Mn[O(O)C(CH$_2$)$_{16}$CH$_3$]$_2$	353–377	75.3	11.45	1.78×10^{-2}	[69]

In acetic acid, the reaction of cobalt ions with ROOH proceeds via two channels: through the mono- and binuclear cobalt complexes.

$$\text{ROOH} + \text{Co}^{2+} \rightleftharpoons \text{ROOH}\cdot\text{Co}^{2+} \qquad K_1$$
$$\text{ROOH}\cdot\text{Co}^{2+} \longrightarrow \text{RO}^\bullet + \text{HOCo}^{2+} \qquad k_{i1}$$
$$\text{ROOH}\cdot\text{Co}^{2+} + \text{Co}^{2+} \rightleftharpoons \text{Co}^{2+}\cdot\text{ROOH}\cdot\text{Co}^{2+} \qquad K_2$$
$$\text{Co}^{2+}\cdot\text{ROOH}\cdot\text{Co}^{2+} \longrightarrow \text{RO}^\bullet + \text{HOCo}^{2+} + \text{Co}^{2+} \qquad k_{i2}$$

Due to this,

$$v = (k_{i1}K_1[\text{Co(II)}] + k_{i2}K_2[\text{Co(II)}]^2)[\text{ROOH}] \qquad (10.10)$$

and for (CH$_3$)$_3$COOH, $k_{i1}K_1 = 2.2 \times 10^{-2}$ L mol^{-1} s^{-1} and $k_{i2}K_2 = 0.62$ L^2 mol^{-2} s^{-1} at 333 K in acetic acid [9]. Similarly, in the case of the Co^{II} complex with ethylenediamine tetraacetate in the mixture AcOH : H$_2$O = 1 : 1 at 308 K, $k_{i1} = 7.2 \times 10^{-5}$ L mol^{-1} s^{-1} and $k_{i2} = 6.1 \times 10^{-3}$ L^2 mol^{-2} s^{-1} [9]. The intermediate complex formation between cumyl hydroperoxide and

Co(II) in a chelate complex (A) was studied and observed by the NMR method [70]. The band of a proton of hydroxyl group absorption was observed to change the frequency of absorption and the height of the peak after addition of the solvent ($CCl_4 + CH_2Cl_2$) to complex A. The calculated values of the equilibrium constant appeared to be very close to those estimated from the kinetic data. The values of the equilibrium constants are given: Co(II) + Me$_2$PhCOOH $\Leftrightarrow$ Complex at $T = 318$ K and rate constants (k_d) of this complex decay to free radicals [71].

Catalyst (Complex A)			
K (L mol^{-1})	28.9	15.0	10.0
k_d (s^{-1})	4.4×10^{-2}	3.5×10^{-2}	5.0×10^{-2}
E_d (kJ mol^{-1})	83.7	77.8	77.4
A_d (s^{-1})	2.0×10^{12}	1.7×10^{14}	3.4×10^{14}
Catalyst (Complex A)			
K (L mol^{-1})	12.0	7.5	1.7
k_d (s^{-1})	0.25	4.1×10^{-2}	0.10
E_d (kJ mol^{-1})	50.6	92.0	64.8
A_d (s^{-1})	2.5×10^{7}	4.1×10^{13}	3.0×10^{9}

In the case of cobalt ions, the inverse reaction of Co^{III} reduction with hydroperoxide occurs also rather rapidly (see Table 10.3). The efficiency of redox catalysis is especially pronounced if we compare the rates of thermal homolysis of hydroperoxide with the rates of its decomposition in the presence of ions, for example, cobalt decomposes 1,1-dimethylethyl hydroperoxide in a chlorobenzene solution with the rate constant $k_d = 3.6 \times 10^{12} \exp(-138.0/RT) = 9.0 \times 10^{-13}$ s^{-1} (293 K). The catalytic decay of hydroperoxide with the concentration [Co^{2+}] = 10^{-4} M occurs with the effective rate constant $k_{eff} = k_{i1}K_1[\text{Co}^{2+}] = 2.2 \times 10^{-6}$ s^{-1}; thus, the specific decomposition rates differ by six orders of magnitude, and this difference can be increased by increasing the catalyst concentration. The kinetic difference between the homolysis of the O—O bond and redox decomposition of ROOH is reasoned by the difference in the thermodynamic routes of decomposition in these cases. The homolysis of ROOH at the O—O bond requires the energy expenditure equal to the strength of this bond (130–150 kJ mol^{-1}). The redox cycle

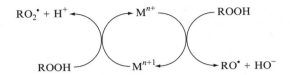

requires an energy expenditure (per ROOH molecule) of only $\Delta H = \frac{1}{2}(D_{O-O} + D_{O-H} - D_{HO-H}) = 10$ kJ mol^{-1}. This results in the great difference in the rates of the catalytic and the noncatalytic decompositions of hydroperoxide.

In real systems (hydrocarbon–O$_2$–catalyst), various oxidation products, such as alcohols, aldehydes, ketones, bifunctional compounds, are formed in the course of oxidation. Many of them readily react with ion-oxidants in oxidative reactions. Therefore, radicals are generated via several routes in the developed oxidative process, and the ratio of rates of these processes changes with the development of the process [5]. The products of hydrocarbon oxidation interact with the catalyst and change the ligand sphere around the transition metal ion. This phenomenon was studied for the decomposition of *sec*-decyl hydroperoxide to free radicals catalyzed by cupric stearate in the presence of alcohol, ketone, and carbon acid [70–74]. The addition of all these compounds was found to lower the effective rate constant of catalytic hydroperoxide decomposition. The experimental data are in agreement with the following scheme of the parallel equilibrium reactions with the formation of Cu-hydroperoxide complexes with a lower activity.

$$CuSt_2 + 2ROOH \rightleftharpoons CuSt_2 \cdot 2ROOH \qquad K_I$$
$$CuSt_2 + ROH \rightleftharpoons CuSt_2 \cdot ROH \qquad K_{II}$$
$$CuSt_2 + 2ROOH + ROH \rightleftharpoons CuSt_2 \cdot 2ROOH \cdot ROH \qquad K_{III}$$
$$CuSt_2 \cdot 2ROOH \longrightarrow \text{Decay} \qquad k_{d1}$$
$$CuSt_2 \cdot 2ROOH \cdot ROH \longrightarrow \text{Decay} \qquad k_{d2}$$

The products of oxidation (alcohol, ketone, acid) lower the concentration of active complexes and, in addition, form complexes with a mixed ligand sphere with lower catalytic activity ($k_{d1} > k_{d2}$). The values of equilibrium constants K_{II} (L mol^{-1}) measured spectrophotometrically in a decane solution for cupric stearate + product are given below [70].

Product	2-Decanol	2-Decanone	RC(O)OH
333 K	6.7	6.2	87
343 K	5.0	4.8	63
353 K	4.0	3.8	55

10.1.4 Competition between Homolytic and Heterolytic Catalytic Decompositions of Hydroperoxides

According to the Haber–Weiss scheme [11], in the framework of which we considered the catalytic decomposition of ROOH, all ROOH molecules decompose under the action of ions only to free radicals, i.e., one-electron redox decomposition occurs. Both the rates of catalytic decomposition of ROOH (from the consumption of ROOH) and the rate of generation of free radicals (from the consumption of the acceptor of free radicals or initiation rate of the chain process of RH oxidation or $CH_2=CHX$ polymerization [53]) were measured for a series of systems (ROOH–catalyst–solvent). The comparison of these two processes showed that there are many systems, indeed, where the rate of ROOH decomposition and radical generation virtually coincide (see Table 10.4).

However, there are known systems in which $v_i \ll v_d$ (see data with Ni, Mo, and V salts). The decrease in v_i compared to v_d can be explained by the cage effect. However, the cage effect of a pair of radicals in low-viscosity liquids are characterized by the ratios $v_i/v_d \approx 0.4$–0.8, so that the v_i/v_d ratios lower than 20% do not agree with the cage effect [80]. In addition, at the cage effect the difference $E_i - E_d = E_D$, and the activation energy of diffusion of a particle in the solvent E_D is 5 to $10 \, \text{kJ} \, \text{mol}^{-1}$. In experiments with these systems (ROOH + metal complex), the difference $E_i - E_d$ often exhibits a very high value of 40–$70 \, \text{kJ} \, \text{mol}^{-1}$. This indicates the parallel occurrence of two different catalytic reactions: homolytic catalysis to form radicals and heterolytic catalysis to form molecular products. Thus, the general scheme of transformations of hydroperoxide in the coordination sphere of the metal includes two routes [80]:

$$M^{n+} + ROOH \longrightarrow M^{n+1} + RO^\bullet + OH^-$$
$$M^{n+} + ROOH \longrightarrow M^{n+} + \text{Molecular products}$$

The cage effect is a component of this scheme. It takes place when the $RO^\bullet$ radical rapidly (within the time of the cage existence) reacts with the metal ion in the oxidized state.

TABLE 10.4
The Yield of Free Radical Generation (e) in Reactions of Hydroperoxide Decomposition Catalyzed by Transition Metal Complexes

Catalyst	ROOH	Solvent	T (K)	e	Ref.
Vanadyl, bis-acetylacetonate	cyclo-C_6H_{11}OOH	cyclo-C_6H_{12}	403	0.08	[75]
Vanadyl, bis-acetylacetonate	cyclo-C_6H_{11}OOH	cyclo-C_6H_{12}	333	0.20	[75]
Vanadyl, bis-acetylacetonate	Me_2PhCOOH	cyclo-C_6H_{12}	333	0.50	[76]
Chromium(III), tris-acetylacetonate	cyclo-C_6H_{11}OOH	cyclo-C_6H_{12}	393	0.08	[75]
Manganese, acetate	Me_3COOH	n-$C_{10}H_{22}$	393	0.017	[77]
Manganese, stearate	sec-$C_{10}H_{11}$OOH	n-$C_{10}H_{22}$	413	0.04	[77]
Cobalt(II), stearate	Me_2PhCOOH	PhCl	323	0.36	[78]
Cobalt(II), stearate	Me_2PhCOOH	n-$C_{10}H_{22}$	323	0.38	[78]
Cobalt(II), stearate	Me_3COOH	C_6H_6	323	1.00	[78]
Cobalt(II), stearate	sec-$C_{10}H_{11}$OOH	n-$C_{10}H_{22}$	323	1.00	[78]
Cobalt(II), bis-acetylacetonate	Me_2PhCOOH	PhCl	318	0.82	[78]
Cobalt(II), bis-acetylacetonate	sec-$C_{10}H_{11}$OOH	n-$C_{10}H_{22}$	318	1.00	[78]
Nickel(II), bis-acetylacetonate	sec-$C_{10}H_{11}$OOH	PhCl	363	0.02	[78]
Nickel(II), bis-acetylacetonate	MePhCHOOH	PhCl	363	0.04	[78]
Nickel(II), stearate	MePhCHOOH	$PhCH_2CH_3$	393	0.012	[78]
Copper(II), stearate	sec-$C_{10}H_{11}$OOH	n-$C_{10}H_{22}$	353	0.016	[79]
Molibdenum dioxide, bis-acetylacetonate	cyclo-C_6H_{11}OOH	cyclo-C_6H_{12}	403	0.017	[79]

The question about the competition between the homolytic and heterolytic catalytic decompositions of ROOH is strongly associated with the products of this decomposition. This can be exemplified by cyclohexyl hydroperoxide, whose decomposition affords cyclohexanol and cyclohexanone [5,6]. When decomposition is catalyzed by cobalt salts, cyclohexanol prevails among the products ([alcohol]:[ketone] > 1) because only homolysis of ROOH occurs under the action of the cobalt ions to form RO$^\bullet$ and RO$_2^\bullet$: the first of them are mainly transformed into alcohol (in the reactions with RH and Co^{2+}), and the second radicals are transformed into alcohol and ketone (ratio 1:1) due to the disproportionation (see Chapter 2). Heterolytic decomposition predominates in catalysis by chromium stearate (see above), and ketone prevails among the decomposition products (ratio [ketone]:[alcohol] = 6 in the catalytic oxidation of cyclohexane at 393 K [81]). These ions, which can exist in more than two different oxidation states (chromium, vanadium, molybdenum), are prone to the heterolytic decomposition of ROOH, and this seems to be mutually related.

10.1.5 Reactions of Transition Metals with Free Radicals

Peroxyl radicals with a strong oxidative effect along with ROOH are continuously generated in oxidized organic compounds. They rapidly react with ion-reducing agents such as transition metal cations. Hydroxyl radicals react with transition metal ions in an aqueous solution extremely rapidly. Alkyl radicals are oxidized by transition metal ions in the higher valence state. The rate constants of these reactions are collected in Table 10.5.

Under the conditions where the chain oxidation process occurs, this reaction results in chain termination. In the presence of ROOH with which the ions react to form radicals, this reaction is disguised. However, in the systems where hydroperoxide is absent and the initiating function of the catalyst is not manifested, the latter has a retarding effect on the process. It was often observed that the introduction of cobalt, manganese, or copper salts into the initial hydrocarbon did not accelerate the process but on the contrary, resulted in the induction period and elongated it [4–6]. The induction period is caused by chain termination in the reaction of RO$_2^\bullet$ with Mn^{n+}, and cessation of retardation is due to the formation of ROOH, which interacts with the catalyst and thus transforms it from the inhibitor into the component of the initiating system.

Some peroxyl radicals (HO$_2^\bullet$, >C(OH)OO$^\bullet$, >C(NHR)OO$^\bullet$) can either oxidize or reduce, for example

$$HO_2^\bullet + Cu^{1+} \longrightarrow HO_2^- + Cu^{2+}$$
$$HO_2^\bullet + Cu^{2+} \longrightarrow O_2 + Cu^{1+} + H^+$$

In systems where such radicals appear (alcohols, amines, some unsaturated compounds), variable-valence metal ions manifest themselves as catalysts for chain termination (see Refs. [150,151] and Chapter 16).

The reaction of ions with peroxyl radicals appears also in the composition of the oxidation products, especially at the early stages of oxidation. For example, the only primary oxidation product of cyclohexane autoxidation is hydroperoxide: the other products, in particular, alcohol and ketone, appear later as the decomposition products of hydroperoxide. In the presence of stearates of metals such as cobalt, iron, and manganese, all three products (ROOH, ROH, and ketone) appear immediately with the beginning of oxidation, and in the initial period (when ROOH decomposition is insignificant) they are formed in parallel with a constant rate [5,6]. The ratio of the rates of their formation is determined by the catalyst. The reason for this behavior is evidently related to the fast reaction of RO$_2^\bullet$ with the

TABLE 10.5
Rate Constants of Free Radical Reactions with Transition Metal Ions and Complexes

Ion	Solvent	T (K)	k (L mol^{-1} s^{-1})	Ref.
	HO•			
Cerium(IV) ion	Water, pH = 0.4	298	2.7×10^6	[82]
Cerium(III) ion	Water	298	3.0×10^8	[83]
Cerium(III) ion	Water	296	2.2×10^8	[84]
Cerium(III) ion	Water	296	3.5×10^8	[85]
Cobalt(III) ion	Water	298	8.0×10^5	[86]
Cobalt(III), pentacyanonitrate	Water	298	1.2×10^8	[87]
Cobalt(III), pentaamminenitride	Water	298	1.8×10^9	[88]
Cobalt(III), tris-acetylacetonate	Water	298	4.9×10^9	[89]
Copper(I) ion	Water	298	2.0×10^{10}	[90]
Copper(II) ion	Water	298	3.5×10^8	[91]
Copper(II) ion	Water, pH = 7	296	3.5×10^8	[92]
Cromium(II) ion	Water, pH = 1.0	298	4.8×10^9	[93]
Cromium(III) ion	Water, pH < 7	298	3.1×10^8	[94]
Cromium(III), pentacyanonitrosyl	Water	298	7.9×10^9	[95]
Europium(II) ion	Water	298	1.1×10^9	[91]
Iron(II) ion	Water, pH = 1.0	298	3.0×10^8	[96]
Iron(II) ion	Water, pH = 1.0	298	3.2×10^8	[97]
Iron(II) ion	Water, pH = 2.1	296	2.5×10^8	[87]
Iron(II) ion	Water, pH = 2.0	296	2.6×10^8	[98]
Iron(II) ion	Water, pH = 0.4	296	2.7×10^8	[99]
Iron(II) ion	Water, pH = 0	288	3.3×10^8	[100]
Iron(II) ion	Water, pH = 3.0	298	4.3×10^8	[101]
Iron(II) ion	Water, pH = 7.0	298	3.2×10^8	[102]
Iron(II) ion	Water, pH = 1.0	298	2.3×10^8	[103]
Iron(II) ion	Water, pH = 4.5–6.2	298	3.5×10^8	[104]
Iron(II) ion	Water	298	1.0×10^9	[105]
Iron(III) ion	Water	298	7.9×10^7	[85]
Ferrocyanide ion	Water, pH = 2.5–10.5	296	2.1×10^9	[106]
Ferrocyanide ion	Water, pH = 7.0	296	1.7×10^{10}	[107]
Ferrocyanide ion	Water, pH = 10.7	296	8.1×10^9	[108]
Ferrocyanide ion	Water, pH = 9.0	296	7.5×10^9	[109]
Ferrocyanide ion	Water, pH = 3–10	296	1.1×10^{10}	[110]
Ferrocyanide ion	Water	298	1.1×10^{10}	[91]
Manganese(II) ion	Water, pH = 7.0	298	$>1.4 \times 10^8$	[96]
Manganese(II) ion	Water, pH = 3.9–6.7	298	3.0×10^7	[91]

continued

TABLE 10.5
Rate Constants of Free Radical Reactions with Transition Metal Ions and Complexes—*continued*

Ion	Solvent	T (K)	k (L mol^{-1} s^{-1})	Ref.
Silver(I) ion	Water	298	1.2×10^{10}	[111]
Stannum(II) ion	Water, pH = 1.0	296	2.0×10^{9}	[112]
Tallium(I) ion	Water, pH = 7.0	296	7.6×10^{9}	[110]
Tallium(I) ion	Water, pH = 1.0	296	8.9×10^{9}	[113]
	$HO_2^{\bullet}$			
Cerium(III) ion	Water, pH = 0.4	298	2.1×10^{5}	[114]
Cerium(IV) ion	Water, pH = 0.4	298	2.7×10^{6}	[82]
Copper(I) ion	Water, pH = 2.3	298	2.3×10^{9}	[115]
Copper(I) ion	Water	296	4.3×10^{9}	[116]
Copper(II) ion	Water, pH = 2.3	296	1.5×10^{7}	[117]
Copper(II) ion	Water, pH = 2.3	298	1.1×10^{8}	[116]
Copper(II) ion	Water, pH = 0.8–2	298	1.0×10^{8}	[118]
Ferrocyanide ion	Water, pH = 0.5–4.4	298	3.0×10^{4}	[119]
Ferrocyanide ion HFe(CN)$_6^{3-}$	Water, pH = 0.5–4.4	298	1.4×10^{5}	[119]
Ferrocyanide ion H$_2$Fe(CN)$_6^{2-}$	Water, pH = 0.5–4.4	298	1.0×10^{4}	[119]
Iron(II) ion	Water, 0.5 M H$_2$SO$_4$	296	7.2×10^{5}	[32]
Iron(II) ion	Water, pH = 1.0	298	1.2×10^{6}	[120]
Iron(II) ion	Water	298	1.5×10^{6}	[121]
Iron(II) ion	Water	298	1.2×10^{6}	[122]
Iron(II) ion	Water, pH = 0.0–2.1	298	2.1×10^{6}	[123]
Iron(III) ion	Water, pH = 1.5	298	2.0×10^{4}	[124]
Iron(III) ion	Water, pH = 2.05	298	3.3×10^{5}	[35]
Iron(III) ion	Water, pH = 2.7	298	3.1×10^{5}	[35]
Iron(III) ion	Water, pH = 1.0	298	4.0×10^{5}	[35]
Iron(III) ion	Water, pH = 2.1	298	1.2×10^{5}	[125]
Iron(III) ion	Water, pH = 2.0	298	1.3×10^{5}	[126]
Iron(III) ion	Water, pH = 2.7	298	3.6×10^{5}	[126]
Iron(III) ion	Water	298	7.3×10^{5}	[127]
Manganese(II) ion	Water, pH = 3–9	298	1.1×10^{8}	[128]
Manganese(II) ion	Water, pH = 2.7–3.4	298	6.0×10^{6}	[129]
Manganese(II) formiate	Water, pH = 2.2–3.0	298	6.0×10^{6}	[129]
Tallium(II) ion	Water, pH = 1.0	298	2.5×10^{9}	[130]
	cyclo-C$_6$H$_{11}$OO$^{\bullet}$			
Cobalt(II),	Cyclohexane	303	2.7×10^{4}	[58]
Cobalt(II),	Cyclohexane	342	9.3×10^{4}	[58]

TABLE 10.5
Rate Constants of Free Radical Reactions with Transition Metal Ions and Complexes—*continued*

Ion	Solvent	T (K)	k (L mol^{-1} s^{-1})	Ref.
Cobalt(III),	Cyclohexane	303	2.0×10^4	[58]
Cobalt(III),	Cyclohexane	342	1.0×10^5	[58]
Copper(II),	Cyclohexane	303	8.6×10^2	[58]
Copper(II),	Cyclohexane	342	1.0×10^4	[58]
Manganese(II)	Cyclohexane	303	9.9×10^2	[58]
Manganese(II)	Cyclohexane	342	6.5×10^3	[58]
Manganese(III)	Cyclohexane	303	1.8×10^3	[58]
Manganese(III)	Cyclohexane	342	2.2×10^4	[58]
	MePhCHOO$^\bullet$			
Cobalt(II), ethylenediamine tetraacetate	Acetic acid/Ethylbenzene	333	3.8×10^5	[131]
Cobalt(II), bis-8-quinolinolate	Acetic acid/Ethylbenzene	333	9.0×10^7	[131]
Copper(II), bis-8-quinolinolate	Acetic acid/Ethylbenzene	333	7.9×10^5	[131]
Ferrocene	Aceticacid/Ethylbenzene	333	2.4×10^5	[132]
Ferrocene, ethyl-	Acetic acid/Ethylbenzene	333	1.8×10^5	[132]
Ferrocene, phenyl-	Acetic acid/Ethylbenzene	333	8.3×10^4	[132]
Ferrocene, acetyl-	Aceticacid/Ethylbenzene	333	3.3×10^4	[132]
Ferrocene, cyano-	Acetic acid/Ethylbenzene	333	4.0×10^3	[132]
Ferrocene, carbonic acid	Acetic acid/Ethylbenzene	333	1.2×10^4	[132]
Manganese, bis-8-quinolinolate	Acetic acid/Ethylbenzene	333	4.2×10^5	[133]
Manganese(II), ethylenediamine tetraacetate	Acetic acid/Ethylbenzene	333	1.0×10^6	[133]
Nickel(II), bis-8-quinolinolate	Acetic acid/Ethylbenzene	333	2.2×10^5	[133]
	Me$_2$(CN)COO$^\bullet$			
Manganese(II), steatrate	Chlorobenzene	333	2.0×10^5	[133]
Manganese(II), steatrate	Chlorobenzene	353	3.0×10^5	[133]
Cobalt(II), bis-acetylacetonate	Chlorobenzene	347	2.3×10^4	[133]
	~CH$_2$C(OO$^\bullet$)HPh			
Cobalt(II), bis-acetylacetonate	Chlorobenzene	328	1.8×10^4	[134]
Cobalt(II), bis-acetylacetonate	Chlorobenzene	353	4.9×10^4	[134]
Iron(II) ion	Water	298	1.7×10^6	[135]
	C$^\bullet$H$_3$			
Copper(II), acetate	Acetonitrile/Aceticacid (2/3)	298	1.5×10^6	[136]
	Me$_2$C$^\bullet$H			
Copper(II) ion	Acetonitrile/Acetic acid (2/3)	298	5.0×10^6	[137]

continued

TABLE 10.5
Rate Constants of Free Radical Reactions with Transition Metal Ions and Complexes—*continued*

Ion	Solvent	T (K)	k (L mol^{-1} s^{-1})	Ref.
	PrC·H$_2$			
Copper(II) ion	Acetonitrile/Acetic acid (2/3)	298	3.1×10^6	[137]
Copper(II), dipyridyl-	Acetonitrile/Acetic acid (2/3)	298	1.7×10^7	[137]
	Me$_3$CC·H$_2$			
Copper(II) ion	Acetonitrile/Acetic acid (2/3)	298	4.5×10^5	[137]
Copper(II), dipyridyl-	Acetonitrile/Acetic acid (2/3)	298	2.5×10^4	[137]
	CH$_2$=CH(CH$_2$)$_3$C·H$_2$			
Copper(II), acetate	Acetonitrile/Aceticacid (2/3)	298	1.5×10^6	[137]
	PhCH$_2$C·H$_2$			
Copper(II) ion	Acetonitrile/Acetic acid (2/3)	298	1.6×10^6	[137]
Copper(II), dipyridyl-	Acetonitrile/Acetic acid (2/3)	298	1.4×10^7	[137]
	C·H$_2$OH			
Iron(III) ion	Water	298	1.0×10^8	[138]
Iron(III) ion	Water	298	8.0×10^7	[139]
Copper(II) ion	Water, pH = 2–5	298	1.6×10^8	[139]
	C·H$_2$CH$_2$OH			
Copper(II) ion	Water, pH = 4.5	298	1.9×10^7	[139]
Copper(II) ion	Water, pH = 2	298	2.2×10^7	[139]
Iron(II) ion	Water	298	1.0×10^6	[105]
	MeC·HOH			
Copper(II) ion	Water, pH = 2–5	298	9.4×10^7	[139]
Copper(II) ion	Water	298	2.7×10^8	[139]
Copper(II) ion	Water	298	3.8×10^8	[139]
Iron(III) ion	Water	298	6.0×10^8	[105]
	Me$_2$C·OH			
Copper(II) ion	Water, pH = 2–5	298	5.2×10^7	[139]
Iron(II) ion	Water	298	2.9×10^6	[140]
Iron(III) ion	Water, pH = 1	298	4.5×10^8	[135]
Iron(III) ion	Water, pH = 1	298	1.8×10^8	[138]
Iron(III) ion	Water	298	5.8×10^8	[139]
	C·H$_2$Me$_2$COH			
Copper(II) ion	Water, pH = 4.5	298	2.7×10^6	[139]
Copper(II) ion	Water, pH = 3.0	298	3.2×10^6	[139]
	~CH$_2$C·HCONH$_2$			
Titan(III) ion	Water, 0.8 N H$_2$SO$_4$	298	6.0×10^2	[141]
Titan(III) ion	D$_2$O, 0.8 N H$_2$SO$_4$	298	8.1×10^2	[141]

TABLE 10.5
Rate Constants of Free Radical Reactions with Transition Metal Ions and Complexes—continued

Ion	Solvent	T (K)	k (L mol^{-1} s^{-1})	Ref.
Vanadyl(IV) ion	Water, 0.8 N H$_2$SO$_4$	298	1.1×10^3	[141]
Vanadium(II) ion	Water, 0.8 N H$_2$SO$_4$	298	1.1×10^5	[141]
Cromium(III) ion	Water, 0.8 N H$_2$SO$_4$	298	2.8×10^5	[141]
Iron(III) ion	Water, pH = 1	298	2.0×10^3	[142]
Iron(III) ion	Water, pH	298	2.0×10^3	[143]
Iron(III) ion	D$_2$O, pH = 0	298	1.9×10^3	[143]
Hydroxoiron(III) ion	Water	298	1.1×10^4	[142]
Hydroxoiron(III) ion	Water	298	2.1×10^4	[141]
Chloroiron(III) ion	Water	298	8.1×10^4	[141]
Dichloroiron(III) ion	Water	298	1.7×10^4	[141]
Trichloroiron(III) ion	Water	298	1.0×10^6	[141]
Bromoiron(III) ion	Water	298	1.7×10^6	[141]
Azidoiron(III) ion	Water	298	1.5×10^6	[141]
Rodanidoiron(III) ion	Water	298	1.3×10^7	[141]
Iron(III), tris-dipyridyl-	Water	298	8.1×10^4	[141]
Iron(III), tris-*ortho*-phenantrolin-	Water	298	3.1×10^5	[141]
Iron(III), tris-(5-methyl-*ortho*-phenantrolin)-	Water	298	2.6×10^5	[141]
Iron(III), tris-(5-phenyl-*ortho*-phenantrolin)-	Water	298	5.1×10^5	[141]
Iron(III), tris-(5-chloro-*ortho*-phenantrolin)-	Water	298	2.3×10^5	[141]
Ferricyanide ion	Water	298	8.5×10^5	[141]
Copper(II) ion	Water, 1 N HClO$_4$	298	1.2×10^5	[143]
Copper(II) ion	D$_2$O, 1 N HClO$_4$	298	1.4×10^3	[143]
Hydroxocerium(IV) ion	Water, pH = 0–1	298	3.3×10^3	[143]
Europium(II) ion	Water, 0.8 N	298	8.4×10^4	[143]
Tallium(III) ion	Water, 0.1 N HClO$_4$	298	21.0	[143]
	p-MeOC$_6$H$_4$CH$_2$C$^\bullet$H$_2$			
Copper(II) ion	Acetonitrile/Aceticacid (2/3)	298	2.1×10^4	[137]
Copper(II), dipyridyl-	Acetonitrile/Aceticacid (2/3)	298	3.0×10^6	[137]
	~CH$_2$C$^\bullet$MeCONH$_2$			
Iron(III) ion	Water	298	53.1	[144]
Hydroxoiron(III) ion	Water	298	1.7×10^3	[144]
Iron(III) methacrylamidyl-	Water	298	6.1×10^3	[144]

continued

TABLE 10.5
Rate Constants of Free Radical Reactions with Transition Metal Ions and Complexes—*continued*

Ion	Solvent	T (K)	k (L mol^{-1} s^{-1})	Ref.
	~$CH_2\overset{\bullet}{C}HCN$			
Ferricyanide ion	Water	298	6.8×10^6	[145]
Trichloroiron(III) ion	Formamide, N,N-dimethyl-	298	6.4×10^3	[146]
Trichloroiron(III) ion	Formamide, N,N-dimethyl-	298	8.1×10^3	[147]
	~$CH_2\overset{\bullet}{C}MeCN$			
Trichloroiron(III) ion	Formamide, N,N-dimethyl-	298	6.2×10^2	[146]
	~$CH_2\overset{\bullet}{C}HPh$			
Trichloroiron(III) ion	Formamide, N,N-dimethyl-	298	5.4×10^4	[146]
	~$CH_2\overset{\bullet}{C}MeCOOMe$			
Dichlorocopper(II) ion	Formamide, N,N-dimethyl-	333	7.8×10^5	[148]
	(2,2,6,6-tetramethyl-4-oxo-piperidine-1-oxyl radical)			
Iron(II) ion	Water, 0.1 N HCl ; 0.4 M NaCl	298	5.9×10^5	[149]
Iron(II) ion	Water, 0.05 N HCl; 0.45 M NaCl	298	1.7×10^5	[149]

catalyst. Thus, the reaction of peroxyl radicals with variable-valence ions manifests itself in the kinetics as well (the induction period appears under certain conditions), and alcohol and ketone are formed in parallel with ROOH from $RO_2^{\bullet}$ among the oxidation products.

The variety of functions of the catalyst is pronounced, in particular, in the technological catalytic oxidation of *n*-paraffins to aliphatic acids [5]. This technology consists of several stages among which the central place is occupied by oxidation. It is conducted at 380–420 K in a series of reactors, with a mixture of salts of aliphatic acids of K^+ and Mn^{2+} or Na^+ and Mn^{2+} as the catalyst. The alkaline metal salt stabilizes (makes it more soluble and stable) the manganese salt [152]. Studies have revealed the multifunctional role of the catalyst (manganese ions) (Mn) [152–154].

First, they (Mn^{2+} and Mn^{3+}) react with the formed hydroperoxide and decompose it to generate radicals and thus initiate the chain oxidation process.

Second, the isomerization of $RO_2^{\bullet}$ of the following type vigorously occurs in oxidized paraffin (without a catalyst]:

$$RCH(OO^{\bullet})CH_2CH_2 \longrightarrow RCH(OOH)CH_2\overset{\bullet}{C}HR$$

due to which polyfunctional compounds are formed and, finally, acid–containing functional groups (oxy acids, keto acids, etc.). This is very unfavorable for obtaining the target product (aliphatic acids). In the presence of the catalysts, the reaction of Mn^{2+} with the peroxyl radical successfully competes with the former reaction

$$RO_2^\bullet + Mn^{2+} \longrightarrow ROO^- + Mn^{3+}$$

due to which the fraction of bifunctional compounds decreases and the yield of aliphatic acids increases.

Third, the Mn^{3+} ions formed in the reactions of Mn^{2+} with ROOH and $RO_2^\bullet$ successfully react rather rapidly and oxidize the carbonyl compounds

$$RCH_2C(O)CH_2R \rightleftharpoons RCH_2C(OH)=CHR$$
$$RCH_2C(OH)=CHR + Mn^{3+} \longrightarrow RCH_2C(O)C^\bullet HR + Mn^{2+} + H^+$$
$$RCH_2C(O)C^\bullet HR + O_2 \longrightarrow RCH_2C(O)CH(OO^\bullet)R$$

Two carboxylic acids are the final products of this ketone oxidation. As a result, the content of ketones in the system decreases, and the yield of aliphatic acids increases.

Fourth, the accumulation of alcohols in the systems sharply retards the developed process, that is, results in the so-called limiting depth of oxidation [154]. The hydroxyperoxyl radical is formed by the attack of $RO_2^\bullet$ at the alcohol group. The Mn^{2+} and Mn^{3+} ions react with these radicals resulting in catalytic chain termination. As alcohol is accumulated, the fraction of hydroperoxyl radicals increases among all peroxyl radicals, and in parallel the rate of chain termination in the reaction of the ions with these radicals increases. The chain process ceases when the termination rate becomes higher than the rate of radical generation involving manganese ions. Thus, there are some limits on the oxidation process (depth, rate) when variable-valence metals used as catalysts.

10.1.6 Reactions of Transition Metal Ions with Dioxygen

Dioxygen oxidizes transition metal ions in the lower valence state generating the hydroxyperoxyl radicals or superoxide ions [155,156]. The thermodynamic characteristics of these reactions are presented in Table 10.6. It is seen that all cited reactions are endothermic, except for the reaction of the cuprous ion with O_2. The reaction of the ferrous ion with dioxygen has a sufficiently low enthalpy (28 kJ mol^{-1}).

The vast majority of the studies in this field relate to the oxidation of iron ions by molecular oxygen. The oxidation of Fe^{2+} is first-order with respect to dioxygen and,

TABLE 10.6
Enthalpies, Entropies, and Gibb's Energies of the Oxidation–Reduction Reactions of the Transition Metal Ions with Dioxygen in Aqueous Solution ($T = 298$ K) [23]

Reaction	ΔG (kJ mol^{-1})	ΔH (kJ mol^{-1})	ΔS (J mol^{-1} K^{-1})
$Ce^{3+} + O_2 + H^+ = Ce^{4+} + HO_2^\bullet$	170.8	101.9	−231
$Co^{2+} + O_2 + H^+ = Co^{3+} + HO_2^\bullet$	190.0	117.5	−243
$Cu^+ + O_2 + H^+ = Cu^{2+} + HO_2^\bullet$	19.6	−19.0	−130
$Fe^{2+} + O_2 + H^+ = Fe^{3+} + HO_2^\bullet$	79.2	28.3	−171
$Mn^{2+} + O_2 + H^+ = Mn^{3+} + HO_2^\bullet$	153.5	118.6	−117

depending on conditions, first- or second-order with respect to Fe^{2+}. The limiting step includes the transfer of an electron from Fe^{2+} to dioxygen. The postulated reaction mechanism is the following [157]:

$$Fe_{aq}^{2+} + O_2 \rightleftharpoons FeO_2^{2+}$$
$$FeO_2^{2+} \longrightarrow Fe^{3+} + O_2^{\bullet-} \text{ (limiting stage)}$$
$$O_2^{\bullet-} + Fe^{2+} \longrightarrow O_2^{2-} + Fe^{3+}$$
$$O_2^{2-} + 2H^+ \rightleftharpoons H_2O_2$$
$$H_2O_2 + Fe^{2+} \longrightarrow Fe^{3+} + HO^- + HO^{\bullet}$$
$$HO^{\bullet} + Fe^{2+} \longrightarrow HO^- + Fe^{3+}$$

The second order with respect to Fe^{2+} is explained by the formation of a binuclear complex and its faster oxidation:

$$FeO_2^{2+} + Fe^{2+} \rightleftharpoons Fe^{2+}O_2Fe^{2+}$$
$$Fe^{2+}O_2Fe^{2+} \longrightarrow Fe^{3+} + FeO_2^+ \text{ (limiting step)}$$
$$FeO_2^+ \longrightarrow Fe^{3+} + O_2^{2-}$$

With increasing hydrogen-ion concentration, the rate of Fe^{2+} oxidation decreases; this is related to the hydrolysis of Fe^{2+}:

$$Fe_{aq}^{2+} \rightleftharpoons FeOH^+ + H^+$$

and a faster oxidation of $FeOH^+$. However, in concentrated hydrochloric acid, an increase in the oxidation rate is observed as the acid concentration increases. This is explained by the formation and rapid oxidation of $Fe^{2+} \cdot HCl$ complexes [155]:

$$Fe_{aq}^{2+} + HCl \rightleftharpoons Fe^{2+}ClH$$
$$Fe^{2+}ClH + O_2 \longrightarrow FeCl^{2+} + HO_2^{\bullet}$$

Tin(II) was found to be oxidized by dioxygen via the chain branching mechanism [156–162]. The oxidation rate is $v = k[O_2]^2$ in organic solvents and $v = k[Sn(II)]^{1/2}[O_2]^{1/2}$ in aqueous solutions. The reaction, under certain conditions, has an induction period. Free radical acceptors retard this reaction. The following kinetic scheme was proposed for tin(II) oxidation by dioxygen.

$$Sn(II) + O_2 \longrightarrow Sn(III) + HO_2^{\bullet}$$
$$Sn(III) + O_2 \longrightarrow Sn(IV) + HO_2^{\bullet}$$
$$Sn(II) + HO_2^{\bullet} \longrightarrow Sn(III) + H_2O_2$$
$$Sn(II) + HO_2^{\bullet} \longrightarrow Sn(IV) + HO^{\bullet}$$
$$Sn(II) + H_2O_2 \longrightarrow Sn(III) + HO^{\bullet}$$
$$Sn(II) + H_2O_2 \longrightarrow Sn(IV) + H_2O$$
$$Sn(II) + HO^{\bullet} \longrightarrow Sn(III)$$
$$Sn(III) + Sn(III) \longrightarrow Sn(II) + Sn(IV)$$

The very active unstable tin(III) ion is supposed to play an important role in this chain mechanism of tin(II) oxidation. Cyclohexane, introduced in the system Sn(II) + dioxygen, is oxidized to cyclohexanol as the result of the coupled oxidation of tin and RH. Hydroxyl radicals, which are very strong hydrogen atom acceptors, attack cyclohexane (RH) with the formation of cyclohexyl radicals that participate in the following transformations:

$$R^{\bullet} + O_2 \longrightarrow RO_2^{\bullet}$$
$$RO_2^{\bullet} + Sn(II) \longrightarrow Sn(III) + ROOH$$
$$RO_2^{\bullet} + Sn(II) \longrightarrow Sn(IV) + RO^{\bullet}$$
$$ROOH + Sn(II) \longrightarrow Sn(III) + RO^{\bullet}$$
$$ROOH + Sn(II) \longrightarrow Sn(IV) + ROH$$
$$RO^{\bullet} + RH \longrightarrow ROH + R^{\bullet}$$
$$RO^{\bullet} + Sn(II) \longrightarrow ROH + Sn(III)$$
$$Sn(III) + Sn(III) \longrightarrow Sn(II) + Sn(IV)$$

Very fast reactions of $RO_2^{\bullet}$ and ROOH with tin(II) and the fast reaction of $RO^{\bullet}$ with cyclohexane result in the formation of cyclohexanol as the main oxidation product.

The preliminary formation of metal–dioxygen complex was postulated in the mechanisms discussed earlier. However, a series of metal complexes were synthesized that form stable complexes with dioxygen. These complexes were studied as individual compounds. A few of them are given in Table 10.7.

Cobalt catalysis of chain generation was found in the system styrene–dioxygen–acetylacetonate cobalt(II) [169]. Free radicals are generated in this system with the rate

$$v = k_1[CH_2{=}CHPh][O_2] + k_2[CH_2{=}CHPh][Co(II)][O_2] \qquad (10.11)$$

The first term on the right-hand side denotes the rate of dioxygen reaction with styrene (see Chapter 4) and the second term is the rate of catalytic free radical generation via the reaction of styrene with dioxygen catalyzed by cobaltous stearate or cobaltous acetylacetonate. The rate constants were found to be $k_1 = 7.45 \times 10^{-6}$ L mol^{-1} s^{-1}, $k_2 = 6.30 \times 10^{-2}$ L^2 mol^{-2} s^{-1} (cobaltous acetylacetonate), and $k_2 = 0.31$ L^2 mol^{-2} s^{-2} (cobaltous stearate) ($T = 388$ K, solvent = PhCl [169]). The mechanism with intermediate complex formation was proposed.

$$Co(II) + O_2 \rightleftharpoons Co(II){\cdot}O_2$$
$$Co(II){\cdot}O_2 + CH_2{=}CHPh \longrightarrow Co(II) + {}^{\bullet}OOCH_2C^{\bullet}HPh$$

Oxidation of ethylbenzene catalyzed by the Cu(II) complex with o-phenanthroline was found to occur with the rate depending on the dioxygen concentration [170].

$$v^2 = a + b[O_2] \qquad (10.12)$$

Since the rate of chain oxidation of hydrocarbon is $v \sim v_i^{1/2}$ and does not depend on $[O_2]$ (see Chapter 2), the catalyst initiates the chains via two parallel reactions: by the reaction with ROOH and by the reaction with dioxygen. The following mechanism was proposed:

$$Cu(II)(o\text{-phen})_2 + O_2 \rightleftharpoons Cu(II)(o\text{-phen})_2{\cdot}O_2$$
$$Cu(II)(o\text{-phen})_2{\cdot}O_2 + RH \longrightarrow Cu(II)(o\text{-phen})_2 + HO_2^{\bullet} + R^{\bullet}$$

TABLE 10.7
Chelate Complexes that form Complexes with Dioxygen

Complex	Me/O$_2$	Absorbtion O$_2$	Ref.
Co salen (bis-salicylaldehyde ethylenediimine)	2	Reversible	[163]
Co salen with additional amine (N-H)	1	Reversible	[163]
Co glycinate (H$_2$N, O, O-NH)	2	Reversible	[164]
Ir(Cl)(CO)(PPh$_3$)$_2$	1	Reversible	[165]
Ir(I)(CO)(PPh$_3$)$_2$	1	Irreversible	[166]
Ru(Cl)(NO)(CO)(PPh$_3$)$_2$	1	Irreversible	[167]
(t-Bu-C≡N)$_2$Ni(N≡C-t-Bu)	1	Irreversible	[168]
(t-Bu-C≡N)$_2$Pd(N≡C-t-Bu)	1	Irreversible	[168]
(t-Bu-C≡N)$_2$Pt(N≡C-t-Bu)	1	Irreversible	[168]

The measurement of the chain generation rate v_{i0} in oxidized ethylbenzene with bis(acetylacetonate) nickel showed that v_{i0} is sufficiently higher in the presence of the catalyst than in the noncatalyzed reaction ($T = 393$ K, [171]).

Ni(acac)$_2$ (mol L^{-1})	0.0	3.0 × 10^{-3}	2.0 × 10^{-2}
v_{i0} (mol L^{-1} s^{-1})	1.5 × 10^{-9}	4.0 × 10^{-7}	9.0 × 10^{-6}

The activation of dioxygen by the nickel complex and the generation of radicals by the reaction of the Ni(II).O$_2$ complex with ethylbenzene were proposed. Examples of reactions of metal–dioxygen complexes with phenols and aromatic amines with the production of phenoxyl and aminyl radicals are known [172–174].

The kinetic parameters characterizing the oxidation of transition metal ions by dioxygen are collected in Table 10.8.

TABLE 10.8
Rate Constants of Oxidation of Transition Metal Ions by Dioxygen

Equation of the Reaction Rate	Conditions	T (K)	E (kJ mol^{-1})	log A, A (L mol^{-1} s^{-1})	k(298 K) (L mol^{-1} s^{-1})	Ref.
		Co(L-histidine)				
k[Co(II)][O$_2$] (L mol^{-1} s^{-1})	pH = 8–11, κ = 0 to 1 M	298	20.9	7.22	3.60 × 10^3	[175]
		Co(D-Histidine)$_2$				
k[Co(II)][O$_2$] (L mol^{-1} s^{-1})	pH 8–11, κ = 0 to 1 M	298	20.9	7.08	2.61 × 10^3	[175]
		[Co(Glycylglycine)$_2$(OH)$_2$]$^{2-}$				
k[Co(II)][O$_2$] (L mol^{-1} s^{-1})	H$_2$O	298			1.0 × 10^3	[176]
		Cu$^+$				
k[Cu$^+$][O$_2$] (L mol^{-1} s^{-1})	HCl	298			1.0 × 10^3	[177]
		CuCl$_2^-$				
k[CuCl$_2^-$][O$_2$] (L mol^{-1} s^{-1})	1 M HCl + KCl	298			2.51 × 10^2	[177]
		Cu(NH$_3$)$_2^+$				
k[Cu(I)][NH$_3$][O$_2$] (L^2 mol^{-2} s^{-1})	κ = 1	298			1.62 × 10^4	[178]
		Cu(Imidazole)$_2$				
k[Cu(I)][imidazole][O$_2$] (L^2 mol^{-2} s^{-1})	κ = 1	298			6.61 × 10^3	[178]
		Cu(dipy)$_2^+$				
k[Cu(Dipy)$_2^+$][O$_2$] (L mol^{-1} s^{-1})	pH = 5.0	298			6.46 × 10^3	[179]
		Fe^{2+} ion				
k[Fe^{2+}][H$_2$PO$_4^-$]$^2 p_{O_2}$ (L^2 mol^{-2} atm^{-1} s^{-1})	pH = 2–3	293–303	83.7	11.55	7.57 × 10^{-4}	[180]
k[Fe^{2+}]$^2 p_{O_2}$ (L mol^{-1} atm^{-1} s^{-1})	[FeSO$_4$] = 0.5 M	333			6.17 × 10^{-5}	[155]
k[Fe^{2+}]$^2 p_{O_2}$ (L mol^{-1} atm^{-1} s^{-1})	[FeSO$_4$] = 1 M	293–333	68.2	6.38	2.66 × 10^{-6}	[155]
k[Fe^{2+}]$^2 p_{O_2}$ (L mol^{-1} atm^{-1} s^{-1})	[FeSO$_4$] = 1 M, [H$_2$SO$_4$] = 1 M	293–333	61.9	5.14	1.95 × 10^{-6}	[155]

continued

TABLE 10.8
Rate Constants of Oxidation of Transition Metal Ions by Dioxygen—*continued*

Equation of the Reaction Rate	Conditions	T (K)	E (kJ mol^{-1})	log A, A (L mol^{-1} s^{-1})	k(298 K) (L mol^{-1} s^{-1})	Ref.
$k[Fe^{2+}]^2 p_{O_2}$ (L mol^{-1} atm^{-1} s^{-1})	[FeCl$_2$] = 0.5 M	333			8.71×10^{-6}	[155]
$k[Fe^{2+}]^2 p_{O_2}$ (L mol^{-1} atm^{-1} s^{-1})	[FeCl$_2$] = 1 M	293–333	56.5	4.01	1.28×10^{-6}	[155]
$k[Fe^{2+}]^2 p_{O_2}$ (L mol^{-1} atm^{-1} s^{-1})	[FeCl$_2$] = 1 M, [HCl] = 2 M	293–333	61.5	5.39	4.07×10^{-6}	[155]
$k[Fe^{2+}][O_2]$ (L mol^{-1} s^{-1})	Fe(NH$_4$)$_2$(SO$_4$)$_2$, pH = 0.76–1.34	293			4.79×10^{-3}	[181]
$k[Fe^{2+}][O_2]$ (L mol^{-1} s^{-1})	[HCl] = 8 M	272–308	61.1	10.41	0.50	[182]
$k[Fe^{2+}]^2 p_{O_2}$ (L mol^{-1} atm^{-1} s^{-1})	[HClO$_4$] = 0.51 M	298–313	72.8	6.36	3.98×10^{-7}	[183]
$k[Fe^{2+}]^2 [H_2P_2O_7^{2-}][O_2]$ (L^2 mol^{-2} s^{-1})	HClO$_4$, pH = 2–3	303			5.13	[184]
$k[Fe^{2+}] p_{O_2}$ (atm^{-1} s^{-1})	[H$_2$SO$_4$] = 1 N	413–453		2.08	1.76×10^{-8}	[185]
$k[Fe^{2+}] p_{O_2}$ (atm^{-1} s^{-1})	[H$_2$SO$_4$] = 0.5 M	303			3.80×10^{-6}	[186]
$k[Fe^{2+}] p_{O_2}$ (L mol^{-1} atm^{-1} s^{-1})	CH$_3$OH	303			1.20×10^{-3}	[187]
	Ti^{3+}					
$k[Ti^{3+}] p_{O_2}$ (atm^{-1} s^{-1})	[HCl] = 1 M	293–323	71.1	8.08	4.14×10^{-5}	[188]
	V^{3+}					
$k([V^{3+}] p_{O_2}/[H^+])$ (mol L^{-1} atm^{-1} s^{-1})	[HClO$_4$] = 1.6–8.3	298	84.1	7.91	1.47×10^{-7}	[189]

10.1.7 Catalytic Oxidation of Ketones

Ketones are resistant to oxidation by dioxygen in aqueous solutions at $T = 300$–350 K. Transition metal ions and complexes catalyze their oxidation under mild conditions. The detailed kinetic study of butanone-2 oxidation catalyzed by ferric, cupric, and manganese complexes proved the important role of ketone enolization and one-electron transfer reactions with metal ions in the catalytic oxidation of ketones [190–194].

Ferric ions catalyze the oxidation of butanone-2 in aqueous solutions [190]. Only two oxidation products, acetic acid and actaldehyde, are formed. The rate of butanone-2 oxidation linearly increases with increasing ketone concentration, does not depend on the dioxygen concentration, beginning from $pO_2 = 58$ kPa, goes through a maximum with a change in pH, and does not depend on the ionic strength of the solution (in interval of 0.2–1.4 mol L^{-1}, 333 K). Anions, such as Cl$^-$, F$^-$, H$_2$PO$_4^-$, AcO$^-$, and SO$_4^{-2}$, lower the activity of ferric ions [190]. The rate of enolization was measured and appeared to be less than or equal to the rate of ketone oxidation. The kinetics of the reaction of ferric ions with butanone-2 in an aqueous solution under anaerobic condition was studied. The following mechanism of oxidation was put forward [190]. Ketone is oxidized by FeOH^{2+} ions in the enolic form. As a result of oxidation, free alkyl radicals are formed, which initiate the polymerization of added methylmethacrylate.

Catalysis in Liquid-Phase Hydrocarbon Oxidation

$$\text{MeC(O)CH}_2\text{Me} + \text{H}_3\text{O}^+ \rightleftharpoons \text{MeC(OH)}=\text{CHMe} + \text{H}_3\text{O}^+ \qquad K_1$$
$$\text{MeC(OH)}=\text{CHMe} + \text{FeOH}^{2+} \rightleftharpoons \text{MeC(OFeOH}^+)=\text{CHMe} + \text{H}^+ \qquad K_2$$
$$\text{MeC(OFeOH}^+)=\text{CHMe} \longrightarrow \text{MeC(O)C}^\bullet\text{HMe} + \text{FeOH}^+ \qquad k_3$$
$$\text{MeC(O)C}^\bullet\text{HMe} + \text{Fe}^{3+} + \text{H}_2\text{O} \longrightarrow \text{MeC(O)CH(OH)Me} + \text{Fe}^{2+} + \text{H}^+ \qquad \text{Fast}$$

The equation for the reaction rate has the following form:

$$\frac{[\text{Ketone}]}{v} = \frac{1}{k_1[\text{H}_3\text{O}^+]} + \frac{1}{[\text{FeOH}^{2+}]}\left(\frac{1}{K_1 k_2} + \frac{[\text{H}_3\text{O}^+]}{K_1 K_2 k_3}\right) \qquad (10.13)$$

The experimental data are in agreement with this equation. In the presence of dioxygen, the alkyl radicals formed from enol rapidly react with dioxygen and thus the formed peroxyl radicals react with Fe^{2+} with the formation of hydroperoxide. The formed hydroperoxide is decomposed catalytically to molecular products (AcOH and AcH) as well as to free radicals. The free radicals initiate the chain reaction resulting in the increase of the oxidation rate.

The oxidation of butanone-2, catalyzed by complexes of pyridine with cupric salts, appeared to be similar in its main features [191]. Butanone-2 catalytically oxidizes to acetic acid and acetaldehyde. The reaction proceeds through the enolization of ketone. Pyridine catalyzes the enolization of ketone. Enole is oxidized by complexes of Cu(II) with pyridine. The complexes Cu(II).Py$_n$ with $n = 2,3$ are the most reactive. Similar results were provided by the study of butanone-2 catalytic oxidation with o-phenanthroline complexes, where Fe(III) and Mn(II) were used as catalysts [192–194].

10.2 COBALT BROMIDE CATALYSIS

Cobalt bromide is used as a catalyst in the technology of production of arylcarboxylic acids by the oxidation of methylaromatic hydrocarbons (toluene, p-xylene, o-xylene, polymethylbenzenes). A cobalt bromide catalyst is a mixture of cobaltous and bromide salts in the presence of which hydrocarbons are oxidized with dioxygen. Acetic acid or a mixture of carboxylic acids serves as the solvent. The catalyst was discovered as early as in the 1950s, and the mechanism of catalysis was studied by many researchers [195–214].

It was shown in the previous section that hydrocarbon oxidation catalyzed by cobalt salts occurs under the quasistationary conditions with the rate proportional to the square of the hydrocarbon concentration and independent of the catalyst (Equation [10.9]). This limit with respect to the rate is caused by the fact that at the fast catalytic decomposition of the formed hydroperoxide, the process is limited by the reaction of $\text{RO}_2^\bullet$ with RH. The introduction of the bromide ions into the system makes it possible to surmount this limit because these ions create a new additional route of hydrocarbon oxidation. In the reactions with ROOH and $\text{RO}_2^\bullet$, the Co^{2+} ions are oxidized into Co^{3+}, which in the reaction with ROOH are reduced to Co^{2+} and do not participate in initiation.

$$\text{RO}_2^\bullet + \text{Co}^{2+} + \text{AcOH} \rightleftharpoons \text{ROOH} + \text{Co}^{3+} + \text{AcO}^-$$

However, in the presence of the bromide ions, the Co^{3+} ions are reduced to Co^{2+}, and the $\text{Br}^\bullet$ atoms that are formed participate in chain propagation, which is very important for accelerating the reaction.

$$Co^{3+} + Br^- \rightleftharpoons CoBr^{2+}$$
$$CoBr^{2+} \longrightarrow Co^{2+} + Br^{\bullet}$$
$$Br^{\bullet} + RH \longrightarrow HBr + R^{\bullet}$$

These reactions result in an additional route of chain propagation, which allows one to exceed the rate limit due to the mechanism of action of only variable-valence ions. In fact, the initial rate of RH transformation in the presence of the cobalt bromide catalyst is determined by the rate of two reactions, namely, $RO_2^{\bullet}$ with RH (k_p) and $RO_2^{\bullet}$ with Co^{2+} (k_p'), followed by the reactions of Co^{3+} with Br^- and $Br^{\bullet}$ with RH. The general scheme proposed by Zakharov includes the following steps (written in the simplified form) [206]:

$$R^{\bullet} + O_2 \longrightarrow RO_2^{\bullet}$$
$$RO_2^{\bullet} + RH \longrightarrow ROOH + R^{\bullet} \qquad k_p$$
$$RO_2^{\bullet} + Co^{2+} + AcOH \longrightarrow ROOH + Co^{3+} + AcO^- \qquad k_p'$$
$$ROOH + Co^{2+} \longrightarrow RO^{\bullet} + Co^{3+} + OH^-$$
$$RO^{\bullet} + RH \longrightarrow ROH + R^{\bullet} \qquad \text{Fast}$$
$$Co^{2+} + Br^- \rightleftharpoons CoBr^+ \qquad K$$
$$RO_2^{\bullet} + CoBr^+ + AcOH \longrightarrow ROOH + CoBr^{2+} + AcO^- \qquad k_p''$$
$$Co^{3+} + Br^- \longrightarrow Co^{2+} + Br^{\bullet}$$
$$Br^{\bullet} + RH \longrightarrow BrH + R^{\bullet}$$
$$RO_2^{\bullet} + RO_2^{\bullet} \longrightarrow \text{Molecular products} \qquad 2k_t$$

The two reactions proposed above limit this process under the conditions of fast ROOH decomposition, so that the oxidation rate of RH is

$$v = (k_p[RH] + k_p'[Co^{2+}] + Kk_p''[Co^{2+}][Br^-])/2k_t \qquad (10.14)$$

This expression is valid for oxidation with the excess of bromide ions over cobalt ions (the conditions of fast oxidation of Co^{3+}). The experimental data agree with this dependence. The k_p, k_p', and Kk_p'' values for three hydrocarbons (343 K, acetic acid) are presented below [206]

Hydrocarbon	k_p (L mol^{-1} s^{-1})	k_p' (L mol^{-1} s^{-1})	Kk_p'' (L mol^{-1} s^{-1})
Toluene	1.6	6.1×10^2	2.1×10^3
o-Xylene	5.5	3.2×10^2	2.0×10^3
m-Xylene	4.0	6.6×10^2	2.2×10^3
p-Xylene	4.5	7.2×10^2	2.3×10^3
Diphenylmethane	19.7	6.0×10^2	1.8×10^3
Ethylbenzene	5.4	1.9×10^2	9.0×10^2

It is seen that $k_p'' \gg k_p$, and the dependence of the reaction rate on the catalyst concentration makes it possible to increase the rate by increasing the catalyst (Co(II) and Br^-) concentration.

Still higher oxidation rates are achieved when hydrocarbons are oxidized in the presence of the catalytic system including the Co, Mn, and Br^- ions. The Mn–Br binary system is less efficient than the cobalt bromide catalyst. Synergism of the mutual catalytic effect of the Co and Mn ions is due to the fact that the Co^{2+} ions rapidly decompose hydroperoxide, and bivalent manganese ions very rapidly react with peroxyl radicals, so that in the presence of Mn^{2+} the chain is more efficiently propagated in the reactions [204]

$$RO_2^{\bullet} + Mn^{2+} + AcOH \longrightarrow ROOH + Mn^{3+} + AcO^-$$
$$Mn^{3+} + Br^- \longrightarrow Mn^{2+} + Br^{\bullet}$$
$$Br^{\bullet} + RH \longrightarrow HBr + R^{\bullet}$$
$$R^{\bullet} + O_2 \longrightarrow RO_2^{\bullet}$$

The peroxyl radical of ethylbenzene reacts with Mn^{2+} in acetic acid with the rate constant $k = 9.7 \times 10^4$ L mol^{-1} s^{-1} (347 K) [204], which is by approximately two orders of magnitude higher than that with the Co^{2+} ions.

As alkylaromatic hydrocarbon (toluene, *p*-xylene, etc.) is oxidized, aldehydes appear; radicals and peracids formed from them play an important role. First, aldehydes react rapidly with the Co^{3+} and Mn^{3+} ions, which intensifies oxidation. Second, acylperoxyl radicals formed from aldehydes are very reactive and rapidly react with the initial hydrocarbon. Third, aldehydes form an adduct with primary hydroperoxide, which decomposes to form aldehyde and acid.

$$ArCH_2OOH + ArCH(O) \rightleftharpoons ArCH_2OOCH(OH)Ar$$
$$ArCH_2OOCH(OH)Ar \longrightarrow ArCH(O) + ArC(O)OH + H_2$$

This creates the possibility of transforming hydroperoxide into aldehydes omitting the stage of alcohol.

Generalizing the known data and established experimental peculiarities of the action of the cobalt bromide catalyst, we have to emphasize the following advantages of cobalt bromide catalysis:

1. This catalyst makes the increase in the oxidation rate of alkylaromatic hydrocarbons possible due to the intense participation of the catalyst itself (Co^{2+}, Co^{3+}, Br^-, and $Br^{\bullet}$) in chain propagation.
2. It provides the fast transformation of intermediate products (hydroperoxide, aldehydes) into the final product, viz., acid.
3. Finally, it makes possible the oxidation of hydrocarbon to a significant depth, and when the RH molecule contains several methyl groups, the catalyst allows all these groups to be transformed into carboxyls. This last specific feature is insufficiently studied so far. Perhaps, it is associated with the following specific features of oxidation of alkylaromatic hydrocarbons. The thermal decomposition of formed hydroperoxide affords hydroxyl radicals, which give phenols after their addition at the aromatic ring

$$ArH + HO^{\bullet} \longrightarrow Ar^{\bullet}H(OH)$$
$$Ar^{\bullet}H(OH) + O_2 \longrightarrow ArOH + HO_2^{\bullet}$$

In addition, alkylaromatic hydroperoxide $ArCH_2OOH$ under the action of acid is heterolytically transformed into phenol and formaldehyde. Phenols are accumulated and retard the oxidation process at early stages when the amount of methylcarboxylic acids (intermediate products) is low and they have no time to be oxidized further. In the presence of the catalyst, during the intense generation of $RO_2^{\bullet}$ and Co^{3+}, phenols are rapidly oxidized and do not retard the process, which makes the complete oxidation of all methyl groups in the hydrocarbon molecule to carboxyl groups possible.

10.3 OSCILLATING OXIDATION REACTIONS

The rate of the chemical process usually changes with time smoothly: decreases with the consumption of reactants or increases and passes through a maximum in autocatalytic processes. According to this, the concentration of the reaction products smoothly changes in time, and the kinetic curves have only one maximum or minimum, if any. However, systems were discovered in which the concentration of intermediate products oscillates, i.e., periodically passes through a maximum and a minimum. The amplitude of oscillations of the product concentration can decrease, increase, or remain unchanged for a long time. This regime was first observed in the decomposition of hydrogen peroxide catalyzed by iodine [215]. Hydrocarbon oxidation in the gas phase is accompanied by the appearance of cold flame, which often flashes periodically (Townend, 1933, propane oxidation [216]). Belousov [217] was first to observe the oscillation of concentrations of ion-oxidants (Ce^{4+} and Ce^{3+}) for the oxidation of malonic acid catalyzed by cerium ions. Later Zhabotinsky [218] showed that the oscillation regime is retained if malonic acid is replaced by another substrate with the activated CH_2 group and cerium ions are replaced by manganese ions. He proposed the mechanism explaining this phenomenon. The more detailed and quantitatively substantiated mechanism of this reaction was proposed and proved in the literature [219–223].

The main processes occurring in this system are the following [219]: bromate oxidizes trivalent cerium to tetravalent cerium; Ce^{4+} oxidizes bromomalonic acid, and is reduced to Ce^{3+}. The bromide ion, which inhibits the reaction, is isolated from the oxidation products of bromomalonic acid. During the reaction, the concentration of the Ce^{4+} ions (and Ce^{3+}) oscillates several times, passing through a maximum and a minimum. The shape of the peaks of concentrations and the frequency depend on the reaction conditions. The autooscillation character of the kinetics of the cerium ions disappears if Ce^{4+} or Br^- are continuously introduced with a low rate into the reaction mixture. The autooscillation regime of the reaction takes place only in a certain interval of concentrations of the reactants: [malonic acid] = 0.013–0.50 M; [$KBrO_3$] = 0.013–0.063; [Ce^{3+}] + [Ce^{4+}] = 10^{-4}–10^{-2}; [H_2SO_4] = 0.5–2.4; [KBr] in traces (2×10^{-5}). The oscillation regime begins after some time after mixing of the reactants; however, if a mixture of Ce^{3+} and Ce^{4+} corresponding to the established regime is introduced, autooscillations begin immediately. The period of vibrations ranges from 5 to 500 s depending on the conditions. The following data are known concerning the mechanism of particular stages. Hydroxybromomalonic acid (ROH) is brominated by BrOH and Br_2, and the formed dibromomalonic acid decomposes to form the bromide ion, which inhibits the oxidation of Ce^{3+} with bromate. The autooscillation regime is observed in the BrO_3^-–cerium ions–reducing agent systems where the reducing agent is oxaloacetic, acetonedicarboxylic, citric, or malic acids; acetoacetic ester, and acetylacetone (all compounds contain the β-diketone group and are readily brominated in the enolic form).

The oscillation regime is observed in the oxidation of the iodide ions by the BrO_3^- ions. The kinetics of this reaction and its mechanism were studied in detail by Citri and Epstein [223]. The process was studied in a jet reactor. The oscillating regime is observed when the concentration of iodide ions changes in an interval of 5×10^{-7} to 4×10^{-2} M (bromate was introduced in excess with respect to the iodide ions). The example of the oscillating kinetic curve can be seen in Figure 10.1.

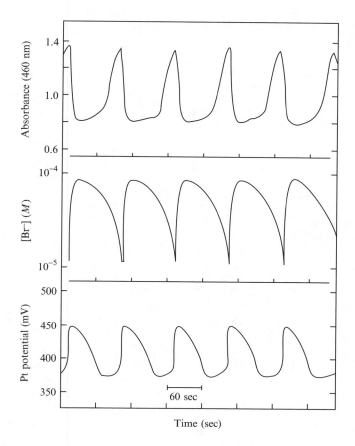

FIGURE 10.1 Kinetics of changing the light absorption by I_2 ($\lambda = 460$ nm), concentration of Br^-, and potential of the Pt electrode in the system where the oscillating reaction of BrO_3^- with I^- occurs ($[BrO_3^-]_0 = 5 \times 10^{-3}$ mol L^{-1}, $[I^-]_0 = 2.5 \times 10^{-3}$ mol L^{-1}, $[H^+] = 1.5$ mol L^{-1}, $T = 298$ K [223]).

The process occurs in two stages. The first stage is completed by the oxidation of I^- to I_2 and proceeds according to the stoichiometric equation

$$6I^- + 6H^+ + BrO_3^- = 3I_2 + 3H_2O + Br^-$$

with the rate ($T = 298$ K)

$$v\,(\text{mol}\,L^{-1}\,s^{-1}) = 45\,(L^3\,\text{mol}^{-3}\,s^{-1})[I^-][BrO_3^-][H_3O^+] \tag{10.15}$$

The second stage is described by the stoichiometric equation

$$I_2 + 2BrO_3^- = Br_2 + 2IO_3^-$$

It proceeds autocatalytically: first I_2 is transformed into BrI, which then is slowly transformed into Br_2 and IO_3^-. In the oscillation regime the concentration of Br^- changes from minimum to its maximum by almost ten times, and the oscillation period is about 90 s ($T = 298$ K, $[BrO_3^-] = 5 \times 10^{-3}$ mol L^{-1}, $[I^-] = 2.5 \times 10^{-3}$ mol L^{-1}, $[H_3O^+] = 1.5$ mol L^{-1}).

The process includes the following stages (the rate constant is measured at $T = 298$ K, water is the solvent):

Reaction	k (298 K)
$BrO_3^- + I^- + 2 H_3O^+ \longrightarrow HBrO_2 + HOI + 2 H_2O$	45 L^3 mol^{-3} s^{-1}
$HBrO_2 + HOI \longrightarrow HIO_2 + HOBr$	10^9 L mol^{-1} s^{-1}
$I^- + HOI + H_3O^+ \longrightarrow I_2 + 2 H_2O$	3.1×10^{12} L^2 mol^{-2} s^{-1}
$I_2 + 2 H_2O \longrightarrow I^- + HOI + H_3O^+$	22.0 L^2 mol^{-2} s^{-1}
$BrO_3^- + HOI + H_3O^+ \longrightarrow HBrO_2 + HIO_2 + H_2O$	8.0×10^2 L^2 mol^{-2} s^{-1}
$BrO_3^- + HIO_2 \longrightarrow IO_3^- + HBrO_2$	1.6×10^{-4} L mol^{-1} s^{-1}
$HOBr + I_2 \longrightarrow HOI + IBr$	8.0×10^{-7} L mol^{-1} s^{-1}
$HOI + IBr \longrightarrow HOBr + I_2$	1.0×10^{-2} L mol^{-1} s^{-1}
$IBr + 2 H_2O \longrightarrow HOI + Br^- + H_3O^+$	30 s^{-1}
$HOI + Br^- + H_3O^+ \longrightarrow IBr + 2 H_2O$	1.0×10^{-8} L^2 mol^{-2} s^{-1}
$Br^- + HbrO_2 \longrightarrow HOBr + BrO^-$	2.0×10^{-6} L mol^{-1} s^{-1}
$HOBr + Br^- + H_3O^+ \longrightarrow Br_2 + 2 H_2O$	8.0×10^{-9} L^2 mol^{-2} s^{-1}
$Br_2 + 2 H_2O \longrightarrow HOBr + Br^- + H_3O^+$	1.1×10^{-2} s^{-1}
$Br^- + 2 H_3O^+ + BrO_3^- \longrightarrow HBrO_2 + HOBr + 2 H_2O$	2.1 L^3 mol^{-3} s^{-1}
$HBrO_2 + HOBr + 2 H_2O \longrightarrow BrO_3^- + Br^- + 2 H_3O^+$	1.0×10^{-4} L mol^{-1} s^{-1}
$Br^- + HIO_2 + H_3O^+ \longrightarrow HOI + HOBr + H_2O$	6.0×10^{-5} L^2 mol^{-2} s^{-1}
$HOI + HOBr + H_2O \longrightarrow Br^- + HIO_2 + H_3O^+$	2.0×10^{-8} L mol^{-1} s^{-1}
$HIO_2 + HOBr \longrightarrow IO_2^- + Br^- + 2 H_3O^+$	2.2×10^{-8} L mol^{-1} s^{-1}
$BrO_3^- + IBr + 2 H_2O \longrightarrow IO_3^- + Br^- + HOBr + H_3O^+$	0.8 L mol^{-1} s^{-1}

The system of differential equations, which describes the process on the basis of all stages presented above, agree well with experiment and reproduces the oscillation regime of the process [223].

One of the simplest schemes, which describes the autooscillation regime, was considered by Lotka [224].

$$A + X \longrightarrow 2X \quad k_1$$
$$X + Y \longrightarrow 2Y \quad k_2$$
$$Y \longrightarrow Z \quad k_3$$

The quasistationary state (if the system has achieved this state) would have the form $[X]_{st} = k_3/k_2$ and $[Y]_{st} = k_1[A]_0/k_2$, where $[A] = [A]_0$ in the course of the whole process. To describe the behavior of the system near the quasistationary state, the following variables are introduced:

$$\xi = [X]_{st} - [X] \quad \text{and} \quad \eta = [Y]_{st} - [Y]$$

Changes in ξ and η are described by the equations

$$d\xi/dt = -k_2\eta(\xi + [X]_{st}) \approx -k_2\eta[X]_{st} \tag{10.16}$$

$$d\eta/dt = -k_2\xi(\eta + [Y]_{st}) \approx -k_2\xi[Y]_{st} \tag{10.17}$$

whose solution has the form $c_1 \exp(-\lambda_1 t) + c_2 \exp(-\lambda_2 t)$, where λ_1 and λ_2 are the roots of the equation $\lambda^2 = k_2^2[X]_{st}[Y]_{st}$, i.e., $\lambda = \pm ik_2([X]_{st}[Y]_{st})^{1/2}$.

The imaginary roots imply that [X] and [Y] undergo oscillation changes but never become equal to $[X]_{st}$ and $[Y]_{st}$ and at $[A]=[A]_0$ they remain unattainable. It can be shown that the system in the coordinates $[X]-[Y]$ executes the cyclic motion around zero (zero has the coordinates $[X]=[X]_{st}$, $[Y]=[Y]_{st}$).

10.4 ACID CATALYSIS IN LIQUID-PHASE OXIDATION OF HYDROCARBONS AND ALCOHOLS

Acids are well known as efficient catalysts of various heterolytic reactions (hydrolysis, esterification, enolyzation, etc. [225,226]). They catalyze the heterolytic decay of hydroperoxides formed during oxidation. For example, they catalyze the decomposition of cumyl hydroperoxide into phenol and acetone (important technological reaction) [5].

$$Me_2PhCOOH \longrightarrow Me_2C(O) + PhOH$$

In addition to heterolysis of hydroperoxides, strong mineral acids were found to catalyze the homolytic splitting of hydrogen peroxide and hydroperoxides into free radicals. This new catalytic ability of acids was found in the kinetic study of isopropanol oxidation [227,228]. Mineral acids (AH), such as H_2SO_4, $HClO_4$, HCl, and HNO_3, were found to accelerate the chain oxidation of 2-propanol due to catalysis of hydrogen peroxide decomposition with free radical generation. The initiation rate was found to be proportional to the product $[H_2O_2]^2 \times [AH]$. Electroconductivity studies proved the participation of hydrogen ions in free radical initiation.

The decay of 1,1-dimethylethyl hydroperoxide into free radicals under action of mineral acids was also established [229]. The similar kinetic equation was observed in this system and the rate of initiation was found to be propotional to the electroconductivity of the solution. The following mechanism of free radical generation was proposed [229].

$$Me_3COOH + HA \rightleftharpoons Me_3COOH_2^+ + A^-$$
$$Me_3COOH + Me_3COOH_2^+ \longrightarrow Me_3CO^\bullet + H_3O^+ + Me_3CO_2^\bullet$$

The homolytic decomposition of hydroperoxides was proved to be catalyzed by Bronsted as well as Lewis acids (for example, BF_3, $AlCl_3$, $SbCl_5$) [230]. Experimental data on acid catalysis of the homolytic decomposition of hydroperoxides are collected in Table 10.9.

Earlier (see Chapter 4) a great discrepancy was noticed between the calculated rate constants for the reactions

$$ROOH + HOR^1 \longrightarrow RO^\bullet + H_2O + R^1O^\bullet$$
$$ROOH + HOC(O)R^1 \longrightarrow RO^\bullet + H_2O + R^1CO_2^\bullet$$

and their experimental values for chain generation by the interaction of hydroperoxides with alcohols and acids [k_i (calculated) $\ll k$ (experimental)]. The probable explanation for such a great difference lies in the participation of the more active protonated form of hydroperoxide in free radical generation. For example, the reaction

$$H_3O_2^+ + H_2O \longrightarrow H_2O_2 + H_3O^+$$

occurs with the enthalpy $\Delta H = 10\,kJ\,mol^{-1}$. Alcohols and carbon acids formed in hydrocarbon oxidation can be donors of a proton.

TABLE 10.9
Acid Catalysis of the Homolytic Decomposition of Hydroperoxides (Experimental Data)

ROOH	Acid	Solvent	T (K)	E (kJ mol^{-1})	log A, A (L^2 mol^{-2} s^{-1})	k (350 K) (L^2 mol^{-2} s^{-1})	Ref.
			$v_i = k_i$ [ROOH]^2[Acid]				
H$_2$O$_2$	H$_2$SO$_4$	Me$_2$CHOH	333–348	108.8	14.93	4.93 × 10^{-2}	[231]
H$_2$O$_2$	HClO$_4$	Me$_2$CHOH	333–348	102.1	14.01	5.92 × 10^{-2}	[228]
H$_2$O$_2$	HClO$_4$	Me$_2$CHOH	333–348	113.0	15.52	4.53 × 10^{-2}	[231]
Me$_3$COOH	HClO$_4$	Me$_2$CHOH	333–348	117.1	15.60	1.33 × 10^{-2}	[229]
Me$_3$COOH	SbCl$_5$	MeCN	313–343	73.2	11.27	2.21	[232]
			$v_i = k_i$ [ROOH][Acid]		log A, A (L mol^{-1} s^{-1})	k (350 K) (L mol^{-1} s^{-1})	
Me$_3$COOH	BF$_3$	Me$_2$CHOH	343			0.10	[230]
Me$_3$COOH	AlCl$_3$	Me$_2$CHOH	343			3.6 × 10^{-2}	[230]
Me$_3$COOH	SbCl$_5$	Me$_2$CHOH	343			1.0 × 10^{-3}	[230]
Me$_3$COOH	LiCl	MeCN/CH$_2$=CHPh	343–353	77.4	11.75	1.58	[233]

10.5 CATALYTIC EPOXIDATION OF OLEFINS BY HYDROPEROXIDES

Olefin epoxidation by alkyl hydroperoxides catalyzed by transition metal compounds occupies an important place among modern catalytic oxidation reactions. This process occurs according to the following stoichiometric equation:

$$\text{ROOH} + \text{RCH=CHR} \longrightarrow \text{ROH} + \underset{H\quad H}{\overset{R\quad O\quad R}{\triangle}}$$

The catalysts of this process are vanadium, molybdenum, tungsten, niobium, chromium, and titanium compounds. The yield of oxide calculated per olefin is close to 100% and that per hydroperoxide reaches 85–95%, and the yield depends on the catalyst, temperature, solvent, and depth of conversion [234]. The following compounds were used and studied as catalysts of epoxidation: Ti(OR)$_4$, VO-acetylacetonate, MoO$_3$, Mo(CO)$_6$, MoCl$_5$, Mo-naphthenate, H$_2$MoO$_4$, Na$_2$MoO$_3$ + Na$_2$PMo$_{12}$O$_{10}$, MoO$_2$-acetylacetonate, WO$_3$, W(CO)$_6$, H$_2$WO$_4$ [234].

The reaction of olefin epoxidation by peracids was discovered by Prilezhaev [235]. The first observation concerning catalytic olefin epoxidation was made in 1950 by Hawkins [236]. He discovered oxide formation from cyclohexene and 1-octane during the decomposition of cumyl hydroperoxide in the medium of these hydrocarbons in the presence of vanadium pentaoxide. From 1963 to 1965, the Halcon Co. developed and patented the process of preparation of propylene oxide and styrene from propylene and ethylbenzene in which the key stage is the catalytic epoxidation of propylene by ethylbenzene hydroperoxide [237,238]. In 1965, Indictor and Brill [239] published studies on the epoxidation of several olefins by 1,1-dimethylethyl hydroperoxide catalyzed by acetylacetonates of several metals. They observed the high yield of oxide (close to 100% with respect to hydroperoxide) for catalysis by molybdenum, vanadium, and chromium acetylacetonates. The low yield of oxide (15–28%) was observed in the case of catalysis by manganese, cobalt, iron, and copper acetylacetonates. The further studies showed that molybdenum, vanadium, and tungsten compounds are the most efficient as catalysts for epoxidation. Atoms and ions of these metals possess vacant orbitals and easily form complexes due to the interaction with electron pairs of other molecules, in particular, with olefins and hydroperoxides. The epoxidation of the double bond by hydroperoxide with these catalysts is heterolytic. It is not accompanied by the formation of free radicals (the initiating effect of this reaction on oxidation is absent, inhibitors do not influence catalytic epoxidation). The catalyst forms a complex with hydroperoxide, and this complex epoxidizes olefin. Therefore, the initial epoxidation period is described by the seemingly simple kinetic scheme (Cat = catalyst)

$$\text{Cat} + \text{ROOH} \rightleftharpoons \text{Cat·ROOH} \qquad K$$
$$\text{Cat·ROOH} + \text{Olefin} \longrightarrow \text{Cat} + \text{ROH} + \text{Epoxide} \qquad k$$

The dependence of the epoxidation rate on the concentrations of olefin and hydroperoxide is described by the Michaelis–Menten equation

$$v = kK[\text{Cat}][\text{ROOH}][\text{Olefin}] \qquad (10.18)$$

The catalyst is preliminarily oxidized to the state of the highest valence (vanadium to V^{5+}; molybdenum to Mo^{6+}). Only the complex of hydroperoxide with the metal in its highest valence state is catalytically active. Alcohol formed upon epoxidation is complexed with the catalyst. As a result, competitive inhibition appears, and the effective reaction rate constant, i.e., $v/[\text{olefin}][\text{ROOH}]$, decreases in the course of the process due to the accumulation of alcohol. Water, which acts by the same mechanism, is still more efficient inhibitor. Several hypothetical variants were proposed for the detailed mechanism of epoxidation.

1. The simplest mechanism includes stages of catalyst oxidation to its highest valence state, the formation of a complex with ROOH, and the reaction (bimolecular) of this complex with olefin [240].

$$VO_2 + \text{ROOH} \longrightarrow V^{5+} + RO^\bullet$$
$$V^{5+} + \text{ROOH} \rightleftharpoons V^{5+}\text{·ROOH}$$

$$V^{5+}\text{·ROOH} + R'CH=CHR'' \longrightarrow ROV^{4+} + \underset{H\quad R''}{\overset{R'\quad\overset{H\oplus}{O}\quad H}{\diagup\!\!\diagdown}}$$

$$ROV^{4+} + \underset{H\quad R''}{\overset{R'\quad\overset{H\oplus}{O}\quad H}{\diagup\!\!\diagdown}} \longrightarrow V^{5+}\text{·ROH} + \underset{H\quad R''}{\overset{R'\quad O\quad H}{\diagup\!\!\diagdown}}$$

The first two stages (oxidation of V^{IV} to V^V and complex formation) were proved but the third stage predicting the heterolysis of the O—O bond with the addition of OH^+ at the double bond is speculative. Heterolysis of this type is doubtful because of the very high energy of the heterolytic cleavage of RO—OH to RO^- and OH^+.

2. From the energetics point of view, the epoxidation act should occur more easily (with a lower activation energy) in the coordination sphere of the metal when the cleavage of one bond is simultaneously compensated by the formation of another bond. For example, Gould proposed the following (schematic) mechanism for olefin epoxidation on molybdenum complexes [240]:

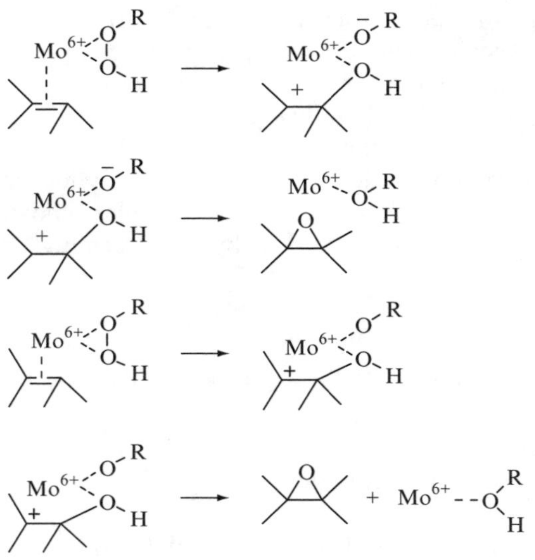

[Scheme showing four sequential steps of Mo^{6+}-mediated epoxidation mechanism with hydroperoxide and olefin]

This scheme differs from the previous one with respect to the heterolysis of the O—O bond, which occurs in the internal coordination sphere of the complex and is compensated by the formation of the C—O bond.

3. The complex in which olefin is bound by the donor–acceptor bond to the oxygen atom of hydroperoxide and exists in the secondary coordination sphere is also considered in the literature [241].
4. Proofs were obtained that the Mo=O group as a proton acceptor participates in epoxidation. In this connection, the following scheme taking into account this circumstance was proposed [242]:

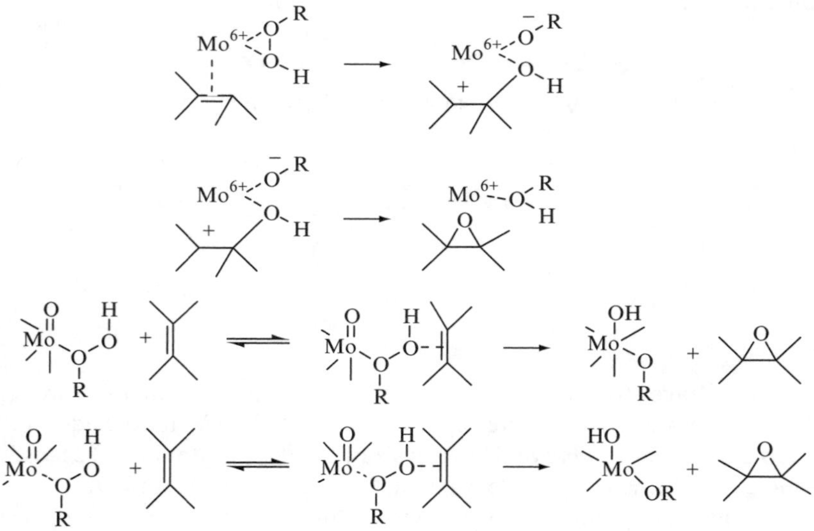

[Scheme showing Mo=O group participating as proton acceptor in epoxidation mechanism, with equilibria leading to epoxide formation and Mo–OH or Mo–OR complexes]

The formation of molybdenum complexes with diols (formed by olefin oxidation) was proved for the use of the molybdenum catalysts. Therefore, the participation of these complexes in the developed epoxidation reaction was assumed [242].

All schemes presented are similar and conventional to a great extent. It is characteristic that the epoxidation catalysis also results in the heterolytic decomposition of hydroperoxides (see Section 10.1.4) during which heterolysis of the O—O bond also occurs. Thus, there are no serious doubts that it occurs in the internal coordination sphere of the metal catalyst. However, its specific mechanism and the structure of the unstable catalyst complexes that formed are unclear. The activation energy of epoxidation is lower than that of the catalytic decomposition of hydroperoxides; therefore, the yield of oxide per consumed hydroperoxide decreases with the increase in temperature.

Sajus et al. [243,244] synthesized the peroxo complex of molybdenum(VI) and studied its reaction with a series of olefins. This peroxo complex MoO_5 was proved to react with olefins with epoxide formation. The selectivity of the reaction increases with a decrease in the complex concentration. It was found to be as much as 95% at epoxidation of cyclohexene by MoO_3 in a concentration 0.06 mol L^{-1} at 288 K in dichloroethylene [244]. The rate of the reaction was found to be

$$v = kK[MoO_3][\text{Olefin}](1 + K[\text{Olefin}])^{-1} \qquad (10.19)$$

This scheme is in accordance with the consecutive two-stage reaction with the first equilibrium stage.

$$MoO_3 + \text{Olefin} \Leftrightarrow MoO_3 \cdot \text{Olefin} \qquad K$$
$$MoO_3 \cdot \text{Olefin} \longrightarrow MoO_2 + \text{Epoxide} \qquad k$$

The reaction proceeds as pseudomonomolecular reaction with a rate k_1 at the constant olefin concetration. The values of k_1 for epoxidation of olefins in dichloroethylene solutions at 288 K and an olefin concentration of 1.96 mol L^{-1} is given below [244].

Olefin	$CH_2=CHC_6H_{13}$	$CH_2=CHPh$	$Me_2C=CHCMe_3$	$MeCH=CHC_5H_{11}$
k_1 (s^{-1})	4.9×10^{-5}	7.3×10^{-5}	1.02×10^{-4}	2.11×10^{-4}
Olefin	$CH_2=CEtMe$	cyclohexene	$Me_2C=CHMe$	$Me_2C=CMe_2$
k_1 (s^{-1})	2.76×10^{-4}	2.78×10^{-4}	6.65×10^{-4}	9.5×10^{-4}

It seems very probable that the epoxidation reaction proceeds through a two-stage mechanism. Hydroperoxide oxidizes the catalyst to peroxo complex and the this complex performs epoxidation of olefins.

The following two processes find technological application:

1. The production of styrene and propylene epoxide from ethylbenzene (Halcon process) that includes the following three technological stages [234].

$$PhCH_2CH_3 + O_2 \longrightarrow PhCH(OOH)CH_3$$
$$PhCH(OOH)CH_3 + CH_2{=}CHCH_3 \longrightarrow PhCH(OH)CH_3 + \underset{}{\triangle}$$
$$PhCH(OH)CH_3 \longrightarrow H_2O + CH_2{=}CHPh$$

2. The production of 2-methylpropylene and propylene epoxide from methylpropane with the following technological stages [234].

$$Me_3CH + O_2 \longrightarrow Me_3COOH$$
$$Me_3COOH + CH_2{=}CHCH_3 \longrightarrow Me_3COH + \underset{}{\triangle}$$
$$Me_3COH \longrightarrow H_2O + CH_2{=}C(CH_3)_2$$

10.6 CATALYTIC OXIDATION OF OLEFINS TO ALDEHYDES

10.6.1 CATALYSIS BY PALLADIUM SALTS

The oxidation of ethylene to acetaldehyde by dioxygen catalyzed by palladium and cupric salts found important technological application. The systematic study of this process was started by Smidt [245] and Moiseev [246]. The process includes the following stoichiometric stages [247,248]:

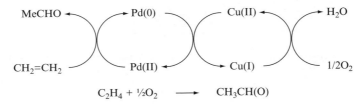

The reaction occurs rapidly in an aqueous solution at mild temperatures (290–350 K) with a high yield of acetaldehyde and evolution of 220 kJ mol^{-1} of heat. Other olefins can be oxidized to subsequent aldehyde or ketone (see Table 10.10).

The di-µ-tetrachloro-bis(ethylene)dipalladium complex rapidly reacts with alcohol (MeOH, EtOH, Me$_2$CHOH, PhCH$_2$OH) with formation of diacetal [247].

$$(C_2H_4PdCl_2)_2 + 4ROH \longrightarrow 2Pd + 4HCl + 2CH_3CH(OR)_2$$

This reaction is transformed into the catalytic process in the presence of cuprous chloride and dioxygen [247,248]. The same complex was found to be oxidized by acetic acid with sodium acetate to vinyl acetate [247,249].

$$(C_2H_4PdCl_2)_2 + 4AcONa \longrightarrow 2Pd + CH_2{=}CHOAc + 4NaCl + AcOH$$

TABLE 10.10
Products of the Catalytic Oxidation ($PdCl_2 + CuCl_2 + O_2$) of Olefins in Water [247]

Olefin	T (K)	Time of Reaction (s)	Product	Yield (%)
$CH_2=CH_2$	293	300	MeCH(O)	85
$CH_2=CHMe$	293	300	MeC(O)Me	90
$CH_2=CHEt$	293	600	EtC(O)Me	80
$CH_2=CHPr$	293	1200	PrC(O)Me	81
$CH_2=CHBu$	303	1800	BuC(O)Me	75
$CH_2=CHCH_2CH=CH_2$	293	900	EtCH=CHCH(O)	91
⬡	303	1800	⬡=O	65
$PhCH=CH_2$	323	10800	PhC(O)Me	57
$CH_2=CHCOOH$	323	11160	MeCH(O)	50
$MeCH=CMeCOOH$	323	3600	MeC(O)Me	75
$CH_2=CMeCOOH$	313	3600	EtCH(O)	61
$CH_2=CHCH_2OH$	298	300	$CH_2=CHCH(O)$	75
$CH_2=CHCN$	303	1800	$CNCH_2CH(O)$	88

10.6.2 Kinetics and Mechanism of Reaction

Olefins and palladium ions $PdCl_4^{2-}$ form the following two types of complexes in an aqueous solution [247].

$$C_2H_4 + PdCl_4^{2-} \rightleftharpoons C_2H_4PdCl_3^- + Cl^- \quad K_1$$

$$C_2H_4 + PdCl_4^{2-} + H_2O \rightleftharpoons C_2H_4PdCl_2OH_2 + 2Cl^- \quad K_2$$

The equilibrium constants have the following values (water, $T = 293$ K, ionic force is 4 g ion L^{-1} [250])

Olefin	Ethylene	Propylene	1-Butene
K_1	15.2	7.9	12.4
K_2 (mol L^{-1})	4.3	4.6	3.4

The kinetic study of the reaction of ethylene with $PdCl_4^{2-}$ proved that this reaction is bimolecular and empirical rate constant k_{exp} depends on the concentration of acid (H_3O^+) and the chloride anion [251].

$$v = k_{exp}[PdCl_4^{2-}][C_2H_4] \quad (10.20)$$

$$k_{exp} = k[H_3O^+]^{-1}[Cl^-]^{-2} \quad (10.21)$$

The following many-stage mechanism was proposed for this reaction [252].

$$C_2H_4 + PdCl_4^{2-} \rightleftharpoons C_2H_4PdCl_3^- + Cl^-$$
$$C_2H_4PdCl_3^- + H_2O \rightleftharpoons C_2H_4PdCl_2OH_2 + Cl^-$$
$$C_2H_4PdCl_2OH_2 + H_2O \rightleftharpoons C_2H_4PdCl_2OH^- + H_3O^+$$
$$C_2H_4PdCl_2OH^- + H_2O \rightleftharpoons H_2OCl_2Pd^- \text{—} CH_2CH_2OH$$
$$H_2OCl_2Pd^- \text{—} CH_2CH_2OH \longrightarrow Cl^- + PdCl^-(\text{solid}) + MeCH(O) + H_3O^+$$

The last stage is supposed to be limiting. However, the limiting stage can be the transformation of the π-complex into the σ-complex of ethylene with palladium (preliminary stage). The rate constant k (water, $T = 298$ K, ion force = 3 g ion L^{-1}) has the following values [246].

Olefin	Ethylene	Propylene	1-Butene
k (mol^2 L^{-2} s^{-1})	1.13×10^{-3}	3.60×10^{-4}	5.30×10^{-4}

The transformation of the π-complex into the σ-complex is supposed to be intramolecular.

10.7 HETEROGENEOUS CATALYSIS IN LIQUID-PHASE OXIDATION

10.7.1 INTRODUCTION

Heterogeneous catalysis is widely used in technology for gas-phase oxidation of hydrocarbons to alcohols, aldehydes, epoxides, anhydrides, etc. Homogeneous catalysis predominates in the liquid-phase oxidation technology. Nevertheless, a series of experimental studies was devoted in the 1970s to 1990s to heterogeneous catalysis. The main objects of study were metal oxides and metals as catalysts.

The obvious technological advantage of a heterogeneous catalyst is that it can be easily separated from reactants and products. However, the serious physical problem is diffusion of reactants to active centers on the surface of the catalyst and back diffusion of the formed intermediate and final products from the surface into the solution. This duffusion occurs much more slowly in the liquid phase compared to the gas phase. The problem of effectiveness of the heterogeneous catalyst in comparison with the homogeneous catalyst is closely connected with the problem of diffusion and sorption on the surface in the liquid phase.

The following four types of heterogeneous catalysis in the liquid-phase oxidation of organic compounds were observed.

1. Catalyst accelerates the decomposition of hydroperoxide to free radicals on the surface. The free radicals then diffuse into the reactant bulk and initiate the chain oxidation of the oxidized substance.
2. Catalyst absorbs dissolved dioxygen. Sorbed dioxygen reacts with the oxidized substance with production of free radicals. The free radicals diffuse into solution and initiate the chain oxidation of hydrocarbon or other substances.
3. Heterogeneous substance inhibits the oxidation of the oxidized substance. The mechanisms of oxidation retardation will be discussed in Chapter 20.
4. Dioxygen and oxidized substances react on the surface of the catalyst only. The pure heterogeneous reaction occurs only after diffusion of reactants to the catalytic surface and back diffusion of products from the surface into the solution. A combination of a few mechanisms of such types are possible.

10.7.2 Decomposition of Hydroperoxides on Catalyst Surface

Hock and Kropf [253] studied cumene oxidation catalyzed by PbO_2. They proposed that PbO_2 decomposed cumyl hydroperoxide (ROOH) into free radicals (RO$^{\bullet}$, RO$_2^{\bullet}$). The free radicals started the chain oxidation of cumene in the liquid phase. Lead dioxide introduced into cumene was found to be reduced to lead oxide. The reduction product lead oxide was found to possess catalytic activity. The following tentative mechanism was proposed.

$$ROOH + PbO_2 \longrightarrow RO_2^{\bullet} + OPbOH$$
$$ROOH + HOPbO \longrightarrow RO^{\bullet} + H_2O + PbO$$
$$ROOH + PbO \longrightarrow RO_2^{\bullet} + PbOH$$
$$ROOH + PbOH \longrightarrow RO^{\bullet} + H_2O + PbO$$

The increase in the amount of catalyst introduced in oxidized cumene (353 K) increases the oxidation rate, decreases the amount of the formed hydroperoxide, and increases the yield of the products of hydroperoxide decomposition: methylphenyl ethanol and acetophenone. Similar mechanism was proposed for catalysis by copper phthalocyanine in cumene oxidation [254].

The generation of free radicals in the catalyzed oxidation of hydroperoxides was observed [255,256]:

1. By the free radical acceptor method in the decomposition of cumyl hydroperoxide catalyzed by Fe_2O_3 (343 K, 0.25 g cm^{-3} Fe_2O_3, ionol as acceptor of free radicals [257])
2. By the CL method in the oxidation catalyzed by Fe_2O_3 decomposition of cumyl hydroperoxide [258]
3. By the EPR method in oxidized cumene with metal oxides (MnO_2, Co_2O_3, CuO, and NiO [259]); there were fixed cumylperoxyl radicals.

The following metal oxides were found to accelerate cumene oxidation and cumyl hydroperoxide decomposition: MnO_2, Fe_2O_3, Co_2O_3, Ni_2O_3, NiO [260]. However, a few heterogeneous catalysts that catalyze the decomposition of cumyl hydroperoxide but inhibit oxidation were found (TiO_2, V_2O_5, WO_3, and MoO_3). These catalysts were assumed to catalyze the heterolytic decomposition of cumyl hydroperoxide with phenol formation. The formed phenol terminates the chains reacting with peroxyl radicals. Metal borides and carbides were found to have the catalytic activity in 2-methyl propane liquid-phase oxidation [261,262]. The kinetic scheme of hydroperoxide (ROOH) homolytic decomposition on the surface of catalyst (S) is supposed to be similar to ROOH decomposition under the action of transition metal ions.

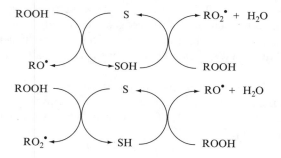

Depending on the electron affinity of the catalyst, one of these two routes predominates. The dependence of the hydroperoxide decomposition rate on [ROOH] is in agreement with the conception of preliminary equilibrium sorption of hydroperoxide on the catalyst surface ($Me_2PhCOOH$, AgO, $16\,m^2\,L^{-1}$, 343 K) [263]). The equilibrium constant was estimated to be $K = 7\,mol\,L^{-1}$, and effective rate constant of described ROOH decomposition is $k_{is} = 70\,s^{-1}$ [263].

The activity of the surface of AgO as a catalyst of hydroperoxide decomposition depends on the gas preliminarily absorbed on the catalyst surface. The preliminary absorption of H_2, NH_3, and CO activates the catalyst. However, the absorption of O_2 and CO_2, on the contrary, lowers the catalyst activity of AgO (343 K, solvent = PhCl, ROOH = $Me_2PhCOOH$, $[ROOH]_0 = 4.4 \times 10^{-1}\,mol\,L^{-1}$, [AgO] = 20 g L^{-1} [264]).

Absorbed Gas	Ar	H_2	NH_3	CO	O_2	CO_2
v_0^{ROOH} ($\times 10^5\,mol\,L^{-1}\,s^{-1}$)	6.7	21.6	12.5	12.6	4.1	0.1

The physical properties of the catalyst surface were found to change during the reaction (for detailed information, see Ref. [255]).

10.7.3 ACTIVATION OF DIOXYGEN

Free radicals were found to be generated on the catalyst surface in hydrocarbon oxidation in the absence of hydroperoxide. The activation of absorbed dioxygen was supposed to be the source of radicals [255]. The catalytic action of the silver surface on cumene oxidation was supposed to be the result of activation of sorbed dioxygen [265].

$$Ag + O_2 \rightleftharpoons Ag.O_2$$
$$Ag.O_2 \longrightarrow Ag^+.O_2^-$$
$$Ag^+.O_2^- + RH \longrightarrow AgOOH + R^\bullet$$
$$R^\bullet + O_2 \longrightarrow RO_2^\bullet$$
$$AgOOH + RH \longrightarrow AgO + H_2O + R^\bullet$$

Dioxygen activation was observed in oxidized cumene by such catalysts as NiO and Bi_2O_3. These catalysts do not decompose hydroperoxide but accelerate cumene oxidation [255]. The formation of alkyl radicals was fixed by the free radical acceptor method with tetramethylpyparidineoxyl in ethylbenzene with AgO only in the presence of dioxygen [266]. The same active centers of the catalyst can be active toward O_2 and ROOH as absorbents. Such competing scheme of heterogeneous catalyst action was discussed in the literature [267]. Dioxygen retards the catalytic decay of hydroperoxide on the surface of such catalysts as NiO, Cr_2O_3, and AgO, as well as phthalocyanine metal complexes [268,269].

The activation of hydrocarbons on the catalyst surface was also discussed in the literature [255]. There are no clear experimental evidences of this activation with free radical generation [270]. However, examples of dimer (RR) formation as a result of oxidation of RH on the surface of MnO_2 are known [270].

Catalysis in Liquid-Phase Hydrocarbon Oxidation

10.7.4 CRITICAL PHENOMENA IN HETEROGENEOUS CATALYSIS

The heterogeneous catalyst accelerates hydrocarbon oxidation. The rate of oxidation increases with increasing concentration of the catalyst. However, this increase in the oxidation rate with the catalyst concentration is not unlimited. The oxidation rate reaches a maximum value and does not increase thereafter. Moreover, the cessation of the reaction was observed and very often at a very small increase in the catalyst concentration. Such phenomenon was named *critical phenomenon*. The basis of critical phenomenon lies in the chain mechanism of oxidation and the dual ability of the catalyst surface to initiate and terminate chains. Numerous observations and studies of critical phenomenon in catalytic liquid-phase oxidations were performed [271–283]. Here are a few examples.

RH	Cumene	Tetraline	Cumene	Cyclohexene
Catalyst	Co_2O_3, MnO	NiO	MnO_2	MnO_2
Ref.	[275]	[259]	[259]	[280]
RH	Cyclohexene	Ethylbenzene	Hexadecane	Ethylbenzene
Catalyst	Mo, Cr/Al_2O_3	$Cu_3[Fe(CN)_6]_2$	Cu/CuO	MnO_2, Co_2O_3
Ref.	[286]	[278]	[277]	[279]

Critical phenomenon results from a series of peculiarites of chain catalytic liquid-phase oxidations:

1. Catalytic surface is active toward hydroperoxide and decomposes it to free radicals. The free radicals initiate the chain oxidation of RH in the liquid phase.
2. Hydroperoxide is produced in the chain reaction with bimolecular chain termination. The rate of ROOH formation is proportional to (initiation rate)$^{1/2}$ and the shorter the chain length higher the initiation rate. With increasing concentration of the catalyst this situation is encountered when the reaction occurs with $\nu = 1$.
3. Catalyst reacts with peroxyl radicals. So, an increase in the catalyst concentration increases the rate of chain reaction proportional to [catalyst]$^{1/2}$ and at the same time it increases the chain termination proportional to the catalyst concentration. Therefore the retardation or cessation of the chain reaction is observed at a high concentration of the catalyst. This is called *critical phenomenon*.
4. In addition, two kinds of active centers are supposed on the catalyst surface: S_1 (reaction with ROOH) and S_2 (reaction with $RO_2^\bullet$). The following kinetic scheme was proposed for the explanation of the critical phenomenon [274].

$$R^\bullet + O_2 \longrightarrow RO_2^\bullet \qquad \text{Fast}$$
$$RO_2^\bullet + RH \longrightarrow ROOH + R^\bullet \qquad k_p$$
$$ROOH + S_1 \longrightarrow RO^\bullet + S_1 \qquad k_{ic}$$
$$ROOH + S_1 \longrightarrow \text{Mol. Products} + S_1 \qquad k_{mc}$$
$$2RO_2^\bullet \longrightarrow ROH + O_2 + \text{Ketone} \qquad 2k_t$$
$$RO_2^\bullet + S_2 \longrightarrow \text{Mol. Products} + S_2 \qquad k_{tc}$$

In such a system, two regimes of hydrocarbon oxidation exist. When $[S] < [S]_{cr}$, the reaction occurs with acceleration. When $[S] > [S]_{cr}$, the reaction occurs with the stationary rate, and higher the amount of the catalyst lower the rate. The critical catalyst concentration

$$[S]_{cr} = \frac{k_p k_{ic}[RH]}{(k_{mc}/k_{ic})(k_{ic} + k_{im})k_{tc}} \qquad (10.22)$$

The $[S]_{cr}$ value depends on the nature of the catalyst, its surface area per unit of weight, the ratio of the rates of hydroperoxide decomposition into free radicals and molecular products, hydrocarbon and dioxygen concentrations, the method of catalyst preparation, and the chemical treatment of the surface.

10.7.5 Selectivity of Catalytic Oxidation

High selectivity of hydrocarbon oxidation to hydroperoxide can be performed with the heterogenous catalyst, which possesses

(a) High activity of hydroperoxide decomposition into free radicals
(b) Low activity in hydroperoxide decay into molecular products ($k_{mc}/k_{ic} \to 0$), especially to phenols from alkylaryl hydroperoxide
(c) Very low activity in chain termination ($S_2/S_1 \to 0$).

Empically found catalysts with high selectivity are known. The results on ethylbenzene oxidation to hydroperoxide with the Ag catalyst with different modifications of the surface (346 K, [catalyst] = 15 g L^{-1}, $pO_2 = 10^5$ Pa [255]) are given below.

Catalyst	Modification	Yield ROOH	Conversion
Ag$_2$O		20%	1.3%
Ag$_2$O	Me$_2$PhCOONa	81%	0.5%
Ag/Al$_2$O$_3$		20%	2.5%
Ag/Al$_2$O$_3$	PhC(O)Me + Me$_2$PhCOH	77%	1.8%
Ag/Al$_2$O$_3$	0.2% Ca	54.5	0.6
Ag/Al$_2$O$_3$	0.2% Ca + PhC(O)Me + Me$_2$PhCOH	96.0	0.6

10.8 OXIDATION IN BASIC SOLUTIONS

The catalysis of hydrocarbon liquid-phase autoxidation by transition metals is the result of the substitution of the slow bimolecular reaction ($E \approx 100$ kJ mol^{-1}, see Chapter 4)

$$ROOH + ROOH \longrightarrow RO^\bullet + H_2O + RO_2^\bullet$$

by the acceleration of the two reactions with rapid electron transfer from hydroperoxide to transition metal ion with activation energy $E \approx 30–40$ kJ mol^{-1} (see Table 10.2).

Catalysis by strong bases in the oxidation of organic compounds in aprotic solvents, principally, is similar. The slow reaction of chain propagation of the type $RO_2^\bullet + RH$

($E_0 = 40$ kJ mol^{-1} at $\Delta H = 0$) is substituted by a few consecutive fast reactions with electron transfer. Russel [284–291] studied a few reactions of oxidation of alkylaromatic hydrocarbons in the presence of strong bases. He proved the chain mechanisms of these reactions. One of them includes a few stages with addition of dioxygen to carbanion. Another includes the electron transfer from carbanion to dioxygen.

Diphenylmethane reacts with dioxygen in the presence of potassium 1,1-dimethylethoxide in various solvents (dimethylformamide [DMF], hexamethylphosphoramide [HMPA], pyridine) to produce nearly 100% yields of benzophenone [284]. The adduct of benzophenone with dimethylsulfoxide (DMSO) [1,1-diphenyl-2-(methylsulfinyl)ethanol] is formed as the final product of the reaction. The stoichiometry of the reaction and the initial rate depends on the solvent (conditions: 300 K, [Ph$_2$CH$_2$] = 0.1 mol L^{-1}, [Me$_3$COK] = 0.2 mol L^{-1}, pO_2 = 97 kPa).

Solvent	C$_5$H$_5$N	DMF/Me$_3$COH	DMSO/Me$_3$COH	HMPA
		4:1 (v/v)	4:1 (v/v)	
$\Delta[O_2]/\Delta[Ph_2CH_2]$	1.4	1.0	2.7	2.7
v_0 (mol L^{-1} s^{-1})	2.5×10^{-3}	6.2×10^{-3}	1.7×10^{-2}	4.8×10^{-2}

The first stage of the reaction is deprotonation of diphenylmethane. This stage is limiting. The formation of benzophenone includes the following stages [284]:

$$Ph_2CH_2 + Me_3CO^- \longrightarrow Ph_2CH^- + Me_3COH$$
$$Ph_2CH^- + O_2 \longrightarrow Ph_2CHOO^-$$
$$Ph_2CHOO^- \longrightarrow Ph_2C(O) + OH^-$$
$$Ph_2CHOO^- + Me_3CO^- \longrightarrow Ph_2C(OO^-)^- + Me_3COH$$
$$Ph_2C(OO^-)^- + O_2 \longrightarrow Ph_2C(OO^-)_2$$
$$Ph_2C(OO^-)_2 + 2Me_3COH \longrightarrow Ph_2C(O) + O_2 + 2Me_3CO^-$$

The rate of the reaction is limited by the first stage and is given by (298 K, DMSO/Me$_3$COK = 4:1 [284]):

$$v \text{(mol L}^{-1}\text{ s}^{-1}) = 1.42 \times 10^{-3} \text{(L mol}^{-1}\text{s}^{-1})[Ph_2CH_2][Me_3COK] \quad (10.23)$$

On the contrary, the oxidation of fluorene in a basic solution is not limited by the deprotonation of hydrocarbon [284]. This is in agreement with the oxidation of fluorene and 9,9-dideuterofluorene at the same rate in DMSO and 1,1-dimethylethanol solution. The stoichiometry of fluorene oxidation is close to unity (except oxidation in HMPA) and the main product of the reaction is fluorenone. The stoichiometry and the initial rate of the reaction depends on the solvent (conditions: 300 K, [fluorene] = 0.1 mol L^{-1}, [Me$_3$COK] = 0.2 mol L^{-1}, pO_2 = 97 kPa).

Solvent	Me$_3$COH	DMF/Me$_3$COH (4:1)	HMPA/Me$_3$COH	HMPA
$\Delta[O_2]/\Delta[RH]$	1.09	1.16	1.22	2.5
v (mol L^{-1} s^{-1})	5.8×10^{-4}	4.5×10^{-2}	4.8×10^{-2}	6.2×10^{-2}

The reaction proceeds as a chain process involving the peroxyl radical and superoxide ion [284].

$$Ar_2CH_2 + B^- \longrightarrow Ar_2CH^- + BH$$
$$Ar_2CH^- + O_2 \longrightarrow Ar_2C^\bullet H + O_2^{-\bullet}$$
$$Ar_2C^\bullet H + O_2 \longrightarrow Ar_2CHOO^\bullet$$
$$A_2CH^- + Ar_2CHOO^\bullet \longrightarrow Ar_2C^\bullet H + Ar_2CHOO^-$$
$$Ar_2CHOO^- + BH \longrightarrow Ar_2CHOOH + B^-$$
$$Ar_2CHOOH + B^- \longrightarrow Ar_2C(O) + BH + HO^-$$
$$Ar_2C^\bullet H + O_2^{-\bullet} \longrightarrow Ar_2CHOO^- + O_2$$

This reaction was found to be accelerated by the addition of electron acceptors such as nitrobenzene and m-trifluoromethylnitrobenzene. These electron acceptors accelerate the electron transfer from the carbanion to dioxygen.

$$Ar_2CH^- + PhNO_2 \longrightarrow Ar_2C^\bullet H + PhNO_2^{-\bullet}$$
$$PhNO_2^{-\bullet} + O_2 \longrightarrow PhNO_2 + O_2^{-\bullet}$$

The increase in the electron acceptor concentration in a solution transforms the reaction of fluorene oxidation into regime when the first stage (deprotonation of fluorene) limits the chain process. For the oxidation of other alkylaromatic hydrocarbons, see Refs. [292–304].

The reaction of benzhydrol with dioxygen in a basic solution (Me_3COK) results in the formation of benzophenone and potassium superoxide KO_2 [288]. The stoichiometry of oxidation depends on the Me_3COK/Ph_2CHOH ratio and increases from 0.73 at $[Me_3COK]/[Ph_2CHOH] = 1$ to 1.45 at $[Me_3COK]/[Ph_2CHOH] = 4$ (solvent Me_3COH, $T = 300\,K$, $[Ph_2CHOH] = 0.12\,mol\,L^{-1}$, $pO_2 = 98\,kPa$). The formation of the ketyl radical-anion was fixed by the EPR method [284]. The reaction was supposed to proceed in two stages. The first one:

$$Ph_2CHOH + B^- \rightleftharpoons Ph_2CHO^- + BH$$
$$Ph_2CHO^- + B^- \rightleftharpoons Ph_2C(O^-)^- + BH$$
$$Ph_2C(O^-)^- + O_2 \longrightarrow Ph_2C(O) + O_2^{2-}$$

The second stage:

$$K_2O_2 + O_2 \rightleftharpoons 2KO_2$$

The oxidation of alcohols in a basic solution catalyzed by Cu(II) o-phenanthroline complexes has been recently studied by Sakharov and Skibida [305–309]. The copper–phenanthroline complex is stable in a basic solution and appears to be a very efficient catalyst for the oxidation of alcohols to carbonyl compounds. The reaction rate increases with an increase in the partial pressure of dioxygen. The solvent dramatically influences the reaction rate (conditions: 348 K, [MeOH] = 20%vol, [Cu—(o—phm)] = 0.01 mol L^{-1}).

Solvent	H_2O	MeOH	C_6H_6	MeCN	$HC(O)NMe_2$
v (mol L^{-1} s^{-1})	1.0×10^{-6}	7.0×10^{-6}	5.1×10^{-5}	4.0×10^{-4}	7.0×10^{-4}

The authors proposed the one-step oxidation of alcohol in the ligand sphere of the Cu(II)·(o-phen)$_2$ complex [309].

$$R_2CHOH + B^- \rightleftharpoons R_2CHO^- + B$$

$$R_2CHO^- + [\text{Cu(o-phen)}_2] \rightleftharpoons \text{Complex I}$$

$$R_2CHO^- + [\text{Cu(o-phen)}_2] \rightleftharpoons \text{Complex (I)}$$

$$\text{Complex(I)} + O_2 \rightleftharpoons \text{Complex(II)}$$
$$\text{Complex(II)} \longrightarrow R_2C(O) + HO_2^- + Cu(I)$$

$$R_2CHO-Cu-O_2 \longrightarrow [\text{Cu(o-phen)}_2] + R_2C(O) + HO_2^-$$

However, consecutive multistage oxidation seems to be the most probable. It can include, for example, the stages:

$$R_2CHO-Cu(II) \longrightarrow R_2C(O) + H^+ + Cu(I)$$
$$Cu(I) + O_2 \longrightarrow Cu(II) + O_2^{-\bullet}$$
$$2O_2^{-\bullet} \longrightarrow O_2 + O_2^{2-}$$
$$BH + O_2^{2-} \longrightarrow B^- + HO_2^-$$

The disproportionation of the superoxide ion occurs very rapidly, and the formed hydrogen peroxide is promptly decomposed in a basic solution.

REFERENCES

1. C Walling. *Free Radicals in Solution*, New York: Wiley, 1962.
2. NN Semenov. *Some Problems of Chemical Kinetics and Reactivity*, vols 1 and 2. London: Pergamon Press, 1958–1959.
3. N Uri. In: WO Lundberg (eds.). *Autoxidation and Antioxidants*, vol 1. New York: Wiley-Interscience, 1961, pp 55–106.

4. ET Denisov, NM Emanuel. *Usp Khim* 29:1409–1438, 1960.
5. NM Emanuel, ET Denisov, ZK Maizus. *Liquid-Phase Oxidation of Hydrocarbons*. New York: Plenum Press, 1967.
6. IV Berezin, ET Denisov, NM Emanuel. *The Oxidation of Cyclohexane*. Oxford: Pergamon Press, 1966.
7. AE Semenchenko, VM Solyanikov, ET Denisov. *Zh Fiz Khim* 47:1148–1151, 1973.
8. NM Emanuel, ZK Maizus, IP Skibida. *Angew Chem* 8:91–101, 1969.
9. IP Skibida. *Usp Khim* 44:1729–1747, 1975.
10. F Haber, R Willstatter. *Chem Ber* 64B:2844–2852, 1931.
11. F Haber, J Weiss. *Naturwissenschaften* 20:948–954, 1932.
12. F Haber, J Weiss. *J Proc Roy Soc* (*Lond*), A147:332–341, 1934.
13. JH Baxendale, MG Evans, GS Park. *Trans Faraday Soc* 42:155, 1946.
14. JH Baxendale. In: WG Frankenburg, VI Komarewsky, EK Rideal (eds). *Advances of Catalysis*, vol 4. London: Academic Press, 1952, p 31.
15. J Weiss. In: WG Frankenburg, VI Komarewsky, EK Rideal (eds). *Advances of Catalysis*, vol 4. London: Academic Press, 1952, p 343.
16. AYa Sychev, SO Travkin, GG Duka, YuI Skurlatov. *Catalytic Reactions and Protection of Environment*. Kishinev: Shtiintsa, 1983 [in Russian].
17. AYa Sychev, VG Isak. *Usp Khim* 64:1183–1209, 1995.
18. B Halliwell, JMC Gutteridge. *Free Radicals in Biology and Medicine*, 2nd ed. Oxford: Clarendon Press, 1989.
19. W Ando, Y Moro-Oka eds. *The Role of Oxygen in Chemistry and Biochemistry*. Amsterdam: Elsevier, 1988.
20. C Reddy, GA Hamilton, KM Madyastha (eds). *Biological Oxidation Systems*. New York: Academic Press, 1990.
21. H Sigel, A Sigel (eds). *Bioinorganic Catalysis*. New York: Marcel Dekker, 1992.
22. T Funabiki (ed.). *Oxygenases and Model Systems*. Dordrecht: Kluwer, 1997.
23. DR Lide (ed.). *Handbook of Chemistry and Physics*, 72nd ed. Boca Raton: CRC Press, 1991–1992.
24. B Chance. *J Franklin Inst* 229:737–742, 1940.
25. G Czapski, BHJ Bielski, N Sutin. *J Phys Chem* 67:201–203, 1963.
26. VM Berdnikov, YuM Kozlov, AP Purmal. *Khim Vysokikh Energiy* 3:321–324, 1969.
27. IM Kolthoff, R Woods. *J Electroanal Chem Interf Chem* 12:385, 1966.
28. R Woods, I M Kolthoff, E J Meehan. *J Am Chem Soc* 85:3334–3337, 1963.
29. WC Barb, JH Baxendale, P George, KR Hargrave. *Trans Faraday Soc* 47:462, 1951.
30. CF Wells. *J Chem Soc* 2741–248, 1969.
31. FS Dainton, HC Sutton. *Trans Faraday Soc* 49:1011–1019, 1953.
32. JP Keene. *Rad Res* 22:14–21, 1964.
33. Z Pospisil. *Coll Czech Chem Commun* 18:337–342, 1953.
34. T Rigg, W Taylor, J Weiss. *J Chem Phys* 22:575–581, 1954.
35. T J Hardwick. *Can J Chem* 35:428–235, 1957.
36. CF Wells, MA Salam. *Trans Faraday Soc* 63:620, 1967.
37. IA Kulikov, VS Koltunov, VI Marchenko, AS Milovranova, LK Nikishova. *Zh Fiz Khim* 53:647–651, 1979.
38. H Po, R Sutin. *Inorg Chem* 7:621, 1968.
39. TL Conoechioli, EJ Hamilton, N Sutin. *J Am Chem Soc* 87:926–927, 1965.
40. CF Wells, MA Salam. *Nature* (*Lond*) 203:751–754, 1964.
41. CF Wells, MA Salam. *Nature* (*Lond*) 205:690–694, 1965.
42. C Walling. *Acc Chem Res* 8:125–133, 1975.
43. WC Barb, JH Baxendale, P George, KR Hargrave. *Trans Faraday Soc* 51:935–944, 1955.
44. G Davies, LJ Kirschenbaum, K Kustin. *Inorg Chem* 7:146–152, 1968.
45. FB Baker, TW Newton. *J Phys Chem* 65:1897–1909, 1961.
46. C Bawn, J Williamson. *Trans Faraday Soc* 47:735–743, 1951.
47. C Bawn, J Jolley. *Proc Roy Soc* A273:297–305, 1956.
48. C Bawn, TP Hobin, L Rafphael. *Proc Roy Soc* A273:313–318, 1956.

49. C Bawn. *Dis Faraday Soc N* 14:181–194, 1953.
50. ET Denisov, NI Mitskevich, VE Agabekov. *Liquid-Phase Oxidation of Oxygen-Containing Compounds*. New York: Consultans Bureau, 1977.
51. L Sajus, I Seree de Roch. In: CH Bamford and CFH Tipper (eds). *Comprehensive Chemical Kinetics*, vol 16. Amsterdam: Elsevier, 1980, pp 89–124.
52. WH Richardson. *J Am Chem Soc* 87:247–255, 1965.
53. K Omuta, K Wada, J Yamashita, H Hashimoto. *Bull Chem Soc Jpn* 40:2000–2010, 1967.
54. R Hiatt, KC Irwin, CW Gould. *J Org Chem* 33:1430–1435, 1968.
55. WH Richardson. *J Am Chem Soc* 87:1096–1102, 1965.
56. H Hock, H Kropf. *J Pract Chem* 16:113–122, 1962.
57. H Kropf. *Ann* 637:111–120, 1960.
58. MT Lisovska, VI Timokhin, AP Pokutsa, VI Kopylets. *Kinet Katal* 41:223–232, 2000.
59. Y Kamiya, S Beaton, A Lafortune, KU Ingold. *Can J Chem* 41:2034–2053, 1963.
60. AG Korsun, VYa Shlyapintoch, NM Emanuel. *Izv AN SSSR Okhn* 788–794, 1961.
61. J Tomiska. *Coll Czech Chem Commun* 27:1549–1558, 1962.
62. WL Reynolds, RJW Lumry. *J Chem Phys* 23:2460–2459, 1955.
63. IM Kolthoff, WL Reynolds. *Disc Faraday Soc* 17:167–175, 1954.
64. WL Reynolds, IM Kolthoff. *J Phys Chem* 60:969–976, 1956.
65. JWL Fordham, HL Williams. *J Am Chem Soc* 73:1634–1637, 1951.
66. JWL Fordham, HL Williams. *J Am Chem Soc* 72:4465–4469, 1950.
67. RJ Orr, HL Williams. *Can J Chem* 30:985–992, 1952.
68. RJ Orr, HL Williams. *J Phys Chem* 57:925–931, 1953.
69. DG Knorre, ZK Maizus, MI Markin, NM Emanuel. *Zh Fiz Khim* 33:398–405, 1959.
70. IP Skibida. Homogeneous Catalysis by Compounds of Transition Metals in Liquid-Phase Oxidation by Molecular Oxygen. Doctoral Thesis Dissertation. RUDN, Moscow, 1997.
71. GM Bulgakova, IP Skibida, EG Rukhadze, GP Talyzenkova. *Kinet Katal* 12:359–363, 1971.
72. BG Balkov, IP Skibida, ZK Maizus. *Izv AN SSSR Ser Khim N* 7:1470–1474, 1969.
73. BG Balkov, IP Skibida, ZK Maizus. *Izv AN SSSR Ser Khim N* 7:1475–1480, 1969.
74. BG Balkov, IP Skibida, ZK Maizus. *Izv AN SSSR Ser Khim N* 8:1780–1785, 1970.
75. GF Pustarnakova, VM Solyanikov. *Neftekhimiya* 15:124–129, 1975.
76. GF Pustarnakova, VM Solyanikov. *React Kinet Lett* 2:313–318, 1975.
77. VM Goldberg, LK Obukhova. *Progress in Chemistry of Organic Peroxides and Autoxidation*. Moscow: Khimiya, 1969, p.160 [in Russian].
78. LI Matienko, LA Goldina, IP Skibida, ZK Maizus. *Izv AN SSSR Ser Khim* 287–292, 1975.
79. ZK Maizus, IP Skibida, NM Emanuel. *Dokl AN SSSR* 164:374–377, 1965.
80. ET Denisov. *Mechanisms of Homolytic Decomposition of Molecules in the Liquid Phase* [*Itogi Nauki i Tekhniki, Kinetika i Kataliz*], vol 9. Moscow: VINITI, 1981, pp 1–158 [in Russian].
81. GF Pustarnakova, VM Solyanikov, ET Denisov. *Izv AN SSSR Ser Khim* 547–552, 1975.
82. G Czapski, BHJ Bielski, N Sutin. *J Phys Chem* 67:201–203, 1963.
83. RW Matthews. *Aust J Chem* 37:475–488, 1984.
84. AE Cahill, H Taube. *J Am Chem Soc* 74:2312–2318, 1952.
85. M Anbar, P Neta. *Int J Appl Radiat Isot* 18:493–552, 1967.
86. GV Buxton, RM Sellers, DR McCracken. *J Chem Soc Faraday Trans 1*, 72:1464–1476, 1976.
87. B Jezowska-Trzebiatowska, E Kalecinska, J Kalecinski. *Bull Acad Pol Sci Ser Sci Chim* 19:265–275, 1971.
88. H Boncher, AM Sargeson, DF Sangster, JC Sullivan. *Inorg Chem* 20:3719–3726, 1981.
89. KVS Rama Rao, LV Shastry, J Shankar. *Radiat Eff* 2:193–202, 1970.
90. NL Sukhov, MA Akinshin, BG Ershov. *Khim Vysokikh Energiy* 20:392–396, 1986.
91. GV Buxton, CL Greenstock, WP Helman, AB Ross. *J Phys Chem Ref Data* 17:513–886, 1988.
92. JH Baxendale, EM Fielden, JP Keene. *Pulse Radiolysis*. London: Academic Press, 1965, p 217.
93. A Samumi, D Meisel, G Czapski. *J Chem Soc Dalton Trans* 1273–1282, 1972.
94. E Hayon, M Moreau. *J Chim Phys* 391–394, 1965.
95. B Jezowska-Trzebiatowska, E Kalecinska, J Kalecinski. *Bull Acad Pol Sci Ser Sci Chim* 17:225–231, 1969.

96. DM Brown, ES Dainton, DC Walker, JP Keene. *Pulse Radiolysis.* London: Academic Press, 1965, p 221
97. D Bunn, FS Dainton, GA Salmon, TJ Hardwick. *Trans Faraday Soc* 55:1760–1767, 1959.
98. HA Schwarz. *J Phys Chem* 66:255–261, 1962.
99. AK Pikaev, PYa Glazunov, VI Spitsyn. *Izv AN SSSR Ser Khim* 401–408, 1965.
100. B Chutny. *Coll Czech Chem Commun* 31:358–361, 1966.
101. H Christensen, K Schested. *Radiat Phys Chem* 18:723–731, 1981.
102. Z Stuglik, ZP Zagorski. *Radiat Phys Chem* 17:229–233, 1981.
103. GG Jayson, BJ Parsons, AJ Swallow. *J Chem Soc Faraday Trans 1*, 68:2053–2058, 1972.
104. D Zehavi, J Rabani. *J Phys Chem* 75:1738–1744, 1971.
105. JSB Park, PM Wood, BC Gilbert, AC Whitwood. *J Chem Soc Perkin Trans 2*, 923–931, 1999.
106. J Rabani, G Stein. *Trans Faraday Soc* 58:2150–2159, 1962.
107. JK Thomas. *Trans Faraday Soc* 61:702–707, 1965.
108. RW Mattews, DF Sangster. *J Phys Chem* 69:1938–1944, 1965.
109. I Kraljic, CN Trumbore. *J Am Chem Soc* 87:2547–2550, 1965.
110. J Rabani, MS Matheson. *J Am Chem Soc* 86:3175–3176, 1964.
111. PC Beaumont, EL Powers. *Int J Radiat Biol Relat Stud Phys Chem Med* 43:485–494, 1983.
112. JW Boyles, S Weiner, CJ Hochanadel. *J Phys Chem* 63:892–895, 1959.
113. TJ Sworski. *J Am Chem Soc* 79:3655–3657, 1957.
114. D Meisel, YA Ilan, GJ Czapski. *J Phys Chem* 78:2330–2334, 1974.
115. YuN Kozlov, VM Berdnikov. *Russ J Phys Chem* 47:338–340, 1973.
116. VM Berdnikov, YuM Kozlov, AP Purmal. *Khim Vysokikh Energiy* 3:321–324, 1969.
117. JH Baxendale. *Rad Res* 17:312–319, 1962.
118. J Rabani, D Klug-Roth, JJ Lilic. *J Phys Chem* 77:1169–1177, 1973.
119. D Zehavi, J Rabani. *J Phys Chem* 76:3703–3709, 1972.
120. GG Jayson, BJ Parsons, AJ Swallow. *J Chem Soc Faraday Trans 1*, 69:236–242, 1973.
121. AD Nadezhdin, YuN Kozlov, AP Purmal. *Zh Fiz Khim* 50:910–912, 1976.
122. JD Rush, BHJ Bielski. *J Phys Chem* 89:5062–5066, 1985.
123. GG Jayson, JP Keene, DA Stirling, AJ Swallow. *Trans Faraday Soc* 65:2453–2462, 1969.
124. K Sehested, E Bjergbukke, OL Rasmussen, H Fricke. *J Chem Phys* 51:3159–3166, 1969.
125. J Pucheault, C Ferradini, A Buu. *Int J Radiat Phys Chem* 1:209–218, 1969.
126. AO Allen, VD Hogan, WG Rotschild. *Rad Res* 7:603–608, 1957.
127. JP Keene. *Rad Res* 17:14–22, 1962.
128. M Pic-Kaplan, J Rabani. *J Phys Chem* 80:1840–1843, 1976.
129. DE Cabelli, BHJ Bielski. *J Phys Chem* 88:3111–3115, 1984.
130. B Cercek, M Ebert, AJ Swallow. *J Chem Soc A*, 612–615, 1966.
131. EM Tochina, LM Postnikov, VYa Shlyapintokh. *Izv AN SSSR Ser Khim* 1489–1493, 1968.
132. LM Postnikov, EM Tochina, VYa Shlyapintokh. *Dokl AN SSSR* 172:651–654, 1967.
133. VM Goldberg, LK Obukhova. *Dokl AN SSSR* 165:860–863, 1965.
134. VP Scheredin, ET Denisov. *Izv AN SSSR Ser Khim* 1428–1431, 1967.
135. J Butler, GG Jayson, AJ Swallow. *J Chem Soc Faraday Trans 1*, 70:1394–1401, 1974.
136. CL Jenkins, JK Kochi. *J Am Chem Soc* 94:843–855, 1972.
137. JK Kochi, A Bemis, CL Jenkins. *J Am Chem Soc* 90:4616–4625, 1968.
138. VM Berdnikov, OS Zhuravleva, LA Terentyeva. *Bull Acad Sci USSR Div Chem Sci* 26:2050, 1977.
139. GV Buxton, JC Green. *J Chem Soc Faraday Trans 1*, 74:697–714, 1978.
140. AG Pribush, SA Brusentseva, VN Shubin, PI Dolin. *High Energy Chem* 9:206–208, 1975.
141. E Collinson, FS Dainton, B Mile, S Tazuke, DR Smith. *Nature* 198:26–29, 1963.
142. E Collinson, FS Dainton, GS MeNaughton. *Trans Faraday Soc* 53:189–196, 1957.
143. E Collinson, FS Dainton, DR Smith, GJ Trudel, S Tazuke. *Dis Faraday Soc* 29:188–199, 1960.
144. FS Dainton, WD Sisley. *Trans Faraday Soc* 59:1377–1388, 1963.
145. KW Chambers, E Collinson, FS Dainton, WA Seddon, F Wilkiason. *Trans Faraday Soc* 63:1699–1708, 1967.
146. CH Bamford, AD Jenkina, R Johnston. *Proc Roy Soc* A239:214–223, 1957.
147. N Colebourne, E Collinson, DJ Currie, FS Dainton. *Trans Faraday Soc* 59:1357–1364, 1963.

148. WI Bengough, WH Fairservice. *Trans Faraday Soc* 61:1206–1213, 1965.
149. RM Davydov. *Zh Fiz Khim* 42:2639–2645, 1968.
150. ET Denisov. *Russ Chem Rev* 65:505–520, 1996.
151. ET Denisov, TG Denisova. *Handbook of Antioxidants*. Boca Raton: CRC Press, 2000.
152. VM Goldberg, LK Obukhova, NM Emanuel. *Neftekhimiya* 2:220–228, 1962.
153. VM Goldberg, LK Obukhova. *Neftekhimiya* 4:466–472, 1964.
154. AV Oberemko, AA Perchenko, ET Denisov, AL Aleksandrov. *Neftekhimiya* 11:229–235, 1971.
155. ET Denisov. *Liquid-Phase Reaction Rate Constants*. New York: IFI/Plenum, 1974.
156. AE Shilov, GB Shulpin. *Activation and Catalytic Reactions of Saturated Hydrocarbons in the Presence of Metal Complexes*. Dordrecht: Kluwer Academic, 2000.
157. LK Lepin, BP Matseevskii. *Dokl AN SSSR* 173:1336–1338, 1967.
158. EA Kutner, BP Matseevskii. *Kinet Katal* 10:997–1003, 1969.
159. H Mimoun, I Seree de Roch. *Tetrahedron* 31:777–786, 1975.
160. IV Zakharov, EI Karasevich, AE Shilov, AA Shteinman. *Kinet Katal* 16:1151–1157, 1975.
161. YuV Geletii, IV Zakharov, EI Karasevich, AA Shteinman. *Kinet Katal* 20:1124–1130, 1979.
162. VV Lavrushko, AM Khenkin, AE Shilov. *Kinet Katal* 21:276–278, 1980.
163. AE Martell, M Calvin. *Chemistry of the Metal Chelate Compounds*. New York: Prentice Hall, 1962.
164. D Burk, J Hearon, L Caroline. *J Biol Chem* 165:723–729, 1946.
165. L Vaska. *Science* 140:809–813, 1963.
166. JA McGinnety, RJ Doedens, JA Ibers. *Inorg Chem* 6:2243–2250, 1967.
167. KR Laing, WR Roper. *Chem Commun* 1556–1562, 1968.
168. S Otsuka, A Nakamura, Y Tatsumo. *Chem Commun* 836–839, 1967.
169. LN Denisova, ET Denisov, TG Degtyareva. *Izv AN SSSR Ser Khim* 1095–1097, 1966.
170. IP Skibida. *Usp Khim* 54:1487–1504, 1985.
171. LI Matienko, ZK Maizus. *Kinet Katal* 15:317–322, 1974.
172. A Nishinaga, H Tomita. *J Molek Catal* 7:179–186, 1980.
173. BM Berdnikov, LI Makarshina, LS Ryvkina. *React Kinet Catal Lett* 9:281–285, 1978.
174. LS Ryvkina, BM Berdnikov. *React Kinet Catal Lett* 21:409–416, 1982.
175. J Simplicio, RG Wilkins. *J Am Chem Soc* 89:6092–6095, 1967.
176. C Tanford, DC Kirk, MK Chantoosi. *J Am Chem Soc* 76:5325–5332, 1954.
177. H Nord. *Acta Chem Scand* 9:430–438, 1955.
178. A Zuberbuhler. *Helv Chim Acta* 50:466–471, 1967.
179. J Pecht, M Anbar. *J Chem Soc A*, 1902–1904, 1968.
180. M Cher, N Davidson. *J Am Chem Soc* 77:793–798, 1955.
181. DI Metelitsa, ET Denisov. *Kinet Katal* 9:733–741, 1968.
182. AM Posner. *Trans Faraday Soc* 49:382–390, 1953.
183. P George. *J Chem Soc* 4349–4359, 1954.
184. S Utsumi, K Murosima. *J Chem Soc Jpn Pure Chem Soc* 86:503–511, 1965.
185. RE Huffman, N Davidson. *J Am Chem Soc* 78:4836–4842, 1956.
186. AB Lamb, LW Elder. *J Am Chem Soc* 53:137–144, 1931.
187. GS Hammond, C-HS Wu. International Oxidations Symposium. San Francisco 2:413, 1967.
188. EA Kutner. PhD Thesis Dissertation. Riga, 1968.
189. JB Ramsey, R Sugimoto, HDe Vorkin. *J Am Chem Soc* 63:3480–3488, 1941.
190. VD Komissarov, ET Denisov. *Neftekhimiya* 8:595–603, 1968.
191. VD Komissarov, ET Denisov. *Neftekhimiya* 7:420–426, 1967.
192. VD Komissarov, ET Denisov. *Zh Fiz Khim* 43:769–771, 1969.
193. VD Komissarov, ET Denisov. *Zh Fiz Khim* 44:390–395, 1970.
194. VD Komissarov, ET Denisov. *Kinet Katal* 10:513–518, 1969.
195. RW Fisher, F Rohrscheid. In: B Cornils, WA Herrmann (eds). *Applied Homogeneous Catalysis with Organometallic Compounds*. Weinheim: VCH, 1996, p 439.
196. FF Rust, WE Vaugan. *Ind Eng Chem* 41:2595–2602, 1949.
197. DAJ Ravens. *Trans Faraday Soc* 55:1768–1776, 1959.
198. AS Hay, HS Blanchard. *Can J Chem* 43:1306–1317, 1965.
199. Y Kamiya. *Tetrahedron* 22:2029–2036, 1966.

200. Y Kamiya, T Naakayama, K Sakoda. *Bull Chem Soc Jpn* 39:2211–2218, 1966.
201. IV Zakharov, LA Balanov, OG Popova. *Dokl AN SSSR* 190:1132–1135, 1970.
202. IV Zakharov, LA Balanov. *Dokl AN SSSR* 193:851–854, 1970.
203. IV Zakharov, VM Muratov. *Dokl AN SSSR* 196:1125–1128, 1971.
204. IV Zakharov, VM Muratov. *Dokl AN SSSR* 200:371–374, 1971.
205. IV Zakharov, VA Sukharev. *Dokl AN SSSR* 204:626–629, 1972.
206. IV Zakharov. *Kinet Katal* 15:1457–1465, 1974.
207. IV Zakharov, YuV Geletii. *Dokl AN SSSR* 217:852–855, 1974.
208. IV Zakharov, YuV Geletii. *Neftekhimiya* 18:615–621, 1978.
209. EJY Scott, AW Chester. *J Phys Chem* 76:1520–1523, 1972.
210. GS Bezhanishvili, NG Digurov, NN Lebedev. *Kinet Katal* 24:1000–1002, 1983.
211. NM Emanuel, D Gal. *Oxidation of Ethylbenzene*. Moscow: Nauka, 1984 [in Russian].
212. M Herustiak, M Hronec, J Ilavsky. *J Mol Catal* 53:209–218, 1989.
213. GM Dugmore, GJ Powels, B Zeelie. *J Mol Catal A: Chem* 99:1–8, 1995.
214. MFT Gomez, OAC Antunes. *J Mol Catal A: Chem* 111:145–152, 1997.
215. WC Bray, AL Caulkins. *J Am Chem Soc* 43:1262–1269, 1921.
216. DTA Townend, M Mandlekar. *Proc Roy Soc A* 141:484, 1933.
217. BP Belousov. *Sbornik Ref Radiats Meditsin*. Moscow: Medgiz, 1958:145, 1959.
218. AM Zhabotinsky. *Dokl AN SSSR* 157:392–395, 1964.
219. RJ Field, E Koros, RM Noyes. *J Am Chem Soc* 94:8649–8664, 1972.
220. KR Sharma, RM Noyes. *J Am Chem Soc* 98:4345–4360, 1976.
221. RM Noyes, SD Furrow. *J Am Chem Soc* 104:45–49, 1982.
222. P DeKepper, IR Epstein. *J Am Chem Soc* 104:49–55, 1982.
223. O Citri, IR Epstein. *J Am Chem Soc* 108:357–363, 1986.
224. AJ Lotka. *J Am Chem Soc* 42:1595, 1920.
225. ET Denisov, OM Sarkisov, GI Likhtenshtein. *Chemical Kinetics*. Amsterdam: Elsevier, 2003.
226. JM Tedder, A Nechvatal, AH Jubb. *Basic Organic Chemistry*. London: Wiley, 1975.
227. VM Solyanikov, ET Denisov. *Dokl AN SSSR* 173:1106–1109, 1967.
228. ET Denisov, VM Solyanikov, AL Aleksandrov. *Adv Chem Ser* 75:112–119, 1967.
229. VM Solyanikov, ET Denisov. *Izv AN SSSR Ser Khim* 1391–1393, 1968.
230. VM Solyanikov, LV Petrov, KhE Kharlampidi. *Dokl AN SSSR* 223:1412–1415, 1975.
231. VM Solyanikov, ET Denisov. *Neftekhimiya* 9:116–123, 1969.
232. LV Petrov, VM Solyanikov. *Izv AN SSSR Ser Khim* 1535–1541, 1980.
233. LV Petrov, VM Solyanikov, ZYa Petrova. *Dokl AN SSSR* 230:366–369, 1976.
234. GA Tolstikov. *Reactions of Hydroperoxide Oxidation*. Moscow: Nauka, 1970 [in Russian].
235. NA Prilezhaev. *Organic Peroxides and their Application for Oxidation of Unsaturated Compounds*. Warsaw, 1912 [in Russian].
236. E Hawkins. *J Chem Soc* 2169–2177, 1950.
237. British Patent 1074330, 1963, Halcon Co; C A 62:16192, 1965.
238. British Patent 1130231, 1964, Halcon Co; C A 65:10565, 1966.
239. N Indictor, W Brill. *J Org Chem* 30:2074–2082, 1965.
240. ES Gould, RR Hiatt, KC Irvin. *J Am Chem Soc* 90:4573–4579, 1968.
241. VI Sapunov, I Margitfalvi, NN Lebedev. *Kinet Katal* 15:1178–1186, 1974.
242. R Sheldon. *Rec Trav Chim* 92:253–262, 1973.
243. H Mimoun, I Seree de Roch, L Sajus. *Bull Soc Chim Fr* 1481–1492, 1969.
244. H Mimoun, I Seree de Roch, L Sajus. *Tetrahedron* 26:37–40, 1970.
245. J Smidt, W Hafner, R Jira, J Sedlmeier, R Sieber, R Ruttinger, H Kojer. *Angew Chem* 71:176–182, 1952.
246. II Moiseev, MN Vargaftik, YaK Syrkin. *Dokl AN SSSR* 130:820–823, 1960.
247. II Moiseev, π-*Complexes in Liquid-Phase Oxidation of Olefins*. Moscow: Nauka, 1970 [in Russian].
248. J Smidt, R Sieber. *Angew Chem* 71:626–627, 1959.
249. II Moiseev, MN Vargaftik, YaK Syrkin. *Dokl AN SSSR* 133:377–380, 1960.
250. GT Van Gemert, PR Wilkinson. *J Phys Chem* 68:645–647, 1964.
251. MN Vargaftik, II Moiseev, YaK Syrkin. *Dokl AN SSSR* 147:399–402, 1962.

252. II Moiseev, MN Vargaftic, YaK Syrkin. *Dokl AN SSSR* 153:140–143, 1963.
253. H Hock, H Kroph. *J Prakt Chem* 6(2):123–132, 1958.
254. H Hock, H Kroph. *J Prakt Chem* 9 (3):173–181, 1959.
255. EA Blumberg, YuD Norikov. *Heterogeneous Catalysis and Inhibition of Reactions of Liquid-Phase Oxidation of Organic Compounds* [Itogi Nauki i Tekhniki, Kinetika i Kataliz], vol 12. Moscow: VINITI, 1984, pp 1–143 [in Russian].
256. YaB Gorokhovatsky, TP Kornienko, VVShalya. *Heterogeneous–Homogeneous Reactions*. Kiev: Tekhnika, 1972, pp 1–22 [in Russian].
257. AYa Valendo, YuD Norikov, EA Blumberg. *Neftekhimiya* 9:866–872, 1969.
258. AYa Valendo, YuD Norikov, EA Blumberg, NM Emanuel. *Dokl AN SSSR* 201:1378–1381, 1971.
259. YaB Gorokhovatsky, NP Evmenenko, MV Kost, VA Khizhnyi. *Zh Teoret Experim Khimii* 9:373–375, 1973.
260. AYa Valendo, YuD Norikov. *Izv AN SSSR Ser Khim* 1972:275–279.
261. I Labody, LI Korablev, LA Tavadyan, EA Blumberg. *Kinet Katal* 23:371–375, 1982.
262. GP Khirnova, MG Bulygin, EA Blumberg. *Neftelhimiya* 21:250–253, 1981.
263. LV Salukvadze, YuD Norikov, VI Naumenko. *Zh Fiz Khim* 46:3113–3117, 1972.
264. YuD Norikov, LV Salukvadze. *Dokl AN SSSR* 203:632–635, 1972.
265. JHR Gasemier, BE Neiuwenhuys, WMH Sachtler. *J Catal* 29:367–373, 1973.
266. YuD Norikov, EA Blumberg, ShK Bochorishvili, MD Irmatov. *Dokl AN SSSR* 223:1187–1190, 1975.
267. YuD Norikov, LV Salukvadze. NM Emanuel (eds). *Theory and Technology of Liquid-Phase Oxidation*. Moscow: Nauka, 1974, p. 197.
268. H Kropf, SK Ivanov, SK Diereks. *Liebigs Ann Chem* 2046–2052, 1974.
269. PK Tarasenko, SA Borisenkova, SA Novikov, GL Popova, AP Rudenko. *Vestnik MGU Ser Khim* 20:380–381, 1979.
270. EF Pratt, SP Suskind. *J Org Chem* 28:638–642, 1963.
271. NP Evmenenko, YaB Gorokhovatskii, VF Tsepalov. *Neftekhimiya* 10:226–229, 1970.
272. AK Agarwal, RD Srivastava. *J Catal* 45:86–99, 1976.
273. DL Allara, RF Roberts. *J Catal* 45:54–67, 1976.
274. NT Silakhtaryan, LV Salukvadze, YuD Norikov, EA Blumberg, NM Emanuel. *Kinet Katal* 23:77–83, 1982.
275. SV Mikhalovsky, NP Evmenenko, YaB Gorokhovatsky. *Teoret Exp Khim* 17:337–347, 1981.
276. HJ Neuburg, MJ Philips, WF Graydon. *J Catal* 38:33–46, 1975.
277. HJ Neuburg, JM Bassel, WF Graydon. *J Catal* 25:425–427, 1972.
278. GR Varma, WF Graydon. *J Catal* 28:236–244, 1979.
279. A Sadana. *Und Eng Chem Process Dev* 18:50–56, 1979.
280. S Fukuzumi, Y Ono. *Bull Chem Soc Jpn* 52:2255–2260, 1979.
281. DL Allara, D Edelson. *J Phys Chem* 81:2443–2451, 1977.
282. K Takehira, T Hayakawa, T Izkikawa. *J Catal* 66:267–280, 1980.
283. T Fukushima, A Ozaki. *J Catal* 59:460–464, 1979.
284. GA Russell, AG Bemis, EJ Geels,EG Janzen, AJ Moyc. *Adv Chem Ser* 75:174–202, 1968.
285. GA Russell, H-D Becker, J Schoeb. *J Org Chem* 28:3584–3593, 1964.
286. GA Russell, AG Bemis. *Inorg Chem* 6:403–409, 1967.
287. GA Russell, AG Bemis. *J Am Chem Soc* 88:5491–5497, 1966.
288. GA Russell, EG Janzen, H-D Becker, EJ Smentowski. *J Am Chem Soc* 84:2652–2653, 1962.
289. GA Russell, EG Janzen, AG Bemis, EJ Geels, AJ Moye, S Mak, ET Strom. *J Am Chem Soc* 84:4155–4157, 1962.
290. GA Russell, EG Janzen, AG Bemis, EJ Geels, AJ Moye, S Mak, ET Strom. *J Am Chem Soc* 86:1807–1814, 1964.
291. GA Russell, AJ Moye,EG Janzen, S Mak, ER Talaty. *J Org Chem* 32:137–146, 1967.
292. CB Wooster. *Chem Rev* 11:1–14, 1932.
293. Y Spinzak. *J Am Chem Soc* 80:5449–5455, 1958.
294. MN Shchukina, VG Ermolova, AE Kolmanson. *Dokl AN SSSR* 158:436–440, 1964.
295. EF Pratt, LE Trapasso. *J Am Chem Soc* 82:6405–6408, 1960.

296. PL Pauson, BJ Williams. *J Chem Soc* 4153–4157, 1961.
297. A LeBerre. *Compt Rend* 252:1341, 1961.
298. A LeBerre. *Bull Soc Chim Fr* 1198, 1543–1551, 1961.
299. A LeBerre. *Bull Soc Chim Fr* 1682–1690, 1962.
300. A Etienne, A LeBerre. *Compt Rend* 252:1166–1172, 1961.
301. A Etieme, Y Fellion. *Compt Rend* 238:1429–1435, 1957.
302. T.L. Cairns, B.C. McKusick, V. Weinmayr. *J Am Chem Soc* 73:1270–1281, 1951.
303. JO Howthorne, KA Schowalter, AW Simon, MH Wilt, MS Morgan. *Adv Chem Ser* 75:203–215, 1968.
304. DHR Barton, DW Jones. *J Chem Soc* 3563–3570, 1965.
305. AM Sakharov, IP Skibida. *Oxid Commun* 1:113–120, 1979.
306. AM Sakharov, IP Skibida. *Kinet Katal* 29:110–117, 1989.
307. AM Sakharov, IP Skibida, *J Molek Catal* 48:157–174, 1988.
308. AM Sakharov, IP Skibida. *Stud Surf Sci Catal* 55:221–228, 1990.
309. AM Sakharov, IP Skibida. *J Chem Biochem Kinet* 1:85–92, 1991.

11 Oxidation of Hydrocarbons in Microheterogeneous Systems

11.1 EMULSION OXIDATION OF HYDROCARBONS

Emulsion oxidation of alkylaromatic compounds appeared to be more efficient for the production of hydroperoxides. The first paper devoted to emulsion oxidation of cumene appeared in 1950 [1]. The kinetics of emulsion oxidation of cumene was intensely studied by Kucher et al. [2–16]. Autoxidation of cumene in the bulk and emulsion occurs with an induction period and autoacceleration. The simple addition of water inhibits the reaction [6]. However, the addition of an aqueous solution of Na_2CO_3 or NaOH in combination with vigorous agitation of this system accelerates the oxidation process [1–17]. The addition of an aqueous phase accelerates the oxidation and withdrawal of water retards it [6]. The addition of surfactants such as salts of fatty acids accelerates the oxidation of cumene in emulsion [3]. The higher the surfactant concentration the faster the cumene autoxidation in emulsion [17]. The rates of cumene emulsion oxidation after an induction period are given below ($T = 353$ K, [RH]:[H_2O] = 2:3 (v/v), $pO_2 = 98$ kPa [17]).

System	$v \times 10^5$ (mol L^{-1} s^{-1})
Cumene (pure)	0.45
Cumene + H_2O_2	1.50
Cumene + 0.1 N Na_2CO_3	2.40
Cumene + 0.1 N Na_2CO_3 + 0.001% $C_{15}H_{31}COOH$	2.80
Cumene + 0.1 N Na_2CO_3 + 0.01% $C_{15}H_{31}COOH$	2.90
Cumene + 0.1 N Na_2CO_3 + 0.2% $C_{15}H_{31}COOH$	3.30
Cumene + 0.1 N Na_2CO_3 + 1% $C_{15}H_{31}COOH$	4.30
Cumene + 0.1 N Na_2CO_3 + 4% $C_{15}H_{31}COOH$	9.60

The optimal pH of an aqueous phase was found to be about pH = 10 with 0.1 N Na_2CO_3. Oxidation of cumene occurs more slowly in H_2O (pH = 7) and in 0.1 N NaOH (pH = 13). The experiments on emulsion oxidation of cumene with the addition of salts of various fatty acids Me(CH_2)$_n$COONa ($n = 2$–16) showed that the optimal number of methylene groups of the fatty acid tail is equal to 14 ($T = 348$ K, [H_2O]:[RH] = 1:1 (v/v), $pO_2 = 98$ kPa) [18]. The addition of a polar solvent to the system RH–H_2O–surfactant substance (sodium laurate) accelerates the emulsion oxidation of cumene ($T = 348$ K, [H_2O]:[RH] = 1:1 (v/v), $pO_2 = 98$ kPa, [RH]:[solvent] = 4:1) [18].

Solvent	MeCN	PhCl	PhH	CCl$_4$	C$_8$H$_{18}$
$v \times 10^5$ (mol L^{-1} s^{-1})	2.53	1.40	1.25	0.93	0.87

The emulsion oxidation of cumene proceeds via the chain mechanism. Inhibitors (phenols) retard the oxidation [11], and initiators accelerate the oxidation [17]. Cobaltous stearate accelerates the emulsion oxidation of cumene ($T = 363$ K, [RH]:[H$_2$O] = 1:1 (v/v), 0.1 N Na$_2$CO$_3$ [4]). The oxidation of cumene starts without an induction period. The accelerating action of St$_2$Co on cumene oxidation is sufficiently stronger for homogeneous oxidation. Ozone accelerates the homogeneous cumene oxidation, however, it does not virtually influence emulsion oxidation [4]. The increase in the dioxygen partial pressure higher than 1 kPa does not influence homogeneous cumene oxidation (see Chapter 2). On the contrary, the rate of emulsion oxidation of cumene is accelerated by an increase in pO_2 from 100 to 1000 kPa ($T = 353$–393 K, [RH]:[H$_2$O] = 3:1 (v/v), 0.1 N Na$_2$CO$_3$ [10]). The action of the hydrogen peroxide on the emulsion oxidation of cumene is interesting and unusual [8]. Hydrogen peroxide introduced in the very beginning of cumene oxidation accelerates oxidation very weakly. However, the introduction of 0.015 mol L^{-1} of hydrogen peroxide the moment when the formed hydroperoxide achieved its maximal concentration gives impetus to the new hydroperoxide formation, and its new concentration maximum appears to be sufficiently high (see Figure 11.1, $T = 358$ K, [H$_2$O]:[RH] = 3:1 (v/v), 0.1 N Na$_2$CO$_3$ [8]).

The effect of jumping of the maximal hydroperoxide concentration after the introduction of hydrogen peroxide is caused by the following processes. The cumyl hydroperoxide formed during the cumene oxidation is hydrolyzed slowly to produce phenol. The concentration of phenol increases in time and phenol retards the oxidation. The concentration of hydroperoxide achieves its maximum when the rate of cumene oxidation inhibited by phenol becomes equal to the rate of hydroperoxide decomposition. The lower the rate of oxidation the higher the phenol concentration. Hydrogen peroxide efficiently oxidizes phenol, which was shown in special experiments [8]. Therefore, the introduction of hydrogen peroxide accelerates cumene oxidation and increases the yield of hydroperoxide.

The experiments on emulsion cumene oxidation with AIBN as initiator proved that oxidation proceeds via the chain mechanism inside hydrocarbon drops [17]. The presence of an aqueous phase and surfactants compounds does not change the rate constants of chain propagation and termination: the ratio $(k_p(2k_t)^{-1/2} = const$ in homogeneous and emulsion oxidation (see Chapter 2). Experiments on emulsion cumene oxidation with cumyl hydroperoxide as the single initiator evidenced that the main reason for acceleration of emulsion oxidation versus homogeneous oxidation is the rapid decomposition of hydroperoxide on the surface of the hydrocarbon and water drops. Therefore, the increase in the aqueous phase and introduction of surfactants accelerate cumene oxidation.

The kinetic study of cumyl hydroperoxide decomposition in emulsion showed that (a) hydroperoxide decomposes in emulsion by 2.5 times more rapidly than in cumene (368 K, [RH]:[H$_2$O] = 2:3 (v/v), 0.1 N Na$_2$CO$_3$) and (b) the yield of radicals from the cage in emulsion is higher and close to unity [19]. The activation energy of ROOH decomposition in cumene is $E_d = 105$ kJ mol^{-1} and in emulsion it is lower and equals $E_d = 74$ kJ mol^{-1} [17].

Hence, the peculiarities of emulsion oxidation of alkylaromatic hydrocarbons can be formulated as follows.

1. Oxidation as a chain reaction occurs inside the hydrocarbon drop. Dioxygen should diffuse first in an aqueous phase and, second, inside the hydrocarbon drop through the dense layer of the polar surface of the drop. Therefore, the dioxygen partial pressure should be higher in emulsion than in homogeneous oxidation.

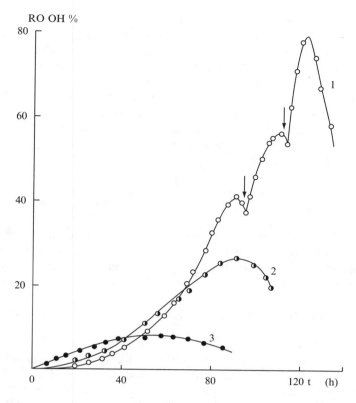

FIGURE 11.1 The kinetic curves of cumyl hydroperoxide formation in emulsion oxidation of cumene [8] at $T = 358\,K$, $H_2O:RH = 3:1$ (v/v) $1\,N$ Na_2CO_3 with input of $0.015\,mol\,L^{-1}$ H_2O_2 in the moments designated by arrows (curve 1), after 8 h (curve 2), and after 4 h (curve 3).

2. The decomposition of hydroperoxides occurs preferentially in the surface layer of water and hydrocarbon. The larger the surface per unit volume of hydrocarbon the faster the decomposition of hydroperoxide. Therefore, the increase in an aqueous phase accelerates hydrocarbon oxidation. The optimal $RH:H_2O$ ratio was found to be nearly 1:1 (v/v) [19], if the calculation of the reaction rate per unit volume of the whole mixture is done. The introduction of surfactants that creates the smaller drops of hydrocarbons increases the surface and, therefore, accelerates the oxidation.
3. Phenol formed in the system due to acid-catalyzed decomposition of hydroperoxide retards the cumene oxidation. The aqueous phase withdraws phenol from the hydrocarbon phase. This is the reason why the emulsion oxidation of cumene helps to increase the yield of hydroperoxide. The addition of hydrogen peroxide into the system helps to increase the yield of hydroperoxide.

11.2 HYDROCARBON OXIDATION IN MICROHETEROGENEOUS SYSTEMS

Hydrocarbon oxidation produces polar oxygen-containing compounds. These compounds tend to associate in nonpolar hydrocarbon and form some aggregates. The formation of such a microheterogeneous polar system can influence the rate and mechanism of oxidation [20]. The study of hydrocarbon oxidation in the systems with reversed micelles has its special interest for the membrane mimetic chemistry [21]. Kasaikina et al. [21–28] have studied the peculiarities of hydrocarbon oxidation in microheterogeneous systems during the last few years [21–28]. They were faced with all possible variants of micelle influence on hydrocarbon oxidation: acceleration, retardation, and neutral behavior (see Table 11.1).

The solubility of water is extremely low in hydrocarbons. For example, as low as 0.7% of water form the separate phase in decane ($T = 298$ K) [22]. The surfactants create a small water micelle in hydrocarbon. For example, sodium bis-2-ethylhexyl sulfosuccinate (AOT) in the ratio $H_2O{:}AOT = 20$ creates stable micelles in decane with the diameter $d = 85 \times 10^{-10}$ m (room temperature) [22]. The radii of a micelle r depends on the ratio $H_2O{:}AOT$; the dependence has the following form [29]:

$$r \times 10^{10} (\text{m}) = 11 + 1.5 \frac{[H_2O]}{[AOT]} \tag{11.1}$$

It is seen from Table 11.1 that surfactant cetyltrimethylammonium bromide (CTAB, $RN^+Me_3Br^-$) exerts a positive catalytic effect on ethylbenzene autoxidation. The kinetic study of this phenomenon [21,27] showed that the acceleration was caused by the additional reaction of hydroperoxide with the bromide ion of CTAB to form free radicals [30].

$$R_1Me_3N^+Br^- + ROOH \longrightarrow R^1N^+Me_3 + OH^- + Br^\bullet + RO^\bullet$$

The active alkoxyl radicals formed by this reaction start new chains. Apparently, the hydroperoxide group penetrates in the polar layer of the micelle and reacts with the bromide anion. The formed hydroxyl ion remains in the aqueous phase, and the $MePhCHO^\bullet$ radical diffuses into the hydrocarbon phase and reacts with ethylbenzene. The inverse emulsion of CTAB accelerates the decay of hydroperoxide MePhCHOOH. The decomposition of hydroperoxide occurs with the rate constant $k_i = 7.2 \times 10^{11} \exp(-91.0/RT)$ L mol^{-1} s^{-1} ($T = 323$–353 K, CTAB, ethylbenzene [28]). The decay of hydroperoxide occurs more rapidly in an O_2 atmosphere, than in an N_2 atmosphere.

The influence of CTAB micelles on dodecane oxidation is opposite: it retards the oxidation of dodecane [24]. The study of the decomposition of hydroperoxide evidenced that

TABLE 11.1
Influence of Reverse Micelles and Surfactant (SA) on the Hydrocarbon Autoxidation

Hydrocarbon	Solvent	T (K)	SA	SA (mmol L^{-1})	Influence	Ref.
Limonene	Decane/limonene	333	AOT	0.92	+	[22]
β-Carotene	$C_{10}H_{24}$	333	AOT	10	0	[21]
Dodecane	$C_{12}H_{26}$	413	AOT	20	0	[24]
Dodecane	$C_{12}H_{26}$	413	SDS	20	−	[24]
Dodecane	$C_{12}H_{26}$	413	CTAB	20	−	[24]
Ethylbenzene	MePhCH$_2$	393	AOT	20	0	[21]
Ethylbenzene	MePhCH$_2$	393	CTAB	10	+	[21]
Ethylbenzene	MePhCH$_2$	393	SDS	10	−	[21]
Ethylbenzene	MePhCH$_2$	393	AOT	20	0	[26]
Ethylbenzene	MePhCH$_2$	393	SDS	10	−	[26]
Ethylbenzene	MePhCH$_2$	393	AOT	20	0	[27]
Ethylbenzene	MePhCH$_2$	393	SDS	10	−	[27]
Ethylbenzene	MePhCH$_2$	393	CTAB	10	+	[27]

CTAB micelles accelerate its decay. However, this decay of hydroperoxide occurs preferentially due to the formation of the molecular product with an extremely low e value (see Chapter 3). This difference depends on the hydroperoxide structure.

The reverse emulsion stabilized by sodium dodecylsulfate (SDS, $ROSO_3^-$ Na^+) retards the autoxidation of dodecane [24] and ethylbenzene [21,26,27]. The basis for this influence lies in the catalytic decomposition of hydroperoxides via the heterolytic mechanism. The decay of hydroperoxides under the action of SDS reverse micelles produces olefins with a yield of 24% ($T = 413$ K, 0.02 mol L^{-1} SDS, dodecane, $[ROOH]_0 = 0.08$ mol L^{-1}) [27]. The thermal decay gives olefins in negligible amounts. The decay of hydroperoxides apparently occurs in the ionic layer of a micelle. Probably, it proceeds via the reaction of nucleophilic substitution in the polar layer of a micelle.

$$R^1OSO_3^- + ROOH \rightleftharpoons R^1OSO_3R + HO_2^-$$
$$R^1OSO_3R \longrightarrow R^1OSO_3H + \text{olefin}$$
$$R^1OSO_3H + H_2O \rightleftharpoons R^1OSO_3^- + H_3O^+$$

The formed hydrogen peroxide decomposes rapidly.

The reverse micelles stabilized by SDS retard the autoxidation of ethylbenzene [27]. It was proved that the SDS micelles catalyze hydroperoxide decomposition without the formation of free radicals. The introduction of cyclohexanol and cyclohexanone in the system decreases the rate of hydroperoxide decay (ethylbenzene, 363 K, $[SDS] = 10^{-3}$ mol L^{-1}, [cyclohexanol] = 0.03 mol L^{-1}, and [cyclohexanone] = 0.01 mol L^{-1} [27]). Such an effect proves that the decay of MePhCHOOH proceeds in the layer of polar molecules surrounding the micelle. The addition of alcohol or ketone lowers the hydroperoxide concentration in such a layer and, therefore, retards hydroperoxide decomposition. The surfactant AOT apparently creates such a layer around water moleculesthat is very thick and creates difficulties for the penetration of hydroperoxide molecules close to polar water. The phenomenology of micellar catalysis is close to that of heterogeneous catalysis and inhibition (see Chapters 10 and 20).

REFERENCES

1. GP Armstrong, RH Hall, DC Quinn. *J Chem Soc* 666–672, 1950.
2. RV Kucher, VD Enaliev, AI Yurzhenko. *Zh Obsch Khim* 27:1774–1779, 1954.
3. RV Kucher, AI Yurzhenko, MA Kovbuz. *Dokl AN SSSR* 117:638–640, 1957.
4. RV Kucher. *Zh Fiz Khim* 33:617–626, 1959.
5. RV Kucher, AI Yurzhenko, MA Kovbuz. *Kolloid Zh* 21:309–314, 1959.
6. RV Kucher, MA Kovbuz. *Zh Fiz Khim* 33:429–436, 1959.
7. RV Kucher, SD Kaz'min, VD Enaliev. *Dokl AN SSSR* 132:1348–1351, 1960.
8. RV Kucher, SD Kaz'min, VD Enaliev. *Zh Fiz Khim* 35:2322–2327, 1961.
9. RV Kucher, SD Kaz'min. *Dokl AN SSSR* 139:1114–1116, 1961.
10. RV Kucher, SD Kaz'min. *Kinet Katal* 2:263–266, 1961.
11. SD Kaz'min, RV Kucher. *Kinet Katal* 2:422–428, 1961.
12. RV Kucher, TM Yuschenko. *Zh Prikl Khim* 35:2068–2073, 1962.
13. NYa Ivanova, AI Yurzhenko, RV Kucher. *Kolloid Zh* 24:178–184, 1962.
14. RV Kucher, SD Kaz'min. *Kinet Katal* 3:31–35, 1962.
15. SD Kaz'min, RV Kucher. *Neftekhimiya* 3:371–375, 1963.
16. RV Kucher., VI Karban. *Chemical Reactions in Emulsion*. Kiev: Naukova Dumka, 1973 [in Russian].
17. NI Solomko, VF Tsepalov, AI Yurzhenko. *Kinet Katal* 9:985–991, 1968.
18. LP Panicheva, EA Turnaeva, SA Panchev, AYa Yuffa. *Neftekhimiya* 38:179–184, 1998.

19. NI Solomko, VF Tsepalov, AI Yurzhenko, AD Shaposhnikova. *Kinet Katal* 10:735–739, 1969.
20. VN Bakunin, GN Kuzmina, OP Parenago. *Neftekhimiya* 37:99–104, 1997.
21. OT Kasaikina, VD Kortenska, ZS Kartasheva, GM Kuznetsova, TV Maksimova, TV Sirota, NV Yanishlieva. *Colloid Surface A: Physicochem Eng Aspects* 149:29–38, 1999.
22. GM Kuznetsova, ZS Kartasheva, OT Kasaikina. *Izv AN Ser Khim* 1682–1685, 1996.
23. ZS Kartasheva, OT Kasaikina. *Izv AN Ser Khim* 1752–1761, 1994.
24. TV Sirota, OT Kasaikina. *Neftekhimiya* 34:467–472, 1994.
25. TV Sirota, NM Evteeva, OT Kasaikina. *Neftekhimiya* 36:169–174, 1996.
26. ZS Kartasheva, TV Maksimova, EV Koverzanova, OT Kasaikina. *Neftekhimiya* 37:153–158, 1997.
27. ZS Kartasheva, TV Maksimova, TV Sirota, EV Koverzneve, OT Kasaikina. *Neftekhimiya* 37:249–253, 1997.
28. LM Pisarenko, TV Maksimova, ZS Kartashova, OT Kasaikina. *Izv AN Ser Khim* 1419–1422, 2003.
29. CJ O'Connor, ED Lomax, RE Ramage. *Adv Colloid Interface Sci* 20:21–28, 1984.
30. ET Denisov. *Izv AN SSSR Ser Khim* 1608–1610, 1967.

12 Sulfoxidation of Hydrocarbons

12.1 INTRODUCTION

The sulfoxidation of aliphatic hydrocarbons is the easiest method for the synthesis of alkylsulfonic acids. Their sodium salts are widely used as surfactive reactants in technology and housekeeping. Platz and Schimmelschmidt [1] were the first to invent this synthetic method. Normal paraffins (C_{14}–C_{18}) are used for the industrial production of alkylsulfonic acids [2–4]. Olefins and alkylaromatic hydrocarbons do not produce sulfonic acids under the action of sulfur dioxide and dioxygen and retard the sulfoxidation of alkanes [5–9].

The reaction occurs according to the stoichiometric equation

$$RH + SO_2 + 0.5O_2 = RSO_3H$$

Sulfoxidation is usually carried out at an atmospheric pressure and the ratio of gaseous reactants $SO_2:O_2 = 2:1$ [1–5]. The temperature of sulfoxidation depends upon the source of initiation. When the reaction is initiated by the UV light or γ-radiation [5,8–11], ozone [12–14], and dichlorine [5], it occurs at room temperature. Sulfoxidation initiated by peroxides or azo-compounds [13,15] occurs at elevated temperatures (320–360 K).

The main products of sulfoxidation of alkanes are alkylsulfonic acids, sulfuric acid, and alkylpolysulfonic acids. The primary unstable product of alkane sulfoxidation is alkylsulfonic peracid (RSO_2OOH). Graf [3] isolated cyclohexyl sulfonic peracid as the product of cyclohexane sulfoxidation. Secondary sulfonic acids are formed preferentially from the normal alkane sulfoxidation [16,17]. Sulfuric acid is formed in parallel with sulfonic acid in the ratio $[H_2SO_4]:[RSO_3H] \approx 0.5$ [2]. The yield of H_2SO_4 decreases and that of polysulfonic acids increases with the depth of the reaction [2]. The sulfoxidation of alkanes proceeds at room temperature via the chain mechanism with short chains. For example, heptane is sulfoxidized at $T = 303$ K and $SO_2:O_2 = 2:1$ with $\nu = 9.0$ [18]. The rate of chain reaction v is proportional to the initiation rate v_i ($v_t \sim$ [radical]), i.e., the chains are terminated linearly [18–22]. The yield of sulfonic acids increases with an increase in pSO_2. The increase of pO_2 increases the rate of sulfoxidation at low pO_2 and decreases at high pO_2 [13,19,21,23]. The temperature dependence of the sulfoxidation rate has a maximum. The sulfoxidation of alkanes initiated photochemically occurs with the maximum rate at $T = 305$–333 K [3]. The sulfoxidation is an exothermic reaction. The heat of decane sulfoxidation is equal to 470 kJ mol^{-1} [15]. For example, the sulfoxidation of decane at $T = 323$ K and $[SO_2] = 2.6$ mol L^{-1} occurs with the following rates at various pO_2:

$[O_2] \times 10^2$ (mol L^{-1})	1.06	1.8	2.7	4.4	5.8	7.3
$v \times 10^5$ (mol L^{-1} s^{-1})	8.2	9.5	8.6	5.7	4.4	3.6

12.2 CHAIN MECHANISM OF SULFOXIDATION

The kinetics of chain sulfoxidation of decane was studied in detail by Komissarov and Saitova [13,21–24]. The reaction was initiated by AIBN and dicyclohexyl-peroxidicarbonate DCHP, $T = 343$ K, $p(SO_2 + O_2) = 97$ kPa, and CCl_4 was used as the solvent. The rate of sulfoxidation was measured volumetrically and found to grow linearly with an increase in the decane concentration for $[RH] \leq 2$ mol L^{-1} and be independent of $[RH]$ for $[RH] \geq 2$ mol L^{-1} in two sets of experiments: (i). $T = 343$ K, $[AIBN] = 2.5 \times 10^{-2}$ mol L^{-1}, $pSO_2 = 32$ kPa, $pO_2 = 32$ kPa, $v_\infty = 4.8 \times 10^{-6}$ mol L^{-1} s^{-1} and (ii). $T = 343$ K, $[AIBN] = 2.5 \times 10^{-2}$ mol L^{-1}, $pSO_2 = 32$ kPa, $pO_2 = 5.1$ kPa, $v_\infty = 2.9 \times 10^{-6}$ mol L^{-1} s^{-1}. The dependence of sulfoxidation on pSO_2 was found to be nonlinear and that on pO_2 is a curve with an extreme v_{max}. The results of three sets of experiments ($T = 343$ K, $pSO_2 = 32$ kPa [13]) are given below.

Initiator	[I] (mol L^{-1})	[RH] (mol L^{-1})	$v_{max} \times 10^6$ (mol L^{-1} s^{-1})
AIBN	2.5×10^{-2}	2.6	8.75
AIBN	1.6×10^{-2}	5.2	6.10
DCHP	1.0×10^{-3}	5.2	2.90

The empirical dependence of the sulfoxidation rate (v) on pO_2 at $[RH] > 2$ (mol L^{-1}) obeys the following equation [13]:

$$\frac{2v_i pO_2}{v} = A + B\frac{(pO_2)^2}{pSO_2} \tag{12.1}$$

Experiments on decane sulfoxidation with the variable glass surface showed that the chains were partly terminated on the surface [13]. The results of the experiments (decane, $T = 343$ K, $[DCHP] = 1.0 \times 10^{-3}$ mol L^{-1}, $v = 5.5 \times 10^{-7}$ mol L^{-1} s^{-1} [13]) are given below.

S (cm^2 L^{-1})	55	98	140
v (mol L^{-1} s^{-1})	5.8×10^{-6}	4.7×10^{-6}	3.4×10^{-6}
ν	11.6	9.4	6.8

The very important factor for the efficient sulfoxidation of alkanes is the partial pressure of sulfur dioxide. The study of alkanes sulfoxidation under SO_2 pressures higher than 1 atm proves that an optimal temperature exists for alkane sulfoxidation. The data (v_{max}) on different alkane sulfoxidation are listed below [25]. The reaction rate v was the maximal rate measured in the experiment.

T (K)	313	323	333	343	353
Decane, $pO_2 = 530$ kPa, $pSO_2 = 675$ kPa, $v_i = 5.0 \times 10^{-7}$ mol L^{-1} s^{-1}					
v_{max} (mol L^{-1} s^{-1})	1.6×10^{-3}	3.3×10^{-3}	4.9×10^{-3}	3.2×10^{-3}	9.0×10^{-4}
Tetradecane, $pO_2 = 750$ kPa, $pSO_2 = 745$ kPa, $v_i = 5.0 \times 10^{-7}$ mol L^{-1} s^{-1}					
v_{max} (mol L^{-1} s^{-1})	1.3×10^{-3}	2.9×10^{-3}	3.6×10^{-3}	3.1×10^{-3}	8.0×10^{-4}
Paraffin C_{14}–C_{19}, $pO_2 = 857$ kPa, $pSO_2 = 1810$ kPa, $v_i = 1.0 \times 10^{-6}$ mol L^{-1} s^{-1}					
v_{max} (mol L^{-1} s^{-1})	8.0×10^{-4}	3.1×10^{-3}	5.9×10^{-3}	1.9×10^{-3}	9.0×10^{-4}

Water retards the sulfoxidation of alkanes [25]. The results of experiments on decane sulfoxidation with addition of water ($T = 333$ K, $pO_2 = 980$ kPa, $[SO_2]_0 = 6.2$ mol L^{-1}, $v_{i0} = 5.0 \times 10^{-7}$ mol L^{-1} s^{-1}) are given in the following table. The rate of sulfoxidation was observed to increase in time due to the accumulation of alkylsulfonic peracid and, hence, $v_0 \ll v_{max}$.

H$_2$O (mol L^{-1})	0.002	0.02	0.06	0.10	1.0
v_0 (mol L^{-1} s^{-1})	1.0×10^{-4}	1.7×10^{-5}	6.0×10^{-4}	6.0×10^{-4}	6.0×10^{-4}
v_{max} (mol L^{-1} s^{-1})	7.0×10^{-3}	4.8×10^{-3}	2.4×10^{-3}	1.8×10^{-3}	~0

Sulfoxidation is accompanied by CL [23,25,26]. Two different sources of CL were found: the first source disappears within 20–40 min, the second source is seen during the whole time of the experiment and is proportional to v_i.

The sulfoxidation of alkanes occurs with heat evolution. This is the basis for rate of oscillation of rapid sulfoxidation at a relatively high pressure when the feedback arises between reaction rate, diffusion of reactants into liquid phase, and heat evolution [27].

The mechanism of alkane sulfoxidation includes the following elementary steps [13,21–24]: initiation, three steps of chain propagation, and a few steps of chain termination.

$$I \longrightarrow r^{\bullet} \xrightarrow{RH} R^{\bullet} \qquad k_i$$
$$R^{\bullet} + SO_2 \longrightarrow RSO_2^{\bullet} \qquad k_{p1}$$
$$RSO_2^{\bullet} \longrightarrow R^{\bullet} + SO_2 \qquad k'_{p1}$$
$$RSO_2^{\bullet} + O_2 \longrightarrow RSO_2OO^{\bullet} \qquad k_{p2}$$
$$RSO_2OO^{\bullet} + RH \longrightarrow RSO_2OOH + R^{\bullet} \qquad k_{p3}$$
$$R^{\bullet} + O_2 \longrightarrow RO_2^{\bullet} \qquad k_{t1}$$
$$RO_2^{\bullet} + RO_2^{\bullet} \longrightarrow ROH + O_2 + R'{=}O \qquad \text{Fast}$$
$$RSO_2^{\bullet} + RSO_2^{\bullet} \longrightarrow \text{Products} \qquad k_{t2}$$
$$RO_2^{\bullet} + RSO_2^{\bullet} \longrightarrow \text{Products} \qquad k_{t13}$$

The chain mechanism of RH sulfoxidation has several peculiarities.

1. The chain propagation proceeds via three consecutive steps. Hence, the different regimes of sulfoxidation are possible when the first or second propagation step limits the chain reaction. The third step proceeds extremely rapidly due to the high activity of alkylsufonylperoxyl radicals.
2. The alkylsulfonylperoxyl radical has a high reactivity in comparison with the alkylperoxyl radical. Therefore, the reaction of the alkyl radical with dioxygen breaks the chain in sulfoxidation. Since dioxygen participates in reactions of chain propagation via the reaction $RSO_2^{\bullet} + O_2$ and chain termination through the reaction $R^{\bullet} + O_2$, the dependence of the sulfoxidation rate on pO_2 has a maximum (sea earlier).
3. The C—S bond in the sulfonyl radical $RSO_2^{\bullet}$ is weak and therefore the reaction of the alkyl radical with the sulfonyl radical is reversible. The decay of the sulfonyl radical is an endothermic reaction. This peculiarity explains the existence of the optimal temperature for sulfoxidation. The increase in temperature lowers the steady-state concentration of sulfonyl radicals and, therefore, increases the chain termination by the reaction of the alkyl radical with dioxygen.

When the hydrocarbon concentration is high enough for the reaction of the sulfonyl-peroxyl radical with hydrocarbon not to limit the chain propagation, the following equation describes the dependence of the initial rate v of initiated (v_i) chain sulfoxidation of alkane [25]:

$$\frac{v_i}{v} = \frac{2k_{t1}v}{k_{p2}^2[O_2]^2} + \frac{(k'_{p1} + k_{p2}[O_2])k_{t1}}{k_{p1}k_{p2}[SO_2]} \quad (12.2)$$

This equation was found to be in good agreement with experimental data [13,21–23,25] and helps to estimate the rate constants of a few stages from the kinetic data.

12.3 ELEMENTARY STEPS OF SULFOXIDATION

The comprehensive study of the elementary steps of chain sulfoxidation was performed by Komissarov and coworkers [21–27]. The addition of sulfur dioxide to the alkyl radical occurs very rapidly. The rate constant of the reaction

was calculated from the kinetic data on cyclohexylsulfonyl chloride chain decomposition and found to be $k = 1.0 \times 10^8$ L mol^{-1} s^{-1} (cyclohexane, $T = 333$ K) [28,29]. The rate of this reaction is close to that controlled by the diffusion of reactants.

The addition of dioxygen to sulfonyl radicals occurs very rapidly and is also limited by the diffusion of reactants in the solvent. The rate constant of the reaction

measured by flash photolysis technique in cyclohexane at different temperatures ($T = 293$–323 K) was found to be equal to $k = 2.0 \times 10^9 \exp(-19.2/RT) = 9.1 \times 10^5$ L mol^{-1} s^{-1} (300 K) [15,30].

The sulfonyl radical is unstable and dissociates via C—S bond back to the alkyl radical and sulfur dioxide. The rate constant of this reaction for the cyclohexylsulfonyl radical was calculated from the kinetic data on the chain decomposition of cyclohexylsulfonyl chloride [2]. This decay of cyclohexylsulfonyl chloride initiated by DCHP occurs according to the following chain mechanism [29,31]:

$$I \longrightarrow r^{\bullet} \xrightarrow{RH} R^{\bullet}$$
$$R^{\bullet} + RSO_2Cl \longrightarrow RCl + RSO_2^{\bullet}$$
$$RSO_2^{\bullet} \longrightarrow R^{\bullet} + SO_2 \qquad k_p$$
$$R^{\bullet} + SO_2 \longrightarrow RSO_2^{\bullet}$$
$$RSO_2^{\bullet} + RSO_2^{\bullet} \longrightarrow \text{Products} \qquad 2k_t$$

The rate of this chain reaction v, measured by SO$_2$ evolution, was found to obey the equation

$$v = \frac{k_p}{\sqrt{2k_t}} \sqrt{v_i} \qquad (12.3)$$

The ratio of the rate constants $k_p(2k_t)^{-1/2}$ can be calculated from experiments with different v_i and then the value of k_p was estimated (for measurement of $2k_t$, see later). These experiments gave $k_p = 9.1 \times 10^{12} \exp(-62.0/RT) = 146 \, s^{-1}$ (300 K) [28]. The activation energy of this reaction seems to be close to the BDE of the C—S bond in the cyclohexylsulfonyl radical. The values of decay rate constants of various alkylsulfonyl radicals are listed in Table 12.1.

The rate constant of $RS^•O_2$ decay increases in the order: *prim*-R< *sec*-R< *tert*-R. The combination of sulfonyl radicals occurs rapidly and is limited by the rate of diffusion of these radicals in most cases (see Table 12.2).

The cross-recombination of alkyl and alkylsulfonyl radicals was studied by the pulse photolysis technique in several special experiments on cyclohexylsulfonyl chloride photolysis [25,37]. The value $k = 6.0 \times 10^{11} \exp(-14.6/RT) = 1.7 \times 10^9$ (300 K) L mol^{-1} s^{-1} was found for the reaction:

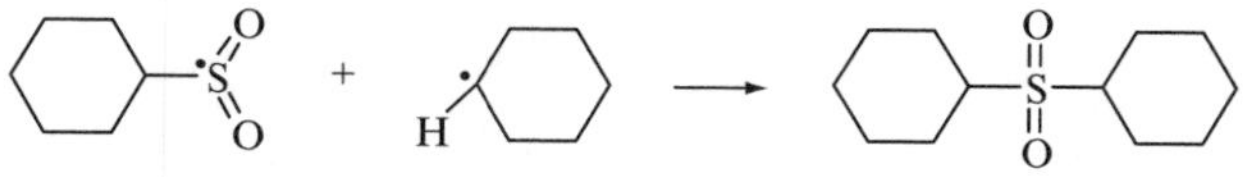

The reaction of the peroxyl radical with the sulfonyl radical was studied by pulse photolysis technique [38]. Both radicals were generated photochemically by a light pulse ($\lambda = 270$–380 nm) in the system: DBP–*cyclo*-$C_6H_{11}SO_2Cl$–*cyclo*-C_6H_{12}(RH)–air ($T = 293$ K). The reactions of free radical formation were the following:

TABLE 12.1
Rate Constant of Decay of Alkylsulfonyl Radicals [25,32]

$RS^•O_2$	Solvent	E (kJ mol^{-1})	log A, A (s^{-1})	k (s^{-1}) T (343 K)
$CH_3S^•O_2$	Gas phase	94.0	13.00	4.8×10^{-2}
$EtS^•O_2$	Gas phase	83.0	14.40	58
$Me_2CHS^•O_2$	$C_{13}H_{28}$	42.0	10.20	6.4×10^3
$Me_2CHCH_2S^•O_2$	$C_{13}H_{28}$	35.8	8.50	1.1×10^3
$EtMeCHS^•O_2$	$C_{13}H_{28}$	51.9	11.50	3.9×10^3
$Me(CH_2)_4CHMeS^•O_2$	$C_{13}H_{28}$	56.6	12.20	3.8×10^3
$Me(CH_2)_6CHMeS^•O_2$	$C_{13}H_{28}$	53.6	11.20	1.1×10^3
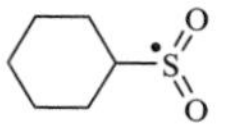 cyclohexyl-$S^•O_2$	 cyclohexane	51.0	11.20	3.2×10^3
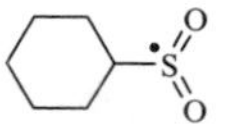 cyclohexyl-$S^•O_2$	$PhCH_3$	39.2	9.3	2.1×10^3
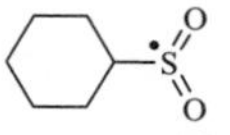 cyclohexyl-$S^•O_2$	$C_{13}H_{28}$			2.7×10^3

TABLE 12.2
Rate Constants of Recombination of Alkylsulfonyl Radicals

RS$^•$O$_2$	Solvent	T (K)	E (kJ mol^{-1})	log A, A (L mol^{-1} s^{-1})	k (355K) (L mol^{-1} s^{-1})	Ref.
MeS$^•$O$_2$	△	163			1.0×10^9	[33]
MeS$^•$O$_2$	△	223			4.5×10^9	[34]
EtS$^•$O$_2$	△	223			4.5×10^9	[34]
EtMeCHS$^•$O$_2$	C$_{13}$H$_{28}$	276–311	19.6	13.0	1.3×10^9	[25]
Me$_2$CHCH$_2$S$^•$O$_2$	C$_{13}$H$_{28}$	282–322	23.1	13.6	1.6×10^9	[25]
BuCH$_2$S$^•$O$_2$	C$_6$H$_{14}$	243–283	6.5	10.8	7.0×10^9	[25]
BuCH$_2$S$^•$O$_2$	C$_{13}$H$_{28}$	277–314	13.6	11.6	4.0×10^9	[25]
Me(CH$_2$)$_4$CHS$^•$O$_2$Me	C$_6$H$_{14}$	274–312	7.0	10.8	5.9×10^9	[25]
Me(CH$_2$)$_4$CHS$^•$O$_2$Me	C$_{13}$H$_{28}$	277–311	18.0	12.4	5.6×10^9	[25]
Me(CH$_2$)$_6$CHS$^•$O$_2$Me	C$_6$H$_{14}$	281–313	10.7	11.2	4.2×10^9	[25]
Me(CH$_2$)$_6$CHS$^•$O$_2$Me	C$_{13}$H$_{28}$	283–320	22.0	12.5	1.8×10^9	[25]
Me(CH$_2$)$_7$CHS$^•$O$_2$Me	C$_6$H$_{14}$	280–311	16.7	12.0	3.5×10^9	[25]
Me(CH$_2$)$_7$CHS$^•$O$_2$Me	C$_{13}$H$_{28}$	281–313	18.3	11.8	1.3×10^9	[25]
cyclohexyl-S$^•$O$_2$	cyclohexane	355			1.4×10^9	[25]
cyclohexyl-S$^•$O$_2$	cyclohexane	350			1.2×10^9	[35]
cyclohexyl-S$^•$O$_2$	cyclohexane	282–313	20.0	12.6	4.5×10^9	[29]
cyclohexyl-S$^•$O$_2$	cyclohexane	285–317	17.6	12.4	6.5×10^9	[29]
cyclohexyl-S$^•$O$_2$	C$_6$H$_{14}$	275–314	5.7	10.6	5.8×10^9	[29]
cyclohexyl-S$^•$O$_2$	C$_7$H$_{16}$	274–313	12.0	11.6	6.8×10^9	[29]
cyclohexyl-S$^•$O$_2$	C$_{10}$H$_{22}$	274–318	16.2	12.2	6.5×10^9	[29]
cyclohexyl-S$^•$O$_2$	C$_{12}$H$_{24}$	274–308	11.0	11.2	3.8×10^9	[29]
cyclohexyl-S$^•$O$_2$	C$_{13}$H$_{28}$	276–313	14.9	11.9	5.6×10^9	[29]
cyclohexyl-S$^•$O$_2$	PhCH$_3$	256–305	6.6	10.8	6.7×10^9	[29]
PhS$^•$O$_2$	△	223			4.5×10^9	[34]

TABLE 12.2
Rate Constants of Recombination of Alkylsulfonyl Radicals—continued

$RS^\bullet O_2$	Solvent	T (K)	E (kJ mol^{-1})	log A, A (L mol^{-1} s^{-1})	k (355K) (L mol^{-1} s^{-1})	Ref.
$PhS^\bullet O_2$	CCl$_4$	296			3.4×10^8	[36]
(methyl-C$_6$H$_4$-SO$_2^\bullet$)	CCl$_4$	296			5.0×10^8	[36]
(Cl-C$_6$H$_4$-SO$_2^\bullet$)	CCl$_4$	296			4.7×10^8	[36]
(Cl-C$_6$H$_4$-SO$_2^\bullet$)	CCl$_4$	296			8.0×10^7	[36]
(Cl-C$_6$H$_4$-SO$_2^\bullet$)	▷	223			4.5×10^9	[36]

$$Me_3COOCMe_3 + h\nu \longrightarrow 2Me_3CO^\bullet$$
$$Me_3CO^\bullet + RH \longrightarrow Me_3COH + R^\bullet$$
$$R^\bullet + O_2 \longrightarrow RO_2^\bullet$$
$$R^\bullet + RSO_2Cl \longrightarrow RCl + RS^\bullet O_2$$

The peroxyl and sulfonyl radicals absorb light in different regions of UV and visible light: $\lambda_{max}(RO_2^\bullet) = 260$ nm and $\lambda_{max}(RS^\bullet O_2) = 360$ nm. Such concentrations of the reactants were chosen to create the ratio of initial concentrations of radicals $[RO_2^\bullet]_0 \gg [RS^\bullet O_2]_0$. Therefore, the kinetics of sulfonyl radicals decay obeys the first-order equation:

$$-\frac{d[RS^\bullet O_2]}{dt} = k_{obs}[RS^\bullet O_2] \tag{12.4}$$

where $k_{obs} \cong k[RO_2^\bullet]_0$.

The same experiments were performed in tridecane ($T = 293$ K). The following rate constants were calculated from the experimental data [25]:

Reaction	Rate Constant (293 K)
cyclohexyl-SO$_2^\bullet$ + $^\bullet$O-O-cyclohexyl(H)	1.5×10^8 L mol^{-1} s^{-1}
cyclohexyl-SO$_2^\bullet$ + C$_{13}$H$_{27}$O$_2^\bullet$	3.0×10^8 L mol^{-1} s^{-1}

We see that such reactions occur very rapidly. One can expect that the peroxyl radical formed in sulfoxidation by the reaction $R^\bullet + O_2$ then reacts with $RO_2^\bullet$ and $RS^\bullet O_2$ as well.

12.4 DECOMPOSITION OF ALKYLSULFONYL PERACIDS

The NMR study of RSO_2OOH formed in the sulfoxidation of decane proved that these peracids are secondary with the $>CHSO_2OOH$ groups. They exist in monomeric and dimeric forms in hydrocarbon and CCl_4 solutions [25,28]. The products of decay of sec-decylsulfonic peracids are sec-decylsulfonic acids (100%), alcohols (75%), ketones (22%), and water (negligible amounts) [39].

The decay of decylsulfonic peracids (RSO_2OOH) occurs as the first-order reaction with $k_{obs} = \Delta \ln[RSO_4H]/t$. However, the rate constant k_{obs} depends on the initial concentration of peracid [25]. For example ($T = 323$ K, decane, $pO_2 = 98$ kPa).

[RSO_4H]$_0$ (mol L^{-1})	2.3×10^{-3}	1.02×10^{-2}	7.6×10^{-2}	0.22
$k_{obs} \times 10^4$ (s^{-1})	2.6	3.6	5.5	6.2

Dioxygen retards the decay of peracids. For example, $k_{obs} = 4.0 \times 10^{-6}$ s^{-1} in a dioxygen atmosphere and $k_{obs} = 1.3 \times 10^{-5}$ s^{-1} in an Ar atmosphere ($T = 345$ K, decane, [RSO_4H]$_0$ = 9.0×10^{-3} mol L^{-1} [25,39]). The decay of peracid is accompanied by the consumption of dioxygen. The ratio of $v_{O_2}/v_d > 1$ and decreases with an increase in the initial peracid concentration. All these facts prove that peracid decomposes with free radical formation and radicals R$^{\bullet}$ formed from the solvent (RH) induce the chain decomposition of peracid with alcohol formation. The decay of peracid to free radicals involves hydrocarbon and bimolecular peracid associates [25,28].

$$RSO_2OOH \longrightarrow RSO_2O^{\bullet} + HO^{\bullet}$$
$$RSO_2OOH + RH \longrightarrow RSO_2O^{\bullet} + H_2O + R^{\bullet}$$
$$2RSO_2OOH \rightleftharpoons RSO_2OO(H) \cdots HOOSO_2R$$
$$RSO_2OO(H) \cdots HOOSO_2R \longrightarrow RSO_2OO^{\bullet} + H_2O + RSO_2O^{\bullet}$$
$$RSO_2O^{\bullet} + RH \longrightarrow RSO_2OH + R^{\bullet}$$
$$R^{\bullet} + RSO_2OOH \longrightarrow ROH + RSO_2O^{\bullet}$$
$$R^{\bullet} + R^{\bullet} \longrightarrow RR$$

The additional reactions occur in the presence of dioxygen:

$$R^{\bullet} + O_2 \longrightarrow RO_2^{\bullet}$$
$$RO_2^{\bullet} + RO_2^{\bullet} \longrightarrow ROH + O_2 + R'{=}O$$

In addition, the peracid dimer decomposes heterolytically to molecular products.

$$RSO_2OO(H) \cdots HOOSO_2R \longrightarrow 2RSO_2OH + O_2$$

Therefore, the yield of free radicals e decreases with an increase in the peracid concentration ($T = 323$ K, decane, [O_2] $= 1.7 \times 10^{-3}$ mol L^{-1} [25]).

[$C_{10}H_{23}SO_2OOH$] (mol L^{-1})	1.4×10^{-3}	1.3×10^{-2}	2.6×10^{-2}	6.4×10^{-2}
e	0.11	0.043	0.023	0.005

Peracid reacts with sulfur dioxide [40]. The reaction proceeds as bimolecular in decane solution.

$$v = k[RSO_2OOH][SO_2] \qquad (12.5)$$

The rate constant values of this reaction at different temperatures are as follows.

T (K)	285	295	303	313	323
k (L mol^{-1} s^{-1})	5.0×10^{-4}	1.6×10^{-3}	2.6×10^{-3}	7.6×10^{-3}	1.65×10^{-2}

The rate constant is $k = 5.0 \times 10^9 \exp(-70.7/RT) = 2.45 \times 10^{-3}$ L mol^{-1} s^{-1} (300 K).

Water accelerates this reaction and acids retard it. The study of the reaction of peracid with SO$_2$ by stop-flow technique [40] showed the following empirical equation for the reaction rate:

$$v = k[RSO_2OOH][SO_2][H_2O]^3[H_2SO_4]^{-1} \qquad (12.6)$$

The reaction proceeds according to the stoichiometric equation:

$$RSO_2OOH + SO_2 + H_2O = RSO_2OH + H_2SO_4$$

The following consequences of the equilibrium reactions were supposed as predecessors of the final step of S(IV) oxidation by protonated peracid [25].

$$SO_2 + H_2O \longrightarrow H_2SO_3$$
$$H_2SO_3 + H_2O \rightleftharpoons HOSO_2^- + H_3O^+$$
$$RSO_2OOH + H_3O^+ \rightleftharpoons RSO_2OOH_2^+ + H_2O$$
$$RSO_2OOH_2^+ + HOSO_2^- \longrightarrow H_2SO_4 + RSO_3H$$

12.5 OXIDATION BY ALKYLSULFONIC PERACIDS

12.5.1 Oxidation of Aromatic Hydrocarbons

Alkylsulfonic acids are active oxidative agents like other organic peracids. Several oxidative reactions of *sec*-decylsulfonic peracid were studied by Safiullin et al. [41]. Peracid was found to oxidize benzene to phenol as the first intermediate product. The formed sulfonic acid accelerates the reaction. Oxidation occurs according to the stoichiometric equation

$$RSO_2OOH + C_6H_6 = RSO_3H + C_6H_5OH$$

with the rate

$$v = (k_1 + k_2[RSO_2OH])[RSO_2OOH][C_6H_6] \qquad (12.7)$$

and rate constants $k_1 = 3.3 \times 10^{-6}$ L mol^{-1} s^{-1}, $k_2 = 1.9 \times 10^{-2}$ L^2 mol^{-2} s^{-1} at $T = 295$ K [41]. The Arrhenius form of k_1 is equal to $k_1 = 1.48 \times 10^8 \exp(-76.5/RT)$ L mol^{-1} s^{-1}. The rate constants of oxidation of the substituted benzenes are listed in Table 12.3. The values of rate constants are correlated with Brawn's σ_p^+ ($T = 295$ K) as

TABLE 12.3
Rate Constants of Aromatic Compounds Oxidation by *sec*-Decylsulfonyl Peracid [41]

Aromatic Compound	$[ArH]_0$ (mol L^{-1})	$[RSO_2OOH]_0$ (mol L^{-1})	k (295 K) (L mol^{-1} s^{-1})
PhH	10.7	7.7×10^{-3}	3.3×10^{-6}
PhMe	9.3	1.0×10^{-3}	6.2×10^{-5}
PhMe	8.7	1.3×10^{-2}	6.1×10^{-5}
PhEt	8.0	1.2×10^{-3}	5.5×10^{-5}
Me$_2$PhCH	6.8	6.0×10^{-3}	1.2×10^{-4}
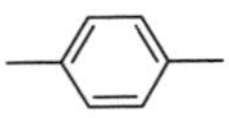 *p*-xylene	8.1	1.7×10^{-3}	2.8×10^{-4}
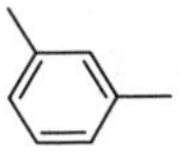 *m*-xylene	8.1	9.0×10^{-4}	7.2×10^{-4}
EtOPh	6.6	7.0×10^{-4}	1.3×10^{-3}
PhOH	8.2×10^{-3}	9.0×10^{-4}	3.2×10^{-2}
PhCl	9.3	8.2×10^{-3}	$>3.4 \times 10^{-6}$

$$\log k(YC_6H_5) = \log k(C_6H_6) - 4.48\sigma_p^+ \quad (12.8)$$

The negative value of the ρ-coefficient means that this reaction proceeds via the electrophilic mechanism.

12.5.2 Oxidation of Olefins

Alkylsulfonic peracids oxidize olefins to epoxides. The formed sulfonic acid reacts with epoxide to form diols and esters. The yields of epoxides in the reactions of oxidation of two cycloolefins are given in Table 12.4.

The detailed kinetic study of octene-1 epoxidation by *sec*-decylsulfonic peracid was performed [25,42]. The 1,2-octanediol monodecylsulfonate was identified as the main product of the reaction. The kinetic dependence of the reaction rate (v) on the reactants concentration obeys the equation

$$v = k_1[RSO_2OOH][C_8H_{16}] + k_2[RSO_2OOH]^2[C_8H_{16}] \quad (12.9)$$

with the following values of the rate constants (CCl$_4$ as solvent, $T = 272$–313 K):

$$k_1 = 5.62 \times 10^7 \exp(-49.3/RT) = 0.15 \, \text{L mol}^{-1} \, \text{s}^{-1} (300 \, \text{K}),$$
$$k_2 = 8.13 \times 10^5 \exp(-20.4/RT) = 228 \, \text{L}^2 \, \text{mol}^{-2} \, \text{s}^{-1} (300 \, \text{K})$$

One can expect that epoxidation occurs as electrophilic reaction. Peracid oxidizes olefin in two forms: monomeric and dimeric. The following scheme of epoxidation was proposed [42]:

Sulfoxidation of Hydrocarbons

$$RSO_2OOH + CH_2{=}CHR^1 \longrightarrow RSO_2OH + cyclo\text{-}[OCH_2CHR^1]$$
$$RSO_2OOH + RSO_2OOH \rightleftharpoons RSO_2OO(H) \cdots HOOSO_2R$$
$$RSO_2OO(H) \cdots HOOSO_2R + CH_2{=}CHR^1 \longrightarrow RSO_2OH + cyclo\text{-}[OCH_2CHR^1] + RSO_2OOH$$

The solvent influences the value of rate constant of these reactions [42].

Solvent	CHCl$_3$	CCl$_4$	Et$_2$O
k_1 (L mol^{-1} s^{-1})	0.18	0.10	1.5×10^{-3}
k_2 (L^2 mol^{-2} s^{-1})	4.34×10^2	1.94×10^2	0.32

The reactivities of different olefins toward sec-decylsulfonic peracid (values of k_1 and k_2) are listed in Table 12.5.

Linear correlation was observed between log k_1 and Taft σ^* function [24]:

$$\log k_1 = 0.76 - 1.20\Sigma\sigma^* \tag{12.10}$$

TABLE 12.4
Epoxide Yields in Reactions of Olefins Oxidation by sec-Decylsulfonic Peracid [40]

Olefin	T (K)	[Olefin]$_0$ (mol L^{-1})	[RSO$_4$H]$_0$ (mol L^{-1})	[Epoxide] (mol L^{-1})	Epoxide (mol %)
 cyclohexane (in CHCl$_3$)	299	9.0×10^{-3}	1.16×10^{-2}	5.0×10^{-3}	56
 cyclohexene (in CHCl$_3$)	299	1.8×10^{-2}	4.8×10^{-2}	1.4×10^{-2}	78
 cyclohexadiene (in CHCl$_3$)	299	5.1×10^{-2}	1.0×10^{-1}	4.6×10^{-2}	90
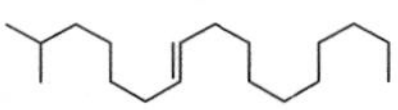 olefin*	296	2.1×10^{-3}	2.3×10^{-2}	1.2×10^{-3}	57
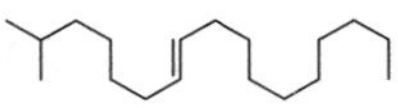 olefin*	300	2.73×10^{-2}	3.82×10^{-2}	1.5×10^{-2}	56
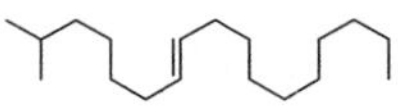 olefin*	300	3.7×10^{-2}	3.9×10^{-2}	1.7×10^{-2}	46
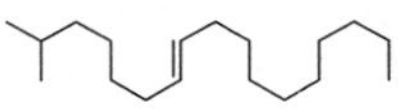 olefin*	300	2.4×10^{-2}	3.1×10^{-2}	1.0×10^{-2}	42
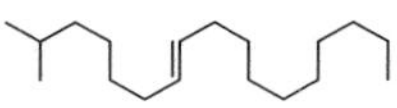 olefin*	273	2.4×10^{-2}	2.5×10^{-2}	9.0×10^{-3}	37
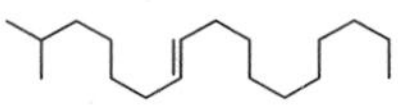 olefin*	296	2.7×10^{-2}	3.1×10^{-2}	1.4×10^{-2}	52

*in CCl$_4$ solution

TABLE 12.5
The Values of Rate Constants k_1 and k_2 of the Epoxidation of Different Olefins by sec-Decylsulfonic Peracid ($T = 297$ K, Equation [12.7]) [40]

Olefin	Solvent	[Olefin]₀ (mol L⁻¹)	[RSO₄H]₀ (mol L⁻¹)	$k_1 \times 10^2$ (L mol⁻¹ s⁻¹)	k_2 (L² mol⁻² s⁻¹)
cyclohexene	Et₂O	0.33	4.5×10^{-3}	1.8×10^{-2}	6.3
$C_{17}H_{35}CH=CH_2$	CCl₄	0.010	1.0×10^{-3}	1.03×10^{-4}	1.41×10^{-3}
cyclohexene	CCl₄	0.010	1.1×10^{-3}	0.21	1.02×10^2
$C_6H_{13}CH=CH_2$	CCl₄	0.10	1.9×10^{-2}	0.12	1.94×10^2
$C_6H_{13}CH=CH_2$	CHCl₃	5.0×10^{-3}	4.5×10^{-4}	0.18	4.34×10^2
$C_6H_{13}CH=CH_2$	Et₂O	0.10	6.2×10^{-3}	1.4×10^{-3}	0.34
$CH_2ClCH=CH_2$	CCl₄	0.49	1.0×10^{-2}	7.0×10^{-3}	0.30
$CH_2BrCH=CH_2$	CCl₄	9.2×10^{-2}	1.0×10^{-2}	6.6×10^{-3}	0.25
$HOCH_2CH=CH_2$	MeCN	0.10	1.0×10^{-2}	5.2×10^{-2}	8.0
$CH_2ClCH=CHCH_2Cl$	CCl₄	0.10	1.0×10^{-3}	1.1×10^{-3}	2.8
$Ph_2C=CH_2$	CCl₄	0.10	9.0×10^{-4}	0.99	1.05×10^3
$EtOCH_2CH=CH_2$	CCl₄	0.10	8.0×10^{-4}	3.0×10^{-2}	33.0
$MeOC(O)CH=CH_2$	CCl₄	4.90	1.0×10^{-3}	7.0×10^{-5}	
cyclooctene	CCl₄	6.0×10^{-3}	3.0×10^{-4}	0.76	
cyclooctene	Et₂O	0.16	6.0×10^{-3}	8.4×10^{-2}	24.5
$C_{10}H_{21}CH=CH_2$	CCl₄	1.0×10^{-3}	1.0×10^{-4}	5.4	

12.5.3 Oxidation of Ketones

Sulfonic peracids oxidize ketones to lactones. The yields of the oxidation products are listed in Table 12.6.

The reaction of sec-decylsulfonic acid with cyclopentanone was studied kinetically [43]. This reaction proceeds bimolecularly with the rate constant equal to $k = 3.98 \times 10^7 \exp(-42.1/RT) = 1.86$ L mol⁻¹ s⁻¹ (300 K) in CCl₄ at $T = 291$–323 K.

This Bayer–Villiger reaction was supposed to proceed via the intermediate formation of the peroxyl adduct between peracid and the carbonyl group [25,43].

$$\text{cyclopentanone} + RSO_2OOH \rightleftharpoons \text{adduct(OH)(OOSO}_2R) \qquad K$$

$$\text{adduct(OH)(OOSO}_2R) \rightarrow \text{lactone} + RSO_2OH \qquad k$$

TABLE 12.6
Yields of Lactones and Rate Constants (k_{obs}) of the Oxidation of Ketones by sec-Decylsulfonic Acid ($T = 290$ K, [43,44])

Ketone	Product	Solvent	k (CCl$_4$) (L mol^{-1} s^{-1})	Time (s)	Yield %
EtC(O)Me	MeC(O)OEt	MeCN	0.17	1.8×10^3	88
MeC(O)Pr	MeC(O)OPr	MeCN	0.19		
cyclopentanone	δ-valerolactone	CCl$_4$	0.94	3.0×10^2	95
cyclohexanone	ε-caprolactone	CCl$_4$	1.80		
cycloheptanone	lactone	MeCN	0.18	1.8×10^3	90
cyclooctanone	lactone	MeCN	0.20	1.8×10^3	82
menthone	lactone	MeCN	0.35	6.0×10^2	98
ADAMANTANONE	lactone	CCl$_4$		3.0×10^2	84

The observed rate constant is $k_{obs} = kK$. Alkylsulfonic peracid was successfully used as an efficient regioselective oxidant of the carbonyl group [25].

REFERENCES

1. C Platz, K Schimmelschmidt. Sulfonic acids. Pat 735096, Germany, (*Chem Abstr* 38, 1249, 1940).
2. F Azinger. *Chemie und Technologi der Paraffin-Kohlenwasserstoffe*. Berlin: Akademie-Verlag, 1956.
3. R Graf. *Ann Chem* 578, 50–82, 1952.
4. L. Orthner. *Angew Chem* N13/14, 302–305, 1950.
5. H Weghofer. *Fette Seifen* 54:260–265, 1952.
6. JH Black, EF Baxter. *Soap Chem Specialities* 34:3–10, 1958.
7. C Beerman. Proc Symp Normal Paraffins. Issue European Chem News. November 16, 1966.
8. AI Popov, PA Zagorets, RV Dzhagatspanyan. *Radiation Induced Sulfoxidation of Paraffins*, vol 2. Moscow: Nauka, 1972, pp. 122–127 [Russian].
9. LN Khmelnitskii, VV Nesterovskii, II Mekhekhanova. *Radiation Induced Sulfoxidation of Paraffins*, vol 2. Moscow: Nauka, 1972, pp. 127–130 [Russian].
10. J. Perwyski, JS Miller. *Tenside Detergents* 21:7–9, 1984.
11. AS Drozdov, V Mukhin, ZV Didenko. *Neftepererabotka Neftekhimiya* 1:28–30, 1973.
12. VD Komissarov, MA Saitova, EM Kuramshin, RG Timirova. *Khim Prom* 738–740, 1973.
13. MA Saitova. Kinetics and Mechanism of Sulfoxidation of Alkanes. Ph.D. thesis, Institute of Chemical Physics, Chernogolovka, 1975.

14. MA Saitova, VD Komissarov. *Izv AN SSSR Ser Khim* 436–438, 1974.
15. RL Safiullin. Reactions of Alkyl and Alkylsulfonyl Radicals in Liquid-Phase Sulfoxidation. Ph.D. thesis, Institute of Chemical Physics, Chernogolovka, 1981.
16. F Azinger, V Fell, S Pottkamper. *Chem Ber* 97:3092–3097, 1964.
17. F Azinger, V Fell, A Commichan. *Chem Ber* 98:2154–2158, 1965.
18. O Cerny. *Coll Czech Chem Commun* 33:257–263, 1968.
19. D Bertram. *Strahlensulfonirungen von Kohlenwasserstofen.* Berlin: Dechema Monographie, vol 42:197–201, 1962.
20. D Bertram. *Radiat Chem Proc Tihany Symp Tihany Hungary* 23–30, 1962.
21. VD Komissarov, MA Saitova. *Dokl AN SSSR* 221:123–125, 1975.
22. VD Komissarov, MA Saitova, NM Vlasova. *React Kinet Catal Lett* 2:105–110, 1975.
23. VD Komissarov, MA Saitova, RF Khalimov. *Neftekhimiya* 14:270–274, 1974.
24. RN Zaripov, RL Safiullin, ShR Rameev, VD Komissarov. *Kinet Katal* 31:1086–1091, 1990.
25. RL Safiullin. Elementary Reactions of Sulfoxidation. Doctoral Dissertation, Institute of Organic Chemistry, Ufa, 2001.
26. RL Safiullin, RN Zaripov, VD Komissarov. *Izv AN SSSR Ser Khim* 1447–1448, 1986.
27. VD Komissarov, RN Zaripov, RL Safiullin. *Izv AN SSSR Ser Khim* 1673–1674, 1984.
28. VD Komissarov, RL Safiullin. *Kinet Katal* 21:594–599, 1980.
29. VD Komissarov, RL Safiullin. *React Kinet Catal Lett* 14:67–72, 1980.
30. RL Safiullin, AI Nikolaev, VD Komissarov, ET Denisov. *Khim Fiz* 1:642–648, 1982.
31. VD Komissarov, RL Safiullin, ET Denisov. *Dokl AN SSSR* 252:1177–1179, 1980.
32. A Good, JCJ Thynne. *Trans Faraday Soc* 63:2708–2719, 1967.
33. C Chatgilialoglu, L Lunnazzi, KU Ingold. *J Org Chem* 48:3588–3589, 1983.
34. AI Nikolaev, RL Safiullin, VD Komissarov. *Izv AN SSSR Ser Khim* 1258–1263, 1986.
35. RL Mc Carthy, A Maclachlan. *Trans Faraday Soc* 57:1107–1116, 1961.
36. S Correa, WA Waters. *J Chem Soc C* 1874–1879, 1968.
37. RL Safiullin, VD Komissarov, Z Akhmadishin, SI Spivak. *React Kinet Catal Lett* 19:65–69, 1982.
38. AI Nikolaev, RL Safiullin, VD Komissarov. *Khim Fiz* 3:257–261, 1984.
39. RL Safiullin, RN Zaripov, AA Elichev, SYu Serenko. *Kinet Katal* 31:808–812, 1990.
40. RL Safiullin, RN Zaripov, GG Savel'eva, VD Komissarov. *Izv AN SSSR Ser Khim* 546–548, 1991.
41. RL Safiullin, LR Enikeeva, VD Komissarov. *Kinet Katal* 30:1040–1044, 1989.
42. RL Safiullin, LR Enikeeva, SYu Serenko, VD Komissarov. *Izv AN SSSR Ser Khim* 333–337, 1991.
43. RL Safiullin, AN Volgarev, VD Komissarov, GA Tolstikov. *Izv AN SSSR Ser Khim* 2188–2189, 1990.
44. RL Safiullin, AN Volgarev, VD Komissarov. *Izv AN SSSR Ser Khim* 1827–1829, 1993.

13 Oxidation of Polymers

13.1 INITIATED OXIDATION OF POLYMERS

13.1.1 CAGE EFFECT IN SOLID POLYMERS

Polyolefins are semicrystalline polymers, so the low-molecular-weight substances penetrate and diffuse only in the amorphous phase. Each pair of particles in the polymer is surrounded by the segments of a macromolecule. The cage formed by the polymer segments is more tight than that of the liquid. The probability e for a pair of radicals to go out of the cage is sufficiently less in polymer than in the liquid. The study of the cage effect and molecular mobility in plasticized polypropylene (PP) proved the important role of not only the translational diffusion but also that of the rotational diffusion in the fate of a radical pair in the polymer cage [1–3] (see Chapter 3). The initiation rate constant is $k_i = 2ek_d$, where k_d is the decomposition rate constant (for the values of k_d of initiators, see Chapter 3). The e values for the most popular initiators are given in Table 13.1. They are sufficiently lower than those in the liquid phase (see Chapter 3 and Ref. [4]).

A solid polymer principally differs from a liquid polymer by its ability to have and maintain a form. In particular, the sample of solid polymer can be stretched. Elongation changes the polymer structure: polymer acquires the microfibrillar structure. This influences the form and elasticity of the polymer cage created by the segments of a macromolecule. The elongation of macromolecules promotes an elongated (cylindric) form of the cage in the polymer. The cage effect in PP films of different oriented elongations was studied for AIBN decay [10]. The rate constant of decay was measured by the kinetics of dinitrogen evolution. The rate of free radical generation was measured using EPR spectroscopic control for decay of the stable nitroxyl radical (2,2,6,6-tetramethyl-4-oxybenzoylpiperidine-N-oxyl). The rate constant of AIBN decay appeared to be the same as in the liquid phase: $k_d = 1.58 \times 10^{15} \exp(-129/RT)$ s^{-1}. The following table gives the values of e measured in the PP powder and films of different degrees of elongation γ:

γ	e (323 K)	e (353 K)	$E_i - E_d$ (kJ mol^{-1})
1	0.013	0.020	12
4	0.056	0.025	−15
8	0.031	0.031	−33

The oriented elongation of the polymer increases the packing of macromolecules and decreases the molecular mobility in the polymer. This was observed by the EPR spectra of the nitroxyl radical in these films. Therefore, one can expect an increase in radical pair recombination in the cage with an increase in γ. However, experiment showed an opposite pattern: the more the γ, the higher the e value. These results found explanation within the scope of the

TABLE 13.1
Probability (e) of Free Radical Escaping from the Cage into Bulk Volume in the Decomposition of Initiators in Polymer Matrix

Polymer	Initiator	T (K)	e	Ref.
PE	AIBN	343	0.006	[1]
PE	AIBN	353	0.011	[1]
PE	AIBN	363	0.020	[1]
IPP	AIBN	333	0.016	[5]
IPP	AIBN	353	0.027	[5]
IPP	DBP	365	0.38	[6]
IPP	AIBN	333	0.013	[2]
IPP	AIBN	343	0.017	[2]
IPP	AIBN	353	0.025	[2]
IPP	AIBN	344	0.14	[7]
IPP	AIBN	378	0.62	[7]
PBD	AIBN	349	0.018	[7]
PBD	AIBN	383	0.59	[7]
PMP	AIBN	358	0.012	[7]
PMP	AIBN	368	0.05	[7]
PS	AIBN	353	0.05	[8]
APP	DBP	298	0.014	[9]
APP	DBP	318	0.061	[9]
APP	DBP	328	0.15	[9]
PP	DLP	353	0.05	[3]

following simple geometric model [10]. Oriented elongation of the polymer film makes the cage to gain a cylindrical shape with rigid walls. The decomposition of AIBN produces the pair of cyanoisopropyl radicals divided by the molecule of dinitrogen in the cage. The cylindrical shape of the cage in combination with rigid cage walls increases the probability of radicals to escape out of the cage. Therefore more the γ, higher the e. Heating of the polymer increases the molecular mobility of cage walls and weakens this effect. Hence, higher the temperature, lesser the influence of elongation (γ) on the cage effect, and the difference $E_i - E_d$ is negative in the stretched polymer. These experiments demonstrate the high importance of the cage shape in a polymer matrix (for the influence of the cage shape on bimolecular reactions, see Chapter 19).

13.1.2 Migration of Free Valence in Solid Polymers

The motion of a free valence in oxidizing solid polymer is a complex process. The following three mechanisms were discussed [11,12]:

1. The encounter of two free valences as a result only of diffusion of segments of a macromolecule. Since the radius of segmental diffusion is limited in the real time, this mechanism can be efficient at the high initiation rate and intense mobility of the polymer segments. Under the conditions of polymer oxidation, this mechanism is possible at the chain length close to unity. Some examples are given in Table 13.2.
2. The combination of segmental diffusion with the transfer of the free valence to another segment due to the chemical reaction, for example POO$^{\bullet}$ + PH or POO$^{\bullet}$ + POOH. The

TABLE 13.2
Kinetic Parameters of Free Valence Migration in Polymers [13,14]: Effective Diffusion Coefficient D and Average Distance of Diffusion r

Polymer, Initiation	Macroradical	T (K)	$[P^\bullet]_0 \times 10^6$ or $[PO_2^\bullet]_0 \times 10^6$ (mol kg^{-1})	$D \times 10^{18}$ (cm^2 s^{-1})	$r \times 10^{10}$ (m)
PE, γ	P$^\bullet$	363	8.3	2.0	50
PE, γ	P$^\bullet$	343	3.5	3.6	23
PE, γ	P$^\bullet$	343	3.7	2.8	8
PE, γ	PO$_2^\bullet$	363	8.3	10	35
PE, γ	PO$_2^\bullet$	363	8.3	90	15
IPP, mechan.	PO$_2^\bullet$	273	1.7	5.0	75
IPP, mechan.	PO$_2^\bullet$	273	3.3	7.0	65
IPP, mechan.	PO$_2^\bullet$	292	5.0	90	75
IPP, mechan.	PO$_2^\bullet$	292	3.3	80	80
PMMA, γ	P$^\bullet$	301	0.5	0.2	55
PMMA, γ	P$^\bullet$	311	0.5	3.0	55
PMMA, γ	P$^\bullet$	318	0.5	30	20
PMMA, γ	P$^\bullet$	328	0.5	20	15
PMMA γ	PO$_2^\bullet$	251	5.0	3.0	60
PMMA, γ	PO$_2^\bullet$	273	6.7	50	55

movement of the free valence occurs as the interchange of segmental diffusion with chemical reactions [12]. The rate of the free valence transfer depends on the velocity of segmental diffusion and the rate of chemical reactions. The decay of free radicals in polymer in an oxygen atmosphere is accompanied by dioxygen consumption (four to ten molecules of dioxygen per radical at room temperature) [15–20]. This mechanism seems to be the most probable for the chain oxidation of solid polymers in the amorphous phase. The values of the rate constants ($2k_t$) of peroxyl radical disproportionation in solid polymers are collected in Table 13.3.

3. In the presence of low-molecular-weight additive rH sufficiently active towards peroxyl radicals, the moving of free valence proceeds as diffusion of free radicals r$^\bullet$ formed in reactions [11,12]:

$$POO^\bullet + rH \longrightarrow POOH + r^\bullet$$

This additive rH should be very reactive toward peroxyl radicals. It can be hydroperoxide ROOH or antioxidant InH.

13.1.3 Initiated Polymer Oxidation

Polymer oxidation is similar to oxidation of low-molecular-weight analogs in the liquid phase and has several peculiarities caused by the specificity of solid-phase free radical reactions of macromolecules. Several monographs are devoted to this field of chemistry [11,12,33–41].

The important characteristics of polymers oxidation were obtained as a result of the study of their initiated oxidation. In the presence of initiator (I) which generates the chains with the rate $v_i = k_i[I]$, the oxidation of polymer PH occurs with the constant rate v. When the macroradical P$^\bullet$ of the oxidized polymer reacts with dioxygen very rapidly (at [O$_2$]

TABLE 13.3
Rate Constants of Macroperoxyradical Disproportionation in Solid Polymers

Polymer	T (K)	E (kJ mol^{-1})	log A, A (kg mol^{-1} s^{-1})	$2k_t$ (293 K) (kg mol^{-1} s^{-1})	Ref.
PE	293	77	16	300	[21]
PE	283–303	90	16.4	3.8	[21]
PE	364			1.0×10^{-2}	[15]
APP	318–336	52	11.0	75	[9]
APP	363–378	48.5	13.0	3×10^4	[7]
IPP	273			3×10^{-3}	[22]
IPP	292			5×10^{-2}	[22]
IPP, [POOH] = 0.1*	328–390	119	21.5	4	[23]
IPP, [POOH] = 2.5×10^{-3}*	298			170	[24]
IPP, [POOH] = 2.5×10^{-2}*	298			12	[24]
IPP, [POOH] = 0.10*	298			5	[24]
IPP	299–320	109	20.0	7.52	[25]
IPP	299–320	110	19.3	1.0	[25]
IPP	298			2.2×10^2	[26]
IPP, [POOH] = 0.13*	298			6	[26]
IPP	363			2.3	[27]
IPP	303			5.4×10^{-3}	[28]
IPP	291			7.2	[21]
IPP	323–636	103	15.7	4×10^{-3}	[29]
PB	373			1.9×10^{-6}	[30]
PS	265–283	73	14.0	16	[31]
PS + 5% C$_6$H$_6$	228–283	26	7.5	870	[31]
PS	248–413	75	14	7	[32]
PIB	210–245	75	17.3	7×10^3	[32]
PMMA	278–310	75	12	7×10^{-2}	[32]
PMMA	293			0.36	[33]
PVA	347–310	75	13	0.7	[32]
PFE	293	42	5	4×10^{-3}	[29]
PMP	295–313	92	18.2	107.5	[32]

* [POOH] in mol kg^{-1}.

$> 10^{-4}$ mol L^{-1}), the chain termination proceeds mainly via reaction of peroxyl radical disproportionation, and the oxidation rate has the following form (see Chapter 2).

$$v = v_i + k_p(2k_t)^{-1/2}[\text{PH}]v_i^{1/2} \quad (13.1)$$

The oxidation of PH ([PH] is the concentration of the monomer fragments (in mol kg^{-1})) proceeds by the chain mechanism when the initiation rate is not very high, i.e., when $2k_t v_i < k_p^2[\text{PH}]^2$ (see Chapter 2). The oxidation rate for very long chains is the following:

$$v = k_p(2k_t)^{-1/2}[\text{PH}]v_i^{1/2} = av_i^{1/2} \quad (13.2)$$

The important peculiarity of semicrystalline polymer oxidation is that only the amorphous phase is oxidized [12,13,33,34,42]. Hence, the oxidation rate and the parameter $a = vv_i^{-1/2}$

TABLE 13.4
Values of Kinetic Parameter $a = v v_i^{-1/2} = k_p[PH](2k_t)^{-1/2}$ for Oxidation of Solid Polymers

Polymer	α (%)	Initiator	T (K)	a (mol$^{1/2}$ kg$^{-1/2}$ s$^{-1/2}$)	Ref.
LDPE	65	DCP	389–402	$1.0 \times 10^{10} \exp(-88/RT)$	[43]
HDPE	75	^{60}Co	318	8.1×10^{-4}	[44]
HDPE	75	^{60}Co	295	8.5×10^{-4}	[45]
HDPE	40	DBP	365	5.4×10^{-3}	[43]
IPP	0	DBP	358–378	$8.5 \times 10^{8} \exp(-71/RT)$	[46]
IPP	65	DBP	349–401	$3.0 \times 10^{6} \exp(-57/RT)$	[43,47]
IPP	60	AIBN	317–365	$4.3 \times 10^{2} \exp(-27/RT)$	[48,49]
IPP	0	DBP	344–378	$1.1 \times 10^{2} \exp(-26/RT)$	[7]
IPP	70	^{60}Co	295	9.0×10^{-3}	[50]
IPP	70	^{60}Co	353	1.9×10^{-2}	[25]
IPP	70	^{60}Co	343	1.4×10^{-2}	[25]
IPP	49	DCP	383	6.3×10^{-2}	[51]
IPP	49	DCP	388	8.0×10^{-2}	[51]
IPP	50	POOH	383–413	$1.2 \times 10^{4} \exp(-38/RT)$	[52]
APP	0	DBP	387	0.21	[43]
APP	0	DBPO	295–318	$2.2 \times 10^{5} \exp(-44/RT)$	[9]
CEP (98/2)	60	DCP	390	2.2×10^{-2}	[47]
CEP (96/4)	60	DCP	382–400	$2 \times 10^{9} \exp(-80/RT)$	[47]
CEP (87/13)	55	DCP	388	3.4×10^{-2}	[47]
CEP (65/35)	50	DBP	353–373	$3 \times 10^{6} \exp(-58/RT)$	[47]
CEP (86/14)	14	^{60}Co	318	1.5×10^{-3}	[44]
CEP (73/27)	5	^{60}Co	318	1.6×10^{-3}	[44]
CEP (37/63)	0	^{60}Co	318	4.4×10^{-3}	[44]
PB		DBP	349–383	$7.2 \times 10^{9} \exp(-80/RT)$	[7]
PMP		DBP	358	7.8×10^{-3}	[7]
PMP		DBP	368	1.3×10^{-2}	[7]
PEA		DBP	358–378	$7.6 \times 10^{5} \exp(-54/RT)$	[46]
PBD		$h\nu$	293	9.9×10^{-2}	[53]
NR		AIBN	353	0.64	[54]
CBDS		$h\nu$	293	3.2×10^{-2}	[53]
PVM		DBP	348–373	$2.5 \times 10^{8} \exp(-67/RT)$	[55]

should expediently be referred to the amorphous phase. This can be done by dividing parameter a into $(1-\alpha)^{1/2}$, where α is the polymer crystallinity. Table 13.4 summarizes the α and a values for some commercially available solid-state polymers.

Let us compare the values of ratio $k_p(2k_t)^{-1/2}$ (kg$^{1/2}$ mol$^{-1/2}$ s^{-1}) of polymers and model hydrocarbons (Table 2.1 and Table 13.4).

Reactant	Phase	E (kJ mol^{-1})	A (kg$^{1/2}$ mol$^{-1/2}$ s^{-1})	$k_p(2k_t)^{-1/2}$ (350 K) (kg mol^{-1} s^{-1})$^{1/2}$
Decane	Liquid	58.3	2.5×10^{5}	5.0×10^{-4}
LDPE	Solid	88.0	1.0×10^{10}	2.1×10^{-5}
Pentane, 2,4-dimethyl	Liquid	44.8	8.1×10^{3}	1.7×10^{-3}
APP	Solid	44.0	2.2×10^{5}	2.5×10^{-3}
IPP	Solid	57.0	3.0×10^{6}	3.9×10^{-4}
PB	Solid	80.0	7.2×10^{9}	4.6×10^{-4}

TABLE 13.5
Rate Constants of Peroxyl Radicals Reactions with C—H bonds of Polymers and Model Compounds in Liquid Phase

Oxidizing Compound	Solvent	Radical	T (K)	k (L mol^{-1} s^{-1})	Ref.
Me(CH$_2$)$_8$Me	PhCl	PhMe$_2$CO$_2^\bullet$	388	6.3×10^{-2}	[56]
PE	PhMe$_2$CH	PhMe$_2$CO$_2^\bullet$	388	0.12	[57]
PE	Solid	PO$_2^\bullet$	388	1.2×10^{-3}	[58]
Me$_2$CH(CH$_2$)$_3$MeCH(CH$_2$)$_3$ Me$_2$CH(CH$_2$)$_3$MeCH(CH$_2$)$_3$ MeCH(CH$_2$)$_3$CHMe$_2$	PhCl	Me$_3$CO$_2^\bullet$	363	0.58	[59]
IPP	PhCl	Me$_3$CO$_2^\bullet$	363	0.98	[59]
EtMe$_2$CH	PhMe$_2$CH	EtMe$_2$CO$_2^\bullet$	363	0.35	[60]
Me$_2$CH(CH$_2$)$_3$MeCH(CH$_2$)$_3$ Me$_2$CH(CH$_2$)$_3$MeCH(CH$_2$)$_3$ MeCH(CH$_2$)$_3$CHMe$_2$	PhCl	tetralyl-OO$^\bullet$	363	2.1	[59]
IPP	PhCl	tetralyl-OO$^\bullet$	363	3.3	[59]
IPP	Solid	PO$_2^\bullet$	363	2.6×10^{-4}	[58]
PhMe$_2$CH	PhMe$_2$CH	PhMe$_2$CO$_2^\bullet$	353	4.2	[30]
PS	PhMe$_2$CH	PhMe$_2$CO$_2^\bullet$	353	0.10	[61]
PS	Solid	PO$_2^\bullet$	353	0.12	[62]

It is seen from comparison that these ratios do not differ much. For 2,4-dimethylpentane and APP, they are practically the same. We observe something like compensation here: the retardation of chain propagation and termination in polymers makes the ratios $k_p(2k_t)^{-1/2}$ close in subsequent hydrocarbons and polymers.

Indeed, the comparison of the absolute rate constants of the peroxyl radical reactions (see Table 13.5) demonstrates the great retarding effect of the solid polymer matrix on the rate of this reaction.

The rate constants of this reaction per reacting bond are close for subsequent hydrocarbon and polymer in solution, however, much different in the liquid and solid phases. Two factors are important for this difference: the rigid polymer cage (see Chapter 19) and the additional activation of adjacent segments to change the C—C bond angles in the PO$_2^\bullet$ + PH elementary act. The absolute values of k_p per reacting C—H bond for solid polymers are collected in Table 13.6.

13.1.4 DIFFUSION OF DIOXYGEN IN POLYMER

The dioxygen solubility λ in hydrocarbons depends on the molecular weight: the higher the molecular weight lower the solubility. The empirical dependence has the following form ($T = 300$ K) [12]:

Oxidation of Polymers

$$\lambda \times 10^8 \text{ mol L}^{-1}\text{Pa}^{-1} = 3.0 + 11 n_C^{-1} \quad (13.3)$$

where n_C is the number of carbon atoms in the hydrocarbon. This is explained by an increase in the intermolecular interaction with an increase in the number of carbon atoms of the hydrocarbon. The greater the forces of intermolecular interaction between the solvent molecules, the smaller the free volume and the lower the oxygen solubility. Oxygen diffuses practically in the amorphous phase of polymer (PE, PP) [42]. Therefore, Henry's coefficient λ for the solubility of oxygen can be represented in the form

$$\lambda = \lambda_{am}(1-\alpha)^{-1} \quad (13.4)$$

where λ_{am} is the solubility of O_2 in the amorphous phase of the polymer, and α is a part of the crystalline phase. The values of λ_{am} for several polymers are given below [12].

Polymer	PE	IPP	CEP	PIB
λ_{am} (298 K) ($\times 10^8$ mol L^{-1} Pa^{-1})	3.0	3.8	4.2	4.9

It is seen that the λ_{am} values are close to λ of hydrocarbons extrapolated to $n_C = \infty$ (Equation [13.3]). The O_2 solubility in elastomers is higher than that in polyolefins [67].

Polymer	NR	CBDS (70/30)	PS*
λ_{am} (298 K) ($\times 10^8$ mol L^{-1} Pa^{-1})	6.8	38	14.7

* Polystyrene.

TABLE 13.6
Rate Constants of the Reaction PO$_2$• + PH Per One C—H Bond in Solid Phase

Polymer	T (K)	A (kg mol^{-1} s^{-1})	E (kJ mol^{-1})	k_p (300 K) (kg mol^{-1} s^{-1})	Ref.
PE	230–270	2.1×10^2	39	3.4×10^{-5}	[58]
DPE	270–300	2.4×10^2	43	7.7×10^{-6}	[58]
APP	363–378	1.6×10^7	50.5	2.5×10^{-2}	[7]
APP	318–336	1.7×10^8	62.8	1.9×10^{-3}	[63]
IPP	383–413	3.0×10^5	38	2.8×10^{-2}	[52]
IPP	317–365	9.6×10^{11}	87.8	4.9×10^{-4}	[64]
IPP	270–310	1.1×10^4	53	6.4×10^{-6}	[58]
IPP	303			1.9×10^{-5}	[28]
DIPP	270–310	1.4×10^4	58	1.9×10^{-6}	[58]
PB	363–378	1.1×10^{11}	75	9.5×10^{-3}	[7]
PS	210–293	3.5×10^8	43	11	[62]
PS	210–293	1.3×10^{12}	88	6.1×10^{-4}	[62]
PMMA	250–360	1.0×10^6	50	1.9×10^{-3}	[65]
PMP	298			1.8×10^{-3}	[66]

The diffusion of oxygen in polymers at temperatures higher than T_g occurs by two to four orders of magnitude more slowly than in standard organic solvents [67]. Diffusion of O_2 occurs with the activation energies as high as 30–50 kJ mol^{-1} (see Table 13.7).

The compensation effect was observed between the pre-exponential factor D_0 ($D = D_0 \exp(-E_D/RT)$) and activation energy of diffusion E_D [12]:

$$\log D_0 \, (\text{cm}^2 \, \text{s}^{-1}) = 3.60 + 0.033 E_D T^{-1} \tag{13.5}$$

Oxygen diffuses inside the amorphous phase of amorphous–crystalline polymers. The diffusion coefficient can be expressed for such polymers in the form [68]

$$D = D_{am}/(A \times B) \tag{13.6}$$

where $D_{am} = D$ in the amorphous phase, A is the factor of geometric resistance, B is the factor of lowering the segmental mobility in the amorphous phase near the crystalline phase.

TABLE 13.7
Diffusion Coefficients D of Dioxygen in Polymers

Polymer	T (K)	D (cm^2 s^{-1})	$D \times 10^7$ (298 K) (cm^2 s^{-1})	Ref.
LDPE	278–328	0.43 exp(−35/RT)	1.7	[68]
LDPE		0.53 exp(−37/RT)	1.7	[69]
HDEP	278–328	5.25 exp(−40.3/RT)	4.6	[68]
HDEP	298–340	2.0 exp(−38.5/RT)	5.3	[70]
PE	278–328	0.83 exp(−39/RT)	12	[68]
IPP	298–340	42 exp(−46.2/RT)	4.7	[71]
IPP	298–340	1.5 exp(−36.4/RT)	6.6	[70]
IPP	403	6 × 10^{-6}		[71]
IPP	366	7 × 10^{-7}		[72]
PS		4.8 × 10^{-4} exp(−26/RT)	0.13	[73]
PS		0.12 exp(−34.9/RT)	1.11	[74]
NR	298–323	1.9 exp(−34.9/RT)	14	[75]
NR	298–323	0.3 exp(−29.7/RT)	18	[75]
SSR	298	17.3	17.3	[76]
PBD	298–323	0.095 exp(−27.2/RT)	16	[75,76]
PBD		0.14 exp(−28.4/RT)	16	[74]
PDMB	298–323	9.2 exp(−44.3/RT)	1.5	[75,76]
PDMB		19 exp(−46.4/RT)	1.4	[74]
CBDS	298–323	0.23 exp(−29.7/RT)	14	[71]
CBA (61/39)	298–323	42.5 exp(−48.6/RT)	1.3	[75,76]
CBA (61/39)		14 exp(−45.6/RT)	1.3	[74]
CBA (68/32)	298–323	7.5 exp(−42.2/RT)	3.0	[76]
CBA (68/32)	298–323	9.8 exp(−43/RT)	36	[75]
CBA (80/20)	298–323	4.8 exp(−34/RT)	52	[75,76]
CBA (80/20)	298–323	0.7 exp(−33.9/RT)	8.1	[74]
PEMA	298–358	0.039 exp(−31.8/RT)	1.0	[77]
PMMA	298	0.03	0.03	[78]
PVC		41 exp(−54/RT)	0.12	[79]
PMP	298	1.4 × 10^{-6}	14	[80]

Polymer	LDPE	HDPE*	PIB (Hydrogenated)
α	0.77	0.43	0.29
A	6.4	3.2	1.4
B	1.6	1.2	1.0

*High-density polyethylene.

The theory of diffusion in polymers as heterogeneous media was discussed in Refs. [68,74,81–85]. The correlation between the frequency of rotation ν_r of the nitroxyl radical (TEMPO) and diffusion coefficient of oxygen D (298 K) was found [86].

$$\log \nu_r (\text{rad s}^{-1}) = 15.2 + 0.87 \log D(\text{cm}^2\ \text{s}^{-1}) \tag{13.7}$$

13.1.5 Diffusion Regime of Polymer Oxidation

Let us consider the easiest example of oxygen diffusion and reaction in polymer film having the thickness $2l$. When oxygen pressure is low (see Chapter 2), the rate of chain oxidation of polymer PH is proportional to pO_2 ($v = k'_p[P^\bullet][O_2] = k'_p(2k'_t)^{-1/2}v_i^{1/2}[O_2] = k'[O_2]$) and proceeds as the reaction of the formally first order. The dependence of the oxygen concentration on the distance x from the surface is described by the diffusion equation in the quasistationary regime.

$$D\frac{d^2[O_2]}{dx^2} = k'[O_2] \tag{13.8}$$

and obeys the following formula [12]:

$$[O_2] = [O_2]_0 \frac{ch(x\sqrt{k/D})}{ch(l\sqrt{k/D})} \tag{13.9}$$

After the differentiation of this equation, one comes to the following reaction rate $v(S)$ per surface unit of the film:

$$v(S)\ (\text{mol cm}^{-2}\ \text{s}^{-1}) = 2D\left(\frac{dC}{dx}\right)_{x=l} = 2 \times 10^{-3}[O_2]_0\sqrt{Dk'} \times \text{th}(l\sqrt{k'/D}) \tag{13.10}$$

This rate $v(S)$ is proportional to the film thickness for thin films ($l < 0.3\sqrt{D/k'}$) and does not depend on l for thick films ($l > 2\sqrt{D/k'}$). The reaction rate calculated per volume unity ($v = v(S)/2l$) is the following:

$$v = [O_2]_0 l^{-1}\sqrt{D/k'} \times \text{th}(l\sqrt{D/k'}) \tag{13.11}$$

and is equal to $k'[O_2]_0$ for thin films and to $[O_2]_0 l^{-2}\sqrt{D/k'}$ for thick films. The time τ of the statement of stationary regime is equal to [12]:

$$\tau = \frac{l^2}{l^2 k' + \pi^2 D} \tag{13.12}$$

and $\tau \approx 1/k'$ at $k' \gg \pi^2 D/l^2$. When oxygen partial pressure is high, the oxidation rate does not depend on pO_2 (see Chapter 2) and is equal to $v = k_p(2k_t)^{-1/2}[\text{PH}]\ v_i^{1/2} = v_\infty$.

Solving the equation

$$D\frac{\partial^2[O_2]}{\partial x^2} = v_\infty \qquad (13.13)$$

we come to the equations

$$v(S) = 2lv \quad \text{for } l < \sqrt{2D[O_2]_0/v_\infty} \qquad (13.14)$$

$$v(S) = \sqrt{2D[O_2]_0 v_\infty} \quad \text{for } l \geq \sqrt{2D[O_2]_0/v_\infty} \qquad (13.15)$$

Two different approaches are possible for the description of polymer oxidation in the diffusion regime at variable pO_2:

1. The first approach deals with the average oxygen concentration in the oxidized sample. Let us suppose that the oxidation rate of polymer depends on the oxygen concentration by the following way [12]:

$$v = \frac{a\sqrt{v_i}}{1 + d/[O_2]}, \qquad (13.16)$$

where $[O_2]$ is the average oxygen concentration in the powder or film of the oxidized polymer. Oxygen penetrates into the amorphous phase of the polymer with the apparent rate κ (s^{-1}). In addition to dioxygen diffusing into and out of the polymer sample, its average rate of dissolution should depend on the difference between λpO_2 and $[O_2]$ in polymer. In the steady-state regime of diffusion and oxidation, the rates of oxygen dissolution and consumption are equal. The following equation results from this hypothesis [87]:

$$\frac{pO_2}{v} = \frac{d}{\lambda} \times \frac{1}{a\sqrt{v_i} - v} + \frac{1}{\lambda \kappa} \qquad (13.17)$$

When diffusion of oxygen in the polymer occurs rapidly ($\lambda \kappa \gg 1$), we observe the kinetic regime of polymer oxidation (see Equation [13.16]). When penetration of oxygen into polymer is slow, the reaction rate $v \approx \kappa \lambda pO_2$.

2. Another approach lies in solving of the diffusion equation mathematically as [12]

$$D\frac{\partial^2[O_2]}{\partial x^2} - \frac{a\sqrt{v_i}}{1 + d/[O_2]} = 0 \qquad (13.18)$$

This equation was calculated numerically for the different values of coefficients D, a, $\sqrt{v_i}$, and d. The theoretical dependence was found to coincide with Equation (13.17) for the coefficient $\kappa = 8D/3l^2$ at $v/v_\infty \geq 0.5$. The numerical calculation gave the following equation for the mixed (kinetic and diffusion) regime of oxidation [87]:

$$\frac{pO_2}{vl^2} = \frac{1}{6D\lambda} + \frac{4pO_2}{5al^2\sqrt{v_i}} \qquad (13.19)$$

Experimental data on the oxidation of the IPP films with different thicknesses at different pO_2 appeared to be in good agreement with this equation ($T = 366$ K, initiator is dibenzoylperoxide) [72]. The following parameters characterize the mixed (kinetic and diffusion) oxidation of IPP films: $a = 1.3 \times 10^{-2}$ mol$^{1/2}$ kg$^{-1/2}$ s$^{-1/2}$, $D\lambda = 2.2 \times 10^{-9}$ cm^2 mol kg^{-1} atm^{-1} s^{-1}, $D = 7.0 \times 10^{-7}$ cm^2 s^{-1} at $\lambda = 3 \times 10^{-3}$ mol kg^{-1} atm^{-1}.

13.1.6 Isomerization of Alkyl and Peroxyl Radicals of Polypropylene

The peculiarities of the oxidation of PP, whose molecules have alternating tertiary C—H bonds in the β-position, are of special interest. Such branched alkanes are oxidized with the formation of polyatomic hydroperoxides produced by the intramolecular isomerization of the peroxyl radical [88].

The greater the number of adjacent tertiary C—H bonds the higher the ratio k_{pis}/k_p [12, 89–91].

Compound	CH$_2$(CHMe$_2$)$_2$	CH(MeCH$_2$CHMe$_2$)$_2$	PP
k_{pis}/k_p (373 K) (mol L^{-1})	40	67	80

The rise in this ratio with the increasing number of tertiary C—H bonds in the molecule is explained by the increased probability of peroxyl radical undergoing isomerization. The experiments indicate that oxidized PP contains mainly block hydroperoxyl groups [12,88]. Hydrocarbons with tertiary C—H bonds (for example, isobutane, isopentane, and cumene) are oxidized in the liquid phase to stable molecular products, mainly hydroperoxides and $\Delta[O_2] = [ROOH]$. The recombination of tertiary peroxyl radicals gives rise to small amounts of dialkyl peroxide and alcohol (see Chapter 2).

During the oxidation of PP, the amount of oxygen consumed is greater than the amount of the hydroperoxy groups formed. The difference could be ascribed to the decomposition of hydroperoxy groups; however, the extrapolation of $[POOH]/\Delta[O_2]$ to the initial conditions, when $\Delta[O_2] \to 0$, shows that $[POOH]/\Delta[O_2] < 1$ beginning from the very onset of oxidation. The difference $1 - ([POOH]/\Delta[O_2])$ exceeds $1/\nu$ and, therefore, cannot be explained by the formation of alternative products in chain termination reactions. At the same time, alternative products (that is, hydroxyl and carbonyl groups) were revealed in chain propagation reactions by measuring the radiochemical yield of these products, G, versus intensity of radiation, I, which appears to be related as [91]:

$$G = \alpha + \beta I^{1/2} \qquad (13.20)$$

where α and β are the empirical characteristics of product formation in the chain termination and propagation reactions, respectively. At $T = 318$ K, these parameters are the following.

Product	$\Delta[O_2]$	POOH	POH	>C=O
α (molecule/100 eV)	9.0	0	6.0	4.6
β	3660	2900	360	210

As is evident from these data, mainly hydroxyl and carbonyl groups are formed in the chain propagation reactions.

During PP oxidation, hydroxyl groups are formed by the intramolecular isomerization of alkyl radicals. Since PP oxidizes through an intense intramolecular chain transfer, many of the alkyl radicals containing hydroperoxy groups in the β-position to an available bond can undergo this reaction. An isomerization reaction has also been demonstrated for the liquid-phase oxidation of 2,4-dimethylpentane [89]. Oxidation products contain, in addition to hydroperoxides, oxide or diol.

At 373 K, the ratio of the rate constants is $k_{is}/k_p' = 8 \times 10^{-5}$ L^{-1} mol, where k_{is} and k_p' refer to the reactions of isomerization and addition of oxygen. Isomerization competes with the reaction of the macroradical P· with O$_2$; therefore, intense hydroxylation during the oxidation of PP may imply that isomerization in the solid phase is slower than in the liquid phase. The experimentally measured ratio Δ[O$_2$]/[ROOH] at different partial pressure of oxygen helps to estimate the ratio k_{is}/k_p' in the oxidized polymer. Since the kinetics of chain PP oxidation are characterized by the rate v

$$v = k_p'[O_2][P^\bullet] + k_{is}[P^\bullet] = v_{POOH} + v_{POH} \tag{13.21}$$

and then

$$\frac{\Delta[O_2]}{[POOH]} = 1 + \frac{k_{is}}{k_p' \lambda pO_2} \tag{13.22}$$

The following values of the k_{is}/k_p' ratio were estimated from the experimental data.

Polymer	T (K)	k_{is}/k_p' (kg mol^{-1})	Ref.
PP	365	1.6×10^{-4}	[92]
PP	298	2.6×10^{-4}	[93]
PS	463	9.1×10^{-4}	[94]

The γ-radiation-initiated oxidation of PE at 295 K was accompanied by the formation of both hydroperoxy (1.4×10^{-2} mol kg^{-1}) and carbonyl (1.6×10^{-2} mol kg^{-1}) groups [95]. The most likely mechanism of their formation is through the consecutive intramolecular isomerization of secondary peroxyl radicals (see Chapter 2).

Oxidation of Polymers

For pentane, the ratio is $k_p'/k_p = 1.4\,\text{L}^{-1}$ mol and $k_p'/k_p[\text{RH}] = 0.16$ [96]. Of the three pentane peroxyl radicals, only two have CH_2 groups in the α-position. Therefore, when considering the behavior of peroxyl radicals of PE, the ratio $k_p'/k_p[\text{RH}]$ must be increased by three times.

The reactions of intramolecular isomerization occur and are important in the oxidation of natural and synthetic rubbers. The peroxyl radical addition to the double bond occurs very rapidly. For example, the peroxyl radical adds to the double bond of 2-methylpropene by 25 times more rapidly than abstraction of hydrogen atom from this hydrocarbon (see Chapter 4). Therefore, the oxidation of polymers having double bonds proceeds as a chain process with parallel reactions of $PO_2^\bullet$ with double and C—H bonds including the intramolecular isomerization of the type [12]:

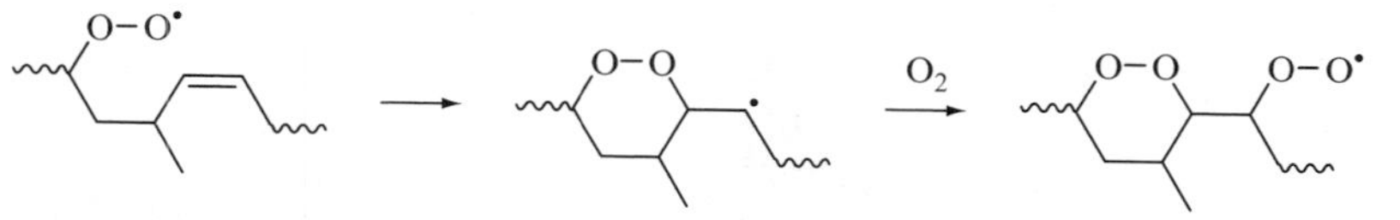

The formation of peroxyl radicals with peroxide bridges are very important for the degradation of this polymer (see later).

The data described above proved that isomerization of alkyl and peroxyl radicals plays a very important role in polymer oxidation. They influence the composition of products of polymer oxidation including the structure of hydroperoxy groups. The competition between reactions of alkyl radical isomerization and addition of dioxygen appeared to be very important for the self-initiation and, hence, autoxidation of PP (see later).

13.2 AUTOXIDATION OF POLYMERS

13.2.1 Chain Generation by Dioxygen

Free radical formation in oxidized organic compounds occurs through a few reactions of oxygen: bimolecular and trimolecular reactions with the weakest C—H bond and double bond (see Chapter 4). The study of free radical generation in polymers (PE, PP) proved that free radicals are produced by the reaction with dioxygen. The rate of initiation was found to be proportional to the partial pressure of oxygen [6,97]. This rate in a polymer solution is proportional to the product $[\text{PH}] \times [O_2]$. The values of the apparent rate constants (k_{i0}) of free radical formation by the reaction of dioxygen ($v_{i0} = k_{i0}[\text{PH}][O_2]$) are collected in Table 13.8.

Such a chain generation could be expected to proceed by the reaction of polymer C—H bonds with dioxygen. The values of k_{i0} per C—H bond should be close in the polymer and the model hydrocarbon. The results of comparison at $T = 403$ K are given below [12].

Compound	PE	Heptane	PP	PP(solution)	Isooctane
k_{i0}(C—H) (L mol^{-1} s^{-1})	2.3×10^{-6}	2.0×10^{-10}	2.5×10^{-6}	4.4×10^{-7}	2.2×10^{-9}

The chain generation in polymers in the solid phase and solution is by several orders magnitude higher than that in pattern hydrocarbon. Hence, other fast reactions of dioxygen generate the chains more rapidly than that in hydrocarbons. The comparison of the k_{i0} values in PE with the percentage of ash showed the correlation ($T = 391$ K, LDPE [6]).

TABLE 13.8
Free Radical Generation on Reaction of Polymer with Dioxygen [6,97]

Polymer	Crystallinity (% of ash)	T (K)	$v_{i0} \times 10^7$ (mol kg^{-1} s^{-1})	$k_{i0} \times 10^6$ (L mol^{-1} s^{-1})	A (L mol^{-1} s^{-1})	E (kJ mol^{-1})
LDPE	65 (0.7)	391	0.67	0.70		
LDPE	65 (0.4)	391	0.20	0.21		
LDPE	65 (0.14)	391	0.13	0.14		
LDPE	55 (0.45)	378	0.54	0.57	2.2×10^{10}	117
LDPE	55 (0.45)	386	2.07	2.18	2.2×10^{10}	117
LDPE	55 (0.45)	388	3.60	3.80	2.2×10^{10}	117
LDPE	55 (0.45)	398	8.01	8.43	2.2×10^{10}	117
LDPE	55 (0.45)	404	12.6	13.3	2.2×10^{10}	117
HDPE	40	362	0.51	0.54	6.8×10^{14}	146
HDPE	40	367.5	1.19	1.25	6.8×10^{14}	146
HDPE	40	372	1.65	1.73	6.8×10^{14}	146
HDPE	40	377	3.94	4.14	6.8×10^{14}	146
HDPE	40	377	33.45	3.62	6.8×10^{14}	146
IPP	65 (0.048)	387	0.10	0.12	2.3×10^7	92
IPP	65 (0.4)	380	4.3	5.33	2.3×10^7	92
IPP	65 (0.4)	389	8.2	10.2	2.3×10^7	92
IPP	65 (0.4)	398	11.2	13.9	2.3×10^7	92
IPP	65 (0.45)	405	20.6	25.5	2.3×10^7	92
IPP*	PhCl (0.4)	385	0.028	0.014	3.4×10^5	98
IPP*	PhCl (0.4)	389	0.47	0.023	3.4×10^5	98
IPP*	PhCl (0.4)	394	0.050	0.025	3.4×10^5	98
IPP*	PhCl (0.4)	399	0.063	0.031	3.4×10^5	98
IPP*	PhCl (0.4)	403	0.089	0.44	3.4×10^5	98

*In chlorobenzene solution.

% Ash	0.14	0.40	0.70
k_{i0} (L mol^{-1} s^{-1})	1.4×10^{-7}	2.1×10^{-7}	7.0×10^{-7}

This means that the main source of free radicals are reactions of dioxygen with impurities and residues of the technological catalyst.

13.2.2 Decomposition of Hydroperoxyl Groups

Like the oxidation of hydrocarbons, the autocatalytic oxidation of polymers is induced by radicals produced by the decomposition of the hydroperoxyl groups. The rate constants of POOH decomposition can be determined from the induction period of polymer-inhibited oxidation, as well as from the kinetics of polymer autoxidation and oxygen uptake. The initial period of polymer oxidation obeys the parabolic equation [12]

$$\Delta[O_2]^{1/2} = bt = ak_i^{1/2}t, \qquad (13.23)$$

or

$$v^2 = a^2 v_i = a^2 k_i [\text{POOH}] \quad (13.24)$$

where k_i is the apparent rate constant of hydroperoxyl group decomposition to free radicals. The values for k_i measured by the inhibition technique and from the kinetics of oxidation are summarized in Table 13.9.

The decomposition of the hydroperoxyl group into molecular products occurs much more rapidly (see Table 13.10).

This is the result of cage effect. The cage of a solid polymer matrix is rigid (see earlier) and the most part of the forming radical pairs recombine in the cage. Hence, the probability of

TABLE 13.9
Rate Constants of Polymer Hydroperoxide Groups Decay

Polymer	T (K)	Medium	k_d (s^{-1})	Ref.
HDPE	373–408	PhCl	$1.1 \times 10^8 \exp(-104/RT)$	[98]
HDPE	393–408	PhCl	$1.2 \times 10^8 \exp(-113/RT)$	[98]
LDPE	383–403	PhCl	$8.5 \times 10^{15} \exp(-146/RT)$	[99]
LDPE	373–393	PhCl + InH	$1.9 \times 10^8 \exp(-84/RT)$	[99]
HDPE	333–373	Solid phase	$2.2 \times 10^8 \exp(-92/RT)$	[100]
HDPE	373–393	Solid phase	$6.2 \times 10^{17} \exp(-162.5/RT)$	[101]
HDPE	413–443	Solid phase	$2.5 \times 10^{14} \exp(-146/RT)$	[102]
HDPE	430	Solid phase, O_2	2.6×10^{-3}	[103]
PE	403	Solid phase	4.7×10^{-4}	[104]
PE	403	Solid phase, O_2	5.3×10^{-3}	[104]
APP	385–395	PhCl	$2.5 \times 10^{15} \exp(-113/RT)$	[105]
APP	363–393	Solid phase	$9.5 \times 10^{11} \exp(-113/RT)$	[106]
APP	363–393	Solid phase	$2.3 \times 10^{12} \exp(-115/RT)$	[107]
IPP	398	PhCl	6×10^{-4}	[99]
IPP	392–407	C_6H_6	$3.8 \times 10^{10} \exp(-109/RT)$	[99]
IPP	393–413	Solid phase	$7.5 \times 10^{10} \exp(-109/RT)$	[108]
IPP	393	Solid phase	$(2.6 \div 3.1) \times 10^{-4}$	[109]
IPP	403	Solid phase	$(6.7 \div 8.7) \times 10^{-4}$	[110]
IPP	403	Solid phase	11.5×10^{-4}	[111]
IPP		Solid phase	$1.3 \times 10^{10} \exp(-100/RT)$	[112]
IPP		Solid phase	$1.7 \times 10^{12} \exp(-129/RT)$	[112]
IPP		Solid phase	$2.2 \times 10^{10} \exp(-104/RT)$	[113]
IPP	403	Solid phase, O_2	8.7×10^{-4}	[114]
IPP	398	Solid phase	1.8×10^{-4}	[99]
PMP		Solid phase	$2 \times 10^7 \exp(-84/RT)$	[115]
PMP		Solid phase	$2.4 \times 10^8 \exp(-96/RT)$	[115]
PMP	403	Solid phase	7.0×10^{-4}	[116]
CEP (95/5)	403	Solid phase	3.7×10^{-4}	[104]
CEP (76/24)	403	Solid phase	1.2×10^{-4}	[104]
PS	353–413	PhCl	$1.2 \times 10^{21} \exp(-82.5/RT)$	[117]
PS	463	Solid phase	6.4×10^{-4}	[94]
PS	463	Solid phase	6.8×10^{-3}	[94]
PS	453–473	Solid phase	$8.1 \times 10^{13} \exp(-150.7/RT)$	[102]
PBD	353–403	Solid phase	$1.1 \times 10^3 \exp(-46.5/RT)$	[118]
SSR	353–403	Solid phase	$2.2 \times 10^7 \exp(-77/RT)$	[118]

TABLE 13.10
Rate Constants of Hydroperoxide Group Decomposition into Free Radicals Measured by Kinetics of Autoxidation and by Free Radical Acceptor Method

Polymer	Conditions of Oxidation	T (K)	k_i (s^{-1})	Ref.
PE	PhCl, 383 K	351–403	$1.6 \times 10^{14} \exp(-146/RT)$	[99]
PE	Solid phase, 363 K	365	5×10^{-6}	[119]
IPP	Me$_2$PhCH, 383 K	392–403	$2.5 \times 10^{12} \exp(-134/RT)$	[99]
IPP	PhCl, 383 K	365–387	$2.4 \times 10^6 \exp(-79/RT)$	[99]
IPP	Solid phase, 363 K	365–387	$7 \times 10^7 \exp(-92/RT)$	[119]
IPP	Solid phase, 358 K	322–370	$3.1 \times 10^{11} \exp(-119/RT)$	[99]
IPP	Solid phase	393	4.6×10^{-5}	[114]
IPP	Solid phase	403	8.1×10^{-5}	[114]
IPP	Solid phase	413	1.4×10^{-4}	[114]
IPP	Solid phase	383	1.2×10^{-4}	[52]
IPP	Solid phase	393	2.5×10^{-4}	[52]
IPP	Solid phase	403	3.5×10^{-4}	[52]
IPP	Solid phase	413	6.0×10^{-4}	[52]
IPP	Solid phase	365	1.4×10^{-5}	[120]
IPP	Solid phase	383	2.6×10^{-6}	[99]
IPP	Solid phase	393	8.9×10^{-6}	[99]
IPP	Solid phase	383	2.0×10^{-5}	[99]
IPP	Solid phase	393	4.6×10^{-5}	[99]
IPP	Solid phase	403	6.2×10^{-5}	[99]
IPP	Solid phase	348	1.3×10^{-6}	[119]
IPP	Solid phase	365	5.0×10^{-6}	[119]
IPP	Solid phase	387	2.8×10^{-5}	[119]
IPP	Solid phase	403	6.0×10^{-5}	[121]
PMP	Solid phase	430	1.2×10^{-5}	[116]
HDPE	Melt	430	1.0×10^{-4}	[103]

radical pairs to escape cage recombination is sufficiently lower than in organic solvents (see Chapter 3). The values of e for decay of hydroperoxyl group are given below.

Polymer	T (K)	e	Ref.
IPP	403	0.14	[113]
	403	0.019	[114]
	398	0.03	[99]
PMP	403	0.04	[115]
	403	0.034	[116]

The decomposition of a single hydroperoxyl group in the absolutely pure polymer can proceed by two mechanisms: by the reaction of POOH with the C—H bond of a macromolecule and by the monomolecular cleavage of the O—O bond (see Chapter 4). If the first reaction prevails, the single POOH group of PP should break down more rapidly than those of PE, because the latter polymer has stronger C—H bonds (the BDE difference between the secondary and tertiary C—H bonds is as much as 22.5 kJ mol^{-1}). The k_d values of the single hydroperoxyl group decay in PE and PP in solution and solid state ($T = 398$ K [99]) are given in the following table:

Polymer	LDPE in PhCl	IPP in PhCl	PE	IPP
k_d (s^{-1})	5.4×10^{-4}	1.8×10^{-4}	5.4×10^{-4}	1.8×10^{-4}

It is seen that the values of k_d are very close. Hence, the reaction of POOH with the C—H bond is not the main initiation reaction. If the breakdown is a monomolecular process, the rate of O—O bond homolysis in polymer must be close to that in the gas phase. 2,2-Dimethylethyl hydroperoxide breaks down in the gas phase with a rate constant of $1.6 \times 10^{13} \exp(-158/RT) = 5.3 \times 10^{-8}$ s^{-1} (398 K, [4]), that is, by four orders of magnitude more slowly than in polymer. Hence, the decomposition reactions in the polymers are much faster than the monomolecular homolysis of peroxide. Decomposition reactions may be of three types (see Chapter 4), such as the reaction of POOH with a double bond

$$POOH + RCH=CHR \longrightarrow PO^{\bullet} + HORCHC^{\bullet}HR$$

the reaction of POOH with the hydroxyl group

$$POOH + POH \longrightarrow PO^{\bullet} + H_2O + PO^{\bullet}$$

and the reaction with the carbonyl group

$$POOH + R_2C=O \longrightarrow POOCR_2OH \longrightarrow PO^{\bullet} + {^{\bullet}}OCR_2OH$$

The mechanism of POOH decomposition in PP is the subject of special discussion.

13.2.3 SPECIFICITY OF FORMATION OF THE HYDROPEROXYL GROUP AND ITS DECAY IN PP

Two different hydroperoxyl groups are formed in oxidized PP: single and adjacent [52]. This is the result of the competition of intramolecular and intermolecular reactions of PP peroxyl radicals (see earlier). According to Mayo's estimation, the first occurs five times more rapidly than the second [122]. The decomposition of PP hydroperoxyl groups depends on the structure [12]. PP with a single hyroperoxyl group was produced by the co-oxidation of PP (PH), with cumene (RH) where oxidation proceeds via the following chain propagation reactions [57]:

$$RO_2^{\bullet} + PH \longrightarrow ROOH + P^{\bullet}$$
$$P^{\bullet} + O_2 \longrightarrow PO_2^{\bullet}$$
$$PO_2^{\bullet} + RH \longrightarrow POOH + R^{\bullet}$$
$$R^{\bullet} + O_2 \longrightarrow RO_2^{\bullet}$$

Due to the high reactivity of cumene, the reaction of the peroxyl macroradical with cumene occurs more rapidly than the intramolecular reaction and the formed POOH is only from the single hydroperoxyl groups. Such POOH decomposes with free radical formation much more slowly than POOH produced in PP oxidation in the solution and solid state.

T (K)	E (kJ mol^{-1})	log A, A (L mol^{-1} s^{-1})	k_i (400 K) (L mol^{-1} s^{-1})	Ref.
	PP was Oxidized in Cumene at 383 K			
392–403	134.0	12.40	7.93×10^{-6}	[99]
	PP was Oxidized in Chlorobenzene at 383 K			
383–431	79.0	6.38	1.16×10^{-4}	[99]
	PP was Oxidized in Solid Phase with Benzoyl Peroxide at 363 K at $pO_2 = 10^5$ Pa			
365–387	92.0	7.84	6.78×10^{-5}	[119]
	PP was Oxidized in Solid Phase with AIBN at 358 K at $pO_2 = 10^5$ Pa			
322–370	119.0	11.49	8.95×10^{-5}	[99]

The rapid decomposition of adjacent POOH groups is the result of decomposition similar to the bimolecular decomposition of two ROOH (see Chapter 4).

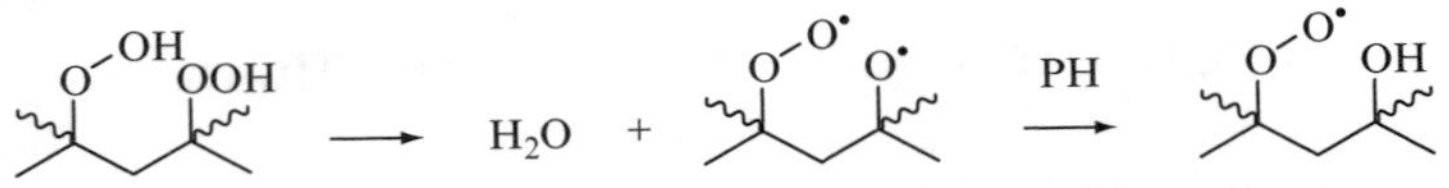

The formed biradical can form trioxide. This trioxide is unstable and rapidly decomposes back to the biradical [4]. Then the biradical rapidly reacts with the C—H bond and forms the peroxyl monoradical.

As noted above, the kinetics of autoxidation of PP depends on the partial pressure of oxygen. This is the result of the mechanistic peculiarity of PP oxidation. This peculiarity lies in competition between two reactions: the addition of oxygen to the 2-hydroperoxyalkyl macroradical and the decomposition of this radical.

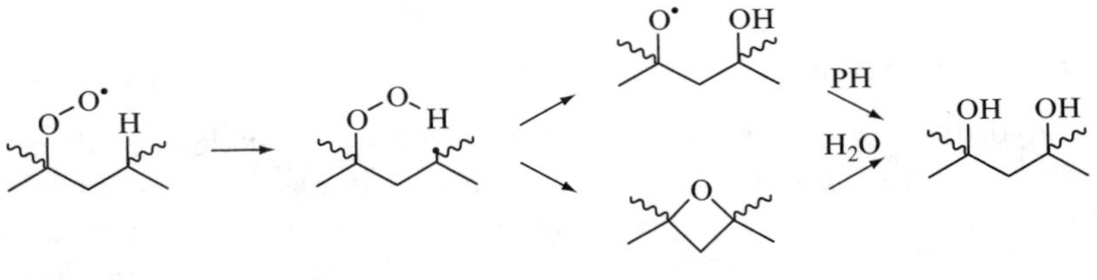

The higher the concentration of O_2 in the oxidized PP, the greater the yield of block hydroperoxyl groups per oxygen molecule consumed and the proportion of block hydroperoxyl groups that break up more rapidly than single groups. This was proved experimentally [99]. IPP was oxidized at different partial pressures of oxygen. Then the initiation rate constant of the formed POOH was measured by the free radical acceptor method at $T = 393$ K. These results demonstrate that the the dioxygen partial pressure in PP oxidation increases with the rate constant k_i.

$pO_2 \times 10^5$ (Pa)	0.21	1.0	10.0
$k_i \times 10^5$ (s^{-1}) ($T = 393$ K)	1.4	2.7	5.1

Hence, the apparent initiation rate constant of POOH in PP depends on the following reactions: bimolecular chain propagation (k_p), intramolecular chain propagation (k_{pis}), isomerization of the 2-hydroperoxylalkyl macroradical (k_{is}), and addition of dioxygen to the alkyl macroradical, as well as on the concentration of dioxygen and cage effect ($k_i = 2ek_d$). The analysis of the full kinetic scheme leads to the following equation for the apparent rate constant of initiation by POOH groups of PP [12].

$$k_i = k_i(\text{single}) \times \left\{ g + \frac{(1-g)h}{h + [O_2]} \right\} + k_i(\text{block}) \frac{(1-g)[O_2]}{h + [O_2]} \quad (13.25)$$

where $g = k_p[PH](k_p[PH] + k_{pis})^{-1}$ and $h = k_p[PH]/k'_p$. This equation was found to agree with experiment [99]. The hypothesis was put forward that dioxygen influences the cage effect in solid polymers [123]. The experimental measurements proved that the partial pressure of dioxygen does not influence the initiation rate constant of POOH in the solid phase. As an example, the values of k_i for POOH of IPP calculated from the experimental values of dioxygen consumption rates expressed in (mol kg^{-1} s^{-1}) and measured at different pO_2 values at $T = 387$ K [6] are given below:

[POOH] (mol kg^{-1})	pO_2 (kPa)				
	40	56	60	70	100
0.040	3.01 × 10^{-5}				3.05 × 10^{-5}
0.095	2.62 × 10^{-5}		2.83 × 10^{-5}	2.85 × 10^{-5}	2.82 × 10^{-5}
0121	2.35 × 10^{-5}	2.95 × 10^{-5}			2.95 × 10^{-5}

Similar to that of hydrocarbons, the autoxidation of PP occurs with acceleration and obeys the following equation [12] because the decay of POOH is negligible.

$$\sqrt{\Delta[O_2]} = \alpha a \sqrt{k_i} \times t \quad (13.26)$$

where α is the yield of POOH groups, $a = (k_p[PH] + k_{pis})(2k_t)^{-1/2}$ and k_i is the apparent rate constant of initiation by POOH groups. As we see, two parameters depend on the concentration of dioxygen, namely, the yield of hydroperoxyl groups α and the rate constant of self-initiation k_i. Besides this, the diffusion of dioxygen in the polymer bulk should also be taken into account too (see Equation [13.19]). Taking into account all these factors we came to the following equation for the initial step of PP autoxidation [12]:

$$\sqrt{\Delta[O_2]} = \frac{1}{2} \frac{a[O_2]}{2d + [O_2]} \sqrt{\frac{[O_2]}{[O_2] + q}} \sqrt{\frac{k_i(\text{block})(1-g)[O_2]}{h + [O_2]}} \, t, \quad (13.27)$$

where a, g, and h are given earlier, $d = k_p[PH]/k'_p$, and $q = k_{is}/k'_p$. These parameters were found to be the following for IPP oxidation at 365 K at different pO_2 values: $a = 1.6 \times 10^{-2}$ (mol kg^{-1} s^{-1})$^{1/2}$, $d = 2.7 \times 10^{-4}$ mol kg^{-1}, $g = 0.8$, $q = 6.0 \times 10^{-5}$ mol kg^{-1}, $k_i(\text{block}) = 5.0 \times 10^{-6}$ s^{-1}.

13.2.4 Chain Decay of Hydroperoxyl Groups of PP

In addition to homolytic decomposition, hydroperoxyl groups of polymer are decomposed in reactions with alkyl radicals (see Chapter 4). Such induced decomposition of POOH groups was found and studied in PE and PP [6,119]. Three methods of study of POOH chain decomposition in the absence of dioxygen and in oxidized polymer were developed.

1. The kinetic study of POOH decay in an inert atmosphere in the presence of another initiator (I). The initiator increases the concentration of macroradicals in the polymer media. If free radicals react with hydroperoxyl groups, one observes the acceleration of POOH decay. The rate of POOH decay in the case of induced decomposition obeys the equation

$$v_{POOH} = k_d[POOH] + k_{ind}[POOH]v_i^{1/2} \qquad (13.28)$$

The experimentally measured rate constant of POOH decay k_Σ is given by

$$k_\Sigma = k_d + k_{ind}(k_I[I] + k_i[POOH])^{1/2} \qquad (13.29)$$

The extrapolation of v_{POOH} to $v_i = 0$ gives $v_d = k_d[ROOH]$ and the slope $\Delta k_\Sigma/\Delta k_I[I]$ is equal to k_{ind}. For example, the study of benzoyl peroxide initiated decay of POOH in IPP gave $k_d = 1.8 \times 10^{-5}\,s^{-1}$ and $k_{ind} = 3.0 \times 10^{-2}\,kg^{1/2}\,mol^{-1/2}\,s^{-1/2}$ ($T = 365\,K$, N_2 [119]).

2. The study of POOH decay in the presence and absence of a free radical acceptor. In the absence of an inhibitor, the decay of POOH proceeds homolytically and under the action of radicals.

$$v_{POOH} = k_d[POOH] + k_{ind}[POOH](k_i[POOH])^{1/2} \qquad (13.30)$$

The addition of an acceptor decreases the rate of POOH decomposition. The increase of added [InH] creates a tendency for k_Σ to decrease to the k_d value, i.e., $k_\Sigma \to k_d$ at [InH] $\to \infty$. Acceptors, which do not react with hydroperoxide groups, were used; sterically hindered phenols and stable nitroxyl radicals (TEMPO) were found to be efficient acceptors. The ratio $k_{ind}(2k_t)^{1/2}$ can be calculated from the values k_Σ and k_d according to the formula:

$$k_{ind} = \frac{k_\Sigma - k_d}{\sqrt{k_i[POOH]}} \qquad (13.31)$$

3. The value of k_Σ can be estimated for POOH decomposition in oxidized polymer from a comparison of POOH produced by oxidation and that analytically found in polymer:
   a. from the values of $[POOH]_{max}$ and rate of oxidation in this moment v_{max}:

$$k_\Sigma = \frac{\alpha v_{max}}{[POOH]_{max}} \qquad (13.32)$$

   b. from the values of $\Delta[O_2]$ consumed and the kinetic curve of POOH produced during oxidation:

$$k_\Sigma = \frac{\alpha\Delta[O_2] - [POOH]}{\int [POOH]dt} \qquad (13.33)$$

The results of k_Σ, k_d, and k_{ind} measurements are presented in Table 13.11.

It is seen that POOH of PP and PE are decomposed under the action of free radicals in an inert atmosphere and in dioxygen. The induced decomposition of POOH in PP proceeds by

TABLE 13.11
Kinetic Parameters of Induced Decomposition of POOH in Solid Phase [119]

Atmosphere	Polymer	T (K)	$k_d \times 10^4$ (s^{-1})	$k_i \times 10^5$ (s^{-1})	$k_{ind} \times 10^2$ (kg$^{1/2}$ mol$^{-1/2}$ s$^{-1/2}$)
N$_2$	IPP, OOH group (adjacent)	387	1.4	2.8	13.0
N$_2$	IPP, OOH group (adjacent)	365	0.21	0.5	4.5
O$_2$, 10^5 Pa	IPP, OOH group (adjacent)	387	1.4	2.8	13.6
O$_2$, 2.6 × 10^4 Pa	IPP, OOH group (adjacent)	365	0.19	0.5	12.0
O$_2$, 10^5 Pa	IPP, OOH group (adjacent)	365	0.21	0.5	2.9
O$_2$, 10^5 Pa	IPP, OOH group (adjacent)	365			3.0
N$_2$	IPP, OOH group (single)	387	0.10	0.21	9.6
N$_2$	IPP, OOH group (single)	365	0.039	0.017	3.0
N$_2$	PE	365	0.05	0.5	0.36
O$_2$, 10^5 Pa	PE	365	0.05	0.5	0.81
N$_2$	PE	393	4.6	33	11.0

the chain mechanism ($\nu = 6$–17). The percentage of hydroperoxyl groups decomposed by free radicals varies from 30 to 80%.

The induced decomposition of POOH in an inert atmosphere is the result of the alkyl macroradical reaction with POOH.

$$P^\bullet + POOH \longrightarrow POH + PO^\bullet$$
$$PO^\bullet + PH \longrightarrow POH + P^\bullet$$

In an oxygen atmosphere, POOH is attacked by peroxyl radicals. For the decay of the secondary hydroperoxyl groups, the following reaction seems to be the most probable:

$$PO_2^\bullet + H-\overset{|}{\underset{|}{C}}-OOH \longrightarrow =O + POOH + HO^\bullet$$

It is rather surprising that the induced decomposition of POOH of PP occurs with long chains in a dioxygen atmosphere. The most probable reaction of tertiary hydroperoxyl groups of PP decay is that of hydroperoxyl radicals formed from POOH (see Chapter 19):

$$PO_2^\bullet + H_2O_2 \longrightarrow POOH + HO_2^\bullet$$
$$HO_2^\bullet + POOH \longrightarrow PO^\bullet + H_2O + O_2$$

13.3 OXIDATIVE DEGRADATION OF MACROMOLECULES

Oxidation of polyalkenes is accompanied by the degradation of macromolecules [11,12,33–41,122,123]. Thus, when HDPE is oxidized in chlorobenzene to a concentration of hydroperoxyl groups of 0.12 mol L^{-1} (378 K, 1 h) and 0.20 mol L^{-1}, the mean molecular weight of polymer decreases from 35,000 to 11,000 and 4,000 (i.e., by three and nine times, respectively) [124]. Various polymers may differ greatly in their susceptibility to oxidative destruction; at elevated temperatures, however, the difference becomes smaller. For instance, the rates of destruction of PIB and PS at 373 K differ by two orders of magnitude, and by only a factor of three to four at 423 K [37].

As follows from the comparison of the rates of oxidation and destruction of linear and branched polyalkenes, the branched form possesses more susceptible tertiary C—H bonds and oxidizes more rapidly (but degrade more slowly) than the linear form. This was observed for PS, PE, PP, poly(n-isopropyl styrene), polymethylene, and polymers produced by the decomposition of diazomethane–diazoethane and diazomethane–diazobutane mixtures [125]. The number of cleavages per oxygen molecule consumed at 403 K amounts to 0.25 for PE and to 0.04 for PP.

Investigations into the molecular weight distribution showed that C—C bond break down randomly [126]. Thus, in the PS oxidized at 473 K, the ratio M_w/M_n initially increased from 1.06 to 1.50 (in this case, M_n decreased from 400,000 to 90,000) and remained virtually unchanged during further oxidation. The situation was different with PE: residual oxygen at concentrations as low as 0.3% promoted the destruction of this polymer [127]. Moreover, anaerobic conditions at 588 K induced cross-linking processes in PE, while in the presence of 0.34% O_2 the destructive processes became predominant. The decrease in M_n at 628 K under anaerobic conditions was not accompanied by changes in the ratio $M_w/M_n = 3$. Conversely, this ratio increased in the course of the oxidative destruction of PE.

The mechanism of oxidative destruction of alkenes is complex and has been well understood for PE and PP only. Because of the formation of hydroperoxyl groups in oxidized polymers, the decomposition of relatively unstable groups has been primarily considered to cause polymer destruction. The two facts in favor of this suggestion are: (i) the generation of alkoxy radicals from hydroperoxyl groups and their rapid breakdown during polymer oxidation; and (ii) the formation of low-molecular-weight products at a rate proportional to the concentration of hydroperoxyl groups [108]. Yet, some experimental data are at variance with this assumption. Indeed, if a polymer degrades by the mechanism

$$POOH \longrightarrow PO^\bullet \longrightarrow \text{cleavage of C—C bond}$$

the rate of destruction v_S should be proportional to [POOH]. At the initial stage of autoxidation, $[POOH] \sim t^2$ (see Chapter 4) and, therefore, the number of chain cleavages [S] must increase proportionally to t^3. However, in the PP oxidized at 370–410 K, $[S] \sim t$ [128], which is inconsistent with the mechanism written above.

Compelling evidence suggesting that the breakdown of hydroperoxyl groups is not related to polymer destruction, at least in the initial period of oxidation at temperatures below 400 K, comes from experiments on the initiated oxidation of polymers. It was found that the destruction of polymers develops in parallel with their oxidation from the very onset of the process, but not after a delay related to the accumulation of a sufficient amount of hydroperoxyl groups [129]. These experiments also demonstrated that it is free macroradicals that undergo destruction. Oxidation of polymers gives rise to alkyl, alkoxyl, and peroxyl macroradicals. Which radicals undergo destruction can be decided based on the kinetics of initiated destructive oxidation.

(1) Alkyl radicals can break down through the cleavage of C—C bonds. Since these radicals are reactive to dioxygen, their concentration must decline with increasing pO_2 and, hence, the rate of alkyl radical destruction must decrease. Experiments show, however, that the rate of destruction of dissolved PE at 388 K [130] and PP at 393 K [131] increases

with pO_2 to reach a saturation level. Therefore, at moderate temperatures (below 400 K), the breakdown of alkyl radicals makes almost no contribution to the oxidative destruction of polymers. Above 500 K, the contribution may be high (see below).

(2) Alkoxyl radicals can result from the isomerization of peroxyl radicals of oxidized PP (see above 13.1.6). If alkoxyl radicals cause polymer destruction, then, as they are produced from alkyl radicals, their accumulation and quasistationary concentration must decrease with increasing pO_2. However, despite varying pO_2, $v_S = const.$ in the oxidized PE and PP and, therefore, alkoxyl radicals essentially do not contribute to the oxidative destruction of polymers. At moderate temperatures, alkoxyl radicals eliminate hydrogen atoms from PH more rapidly than they undergo degradation.

(3) There is ample evidence that oxidized polyalkenes degrade through a monomolecular destruction of peroxyl macroradicals. Indeed, it was demonstrated experimentally that $v_S \sim v_i^{1/2}$ and $v_S \sim v_0$ [129–132], from when $v_S \sim [PO_2^\bullet]$. The constant v_S/v_0 ratio under varying initiation rates and partial dioxygen pressures suggests that if the $PO_2^\bullet$ is a macroradical, it breaks down, but not the products of its destruction. A similar inference was drawn by Ivanov et al. [128], who studied the destruction of autoxidized PP and showed, by extrapolating the ratio v_S/v_0 to $t = 0$ (where $v_0 = 0$), that this ratio depends on the temperature but not on pO_2. The following mechanism of the decomposition of a peroxyl macroradical with coordinated cleavage of three bonds has been proposed [133].

This reaction is very exothermic ($\Delta H \approx -180$ to -200 kJ mol^{-1}) and, therefore, seems to be very probable from the thermochemical point of estimation. The pre-exponential factor is expected to be low due to the concentration of the energy on three bonds at the moment of TS formation (see Chapter 3). To demonstrate that this reaction is responsible for the oxidative destruction of polymers, PP and PE were oxidized in chlorobenzene with an initiator and analyzed for the rates of oxidation, destruction (viscosimetrically), and double bond formation (by the reaction with ozone) [131]. It was found that: (i) polymer degradation and formation of double bonds occur concurrently with oxidation; (ii) the rates of all three processes are proportional to $v_i^{1/2}$, (iii) independent of pO_2, and (iv) $v_S = v_{dbf}$ in PE and $v_S = 1.6 v_{dbf}$ in PP (v_{dbf} is the rate of double bond formation). Thus, the rates of destruction and formation of double bonds, as well as the kinetic parameters of these reactions, are close, which corroborates with the proposed mechanism of polymer destruction. Therefore, the rate of peroxyl macromolecules degradation obeys the kinetic equation:

$$v_S = k_S[PO_2^\bullet] = k_S(v_i/2k_t)^{1/2} \tag{13.34}$$

The rate constant of decomposition of PE peroxyl macroradical is $k_S = 3.2 \times 10^{11} \exp(-73/RT)$ s^{-1} [131]. The values of k_S calculated from experimental data are collected in Table 13.12.

The radiochemical oxidation of PS in a chloroform solution is accompanied by its destruction and formation of products of styrene oxidation, namely, benzaldehyde and styrene oxide [136]. The radiochemical yield of these products was equal to the radiochemical yield of PS macromolecule cleavages. Butyagin [137] analyzed the products of decomposition of the peroxyl radicals of PS and polyvinylcyclohexane. Alkyl macroradicals were produced mechano- or photochemically, volatile products were evaporated in vacuum, and alkyl radicals were converted into peroxyl radicals using labeled ^{18}O. Peroxyl radicals were then

TABLE 13.12
Kinetic Parameters of Macromolecules Degradation in Oxidized Polymers

Polymer	T (K)	Conditions of Oxidation	$(v_S/v) \times 100$	$k_S \times 10^3$ (s^{-1})	Ref.
LDPE	378	0.36 mol L^{-1} in PhCl	2.0	3.2	[134]
LDPE	388	0.36 mol L^{-1} in PhCl	1.7	5.0	[134]
LDPE	398	0.36 mol L^{-1} in PhCl	1.3	8.1	[134]
LDPE	388	0.36 mol L^{-1} in PhCl	1.6	4.8	[132]
LDPE	388	Solid phase	1.1		[135]
IPP	393	0.12 mol L^{-1} in PhCl	1.6	3.8	[129]
IPP	393	0.12 mol L^{-1} in PhCl	1.3	3.1	[132]

eliminated by an increasing temperature, and the polymer was analyzed for low-molecular-weight products. The decay of peroxyl macroradicals was found to occur concurrently with the formation of isotopically labeled water, probably resulting from the destruction of peroxyl radicals, which were from hydroxyl radicals. The decomposition of PO$_2^\bullet$ radicals was accompanied by the formation of water and benzaldehyde with a total yield of 10% of the oxygen consumed. On the other hand, the decomposition of polyvinylcyclohexane PO$_2^\bullet$ radicals was accompanied by the formation of water, cyclohexanone, and cyclohexene with a total yield of 30% of the oxygen consumed. These results are consistent with the following mechanism that is similar to the mechanism described above.

At 300 K and below, when hydroperoxides are stable, the decay of PMP peroxyl radicals gives rise to low-molecular-weight products, namely, water, acetone, and isobutyric aldehyde. The formation of these products can be explained by the breakdown of various peroxyl radicals with production of hydroxyl ion and cleavage of the C—C bond.

Hence, the experimental data on polymer degradation and formation of low-molecular-weight products due to the decomposition of peroxyl macrorodicals are in good agreement with the proposed [131] mechanism of PO$_2^\bullet$ degradation with the formation of a double bond and hydroxyl ion. At high temperatures, the cleavage of alkoxyl radicals can lead to polymer degradation also. Let us compare the rates of PO$_2^\bullet$ and PO$^\bullet$ degradation in the process of PE oxidation.

(A) Initial stage of oxidation at moderate temperatures when hydroperoxyl groups are relatively stable. The rates of C—C bonds scission by decomposition of alkoxyl and peroxyl radicals are the following:

$$v_S(PO_2^\bullet) = k_{SPO_2}[PO_2^\bullet] \text{ and } [PO_2^\bullet] = \sqrt{\frac{k_i[POOH]}{2k_t}} \quad (13.35)$$

$$v_S(PO^\bullet) = k_{SPO}[PO^\bullet] \text{ and } [PO^\bullet] = \frac{k_i[POOH]}{k_{SPO} + k_a[PH]} \quad (13.36)$$

The ratio of polymer degradation via these two reactions is:

$$\frac{v_{S\,PO_2}}{v_{S\,PO}} = \frac{k_{S\,PO_2}}{k_{S\,PO}} \times \frac{k_{S\,PO} + k_a[PH]}{\sqrt{2k_t k_i[POOH]}} \quad (13.37)$$

The results of the calculation of this ratio for PE oxidation at $T = 388$ K are listed here. The following values of the rate constants were considered: $k_{S\,PO_2} = 4.5 \times 10^{-3}\,s^{-1}$, $k_{S\,PO} = 1.9 \times 10^6\,s^{-1}$, $k_i = 3.5 \times 10^{-6}\,s^{-1}$, $2k_t = 1.9 \times 10^4$ kg mol^{-1} s^{-1}, and $k_a[PH] = 8.8 \times 10^7\,s^{-1}$ [12].

[POOH] (mol kg^{-1})	0.001	0.005	0.01	0.1	0.66
$v_S(PO_2^\bullet)/v_S(PO^\bullet)$	26.6	11.4	8.1	2.6	1.0

At elevated temperatures, the oxidation of PE occurs as a quasistationary process with the kinetic equilibrium between the formation and decay of hydroperoxyl groups (see Chapter 4). The ratio of the two discussed reactions of polymer degradation is:

$$\frac{v_{S\,PO_2}}{v_{S\,PO}} = \frac{k_{S\,PO_2}}{k_p[PH]} \times \frac{k_\Sigma}{k_i} \left\{ 1 + \frac{k_a[PH]}{k_{S\,PO}} \right\} \quad (13.38)$$

The ratios of rate constants in their temperature dependence are the following: $k_{S\,PO_2}/k_p[PH] = 1.2 \exp(-15/RT)$, $k_{S\,PO}/k_a[PH] = 2.8 \times 10^3 \exp(-38/RT)$, and $k_\Sigma/k_i = 50$ [12].

T (K)	400	430	460	480	500
$v_S(PO_2^\bullet)/v_S(PO^\bullet)$	21	14	10	8	7

These calculations show that the degradation via decomposition of peroxyl macroradicals prevails inside the tight range of oxidation conditions.

The degradation of alkyl radicals can participate in polymer degradation in the absence of dioxygen or at oxidation in the diffusion regime.

The mechanism of PIP degradation appeared to be principally different. PIP has double bonds and oxidizes through intramolecular peroxyl radical addition to the double bond with formation of peroxide bridges.

Mayo [138] proposed that PIP degradation occurs as a result of unstable alkoxyl radicals formed in oxidized polymer. Tobolsky [139] cleared that the degradation rate v_S in experiments on PIP oxidation in a benzene solution is proportional to the initiation rate: $v_S \sim v_i$.

He supposed that the decay proceeds as the decomposition of alkoxyl radicals formed in the bimolecular reaction of two tertiary peroxyl radicals (see Chapter 2).

$$2PO_2^{\bullet} \longrightarrow 2PO^{\bullet} + O_2$$
$$PO^{\bullet} \longrightarrow \text{Degradation}$$

The kinetics of PIP oxidation and degradation was studied by Pchelintsev and Denisov [54]. Films of PIP with AIBN as an initiator were oxidized at $T = 353$ K in the kinetic regime. Oxidation proceeded by the chain mechanism with long chains. The molecular weights of the initial and the oxidized PIP were measured osmometrically. It was stated that the oxidation rate $v \sim v_i$ but the rate of degradation $v_S \sim v_i$. The v_S/v_i ratio was found to be very high and equal to 15. This result is in good agreement with the ratio of two parallel reactions between tertiary peroxyl radicals (see Chapter 2).

$$2PO_2^{\bullet} \longrightarrow 2PO^{\bullet} + O_2$$
$$2PO_2^{\bullet} \longrightarrow POOP + O_2$$

The formed alkoxyl macroradical is very unstable and decomposes with the formation of three carbonyl compounds.

The degradation of polystyrene in a chlorobenzene solution in the presence of dioxygen and initiator also occurs with the rate $v_S \sim v_i$ [140,141]. However, the v_S/v_i ratio was found to be much less than unity.

Polymer, Copolymer	T (K)	[Polymer] (mol L^{-1})	$(v_S/v_i) \times 10^2$
PS	348	0.48	7.2
PS	393	0.70	7.5
Isobutylene–styrene	393	0.91	7.5
Isobutylene–p-chlorostyrene	393	0.56	6.2
Isobutylene–o-chlorostyrene	393	0.56	5.6
Isobutylene–p-methylstyrene	393	0.69	7.1
Isobutylene–o-methylstyrene	393	0.69	5.7

13.4 HETEROGENEOUS CHARACTER OF POLYMER OXIDATION

In a liquid hydrocarbon subjected to vigorous stirring, the active species, peroxyl radicals, in particular, diffuse with the effective coefficient of diffusion $D_{\text{eff}} = 10^{-3}$–10^{-2} cm^2 s^{-1}. At a rate of initiation of 10^{-6} to 10^{-8} mol L^{-1} s^{-1} and $2k_t = 10^4$–10^6 L mol^{-1} s^{-1}, the lifetime of peroxyl radicals given by $(2v_i k_t)^{-1/2}$ ranges from 1 to 100 s. Within its lifetime, the peroxyl radical diffuses over the distance $x = (2D't)^{1/2} \sim 0.1$–1 cm, which is commensurate with the reactor size.

In a solid polymer, low-molecular-weight species diffuse with $D \sim 10^{-7}$–10^{-9} cm^2 s^{-1} [67]. Segments of a macromolecule with the free valence diffuse still more slowly ($D_{\text{eff}} \approx 10^{-9}$–$10^{-12}$ cm^2 s^{-1}). With the lifetime of PO$_2^{\bullet}$ of 10–100 s, the distance covered by this radical comprises 10^{-4}–10^{-6} cm, which is 10^3 to 10^4 times less than in a liquid. It is of interest to compare this distance with the mean size of the macromolecules. At the molecular weight between 10^4 and 10^5 g mol^{-1} and a density of about 1000 kg m^{-3}, the radius of a globular macromolecule varies from 3×10^{-9} to 9×10^{-9} m, which is only ten times smaller than the mean distance of free valence migration. Experimental data on the initiated oxidation of ethylene–propylene copolymer agree well with these estimates [47]. In particular, it was shown that the copolymer oxidizes at a rate, which is the sum of the rates of oxidation of particular homopolymers. At the same time, the overall rate of the copolymer oxidation nonlinearly depends on the propylene content, as much as the rate of the concurrent oxidation of isooctane and decane depends on the isooctane content of the mixture [47]. Thus, because of the slow motion of available bonds in polymers, their chain oxidation occurs in a small volume, which is 10^{10} to 10^{13} times smaller than in the case of liquid hydrocarbons.

Another specific feature of solid polyalkenes is their heterogeneous structure representing a mosaic of crystalline and amorphous regions. The crystalline phase of PE is rather dense (1000 kg m^{-3}), while the density of the amorphous phase is 850 kg m^{-3}. In IPP, the densities of crystalline and amorphous phases are 940 and 850 kg m^{-3}, respectively. In the dense crystalline regions of polymers, diffusion of oxygen is very slow. Thus, the coefficients of diffusion of dioxygen in the crystalline phases of PE and PP are 10^8 times lower than in their amorphous phases, so that oxygen molecules hardly have time to penetrate into the crystalline regions of polymers in the course of the experiment [42]. Upon the swelling of polymers in solvents, solvent and solute molecules can penetrate only into the amorphous phase of polymers. Thus, stable phenoxyl radicals in a swelling solvent were revealed only in the amorphous phase of polymer (this was shown by the concentration-dependent broadening of the EPR signals of this radical). In experiments with various initiators (AIBN, bis(1,1-dimethylethyl) peroxide, and dicumyl peroxide), the initiated oxidation of polymers was observed only in the amorphous regions. Decker et al. [44] studied the γ-radiation-initiated oxidation of PEs of various crystallinities at 318 K and found that the parameter a ($v = av_i^{1/2}$) grew linearly with the parameter α_{am} characterizing the polymer amorphism given by the equation

$$a \text{ (mol}^{1/2}\text{kg}^{-1/2}\text{s}^{-1/2}) = 1.2 \times 10^{-3} + 1.87 \times 10^{-2} \alpha_{am} \tag{13.39}$$

The oxidation of the crystalline phase occurs 15 times more slowly than the oxidation of the amorphous phase and is likely localized in the boundary regions. The autoxidation of PE and PP occurs at the rate, which is slower at higher crystallinity of polymers and vice versa, as indicated by the amount of the oxygen consumed. Based on the evidence accumulated, it is safe to say that PE and PP are oxidized in their amorphous regions [12,33,34,42,67].

The localization of solid PP oxidation has been investigated by a combination of specific staining techniques with UV spectroscopy. The formed carbonyl groups were reacted with 2,4-dinitrophenylhydrazine to visualize localized oxidation [142–144] and the formed hydroperoxyl groups reacted with SO_2, forming alkyl sulfates with a strong IR absorbance band [145]. Localized regions of extensive oxidation around iron impurities were identified. Billingham [146] assumed that the oxidation of polymer was localized around the initiator center. This center does not diffuse, and degradation takes place in a region of high concentrations of oxidation products around this center. Such active centers appeared to be residues of the catalyst used in polyolefins production [147]. When the Ziegler–Natta catalytic system is used for polymer production, the produced polymer contains 1–20 ppm of titanium [148]. Residues of the Ziegler–Natta catalyst were found to consist of Ti(III) and small amounts Ti(II) bounded with aluminum alkyls [148].

New experimental evidence of heterogeneity of PP film oxidation and photooxidation were obtained using the special resolution of the CL-imaging system [148–151]. George and coworkers [151,152] formulated the "epidemic model" of heterogeneous polymer oxidation. Three distinct populations were postulated in the amorphous phase of oxidized polymer after some time of oxidation: (a) the unoxidized fraction p_r, (b) oxidizing fraction p_i, and (c) oxidized fraction p_d. Oxidation is treated as spreading of the oxidizing fraction with the rate equal to αp_i where α is the rate coefficient for the formation of the oxidized material from the oxidized fraction. Oxidation can spread if the nonoxidized polymer is in the contact zone with the rate equal to $bp_r(1-p_r)$, which represents the efficient rate coefficient for spreading and $p_d \ll p_r$. The solvation of coupled differential equations leads to the following formula for the kinetics of polymer oxidation [152]:

$$p_i = p_0 p_r \exp(b-\alpha)t \qquad (13.40)$$

where p_0 is the initial fraction of active centers in the oxidized polymer. The CL kinetic curves of PP oxidation were described by this equation. The values of parameters calculated from these kinetic curves of single PP particles oxidation are listed below [152].

No.	$p_0 \times 10^3$	$\alpha \times 10^3$ (s^{-1})	$b \times 10^4$ (s^{-1})
1	11.6	2.2	3.6
2	23.2	1.5	3.3
3	11.3	3.2	3.8
4	0.8	9.1	5.2
5	1.8	9.6	4.0
6	0.2	12.4	5.3

A wide variation exists for the number of active centers of oxidation in polymer samples. This reflects the statistical nature of the catalyst residues in polymer particles. On the contrary, the spreading rate coefficient b is approximately constant for the studied samples of PP. The coefficient α is probably sensitive to the morphology of the particles.

REFERENCES

1. AP Griva, LN Denisova, ET Denisov. *Vysokomol Soed* A18:219–224, 1977.
2. AP Griva, LN Denisova, ET Denisov. *Zh Fiz Khim* 59:2944–2951, 1985.
3. AP Griva, LN Denisova, ET Denisov. *Zh Fiz Khim* 58:557–583, 1984.
4. ET Denisov, TG Denisova, TS Pokidova. *Handbook of Free Radical Initiators*. New York: Wiley, 2003.
5. VA Roginskii, VA Shanina, VB Miller. *Dokl AN SSSR* 227:1167–1170, 1976.
6. YuB Shilov. Kinetics and Mechanism of Initiated Oxidation of Polypropylene. Ph.D. thesis Dissertation, Institute Chemical Physics, Chernogolovka, 1979.
7. JCW Chien, DSF Wang. *Macromolecules* 8:920–928, 1975.
8. NV Zolotova, ET Denisov. *Vysokomol Soed* A11:946–950, 1969.
9. E Niki, C Decker, F Mayo. *J Polym Sci Polymer Chem Ed* 11:2813–2844, 1973.
10. BE Krisyuk, EN Ushakov, AP Griva, ET Denisov. *Dokl AN SSSR* 277:630–633, 1984.
11. NM Emanuel, AL Buchachenko. *Chemical Physics of Polymer Degradation and Stabilization*. Utrecht: VNU Science, 1987.
12. ET Denisov. *Oxidation and Degradation of Carbon chain Polymers*. Leningrad: Khimiya, 1990, [in Russian].
13. WY Wen, DR Johnson, M Dole. *J Phys Chem* 78:1798–1804, 1974.
14. PYu Butyagin. *Vysokomol Soed* A16:63–70, 1974.

15. E Lawton, R Powell, J Balwitt. *J Polym Sci* 32:257–263, 1958.
16. P Neudorf. *Kolloid Ztschr* 224:132–136, 1968.
17. VA Radtsig. *Vysokomol Soed* A28:777–784, 1986.
18. B Loy. *J Phys Chem* 65:58–61, 1961.
19. H Fischer, KH Hellwege, P Neudorf. *J Polym Sci* A1:2109–2117, 1963.
20. P Neudorf. *Kolloid Ztschr* 224:25–32, 1968.
21. L Davis, C Pampilo, T Chiang. *J Polym Sci Polym Phys Ed* 11:841–853, 1973.
22. ER Klinshpont, VK Milinchuk. *Khim Vysok Energii* 3:74–79, 1969.
23. VA Roginskii, VB Miller. *Dokl AN SSSR* 215:1164–1167, 1974.
24. VA Roginskii, EL Shanina, VB Miller. *Vysokomol Soedin* A24:189–195, 1982.
25. NYa Rapoport, ASh Goniashvili, MS Akutin, VB Miller. *Vysokomol Soed* A20:1432–1437, 1978.
26. AL Margolin, AE Kordonskii, YuV Makedonov, VYa Shlyapintokh. Vysokomol Soed A29:1067–1073, 1987.
27. F Czocs. *J Appl Polym Sci* 27:1865–1871, 1982.
28. LM Postnikov, G Geskens. *Vysokomol Soed* B28:89–93, 1986.
29. YaS Lebedev, YuD Tsvetkov, VV Voevodskii. *Kinet Katal* 1:496–502, 1960.
30. IS Gaponova, TV Fedorova, VF Tsepalov, VF Shuvalov, YaS Lebedev. *Kinet Katal* 12:1137–1143, 1971.
31. GP Vskretchyan, YaS Lebedev. *Izv AN SSSR Ser Khim* 1378–1380, 1976.
32. PYu Butyagin, AM Dubinskaya, VA Radtsig. *Usp Khim* 38:593–623, 1969.
33. SH Hamid (ed.). *Handbook of Polymer Degradation*. New York: Marcel Dekker, 2000.
34. G Scott. *Atmospheric Oxidation and Antioxidants*. Amsterdam: Elsevier, 1965.
35. G Geiskens (ed.). *Degradation and Stabilisation of Polymers*. London: Applied Science Publishers, 1975.
36. B Ranby, JF Rabek. *Photodegradation, Photooxidation and Photostabilization of Polymers*. London: John Wiley, 1975.
37. HHG Jellinec (ed.). *Aspects of Degradation and Stabilisation of Polymers*. Amsterdam: Elsevier, 1978.
38. JF McKellar, NS Allen. *Photochemistry of Man-Made Polymers*. London: Applied Science Publishers, 1979.
39. AA Popov. NYa Rapoport, GE Zaikov. *Oxidation of Oriented and Strained Polymers*. Moscow: Khimiya, 1987, p 25 [in Russian].
40. G Scott (ed.), *Mechanisms of Polymer Degradation and Stabilisation*. London: Elsevier Applied Science, 1990.
41. W Hawkins. *Polymer Degradation and Stabilization*. Berlin: Springer-Verlag, 1984.
42. NC Billingham, PD Calvert. In: G Scott (ed.), *Developments in Polymer Stabilisation-3*. London: Applied Science Publishers 1980, pp 139–190.
43. YuB Shilov, ET Denisov. *Vysokomol Soed* A11:1812–12816, 1969.
44. C Decker, F Mayo, H Richardson. *J Polym Sci* 11:2875–2898, 1973.
45. NYa Rapoport, ASh Goniashvili, MS Akutin. *Vysokomol Soed* A23:393–399, 1981.
46. AV Tobolsky, PM Norling, NH Frick, M Yu. *J Am Chem Soc* 86:3925–3930, 1964.
47. YuB Shilov, ET Denisov. *Vysokomol Soed* A14:2385–2390, 1972.
48. VA Roginskii. Oxidation of polyolefines inhibited by sterically hindered phenols, Doctoral Thesis Dissertation, Institute of Chemical Physics, Chernogolovka, 1983.
49. EL Shanina, VA Roginskii, VB Miller. *Vysokomol Soed* A18:1160–1164, 1976.
50. NYa Rapoport, ASh Goniashvili, MS Akutin, VB Miller. *Vysokomol Soed* A19:2211–2216, 1977.
51. AV Kirgin, YuB Shilov, ET Denisov, AA Efimov. *Vysokomol Soed* A28:2236–2241, 1986.
52. JCW Chien, CR Boss. *J Polym Sci* A-1, No 12: 3091–3101, 1967.
53. VB Ivanov, SG Burkova, YuL Morozov, VYa Shlyapintokh. *Vysokomol Soed* B20:852–855, 1978.
54. VV Pchelintsev, ET Denisov. *Vysokomol Soed* A25:781–786, 1983.
55. PM Norling, AV Tobolsky. *Rubber Chem Technol* 39:278–283, 1966.
56. ET Denisov, NI Mitskevich, VE Agabekov. *Liquid-Phase Oxidation of Oxygen-Containing Compounds*. New York: Consultants Bureau, 1977, p 7.
57. PA Ivanchenko, ET Denisov, VV Kharitonov. *Kinet Katal* 12:492–495, 1971.

58. ER Klinshpont, BK Milinchuk. *Vysokomol Soed* B17:358–361, 1975.
59. E Niki, Y Kamiya. *Bull Chem Soc Jpn* 48:3226–3229, 1975.
60. TG Degtyareva. Mechanism of Oxidation of 2-Methylbutane. PhD Thesis Dissertation. Institute of Chemical Physics, Chernogolovka, 1972 [in Russian].
61. AF Guk, SP Erminov, VF Tsepalov. *Kinet Katal* 13:86–91, 1972.
62. VA Radzig, MM Rainov. *Vysokomol Soed* A18:2022–2030, 1976.
63. E Niki, C Dekker, F Mayo. *J Polym Sci Polym Chem Ed* 11:2813–2844, 1973.
64. VA Roginskii, EL Shanina, VB Miller. *Vysokomol Soed* A20:265–272, 1978.
65. PYu Butyagin, IV Kolbanov, AM Dubinskaya, MU Kislyuk. *Vysokomol Soed* A26:2265–2277, 1968.
66. LS Shibryaeva, NYa Rapoport, GE Zaikov. *Vysokomol Soed* A28:1230–1239, 1986.
67. ET Denisov. In: SH Hamid (ed.), *Handbook of Polymer Degradation*. New York: Marcel Dekker, 2000, pp 383–420.
68. AS Michaels, HJ Bixler. *J Polym Sci* 50:393–439, 1961.
69. *Polymer Handbook*, New York: Wiley, 1979, vol 6, p 43.
70. SG Kiryushkin, BA Gromov. *Vysokomol Soed* A14:1715–1717, 1972.
71. SG Kiryushkin, VP Filipenko, VT Gontkovskaya. *Carbonchain Polymers*. Moscow: Nauka, 1971, p 182 [in Russian].
72. ET Denisov, YuB Shilov. *Vysokomol Soed* A15:1196–1203, 1983.
73. Y Watanade, T Shirota. *Bull Agency Ind Tech* 2:149–154, 1981.
74. J Crank (ed.), *Diffusion in Polymers*. New York: Academic Press, 1968.
75. R Kosiyanov, R Gregor. *J Appl Polym Sci* 26:629–641, 1981.
76. G Amerongen. *J Polym Sci* 5:307–312, 1950.
77. V Stannet, JL Williams. *J Polym Sci Part C*, 10: 45–59, 1965.
78. G Shaw. *Trans Faraday Soc* 63:2181–2189, 1967.
79. BP Tikhomirov, HB Hopfenberg, V Stannets, JL Williams. *Makromol Chem* 118:177–188, 1968.
80. H Yasuda, K Rosengren. *J Appl Polym Sci* 14:2834–2877, 1970.
81. YaI Frenkel. *Kinetic Theory of Liquids*. Leningrad: Nauka, 1975 [in Russian].
82. HL Frish, JB Bdzil. *J Phys Chem* 62:4804–4808, 1975.
83. MN Cohen, D Turbull. *J Chem Phys* 31:1164–1169, 1959.
84. A Peterlin, FL Mc Grackin. *J Polym Sci Polym Phys Ed* 19:1003–1006, 1981.
85. AT Di Benedetto, DR Paul. *J Polym Sci Part C*, 10, 17–44, 1965.
86. YuP Yampolskii, SG Durgarian, NS Nametkin. *Dokl AN SSSR* 261:708–710, 1981.
87. ET Denisov, AI Volpert, VP Filippenko. *Vysokomol Soed* A28:2083–2089, 1981.
88. JCW Chien, EJ Vandenberg, H Jabloner. *J Polym Sci Part A*-1, 6:381–392, 1968.
89. T Mill, G Montorsi. *Int J Chem Kinet* 5:119–136, 1973.
90. DE Van Sickle. *J Org Chem* 37:755–760, 1972.
91. C Decker, F Mayo. *J Polym Sci Polym Chem Ed* 11:2847–2874, 1973.
92. AV Kirgin, YuB Shilov, ET Denisov, AA Efimov. *Vysokomol Soed* A28:2236–2241, 1986.
93. NYa Rapoport, ASh Goniashvili, MS Akutin, VB Miller. *Vysokomol Soed* A20:1652–1659, 1978.
94. VM Goldberg, VN Esenin, GE Zaikov. *Vysokomol Soed* A28:1634–1639, 1986.
95. NYa Rapoport, ASh Goniashvili, MS Akutin. *Vysokomol Soed* A23:393–399, 1981.
96. DE Van Sickle, T Mill, F Mayo. *J Org Chem* 38:4435–4440, 1973.
97. LN Denisova, ET Denisov. *Kinet Katal* 17:596–600, 1976.
98. JCW Chien. *J Polym Sci Part A-1*, 6:375–380, 1968.
99. NV Zolotova, ET Denisov. *J Polym Sci* 9:3311–3320, 1971.
100. NM Emanuel, SG Kiryushkin, AP Mar'in. *Dokl AN SSSR* 275:408–411, 1984.
101. AM Tolks. Kinetics and Mechanism of Hydroperoxide Groups Decay in Polymers. PhD Thesis, Polytechnic Institute, Riga, 1973 [in Russian].
102. VM Goldberg, VI Esenin, IA Krasochkina. *Vysokomol Soed* A19:1720–1727, 1977.
103. M Iring, T Kelen, F Tudos. *Makromol Chem* 175:467–481, 1974.
104. IG Latyshkaeva, GN Belov, TA Bogaevskaya, YuA Shlyapnikov. *Vysokomol Soed* B24:70–72, 1982.
105. SS Stivala, V Chavla, L Reich. *J Appl Polym Sci* 17:2739–2748, 1973.

106. Z Manyasek, D Beren, M Michko. *Vysokomol Soed* 3:1104–1115, 1961.
107. VV Dudorov, AL Samvelyan, AF Lukovnikov. *Izv AN ArmSSR Ser Khim* 15:1104–1112, 1962.
108. AS Kuzminskii (ed.), *Ageing and Stabilization of Polymers*. Moscow: Khimiya, 1966 [in Russian].
109. TV Monakhova, TA Bogaevskaya, BA Gromov, YuA Shlyapnikov. *Vysokomol Soed* B16:91–94, 1974.
110. SG Kiryushkin, YuA Shlyapnikov. *Dokl AN SSSR* 220:1364–1367, 1975.
111. SG Kiryushkin, YuA Shlyapnikov. *Vysokomol Soed* B16:702–707, 1974.
112. JCW Chien, H Jabloner. *J Polym Sci Part A-1*, 6:393–402, 1968.
113. VS Pudov, MB Neiman. *Neftekhimiya* 2:918–923, 1962.
114. VS Pudov, BA Gromov, ET Sklyarova, MB Neiman. *Neftekhimiya* 3:743–749, 1963.
115. VS Pudov. Kinetics and Mechanism of Solid Polymer Degradation and Stabilization, Doctoral Thesis Dissertation, Institute Chemical Physics, Moscow, 1980.
116. NYa Rapoport, LS Shibryaeva, VB Miller. *Vysokomol Soed* A25: 831–839, 1983.
117. L Dulog, KH David. *Makromol Chem* 53B:72–83, 1962.
118. EA Il'ina, SM Kavun, ZI Tarasova. *Vysokomol Soed* B17:388–390, 1975.
119. YuB Shilov, ET Denisov. *Vysokomol Soed* A19:1244–1253, 1977.
120. EL Shanina, VA Roginskii. *Vysokomol Soed* B20:145–148, 1978.
121. NYa Rapoport, VB Miller. *Vysokomol Soed* A18:2343–2347, 1976.
122. F Mayo. *Macromolecules* 11:942–946, 1978.
123. TV Monakhova, TA Bogaevskaya, YuA Shlyapnikov. *Vysokomol Soed* A18:808–811, 1976.
124. M Iring, T Kelen, F Tudos, ZS Laslo-Hedvig. *J Polym Sci Polym Symp* 57:89–99, 1976.
125. M Iring, ZS Laslo-Hedvig, K Bavabas. *Eur Polym J* 14:439–442, 1978.
126. IM Bel'govskii, VM Goldberg, IA Krasotkina, DYa Toptygin. *Dokl AN SSSR* 192:121–122, 1970.
127. A Holmstrom, EM Sorvik. *J Appl Polym Sci* 18:779–804, 1974.
128. AA Ivanov, AP Firsov, VD Grigoryan, AA Berlin. *Vysokomol Soed* A11:1390–1397, 1969.
129. PA Ivanchenko, VV Kharitonov, ET Denisov. *Kinet Katal* 13:218–221, 1972.
130. PA Ivanchenko. Mechanism of C—C Bond Scission in Oxidized Polyethylene. PhD Thesis Dissertation, Institute Chemical Physics, Chernogolovka, 1979, [in Russian].
131. NF Trofimova, VV Kharitonov, ET Denisov. *Dokl AN SSSR* 253:651–653, 1980.
132. TG Degtyareva, NF Trofimova, VV Kharitonov. *Vysokomol Soed* A20:1873–1878, 1978.
133. ET Denisov. *Dokl AN SSSR* 235:615–617, 1977.
134. J Lacoste, DJ Carlsson, S Falicki, DM Wiles. *Polym Degrad Stabil* 34:309–323, 1991.
135. PA Ivanchenko, VV Kharitonov, ET Denisov. *Vysokomol Soed* A11:1622–1630, 1969.
136. M Abadie, J Marchal. *Makromol Chem* 141:299–316, 1971.
137. GP Butyagin, PYu Butyagin, VYa Shlyapintokh. *Vysokomol Soed* A24:165–172, 1971.
138. FR Mayo. *Ind Eng Chem* 52:614–627, 1960.
139. AV Tobolsky, A Mercurio. *J Am Chem Soc* 81:5535–5538, 1959.
140. NM Zalevskaya, IA Opeyda, RV Kucher. *Vysokomol Soed* B20:493–496, 1978.
141. ZA Sadykov, SR Kusheva. *Vysokomol Soed* B22:403–406, 1980.
142. JB Knight, PD Calvert, NC Billingham. *Polymer* 26:1713–1722, 1985.
143. NC Billingham, PD Calvert, JB Knight, G Ryan. *Brit Polym J* 11:155–159, 1979.
144. NC Billingham, PD Calvert. *Pure Appl Chem* 57:1727–1732, 1985.
145. P Richters. *Macromolecules* 3:362–371, 1970.
146. NC Billingham. *Macromol Chem Macromol Symp* 28:145–152, 1989.
147. KY Choi, WH Ray. *J Macromol Sci Rev Macromol Chem Phys* C25:57–64, 1985.
148. TE Nowlin. *Prog Polym Sci* 11:29–41, 1985.
149. M Selina, GA George, DJ Lacey, NC Billingham. *Polym Degrad Stabil* 47:311–320, 1995.
150. M Selina, GA George, NC Billingham. *Polym Degrad Stabil* 42:335–342, 1993.
151. GA George, M Selina, G Cash, D Weddell. 11th Bratislava Int Conf Polymers. 2–3, Stara Lesna, 1996.
152. GA George, M Selina. In: SH Hamid (ed.), *Handbook of Polymer Degradation*. New York: Marcel Dekker, 2000, pp 277–313.

Part II

Chemistry of Antioxidants

14 Theory of Inhibition of Chain Oxidation of Organic Compounds

14.1 HISTORICAL INTRODUCTION

Oxidation of organic compounds is one of the efficient methods of organic synthesis in modern petrochemical industry. On the other side, the autoxidation of organic compounds, their mixtures, and products promotes their rapid deterioration due to the action of atmospheric oxygen. Products such as rubber, polymers, hydrocarbon fuels, lubricants, organic solvents, semiproducts, food, drags, etc. are spoiled due to oxidation by oxygen. Antioxidants prevent the rapid development of these undesirable processes. They were the object of intensive study during the last 40 years. The practical use of antioxidants began in the end of the 19th century.

The early practical application of antioxidants was connected with the development of rubber production. The rubber is easily oxidized in air, and the first antioxidants were empirically found and used to stabilize it [1]. Empirical search for antioxidants was performed by Moureu and Dufresse [2] during the First World War. These researchers successfully solved the problem of acrolein stabilization by the addition of hydroquinone. They explained the retarding action of the antioxidant in the scope of peroxide conception of Bach and Engler (see Chapter 1). They proposed that the antioxidant rapidly reacts with the formed hypothetical moloxide and in such a way prevents the autoxidation of the substrate.

It was in 1924 when Christiansen [3] put forward the conception of the chain mechanism of oxidation and explained the action of antioxidants via chain termination by the antioxidant [3]. Three years later, Backstrom and coworkers [4–6] experimentally proved the chain mechanism of benzaldehyde oxidation (see Chapter 1) and the mechanism of antioxidant (hydroquinone) action via *chain termination*. The systematic study of the oxidation kinetics of esters of nonsaturated acids was performed by Bolland and ten Have [7,8]. They observed in the kinetic experiments that substrates are oxidized by the chain mechanism with chain propagation via the cycle of reactions (see Chapter 2).

$$ROOH \leftarrow \quad R^\bullet \quad \rightarrow O_2$$
$$RH \rightarrow \quad RO_2^\bullet \leftarrow$$

The retarding action of antioxidants (InH), such as phenols and aromatic amines, was proved to be the result of chain termination by accepting the peroxyl radicals.

$$RO_2^\bullet + InH \longrightarrow ROOH + In^\bullet$$
$$RO_2^\bullet + In^\bullet \longrightarrow Products$$

In a number of kinetic investigations that followed, it was clarified that the phenoxyl radical formed from phenol reacts with the peroxyl radical with the formation of quinone or quinolide peroxide [9]. The stoichiometry of chain termination by antioxidants was studied by Boozer et al. [10] and it was shown that the stoichiometric coefficient is $f \approx 2$. Howard and Ingold [11] studied the reactions of peroxyl radicals with phenols (ArOH) and deuteriophenols (ArOD) and evidenced the existence of the kinetic isotope effect ($k_H/k_D > 1$) [11]. They proved the reaction of hydrogen atom abstraction by peroxyl radicals from such antioxidants as phenols and aromatic amines. This agrees with the effect of hydrogen bond formation between the antioxidant and oxygen-containing substrates (alcohols, ethers, esters, etc) [12]. The phenomena of *cyclic chain termination* was discovered in experiments on cyclohexanol oxidation with 1-naphthylamine in 1963 [13]. The interpretation of this phenomenon was given in 1967 [14]. The full kinetic scheme of retarding action of phenols and aromatic amines was developed in several excellent experimental studies during 1970–1980 (see Chapter 17). Besides peroxyl radical acceptors, alkyl radical acceptors (nitroxyl radicals, quinones, nitrones) can also retard the oxidation of polymers [15]. Autoxidation is accelerated by forming hydroperoxide (see Chapter 4) and can be retarded by hydroperoxide decomposers (S, P-containing compounds, see Chapter 18) [16]. The common action of a few antioxidants often gives a *synergistic* effect which is widely used for efficient stabilization (see Chapter 19). The history of the study of antioxidants and their applications for the stabilization of organic substances in the years prior to the Second Word War was comprehensively accounted in Bailey's monograph [1]. The chemistry and kinetics of the antioxidant action are presented in handbooks [17–19], monographs [20–34], and reviews [35–40].

14.2 KINETIC CLASSIFICATION OF ANTIOXIDANTS

14.2.1 CLASSIFICATION

Oxidation of organic compounds occurs by the chain mechanism via alternating reactions of alkyl and peroxyl radicals (see Chapter 2). The accumulated hydroperoxide decomposes into radicals, thereby increasing the rate of oxidation (see Chapter 4). The oxidation of an organic compound may be retarded by one of the following three ways:

1. Breaking the chains perpetuated by the acceptor reaction with peroxyl radicals.
2. Breaking the chains by the reaction of an acceptor with alkyl radicals.
3. If hydroperoxide is the main autoinitiator, oxidation can be retarded by the introduction of chemical compounds capable to decompose hydroperoxide without the formation of free radicals.

Polyfunctional inhibitors can exert a combined action by participating in different inhibitory reactions. The mixture of antioxidants can act similarly. With respect to the mechanism of action, antioxidants can be divided into the following seven groups [41].

1. *Antioxidants that break chains by reactions with peroxyl radicals.* These are reductive compounds with relatively weak O—H and N—H bonds (phenols, naphthols, hydroquinones, aromatic amines, aminophenols, diamines), which readily react with peroxyl radicals forming intermediate radicals of low activity.

2. *Antioxidants that break chains by reactions with alkyl radicals.* These are compounds, such as quinones, nitrones, iminoquinones, methylenequinones, stable nitroxyl radicals, and nitrocompounds that readily accept alkyl radicals. Such antioxidants are efficient at very low concentrations of dioxygen and in solid polymers.
3. *Hydroperoxide decomposing antioxidants.* These are compounds that react with hydroperoxides without forming free radicals: sulfides, phosphites, arsenites, thiophosphates, carbamates, and some metal complexes. Reactions with hydroperoxides can be either stoichiometric (typical of, for example, sulfides and phosphites) or catalytic (typical of chelate metal complexes).
4. *Metal-deactivating antioxidants.* Transition metal compounds decompose hydroperoxides with the formation of free radicals, thereby increasing the rate of oxidation. Such an enhanced oxidation can be slowed down by the addition of a compound that interacts with metal ions to form complexes that are inactive with respect to hydroperoxides. Diamines, hydroxy acids, and other bifunctional compounds exemplify this type of antioxidants.
5. *Cyclic chain termination by antioxidants.* Oxidation of some substances, such as alcohols or aliphatic amines, gives rise to peroxyl radicals of multiple (oxidative and reductive) activity (see Chapters 7 and 9). In the systems containing such substances, antioxidants are regenerated in the reactions of chain termination. In other words, chain termination occurs as a catalytic cyclic process. The number of chain termination events depends on the proportion between the rates of inhibitor consumption and regeneration reactions. Multiple chain termination may take place, for instance, in polymers. Inhibitors of multiple chain termination are aromatic amines, nitroxyl radicals, and variable-valence metal compounds.
6. *Inhibitors of combined action.* Some antioxidants inhibit oxidation through various reactions. For instance, anthracene and methylenequinone can react with alkyl radicals, as well as with peroxyl radicals, whereas carbonates and thiophosphates can decompose hydroperoxides and break chains through the reaction with peroxyl radicals. Moreover, the same reaction center of an inhibitor (for example, the double bond of methylenequinone) may interact with the $R^{\bullet}$ and $RO_2^{\bullet}$ radicals. However, an inhibitor molecule may have two and more functional groups, each of which can undergo its own reaction. For example, the phenolic group of phenolsulfide reacts only with the peroxyl radical, and sulfide group reacts with hydroperoxide. Furthermore, the original inhibitor and its conversion products can exert their inhibitory action through various reactions.
7. *Synergism of action of several antioxidants.* When two inhibitors mutually enhance their inhibitory effects, it is a *synergistic* action. If the inhibitory effects of two inhibitors are simply added, it is *additive* inhibition. However, if the inhibitory effect of a mixture of inhibitors is lower than the sum of effects of individual inhibitors, it is known as the so-called *antagonism* of inhibitors. Thus, phenol and sulfide added to a hydrocarbon inhibit its oxidation through the following mechanisms: phenol terminates chains by reacting with the peroxyl radical, whereas sulfide slows the degenerate chain branching by breaking down hydroperoxide.

In addition to the classification of inhibitors according to their mechanisms of the action on oxidation, they can be classified into consumable and long-lived inhibitors. A consumable inhibitor is irreversibly consumed in its reactions with free radicals ($R^{\bullet}$ or $RO_2^{\bullet}$) or hydroperoxide. The stoichiometric coefficient of inhibition of such inhibitors is typically equal to one or two per inhibitory functional group. However, in some systems (for example $RH-O_2-InH$), an inhibitor can act cyclically so that, getting repeatedly regenerated, the inhibitor molecule may participate in more than one of the inhibitory events, thereby causing multiple chain termination or the catalytic decomposition of ROOH. In this case the coefficient of inhibition is $f \gg 2$.

14.2.2 Capacity, Strength, and Efficiency of Antioxidants

Inhibitors slow down oxidation by breaking chains or breaking down hydroperoxide. The inhibitory action of an antioxidant ceases when it is completely consumed. The duration of inhibitory effect depends on the mechanism of action of the antioxidant, the nature of inhibitory reactions, and the occurrence of side reactions in which the inhibitor is uselessly consumed. The action of the antioxidant in a given system can expediently be characterized in terms of *inhibitory capacity*. The capacity of a chain-breaking inhibitor can be characterized by the *inhibition stoichiometric coefficient f* [18].

For an antioxidant that breaks chains by the reaction with peroxyl radical according to the reactions:

$$RO_2^{\bullet} + InH \longrightarrow ROOH + In^{\bullet} \qquad k_7$$
$$RO_2^{\bullet} + In^{\bullet} \longrightarrow Products \qquad k_8$$
$$In^{\bullet} + In^{\bullet} \longrightarrow Products \qquad k_9$$

the inhibitory capacity is equal to the inhibition stoichiometric coefficient $f = 1-2$. In addition to the mechanism of chain termination, the capacity of inhibitor depends on the side reactions in which it is inefficiently consumed. In this case, the inhibitory capacity and the intensity of side reactions are related inversely. For instance, if an inhibitor is oxidized by dioxygen through a bimolecular reaction, its inhibitory capacity depends on the concentration of O_2 and the rate constant of this reaction (see later).

For antioxidant S, decomposing hydroperoxide in the stoichiometric reaction

$$n_{ROOH} ROOH + n_S S \longrightarrow Products$$

its inhibitory capacity is equal to n_{ROOH}/n_S, that is, the number of hydroperoxide molecules broken per antioxidant molecule. In the case of the catalytic decomposition of ROOH, the inhibitory capacity is defined as the number of the catalytic cycles of decomposition reactions.

The degree of inhibition by the antioxidant can be estimated by the ratio v_0/v, where v and v_0 are the rates of oxidation in the presence and absence of the inhibitor, respectively. With the increasing concentration of the inhibitor, the rate of oxidation tends to a nonzero value, v_∞. This can be explained by the ability of the intermediate products of inhibitor conversion or inhibitor itself to contribute to the chain propagation or initiation. Even an almost total replacement of the inactive radical $In^{\bullet}$ for active $RO_2^{\bullet}$ does not lead to complete inhibition, since the $In^{\bullet}$ radical can, to a certain extent, propagate chains. The lower the reactivity of $In^{\bullet}$, the greater the extent to which it is able to slow down the oxidation, provided its concentration is sufficiently high. On the other hand, an antioxidant can participate in chain initiation by reactions with dioxygen and hydroperoxide (see later). Therefore, the *strength* of the inhibitor can be defined as its ultimate inhibitory capacity expressed through the ratio v_0/v_∞ where $v_\infty = v$ at $[InH] \to \infty$.

If the radicals formed from inhibitor are inactive with respect to chain propagation, the chain length is $\nu \to 0$ and oxidation rate is $v \to v_\infty = v_i$ with increasing $[InH]_0$. Therefore, *strong inhibitors* (under given conditions) are those for which $v_0/v_\infty \geq \nu$ and $v_\infty \leq v_i$. *Weak inhibitors* can be defined as inhibitors, which form sufficiently active intermediate radicals with respect to chain propagation. For them, the ratio $v_0/v_\infty < v$ and $v_\infty > v_i$ so that the weaker the inhibitor the higher the ratio v_i/v_∞ [42].

The extent to which chain oxidation is inhibited depends on the activity and concentration of the antioxidant. A specific activity of an antioxidant as a retarding agent should be expressed per unit concentration of the inhibitor. If the antioxidant terminates chains, chain self-termination by the reaction of peroxyl radical disproportionation should be taken into account. As a result, one obtains the following expression for estimation of the activity F of the introduced amount of the antioxidant [18]:

$$F = \frac{v_0}{v}\left(1 - \frac{v^2}{v_0^2}\right) \quad (14.1)$$

The *efficiency* of the chosen inhibitor can be expressed by the ratio $F/[\text{In H}]$. This ratio does not depend on the antioxidant concentration if the latter terminates the chains and the intermediate radical In$^\bullet$ does not propagate through the chains (see later).

14.3 MECHANISMS OF INHIBITION OF INITIATED CHAIN OXIDATION

Depending on the oxidation conditions and its reactivity, the inhibitor InH and the formed radical In$^\bullet$ can participate in various reactions determining particular mechanisms of inhibited oxidation. Of the various mechanisms, one can distinguish 13 basic mechanisms, each of which is characterized by a minimal set of elementary steps and kinetic parameters [38,43–45]. These mechanisms are described for the case of initiated chain oxidation when the initiation rate $v_i = const$, autoinitiation rate $k_3[\text{ROOH}] \ll v_i$, and the concentration of dissolved dioxygen is sufficiently high for the efficient conversion of alkyl radicals into peroxyl radicals. The initiated oxidation of organic compounds includes the following steps (see Chapter 2).

$$\begin{array}{ll}
\text{Initiator} \longrightarrow \text{r}^\bullet & \text{i} \\
\text{r}^\bullet + \text{RH} \longrightarrow \text{rH} + \text{R}^\bullet & 2\text{i} \\
\text{R}^\bullet + \text{O}_2 \longrightarrow \text{RO}_2^\bullet & 1 \\
\text{RO}_2^\bullet + \text{RH} \longrightarrow \text{ROOH} + \text{R}^\bullet & 2 \\
\text{ROOH} + \text{RH} \longrightarrow \text{RO}^\bullet + \text{H}_2\text{O} + \text{R}^\bullet & 3 \\
2\text{R}^\bullet \longrightarrow \text{RR} & 4 \\
\text{R}^\bullet + \text{RO}_2^\bullet \longrightarrow \text{ROOR} & 5 \\
\text{RO}_2^\bullet + \text{RO}_2^\bullet \longrightarrow \text{ROOR} + \text{O}_2 & 6
\end{array}$$

The rate of initiated oxidation is the following:

$$v = k_2[\text{RH}][\text{RO}_2^\bullet] = k_2[\text{RH}](v_i/2k_6)^{1/2} \quad (14.2)$$

Mechanism I: Initiatory Radicals React Primarily with the Inhibitor
If the concentration of an antioxidant is such that initiatory radicals react mainly with InH, the following reactions occur additionally in the system.

$$\begin{array}{ll}
\text{r}^\bullet + \text{InH} \longrightarrow \text{rH} + \text{In}^\bullet & 7\text{i} \\
\text{RO}_2^\bullet + \text{In}^\bullet \longrightarrow \text{ROOIn} & 8 \\
\text{In}^\bullet + \text{In}^\bullet \longrightarrow \text{Products} & 9
\end{array}$$

The rate of oxidation is

$$v = k_{2i}[\text{RH}][\text{r}^\bullet] = v_i(1 + fk_{7i}[\text{InH}]/k_{2i}[\text{RH}])^{-1} \tag{14.3}$$

If the initiator is a peroxide compound capable of generating very active but nonselective alkoxyl radicals, the concentration of the inhibitor must be sufficiently high to provide the above conditions. For instance, the ratio $k_{7i}/k_{2i} = 50$ (408 K) for the reaction of the *tert*-butoxyl radical with ionol and cyclohexane [46]. In this case, alkoxyl radicals will react with InH more readily than with RH, provided $[\text{InH}] > 0.2 \text{ mol L}^{-1}$.

Mechanism II: Inhibited Nonchain Oxidation
If virtually all the initiatory radicals react with RH, while the peroxyl radicals produced in reaction (2i) react primarily with the inhibitor, oxidation proceeds as a straight chain reaction (in the RH–O$_2$–InH–I system, reactions (i), (1), (7), and (8) occur). In this case

$$v = k_7[\text{InH}][\text{RO}_2^\bullet] = f^{-1}v_i. \tag{14.4}$$

Inhibitory radicals can react not only with $\text{RO}_2^\bullet$ according to reaction (8), but also with one another, according to reaction (9). The proportion between the rates of the reactions (8) and (9) determines the stoichiometric coefficient of inhibition: $f = 2$ when $v_8 \gg v_9$, and $f = 1$ when $v_9 \gg v_8$. The value of f coefficient depends on the competition of both reactions and, in the general case, is equal to

$$f = 1 + (1 + k_9[\text{In}^\bullet]/k_8[\text{RO}_2^\bullet])^{-1} \tag{14.5}$$

The concentrations of the inhibitor, at the level of which this type of oxidation is possible, depend on the oxidizability of RH and reactivity of InH. In the case of oxidized cyclohexane, for which $k_2[\text{RH}] = 34 \text{ s}^{-1}$ (293 K), this type of oxidation takes place at ionol concentrations higher than $5.5 \times 10^{-4} \text{ mol L}^{-1}$, whereas in the case of ethylbenzene ($k_2[\text{RH}] = 148 \text{ s}^{-1}$ at 393 K) this oxidation is observed at $[\text{InH}] > 2.4 \times 10^{-3} \text{ mol L}^{-1}$ [47].

Mechanism III: Inhibited Chain Oxidation
When the concentration of an inhibitor is not high, oxidation occurs as a chain process (reactions (i), (1), (2), (7), and (8)) and, the oxidation rate is

$$v = v_i + k_2[\text{RH}][\text{RO}_2^\bullet] = v_i(1 + k_2[\text{RH}]/2k_7[\text{InH}]) \tag{14.6}$$

Since peroxyl radicals are also removed as a result of disproportionation, the reaction conditions are quasi-stationary for which the equality $v_i = 2k_7[\text{InH}][\text{RO}_2^\bullet] + 2k_6[\text{RO}_2^\bullet]^2$ is more appropriate. In this case, the rate of initiated chain oxidation of RH is equal to:

$$v = v_i + \frac{k_2 k_7 [\text{RH}][\text{InH}]}{2k_6} \left\{ \sqrt{1 + \frac{2v_i k_6}{k_7^2 [\text{InH}]^2}} - 1 \right\} \tag{14.7}$$

If $v \ll v_0 = k_2(2k_6)^{-1/2}[\text{RH}]v_i^{1/2}$ where v_0 is the rate of noninhibited chain oxidation, virtually all the chains are broken by the inhibitor molecules and radicals. Then antioxidant is consumed at the constant rate $v_{\text{In H}} = f^{-1}v_i$ and induction period $\tau = f[\text{InH}]_0/v_i$. The kinetics of oxygen consumption is now described for $t < \tau$ by the expression [48]

$$\Delta[O_2] = v_i t - \frac{k_2}{k_7}[RH]\ln(1 - t/\tau) \qquad (14.8)$$

This equation helps to use the whole experimental curve for estimation of the k_2/k_7 ratio.

Mechanism IV: Inhibited Chain Oxidation when In• Propagates the Chain by Reaction with RH
If the radicals In• formed from an antioxidant are active toward RH, the chain termination is limited by reactions (8) and (9) rather than by reaction (7). Inhibited oxidation also involves the following reaction [18,23,31,32,38]:

$$In^\bullet + RH \longrightarrow InH + R^\bullet \qquad (10)$$

If $v_8 \gg v_9$ and $v_{10} > v_i$, the rate of inhibited oxidation is described by the following equation [18]:

$$v = k_2[RH]^{3/2}\sqrt{\frac{k_{10}v_i}{fk_7k_8[InH]}} \qquad (14.9)$$

Let us compare the equations of oxidation rate mechanisms III (Equation (14.6)) and IV (Equation (14.9)).

Parameter	v_i	[InH]	[RH]
Mechanism III	$v \sim v_i$	$v \sim [InH]^{-1}$	$v \sim [RH]$
Mechanism IV	$v \sim v_i^{1/2}$	$v \sim [InH]^{-1/2}$	$v \sim [RH]^{3/2}$

We see that the participation of the inhibitor radical in chain propagation decreases the reaction order of the initiation rate, i.e., initiator diminishes the dependence of the reaction rate on the inhibitor concentration. The retardation of oxidation can be performed with the antioxidant concentration $[InH] > 2k_6k_{10}[RH]/fk_7k_8$. The kinetics of chain oxidation obeys the equation (for $t < \tau$):

$$\Delta[O_2] = 2v_0\tau\left\{1 - \sqrt{1 - t/\tau}\right\}, \qquad (14.10)$$

where the induction period $\tau = f[InH]_0/v_i$ and for the uninhibited oxidation rate, see Equation (14.2). Mahoney studied this kinetics by the oxidation of 9,10-dihydroanthracene inhibited by several substituted phenols [23,31,32,37,38,49]. 9,10-Dihydroanthracene possesses weak C—H bonds that are easily attacked not only by peroxyl radicals but also by phenoxyl radicals as well (for the rate constants of reaction (10), see Chapter 15).

Mechanism V: Inhibited Chain Oxidation When In• Propagates the Chain by Reaction with Hydroperoxide
If an oxidized RH substance is partially oxidized and contains hydroperoxide, then sufficiently active In• radicals react with ROOH according to the reaction [18,23,31,32,38,50]:

$$In^\bullet + ROOH \longrightarrow InH + RO_2^\bullet, \qquad (-7)$$

which results in chain propagation. If reaction (−7) is vigorous ($v_{-7} > v_i$), then chains are terminated by reaction (8) rather than by reaction (7), whereas reaction (9) is insignificant. This affects the kinetics of inhibited oxidation, so the latter becomes like that for mechanism

IV (see earlier). The rate of oxidation inhibited by mechanism V is described by the following equation [23,31,32,38,50]:

$$v = k_2[RH]\sqrt{\frac{k_{-7}v_i[ROOH]}{fk_7k_8[InH]}}. \qquad (14.11)$$

The retardation of oxidation can be performed with the antioxidant concentration $[InH] > 2k_6k_{-7}[ROOH]/fk_7k_8$. The kinetics of chain oxidation obeys the equation (for $t < \tau$):

$$\Delta[O_2] = \left(1 - \sqrt{1 - t/\tau}\right)\left(\frac{v_0^2\tau^2}{[ROOH]_0} + 2v_0\tau\right), \qquad (14.12)$$

where the induction period $\tau = f[InH]_0/v_i$ and for the uninhibited oxidation rate v_0, see Equation (14.2).

Mechanism VI: Reaction of Inhibitor with Hydroperoxide
Antioxidants possess the reducing activity and can be oxidized by hydroperoxide formed in the oxidizing substance [23,31,32,38,51]. This reaction produces an active radical

$$InH + ROOH \longrightarrow In^\bullet + H_2O + RO^\bullet \qquad (11)$$

This reaction is slow due to a high activation energy (see Chapter 15). However, at elevated temperatures and sufficiently high concentrations of antioxidant and hydroperoxide, this reaction becomes fast and, hence, can accelerate the rate of oxidation. As a result, the rate of initiation increases: $v_i = v_{i0} + e_{11}k_{11}[ROOH][InH]$ (e_{11} is the probability of the appearance of active radicals in the bulk). From the other side, this reaction shortens the induction period ($\tau_0 = f[InH]_0/v_{i0}$).

$$\tau = \frac{\ln(1 + 2k_{11}\tau_0[ROOH]_0)}{2k_{11}[ROOH]_0} \qquad (14.13)$$

The rate of oxidation and kinetics of dioxygen consumption are the following:

$$v = \frac{k_2[RH](v_i + e_{11}k_{11}[ROOH][InH])}{fk_7[InH]} \qquad (14.14)$$

and

$$\Delta[O_2] = \frac{k_2[RH]}{fk_7}\left(bt + \ln\frac{2b\tau_0}{1 + 2b\tau_0 - e^{-bt}}\right) \qquad (14.15)$$

where coefficient $b = e_{11}k_{11}[ROOH]$.

Mechanism VII: Reaction of Antioxidant with Dioxygen
Dioxygen is a weak oxidant. Nevertheless, phenols and aromatic amines interact with dioxygen at elevated temperatures according to the reaction [23,31,32,38,52]

$$InH + O_2 \longrightarrow In^\bullet + HO_2^\bullet \qquad (12)$$

The rate of initiation via this reaction is $v_i(InH + O_2) = e_{12}k_{12}[InH][O_2]$, where the coefficient e_{12} describes the effectiveness of initiation by the reaction. Normally, such reaction occurs slowly, however, at a sufficiently high temperature and high concentrations of antioxidant and dioxygen, the process of inhibited oxidation is characterized by increasing v_i and decreasing induction period ($\tau_0 = f[\text{InH}]_0/v_{i0}$).

$$\tau = \frac{\ln(1 + \tau_0 e_{12} k_{12}[O_2])}{2 e_{12} k_{12}[O_2]} \tag{14.16}$$

The rate of oxidation and dioxygen consumption is

$$v = \frac{k_2[\text{RH}](v_i + e_{12} k_{12}[O_2][\text{InH}])}{f k_7[\text{InH}]} \tag{14.17}$$

$$\Delta[O_2] = \frac{k_2[\text{RH}]}{f k_7} \left(ct + \ln \frac{2c\tau_0}{1 + 2c\tau_0 - e^{-ct}} \right) \tag{14.18}$$

where coefficient $c = e_{12} k_{12}[O_2]$.

Mechanism VIII: Decomposition of Quinolide Peroxides
With 2,4,6-trialkylphenols used as inhibitors, the formed phenoxyl radicals produce quinolide peroxides by the reactions with peroxyl radicals. At sufficiently high temperatures, quinolide peroxides decompose giving rise to free radicals [18,31,32,53,54]:

$$\text{In}^\bullet + \text{RO}_2^\bullet \longrightarrow \text{InOOR} \tag{8}$$
$$\text{InOOR} \longrightarrow \text{InO}^\bullet + \text{RO}^\bullet \tag{13}$$

which can affect the rate of inhibited oxidation and induction period. When the temperature and inhibitor concentrations are such that $f[\text{InH}]_0/v_i \gg k_{13}^{-1}$, the quasi-stationary concentration of quinolide peroxide is reached over a time period shorter than $\tau_0 = f[\text{InH}]_0/v_i$. In this case, the induction period is shorter than usual.

$$\tau = \frac{f[\text{InH}]_0}{v_i(1 + 0.5 e_{13})} = \frac{\tau_0}{1 + 0.5 e_{13}} \tag{14.19}$$

The rate of oxidation and dioxygen consumption are [18,31–33,38]

$$v = \frac{k_2[\text{RH}] v_i (1 + 0.5 e_{13})}{f k_7[\text{InH}]} \tag{14.20}$$

$$\Delta[O_2] = v_i t - \frac{k_2}{k_7}[\text{RH}] \ln(1 - t/\tau) \tag{14.21}$$

It is seen that the decomposition of quinolide peroxides shortens the induction period.

Mechanism IX: Decomposition of Phenoxyl Radicals
Phenoxyl radicals with alkoxyl substituents in the *ortho-* and *para-*positions decompose with the formation of alkyl radicals [55].

(14)

If the decomposition is intense ($v_{14} > v_i$), reaction (14) contributes to chain propagation. This reaction shortens the induction period

$$\tau = 2\sqrt{\frac{2k_8[\text{InH}]_0}{k_7 k_{14} v_i}} \tag{14.22}$$

and lowers the effectiveness of chain termination.

$$v = k_2[\text{RH}]\sqrt{\frac{k_{14} v_i}{f k_7 k_8 [\text{InH}]}} \tag{14.23}$$

$$\Delta[\text{O}_2] = 2 v_0 \tau \left\{1 - \sqrt{1 - t/\tau}\right\} \tag{14.24}$$

The decomposition of alkoxyphenoxyl radicals occurs with the activation energy and plays an important role in the inhibited oxidation at elevated temperature.

Mechanism X: Inhibitor Reacts with $RO_2^{\bullet}$ and Radical $In^{\bullet}$ Reacts with Dioxygen
Inhibitors such as diatomic phenols (hydroquinone, pyrocatechol), aminophenols, and aromatic diamines produce phenoxyl and aminyl radicals, which are efficient hydrogen donors rapidly reacting with dioxygen [56], for example,

$$\text{O}^{\bullet}\text{-C}_6\text{H}_4\text{-OH} + \text{O}_2 \longrightarrow \text{O=C}_6\text{H}_4\text{=O} + \text{HO}_2^{\bullet} \tag{15}$$

Reaction (15) contributes to the chain propagation. The higher rates of this reaction correspond to lower inhibitory efficiencies and lower stoichiometric coefficients of inhibition. The equations describing the induction period, oxidation rate, and kinetics of oxygen consumption for this mechanism when $k_{15}[\text{O}_2][\text{In}] > v_i$ are given here.

$$\tau = \frac{2v_i}{b^2}\left\{\ln\left[1 + \frac{b\sqrt{[\text{InH}]_0}}{v_i}\right] - 1\right\} - \frac{2\sqrt{[\text{InH}]_0}}{b} \tag{14.25}$$

$$v = \frac{k_2 b[\text{RH}]/k_7}{\sqrt{[\text{InH}]_0} - 0.5bt} + b\left\{\sqrt{[\text{InH}]_0} - 0.5bt\right\} \tag{14.26}$$

$$\Delta[\text{O}_2] = -\frac{2k_2[\text{RH}]}{k_7}\ln\left(1 - \frac{bt}{2\sqrt{[\text{InH}]_0}}\right) - \frac{1}{4}b^2 t^2, \tag{14.27}$$

where coefficient $b = (k_7 k_{15}[\text{O}_2] v_i / k_8)^{1/2}$. The induction period depends on the dioxygen concentration and rate constant of dioxygen reaction with the semiquinone radical.

Mechanism XI: Acceptors of Alkyl Radicals as Antioxidants
Some compounds, for example, quinones Q and nitroxyl radicals $AmO^{\bullet}$, inhibit oxidation by accepting alkyl radicals [15].

$$R^{\bullet} + Q \longrightarrow RQ^{\bullet} \tag{16}$$
$$RQ^{\bullet} + R^{\bullet} \longrightarrow RQR \quad \text{Fast}$$

This greatly alters the kinetics of inhibited oxidation by making it dependent on pO_2, since chain termination by reaction (16) competes with reaction (1) of chain propagation. The rate of oxidation becomes dependent on the dioxygen partial pressure. If the radical formed in the reaction of an acceptor with the alkyl radical does not participate in chain propagation, the following equations are valid for description of τ, v, and $\Delta[O_2]$:

$$\tau = f[Q]_0/v_i \tag{14.28}$$

$$v = v_i + k_1[R^\bullet][O_2] = v_i(1 + k_1[O_2]/fk_{16}[Q]) \tag{14.29}$$

$$\Delta[O_2] = v_i t - \frac{k_1}{k_{16}}[O_2]\ln(1 - t/\tau) \tag{14.30}$$

Acceptors of alkyl radicals showed their antioxidant activity in solid polymers where dioxygen reacts relatively slowly with macroradicals (see Chapters 13 and 19).

Mechanism XII: Inhibitor Reacts with $R^\bullet$ and its Radical Reacts with RH
In this case, the chain spreads through the reaction of formed $RQ^\bullet$ with RH [17]. The kinetics of oxidation is characterized by the following equations:

$$v = k_1[O_2]\sqrt{\frac{k_{17}v_i[RH]}{fk_{16}k_8[Q]}} \tag{14.31}$$

$$\Delta[O_2] = 2v_0\tau\left\{1 - \sqrt{1 - t/\tau}\right\}, \tag{14.32}$$

where the rate of noninhibited oxidation is $v_0 = k_1[O_2](v_i/2k_4)^{1/2}$ (see Chapter 2).

Mechanism XIII: Acceptor Reacts with $R^\bullet$ and its Radical Reacts with ROOH
The effectiveness of the acceptor decreases with increasing hydroperoxide concentration due to reaction (-7) (see mechanism V).

$$RQ^\bullet + HOOR \longrightarrow RQH + RO_2^\bullet \tag{-7}$$

This reaction does not change the induction period and accelerates inhibited oxidation only. The equations for v and $\Delta[O_2]$ are the following:

$$v = k_1[O_2]\sqrt{\frac{k_{-7}v_i[ROOH]}{fk_{16}k_8[Q]}} \tag{14.33}$$

$$\Delta[O_2] = \left(1 - \sqrt{1 - t/\tau}\right)\left(\frac{v_0^2\tau^2}{[ROOH]_0} + 2v_0\tau\right). \tag{14.34}$$

The expression for noninhibited oxidation rate is given earlier.

14.3.1 CHARACTERISTICS OF ANTIOXIDANT EFFICIENCY

The introduction of an inhibitor to oxidized hydrocarbon ($v_i = const.$) slows down the chain process due to the intensification of chain termination. The inhibitory efficiency can be characterized empirically by the ratio of the rates of chain termination with and without inhibitor (see earlier). In the absence of the inhibitor, chains are terminated via the disproportionation reaction of peroxyl radicals with a rate of $k_6[RO_2^\bullet]^2$. When an inhibitor is present, the chains involved in the oxidized RH can be terminated either by reaction (6) or additionally by reactions (7) and (8). Taking into account that the rates of chain initiation and termination are equal ($v_i = fk_7[\text{InH}][RO_2^\bullet] + k_6[RO_2^\bullet]^2$), we arrive at the expression (Equation (14.35)) for the retarding effectiveness of the inhibitor. When the inhibitor is introduced in a sufficiently high concentration chains are terminated mainly by reactions (7) and (8) (see scheme) and the function F takes the very simple form: $F = v_0/v$. This approach allows the comparison of inhibitors with respect to their effectiveness provided that all experimental conditions, including the concentrations of inhibitors and the rate of initiation, are identical. The function F is convenient for elucidating the mechanism of action of inhibitors. It can be written in the form [41]:

$$F = g[\text{RH}]^{n_{\text{RH}}}[O_2]^{n_{O_2}}[\text{InH}]^{n_{\text{InH}}}[\text{ROOH}]^{n_{\text{ROOH}}} v_i^{n_i} \qquad (14.35)$$

and the indices of powers depend on the mechanism of inhibition. Relevant parameters, together with the overall parameter $n = n_{\text{RH}} + n_{O_2} + n_{\text{InH}} + n_{\text{ROOH}} + n_i$, are summarized in Table 14.1.

Some inferences that can be deduced from this table are the following:

(i) If radicals In$^\bullet$ are not involved in chain propagation and, consequently, chain termination is linear, then $n_{\text{InH}} = 1$, $n_i = -0.5$, and $n = 0.5$.
(ii) If chains are terminated by only peroxyl radicals, then $n_{O_2} = 0$; but participating in chain termination makes $n_{O_2} \neq 0$.
(iii) If inhibitory radicals are involved in chain propagation, then $n_{\text{InH}} = 0.5$, $n_{\text{InH}} = 0$, and $n = 0$.
(iv) If inhibitory radicals and hydroperoxide are involved in chain propagation $n_{\text{ROOH}} = -0.5$.

It should be noted that Table 14.1 presents the most simple variants of inhibitory mechanisms with a minimal set of basic reactions. In practice, the dependancies of F may be more complex. However, the function F allows the mechanism of the inhibitory action of antioxi-

---

TABLE 14.1
Parameters of the Efficiency of Inhibition, $F = v_0/v - v/v_0 = \alpha[\text{RH}]^{n_1}[O_2]^{n_2}[\text{InH}]^{n_3}[\text{ROOH}]^{n_4} v_i^{n_5}$, for Various Inhibitory Mechanisms [41]

Mechanism	Key Steps	α	n_1	n_2	n_3	n_4	n_5	n
		Linear Chain Termination on $RO_2^\bullet$ and $R^\bullet$						
III	(2), (7)	$fk_7(2k_6)^{-1/2}$	0	0	1	0	$-1/2$	$1/2$
X	(1), (7)	$fk_1^{-1}k_2(2k_6)^{-1/2}k_7$	1	-1	1	0	$-1/2$	$1/2$
		Quadratic Chain Termination on $RO_2^\bullet$						
IV	(2), (7), (8), (10)	$(fk_7k_8/2k_6k_{10})^{-1/2}$	$-1/2$	0	$1/2$	0	0	0
V	(2), (7), (−7), (8)	$(fk_7k_8/2k_6k_{-7})^{-1/2}$	0	0	$1/2$	$-1/2$	0	0
IX	(2), (7), (8), (14)	$(fk_7k_8/2k_6k_{14})^{1/2}$	0	0	$1/2$	0	0	$1/2$
		Quadratic Chain Termination on $R^\bullet$						
XI	(1), (7), (8), (10)	$(fk_2k_7k_8/2k_1k_6k_{10})^{1/2}$	$1/2$	-1	$1/2$	0	0	0
XII	(1), (7), (−7), (8)	$(fk_2k_7k_8/2k_1k_6k_{-7})^{1/2}$	1	-1	$1/2$	$-1/2$	0	0

Theory of Inhibition of Chain Oxidation of Organic Compounds

dant to be elucidated. The coefficient g can be considered as a quantitative characteristic of the efficiency of a given type of inhibitor under given experimental conditions, including temperature, solvent, and the nature of oxidized RH.

14.4 KINETICS OF AUTOXIDATION OF ORGANIC COMPOUNDS INHIBITED BY ACCEPTORS OF PEROXYL RADICALS

As already noted (see Chapter 4), autoxidation is a degenerate branching chain reaction with a positive feedback via hydroperoxide: the oxidation of RH produces ROOH that acts as an initiator of oxidation. The characteristic features of inhibited autoxidation, which are primarily due to this feedback, are the following [18,21,23,26,31–33]:

1. The duration of the inhibition period of a chain-breaking inhibitor of autoxidation is proportional to its efficiency. Indeed, with an increasing rate of chain termination, the rates of hydroperoxide formation and, hence, chain initiation decrease, which results in the lengthening of the induction period (this problem will be considered in a more detailed manner later). It should be noted that when initiated oxidation occurs as a straight chain reaction, the induction period depends on the concentration of the inhibitor, its inhibitory capacity, and the rate of initiation, but does not depend on the inhibitor efficiency.
2. A feedback between the chain length and initiation rate results in critical phenomena, when small changes in the concentration of the inhibitor (by 0.1–1.0%) can greatly influence the duration of the induction period.
3. Since autoxidation is mainly initiated by hydroperoxide, it can be inhibited not only by scavengers of peroxyl and alkyl radicals, but also by compounds reactive to hydroperoxide (see Chapter 12).
4. Since variable-valence metals catalyze the decomposition of ROOH into radicals, autoxidation in the presence of these metals is inhibited by the respective complexing agents.

14.4.1 NONSTATIONARY KINETICS OF INHIBITED AUTOXIDATION

As shown above (see earlier) for straight chain reactions, the inhibitor is consumed at a constant rate v_i/f. Similarly, during the inhibited autoxidation of RH, the inhibitor is initially consumed at a constant rate v_{i0}/f, but then the rate of inhibitor consumption drastically increases [57,58], which leads to a rapid accumulation of hydroperoxide and the enhancement of initiation (see Figure 14.1).

This problem was first approached in the work of Denisov [59] dealing with the autoxidation of hydrocarbon in the presence of an inhibitor, which was able to break chains in reactions with peroxyl radicals, while the radicals produced failed to contribute to chain propagation (see Chapter 5). The kinetics of inhibitor consumption and hydroperoxide accumulation were elucidated by a computer-aided numerical solution of a set of differential equations. In full agreement with the experiment, the induction period increased with the efficiency of the inhibitor characterized by the ratio of rate constants [59]. An initiated inhibited reaction ($v_i = v_{i0} = const.$) transforms into the autoinitiated chain reaction ($v_i = v_{i0} + k_3[\text{ROOH}] \gg v_{i0}$) if the following condition is satisfied.

$$k_7 v_{i0} < 5 k_2 k_3 [\text{RH}] \qquad (14.36)$$

At a sufficiently high concentration of an efficient inhibitor the chain length is equal to 1, and RH converts at the rate v_{i0} whatever the concentration of the inhibitor. The induction period

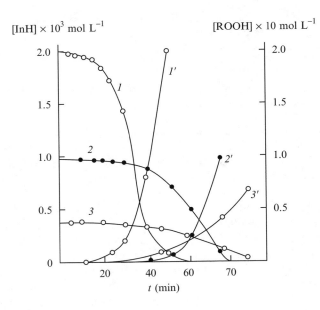

FIGURE 14.1 Oxidation of cyclohexanone: kinetic curves of 1-naphtol decay (curves without primes) and hydroperoxide formation (curves with primes) at $T = 413\,\text{K}$ (1, 1′), 403 K (2, 2′), and 393 K (3, 3′).

becomes independent of k_7/k_2, provided it is defined as the time necessary for the oxidation of a certain portion of RH (for example, 0.01%) [60].

The mechanisms responsible for inhibited oxidation depend on the experimental conditions and particular properties of RH and antioxidant (see earlier). Let us assume that hydroperoxide is relatively stable, so that it virtually does not decompose during the induction period ($k_d\tau \ll 1$). Actually, this means that the rate of ROOH formation is much higher than the rate of its decomposition, $k_2[\text{RH}][\text{RO}_2^\bullet] \gg k_d[\text{ROOH}]$. For each of the mechanisms of inhibited autoxidation, there is a relationship between the amounts of the inhibitor consumed and hydroperoxide produced (see Table 14.2). For example, for mechanism V with key reactions (2), (7), (−7), and (8), we can get (by dividing the oxidation rate v into the rate of inhibitor consumption) the following equation:

$$\frac{d[\text{ROOH}]}{d[\text{InH}]} = k_2[\text{RH}]\sqrt{\frac{fk_{-7}}{k_3 k_7 k_8 [\text{RH}][\text{InH}]}} \qquad (14.37)$$

After integrating it, we arrive at the formula

$$[\text{ROOH}] = 2k_2[\text{RH}]\sqrt{\frac{fk_{-7}}{k_3 k_7 k_8}}\left\{\sqrt{[\text{InH}]_0} - \sqrt{[\text{InH}]}\right\} \qquad (14.38)$$

14.4.2 Quasistationary Kinetics of Inhibited Autoxidation

At high temperatures or in the presence of catalysts, hydroperoxide decomposes at a high rate, so that, after $t \ll \tau$, inhibited oxidation becomes a quasi-stationary process with balanced rates of ROOH formation and decomposition. In this case, $k_d\tau \gg 1$, where k_d is the overall specific rate of ROOH decomposition with allowance made for its decomposition

TABLE 14.2
Equations for [ROOH] = F ([InH]) in the Nonstationary Regime of Inhibited Hydrocarbon Oxidation; $x = $ [InH]/[InH]$_0$, [InH] = [InH]$_0$ at $t = 0$ [33,38,45]

Key Steps	[ROOH]	a
(2), (7)	$[InH]_0(1-x) + a \ln \frac{1}{x}$	$\frac{k_2[RH]}{k_7}$
(2), (3), (7), (−7), (8)	$a(1 - \sqrt{x})$	$2k_2[RH]\sqrt{\frac{fk_{-7}[InH]_0}{k_3 k_7 k_8}}$
(2), (3), (7), (8), (10)	$a(1 - \sqrt{x})^{2/3}$	$\sqrt[3]{\frac{9fk_2^2 k_{10}[InH]_0}{k_3 k_7 k_8}}[RH]$
(2), (3), (7), (11)	$b(1-x) + a \ln \frac{1}{x}$	$\frac{k_2[RH]}{k_7}$; $b = \frac{2k_2 k_{11}[RH][InH]_0}{fk_3 k_7}$
(2), (3), (7), (12)	$\frac{2a}{b} \ln \frac{1}{x}$	$\frac{k_2[RH]}{k_7}$; $b = 1 + \frac{0.25 fk_{12}[O_2]}{k_3 + k_{12}[O_2] + 2k_2 k_3[RH]/k_7[InH]_0}$
(2), (3), (7), (13)	$[InH]_0(1-x) + a \ln \frac{1}{x}$	$\frac{k_2[RH]}{k_7}$
(2), (3), (8), (14)	$a(1 - \sqrt{x})^{2/3}$	$\sqrt[3]{\frac{9fk_2^2 k_{14}[InH]_0}{k_3 k_7 k_8}}[RH]$
(2), (3), (8), (15)	$a(1 - \sqrt{x})^{2/3}$	$\sqrt[3]{\frac{9fk_2^2 k_{15}[O_2][InH]_0}{k_3 k_7 k_8}}[RH]$

into radicals and molecular products. The transition from nonstationary to quasistationary patterns of inhibited autoxidation is related to the induction period τ and depends on the nature of the inhibitor and its concentration.

This transition may exhibit a critical behavior when, at a certain concentration of inhibitor known as the critical concentration [InH]$_{cr}$, the dependence of the induction period on [InH] drastically changes, so that $d\tau/d[InH]$ at [InH] > [InH]$_{cr}$ becomes much higher than $d\tau/d[InH]$ at [InH] < [InH]$_{cr}$. In the literature this problem has been treated only with reference to mechanisms II, III, and VIII [61–68], while all the known mechanisms of inhibited oxidation of RH will be envisaged here (see earlier) [69]. The equations for the chain length, critical antioxidant concentration [InH]$_{cr}$, stationary concentration of hydroperoxide [ROOH]$_{st}$, and induction period are given in Table 14.3 and Table 14.4.

The theoretical analysis presented in Table 14.3 and Table 14.4 allows the following inferences to be drawn:

1. No critical phenomenon is observed for mechanisms IV, IX, and XI, that is, free radicals In$^\bullet$ cancel out in the reactions RO$_2^\bullet$ + In$^\bullet$ or In$^\bullet$ + In$^\bullet$.
2. Linear chain termination is not, however, a necessary condition for the critical behavior. Indeed, with mechanisms V and XII, chain termination is quadratic ($v \sim v_i^{1/2}$), but critical transition does take place because hydroperoxide decomposes into radicals that contribute to chain propagation. As a result, $v \sim (v_i[ROOH])^{1/2}$ $v_i \sim [ROOH]^{1/2}$, and $v \sim [ROOH]$ (see Equation (14.11)) which explains the critical behavior.
3. Quasistationary (with respect to hydroperoxide) oxidation is possible for the mechanisms with a critical behavior (at [InH] > [InH]$_{cr}$) as well as for ordinary mechanisms (at $k_d \tau > 1$).

TABLE 14.3
Formulas for Kinetic Parameters of Hydrocarbon Autoxidation as Chain Reaction in Quasistationary Regime: Chain Length ν, Critical Concentration of Inhibitor $[InH]_{cr}$, and Quasistationary Concentration of Hydroperoxide $[ROOH]_s$. The Following Symbols are Used: $\beta = k_3/k_d$ and ν_{i0} is the Rate of Free Radical Generation on Reaction of RH with Dioxygen [33,38,45]

Key Steps	ν	$[InH]_{cr}$	$[ROOH]_s$
(2), (3), (7) (mechanism III)	$\dfrac{k_2[RH]}{fk_7[InH]}$	$\dfrac{\beta k_2[RH]}{fk_7}$	$\dfrac{\beta w_{i0}}{k_3(1-\beta\nu)}$
(2), (3), (7), (8), (10) (mechanism IV)	β^{-1}	—	$\dfrac{\beta^2 k_2 k_{10}[RH]^3}{fk_3 k_7 k_8[InH]}$
(2), (3), (7), (−7), (8) (mechanism V)	$\dfrac{k_2 k_{-7}^{1/2}[RH]}{\sqrt{fk_3 k_7 k_8[InH]}}$	$\dfrac{\beta^2 k_2^2 k_{-7}[RH]^2}{fk_3 k_7 k_8}$	$\dfrac{\nu_{i0}[InH]_{cr}}{fk_3([InH]-[InH]_{cr})}$
(2), (3), (7), (11) (mechanism VI)	$\dfrac{k_2[RH]}{fk_7[InH]}$	$\dfrac{\beta k_2[RH]}{fk_7}$	$\dfrac{\beta w_{i0}}{(k_3+k_{11}[InH])(1-\beta\nu)}$
(2), (3), (7), (12) (mechanism VII)	$\dfrac{k_2[RH]}{fk_7[InH]}$	$\dfrac{\beta k_2[RH]}{fk_7}$	$\dfrac{\beta\nu(\nu_{i0}+k_{12}[O_2][InH])}{k_3(1-\beta\nu)}$
(2), (3), (7), (13) (mechanism VIII)	$\dfrac{k_2[RH]}{fk_7[InH]}$	$\dfrac{\beta k_2[RH]}{fk_7}$	$\dfrac{2\beta w_{i0}}{k_3(1-2\beta\nu)}$
(2), (3), (7), (14) (mechanism IX)	β^{-1}	—	$\dfrac{k_{14}(\beta k_2[RH])^2}{fk_3 k_7 k_8[InH]}$
(2), (3), (7), (15) (mechanism X)	β^{-1}	—	$\dfrac{k_{15}[O_2](\beta k_2[RH])^2}{fk_3 k_7 k_8[InH]}$

4. When the inhibitor is able to interact with hydroperoxide (mechanism VI), the necessary conditions for critical behavior are $fk_7\,k_d > 2k_2 e_{11}\,k_{11}[RH]$ or $k_d > 2e_{11}k_{11}[InH]\nu$, that is, the reaction of inhibitor with hydroperoxide must not be too fast.
5. Critical behavior is profound only if the rate of chain initiation is sufficiently low, that is, $\nu_{i0} \ll k_3[RH]$ [62].

14.5 TOPOLOGICAL KINETICS OF INHIBITED OXIDATION OF HYDROCARBONS

The mechanisms by which an inhibitor adds to an oxidized hydrocarbon exerts its influence may differ depending on the reaction conditions. If the rate constants of the elementary reactions of RH, InH, $RO_2^\bullet$, $In^\bullet$, ROOH, and O_2 are known, the kinetics of the inhibited oxidation of RH can mathematically be described for any conditions. However, such an approach fails to answer questions how the mechanism of inhibited oxidation is related to the structure and reactivity of InH, RH, and $RO_2^\bullet$ or what inhibitor appears the most efficient under the given conditions, and so on. At the same time, these questions can easily be clarified in terms of a topological approach whose basic ideas are the following [43–45,70–72]:

1. Inhibited oxidation of RH occurs through a relatively large number of reactions (see Chapter 4), but the primary mechanism of inhibited oxidation is determined by a few key reactions. For example, cumene is oxidized in the presence of p-cresol at 320–380 K

TABLE 14.4
Formulas for Induction Period τ of Inhibited Oxidation of Hydrocarbons in Quasistationary Regime. Symbols are the Following: $\tau_0 = f[\text{InH}]_0 v_{i0}^{-1}$, $\beta = k_3/k_d$, v_{i0} is the Rate of Free Radical Generation on Reaction of RH with Dioxygen [33,38,45]

Key Steps	τ/τ_0	x
(2), (7), (8)	$1 - \dfrac{1+\ln x}{x}$	$\dfrac{fk_7[\text{InH}]_0}{\beta k_2[\text{RH}]}$
(2), (7), (8), (10)	$1 - \dfrac{1+\ln x}{x}$	$\dfrac{fk_7 k_8[\text{InH}]_0 v_{i0}}{(\beta k_2[\text{RH}])^2 k_{10}[\text{RH}]}$
(2), (−7), (8)	$1 - \dfrac{1+\ln(2 - 2\sqrt{x} + x)}{x} + \arctan(\sqrt{x}-1)$	$\dfrac{fk_3 k_7 k_8 [\text{InH}]_0}{(\beta k_2[\text{RH}])^2 k_{-7}}$
(2), (7), (11)	$[\text{InH}]_0/[\text{InH}]_{cr} - a \ln x;$ $a = \dfrac{\beta[\text{RH}]}{f[\text{InH}]_0 (k_2^{-1} k_7 + \beta k_3^{-1} k_{11}(\text{RH}))}$	$\dfrac{(k_7 k_{11}[\text{InH}]_{cr} + b)[\text{InH}]_0}{(k_7 k_{11}[\text{InH}]_0 + b)[\text{InH}]_{cr}}$; $b = k_3 k_7 + \beta k_2 k_{11}[\text{RH}]$
(2), (7), (12)	$cx^{-1}(2 - b/a)\{\arctan[c(2ax+b)] -$ $\arctan[c(2a+b)]\}$	$fk_7[\text{InH}]_0/\beta k_2[\text{RH}]$; $a = fk_7 k_{12}[\text{O}_2][]$
(2), (8), (14)	$1 - \dfrac{1+\ln x}{x}$	$\dfrac{fk_7 k_8[\text{InH}]_0 v_{i0}}{(\beta k_2[\text{RH}])^2 k_{14}}$
(2), (8), (15)	$1 - \dfrac{1+\ln x}{x}$	$\dfrac{fk_7 k_8[\text{InH}]_0 v_{i0}}{(\beta k_2[\text{RH}])^2 k_{15}[\text{O}_2]}$

by the mechanism that includes reactions (2), (7), (−7), (8), and (10). The rates of the reactions of In$^{\bullet}$ with ROOH (−7) and In$^{\bullet}$ with RH (10) are considerably lower than that of the reaction RO$_2^{\bullet}$ + In$^{\bullet}$ (8). If $v_2 > v_i$, the key reactions of inhibited oxidation are RO$_2^{\bullet}$ + RH (2) and RO$_2^{\bullet}$ + InH (7) and the rate of the inhibited oxidation is given by the formula (see earlier) $v = k_2[\text{RH}]v_i/2k_7[\text{InH}]$. The efficiency of the inhibitor is characterized by the ratio of the rate constants of these reactions, k_7/k_2. For inhibitors that terminate chains by reacting with RO$_2^{\bullet}$, we can distinguish nine primary inhibitory mechanisms (see earlier).

2. The region of implementation of a particular mechanism of inhibited oxidation in the coordinates: T—[RH]—[InH] is a combination of oxidized substances, inhibitors, and reaction conditions for which this mechanism is appropriate.
3. This region is determined by proportions between the rates of elementary reactions. For instance, when RH oxidation is chain-like and chains are terminated by the reaction RO$_2^{\bullet}$ of with InH (mechanism III), the following six inequalities must be satisfied.

$$k_2[\text{RH}] > 2k_7[\text{InH}] \qquad k_{-7}[\text{ROOH}] \ll k_8[\text{RO}_2^{\bullet}]$$
$$k_{10}[\text{RH}] \ll k_8[\text{RO}_2^{\bullet}] \qquad v_i \gg (k_{11}[\text{ROOH}] + k_{12}[\text{O}_2])[\text{InH}]$$
$$k_{13}[\text{InOOR}] \ll v_i k_{14} \qquad v_i k_{14} \ll k_8[\text{RO}_2^{\bullet}]$$

This multiparametric dependence can be simplified by using correlation equations, each of which describes a particular group of inhibitors. For instance, phenols can be divided into three groups (see Chapter 15).

(A) Phenols of this group react with peroxyl radicals, hydroperoxide, and dioxygen, while respective phenoxyl radicals can react with RH and ROOH. Reactions of these phenols with $RO_2^\bullet$ most commonly give rise to quinones; the breakdown of phenoxyls does not produce active radicals. This group includes all phenols, except 2,6-di-*tert*-alkylphenols and alkoxy-substituted phenols. Phenols of this group can inhibit oxidation by mechanisms I–VII.

(B) Phenols of this group slowly react with hydroperoxide and dioxygen. Respective phenoxyl radicals are relatively unreactive toward RH and ROOH, but can react with $RO_2^\bullet$ giving rise to peroxides and then to free radicals. For these phenols, appropriate inhibitory mechanisms are I–III and VI–VIII.

(C) This group includes phenols with alkoxy substituents. Respective phenoxyl radicals decompose with the formation of chain-propagating alkyl radicals. In addition to inhibitory mechanisms determined by substituents, these phenols can realize mechanism IX.

4. Reactions with the participation of phenols and phenoxyl radicals depend on the strength of the phenolic O—H bond. Therefore, in each group of phenols, the activation energies of elementary reactions can be expressed through the energy of the In—H bond (Table 14.5), while the activation energies of reactions with RH can be expressed through the strength of the R—H bond (one should take into account that tertiary $RO_2^\bullet$ is less reactive than secondary or primary $RO_2^\bullet$). Thus, the effect of structural factors accounted for the rate constants of nine reactions can be reduced to only two parameters, $D(R—H)$ and $D(In—H)$.

In spite of the paucity of data on the energy of R—H and In—H bonds, the rate constants of the reactions $RO_2^\bullet + RH$ (2) and $RO_2^\bullet + InH$ (7) have been measured for a great number of compounds (see Database [73]). This explains why these are the parameters that were taken as the kinetic characteristics of $RO_2^\bullet$, $In^\bullet$, RH, and InH (Table 14.5). The symbol "*" denotes that these rate constants (k_2 and k_7) refer to a reaction temperature of 333 K.

14.5.1 Initiated Oxidation

The region of each inhibitory mechanism can be presented as a domain in the three-dimensional space: $\log k_2 - \log k_7$ – ambient conditions (T, v_i, [InH], [RH], [ROOH], and [O_2], Figure 14.2).

The boundary between two neighboring domains is characterized by a set of the k_2 and k_7 values and ambient conditions at which the inhibitory mechanism possesses the features of two basic mechanisms. These boundary quantities and conditions can be described by respective parametric expressions (Table 14.6). Since boundaries have a finite width, rate constants change continuously between domains. Conventionally, the boundary width is taken such that the ratio of the rate constants of the key reactions changes across the boundary e times, which corresponds to a threefold change in the boundary parameters.

The analysis of the expressions defining boundaries between various inhibitory mechanisms made it possible to distinguish three groups of parameters.

(a) Parameters determining inhibitory mechanisms and respective three-dimensional domains: $D(R—H)$, $D(In—H)$, and T.
(b) Parameters exerting a weak influence on inhibitory mechanisms: [RH] and [InH]. These parameters enter the boundary equations as logarithms. Since log[RH] usually changes by at most unity and log[InH] by at most two to three times, the effect of

TABLE 14.5
Linear Relationship between E and D in Reactions of $RO^{\bullet}$, $RO_2^{\bullet}$, and $In^{\bullet}$ with InH, RH, and ROOH [33,38,45]

Reaction	log A, A (L mol^{-1} s^{-1})	E (kJ mol^{-1})	log k, k (L mol^{-1} s^{-1})
$RO^{\bullet}$ + RH (2′)	9.2	$0.30D(R-H) - 92.5$	
sec-$RO_2^{\bullet}$ + RH (2)	9.0	$0.55D(R-H) - 144$	$9.0 - 3000/T + 330/T \log k_2^{*}$
$tert$-$RO_2^{\bullet}$ + RH (2)	8.2	$0.55D(R-H) - 144$	$8.2 - 3000/T + 330/T \log k_2^{*}$
	Sterically Nonhindered Phenol (Ar_1OH)		
$RO^{\bullet}$ + InH (7′)	11 0	$16D(In-H) - 42$	
sec-$RO_2^{\bullet}$ + InH (7)	7.2	$0.32D(In-H) - 94$	$7.2 - 2400/T + (330/T)\log k_7^{*}$
sec-$RO_2^{\bullet}$ + $In^{\bullet}$ (8)	8.5	0	8.5
$In^{\bullet}$ + $In^{\bullet}$ (8)	8.5	0	8.5
$In^{\bullet}$ + sec-ROOH (−7)	7.2	$274 - 0.68D(In-H)$	$7.2 + 1260/T - (710/T)\log k_7^{*}$
$In^{\bullet}$ + RH (10)	9.2	$0.69\{D(R-H) - D(In-H)\} + 6.5$	$9.2 - 1800/T - (520/T)\log k_7^{*} + (530/T) \log k_2^{*}$
InH + ROOH (11)	10.0	$D(In-H) - 254$	$10.0 - 8100/T + (570/T) \log k_7^{*}$
InH + O_2 (12)	9.9	$D(In-H) - 218$	$9.9 - 8700/T - (570/T)\log k_7^{*}$
	Sterically Hindered Phenol (Ar_2OH)		
sec-$RO_2^{\bullet}$ + InH (7)	7.2	$0.37D(In-H) - 107$	$7.2 - 2410/T + (330/T)\log k_7^{*}$
$In^{\bullet}$ + sec-ROOH (−7)	7.2	$266 - 0.63D(In-H)$	$7.2 + 530/T - (710/T)\log k_7^{*}$
	Phenols of Group C (Alcoxy-Substituted)		
$In^{\bullet} \to Q + R^{\bullet}$	12.7		$12.7 - 5810/T$

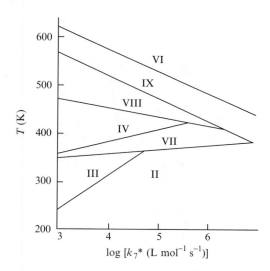

FIGURE 14.2 Domains of realization of various mechanisms of phenol-inhibited hydrocarbon oxidation in coordinates: T versus $\log k_7^{*}$ for conditions: $k_2^{*} = 1$ (L mol^{-1} s^{-1}), [RH] = 10 (mol L^{-1}), $f = 2$, [InH] = [ROOH] = [O_2] = 10^{-3} mol L^{-1} (see Table 14.5).

TABLE 14.6
Equations (in the form $T = A/B$) for the Bounday Mechanisms of the Phenol-Inhibited Oxidation of RH [69]

Mechanisms	Boundary Equation	A	B
		Phenols of Group A	
II–III	$k_2[RH] = fk_7[InH]$	$600 + 330 (\log k_7^* - \log k_2^\bullet)$	$1.8 + \log([RH]/f[InH])$
III–IV	$k_8\nu_i = k_7 k_{10}[RH][InH]$	$4200 + 90 \log k_7^\bullet - 530 \log k_2^\bullet$	$7.9 + \log([RH][InH]/\nu_i)$
III–V	$k_8\nu_i = k_7 k_{-7}[ROOH][InH]$	$1140 + 380 \log k_7^\bullet$	$5.9 + \log([ROOH][InH]/\nu_i)$
III, IV–VII	$fk_{12}[InH][O_2] = \nu_i$	$8700 - 570 \log k_7^\bullet$	$9.9 + \log(f[InH][O_2]/\nu_i)$
III–VI	$fk_{11}[InH][ROOH] = \nu_i$	$8100 - 570 \log k_7^\bullet$	$10 + \log(f[InH][ROOH]/\nu_i)$
III–IX	$k_4\nu_i = k_7 k_{14}[InH]$	$8210 - 330 \log k_7^\bullet$	$11.4 + \log([InH]/\nu_i)$
		Phenols of Group B	
II–III	$k_2[RH] = fk_7[InH]$	$600 + 330 (\log k_7^\bullet - \log k_2^\bullet)$	$1.8 + \log([RH]/f[InH])$
III–VIII	$k_{13}[InH] = \nu_i$	7225	$14 + \log([InH]/\nu_i)$

[RH] and [InH] on inhibitory mechanisms is relatively weaker than that of the parameters of the first group ($D(R{-}H)$, $D(In{-}H)$, and T).

(c) Initiation rate, $[O_2]$, and [ROOH] may affect inhibitory mechanisms only under certain conditions. In particular, ν_i influences the boundaries between mechanism III and mechanisms V, VIII, and IX, whereas $[O_2]$ affects the domain of mechanism VII, and [ROOH] affects the domains of mechanisms V and VI.

It is apparent that the mechanism of inhibited oxidation can be changed by varying the reaction conditions (T, ν_i, [InH], etc.). The question arises whether this is possible for any RH and InH. The formulae for the mechanism boundaries provide a clearly defined answer to this question. For instance, the region of mechanism II is limited from the left and from the right. By invoking equations for boundaries I–II and II–VII, we get for the upper limit of $D(R{-}H)$, (under fixed conditions)

$$D(R{-}H) < 1.15 D(In{-}H) + 118. \tag{14.39}$$

Similarly, using the equations for boundaries II–III and II–IV, one can derive expressions for the lower limit.

$$D(R{-}H) > 0.58 D(In{-}H) + 0.19T + 91 \text{ at } T < 268 + 0.26 D(In{-}H) \tag{14.40}$$

$$D(R{-}H) > 0.53 D(In{-}H) + 43 \text{ at } T > 268 + 0.26 D(In{-}H) \tag{14.41}$$

By combining the respective equations for bis-2,6(1,1-dimethylethyl)phenols, we arrive at the inequality describing the domain of existence of mechanism II.

$$D(R{-}H) > 0.67 D(In{-}H) + 145 \tag{14.42}$$

In other words, mechanism II is possible only for inhibitors and hydrocarbons with $D(R{-}H)$ and $D(In{-}H)$ that conform to the inequality (Equation (14.42)).

Similarly, by using equations for boundaries II–IV and IV–VII, we obtain the following equation for region III:

$$D(R-H) > 0.63D(In-H) + 142 \tag{14.43}$$

and the following equation for region IV

$$D(R-H) < 1.42D(In-H) + 24 \tag{14.44}$$

Thus, structural restrictions exist that make a given inhibitory mechanism hardly realizable for all RH–InH pairs.

Assuming that $k_8 = k_9 = 3 \times 10^8 \, L\,mol^{-1}\,s^{-1}$ (the key value), the diversity of the rate constants of the reactions of phenols and phenoxyl radicals (7, −7, 10, 11, and 12) can be reduced to only two parameters, k_7 and T. This allows one to get the universal formulae for the oxidation rate v, into which these parameters enter as functions of k_2^*, k_7^*, T, and ambient conditions (Table 14.7). When considering this table, it should be taken into account that mechanism VII is possible only for 2,4,6-tris-alkylphenols, while mechanism IX holds only for o- and p-alkoxyphenols.

The chain-terminating efficiency of inhibitors can be characterized by the chain length $\nu = v/v_i$ (more potent inhibitors provide shorter chains). From Table 14.7, the inhibitory efficiency depends on the mechanism of oxidation of the reactants RH and InH at the ambient conditions (T, v_i, and the reactant concentrations). It can be seen from Table 14.7 that higher k_7^* values always correspond to lower ν. Therefore, the most potent inhibitors are those that possess maximal k_7^*.

Another important characteristic of inhibitors is the time of their inhibition action. If an inhibitor is consumed only in chain termination reactions, this time is determined by the initial concentration $[InH]_0$, stoichiometric coefficient of inhibition f and v_i. In this case, the rate of inhibitor consumption is $v_{InH} = v_i/f$. Side reactions of InH with dioxygen and hydroperoxide shorten the inhibitory period and increase the rate of inhibitor consumption. Therefore, an inhibitor is efficient when it provides a minimal chain length ν and its own loss in side reactions w is low. Assuming that an efficient inhibitor has $w < 0.25$, we get the inequality $4k_{12}[InH][O_2] < v_i$ which can be transformed, by substituting the correlation equation from Table 14.7, into the following equation

$$\log k_7 \leq 15.3 - \{18.4 - 1.81 \log(v_i/f[InH][O_2])\} \times 10^{-3} T. \tag{14.45}$$

It is apparent that the efficient inhibitor has maximal k_7 and conforms to this inequality. The optimal value for k_7 is lower, higher are the reaction temperature and the concentrations of inhibitor and dioxygen. At $v_i = 10^{-7}\,mol\,L^{-1}\,s^{-1}$, $[InH] = [O_2] = 10^{-7}\,mol\,L^{-1}$, and $f = 2$, the

---

TABLE 14.7
Parameters of Inhibited Oxidation v Expressed Through k_2^* (*tert*-RO$_2^\bullet$) and k_7^* (Phenols of Group A) [69]

Mechanism	$\log(v/mol\,L^{-1}\,s^{-1})$
III (2, 7)	$1.8 - 600/T + (330/T)(\log k_2^* - \log k_7^*) + \log([RH]v_i/f[InH])$
IV (2, 7, 8, 10)	$5.8 - 2700/T + (600/T) \log k_2^* - (420/T) \log k_7^* + 1.5 \log[RH] + (1/2) \log(v_i/f[InH])$
IV (2, 7, 9, 10)	$5.0 - 1800/T + (520/T) \log k_7^* - (530/T) \log k_2^* + \log[RH] + 0.5 \log v_i$
V (2, 7, 8, −7)	$4.8 - 1170/T + (330/T) \log k_2^* - (520/T) \log k_7^* + \log[RH] + (1/2) \log([ROOH]v_i/f[InH])$
IX (2, 7, 8, 14)	$7.5 - 4700/T + (330/T) \log k_2^* - (170/T) \log k_7^* + (1/2) \log([RH]^2 v_i/f[InH])$

optimal values of $\log k_7$ are equal to 8.0 ($T=350$ K), 7.0 ($T=400$ K), 6.0 ($T=450$ K), and 5.0 ($T=500$ K).

If the oxidation of the inhibitor by hydroperoxide is the main side reaction, the inequality $w<0.25$ can be transformed, by substituting the correlation equation, into the following.

$$\log k_7 \leq 14.2 - \{18.6 - 1.81 \log(v_i/f[\text{InH}][\text{ROOH}])\} \times 10^{-3} T \tag{14.46}$$

It can be seen that the elevated temperatures narrow the range of promising phenolic inhibitors. Indeed, at $v_i = 10^{-7}$ mol L^{-1} s^{-1}, $[\text{InH}] = [\text{ROOH}] = 10^{-3}$ mol L^{-1}, and $f=2$, the optimal values of $\log k_7$ are equal to 6.9 ($T=350$ K), 5.8 ($T=400$ K), 4.5 ($T=450$ K), and 3.8 ($T=500$ K).

Generally, an inhibitor reacts with both dioxygen and hydroperoxide. For the inhibitor to be efficient, the overall rate of these reactions must be low. The competition between reactions (11) and (12) depends on the temperature and concentrations of O_2 and ROOH. At typical temperatures of oxidation (350–450 K) and [ROOH] ~ 0.1[O_2], the reaction of phenol with hydroperoxide is predominant.

14.5.2 Inhibited Autoxidation

As described earlier, the mechanism of inhibited chain oxidation depends on the structural features of RH and InH, as well as on the reaction conditions (T, v_i[RH], [InH], [O_2], and [ROOH]). In this section we present data illustrating this approach with reference to the autoxidation of hydrocarbons inhibited by sterically nonhindered phenols of group A.

When inhibited oxidation is nonstationary with respect to hydroperoxide, there is a definite, mechanism-dependent correlation between the amounts of the inhibitor consumed and hydroperoxide produced (see Table 14.2). The values of parameter a can be expressed through the rate constants k_2^* and k_7^* (the symbol * denotes that these constants are measured at 333 K) from Table 14.5.

When inhibited oxidation is quasistationary with respect to hydroperoxide, the induction period τ can be expressed through $[\text{InH}]_0$, [RH], v_i, and the rate constants of key reactions (see Equations [8.8]–[8.14]). Parametric equations make it possible to derive simple expressions for τ. Table 14.8 summarizes expressions for $\log x$ in terms of k_2^* and k_7^*, T, f, and $\beta = k_3/k_d$.

The kinetic topology provides an opportunity of choosing an optimal inhibitor by estimating k_7 for the minimal consumption of inhibitor at efficient chain termination. This can be illustrated with reference to oxidation by mechanism VI. Taking into account the initiation reaction, the consumption of an inhibitor can be described by the expressions

TABLE 14.8
Formulae for Induction Period $\tau = \tau_0 f(x)$ of Hydrocarbon Oxidation (Table 14.4) Inhibited by Phenols of Group A: Reactions are Nonstationary with Respect to ROOH; $\tau_0 = f[\text{InH}]_0/v_{i0}$, $\beta = k_3/k_d$, v_{i0} is the Rate of Free Radical Generation on Reaction of RH with Dioxygen [33,38,45]

Mechanisms	$\log(x/(\text{L}^{1/2}\text{ mol}^{-1/2}\text{ s}^{-1/2}))$
III	$-1.8 - 600/T - (330/T) \log(k_2^*/k_7^*) + \log(f[\text{InH}]_0/\beta[\text{RH}])$
IV	$-11.5 + 5400/T - (1200/T) \log k_2^* + (850/T) k_7^* + \log(f[\text{InH}]_0 v_{i0}/\beta^2[\text{RH}]^3)$
V	$-9.7 + 2340/T + (1040/T) \log k_7^* - (670/T) \log k_2^* + \log(fk_3[\text{InH}]0/\beta^2[\text{RH}]^2)$
VIII	$-15 + 9400/T - (670/T) \log k_2^* + (330/T) k_7^* + \log(f[\text{InH}]_0 v_{i0}/\beta^2[\text{RH}]^2)$

$$v_{\text{InH}} = v_i f^{-1} + k_{11}[\text{InH}][\text{ROOH}]_s \tag{14.47}$$

$$v_i = v_{i0} + (k_3 + 2e_{11}k_{11}[\text{InH}])[\text{ROOH}]_s \tag{14.48}$$

$$v_{\text{InH}} = v_{i0}f^{-1} + \{k_3 f^{-1} + k_{11}(1 + 2e_{11}f^{-1})[\text{InH}]\}[\text{ROOH}]_s \tag{14.49}$$

The probability of the appearance of radicals in the bulk (e_{11}) by reaction (11) is 0.5; therefore, $2e_{11} = 1$ [51]. Substituting $[\text{ROOH}]_s$ and assuming that $[\text{InH}] = 0.5[\text{InH}]_0$ and $\beta v \ll 1$, we get the expression

$$v_{\text{InH}} = v_{i0}f^{-1} + \beta v v_{i0}f^{-1}\{1 + 0.5k_3^{-1}k_{11}(1+f)[\text{InH}]_0\} \tag{14.50}$$

which, by substituting correlation equations for k_7 and k_{11} and differentiating with respect to $z = k_7^*$, can be transformed into the following differential equation:

$$\frac{dv_{\text{InH}}}{dz} = const\left\{\frac{570 A_{11}(1+f)[\text{InH}]_0}{3k_3 T}z^{(570/T)-1} - \frac{333}{T}z^{-(330/T)-1}\right\} \tag{14.51}$$

From the condition of the minimum ($dv_{\text{InH}}/dz = 0$), we arrive at the formula for k_7^* corresponding to the minimal value of v_{InH}.

$$\log k_7^* = 14.2 - 17.5 \times 10^{-3}T + 1.8 \times 10^{-3}T \log(k_3/(1+f)[\text{InH}]_0) \tag{14.52}$$

Table 14.9 summarizes respective formulae for k_7^* of optimal inhibitors as functions of T, $[\text{InH}]_0$, f, and k_3. At $v_i = const$, the k_7^* value of optimal inhibitor decreases with increasing temperature. But during autoxidation, k_7^* and T change unidirectionally. Such an inconsistency is due to an inverse relation between the efficiency of inhibitor and the temperature dependence of v_{i0}. The temperature-dependent rate constant k_3 may also contribute to this inconsistency, with the contribution depending on the ratio $k_3/(1+f)[\text{InH}]_0$.

TABLE 14.9
Formulae for Finding Optimal Inhibitors among the Phenols of Group A [69]

Mechanism	$\log(k_7/(\text{L mol}^{-1}\text{s}^{-1}))$ at $T = 333$ K
	Quasi-Stationary Reactions
III	$14.2 - 17.5 \times 10^{-3}T + 1.8 \times 10^{-3}T \log[k_3/(1+f)[\text{InH}]_0]$
IV	Maximal k_7
V	Maximal k_7
IX	$14.2 - 17.5 \times 10^{-3}T + 1.8 \times 10^{-3}T \log[k_3/(1+f)[\text{InH}]_0]$
	Nonstationary Reactions
III	$14.2 - 17.5 \times 10^{-3}T + 1.8 \times 10^{-3}T \log[k_3/(1+f)[\text{InH}]_0]$
IV	$14.2 - 17.5 \times 10^{-3}T + 1.8 \times 10^{-3}T \log[k_3/(1+f)[\text{InH}]_0]$
V	$15.2 - 17.5 \times 10^{-3}T + 1.8 \times 10^{-3}T \log[k_3/(1+f)[\text{InH}]_0]$
IX	$13.6 - 17.5 \times 10^{-3}T + 1.8 \times 10^{-3}T \log[k_3/(1+f)[\text{InH}]_0]$

REFERENCES

1. K Bailey. *The Retardation of Chemical Reactions.* London: Edward Arnold, 1937.
2. C Moureu, C Dufresse. *Chem Rev* 3:113, 1929.
3. JA Christiansen. *J Phys Chem* 28:145, 1924.
4. HLJ Backstrom. *J Am Chem Soc* 49:1460–1491, 1927.
5. H Backstrom. *Trans Faraday Soc* 24:601–612, 1928.
6. H Alea, H Backstrom. *J Am Chem Soc* 51:90–109, 1929.
7. JL Bolland, P ten Have. *Trans Faraday Soc* 43:201–210, 1947.
8. JL Bolland, P ten Have. *Discuss Faraday Soc* 2:252–263, 1947.
9. WA Waters, A Wilkham-Jones. *J Chem Soc* 812–820, 1951.
10. CE Boozer, GS Hammond, CE Hamilton, JN Sen. *J Am Chem Soc* 77:3233–3237, 1955.
11. JA Howard, KU Ingold. *Can J Chem* 40:1851–1864, 1962.
12. ET Denisov, AL Aleksandrov, VP Scheredin. *Izv AN SSSR Ser Khim* 1583–1590, 1964.
13. VV Kharitonov, ET Denisov. *Izv AN SSSR Ser Khim* 2222–2225, 1963.
14. VV Kharitonov, ET Denisov. *Izv AN SSSR Ser Khim* 2764–2766, 1967.
15. ET Denisov. In: G Scott (ed.) *Developments in Polymer Stabilisation-5.* London: Applied Science Publishers, 1982, pp. 23–40.
16. GH Denison, PC Condit. *Ind Eng Chem* 37:1102–1112, 1945.
17. ET Denisov. *Handbook of Antioxidants.* Boca Raton, FL: CRC Press, 1995.
18. ET Denisov, TG Denisova. *Handbook of Antioxidants.* Boca Raton, FL: CRC Press, 2000.
19. SH Hamid (ed.), *Handbook of Polymer Degradation.* New York: Marcel Dekker, 2000.
20. N Uri. In: WO Lundberg (ed.), *Autoxidation and Antioxidants*, vol 1. New York: Wiley Interscience, 1961, pp. 133–169.
21. G Scott. *Atmospheric Oxidation and Antioxidants.* Amsterdam: Elsevier, 1965.
22. NM Emanuel, YuN Lyaskovskaya. *Inhibition of Fat Oxidation.* Moscow: Pischepromizdat, 1961 [in Russian].
23. NM Emanuel, ET Denisov, ZK Maizus. *Liquid-Phase Oxidation of Hydrocarbons.* New York: Plenum Press, 1967.
24. ET Denisov. In: CH Bamford and CFH Tipper (eds.), *Comprehensive Chemical Kinetics*, vol 16. Amsterdam: Elsevier, 1980, pp. 125–204.
25. AM Kuliev. *Chemistry and Technology of Additives to Oils and Fuels.* Moscow: Khimiya 1972 [in Russian].
26. ET Denisov, GI Kovalev. *Oxidation and Stabilization of Jet Fuels.* Moscow: Khimiya, 1983 [in Russian].
27. ET Denisov, NI Mitskevich, VE Agabekov. *Liquid-Phase Oxidation of Oxygen-Containing Compounds.* New York: Consultants Bureau, 1977.
28. NM Emanuel, GE Zaikov, ZK Maizus. *Oxidation of Organic Compounds. Medium Effect in Radical Reaction.* Oxford: Pergamon Press, 1984.
29. G Scott (ed.), *Atmospheric Oxidation and Antioxidants*, vols 1–3. Amsterdam: Elsevier, 1993.
30. J Pospisil, PP Klemchuk (eds.), *Oxidation Inhibition in Organic Materials*, vols 1, 2. Boca Raton, FL: CRC Press, 1990.
31. ED Denisov, VV Azatyan. *Inhibition of Chain Reactions.* London: Gordon and Breach. 2000.
32. VA Roginskii. *Phenolic Antioxidants.* Moscow: Nauka, 1988 [in Russian].
33. ET Denisov. *Reactions of Inhibitor Radicals and Mechanism of Inhibited Oxidation of Hydrocarbons*, vol 17. Moscow: VINITI, Itogi Nauki i Tekhniki, 1987, pp. 3–115 [in Russian].
34. MM Mogilevich, EM Pliss. *Oxidation and Oxidative Polymerization of Nonsaturated Compounds.* Moscow: Khimiya, 1990 [in Russian].
35. KU Ingold. *Chem Rev* 61:563–589, 1961.
36. JA Howard. *Adv Free Radical Chem* 4:49–173, 1972.
37. LR Mahoney. *Angew Chem Int Ed* 8:547–555, 1969.
38. ET Denisov, IV Khudyakov. *Chem Rev* 87:1313–1357, 1987.
39. ET Denisov. *Russ Chem Rev* 65:505–520, 1996.
40. ET Denisov. *Usp Khim* 42:361–390, 1973.

41. NM Emanuel, ET Denisov. *Neftekhimiya* 16:366–382, 1976.
42. ET Denisov. *Zh Fiz Khim* 32:99–108, 1958.
43. ET Denisov. *Oxid Commun* 6:309–317, 1984.
44. ET Denisov. *Khim Fiz* 3:1114–1120, 1984.
45. ET Denisov. In: G Scott (ed.), *Mechanism of Polymer Degradation and Stabilization*. London: Elsevier, 1990, pp. 1–22.
46. ET Denisov. *Liquid-Phase Reactions Rate Constants*. New York: Plenum Press, 1974.
47. ET Denisov. *Usp Khim* 39:62–93, 1970.
48. VF Tsepalov. *The Study of Elementary Stages of Liquid-Phase Oxidation of Alkylaromatic Hydrocarbons*. Doctoral Thesis, Institute of Chemical Physics, Chernogolovka, 1975, pp. 1–42 [in Russian].
49. LR Mahoney, FC Ferris. *J Am Chem Soc* 85:2345–2346, 1963.
50. JR Thomas. *J Am Chem Soc* 85:2166–2167, 1963.
51. VS Martemiaynov, ET Denisov, LA Samoylova. *Izv AN SSSR Ser Khim* 1039–1042, 1972.
52. LN Denisova, ET Denisov, DI Metelitsa. *Izv AN SSSR Ser Khim* 1657–1662, 1969.
53. IA Shlyapnikova, VB Dubinskii, VA Roginskii, VB Miller. *Izv AN SSSR Ser Khim* 57–60, 1977.
54. VA Roginskii, VB Dubinskii, IA Shlyapnikova, VB Miller. *Eur Polym J* 13:1043–1051, 1977.
55. VA Roginskii, VB Dubinskii, VB Miller. *Izv AN SSSR Ser Khim* 2808–2812, 1981.
56. NN Pozdeeva, IK Yakuschenko, AL Aleksandrov, ET Denisov. *Kinet Catal* 32:1162–1169, 1991.
57. ET Denisov. In: G Scott (ed.), *Developments in Polymer Stabilization—5*. London: Applied Science, 1982, pp. 23–40.
58. ET Denisov. *Dokl AN SSSR* 141:131–134, 1961.
59. ET Denisov. *Kinet Katal* 4:53–59, 1963.
60. ET Denisov. *Kinet Katal* 4:508–516, 1963.
61. AB Gagarina, TD Nekipelova, NM Emanuel. *Dokl AN SSSR* 264:1412–1417, 1982.
62. NM Emanuel, AB Gagarina. *Usp Khim* 35:619–655, 1966.
63. AB Gagarina, ZK Maizus, NM Emanuel. *Dokl AN SSSR* 135:354–356, 1982.
64. AB Gagarina, ZK Maizus, NM Emanuel. *Dokl AN SSSR* 140:153–156, 1961.
65. AB Gagarina. *Izv AN SSSR Ser Khim* 444–450, 1964.
66. YuA Shlyapnikov, VB Miller. *Zh Fiz Khim* 39:2418–2424, 1965.
67. YuA Ershov. *Vysokomol Soed Ser A* 18:2706–2711, 1976.
68. VA Roginskii. *Kinet Katal* 26:571–577, 1985.
69. AB Gagarina. *Zh Fiz Khim* 39:2503–2509, 1965.
70. ET Denisov. *Khim Fiz* 4:67–74, 1985.
71. ET Denisov. *Khim Fiz* 2:229–238, 1983.
72. ET Denisov. *Neftekhimiya* 22:448–453, 1982.
73. ET Denisov, TG Denisova, SV Trepalin, TI Drozdova. *Database: Oxidation and Antioxidants in Organic Chemistry and Biology*. New York: CRC Press, 2005.

15 Antioxidants Reacting with Peroxyl Radicals

15.1 REACTION OF PHENOLS WITH PEROXYL RADICALS

Phenols rapidly react with peroxyl radicals terminating the chains by reaction [1–5]:

$$RO_2^{\bullet} + ArOH \longrightarrow ROOH + ArO^{\bullet}. \tag{7}$$

Several experimental evidences of this mechanism were obtained. The rate of chain oxidation of a substrate (RH) was proved to be proportional to $[ArOH]^{-1}$ (see Chapter 14) [1–6]. The phenoxyl radicals were identified by the EPR method as intermediate products of the reaction [6]

Phenols decrease the intensity of CL I_{chl} in oxidized hydrocarbons as a result of chain termination by the reaction with peroxyl radicals. Since $I_{chl} \sim [RO_2^{\bullet}]^2$ (see Chapter 2), the ratio $(I_0/I)^{1/2}$ was found to be proportional to [ArOH] [7]. The kinetic isotope effect ($k_{OH}/k_{OD} \gg 1$) proves that the peroxyl radical abstracts a hydrogen atom from the O—H bond of phenol [2,8].

Polar solvents block the O—H bond of phenols in the reaction with peroxyl radicals due to the formation of hydrogen bond and decrease the activity of phenols as chain terminating agents [1,9,10].

$$ArOH + O{=}CR^1R^2 \rightleftharpoons ArOH \cdots O{=}CR^1R^2$$

The reactivity of phenols depends on several structural factors, as well as on the conditions of oxidation (solvent, temperature). Let us discuss the main factors that determine phenol reactivity. This analysis will be performed within the scope of IPM (see Chapter 6).

15.1.1 REACTION ENTHALPY

In the reaction of the peroxyl radical with the O—H bond of phenol is cleaved, and hydroperoxide ROOH is formed. The reaction enthalpy ΔH is

$$\Delta H = D(ArO{-}H) - D(ROO{-}H) \tag{15.1}$$

and depends on the strength of two reacting O—H bonds.

The BDE of O—H bonds of phenols were studied intensively during the last 20 years [11–31]. They are collected in handbooks [4,32]. Substituents in the aromatic ring have influence on the BDE of the O—H bond of phenol. A few examples for monosubstituted phenols YC_6H_4OH are given below (values of BDE are given in kJ mol^{-1}, for PhOH $D(O—H) = 369$ kJ mol^{-1} [4].

Y	Ac	NH$_2$	CH$_3$	CH$_3$O	OH	Ph
o-	363.0	338.1	359.9	352.1	339.6	361.9
m-	376.5	367.2	366.7	369.8	369.1	
p-	371.6	356.7	362.2	345.8	352.0	368.7
X	Br	Cl	F	CN	NO$_2$	CF$_3$
o-	361.8	359.9	361.0	368.1	363.4	
m-		369.9	372.7	373.6	366.9	385.7
p-	372.3	370.5	365.0	370.2	372.8	382.4

We see that the BDE values of monosubstituted phenols vary from 338 kJ mol^{-1} (Y = o-NH$_2$) to 382 kJ mol^{-1} (Y = p-CF$_3$). These values are sufficiently lower for sterically hindered phenols (2,6-bis(1,1-dimethylethyl)-4-Y-phenols). A few examples are as follows [4].

Y	H	NH$_2$	CH$_3$	CH$_3$O	Ph	PhCH$_2$
D (kJ mol^{-1})	346.4	322.9	339.0	327.1	337.4	339.7
Y	Cl	NO	NO$_2$	SH	PhS	Me$_3$C
D (kJ mol^{-1})	344.5	346.0	358.0	340.1	346.4	339.7

The BDE values of O—H bond of hydroperoxide depend on the substituent near the hydroperoxide group (see Chapters 2, 7, 8, and 9). The higher the value of D(ROO—H) the faster the exothermic reaction of the peroxyl radical with phenol. The values of ΔH of reactions of different RO$_2^\bullet$ with several of the monosubstituted phenols (Ar$_1$OH) and sterically hindered phenols (Ar$_2$OH) are collected in Table 15.1.

These data illustrate that the reaction enthalpy ΔH varies in an interval from 45 kJ mol^{-1} due to different BDE of hydroperoxides to 44 kJ mol^{-1} due to different BDE of chosen phenols. Recently Luo [31] performed a correlation of BDE of O—H bonds of phenols YC_6H_4OH with $\delta^+(Y)$ Brown–Okamoto constants and arrived at the following equation:

$$\Delta D = D(YC_6H_4O—H) - D(PhO—H) = -3.18 + 29.75\Sigma(\sigma_0^+ + \sigma_m^+ + \sigma_p^+) \quad (15.2)$$

The influence of the reaction enthalpy on the activation energy can be estimated using the IPM method (see Chapter 6). The parameters of free radical reactions with phenols are collected in Table 15.2. They are different for sterically nonhindered (Ar$_1$OH) and sterically hindered (Ar$_2$OH) phenols. The values of coefficients α, b, and zero-point vibration energy of the phenolic O—H bond are the following: $\alpha = 1.014$, $b = 4.665 \times 10^{11}$ (kJ mol^{-1})$^{1/2}$ m^{-1}, and $0.5hLv = 21.5$ kJ mol^{-1} [33].

TABLE 15.1
Enthalpies of Reactions of Various Peroxyl Radicals with Phenols. BDE of O—H Bonds of Phenols were Taken from Ref. [32], BDE of O—H Bonds of ROOH (See Chapters 2, 7, 8, and 9)

Substituent	$R_3CO_2^\bullet$	$R_2CHO_2^\bullet$	$R(RO)CHO_2^\bullet$	$AcOPhCHO_2^\bullet$	$RC(O)O_2^\bullet$	$PhC(O)O_2^\bullet$
			Substituted Phenols			
C_6H_5OH	10.4	3.5	1.7	−7.4	−18.0	−35.0
2-Ac	4.4	−2.5	−4.3	−13.4	−24.0	−41.0
4-Ac	13.0	6.1	4.3	−4.8	−15.4	−32.4
2-NH_2	−20.5	−27.4	−29.3	−38.3	−48.9	−65.9
4-NH_2	−1.9	−8.8	−10.6	−19.7	−30.3	−47.3
2-OH	−19.0	−25.9	−27.7	−36.8	−47.4	−64.4
4-OH	−6.6	−13.5	−15.3	−24.4	−35.0	−52.0
4-COOH	13.1	6.2	4.4	−4.7	−15.3	−32.3
2-CN	9.5	2.6	0.8	−8.3	−18.9	−35.9
3-CN	15.0	8.1	6.3	−2.8	−13.4	−30.4
4-CN	11.6	4.7	2.9	−6.2	−16.8	−33.8
2-Ph	3.3	−3.6	−5.4	−14.5	−25.1	−42.1
4-Ph	10.1	3.2	1.4	−7.7	−18.3	−35.3
2-Br	3.2	−3.7	−5.5	−14.6	−25.2	−42.2
4-Br	13.7	6.8	5.0	−4.1	−14.7	−31.7
2-Me_3C	−4.8	−11.7	−13.5	−23.3	−33.2	−50.2
4-Me_3C	1.5	−5.4	−7.2	−16.3	−26.9	−43.9
2-Cl	1.3	−5.6	−7.4	−16.5	−27.1	−44.1
4-Cl	11.9	5.0	3.2	−5.9	−16.5	−33.5
2-F	2.4	−4.5	−6.3	−15.4	−26.0	−43.0
4-F	6.4	−0.5	−2.3	−11.4	−22.0	−39.0
4-CF_3	23.8	16.9	15.1	6.0	−4.6	−21.6
2-CH_3O	−6.5	−13.4	−15.2	−24.3	−34.9	−51.9
4-CH_3O	−12.8	−19.7	−21.5	−30.6	−41.2	−58.2
2-CH_3	1.3	−5.6	−7.4	−16.5	−27.1	−44.1
3-CH_3	8.1	1.2	−0.6	−9.7	−20.3	−37.3
4-CH_3	3.6	−3.3	−5.1	−14.2	−24.8	−42.1
2-NO_2	4.8	−2.1	−3.9	−13.0	−23.6	−40.6
3-NO_2	8.3	1.4	−0.4	−9.5	−20.1	−37.1
4-NO_2	14.2	7.3	5.5	−3.6	−14.2	−31.2
			4-Substituted 2,6-bis(1,1-dimethylethyl)phenol			
4-H	−12.1	−19.1	−20.9	−30.0	−40.6	−57.6
4-$PhCH_2$	−18.9	−25.8	−27.6	−36.7	−47.3	−64.3
Galvinol	−29.5	−36.4	−38.2	−47.9	−57.9	−74.9
4-Cl	−14.1	−21.0	−22.8	−31.9	−42.5	−59.5
4-Me_3CO	−27.3	−34.2	−36.0	−45.1	−55.7	−72.7
4-SH	−18.5	−25.4	−27.2	−36.3	−46.9	−63.9
4-CH_3O	−31.5	−38.4	−40.2	−49.3	−59.9	−77.0
4-CH_3	−19.6	−26.5	−28.3	−37.4	−48.0	−65.0
4-NO_2	−0.6	−7.5	−9.3	−18.4	−29.0	−46.0
4-NO	−12.6	−19.5	−21.3	−30.4	−41.0	−58.0
4-Ph	−21.2	−28.1	−29.9	−39.0	−49.6	−66.6
4-PhS	−12.2	−19.1	−20.9	−30.0	−40.6	−57.6
4-Me_3C	−18.9	−25.8	−27.6	−36.7	−47.3	−64.3
4-NH_2	−35.7	−42.6	−44.4	−53.5	−64.1	−81.1
4-CH_3O	−30.6	−37.5	−39.3	−48.4	−59.0	−76.0

TABLE 15.2
Parameters of Phenols (Ar$_1$OH) and Sterically Hindered Phenols (Ar$_2$OH) Reactions with Free Radicals in IPM Model [33,34]

Radical	α	br_e (kJ mol^{-1})$^{1/2}$	E_{e0} (kJ mol^{-1})	$0.5hL(\nu - \nu_f)$ (kJ mol^{-1})	A (L mol^{-1} s^{-1})	$\Delta H_{e\ min}$ (kJ mol^{-1})	$\Delta H_{e\ max}$ (kJ mol^{-1})
				Ar$_1$OH			
R$_1^\bullet$	1.247	17.61	61.5	4.1	1.0×10^8	86.9	136.7
R$_2^\bullet$	1.247	19.89	78.4	4.1	1.0×10^8	128.1	198.3
R$_3^\bullet$	1.247	18.73	69.5	4.1	1.0×10^8	106.2	165.6
RO$^\bullet$	0.992	14.16	50.5	−0.2	1.0×10^9	70.6	69.2
RO$_2^\bullet$	1.014	13.16	42.7	0.3	3.2×10^7	49.1	53.6
				Ar$_2$OH			
R$_1^\bullet$	1.246	20.58	84.0	4.1	1.0×10^8	142.2	219.2
R$_2^\bullet$	1.246	23.17	106.4	4.1	1.0×10^8	199.3	305.4
R$_3^\bullet$	1.246	21.82	94.34	4.1	1.0×10^8	168.3	258.8
RO$^\bullet$	0.992	15.84	63.2	−0.2	1.0×10^9	106.0	104.2
RO$_2^\bullet$	1.014	14.30	50.4	0.3	3.2×10^7	69.3	71.6

The increment of ΔH influence on the activation energy ΔE_H was calculated according to Equation (6.21). The activation energy was calculated by the IPM model via Equation (15.3). The geometrical parameters of TS of this reaction was calculated by the following equations [4,33–35]:

$$E = \left\{ 0.4965 br_e + 0.507 \frac{\Delta H + 0.3}{br_e} \right\}^2 - 21.5 + \frac{1}{2} RT \tag{15.3}$$

$$r(\text{O—O}) \times 10^{10} (\text{m}) = 1.92 + 3.09 \times 10^{-2} \left(\sqrt{E_e} + \alpha \sqrt{E_e - \Delta H_e} \right) \tag{15.4}$$

$$r(\text{ArO—H}) \times 10^{10} (\text{m}) = 0.955 + 3.09 \times 10^{-2} \sqrt{E_e} \tag{15.5}$$

$$r(\text{ROO—H}) \times 10^{10} (\text{m}) = 0.965 + 3.13 \times 10^{-2} \sqrt{E_e - \Delta H_e} \tag{15.6}$$

The results of calculation are given in Table 15.3.

15.1.2 Triplet Repulsion in Transition State

The reaction enthalpy is not the single factor that influences the activation energy of this reaction. The parameters of the peroxyl radical reaction with hydrocarbon (ethylbenzene) and phenol (p-cresol) are compared in the following table [34].

RO$_2^\bullet$ + YH, YH=	PhCH$_2$CH$_3$	4-MeC$_6$H$_4$OH
k (333 K) (L mol^{-1} s^{-1})	2.0	1.4×10^4
ΔH (kJ mol^{-1})	−8.5	−7.7
A (L mol^{-1} s^{-1})	2×10^7	3×10^7
E (kJ mol^{-1})	44.6	21.4

TABLE 15.3
The Values of ΔH, E, ΔE_H, and Geometric Parameters of Secondary Alkylperoxyl Radicals Reactions with Phenols, Calculated by IPM Method (Equations (15.3)–(15.6))

Substituent of Phenol	D (kJ mol^{-1})	ΔH (kJ mol^{-1})	E (kJ mol^{-1})	r(ArO—H) × 10^{11} (m)	r(H—OOR) × 10^{11} (m)	ΔE_H (kJ mol^{-1})
			Ar$_1$OH			
H	369.0	3.8	25.2	11.11	12.27	2.2
2-Ac	363.0	−2.2	22.2	11.06	12.32	−0.8
4-Ac	371.6	6.4	26.6	11.13	12.25	3.6
2-NH$_2$	338.1	−27.1	10.8	10.86	12.52	−12.2
4-NH$_2$	356.7	−8.5	19.2	11.01	12.37	−3.8
2-OH	339.6	−25.6	11.4	10.87	12.51	−11.6
4-OH	352.0	−13.2	16.9	10.97	12.41	−6.1
4-COOH	371.7	6.5	26.6	11.13	12.25	3.6
2-CN	368.1	2.9	24.8	11.10	12.28	1.8
3-CN	373.6	8.4	27.6	11.15	12.23	4.6
4-CN	370.2	5.0	25.8	11.12	12.26	2.8
2-Ph	361.9	−3.3	21.7	11.05	12.333	−1.3
4-Ph	368.7	3.5	25.1	11.11	12.27	2.1
2-Br	361.8	−3.4	21.6	11.05	12.33	−1.4
4-Br	372.3	7.1	27.0	11.14	12.24	4.0
2-Me$_3$C	353.8	−11.4	17.8	10.99	12.39	−5.2
4-Me$_3$C	360.1	5.1	20.8	11.04	12.34	−2.2
2-Cl	359.9	−5.3	20.7	11.04	12.34	−2.3
4-Cl	370.5	5.3	26.0	11.12	12.26	3.0
2-F	361.0	−4.2	21.2	11.05	12.33	−1.8
4-F	365.0	−0.2	22.9	11.08	12.30	−0.1
4-CF$_3$	382.4	17.2	32.3	11.22	12.16	9.3
2-MeO	352.1	−13.1	17.0	10.98	12.41	−6.0
4-MeO	345.8	−19.4	14.1	10.92	12.46	−8.9
2-Me	359.9	−5.3	21.0	11.04	12.34	−2.0
3-Me	366.7	1.5	24.0	11.09	12.29	1.0
4-Me	362.2	−3.0	21.8	11.06	12.32	−1.2
2-NO$_2$	363.4	−1.8	22.4	11.07	12.31	−0.6
3-NO$_2$	366.9	1.7	24.2	11.09	12.29	1.2
4-NO$_2$	372.8	7.6	27.2	11.14	12.24	4.2
			Ar$_2$OH			
4-H	346.4	−18.8	22.2	11.066	12.67	−8.6
PhCH$_2$	339.7	−25.5	19.2	11.01	12.72	−11.6
Galvinol	329.1	−36.1	15.0	10.93	12.80	−15.8
Cl	344.5	−20.7	21.3	11.05	12.68	−9.5
Me$_3$CO	331.3	−33.9	16.0	10.95	12.78	−14.8
SH	340.1	−25.1	19.3	11.01	12.72	−11.5
MeO	327.1	−38.1	14.0	10.92	12.81	−16.8
Me	339.0	−26.2	19.0	11.01	12.72	−11.8
NO$_2$	358.0	−7.2	28.0	11.15	12.58	−2.8
NO	346.0	−19.2	22.0	11.06	12.67	−8.8
Ph	337.4	−27.8	18.2	10.99	12.74	−12.6
PhS	346.4	−18.8	22.2	11.06	12.67	−8.6
Me$_3$C	339.7	−25.5	19.2	11.01	12.72	−11.6
NH$_2$	322.9	−42.3	12.2	10.89	12.84	−18.6
MeO	328.0	−37.2	14.2	10.92	12.81	−16.6

It is apparent that these reactions are close in their enthalpies, but greatly differ in the rate constants. The peroxyl radical reacts with p-cresol by four orders of magnitude more rapidly than with ethylbenzene. Such a great difference in the reactivities of RH and ArOH is due to the different activation energies of these reactions, while their pre-exponential factors are close. This situation was analyzed within the scope of the parabolic model of transition states (see Chapter 6 and Refs. [33–38]).

If the peroxyl radical abstracts the hydrogen atom from the YH molecule, then in the transition state $RO_2^\bullet \cdots H \cdots Y$ in accordance with the Pauli exclusion principle, two electrons occupy the bonding orbital of the ROO—Y bond, whereas one electron occupies the nonbonding orbital [39]. The stronger the ROO—Y bond, the higher the nonbonding orbital energy. Consequently, the higher the TS energy and the activation energy of reaction. This leads to a correlation between the parameter r_e (see Chapter 6) and the energy of the ROO—Y bond [40–42]. The stronger ROO—Y bond also corresponds to higher energies of thermally neutral reactions (that is, reactions with $\Delta H_e = 0$). The O—O bond in the hypothetical compound ROO—OAr is weak and, hence, the triplet repulsion is close to zero. The O—C bond in the compound ROO—R (for example, hydrocarbon) is relatively strong (its energy is about 270 kJ mol^{-1}); therefore, a strong triplet repulsion does not allow this reaction to occur with a high rate. The contribution of the triplet repulsion to the activation energy of the thermally neutral reaction $RO_2^\bullet + RH$ is 37 kJ mol^{-1} [34], which explains why peroxyl radicals react with the O—H bonds of organic compounds with relatively low activation energies E_{e0} at $\Delta H_e = 0$ [36].

YH	RH	ArOH	AmOH	ROOH
E_{e0} (kJ mol^{-1})	61.5	45.3	45.6	43.1

Alkoxyl radicals react with phenols extremely rapidly. The rate constants of the reactions of *tert*-butoxyl radicals with some phenols in benzene are given below (Y is the p-substituent in the phenol molecule) [43,44].

p-YC$_6$H$_4$OH, Y=	H	Br	HO	CN	MeO
k (295 K)	3.3×10^8	3.0×10^8	3.2×10^9	9.7×10^7	1.6×10^7
E (kJ mol^{-1})	11.7	10.9			16.7
log A, A (L mol^{-1} s^{-1})	10.6	13.4			12.1

15.1.3 STERIC FACTOR

Tertiary alkyl substituents in the *ortho*-position to the phenolic group cause an additional repulsion in TS and, hence, diminish the reactivity of phenols. Using values of E_{e0} for Ar$_1$OH and sterically hindered Ar$_2$OH, we can estimate the contribution from such a steric repulsion ΔE_S to the activation energy E [33,34].

$$\Delta E_S = E_{e0}(Ar_2OH) - E_{e0}(Ar_1OH) \tag{15.7}$$

The results of such estimations are given below [34,38]:

Reaction	br_e (kJ mol^{-1})$^{1/2}$	E_{e0} (kJ mol^{-1})	ΔE_S (kJ mol^{-1})
$RO_2^{\bullet} + Ar_1OH$	13.76	45.3	0
$RO_2^{\bullet} + Ar_2OH$	14.40	51.8	6.5
$CH_3^{\bullet} + Ar_1OH$	17.38	59.8	0.0
$CH_3^{\bullet} + Ar_2OH$	18.92	84.2	24.4

It can be seen that the steric effect is profound in radical reactions of Ar_2OH with peroxyl and methyl radicals. It will be shown later that the steric effect exists in other free radical reactions of Ar_2OH. The ΔE_S values of the reactions of alkyl radicals with Ar_2OH are considerably higher than those for phenols reacting with oxygen-centered radicals. The steric effect can also manifest itself in the inverse reactions of sterically hindered phenoxyl radicals $Ar_2O^{\bullet}$ with various molecules (see later).

15.1.4 Effect of Hydrogen Bonding on Activity of Inhibitors

Phenols and aromatic amines form hydrogen bonds with polar molecules Y containing heteroatoms or π-bonds, for example, with ketones [45]:

$$ArOH + O=CR^1R^2 \rightleftharpoons ArOH \cdots O=CR^1R^2$$

Therefore, in such systems an inhibitor may occur either in the free form or as a complex held together by hydrogen bonds. The hydrogen bonding between phenols and methylethylketone is characterized by the following equilibrium constants K_H parameters [46].

Phenol			
K_H (295 K) (L mol^{-1})	7.5	9.3	2.9
K_H (333 K) (L mol^{-1})	3.0	3.6	1.3

As the reaction temperature increases, the equilibrium constant diminishes, since complex formation is accompanied by heat liberation. Sterically hindered phenols form loose complexes because of the impeding effect of voluminous alkyl substituents in the *ortho*-position. Hydrogen bonding reduces the activity of phenols, which was first observed in the studies of the effects of cyclohexanol and butanol on the inhibitory activity of α-naphthol in cyclohexane [9]. This phenomenon was investigated in detail with reference to the oxidation of methylethylketone [10]. The k_7 values for some inhibitors of the oxidation of ethylbenzene and methylethylketone are given below (333 K) [10,46]:

	Phenol	(1-naphthol)	(2,4,6-trimethylphenol)	(2,4,6-tri-tert-butylphenol)
k_7 (PhEt) (L mol^{-1} s^{-1})		4.0×10^5	2.2×10^5	4.2×10^4
k_7 (MeC(O)Et) (L mol^{-1} s^{-1})		7.5×10^2	1.0×10^3	5.7×10^2

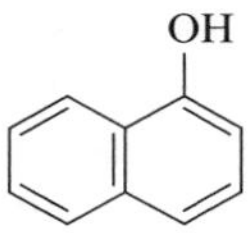

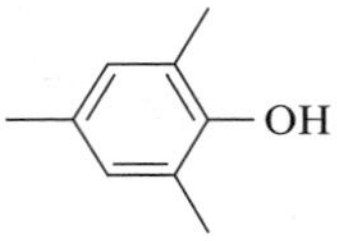

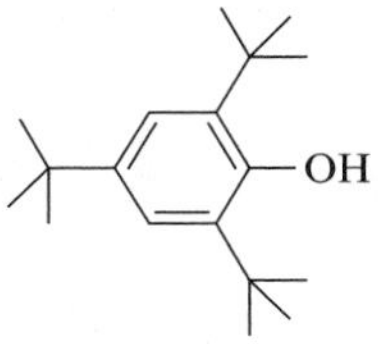

Since inhibitor molecules in a polar solvent may exist in the free (InH) or bound (InH $\cdots$ Y) form, and peroxyl radicals attack preferentially the free In—H bonds that are not involved in complex formation, the decrease in k_7 in such solvents is due to the decreasing concentration of free and, hence, more reactive phenol molecules. The concentrations of phenol and the complex are related as [InH $\cdots$ Y] = K_H[InH][Y]. An inhibitor occurring in the complex is unlikely to react with the peroxyl radical by virtue of this reaction. Therefore, the empirical rate constant $k_{7\text{emp}}$ is related to [Y] in the following way:

$$k_{7\text{emp}} = k_7(1 + K_H[Y])^{-1} \tag{15.8}$$

The K_H values derived from kinetic data are close to those obtained by IR and NMR spectroscopies. An empirical formula for the evaluation of the hydrogen bond enthalpy and Gibbs potential was derived [47]:

$$\Delta H_H(\text{kJ mol}^{-1}) = 4.96 E_{\text{ArOH}} E_Y, \tag{15.9}$$

$$\Delta G_H(\text{kJ mol}^{-1}) = 5.70 + 2.43 C_{\text{ArOH}} C_Y, \tag{15.10}$$

where E_{ArOH}, E_Y, C_{ArOH}, and C_Y are empirical increments. The entropy of hydrogen bonding is about -50 J mol^{-1} K^{-1}. The calculated values of ΔH_H for several phenols and oxygen-containing compounds are presented in Table 15.4.

Since phenol forms the hydrogen bond with any O-containing compound, it can form hydrogen bond with the peroxyl radical too [48] and hence the reaction of the peroxyl radical with phenol should be treated as a two-stage process:

$$RO_2^\bullet + ArOH \rightleftharpoons RO_2^\bullet \cdots HOAr \longrightarrow ROOH + ArO^\bullet$$

Peroxyl radicals are highly polar species; therefore, they can form complexes with water, alcohols, and acids according to the following reaction (see Chapter 6 and [49]):

$$RO_2^\bullet + HOX \rightleftharpoons RO_2^\bullet \cdots HOX$$

Generally, hydroxyl-containing compounds form with peroxyl radical an equilibrium system

$$ArOH + HOX \rightleftharpoons ArOH \cdots O(H)X \qquad K_{H1}$$
$$RO_2^\bullet + HOX \rightleftharpoons RO_2^\bullet \cdots HOX \qquad K_{H2}$$
$$RO_2^\bullet + HOAr \longrightarrow ROOH + ArO^\bullet \qquad k_7$$
$$RO_2^\bullet \cdots HOX + ArOH \longrightarrow ROOH + ArO^\bullet + HOX \qquad k_7'$$

TABLE 15.4
The Values of Gibbs Potential ($-\Delta G$(298 K)) (kJ mol^{-1}) for Hydrogen Bonding ArOH $\cdots$ Y, Calculated by Equation (15.9), for Increments see Handbook [4]

Phenol	EtOH	EtOEt	MeC(O)Et	PhCH(O)	cyclohexanone	tetrahydrofuran
PhOH	17.0	16.1	17.4	15.9	18.2	18.4
2-Me-C$_6$H$_4$-OH	16.2	15.4	16.6	15.2	17.4	17.5
4-Me-C$_6$H$_4$-OH	16.5	15.7	17.0	15.5	17.7	17.9
4-MeO-C$_6$H$_4$-OH	16.6	15.8	17.0	15.3	17.8	17.9
3-HO-C$_6$H$_4$-OH	19.8	18.8	20.4	18.5	21.4	21.6
4-Br-C$_6$H$_4$-OH	18.5	17.5	19.0	17.2	19.9	20.1
4-Cl-C$_6$H$_4$-OH	18.4	17.4	18.9	17.2	19.8	20.0
4-F-C$_6$H$_4$-OH	17.8	16.9	18.3	16.6	19.1	19.3
4-NC-C$_6$H$_4$-OH	20.9	19.8	21.5	19.5	22.6	22.8
4-O$_2$N-C$_6$H$_4$-OH	21.5	20.3	22.1	20.0	23.2	23.5
2,6-diMe-C$_6$H$_3$-OH	13.5	12.9	13.8	12.7	14.3	14.4
2,4-diMe-C$_6$H$_3$-OH	16.0	15.2	16.3	15.0	17.1	17.2
1-naphthol	17.4	16.6	17.8	16.3	18.7	18.9
2-naphthol	17.6	17.0	18.0	16.4	18.9	19.1
2-Me-6-tBu-C$_6$H$_3$-OH	12.7	12.2	13.0	12.0	13.5	13.6
2,6-ditBu-C$_6$H$_3$-OH	9.9	9.6	10.1	9.5	10.4	10.4
2,4,6-tritBu-C$_6$H$_2$-OH	8.0	7.9	8.1	7.8	8.3	8.3
C$_6$F$_5$-OH	22.6	21.3	23.2	21.0	24.4	24.7

The complexed peroxyl radical possesses a lower activity, so that the empirical rate constant $k_{7\text{emp}}$ in the presence of HOX has the following form [10]:

$$k_{7\text{emp}} = \frac{k_7 + k'_7 K_{H2}[\text{HOX}]}{1 + K_{H1}[\text{HOX}] + K_{H2}[\text{HOX}]} \qquad (15.11)$$

The experimental rates and equilibrium constants (333 K) of the peroxyl radicals of methylethylketone reacting in 1,1-dimethylethanol are given below [10].

Phenol	HO–C$_6$H$_4$–OH	naphthol	dimethylphenol	di-tert-butylphenol
K_{H1} (L mol^{-1})	3.2	3.6	3.0	1.3
K_{H2} (L mol^{-1})	1.2	43	15.8	11.8
k_7 (L mol^{-1} s^{-1})	1.4×10^5	3.1×10^4	3.6×10^4	9.5×10^3
k'_7 (L mol^{-1} s^{-1})	1.7×10^4	1.6×10^3	3.0×10^3	3.3×10^3

The same effect is typical of the reactions of alkoxyl radicals with phenols, that is, these reactions are much slower in solvents capable of forming hydrogen bonds with O—H and N—H groups [50]. MacFaul et al. [50] proposed a universal scale for correlating the reactivities of phenols and the hydrogen-bonding abilities of solvents.

Intramolecular hydrogen bond is much stronger than the intermolecular hydrogen bond. The reactivity of such phenolic groups was studied by Pozdeeva et al. [51]. Crown phenol A was synthesized, and the reactivity as chain terminating antioxidant in oxidized styrene (323 K) was studied. The chain terminating activity of this crown phenol A was compared with that of ionol.

$k_7 = 56$ L mol^{-1} s^{-1} and $k_7 = 1.2 \times 10^4$ L mol^{-1} s^{-1}.

Such a great difference (2.5 orders of magnitude) in chain terminating activity of crown phenol and ionol demonstrates the influence of the intermolecular hydrogen bond on the reaction of peroxyl radical with the phenolic group. The spectroscopic study shows that 99.9% phenolic groups in crown phenol are bound by the hydrogen bond. Therefore, only a small portion of the free phenolic groups react. The thermodynamic parameters of hydrogen bonding in crown phenol was found to be the following: $\Delta H = -21$ kJ mol^{-1} and $\Delta S = -5.7$ J mol^{-1} K^{-1} [51].

Another factor influencing the reactivities of polar particles is their nonspecific solvation. Since both the individual particles, namely phenol and peroxyl radicals and their complex are polar, rate constants must depend on the polarity of the medium, its permittivity ε, in particular. This was confirmed in experiments with mixtures of benzene and methylethylketone, which showed that k_7 diminishes as the concentration of methylethylketone decreases provided the hydrogen bonding between the benzene and methylethylketone molecules are taken into account [10]. The dependence of $\log k_7$ on the medium permittivity ε is described by the formula

$$\log k_7 = \log k_7(\varepsilon = 1) + d\frac{\varepsilon - 1}{2\varepsilon + 1}, \tag{15.12}$$

which accounts for the effect of nonspecific formation. The values of rate constants of peroxyl radicals with phenols are given in Database [52].

15.2 REACTION OF AROMATIC AMINES WITH PEROXYL RADICALS

15.2.1 Dissociation Energies of N—H Bonds of Aromatic Amines

Aromatic amines possess weak N—H bonds and the latter are attacked by peroxyl radicals when amines are used as antioxidants [1–9].

$$RO_2^\bullet + Ar_2NH \longrightarrow ROOH + Ar_2N^\bullet \tag{7}$$

The experimental evidences for this reaction are the same as in the case of phenols (see earlier). The BDE of N—H bonds of aromatic amines are collected in Table 15.5. They vary for the known amines from 387 kJ mol^{-1} (aniline) to 331 kJ mol^{-1} (phenothiazine).

15.2.2 Rate Constants of Reaction of $RO_2^\bullet$ with Aromatic Amines

The experimental values of rate constants of $RO_2^\bullet$ reactions with aromatic amines (AmH) are given in Database [52]. The experimental measurement of the rate constant k_7 for aromatic amines from kinetics of oxidation faced with great difficulties. These difficulties arise due to the extremely high activity of aminyl radicals toward hydroperoxide [53–56]. The reaction

$$Am^\bullet + HOOR \longrightarrow AmH + RO_2^\bullet \tag{-7}$$

occurs so rapidly that an equilibrium exists between $Am^\bullet$ and $RO_2^\bullet$ in the presence of amine and hydroperoxide. Chain termination proceeds via the reaction of aminyl with peroxyl radicals. As a result, reaction (−7) affects the mechanism and decreases the value of k_7 rate constant. The precise measurement of the rate constant of the reaction of $RO_2^\bullet$ with aromatic amine was performed after a detailed study of the equilibrium between $Am^\bullet$ and $RO_2^\bullet$ in the presence of amine and hydroperoxide [55]. The parameters of free radical reactions with aromatic amines calculated from experimental data in the IPM model [54] are presented in Table 15.6.

According to these parameters, the activation energy of the peroxyl radical reaction with aromatic amine can be calculated using the reaction enthalpy by the equation [54]

$$E = \{6.247 + 3.88 \times 10^{-2}(\Delta H - 1.2)\}^2 - 20.0 + 0.5RT \tag{15.13}$$

TABLE 15.5
Dissociation Energies of N—H Bonds of Aromatic Amines

Name	Structural Formula	D(N—H) (kJ mol^{-1})	Ref.
Benzenamine, 4-bromo-N-(4-bromophenyl)-		364.2	[53]
Benzenamine, 4-(1,1-dimethylethyl)-N-[4-(1,1-dimethylethyl)phenyl]-		358.8	[53]
Benzenamine, N-1-[3,7-bis (1,1-dimethylethyl)naphthalenyl]-		344.9	[54]
Benzenamine, 4-(1,1-dimethylethyl)-N-phenyl-		360.3	[53]
Benzenamine, 4-methoxy-N-4-methoxyphenyl-		348.6	[53]
Benzenamine, 4-methoxy-N-phenyl-		355.9	[53]
Benzenamine, 4-methyl-N-(4-methylphenyl)-		357.5	[53]
Benzenamine, 4-nitro-N-phenyl-		372.9	[53]
Benzenamine, N-phenyl-, (Diphenylamine)		364.7	[53]
1,4-Benzenediamine, N,N'-bis[4-(1-methylethyl)phenyl]-, (p-Phenylendiamine, N,N'-di (4-isopropylphenyl-)		333.6	[54]

TABLE 15.5
Dissociation Energies of N—H Bonds of Aromatic Amines—*continued*

Name	Structural Formula	D(N—H) (kJ mol^{-1})	Ref.
1,4-Benzenediamine, N,N'-di-2-naphthalenyl-, (*p*-Phenylendiamine, N,N'-di-2-naphthyl-)		346.6	[54]
1,4-Benzenediamine, N,N'-dioctyl-, (*p*-Phenylendiamine, N,N'-dioctyl-)		352.9	[54]
1,4-Benzenediamine, N,N'-diphenyl-, (*p*-Phenylendiamine, N,N'-diphenyl-)		346.9	[54]
1,4-Benzenediamine, N-phenyl-N'-(1-methylethyl)-, (*p*-Phenylendiamine, N-phenyl-N'-isopropyl-)		340.2	[54]
9H-Carbazole, (Dibenzo[*b,d*]pyrrole)		371.6	[54]
1-Naphthalenamine (1-Naphtylamine)		374.7	[54]
2-Naphthalenamine (2-Naphtylamine)		379.5	[54]
1-Naphthalenamine, 4-(1,1-dimethylethyl)-N-[4-(1,1-dimethylethyl) phenyl]-		352.1	[54]
2-Naphthalenamine, N-2-naphthalenyl- (bis-2-naphtylamine)		360.2	[53]
1-Naphthalenamine, N-phenyl-, (Phenyl-1-naphthylamine)		352.5	[54]
2-Naphthalenamine, N-phenyl- (Phenyl-2-naphthylamine)		362.9	[53]

continued

TABLE 15.5
Dissociation Energies of N—H Bonds of Aromatic Amines—*continued*

Name	Structural Formula	D(N—H) (kJ mol^{-1})	Ref.
2-Naphthalenamine, *N*-(4-phenoxy)phenyl-		349.5	[54]
1,2,3,4-Tetrahydroquinoline, 2-spirocyclohexyl-4-(spirotetrahydrofuran-2-)-		360.8	[54]
1,2,3,4-Tetrahydroquinoline, 2,2,4-trimethyl-		359.5	[54]
Phenothiazine		331.4	[21]

TABLE 15.6
Kinetic Parameters of Free Radical Reactions with Aromatic Amines in IPM Model [34,54]

Reaction	α	$b \times 10^{-11}$ (kJ mol^{-1})$^{1/2}$ (m)	$0.5hL\nu$ (kJ mol^{-1})	$0.5hL(\nu - \nu_f)$ (kJ mol^{-1})
R$^{\bullet}$ + AmH	1.246	4.306	20.0	2.6
RO$^{\bullet}$ + AmH	0.916	4.306	20.0	−1.7
RO$_2^{\bullet}$ + AmH	0.936	4.306	20.0	−1.2

Reaction	br_e (kJ mol^{-1})$^{1/2}$	E_{e0} (kJ mol^{-1})	A (L mol^{-1} s^{-1})	$\Delta H_{e\ min}$ (kJ mol^{-1})	$\Delta H_{e\ max}$ (kJ mol^{-1})
R$_1^{\bullet}$ + AmH	18.46	67.5	1.0×10^9	129.7	169.3
R$_2^{\bullet}$ + AmH	20.21	81.0	1.0×10^9	167.3	222.2
R$_3^{\bullet}$ + AmH	19.41	74.7	1.0×10^9	148.7	196.1
RO$^{\bullet}$ + AmH	13.27	48.0	5.0×10^{10}	72.2	59.3
RO$_2^{\bullet}$ + AmH	12.12	39.2	1.0×10^8	46.8	39.8

The geometric parameters of TS of this reaction can be calculated using the values of ΔH_e and E_e according to the following equations [35]:

$$r(\text{N—H})[10^{-10}(\text{m})] = 1.01 + 3.30 \times 10^{-2}\sqrt{E_e} \qquad (15.14)$$

$$r(\text{O—H})[10^{-10}(\text{m})] = 0.97 + 3.13 \times 10^{-2}\sqrt{E_e - \Delta H_e} \qquad (15.15)$$

The values of rate constants, activation energies, and geometric parameters of secondary alkyl peroxyl radical reaction with several aromatic amines are presented in Table 15.7.

TABLE 15.7
Enthalpies, Activation Energies, Rate Constants, and Geometric Parameters of TS of Reaction RO$_2^\bullet$ with Aromatic Amines Calculated by Equations (15.13)–(15.15)

Amine	ΔH (kJ mol^{-1})	E (kJ mol^{-1})	r(N—H) × 10^{10} (m)	r(O—H) × 10^{10} (m)	k (350 K) (L mol^{-1} s^{-1})
(4,4'-dibromodiphenylamine)	−1.3	19.2	1.21	1.17	1.35 × 10^5
(4,4'-di-tert-butyldiphenylamine)	−6.7	16.7	1.20	1.18	3.23 × 10^5
(2,4-di-tert-butyl-N-phenyl-1-naphthylamine)	−20.6	10.6	1.19	1.19	2.64 × 10^6
(4-tert-butyldiphenylamine)	−5.2	14.0	1.21	1.17	2.54 × 10^5
(4,4'-dimethoxydiphenylamine)	−16.9	12.1	1.19	1.19	1.54 × 10^6
(4-methoxydiphenylamine)	−9.6	15.4	1.20	1.18	5.09 × 10^5
(4,4'-dimethyldiphenylamine)	−8.0	16.1	1.20	1.18	3.97 × 10^5
(4-nitrodiphenylamine)	7.4	23.5	1.22	1.16	3.12 × 10^4
(diphenylamine)	−0.8	19.5	1.21	1.17	1.25 × 10^5
(N,N'-bis(4-isopropylphenyl)-p-phenylenediamine)	−31.9	6.0	1.17	1.21	2.52 × 10^7
(N,N'-di-2-naphthyl-p-phenylenediamine)	−18.9	11.3	1.19	1.19	4.14 × 10^6
(N,N'-dialkyl-p-phenylenediamine)	−13.3	12.5	1.19	1.19	2.73 × 10^6

continued

TABLE 15.7
Enthalpies, Activation Energies, Rate Constants, and Geometric Parameters of TS of Reaction $RO_2^\bullet$ with Aromatic Amines Calculated by Equations (15.13)–(15.15)—*continued*

Amine	ΔH (kJ mol^{-1})	E (kJ mol^{-1})	r(N—H) × 10^{10} (m)	r(O—H) × 10^{10} (m)	k (350 K) (L mol^{-1} s^{-1})
	−18.6	11.3	1.19	1.19	4.12 × 10^6
	−25.3	8.3	1.18	1.20	1.51 × 10^7
	6.1	22.8	1.22	1.16	3.91 × 10^4
	9.2	24.4	1.23	1.16	4.56 × 10^4
	14	26.9	1.23	1.15	1.95 × 10^4
	−13.4	13.7	1.20	1.18	9.12 × 10^5
	−5.3	17.3	1.21	1.17	2.58 × 10^5
	−13.0	13.8	1.21	1.17	8.72 × 10^5
	−2.6	18.6	1.21	1.17	1.67 × 10^5
	−16.0	12.5	1.19	1.19	1.35 × 10^6
	−4.7	17.6	1.21	1.17	2.35 × 10^5

TABLE 15.7
Enthalpies, Activation Energies, Rate Constants, and Geometric Parameters of TS of Reaction $RO_2^{\bullet}$ with Aromatic Amines Calculated by Equations (15.13)–(15.15)—continued

Amine	ΔH (kJ mol^{-1})	E (kJ mol^{-1})	r(N—H) × 10^{10} (m)	r(O—H) × 10^{10} (m)	k (350 K) (L mol^{-1} s^{-1})
(tetrahydroquinoline)	−6.0	17.2	1.20	1.18	2.71 × 10^5
(phenothiazine)	−34.1	5.21	1.17	1.21	1.68 × 10^7

15.2.3 Electronegativity of Atoms in the Transition State

Let us compare the rate constants and activation energies of sec-$RO_2^{\bullet}$ with phenol and aromatic amine with close reaction enthalpies.

$RO_2^{\bullet}$ + YH, YH=	(phenol)	(diphenylamine)
k (333 K) (L mol^{-1} s^{-1})	1.4 × 10^4	3.0 × 10^5
ΔH (kJ mol^{-1})	−7.7	−8.0
A (L mol^{-1} s^{-1})	3.0 × 10^7	1.0 × 10^8
E (kJ mol^{-1})	21.4	16.1

Both reactions have practically the same enthalpy, however, amine reacts by one order of magnitude more rapidly than phenol. Comparison of the activation energies of the thermally neutral reactions $RO_2^{\bullet}$ + Ar_1OH and $RO_2^{\bullet}$ + AmH also proves that the hydrogen atom can be separated from the N—H bond still more easily than from the O—H bond, since E_{e0} is 45 kJ mol^{-1} for the first and 38 kJ mol^{-1} for the second reactions, respectively [54]. The strength of the ROO—NAr$_2$ bond is unknown but is, undoubtedly, positive and greater than that of the ROO—OAr bond. Accordingly, the triplet repulsion in the transition state ROO$^{\bullet}$ ··· H ··· N is greater than in the transition state ROO$^{\bullet}$ ··· H ··· OAr. Nevertheless, E_{e0}(O ··· H ··· N) < E_{e0}(O ··· H ··· O), which is due to the fact that oxygen and nitrogen atoms possess different electron affinities and are, therefore, attracted in the TS with a concurrent decrease in the activation energy [40]. A similar phenomenon is typical of the reactions of aminyl radicals with phenols and hydroperoxides (see later).

The empirical correlation equation for r_e as a function of the bond dissociation energy D_{XY}, difference in electronegativity ΔEA_{XY} for the radical reaction abstraction of the type

$$X^{\bullet} + HY \longrightarrow XH + Y^{\bullet}$$

has the following form [40]:

$$r_e^2 \times 10^{22}(m) = const + 13.7(D_{X-Y}/D_{H-H}) - 22.4(\Delta EA_{XY}/D_{H-H}) \qquad (15.16)$$

where D_{H-H} is the bond strength of the hydrogen molecule. This equation allows the contributions from each of the factors influencing the E_{e0} values to be calculated. Given below are the contributions from the orbital repulsion of the atoms X and Y (ΔE_R), triplet repulsion (ΔE_T), and the difference in electronegativity of the atoms X and Y (ΔEA_{XY}) in the activation energies E_{e0} of various reactions [34].

Reaction	E_{e0} (kJ mol^{-1})	ΔE_R (kJ mol^{-1})	ΔE_T (kJ mol^{-1})	ΔE_{EA} (kJ mol^{-1})
$RO_2^{\bullet} + RH$	61.5	32	37	-7
$RO_2^{\bullet} + Ar_1OH$	45.3	45	0	0
$RO_2^{\bullet} + AmH$	39.0	44	15	-20

It is apparent that the high reactivity of phenols is due to the absence of triplet repulsion ($\Delta E_T = 0$), whereas that of aromatic amines results from a large difference in the electronegativities of nitrogen and oxygen atoms ($\Delta E_{EA} = -20$ kJ mol^{-1}).

Amines, similar to phenols, form hydrogen bonds with polar compounds. Hydrogen bonding decreases the amines activity in reactions with peroxyl radicals. For example, 1-naphthylamine reacts with the peroxyl radical with $k_7 = 2.0 \times 10^5$ L mol^{-1} s^{-1} in cyclohexane and $k_7 = 1.6 \times 10^4$ L mol^{-1} s^{-1} in cyclohexanone at 347 K [9]. The values of hydrogen bonding enthalpies of a few amines and several O-containing compounds are given in Table 15.8. These values are calculated via Equation (15.9) using the increments suggested in Ref. [47].

One can suppose the reaction of the peroxyl radical with amine to proceed in two stages via the preliminary equilibrium formation of a complex.

$$RO_2^{\bullet} + HNAr_2 \rightleftharpoons RO_2^{\bullet} \cdots HNAr_2 \longrightarrow ROOH + Ar_2N_2^{\bullet}$$

The reaction between $RO_2^{\bullet}$ and amine can proceed through the electron transfer in the very polar media (for example, water).

15.3 REACTIONS OF PHENOXYL AND AMINYL RADICALS WITH $RO_2^{\bullet}$

15.3.1 RECOMBINATION OF PEROXYL WITH PHENOXYL RADICALS

Mahoney and DaRooge [57] studied the kinetics of oxidation of 9,10-dihydroanthracene at 333 K in the presence of some phenols and estimated the ratio of rate constants k_8/k_2. From these ratios, the rate constants k_8 were calculated for several *para*-substituent phenols 4-YC$_6$H$_4$OH using $k_2 = 850$ L mol^{-1} s^{-1}.

Substituent Y	OMe	Ph	Me	CMe$_3$	H
$k_8 \times 10^{-8}$ (L mol^{-1} s^{-1})	7.8	3.9	2.4	1.1	1.6

It can be seen that the reaction of radical recombination occurs rapidly, rate constants are substituent-dependent and range from 10^8 to 10^9 L mol^{-1} s^{-1}. Peroxyl radicals appeared to be

TABLE 15.8
The Values of Gibbs Potential ($-\Delta G$ (298 K)) (kJ mol^{-1})) for Hydrogen Bonding Ar$_2$NH $\cdots$ Y, Calculated by Equation (15.9), Increments See in Refs. [4,47]

Amine	EtOH	EtOEt	MeC(O)Et	PhCH(O)	cyclohexanone	tetrahydrofuran
PhNH$_2$	7.8	7.6	7.9	7.6	8.0	8.0
Ph$_2$NH	10.5	10.2	10.7	10.1	11.1	11.1
(4-Br-C$_6$H$_4$)$_2$NH	12.7	12.1	12.9	12.0	13.4	13.5
4-MeO-C$_6$H$_4$-NH-Ph	9.8	9.5	10.0	9.4	10.2	10.3
(4-IO-C$_6$H$_4$)$_2$NH	18.0	17.1	18.5	16.8	19.4	19.6
(4-Me-C$_6$H$_4$)$_2$NH	9.5	9.2	9.7	9.2	9.9	10.0
carbazole	14.9	14.2	15.3	14.0	15.9	16.1
1-naphthyl-NH-Ph	7.5	7.3	7.5	7.3	7.6	7.7
2-naphthyl-NH-Ph	10.8	10.5	11.0	10.3	11.4	11.5

more reactive toward phenoxyl radicals with electropositive substituents. This reaction has not been studied in depth; however, by analogy with the 2,4,6-substituted phenoxyl radicals (see below), it can be assumed that RO$_2^\bullet$ adds to the benzene ring in the *ortho-* and *para-*positions with respect to the O—H group to produce alcohol and *ortho-* and *para-*quinones.

The 2,4,6-substituted phenoxyl radicals recombine slowly and selectively react with peroxyl radicals, producing quinolidic peroxides [57].

ROO• + [2,4,6-tri-tert-butylphenoxyl radical] → [quinolidic peroxide with ROO group]

At moderate temperatures ($T < 350$ K), these peroxides are stable but readily decompose at elevated temperatures to form radicals [58–60]. Peroxyl radicals can add to phenoxyl radicals at both *para-* and *ortho-*positions with the proportion dependent on the substituent. Thus, the peroxyl radical adds to 2,4,6-tris(1,1-dimethylethyl)phenoxyl in the *para-*position by eight times more rapidly than in the *ortho-*position, whereas it adds to 2-methyl-4,6-bis(1,1-dimethylethyl)phenoxyl preferentially in the *ortho-*position [59]. The rate constants of the reactions of the peroxyl radical with 2,6-bis(1,1-dimethylethyl)-4-substituted phenoxyl radicals in benzene were measured in the presence of respective phenol and AIBN as a source of 1-cyano-1-methylethylperoxyl radicals [60,61]. The concentration of the formed phenoxyl radical in these experiments peaked at $[ArO^•]_{max} = (k_7/k_8)[ArOH]$. Using this expression and taking the known k_7 values, the values of k_8 for various phenoxyl radicals 4-Y-2,6-$(Me_3C)_2$-$C_6H_2O^•$ were estimated at $T = 353$ K (see Database [52]):

Substituent Y	OMe	Ph	CMe$_3$	CN	PhC(O)	Cl
$k_8 \times 10^{-8}$ (L mol^{-1} s^{-1})	7.2	6.0	3.2	2.7	1.9	1.7

It is seen that the rate constant k_8 is lower for compounds with electron-accepting substituents than with electron-donating substituents, which implies a dependence of the rate of $Ar_2O^•$ and $RO_2^•$ recombination on the electron density at the *para-* and *ortho-*positions of the benzene ring of the phenoxyl radical. The activation energies of this reaction vary from −33 to 10 kJ mol^{-1}; however, the concurrent variation in the pre-exponential factor from 10^3 to 10^{10} L mol^{-1} s^{-1} causes a strong compensatory effect. It can also be seen that phenoxyl radicals readily react with peroxyl radicals ($k = 10^8$–10^9 L mol^{-1} s^{-1}), whereas the disproportionation of peroxyl radicals is sufficiently slower (see Chapter 2). Hence, during the oxidation of hydrocarbons in the presence of phenols when $k_7[ArOH] > k_2[RH]$, the recombination reaction of $ArO^•$ with $RO_2^•$ is always faster than the reaction of disproportionation of peroxyl radicals.

The formed quinolide peroxides become unstable at elevated temperatures and decompose into free radicals. The rate constants of decomposition for two peroxides are given below [4]:

$k = 10^{14} \exp(-127.5/RT)$ s^{-1} $k = 10^{14} \exp(-137.9/RT)$ s^{-1}

The full list of the data are given in handbooks [4,62].

15.3.2 REACTION OF PEROXYL WITH AMINYL RADICALS

Aminyl radicals react with peroxyl radicals by two ways with the formation of either N—O or C—O bonds [63]. The decomposition of the resulting unstable peroxides gives rise to the nitroxyl radical and iminoquinone, respectively.

$$RO_2^\bullet + {}^\bullet NAr_2 \longrightarrow ROONAr_2 \longrightarrow RO^\bullet + Ar_2NO^\bullet$$

$$RO_2^\bullet + \text{C}_6H_5\text{-N}^\bullet Ph \longrightarrow \underset{ROO}{\overset{H}{\diagdown}}\!\!=\!\!NPh$$

$$\underset{ROO}{\overset{H}{\diagdown}}\!\!=\!\!NPh \longrightarrow ROH + O=\!\!\diagdown\!\!=\!\!NPh$$

In the reactions with the diphenylaminyl radical, the relative yield of nitroxyl radicals is in the range 0.11 to 0.19 and 0.20 to 0.33 for the secondary and tertiary peroxyl radicals, respectively [63]. The rate constant of this reaction was measured by flash photolysis with tetraphenylhydrazine as a source of diphenylaminyl radicals [64]. Peroxyl radicals were generated through the photodecay of bis(1,1-dimethylethyl) peroxide followed by the rapid reaction of formed alkoxyl radicals with cyclohexane used as a solvent; the reaction conditions were chosen in such a way that $[RO_2^\bullet]_0 > [Ph_2N^\bullet]_0$. Under these conditions, aminyl radicals were destroyed only in the reaction with peroxyl radicals and, hence, the concentration of peroxyl radicals during the measurement of the kinetics of $Ph_2N^\bullet$ consumption varied insignificantly. The rate constant of the recombination of peroxyl and diphenylamine radicals was found to be equal to 6×10^8 L mol^{-1} s^{-1} (cyclohexane, 283–303 K).

Vardanyan [65,66] discovered the phenomenon of CL in the reaction of peroxyl radicals with the aminyl radical. In the process of liquid-phase oxidation, CL results from the disproportionation reactions of primary and secondary peroxyl radicals, giving rise to triplet-excited carbonyl compounds (see Chapter 2). The addition of an inhibitor reduces the concentration of peroxyl radicals and, hence, the rate of $RO_2^\bullet$ disproportionation and the intensity of CL. As the inhibitor is consumed in the oxidized hydrocarbon the initial level of CL is recovered. On the other hand, the addition of primary and secondary aromatic amines to chlorobenzene containing some amounts of alcohols, esters, ethers, or water enhances the CL by 1.5 to 7 times [66]. This effect is probably due to the reaction of peroxyl radicals with the aminyl radical, since the addition of phenol to the reaction mixture under these conditions must extinguish CL. Indeed, the fast exchange reaction

$$ArOH + Am^\bullet \rightleftharpoons ArO^\bullet + AmH$$

gives rise to $ArO^\bullet$ radicals. As a result, instead of emitting light in the reaction with aminyl radicals, peroxyl radicals react with phenoxyl radicals without CL. After the phenolic inhibitor has been exhausted, light emission again increases due to the reaction of $RO_2^\bullet$ with $Am^\bullet$. A significant role in the excitation of light-emitting particles is played by the polar environment, so that CL is absent in a nonpolar (for example, hydrocarbon) medium.

Nitroxyl radicals produced in the reactions of $RO_2^\bullet$ with aminyl radicals react with peroxyl radicals. The latter reaction is considerably slower than the reaction of peroxyl with aminyl radicals, which can be seen from the following data derived by flash photolysis (the photolysis of bis(1,1-dimethylethyl)peroxide in toluene was performed in the presence of methylcyclohexyl hydroperoxide at 253 K [67]). The values of the rate constants of recombination $RO_2^\bullet + AmO^\bullet$ [67,68] are given below.

AmO•	(1-naphthyl)(phenyl)nitroxyl	(2-naphthyl)(phenyl)nitroxyl	(1-naphthyl)(mesityl)nitroxyl	diphenylnitroxyl
$RO_2^\bullet$	cyclohexyl-OO•	cyclohexyl-OO•	cyclohexyl-OO•	cumyl-OO•
k (L mol^{-1} s^{-1})	3.4×10^3	3.3×10^3	1.4×10^3	5.0×10^3
Solvent	PhMe	PhMe	PhMe	PhCMe$_3$
T (K)	253	253	253	341
Ref.	[67]	[67]	[67]	[68]

p-Methoxydiphenylnitroxyl reacts with cumylperoxyl radicals considerably more rapidly, with $k = 6.0 \times 10^5$ L mol^{-1} s^{-1} (ethylbenzene, 333 K, CL method [69]).

15.4 RECOMBINATION AND DISPROPORTIONATION OF PHENOXYL AND AMINYL RADICALS

15.4.1 Recombination and Disproportionation of Phenoxyl Radicals

The intensive mechanistic studies of phenoxyl self-reactions proved a great variety of mechanisms and rate constants of these reactions [2,3,6]. The substituents can dramatically influence the mechanism and kinetics of self-reactions. Due to free valence delocalization the phenoxyl radical possesses an excess of the electron density in the *ortho-* and *para*-positions. Mono- and disubstituted phenoxyls recombine with the formation of labile dimers that after enolization form bisphenols [3,6].

Phenoxyl with such a structure recombines with the rate constant close to that of the diffusionally controlled reaction. 2,4,6-Trisubstituted phenoxyls form unstable dimers. The latter dissociate back to phenoxyls. The values of the formed bonds lie between 30 and 120 kJ mol^{-1} [3]. The rate constants and equilibrium constants of dimerization for a few phenoxyls are presented in Table 15.9.

TABLE 15.9
Rate Constants, Equilibrium Constants, Enthalpies, and Entropies of Aroxyl Radicals Dimerization in Hexane at $T = 293\,\text{K}$

Phenoxyl	k_{dim} (L mol^{-1} s^{-1})	K_{eq} (mol^{-1})	ΔH (kJ mol^{-1})	ΔS (J mol^{-1} K^{-1})	Ref.
	1.1×10^8	4.5×10^{-7}	96.0	205	[70]
	3.2×10^9	2.5×10^{-8}	51.5	30	[71]
	1.8×10^9	7.4×10^{-10}	92.0	138	[72]
	8.0×10^8				[72]
	7.5×10^8	4.0×10^{-9}	73.0	88	[73]
	1.1×10^9	2.0×10^{-10}	105.0	188	[70]
	1.3×10^9	2.0×10^{-11}	117.0	209	[70]
	5.0×10^7	6.0×10^{-7}	65.0	109	[70]
	2.2×10^9	1.4×10^{-7}	71.0	117	[70]

A solvent can have influence on the structure of dimer. For example, 2,6-dimethoxyphenoxyl radicals recombine in a polar solvent (dimethylformamide) with the formation of C—C dimer [74].

The same radicals form the C—O dimer in a nonpolar (benzene) solvent [74].

2,4,6-Trisubstituted phenoxyl radicals disproportionate if the *para*-substituent bears the α-C—H bond. Two ways were proposed [3]: the direct disproportionation and the formation of labile dimer followed by the decay.

The values of rate constants of phenoxyl radical disproportionation are given in Table 15.10.

The rapid recombination of phenoxyl radicals occurs with rate constants close to the diffusionally controlled rate. The rate constant of diffusionally controlled reaction depends on the viscosity ($k_D = 8RT/3000\eta$ L mol^{-1} s^{-1} and varies in organic solvents in the limits of 3×10^8 to 5×10^9 L mol^{-1} s^{-1}).

TABLE 15.10
Rate Constants of Disproportionation of 4-Y-2,6-bis(1,1-dimethylethyl) Phenoxyls in Hydrocarbon Solutions at $T = 323\,K$

4-Y	k_9 (L mol^{-1} s^{-1})	log A, A (L mol^{-1} s^{-1})	E (kJ mol^{-1})	Ref.
Me	8.7×10^3	7.17	20.0	[75]
Et	2.4×10^3	7.58	25.9	[76]
MeOCH$_2$	1.1×10^2	7.18	31.8	[76]
EtMeCH	2.9	5.22	29.4	[77]
HOCH$_2$CH$_2$	1.1×10^3	7.27	26.4	[77]
MeOC(O)CH$_2$CH$_2$	7.4×10^2	7.27	27.2	[76]
C$_{18}$H$_{37}$OC(O)CH$_2$CH$_2$	1.5×10^3	6.48	20.5	[76]
Me$_2$CH	7.7	6.23	33.0	[77]
cyclo-C$_6$H$_{11}$	11.0	5.24	25.9	[77]

15.4.2 REACTIONS CONTROLLED BY ROTATIONAL DIFFUSION

Khudyakov et al. [78] faced with unusual phenomena in his study of these reactions. He found that the rate constant of recombination is much lower (10^7–10^8 L mol^{-1} s^{-1}) than k_D; however, it depends on the viscosity of the solvent. It was found for the recombination of the following phenoxyl radical:

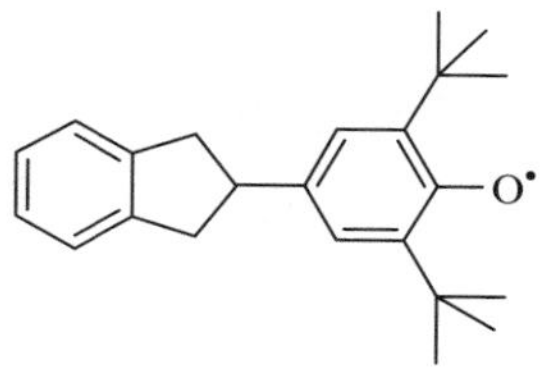

This phenoxyl radical recombines with the rate constant $k = 7.5 \times 10^7$ L mol^{-1} s^{-1} in toluene where its $k_D \approx 2$–4×10^9 L mol^{-1} s^{-1}. This phenomenon was explained within the scope of conception of bimolecular reaction as the interaction of two spheres with black spots [79–85]. Two different situations are possible for reactions that occur without an activation energy:

1. The steric factor P is high enough for the bimolecular reaction to occur when two radicals met in the cage. This reaction is limited only by translational diffusion of the reacting radicals and depends on the viscosity of the solvent. The rate constant of such reaction is close to the frequency of encounters of radicals, namely, $k = \sigma$, $k_D = 0.25 \times 4\pi r_{AB} D_{AB} = RT/6000\eta$ where $\sigma = 0.25$ is the spin-statistical factor, r_{AB} is the sum of radii of reactants A and B, and D_{AB} is the sum of their diffusion coefficients.
2. The steric factor of the reaction is very low. As a result, both reactants can meet and separate in the cage many times before the performance of the reaction due to a low probability to reach the configuration geometry of the reactants. This reaction does not depend on translational diffusion but on rotational diffusion of reacting particles. The rate constant of this reaction is $k \ll k_D$ but depends on the viscosity of the solvent. These two types of reactions without activation energies are illustrated by the following scheme (Figure 15.1).

The detailed analysis of the phenoxyl radical self-reactions are described in reviews [2–6].

15.4.3 Disproportionation of Semiquinone Radicals

Semiquinone radicals disproportionate with the formation of hydroquinone and quinone [2].

$$2\ \text{HO-C}_6\text{H}_4\text{-O}^\bullet \longrightarrow \text{HO-C}_6\text{H}_4\text{-OH} + \text{O=C}_6\text{H}_4\text{=O}$$

Disproportionation of sterically nonhindered semiquinone radicals occurs with diffusion rate constants. On the other hand, sterically hindered semiquinone radicals react by several orders of magnitude more slowly (see Table 15.11).

The semiquinone radical dissociates to a proton and semiquinone radical anion in the polar solvent (water).

$$\text{HO-C}_6\text{H}_4\text{-O}^\bullet + \text{H}_2\text{O} \rightleftharpoons {}^-\text{O-C}_6\text{H}_4\text{-O}^\bullet + \text{H}_3\text{O}^+$$

Hydroxyphenoxyl radicals have the following pK_a values [92]:

Radical	*ortho*-HO-C$_6$H$_4$-O$^\bullet$	*meta*-HO-C$_6$H$_4$-O$^\bullet$	*para*-HO-C$_6$H$_4$-O$^\bullet$
pK_a	−1.65	−1.45	−0.77

Self-reactions of semiquinone radical anions proceed via electron transfer reaction [2].

15.4.4 Recombination and Disproportionation of Aminyl Radicals

Diphenylaminyl radicals recombine with the formation of either N—N or C—N bonds [93].

$$\text{Ph}_2\text{N}^\bullet + \text{Ph}_2\text{N}^\bullet \longrightarrow \text{Ph}_2\text{N-NPh}_2 \qquad (A)$$

$$2\ \text{Ph}_2\text{N}^\bullet \longrightarrow \text{Ph(H)N-C}_6\text{H}_4\text{(Ph)=NPh} \qquad (B)$$

$$2\ \text{Ph}_2\text{N}^\bullet \longrightarrow \text{Ph(Ph)N-C}_6\text{H}_4\text{(H)=NPh} \qquad (C)$$

TABLE 15.11
Rate Constants of Disproportionation of Semiquinone Radicals at 298 K

Semiquinone Radical	Solvent	k_{dis} (L mol^{-1} s^{-1})	log A, A (L mol^{-1} s^{-1})	E (kJ mol^{-1})	Ref.
HO–⟨⟩–O•	Me$_2$CHOH	1.5×10^9	11.8	15.5	[86]
tetrachloro HO–⟨⟩–O•	Me$_2$CHOH	1.7×10^8			[86]
2,6-dimethyl HO–⟨⟩–O•	Me$_2$CHOH	3.0×10^8	10.9	13.8	[87]
tetramethyl HO–⟨⟩–O•	Me$_2$CHOH	7.0×10^8	11.0	13.8	[86]
2,6-di-tert-butyl HO–⟨⟩–O•	Me$_2$CHOH	1.7×10^6	10.9	25.9	[87]
2,6-di-tert-butyl ⟨⟩–OH, O•	PhMe	2.2×10^6	6.3	0	[88]
3,5-di-tert-butyl ⟨⟩–OH, O•	PhMe	6.0×10^6	6.8	0	[88]
Cl, di-tert-butyl ⟨⟩–OH, O•	PhMe	5.6×10^6	6.8	0	[88]
2,6-diphenyl ⟨⟩–OH, O•	PhMe	5.5×10^6	6.7	0	[88]

continued

TABLE 15.11
Rate Constants of Disproportionation of Semiquinone Radicals at 298 K—*continued*

Semiquinone Radical	Solvent	k_{dis} (L mol^{-1} s^{-1})	log A, A (L mol^{-1} s^{-1})	E (kJ mol^{-1})	Ref.
	PhMe	1.3×10^6	6.1	0	[88]
	C_6H_6	9.5	8.9	38.9	[89]
	Me$_2$CHOH	2.3×10^8			[86]
	Me$_2$CHOH	1.2×10^9			[86]
	PhMe/Me$_2$CHOH	1.4×10^8	10.0	11.0	[90]
	AcOH	1.2×10^8			[91]

The most rapid reaction is N—N-dimerization (the rates of reactions A, B, C are related as 1:0.15:0.02 [94]. Naphthylaminyl radicals recombine with the formation of N—C-dimers only [95], probably because voluminous naphthalene rings sterically hinder N—N-dimerization. A correlation between the rate constant of hyperfine splitting on the nitrogen atom of the aminyl radical and the rate constant of recombination of substituted ($(YC_6H_4)_2N^\bullet$) diphenyl-aminyl radicals was observed [95].

Y	OMe	CMe$_3$	H	Br
$2k_9 \times 10^6$ (L mol^{-1} s^{-1})	2.3	3.4	2.7	60
$a^N \times 10^{-3}$ (A m^{-1})	8.49	8.69	8.80	9.0

The aminyl radicals of p-phenylenediarylamines readily undergo disproportionation.

The literature data on the recombination of diphenylaminyl radicals correlate well (see Table 15.12). The rate constants of recombination of various aminyl radicals measured by flash photolysis vary from 10^5 to 10^9 L mol^{-1} s^{-1}. For substituted diphenylaminyl radicals, a linear dependence was observed between log k_9 and σ-Hammett [96]:

$$\log k_9 = 7.24 + 1.50 n\sigma, \tag{15.17}$$

where $n = 1$ or 2 is the number of *para*-substituents in the benzene rings.

Two aminyl radicals combine to form the N—N-dimer with the rate constant proportional to the spin density on the nitrogen atom [96]. This recombination also depends on the steric factor. Thus, diethylaminyl radicals recombine with a rate constant of 10^9 L mol^{-1} s^{-1}, while diisopropylaminyl radicals are only able to disproportionate with a rate constant of 4.5×10^5 L mol^{-1} s^{-1}, probably because of the voluminous isopropyl substituent [5]. The formation of N—N-dimers is accompanied by the turn of phenyl groups toward establishing an N—N bond. In this case, the phenyl groups lose their capacity for free rotation, which results in an additional entropy loss in TS. The lower the electron density on the nitrogen atom, the closer the aminyl fragments approach each other in TS. This leads to a stronger repulsion of the phenolic rings and, hence, to a greater entropy loss.

15.5 REACTIONS OF PHENOXYL AND AMINYL RADICALS WITH HYDROCARBONS, HYDROPEROXIDES, AND DIOXYGEN

Reactions of phenoxyl and aminyl radicals with RH and ROOH are chain propagation steps in oxidation inhibited by phenols and amines (see Chapter 14). Both reactions become important when their rates are close to the initiation rate (see Chapter 14). Mahoney and DaRooge [57] studied the oxidation of 9,10-dihydroanthracene inhibited by different phenols. He went on to estimate the values of rate constants ratio of the reaction of ArO$^\bullet$ with RH and the reaction In$^\bullet$ + In$^\bullet$ (reactions (9) and (10), see Chapter 14) by the kinetic study. The values of k_{10} for the reaction

$$YC_6H_4O^\bullet + RH \longrightarrow YC_6H_4OH + R^\bullet \tag{10}$$

are given below (RH is 9,10-dihydroanthracene, $T = 333$ K).

Y	4-Ph	4-MeO	4-CMe$_3$	4-Me	H
k_{10} (L mol^{-1} s^{-1})	12	37	87	99	110

A linear dependence between log k_{10} and the reaction enthalpy ΔH ($T = 333$ K) is observed.

$$\log k_{10} (\text{L mol}^{-1} \text{s}^{-1}) = -1.7 - 0.56 \Delta H (\text{kJ mol}^{-1}) \tag{15.18}$$

TABLE 15.12
Rate Constants of Recombination and Disproportionation of Aminyl Radicals in Hydrocarbon Solutions Measured by the Flash Photolysis Technique

Aminyl Radical	T (K)	k (L mol^{-1} s^{-1})	Ref.
(2,2'-dinaphthylaminyl)	298	4.5×10^8	[96]
(diphenylaminyl)	298	2.7×10^7	[96]
(diphenylaminyl)	293	1.8×10^7	[97]
(diphenylaminyl)	298	2.5×10^7	[93]
(4,4'-dibromodiphenylaminyl)	298	6.0×10^7	[96]
(4,4'-dimethoxydiphenylaminyl)	298	2.3×10^6	[96]
(4,4'-dimethyldiphenylaminyl)	298	2.3×10^6	[96]
(4,4'-di-tert-butyldiphenylaminyl)	298	3.4×10^6	[96]
(4-methoxydiphenylaminyl)	298	6.0×10^6	[96]
(4-tert-butyldiphenylaminyl)	298	1.1×10^7	[96]
(N,N'-di(2-naphthyl)-p-phenylenediaminyl)	298	1.9×10^8	[96]
(N,N'-diphenyl-p-phenylenediaminyl)	298	1.6×10^9	[96]
(1-naphthylphenylaminyl)	298	6.2×10^8	[96]
(2-naphthylphenylaminyl)	298	3.8×10^8	[96]

TABLE 15.12
Rate Constants of Recombination and Disproportionation of Aminyl Radicals in Hydrocarbon Solutions Measured by the Flash Photolysis Technique—*continued*

Aminyl Radical	T (K)	k (L mol^{-1} s^{-1})	Ref.
	298	1.1×10^5	[96]
	298	2.4×10^6	[96]
	298	4.5×10^8	[96]
	298	5.4×10^8	[96]
	298	4.3×10^8	[96]
	298	1.2×10^9	[96]
	208–233	1.6×10^{12} exp(−65.2/RT)	[98]

15.5.1 Triplet Repulsion

The comparison of ArO$^\bullet$ reactions with RH and ROOH illustrates a great role of the triplet repulsion in free radical abstraction reactions. The IPM method helps to clarify this important factor (see Ref. [33] and Chapter 6). The parameters of reactions of Ar$_1$O$^\bullet$ and sterically hindered phenoxyls Ar$_2$O$^\bullet$ with hydrocarbons (R^{1}H, R^{2}H, and R^{3}H) and hydroperoxides ROOH are collected in Table 15.13.

Since the reaction enthalpy has influence on the activation energy (see earlier), we compared the activation energies of the reactions Ar$_1$O$^\bullet$ + R^{1}H and Ar$_1$O$^\bullet$ + ROOH at $\Delta H_e = 0$. The reaction Ar$_1$O$^\bullet$ + R^{1}H occurs with the activation energy of thermoneutral reaction $E_{e0} = 62.9$ kJ mol^{-1} and $E_0(T=0) = 62.9 - 17.4 = 45.5$ kJ mol^{-1}. The reaction Ar$_1$O$^\bullet$ + ROOH occurs with $E_{e0} = 43.3$ kJ mol^{-1} and $E_0(T=0) = 43.3 - 21.5 = 21.8$ kJ mol^{-1}. We observed the great difference $\Delta E = E_0(R^1H) - E_0(ROOH) = 23.7$ kJ mol^{-1}. This difference in activation energies E_0 of reactions (10) and (−7) illustrates, first of all, the influence of the triplet repulsion on the energy of the TS [33]. In the reaction of $Ar_1O^{\bullet}$ with R^1H, the TS has the structure $O \cdots H \cdots C$, and the hydrogen atom is moving along the pseudobond $Ar_1O - R^1$. This bond has the BDE of about 268–280 kJ mol^{-1}[32]. The TS of the reaction $Ar_1O^{\bullet} + ROOH$ has the structure: $-O \cdots H \cdots OO-$, and the hydrogen atom is moving along the pseudobond $O \cdots OO-$ that has practically the zero energy. Hence, there is zero triplet repulsion in such TS and as a result, a sufficiently lower activation energy. When the aroxyl radical reacts with α-C—H bonds of alkylaromatic (R^3H) and olefinic (R^2H) hydrocarbons, the energy of TS increases due to the interaction of π-electrons with electrons of TS. The values of E_{e0} illustrate this influence [33,34,38].

Reaction	$Ar_1O^{\bullet} + R^1H$	$Ar_1O^{\bullet} + R^2H$	$Ar_1O^{\bullet} + R^3H$
E_{e0} (kJ mol^{-1})	62.9	74.3	69.5

In addition to the repulsion, the reaction enthalpy has influence on the activation energy. Table 15.14 illustrates the increments ΔE_H of this influence, as well as the geometrical

TABLE 15.13
Kinetic Parameters of Phenoxyl and Aminyl Radical Reactions with Hydrocarbons and Hydroperoxides in IPM Model [4,34,38]

Reaction	α	$b \times 10^{-11}$ (kJ mol^{-1})$^{1/2}$ (m)	$0.5hL\nu$ (kJ mol^{-1})	$0.5hL(\nu - \nu_f)$ (kJ mol^{-1})
$ArO^{\bullet} + RH$	0.802	3.743	17.4	−4.1
$ArO^{\bullet} + ROOH$	0.986	4.600	21.2	−0.3
$AmO^{\bullet} + RH$	0.802	3.743	17.4	−4.1
$AmO^{\bullet} + ROOH$	0.986	4.600	21.2	−0.3
$Am^{\bullet} + RH$	0.866	3.743	17.4	−2.6
$Am^{\bullet} + ROOH$	1.064	4.600	21.2	1.2

Reaction	br_e (kJ mol^{-1})$^{1/2}$	E_{e0} (kJ mol^{-1})	A (L mol^{-1} s^{-1})	$-\Delta H_{e\ min}$ (kJ mol^{-1})	$\Delta H_{e\ max}$ (kJ mol^{-1})
$Ar_1O^{\bullet} + R^1H$	14.29	62.9	1.0×10^9	141.8	90.2
$Ar_1O^{\bullet} + R^2H$	15.95	78.3	1.0×10^8	198.3	128.1
$Ar_1O^{\bullet} + R^3H$	15.02	69.5	1.0×10^8	165.6	106.2
$Ar_1O^{\bullet} + ROOH$	13.00	43.3	3.2×10^7	51.3	52.3
$Ar_2O^{\bullet} + R^1H$	16.65	85.4	1.0×10^9	224.7	145.7
$Ar_2O^{\bullet} + R^2H$	18.58	106.3	1.0×10^8	305.4	199.3
$Ar_2O^{\bullet} + R^3H$	17.50	94.3	1.0×10^8	258.8	168.3
$Ar_2O^{\bullet} + ROOH$	14.10	51.1	3.2×10^7	71.5	72.8
$AmO^{\bullet} + R^1H$	14.30	63.0	1.0×10^9	170.7	128.0
$AmO^{\bullet} + R^2H$	16.13	80.1	1.0×10^8	222.0	167.6
$AmO^{\bullet} + R^3H$	15.20	71.1	1.0×10^8	197.6	148.9
$AmO^{\bullet} + ROOH$	12.89	39.0	1.0×10^7	44.5	46.1
$Am^{\bullet} + R^3H$	16.87	81.7	1.0×10^8	196.1	148.7
$Am^{\bullet} + ROOH$	12.95	39.4	1.0×10^7	39.8	46.8

TABLE 15.14
Enthalpies, Activation Energies, Rate Constants, Increment ΔE_H, and Geometrical Parameters of TS Reactions of Phenoxyl Radicals with Cumene (Reaction 10) Calculated by IPM Method

Phenoxyl	ΔH (kJ mol^{-1})	E (kJ mol^{-1})	k_{10} (400 K) (L mol^{-1} s^{-1})	ΔE_H (kJ mol^{-1})	$r(C-H) \times 10^{10}$ (m)	$r(O-H) \times 10^{10}$ (m)
(acetyl-phenoxyl)	−16.9	44.8	1.41×10^2	−7.1	1.39	1.25
(amino-phenoxyl)	−2.0	51.1	21.4	−0.8	1.40	1.24
(hydroxy-phenoxyl)	2.7	53.1	15.5	1.2	1.41	1.23
(cyano-phenoxyl)	−15.5	45.4	1.19×10^2	−6.5	1.39	1.25
(chromanoxyl)	23.9	62.9	0.60	11.0	1.43	1.21
(dihydrobenzofuranoxyl)	28.7	65.2	0.30	13.3	1.44	1.20
(hydroxynaphthoxyl)	21.2	61.6	0.89	9.7	1.43	1.21
(1-naphthoxyl)	11.3	57.0	3.58	5.1	1.42	1.22
(2-naphthoxyl)	0.9	52.3	14.6	0.4	1.41	1.23
(phenanthroxyl)	0.0	51.9	16.4	0.0	1.41	1.23
PhO•	−14.3	45.8	1.02×10^2	−6.1	1.39	1.25
(t-butyl-phenoxyl)	−5.4	49.6	33.2	−2.3	1.40	1.24
(dichloro-phenoxyl)	−31.5	39.0	8.05×10^2	−12.9	1.38	1.26
(fluoro-phenoxyl)	−10.3	47.5	62.0	−4.4	1.40	1.24
(methoxy-phenoxyl)	8.9	55.9	5.0	4.0	1.42	1.22

continued

TABLE 15.14
Enthalpies, Activation Energies, Rate Constants, Increment ΔE_H, and Geometrical Parameters of TS Reactions of Phenoxyl Radicals with Cumene (Reaction 10) Calculated by IPM Method—*continued*

Phenoxyl	ΔH (kJ mol^{-1})	E (kJ mol^{-1})	k_{10} (400 K) (L mol^{-1} s^{-1})	ΔE_H (kJ mol^{-1})	r(C—H) × 10^{10} (m)	r(O—H) × 10^{10} (m)
(4-MeO-C$_6$H$_4$-O•)	−7.5	48.7	43.5	−3.2	1.40	1.24
(4-Me$_2$N-C$_6$H$_4$-O•)	25.9	63.9	45.2	12.0	1.09	1.55

X	ΔH	E	k_{10} (400 K)	ΔE_H	r(C—H) × 10^{10}	r(O—H) × 10^{10}
CH$_2$COOH	17.8	84.9	8.34 × 10^{-4}	8.1	1.48	1.25
Ac	6.9	79.8	3.72 × 10^{-3}	3.0	1.47	1.26
CN	2.3	77.8	6.89 × 10^{-3}	1.0	1.46	1.27
OCMe$_3$	23.4	87.4	3.80 × 10^{-4}	10.6	1.48	1.25
OMe	27.6	89.4	2.09 × 10^{-4}	12.6	1.49	1.24
Me	15.7	83.8	1.12 × 10^{-3}	7.0	1.47	1.26
NO$_2$	−3.3	75.3	1.44 × 10^{-2}	−1.5	1.46	1.27
Ph	17.0	84.4	9.32 × 10^{-4}	7.6	1.48	1.25

parameters of TS for several reactions $Ar_1O^\bullet$ with R^1H. Calculation was performed according to the following equations:

$$E = (2.80\, br_e)^2 \left\{ 1 - 0.802\sqrt{1 - \frac{0.357\Delta H_e}{(br_e)^2}} \right\}^2 - 17.4 + 0.5RT \qquad (15.19)$$

$$r(\text{C—H})[10^{-10}\,\text{m}] = 1.096 + 3.85 \times 10^{-2}\sqrt{E_e} \qquad (15.20)$$

$$r(\text{O—H})[10^{-10}\,\text{m}] = 0.970 + 3.09 \times 10^{-2}\sqrt{E_e - \Delta H_e} \qquad (15.21)$$

15.5.2 Steric Factor

We came across the influence of the steric factor earlier when discussing the difference in the reactivities of sterically nonhindered phenols (Ar_1OH) and sterically hindered phenols (Ar_2OH) in their reactions with peroxyl radicals. The same factor, namely, the influence of two voluminous *ortho–tert*-alkyl groups on the reactivity of sterically hindered phenoxyls ($Ar_2O^{\bullet}$) appeared in the reactions of phenoxyl radicals with hydrocarbons. Let us compare the values of E_{e0} for reactions of $Ar_1O^{\bullet}$ and $Ar_2O^{\bullet}$ with RH and ROOH [33,34]. Calculation was performed according to the following equations:

Substrate	R^1H	R^2H	R^3H	ROOH
$E_{e0}(Ar_1O^{\bullet})$ (kJ mol^{-1})	61.5	78.4	69.5	42.7
$E_{e0}(Ar_2O^{\bullet})$ (kJ mol^{-1})	84.0	106.5	94.3	50.4
ΔE_S (kJ mol^{-1})	22.5	28.1	24.8	7.7

We see that the increment of the steric influence ΔE_S ranges from 22 to 28 kJ mol^{-1} for the $Ar_2O^{\bullet}$ reaction with the C—H bond and amounts to 8 kJ mol^{-1} for the reaction of $Ar_2O^{\bullet}$ with hydroperoxide. Due to the steric effect, sterically hindered phenoxyl practically does not participate in the chain propagation. The same effect is observed in the comparison of diphenylaminyl and picrylhydrazyl reactions with R^3H [34]: $E_{e0}(Ph_2N^{\bullet}) = 81.7$ kJ mol^{-1}, $E_{e0}(DPPH^{\bullet}) = 99.3$ kJ mol^{-1}, and $\Delta E_S = 17.6$ kJ mol^{-1}.

15.5.3 Electron Affinity of Atoms in the TS

Aminyl radicals react with C—H bonds of hydrocarbons more slowly than phenoxyl radical when compared to the activation energies of their thermoneutral reactions [33,34,38].

Reactant	R^1H	R^2H	R^3H	ROOH
$E_{e0}(Ar_1O^{\bullet})$ (kJ mol^{-1})	61.5	74.8	69.5	42.7
$E_{e0}(Am^{\bullet})$ (kJ mol^{-1})	73.7	88.4	81.5	39.2
ΔE (kJ mol^{-1})	12.5	13.6	12.0	−3.5

The difference $E_{e0}(Am^{\bullet} + RH) - E_{e0}(Ar_1O^{\bullet} + RH) \approx 12.5$ kJ mol^{-1} can be explained by the additional steric hindrance in the reaction of the aminyl radical due to its pyramidal geometry and additional repulsion of phenyl groups. One can expect the same effect in reactions of phenoxyl and aminyl radicals with hydroperoxide. However, the opposite effect is observed in this reaction, where ΔE is negative. The high reactivity of aminyl radicals in reaction with hydroperoxide has the same principle as the high reactivity of $RO_2^{\bullet}$ in reaction with AmH (see earlier). Aminyl radicals react with the O—H bond of hydroperoxide with the low activation energy (E_{e0}) due to a great difference in electron affinity of N and O atoms in the TS of the N $\cdots$ H $\cdots$ OR type [33]. The values of ΔH, E, and ΔE_H and geometric parameters of TS for some reactions of aminyl radicals with hydrocarbons and hydroperoxides are given in Table 15.15.

The flash photolysis study of the reaction of $Ph_2N^{\bullet}$ with cumyl hydroperoxide showed the more sophisticated mechanism of this reaction [99]. When [ROOH] is low (less than 0.01 mol L^{-1}), the reaction proceeds as bimolecular. The mechanism changes at the hydroperoxide concentration greater than 0.02 mol L^{-1}. The diphenylaminyl radical forms complex with hydroperoxide, and the reaction proceeds through the electron transfer.

TABLE 15.15
Enthalpies, Activation Energies, Rate Constants, Increment ΔE_H, and Geometrical Parameters of TS Reactions of Aminyl Radicals with Cumene (Reaction 10) Calculated by IPM Method

Aminyl	ΔH (kJ mol⁻¹)	E (kJ mol⁻¹)	k (400K) (L mol⁻¹ s⁻¹)	ΔE_H (kJ mol⁻¹)	r(C—H) × 10¹⁰ (m)	r(N—H) × 10¹⁰ (m)
PhN·H	−31.9	50.9	22.5	−14.1	1.40	1.23
(di-tert-butyl diphenylamine)	−4.1	62.9	0.60	−1.9	1.43	1.20
(di-methoxy diphenylamine)	6.1	67.6	0.15	2.8	1.44	1.19
(methoxy diphenylamine)	−1.2	64.2	0.40	−0.6	1.43	1.20
(di-methyl diphenylamine)	−2.8	63.5	0.50	−1.3	1.43	1.20
Ph₂N·	−10.0	60.3	1.34	−4.5	1.43	1.20
(PhHN-C₆H₄-N·Ph)	−1.2	64.2	0.40	−0.6	1.43	1.20
(Ph-N·-C₆H₄-N-iPr)	5.5	67.3	0.16	2.5	1.44	1.19
(carbazolyl)	−16.9	57.2	3.33	−7.6	1.42	1.21
(1-naphthyl-N·H)	−20.0	55.9	4.98	−8.9	1.42	1.21
(2-naphthyl-N·H)	−24.8	53.8	9.22	−11.0	1.41	1.22
(1-naphthyl-N·-Ph)	−2.4	63.7	0.48	−1.1	1.43	1.20
(2-naphthyl-N·-Ph)	−8.2	61.1	1.05	−3.7	1.43	1.20
(spiro compound)	−15.1	58.0	2.63	−6.8	−1.42	1.21

Antioxidants Reacting with Peroxyl Radicals

$$Am^\bullet + ROOH \longrightarrow AmH + RO_2^\bullet \qquad k_7$$
$$Am^\bullet + ROOH \rightleftharpoons AmH^{\bullet+} \cdots RO_2^- \qquad K$$
$$AmH^{\bullet+} \cdots RO_2^- \longrightarrow AmH + RO_2^\bullet \qquad k$$

The values of rate constants were estimated as $k_7 = 1.1 \times 10^5$ L mol^{-1} s^{-1}, $k = 1.0 \times 10^3$ s^{-1}, and $K = 42$ L mol^{-1} (348 K).

15.5.4 Reactions of Phenoxyl Radicals with Dioxygen

The phenoxyl radical has an increased electron density in the *ortho-* and *para-*positions and adds dioxygen similar to alkyl radicals. However, the C—OO bond is weak in this peroxyl radical and back dissociation occurs rapidly. Therefore, the formation of quinolide peroxide occurs in two steps, which was studied for the 2,4,6-tris(1,1-dimethylethyl)phenoxyl radical [100,101].

The phenoxyl radical is consumed with the rate

$$v = 2kK[Ar_2O^\bullet]^2[O_2] \qquad (15.22)$$

The coefficient kK was found to decrease with the increasing temperature, which can be explained by the temperature-dependent equilibrium between dioxygen, phenoxyl radical, and the formed peroxyl radical [100].

$$kK = 3.2 \times 10^{-5} \exp(114/RT) \text{ (L}^2 \text{ mol}^{-2} \text{ s}^{-1}) \qquad (15.23)$$

The reaction between the semiquinone radical and dioxygen was found recently in the kinetic study of styrene oxidation inhibited by hydroquinone [102].

$$HO-C_6H_4-O^\bullet + O_2 \longrightarrow O=C_6H_4=O + HO_2^\bullet \qquad (10)$$

The rate constant of this reaction is $k = 4.7 \times 10^2$ L mol^{-1} s^{-1} (styrene, $T = 323$ K). This reaction decreases the value of f coefficient for hydroquinone as inhibitor. The latter depends on pO_2 and is lower, the greater the pO_2.

The chain mechanism of oxidation of N,N-diphenyl-1,4-phenylenediamine by dioxygen was proved and studied by Varlamov and Denisova [103]. The introduction of an initiator accelerates this reaction. The following chain mechanism was proposed [103].

The rate constant of the aminyl radical reaction with dioxygen was estimated as $k_{10} = 1.1 \times 10^6 \exp(-26.3/RT)$ L mol^{-1} s^{-1}.

15.6 DIRECT OXIDATION OF PHENOLS AND AMINES BY DIOXYGEN AND HYDROPEROXIDES

Dioxygen is a weak hydrogen atom acceptor due to a low strength of the formed O_2—H bond (220 kJ mol^{-1}). However, at high temperature ($T > 450$ K) the reaction

$$\text{InH} + O_2 \longrightarrow \text{In}^\bullet + HO_2^\bullet \tag{12}$$

occurs with a measurable rate. The kinetics of direct oxidation of phenols and aromatic amines by dioxygen was studied [103–108]. The comparison of the experimental and theoretical values of rate constants later showed that the studied reactions occur apparently by the chain mechanism via intermediate participation of $HO_2^\bullet$ and $HO^\bullet$ radicals.

Such reactions are endothermic, since the O—H bond of phenols and the N—H bond of amines are considerably stronger (see Table 5.1 and Table 5.3) than the O—H bond of the resulting $HO_2^\bullet$ radical ($D_{O-H} = 220$ kJ mol^{-1}). The activation energies of these reactions are virtually equal to their enthalpies ($E = \Delta H + 0.5RT$). For highly endothermic reactions, the pre-exponential factor A and reaction enthalpy ΔH change is unidirectional, since higher reaction enthalpies correspond to higher vibrational amplitudes of breaking bonds in the TS and larger collision cross sections of interacting particles [109]. In terms of the parabolic model, the factor A depends on ΔH as

$$(A/A_0)^{1/2} = 1 + 1.3\left(\Delta H_e^{1/2} - \Delta H_{e\,\max}^{1/2}\right), \tag{15.24}$$

where A_0 is the pre-exponential factor typical of a given class of reactions (for example, $A_0 = 3.2 \times 10^7$ L mol^{-1} s^{-1} for the reactions $RO_2^\bullet + \text{ArOH}$). The limiting values of $\Delta H_{e\,\max}$ are given here [110].

InH	Ar$_1$OH	Ar$_2$OH	AmH
$\Delta H_{e\,\max}$ (kJ mol^{-1})	57.2	74.8	42.7

The calculated ΔH, A, and k values are summarized in Table 15.16. It can be seen that, normally, the rate constants of these reactions are low, however, at elevated temperatures (500 K and above) the rate of antioxidant consumption by reaction (11) can noticeably increase.

Similar calculations were performed for the reactions of antioxidants with NO_2, assuming that the energy of the O—H bond of the formed ONOH is equal to 327.6 kJ mol^{-1} [111]. The parameters used for calculations are summarized in Table 15.17. The calculated values of activation energies and rate constants of the reactions of nitrogen dioxide with antioxidants are given in handbook [4].

Ozone vigorously reacts with the double bonds of alkenes and quite rapidly with the π-bonds of aromatic rings. Its reaction with the C—H bonds of hydrocarbons occurs relatively slowly and is accompanied by the formation of free R$^\bullet$ and $HO_2^\bullet$ radicals followed by the decomposition of the unstable radical $HO_3^\bullet$ (see Chapter 3).

$$RH + O_3 \longrightarrow R^\bullet + HO_3^\bullet$$
$$HO_3^\bullet \longrightarrow HO^\bullet + O_2$$

The dissociation energy of the O—H bond of $HO_3^\bullet$ is 350.4 kJ mol^{-1} [112]. It can be anticipated that, like peroxyl radicals, ozone reacts with inhibitors (phenols) by the reaction [113]:

$$O_3 + ArOH \longrightarrow HO_3^\bullet + ArO^\bullet$$

with an insignificant side reaction of addition to the aromatic ring. The activation energy of such reactions can be calculated by the following simple formula of the parabolic model [113,114].

TABLE 15.16
Kinetic Parameters of Phenols and Amines Reactions with Hydroperoxides, Dioxygen, Ozone, and Nitrogen Dioxide in IPM Model [4,110–114]

Reaction	α	$b \times 10^{-11}$ (kJ mol^{-1})$^{1/2}$ m	$0.5hL\nu$ (kJ mol^{-1})	$0.5hL(\nu - \nu_f)$ (kJ mol^{-1})
ArOH + O_2 (O_3)	1.014	4.665	21.5	0.3
ArOH + NO_2	1.000	4.665	21.5	0.0
ArOH + ROOH	0.982	4.665	21.5	−0.7
AmH + O_2 (O_3)	0.940	4.324	20.0	−1.2
AmH + NO_2	0.927	4.324	20.0	−1.5
AmH + ROOH	0.907	4.324	20.0	−2.2

Reaction	br_e (kJ mol^{-1})$^{1/2}$	E_{e0} (kJ mol^{-1})	A (L mol^{-1} s^{-1})	$-\Delta H_{e\ min}$ (kJ mol^{-1})	$\Delta H_{e\ max}$ (kJ mol^{-1})
Ar$_1$OH + O_2 (O_3)	13.16	45.3	6.4×10^8	49.1	53.6
Ar$_1$OH + NO_2	13.54	45.8	2.0×10^8	57.8	57.8
Ar$_1$OH + ROOH	20.45	107.8	1.0×10^8	228.6	228.6
Ar$_2$OH + O_2 (O_3)	14.30	51.8	6.4×10^8	69.3	71.6
Ar$_2$OH + NO_2	14.49	52.5	2.0×10^8	75.6	75.6
Ar$_2$OH + ROOH	21.60	120.2	1.0×10^8	251.2	251.2
AmH + O_2 (O_3)	12.12	39.0	2.0×10^9	46.2	39.5
AmH + NO_2	11.93	38.3	2.0×10^9	44.7	36.7
AmH + ROOH	18.87	99.2	1.0×10^8	238.8	192.5

TABLE 15.17
Enthalpies, Pre-exponential Factors, and Rate Constants of Reaction InH + $O_2 \rightarrow$ In' + $HO_2^{\bullet}$ Calculated by IPM method [110]

InH	D (kJ mol^{-1})	ΔH (kJ mol^{-1})	A (L mol^{-1} s^{-1})	k (400) (L mol^{-1} s^{-1})
		Phenols		
PhOH	369.0	1.490×10^2	3.473×10^{10}	1.22×10^{-9}
H$_2$N-C$_6$H$_4$-OH	356.7	1.367×10^2	2.872×10^{10}	4.06×10^{-8}
HO-C$_6$H$_4$-OH	352.0	1.320×10^2	2.650×10^{10}	1.54×10^{-7}
HO-C$_6$H$_3$(CH$_3$)-OH	349.8	1.298×10^2	2.549×10^{10}	2.87×10^{-7}
NC-C$_6$H$_4$-OH	370.2	1.502×10^2	3.534×10^{10}	8.63×10^{-10}
1-phenanthrol	338.3	1.183×10^2	2.036×10^{10}	7.27×10^{-6}
1-naphthol	343.4	1.234×10^2	2.259×10^{10}	1.74×10^{-6}
Cl-C$_6$H$_4$-OH	370.5	1.505×10^2	2.549×10^{10}	7.92×10^{-10}
CH$_3$O-C$_6$H$_4$-OH	345.8	1.258×10^2	2.366×10^{10}	8.87×10^{-7}
CH$_3$O-C$_6$H$_2$(CH$_3$)$_2$-OH	331.4	1.114×10^2	1.745×10^{10}	4.96×10^{-5}
CH$_3$-C$_6$H$_4$-OH	362.2	1.422×10^2	3.137×10^{10}	8.49×10^{-9}
2,6-dimethylphenol	354.6	1.346×10^2	2.772×10^{10}	7.37×10^{-8}
(CH$_3$)$_2$N-C$_6$H$_4$-OH	328.8	1.088×10^2	1.639×10^{10}	1.02×10^{-4}
dihydroxypyrene	315.9	9.590×10^2	1.147×10^{10}	3.45×10^{-3}
		Sterically Hindered Phenols		
2,6-di-t-butylphenol	346.4	1.264×10^2	1.374×10^{10}	4.30×10^{-7}

TABLE 15.17
Enthalpies, Pre-exponential Factors, and Rate Constants of Reaction InH + $O_2 \rightarrow$ In' + $HO_2^{\bullet}$ Calculated by IPM method [110]—*continued*

InH	D (kJ mol^{-1})	ΔH (kJ mol^{-1})	A (L mol^{-1} s^{-1})	k (400) (L mol^{-1} s^{-1})
(2,6-di-t-butyl-4-methoxyphenol)	327.1	1.071×10^2	7.716×10^{10}	7.99×10^{-5}
(2,6-di-t-butyl-4-methylphenol)	339.0	1.190×10^2	1.128×10^{10}	3.27×10^{-6}
(bis-phenol methylene)	340.9	1.209×10^2	1.190×10^{10}	1.95×10^{-6}
(2,6-diphenyl-4-methoxyphenol)	328.0	1.080×10^2	7.968×10^{10}	6.30×10^{-5}
(thiobisphenol derivative)	335.2	1.152×10^2	1.009×10^{10}	9.16×10^{-6}
(CN-thio-phenol derivative)	345.5	1.255×10^2	1.343×10^{10}	5.51×10^{-7}
Amines				
Ph$_2$NH	364.7	1.447×10^2	4.606×10^{10}	5.88×10^{-9}
(4,4'-dimethoxydiphenylamine)	348.6	1.286×10^2	3.686×10^{10}	5.95×10^{-7}
(4-methoxydiphenylamine)	355.9	1.359×10^2	4.097×10^{10}	7.37×10^{-8}
(4,4'-dimethyldiphenylamine)	357.5	1.375×10^2	4.189×10^{10}	4.66×10^{-8}
O_2N—C$_6$H$_4$—NHPh	372.9	1.529×10^2	5.093×10^{10}	5.52×10^{-10}
(N,N'-bis(4-isopropylphenyl)-p-phenylenediamine)	333.6	1.136×10^2	2.876×10^{10}	4.22×10^{-5}

continued

TABLE 15.17
Enthalpies, Preexponential Factors, and Rate Constants of Reaction InH + $O_2 \to$ In' + $HO_2^\bullet$ Calculated by IPM method [110]—*continued*

InH	D (kJ mol^{-1})	ΔH (kJ mol^{-1})	A (L mol^{-1} s^{-1})	k (400) (L mol^{-1} s^{-1})
naphthyl-NH-C6H4-NH-naphthyl	346.6	1.266×10^2	3.575×10^{10}	1.05×10^{-6}
naphthyl-NH-naphthyl	360.2	1.402×10^2	4.344×10^{10}	2.14×10^{-8}
naphthyl-NH-Ph	357.1	1.371×10^2	4.166×10^{10}	5.22×10^{-8}
naphthyl-NH-Ph (isomer)	362.9	1.429×10^2	4.501×10^{10}	9.86×10^{-9}
naphthyl-NH-C6H4-O-Ph	349.5	1.295×10^2	3.736×10^{10}	4.60×10^{-7}

$$E = \{0.5br_e + \Delta H(2br_e)^{-1}\}^2 - 20.1 + 0.5RT \tag{15.25}$$

The parameters used for calculations of activation energies and rate constants are summarized in Table 15.16. The calculated values of E and k for the reaction of ozone with antioxidants are given in the handbook [4]. They are in good agreement with the experimental data [113].

Phenols react with hydroperoxides by the bimolecular reaction [5,115]

$$ArOH + ROOH \longrightarrow ArO^\bullet + H_2O + RO^\bullet$$

with the rate $v = k_{11}$[PhOH][ROOH]. The thermal effect of this reaction is $q = 347 - D_{O-H}$ (kJ mol^{-1}). For most phenols, this reaction is exothermic. The energy of the phenolic O—H bond and reaction rate are related inversely. Thus, *p*-methoxyphenol ($D_{O-H} = 351$ kJ mol^{-1}) and *p*-cresol ($D_{O-H} = 362$ kJ mol^{-1}) react at 413 K with $k_{11} = 4.3 \times 10^{-3}$ and 2×10^{-4} L mol^{-1} s^{-1}, respectively [115]. The reaction occurs homolytically with the formation of radicals, as demonstrated with reference to *p*-methoxyphenol in oxidized cumene. The yield of radicals is 0.27 (cumene, 293 K, [115]). The activation energy of this reaction is related to the energy of the phenolic O—H bond as $E_{11} = D_{O-H} - 254$ kJ mol^{-1} with the mean pre-exponential factor equal to 10^{10} L mol^{-1} s^{-1} [115]. Similar to the interaction of $RO_2^\bullet$ with phenol, the reaction proceeds via an intermediate step of complexing by hydrogen bonds:

$$PhOH + ROOH \Longleftrightarrow PhOH \cdots OOR \to PhO^\bullet + H_2O + RO^\bullet$$

The experimental data on antioxidant oxidation by hydroperoxide are found in the literature [115–124]. The calculated values of ΔH, E, and k of reaction ROOH with phenols are collected in Table 15.18.

TABLE 15.18
Enthalpies, Activation Energies, and Rate Constants of Reactions of 1,1-dimethylethyl Hydroperoxide with Phenols (Ar₁OH) in Hydrocarbon Solutions: ROOH + Ar₁OH → RO• + H₂O + Ar₁O• Calculated via IPM Method [114]

InH	ΔH (kJ mol⁻¹)	E (kJ mol⁻¹)	A (L mol⁻¹ s⁻¹)	k (400 K) (L mol⁻¹ s⁻¹)
		Phenols		
4-acetylphenol	49.9	111.0	1.4×10^8	4.5×10^{-7}
4-aminophenol	35	102.8	1.4×10^8	5.5×10^{-6}
4-hydroxyanisole (hydroquinone derivative)	30.0	100.2	2.9×10^8	2.4×10^{-5}
2,5-dimethylhydroquinone	23	96.4	3.0×10^8	7.8×10^{-5}
4-cyanophenol	48.5	110.2	1.4×10^8	5.6×10^{-7}
1-hydroxyphenanthrene	16.6	93.0	1.5×10^8	1.1×10^{-4}
1,5-dihydroxynaphthalene	11.8	90.5	3.1×10^8	4.6×10^{-4}
1-naphthol	21.7	95.7	1.5×10^8	4.8×10^{-5}
hydroxyphenanthrene	33	101.7	1.4×10^8	7.6×10^{-6}
PhOH	47.3	109.5	1.4×10^8	6.9×10^{-7}
4-tert-butylphenol	38.4	104.6	1.4×10^8	3.1×10^{-6}
4-(trifluoromethyl)phenol	60.7	117.1	1.4×10^8	6.9×10^{-8}
4-methoxyphenol	24.1	96.9	1.5×10^8	3.2×10^{-5}
2,3-dimethyl-4-methoxyphenol	9.7	89.5	1.5×10^8	3.2×10^{-4}
4-methylphenol	40.5	105.8	1.4×10^8	2.2×10^{-6}

TABLE 15.18
Enthalpies, Activation Energies, and Rate Constants of Reactions of 1,1-dimethylethyl Hydroperoxide with Phenols (Ar$_1$OH) in Hydrocarbon Solutions: ROOH + Ar$_1$OH → RO$^\bullet$ + H$_2$O + Ar$_1$O$^\bullet$ Calculated via IPM Method [114]—*continued*

InH	ΔH (kJ mol^{-1})	E (kJ mol^{-1})	A (L mol^{-1} s^{-1})	k (400 K) (L mol^{-1} s^{-1})
2,6-dimethylphenol	32.9	101.6	1.4×10^8	7.7×10^{-6}
4-(dimethylamino)phenol	7.1	88.1	1.6×10^8	4.8×10^{-4}
1,6-dihydroxypyrene	−5.8	81.7	3.2×10^8	7.0×10^{-3}
N-acetyl tetrahydroquinoline derivative	14	91.7	1.5×10^8	1.6×10^{-4}
Sterically Hindered Phenols				
2,6-diphenyl-4-(cyanomethyl)phenol	−10.6	86.6	1.6×10^8	7.7×10^{-4}
3,5-di-tert-butyl-4-hydroxyacetophenone	26.1	105.3	1.4×10^8	2.6×10^{-6}
2,6-di-tert-butyl-4-methylphenol	24.7	104.5	1.4×10^8	3.2×10^{-6}
2,6-di-tert-butyl-4-methoxyphenol	5.4	94.5	1.5×10^8	6.8×10^{-5}
2,6-di-tert-butyl-4-methylphenol (alt)	17.3	100.6	1.5×10^8	1.0×10^{-5}

TABLE 15.18
Enthalpies, Activation Energies, and Rate Constants of Reactions of 1,1-dimethylethyl Hydroperoxide with Phenols (Ar$_1$OH) in Hydrocarbon Solutions: ROOH + AmH → RO$^\bullet$ + H$_2$O + Am$^\bullet$ Calculated via IPM Method [114]—*continued*

InH	ΔH (kJ mol^{-1})	E (kJ mol^{-1})	A (L mol^{-1} s^{-1})	k (400 K) (L mol^{-1} s^{-1})
		Amines		
(4,4'-dimethoxydiphenylamine)	26.9	92.3	1.5×10^8	1.3×10^{-4}
(4-methoxydiphenylamine)	34.2	100.5	1.5×10^8	1.1×10^{-5}
(4,4'-dimethyldiphenylamine)	35.8	102.3	1.4×10^8	6.3×10^{-6}
O$_2$N–C$_6$H$_4$–NHPh	51.2	120.5	1.3×10^8	2.5×10^{-8}
Ph$_2$NH	43.0	110.6	1.4×10^8	4.4×10^{-7}
(N,N'-bis(4-isopropylphenyl)-p-phenylenediamine)	11.9	76.5	3.3×10^8	3.4×10^{-2}
(N,N'-di-2-naphthyl-p-phenylenediamine)	24.9	90.2	3.1×10^8	5.2×10^{-4}
(N,N'-diphenyl-p-phenylenediamine)	34.2	100.5	2.9×10^8	2.2×10^{-5}
(2,2'-dinaphthylamine)	38.5	105.4	1.4×10^8	2.5×10^{-6}
(N-phenyl-1-naphthylamine)	35.4	101.8	1.4×10^8	7.3×10^{-6}
(N-phenyl-2-naphthylamine)	41.2	108.5	1.4×10^8	9.4×10^{-7}
(N-(4-phenoxyphenyl)-2-naphthylamine)	27.8	93.3	1.5×10^8	9.8×10^{-5}

In polar solvents, this reaction is fast. For instance, at 333 K *p*-methoxyphenol is oxidized by cumyl hydroperoxide in chlorobenzene and a mixture of chlorobenzene: *tert*-butanol = 4:1 with $k_{11} = 3.5 \times 10^{-6}$ and 2.5×10^{-4} L mol^{-1} s^{-1}, respectively [124]. The acceleration of this reaction is due to phenol ionization and the rapid oxidation of the phenolate ion by hydroperoxide.

The introduction of 0.5 mol L^{-1} pyridine into 1,1-dimethylethanol containing phenol and hydroperoxide increases the rate constant k_{11} from 10^{-4} to 1.2×10^{-3} L mol^{-1} s^{-1} (353 K) [121]. At concentrations lower than 0.1 mol L^{-1}, pyridine enhances the rate of the reaction of p-methoxyphenol with hydroperoxide in benzene at 353 K from almost zero to $v = 3.7 \times 10^{-2}$[ArOH][ROOH][C$_5$H$_5$N] [121]. The addition of pyridine to tert-butanol with p-methoxyphenol increases both the reaction rate and the electroconductivity of an ArOH solution. All these results are in agreement with the following reaction mechanism:

$$ArOH + ROH \rightleftharpoons ArO^- + ROH_2^+$$
$$ArOH + C_6H_5N \rightleftharpoons ArO^- + C_6H_5NH^+$$
$$ArO^- + ROOH \longrightarrow ArO^\bullet + RO^\bullet + OH^-$$
$$RO^\bullet + ArOH \longrightarrow ROH + ArO^\bullet$$
$$2ArO^\bullet \longrightarrow \text{Recombination products}$$

In aqueous solutions, ROOH can react with p-methoxyphenol at pH > 8; however, at pH > 10 the reaction slows down due to ROOH dissociation and a weak oxidative ability of the RO$_2^-$ anion.

Hydroperoxides oxidize aromatic amines more readily than analogous phenols. Thus, at 368 K cumyl hydroperoxide oxidizes α-naphthylamine and α-naphthol with $k_{11} = 1.4 \times 10^{-4}$ and 1.7×10^{-5} L mol^{-1} s^{-1}, respectively [115,118]. The oxidation of amines with hydroperoxides occurs apparently by chain mechanism, since the step of free radical generation proceeds much more slowly. This was proved in experiments on amines oxidation by cumyl hydroperoxide in the presence of N,N'-diphenyl-1,4-phenylenediamine (QH$_2$) as a radical acceptor [125]. The following reactions were supposed to occur in solution (80% decane and 20% chlorobenzene):

$$ROOH + \text{Solvent} \longrightarrow RO^\bullet + \text{Products} \quad (k_i)$$
$$AmH + ROOH \longrightarrow Am^\bullet + H_2O + RO^\bullet \quad (k_{11})$$
$$RO^\bullet + QH_2 \longrightarrow ROH + {}^\bullet QH$$
$$Am^\bullet + QH_2 \rightleftharpoons AmH + {}^\bullet QH$$
$${}^\bullet QH + {}^\bullet QH \longrightarrow QH_2 + Q$$

The k_{11}/k_i ratio was measured in kinetic experiments and k_{11} was estimated at $k_i = 2.5 \times 10^{-6}$ s^{-1} (403 K). The values of k_{11} (403 K) are given below [125]:

AmH	Ph$_2$NH	(MeO-C$_6$H$_4$)$_2$NH	MeO-C$_6$H$_4$-NH-Ph	(4-Me-C$_6$H$_4$)$_2$NH
k_{11} (L mol^{-1} s^{-1})	1.9×10^{-5}	3.4×10^{-4}	1.1×10^{-4}	8.8×10^{-5}

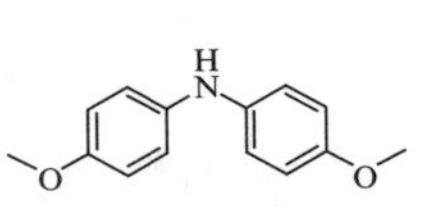

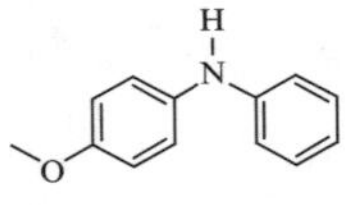

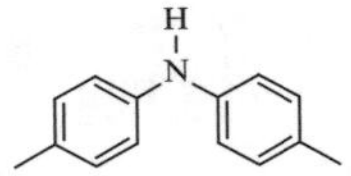

AmH	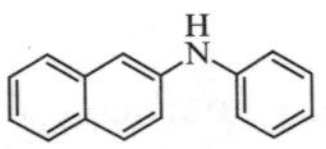 2-naphthyl-NH-Ph	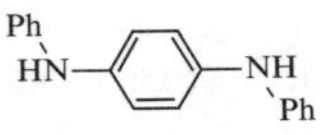 Ph-HN-C$_6$H$_4$-NH-Ph	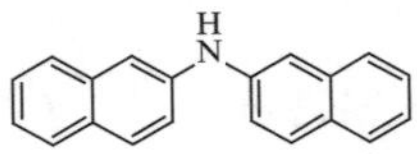 di(2-naphthyl)NH	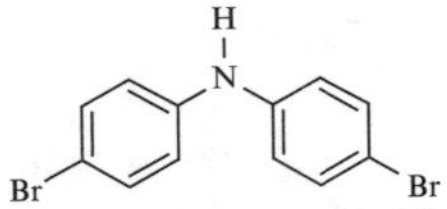 (4-Br-C$_6$H$_4$)$_2$NH
k_{11} (L mol^{-1} s^{-1})	4.2×10^{-5}	2.9×10^{-4}	7.1×10^{-5}	3.0×10^{-5}

The comparison of the rate constants of these reactions with BDE of N—H bonds proves that the higher the BDE, the lower rate the constant. The rate constants of the reaction of amine with hydroperoxide calculated by the IPM method are listed in Table 15.18. The parameters of these reactions are given in Table 15.18. The values of activation energies for all reactions of the type

$$ROOH + HIn \rightarrow RO^\bullet + H_2O + In^\bullet$$

are very high (90–120 kJ mol^{-1}) due to intensive repulsion of four atoms of the transition center.

The decay of amine oxidized by hydroperoxide occurs much more rapidly than free radical generation. Apparently, these reactions proceed by chain mechanism. The diatomic phenols and aryldiamines (QH$_2$) must react with ROOH by the chain mechanism in which the semiquinone radical $^\bullet$QH that reduces hydroperoxide plays the key role. The following chain mechanism can be supposed [122]:

$$ROOH + QH_2 \longrightarrow RO^\bullet + H_2O + {}^\bullet QH$$
$$ROOH + {}^\bullet QH \longrightarrow RO^\bullet + H_2O + Q$$
$$RO^\bullet + QH_2 \longrightarrow ROH + {}^\bullet QH$$
$${}^\bullet QH + {}^\bullet QH \longrightarrow QH_2 + Q$$

REFERENCES

1. ET Denisov. *Usp Khim* 42:361–390, 1973.
2. JA Howard. *Adv Free Radical Chem* 4:49–173, 1972.
3. ET Denisov, IV Khudyakov. *Chem Rev* 87:1313–1357, 1987.
4. ET Denisov, TG Denisova. *Handbook of Antioxidants*. Boca Raton: CRC Press, 2000.
5. ET Denisov, VV Azatyan. *Inhibition of Chain Reactions*. London: Gordon and Breach, 2000.
6. VA Roginskii. *Phenolic Antioxidants*. Moscow: Nauka, 1988 [in Russian].
7. VYa Shlyapintokh, ON Karpukhin, LM Postnikov, VF Tsepalov, AA Vichutinskii, IV Zakharov. *Chemiluminescence Techniques in Chemical Reactions*. New York: Consultants Bureau, 1968.
8. JA Howard, KU Ingold. *Can J Chem* 43:2724–2728, 1965.
9. ET Denisov, AL Aleksandrov, VP Scheredin. *Izv AN SSSR* 1583–1590, 1964.
10. NM Emanuel, GE Zaikov, ZK Maizus. *Oxidation of Organic Compounds. Medium Effect in Radical Reaction*. Oxford: Pergamon Press, 1984.
11. P Mulder, OW Saastad, D Griller. *J Am Chem Soc* 110:4090–4092, 1988.
12. VD Parker. *J Am Chem Soc* 114:7458–7462, 1992.
13. J Lind, X Shen, TE Eriksen, G Merenyi. *J Am Chem Soc* 112:479–482, 1990.
14. MM Suryan, SA Kafafi, SE Stein. *J Am Chem Soc* 111:4594–4600, 1989.
15. LR Mahoney, MA DaRooge. *J Am Chem Soc* 92:890–899, 1970.
16. LR Mahoney, MA DaRooge. *J Am Chem Soc* 97:4722–4731, 1975.
17. FG Bordwell, JP Cheng. *J Am Chem Soc* 113:1736–1743, 1991.
18. MEJ Coronel, AJ Colussi. *J Chem Soc Perkin Trans* 2:785–787, 1994.
19. XM Zhang, FG Borwell. *J Am Chem Soc* 116:4251–4254, 1994.
20. FG Bordwell, WZ Liu. *J Am Chem Soc* 118:10819–10823, 1996.
21. M Lucarini, P Pedrielli, GF Pedulli. *J Org Chem* 61:9259–9269, 1996.
22. MI de Heer, HG Korth, P Mulder. *J Org Chem* 64:6969–6975, 1999.
23. VF DeTuri, KM Ervin. *J Phys Chem* A103:6911–6920, 1999.
24. RMB dos Santos, JA Simoes. *J Phys Chem Ref Data* 27:707, 1998.

25. DDM Wayner, E Lusztyk, KU Ingold. *J Org Chem* 61:6430–6433, 1996.
26. FG Bordwell, XM Zhang. *J Phys Org Chem* 8:529–534, 1995.
27. SH Hoke, SS Yang, RG Cooks, DA Hrovat, WT Borden. *J Am Chem Soc* 116:4888–4892, 1994.
28. ET Denisov. *Russ J Phys Chem* 69:563–574, 1995.
29. DDM Wayner, E Lusztyk, D Page, KU Ingold, P Mulder, LJJ Laarkoven, HS Aldrich. *J Am Chem Soc* 117:8737–8744, 1995.
30. ET Denisov. *Polym Deg Stab* 49:71–75, 1995.
31. Y-R Luo. *Int J Chem Kinet* 34:453–466, 2002.
32. Y-R Luo. *Handbook of Bond Dissociation Energies in Organic Compounds.* Boca Raton: CRC Press, 2003.
33. ET Denisov, TI Drozdova. *Kinet Catal* 35:155–162, 1994.
34. ET Denisov. *Russ Chem Rev* 66:859–876, 1997.
35. AF Shestakov, ET Denisov. *Russ Chem Bull* 320–329, 2003.
36. ET Denisov. *Chem Phys Rep* 14:1513–1520, 1995.
37. VE Tumanov, ET Denisov. *Kinet Catal* 38:339–344, 1997.
38. ET Denisov. In: ZB Alfassi, (ed.) *General Aspects of the Chemistry of Radicals.* Chichester: Wiley, 1999, pp. 79–137.
39. HS Johnston. *Gas Phase Reaction Rate Theory.* New York: Ronald, 1966.
40. ET Denisov. *Kinet Catal* 35:293–298, 1994.
41. ET Denisov. *Russ J Phys Chem* 68:1089–1093, 1994.
42. ET Denisov. *Kinet Catal* 35:617–635, 1994.
43. PK Das, MV Encinas, S Steenken, JC Scaiano. *J Am Chem Soc* 103:4162–4166, 1981.
44. MV Encinas, JC Scaiano. *J Am Chem Soc* 103:6393–6397, 1981.
45. GC Pimentel, AL McClelan. *The Hydrogen Bond.* San Francisco: Pergamon Press, 1960.
46. GI Zaikov, VP Lezina, ZK Maizus. *Zh Fiz Khim* 42:1273–1275, 1968.
47. OA Raevskii, VYu Grigor'ev, DB Kireev, NS Zefirov. *Quant Struct Act Relat* 11:49–63, 1992.
48. ET Denisov, RL Vardanyan. *Izv AN SSSR Ser Khim* 2463–2467, 1972.
49. ET Denisov. *Izv AN SSSR Ser Khim* 53–58, 1960.
50. PA MacFaul, KU Ingold, J Lusztyk. *J Org Chem* 61:1316–1321, 1996.
51. NN Pozdeeva, IK Yakuschenko, AL Aleksandrov, VB Luzhkov, ET Denisov. *Kinet Katal* 30:31–37, 1989.
52. ET Denisov, TG Denisova, SV Trepalin, TI Drozdova. *Database: Oxidation and Antioxidants in Organic Chemistry and Biology.* New York: Marcel Dekker, 2003.
53. VT Varlamov, ET Denisov. *Bull Acad Sci USSR* 39:657–662, 1990.
54. ET Denisov. *Kinet Catal* 36:345–350, 1995.
55. VT Varlamov, ET Denisov. *Dokl AN SSSR* 293:126–128, 1987.
56. VT Varlamov, ET Denisov. *Bull Acad Sci USSR* 36:1607–1612, 1987.
57. IR Mahoney, MA DaRooge. *J Am Chem Soc* 97:4722–4731, 1975.
58. VA Roginskii, VZ Dubinskii, IA Shlyapnikova, VB Miller. *Eur Polym J* 13:1043–1051, 1977.
59. VA Roginskii, VZ Dubinskii, VB Miller. *Izv AN SSSR Ser Khim* 2808–2812, 1981.
60. VI Rubtsov, VA Roginskii, VB Miller, GE Zaikov. *Kinet Katal* 21:612–615, 1980.
61. AP Griva, ET Denisov. *Int J Chem Kinet* 5:869–877, 1973.
62. ET Denisov, TG Denisova, TS Pokidova. *Handbook of Free Radical Initiators.* New York: Wiley, 2003.
63. K Adamic, KU Ingold. *Can J Chem* 47:295–299, 1968.
64. VT Varlamov, RL Safiulin, ET Denisov. *Khim Fiz* 2:408–412, 1983.
65. RL Vardanyan, VV Kharitonov, ET Denisov. *Kinet Katal* 12:903–909, 1971.
66. RL Vardanyan. *Kinet Katal* 15:794–796, 1974.
67. JE Bennett, G Brunton, AR Forrester, JD Fullerton. *J Chem Soc Perkin Trans 2* 1477–1480, 1983.
68. JR Thomas, CA Tolman. *J Am Chem Soc* 84:2930–2935, 1962.
69. VA Roginskii. *Khim Fizika* 4:1244–1249, 1985.
70. IV Khudyakov, PP Levin, VA Kuzmin. *Usp Khim* 49:1990–2033, 1980.
71. IV Khudyakov, PP Levin, VA Kuzmin, CRHL de Jonge. *Int J Chem Kinet* 11:357–364, 1979.
72. PP Levin, IV Khudyakov, VA Kuzmin. *Izv AN SSSR Ser Khim* 255–261, 1980.

73. IV Khudyakov, CRH de Jonge, PP Levin, VA Kuzmin. *Izv AN SSSR Ser Khim* 1492–1498, 1978.
74. CRH de Jonge, EA Giezen, FPB van der Maeden. *Adv Chem Ser* 169:31, 399, 1978.
75. HLJ Backstrom. *Trans Faraday Soc* 49:1460–1491, 1927.
76. EJ Land. *Prog React Kinet* 3:369–402, 1965.
77. IV Khudyakov, VA Kuzmin. *Usp Khim* 47:39–82, 1978.
78. AL Yasmenko, IV Khudyakov, VA Kuzmin, AP Khardin. *Kinet Katal* 22:122–126, 1981.
79. S Lee, MJ Karpius. *J Chem Phys* 86:1904–1921, 1987.
80. AB Doktorov, PA Purtov. *Khim Fiz* 6:484–491, 1987.
81. SI Temkin, BI Yakobson. *J Phys Chem* 88:2679–2681, 1984.
82. AB Doktorov, NN Lukzen. *Khim Fiz* 4:616–623, 1985.
83. AB Doktorov. *Khim Fiz* 4:800–808, 1985.
84. IA Pritchin, KM Salikhov. *J Phys Chem* 89:5212–5217, 1985.
85. IV Khudyakov, LI Koroli. *Chem Phys Lett* 103:383–385, 1984.
86. SK Wong, W Sytnnjk, IKS Wan. *Can J Chem* 50:3052–3057, 1972.
87. T Forster, AJ Elliot, BB Adeleke. *Can J Chem* 56:869–877, 1978.
88. EL Tumanskii, AI Prokof'ev, NN Bubnov SP Solodovnikov, AA Khodak. *Izv AN SSSR Ser Khim* 268–274, 1983.
89. SP Yarkov, VA Roginskii, GE Zaikov. *Khim Fiz* 1410–1415, 1983.
90. AJ Elliot, KL Egan, JKS Wan. *J Chem Soc Faraday Trans 1*, 2111–2120, 1978.
91. MV Voevodskaya, IV Khudyakov, PP Levin, VA Kuzmin. *Izv AN SSSR Ser Khim* 1925–1927, 1980.
92. WT Dixon, D Murphy. *J Chem Soc Faraday Trans 2*, 1221–1230, 1976.
93. T Shida, A Kira. *J Phys Chem* 73:4315–4320, 1969.
94. VT Varlamov. Elementary Reactions and Products of the Liquid-Phase Conversion of Diphenylaminyl Radical. Ph.D. Thesis, Institute of Chemical Physics, Chernogolovka, 1986 [in Russian].
95. RF Bridger. *J Am Chem Soc* 94:3124–3132, 1972.
96. EA Efremkina, IV Khudyakov, ET Denisov. *Khim Fiz* 6:1289–1291, 1987.
97. VT Varlamov, RL Safiullin, ET Denisov. *Khim Fiz* 2:408–412, 1983.
98. Landolt-Bornstein. In: H. Fischer, (ed.) *Numerical Data and Functional Relationships in Science and Technology, New Series, Group II: Atomic and Molecular Physics*, vol 13, subvol c, Radical Reaction Rates in Liquids, Berlin: Springer-Verlag, 1984.
99. VT Varlamov, RL Safiullin, ET Denisov. *Khim Fiz* 4:789–793, 1985.
100. AP Griva, ET Denisov. *Int J Chem Kinet* 5:869–877, 1973.
101. VI Rubtsov, VA Roginskii, VB Miller. *Vysokomol Soed* B22:446–449, 1980.
102. NN Pozdeeva, IK Yakuschenko, AL Alexandrov, ET Denisov. *Kinet Catal* 32:1162–1169, 1991.
103. VT Varlamov, LN Denisova. *Dokl AN SSSR* 320:1156–1159, 1991.
104. LN Denisova, ET Denisov, DI Metelitsa. *Izv ANSSR Ser Khim* 1657–1663, 1969.
105. AN Nikolaevskii, TA Filippenko, RV Kucher. *Zh Org Khim* 16:331–336, 1980.
106. LI Mazaletskaya, GV Karpukhina, NL Komissarova. *Izv AN SSSR Ser Khim* 505–508, 1982.
107. EL Shanina, VA Roginskii, GE Zaikov. *Vysokomol Soed* A28:1971–1976, 1986.
108. LN Denisova, ET Denisov, DI Metelitza. *Izv AN SSSR Ser Khim* 44:1670–1675, 1970.
109. ET Denisov. *Kinet Catal* 37:519–523, 1996.
110. ET Denisov. *Kinet Catal* 39:17–23, 1998.
111. ET Denisov, TG Denisova, YuV Geletii, J Balavoine. *Kinet Catal* 39:312–319, 1998.
112. ET Denisov, TG Denisova. *Kinet Catal* 37:46–50, 1996.
113. TG Denisova, ET Denisov. *Polym Degrad Stabil* 60:345–350, 1998.
114. TG Denisova, ET Denisov. *Petrol Chem* 40:141–147, 2000.
115. VS Martem'yanov, ET Denisov, LA Samoilova. *Izv AN SSSR Ser Khim* 1039–1042, 1972.
116. K Heberger. *Int J Chem Kinet* 17:271–280, 1985.
117. L Zikmund, J Bradilova, J Pospisil. *J Polym Sci Polym Symposia* 271–279, 1973.
118. AN Nikolaevskii, TA Filipenko, AI Peicheva, RV Kucher. *Neftekhimiya* 16:758–761, 1976.
119. AV Gerasimova, GI Kovalev, LD Gogitidze, VI Kuranova, NS Zvereva, ET Denisov. *Neftekhimiya* 25:555–561, 1985.
120. VL Antonovskii, ET Denisov, V Solntseva. *Kinet Katal* 6:815–819, 1965.
121. NV Zolotova, ET Denisov. *Izv AN SSR Ser Khim* 767–768, 1966.

122. NV Zolotova, ET Denisov. *Zh Fiz Khim* 46:2008–2011, 1973.
123. VS Martem'yanov, ET Denisov. *Izv AN SSSR Ser Khim* 2191–2194, 1972.
124. VS Martem'yanov, ET Denisov, VV Fedorova. *Kinet Katal* 13:303–307, 1972.
125. LN Denisova, VT Varlamov. *Kinet Katal* 32:845–851, 1991.

16 Cyclic Chain Termination in Oxidation of Organic Compounds

16.1 CYCLIC CHAIN TERMINATION BY AROMATIC AMINES AND AMINYL RADICALS

Aromatic amines terminate chains in oxidizing hydrocarbons with the stoichiometric coefficient between 1 and 2 as a result of the consecutive reactions (see Chapter 15).

$$RO_2^\bullet + AmH \longrightarrow ROOH + Am^\bullet$$
$$Am^\bullet + RO_2^\bullet \longrightarrow Products$$
$$Am^\bullet + Am^\bullet \longrightarrow Products$$

High values of the inhibition coefficient ($f = 12-28$) were detected for the first time in the oxidation of cyclohexanol [1] and butanol [2] inhibited by 1-naphthylamine. For the oxidation of decane under the same conditions, $f = 2.5$. In the case of oxidation of the decane–cyclohexanol mixtures, the coefficient f increases with an increase in the cyclohexanol concentration from 2.5 (in pure decane) to 28 (in pure alcohol). When the oxidation of cyclohexanol was carried out in the presence of tetraphenylhydrazine, the diphenylaminyl radicals produced from tetraphenylhydrazine were found to be reduced to diphenylamine [3]. This conclusion has been confirmed later in another study [4]. Diphenylamine was formed only in the presence of the initiator, regardless of whether the process was conducted under an oxygen atmosphere or under an inert atmosphere. In the former case, the aminyl radical was reduced by the hydroperoxyl radical derived from the alcohol (see Chapter 6), and in the latter case, it was reduced by the hydroxyalkyl radical.

On the basis of these results, the following general mechanism was suggested for the *cyclic chain termination* by inhibitor InH in oxidized alcohol [3]:

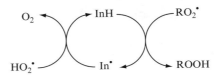

The oxidation of primary and secondary alcohols in the presence of 1-naphthylamine, 2-naphthylamine, or phenyl-1-naphthylamine is characterized by the high values of the inhibition coefficient $f \gg 10$ [1–7]. Alkylperoxyl, α-ketoperoxyl radicals, and β-hydroxyperoxyl radicals, like the peroxyl radicals derived from tertiary alcohols, appeared to be incapable of reducing the aminyl radicals formed from aromatic amines. For example, when the oxidation of *tert*-butanol is inhibited by 1-naphthylamine, the coefficient f is equal to 2, which coincides with the value found in the inhibited oxidation of alkanes [3]. However, the addition of hydrogen peroxide to the *tert*-butanol getting oxidized helps to perform the cyclic chain termination mechanism (1-naphthylamine as the inhibitor, $T = 393\,K$, cumyl peroxide as initiator, $pO_2 = 98\,kPa$ [8]). This is due to the participation of the formed hydroperoxyl radical in the chain termination:

$$RO_2^{\bullet} + AmH \longrightarrow ROOH + Am^{\bullet}$$
$$RO_2^{\bullet} + H_2O_2 \longrightarrow ROOH + HO_2^{\bullet}$$
$$Am^{\bullet} + HO_2^{\bullet} \longrightarrow AmH + O_2$$

Since the peroxyl radicals derived from alcohol dissociate to a carbonyl compound and $HO_2^{\bullet}$ (see Chapter 8), two reactions in which the aminyl radicals formed from amine are reduced, occur in parallel under the conditions of the alcohol oxidation:

$$>C(OH)OO^{\bullet} \longrightarrow >C=O + HO_2^{\bullet}$$
$$RO_2^{\bullet} + H_2O_2 \longrightarrow ROOH + HO_2^{\bullet}$$
$$Am^{\bullet} + HO_2^{\bullet} \longrightarrow AmH + O_2$$
$$>C(OH)OO^{\bullet} + Am^{\bullet} \longrightarrow >C=O + O_2 + AmH$$

In the presence of dissolved dioxygen, the hydroxyalkyl radicals are converted into the hydroxyperoxyl radicals very rapidly; therefore, only hydroperoxyl and hydroxyalkylperoxyl radicals participate in the reduction of the aminyl radicals. The higher the temperature, the more effective the decomposition of the hydroxyperoxyl radicals and the higher the proportion of the $HO_2^{\bullet}$ radicals participating in the regeneration of the amine.

Later it was shown [9] that in the case of repeated chain termination with aromatic amines in the oxidation of alcohols the situation is more complicated. In parallel with the reaction of disproportionation with the aminyl radical, the following reactions occur:

$$>C(OH)OO^{\bullet} + Am^{\bullet} \longrightarrow >C(OH)O^{\bullet} + AmO^{\bullet}$$
$$AmO^{\bullet} + >C(OH)OO^{\bullet} \longrightarrow AmOH + O_2 + >C=O$$
$$>C(OH)OO^{\bullet} + AmOH \longrightarrow >C(OH)OOH + AmO^{\bullet}$$

The intermediate formation of the nitroxyl radical was detected in the oxidation of 2-propanol retarded by diphenylamine; chain termination occurs by cyclic mechanisms involving both aminyl and nitroxyl radicals [9]. In the case where the oxidation of 2-propanol is decelerated by bis(p-methoxyphenyl)amine, only the cycle involving nitroxyl radicals is realized [9].

Organic acids retard the formation of nitroxyl radicals via the reaction of the peroxyl radical with the aminyl radical [10]. Apparently, the formation of a hydrogen bond of the $>N \cdots HOC(O)R$ type leads to the shielding of nitrogen, which precludes the addition of dioxygen to it, yielding the nitroxyl radical. Thus, the products of the oxidation of alcohols, namely, acids have an influence on the mechanism of the cyclic chain termination.

As noted above, the duration of the retarding action of an inhibitor is directly proportional to the f value. In systems with a cyclic chain termination mechanism, the f coefficient depends on the ratio of the rate constants for two reactions, in which the inhibitor is regenerated and irreversibly consumed. In the oxidation of alcohols, aminyl radicals are consumed irreversibly via the reaction with nitroxyl radical formation (see earlier) and via the following reaction [11]:

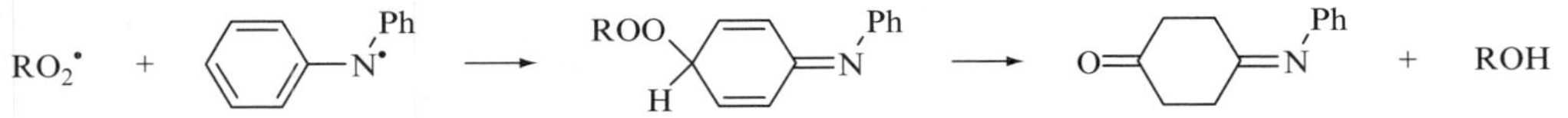

The formation of the nitroxyl radical and quinone imine precludes the possibility of the recovery of amine and, hence, any of the above reactions interrupts the cycle at the aminyl radical. Taking these reactions into account, we come to the following expression for the coefficient f:

$$f = 1 + \frac{k(Am^\bullet + HO_2^\bullet \longrightarrow AmH + O_2)}{k(Am^\bullet + HO_2^\bullet \longrightarrow AmO^\bullet + HO^\bullet) + k(Am^\bullet + HO_2^\bullet \longrightarrow Q + H_2O)} \qquad (16.1)$$

Table 16.1 presents the inhibition coefficients f and the termination rate constants k_7 in systems with the cyclic chain termination mechanism with aromatic amines. Naturally, these are apparent rate constants, which characterize primarily the rate-limiting step of the chain termination process.

The question why the aminyl radicals ensure cyclic chain termination in those systems in which the hydroperoxyl and hydroxyalkylperoxyl radicals are formed, but not in the oxidation of hydrocarbons where alkylperoxyl radicals are the chain-propagating species deserves special attention [22–24]. Indeed, the disproportionation of the aminyl and peroxyl radicals that involve both the O—H and C—H bonds is, in principle, possible

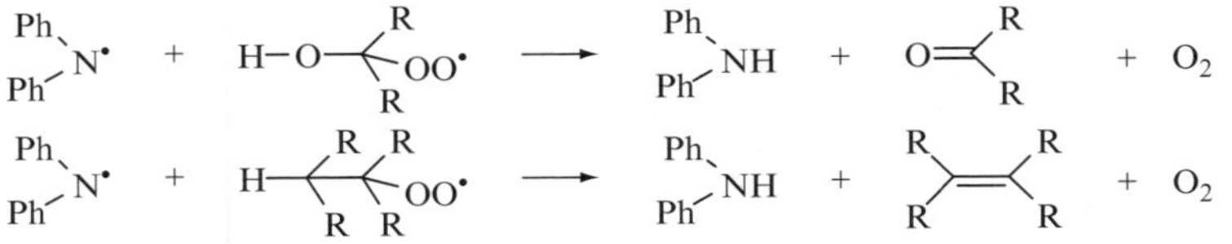

We present the enthalpies of disproportionation of aminyl radicals with hydroxyalkyl and alkyl peroxyl radicals in Table 16.2 [22–24].

All these reactions are exothermic, and the ΔH values are negative. All these reactions should seemingly occur equally rapidly. The question to how easily the aminyl radicals react with the H—O and H—C bonds of the peroxyl radicals can be answered by analyzing these reactions in terms of the IPM model of free radical reaction (see Chapter 6). This model gives a tool to perform the calculation of the activation energy for a thermally neutral reaction of each class. Analysis of experimental data has shown (see Chapter 15) that, when aminyl

TABLE 16.1
Inhibition Coefficients f and Rate Constants k for the Reactions of Peroxyl Radicals with Aromatic Amines in Systems with a Cyclic Mechanism of Chain Termination (Experimental Data)

Amine	Oxidizing Substrate	T (K)	f	k (L mol^{-1} s^{-1})	Ref.
Ph_2NH	cyclohexadiene	348	200		[12]
Ph_2NH	trans-PhCH=CHCOOEt	323	20	3.6×10^5	[13]
Ph_2NH	trans-PhCH=CHCOOMe	323	20	2.8×10^5	[13]
Ph_2NH	trans-PhCH=CHCOOPh	323	20	1.1×10^4	[13]
Ph_2NH	Me_2CHOH	343	>23		[9]
Ph_2NH	cyclohexanol	393	56	5.0×10^3	[14]
Ph_2NH	$Me_2N(CH_2)_2OCOMeC=CH_2$	323	10	6.0×10^3	[15]
diphenylamine-diol	$Me_2NCH_2CH_2NMe_2$	313	26	6.5×10^4	[16]
diphenylamine-diol	Me_2CHOH	343	22		[9]
diphenylamine-diol	N-methylpiperidine	323	52	7.3×10^4	[17]
diphenylamine-diol	$N(CH_2Me)_3$	313	80	4.4×10^5	[18]
diphenylamine-diol	N,N-dimethylcyclohexylamine	323	70	1.5×10^6	[18]
diphenylamine-diol	$MeCON(CH_2Me)_2$	348	30	$>1.0 \times 10^5$	[19]
diphenylamine-diol	cyclohexylamine	348	18	2.2×10^4	[20]
diphenylamine-diol	$(C_4H_9)_2NH$	348	29	4.2×10^4	[20]
diphenylamine-diol	$PhCH_2NH_2$	338	25	8.2×10^3	[20]
diphenylamine-diol	$PhCH_2NHMe$	333	>10	3.7×10^4	[18]
diphenylamine-diol	$Me_2N(CH_2)_2OCOMeC=CH_2$	323	18	1.2×10^4	[15]
diphenylamine-diol	$Me_2NC_4H_9$	323	16	1.1×10^5	[21]

TABLE 16.1
Inhibition Coefficients f and Rate Constants k for the Reactions of Peroxyl Radicals with Aromatic Amines in Systems with a Cyclic Mechanism of Chain Termination (Experimental Data)—*continued*

Amine	Oxidizing Substrate	T (K)	f	k (L mol^{-1} s^{-1})	Ref.
4,4′-dimethoxydiphenylamine	Me$_2$N(CH$_2$)$_2$OCOCEt	323	26	8.0×10^4	[21]
1-naphthylamine	cyclohexene	348	28		[12]
1-naphthylamine	cyclohexanol	393	56	1.3×10^3	[14]
1-naphthylamine	cyclohexanol	393	28	3.2×10^4	[1]
1-naphthylamine	cyclohexanol	413	30		[1]
1-naphthylamine	cyclohexanol	393	28		[2]
1-naphthylamine	cyclohexanol	348	15		[3]
1-naphthylamine	cyclohexanol	348	90		[8]
1-naphthylamine	Me$_2$CHOH	348	90		[8]
1-naphthylamine	cyclohexanol + H$_2$O$_2$	398	22		[8]
1-naphthylamine	N,N-dimethylcyclohexylamine	323	>4	4.2×10^4	[18]

TABLE 16.1
Inhibition Coefficients *f* and Rate Constants *k* for the Reactions of Peroxyl Radicals with Aromatic Amines in Systems with a Cyclic Mechanism of Chain Termination (Experimental Data)—*continued*

Amine	Oxidizing Substrate	T (K)	f	k (L mol^{-1} s^{-1})	Ref.
1-naphthylamine	Me(CH$_2$)$_3$OH	347	12		[2]
1-naphthylamine	Me(CH$_2$)$_3$OH	383	17		[2]
1-naphthylamine	MeCH(OH)Et	347	12		[2]
1-naphthylamine	Me$_3$COH + H$_2$O$_2$	358	9		[2]
1-naphthylamine	Me$_3$COH + H$_2$O$_2$	348	22		[2]
1-naphthylamine	Me$_2$CHOH	348	90		[8]
1-naphthylamine	Me$_2$NCH$_2$CH$_2$NMe$_2$	313		1.6×10^3	[16]
1-naphthylamine	N-methylpiperidine	323		2.0×10^3	[17]
1-naphthylamine	N(CH$_2$CH$_3$)$_3$	313		1.3×10^4	[18]
1-naphthylamine	(C$_4$H$_9$)$_2$NH	348	26	1.5×10^3	[20]
1-naphthylamine	Me$_2$N(CH$_2$)$_2$OCO MeC=CH$_2$	323	10	6.0×10^3	[15]

TABLE 16.1
Inhibition Coefficients *f* and Rate Constants *k* for the Reactions of Peroxyl Radicals with Aromatic Amines in Systems with a Cyclic Mechanism of Chain Termination (Experimental Data)—*continued*

Amine	Oxidizing Substrate	T (K)	f	k (L mol^{-1} s^{-1})	Ref.
1-naphthylamine	cyclohexyl-NH$_2$	348	16	8.0×10^2	[20]
1-naphthylamine	PhCH$_2$NH$_2$	338	30	1.6×10^3	[20]
2-naphthylamine	Me(CH$_2$)$_3$OH	347	6		[2]
2-naphthylamine	cyclohexanol	393	28	2.0×10^3	[14]
N,N'-di(2-naphthyl)-p-phenylenediamine	Me$_2$N(CH$_2$)$_2$OCOMeC=CH$_2$	323	26	1.2×10^5	[15]
N,N'-di(2-naphthyl)-p-phenylenediamine	cyclohexene	348	6.5		[12]
N,N'-di(2-naphthyl)-p-phenylenediamine	cyclohexene	313	22		[12]
N,N'-di(2-naphthyl)-p-phenylenediamine	N-methylpiperidine	323	43	8.2×10^4	[17]
N,N'-di(2-naphthyl)-p-phenylenediamine	N,N-dimethylcyclohexylamine	323	17	2.1×10^5	[18]

continued

TABLE 16.1
Inhibition Coefficients *f* and Rate Constants *k* for the Reactions of Peroxyl Radicals with Aromatic Amines in Systems with a Cyclic Mechanism of Chain Termination (Experimental Data)—*continued*

Amine	Oxidizing Substrate	T (K)	f	k (L mol^{-1} s^{-1})	Ref.
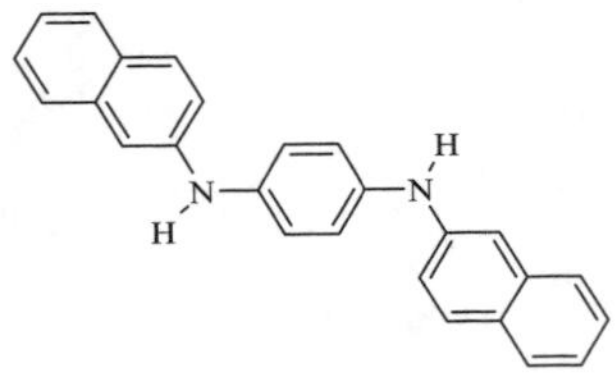 *N,N'*-di(2-naphthyl)-*p*-phenylenediamine	*trans*-PhCH=CHCOOEt	323	40	2.6×10^5	[13]
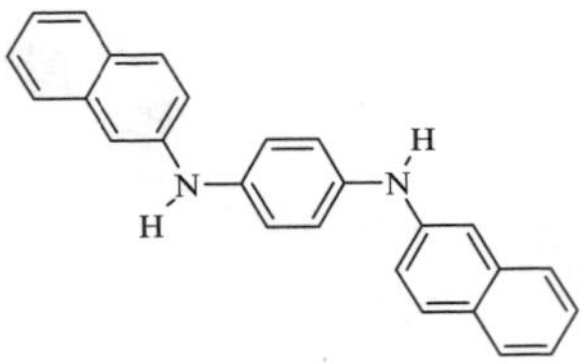 *N,N'*-di(2-naphthyl)-*p*-phenylenediamine	*trans*-PhCH=CHPh	323	40	1.7×10^5	[13]
N,N'-di(2-naphthyl)-*p*-phenylenediamine	Me$_2$NCH$_2$CH$_2$NMe$_2$	313	22	4.0×10^4	[16]
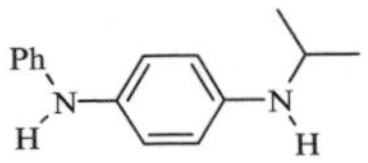 *N*-phenyl-*N'*-isopropyl-*p*-phenylenediamine	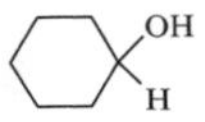 cyclohexanol	393	200		[22]
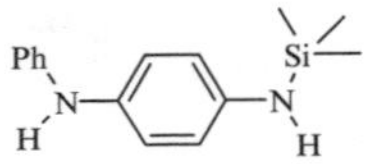 *N*-phenyl-*N'*-trimethylsilyl-*p*-phenylenediamine	Me$_2$CHOH	343	18		[9]
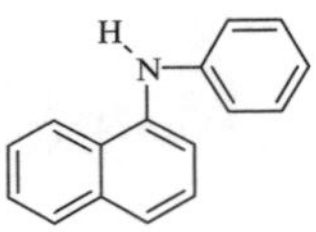 *N*-phenyl-1-naphthylamine	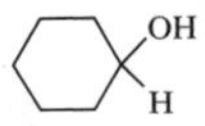 cyclohexanol	393	15		[14]
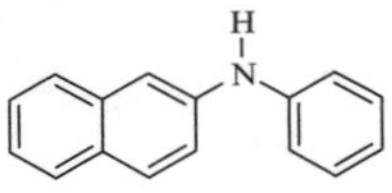 *N*-phenyl-2-naphthylamine	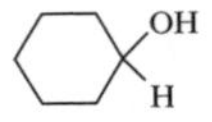 cyclohexanol	393	26	9.0×10^3	[14]
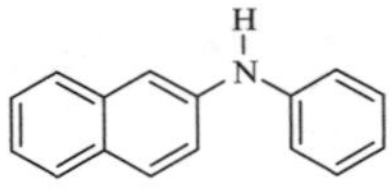 *N*-phenyl-2-naphthylamine	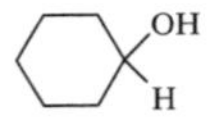 cyclohexanol	348	36	1.2×10^4	[20]
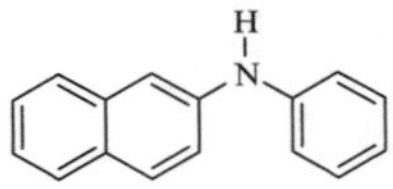 *N*-phenyl-2-naphthylamine	PhCH$_2$NH$_2$	338	52	1.2×10^4	[20]
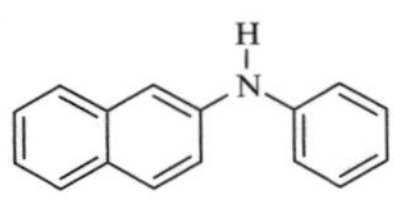 *N*-phenyl-2-naphthylamine	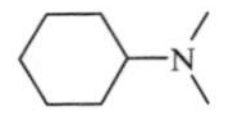 *N*,*N*-dimethylcyclohexylamine	323	20	1.0×10^4	[18]

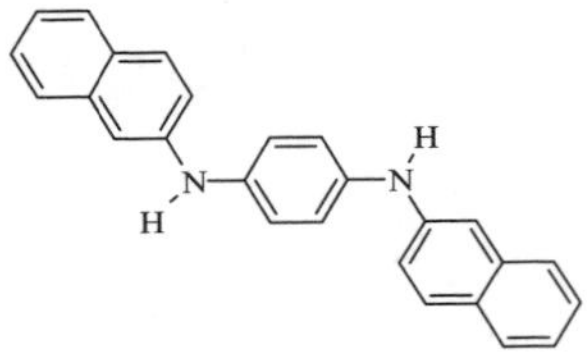

TABLE 16.2
Enthalpies of Disproportionation Reactions: Am$^\bullet$ + RO$_2^\bullet$ $\longrightarrow$ AmH + O$_2$ + Molecule (olefin or ketone)

Aminyl Radical	$-\Delta H$ (kJ mol^{-1})			
	HO$_2^\bullet$	HO–C$_6$H$_{10}$–OO$^\bullet$	H$_3$C(OO$^\bullet$)PhMe	$^\bullet$OO–C$_6$H$_{10}$–H,H
Ph$_2$N$^\bullet$	144.7	134.9	94.7	84.4
(4-MeO-C$_6$H$_4$)$_2$N$^\bullet$	128.6	118.8	78.6	68.3
1-Naphthyl-N$^\bullet$-Ph	137.1	127.3	87.1	76.8
2-Naphthyl-N$^\bullet$-Ph	142.9	127.3	87.1	76.8
2-Naphthyl-N$^\bullet$H	159.5	144.9	104.7	94.4

radicals react with hydrocarbons, phenols, and hydroperoxides, they are extremely reactive toward specifically O—H bonds [25,26]. The values of E_{e0} for the reactions of aminyl radicals with various compounds are presented below.

Compound	R^1—H	R^3—H	ROO—H	Ar$_1$O—H
E_{e0} (kJ mol^{-1})	69.7	81.7	39.2	29.4
$-\Delta H_{e\,min}$ (kJ mol^{-1})	169.3	196.1	39.8	29.4

The reactions of aminyl radicals with O—H and C—H bonds are very different. The first have low triplet repulsion and high difference in electron affinity of atoms in the TS reaction center N $\cdots$ H $\cdots$ O and, therefore, a low value of E_{e0}. The second has high triplet repulsion and moderate difference in electron affinity of atoms in the TS reaction center N $\cdots$ H $\cdots$ C and, therefore, high value of E_{e0}. According to the parabolic model (see Chapter 6) for highly exothermic reactions with fairly high $-\Delta H$ values the activation energies $E = 0.5RT$. The transition from higher activation energies to $E = 0.5RT$ occurs at a certain critical value of $-\Delta H = \Delta H_{e\,min}$, which depends on E_{e0} for the class of reactions considered. The higher E_{e0}, the higher $|\Delta H_{e\,min}|$ and the broader the ΔH range in which $E > 0.5RT$.

For the reaction of aminyl radicals with the O—H bond in ROOH, $\Delta H_{e\,min} = -39.8$ kJ mol^{-1}, which is substantially smaller by magnitude than $|\Delta H|$ for the reactions of diphenyl-aminyl radicals with H—O bonds of subsequent peroxyl radicals (from -161 to -135 kJ mol^{-1}, see above). Thus, all these reactions require virtually no activation energy, i.e., occurs fairly rapidly. Quite another situation arises with disproportionation of aminyl and alkylperoxyl radicals. The $|\Delta H|$ value for the reactions of the diphenylaminyl radical with the C—H bond is markedly lower than the value of $|\Delta H_{e\ min}|$ (see above). Hence, these reactions should proceed with positive activation energies $E > 0.5RT$, i.e., their rates are much lower than the rate of interaction of aminyl radical with the O—H bond in the corresponding peroxyl radical. This is illustrated in Figure 16.1. It is clearly seen that the disproportionation reactions of all known aminyl radicals with $HO_2^•$ occur with $E = 0.5RT$. The reactions of aminyl radicals with alkylperoxyl radicals, on the contrary, should occur with a sufficiently high activation energy.

To give the ultimate answer to the question why the cyclic chain termination mechanism is realized in the systems where $HO_2^•$ and $>C(OH)OO^•$ radicals act as chain-propagating species but is not realized during the oxidation of hydrocarbons, one should compare the rate constants for disproportionation reactions in which $Am^•$ affords the starting AmH with the rate constant for the irreversible destruction of $Am^•$. The diphenylaminyl radical reacts with the peroxyl radical via two parallel routes with the formation of quinone imine and nitroxyl radical with approximately identical rate [9]. The rate constant for the overall reaction k is known to be 6×10^8 L mol^{-1} s^{-1} [27]. This implies that the rate constant for the addition of the peroxyl radical to the aromatic ring of the diphenylaminyl radical is 3×10^8 L mol^{-1} s^{-1}. The cyclic mechanism is realized if the inequality $k(\text{regeneration}) > k(\text{decay})$ holds. Therefore, it is necessary to estimate the rate constants for the disproportionation of $Am^•$ and $RO_2^•$ and for the regeneration of AmH by this reaction. It should be noted that, according to the parabolic model, in the case of highly exothermic reactions the pre-exponential factor A is related to the enthalpy of the reaction at $|\Delta H_e| > |\Delta H_{e\ min}|$ by the following relationship [28]:

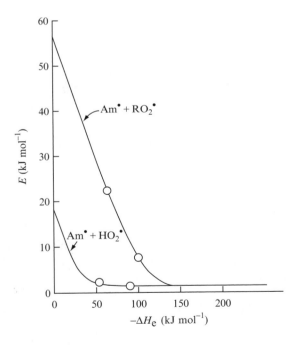

FIGURE 16.1 The dependence of activation energy E on reaction enthalpy ΔH_e for reaction of hydrogen atom abstraction by aminyl radical from the C—H bond of alkylperoxyl radical and O—H bond of hydroperoxyl radical calculated by IPM method (see Chapter 6). The points fix the reactions with minimum and maximum enthalpy among known aromatic aminyl radicals.

$$\sqrt{\frac{A}{A_0}} = 1 + 1.3\left\{\sqrt{|\Delta H_e|} - \sqrt{|\Delta H_{e\,\min}|}\right\} \qquad (16.2)$$

The activation energies and rate constants for the discussed reactions of the diphenylaminyl radical with various peroxyl radicals calculated in this way are presented below [22].

Reaction	$-\Delta H$ (kJ mol^{-1})	E (kJ mol^{-1})	k (L mol^{-1} s^{-1})
$Ph_2N^\bullet + H{-}O_2^\bullet$	145	1.4	8.4×10^9
$Ph_2N^\bullet\ +\ ^\bullet OO{-}C_6H_{10}{-}OH$	135	1.4	7.6×10^9
$Ph_2N^\bullet + H{-}C(OO^\bullet)PhMe_2$	95	16.6	1.5×10^7
$Ph_2N^\bullet\ +\ H{-}C_6H_{10}{-}OO^\bullet$	34	19.8	3.1×10^6

One can see that the rate constants for the disproportionation of the aminyl and peroxyl radicals involving the C—H bond are substantially lower than the rate constants for the addition of RO$_2^\bullet$ to the aromatic ring of the aminyl radical ($k = 3 \times 10^8$ L mol^{-1} s^{-1}). Conversely, the reaction of Am$^\bullet$ with HO$_2^\bullet$ occurs very rapidly and can compete successfully with the addition reaction. Thus, the result obtained within the scope of the parabolic model is in good agreement with the experiment.

The reason for the different reactivities of the aminyl radical with respect to alkylperoxyl and hydroperoxyl radicals lies in the energy characteristics of the TS of the C $\cdots$ H $\cdots$ N and O $\cdots$ H $\cdots$ N types (see Chapter 15). In the former case, the triplet repulsion makes a substantial contribution to the energy of the TS, because the C—N bond is fairly strong [29] and the triplet repulsion is high [22]. In the latter case, the hypothetical compound ROOAm incorporates a weak N—O bond, and the contribution of the triplet repulsion to the activation energy is low. Apart from the triplet repulsion, the electronegativities of the atoms forming the reaction center of the TS are also significant: the higher the difference between the electronegativities of the atoms, the lower the E_{e0} activation barrier. In this respect, the relationship between the C $\cdots$ H $\cdots$ N and O $\cdots$ H $\cdots$ N reaction centers counts in favor of the latter: the electron affinities of the C and N atoms are close to each other, whereas the electronegativities of the O and N atoms are substantially different.

The estimation based on the equations of the parabolic model indicates that a reaction of the type (ArO$^\bullet$ + HO$_2^\bullet$ → ArOH + O$_2$) involving phenoxyl radicals also requires no activation energy (in this case, $\Delta H > \Delta H_{e\,\min} = 57$ kJ mol^{-1}). However, the addition of the peroxyl radical to the aromatic ring of the phenoxyl radical occurs very rapidly. Hence, the rate constant for this reaction is determined by diffusion processes. The data on the E_{e0} values are also consistent with this. For the ArO$^\bullet$ + HOOR reactions with the O $\cdots$ H $\cdots$ O reaction center and for Am$^\bullet$ + HOOR reactions with the N $\cdots$ H $\cdots$ O reaction center, these values are 45.3 and 39.8 kJ mol^{-1}, respectively [23]. At the same time, the calculation of the pre-exponential factor in terms of the parabolic model indicates that the rate constant k_{-7} for the reaction of ROOH with the participation of the aminyl radical is several times higher than that for the reaction involving the phenoxyl radical, where the enthalpies of these reactions are identical. Thus, the irreversible destruction of ArO• in its reactions with $HO_2^•$ and >C(OH)$O_2^•$ occurs more rapidly than the conversion of ArO• into ArOH, and the cyclic mechanism of chain termination is not realized with phenols in such systems. An exception is provided by hydroquinones, which are considered below. Aromatic amines also cause cyclic chain termination in the oxidation of primary and secondary aliphatic amines (Table 16.1). Amine is recovered from Am• via the reaction of Am• with the O—H bond of the peroxyl radical. The latter is formed as a result of decomposition of the aminoperoxyl radical RCH(OO•)NHR with the formation of imine RCH=NHR and $HO_2^•$ (see Chapter 9).

Cyclic chain termination with aromatic amines also occurs in the oxidation of tertiary aliphatic amines (see Table 16.1). To explain this fact, a mechanism of the conversion of the aminyl radical into AmH involving the β-C—H bonds was suggested [30]. However, its realization is hampered because this reaction due to high triplet repulsion should have high activation energy and low rate constant. Since tertiary amines have low ionization potentials and readily participate in electron transfer reactions, the cyclic mechanism in systems of this type is realized apparently as a sequence of such reactions, similar to that occurring in the systems containing transition metal complexes (see below).

16.2 CYCLIC CHAIN TERMINATION BY QUINONES

Quinones (Q) are well known as inhibitors of radical polymerization; they terminate chains by the addition of alkyl radicals via the following reactions [7]:

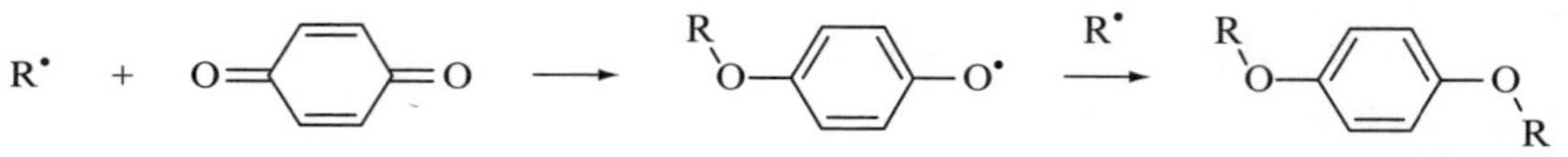

At the same time, quinones do not practically retard oxidation of hydrocarbons, since alkyl radicals react very rapidly with dioxygen (see Chapter 4) to give alkylperoxyl radicals, which scarcely react with quinones. Quinones exhibit their inhibiting properties as alkyl radical acceptors only in the oxidation of polymers (see Chapter 19). However, quinones were found to decelerate the oxidation of alcohols very efficiently and for long periods by ensuring cyclic chain termination via the following reactions [31–34]:

$$R_2C^•OH + O_2 \longrightarrow R_2C(OO^•)OH$$
$$R_2C(OO^•)OH \longrightarrow R_2C(O) + HO_2^•$$

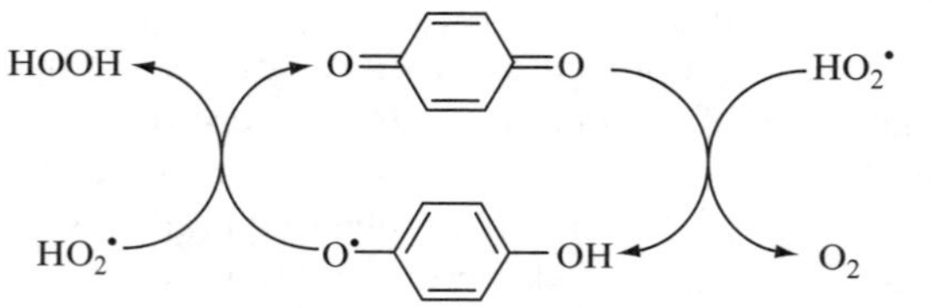

The processes of oxidation of cyclohexadiene, 1,2-substituted ethenes, and aliphatic amines are decelerated by quinones, hydroquinones, and quinone imines by a similar mechanism. The values of stoichiometric inhibition coefficients f and the rate constants k for the corresponding reactions involving peroxyl radicals ($HO_2^•$ and >C(OH)OO•) are presented in Table 16.3. The f coefficients in these reactions are relatively high, varying from 8 to 70. Evidently, the irreversible consumption of quinone in these systems is due to the addition of peroxyl radicals to the double bond of quinone and alkyl radicals to the carbonyl group of quinone.

TABLE 16.3
Rate Constants and Inhibition Coefficients for Chain Termination by Quinones and Hydroquinones in Oxidized Alcohols, Amines, and Polypropylene

Antioxidant	RH	T (K)	f	k (L mol^{-1} s^{-1})	Ref.
benzoquinone	cyclohexene	348	70		[12]
benzoquinone	Me$_2$CHOH	344	23	2.9×10^5	[31]
benzoquinone	NEt$_3$	313	10	2.9×10^5	[18]
benzoquinone	BuNMe$_2$	323	12	4.2×10^4	[21]
benzoquinone	C$_6$H$_{11}$NMe$_2$	323	16	2.7×10^4	[18]
benzoquinone	N-methylpiperidine	323	8	4.0×10^4	[18]
benzoquinone	Me$_2$NCH$_2$CH$_2$NMe$_2$	313		1.0×10^4	[16]
benzoquinone	Me$_2$NCH$_2$CH$_2$OC(O)MeC=CH$_2$	323	34	1.6×10^4	[15]
hydroquinone	trans-PhCH=CHPh	323	15	3.0×10^3	[35]
hydroquinone	trans-PhCH=CHC(O)OEt	323	15	1.4×10^5	[35]
hydroquinone	NEt$_3$	313	>20	1.4×10^5	[18]
hydroquinone	C$_6$H$_{11}$NMe$_2$	323	20	7.2×10^4	[18]
hydroquinone	N-methylpiperidine	323	14	3.2×10^4	[18]
hydroquinone	Me$_2$NCH$_2$CH$_2$NMe$_2$	313		7.0×10^3	[33]
hydroquinone	AcC(O)NEt$_2$	348	10	9.4×10^3	[19]
PhN=quinone	Me$_2$CHOH	343	12	2.0×10^5	[9]
hindered phenol quinone methide	Polypropylene	366	36		[32]
bis-phenol	Polypropylene	366	57		[36]

It is evident from Table 16.3 that quinones react rapidly with the hydroperoxyl radicals. Calculation shows that the change in the enthalpy of this reaction is relatively small. The dissociation energy of the O—H bond in the semiquinone radical 4-HOC$_6$H$_4$O$^\bullet$ is 228.1 kJ mol^{-1} and the enthalpy of the reaction Q + HO$_2^\bullet$ is $\Delta H = 220.0 - 228.1 = -8.1$ kJ mol^{-1}. Calculation of the kinetic parameters of the quinone reaction with hydroperoxyl radical by the IPM model [31–34] gives the following values: $\Delta H = -8.1$ kJ mol^{-1}, $E = 27.3$ kJ mol^{-1}, $A = 2.0 \times 10^8$ L mol^{-1} s^{-1}, and $k(344 \text{ K}) = 1.4 \times 10^3$ L mol^{-1} s^{-1}.

Reaction	α	$b \times 10^{-11}$ ((kJ mol^{-1})$^{1/2}$m)	$0.5\, hL\nu$ (kJ mol^{-1})	$0.5\, hL(\nu - \nu_f)$ (kJ mol^{-1})
Q + R^{1}H	0.802	3.743	17.4	−4.1
Q + ROOH	0.986	4.600	21.2	−0.3

	br_e (kJ mol^{-1})$^{1/2}$	E_{e0} (kJ mol^{-1})	A (L mol^{-1} s^{-1})	$\Delta H_{e\,\min}$ (kJ mol^{-1})	$\Delta H_{e\,\max}$ (kJ mol^{-1})
Q + R^{1}H	14.29	62.9	2.0×10^9	141.8	90.2
Q + ROOH	13.00	43.3	1.3×10^8	51.3	52.3

This value is close to that estimated from experiments (see Table 16.3). The activation energy of this reaction seems to be relatively low. The low activation barriers to these reactions are accounted for the fact that the transition state of the O $\cdots$ H $\cdots$ O type arising in the reaction of Q with HO$_2^\bullet$ is characterized by the absence of the triplet repulsion, which makes a substantial contribution to the activation energy in other reactions.

The question arises why quinones terminate chains via the reaction with hydroperoxyl and hydroxyalkylperoxyl radicals and do not via the reaction with the β-C—H bond of the alkylperoxyl radical.

The cleavage of the C—H and C—O bonds in this reaction is compensated to a substantial extent by the formation of the π-C—C and O—H bonds. The ΔH values for the discussing reactions involving HO$_2^\bullet$ and two hydrocarbon peroxyl radicals PhMe$_2$CO$_2^\bullet$ and cyclo-C$_6$H$_{11}$O$_2^\bullet$ are given below [31].

Radical	HO$_2^\bullet$	cyclo-C$_6$H$_{11}$O$_2^\bullet$	PhMe$_2$CO$_2^\bullet$
ΔH (kJ mol^{-1})	−8.1	52.9	41.9
E_{e0} (kJ mol^{-1})	42.8	62.9	62.9
E (kJ mol^{-1})	27.3	78.9	73.6
k (333 K) (L mol^{-1} s^{-1})	1.0×10^4	1.7×10^{-3}	3.4×10^{-2}

We see, at first, that the reaction enthalpy for quinone abstraction reactions with the C—H bond of alkylperoxyl radicals is higher than with the O—H bond of the hydroperoxyl radical. The second important factor is different triplet repulsions in these two types of abstraction reactions. Indeed, the reaction with $RO_2^{\bullet}$ proceeds via TS of the C $\cdots$ H $\cdots$ O type. Such reaction is characterized by the high thermally neutral activation energies $E_{e0} = 62.9$ kJ mol^{-1}. The value of E_{e0} for the reaction involving the O $\cdots$ H $\cdots$ O TS reaction center is much lower (27.3 kJ mol^{-1}). With the rate constants have a very low value, the reaction $Q + RO_2^{\bullet}$ cannot influence the oxidative chain termination in comparison with the interaction of two $RO_2^{\bullet}$ radicals. Indeed, the rate constant for the latter is 10^5-10^7 L mol^{-1} s^{-1} and, in these cases, the inequality $(2k_6v_i)^{1/2} \gg 2k[Q]$ always holds. The reason for such high E_{e0} values and, hence, for the high activation energies characteristic of the reactions involving TS of the O $\cdots$ H $\cdots$ C type lies in the high triplet repulsion (see Chapters 6 and 15), as well as in the reactions involving aminyl radicals. In the case of oxidation of primary and secondary amines, the high rate of chain termination by quinones occurs as a result of the fast decomposition of aminoperoxyl radicals.

$$R_1NHCH(OO^{\bullet})R_2 \longrightarrow R_1N{=}CHR_2 + HO_2^{\bullet}$$

The formed hydroperoxyl radicals react with quinone. In addition to $HO_2^{\bullet}$, the aminoperoxyl radicals can apparently reduce quinones also.

The high selectivity of hydroperoxyl and alkylhydroxyperoxyl radicals toward quinone has been used for the development of an original method for measuring the rate constant of peroxyl radicals with alcohols (see Chapter 7).

Nitro compounds, like quinones, terminate chains in oxidizing compounds where hydroperoxyl radicals are formed. This was proved for the oxidation of polyatomic esters [37] and PP [38]. Nitrobenzene retards the initiated oxidation of the following esters: tetrapropionate of pentaerythritol, propionate of 2,2-dimethylbutanol, and dipropionate of 2,2-dimethylpropanediol terminating chains by the reaction with peroxyl radicals [37]. The hydroperoxyl radicals were supposed to be formed as a result of the following reactions:

1. Hydrolysis of formed hydroperoxide and exchange reaction with $HO_2^{\bullet}$ formation:

$$ROOH + H_2O \longrightarrow ROH + H_2O_2$$
$$RO_2^{\bullet} + H_2O_2 \longrightarrow ROOH + HO_2^{\bullet}$$

2. Oxidation of the formed alcohol with production of hydroperoxyl radicals (see Chapter 7).

The addition of nitrobenzene into oxidized ester decreases the oxidation rate to some limit ($v \to v_{\infty}$ at $[PhNO_2] \to \infty$) [37]. Such a limitation of the retarding action occurs as a result of the participation of two types of peroxyl radicals, namely, $RO_2^{\bullet}$ and $HO_2^{\bullet}$.

16.3 CYCLIC CHAIN TERMINATION BY NITROXYL RADICALS

Nitroxyl radicals (AmO$^{\bullet}$) are known to react rapidly with alkyl radicals and efficiently retard the radical polymerization of hydrocarbons [7]. At the same time, only aromatic nitroxyls are capable of reacting with alkylperoxyl radicals [10,39] and in this case the chain termination in the oxidation of saturated hydrocarbons occurs stoichiometrically. However, in the processes of oxidation of alcohols, alkenes, and primary and secondary aliphatic amines in which the chain reaction involves the $HO_2^{\bullet}$, $>C(OH)O_2^{\bullet}$, and $>C(NHR)O_2^{\bullet}$ radicals, possessing the high reducing activity, nitroxyl radicals are capable of terminating the chains repeatedly by a cyclic mechanism [9,11,19,39]. For example, chain termination in the oxidation of alcohols occurs via the following cyclic reactions:

$$ROOH \leftarrow \quad \nearrow AmO^\bullet \quad \leftarrow \quad HO_2^\bullet$$
$$RO_2^\bullet \nearrow \quad \searrow AmOH \leftarrow \quad \searrow O_2$$

The first reaction is the rate-determining step. During oxidation a kinetic equilibrium is established between the two forms of the inhibitor: $AmO^\bullet$ and $AmOH$ [38]. The kinetic characteristics of this mechanism are presented in Table 16.4.

The cross-disproportionation of nitroxyl and hydroperoxyl radicals is an exothermic reaction. For example, the enthalpies of disproportionation of TEMPO radical with $HO_2^\bullet$, $Me_2C(OH)O_2^\bullet$, and $cyclo$-$C_6H_{10}(OH)O_2^\bullet$ radicals are equal to -109, -92, and $-82\,kJ\,mol^{-1}$, respectively. The E_{e0} value for the abstraction of an H atom from the O—H bond in ROOH by a nitroxyl radical is $45.6\,kJ\,mol^{-1}$ and $\Delta H_{e\,min} = -58\,kJ\,mol^{-1}$. Since $\Delta H_e < \Delta H_{e\,min}$, (see Chapter 6), the activation energy of such exothermic reactions for these reactions is low ($E \sim 0.5RT$), and the rate constant correspondingly is high [31–34]. Therefore, in the systems in which hydroperoxyl, hydroxyperoxyl, and aminoperoxyl radicals participate in chain propagation, the cyclic chain termination mechanism should be realized.

Principally, nitroxyl radicals can disproportionate with alkylperoxyl radicals.

$$AmO^\bullet + H\text{—}CHRCH(O_2^\bullet)R \longrightarrow AmOH + O_2 + RCH\text{=}CHR$$

Such reactions are exothermic. For example, the TEMPO radical abstracts an H atom from the β-C—H bond of cumylperoxyl and cyclohexylperoxyl radicals with enthalpies equal to -42 and $-32\,kJ\,mol^{-1}$, respectively. However, such reactions should occur with relatively high activation energy as it is seen from the data calculated by the IPM method [31].

Radical	$HO_2^\bullet$	cyclohexyl-O-O$^\bullet$	$PhMe_2CO_2^\bullet$
ΔH (kJ mol^{-1})	−77.7	−32.0	−42.0
$\Delta H_{e\,min}$ (kJ mol^{-1})	−60.2	−142.1	−142.1
E_{e0} (kJ mol^{-1})	42.8	63.0	63.0
E (kJ mol^{-1})	0.5 RT	27.4	23.7
k (333K) (L mol^{-1} s^{-1})	1.1×10^8	1.0×10^5	1.1×10^5

The reaction of $AmO^\bullet$ with $HO_2^\bullet$ occurs with $\Delta H < \Delta H_{e\,min}$ and, subsequently, with a low activation energy ($E = 0.5RT$) and a high rate constant. The latter is higher than $2k_t$ for peroxyl radicals (see Chapter 6), which is important for cyclic chain termination. The inverse situation takes place in reactions of nitroxyl radical disproportionation with alkylperoxyl radicals. For these reactions we observe inequality $\Delta H > \Delta H_{e\,min}$ and, subsequently, relatively a high activation energy ($E > 0.5RT$) and a low rate constant. The latter are lower than $2k_t$ for

TABLE 16.4
Kinetic Characteristics of the Cyclic Mechanism of the Chain Termination on Nitroxyl Radicals in the Oxidation of Alcohols and Amines (Experimental Data)

Inhibitor	Substrate	T (K)	f	k (L mol^{-1} s^{-1})	Ref.
(MeO-C$_6$H$_4$)$_2$N-O•	Me$_2$CHOH	343	50		[9]
(MeO-C$_6$H$_4$)$_2$N-O•	cyclohexyl-OH	348	50	8.4×10^4	[19]
(MeO-C$_6$H$_4$)$_2$N-O•	MeC(O)NEt$_2$	348	30	3.5×10^4	[19]
(MeO-C$_6$H$_4$)$_2$N-O•	Bu$_2$NH	348	100	2.1×10^5	[19]
(MeO-C$_6$H$_4$)$_2$N-O•	cyclohexyl-NH$_2$	348	100	8.1×10^4	[19]
(MeO-C$_6$H$_4$)$_2$N-O•	PhCH$_2$NH$_2$	338	140	3.4×10^5	[19]
TEMPO	trans-PhCH=CHCOOEt	323	>100	1.9×10^5	[13]
TEMPO	trans-PhCH=CHCOOMe	323	>100	1.2×10^5	[13]
TEMPO	trans-PhCH=CHC(O)OEt	323	>100	1.4×10^5	[13]
TEMPO	trans-PhCH=CHPh	323	>100	3.2×10^4	[13]
TEMPO	Me$_2$C=CHC(O)Me	323	>100	2.2×10^4	[13]

continued

TABLE 16.4
Kinetic Characteristics of the Cyclic Mechanism of the Chain Termination on Nitroxyl Radicals in the Oxidation of Alcohols and Amines (Experimental Data)—*continued*

Inhibitor	Substrate	T (K)	f	k (L mol^{-1} s^{-1})	Ref.
TEMPO	EtOCOCH=CHC(O)OEt	323	>100	1.2×10^4	[13]
TEMPO	CH_2=CMeC(O)OCH$_2$CH$_2$NMe$_2$	323	30	2.6×10^4	[15]
TEMPO	MeC(O)NEt$_2$	348	50	1.3×10^4	[19]
TEMPO	C$_6$H$_{11}$NH$_2$ (cyclohexylamine)	348	200	2.4×10^4	[19]
TEMPO	Bu$_2$NH	348	160	1.4×10^5	[19]
TEMPO	EtC(O)OCH$_2$CH$_2$NMe$_2$	323	10	2.0×10^3	[13]
TEMPO	N-methylpiperidine	323	90	7.2×10^4	[19]
TEMPO	Et$_3$N	313	25	5.7×10^5	[19]
TEMPO	C$_6$H$_{11}$NMe$_2$	323	92	1.8×10^5	[19]
TEMPO	PhCH$_2$NHCH$_3$	323	16	4.2×10^4	[19]
4-hydroxy-TEMPO	cyclohexanol	348	120	2.1×10^5	[20]

TABLE 16.4
Kinetic Characteristics of the Cyclic Mechanism of the Chain Termination on Nitroxyl Radicals in the Oxidation of Alcohols and Amines (Experimental Data)—*continued*

Inhibitor	Substrate	T (K)	f	k (L mol^{-1} s^{-1})	Ref.
HO-TEMPO	C$_6$H$_{11}$NH$_2$ (cyclohexyl-NH-H)	348	100	2.8×10^4	[19]
HO-TEMPO	Bu$_2$NH	348	100	1.5×10^5	[19]
HO-TEMPO	Me$_2$NCH$_2$CH$_2$NMe$_2$	313	>4	1.3×10^4	[16]
HO-TEMPO	PhCH$_2$NH$_2$	338	100	7.9×10^4	[19]
HO-TEMPO	CH$_2$=CMeCO$_2$CH$_2$CH$_2$NMe$_2$	323	70	8.0×10^3	[15]
HO-TEMPO	BuNMe$_2$	323	80	5.6×10^4	[19]
HO-TEMPO	EtCO$_2$CH$_2$CH$_2$NMe$_2$	323	10	3.0×10^3	[13]
O=TEMPO	C$_6$H$_{11}$NH$_2$	348		3.0×10^4	[20]
O=TEMPO	Bu$_2$NH	348		1.3×10^5	[20]
O=TEMPO	PhCH$_2$NH$_2$	338		1.9×10^5	[20]
PhC(O)O-TEMPO	C$_6$H$_{11}$NH$_2$	348		2.5×10^4	[20]

continued

TABLE 16.4
Kinetic Characteristics of the Cyclic Mechanism of the Chain Termination on Nitroxyl Radicals in the Oxidation of Alcohols and Amines (Experimental Data)—*continued*

Inhibitor	Substrate	T (K)	f	k (L mol^{-1} s^{-1})	Ref.
Ph-C(O)-O-CH$_2$-(tetramethylpiperidine-N-O•)	Bu$_2$NH	348		1.7×10^5	[20]
Ph-C(O)-O-CH$_2$-(tetramethylpiperidine-N-O•)	Polypropylene	387	14		[40]
C$_{17}$H$_{35}$-C(O)-O-CH$_2$-(tetramethylpiperidine-N-O•)	Oxidized polypropylene	388	>40		[41]
Ph-substituted imidazolidine nitroxide (with N=O)	cyclohexanol (C$_6$H$_{11}$OH)	348		5.9×10^4	[20]
Ph-substituted imidazolidine nitroxide	PhCH$_2$NH$_2$	338		7.2×10^4	[20]
Ph-substituted imidazolidine nitroxide	cyclohexylamine (C$_6$H$_{11}$NH$_2$)	348		7.0×10^3	[20]
Ph-substituted imidazolidine nitroxide	Bu$_2$NH	348		6.6×10^4	[20]

peroxyl radicals (see Chapter 6) and cyclic chain termination cannot be performed. The values of activation energies as functions of the enthalpy are shown in Figure 16.2 for both types of disproportionation.

It is clearly seen that for the known nitroxyl radicals $E \approx 0$ for reactions of nitroxyls with the hydroxyl group containing peroxyl radicals and $E > 0$ for all nitroxyl reactions with alkylperoxyl radicals.

Why are the activation energies of the reactions of nitroxyl radicals with O—H bonds lower than those in their reactions with C—H bonds? As in the case of the reaction of RO$_2^\bullet$ with quinones, the difference in E values occurs as a result of the different triplet repulsions in TS [23]. When a TS of the O $\cdots$ H $\cdots$ O type is formed (the AmO$^\bullet$ + HO$_2^\bullet$ reaction), the triplet repulsion is close to zero because the O—O bond in the labile compound AmOOH is very weak. Conversely, the triplet repulsion in the reaction of AmO$^\bullet$ with the C—H bond is fairly great, due to the high dissociation energy of the AmO—R bond. This accounts for the difference between the activation energies and between the rate constants for the reactions considered above. Thus, the possibility of the realization of a cyclic chain termination mechanism in the reactions of nitroxyl radicals with peroxyl radicals, incorporating O—H groups, is caused by the weak triplet repulsion in the TS of such disproportionation reactions

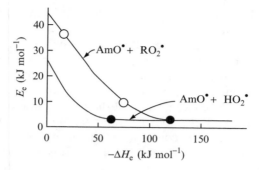

FIGURE 16.2 The dependence of activation energy E on reaction enthalpy ΔH_e for reaction of hydrogen atom abstraction by nitroxyl radical from the C—H bond of alkylperoxyl radical and the O—H bond of hydroperoxyl radical calculated by IPM method (see Chapter 6). The points fix the reactions with minimum and maximum enthalpy among known aromatic aminyl radicals.

of peroxyl radicals. The mechanisms of cyclic chain termination in polymer oxidation will be discussed in Chapter 19.

A new cyclic mechanism of chain termination by nitroxyl radicals, including the formation of aminyl radicals as intermediate species, has been proposed by Korcek and coworkers [42,43]. It was shown that the addition of 4,4′-dioctyldiphenylnitroxyl radical to the hexadecane that is oxidized ($T = 433$ K) leads to the formation of the corresponding diphenylamine as an intermediate compound during its transformations. The following cyclic mechanism of chain termination was suggested:

$$AmO^\bullet + R^\bullet \longrightarrow AmOR$$
$$AmOR \longrightarrow [Am^\bullet + RO^\bullet]$$
$$[Am^\bullet + RO^\bullet] \longrightarrow AmH + Y$$
$$RO_2^\bullet + AmH \longrightarrow ROOH + Am^\bullet$$
$$RO_2^\bullet + Am^\bullet \longrightarrow RO^\bullet + AmO^\bullet$$

An alternative mechanism of the regeneration of the amino radical from the nitroxyl radical is also possible:

$$AmO^\bullet + RO_2^\bullet \longrightarrow [Am^\bullet + O_2 + RO^\bullet]$$
$$[Am^\bullet + O_2 + RO^\bullet] \longrightarrow AmH + R'\!=\!O + O_2$$
$$RO_2^\bullet + AmH \longrightarrow ROOH + Am^\bullet$$
$$RO_2^\bullet + Am^\bullet \longrightarrow RO^\bullet + AmO^\bullet$$

Thus, nitroxyl radicals can participate in various cyclic mechanisms of chain termination. Additional information about cyclic chain termination is described in Chapter 19.

16.4 ACID CATALYSIS IN CYCLIC CHAIN TERMINATION

A new interesting branch of the modern antioxidant chemistry deals with the cyclic mechanisms involving acid catalysis. The first inhibiting system of this type was discovered in 1988 [44]. It consisted of an alcohol (primary or secondary), a stable nitroxyl radical TEMPO, and an organic (citric) acid. This three-component system efficiently retards the initiated oxidation of ethylbenzene. Each component taken by itself, as well as binary combinations of these components, does not retard oxidation of ethylbenzene or retards it only slightly. Intense chain termination was ensured only by the presence of three components of this system. The important role of the acid was established in experiments on the oxidation of ethylbenzene containing an alcohol and nitroxyl radicals, when air or air mixed with gaseous HCl was passed through the reactor. The concentration of the peroxyl radicals in the oxidized hydrocarbon was determined by the CL method. The intensity of CL was evaluated in terms of $I_{chl} \sim [RO_2^\bullet] \sim v_i/2k_t$ where k_t is the rate constant for the disproportionation of $RO_2^\bullet$. When a mixture of air with HCl was used, the intensity of the CL decreases by approximately two orders of magnitude [44].

The $H_2O_2 + AmO^\bullet$ (TEMPO) + HA (citric acid) system also exhibits an extremely high inhibiting effect [45,46]. The data presented below demonstrate how the individual components of this system and the whole system influences the rate of the chain oxidation of ethylbenzene (343 K, $pO_2 = 10^5$ Pa, $v_i = 5.21 \times 10^{-7}$ mol L^{-1} s^{-1} [44]).

[AmO$^\bullet$] × 10^4 (mol L^{-1})	[HA] × 10^4 (mol L^{-1})	[H$_2$O$_2$] × 10^3 (mol L^{-1})	v × 10^6 (mol L^{-1} s^{-1})
0.0	0.0	0.0	4.92
5.0	0.0	0.0	3.78
5.0	5.0	0.0	3.00
5.0	5.0	2.5	0.22

It can be seen that complete termination of the chain process is ensured only by the three-component system. The very efficient retardation of hydrocarbon oxidation by such a triple system can be explained by the following mechanism [45,46]:

$$AmO^\bullet + HA \rightleftharpoons [AmOH^{\bullet+}, A^-] \qquad K_a$$
$$[AmOH^{\bullet+}, A^-] + RO_2^\bullet \longrightarrow ROOH + [AmO^{\bullet+}, A^-] \qquad k_b$$
$$[AmO^{\bullet+}, A^-] + H_2O_2 \longrightarrow AmOH + HA + O_2 \qquad k_c$$
$$RO_2^\bullet + AmOH \longrightarrow ROOH + AmO^\bullet \qquad \text{Fast}$$

According to this mechanism, hydrogen peroxide is consumed during oxidation of hydrocarbons with a rate of $0.5v_i$, whereas the rate at which nitroxyl radicals are consumed via the reaction with alkyl radicals is much lower. Acid and the nitroxyl radical act as catalysts of cyclic chain termination. Only a small portion of the R$^\bullet$ radicals react with AmO$^\bullet$, since the latter reacts very rapidly with dioxygen. The mechanism of chain termination consisting of the above described reactions is notable because the acid protonates nitroxyl and thus converts it into a hydrogen atom donor for the peroxyl radical, while hydrogen peroxide reduces the nitroxyl radical cation to hydroxylamine, which reacts rapidly with the peroxyl radical. Thus, the acid and hydrogen peroxide transform AmO$^\bullet$, which is unreactive with respect to RO$_2^\bullet$, into AmOH$^{\bullet+}$ and AmOH, and the latter inhibit efficiently the oxidation of hydrocarbons. The kinetic characteristics of the inhibited oxidation are fully consistent with the suggested scheme for the cyclic chain termination. When the concentration of hydrogen peroxide is relatively low, the chain termination rate is limited by the stage ([AmO$^{\bullet+}$, A$^-$] + H$_2$O$_2$), and then the function F (see chapter 14) has the following form [45]:

$$F[H_2O_2] = 2k_b K_a (2k_t v_i)^{-1/2} [AmO^\bullet][HA][H_2O_2] \qquad (16.3)$$

If the hydrogen peroxide concentration is large, the exchange reaction between $RO_2^\bullet$ and H_2O_2 occurs rapidly, and this reaction becomes the rate-limiting stage of cyclic chain termination.

Recently an analogous mechanism for cyclic chain termination has been established for quinones [47]. Quinones, which can act as acceptors of alkyl radicals, do not practically retard the oxidation of hydrocarbons at concentrations of up to 5×10^{-3} mol L^{-1}, because the alkyl radicals react very rapidly with dioxygen. However, the ternary system, N-phenylquinone imine (Q) + H_2O_2 + acid (HA), efficiently retards the initiated oxidation of methyl oleate and ethylbenzene [47]. This is indicated by the following results obtained for the oxidation of ethylbenzene (343 K, $pO_2 = 98$ kPa, $v_i = 5.21 \times 10^{-7}$ mol L^{-1} s^{-1}).

[Q] × 10^4 (mol L^{-1})	[HA] × 10^4 (mol L^{-1})	[H_2O_2] × 10^3 (mol L^{-1})	v × 10^7 (mol L^{-1} s^{-1})
0.0	0.0	0.0	4.92
5.0	0.0	0.0	3.01
0.0	0.0	1.25	3.27
0.0	5.0	0.0	3.93
5.0	5.0	1.25	2.39
5.0	5.0	12.5	0.69

The following mechanism, analogous to the mechanism involving nitroxyl radicals, was suggested for the retarding influence of this system [47]:

$$Q + HA \rightleftharpoons [QH^+, A^-] \qquad K_a$$
$$RO_2^\bullet + [QH^+, A^-] \longrightarrow ROOH + [Q^{\bullet+}, A^-] \qquad k_b$$
$$[Q^{\bullet+}, A^-] + H_2O_2 \longrightarrow {}^\bullet QH + HA + O_2 \qquad k_c$$
$$RO_2^\bullet + HQ^\bullet \longrightarrow ROOH + Q \qquad \text{Fast}$$
$$[Q^{\bullet+}, A^-] + HR \longrightarrow R^\bullet + HA + Q \qquad k_p$$

In conformity with this mechanism, the function F depends on the concentrations of the components of the inhibiting mixture in the following way:

$$F[H_2O_2] = \frac{2k_b k_c K_a [Q][HA]}{(k_c[H_2O_2] + k_p[RH])\sqrt{2k_t v_i}} \qquad (16.4)$$

Experimental data on the kinetics of oxidation of methyl oleate and ethylbenzene are in good agreement with the proposed scheme. The cyclic mechanism of chain termination with quinone imines involving acid catalysis is very important for an understanding of the retarding influence of aromatic amines on the oxidation of hydrocarbons and of other organic compounds via the reactions with hydrogen peroxide. Hydrogen peroxide is formed during the oxidation of alcohols due to the decomposition of the alkylhydroxyperoxyl radical to yield $HO_2^\bullet$ and ketone and due to the decomposition of hydroperoxide $>C(OH)OOH$ giving ketone and hydrogen peroxide. In addition, hydrogen peroxide is formed in the hydrolysis of alkyl hydroperoxide through the action of water or acids, which also arise during oxidation of hydrocarbons. Thus, two of the three components needed for the realization of the cyclic chain termination mechanism involving acid catalysis result from the oxidation of the hydrocarbon, while the third component, namely, quinone imine results from the oxidation of the starting inhibitor (aromatic amine). It is possibly owing to this mechanism that aromatic amines exhibit high inhibiting capacity during oxidation at elevated temperatures (>400 K).

16.5 TRANSITION METAL IONS AS CATALYSTS FOR CYCLIC CHAIN TERMINATION

Compounds of transition metals (Mn, Cu, Fe, Co, Ce) are well known as catalysts for the oxidation of hydrocarbons and aldehydes (see Chapter 10). They accelerate oxidation by destroying hydroperoxides and initiating the formation of free radicals. Salts and complexes containing transition metals in a lower-valence state react rapidly with peroxyl radicals and so when these compounds are added to a hydrocarbon prior to its oxidation an induction period arises [48]. Chain termination occurs stoichiometrically $(f \sim 1)$ and stops when the metal passes to a higher-valence state due to oxidation. On the addition of an initiator or hydroperoxide, the induction period disappears.

Another situation is observed when salts or transition metal complexes are added to an alcohol (primary or secondary) or alkylamine subjected to oxidation: in this case, a prolonged retardation of the initiated oxidation occurs, owing to repeated chain termination. This was discovered for the first time in the study of cyclohexanol oxidation in the presence of copper salt [49]. Copper and manganese ions also exert an inhibiting effect on the initiated oxidation of 1,2-cyclohexadiene [12], aliphatic amines [19], and 1,2-disubstituted ethenes [13]. This is accounted for, first, by the dual redox nature of the peroxyl radicals $HO_2^\bullet$, $>C(OH)O_2^\bullet$, and $>C(NHR)O_2^\bullet$, and, second, for the ability of ions and complexes of transition metals to accept and release an electron when they are in an higher- and lower-valence state.

$$ROOH \xleftarrow{H^+} RO\bar{O} \quad Cu^{2+} \quad HO_2^\bullet$$
$$RO_2^\bullet \quad Cu^{1+} \quad O_2 + H^+$$

Table 16.5 presents the results of the kinetic study of such reactions. The measured rate constants are effective values, since both forms of transition metal react with the peroxyl radical.

$$k = \frac{k_1[Me^{n+}] + k_2[Me^{n+1}]}{[Me^{n+}] + [Me^{n+1}]} \qquad (16.5)$$

Copper and manganese compounds exhibit the highest inhibiting activity.

The chain termination on variable-valence metals in the systems in which hydroperoxyl and hydroxyperoxyl radicals act as chain-propagating species is characterized by very high coefficients: $f \sim 10^4 - 10^6$. Both reactions may include several stages. If an aqueous solution of a copper salt, for example, sulfate, is used, the following mechanism involving the incorporation of $HO_2^\bullet$ into the inner coordination sphere of the metal ion is possible:

$$HO_2^\bullet + Cu^{2+}(H_2O)_6 \rightleftharpoons Cu^{2+}(H_2O)_5(HO_2^\bullet) + H_2O$$
$$Cu^{2+}(H_2O)_5(HO_2^\bullet) + H_2O \rightleftharpoons Cu^{2+}(H_2O)_5(O_2^{\bullet-}) + H_3O^+$$
$$Cu^{2+}(H_2O)_5(O_2^{\bullet-}) \longrightarrow Cu^+(H_2O)_5(O_2)$$
$$Cu^+(H_2O)_5(O_2) + H_2O \longrightarrow Cu^+(H_2O)_6 + O_2$$

TABLE 16.5
Rate Constants for the Reactions of Transition Metal Salts and Complexes with Peroxyl Radicals in Systems with a Cyclic Mechanism of Chain Termination (Experimental Data)

Metal Complex	Oxidized Substrate	T (K)	k (L mol^{-1} s^{-1})	Ref.
Mn(OAc)$_2$	Bu$_2$NH	348	2.5×10^6	[19]
Mn(OAc)$_2$	Me$_2$NCH$_2$CH$_2$NMe$_2$	313	1.1×10^6	[16]
Mn(OAc)$_2$	cyclohexylamine	348	7.3×10^7	[18,20]
Mn(OAc)$_2$	PhCH$_2$NH$_2$	338	2.8×10^8	[20]
Mn(OAc)$_2$	CH=CMeC(O)OC$_2$H$_4$NMe$_2$	323	1.2×10^7	[15]
Mn(OAc)$_2$	PrOC(O)C$_2$H$_4$NMe$_2$	323	1.5×10^7	[15]
Mn(OC(O)C$_{17}$H$_{35}$)$_2$	cyclohexene	348	1.9×10^6	[12]
Mn(OC(O)C$_{17}$H$_{35}$)$_2$	Bu$_2$NH	348	3.5×10^6	[20,50]
Mn(OC(O)C$_{17}$H$_{35}$)$_2$	cyclohexylamine	348	1.6×10^8	[20,50]
Mn(OC(O)C$_{17}$H$_{35}$)$_2$	cyclohexanol	348	6.8×10^6	[51]
Mn(OC(O)C$_{17}$H$_{35}$)$_2$	MeC(O)NEt$_2$	348	7.8×10^5	[19]
Mn(acac)$_2$	cyclohexylamine	363	9.8×10^7	[52]
Mn(salen)	cyclohexylamine	348	3.1×10^8	[52]
MnSt$_2$ with 2,4-dioxo-1,5,8,12-tetraazatetradecane	cyclohexanol	343	5.4×10^7	[53]
MnSt$_2$ with 2,4-dioxo-1,5,8,12-tetraazatetradecane	cyclohexanol	363	4.3×10^7	[53]
Fe(OC(O)C$_{17}$H$_{35}$)$_3$	cyclohexanol	348	1.3×10^4	[52]
Fe(OC(O)C$_{17}$H$_{35}$)$_3$	cyclohexylamine	348	1.2×10^6	[19,20]
Fe(acac)$_3$	cyclohexylamine	348	1.0×10^5	[19,20]
Fe(DMG)$_2$Py$_2$	cyclohexene	348	2.3×10^3	[54]
Fe(DMG)$_2$Py$_2$	Me$_2$CHOH	344	1.0×10^3	[55]
CoCl$_2$	cyclohexanol	348	6.1×10^4	[19]
Co(OAc)$_2$	cyclohexylamine	348	1.2×10^5	[19,20]

continued

TABLE 16.5
Rate Constants for the Reactions of Transition Metal Salts and Complexes with Peroxyl Radicals in Systems with a Cyclic Mechanism of Chain Termination (Experimental Data)—*continued*

Metal Complex	Oxidized Substrate	T (K)	k (L mol^{-1} s^{-1})	Ref.
Co(OC(O)C$_{17}$H$_{35}$)$_2$	cyclohexanol	348	2.8×10^5	[51]
Co(OC(O)C$_{17}$H$_{35}$)$_2$	cyclohexylamine (N-H)	348	3.2×10^4	[20]
Co(acac)$_2$ dimer structure	Me$_2$CHOH	344	2.9×10^2	[55]
Co(acac)$_2$	cyclohexanol	348	6.4×10^6	[20]
Co(acac)$_2$	cyclohexylamine	348	6.4×10^4	[20]
Co(OC(O)C$_6$H$_{11}$)$_2$	cyclohexanol	348	8.7×10^4	[19]
Co(II)(DMG)$_2$NH$_3$Cl	cyclohexene	348	8.4×10^2	[54]
Co(II)(DMG)$_2$NH$_3$Cl	Me$_2$CHOH	344	5.0×10^3	[55]
Co(II)(DMG)$_2$NH$_3$I	cyclohexene	348	2.3×10^4	[54]
Co(II)(DMG)$_2$NH$_3$I	Me$_2$CHOH	344	3.1×10^2	[55]
Co(II)(DMG)$_2$Ipy	Me$_2$CHOH	344	7.0×10^4	[55]
Co(II)(DMG)$_2$Py$_2$	Me$_2$CHOH	344	1.2×10^4	[55]
Co(salen)	cyclohexanol	348	3.3×10^5	[52]
Co(II)Porphirine	cyclohexanol	348	3.6×10^4	[52]
CoSt$_2$ with 2,4-dioxo-1,5,8,12-tetraazatetradecane	cyclohexanol	343	1.5×10^5	[53]
CoSt$_2$ with 2,4-dioxo-1,5,8,12-tetraazatetradecane	cyclohexanol	363	1.1×10^5	[53]
Ni(salicylate) complex	C$_7$H$_{15}$CH$_2$OH	363	1.6×10^4	[52]
Ni(salicylate) complex	cyclohexylamine	353	2.8×10^4	[52]

TABLE 16.5
Rate Constants for the Reactions of Transition Metal Salts and Complexes with Peroxyl Radicals in Systems with a Cyclic Mechanism of Chain Termination (Experimental Data)—continued

Metal Complex	Oxidized Substrate	T (K)	k (L mol^{-1} s^{-1})	Ref.
Ni(salen) complex	$C_8H_{17}OH$	363	7.9×10^3	[52]
Ni(salen) complex	$C_6H_{11}NH_2$ (cyclohexylamine)	353	1.5×10^4	[52]
Ni(salen)·H$_2$ complex	$C_6H_{11}NH_2$	353	1.7×10^5	[52]
$CuSO_4$	cyclohexanol	353	8.8×10^6	[49]
$CuCl_2$	trans-PhCH=CHC(O)OEt	348	5.9×10^5	[13]
$CuCl_2$	trans-PhCH=CHC(O)OMe	323	5.6×10^5	[13]
$CuCl_2$	trans-PhCH=CHPh	323	2.6×10^5	[13]
$Cu(OAc)_2$	$PhCH_2NH_2$	323	1.5×10^8	[19]
$Cu(OAc)_2$	$PrC(O)OC_2H_4NMe_2$	338	1.3×10^7	[56]
$Cu(OAc)_2$	$BuNMe_2$	323	1.0×10^7	[56]
$Cu(OAc)_2$	$C_6H_5CH=CHCOOC_2H_5$	323	2.2×10^5	[13]
$Cu(OAc)_2$	PhCH=CHC(O)OMe	323	4.5×10^5	[13]
$Cu(OAc)_2$	PhCH=CHPh	323	5.3×10^5	[13]
$Cu(OC(O)C_{17}H_{35})_2$	cyclohexene	348	4.5×10^6	[12]
$Cu(OC(O)C_{17}H_{35})_2$	cyclohexanol	348	2.9×10^6	[51]
$Cu(OC(O)C_{17}H_{35})_2$	Bu_2NH	348	4.1×10^6	[19]
$Cu(OC(O)C_{17}H_{35})_2$	cyclohexylamine	348	9.5×10^7	[20]

DMG, dimethyglionime; StH, stearic acid.

In the case of metal complexes with strongly bound ligands, the out-of-sphere transfer of an electron from the radical to the ion and back is more probable.

When variable-valence metals are used as catalysts in the oxidation of hydrocarbons, the chain termination via such reactions manifests itself later in the process. This case has specially been studied in relation to the oxidation of paraffins to fatty acids in the presence of the K–Mn catalyst [57], which ensures a high oxidation rate and a high selectivity of formation of the target product (carboxylic acids). As the reaction occurs, alcohols are accumulated in the reaction mixture, and their oxidation is accompanied by the formation of hydroxyperoxyl radicals. The more extensively the oxidation occurs, the higher the concentration of alcohols in the oxidized paraffin, and, hence, the higher is the kinetic

TABLE 16.6
Cyclic mechanisms of Chain Termination in the Liquid-Phase Oxidation of Organic Compounds

Inhibiting System	Oxidized Reactant	Mechanism	Ref.
AmH, Am$^\bullet$	Olefins	$HO_2^\bullet + AmH \longrightarrow H_2O_2 + Am^\bullet$ $Am^\bullet + HO_2^\bullet \longrightarrow AmH + O_2$	[8,12,13]
AmH, Am$^\bullet$	Alcohols	$>C(OH)OO^\bullet + AmH \longrightarrow >C(OH)OOH + Am^\bullet$ $>C(OH)OO^\bullet + Am^\bullet \longrightarrow >C=O + O_2 + AmH$	[3,8,9,14,44,61]
AmH, Am$^\bullet$	Amines	$>C(NHR)OO^\bullet + AmH \longrightarrow >C(NHR)OOH + Am^\bullet$ $>C(NHR)OO^\bullet + Am^\bullet \longrightarrow >C=NR + O_2 + AmH$	[15–21]
AmO$^\bullet$, AmOH	Polypropylene	$AmO^\bullet + HO_2^\bullet \longrightarrow AmOH + O_2$ $AmOH + HO_2^\bullet \longrightarrow AmO^\bullet + H_2O_2$	[12,13,43]
AmO$^\bullet$, AmOH	Alcohols	$>C(OH)OO^\bullet + AmO^\bullet \longrightarrow >C=O + O_2 + AmOH$ $>C(OH)OO^\bullet + AmOH \longrightarrow >C(OH)OOH + AmO^\bullet$	[9,62]
AmO$^\bullet$, AmOH	Amines	$>C(NHR)OO^\bullet + AmO^\bullet \longrightarrow >C=NR + AmOH + O_2$ $>C(NHR)OO^\bullet + AmOH \longrightarrow >C(NHR)OOH + AmO^\bullet$	[15,19,20,24]
Q, HQ$^\bullet$	Olefins	$Q + HO_2^\bullet \longrightarrow HQ^\bullet + O_2$ $HQ^\bullet + HO_2^\bullet \longrightarrow Q + H_2O_2$	[9,12,13,63] [9,63]
Q, HQ$^\bullet$	Alcohols	$>C(OH)OO^\bullet + Q \longrightarrow >C=O + O_2 + HQ^\bullet$ $>C(OH)OO^\bullet + HQ^\bullet \longrightarrow >C(OH)OOH + Q$	[31–34,63]
Q, HQ$^\bullet$	Amines	$>C(NHR)OO^\bullet + Q \longrightarrow >C=NR + O_2 + HQ^\bullet$ $>C(NHR)OO^\bullet + HQ^\bullet \longrightarrow >C(NHR)OOH + Q$	[16,18–21]
RNO_2, $RN(O)O^\bullet$	Ester	$RNO_2 + HO_2^\bullet \longrightarrow RN(OH)O^\bullet + O_2$ $RN(OH)O^\bullet + HO_2^\bullet \longrightarrow RNO_2 + HOOH$	[37]
RNO_2, $RN(O)O^\bullet$	Polypropylene	$RNO_2 + HO_2^\bullet \longrightarrow RN(OH)O^\bullet + O_2$ $RN(OH)O^\bullet + HO_2^\bullet \longrightarrow RNO_2 + HOOH$	[38]
Me^{n+}, Me^{n+1}	Olefins	$HO_2^\bullet + Me^{n+} \longrightarrow Me^{n+1} + HO_2^-$ $HO_2^\bullet + Me^{n+1} \longrightarrow H^+ + O_2 + Me^{n+}$	[12,13,54]
Me^{n+}, Me^{n+1}	Alcohols	$>C(OH)OO^\bullet + Me^{n+} \longrightarrow >C(OH)\,O_2^- + Me^{n+1}$ $>C(OH)OO^\bullet + Me^{n+1} \longrightarrow >C=O + O_2 + H^+ + Me^{n+}$	[19,49–56]
Me^{n+}, Me^{n+1}	Amines	$>C(NHR)OO^\bullet + Me^{n+} \longrightarrow >C(NHR)\,O_2^- + Me^{n+1}$ $>C(NHR)OO^\bullet + Me^{n+1} \longrightarrow >C=NR + O_2 + H^+ + Me^{n+}$	[15,19,20,50,51,54]
AmO$^\bullet$, AmOR	Polymer	$AmO^\bullet + R^\bullet \longrightarrow AmOR$ $AmOR + RO_2^\bullet \longrightarrow ROOH + Alkene + AmO^\bullet$	[40–43,63–65]
AmO$^\bullet$, AmOR	Polymer	$AmO^\bullet + R^\bullet \longrightarrow AmOR$ $AmOR \longrightarrow [AmO^\bullet + R^\bullet] \longrightarrow AmOH + Ketone$ $AmOH + RO_2^\bullet \longrightarrow ROOH + AmO^\bullet$	[66–70]
AmO$^\bullet$, AmOR	Polymer	$AmO^\bullet + R^\bullet \longrightarrow AmOR$ $AmOR \longrightarrow [Am^\bullet + RO^\bullet] \longrightarrow AmH + Ketone$ $AmH + RO_2^\bullet \longrightarrow ROOH + AmO^\bullet$ $Am^\bullet + RO_2^\bullet \longrightarrow AmO^\bullet + RO^\bullet$	[42,43]
I_2	Hydrocarbon	$R^\bullet + I_2 \longrightarrow RI + I^\bullet$ $RI + RO_2^\bullet \longrightarrow R^\bullet + ROOI$ $2ROOI \longrightarrow I_2 + O_2 + ROOR$	[71]
AmO$^\bullet$, H_2O_2, HA	Hydrocarbon	$AmO^\bullet + HA \leftrightarrow [AmOH^{+\bullet}, A^-]$ $RO_2^\bullet + [AmOH+^\bullet, A^-] \longrightarrow ROOH + [AmO^+, A^-]$ $[AmO^+, A^-] + H_2O_2 \longrightarrow AmOH + HA + O_2$ $RO_2^\bullet + AmOH \longrightarrow ROOH + AmO^\bullet$	[10,45,46]
AmO$^\bullet$, HA, $>CH(OOH)OH$	Hydrocarbon + Alcohol	$AmO^\bullet + HA \leftrightarrow [AmOH^{+\bullet}, A^-]$ $RO_2^\bullet + [AmOH^{+\bullet}, A^-] \longrightarrow ROOH + [AmO^+, A^-]$ $[AmO^+, A^-] + >C(OH)OOH \longrightarrow AmOH + HA + O_2 + >C=O$ $RO_2^\bullet + AmOH \longrightarrow ROOH + AmO^\bullet$	[47]

TABLE 16.6
Cyclic mechanisms of Chain Termination in the Liquid-Phase Oxidation of Organic Compounds—*continued*

Inhibiting System	Oxidized Reactant	Mechanism	Ref.
Q, H_2O_2, HA	Hydrocarbon	$Q + HA \leftrightarrow (QH^+, A^-)$ $(QH^+, A^-) + RO_2^\bullet \longrightarrow (Q^{+\bullet}, A^-) + ROOH$ $(Q^{+\bullet}, A^-) + H_2O_2 \longrightarrow HQ^\bullet + HA + O_2$ $RO_2^\bullet + HQ^\bullet \longrightarrow ROOH + Q$	[47]
B	Alcohol	$>C(OH)O_2^\bullet + B \longrightarrow >C=O + O_2^{\bullet -} + BH^+$ $>C(OH)O_2^\bullet + O_2^{\bullet -} \longrightarrow >C(O) + O_2 + HO_2^-$	[58–60]

equilibrium concentration of $HO_2^\bullet$ radicals. The latter enter into the cyclic reactions of chain termination with Mn^{2+} and Mn^{3+} ions and thus cause chain termination. When the rate of chain termination with the hydroxyperoxyl radicals becomes equal to the rate of generation of the chain-propagating radicals, the oxidation virtually ceases.

The oxidation of alcohols is also retarded by bases [58–60], which is apparently due to the decomposition of the hydroxyperoxyl radical giving ketone and ion $O_2^{\bullet -}$.

$$B + HO_2^\bullet \longrightarrow BH^+ + O_2^{\bullet -}$$

The superoxide ion is a very weak hydrogen atom abstractor, which cannot continue the chain, and is destroyed via disproportionation with any peroxyl radical. So, the studies of the mechanisms of cyclic chain termination in oxidation processes demonstrate that they, on the one hand, are extremely diverse and, on the other, that they are highly structurally selective. The 20 currently known mechanisms are presented in Table 16.6.

REFERENCES

1. ET Denisov, VV Kharitonov. *Izv AN SSSR Ser Khim* 2222–2225, 1963.
2. ET Denisov, VP Scheredin. *Izv AN SSSR Ser Khim* 919–921, 1964.
3. VV Kharitonov, ET Denisov. *Izv AN SSSR Ser Khim* 2764–2766, 1967.
4. VV Shalya, BI Kolotusha, FA Yampol'skaya, YaB Gorokhovatskiy. *Kinet Katal* 10:1090–1096, 1969.
5. ET Denisov, IV Khudyakov. *Chem Rev* 1313–1357, 1987.
6. ET Denisov, TG Denisova. *Handbook of Antioxidants*. Boca Raton: CRC Press, 2000.
7. ET Denisov, VV Azatyan. *Inhibition of Chain Reactions*. London: Gordon and Breach, 2000.
8. RL Vardanyan, VV Kharitonov, ET Denisov. *Izv AN SSSR Ser Khim* 1536–1542, 1970.
9. ET Denisov, VI Goldenberg, LG Verba. *Izv AN SSSR Ser Khim* 2217–2223, 1988.
10. VI Goldenberg, ET Denisov, LG Verba. *Izv AN SSSR Ser Khim* 2223–2226, 1988.
11. VT Varlamov. Elementary Reactions and Products of the Liquid-Phase Conversion of Diphenylaminyl Radical. Ph.D. Thesis, Institute of Chemical Physics, Chernogolovka, 1986, pp. 1–21 [in Russian].
12. RL Vardanyan, ET Denisov. *Izv Akad Nauk SSSR Ser Khim* 2818–2820, 1971.
13. EM Pliss. Oxidation of vinyl compounds: mechanism, elementary reactions, structure and reactivity, Doctoral Thesis, Institute of Chemical Physics, Chernogolovka, 1990, pp. 1–40 [in Russian].

14. RL Vardanyan, VV Kharitonov, ET Denisov. *Neftekhimiya* 11:247–252, 1971.
15. EM Pliss, AL Aleksandrov, MM Mogilevich. *Izv AN SSSR Ser Khim* 1441–1443, 1977.
16. UYa Samatov, AL Aleksandrov, IR Akhyndov. *Izv AN SSSR Ser Khim* 2254–2258, 1978.
17. AL Aleksandrov, LD Krisanova. *Izv AN SSSR Ser Khim* 2469–2474, 1980.
18. GA Kovtun, AL Aleksandrov. *Izv AN SSSR Ser Khim* 1274–1279, 1974.
19. AL Aleksandrov. Negative Catalysis in Radical-chain Reactions of Oxidation of N- and O-Containing Compounds, Doctoral Dissertation, Institute of Chemical Physics, Chernogolovka, 1987, pp. 1–40 [in Russian].
20. GA Kovtun. Mechanism of Oxidation of Aliphatic Amines and Regeneration of Antioxidants, Cand. Sci. (Chem.) Thesis Dissertation, Institute of Chemical Physics, Chernogolovka, 1974, pp. 1–23 [in Russian].
21. EM Pliss, AL Aleksandrov, VS Mikhlin, MM Mogilevich. *Izv AN SSSR Ser Khim* 2259–2262, 1978.
22. ET Denisov. *Russ Chem Bull* 45:1870–1874, 1996.
23. ET Denisov. *Russ Chem Rev* 65:505–520, 1996.
24. ET Denisov. *Kinet Catal* 38:236–244, 1997.
25. ET Denisov. *Kinet Catal* 36:345–350, 1995.
26. ET Denisov. *Kinet Catal* 38:28–34, 1997.
27. VT Varlamov, RL Safiulin, ET Denisov. *Khim Fiz* 2:408–412, 1983.
28. ET Denisov. *Kinet Catal* 37:519–523, 1996.
29. DR Lide. *Handbook of Chemistry and Physics*. Boca Raton: CRC Press, 1992.
30. AL Aleksandrov. *Izv AN SSSR Ser Khim* 2720–2726, 1980.
31. ET Denisov. *Izv AN SSSR Ser Khim* 328–331, 1969.
32. YuB Shilov, ET Denisov. *Vysokomol Soed A*26:1753–1758, 1984.
33. RL Vardanyan, ET Denisov, VI Zozulya. *Izv AN SSSR Ser Khim* 611–613, 1972.
34. TG Degtyareva, ET Denisov, VS Martemyanov, LYa Badretdinova. *Izv AN SSSR Ser Khim* 1219–1225, 1979.
35. MM Mogilevich, EM Pliss. *Oxidation and Oxidative Polymerization of Nonsaturated Compounds*. Moscow: Khimiya, 1990 [in Russian].
36. YuB Shilov, ET Denisov. *Vysokomol Soed A*24:837–842, 1984.
37. MI Borisov, ET Denisov, ZF Sharafuddinova. *Petrol Chem* 40:166–168, 2000.
38. YuB Shilov, ET Denisov. *Kinet Catal* 42:238–242, 2001.
39. YuB Shilov, ET Denisov. *Vysokomol Soed A*16:2313–2316, 1974.
40. YuB Shilov, RM Battalova, ET Denisov. *Dokl AN SSSR* 207:388–389, 1972.
41. AV Kirgin, YuB Shilov, BE Krisyuk, AA Efimov, VV Pavlikov. *Kinet Catal* 31:52–57, 1990.
42. S Korcek, RK Jensen, M Zinbo, JL Gerlock. In: H Fischer (ed.) *Organic Free Radicals*. Berlin: Springer-Verlag, 1988, p. 95.
43. RK Jensen, S Korcek, M Zinbo, JL Gerlock. *J Org Chem* 60:5396–5400, 1995.
44. VI Goldenberg, NV Katkova, ET Denisov. *Bull Acad Sci USSR* 37:214–219, 1988.
45. VI Goldenberg, ET Denisov, NA Ermakova. *Bull Acad Sci USSR* 39:651–656, 1990.
46. ET Denisov. In: Thesis of 12th International Conference on Advances in the Stabilization and Controlled Degradation of Polymers. Luzern, Switzerland, 1990, pp. 77–82.
47. VI Goldenberg, NA Ermakova, ET Denisov. *Russ Chem Bull* 44:74–77, 1995.
48. NM Emanuel, ET Denisov, ZK Maizus. *Liquid-Phase Oxidation of Hydrocarbons*. New York: Plenum Press, 1967.
49. AL Aleksandrov, ET Denisov. *Izv AN SSSR Ser Khim* 652–1657, 1969.
50. GA Kovtun, AL Aleksandrov. *Izv AN SSSR Ser Khim* 2611–2613, 1973.
51. AL Aleksandrov, GI Solov'ev, ET Denisov. *Izv AN SSSR Ser Khim* 527–1533, 1972.
52. GA Kovtun. Complexes of Transition Metals as Catalysts of Chain Termination in Oxidation, Doct. Sci. (Chem.) Thesis Dissertation, Institute of General and Inorganic Chemistry, Moscow, 1984, pp. 1–49 [in Russian].
53. AL Aleksandrov, EP Marchenko, TS Pokidova. *Kinet Katal* 36:825–830, 1995.
54. NG Zubareva, ET Denisov, AV Ablov. *Kinet Katal* 14:579–583, 1973.
55. NG Zubareva, ET Denisov, AV Ablov. *Kinet Katal* 14:346–351, 1973.
56. EM Pliss, AL Aleksandrov. *Izv AN SSSR Ser Khim* 214–216, 1978.

57. AV Oberemko, AA Perchenko, ET Denisov, AL Aleksandrov. *Neftekhimiya* 11:229–235, 1971.
58. ET Denisov, VM Solyanikov, AL Aleksandrov. *Adv Chem Ser* 75(1):112–119, 1968.
59. AL Aleksandrov, ET Denisov. *Dokl AN SSSR* 178:379–382, 1968.
60. AL Aleksandrov, ET Denisov. *Izv AN SSSR Ser Khim* 2322–2324, 1969.
61. ET Denisov. *Kinet Katal* 11:312–320, 1970.
62. GA Kovtun, VA Golubev, AL Aleksandrov. *Izv AN SSSR Ser Khim* 793–799, 1974.
63. YuB Shilov, ET Denisov. *Vysokomol Soed A*29:1359–1363, 1987.
64. ET Denisov. *Polym Degrad Stabil* 25:209–215, 1989.
65. ET Denisov. *Polym Degrad Stabil* 34:325–332, 1991.
66. DJ Carlsson, A Garton, DM Willes. In: G. Scott (ed.) *Developments in Polymer Stabilisation-1*. London: Applied Science, 1979, pp. 219–246.
67. KB Chakraborty, G Scott. *Polymer* 21:252–253, 1980.
68. H Berger, TA Bolsman, DM Brower. In: G Scott (ed.) *Developments in Polymer Stabilisation-6*. London: Applied Science, 1983, pp. 1–27.
69. G Scott. *Br Polym J* 15:208–223, 1983.
70. DJ Carlsson, DM Willes. *Polym Degrad Stabil* 6:1–12, 1984.
71. AL Aleksandrov, TI Sapacheva, ET Denisov. *Neftekhimiya* 10:711–716, 1970.

17 Hydroperoxide Decomposing Antioxidants

17.1 INTRODUCTION

Autoxidation of organic compounds occurs as an autoinitiated process where hydroperoxides are the major source of free radicals (see Chapter 6). This fact determines the kinetic pattern of autoxidation, whose rate is proportional to either $[ROOH]^{1/2}$ or $[ROOH]$, depending on the mechanism of chain initiation and termination. One can anticipate that autoxidation can be affected by factors influencing the concentration and decay of hydroperoxides. It should be taken into account here that hydroperoxide can decompose either homolytically or heterolytically. In the former case, the resulting free radicals accelerate oxidation (see Chapter 6). On the other hand, the heterolytic decomposition of hydroperoxide into molecular products (the so-called molecular decomposition) decreases the concentration of hydroperoxide and, hence, reduces the rate of autoinitiation (see Chapter 10). Namely, heterolytic decomposers of hydroperoxides can retard the autoxidation of organic compounds. Since as hydroperoxides are oxidizing reactants the substances with reducing activity should be good hydroperoxide decomposers. Among them, phosphorus and sulfur compounds are used as antioxidants of organic compounds. The detailed kinetic study of their action evidenced that they are antioxidants of a complex mechanism of inhibiting action. They react with hydroperoxide, as well as with peroxyl radicals, and, in addition, produce products with antioxidant properties.

17.2 ORGANOPHOSPHORUS ANTIOXIDANTS

17.2.1 REACTION WITH HYDROPEROXIDES

Aryl phosphites are widely used as inhibitors of oxidation. Phosphites react with hydroperoxides and are oxidized into phosphates by the following stoichiometric reaction [1–10]:

$$R'OOH + P(OR)_3 \longrightarrow R'OH + O{=}P(OR)_3$$

The stoichiometry of this reaction is usually close to unity [6–9]. Thus, cumyl hydroperoxide oxidizes triphenyl phosphite in the stoichiometry $\Delta[ROOH]/\Delta[Ph_3P]$ from 1.02:1 to 1.07:1, depending on the proportion between the reactants [6]. The reaction proceeds as bimolecular. The oxidation of phosphite by hydroperoxide proceeds mainly as a heterolytic reaction (as follows from conservation of the optical activity of reaction products [5,11]). Oxidation is faster in more polar solvents, as evident from the comparison of k values for benzene and chlorobenzene. Heterolysis can occur via two alternative mechanisms [1–7].

$$R^1OOH + P(OR)_3 \longrightarrow [R^1OP^+(OR)_3, HO^-]$$
$$[R^1OP^+(OR)_3, HO^-] \longrightarrow R^1OH + O{=}P(OR)_3$$
$$R^1OOH + P(OR)_3 \longrightarrow [HOP^+(OR)_3, R^1O^-]$$
$$[HOP^+(OR)_3, R^1O^-] \longrightarrow R^1OH + O{=}P(OR)_3,$$

as well as via the coordinated heterolysis of the O—H and O—O bonds in TS.

$$ROOH + P{\diagup}^{\diagdown} \longrightarrow RO{\cdots}\overset{H}{O}{\cdots}P{\diagup}^{\diagdown} \longrightarrow ROH + O{=}P{\diagup}^{\diagdown}$$

The rate of the reaction is proportional to the product of concentrations of two reactants.

$$v = k[ROOH][P(OR^1)_3] \tag{17.1}$$

Trialkyl phosphites were found to be more reactive to hydroperoxide than aryl phosphites; the activation energy depends on a particular phosphite and ranges from 25 to 77 kJ mol^{-1}. At the same time, the reaction depends weakly on the type of hydroperoxide, which can be seen from the comparison of rate constants for the reactions of two hydroperoxides (solvent was PhCl, $T = 303$ K [12]).

R_1	R_2	PhMe$_2$COOH, k (L mol^{-1} s^{-1})	Me$_3$COOH, k (L mol^{-1} s^{-1})
H	n-C$_6$H$_{13}$	1.36	0.85
H	Ph	0.22	0.13
Ph	Ph	6.0×10^{-2}	5.1×10^{-2}
cyclo-C$_6$H$_{11}$	cyclo-C$_6$H$_{11}$	5.3×10^{-2}	2.6×10^{-2}

The values of rate constants of the hydroperoxide reaction with phosphites are given in Table 17.1.

The concurrent slow homolytic reaction gives rise to free radicals [14]. The occurrence of the homolytic reaction can be revealed by the consumption of free radical acceptors [8,15], CL [16], or NMR spectroscopy [17,18]. The introduction of phosphite into the hydroperoxide-containing cumene causes an initiation, pro-oxidative effect related to the formation of free radicals [6]. The yield of radicals from aliphatic phosphites is much lower (0.01–0.02%) than that from aromatic phosphites (up to 5%) [17]. The homolytic reaction of phosphites with hydroperoxide has a higher activation energy than the heterolytic reaction, which results in the predominance of the former reaction at elevated temperatures.

Once aryl phosphite has oxidized into phosphate followed by the catalytic decomposition of hydroperoxide [6,10,19,20]. The catalyst is obviously stable, since additional hydroperoxide introduced into the system immediately breaks down [10]. Repeatedly, ROOH decomposes via the acid-catalyzed reaction with the products of phosphate hydrolysis acting as catalysts [6,10].

TABLE 17.1
Rate Constants of Reactions of Cumylhydroperoxide with Phosphites

Phosphite	Solvent	E (kJ mol^{-1})	log A, A (L mol^{-1} s^{-1})	k (303 K) (L mol^{-1} s^{-1})	Ref.
$(C_4H_9O)_3P$	PhCl	24.7	3.50	0.18	[7]
(tricyclohexyl phosphite)	PhH	25.5	3.58	0.15	[7]
(benzodioxaphosphole, isopropoxy)	PhCl			0.19	[12]
(benzodioxaphosphole, phenoxy)	PhCl			9.9×10^{-2}	[12]
(naphthodioxaphosphole, aryloxy)	PhCl	77.0	11.62	2.23×10^{-2}	[12]
(naphthodioxaphosphole, aryloxy)	PhCl	67.0	9.73	1.52×10^{-2}	[12]
(dioxaphosphorinane, aryloxy)	PhCl	54.0	7.88	3.73×10^{-2}	[12]
(dibenzyl phosphite)	PhH	57.7	7.58	4.29×10^{-3}	[7]
(dibenzyl phosphite, phenoxy)	PhCl	77.0	10.94	4.63×10^{-3}	[12]

continued

TABLE 17.1
Rate Constants of Reactions of Cumylhydroperoxide with Phosphites—continued

Phosphite	Solvent	E (kJ mol^{-1})	log A, A (L mol^{-1} s^{-1})	k (303 K) (L mol^{-1} s^{-1})	Ref.
[structure: 2,2'-biphenylenedioxy phosphite with 2,6-di-tert-butyl-4-methylphenoxy group]	PhH	45.6	5.50	4.41×10^{-3}	[7]
[structure: 2,2'-biphenylenedioxy phosphite with 2,6-di-tert-butyl-4-methylphenoxy group]	PhCl	72.0	10.0	3.87×10^{-3}	[13]
[structure: tri(1-naphthyl) phosphite]	PhH	37.6	4.69	1.62×10^{-2}	[7]

Phosphines react readily with hydroperoxides with a stoichiometric ratio of 1:1 in the following manner [15]:

$$R'OOH + P(C_6H_5)_3 \longrightarrow R'OH + O{=}P(C_6H_5)_3$$

Since this reaction is not affected by hydroquinone and galvinoxyl and does not initiate polymerization of styrene, it obviously occurs without the formation of free radicals. The kinetic parameters of the reactions of three hydroperoxides with triphenyl phosphite in different solvents are given in Table 17.2 [21].

We observe that both the activation energy and preexponential factor increase with the polarity of the solvent. That is probably due to the polar structure of reactants and TS. In hexane, this reaction proceeds via two routes at the total rate v.

$$v = k_1[\text{ROOH}][\text{P(OPh)}_3] + k_2[\text{ROOH}]^2[\text{P(OPh)}_3], \tag{17.2}$$

where $k_2 = 38$ L^2 mol^{-2} s^{-1} at $T = 295$ K [21]. The formal trimolecular reaction is probably the result of hydroperoxide autoprotolysis and reaction of hydroperoxide with triphenylphosphine.

$$\text{ROOH} + \text{ROOH} \longrightarrow \text{ROOH}_2^+ + \text{ROO}^-$$
$$\text{ROOH}_2^+ + \text{P(C}_6\text{H}_5)_3 \longrightarrow \text{ROH}_2^+ + \text{O}{=}\text{P(C}_6\text{H}_5)_3$$

TABLE 17.2
Rate Constants and Activation Energies of Triphenyl Phosphite Reaction with Different Hydroperoxide [21]

Hydroperoxide	Solvent	E (kJ mol^{-1})	log A, A (L mol^{-1} s^{-1})	k (303 K) (L mol^{-1} s^{-1})
PrCH$_2$OOH	Hexane	19.7	4.90	31.9
PrCH$_2$OOH	CH$_2$Cl$_2$	35.1	7.40	22.3
PrCH$_2$OOH	Ethanol	35.1	7.20	14.1
EtMeCHCOOH	Hexane	27.2	5.80	12.9
EtMeCHCOOH	CH$_2$Cl$_2$	42.3	8.30	10.2
EtMeCHCOOH	Ethanol	45.2	9.00	16.1
Me$_3$COOH	Hexane	25.1	5.30	9.42
Me$_3$COOH	CH$_2$Cl$_2$	41.0	7.80	5.39
Me$_3$COOH	Ethanol	46.9	8.80	5.19

This mechanism agrees with the acid catalysis of this reaction when the overall reaction rate in the presence of HClO$_4$ obeys the equation

$$v = k_1[\text{Me}_3\text{COOH}][\text{P(OPh)}_3] + k_3[\text{Me}_3\text{COOH}][\text{P(OPh)}_3][\text{HClO}_4] \quad (17.3)$$

where $k_1 = 1.28$ L mol^{-1} s^{-1} and $k_2 = 24$ L^2 mol^{-2} s^{-1} at $T = 298$ K in ethanol [22]. The complexes of molybdenum, vanadium, and tungsten appeared to be very strong catalysts of this reaction [22]. For example, molybdenyl acetylacetonate catalyses this reaction with $k_3 = 10^5$ L^2 mol^{-2} s^{-1} at $T = 298$ K in ethanol.

17.2.2 Reaction with Peroxyl Radicals

Phosphites can react not only with hydroperoxides but also with alkoxyl and peroxyl radicals [9,14,17,23,24], which explains their susceptibility to a chain-like autoxidation and, on the other hand, their ability to terminate chains. In neutral solvents, alkyl phosphites can be oxidized by dioxygen in the presence of an initiator (e.g., light) by the chain mechanism. Chains may reach 10^4 in length. The rate of oxygen consumption is proportional to $v_i^{1/2}$, thus indicating a bimolecular mechanism of chain termination. The scheme of the reaction

$$\text{RN}{=}\text{NR} \longrightarrow 2\text{R}^\bullet + \text{N}_2$$
$$\text{R}^\bullet + \text{O}_2 \longrightarrow \text{RO}_2^\bullet$$
$$\text{RO}_2^\bullet + \text{P(OR)}_3 \longrightarrow \text{ROOP}^\bullet(\text{OR})_3$$
$$\text{ROOP}^\bullet(\text{OR})_3 \longrightarrow \text{RO}^\bullet + \text{O}{=}\text{P(OR)}_3$$
$$\text{RO}^\bullet + \text{P(OR)}_3 \longrightarrow \text{ROP}^\bullet(\text{OR})_3$$
$$\text{ROP}^\bullet(\text{OR})_3 \longrightarrow \text{R}^\bullet + \text{O}{=}\text{P(OR)}_3$$
$$2\text{RO}_2^\bullet \longrightarrow \text{ROOR} + \text{O}_2$$

is confirmed by both the kinetic patterns of oxidation and experimental data on particular elementary steps [11,14,25]. According to the data of EPR spectroscopy [11], phosphoranyl radicals rapidly decompose.

$$(\text{MeO})_3\text{P}^\bullet\text{OCMe}_3 \longrightarrow (\text{MeO})_3\text{P}{=}\text{O} + {}^\bullet\text{CMe}_3$$
$$k = 3.2 \times 10^{10} \exp(-33.0/RT) = 5.3 \times 10^4 \text{ s}^{-1} (298 \text{ K})$$

$$(EtO)_3P^\bullet OCMe_3 \longrightarrow (EtO)_3P{=}O + {}^\bullet CMe_3$$
$$k = 2.0 \times 10^{10} \exp(-31.4/RT) = 6.3 \times 10^4 \text{ s}^{-1} (298 \text{ K})$$
$$(MeO)_4P^\bullet \longrightarrow (MeO)_3P{=}O + MeO^\bullet$$
$$k = 2.0 \times 10^{11} \exp(-33.0/RT) = 3.3 \times 10^5 \text{ s}^{-1} (298 \text{ K})$$

Of these reactions, the reaction of the peroxyl radical with phosphite is the slowest. The rate constant of this reaction ranges from 10^2 to 10^3 L mol^{-1} s^{-1} which is two to three orders of magnitude lower than the rate constant of similar reactions with phenols and aromatic amines. Namely, this reaction limits chain propagation in the oxidation of phosphites. Therefore, the chain oxidation of trialkyl phosphites involves chain propagation reactions with the participation of both peroxyl and phosphoranylperoxyl radicals:

$$(R'O)_3(RO)POO^\bullet + P(OR')_3 \longrightarrow (R'O)_3(RO)P{-}O^\bullet + O{=}P(R'O)_3$$
$$(R'O)_3(RO)P{-}O^\bullet \longrightarrow (R'O)_3P{=}O + RO^\bullet$$
$$(R'O)_3(RO)P{-}O^\bullet \longrightarrow (R'O)_2(RO)P{=}O + R'O^\bullet$$

According to the NMR spectroscopy data, oxidized trialkyl phosphites give rise to phosphates with a negative polarization of the ^{31}P nucleus [26]. This has been interpreted in terms of a series of the following reactions:

$$(R'O)_3(RO)P^\bullet + O_2 \longrightarrow (R'O)_3(RO)POO^\bullet$$
$$(R'O)_3(RO)P^\bullet + (R'O)_3(RO)POO^\bullet \longrightarrow (R'O)_3P{=}O + \text{Products}$$

The situation is different for aryl phosphites, which give rise to shorter chains than alkyl phosphites do (about 100 elementary reactions). The rate of aryl phosphite oxidation, $v \sim v_i$, suggests that chains are terminated in a linear manner [4,14,27]. Linear chain termination was interpreted in terms of the following mechanism of chain oxidation of aryl phosphites:

$$RN_2R \longrightarrow N_2 + 2R^\bullet$$
$$R^\bullet + O_2 \longrightarrow RO_2^\bullet$$
$$RO_2^\bullet + P(OAr)_3 \longrightarrow ROOP^\bullet(OAr)_3$$
$$ROOP^\bullet(OAr)_3 \longrightarrow RO^\bullet + (ArO)_3P{=}O$$
$$RO^\bullet + P(OAr)_3 \longrightarrow ROP^\bullet(OAr)_3$$
$$ROP^\bullet(OAr)_3 \longrightarrow R^\bullet + (ArO)_3P{=}O$$
$$ROP^\bullet(OAr)_3 \longrightarrow ROP(OAr)_2 + ArO^\bullet$$
$$RO_2^\bullet + ArO^\bullet \longrightarrow ROH + \text{quinone or quinolide peroxide.}$$

Aryl phosphites inhibit the initiated oxidation of hydrocarbons and polymers by breaking chains on the reaction with peroxyl radicals (see Table 17.3). The low values of the inhibition coefficient f for aryl phosphites are explained by their capacity for chain autoxidation [14]. Quantitative investigations of the inhibited oxidation of tetralin and cumene at 338 K showed that with increasing concentration of phosphite f rises tending to 1 [27].

The mechanism of inhibitory action of aryl phosphites seems to be relatively complex. Phosphites reduce hydroperoxide and thus decrease chain autoinitiation. The formed peroxyl and alkoxyl radicals react with phosphites to form aroxyl radicals. The latter terminates the chains by reaction with peroxyl radicals. On the other hand, phosphites are hydrolyzed with

TABLE 17.3
Rate Constants of Reactions of Peroxyl Radicals with Phosphines and Phosphites in Hydrocarbon Solutions (Experimental Data)

Phosphines or Phosphites	$RO_2^\bullet$	E (kJ mol^{-1})	log A, A (L mol^{-1} s^{-1})	k (338 K) (L mol^{-1} s^{-1})	Ref.
		Phosphines			
$(Me_2CH)_3P$	$Me_3CO_2^\bullet$	18.0	6.20	2.64×10^3	[28]
$(Me_2CH)_3P$	$Me_2(CN)CO_2^\bullet$			1.30×10^3	[29]
Ph_2PH	$Me_3CO_2^\bullet$	10.4	5.00	2.47×10^3	[28]
Ph_2PCl	$Me_3CO_2^\bullet$	5.4	3.70	7.32×10^2	[28]
Ph_2POMe	$Me_3CO_2^\bullet$	5.4	3.70	7.32×10^2	[28]
Ph_2PMe	$Me_3CO_2^\bullet$	6.3	5.11	1.38×10^4	[28]
$(MeO)_3P$	$Me_3CO_2^\bullet$	15.9	5.70	1.74×10^3	[28]
Ph_3P	$Me_3CO_2^\bullet$	12.5	6.00	1.17×10^4	[28]
tris(4-fluorophenyl)phosphine	$Me_3CO_2^\bullet$	13.0	5.90	7.74×10^3	[28]
tris(4-methoxyphenyl)phosphine	$Me_3CO_2^\bullet$	13.4	5.90	6.75×10^3	[28]
tris(4-methylphenyl)phosphine	$Me_3CO_2^\bullet$	15.9		6.98×10^3	[28]
tris(4-chlorophenyl)phosphine	$Me_3CO_2^\bullet$	17.1	6.80	1.43×10^4	[28]
		Phosphites			
$(EtO)_3P$	$Me_3CO_2^\bullet$	13.8	5.00	7.37×10^2	[28]
$(EtO)_3P$	$Me_2(CN)CO_2^\bullet$			1.9×10^3	[29]
$(Me_3CO)_3P$	$Me_3CO_2^\bullet$	23.0	6.70	1.39×10^3	[28]
$(Me_2CHO)_3P$	$Me_2(CN)CO_2^\bullet$			7.4×10^2	[29]
$(CH_2=CHCH_2O)_3P$	$Me_3CO_2^\bullet$	15.5	5.70	2.10×10^3	[28]
isopropyl benzodioxaphosphole	$Me_2(CN)CO_2^\bullet$			9.4×10^2	[29]

continued

TABLE 17.3
Rate Constants of Reactions of Peroxyl Radicals with Phosphines and Phosphites in Hydrocarbon Solutions (Experimental Data)—*continued*

Phosphines or phosphites	$RO_2^\bullet$	E (kJ mol^{-1})	log A, A (L mol^{-1} s^{-1})	k (338 K) (L mol^{-1} s^{-1})	Ref.
[structure]	$Me_3CO_2^\bullet$	19.2	6.00	1.08×10^3	[28]
$(PhO)_3P$	$Me_3CO_2^\bullet$	20.1	7.30	1.57×10^4	[28]
	$PhMe_2CO_2^\bullet$			1.50×10^2	[29]
[structure]	$PhMe_2CO_2^\bullet$			80	[29]
[structure]	$PhMe_2CO_2^\bullet$			50	[29]
[structure]	$PhMe_2CO_2^\bullet$			1.4×10^2	[29]

water with phenol formation. This process is accelerated by hydroperoxide and acids. Hence, phosphites themselves are weak inhibitors, however, they can serve as a source of efficient inhibitors, phenols and aroxyl radicals. The formed phosphorous acid catalyzes the decay of hydroperoxide.

17.3 SULFUR-CONTAINING ANTIOXIDANTS

17.3.1 REACTION WITH HYDROPEROXIDE

Sulfur(II)-containing compounds possess the reducing activity and react with hydroperoxides and peroxyl radicals [1–5]. They are employed as components of antioxidant additives to lubricants and polymers [30–35]. Denison and Condit [36] were the first to show that dialkyl sulfides are oxidized by hydroperoxides to sulfoxides and then to sulfones

$$\text{RSR} + \text{R'OOH} \longrightarrow \text{RS(O)R} + \text{R'OH}$$
$$\text{RS(O)R} + \text{R'OOH} \longrightarrow \text{RSO}_2\text{R} + \text{R'OH}$$

whereas hydroperoxides are reduced to alcohols. Later, this reaction was studied in detail [5,37–41]. In aprotic solvents the reaction rate is

$$v = k[\text{R}_2\text{S}][\text{R'OOH}]^2. \tag{17.4}$$

The second order of this reaction with respect to hydroperoxide is due to the second molecule of R'OOH acting as an acid molecule [37].

$$\text{R'OOH} + \text{HOOR} \rightleftharpoons \text{R'OOH}_2^+, \text{ROO}^-$$
$$\text{R'OOH}_2^+, \text{ROO}^- + \text{SR}_2 \longrightarrow \text{R'OH} + \text{R'OOH} + \text{R}_2\text{S(O)}$$

In the presence of alcohols and acids, the reaction is bimolecular and acid-catalyzed [37]:

$$\text{R'OOH} + \text{HA} \rightleftharpoons \text{R'OOH}_2^+\text{A}^-$$
$$\text{R'OOH}_2^+\text{A}^- + \text{SR}_2 \longrightarrow \text{R'OH} + \text{HA} + \text{R}_2\text{S(O)}$$

The rate of the reaction, obviously, depends on the acid concentration.

$$v = k[\text{R}_2\text{S}][\text{R'OOH}][\text{AH}]. \tag{17.5}$$

These reactions produce free radicals, as follows from the fact of consumption of free radical acceptor [42]. The oxidation of ethylbenzene in the presence of thiophenol is accompanied by CL induced by peroxyl radicals of ethylbenzene [43]. Dilauryl dithiopropionate induces the pro-oxidative effect in the oxidation of cumene in the presence of cumyl hydroperoxide [44] provided that the latter is added at a sufficiently high proportion ([sulfide]/[ROOH] $\geq$ 2). By analogy with similar systems, it can be suggested that sulfide should react with ROOH both heterolytically (the major reaction) and homolytically producing free radicals. When dilauryl dithiopropionate reacts with cumyl hydroperoxide in chlorobenzene, the rate constants of these reactions (molecular m and homolytic i) in chlorobenzene are [42]

$$k_\text{m} = 8 \times 10^4 \exp(-50/RT) \text{ L mol}^{-1} \text{ s}^{-1}$$
$$k_\text{i} = 2.5 \times 10^{14} \exp(-121/RT) \text{ L mol}^{-1} \text{ s}^{-1}$$

The ratio $k_\text{i}/k_\text{m} = 5 \times 10^{-4}$ at 300 K increases with increasing temperature, since $E_\text{i} > E_\text{m}$. The rate constants of the sulfide reaction with hydroperoxide are collected in Table 17.4.

The reactions of sulfides with ROOH give rise to products that catalyze the decomposition of hydroperoxides [31,38–47]. The decomposition is acid-catalyzed, as can be seen from the analysis of the resulting products: cumyl hydroperoxide gives rise to phenol and acetone, while 1,1-dimethylethyl hydroperoxide gives rise to 1,1-dimethylethyl peroxide, where all the three are the products of acid-catalyzed decomposition [46–49]. It is generally accepted that the intermediate catalyst is sulfur dioxide, which reacts with ROOH as an acid [31,46–50].

$$\text{SO}_2 + \text{H}_2\text{O} \rightleftharpoons \text{H}_2\text{SO}_3$$
$$\text{PhMe}_2\text{COOH} + \text{H}_2\text{SO}_3 \rightleftharpoons \text{PhMe}_2\text{COOH}_2^+ + \text{HSO}_2^-$$
$$\text{PhMe}_2\text{COOH}_2^+ \longrightarrow \text{PhOH} + \text{MeC(O)Me} + \text{H}^+$$

TABLE 17.4
Rate Constants of Reactions of Sulfides, Thiols, Disulfides, and Sulfoxides with Hydroperoxides (Experimental Data)

R_1SR_2	ROOH	Solvent	T (K)	k (L mol^{-1} s^{-1})	Ref.
2-naphthyl-S-H	PhMe$_2$COOH	Mineral oil	423	1.0	[45]
PhCH$_2$SSCH$_2$C(O)OCH$_2$Ph	Me$_3$COOH	PhCl	313–373	1.0×10^3 exp(−21.2/RT)	[46]
PhCH$_2$SSCH$_2$C(O)OH	Me$_3$COOH	PhCl	353–373	1.3×10^8 exp(−59.5/RT)	[46]
PhC(O)SSC(O)Ph	PhMe$_2$COOH	PhCl	353	1.4×10^{-2}	[42]
BuSSBu	PhMe$_2$COOH	Mineral oil	423	1.0	[45]
PrSSPr	Me$_3$COOH	PhCl	353–383	1.6×10^{11} exp(−82/RT)	[46]
EtS(O)S(O)Et	Me$_3$COOH	PhCl	373	0.39	[46]
4,4′-dihydroxydiphenyl sulfide	PhMe$_2$COOH	PhCl	423	1.0	[45]
PhC(O)SC(O)Ph	PhMe$_2$COOH	PhCl	353	2.3×10^{-2}	[42]
PhCH$_2$SCH$_2$Ph	PhMe$_2$COOH	PhCl	353	1.3×10^{-3}	[42]
BuSBu	PhMe$_2$COOH	Mineral oil	423	0.67	[45]
BuS(O)Bu	PhMe$_2$COOH	Mineral oil	423	3.3×10^{-2}	[45]
(C$_{10}$H$_{21}$)$_2$S	PhMe$_2$COOH	Mineral oil	423	8.3	[45]
PhSPh	PhMe$_2$COOH	Mineral oil	423	1.0×10^{-2}	[45]
4-hydroxyphenyl phenyl sulfide	PhMe$_2$COOH	Mineral oil	423	0.10	[45]
[Me$_2$NC(S)S]$_2$	PhMe$_2$COOH	PhCl	353	2.0×10^{-2}	[42]
[Me$_2$NC(S)]$_2$S	PhMe$_2$COOH	PhCl	353	6.2×10^{-2}	[42]

$$Me_3COOH + H_2SO_3 \rightleftharpoons Me_3COOH_2^+ + HSO_2^-$$
$$Me_3COOH_2^+ \rightleftharpoons Me_3C^+ + H_2O_2$$
$$Me_3C^+ + Me_3COOH \longrightarrow Me_3COOCMe_3 + H^+$$

The formation of free radicals and alcohol (in addition to the products of hydroperoxide heterolysis) implies that the catalytic decomposition of hydroperoxide occurs both heterolytically and homolytically. The mechanism of homolytic hydroperoxide decomposition was proposed by Van Tilborg and Smael [48].

$$ROOH + SO_2 \longrightarrow ROO(H^+)SO_2^- \longrightarrow ROOS(O)OH$$
$$ROOS(O)OH \longrightarrow RO^\bullet + HOS^\bullet O_2$$

The formation of the HOS$^\bullet$O$_2$ radical was demonstrated by EPR spectroscopy [51]. The homolytic decomposition of hydroperoxide catalyzed by SO$_2$ underlies an oscillating pattern of the n-hexylbenzene oxidation inhibited by thiophene or BaSO$_4$ [48]. Radicals can also be produced in the reaction of ROOH with sulfenic acid resulting from the decomposition of sulfoxide [31,33,44]:

$$RS(O)R \longrightarrow \text{olefin} + RSOH$$
$$RSOH + R'OOH \longrightarrow RS^{\bullet}O + H_2O + R'O^{\bullet}$$

The oxidation of sulfides is a complex process involving a number of conversions [32,46].

Disulfides are oxidized by hydroperoxide via the intermediate thiosulfinate RSSOR, which is very reactive to ROOH [32,52–54]. The interaction of ROOH with phenolsulfoxides also gives rise to intermediate catalytic compounds, as a result of which the reaction proceeds as an autocatalytic process [46,55]. The rate of the catalytic decomposition of R'OOH is described by one of the following equations:

$$v = k[\text{RSSR}][R'OOH] \tag{17.6}$$
$$v = k[\text{RSSR}][R'OOH]^2. \tag{17.7}$$

The rate constants are summarized in Table 17.5.

17.3.2 Reaction with Peroxyl Radicals

During the chain oxidation of hydrocarbons, sulfides and disulfides terminate chains by reacting with peroxyl radicals [40,42,44], which, as opposed to phenols, are weak inhibitors (see Table 17.6). The mechanism and stoichiometry of the termination reaction by sulfides remain yet unclear. Since sulfenic acid is an efficient scavenger of free radicals, the oxidation of tetralin in the presence of dialkylsulfoxide occurs only if the initiation rate $v_i > v_{i\min}$ is proportional to the concentration of sulfoxide [5], so that the rate of oxidation is

$$v = k_p(2k_t)^{1/2}[RH](v_i - k[R_1SOR_2]). \tag{17.8}$$

This can be explained by the breakdown of sulfoxide into olefin and sulfenic acid, where the latter is an efficient acceptor of peroxyl radicals.

$$R^1S(O)CH_2CH_2R^2 \longrightarrow R^1S(O)H + CH_2{=}CHR^2$$
$$R^1S(O)H + RO_2^{\bullet} \longrightarrow R^1S(O^{\bullet}) + ROOH$$
$$R^1S(O^{\bullet}) + RO_2^{\bullet} \longrightarrow R^1SO_2OR$$

Oxidation develops only if the rate of initiation is higher than the rate of generation of an efficient inhibitor. For the rate constants of sulfoxides decomposition, see Table 17.7.

The decay of alkylsulfoxides was studied by the quantum-chemical and IPM methods [65]. These reactions are endothermic. The activation energy of the thermoneutral analog, E_{e0}, depends on the structure of the alkyl radical:

$$R^1S(O)CHMe_2 \longrightarrow R^1S(O)H + CH_2{=}CHMe \qquad E_{e0} = 101\,\text{kJ mol}^{-1}$$
$$R^1S(O)CH_2CH_2R \longrightarrow R^1S(O)H + CH_2{=}CHR \qquad E_{e0} = 111\,\text{kJ mol}^{-1}$$

The activation energy of the thermoneutral reaction is lower when the methyl group participates in TS formation. There are two ways how the R substituent can have effect on the reaction. First, it is simple electronic effect that makes the TS less stable due to the repulsion

TABLE 17.5
Rate Constants of Bimolecular Catalytic Decomposition of Hydroperoxides Under Action of Products of Oxidation of S-Containing Compounds in Chlorobenzene (Experimental Data)

S-Containing Compound	Hydroperoxide	T (K)	k (L mol^{-1} s^{-1})	Ref.
	$v = k$ [RSR][ROOH]			
BuOCH$_2$CH(SBu)CH$_2$NHPh	PhMe$_2$COOH	354–383	8.9×10^{12} exp($-96.0/RT$)	[56]
BuOCH$_2$CH(SH)CH$_2$NHPh	PhMe$_2$COOH	353–383	9.1×10^{12} exp($-89.5/RT$)	[56]
PhCH$_2$SSCH$_2$C(O)OH	Me$_3$COOH	353–373	1.3×10^{8} exp($-59.5/RT$)	[57]
(bis-phenol disulfide)	PhMe$_2$COOH	393	0.51	[46]
(bis-phenol disulfide)	PhMe$_2$COOH	353–383	3.2×10^{8} exp($-52.8/RT$)	[58]
PhCH$_2$SSCH$_2$Ph	Me$_3$COOH	353–383	1.7×10^{11} exp($-82.0/RT$)	[59]
PhCH$_2$SSCH$_2$Ph	PhMe$_2$COOH	363–383	1.7×10^{7} exp($-48.0/RT$)	[59]
EtSSEt	PhMe$_2$COOH	363–383	7.9×10^{10} exp($-77.0/RT$)	[59]
(bis-phenol thiomethylene)	PhMe$_2$COOH	353–383	7.1×10^{9} exp($-64.3/RT$)	[58]
(morpholine thiobenzamide)	PhMe$_2$COOH	343–383	5.0×10^{7} exp($-46.6/RT$)	[60]
Me$_3$CSCMe$_3$	PhMe$_2$COOH	393	0.34	[46]
(bis-phenol sulfide)	PhMe$_2$COOH	393	0.38	[46]
(bis-phenol sulfide isomer)	PhMe$_2$COOH	353–383	7.1×10^{7} exp($-50.7/RT$)	[58]
(bis-phenol sulfide isomer)	PhMe$_2$COOH	353–383	1.3×10^{11} exp($-79.0/RT$)	[58]
(phenol thiobenzyl)	PhMe$_2$COOH	343–383	1.6×10^{5} exp($-34.3/RT$)	[60]

TABLE 17.5
Rate Constants of Bimolecular Catalytic Decomposition of Hydroperoxides Under Action of Products of Oxidation of S-Containing Compounds in Chlorobenzene (Experimental Data)—*continued*

S-Containing Compound	Hydroperoxide	T (K)	k (L mol^{-1} s^{-1}; L^2 mol^{-2} s^{-1})	Ref.
PhCH$_2$SS(O)CH$_2$Ph	Me$_3$COOH	353–383	2.4×10^{11} exp($-82.0/RT$)	[59]
PhCH$_2$SO$_2$CH$_2$Ph	Me$_3$COOH	353–383	5.7×10^9 exp($-76.5/RT$)	[59]
[structure: bis(di-tert-butylhydroxyphenyl) sulfone]	PhMe$_2$COOH	—	2.5×10^6 exp($-50.3/RT$)	[46]
[structure: bis(di-tert-butylhydroxyphenyl) sulfoxide]	PhMe$_2$COOH	—	2.4×10^8 exp($-67.3/RT$)	[46]
Me$_3$CSOCMe$_3$	PhMe$_2$COOH	383–403	6.2×10^{12} exp($-100/RT$)	[46]
Me$_3$CSS(O)CMe$_3$	PhMe$_2$COOH	383–403	3.9×10^{13} exp($-106/RT$)	[46]
PhCH$_2$C(S)NHPr	PhMe$_2$COOH	343–383	2.0×10^9 exp($-63/RT$)	[60]
[structure: hydroxyphenyl thioamide]	PhMe$_2$COOH	343–383	4.0×10^{14} exp($-98.7/RT$)	[60]
PhC(S)NHCH$_2$Ph	PhMe$_2$COOH	343–383	2.5×10^{15} exp($-97.6/RT$)	[60]
	$v = k$ [RSR][ROOH]^2			
[structure: 2,3-dihydroxythiophenol]	PhMe$_2$COOH	343–383	6.3×10^{11} exp($-66.5/RT$)	[60]
[structure: 2-hydroxybenzyl alkyl sulfide]	PhMe$_2$COOH	—	1.0×10^{11} exp($-71.0/RT$)	[61]
[structure: hydroxyphenyl phenyl sulfide]	PhMe$_2$COOH	343–383	2.5×10^7 exp($-47.4/RT$)	[60]
[structure: tert-butyl hydroxyphenyl sulfide]	PhMe$_2$COOH	383	0.59	[61]
[structure: hydroxyphenyl polysulfide]	PhMe$_2$COOH	—	4.0×10^{11} exp($-78.6/RT$)	[61]
[structure: hydroxybenzyl phenyl sulfide]	PhMe$_2$COOH	—	3.8×10^9 exp($-72.7/RT$)	[61]
[structure: tert-butyl hydroxybenzyl thiol]	PhMe$_2$COOH	—	9.5×10^3 exp($-19.6/RT$)	[61]

TABLE 17.6
Rate Constants of Reactions of Peroxyl Radicals with Sulfides and Disulfides in Hydrocarbon Solutions (Experimental Data)

Sulfides or Disulfides	$RO_2^{\bullet}$	T (K)	k (L mol^{-1} s^{-1})	Ref.
Sulfides				
$PhCH_2SPh$	$Me_3CO_2^{\bullet}$	303	0.16	[62]
$PhCH_2SPh$	$PhMe_2CO_2^{\bullet}$	303	9.5	[62]
$[C_{12}H_{25}OC(O)CH_2CH_2]_2S$	$PhMe_2CO_2^{\bullet}$	333	40	[45]
tetrahydrothiophene	$Me_3CO_2^{\bullet}$	303	8.0×10^{-2}	[62]
tetrahydrothiophene	tetrahydrothiophen-2-ylperoxyl	303	6.4	[62]
thiane	thian-2-ylperoxyl	303	1.5	[62]
thiane	$Me_3CO_2^{\bullet}$	303	2.0×10^{-2}	[62]
Disulfides				
$PhCH_2SSCH_2C(O)OCH_2Ph$	$PhMe_2CO_2^{\bullet}$	333	21.0	[59]
$PhCH_2SSCH_2C(O)OH$	$PhMe_2CO_2^{\bullet}$	333	50.4	[59]
$PhCH_2SCH_2Ph$	$Me_3CO_2^{\bullet}$	303	0.36	[62]
$PhCH_2SSCH_2Ph$	$Me_3CO_2^{\bullet}$	303	0.15	[62]
$PhCH_2SSCH_2Ph$	$PhMe_2CO_2^{\bullet}$	333	7.8	[59]
$EtSSEt$	$PhMe_2CO_2^{\bullet}$	333	0.35	[59]
$Me_3CS(O)H$	tetralin-1-ylperoxyl	333	1.0×10^7	[41]
$PhCH_2SS(O)CH_2Ph$	$\sim CH_2PhCHO_2^{\bullet}$	333	1.7×10^5	[59]
$PhCH_2SS(O_2)CH_2Ph$	$\sim CH_2PhCHO_2^{\bullet}$	333	8.0×10^2	[59]

of electrons on the occupied energy levels. The second mechanism appears when the system has a possibility for a few TS conformations with different occupations as a result. If R facilitates the occupation of the TS, which has a lower advantage for the reaction, then the reaction demands more energy for crossing the TS.

So, the effect of sulfur compounds on the oxidation of hydrocarbons and polymers is rather complex.

1. Sulfur-containing antioxidants reduce hydroperoxide to alcohols, thereby decreasing the rate of autoinitiation and decreasing the oxidation rate.

TABLE 17.7
Rate Constants of Sulfoxide Decomposition (Experimental Data)

Reaction	Solvent	T (K)	k (s^{-1})	Ref.
$Pr_2SO \longrightarrow PrSOH + MeCH=CH_2$	PhMe	373	6.0×10^{-8}	[63]
$[Me_2CH]_2SO \longrightarrow Me_2CHSOH + MeCH=CH_2$	PhMe	303	2.0×10^{-2}	[63]
$MeSOCMe_3 \longrightarrow MeSOH + Me_2CH=CH_2$	PhMe	373	6.3×10^{-8}	[63]
$EtSOBu \longrightarrow BuSOH + CH_2=CH_2$	Gas	453–473	$2.2 \times 10^{11} \exp(-131.5/RT)$	[64]
$EtSOBu \longrightarrow EtSOH + EtCH=CH_2$	Gas	453–473	$1.4 \times 10^{11} \exp(-132.0/RT)$	[64]
$PrSOBu \longrightarrow BuSOH + MeCH=CH_2$	Gas	453–473	$1.4 \times 10^{11} \exp(-130.9/RT)$	[64]
$PrSOBu \longrightarrow PrSOH + EtCH=CH_2$	Gas	453–473	$1.4 \times 10^{11} \exp(-130.5/RT)$	[64]
$Me_2CHSOBu \longrightarrow BuSOH + MeCH=CH_2$	Gas	453–473	$4.2 \times 10^{11} \exp(-126.3/RT)$	[64]
$Me_2CHSOBu \longrightarrow Me_2CHSOH + EtCH=CH_2$	Gas	453–473	$1.4 \times 10^{11} \exp(-130.0/RT)$	[64]
$Me_2CHSOCMe_3 \longrightarrow Me_2CHSOH + Me_2C=CH_2$	PhMe	373	4.6×10^{-5}	[63]
$EtSOCHMeEt \longrightarrow EtSOH + EtCH=CH_2$	Gas	453–473	$2.1 \times 10^{11} \exp(-124.3/RT)$	[64]
$EtSOCHMeEt \longrightarrow MeEtCHSOH + CH_2=CH_2$	Gas	453–473	$2.1 \times 10^{11} \exp(-129.4/RT)$	[64]
$EtSOCHMeEt \longrightarrow EtSOH + trans\text{-}MeCH=CHMe$	Gas	453–473	$1.4 \times 10^{11} \exp(-127.4/RT)$	[64]
$PrSOCHMeEt \longrightarrow PrSOH + EtCH=CH_2$	Gas	453–473	$2.1 \times 10^{11} \exp(-123.3/RT)$	[64]
$PrSOCHMeEt \longrightarrow MeEtCHSOH + MeCH=CH_2$	Gas	453–473	$1.4 \times 10^{11} \exp(-129.5/RT)$	[64]
$PrSOCHMeEt \longrightarrow PrSOH + trans\text{-}MeCH=CHMe$	Gas	453–473	$1.4 \times 10^{11} \exp(-125.9/RT)$	[64]
$Me_3CSOCMe_3 \longrightarrow Me_3CSOH + Me_2C=CH_2$	PhMe	345	3.3×10^{-5}	[63]
$Me_3CSOCMe_3 \longrightarrow Me_3CSOH + Me_2C=CH_2$	PhMe	352	1.1×10^{-4}	[63]
$Me_3CSOCMe_3 \longrightarrow Me_3CSOH + Me_2C=CH_2$	PhMe	363	3.3×10^{-4}	[63]
$Me_3CSOCMe_3 \longrightarrow Me_3CSOH + Me_2C=CH_2$	PhMe	373	1.8×10^{-3}	[63]
$Me_2CHSOPh \longrightarrow PhSOH + MeCH=CH_2$	PhMe	373	1.3×10^{-6}	[63]
$Me_3CSOPh \longrightarrow PhSOH + Me_2C=CH_2$	PhMe	373	2.1×10^{-4}	[63]
$4\text{-}MeOC_6H_4SOCMe_3 \longrightarrow 4\text{-}MeOC_6H_4SOH + Me_2C=CH_2$	PhMe	374	9.7×10^{-5}	[63]
$4\text{-}MeC_6H_4SOCMe_3 \longrightarrow 4\text{-}MeC_6H_4SOH + Me_2C=CH_2$	PhMe	374	1.2×10^{-4}	[63]
$3\text{-}MeC_6H_4SOCMe_3 \longrightarrow 3\text{-}MeC_6H_4SOH + Me_2C=CH_2$	PhMe	374	1.8×10^{-4}	[63]
$3\text{-}MeOC_6H_4SOCMe_3 \longrightarrow 3\text{-}MeOC_6H_4SOH + Me_2C=CH_2$	PhMe	374	2.3×10^{-4}	[63]
$4\text{-}ClC_6H_4SOCMe_3 \longrightarrow 4\text{-}ClC_6H_4SOH + Me_2C=CH_2$	PhMe	374	2.4×10^{-4}	[63]
$3\text{-}ClC_6H_4SOCMe_3 \longrightarrow 3\text{-}ClC_6H_4SOH + Me_2C=CH_2$	PhMe	374	3.2×10^{-4}	[63]
$3\text{-}NO_2C_6H_4SOCMe_3 \longrightarrow 3\text{-}NO_2C_6H_4SOH + Me_2C=CH_2$	PhMe	374	5.0×10^{-4}	[63]
$4\text{-}NO_2C_6H_4SOCMe_3 \longrightarrow 4\text{-}NO_2C_6H_4SOH + Me_2C=CH_2$	PhMe	374	9.0×10^{-4}	[63]

2. On the other hand, they react with hydroperoxide to produce free radicals and, hence, contribute to the initiation of oxidation.
3. As other antioxidants, they terminate chains, albeit slowly, by reacting with peroxyl radicals.
4. The significant role of intermediates of conversion of sulfides and disulfides should be highlighted. Thus, sulfenic acid is formed by sulfoxide decomposition and terminates chains by reacting with $RO_2^\bullet$ and, on the other hand, initiates chains by reacting with hydroperoxide.
5. Acidic products (SO_2, H_2SO_4, RSO_2H, RSO_3H) break hydroperoxides into molecular products, thereby inhibiting autoinitiation. At the same time, they act as initiators by breaking hydroperoxide with the formation of free radicals.

17.4 METAL THIOPHOSPHATES AND THIOCARBAMATES

Metal complexes, dialkyl thiophosphates and dialkyl thiocarbamates of Zn, Ni, Ba, and Ca, in particular, are widely used for the stabilization of lubricants [30–32,34]. At moderate temperatures (350–400 K), these inhibitors are less efficient than phenols, but they are more potent at higher temperatures (430–480 K). The sophisticated mechanism of action of these antioxidants involves their reactions with hydroperoxide. The interaction of hydroperoxide with metal dialkyl thiophosphates induces a cascade of reactions [5,66–69].

The resulting products, such as sulfenic acid or sulfur dioxide, are reactive and induce an acid-catalyzed breakdown of hydroperoxides. The important role of intermediate molecular sulfur has been reported [68–72]. Zinc (or other metal) forms a precipitate composed of ZnO and $ZnSO_4$. The decomposition of ROOH by dialkyl thiophosphates is an autocatalytic process. The interaction of ROOH with zinc dialkyl thiophosphate gives rise to free radicals, due to which this reaction accelerates oxidation of hydrocarbons, excites CL during oxidation of ethylbenzene, and intensifies the consumption of acceptors, e.g., stable nitroxyl radicals [68]. The induction period is often absent because of the rapid formation of intermediates, and the kinetics of decomposition is described by a simple bimolecular kinetic equation

$$v = k[(RO)_2PS_2ZnS_2P(OR)_2][R'OOH]. \qquad (17.9)$$

The reaction occurs both heterolytically and homolytically, so that the overall rate constant $k_\Sigma = k_m + k_i$. For the rate constants, see Table 17.8.

Dialkyl thiophosphates can also terminate chains by reacting with peroxyl radicals, as follows from their ability to inhibit the initiated oxidation of hydrocarbons and to extinguish CL in oxidizing hydrocarbon [66,68]. The stoichiometric coefficient of inhibition f for various dialkyl thiophosphates is close to 2, as follows from the data for diisopropyl dithiophosphates:

Me:	Zn	Ni	Pb	Fe	Cu	Co
f [66]	1.81	1.94	1.90	1.25	—	—
f [68]	2.04	1.74	—	—	2.14	1.91

The limiting step was supposed to be the electron transfer from the complex to peroxyl radical succeeded by the decomposition of the formed complex [66].

TABLE 17.8
Rate Constants of Reactions of Metal Thiophosphates and Metal Thiocarbamates with Cumylhydroperoxide (Experimental Data)

Metal Complex	T (K)	E (kJ mol^{-1})	log A, A (L mol^{-1} s^{-1})	k (303 K) (L mol^{-1} s^{-1})	Ref.
Thiocarbamates					
[(Me$_2$CHO)$_2$NCS$_2$]$_2$Ni	303			0.16	[73]
(Bu$_2$NCS$_2$)$_2$Ni	293–358	20.9	2.72	0.13	[74]
(Et$_2$NCS$_2$)$_2$Ni	293–358	29.9	2.30	3.50 × 10^{-3}	[74]
[(Me$_3$CCH$_2$MeCH)$_2$NCS$_2$]$_2$Zn	293–358	45.7	4.92	1.10 × 10^{-3}	[75]
(Bu$_2$NCS$_2$)$_2$Zn	318–358	54.0	6.04	5.40 × 10^{-4}	[75]
Thiophosphates					
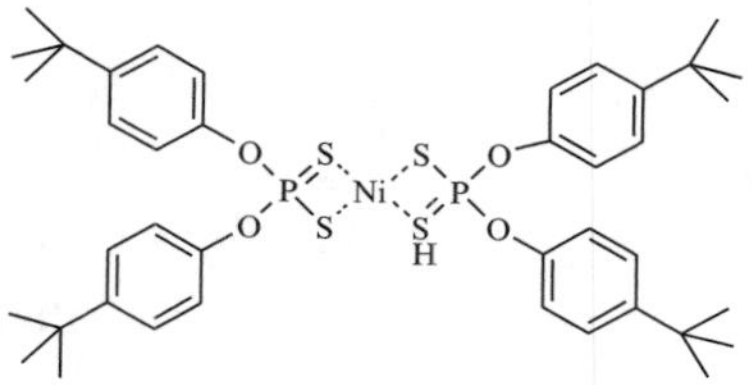 (structure shown)	298–353	39.0	4.00	1.89 × 10^{-3}	[76]
[(Me$_2$CHO)$_2$PS$_2$]$_2$Ni	298–353	55.5	7.54	9.47 × 10^{-3}	[76]
[(Me$_2$CHO)$_2$PS$_2$]$_2$Ni	303			4.0 × 10^{-3}	[73]
[(Me$_2$CHO)$_2$PS$_2$]$_2$Zn	328			3.8 × 10^{-3}	[5]

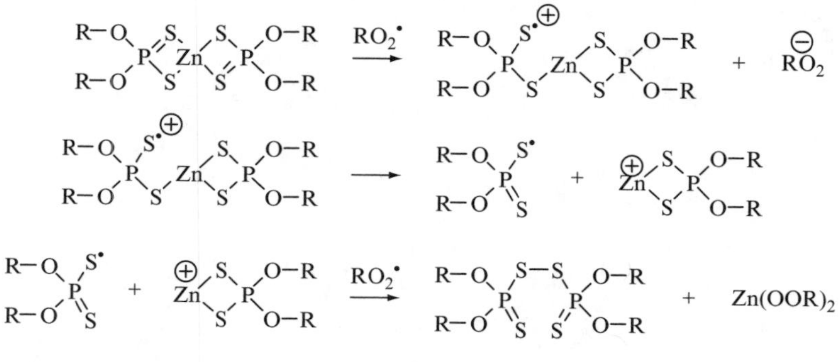

For the rate constants of these reactions, see Table 17.9.

Metal dialkyl dithiocarbamates inhibit the oxidation of hydrocarbons and polymers [25,28,30,76–79]. Like metal dithiophosphates, they are reactive toward hydroperoxides. At room temperature, the reactions of metal dialkyl dithiocarbamates with hydroperoxides occur with an induction period, during which the reaction products are formed that catalyze the breakdown of hydroperoxide [78]. At higher temperatures, the reaction is bimolecular and occurs with the rate $v = k$[ROOH][inhibitor]. The reaction of hydroperoxide with dialkyl dithiocarbamate is accompanied by the formation of radicals [30,76,78]. The bulk yield of radicals in the reaction of nickel diethyl dithiocarbamate with cumyl hydroperoxide is 0.2 at

TABLE 17.9
Rate Constants of Reactions of Peroxyl Radicals with Metal Thiocarbamates and Thiophosphates in Hydrocarbon Solutions (Experimental Data)

Metal Complex	Peroxyl Radical	T (K)	k (L mol^{-1} s^{-1})	Ref.
	Thiocarbamates			
[(Me$_2$CHO)$_2$NCS$_2$]$_2$Ni	tetralinyl-OO•	323	6.0×10^3	[80]
[(Me$_2$CHO)$_2$NCS$_2$]$_2$Ni	PhCH(O$_2$•)CH$_2$~	323	9.0×10^2	[80]
[(Me$_2$CHO)$_2$NCS$_2$]$_2$Ni	HO$_2$•	303	7.0×10^4	[80]
(Bu$_2$NCS$_2$)$_2$Ni	tetralinyl-OO•	323	1.4×10^4	[80]
(Bu$_2$NCS$_2$)$_2$Ni	PhCH(O$_2$•)CH$_2$~	323	5.0×10^3	[80]
(Bu$_2$NCS$_2$)$_2$Ni	PhMe$_2$CO$_2$•	333	1.9×10^3	[73]
(Et$_2$NCS$_2$)$_2$Ni	PhMe$_2$CO$_2$•	348	2.8×10^4	[81]
(Et$_2$NCS$_2$)$_2$Ni	PhMe$_2$CO$_2$•	333	2.3×10^3	[77]
(Et$_2$NCS$_2$)$_2$Bi	PhMe$_2$CO$_2$•	348	1.4×10^4	[81]
(Et$_2$NCS$_2$)$_2$Cu	PhMe$_2$CO$_2$•	348	1.0×10^5	[81]
(Et$_2$NCS$_2$)$_2$Pb	PhMe$_2$CO$_2$•	348	9.0×10^3	[81]
(Bu$_2$NCS$_2$)$_2$Sb	PhMe$_2$CO$_2$•	333	3.8×10^3	[82]
(Me$_3$CCH$_2$CHMeCH$_2$NS$_2$)$_2$Sb	PhMe$_2$CO$_2$•	333	3.5×10^3	[82]
Zn complex with cyclobutyl N-substituents	PhMe$_2$CO$_2$•	333	4.5×10^3	[82]
Zn complex with cyclooctyl N-substituents	PhMe$_2$CO$_2$•	333	8.0×10^2	[82]
Zn complex with cyclohexyl N-substituents	PhMe$_2$CO$_2$•	333	3.5×10^3	[82]
Zn complex with long-chain alkyl N-substituents	PhMe$_2$CO$_2$•	333	4.6×10^3	[82]

TABLE 17.9
Rate Constants of Reactions of Peroxyl Radicals with Metal Thiocarbamates and Thiophosphates in Hydrocarbon Solutions (Experimental Data)—continued

Metal Complex	Peroxyl Radical	T (K)	k (L mol^{-1} s^{-1})	Ref.
[structure: Zn complex with C$_{13}$H$_{27}$ groups]	PhMe$_2$CO$_2^\bullet$	333	4.0×10^2	[82]
(Me$_2$CHNCS$_2$)$_2$Zn	PhMe$_2$CO$_2^\bullet$	333	2.4×10^3	[78]
[(Me$_3$CCH$_2$CHMeCH$_2$)$_2$NCS$_2$]$_2$Zn	PhMe$_2$CO$_2^\bullet$	333	6.1×10^3	[82]
(Bu$_2$NCS$_2$)$_2$Zn	PhMe$_2$CO$_2^\bullet$	333	4.7×10^3	[82]
(Et$_2$NCS$_2$)$_2$Zn	PhMe$_2$CO$_2^\bullet$	348	2.1×10^3	[81]
Thiophosphates				
[(Me$_2$CHO)$_2$PS$_2$]$_2$Zn	Me$_3$CO$_2^\bullet$	200	44	[83]
[(Me$_2$CHO)$_2$PS$_2$]$_2$Zn	PhCH(O$_2^\bullet$)CH$_2$~	303	2.8×10^3	[5]
[(Me$_2$CHO)$_2$PS$_2$]$_2$Zn	cyclohexyl-O$_2^\bullet$	303	4.3×10^3	[5]
[(Me$_2$CHO)$_2$PS$_2$]$_2$Zn	tetralyl-OO$^\bullet$	303	3.0×10^3	[5]
[(Me$_2$CHO)$_2$PS$_2$]$_2$Zn	PhC(O)O$_2^\bullet$	303	4.8×10^5	[5]
[(Me$_2$CHO)$_2$PS$_2$]$_2$Zn	HO$_2^\bullet$	303	2.8×10^4	[5]
[(EtMeCHO)$_2$PS$_2$]$_2$Zn	Me$_3$CO$_2^\bullet$	200	34	[83]
[(EtMeCHO)$_2$PS$_2$]$_2$Zn	Me$_3$CO$_2^\bullet$	200	43	[83]

293 K and 0.7 at 343 K [76]. Carbamates are antioxidants of the complex action: they react with hydroperoxide, as well as with peroxyl radicals [5].

17.5 METAL COMPLEXES WITH PHOSPHITES

Pobedimskii and coworkers [84–92] studied hydroperoxide decomposition under the combined action of a transition metal complex and phosphite. He found that this binary system induces three parallel catalytic reactions of hydroperoxide decomposition.

1. Hydroperoxide forms a complex with the metal complex and is decomposed. The formation of such complex was evidenced by NMR spectroscopy [86,87].
2. Hydroperoxide is reduced by phosphite (see earlier).
3. Metal forms a complex with hydroperoxide and phosphite reacts rapidly with hydroperoxide as a metal ligand.

Besides this, phosphite reacts with metal forming an inactive complex. Hence, the catalytic activity of the metal decreases with an increase in the phosphite concentration. The produced

TABLE 17.10
Kinetic Parameters of Peroxyl Radicals of Oxidized Styrene with Phosphites and their Complexes with CuCl in Chlorobenzene Solution [89]

Complex	T (K)	f	E (kJ mol^{-1})	log A, A (L mol^{-1} s^{-1})	k (393 K) (L mol^{-1} s^{-1})
[Cu$^+$ PEt$_3$]Cl$^-$	393	3.0			6.0×10^4
[Cu$^+$ PEt$_3$]Cl$^-$	428	0.2			5.5×10^5
[Cu$^+$ PEt$_3$]Cl$^-$	393–428		58.1	12.50	6.0×10^4
(PhO)$_3$P	393	0.04			5.0×10^4
(PhO)$_3$P	428	0.003			3.3×10^5
(PhO)$_3$P	393–428		56.4	12.20	5.0×10^4
[Cu$^+$ P(OPh)$_3$]Cl$^-$	393	2.0			1.1×10^5
[Cu$^+$ P(OPh)$_3$]Cl$^-$	428	0.1			1.0×10^6
[Cu$^+$ P(OPh)$_3$]Cl$^-$	393–428		62.9	13.40	1.1×10^5
[Cu$^+$ (P(OPh$_3$)$_2$]Cl$^-$	393	1.0			3.6×10^5
[Cu$^+$ (P(OPh$_3$)$_2$]Cl$^-$	428	0.1			1.6×10^6
[Cu$^+$ (P(OPh$_3$)$_2$]Cl$^-$	393–428		35.7	10.30	3.6×10^5
(structure)	393	0.04			5.0×10^4
(structure)	428	0.003			3.3×10^5
(structure)	393–428		39.0	12.20	1.0×10^7
(structure)	393	1.4			3.4×10^5
(structure)	428	0.03			2.0×10^6
(structure)	393–428		35.6	11.90	1.5×10^7

TABLE 17.10
Kinetic Parameters of Peroxyl Radicals of Oxidized Styrene with Phosphites and their Complexes with CuCl in Chlorobenzene Solution [89]—*continued*

Complex	T (K)	f	E (kJ mol^{-1})	log A, A (L mol^{-1} s^{-1})	k (393 K) (L mol^{-1} s^{-1})
	393	0.9			4.8×10^5
	428	0.01			1.0×10^6
	393–428		12.6	7.90	1.7×10^6
	393	0.2			5.0×10^4
	428	0.02			2.5×10^5
	393–428		29.1	10.0	1.4×10^6
	393	1.1			1.4×10^5

continued

TABLE 17.10
Kinetic Parameters of Peroxyl Radicals of Oxidized Styrene with Phosphites and their Complexes with CuCl in Chlorobenzene Solution [89]—*continued*

Complex	T (K)	f	E (kJ mol^{-1})	log A, A (L mol^{-1} s^{-1})	k (393 K) (L mol^{-1} s^{-1})
(structure)	428	0.1			5.0×10^5
(structure)	393–428		23.8	9.40	1.7×10^6

phosphate also forms an inactive complex with the metal. So, the activity of metal drops in time of the experiment. The resulting kinetic scheme includes the following stages [85]:

$$Me_3COOH + P(OPh)_3 \longrightarrow Me_3COH + (Ph)_3P(O) \qquad k_0$$
$$Me_3COOH + Me^{n+} \rightleftharpoons Me_3COOH \cdot Me^{n+} \qquad K_1$$
$$Me_3COOH \cdot Me^{n+} \longrightarrow Me^{n+} + Products \qquad k_2$$
$$MeCOOH \cdot Me^{n+} + P(OPh)_3 \longrightarrow Me_3COH + (PhO)_3P(O) + Me^{n+} \qquad k_3$$
$$Me^{n+} + P(OPh)_3 \rightleftharpoons Me^{n+} \cdot P(OPh)_3 \qquad K_4$$

The following equilibrium and rate constants were found as a result of the kinetic and NMR studies (CH$_2$Cl$_2$ as solvent, $T = 292$ K [85]).

Catalyst	k_0 (L mol^{-1} s^{-1})	K_1 (L mol^{-1})	k_3 (L mol^{-1} s^{-1})	K_4 (L mol^{-1})
VO(acac)$_2$	5.0×10^{-3}	1.04	1.53×10^4	1.02×10^2
MoO$_2$(acac)$_2$	5.0×10^{-3}	1.00	2.75×10^4	3.05×10^2

In addition to the decay of hydroperoxides, metal complexes accelerate the reaction of phosphite with peroxyl radicals [90,91]. Phosphite forms a complex with the metal ion, and the formed complex terminates the chains more rapidly than phosphite does alone. For example, triphenyl phosphite terminates the chains in oxidized styrene with $fk_7 = 2 \times 10^3$ L mol^{-1} s^{-1} at $T = 393$ K and the complex of this phosphite with CuCl does it with $fk_7 = 2 \times 10^5$ L mol^{-1} s^{-1}, i.e., two orders of magnitude more rapidly. The values of f and k_7 for three phosphites and six complexes (phosphite with CuCl) are given in Table 17.10.

The effectiveness of complexes metal–acetylacetonate with tris(1,1-dimethylethyl-4-methylphenyl) phosphite in their reaction with peroxyl radicals of styrene and tetralin (323 K) decreases in the row: Co$^{2+} \geq$ VO$^{2+} >$ Cr$^{3+} \gg$ Fe^{2+} [88].

So, the action of the system metal ion–phosphite on oxidation is complex. The formed complex terminates the chains and decomposes the formed hydroperoxide more rapidly than phosphite alone.

REFERENCES

1. ET Denisov, TG Denisova. *Handbook of Antioxidants*. Boca Raton: CRC Press, 2000.
2. WD Habicher, I Bauer. In: Hamid, (ed.) *Handbook of Polymer Degradation*. New York: Marcel Dekker, 2000, pp. 81–104.
3. ET Denisov, VV Azatyan. *Inhibition of Chain Reactions*. London: Gordon and Breach, 2000.
4. K Schwetlick. In: G Scott, (ed.) *Mechanisms of Polymer Degradation and Stabilisation*. London: Elsevier Applied Science, 1990, pp. 23–60.
5. DM Shopov, SK Ivanov. *Mechanism of Action of Hydroperoxide-Breaking Inhibitors*. Sofia: Izdat Bulg Akad Nauk, 1988 [in Russian].
6. KJ Humphris, G Scott. *Pure Appl Chem* 36:163–176, 1973.
7. EG Chebotareva, DG Pobedimskii, NS Kolyubakina, NA Mukmeneva, PA Kirpichnikov, AG Akhmadullina. *Kinet Katal* 14:891–895, 1973.
8. DG Pobedimskii, AL Buchachenko. *Izv AN SSSR Ser Khim* 1181–1186, 1968.
9. PA Kirpichnikov, DG Pobedimskii, NA Mukmeneva. *Chemistry and Applications of Organophosphorus Compounds*. Moscow: Nauka, 1974, p. 215 [in Russian].
10. KJ Humphris, G Scott. *J Chem Soc Perkin II* 826–830, 1973.
11. DA Levin, EI Vorkunova. *Homolytic Chemistry of Phosphorus*. Moscow: Nauka, 1978, p. 123, 147, 171, and 176 [in Russian].
12. C Ruger, T Konig, K Schwetlick. *J Prakt Chem* 326:622–632, 1984.
13. K Schwetlick, C Ruger, R Noack. *J Prakt Chem* 324:697–706, 1982.
14. DG Pobedimskii. *Neftekhimiya* 18:701–707, 1978.
15. DG Pobedimskii, PI Levin, ZB Chelnokova. *Izv AN SSSR Ser Khim* 2066–2068, 1969.
16. DG Pobedimskii, VA Belyakov. *Kinet Katal* 10:64–68, 1969.
17. PA Kirpichnikov, NA Mukmeneva, DG Pobedimskii. *Usp Khim* 52:1831–1851, 1983.
18. AD Pershin, DG Pobedimskii, VA Kurbatov, AL Buchachenko. *Izv AN SSSR Ser Khim* 581–586, 1975.
19. LP Zaichenko, VG Babel, VA Proskuryakov. *Zh Prikl Khim* 47:1168–1171, 1974.
20. LP Zaichenko, VG Babel, VA Proskuryakov. *Zh Prikl Khim* 49:465–467, 1976.
21. R Hiatt, RJ Smythe, C McColeman. *Can J Chem* 49:1707–1711, 1971.
22. R Hiatt, C McColeman. *Can J Chem* 49:1712–1715, 1971.
23. C Walling, R Rabinowitz. *J Am Chem Soc* 81:1243–1249, 1950.
24. DG Pobedimskii, NA Mukmeneva, PA Kirpichnikov. *Usp Khim* 41:1242–1259, 1972.
25. VA Kurbatov, GP Gren, LA Pavlova, PA Kirpichnikov, DG Pobedimskii. *Kinet Katal* 17:329–332, 1976.
26. DG Pobedimskii, VA Kurbatov, ID Temyachev, YuYu Samitov, PA Kirpichnikov. *Teoret Experim Khim* 10:492–499, 1974.
27. K Schwetlick, T Konig, J Pionteck, D Sasse, WD Habicher. *Polym Degrad Stabil* 22:357, 1988.
28. E Furimsky, JA Howard. *J Am Chem Soc* 95:369–379, 1973.
29. K Schwetlick, T Konig, C Ruger, J Pionteck, WD Habicher. *Polym Degrad Stabil* 15:97, 1986.
30. J von Voigt. *Die Stabilisierung der Kunststoffe gegen Licht und Warme*. Berlin: Springer-Verlag, 1966.
31. G Scott, (ed.) *Atmospheric Oxidation and Antioxidants*, vols 1–3. Amsterdam: Elsevier, 1993.
32. S Patai, Z Rappoport, C Stirling, (eds.) *The Chemistry of Sulfones and Sulfoxides*. New York: John Wiley & Sons, 1988.
33. JR Shelton, KE Davis. *Int J Sulfur Chem* 8:205–216, 1973.
34. AM Kuliev. *Chemistry and Technology of Additives to Oils and Fuels*. Moscow: Khimiya, 1972 [in Russian].

35. ET Denisov, GI Kovalev. *Oxidation and Stabilization of Jet Fuels.* Moscow: Khimiya, 1983, pp. 12–127 [in Russian].
36. GH Denison, PC Condit. *Ind Eng Chem* 37:1102–1112, 1945.
37. L Bateman, KR Hargrave. *Proc Roy Soc* 224:389–399, 1954.
38. G Scott. *Chem Commun* 1572–1583, 1968.
39. G Scott. PA Shearn. *J Appl Polym Sci* 13:1329–1338, 1969.
40. G Scott. *Pure Appl Chem* 30:267–278, 1972.
41. P Koelewijn, H Berger. *Rec Trav Chim* 93:63–68, 1974.
42. NV Zolotova, LL Gervitz, ET Denisov. *Neftekhimiya* 15:146–150, 1975.
43. GV Karpukhina, ZK Maizus, LI Matienko. *Zh Fiz Khim* 41:733–735, 1967.
44. CA Armstrong, MA Plant, G Scott. *Eur Polym J* 11:161–168, 1975.
45. GW Kennerly, WL Patterson. *Ind Eng Chem* 48:1917–1926, 1956.
46. AJ Bridgewater, MD Sexton. *J Chem Soc Perkin II* 530–536, 1978.
47. JCW Chien, CR Boss. *Polym Sci A-1* 10:1579–1590, 1972.
48. WJM Van Tilborg, P Smael. *Rec Trav Chim* 95:138–149, 1976.
49. ON Grishina, MI Potekhina, VM Baschinova, NP Anoshina, FYa Yusupova. *Neftekhimiya* 17:790–795, 1977.
50. JR Shelton. In: G Scott, (ed.) *Developments in Polymer Stabilisation—4.* London: Applied Science, 1981.
51. BD Flockhart, KJ Ivin, RC Pink, BD Sharma. *Chem Commun* 339–340, 1971.
52. DI Barnard, L Bateman, ER Cole, JI Cunneen. *Chem Ind (L)* 918–925, 1958.
53. AD Aslanov, LV Petrov, ET Denisov, FA Kuliev. *Neftekhimiya* 25:562–565, 1985.
54. S Al-Malaika. *Polym Degrad Stabil* 34:1–36, 1991.
55. VM Farzaliev, WSE Fernand, G Scott. *Eur Polym J* 11:785–794, 1978.
56. MM Akhundova, VM Farzaliev, VM Solyanikov, ET Denisov. *Izv AN SSSR Ser Khim* 741–746, 1981.
57. AD Aslanov, NV Zolotova, ET Denisov, FA Kuliev. *Neftekhimiya* 22:504–509, 1982.
58. VM Farzaliev, FA Kuliev, MM Akhundova, ET Denisov. *Azerb Khim Zh* 117–122, 1981.
59. AD Aslanov. Organic Disulfides as Inhibitors of Oxidative Processes, Cand. Sci. (Chem.) Thesis Dissertation, Inst. Chem. Phys., Chernogolovka, 1985 [in Russian].
60. FA Kuliev. Organic Sulfides, Disulfides and Thioamides as Inhibitors of Oxidation, Doct. Sci. (Chem.) Thesis Dissertation, Petrochem. Inst., Ufa, 1986 (in Russian).
61. AS Aliev. VM Farzaliev. FA Abdullaeva. ET Denisov. *Neftekhimiya* 15:890–895, 1975.
62. JA Howard, S Korcek. *Can J Chem* 49:2178–2182, 1971.
63. JR Shelton, KE Davis. *Int J Sulfur Chem* 8:197–204, 1973.
64. DW Emerson, AP Craig, IW Potts. *J Org Chem* 32:102–105, 1967.
65. ET Denisov, VM Anisimov. *J Molec Struct (Theochem)* 545:49–60, 2001.
66. AJ Burn, R Cecil, VO Young. *J Inst Petrol* 57:319–326, 1971.
67. PI Sanin, VV Sher, IV Shkhiyants, GN Kuzmina. *Neftekhimiya* 18:693–700, 1978.
68. SK Ivanov. In: G Scott, (ed.) *Developments in Polymer Stabilisation-3.* London: Applied Science, 1980, pp. 55–116.
69. G Scott. *Eur Polym J (Suppl)* 189–197, 1969.
70. ON Grishina, VM Bashinova. *Neftekhimiya* 14:142–146, 1974.
71. ON Grishina, MI Potekhina, VM Bashinova, SF Kadyrova, YuYa Efremov. *Neftekhimiya* 24:691–695, 1984.
72. VG Vinogradova, ZK Maizus. *Kinet Katal* 13:298–302, 1972.
73. AN Zverev, VG Vinogradova, EG Rukhadze, ZK Maizus. *Izv AN SSSR Ser Khim* 2437–2440, 1973.
74. VG Vinogradova, AN Zverev, ZK Maizus. *Kinet Katal* 15:323–325, 1974.
75. IV Shkhiyants, MA Dzyubina, VV Sher, PI Sanin. *Neftekhimiya* 13:571–573, 1973.
76. IL Edilashvili, KB Ioseliani, NV Zolotova, GSh Bakhturidze, NF Dzhanibekov. *Neftekhimiya* 29:892–896, 1979.
77. JA Howard, JHB Chenier. *Can J Chem* 54:390–401, 1976.
78. LL Gervits, NV Zolotova, ET Denisov. *Neftekhimiya* 15:135–140, 1975.

79. RG Korenevskaya, GN Kuzmina, EI Markova, PI Sanin. *Neftekhimiya* 22:477–482, 1982.
80. JA Howard, JHB Chenier. *Can J Chem* 54:382–389, 1976.
81. AB Mazaletskii. The Antioxidant Activity of S-Containing Chelates of Heavy Metals in Oxidizing Hydrocarbons. PhD (Chem.) Thesis Dissertation, Institute of Chemical Physics, Chernogolovka, 1979 [in Russian].
82. RG Shelkova. Mechanism Action of Dialkyldithiocarbamates of Metals as Effective Inhibitors of Hydrocarbon Oxidation. PhD (Chem.) Thesis Dissertation, Petrochemical Institute, Moscow, 1990 [in Russian].
83. JA Howard, Y Ohkatsu, JHB Chenier, KU Ingold. *Can J Chem* 51:1543–1553, 1973.
84. DG Pobedimskii, EG Chebotareva, ShA Nasybullin, PA Kirpichnikov, AL Buchachenko. *Dokl AN SSSR* 220, 641–643, 1975.
85. DG Pobedimskii, ShA Nasybullin, EG Chebotareva, AL Buchachenko. *Izv AN SSSR Ser Khim* 1271–1277, 1976.
86. DG Pobedimskii, AD Pershin, ShA Nasybullin, AL Buchachenko. *Izv AN SSSR Ser Khim* 79–81, 1976.
87. DG Pobedimskii, ShA Nasybullin, PA Kirpichnikov, RB Svitych, OP Yablonskii, AL Buchachenko. *Org Magnetic Res* 9:61–63, 1977.
88. DG Pobedimskii, ShA Nasybullin, VKh Kadyrova, PA Kirpichnikov. *Dokl AN SSSR* 226:634–636, 1976.
89. DG Pobedimskii, NN Rzhevskaya, ShA Nasybullin, OP Yablonskii. *Kinet Katal* 18:830–836, 1977.
90. VI Kurashov, DG Pobedimskii, PA Kirpichnikov. *Dokl AN SSSR* 238:1407–1410, 1978.
91. ShA Nasybullin, NM Valeev, DG Pobedimskii. *React Kinet Catal Lett* 7:69–74, 1977.
92. VI Kurashov, DG Pobedimskii, LG Krivileva. *Zh Fiz Khim* 53:501–503, 1979.

18 Synergism of Antioxidant Action

18.1 INTRODUCTION

As shown in Chapters 14–17, the mechanisms of the antioxidant action are very diverse. One inhibitor terminates chains by reacting with the peroxyl radical, another by reacting with the alkyl radical, and the third one decreases the autoxidation rate decomposing the hydroperoxide formed. The resulting radicals of the inhibitor are either nonreactive, or can contribute to chain propagation by reacting with ROOH or RH, or can induce cyclic reactions of chain termination. An inhibitor can either reduce hydroperoxide or catalyze its breakdown. If two or more inhibitors are added to oxidized hydrocarbon (or other substance), their combined inhibitory effect can be either *additive* (the total inhibitory effect is the sum of individual effects), *antagonistic* (the inhibitors cancel out each other's effects), or *synergistic* (the total inhibitory effect is greater than the sum of individual effects).

For initiated oxidation, the inhibitory criterion could be defined as the ratio v_0/v or $(v_0/v - v/v_0)$, where v_0 and v are the rates of initiated oxidation in the absence and presence of the fixed concentration of an inhibitor, respectively. Another criterion could be defined as the ratio of the inhibition coefficient of the combined action of a few antioxidants f_Σ to the sum of the inhibition coefficients of individual antioxidants when the conditions of oxidation are fixed ($f_\Sigma = \Sigma f_i x_i$ where f_i and x_i are the inhibition coefficient and molar fraction of ith antioxidant terminating the chain). It should, however, be noted that synergism during initiated oxidation seldom takes place and is typical of autoxidation, where the main source of radicals is formed hydroperoxide. It is virtually impossible to measure the initial rate in the presence of inhibitors in such experiments. Hence, inhibitory effects of individual inhibitors and their mixtures are usually evaluated from the duration of retardation (induction period), which equals the span of time elapsed from the onset of experiment to the moment of consumption of a certain amount of oxygen or attainment of a certain, well-measurable rate of oxidation. Then three aforementioned cases of autoxidation response to inhibitors can be described by the following inequalities (τ_Σ is the induction period of a mixture of antioxidants).

Synergism	Addition	Antagonism
$\tau_\Sigma > \Sigma x_i \tau_i$	$\tau_\Sigma = \Sigma x_i \tau_i$	$\tau_\Sigma < \Sigma x_i \tau_i$

Synergism of antioxidants was reviewed in the monographs [1–9]. The classification of binary mixtures of inhibitors is based on the mechanisms of action of particular inhibitors [10,11]. One of the classification schemes is given in Ref. [11]. If one takes into account the relationship between the inhibitors and their products, the synergistic system can be divided into the following three groups [9].

1. One inhibitor breaks chains, and another diminishes the rate of autoinitiation by breaking down hydroperoxide into molecular products or deactivating a catalyst that breaks hydroperoxide into radicals.
2. Two substances (either inhibitory or not) react to form an efficient antioxidant.
3. A synergistic effect is exerted by the interaction of intermediate products formed from inhibitors. Actually, inhibitory systems may combine various synergistic mechanisms.

18.2 SYNERGISM OF CHAIN TERMINATION AND HYDROPEROXIDE DECOMPOSING THE ANTIOXIDANTS

A combined addition of a chain-breaking inhibitor and a hydroperoxide-breaking substance is widely used to induce a more efficient inhibition of oxidative processes in polyalkenes, rubbers, lubricants, and other materials [3–8]. Kennerly and Patterson [12] were the first to study the combined action of a mixture, phenol (aromatic amine) + zinc dithiophosphate, on the oxidation of mineral oil. Various phenols and aromatic amines can well serve as peroxyl radical scavengers (see Chapter 15), while arylphosphites, thiopropionic ethers, dialkylthiopropionates, zinc and nickel thiophosphates, and other compounds are used to break down hydroperoxide (see Chapter 17). Efficient inhibitory blends are usually prepared empirically, by choosing such blend compositions that induce maximal inhibitory periods [13].

The addition of an hydroperoxide-breaking substance S to oxidized RH decreases the current concentration of ROOH. This cannot affect the rate of oxidation if the radicals are produced by radiant energy or other sources unreactive to S, but it slows down the oxidation if radicals are produced from hydroperoxide. The initial step of oxidation in the absence of a peroxy-trapping substance is described by the equation (see Chapter 4)

$$\sqrt{\Delta[O_2]} = bt, \qquad (18.1)$$

where coefficient $b = 0.5 k_2 k_3 (2k_6)^{-1/2} [RH]$. The addition of antioxidant S disturbs the relationship between [ROOH] and dioxygen consumption. If S breaks down hydroperoxide by a bimolecular reaction with the rate

$$v_{ROOH} = k_S[ROOH][S] \qquad (18.2)$$

the destruction of one S molecule is accompanied by the destruction of f_S hydroperoxide molecules. The rate of hydroperoxide accumulation is given by the differential equation

$$\frac{d[ROOH]}{dt} = \frac{k_2[RH]}{\sqrt{2k_6}} \sqrt{v_{i0} + k_3[ROOH]} - (k_d + k_S[S])[ROOH] \qquad (18.3)$$

If hydroperoxide decomposition is rapid, then a quasistationary oxidation is established after the time $t \approx (k_S[S]_0)^{-1}$, and the quasistationary concentration $[ROOH]_{st}$ is given by the equation when the rate of chain initiation is very low [9] where k_d is rate constant of ROOH decomposition.

$$[ROOH]_{st} \cong \left(\frac{a[RH]\sqrt{k_3}}{k_d + k_S[S]}\right)^2 \approx \left(\frac{a[RH]\sqrt{k_3}}{k_S[S]}\right)^2 \qquad (18.4)$$

Synergism of Antioxidant Action

providing that [S] is so high that $k_S[S] \gg k_3$, coefficient $a = k_2/(2k_6)^{1/2}$. The induction period determined as the time span during which [S] decreases to the concentration $[S]_\tau = k_3/k_S$ is described by the equation [9]

$$\tau = \frac{f_S k_d}{2 k_S a^2 [RH]^2} \left\{ \left(\frac{k_S[S]_0}{k_d}\right)^2 + \frac{4 k_S[S]_0}{k_d} + 2 \ln \frac{k_S[S]_0}{k_d} \right\} \tag{18.5}$$

At $[S]_0 \gg k_d/k_S$, this expression is reduced to

$$\tau \cong \frac{f_S k_S [S]_0^2}{2 a^2 k_d [RH]^2} \tag{18.6}$$

The greater k_S and f_S, the longer the inhibitory period $\tau \sim [S]_0^2$. Obviously, S slows down oxidation only when added at a sufficiently high concentration, $[S] > k_3/k_S$. If S breaks down hydroperoxide so rapidly that $[ROOH]_{st} < v_{i0}/k_3$, then the amount of formed and broken hydroperoxide during the induction period is given by νv_{i0}. Thus, S is consumed by ν times more rapidly than free radicals are generated. Recall that, under similar conditions, a highly efficient peroxyl radical scavenger is consumed with the rate v_{i0}/f and induction period $\tau = f[\text{InH}]_0/v_{i0}$. In the other words, substance S will inhibit oxidation only if the reaction is chain-like ($\nu \gg 1$), and the basic initiator is hydroperoxide ($k_3[ROOH] > v_{i0}$). Inhibition will be noticeable if $k_S[S]_0 > k_3$. It should also be emphasized that, since hydroperoxide is usually broken both heterolytically (with the rate constant k_S) and homolytically (with the rate constant k_{Si}), the inhibitory action of S will manifest itself at $\beta_S = k_{Si}/k_S \ll 1$ and $k_{Si} \ll k_3$. Hence, the homolysis of ROOH in the reaction with S should be very low.

Consider now the case of a combined action of two inhibitors, one of which (InH) breaks the chains by reacting with peroxyl radicals and the other (S) breaks down ROOH [9]. The radical In• formed in the reaction of InH with $RO_2^\bullet$ is trapped by the peroxyl radical and, therefore, does not contribute to chain propagation (mechanism III, Chapter 14). If $k_{Si} \ll k_3$, the inhibitor InH is consumed with a rate of v_i/f, where the initiation rate is $v_i = v_{i0} + k_3[ROOH]$. After the time $t_s = (k_S[S]_0)^{-1}$, oxidation becomes quasistationary with the quasistationary hydroperoxide concentration (see Chapter 14)

$$[ROOH]_{st} = \frac{k_2[RH]v_{i0}}{fk_7[\text{InH}](k_d + k_S[S]) - k_2 k_3 [RH]} \tag{18.7}$$

It is evident from this expression that the condition of quasistationarity has the following form:

$$fk_7[\text{InH}](k_d + k_S[S]) > k_2 k_3 [RH] \tag{18.8}$$

In the presence of a hydroperoxide decomposer, InH is consumed with the rate:

$$v_{\text{InH}} = \frac{v_{i0}}{f} + \frac{k_2 k_3 [RH] v_{i0}}{f^2 k_7 [\text{InH}](k_d + k_S[S]) - f k_2 k_3 [RH]} \tag{18.9}$$

that is, S actually decreases the rate of InH decay and extends the induction period. Analysis of formulas (18.8) and (18.9) yields the following inferences. For inhibition to be efficient, the concentration [S] should be sufficiently high, so that $[S] > k_2 k_3 [RH]/fk_7 k_S[\text{InH}]$, that is, the lower [InH], the higher should be [S] to attain quasistationarity of the process. In this case, one can speak about a critical value of the product of the concentrations of two inhibitors

$([InH][S])_{cr} = k_2 k_3 [RH]/fk_7 k_S$. At such a sufficiently high [S], that $k_3[ROOH]_{st} < v_{i0}$, the rate of InH decay does not depend on [S]. Therefore, the range of efficient inhibitory concentrations of S lies in the limits

$$\frac{k_d}{k_S} \leq [S] \leq \frac{2k_2 k_3 [RH]}{fk_7 k_S [InH]} \tag{18.10}$$

The induction period is measured experimentally at the constant sum of concentrations of two antioxidants, namely, $C_0 = [S]_0 + [InH]_0 = const$. Theoretically this problem was analyzed in [9] for different mechanisms of chain termination by the peroxyl radical acceptor InH (see Chapter 14). It was supposed that antioxidant S breaks ROOH catalytically and, hence, is not consumed. The induction period was defined as $\tau = (f[InH]_0)/v_{1/2}$, where $v_{1/2}$ is the rate of InH consumption at its concentration equal to $0.5[InH]_0$. The results of calculations are presented in Table 18.1.

It is apparent from this table that the maximum of the induction period depends on the concentration of introduced inhibitors C_0 for mechanisms III and VI and is close to $x = 1/2$ for mechanisms IV and V. The experimental data on the common action of S and InH inhibitors are given in the monographs by von Voigt [6] and Shopov and Ivanov [14].

Thus, the introduction of S into RH oxidized in the presence of InH leads to the following events. The concentration of hydroperoxide and the rate of autoinitiation decrease, whereas the duration of the InH-induced inhibitory period increases. When added at a sufficiently high concentration, S leads to a quasistationary regime of oxidation. If an inhibitory mechanism implies the occurrence of critical phenomena, the addition of S decreases the critical concentration $[InH]_{cr}$ (see Chapter 14). For mechanism III,

$$[InH]_{cr} = k_2 k_3 [RH]/fk_7 k_d \tag{18.11}$$

and in the presence of S,

$$[InH]_{cr} \approx k_2 k_3 [RH]/fk_7 k_S [S]_0. \tag{18.12}$$

The synergistic effect of S diminishes if the InH-induced inhibition is so strong that it suppresses the accumulation of hydroperoxide. In this case, oxidation proceeds as a radical

---

TABLE 18.1
Equations for $[ROOH]_s$, v_{InH}, and x_{max} for Autoxidation of Hydrocarbons Inhibited by Peroxyl Radical Acceptor InH and Hydroperoxide Decomposer S where $x = [InH]_0 ([InH]_0 + [S]_0)^{-1}$, $\lambda = k_s C_0/k_d$, $\delta = fk_{11} C_0/2k_3$; for Kinetic Scheme, See Chapter 14 [9]

Key Reactions	$[ROOH]_{st}$	v_{InH}	x_{max}
(2), (3), (7)	$\dfrac{k_2[RH]v_{i0}}{fk_7 k_s[InH][S]}$	$\dfrac{k_2 k_3[RH]v_{i0}}{f^2 k_7 k_s[InH][S]}$	$1 - \left(\dfrac{2k_2 k_3[RH]}{fk_7 k_s C_0^2}\right)^{1/2}$
(2), (3), (7), (8), (10)	$\dfrac{k_2^2 k_3 k_{10}[RH]^3}{fk_7 k_8 k_s^2[S]^2[InH]}$	$\dfrac{k_2^2 k_3^2 k_{10}[RH]^3}{fk_7 k_8 k_s^2[S]^2[InH]}$	$\dfrac{3(\lambda+1)}{4\lambda} - \dfrac{1}{4}$
(2), (3), (7), (−7), (8)	$\dfrac{k_2^2 k_{-7}[RH]^2 v_{i0}}{f^2 k_7 k_8 k_s^2[S]^2[InH]}$	$\dfrac{k_2^2 k_3 k_{-7}[RH]^2 v_{i0}}{f^3 k_7 k_8 k_s^2[S]^2[InH]}$	$\dfrac{3(\lambda+1)}{4\lambda} - \dfrac{1}{4}$
(2), (3), (7), (11)	$\dfrac{k_2^2[RH]v_{i0}}{fk_7 k_s[InH][S]}$	$\dfrac{k_2[RH]v_{i0}(k_3+fk_{11}[InH])}{f^2 k_7 k_s[InH][S]}$	$\dfrac{3+\delta}{4\delta}\left(\left[1+\dfrac{16\delta}{(\delta+3)^2}\right]^{1/2}-1\right)$

reaction with $\tau = \tau_0 = (f[\text{InH}]_0)/v_{i0}$. Similar effects are observed when oxidation occurs in the presence of a catalyst capable of breaking hydroperoxide into free radicals. The introduction of a deactivator decreases the rate of autoinitiation and lengthens the inhibitory action of InH.

Sulfur compounds in combination with peroxyl radical acceptors are often used for the efficient break of hydroperoxide [14]. The mechanism of action of these inhibitory mixtures can, however, be more complex, as demonstrated with reference to a pair of 2,6-diphenylphenol and distearyl dithiopropionate [15]. The combined addition of these compounds with concentrations of 0.05% and 0.3%, respectively, results in an extended inhibitory period during the oxidation of PP (up to 3000 h at 413 K). Sulfide (for instance, β,β'-diphenylethyl sulfide) or its products not only break down ROOH, but also reduce the phenoxyl radical. Sulfoxide formed in the reaction of the sulfide with ROOH can react with ArO$^\bullet$. Thus, the ability of sulfides and their products to reduce phenoxyl radicals can contribute to their synergistic effect.

18.3 SYNERGISM OF TWO ANTIOXIDANTS TERMINATING CHAINS

18.3.1 COMBINED ACTION OF PHENOL AND AMINE ON HYDROCARBON OXIDATION

The synergistic action of a phenol and aromatic amine mixture on hydrocarbon oxidation was found by Karpukhina et al. [16]. A synergistic effect of binary mixtures of some phenols and aromatic amines in oxidizing hydrocarbon is related to the interaction of inhibitors and their radicals [16–26]. In the case of a combined addition of phenyl-N-2-naphthylamine and 2,6-bis(1,1-dimethylethyl)phenol to oxidizing ethylbenzene ($v_i = const$, 343 K), the consumption of amine begins only after the phenol has been exhausted [16], in spite of the fact that peroxyl radicals interact with amine more rapidly than with phenol (k_7 (amine) $= 1.3 \times 10^5$ and k_7 (phenol) $= 1.3 \times 10^4$ L mol^{-1} s^{-1}, respectively; 333 K). This phenomenon can be explained in terms of the fast equilibrium reaction [27–30]:

$$\text{Am}^\bullet + \text{HOAr} \rightleftharpoons \text{AmH} + \text{ArO}^\bullet$$
$$\text{Am}^\bullet + \text{HOOR} \rightleftharpoons \text{AmH} + \text{RO}_2^\bullet$$

Due to the high difference in electron affinity of the N and O atoms in the TS the reaction between peroxyl radicals and amine and the back reaction of the aminyl radical with hydroperoxide occur very rapidly (see Chapter 15). As a result, the reaction of the peroxyl radical with aromatic amine does not lead to chain termination, as well as the aminyl radical can propagate chains by reacting with ROOH. Sterically hindered phenoxyls formed from phenol, on the contrary, do not participate in chain propagation via the reaction with hydroperoxide and are trapped by RO$_2^\bullet$. The reversible equilibrium acts to decrease the concentration of Am$^\bullet$. The resulting radical ArO$^\bullet$ has low reactivity toward ROOH and reacts preferentially with the peroxyl radical terminating the chain. Hence, the peculiarities of both antioxidants lead to synergism. Synergism of this system appears to be the result of the combination of the high reactivity of amine toward the peroxyl radical, fast equilibrium exchange between two inhibitors and their radicals, and the rapid reaction of phenoxyl with the peroxyl radical. This allows one to relate synergism to the retardation of chain propagation by the reaction of Am$^\bullet$ with ROOH. For example, 2,4,6-tris(1,1-dimethylethyl)phenoxyl reacts with ROOH with the rate constant $k_{-7} = 1.3$ L mol^{-1} s^{-1} (333 K), i.e., five orders of magnitude more slowly than the diphenylaminyl radical ($k_{-7} = 6.7 \times 10^4$ L mol^{-1} s^{-1} (333 K) [4], so that reaction (8) of ArO$^\bullet$ with RO$_2^\bullet$ appears to be much more rapid than the reaction of ArO$^\bullet$ with ROOH. Varlamov and coworkers [28–33] intensively studied the reactions of phenoxyl radical with amines. They found that these reactions occur extremely rapidly. Some of them possess a negative activation energy. The following mechanism was proposed to explain this unusual peculiarity:

$$\text{Am}^\bullet + \text{HOAr} \rightleftharpoons \text{Am}^\bullet \cdots \text{HOAr}$$
$$\text{Am}^\bullet \cdots \text{HOAr} \longrightarrow \text{AmH} + \text{ArO}^\bullet$$

The experimentally measured activation energy of this reaction is the sum of the equilibrium enthalpy ΔH and activation energy E_2 of the second step: $E_{exptl} = \Delta H + E_2$. Since the enthalpy of associate $\text{Am}^\bullet \cdots \text{HOAr}$ formation is negative, $E_{exptl} < 0$ when $E_2 < |\Delta H|$. The low values of E_2 is the result of the very low thermoneutral activation energies for reactions with TS with the $\text{O} \cdots \text{H} \cdots \text{N}$ reaction center due to a great difference in electron affinity of oxygen and nitrogen atoms (see Chapter 15). The values of rate constants of aminyl radical reactions with phenols is presented in Table 18.2. The values of E_{exptl} for a few such reactions are given here [33].

Am$^\bullet$	ArOH	Solvent	E_{exptl} (kJ mol^{-1})	log A, A (L mol^{-1} s^{-1})
Ph$_2$N$^\bullet$	2,6-di-t-Bu-phenol	Toluene	−4.1	5.72
Ph$_2$N$^\bullet$	2,4,6-tri-substituted phenol	Toluene	−5.2	5.70
Ph$_2$N$^\bullet$	2,4,6-tri-t-Bu-phenol	Toluene	−5.7	5.91
Ph$_2$N$^\bullet$	2,4,6-tri-t-Bu-phenol	Decane	−9.5	5.46
(4-MeC$_6$H$_4$)$_2$N$^\bullet$	4-alkoxyphenol	Decane	−12.3	5.63
(4-MeC$_6$H$_4$)$_2$N$^\bullet$	2,4,6-tri-t-Bu-phenol	Decane	−8.1	5.59
aryl aminyl	2,6-di-t-Bu-phenol	Decane	−6.6	5.74

TABLE 18.2
Rate Constants of Reactions of *para*-Disubstituted Diphenylaminyl Radicals with Phenols in Decane and Toluene Estimated by the Laser Photolysis Technique [31–33]

Phenol	k (294 K) (L mol^{-1} s^{-1})			
	4-XC$_6$H$_4$N$^•$C$_6$H$_4$-4Y; Solvent: Decane			
para-Substituents (X, Y):	H, H	Br, Br	CH$_3$, CH$_3$	CH$_3$O, H
PhOH	9.8 × 10^5	7.0 × 10^5	6.5 × 10^5	3.6 × 10^5
4-methylphenol	7.0 × 10^6	9.6 × 10^6	3.2 × 10^6	2.2 × 10^6
hydroquinone	1.3 × 10^7	1.0 × 10^7	2.8 × 10^6	2.0 × 10^6
4-chlorophenol	2.9 × 10^6	3.8 × 10^6	1.1 × 10^6	5.6 × 10^5
4-methoxyphenol	1.5 × 10^8	1.6 × 10^8	5.9 × 10^7	2.6 × 10^7
2,6-dimethoxyphenol	2.9 × 10^7	3.0 × 10^7	1.7 × 10^7	1.3 × 10^7
2,6-dimethylphenol	2.1 × 10^7	1.8 × 10^7	9.0 × 10^6	5.1 × 10^6
2,4,6-trimethylphenol	5.0 × 10^7	4.2 × 10^7	3.1 × 10^7	2.1 × 10^7
2,5-di-*tert*-butylhydroquinone	2.6 × 10^8	2.5 × 10^8	2.3 × 10^8	1.6 × 10^8
2,6-di-*tert*-butylphenol	7.3 × 10^6	6.4 × 10^6	5.2 × 10^6	3.4 × 10^6
2,6-di-*tert*-butyl-4-methylphenol	1.2 × 10^7	7.9 × 10^6	1.1 × 10^7	7.5 × 10^6

continued

TABLE 18.2
Rate Constants of Reactions of *para*-Disubstituted Diphenylaminyl Radicals with Phenols in Decane and Toluene Estimated by the Laser Photolysis Technique [22–24]—*continued*

para-Substituents (X, Y):	H, H	Br, Br	CH$_3$, CH$_3$	CH$_3$O, H
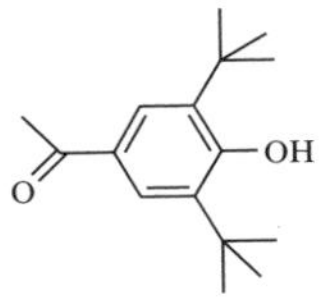 (O=C, 2,6-di-t-Bu phenol)	1.0×10^7	3.4×10^6	6.4×10^6	5.8×10^6
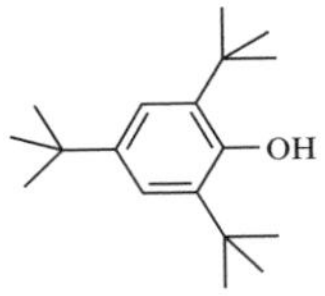 (t-Bu, 2,6-di-t-Bu phenol)	1.3×10^7	7.9×10^6	1.1×10^7	7.9×10^6
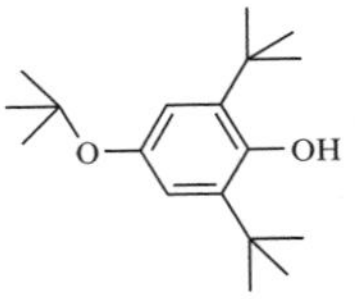 (t-BuO, 2,6-di-t-Bu phenol)	3.5×10^7	2.0×10^7	2.9×10^7	2.6×10^7
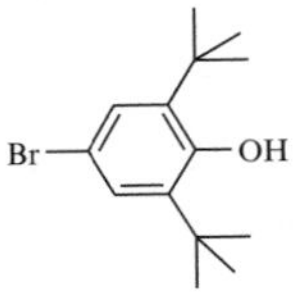 (Br, 2,6-di-t-Bu phenol)	1.6×10^7	8.7×10^7	9.0×10^6	7.8×10^6
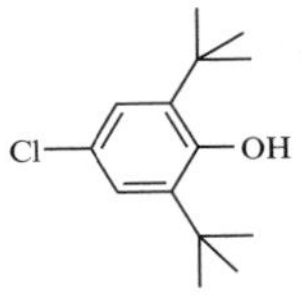 (Cl, 2,6-di-t-Bu phenol)	1.4×10^7	9.1×10^6	1.1×10^7	7.3×10^6
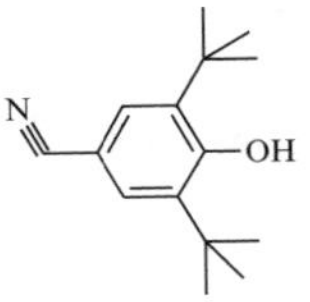 (NC, 2,6-di-t-Bu phenol)	1.1×10^7	3.3×10^6	1.2×10^7	9.8×10^6

Synergism can be observed for mixtures of amines with 2,6-bis(1,1-dimethylethyl)phenol but not with monosubstituted phenols [19]. There are two reasons for this. First, 2,6-dialkylphenols are characterized by $D_{O-H} < D_{N-H}$; therefore, the equilibrium of the above reaction is displaced toward the formation of ArO$^\bullet$. Second, phenoxyls like these are sterically hindered and, hence, must be less reactive in abstraction reactions. Thus, the necessary conditions for synergism to occur are the following.

1. An equilibrium of the reaction between Am$^\bullet$ and ArOH must easily be attainable and displaced to the right. This is the case of $D_{N-H} > D_{O-H}$ when aminyl radicals rapidly react with phenols.
2. During amine-inhibited oxidation, hydroperoxide must be accumulated at concentrations that allow the aminyl radical to participate in chain propagation. This is possible

with the chain-like pattern of inhibited oxidation when $k_2[RH] > fk_7[AmH]$. The more oxidizable the hydrocarbon, the stronger the synergistic effect [23].
3. Radicals cancel out each other mainly in the reaction of peroxyl radicals with phenoxyls to form quinolidic peroxide (see Chapter 6). At temperatures above 420 K, quinolidic peroxide breaks down into radicals, which diminishes the inhibitory effect. Therefore, the synergistic effect of a mixture of AmH and ArOH can be observed in the temperature interval T_{min} to T_{max}.

The rare example of synergistic action of a binary mixture of 1-naphthyl-N-phenylamine and phenol (1-naphthol, 2-(1,1-dimethylethyl)hydroquinone) on the initiated oxidation of cholesterol esters was evidenced by Vardanyan [34]. The mixture of two antioxidants was proved to terminate more chains than both inhibitors can do separately ($f_\Sigma > \Sigma f_i x_i$). For example, 1-naphtol in a concentration of 5×10^{-5} mol L^{-1} creates the induction period $\tau = 170$ s, 1-naphthyl-N-phenylamine in a concentration of 1.0×10^{-4} mol L^{-1} creates the induction period $\tau = 400$ s, and together both antioxidants create the induction period $\tau = 770$ s (oxidation of ester of pelargonic acid cholesterol at $T = 348$ K with AIBN as initiator). Hence, the ratio $f_\Sigma/\Sigma f_i x_i$ was found equal to 2.78. The formation of an efficient intermediate inhibitor as a result of interaction of intermediate free radicals formed from phenol and amine was postulated. This inhibitor was proved to be produced by the interaction of oxidation products of phenol and amine.

18.3.2 COMBINED ACTION OF TWO PHENOLS

The combined addition of two phenols, one of which is sterically hindered, for example, 2,6-bis(1,1-dimethylethyl)phenol, and another is sterically nonhindered also leads to a synergistic effect [35–38]. As found by Mahoney [35], 2,4,6-tris(1,1-dimethylethyl)phenol with a concentration of 10^{-4} L mol^{-1} does not virtually inhibit the initiated oxidation of 9,10-dihydroanthracene (333 K), but p-methoxyphenol, taken in the same concentration, does inhibit oxidation. The induction period doubles if two phenols are added together in equal concentrations, which indicates that both phenols are involved in chain termination. The mechanism of synergistic action can be explained by the following kinetic scheme [35]:

$$RO_2^\bullet + Ar_1OH \longrightarrow ROOH + Ar_1O^\bullet$$
$$RO_2^\bullet + Ar_2OH \longrightarrow ROOH + Ar_2O^\bullet$$
$$Ar_1O^\bullet + Ar_2OH \rightleftharpoons Ar_1OH + Ar_2O^\bullet$$
$$Ar_1O^\bullet + ROOH \longrightarrow Ar_1OH + RO_2^\bullet$$
$$Ar_1O^\bullet + RH \longrightarrow Ar_1OH + R^\bullet$$
$$R^\bullet + O_2 \longrightarrow RO_2^\bullet$$
$$Ar_2O^\bullet + RO_2^\bullet \longrightarrow \text{Quinolide peroxide}$$

The peroxyl radical reacts rapidly with nonhindered phenol. However, the formed phenoxyl radical is active in reactions with RH and ROOH and these reactions decrease the inhibiting activity of lone Ar_1OH. Sterically hindered phenol Ar_2OH possesses lower activity in the reaction with the peroxyl radical due to the steric effect (see Chapter 15). However, the formed sterically hindered phenoxyl $Ar_2O^\bullet$ practically does not participate in chain propagation reactions and reacts only with peroxyl radicals. The rapid equilibrium exchange between phenols and phenoxyls helps to intensify chain termination at the introduction of both phenols. The fast exchange reaction of phenoxyl $Ar_1O^\bullet$ with phenol Ar_2OH is the result of the low thermoneutral energy of the reaction with the TS of the ArO $\cdots$ H $\cdots$ OAr type due to the absence of triplet repulsion in such TS [39]. This is seen from the values of activation energies of thermoneutral reactions E_{e0} (see Chapter 6) of phenoxyl radicals with phenols and hydrocarbons for comparison [39,40].

Reaction	$Ar_1O^\bullet + R^1H$	$Ar_1O^\bullet + Ar_1OH$	$Ar_2O^\bullet + R^1H$	$Ar_2O^\bullet + Ar_1OH$
E_{e0} (kJ mol^{-1})	62.9	39.7	85.4	43.6
ΔE_{e0} (kJ mol^{-1})		−23.2		−41.8

We observe a great difference in the E_{e0} values for the reactions with TS of the type ArO $\cdots$ H $\cdots$ R (high triplet repulsion) and ArO $\cdots$ H $\cdots$ OAr (low triplet repulsion). The parameters of IPM for reactions of phenoxyl with phenols are given in Table 18.3.

The values of rate constants of the reaction of selected phenoxyls with ionol calculated by the IPM method are given in Table 18.4. Experimental data on reactions of phenoxyls with phenols, hydroperoxides, and hydrocarbons can be found in Database [41] and those calculated by the IPM method values in the *Handbook of Antioxidants* [4].

TABLE 18.3
Kinetic Parameters of Radical Abstraction Reactions by Phenoxyl Radicals in IPM [4]

Reaction	α	$b \times 10^{-11}$ (kJ mol^{-1})$^{1/2}$ m^{-1}	$0.5hL\nu_i$ (kJ mol^{-1})	$0.5hL(\nu_i - \nu_f)$ (kJ mol^{-1})	$(r^{\neq}/r_e)_0$
ArO$^\bullet$ + HR	0.802	3.743	17.4	−4.1	0.445
ArO$^\bullet$ + HOOR	0.986	4.600	21.2	−0.3	0.500
ArO$^\bullet$ + HOAr	1.000	4.665	21.5	0.0	0.500
ArO$^\bullet$ + HOAm	1.000	4.665	21.5	0.0	0.500
ArO$^\bullet$ + HAm	0.927	4.324	20.0	−1.5	0.519
ArO$^\bullet$ + HSAr	0.649	3.026	13.8	−7.7	0.606

Reaction	br_e (kJ mol^{-1})$^{1/2}$	E_{e0} (kJ mol^{-1})	A (L mol^{-1} s^{-1})	$-\Delta H_{e\ min}$ (kJ mol^{-1})	$\Delta H_{e\ max}$ (kJ mol^{-1})
$Ar_1O^\bullet + R_1H$	14.29	62.9	1.0×10^9	141.8	90.2
$Ar_1O^\bullet + R_2H$	15.95	78.3	1.0×10^8	198.3	128.1
$Ar_1O^\bullet + R_3H$	15.02	69.5	1.0×10^8	165.6	106.2
$Ar_1O^\bullet + ROOH$	13.00	43.3	3.2×10^7	51.3	52.3
$Ar_1O^\bullet + Ar_1OH$	12.61	39.7	1.0×10^9	42.1	42.1
$Ar_1O^\bullet + Ar_2OH$	13.20	43.6	1.0×10^8	51.8	51.8
$Ar_1O^\bullet + AmH$	10.48	29.6	1.0×10^8	22.0	16.7
$Ar_1O^\bullet + AmOH$	13.54	45.8	1.0×10^8	57.8	57.8
$Ar_1O^\bullet + ArSH$	10.75	42.5	1.0×10^8	103.7	38.4
$Ar_2O^\bullet + R_1H$	16.65	85.4	1.0×10^9	224.7	145.7
$Ar_2O^\bullet + R_2H$	18.58	106.3	1.0×10^8	305.4	199.3
$Ar_2O^\bullet + R_3H$	17.50	94.3	1.0×10^8	258.8	168.3
$Ar_2O^\bullet + ROOH$	14.10	51.1	3.2×10^7	71.5	72.8
$Ar_2O^\bullet + Ar_1OH$	13.20	43.6	1.0×10^8	51.8	51.8
$Ar_2O^\bullet + Ar_2OH$	14.37	51.6	1.0×10^8	73.2	73.2
$Ar_2O^\bullet + AmOH$	14.49	52.5	1.0×10^8	75.6	75.6
$Ar_2O^\bullet + AmH$	11.15	33.5	1.0×10^8	31.9	26.1
$Ar_2O^\bullet + ArSH$	11.25	46.5	1.0×10^8	121.0	46.4

TABLE 18.4
Enthalpies, Activation Energies, and Rate constants of Reactions of Phenoxyl Radicals ($Ar_1O^\bullet$) with 2,6-bis(1,1-dimethylethyl)-4-methylphenol (Ionol) Calculated by IPM Method, Equations See in Chapter 6, for the Values of α, br_e, and A, See Table 18.3

Phenoxyl Radical	ΔH (kJ mol^{-1})	E (kJ mol^{-1})	k (333 K) (L mol^{-1} s^{-1})	k (400 K) (L mol^{-1} s^{-1})
PhO$^\bullet$	−30.0	9.8	2.9×10^6	5.2×10^6
2-Me-C$_6$H$_4$O$^\bullet$	−11.3	5.5	1.4×10^7	1.9×10^7
3-Me-C$_6$H$_4$O$^\bullet$	−18.1	2.7	3.8×10^7	4.5×10^7
4-Me-C$_6$H$_4$O$^\bullet$	−13.6	4.5	1.9×10^7	2.6×10^7
4-t-Bu-C$_6$H$_4$O$^\bullet$	−21.1	13.6	7.2×10^5	1.6×10^6
4-HO-C$_6$H$_4$O$^\bullet$	−3.4	9.0	3.8×10^6	6.6×10^6
4-Me$_2$N-C$_6$H$_4$O$^\bullet$	10.2	28.8	3.0×10^3	1.7×10^4
4-O$_2$N-C$_6$H$_4$O$^\bullet$	−24.2	0.5RT	1.6×10^8	1.6×10^8
4-MeC(O)-C$_6$H$_4$O$^\bullet$	−32.6	8.8	4.2×10^6	7.1×10^6
4-H$_2$N-C$_6$H$_4$O$^\bullet$	−17.7	15.1	4.2×10^5	1.1×10^6
4-NC-C$_6$H$_4$O$^\bullet$	−31.2	9.4	3.4×10^6	6.0×10^6
2,6-(MeO)$_2$-C$_6$H$_3$O$^\bullet$	−8.9	19.2	9.7×10^4	3.1×10^5
2,4-Me$_2$-C$_6$H$_3$O$^\bullet$	−21.5	13.5	7.7×10^5	1.7×10^6
2,6-Me$_2$-C$_6$H$_3$O$^\bullet$	−15.6	16.1	3.0×10^5	7.9×10^5
3-MeO-C$_6$H$_4$O$^\bullet$	−21.2	0.5RT	7.6×10^7	7.6×10^7

continued

TABLE 18.4
Enthalpies, Activation Energies, and Rate Constants of Reactions of Phenoxyl Radicals (Ar$_1$O$^•$) with 2,6-bis(1,1-dimethylethyl)-4-methylphenol (Ionol) Calculated by IPM Method, Equations See in Chapter 6, for the Values of α, br_e, and A, See Table 18.3—*continued*

Phenoxyl Radical	ΔH (kJ mol^{-1})	E (kJ mol^{-1})	k (333 K) (L mol^{-1} s^{-1})	k (400 K) (L mol^{-1} s^{-1})
	2.8	12.0	1.3×10^6	2.7×10^6
	−11.0	18.2	1.4×10^5	4.2×10^5
	−16.8	15.6	3.6×10^5	9.3×10^5
	17.2	19.5	8.8×10^4	2.9×10^5
	13.9	17.7	1.7×10^5	4.9×10^5
	1.1	11.2	1.8×10^6	3.5×10^6

One more mechanism responsible for the synergistic action of two phenols has recently been discovered during the study of disproportionation of phenoxyl radicals [42,43]. This reaction is possible only for phenoxyls possessing C—H groups in the *ortho-* or *para-*position [43]. For instance, 2,4,6-tris(1,1-dimethylethyl)phenoxyl is unable to disproportionate, whereas ionol radicals disproportionate. It was found that the cross-disproportionation of ionol and α-tocopherol radicals occurs much more rapidly than homodisproportionation (323 K [42]):

Phenoxyl	Ionol	α-Tocopherol	Ionol + α-tocopherol
$2k_{dis}$ (L mol^{-1} s^{-1})	8.7×10^3	2.2×10^3	1.8×10^5

In this reaction, α-tocopherol is regenerated at the expense of the ionoxyl radical. Rapid cross-disproportionation diminishes the total concentration of phenoxyl radicals (thereby preventing their participation in chain propagation) and reduces the reactive phenol (α-tocopherol is more reactive toward ROOH than ionol) into methylenequinone, which terminates the chains in the reaction with peroxyl radicals.

18.3.3 SYNERGISM OF PHENOL AND NITROXYL RADICAL ACTION ON HYDROCARBON OXIDATION

A mixture of two antioxidants, one acceptor of the peroxyl radical (phenol or amine) and another alkyl radical acceptor (stable nitroxyl radical), causes the synergistic effect in autoxidation of hydrocarbons (ethylbenzene and nonene-1) [44–46].

Nitroxyl radicals as alkyl radical acceptors are known to be very weak antioxidants due to the extremely fast addition of dioxygen to alkyl radicals (see Chapter 2). They retard the oxidation of solid polymers due to specific features of free radical reactions in the solid polymer matrix (see Chapter 19). However, the combination of two inhibitors, one is the peroxyl radical acceptor (phenol, aromatic amine) and another is the alkyl radical acceptor (nitroxyl radical) showed the synergistic action [44–46]. The results of testing the combination of nitroxyl radical (>NO$^\bullet$) (2,2,6,6-tetramethyl-4-benzoylpiperidine-1-oxyl) + amine (phenol) in the autoxidation of nonene-1 at 393 K are given here ([>NO$^\bullet$]$_0$ + [InH]$_0$ = 1.5 × 10^{-4} mol L^{-1} pO_2 = 98 kPa) [44].

InH	τ (InH) (h)	τ (InH +>NO$^\bullet$) (h)	[InH]/[>NO$^\bullet$]
(tert-butyl phenol structure)	1.8	14.5	1.9
(methoxy phenol structure)	10.0	18.5	5.7
(dimethyl phenol structure)	7.8	18.0	4.0
(di-isopropenyl phenol structure)	5.0	21.5	3.0
Ph$_2$NH	7.0	16.5	5.7
(1-naphthylamine)	2.0	16.0	3.0
(N-phenyl-2-naphthylamine)	9.0	16.0	3.0

Synergism increases when oxidation occurs in air at pO_2 = 20 kPa. The kinetic scheme of autoxidation was analyzed with independent chain termination by the reactions of InH with RO$_2^\bullet$ and >NO$^\bullet$ with R$^\bullet$(A). The result was negative, namely, such a simple mechanism does not show a synergistic action. The explanation of synergism in such system lies in the exchange of the type.

$$>NO^\bullet + HOAr \rightleftharpoons\, >NOH + ArO^\bullet$$
$$>NO^\bullet + HAm \rightleftharpoons\, >NOH + Am^\bullet \qquad (16)$$

and a very rapid reaction of peroxyl radicals with the formed hydroxylamine and alkyl radicals with >NO$^\bullet$, ArO$^\bullet$, and Am$^\bullet$. Nitroxyl radicals, like phenoxyl, react with phenols and amines with a very low activation energy of the thermonentral reaction [47].

Reactant	R¹H	Ar_1OH	Ar_2OH	AmH	AmOH
E_{e0} (kJ mol^{-1})	63.0	45.8	52.5	38.3	45.8

Similar to phenoxyl radical, the nitroxyl radical reacts rapidly with phenol Ar_1OH due to the low triplet repulsion in the TS of the structure >NO $\cdots$ H $\cdots$ OAr and very rapidly with amine due to a high difference in electron affinity in the TS of the structure >NO $\cdots$ H $\cdots$ Am. The IPM parameters for the nitroxyl radical reactions are presented in Table 18.5.

The values of calculated activation energies and rate constants of the >NO$^•$ reactions with chosen phenols and amines are given in Table 18.6. The hydroxylamine formed by the reaction of the nitroxyl radical with InH reacts with peroxyl radicals very rapidly (see Table 18.7). So, two reactions of chain termination occur in oxidized RH in the presence of >NO$^•$ and InH and chain termination includes the following cycles of reactions.

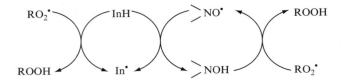

TABLE 18.5
Kinetic Parameters of Radical Abstraction Reactions by Nitroxyl Radicals in IPM [4]

Reaction	α	$b \times 10^{-11}$ (kJ mol^{-1})$^{1/2}$ m^{-1}	$0.5hL\nu_i$ (kJ mol^{-1})	$0.5hL(\nu_i - \nu_f)$ (kJ mol^{-1})	$(r^{\neq}/r_e)_0$
>NO$^•$ + HR	0.802	3.743	17.4	−4.1	0.555
>NO$^•$ + HOOR	0.986	4.600	21.2	−0.3	0.500
>NO$^•$ + HOAr	1.000	4.665	21.5	0.0	0.500
>NO$^•$ + HAm	0.927	4.324	20.0	−1.5	0.519
>NO$^•$ + HOAm	1.000	4.665	21.5	0.0	0.500
>NO$^•$ + HSAr	0.649	3.026	13.8	−7.7	0.606
$RO_2^•$ + HON<	1.014	4.665	21.5	0.3	0.500

Reaction	br_e (kJ mol^{-1})$^{1/2}$	E_{e0} (kJ mol^{-1})	A (L mol^{-1} s^{-1})	$-\Delta H_{e\,min}$ (kJ mol^{-1})	$\Delta H_{e\,max}$ (kJ mol^{-1})
>NO$^•$ + R_1H	14.30	63.0	1.0×10^9	142.1	90.5
>NO$^•$ + R_2H	16.13	80.1	1.0×10^8	204.6	132.3
>NO$^•$ + R_3H	15.20	71.1	1.0×10^8	166.2	106.7
>NO$^•$ + ROOH	12.89	39.0	1.0×10^8	44.5	46.1
>NO$^•$ + Ar_1OH	13.54	45.8	1.0×10^8	57.8	57.8
>NO$^•$ + Ar_2OH	14.49	52.5	1.0×10^8	75.6	75.6
>NO$^•$ + AmH	11.93	38.3	1.0×10^8	37.1	32.6
>NO$^•$ + AmOH	13.53	45.8	1.0×10^8	57.6	57.6
>NO$^•$ + ArSH	10.80	42.9	1.0×10^8	105.4	39.2
$RO_2^•$ + HON<	13.50	45.6	3.2×10^7	54.9	56.8

TABLE 18.6
Enthalpies, Activation Energies, and Rate Constants of Reactions of Nitroxyl Radicals (>NO$^\bullet$) with O—H Bond of *p*-methoxyphenol, Ionol, and 4,4-dimethoxydiphenylamine Calculated by IPM Method; for Equations See Chapter 6 and for the Values of α, br_e, and A, See Table 18.5

Nitroxyl Radical	ΔH (kJ mol^{-1})	E (kJ mol^{-1})	k (333 K) (L mol^{-1} s^{-1})	k (400 K) (L mol^{-1} s^{-1})
		p-methoxyphenol		
(structure)	38.1	46.9	4.5	76
(structure)	29.3	41.6	29	3.6×10^2
(structure)	50.6	54.6	0.27	7.4
(structure)	48.1	53.0	0.48	12
(structure)	45.5	51.4	0.86	19
(structure)	42.4	49.5	1.7	35
(structure)	37.9	46.7	4.7	79
(structure)	43.6	50.2	1.3	28
(structure)	45.4	51.3	0.88	20
(structure)	54.8	57.3	0.10	3.3

continued

TABLE 18.6
Enthalpies, Activation Energies, and Rate Constants of Reactions of Nitroxyl Radicals (>NO•) with O—H Bond of *p*-methoxyphenol, Ionol, and 4,4-dimethoxydiphenylamine Calculated by IPM Method; for Equations See Chapter 6 and for the Values of α, br_e, and A, See Table 18.5—*continued*

Nitroxyl Radical	ΔH (kJ mol^{-1})	E (kJ mol^{-1})	k (333 K) (L mol^{-1} s^{-1})	k (400 K) (L mol^{-1} s^{-1})
	47.1	52.4	0.60	14
	44.9	51.0	0.99	22
	31.3	49.3	1.9	36
	22.5	44.3	11	1.6×10^2
	43.8	56.7	0.13	4.0
	41.3	55.2	0.22	6.3
	38.7	53.6	0.39	10
	35.6	51.8	0.75	17
	31.1	49.2	1.9	38
	36.8	52.5	0.58	14

TABLE 18.6
Enthalpies, Activation Energies, and Rate Constants of Reactions of Nitroxyl Radicals (>NO•) with O—H Bond of *p*-methoxyphenol, Ionol, and 4,4-dimethoxydiphenylamine Calculated by IPM Method; for Equations See Chapter 6 and for the Values of α, br_e, and A, See Table 18.5—*continued*

Nitroxyl Radical	ΔH (kJ mol^{-1})	E (kJ mol^{-1})	k (333 K) (L mol^{-1} s^{-1})	k (400 K) (L mol^{-1} s^{-1})
	38.6	53.6	0.40	10
	48.0	59.2	5.1×10^{-2}	1.8
	40.3	54.6	0.27	7.5
	38.1	53.3	0.44	11
	40.9	42.4	53.6	6.9×10^2
	32.1	36.1	2.17×10^2	1.9×10^3
	53.4	54.9	2.2	61.0
	50.9	52.4	4.5	1.1×10^2
	48.3	49.8	9.1	1.9×10^2

continued

TABLE 18.6
Enthalpies, Activation Energies, and Rate Constants of Reactions of Nitroxyl Radicals (>NO•) with O—H Bond of *p*-methoxyphenol, Ionol, and 4,4-dimethoxydiphenylamine Calculated by IPM Method; for Equations See Chapter 6 and for the Values of α, br_e, and A, See Table 18.5—*continued*

Nitroxyl Radical	ΔH (kJ mol^{-1})	E (kJ mol^{-1})	k (333 K) (L mol^{-1} s^{-1})	k (400 K) (L mol^{-1} s^{-1})
(Br, O=, N—O•)	45.2	46.7	20.3	3.4×10^2
(Cl, O=, N—O•)	40.7	42.2	55.2	7.1×10^2
(HO—, N—O•)	46.4	47.9	15.0	2.7×10^2
(O=, N—O•)	48.2	49.7	9.3	1.9×10^2
(O=CH—NH, N—O•)	57.6	59.1	6.44×10^{-2}	2.3
(O=, N—O•)	49.9	51.4	5.89	1.3×10^2
(HO, N, N—O•, O=)	47.7	49.2	10.7	2.1×10^2

The combined action of InH and >NOH appears more efficient due to two parallel reactions of chain termination via reactions of peroxyl radicals with InH and >NOH and the rapid exchange between InH and the nitroxyl radical. The decay of inhibitors proceed by the reactions of In• with peroxyl and >NO• with alkyl radicals.

18.3.4 SYNERGISTIC ACTION OF QUINONE IN COMBINATION WITH PHENOL OR AMINE

Quinones are formed by the reaction of the peroxyl radical with phenoxyls (see Chapter 15). They are known as inhibitors of free radical polymerization of monomers where they retard the reaction terminating chains by the reaction with macroradicals [9]. Quinones do not react with peroxyl radicals and react with alkyl radicals by a few orders magnitude [5–7] more slowly than dioxygen does. It was a surprising phenomena that quinones appeared to

TABLE 18.7
Enthalpies, Activation Energies, and Rate Constants of Reactions Secondary Peroxyl Radicals with Hydroxylamines (>NOH): $RO_2^• + HON< \longrightarrow ROOH + >NO^•$ in Hydrocarbon Solutions Calcualted by IPM Method; for Equations See Chapter 6 and for the Values of α, br_e, and A, See Table 18.5

Hydroxylamine	ΔH (kJ mol^{-1})	E (kJ mol^{-1})	k (333 K) (L mol^{-1} s^{-1})	k (400 K) (L mol^{-1} s^{-1})
	−42.5	7.0	2.6×10^6	3.9×10^6
	−70.4	$0.5RT$	2.0×10^8	2.0×10^8
	−68.6	$0.5RT$	1.1×10^8	1.1×10^8
	−59.7	$0.5RT$	4.2×10^7	4.2×10^7
	−12.8	19.7	2.7×10^4	8.6×10^4
	−27.0	13.3	2.6×10^5	5.9×10^5
	−29.9	12.1	4.4×10^5	8.3×10^5
	−67.8	$0.5RT$	1.7×10^8	1.7×10^8
	−65.2	$0.5RT$	1.3×10^8	1.3×10^8
	−63.3	$0.5RT$	1.1×10^8	1.1×10^8
	−65.1	$0.5RT$	1.3×10^8	1.3×10^8

be synergistic in combination with phenols and aromatic amines introduced in oxidized hydrocarbons [44–46]. The values of induction period τ of 1-nonene autoxidation at $T = 393$ K with the total inhibitor concentration $[InH]_0 + [Q]_0 = 1.5 \times 10^{-4}$ mol L^{-1} [44] are presented below:

Inhibitor	Q	[InH]:[Q]	pO_2(kPa)	τ (h)
N-phenyl-2-naphthylamine	—	1:0	98	9
N-phenyl-2-naphthylamine	2,6-di-tert-butyl-1,4-benzoquinone	7:3	98	17
N-phenyl-2-naphthylamine	2,6-di-tert-butyl-1,4-benzoquinone	8:2	20	27
2,6-di-tert-butylphenol	—	1:0	20	5
2,6-di-tert-butylphenol	2,6-di-tert-butyl-1,4-benzoquinone	7:3	98	8.5
2,6-di-tert-butylphenol	2,6-di-tert-butyl-1,4-benzoquinone	7:3	20	15.5

It is seen that the substitution of the part of amine or phenol by quinone prolongs the induction period by two or three times. The mechanism of synergistic action of quinone is the same as in the case of nitroxyl radicals. Quinone reacts with InH with production of semiquinone radicals. The latter rapidly reacts with peroxyl radicals and provokes the additional rapid chain termination [47].

$$InOOR \xleftarrow{RO_2^\bullet} In^\bullet \xleftarrow{InH} Q \xrightarrow{} {}^\bullet QH \xrightarrow{RO_2^\bullet} ROOH$$

The equations for the rate of oxidation of organic compound RH inhibited by a mixture of InH and Q and the characteristics of autoxidation when the mechanism includes different combinations of key reactions are presented in Table 18.8.

TABLE 18.8
Equations for the Rate of Oxidation v, Critical Concentration of Antioxidant $[InH]_{cr}$, and Quasistationary Concentration of Formed Hydroperoxide $[ROOH]_s$ at Common Action of Antioxidant and Quinone Q or Nitroxyl (>NO$^\bullet$) [4]

v (mol L^{-1} s^{-1})	$[ROOH]_s$
Key Reactions: (2), (3), (7), (16), $[InH]_{cr} = \beta k_2[RH]/fk_7$	
$\dfrac{k_2[RH]v_i}{fk_7[InH]} - \dfrac{k_7 k_{16}[InH][>NO^\bullet]}{fk_7}$	$\dfrac{\beta k_2[RH](v_{i0} - 2k_{16}[InH][>NO^\bullet])}{fk_7(1-\beta v)}$
Key Reactions: (2), (7), (8), (10), (16)	
$k_2[RH]\sqrt{\dfrac{k_{10}[RH]v_i}{fk_7k_8[InH]}}\left\{1 - \dfrac{k_{16}[>NO^\bullet][InH]}{k_{10}[RH]v_i}\right\}$	$\dfrac{(\beta k_2[RH])^2 k_{10}[RH]}{4k_3 k_7 k_8[RH]}\left(1 - \sqrt{1 - \dfrac{8k_7 k_8 k_{16}[InH][Q]}{\beta^2 k_2^2 k_{10}[RH]^3}}\right)$
Key Reactions: 2, 7, −7, 8, 16; $[InH]_{cr} = \beta^2 k_2^2 k_{-7}[RH]^{2}/f k_3 k_7 k_8$	
$k_2[RH]\sqrt{\dfrac{k_{-7}[ROOH]v_i}{fk_7k_8[InH]}}\left\{1 - \dfrac{k_{16}[Q][InH]}{k_{-7}[ROOH]v_i}\right\}$	$\dfrac{(v_{i0} - k_{16}[InH][Q])[InH]_{cr}}{fk_3([InH] - [InH]_{cr})}$

The reactions of quinones with phenols and amines are endothermic due to low BDE of the formed $^\bullet$Q—H bond. Hence, the synergism of quinones should be noticeable at elevated temperatures. The values of BDE in a few semiquinone radicals are given below:

Radical	(p-benzosemiquinone)	(methyl)	(2,5-dimethyl)	(tetrachloro)
D (kJ mol^{-1})	226.3	229.8	229.5	245.0

Reactions of quinone with phenols and amines are endothermic; however, the rate constants of reactions Q + InH are not low at elevated temperatures due to the low E_{e0} values, as in the case of nitroxyl radicals reactions [4,48,49].

Reaction	Q + R^{1}H	Q + Ar$_1$OH	Q + Ar$_2$OH	Q + AmH
E_{e0} (kJ mol^{-1})	62.9	39.7	43.6	29.6

As in the case of phenoxyl and nitroxyl radical reactions, the value of E_{e0} for the quinone reaction with phenol (Ar$_1$OH) is much lower than that for the reaction of Q with R^{1}H ($\Delta E_{e0} = 23$ kJ mol^{-1}). Such a difference is the result of the high triplet repulsion in TS of the type C $\cdots$ H $\cdots$ O and low in the TS of the type O $\cdots$ H $\cdots$ O, as in the reactions of the nitroxyl radical. The very low value of E_{e0} for the reaction Q with aromatic amine is due to a high difference in electron affinity of N and O atoms in TS of the type O $\cdots$ H $\cdots$ N. The values of rate constants of p-benzoquinone with several inhibitors were calculated by the IPM method. The parameters of the IPM model are collected in Table 18.9.

TABLE 18.9
Kinetic Parameters of Radical Abstraction Reactions by Quinones in IPM [4]

Reaction	α	$b \times 10^{-11}$ (kJ mol^{-1})$^{1/2}$ m^{-1}	$0.5hL\nu_i$ (kJ mol^{-1})	$0.5hL(\nu_i - \nu_f)$ (kJ mol^{-1})	$(r^{\neq}/r_e)_0$
Q + HR	0.802	3.743	17.4	−4.1	0.445
Q + HOOR	0.986	4.600	21.2	−0.3	0.500
Q + HOAr	1.000	4.665	21.5	0.0	0.500
Q + HOAm	1.000	4.665	21.5	0.0	0.500
Q + HAm	0.927	4.324	20.0	−1.5	0.519
Q + HSAr	0.649	3.026	13.8	−7.7	0.606

Reaction	br_e (kJ mol^{-1})$^{1/2}$	E_{e0} (kJ mol^{-1})	A (L mol^{-1} s^{-1})	$-\Delta H_{e\,min}$ (kJ mol^{-1})	$\Delta H_{e\,max}$ (kJ mol^{-1})
Q + R$_1$H	14.29	62.9	2.0×10^9	141.8	90.2
Q + R$_2$H	15.95	78.3	2.0×10^8	198.3	128.1
Q + R$_3$H	15.02	69.5	2.0×10^8	165.6	106.2
Q + ROOH	13.00	43.3	2.0×10^8	51.3	52.3
Q + Ar$_1$OH	12.61	39.7	2.0×10^9	42.1	42.1
Q + Ar$_2$OH	13.20	43.6	2.0×10^8	51.8	51.8
Q + AmH	10.48	29.6	2.0×10^8	22.0	16.7
Q + AmOH	13.54	45.8	2.0×10^8	57.8	57.8
Q + ArSH	10.75	42.5	2.0×10^8	103.7	38.4

The values of activation energies and rate constants of the reactions of p-benzoquinone with a few phenols and amines are presented in Table 18.10. For additional data on this reaction see handbook [4] and Database [41].

18.3.5 THE COMBINED ACTION OF FE AND CU SALTS

Iron salts and complexes soluble in hydrocarbon catalyze oxidation due to the rapid decomposition of forming peroxides into free radicals, for example (see Chapter 10)

$$ROOH + Fe(II) \longrightarrow RO^{\bullet} + Fe(III) + HO^{-}$$

Soluble copper salts were found to possess the outstanding inhibiting activity in the presence of iron salts [50]. This copper antioxidant activity is believed to be the result of the fast oxidation of catalytically active ferrous ions by cupric ions

$$Cu^{2+} + Fe^{2+} \longrightarrow Cu^{1+} + Fe^{3+}$$

followed by fast chain termination in the reactions

$$RO_2^{\bullet} + Cu^{1+} \longrightarrow RO_2^{-} + Cu^{2+}$$
$$R^{\bullet} + Cu^{2+} \longrightarrow olefin + H^{+} + Cu^{1+}$$

Since transition metal salts catalyze the autoxidation of organic compounds, the various deactivators are used to decrease catalyzed hydroperoxide decomposition to free radicals. Different chelate-forming compounds are used as such deactivators [6]. For example, for

TABLE 18.10
Enthalpies, Activation Energies, and Rate Constants of Reactions of p-Benzoquinone with Phenols and Amines: InH + Q ⟶ In• + HQ• in Hydrocarbon Solutions Calculated by IPM Method; for Equations See Chapter 6 and for the Values of α, br_e, and A, See Table 18.9

Phenol	ΔH (kJ mol^{-1})	E (kJ mol^{-1})	A (L mol^{-1} s^{-1})	k (400 K) (L mol^{-1} s^{-1})
		Ar$_1$OH		
PhOH	140.9	142.6	1.8×10^{10}	4.3×10^{-9}
	134.1	135.8	1.6×10^{10}	2.9×10^{-8}
	100.7	102.4	9.0×10^{9}	3.8×10^{-4}
	117.7	119.4	1.1×10^{10}	2.8×10^{-6}
	125.7	127.4	1.5×10^{10}	3.5×10^{-7}
	99.0	100.7	8.7×10^{9}	6.2×10^{-4}
	119.8	121.5	1.3×10^{10}	1.8×10^{-6}
	115.3	117.0	1.2×10^{10}	6.3×10^{-6}
	128.6	130.3	1.5×10^{10}	1.4×10^{-7}
	103.3	105.0	9.0×10^{9}	1.7×10^{-4}
	106.6	108.3	8.0×10^{9}	5.8×10^{-5}
	123.9	125.6	1.4×10^{10}	5.5×10^{-7}
	116.6	118.3	1.2×10^{10}	4.3×10^{-6}
	110.2	111.9	9.0×10^{9}	2.2×10^{-5}

continued

TABLE 18.10
Enthalpies, Activation Energies, and Rate Constants of Reactions of *p*-Benzoquinone with Phenols and Amines: InH + Q ⟶ In• + HQ• in Hydrocarbon Solutions Calculated by IPM Method; for Equations See Chapter 6 and for the Values of α, br_e, and A, See Table 18.9—*continued*

Phenol	ΔH (kJ mol^{-1})	E (kJ mol^{-1})	A (L mol^{-1} s^{-1})	k (400 K) (L mol^{-1} s^{-1})
1-naphthol	115.3	117.0	1.2×10^{10}	6.3×10^{-6}
2-naphthol	125.7	127.4	1.5×10^{10}	3.5×10^{-7}
1,5-dihydroxynaphthalene	105.4	107.1	1.0×10^{10}	1.0×10^{-4}
dihydroxypyrene	87.8	89.5	6.3×10^{9}	1.3×10^{-2}
N-acetyl-hydroxytetrahydroquinoline	107.6	109.3	1.1×10^{10}	5.9×10^{-5}
Ar$_2$OH				
4-amino-2,6-di-tert-butylphenol	106.7	108.4	7.2×10^{9}	5.0×10^{-5}
2,6-di-tert-butylphenol	118.6	120.3	9.2×10^{9}	1.8×10^{-6}
4-diphenylmethyl-2,6-di-tert-butylphenol	114.2	115.9	8.3×10^{9}	6.1×10^{-6}
4-methoxy-2,6-di-tert-butylphenol	99.0	101.7	5.6×10^{9}	2.9×10^{-4}

TABLE 18.10
Enthalpies, Activation Energies, and Rate Constants of Reactions of *p*-Benzoquinone with Phenols and Amines: InH + Q $\longrightarrow$ In$^\bullet$ + HQ$^\bullet$ in Hydrocarbon Solutions Calculated by IPM Method; for Equations See Chapter 6 and for the Values of α, br_e, and A, See Table 18.9—*continued*

Phenol	ΔH (kJ mol^{-1})	E (kJ mol^{-1})	A (L mol^{-1} s^{-1})	k (400 K) (L mol^{-1} s^{-1})
	110.9	112.6	7.7×10^9	1.5×10^{-5}
AmH	130.7	132.4	3.7×10^{10}	1.9×10^{-7}
	116.8	118.5	3.2×10^{10}	1.1×10^{-5}
	132.2	133.9	3.8×10^{10}	1.2×10^{-7}
	120.5	122.2	3.4×10^{10}	3.8×10^{-6}
	127.8	129.5	3.6×10^{10}	4.6×10^{-7}
	129.4	131.1	3.7×10^{10}	2.9×10^{-7}
PhNHPh	136.6	138.3	4.0×10^{10}	3.5×10^{-8}
	105.5	107.2	2.9×10^{10}	2.9×10^{-4}

continued

TABLE 18.10
Enthalpies, Activation Energies, and Rate Constants of Reactions of *p*-Benzoquinone with Phenols and Amines: InH + Q $\longrightarrow$ In$^\bullet$ + HQ$^\bullet$ in Hydrocarbon Solutions Calculated by IPM Method; for Equations See Chapter 6 and for the Values of α, br_e, and A, See Table 18.9—*continued*

Phenol	ΔH (kJ mol^{-1})	E (kJ mol^{-1})	A (L mol^{-1} s^{-1})	k (400 K) (L mol^{-1} s^{-1})
	118.5	120.2	3.3×10^{10}	6.6×10^{-6}
	127.8	129.5	3.6×10^{10}	4.6×10^{-7}
	121.1	122.8	3.4×10^{10}	4.6×10^{-3}
	146.6	148.3	4.3×10^{10}	1.9×10^{-9}

deactivation of cupric salts that catalyze the autoxidation of hydrocarbons and rubber salicylaldoxim and *N,N*-disalicylidenethylenedimine were used. They form very stable complexes with copper ions.

The formation of chelate complexes is supposed to change the redox potential of the transition metal and decrease its ability to catalyze the decay of hydroperoxide into free radicals [6]. However, the general basis for transition metal transformation from a catalyst to inhibitors of hydrocarbon autoxidation can be the following. In general, the metal catalyst decomposes hydroperoxide by two parallel ways, namely, homolytic and heterolytic (see Chapter 10). If the first channel predominates, the metal compound accelerates autoxidation, if the second channel prevails, metal compound retards autoxidation. The formation of a new complex because of the introduction of the ligand in the oxidized system changes the proportion of formed hydroperoxide homolysis and heterolysis as well as the general catalytic activity of the metal. So, the most active antioxidant ligand additive should be the one which transforms the metal into an active heterolytic decomposer of hydroperoxide. The same ligand can strengthen the heterolytic activity of one metal and homolytic activity of another in the catalytic decomposition of hydroperoxides. For example, the two ligands discussed above decrease the catalytic activity of copper salts and strengthen the activity of manganese and ferrous salts in oxidized hydrocarbons [6]. Hence, only the empirical testing is used for identifying efficient transition metal deactivators.

REFERENCES

1. K Bailey. *The Retardation of Chemical Reactions*. London: Edward Arnold and Co, 1937.
2. KU Ingold. *Chem Rev* 61:563–589, 1961.
3. NM Emanuel, YuN Lyaskovskaya. *Inhibition of Fat Oxidation*. Moscow: Pischepromizdat, 1961 [in Russian].
4. ET Denisov, TG Denisova. *Handbook of Antioxidants*. Boca Raton: CRC Press, 2000.
5. G Scott, ed. *Atmospheric Oxidation and Antioxidants*, vols 1–3. Amsterdam: Elsevier Publishers, 1993.
6. J von Voigt. *Die Stabilisierung der Kunststoffe gegen Licht und Warme*. Berlin: Springer-Verlag, 1966.
7. NM Emanuel, ET Denisov, ZK Maizus. *Liquid-Phase Oxidation of Hydrocarbons*. New York: Plenum Press, 1967, pp. 223–281.
8. AM Kuliev. *Chemistry and Technology of Additives to Oils and Fuels*. Moscow: Khimiya 1972 [in Russian].
9. ET Denisov, VV Azatyan. *Inhibition of Chain Reactions*. London: Gordon and Breach, 2000.
10. YuA Ershov, GP Gladyshev. *Vysokomol Soed* A19:1267–1273, 1977.
11. GV Karpukhina, NM. Emanuel. *Dokl AN SSSR* 276:1163–1167, 1984.
12. GW Kennerly, WL Patterson. *Ind Eng Chem* 48:1917–1927, 1956.
13. RI Levin, VV Mikhailov. *Usp Khim* 39:1687–1706, 1970.
14. DM Shopov, SK Ivanov. *Mechanism of Action of Hydroperoxide-Breaking Inhibitors*. Sofia: Izdat Bulg Akad Nauk, 1988 [in Russian].
15. CRHI de Jonge, P Hope. In: G Scott, ed. *Developments in Polymer Stabilisation—3*. London: Applied Science Publishers, 1980, pp 21–54.
16. GV Karpukhina, ZK Maizus, NM Emanuel. *Dokl AN SSSR* 152:110–113, 1963.
17. GV Karpukhina, ZK Maizus, ON Karpukhin. *Zh Fiz Khim* 39:498–500, 1965.
18. GV Karpukhina, ZK Maizus, NM Emanuel. *Dokl AN SSSR* 160:158–161, 1965.
19. GV Karpukhina, ZK Maizus, LI Matienko. *Neftekhimiya* 6:603–611, 1966.
20. GV Karpukhina, ZK Maizus, MYa Meskina. *Kinet Katal* 9:245–249, 1968.
21. GV Karpukhina, ZK Maizus, NM Emanuel. *Dokl AN SSSR* 182:870–873, 1968.
22. MYa Meskina, GV Karpukhina, ZK Maizus. *Neftekhimiya* 11:213–218, 1971.
23. MYa Meskina, GV Karpukhina, ZK Maizus. *Izv AN SSSR Ser Khim* 1481–1488, 1973.
24. MYa Meskina, GV Karpukhina, ZK Maizus, NM Emanuel. *Dokl AN SSSR* 213:1124–1127, 1973.
25. TV Lomteva, GV Karpukhina, ZK Maizus. *Izv AN SSSR Ser Khim* 930–934, 1973.
26. GV Karpukhina, ZK Maizus, NV Zolotova, LI Mazaletskaya, MYa Meskina. *Neftekhimiya* 18:708–715, 1978.
27. GV Karpukhina, ZK Maizus. *Izv AN SSSR Ser Khim* 957–962, 1968.
28. VV Varlamov, ET Denisov, *Bull Acad Sci USSR* 39:657–662, 1990.
29. VV Varlamov, ET Denisov, *Bull Acad Sci USSR* 36:1607–1612, 1987.
30. VV Varlamov, RL Safiullin, ET Denisov. *Khim Fiz* 4:901–904, 1985.
31. VV Varlamov, NN Denisov, VA Nadtochenko, EP Marchenko. *Kinet Katal* 35:833–837, 1994.
32. VV Varlamov, NN Denisov, VA Nadtochenko, EP Marchenko, IV Petrov, LG Plekhanova. *Kinet Katal* 35:838–840, 1994.
33. VV Varlamov, NN Denisov, VA Nadtochenko. *Russ Chem Bull* 44:2282–2286, 1995.
34. RL Vardanyan, AG Vanesyan, TM Ayvazayn, AV Tigranyan. *Dokl AN SSSR* 248:1144–1147, 1979.
35. LR Mahoney, MA da Rooge. *J Am Chem Soc* 89:5619–5629, 1967.
36. NA Azatyan, TV Zolotova, GV Karpukhina, ZK Maizus. *Neftekhimiya* 11:568–573, 1971.

37. NA Azatyan, GV Karpukhina, IS Belostotskaya, NL Komissarova. *Neftekhimiya* 13:435–440, 1973.
38. EA Arakelyan, NA Azatyan, MYa Meskina, ZK Maizus. *Neftekhimiya* 22:464–468, 1982.
39. ET Denisov, TI Drozdova. *Kinet Catal* 35, 155–162, 1994.
40. ET Denisov. *Russ Chem Rev* 66:859–876, 1997.
41. ET Denisov, TG Denisova, SV Trepalin, TI Drozdova. *CD-ROM: Oxidation and Antioxidants in Organic Chemistry and Biology*. New York: Marcel Dekker, 2003.
42. VA Roginskii, GA Krasheninnikova. *Dokl AN SSSR* 293:157–162, 1987.
43. VA Roginskii. *Izv AN SSSR Ser Khim* 1987–1996, 1985.
44. LI Mazaletskaya, GV Karpukhina, ZK Maizus. *Neftekhimiya* 19:214–219, 1979.
45. LI Belova, GV Karpukhina, ZK Maizus,e.g.,Rozantsev, NM Emanuel. *Dokl AN SSSR* 231: 369–372, 1976.
46. LI Mazaletskaya, GV Karpukhina. *Izv AN SSSR Ser Khim* 1741–1747, 1984.
47. ET Denisov. *Kinet Catal* 36:351–355, 1995.
48. ET Denisov. *Kinet Catal* 38:762–768, 1997.
49. ET Denisov. *Macromol Symp* 143:65–74, 1999.
50. T Colclough. *J Chem Soc* 26:1888, 1987.

19 Peculiarities of Antioxidant Action in Polymers

19.1 RIGID CAGE OF POLYMER MATRIX

19.1.1 Comparison of Bimolecular Reactions in Liquid and Solid Phases

Diffusion of particles in the polymer matrix occurs much more slowly than in liquids. Since the rate constant of a diffusionally controlled bimolecular reaction depends on the viscosity, the rate constants of such reactions depend on the molecular mobility of a polymer matrix (see monographs [1–4]). These rapid reactions occur in the polymer matrix much more slowly than in the liquid. For example, recombination and disproportionation reactions of free radicals occur rapidly, and their rate is limited by the rate of the reactant encounter. The reaction with sufficient activation energy is not limited by diffusion. Hence, one can expect that the rate constant of such a reaction will be the same in the liquid and solid polymer matrix. Indeed, the process of a bimolecular reaction in the liquid or solid phase occurs in accordance with the following general scheme [4,5]:

$$A + B \underset{v_D}{\overset{k_D}{\rightleftharpoons}} A \cdots B \overset{k}{\longrightarrow} \text{Products}$$

The observed rate constant is $k_{obs} = kk_D(k + v_D)^{-1}$. For the fast reactions with $k \gg v_D$ the rate constant is $k_{obs} = k_D$. In the case of a slow reaction with $k \ll v_D$ the rate constant is $k_{obs} = k \times K_{AB}$, where $K_{AB} = k_D/v_D$ is the equilibrium constant of formation of cage pairs A and B in the solvent or solid polymer matrix. The equilibrium constant K_{AB} should not depend on the molecular mobility. According to this scheme, the rate constant of a slow bimolecular reaction $k_{obs} = kK_{AB}(k_{obs} \ll k_D)$ should be the same in a hydrocarbon solution and the nonpolar polymer matrix. However, it was found experimentally that several slow free radical reactions occur more slowly in the polymer matrix than in the solvent. A few examples are given in Table 19.1.

It is clearly seen that the rate constants of all the studied reactions are considerably lower in the solid phase of PE and PP than in benzene or chlorobenzene solution. At the same time, these reactions are not limited by translational or rotational diffusion.

The above kinetic scheme of the bimolecular reaction simplifies physical processes that proceed via the elementary bimolecular act. To react, two reactants should (a) meet, (b) be oriented by the way convenient for the elementary act, and (c) be activated to form the TS and then react. Hence, not only translational but also rotational diffusion of particles in the solution and polymer are important for the reaction to be performed. So, the more detailed kinetic scheme of a bimolecular reaction includes the following stages: diffusion and encounter the reactants in the cage, orientation of reactants in the cage due to rotational diffusion, and activation of reactants followed by reaction [5,13].

TABLE 19.1
Comparison of Rate Constants and Activation Energies of Bimolecular Reactions in Solution and Polymer Matrix

Medium	E (kJ mol^{-1})	log A, A (L mol^{-1} s^{-1})	k (300 K) (L mol^{-1} s^{-1})	k_D (300 K) (L mol^{-1} s^{-1})	Ref.
\multicolumn{6}{c}{PhMe$_2$CO$_2^\bullet$ + HO–Ar(tBu)$_2$Me → PhMe$_2$COOH + $^\bullet$O–Ar(tBu)$_2$Me}					
PhEt	17.1	6.88	8.1×10^3	3×10^9	[6]
PP	82.7	15.80	2.7×10^2	2×10^6	[6]
\multicolumn{6}{c}{PhMe$_2$COOH + $^\bullet$O–Ar(tBu)$_2$ → PhMe$_2$COO$^\bullet$ + HO–Ar(tBu)$_2$}					
PhH	46.0	7.11	0.13	3×10^9	[7]
PE	69.0	10.40	2.5×10^{-2}	9×10^6	[7]
PP	67.0	9.40	5.5×10^{-3}	2×10^6	[7]
\multicolumn{6}{c}{TEMPO–N–O$^\bullet$ + HO–Ph → TEMPO–N–OH + $^\bullet$O–Ph}					
PP	41.2	4.88	5.09×10^{-3}	2×10^6	[8]
PhH	57.1	8.00	1.14×10^{-2}	3×10^9	[8]
\multicolumn{6}{c}{TEMPO–N–O$^\bullet$ + HO–C$_6$H$_4$–Cl → TEMPO–N–OH + $^\bullet$O–C$_6$H$_4$–Cl}					
PP	65.0	8.43	1.32×10^{-3}	2×10^6	[8]
PhH	57.4	8.00	1.01×10^{-2}	3×10^9	[8]
\multicolumn{6}{c}{TEMPO–N–O$^\bullet$ + HO–C$_6$H$_4$–NO$_2$ → TEMPO–N–OH + $^\bullet$O–C$_6$H$_4$–NO$_2$}					
PP	76.4	9.56	1.77×10^{-4}	2×10^6	[8]
PhH	55.7	8.00	2.00×10^{-2}	3×10^9	[8]
\multicolumn{6}{c}{PhC(O)O–piperidinyl–N–O$^\bullet$ + HO–Ar(tBu)$_2$ → PhC(O)O–piperidinyl–N–OH + $^\bullet$O–Ar(tBu)$_2$}					
PhH	42.0	4.11	6.5×10^{-4}	3×10^9	[9]
PP	75.0	75.0	4.5×10^{-5}	2×10^6	[9]

TABLE 19.1
Comparison of Rate Constants and Activation Energies of Bimolecular Reactions in Solution and Polymer Matrix—*continued*

Medium	E (kJ mol^{-1})	log A, A (L mol^{-1} s^{-1})	k (300 K) (L mol^{-1} s^{-1})	k_D (300 K) (L mol^{-1} s^{-1})	Ref.

Reaction: PhC(O)O–piperidine–N–O• + HO–naphthyl → PhC(O)O–piperidine–N–OH + •O–naphthyl

Medium	E (kJ mol^{-1})	log A, A	k (300 K)	k_D (300 K)	Ref.
PhH			6.8×10^{-3} (333K)	3×10^9	[9]
PP			4.8×10^{-4} (333K)	2×10^6	[9]

Reaction: piperidine–N–O• + HO–(dimethylphenyl)–N=CH–(hydroxyphenyl) → N–OH + •O–aryl–N=CH–ArOH

Medium	E (kJ mol^{-1})	log A, A	k (300 K)	k_D (300 K)	Ref.
PP	42.6	6.19	5.93×10^{-2}	2×10^6	[10]

Reaction: PhC(O)O–piperidine–N–O• + HO–Ar–N=CH–ArOH → PhC(O)O–piperidine–N–OH + •O–Ar–N=CH–ArOH

Medium	E (kJ mol^{-1})	log A, A	k (300 K)	k_D (300 K)	Ref.
PP	48.2	6.70	2.03×10^{-2}	2×10^6	[10]

Reaction: PhC(O)O–piperidine–N–O• + H–N(Ph)$_2$ → PhC(O)O–piperidine–N–OH + •N(Ph)$_2$

Medium	E (kJ mol^{-1})	log A, A	k (300 K)	k_D (300 K)	Ref.
PhH $T = 333$ K	64.4	8.00	5.25×10^{-3}	6.8×10^9	[11]
PP $T = 333$ K	84.0	10.0	6.40×10^{-4}	1.3×10^7	[11]

Reaction: naphthyl–NH–C$_6$H$_4$–NH–naphthyl + PhMe$_2$COOH → naphthyl–N•–C$_6$H$_4$–NH–naphthyl + H$_2$O + PhMe$_2$CO•

Medium	E (kJ mol^{-1})	log A, A	k (300 K)	k_D (300 K)	Ref.
PhCl	63.0	7.78	6.7×10^{-4}	3×10^9	[12]
PP	75.0	8.83	6.2×10^{-5}	2×10^6	[12]

$$A + B \underset{\nu_D}{\overset{k_D}{\rightleftharpoons}} A \cdots B \underset{\nu_r}{\overset{\nu_{or}}{\rightleftharpoons}} [A \cdots B]_{oriented} \overset{k}{\rightarrow} \text{Products}$$

Rotational diffusion of particles occurs in polymer much slowly than in liquids. Therefore, the observed difference in liquid (k_l) and solid polymer (k_S) rate constants can be explained by the different rates of reactant orientation in the liquid and polymer. The EPR spectra were obtained for the stable nitroxyl radical (2,2,6,6-tetramethyl-4-benzoyloxypiperidine-1-oxyl). The molecular mobility was calculated from the shape of the EPR spectrum of this radical [14,15]. These values were used for the estimation of the orientation rate of reactants in the liquid and polymer cage. The frequency of orientation of the reactant pairs was calculated as $\nu_{or} = P\nu_{rot}$, where P is the steric factor of the reaction, and ν_{rot} is the frequency of particle rotation to the angle equal to 4π. The results of this comparison are given in Table 19.2.

TABLE 19.2
Comparison of Rate Constants and Steric Factors of Bimolecular Reactions with Frequency of Reactant Rotation and Orientation in Polymer Matrix ($T = 300\,K$)

Medium	ν_{rot} (s^{-1})	P	ν_{or} (s^{-1})	k/K_{AB} (s^{-1})
	PhMe$_2$CO$_2^\bullet$ + HO–ArBu$_3$ $\rightarrow$ PhMe$_2$COOH + $^\bullet$O–ArBu$_3$			(6)
PP	8×10^6	6×10^{-6}	48	2.7×10^2
	PhMe$_2$COOH + $^\bullet$O–ArBu$_3$ $\rightarrow$ PhMe$_2$COO$^\bullet$ + HO–ArBu$_3$			(7)
PE	4×10^7	6×10^{-6}	2.4×10^2	2.5×10^{-2}
PP	8×10^6	6×10^{-6}	48	5.5×10^{-3}
	PhC(O)O–pip–N–O$^\bullet$ + HO–Ar $\rightarrow$ PhC(O)O–pip–N–OH + $^\bullet$O–Ar			(9)
PP	1.4×10^7	1.3×10^{-9}	1.8×10^{-2}	4.5×10^{-5}
	PhC(O)O–pip–N–O$^\bullet$ + H–NPh$_2$ $\rightarrow$ PhC(O)O–pip–N–OH + $^\bullet$NPh$_2$			(11)
PP ($T = 333\,K$)	3.9×10^7	1.0×10^{-4}	3.9×10^3	6.4×10^{-4}

TABLE 19.2
Comparison of Rate Constants and Steric Factors of Bimolecular Reactions with Frequency of Reactant Rotation and Orientation in Polymer Matrix ($T = 300$ K)—*continued*

Medium	ν_{rot} (s^{-1})	P	ν_{or} (s^{-1})	k/κ_{AB} (s^{-1})
PP ($T = 333$ K)	3.9×10^7	1.0×10^{-4}	3.9×10^3	8.0×10^{-5} (11)
PP ($T = 333$ K)	3.9×10^7	1.0×10^{-4}	3.9×10^3	9.1×10^{-3} (11)
PP	1.4×10^7	1.3×10^{-10}	1.8×10^{-3}	6.2×10^{-5} (12)

We see that the rate constant of the bimolecular reaction is two to three orders of magnitude lower than the rate of reactant orientation.

19.1.2 CONCEPTION OF RIGID CAGE OF POLYMER MATRIX

Hence, the phenomena of the low reaction rate in the polymer matrix cannot be explained by the limiting rate of reactant orientation (rotational diffusion) in the cage. This result becomes the impetus to formulate the conception of the "rigid" cage of polymer matrix [16–20]. In addition to the experiments with comparison of the rate constants in the liquid phase and polymer matrix, experiments on the kinetic study of radical reactions in polymers with different amounts of introduced plasticizer were carried out [7,9,15,21]. A correlation between the rate constant of the reaction k and the frequency of rotation ν_{or} of the nitroxyl radical (2,2,6,6-tetramethyl-4-benzoyloxypiperidine-N-oxyl) was found. The values of the rate constants for the reaction of this nitroxyl radical with sterically hindered phenol (2,6-bis(1,1-dimethylethyl)phenol) and 1-naphthol in polymers at different concentrations of chlorobenzene as plasticizer together with the values of rotational rate of nitroxyl are given in Table 19.3.

The rate constant of this reaction is lower when rotational diffusion is slower. The experimental data given above prove that the medium of the polymer matrix influences on the bimolecular reaction quite differently than in liquid.

The interaction of two reagents in the liquid phase occurs in the cage formed by small labile molecules. All geometric shapes of such a cage are energetically equivalent due to the high flexibility of molecules surrounding the pair of interacting particles, and such a cage may be regarded as "soft." The formation of the TS in the "soft" cage of nonpolar liquid does not need an additional energy for reorganization of the pair of reactants among the surrounding molecules.

In polymer matrix quite another situation was observed. Each particle or a pair of particles is surrounded in the polymer matrix by segments of the macromolecule. These segments are connected by C—C bonds and form a "rigid" cage. In such a rigid cage, there are geometrically and energetically unequal orientations of particles. This is the reason why a pair of reactants reacting in a polymer matrix needs an additional energy to assume the necessary orientation to react in a rigid cage. Therefore, we should introduce for the reaction in a polymer matrix the additional coefficient F_S incorporated into the steric factor P_S, which describes the influence of the cage walls on the mutual orientation of reagents: $P_S = P_1 \times F_S$ where P_1 is the steric factor of the reaction in the liquid phase. This coefficient F_S should depend on temperature, because it includes the Boltzmann factor equal to $\exp(-E_{or}/RT)$. The activation energy E_{or} is the difference between the energy of the energetically most convenient orientation of interacting particles and the energy necessary for the reaction to occur (see Figure 19.1).

A particle in a polymer cage is regarded as being in the field of forces of intermolecular interaction, which is approximated by a cosine function if the rotation is regarded in one plane.

$$E_{or}(\theta) = 0.5 E_{rot}(1 - \cos n_\theta \theta) \tag{19.1}$$

where E_{rot} is the energy barrier dividing two energetically convenient positions of the reacting particles in the cage and is supposed to be the same as the activation energy of rotational diffusion, and $2n_\theta$ is the number of such positions. The value of E_{rot} can be estimated through the ratio of rotation frequency of the reactant in the liquid and solid phases.

$$E_{rot} = RT \ln(\nu_1/\nu_S). \tag{19.2}$$

For the reaction to occur, particle A must be oriented in the cage at an angle of $\theta_A \pm \Delta\theta_A$. The probability of this orientation is:

$$W(\vartheta_A) = \frac{\Delta\vartheta_A \exp(-E_{rot}/RT)}{2\pi I_0(E_{rot}/2RT) \exp(-E_{rot}/2RT)} \tag{19.3}$$

where E_{or} is the energy of mutual orientation of reacting particles in the rigid cage, and $I_0(x)$ is the modified Bessel function of imaginary argument. The steric factor for the reaction of A with B in a polymer matrix in the scope of such a model is

$$P_S = \frac{\Delta\vartheta_A \Delta\vartheta_B \exp(-E_{rot}/2RT)}{16\pi^2 I_0(E_{rot}/2RT)}. \tag{19.4}$$

TABLE 19.3
Influence of Plastification of Polymer on the Molecular Mobility and Rate Constant of Bimolecular Reaction [7,14,15,21]

PhC(O)O–[piperidine]–N–O• + HO–[2,6-di-tert-butylphenol] → PhC(O)O–[piperidine]–N–OH + O•–[2,6-di-tert-butylphenoxyl]

Medium	T (K)	% PhCl	ν_{rot} (s^{-1})	k (L mol^{-1} s^{-1})	k_l/k_s
Benzene	313		5.4×10^{10}	1.38×10^5	1.0
PP	313	36.0	3.5×10^9	7.0×10^4	1.97
PP	313	9.0	7.5×10^8	3.6×10^4	3.83
PP	313	4.8	5.8×10^8	2.0×10^4	6.90
PP	313	3.0	5.3×10^8	1.4×10^4	9.86
PP	313	<0.4	4.2×10^8	1.0×10^4	13.8
PP	313	2.0	6.0×10^8	3.1×10^4	4.45
PP	313	<0.4	2.3×10^8	1.0×10^4	13.8
Benzene	323		6.5×10^{10}	2.14×10^5	1.00
PP	323	4.0	9.4×10^8	7.8×10^4	2.74
PP	323	<0.4	5.8×10^8	5.6×10^4	7.18
PP	323	1.5	4.7×10^8	4.6×10^4	4.65
PP	323	<0.4	2.8×10^8	2.6×10^4	8.23
Benzene	333		7.2×10^{10}	2.98×10^5	1.00
PP	333	<0.4	8.3×10^8	1.10×10^5	2.71
PP	333	2.0	1.0×10^9	1.26×10^5	2.36
PP	333	1.0	6.9×10^8	1.04×10^5	2.86
PP	333	<0.4	4.5×10^8	7.60×10^4	3.92
PS	333	38.0	3.0×10^9	1.60×10^5	1.86
PS	333	15.0	6.2×10^8	6.20×10^4	4.81
PS	333	9.0	1.5×10^8	4.02×10^4	7.41

PhC(O)O–[piperidine]–N–O• + HO–[naphthyl] → PhC(O)O–[piperidine]–N–OH + O•–[naphthyl]

Medium	T (K)	% PhCl	ν_{rot} (s^{-1})	k (L mol^{-1} s^{-1})	k_l/k_s
Benzene	333		7.2×10^{10}	6.80×10^5	1.00
PS	333	50.0	4.7×10^9	3.90×10^5	1.74
PS	333	18.0	3.3×10^8	1.68×10^5	4.05
PS	333	6.0	1.0×10^8	4.80×10^4	14.17

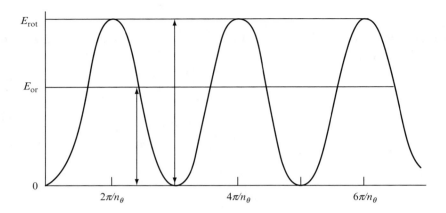

FIGURE 19.1 Cosine type angle dependence of the potential energy of the molecule or radical orientation in the model of anisotropic hard cage of the polymer matrix.

In liquids all orientations of reactants A and B are energetically equivalent and $P_1 = (4\pi)^{-2} \Delta\theta_A \Delta\theta_B$. Therefore, the ratio of rate constants in polymer and liquid is

$$\frac{k_S}{k_l} = \frac{P_S}{P_l} = \frac{\exp(-E_{rot}/2RT)}{I_0(E_{rot}/2RT)} \qquad (19.5)$$

This model explains the above-mentioned peculiarities of free radical reactions in polymers.

First, reaction occurs more slowly in polymer than in liquid because of the reorganization of the surrounding reactant polymer segments to achieve the TS. Additional energy is needed to perform this reorganization.

Second, the correlation between k_S and ν_{rot} finds its natural explanation, because k_S and ν_{rot} depend on the energy E_{or}. After the substitution of the value of E_{rot} for $RT\ln(\nu_{rotl}/\nu_{rotS})$ in Equations (19.4) and (19.5) one obtains ($m = E_{or}/E_{rot}$)

$$\frac{k_S}{k_l}\sqrt{\frac{\nu_l}{\nu_S}} = \left(\frac{\nu_l}{\nu_S}\right)^{1/m} \times I_0\left\{\frac{1}{2}\ln\frac{\nu_l}{\nu_S}\right\}, \qquad (19.6)$$

or in the logarithmic form

$$\ln\left\{\frac{k_l}{k_S}\sqrt{\frac{\nu_S}{\nu_l}} \times I_0\left\{\frac{1}{2}\ln\frac{\nu_l}{\nu_S}\right\}\right\} = m\ln\frac{\nu_l}{\nu_S}. \qquad (19.7)$$

Experimental data are in good agreement with this equation (see Table 19.4 and Figure 19.2). As was shown, the parameter m is constant in the very line of experiments with one reaction in the same polymeric matrix and is always less than unity. The latter means that $E_{rot} > E_{or}$. This agrees with the conception of a rigid cage.

We see that the parameter m is constant in the very line of experiments and lies in the limit 0.5–0.75. The energetic barrier of orientation for the studied reactions in polyethylene and polypropylene amounts 5–12 kJ mol^{-1}.

Third, according to the rigid cage model, the larger the volume of reagents, the more the number of polymer segments surrounding the reactants and the higher should be the potential barrier E_{rot} for appropriate orientation of reactants. This was confirmed in the study of

TABLE 19.4
The Values of E_r and E_{or} for Bimolecular Reactions of Nitroxyl Radicals with Phenols Calculated According to the Rigid Cage Model for Reaction in a Polymer Matrix (Equation (19.7)) [7,9,14,15,21]

Medium	ν_{rotl}/ν_{rots} (s^{-1})	k_l/k_S	E_{rot} (kJ mol^{-1})	m	E_{rot} (kJ mol^{-1})
PP, 303(K)	27	6.09	8.3	0.54	4.5
PP, 363(K)	18	3.18	8.7	0.40	3.5
PP, 303(K)	51	18.3	9.9	0.74	7.3
PP, 363 (K)	10	2.1	7.1	0.31	2.2
PP, 363 (K)	25	4.4	9.8	0.45	4.4
PE, 313(K)	128	13.8	13.1	0.81	10.7
PE + 5.5% PhCl, 313 (K)	93	8.1	11.7	0.74	8.6
PE + 36% PhCl, 313 (K)	15	2.0	10.7	0.69	7.4
PP, 313(K)	235	13.8	14.1	0.73	10.3
PP + 2% PhCl, 313 (K)	90	4.4	11.7	0.61	7.1
PE, 323 (K)	112	3.7	12.6	0.55	6.9
PE + 4% PhCl, 323 (K)	64	2.7	11.1	0.52	5.8
PP, 323 (K)	232	8.0	14.6	0.63	9.2
PP + 1% PhCl, 323 (K)	155	4.8	13.5	0.57	7.6
PE, 333 (K)	90	2.9	12.4	0.51	6.5
PE, 333 (K)	150	4.0	13.8	0.53	7.4
PP + 1% PhCl, 333 (K)	104	2.9	13.0	0.51	6.6
PP + 2% PhCl, 333 (K)	72	2.4	11.8	0.49	5.7

continued

TABLE 19.4
The Values of E_r and E_{or} for Bimolecular Reactions of Nitroxyl Radicals with Phenols Calculated According to the Rigid Cage Model for Reaction in a Polymer Matrix (Equation (19.7)) [7,9,14,15,21]—*continued*

Medium	v_{rotl}/v_{rotS} (s⁻¹)	k_l/k_S	E_{rot} (kJ mol⁻¹)	m	E_{rot} (kJ mol⁻¹)
PS + 9% PhCl, 333 (K)	480	7.5	17.0	0.56	9.5
PS + 38% PhCl, 333 (K)	240	1.9	8.8	0.52	4.6

PhC(O)O—[tetramethylpiperidine]—N–O• + HO—[naphthyl] → PhC(O)O—[tetramethylpiperidine]—N–OH + O•—[naphthyl]

Medium	v_{rotl}/v_{rotS} (s⁻¹)	k_l/k_S	E_{rot} (kJ mol⁻¹)	m	E_{rot} (kJ mol⁻¹)
PE + 6% PhCl, 333 (K)	720	14.2	18.1	0.63	11.4
PS + 18% PhCl, 333 (K)	218	4.0	14.9	0.52	7.9
PS + 50% PhCl, 333 (K)	15	1.7	7.4	0.55	4.1

PhC(O)O—[tetramethylpiperidine]—N–O• + HO—[dimethylphenyl]—N=CH—[salicyl] → PhC(O)O—[tetramethylpiperidine]—N–OH + O•—[dimethylphenyl]—N=CH—[salicyl]

Medium	v_{rotl}/v_{rotS} (s⁻¹)	k_l/k_S	E_{rot} (kJ mol⁻¹)	m	E_{rot} (kJ mol⁻¹)
PP, 303 (K)	36	7.6	9.0	0.55	5.0
PP, 333 (K)	22	3.8	8.6	0.43	3.7
PP, 363 (K)	11	2.2	7.3	0.33	2.4

PhC(O)O—[tetramethylpiperidine]—N–O• + HO—[dimethylphenyl]—N=CH—[salicyl]—Fe complex → PhC(O)O—[tetramethylpiperidine]—N–OH + O•—[dimethylphenyl]—N=CH—[salicyl]—Fe complex

Medium	v_{rotl}/v_{rotS} (s⁻¹)	k_l/k_S	E_{rot} (kJ mol⁻¹)	m	E_{rot} (kJ mol⁻¹)
PP, 303 (K)	16	40	7.0	0.88	6.2
PP, 333 (K)	14	23	7.3	0.78	5.7
PP, 363 (K)	11	13	7.3	0.71	5.2

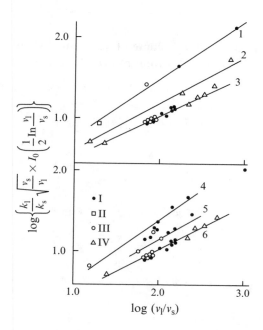

FIGURE 19.2 The correlation of rate constants of various free radical reactions with molecular mobility of nitroxyl radical in the polymer matrix of different polymers with addition of plastificator: I in IPP, II in preliminary oxidized IPP, III in PE, and IV in PS. Line 1 for the reaction of 2,6-bis(1,1-dimethylethyl)phenoxyl radical with hydroperoxide groups at $T = 295$ K; line 2 for the reaction of 2,2,6,6-tetramethyl-4-benzoyloxypiperidine-N-oxyl with 1-naphthol at $T = 333$ K; line 3 for the reaction of 2,2,6,6-tetramethyl-4-benzoyloxypiperidine-N-oxyl with 2,6-bis(1,1-dimethylethyl)phenol at $T = 333$ K; line 4 for the same reaction at $T = 303$ K; line 5 for the same reaction at $T = 313$ K; and line 6 for the same reaction at $T = 323$ K [18].

reactions of stable nitroxyl radicals with different volumes with phenols and aromatic amines (InH).

$$>NO^{\bullet} + HIn \longrightarrow >NOH + In^{\bullet}$$

Substituents change the volume of the reactant. These reactions were studied in the liquid phase and polymer matrix (PP) [7,8,11]. The experiments evidenced that the rate constant of the studied reactions really depends on the volume of reactants in the polymer matrix and is independent in the liquid phase (see Table 19.5).

The results of the experimental estimation of rate constants for all these reactions prove that larger the volume $V^{\#}$ of TS, lower the rate constant and higher the activation energy for reconstruction of the shape of the cage to form an appropriate orientation of polymer segments around TS. An empirical linear correlation between $\Delta E_{or} = RT \ln(k_l/k_S)$ and the volume $V^{\#}$ of TS was found [8] as follows:

$$\Delta E_{or} = RT \ln(k_1/k_S) = a(V^{\#} - V^{\#}_{min}) \qquad (19.8)$$

where $V^{\#}_{min}$ is the minimal volume of TS when E_{or} depends on $V^{\#}$ in the polymer matrix, and a is the empirical coefficient for the reaction in polypropylene at a fixed temperature (a is temperature-dependent [8]).

TABLE 19.5
Dependence of Rate Constant on the Volume of Reactants for Bimolecular Reactions of Nitroxyl Radicals with Phenols in Polymer Matrix [8,10,11]

ArOH	T (K)	k (L mol^{-1} s^{-1})	$V^{\#}$ (cm^3 mol^{-1})	E_{or} (kJ mol^{-1})
PhOH	303	5.8×10^{-3}	157.6	4.5
PhOH	363	8.7×10^{-2}	157.6	3.5
Cl-C$_6$H$_4$-OH	303	1.6×10^{-3}	169.3	7.3
O$_2$N-C$_6$H$_4$-OH	362	7.1×10^{-2}	179.1	4.4
(bis-phenol Schiff base)	303	6.7×10^{-2}	240.2	5.0
(bis-phenol Schiff base)	363	1.15	240.2	2.4
Co-salen complex	303	0.19	378.1	5.3
Co-salen complex	363	1.07	378.1	5.5
Fe-salen complex	303	0.14	376.3	6.2
Fe-salen complex	363	1.05	376.3	5.2

TABLE 19.5
Dependence of Rate Constant on the Volume of Reactants for Bimolecular Reactions of Nitroxyl Radicals with Phenols in Polymer Matrix [8,10,11]—*continued*

ArOH	T (K)	k (L mol^{-1} s^{-1})	$V^{\#}$ (cm^3 mol^{-1})	E_{or} (kJ mol^{-1})
[Ni-salen bis-phenol complex]	303	0.11	379.9	6.3
[Ni-salen bis-phenol complex]	363	0.46	379.9	6.7
[phenol-salicylaldimine]	303	2.4×10^{-2}	304.6	7.6
[Co-salen bis-phenol complex]	303	3.4×10^{-2}	440.6	9.6
[Fe-salen bis-phenol complex]	303	4.3×10^{-2}	442.5	9.3
[Ni-salen bis-phenol complex]	303	3.7×10^{-2}	444.3	9.0

continued

TABLE 19.5
Dependence of Rate Constant on the Volume of Reactants for Bimolecular Reactions of Nitroxyl Radicals with Phenols in Polymer Matrix [8,10,11]—continued

ArOH	T (K)	k (L mol^{-1} s^{-1})	$V^{\#}$ (cm^3 mol^{-1})	E_{or} (kJ mol^{-1})
(Ni complex structure)	363	0.24	444.3	8.7
(trinitro compound structure)	363	0.17	433.4	8.1
(phenol imine structure)	363	0.13	573.1	10.4

T (K)	303	333	363
$a \times 10^2$ (kJ cm^{-3})	2.50	2.23	1.98
$V^{\#}_{min}$ (cm^3 mol^{-1})	127	110	117

A decrease in the coefficient a with an increase in temperature as a result of the intensification of the molecular mobility in the polymer matrix with the increase in temperature. The increase in temperature decreases the energetic barrier E_{or}. In the amorphous–crystalline polymers all these processes occur in the amorphous phase of the polymer where reactants are dissolved.

Fourth, the model of a rigid cage for a bimolecular reaction in the polymer matrix helps to explain another specific feature. This model explains the simultaneous increase in activation energy and preexponential factor on transferring the reaction from the liquid (E_l, A_l) to solid polymer matrix (E_S, A_S). In the nonpolar liquid phase $E_{obs} = E_l = E_{gas}$ but in the polymer matrix [3,21] it is

$$E_{obs} = E_l + dP_S/d(T^{-1}) = E + E_{or} + dE_{or}/d(T^{-1}) \tag{19.9}$$

and, hence, the preexponential factor is

$$A_S = const. + (12\alpha_T E_S - E - mRT)/(1 + 12\alpha_T T)R, \quad (19.10)$$

where α_T is the coefficient of linear temperature expansion of the polymer. This equation can be used for the independent estimation of the value of E_{or}.

$$E_{or} = \frac{E_{rot} + 12\alpha_T RT^2}{1 + 2\alpha_T RT} \quad (19.11)$$

The calculation of E_{or} by the two methods gives close values.

Polymer	T (K)	E_{or} (Eq. 19.2, 19.7) (kJ mol^{-1})	E_{or} (Eq. 19.11) (kJ mol^{-1})
PhC(O)O—[piperidine]—N–O• + HO—[di-tBu-phenyl] → PhC(O)O—[piperidine]—N–OH + •O—[di-tBu-phenyl]			
PE	313	13.1	13.1
PE	323	12.9	12.6
PE	333	12.5	12.5
PhMe$_2$COOH + •O—[di-tBu-phenyl] → PhMe$_2$COO• + HO—[di-tBu-phenyl]			
PE	395	13.3	12.7
PE	395	15.4	16.4

19.1.3 THE PHENOMENA OF REACTIVITY LEVELING IN POLYMERIC MATRIX

Rapid bimolecular reactions are limited by diffusion of reactants in the liquid and solid phases. Diffusion occurs in polymers much more slowly than in liquids. Hence, such rapid reactions as recombination of free radicals occurs in polymers with rate constants of a few order of magnitude more slowly than in solution. For example, the reaction of sterically hindered phenoxyl with the peroxyl radical

$$RO_2^\bullet + {}^\bullet O\text{—[di-tBu-phenyl-Y]} \longrightarrow O={[cyclohexadienyl-Y]}\text{—OOR}$$

occurs approximately 100 times more slowly in the amorphous phase of PP in comparison with benzene [22].

Y	Me$_3$C	PhC(O)	Ph	MeO	Me$_3$CO	CN
$k_8 \times 10^{-8}$ (333 K, C$_6$H$_6$) (L mol^{-1} s^{-1})	3.2	1.9	6.0	7.2	4.1	2.7
$k_8 \times 10^{-6}$ (353 K, PP) (L mol^{-1} s^{-1})	2.9	4.1	1.6	1.2	5.9	24

Slow diffusion of molecules and radicals in polymer contracts the interval of the observed rate constants of bimolecular reactions.

Besides diffusion, there is another underlying reason for leveling of reactivity of reactants in polymer media. The phenomena of leveling of reactivity of slowly reacting reactants was observed in the study of the reactions of peroxyl radicals with phenols and amines in the solid PS [23]. Later, this peculiarity was detected for different free radical reactions in the polymer matrix. All these reactions occur with rate constants much lower than k_D in polymer and cannot be limited by translational diffusion (see Table 19.6).

All reactions collected in Table 19.6 are slow. They occur with rate constants that are sufficiently lower than the rate constants of diffusion in polymer, as well as the frequency of reactant orientation in the cage ($\nu_{or} = \nu_r \times P$). Hence, physical processes are not limited by the rates of these reactions. However, polymer media influences the kinetics of these reactions.

This phenomenon can be explained within the framework of the conception of rigid polymer cage [25]. As described earlier, the polymer cage should be activated for two reactants to be oriented before to form a TS. Hence, the activation energy of the bimolecular reaction in the polymer matrix E_S can be presented as consisting of two terms, namely, the activation energy of the reaction $E = E_l$ (E_l is the activation energy in a nonpolar solvent) and the activation energy of cage reorganization E_{cr}: $E_S = E_l + E_{cr}$. The activation of the molecule is performed due to heat fluctuation. This fluctuation copes not only one molecule, but also an ensemble of polymer segments surrounding the reactant (walls of the cage). The study of density fluctuations in polymer at temperatures close to T_g (temperature of glass formation) showed that one heat fluctuation copes around 20–100 segments of a macromolecule. So, the activation of a molecule is accompanied by the activation of the cage. Consequently, there should be a feedback between the activation of reactants and polymer cage. Due to the collective character of the heat fluctuation, the activation of reactants is followed by the partial activation of the segments that form the cage. Therefore, the activation energy of cage

TABLE 19.6
Limits of Reactivity (k_{max}/k_{min}) for Different Reactions in Liquid Phase and Polymer Media

Reaction	Media	T (K)	k_{max}/k_{min}	Refs.
RO$_2^{\bullet}$ + InH	Benzene	343	413	[24]
RO$_2^{\bullet}$ + InH	PS	343	170	[24]
RO$_2^{\bullet}$ + ArOH	Ethylbenzene	353	71	[25]
RO$_2^{\bullet}$ + ArOH	PP	353	5.4	[26]
ArO$^{\bullet}$ + ROOH	Benzene	353	620	[27]
ArO$^{\bullet}$ + ROOH	PP	353	15.7	[26]
>NO$^{\bullet}$ + ArOH	Benzene	333	68	[7,8,10]
>NO$^{\bullet}$ + ArOH	PP	333	25	[7,8,10]
>NO$^{\bullet}$ + AmH	Benzene	333	369	[11]
>NO$^{\bullet}$ + AmH	PP	333	114	[11]
>NO$^{\bullet}$ + ArOH	Benzene	333	2.3	[7]
>NO$^{\bullet}$ + ArOH	PS	333	1.2	[7]

reorganization can be divided into two parts: $E_{cr} = E_{cr}' + E_{cr}''$ where $E_{cr}' = \beta E_1$ and automatically appears when reactants in the cage are activated and E_{cr}'' is additional activation energy. As a result, we derive the following equations [25]:

$$E_S = E_1 + E_{cr}'' = (1-\beta)E_1 + E_{cr}, \qquad (19.12)$$

$$RT \ln \frac{k_1}{k_S} = E_{cr} + \beta RT \ln \frac{k_1}{A}. \qquad (19.13)$$

Experimental values of k_S and k_1 are in good agreement with this equation [21,24]. The values of E_{cr} and coefficient β for different reactions of peroxyl and nitroxyl (2,2,6,6-tetramethyl-4-benzoyloxypiperidine-N-oxyl) radicals are presented below.

Reaction	T (K)	E_{cr} (kJ mol^{-1})	β	Ref.
$RO_2^\bullet + Ar_2OH \longrightarrow ROOH + Ar_2O^\bullet$	388	27 ± 2	0.53	[27]
$RO_2^\bullet + Ar_2OH \longrightarrow ROOH + Ar_2O^\bullet$	353	34 ± 3	0.48	[26]
$>NO^\bullet + Ar_1OH \longrightarrow >NOH + Ar_1O^\bullet$	333	30 ± 5	0.24	[10]
$>NO^\bullet + AmH \longrightarrow >NOH + Am^\bullet$	333	33 ± 5	0.33	[11]

These reactions were studied in the media of PE, PP, and PS. The mean value of E_{cr} is 32 ± 3 kJ mol^{-1} at 330–350 K. The activation energy decreases with an increase in temperature and becomes zero at $T \approx T_m$, so that at $T > T_m$, $E \approx E_1$. The limits of this phenomenon are obvious. First, all reactions should not be limited by translational or rotational diffusion, i.e., $k_S \ll k_D$ and $k_S \ll \nu_{rot} \times P$. Second, due to the automatic activation of the cage with reactant activation, the value of E_{cr}'' cannot be higher than E_{cr}, i.e., $\beta E_e \leq E_{cr}''$ and Equations (19.12)–(19.14) are valid for the reactions with $E_1 \leq E_{cr}/\beta$. If $E_1 \geq E_{or}/\beta$, $E_S = E_1$, and $k_S = k_1$, i.e., very slow reactions should occur with the same rate constants in the liquid phase and polymer matrix. If $E_{cr} = 32$ kJ mol^{-1} and $\beta = 0.32$, then $E_{1\,max} = 100$ kJ mol^{-1} and $k_S = k_1 = 2 \times 10^{-8}$ L mol^{-1} s^{-1} at $T = 333$ K.

19.1.4 Reactions Limited by Rotational Diffusion in Polymer Matrix

Reactions described earlier were not limited by rotational diffusion of reactants. It is evident that such bimolecular reactions can occur that are limited not by translational diffusion but by the rate of reactant orientation before forming the TS. We discussed the reactions of sterically hindered phenoxyl recombination in viscous liquids (see Chapter 15). We studied the reaction of the type: radical + molecule, which are not limited by translational diffusion in a solution but are limited by the rate of reactant orientation in the polymer matrix [28]. This is the reaction of stable nitroxyl radical addition to the double bond of methylenequinone.

This reaction occurs in solution with an extremely low preexponential factor with the rate constant $k = 1.6 \times 10^2 \exp(-29.0/RT)$ L mol^{-1} s^{-1} (benzene, 293–333 K [28]). Reactants are dissolved and react in the amorphous phase of the polymer. The PP matrix retards the reaction. The experiments with polymer plasticization showed that the rate constant is proportional to the frequency of nitronyl radical rotation measured by the ERP method.

%C_6H_6 ($T = 301$ K)	<0.4	1.0	1.5	2.0	3.0
$\nu_{rot} \times 10^{-8}$ (rad s^{-1})	0.9	2.0	2.6	2.9	4.2
$k \times 10^4$ (L mol^{-1} s^{-1})	1.8	4.3	5.0	5.9	8.0
$(k/\nu_{rot}) \times 10^{12}$ (L mol^{-1} rad^{-1})	2.0	2.1	1.9	2.0	1.9

This reaction occurs with an extremely low steric factor $P = 1.6 \times 10^{-11}$. Therefore, the rate of reactant orientation is very low, and the reaction is limited by reactant orientation in the polymer matrix. This reaction occurs according to the following kinetic scheme:

$$A + B \underset{\nu_D}{\overset{k_D}{\rightleftharpoons}} A \cdots B \overset{\nu_{rot}}{\longrightarrow} \text{Products}$$

The kinetic parameters of this reaction in the liquid and solid (IPP) phases [28] are given below.

Media	E (kJ mol^{-1})	log A, A (L mol^{-1} s^{-1})	k (300 K) (L mol^{-1} s^{-1})
C_6H_6	29.0 ± 4.0	2.2 ± 0.6	1.41×10^{-3}
PP	46.0 ± 4.0	4.2 ± 0.6	1.55×10^{-4}
ν_{or} ($2\pi s^{-1}$)(>NO$^\bullet$)	40.0 ± 3.0	3.3 ± 0.3	2.17×10^{-4}

19.2 ANTIOXIDANTS REACTING WITH PEROXYL RADICALS

The mechanism of antioxidant action on the oxidation of carbon-chain polymers is practically the same as that of hydrocarbon oxidation (see Chapters 14 and 15 and monographs [29–40]). The peculiarities lie in the specificity of diffusion and the cage effect in polymers. As described earlier, the reaction of peroxyl radicals with phenol occurs more slowly in the polymer matrix than in the liquid phase. This is due to the influence of the polymeric rigid cage on a bimolecular reaction (see earlier). The values of rate constants of macromolecular peroxyl radicals with phenols are collected in Table 19.7.

It should be taken into account that the reaction of chain propagation occurs in polymer more slowly than in the liquid phase also. The ratios of rate constants k_2/k_7, which are so important for inhibition (see Chapter 14), are close for polymers and model hydrocarbon compounds (see Table 19.7). The effectiveness of the inhibiting action of phenols depends not only on their reactivity, but also on the reactivity of the formed phenoxyls (see Chapter 15). Reaction 8 (In$^\bullet$ + RO$_2^\bullet$) leads to chain termination and occurs rapidly in hydrocarbons (see Chapter 15). Since this reaction is limited by the diffusion of reactants it occurs in polymers much more slowly (see earlier). Quinolide peroxides produced in this reaction in the case of sterically hindered phenoxyls are unstable at elevated temperatures. The rate constants of their decay are described in Chapter 15. The reaction of sterically hindered phenoxyls with hydroperoxide groups occurs more slowly in the polymer matrix in comparison with hydrocarbon (see Table 19.8).

The effectiveness of the antioxidant action is characterized by the induction period τ under fixed oxidation conditions and inhibitor concentration. Another parameter is the critical

TABLE 19.7
Rate Constants of Reactions of $PO_2^\bullet$ with Phenols in Solid Polymer

Phenol	PH	T (K)	$(k_7/k_2) \times 10^{-3}$	$k_7 \times 10^{-3}$ (L mol^{-1} s^{-1})	Ref.
	IPP	388	3.5	5.6	[41]
	IPP	353	36.5	3.5	[1]
	SSR	353	1.0	13.5	[42]
	IPP	353	84.4	8.1	[1]
	IPP	353	40.6	3.9	[1]
	IPP	353	33.3	3.2	[1]
	IPP	353	18.8	1.8	[1]
	IPP	353	28.1	2.7	[1]
	IPP	353	5.7	0.55	[1]
	IPP	353	9.9	0.95	[1]

continued

TABLE 19.7
Rate Constants of Reactions of PO$_2^\bullet$ with Phenols in Solid Polymer—*continued*

Phenol	PH	T (K)	$(k_7/k_2) \times 10^{-3}$	$k_7 \times 10^{-3}$ (L mol^{-1} s^{-1})	Ref.
	IPP	353	6.4	0.61	[1]
	IPP	353	13.3	1.28	[1]
	IPP	353	10.6	1.02	[1]
	IPP	388	2.6	4.2	[41]
	IPP	388	7.0	11.2	[41]
	IPP	353	1.0	13.3	[42]
	SSR	353	0.89	12.0	[42]
	SSR	353	0.67	9.1	[42]

TABLE 19.7
Rate Constants of Reactions of $PO_2^\bullet$ with Phenols in Solid Polymer—*continued*

Phenol	PH	T (K)	$(k_7/k_2) \times 10^{-3}$	$k_7 \times 10^{-3}$ (L mol^{-1} s^{-1})	Ref.
(2,6-di-tert-butyl-4-isobutylphenol structure)	IPP	388	14.0	22	[41]
(tetrakis phenol structure)	SSR	353	0.89	12.2	[42]
(2,6-dithiol-4-tert-butylphenol structure)	IPP	388	0.77	1.2	[43]

TABLE 19.8
Rate Constants of *Para*-Substituted 2,6-di-*tert*-Butylphenoxyl Reaction with Hydroperoxide Groups of IPP [7,44]

para-Substituent	k_{-7} (353 K) (L mol^{-1} s^{-1})	log A, A (L mol^{-1} s^{-1})	E (kJ mol^{-1})	k_s/k_l (353 K)
H	0.29	9.4	67	0.13
H	1.49	10.4	67	0.67
Me$_3$CO	8.9×10^{-3}	15.0	115	1.5×10^{-2}
Me$_3$C	1.7×10^{-2}	12.8	98	4.3×10^{-3}
PhC(O)	4.4×10^{-2}	12.5	94	2.2×10^{-3}
CN	0.14	16.0	115	4.0×10^{-4}

concentration of antioxidant $[InH]_{cr}$. (see Chapter 14). The value of $[InH]_{cr}$ is found experimentally as the point on the τ–$[InH]_{cr}$ curve, so that τ slightly depends on $[InH]$ at $[InH] < [InH]_{cr}$, and this dependence becomes strong at $[InH] > [InH]_{cr}$. Some values of $[InH]_{cr}$ and τ are given in Table 19.9.

The effectiveness of the antioxidant depends not only on its reactivity, but also on its molecular weight that affects the rate of the antioxidant loss due to evaporation. The following example illustrates this dependence. Antioxidants of the structure 2,6-bis (1, 1-dimethylethyl)phenols with *para*-substituents of the general structure ROCOCH$_2$CH$_2$ were introduced into decalin and polypropylene films that were oxidized by dioxygen at

TABLE 19.9
Critical Concentrations of Phenols and Induction Periods of IPP Oxidation at 473 K and Dioxygen Partial Pressure 10^4 Pa [45–47]

Inhibitor	$[InH]_{cr} \times 10^3$ (mol kg^{-1})	$\tau \times 10^{-3}$ (s)
(bis-phenol methylene, tBu)	1.2	14
(bis-phenol sulfide, tBu)	2.0	13
(dithiol, tBu)	3.0	15
(bis-phenol methylene, tBu, Cl)	4.0	2.7
(bis-phenol sulfide, tBu, Cl)	4.0	6.0
(bis-phenol methylene, tBu, reversed)	4.0	7.5

403 K [29]. The experimentally measured values were the following: the induction period τ_1 of decalin oxidation, induction period τ_2 of PP film oxidation in an atmosphere of dioxygen, induction period τ_3 of PP film oxidation in the dioxygen flow, and time of evaporation of a half of phenol $t_{1/2}$ in a nitrogen atmosphere at $T = 403$ K. The results are given below.

R	CH$_3$	C$_6$H$_{13}$	C$_{12}$H$_{25}$	C$_{18}$H$_{37}$
MW (g mol^{-1})	292	362	446	530
$t_{1/2}$ (s)	1.01×10^3	1.30×10^4	2.29×10^5	2.38×10^6
$\tau_1 \times 10^{-3}$ (s)	90	83	72	72
$\tau_2 \times 10^{-4}$ (s)	34	112	151	72
$\tau_3 \times 10^{-3}$ (s)	7.2	7.2	7.2	594

The following correlation was established for IPP oxidation at $T = 473$ K inhibited by phenols between the induction period τ and the rate constant of the reaction of the same phenols with cumylperoxyl radicals k_7 in cumene at $T = 333$ K [48]:

$$\tau(s) = 1.44 \times 10^4 - 2.88 \times 10^4 \, k_7^{-1} \quad (19.14)$$

The diffusion coefficients and solubility of phenols in polymers play an important role for polymer stabilization also. The values of these parameters can be found in the *Handbook of Polymer Degradation* [34].

19.3 ANTIOXIDANTS REACTING WITH ALKYL RADICALS

Acceptors of alkyl radicals are known to be very weak inhibitors of liquid-phase hydrocarbon oxidation because they compete with dioxygen, which reacts very rapidly with alkyl radicals. The situation dramatically changes in polymers where an alkyl radical acceptor effectively terminates the chains [3,49]. The study of the inhibiting action of *p*-benzoquinone [50], nitroxyl radicals [51–53], and nitro compounds [54] in oxidizing PP showed that these alkyl radical acceptors effectively retard the oxidation of the solid polymer at concentrations (~10^{-3} mol L^{-1}) at which they have no retarding effect on liquid hydrocarbon oxidation. It was proved from experiments on initiated PP oxidation at different pO_2 that these inhibitors terminate chains by the reaction with alkyl macroradicals. The general scheme of such inhibitors action on chain oxidation includes the following steps:

$$\text{Initiator} \longrightarrow R^\bullet \longrightarrow P^\bullet \qquad k_i$$
$$P^\bullet + O_2 \longrightarrow PO_2^\bullet \qquad k_1$$
$$PO_2^\bullet + PH \longrightarrow POOH + P^\bullet \qquad k_2$$
$$PO_2^\bullet + PO_2^\bullet \longrightarrow \text{Molecular products} \qquad k_6$$
$$PO_2^\bullet + InH \longrightarrow POOH + In^\bullet \qquad k_7$$
$$P^\bullet + InH \longrightarrow In^\bullet + PH \qquad k_7'$$
$$PO_2^\bullet(P^\bullet) + In^\bullet \longrightarrow \text{Molecular products} \qquad \text{fast}$$

According to this scheme, function F (see Chapter 14) has the following form:

$$F = \frac{v_i}{v}\left(1 - \frac{v^2}{v_0^2}\right) = c[\text{InH}] + d\frac{[\text{InH}]}{pO_2} \quad (19.15)$$

where v_i is the rate of initiation, v and v_0 are the rates of oxidation with and without inhibitor, $c = fk_7/k_2[\text{PH}]$, and $d = fk_7'/\gamma k_1$. If the inhibitor terminates the chains only by the reaction with dioxygen, the coefficient $d = 0$. If the inhibitor terminates the chains only by the reaction with alkyl macroradicals, the coefficient $c = 0$. The ratios $k_7/k_2[\text{PH}]$ and $k_7'/k_1\gamma$, estimated from the experimental data, are collected in Table 19.10.

Nitroxyl radicals, *p*-benzoquinone, and dinitrotoluene terminate chains only by the reaction with alkyl macroradicals. They form the following series according to their activity: nitroxyl radical > quinone > nitro compound.

Anthracene and 2,6-dinitrophenol terminate chains in oxidizing PP reacting with alkyl as well as with peroxyl radicals [50]. It is important to note that the last two inhibitors retard the liquid phase oxidation of hydrocarbons and aldehydes only by the reaction with peroxyl

TABLE 19.10
Relative Rate Constants of Inhibitors Terminating the Chains by Reactions with Alkyl and Peroxyl Macroradicals

Inhibitor	Polymer	T (K)	fk_7/k_2	$fk_7'/\gamma k_1$	Ref.
Ph-C(O)-O-[TEMP]-N-O•	PP	387	0	9.5×10^{-2}	[50]
Ph-C(O)-O-[TEMP]-N-O•	HDPE	366	0	2.7×10^{-2}	[52]
Ph-C(O)-O-[TEMP]-N-O•	PP	366		4.6×10^{-2}	[52]
O=⟨⟩=O (benzoquinone)	PP	387	0	7.6×10^{-2}	[50]
2,4-dinitrotoluene	PE	389	0	7.8×10^{-3}	[54]
2,4-dinitrotoluene	PP	389	0	4.5×10^{-3}	[54]
anthracene	PP	387	1.3×10^{-1}	3.6×10^{-3}	[50]
2,6-dinitrophenol	PP	387	9.0×10^{-2}	4.8×10^{-3}	[50]

radicals [55]. The values of ratio $k_7'/\gamma k_1$ are close for 2,6-dinitrophenol and 2,4-dinitrotoluene. So, 2,6-dinitrophenol behaves as a bifunctional inhibitor reacting with the peroxyl radical by the phenolic group and with the alkyl radical by its nitro group.

$$PO_2^\bullet + HO\text{-}C_6H_3(NO_2)_2 \longrightarrow POOH + {}^\bullet O\text{-}C_6H_3(NO_2)_2$$

$$P^\bullet + HO\text{-}C_6H_3(NO_2)_2 \longrightarrow \text{[product with P-O-N(O}^\bullet\text{) and HO, NO}_2\text{]}$$

Anthracene apparently adds alkyl as well as peroxyl macroradicals in positions 9 and 10.

$$PO_2^\bullet + \text{anthracene} \longrightarrow POO\text{-anthracenyl}^\bullet$$

$$P^\bullet + \text{anthracene} \longrightarrow P\text{-anthracenyl}^\bullet$$

The phenomena of relatively high activity of alkyl radical acceptors as antioxidants in solid polymer media seems to be the result of a line peculiarities of free radical reactions in the polymer matrix. Let us compare the features of these reactions in solution and polymer media.

Reaction $R^\bullet + O_2$ in Solution

1. Diffusion of dioxygen occurs with the diffusion coefficient $D \sim 10^{-5}$ cm^2 s^{-1}

2. Reaction is controlled by diffusion and occurs with a low activation energy equal to that of dioxygen diffusion in the liquid

3. Solubility of dioxygen in hydrocarbon is about 10^{-2} mol L^{-1} atm^{-1}

4. Liquid phase does not influence the activation energy of the reaction (see earlier)

Reaction $P^\bullet + O_2$ in Polymer

Diffusion of dioxygen occurs 10^2–10^5 times more slowly with the diffusion coefficient $D \sim 10^{-7}$–10^{-10} cm^2 s^{-1}

Carbon-centered atom of $P^\bullet$ changes its orbital hybridization in this reaction and changes the C—C bond angles from 120° to 109°. Since $P^\bullet$ is macroradical and is surrounded by segments of macromolecules, this process occurs with an activation energy

Solubility of dioxygen in the amorphous phase of polymer is about 3×10^{-4}–2×10^{-3} mol L^{-1} atm^{-1}

Polymer media influences the activation energy of the bimolecular reaction and causes the effect of reactivity leveling (see earlier)

As a result, the difference in activity of free radical acceptor and dioxygen is not so great in polymer as in the liquid.

Another important peculiarity of the retarding action of free radical acceptors in oxidized PP is their ability for chain termination by the reaction with peroxyl radicals (see Table 19.9 and refer to the literature [53,54,56,57]). As mentioned earlier, acceptors such as nitroxyls, quinones, and nitro compounds do not react with peroxyl radicals of hydrocarbons (see Chapter 16). Their ability to terminate chains in PP containing hydroperoxide groups is the result of the decomposition of the formed hydroperoxide groups, with the formation of hydrogen peroxide [58]. The latter reacts rapidly with peroxyl radicals. The formed hydroperoxyl radicals possess the high reducing activity and react with acceptors of hydrogen atoms (see later).

19.4 CYCLIC CHAIN TERMINATION IN OXIDIZED POLYMERS

19.4.1 REGENERATION OF NITROXYL RADICALS IN POLYMER OXIDATION

Nitroxyl radicals are formed as intermediates in reactions of polymer stabilization by sterically hindered amines as light stabilizers (HALS) [30,34,39,59]. The very important peculiarity of nitroxyl radicals as antioxidants of polymer degradation is their ability to participate in cyclic mechanisms of chain termination. This mechanism involves alternation of reactions involving alkyl and peroxyl radicals with regeneration of nitroxyl radical [60–64].

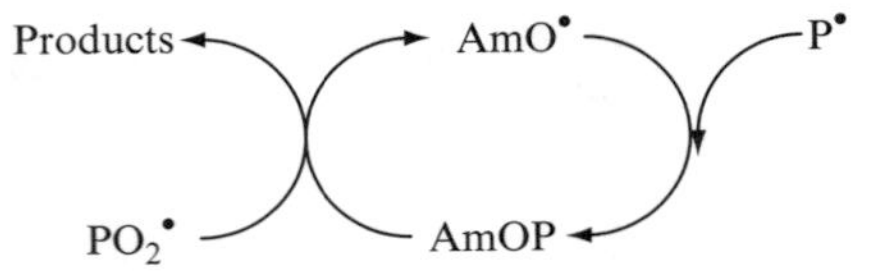

The phenomena of nitroxyl radicals regeneration has been discovered in the study of the retarding effect of 2,2,6,6-tetramethyl-4-benzoyloxypiperidine-N-oxyl on PP initiated oxidation [51]. It has been shown that the limiting step of chain termination by the nitroxyl radical is the reaction with the alkyl macroradical of PP. The resulting compound AmOP is fairly reactive with respect to the peroxyl radical and nitroxyl radical is regenerated in this reaction. Thus, the cycle includes the following two reactions (mechanism I) [60–64]:

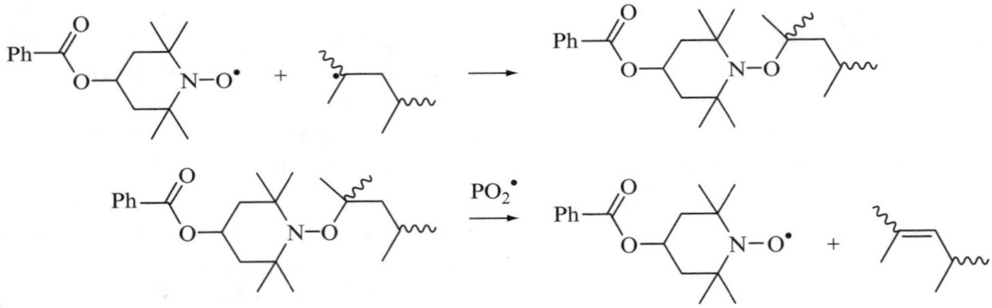

The regeneration of nitroxyl radical from the product of the reaction of nitroxyl radical with the alkyl macroradical was proved in the following experiments [51]. The nitroxyl radical and initiator (dicumyl peroxide) were introduced in a PP powder and this sample was heated to $T = 387\,K$ in an argon atmosphere. The concentration of nitroxyl radical was monitored by the EPR technique. The nitroxyl radical was consumed in PP with the rate of free radical generation by the initiator (see Figure 19.3). Dioxygen was introduced in the reactor after the nitroxyl radical was consumed. The generation of peroxyl radicals induced the formation of nitroxyl radicals from the adduct of the nitroxyl radical with the PP macroradical.

Two products can be formed in the reaction of the nitroxyl radical with the alkyl macroradical.

$$AmO^{\bullet} + P^{\bullet} \longrightarrow AmOP$$
$$AmO^{\bullet} + P^{\bullet} \longrightarrow AmOH + {\sim}MeC{=}CH{\sim}$$

Special experiments with preliminary withdrawal of all low-molecular products formed by the reaction of nitroxyl with alkyl macroradicals from the polymer sample were performed. These

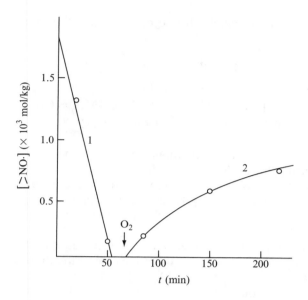

FIGURE 19.3 Kinetics of 2,2,6,6-tetramethyl-4-benzoyloxypiperidine-N-oxyl consumption in IPP containing AIBN ($v_i = 8.0 \times 10^{-7}$ mol kg^{-1} s^{-1}) at 387 K in Ar (1) and dioxygen (2) atmosphere [51].

experiments proved that nitroxyl radicals were formed precisely by the reaction of peroxyl radicals with AmOP. This conclusion was confirmed later in the study of radiation-induced chemical oxidation of octane where alkoxyamine AmOR was identified as the reaction product [65].

The reaction of the peroxyl radical with the product AmOP should occur rapidly to perform cyclic chain termination. Let us estimate the enthalpy of this reaction. In the reaction $RO_2^{\bullet} + RON{<}$ two bonds, namely, C—H and C—ON, are broken, and two bonds are formed, namely, ROO—H and C=C. The values of these bonds are given below [66].

Bond	Me$_3$COO—H	RC$^{\bullet}$HCH—HR	⌬	ΔH
D (kJ mol^{-1})	358.6	139.0	134.0	−85.6

This calculation shows that the discussed reaction is very exothermic. The activation energy of this reaction calculated by the IPM method (see Chapter 6) is equal to 8.7 kJ mol^{-1} and rate constant is $k = 7.3 \times 10^6$ L mol^{-1} s^{-1} at $T = 400$ K. This rate constant is close to that of the acceptance of the alkyl macroradical by the nitroxyl radical. Hence, this reaction is rapid enough to be the efficient step in cyclic chain termination in polymer.

Another mechanism of nitroxyl radical regeneration was proposed and discussed in the literature [67–71]. The alkoxyamine AmOR is thermally unstable. At elevated temperatures it dissociates with cleavage of the R—O bond, which leads to the appearance of an [AmO$^{\bullet}$ + R$^{\bullet}$] radical pair in the cage of polymer. The disproportionation of this radical pair gives hydroxylamine and alkene. The peroxyl radical reacts rapidly with hydroxylamine thus regenerating the nitroxyl radical. Such a cyclic mechanism includes the following steps (mechanism II):

$$AmO^{\bullet} + P^{\bullet} \longrightarrow AmOP$$
$$AmOP \longrightarrow [AmO^{\bullet} + P^{\bullet}]$$
$$[AmO^{\bullet} + P^{\bullet}] \longrightarrow AmOH + \sim MeC{=}CH\sim$$
$$PO_2^{\bullet} + AmOH \longrightarrow POOH + AmO^{\bullet}$$

The fact that hydroxylamine has been found among the products of transformations of nitroxyl radical during the oxidation of hydrocarbons is the evidence in support of this mechanism. Both mechanisms described earlier are realized in parallel and supplement each other. The result of the competition between them depends primarily on the temperature, because the thermal decomposition of alkoxyamine AmOR requires a fairly high activation energy (see Table 19.11).

The problem of competition between these two mechanisms I and II of cyclic chain termination has been considered in Ref. [60]. In the case of the PP oxidation at the initiation rate $v_i = 10^{-7}$ mol kg^{-1} s^{-1} and for $[O_2] = [AmO^{\bullet}] = 10^{-3}$ mol kg^{-1}, the ratio of the rates of chain termination via reactions ($PO_2^{\bullet} + AmOP$ (I)) and ($PO_2^{\bullet} + AmOH$ (II)) varies as a function of the temperature in the following way [60]:

T (K)	350	365	380	388	400	420
v_I/v_{II}	50.0	9.55	2.14	1.00	0.33	0.029

It is obvious that when the temperature is not very high and alkoxyamine AmOR is stable, mechanism I predominates, that is, the regeneration of AmO$^{\bullet}$ from AmOR is ensured by the reaction of AmOH with the peroxyl radical. At higher temperatures, when AmOR becomes unstable, mechanism II predominates. It should be taken into account that only some of the radical pairs formed upon decomposition of AmOR disproportionate giving AmOH and alkene. The remaining radical pairs pass into the bulk, and the radical R$^{\bullet}$ reacts with dioxygen to give peroxyl radicals and thus initiate a new oxidation chain. Hence, the regeneration of AmO$^{\bullet}$ from AmOH is inevitably accompanied by the initiation of the process, which certainly decreases the efficiency of inhibition.

19.4.2 Cyclic Chain Termination in Oxidized Polypropylene

The PP oxidation proceeds with preferential intramolecular chain propagation and formation of the adjacent hydroperoxyl groups (see Chapter 13). These groups decompose to give hydrogen peroxide [58].

The latter rapidly reacts with peroxyl radicals to produce hydroperoxyl radicals possessing the high reducing activity (see Chapter 16).

$$PO_2^{\bullet} + H_2O_2 \longrightarrow POOH + HO_2^{\bullet}$$

TABLE 19.11
Rate Constants of Thermal Decay of N-alkoxyamines

N-Alkoxyamine	Solvent	T (K)	E (kJ mol^{-1})	log A, A (s^{-1})	k_d (350 K) (s^{-1})	Ref.
	Gas phase	503–533	189.5	15.30	1.05×10^{-13}	[72]
	Cyclohexadiene or 9,10-dihydroanthracene	503–533	192.5	15.30	3.72×10^{-14}	[72]
	Cyclohexadiene or 9,10-dihydroanthracene	503–533	179.5	14.10	2.05×10^{-13}	[72]
	Chlorobenzene		102.6	14.80	0.31	[73]
	Isooctane		134.0	13.70	5.03×10^{-7}	[74]
	Cyclohexane		92.1	9.04	1.97×10^{-5}	[73]
	Isooctane		87.0	8.60	4.13×10^{-5}	[74]
	Toluene		133.2	14.04	1.45×10^{-6}	[75]
	Toluene		138.8	14.96	1.76×10^{-6}	[75,76]
	tert-Butylbenzene	343–363	114.4	14.11	1.09×10^{-3}	[77]
	Cyclohexane		114.0	14.00	9.69×10^{-4}	[78]
	Isooctane		121.0	14.90	6.95×10^{-4}	[74]
	Cyclohexane		129.0	13.70	2.81×10^{-6}	[78]
	tert-Butylbenzene		99.0	11.85	1.19×10^{-3}	[79]

continued

TABLE 19.11
Rate Constants of Thermal Decay of N-Alkoxyamines—*continued*

N-Alkoxyamine	Solvent	T (K)	E (kJ mol^{-1})	log A, A (s^{-1})	k_d (350 K) (s^{-1})	Ref.
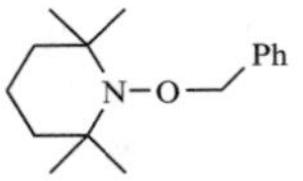	Cyclohexane		137.0	13.60	1.43×10^{-7}	[78]
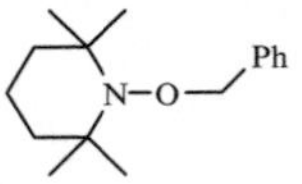	Styrene	396			7.7×10^{-5}	[80]
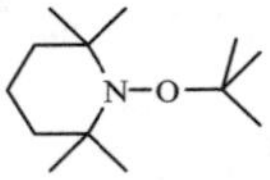	tert-Butylbenzene		97.0	10.76	1.92×10^{-4}	[79]
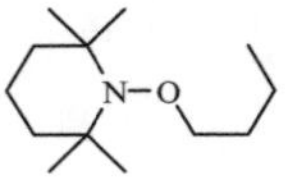	tert-Butylbenzene		99.0	10.40	4.22×10^{-5}	[79]
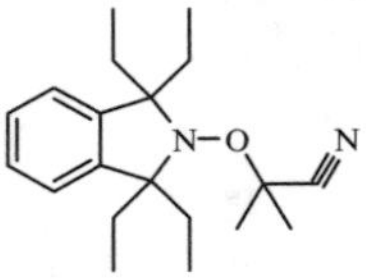	Hexane	333			3.0×10^{-4}	[81]
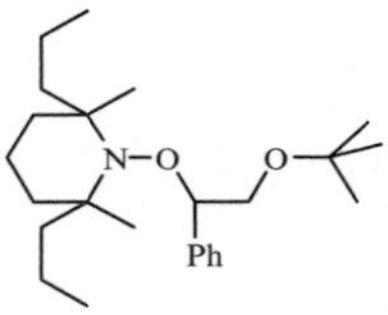	Ethyl acetate	353			2.9×10^{-5}	[81]
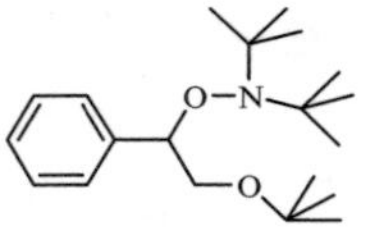	Ethyl acetate	363			1.65×10^{-4}	[81]

Therefore, such alkyl radical acceptors as quinones, nitroxyl radicals, and nitro compounds retard the oxidation of PP according to the following cyclic mechanism of chain termination:

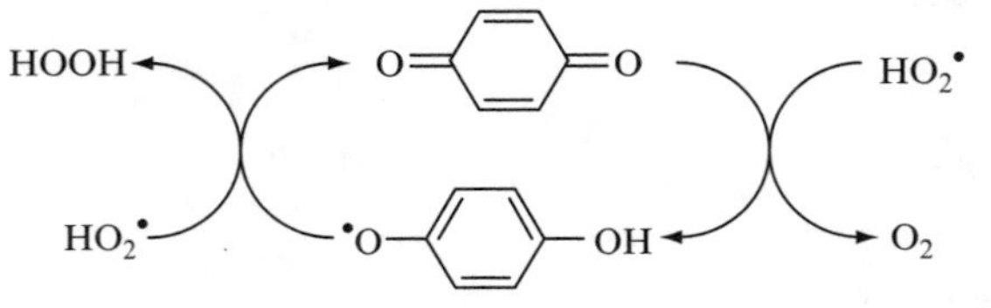

In addition to this reaction, quinones and other alkyl radical acceptors retard polymer oxidation by the reaction with alkyl radicals (see earlier). As a result, effectiveness of these inhibitors increases with the formation of hydroperoxide groups in PP. In addition, the inhibiting capacity of these antioxidants grows with hydroperoxide accumulation. The results illustrating the efficiency of the antioxidants with cyclic chain termination mechanisms in PP containing hydroperoxide groups is presented in Table 19.12. The polyatomic phenols producing quinones also possess the ability to terminate several chains.

TABLE 19.12
Kinetic Parameters of Inhibitors of Cyclic Chain Termination in PP Containing Hydroperoxide Groups and Reacting with $HO_2^\bullet$ Radicals

Inhibitor	T (K)	fk_7/k_2	fk_7/k_1	f	Ref.
(piperidine N–O• benzoate)	387			14	[50]
($C_{17}H_{35}$ piperidine N–O• ester)	388	800	0.78	>40	[56]
($C_{17}H_{35}$ piperidine NH ester)	388	800	2.3		[56]
(bis-quinone methide)	366	6.3×10^{-1}	1.8×10^{-2}	20	[53]
(bisphenol methylene)	366	8.4×10^{-1}	9.3×10^{-2}	>60	[82]
(bisphenol methylene)	366			60	[82]
(hindered phenol piperidine NH ester)	388	460	2.5	30	[57]
(phenol quinone methide)	366			35	[53]
(stilbenequinone)	366	0.84	9.3×10^{-2}		[53]

continued

TABLE 19.12
Kinetic Parameters of Inhibitors of Cyclic Chain Termination in PP Containing Hydroperoxide Groups and Reacting with $HO_2^{\bullet}$ Radicals—continued

Inhibitor	T (K)	fk_7/k_2	fk_7/k_1	f	Ref.
[structure]	366	0.63	1.8×10^{-2}		[53]
[structure]	366			21	[82]

Phenols usually terminate two chains in the oxidation of hydrocarbons and solid polymers (see Chapter 15). The study of the f value dependence on partial dioxygen pressure showed, however, that the stoichiometric coefficient of inhibition has a tendency to increase with decreasing the dioxygen pressure and, in an inert atmosphere, it is markedly higher than in dioxygen [83]. The results of f value estimation ($f = v_i/v_{InH}$, phenol concentration was measured spectroscopically) are given in Table 19.13.

The following explanation accounts for the facts. Alkyl radicals react with phenoxyl radicals by two parallel reactions, namely, recombination and disproportionation.

TABLE 19.13
The Dependence of f Values for Phenols on Dioxygen Pressure in PP at $T = 388$ K [83]

Phenols	f values at		
	$pO_2 = 1$ (atm)	$pO_2 = 0.2$ (atm)	Argon
[structure]	1.0	1.8	2.0
[structure]	1.0	2.0	3.3
[structure]	3.0	5.0	10.0

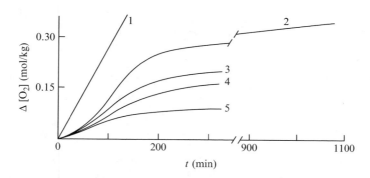

FIGURE 19.4 Kinetics of initiated IPP oxidation ($v_i = 8.5 \times 10^{-7}$ (mol kg^{-1} s^{-1}) $T = 388$ K, $pO_2 = 98$ kPa) (1) without (2) and with various concentrations of 2-hydroxy-5-methylbenzylphenyl sulfide: 2.0×10^{-3} (mol kg^{-1}) (2), 3.0×10^{-3} (mol kg^{-1}) (3), 5.0×10^{-3} (mol kg^{-1}) (4), and 1.0×10^{-2} (mol kg^{-1}) (5) [84].

Consequently, in an inert atmosphere $f = 2(1 + k_{dis}/k_{rec}) > 2$. When phenoxyl radicals react only with peroxyl radicals, $f = 2$ and there is no regeneration. At low dioxygen pressures, phenoxyl radicals react with both peroxyl and alkyl radicals; f ranges between 2 and $2(1 + k_{dis}/k_{rec})$ and increases with decreasing pO_2. In addition to this, the product of phenol oxidation, quinone, becomes the efficient alkyl radical acceptor at low dioxygen pressure (see earlier).

Among the products formed from antioxidants may be inhibitors terminating chains with high f values. Such an example was found for PP oxidation with phenol sulfide (2-hydroxy-5-methylbenzylphenyl sulfide) [84]. When the inhibitor concentration is not high (for example, 10^{-3} mol kg^{-1}), oxidation involves three steps (Figure 19.4): first, with the induction period nearly equal to $2[InH]_0/v_i$ typical of phenol inhibition; second, when oxidation occurs with the rate of a noninhibited reaction; and third, when the second period of the inhibited reaction appears. This period is obviously connected with the formation of a product (or products) with high inhibition activity. It will be noted that the second inhibition period is extremely long. If all phenol sulfide is assumed to be converted into this inhibitor, its f value is higher than 20. It seems very probable that the second period of inhibition is the result of hydroperoxide groups formation and participation of HO$_2^\bullet$ radicals in cyclic chain termination by sulfoxide or disulfone by a mechanism similar to that of quinone.

19.5 INHIBITION OF SYNTHETIC RUBBER DEGRADATION

Oxidative degradation of 1,3-polyisobutylene (PIB) occurs as a result of the rapid degradation of alkoxyl macroradicals with peroxide bridges (see Chapter 13). The rate of PIB oxidative degradation is $v_S \sim [PO_2^\bullet]^2$. If this mechanism is the same in the presence of the peroxyl radical acceptor InH, the retarding action of such antioxidants as phenols and amines would be extremely efficient. For example, a decrease in the oxidation rate by 10 times would be lower than the rate of degradation by 100 times. However, the experiments with measurement of the rate of oxidation v as well as that of degradation v_S gave quite another result. The results of the experiments on PIB oxidation retarded by ionol are given below (PIB, $v_i = 1.9 \times 10^{-6}$ mol $L^{-1} s^{-1}$, [ionol] $= 4.5 \times 10^{-3}$ mol L^{-1}, v and v_S are given in mol $L^{-1} s^{-1}$ [85]).

Conditions of PIB Oxidation	Rate of Oxidation	Rate of Degradation
Without [InH]	$v_0 = 8.8 \times 10^{-4}$	$v_S = 2.8 \times 10^{-5}$
With [ionol] $= 4.5 \times 10^{-3}$ (mol L^{-1})	$v = 9.4 \times 10^{-6}$	$v_S = 4.0 \times 10^{-6}$

The introduction of ionol in oxidized PIB decreases the peroxyl radical concentration by 94 times and rate of degradation via the reaction $PO_2^{\bullet} + PO_2^{\bullet}$ by $94^2 = 8836$ times, but the observed rate of macromolecules degradation decreases by seven times only. This means the appearance of another mechanism of degradation in the presence of the antioxidant. The kinetic study of PIB degradation in the process of initiated oxidation confirmed this hypothesis. The PIB degradation via the reaction $PO_2^{\bullet} + PO_2^{\bullet}$ leads to the proportionality: $v_S \sim [InH]^{-2} v_i^2$. Experiments on PIB oxidation with different antioxidants (phenols, amines, and aminophenols) and variation of concentrations of an initiator, as well as inhibitor, proved the following equation:

$$v_S = const. \times \frac{v_i^2}{[InH]}. \tag{19.16}$$

The results of the experiments are presented in Table 19.14.

The studied inhibitors differ in their ability to retard degradation as well as oxidation of PIB. There is no similarity in their activity to retard oxidation and destruction. The following mechanism of polymer degradation was proposed for PIB [85]:

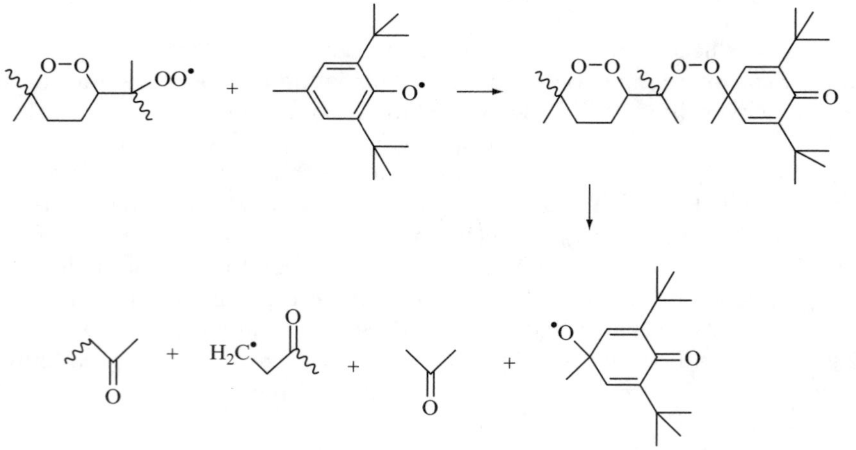

The oxidation of PIB occurs mainly via intramolecular addition of dioxygen to double bonds of polymer. The reaction of peroxyl radical addition to the phenoxyl radical leads to the formation of quinolide peroxide (see Chapter 15). This peroxide is unstable, and its decomposition provokes the degradation of PIB. Another reaction predominates in case of aromatic diamine.

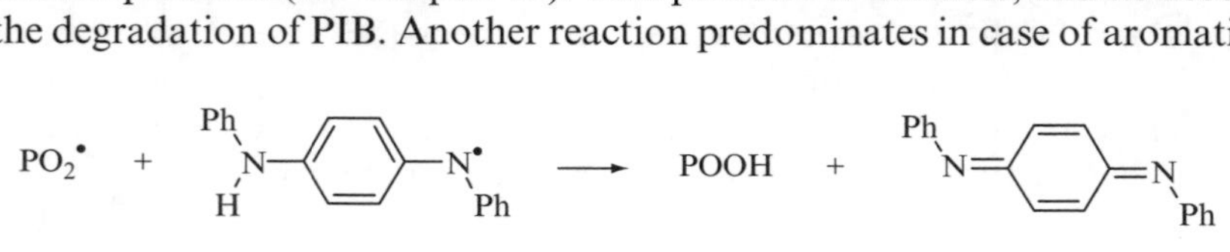

TABLE 19.14
The Kinetic Parmeters of Retarding Action of Phenols and Amines on Oxidation and Degrdation of PIB at $T = 353$ (K) and $pO_2 = 10^5$ Pa [85]

Antioxidant	$k_7(k_p[PH])^{-1}$ (kg mol^{-1})	$v_i^2/v_s[InH]_0$ (s^{-1})
	67.4	1.82×10^{-4}
	56.4	4.00×10^{-4}
	56.4	
	50.2	6.67×10^{-4}
	56.4	2.44×10^{-4}
	81.5	4.76×10^{-4}
	522	3.70×10^{-3}

continued

TABLE 19.14
The Kinetic Parmeters of Retarding Action of Phenols and Amines on Oxidation and Degrdation of PIB at $T = 353$ (K) and $pO_2 = 10^5$ Pa [85]—continued

Antioxidant	$k_7(k_p[PH])^{-1}$ (kg mol^{-1})	$v_i^2/v_s[InH]_0$ (s^{-1})
(phenyl-aminophenol)	6.10	7.14×10^{-4}
(naphthyl-aminophenol)	7.20	7.69×10^{-4}

Hence, the degradation of PIB occurs sufficiently slowly (see Table 19.14). Aminophenols retard oxidation more weakly than phenols but are efficient in retardation of degradation.

REFERENCES

1. VA Roginskii. Oxidation of polyolefines inhibited by sterically hindered phenols, Thesis Doctoral Diss., Institute of Chemical Physics, Chernogolovka, 1983 [in Russian].
2. NM Emanuel, AL Buchachenko. *Chemical Physics of Polymer Degradation and Stabilization.* Utrecht: VNU Science, 1987.
3. ET Denisov. *Oxidation and Degradation of Carbon Chain Polymers.* Leningrad: Khimiya, 1990, pp. 1–287 [in Russian].
4. AM North. *The Collision Theory of Chemical Reaction in Liquids.* London: Methuen, 1964.
5. ET Denisov, OM Sarkisov, GI Likhtenshtein. *Chemical Kinetics.* Amsterdam: Elsevier, 2003.
6. VI Rubtsov, VA Roginskii, VB Miller. *Vysokomol Soed* A22:2506–2513, 1980.
7. AP Griva, ET Denisov. *J Polym Sci Polym Chem Ed* 14:1051–1064, 1976.
8. SD Grinkina, ET Denisov, AP Griva. *Kinet Katal* 35:523–525, 1994.
9. AP Griva, LN Denisova, ET Denisov. *Kinet Katal* 19:309–315, 1978.
10. SD Grinkina, AP Griva, ET Denisov, VD Sen', IK Yakuschenko. *Kinet Katal* 34:245–249, 1993.
11. AP Griva, LN Denisova, ET Denisov. *Zh Fiz Khim* 63:2601–2607, 1989.
12. NV Zolotova, ET Denisov. *Zh Fiz Khim* 46:2008–2011, 1972.
13. ET Denisov, IV Khudyakov. *Chem Rev* 87:1313–1357, 1987.
14. AP Griva, LN Denisova, ET Denisov. *Vysokomol Soed* A21:849–853, 1979.
15. AP Griva, LN Denisova, ET Denisov. *Dokl AN SSSR* 232:1343–1346, 1977.
16. ET Denisov. *Zh Fiz Khim* 49:2473–2410, 1975.
17. ET Denisov. *Macromol Chem Suppl* 8:63–78, 1984.
18. ET Denisov, AP Griva. *Zh Fiz Khim* 53:2417–2437, 1979.
19. ET Denisov. In: ZB Alfassi, (ed.) *General Aspects of the Chemistry of Radicals.* New York: Wiley, 1999, pp. 79–137.
20. ET Denisov. *Reactions of Inhibitor Radicals and Mechanism of Inhibited Oxidation of Hydrocarbons*, vol 17. Moscow: VINITI, Itogi Nauki i Tekhniki, 1987, pp. 3–115 [in Russian].
21. AP Griva, ET Denisov. *Dokl AN SSSR* 219:640–643, 1974.
22. VI Rubtsov, VA Roginskii, VB Miller. *Vysokomol Soed* A22:2506–2513, 1980.
23. ON Karpukhin, TV Pokholok, VYa Shlyapintokh. *Vysokomol Soed* A13:22–28, 1971.

24. ON Karpukhin, TV Pokholok, VYa Shlyapintokh. *J Polymer Sci Polymer Chem Ed* 13:525–548, 1975.
25. ET Denisov. *Zh Fiz Khim* 61:3100–3102, 1987.
26. VI Rubtsov, VA Roginskii, VB Miller. *Vysokomol Soed* A22:2506–2513, 1980.
27. VI Rubtsov, VA Roginskii, VB Miller. *Kinet Katal* 21:612–615, 1980.
28. AP Griva, LN Denisova, ET Denisov. *Vysokomol Soed* A21:849–854, 1979.
29. N Grassi, G Scott. *Polymer Degradation and Stabilization*. Cambridge: University Press, 1985.
30. G Geiskens (ed.), *Degradation and Stabilisation of Polymers*. London: Applied Science, 1975.
31. HHG Jellinec (ed.), *Aspects of Degradation and Stabilisation of Polymers*. Amsterdam: Elsevier, 1978.
32. VYa Shlyapintokh. *Photochemical Conversion and Stabilization of Polymers*. Munich: Hanser Publishers, 1984.
33. AA Popov, NYa Rapoport, GE Zaikov. *Oxidation of Oriented and Strained Polymers*. Moscow: Khimiya, 1987, p. 25 [in Russian].
34. SH Hamid (ed.), *Handbook of Polymer Degradation*. New York: Marcel Dekker, 2000.
35. VJ von Voigt. *Die Stabilisierung der Kunststoffe gegen Licht und Warme*. Berlin: Springer-Verlag, 1966.
36. WL Hawkins. *Polymer Stabilization*. New York: Wiley, 1972.
37. J Pospisil. In: G Scott (ed.), *Developments in Polymer Stabilization-1*. London: Applied Science, 1978, p. 1.
38. G Scott. In: G Scott (ed.), *Developments in Polymer Stabilization-4*. London: Applied Science, 1981, p. 1.
39. P Klemchuk, J Pospisil. *Oxidation Inhibition of Organic Materials*, vols 1 and 2. Boca Raton: CRC Press, 1990.
40. ET Denisov. *Handbook of Antioxidants*. Boca Raton: CRC Press, 1995, pp. 1–174.
41. NV Zolotova, ET Denisov. *Vysokomol Soed* B18:605–608, 1976.
42. VV Pchelintsev, ET Denisov. *Vysokomol Soed* A25:1035–1041, 1983.
43. NV Zolotova, ET Denisov. *Vysokomol Soed* A24:420–424, 1982.
44. VA Roginskii, VB Miller. *Dokl AN SSSR* 213:642–645, 1973.
45. NM Livanova, NS Vasileiskaya, DV Muslin, VB Miller, YuA Shlyapnikov. *Izv AN SSSR Ser Khim* 1074–1079, 1972.
46. NS Vasileiskaya, NM Livanova, VB Miller, LV Samarina, YuA Shlyapnikov. *Izv AN SSSR Ser Khim* 2614–2616, 1972.
47. NM Livanova, NS Vasileiskaya, LV Samarina, VB Miller, YuA Shlyapnikov. *Izv AN SSSR Ser Khim* 1672–1673, 1973.
48. AA Kharitonova, GP Gladyshev, VF Tsepalov. *Kinet Katal* 20:245–247, 1979.
49. ET Denisov. In: G Scott, (ed.) *Developments and Polymer Stabilization-5*. London: Applied Science, 1982, pp. 23–40.
50. YuB Shilov, ET Denisov. *Vysokomol Soed* A16:1736–1741, 1974.
51. YuB Shilov, RT Battalova, ET Denisov. *Dokl AN SSSR* 207:388–389, 1972.
52. YuB Shilov, ET Denisov. *Vysokomol Soed* A16:2313–2316, 1974.
53. YuB Shilov, ET Denisov. *Vysokomol Soed* A26:1753–1758, 1984.
54. YuB Shilov, ET Denisov. *Kinet Catal* 42:238–242, 1984.
55. NM Emanuel, ET Denisov, ZK Maizus. *Liquid-Phase Oxidation of Hydrocarbons*. New York: Plenum Press, 1967.
56. AV Kirgin, YuB Shilov, BE Krisyuk, AA Efimov, VV Pavlikov. *Kinet Katal* 31:58–64, 1990.
57. AV Kirgin, YuB Shilov, ET Denisov, VV Pavlikov, AA Efimov. *Kinet Katal* 31:65–71, 1990.
58. YuB Shilov, ET Denisov. *Vysokomol Soed* A29:1359–1361, 1987.
59. PP Klemchuk (ed.), *Polymer Stabilization and Degradation*. ACS Symposium Series 280. Washington, D.C.: American Chemical Society, 1985.
60. ET Denisov. In: G Scott (ed.), *Developments and Polymer Stabilisation-3*. London: Applied Science, 1980, pp. 1–20.
61. ET Denisov. *Russ Chem Rev* 65: 505–520, 1996.

62. ET Denisov. In: PP Klemchuk (ed.), *Polymer Stabilization and Degradation*. ACS Symposium Series 280. Washington, D.C.: American Chemical Society, 1985, pp. 87–90.
63. ET Denisov. *Polym Degrad Stab* 25:209–215, 1989.
64. ET Denisov. *Polym Degrad Stab* 34:325–332, 1991.
65. MV Sudnik, MF Romantsev, AB Shapiro, EG Rozantsev. *Izv AN SSSR Ser Khim* 2813–2816, 1975.
66. ET Denisov, TG Denisova, TS Pokidova. *Handbook of Free Radical Initiators*. New York: Wiley, 2003.
67. DJ Carlsson, A Garton, DM Wiles. In: G Scott (ed.), *Developments and Polymer Stabilisation-1*, 219, 1978.
68. KB Chakraborty, G Scott. *Polymer* 21:252–261, 1980.
69. H Berger, TABM Bolsman, DM Brower. *Dev Polym Stabil* 6:1–27, 1983.
70. G Scott. *J Polym Sci Polym Lett Ed* 22:553–562, 1984.
71. DJ Carlsson. *Pure Appl Chem* 55:1651–1662, 1983.
72. MV Ciriano, H-G Korth, WB van Scheppingen, P Mulder. *J Am Chem Soc* 121:6375–6381, 1999.
73. DW Grattan, DJ Carlsson, JA Howard, DM Wiles. *Can J Chem* 57:2834–2842, 1979.
74. C Neri, S Costanzi, R Farris, V Malatesta. Thirteenth International Conference on Advances in the Stabilization and Degradation of Polymers. Lucern, 1991.
75. SAF Bon, G Chambard, AL German. *Macromolecules* 32:8269–8278, 1999.
76. SAF Bon, G Chambard, FAC Bergman, EHH Snellen, B Klumperman, AL German. *Polym Prepr (Am Chem Soc Div Polym Chem)* 38(1), 748–759 (1997).
77. T Kothe, S Marque, R Martschke, M Popov, H Fisher. *J Chem Soc Perkin Trans 2*, 1553–1559, 1998.
78. WG Skene, ST Belt, TJ Connolly, P Hahn, JC Scaiano. *Macromolecules* 31:9103–9105, 1998.
79. P Slipa, L Greei, P Carloni, E Damiani. *Polym Degrad Stab* 55:323–331, 1997.
80. CJ Hawker, GG Barelay, A Orellana, J Dao, W Devonport. *Macromolecules* 29:5245–5251, 1996.
81. G Moad, E Rizzardo. *Macromolecules* 28:8722–8729, 1995.
82. YuB Shilov, ET Denisov. *Vysokomol Soed* A24:837–842, 1982.
83. NV Zolotova, ET Denisov. *Vysokomol Soed* B18:605–608, 1976.
84. LL Gervits, NV Zolotova, ET Denisov. *Vysokomol Soed* B18:524–526, 1976.
85. VV Pchelintsev, ET Denisov. *Vysokomol Soed* A26:624–628, 1982.

20 Heterogeneous Inhibition of Oxidation

20.1 RETARDING ACTION OF SOLIDS

Metals and metal oxides, as a rule, accelerate the liquid-phase oxidation of hydrocarbons. This acceleration is produced by the initiation of free radicals via catalytic decomposition of hydroperoxides or catalysis of the reaction of RH with dioxygen (see Chapter 10). In addition to the catalytic action, a solid powder of different compounds gives evidence of the inhibiting action [1–3]. Here are a few examples. The following metals in the form of a powder retard the autoxidation of a hydrocarbon mixture (fuel T-6, at $T = 398$ K): Mg, Mo, Ni, Nb V, W, and Zn [4,5]. The retarding action of the following compounds was described in the literature.

RH	Cyclohexane	Hydrocarbon Mixture	Cumene+1-nonene	Benzaldehyde
Compounds	$CuSe$, $MoSe_2$, $TaSe_2$, WSe_2	MoS_2	$K_3Fe(CN)_6$	NoS_2, TiB_2
Ref.	[6]	[7]	[8]	[2]

Critical phenomena, observed in heterogeneous catalysis (see Chapter 10), proved that the surface of a catalyst possesses two kinds of action on the liquid-phase oxidation: chain initiation and chain termination (see Chapter 10).

The kinetics of a hydrocarbon mixture (fuel T-6) oxidation inhibited by a powder of Mo and MoS_2 was studied by Kovalev et al. [7,9,10]. The results of the study of initiated oxidation of T-6 in the presence of Mo and MoS_2 are presented in Table 20.1.

One observes that the larger the surface of Mo or MoS_2 the shorter the chain length v. The dependence of the function $F = v_0/v - v/v_0$, where $v_0 = v$ at [inhibitor] = 0, on the amount of inhibitor is unusual. The dependence $F \sim$ [inhibitor] is usual for homogeneous chain termination (see Chapter 14). Quite another dependence was observed for the heterogeneous chain termination by powders of Mo and MoS_2 [1,7,9,10], namely

$$F = const \times [\text{solid inhibitor}]^{1/3}[\text{initiator}]^{-1/2}. \qquad (20.1)$$

Such a dependence was interpreted within the scope of the model of chain oxidation with diffusionally controlled chain termination on the surface of solid antioxidant (for example, Mo or MoS_2). According to the Smolukhovsky equation, the diffusion velocity of radical $RO_2^\bullet$ at the distance $l/2$ is $v = 0.2D\kappa S^{1/3}$, where S is the surface of the solid inhibitor and κ is the coefficient of proportionality between the surface and number n of the solid particles ($S = \kappa \times n$). The function F for such diffusionally controlled chain termination is the following:

TABLE 20.1
Retarding Action of Mo and MoS$_2$ Powder on Initiated Oxidation of Fuel T-6 (Mixture of Hydrocarbons) at $T = 398$ K, $pO_2 = 10^5$ (Pa), $v_i = 7.80 \times 10^{-7}$ mol L^{-1} s^{-1}, with Dicumylperoxide as Initiator [5,7,9,10]

S (cm^2 L^{-1})	$v \times 10^5$ (mol L^{-1} s^{-1})	ν	F	$F[I]^{1/2} S^{-1/3}$
		Mo		
0	2.22	28	0	0
30	1.66	21	0.59	1.47
70	1.67	21	0.58	1.09
900	1.33	17	1.07	0.86
150	1.16	15	1.39	0.94
750	0.76	10	2.58	1.02
12,000	0.71	9	2.80	0.95
		MoS$_2$		
0	2.22	28	0	0
2.54	1.11	14	1.20	6.75
5.52	0.98	13	1.31	5.80
8.15	0.81	10	1.72	6.50
8.15	1.57[a]	10	1.23	6.75
8.15	0.50	6	3.30	6.00

[a] Initiation rate is 1.56×10^{-6} mol L^{-1} s^{-1}.

$$F = \frac{0.2 D \kappa}{\sqrt{2 k_t k_i}} \times \frac{S^{1/3}}{[I]^{1/2}} \qquad (20.2)$$

This equation agrees with the experimental data (see Table 20.1). The chain termination on the surface of MoS$_2$ occurs catalytically with a very high inhibition coefficient $f \approx 3 \times 10^5$.

In addition to chain termination, Mo and MoS$_2$ catalytically decompose the formed hydroperoxide according to the following empirical equations [9,10]:

Catalyst	Equation	Rate Constant
Mo	$v = k_d[\text{ROOH}] S_{\text{Mo}}^{0.25}$	$k_d = 4.5 \times 10^4 \exp(-58.5/RT)$ (L cm^{-2})$^{0.25}$ s^{-1}
MoS$_2$	$v = k_d[\text{ROOH}] S_{\text{Mo}}^{0.10}$	$k_d = 3.2 \exp(-29.0/RT)$ (L cm^{-2})$^{0.1}$ s^{-1}

The decomposition of hydroperoxide on the surface of Mo and MoS$_2$ started after some induction period (see Table 20.2). This induction period is the time for the activation of the surface toward hydroperoxide decomposition. It was evidenced in special experiments that the catalyst is not dissolved in hydrocarbon and catalytic hydroperoxide decomposition occurs only heterogeneously.

The catalyst can decompose hydroperoxide homolytically, as well as heterolytically (see Chapters 10 and 17). Special experiments on the oxidation of fuel T-6 were performed in the presence of MoS$_2$ with combined initiation by the initiator (DCP) and hydroperoxide formed in T-6 in the presence and absence of MoS$_2$ [10]. It was found that the rate of the free radical generation and the rate of the hydroperoxide decay proceeds by the equations

TABLE 20.2
Rate Constants k_d and Induction Periods τ of Hydroperoxide Decomposition Catalyzed by Mo and MoS_2 Powder in Hydrocarbon Mixture (Fuel T-6) [10]

[ROOH] × 10³ (mol L⁻¹)	S (cm² L⁻¹)	T (K)	k_d × 10⁴ (s⁻¹)	τ × 10⁻³ (s)
		Mo		
3.25	30	398		7.2
3.25	120	398	4.6	2.34
3.25	200	398	5.1	1.20
3.25	1,200	398	9.1	0.90
3.70	1,200	398	9.2	0.90
2.50	1,200	398	9.2	0.90
1.82	1,200	398	9.2	0.90
3.25	1,200	388	5.3	1.26
3.25	1,200	408	12.3	0.60
3.25	1,200	418	21.8	0.18
		MoS_2		
3.12	1,700	398		12.6
3.12	1,900	398	3.20	4.4
3.12	6,500	398	3.45	3.6
3.12	13,000	398	4.00	2.7
3.12	65,000	398	5.10	1.4
3.12	6,50,000	398	10.50	0.0
3.40	65,000	398	5.10	1.4
2.90	65,000	398	5.10	1.4
2.10	65,000	398	5.10	1.4
1.10	65,000	398	5.10	1.2
3.25	65,000	408	6.50	0.6
3.25	65,000	388	4.05	1.5

$$v_i = k_1[\text{ROOR}] + k_i[\text{ROOH}] + k_{iS}[\text{ROOH}][S]^n \qquad (20.3)$$

$$v_{d\text{ROOH}} = k_d[\text{ROOH}] + k_{dS}[\text{ROOH}][S]^n \qquad (20.4)$$

The rate constants were found to be $k_{dS}[S]^n = 5.1 \times 10^{-4}$ s⁻¹, $k_{iS}[S]^n = 1.0 \times 10^{-4}$ s⁻¹, and $k_d = 2.0 \times 10^{-5}$ s⁻¹ at $T = 398$ K, $pO_2 = 98$ kPa, $[\text{ROOH}]_0 = 1.8 \times 10^{-3}$ mol L⁻¹, and $S_{MoS_2} = 6.5 \times 10^4$ cm² L⁻¹.

The retarding action of metal selenides on cyclohexene oxidation initiated by AIBN was proved in Ref. [11]. The addition of metal selenides decreases the oxidation rate ($T = 343$ K, $pO_2 = 98$ kPa, $[\text{AIBN}] = 2.5 \times 10^{-2}$ mol L⁻¹).

Selenide	Without	WSe_2	$TaSe_2$	$MoSe_2$
$v \times 10^5$ (mol L⁻¹ s⁻¹)	1.7	0.50	0.40	0.20
v/v_0	1.0	0.29	0.23	0.12

An increase in the amounts of $MoSe_2$ and WSe_2 in oxidized cyclohexene decreases the oxidation rate to the limiting value that is nearly 10 times more than the initiation rate. The authors proposed that the surface of these compounds possesses the ability to terminate chains by the reaction with peroxyl radicals as well as the initiating activity.

20.2 INHIBITOR ACTION IN CATALYZED HYDROCARBON OXIDATION

There is specificity of the antioxidant action in the presence of heterogeneous catalyst. The kinetics of ionol retarding action on the oxidation of fuel T-6 catalyzed by the copper powder and homogeneous catalyst copper oleate was studied in Ref. [12]. Copper oleate appeared to be very active homogeneous catalyst: it was found to catalyze the autoxidation of T-6 in such small concentration as 10^{-6} mol L^{-1} ($T = 398$ K). The kinetics of autoxidation catalyzed by copper salt obeys the parabolic law (see Chapter 4):

$$\sqrt{\Delta[O_2]} = b \times t \qquad (20.5)$$

The generation of free radicals occurs by the thermal and copper oleate-catalyzed decomposition of formed hydroperoxides with the rate [12]

$$v_i = k_i[\text{ROOH}] + k_{\text{icat}}[\text{ROOH}][\text{Cu}(\text{OC(O)R}')_2], \quad k_i = 3.5 \times 10^{-5} \text{s}^{-1} \text{ and}$$
$$k_{\text{icat}} = 7.7 \text{ L mol}^{-1}\text{s}^{-1} \qquad (20.6)$$

Similar kinetics of fuel T-6 oxidation was found for the copper powder as a catalyst. The copper powder accelerates fuel T-6 oxidation via the decomposition of formed hydroperoxides on the surface. The rate of this decomposition increases linearly with the amount of introduced powder ($T = 398$ K, $pO_2 = 98$ kPa [13]).

$$v_i = k_i[\text{ROOH}] + k_{iS}[\text{ROOH}]S_{\text{Cu}}, \quad k_i = 3.5 \times 10^{-5} \text{ s}^{-1} \text{ and}$$
$$k_{iS} = 3.4 \text{ L cm}^{-2} \text{ s}^{-1} \qquad (20.7)$$

In addition to hydroperoxide decomposition, the copper surface was found to initiate the chains through activation of dioxygen. The rate of chain initiation in the presence of the copper powder was found to be

$$v_{i0} = v_{i0} + k_{i0S}[O_2]S_{\text{Cu}}, \quad v_{i0} = 3.9 \times 10^{-9} \text{ mol L}^{-1} \text{ s}^{-1} \text{ and}$$
$$k_{i0S} = 1.22 \times 10^{-8} \text{L cm}^{-2} \text{ s}^{-1} \qquad (20.8)$$

Experiments with different ionol concentrations in oxidized T-6 with the copper powder showed another important peculiarity of the metal surface. The reaction of direct ionol oxidation by hydroperoxide on the catalyst surface was found to occur with the rate

$$v_{\text{InH}} = k_{\text{InH}}[\text{ROOH}][\text{InH}]S_{\text{Cu}}, \quad k_{\text{InH}} = 3.5 \times 10^{-6} \text{L}^2 \text{ mol}^{-1} \text{ cm}^{-2} \text{ s}^{-1} (398 \text{ K}) \qquad (20.9)$$

As a result, the decay of antioxidant proceeds via two ways, namely, by the reaction with free radicals and by the oxidation by hydroperoxide on the surface of Cu with the summary rate

$$v_\Sigma = f^{-1}(k_i[\text{ROOH}] + k_{iS}[\text{ROOH}]S_{\text{Cu}}) + k_{\text{InH}}[\text{ROOH}][\text{InH}]S_{\text{Cu}}. \qquad (20.10)$$

The following kinetic scheme was suggested to explain these results on the catalytic action of the copper surface in the presence of antioxidant InH [12].

$$O_2 + S \rightleftharpoons S \cdots O_2$$
$$S \cdots O_2 + RH \longrightarrow S \cdots O_2H + R^\bullet$$
$$R^\bullet + O_2 \longrightarrow RO_2^\bullet$$
$$ROOH + S \rightleftharpoons S \cdots ROOH$$
$$S \cdots ROOH \longrightarrow RO^\bullet + SOH$$
$$RO^\bullet + RH \longrightarrow ROH + R^\bullet$$
$$RO_2^\bullet + InH \longrightarrow ROOH + In^\bullet$$
$$In^\bullet + RO_2^\bullet \longrightarrow ROOIn$$
$$InH + S \cdots ROOH \longrightarrow Products$$

We observed a more complicated behavior in the study of retarding action of amines (N-benzyl-N'-phenyl-1,4-benzenediamine and 4-hydroxyphenyl-2-naphtalenamine) on fuel T-6 oxidation catalyzed by the copper powder [13]. Both antioxidants appeared to retard the autoxidation of T-6 very effectively. They stop chain oxidation during the induction period in concentrations equal to 5×10^{-5} mol L^{-1} and higher. The induction period was found to be the longer, the higher the concentration of the antioxidant and lower the amount of the copper powder introduced in T-6.

$$\tau = f(v_{i0}^{-1} + g/S_{Cu})[InH]_0 \tag{20.11}$$

The kinetic study of antioxidant decay by the reaction with hydroperoxides produced by T-6 oxidation proved the catalytic influence of the copper surface. The rate of this reaction in the absence of dioxygen obeys the equation

$$v(InH + ROOH) = (k_{iInH} + k_{CuROOH} S_{Cu})[ROOH][InH] \tag{20.12}$$

In addition to peroxyl radicals and hydroperoxide, amines are oxidized by dioxygen and this reaction was found to be catalyzed by the copper surface also. This reaction was studied in chlorobenzene and occurs with the rate:

$$v(InH + O_2) = (k_{i0} + k_{CuO_2} S_{Cu})[O_2][InH]. \tag{20.13}$$

Hence, the copper surface catalyzes the following reactions: (a) decomposition of hydroperoxide to free radicals, (b) generation of free radicals by dioxygen, (c) reaction of hydroperoxide with amine, and (d) heterogeneous reaction of dioxygen with amine with free radical formation. All these reactions occur homolytically [13]. The products of amines oxidation additionally retard the oxidation of hydrocarbons after induction period. The kinetic characteristics of these reactions (T-6, $T = 398$ K, [13]) are presented below.

Antioxidant	k_{i0} (L mol^{-1} s^{-1})	k_{CuO_2} (L^2 mol^{-1}cm^{-2} s^{-1})	k_{iInH} (L mol^{-1} s^{-1})	k_{CuROOH} (L^{2}mol^{-1}cm^{-2} s^{-1})
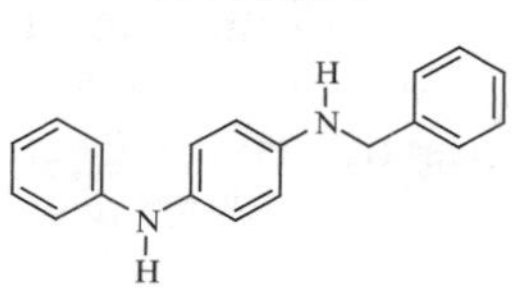 N-benzyl-N'-phenyl-1,4-benzenediamine	1.7×10^{-5}	5.9×10^{-8}	0.16	1.35×10^{-2}
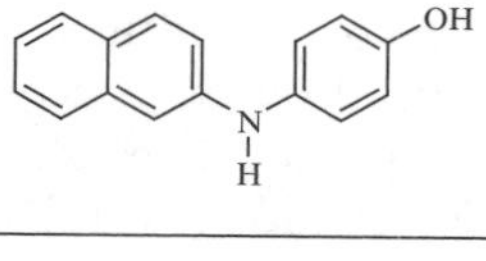 4-hydroxyphenyl-2-naphtalenamine	1.4×10^{-5}	7.2×10^{-8}	0.13	5.07×10^{-3}

Let us compare the rates (mol L^{-1} s^{-1}) of these four reactions under fixed conditions of catalyzed oxidation for the studied amines (T-6, $T = 398$ K, $pO_2 = 98$ kPa, [ROOH] = 1×10^{-3} mol L^{-1}, [InH] = 1×10^{-3} mol L^{-1}, $S_{Cu} = 3000$ cm^2 L^{-1}).

Reaction	InH + O$_2$	InH + O$_2$ (Cu)	InH + ROOH	InH + ROOH (Cu)
(diphenylamine-benzylamine structure)	1.7×10^{-10}	1.8×10^{-9}	1.6×10^{-7}	4.0×10^{-5}
(naphthylamine-phenol structure)	1.4×10^{-10}	2.2×10^{-9}	1.3×10^{-7}	1.5×10^{-5}

20.3 DIOXYGEN ACCEPTORS AS STABILIZERS OF POLYMERS

The oxidative destruction of polymer occurs only in the presence of dioxygen. Principally, one can prevent destruction by introducing an acceptor of dioxygen into the polymer. When the temperature is not high the oxidation of polymer occurs much more slowly than the diffusion of dioxygen into the polymer bulk. Therefore, antioxidants reacting with free radicals are more efficient. At elevated temperatures ($T > 500$ K) oxidation occurs so rapidly that diffusion of dioxygen into the polymer bulk becomes the limiting step of the process. Acceptors of dioxygen can effectively retard the oxidation of polymer under such conditions. However, the introduction of an acceptor of dioxygen beforehand is unreasonable because it will be consumed by dioxygen during the time of storage. This acceptor is needed in the very moment of polymer heating.

The original method of polymers stabilization was invented by Gladyshev and coworkers [14–18]. They proposed to introduce in polymer a metal compound inert toward dioxygen. This compound is decomposed at elevated temperatures with production of a thin metal powder. Formates, oxalates, and carbonyls of metals were suggested as predecessors of an active metal powder. For example, ferrous oxalate decomposes at 600–630 K with the formation of pyrofore iron and ferrous oxide

$$FeC_2O_4 \longrightarrow Fe + 2CO_2$$
$$FeC_2O_4 \longrightarrow FeO + CO + CO_2$$

Oxalates of transition metals (Fe, Co, Mn, and Cu) decompose with the formation of a thin powder of metal and gaseous products, such as CO_2, CO, CH_4, and H_2, at 500–650 K. This is the temperature when degradation of thermostable polymers begins. The formed metal is oxidized and accepting dioxygen prevents the polymer from oxidative degradation. In addition, metals accept alkyl and peroxyl radicals terminating chains. Hence, the retarding action of such antioxidants is complex. The temperatures of metal formates and oxalates degradation are given in Table 20.3.

Oxidation of polymer in the presence of dioxygen acceptors is limited by the diffusion of dioxygen into the polymer bulk. The lifetime of polymer does not depend on the acceptor concentration at [acceptor] $\geq$ [acceptor]$_{min}$. The lifetime of a polymer sample τ depends on pO_2, λ, D, and the thickness of the sample l according to the parabolic equation [14–18]

TABLE 20.3
Temperature Limits for Decay of Metal Formates and Oxalates [16]

Formates	T (K)	Oxalates	T (K)
$Cd(HCOO)_2$	500–580	CoC_2O_4	580–600
$Cu(HCOO)_2$	440–500	NiC_2O_4	590–620
$Co(HCOO)_2$	520–570	MnC_2O_4	610–640
$Fe(HCOO)_2$	550–690	PbC_2O_4	580–640
$Mn(HCOO)_2$	530–680	FeC_2O_4	590–680
$Ni(HCOO)_2$	530–560	AgC_2O_4	370–430
$Pb(HCOO)_2$	530–610	ZnC_2O_4	580–620
$Zn(HCOO)_2$	480–630	CuC_2O_4	540–600

$$\tau = t_{degr} + \frac{fl^2[\text{acceptor}]_0}{8\lambda D p O_2}, \tag{20.14}$$

where t_{degr} is the time of polymer degradation in the absence of acceptor and D is the diffusion coefficient of dioxygen in the polymer.

REFERENCES

1. ET Denisov, GI Kovalev. *Oxidation and Stabilization of Jet Fuels*. Moscow: Khimiya, 1983 [in Russian].
2. EA Blumberg, YuD Norikov. Heterogeneous catalysis and inhibition of liquid-phase oxidation of organic compounds. In: *Itogi Nauki i Tekhniki, Ser Kinetika i Kataliz*, vol. 12. Moscow: VINITI, 1984, pp. 3–143 [in Russian].
3. ET Denisov, VV Azatyan. *Inhibition of Chain Reaction*. London: Gordon and Breach, 2000.
4. GI Kovalev, LD Gogitidze, VI Kuranova, ET Denisov. *Neftekhimiya* 19:237–243, 1979.
5. GI Kovalev. *Neftekhimiya* 18:584–589, 1978.
6. AL Smirnova, LA Tavadyan, EA Blumberg. *Neftekhimiya* 22:513–515, 1982.
7. GI Kovalev, LD Gogitidze, VI Kuranova, YuN Dyshlevskii, ET Denisov. *Neftekhimiya* 17:438–443, 1977.
8. NT Silakhtaryan, LV Salukhvadze, YuD Norikov, EA Blumberg, NM Emanuel. *Kinet Katal* 23:77–82, 1982.
9. GI Kovalev, LD Gogitidze, VI Kuranova, ET Denisov. *Neftekhimiya* 19:88–91, 1979.
10. GI Kovalev, YuN Dyshlevskii, ET Denisov. *Neftekhimiya* 20:446–450, 1980.
11. AM Rubinstein, AA Dulov, AA Slinkin, LA Abramova. *J Catal* 35:80–91, 1974.
12. AV Gerasimova, GI Kovalev, ET Denisov, NS Zvereva. *Neftekhimiya* 22:516–521, 1982.
13. AV Gerasimova, GI Kovalev, LD Gogitidze, VI Kuranova, NS Zvereva, ET Denisov. *Neftekhimiya* 25:555–561, 1985.
14. GP Gladyshev. *Thermodynamics and Kinetics of Nature Hierarchical Processes*. Moscow: Nauka, 1988 [in Russian].
15. GP Gladyshev. Stabilization of polymers, vol 4. In: *Khimicheskaya Encyclopediya*. Moscow: Bol'shaya Rossiyskaya Encyclopediya, 1995, pp. 411–413.
16. GP Gladyshev, OA Vasnetsova. In: G Scott (ed.), *Developments in Polymer Stabilization—6*. London: Applied Science, 1983, pp. 295–334.
17. GP Gladyshev, YuA Ershov, OA Shustova. *Stabilization of Thermostable Polymers*. Moscow: Khimiya, 1979, pp. 1–271 [in Russian].
18. OA Shustova, GP Gladyshev. *Usp Khim* 45:1695–1724, 1976.

Part III

Biological Oxidation and Antioxidants

21 Initiators of Free Radical-Mediated Processes

Numerous initiators of free radical-mediated processes, lipid peroxidation, oxidative destruction of proteins and DNA, cell damage, and others are now well known. Among them are free radicals, transition metals, pollutants, drugs, food components, radiation, and even magnetic field. Despite a great number of initiators, all of them are the producers of free radicals, such as superoxide, hydroxyl radical, perhydroxyl radical, and nitric oxide. We will discuss many of these species and their major reactions excluding the effects of UV, visible, and high-energy radiation, which are mainly relevant to the field of radiation biology.

21.1 SUPEROXIDE

It has already been stressed that the discovery of superoxide as the enzymatically produced diffusion-free dioxygen radical anion [1–3] was a pivotal event in the study of free radical processes in biology. It is not of course that the McCord and Fridovich works were the first ones in free radical biology, but the previous works were more of hypothetical character, and only after the identification of superoxide by physicochemical, spectral, and biochemical analytical methods the enzymatic superoxide production became a proven fact.

Chemical and biochemical properties of superoxide have already been considered earlier [4,5]. It is now understood that in spite of its pompous name superoxide is a relatively innocuous free radical and that its main role is to be a precursor of other much more reactive species (see below). At the same time, many new findings have been obtained concerning biological activity of superoxide. Due to its mainly harmless nature, superoxide nonetheless is able to interact with some biological molecules and affect various biological systems. For example, superoxide produced by stimulated neutrophils is able to damage erythrocytes [6]. Shibanuma et al. [7] has shown that superoxide increases intracellular pH of human leukemia cells. Everett et al. [8] has studied the interaction of superoxide with N^w-hydroxy-L-arginine (NHA), a stable intermediate formed in the oxidation of L-arginine to L-citrulline and NO. It has earlier been suggested that superoxide might mediate the oxidative denitrification of NHA catalyzed by NO synthase and cytochrome P-450. However, the rate constant for reaction of $O_2^{\bullet-}$ with NHA was estimated as $200-500\,l\,mol^{-1}\,s^{-1}$ that is a too small value to be of importance for this enzymatic process. Similarly, it was found [9] that the reaction of superoxide with another important substrate S-nitrosoglutathione has a rate constant of $300\pm100\,l\,mol^{-1}\,s^{-1}$ that is much slower that it has been proposed earlier and therefore, is unlikely to be of biological importance. It has long been known that superoxide is able to release iron from ferritin [10,11]. To exclude possible superoxide-independent contribution of xanthine oxidase-stimulated iron release from ferritin, Paul [12] applied a new chemical source of superoxide, di-(4-carboxybenzyl) hyponitrite (SOTS-1). It was found that prolonged superoxide flux on ferritin stimulated the release of as many as 130 iron atoms from the ferritin molecule.

The most interesting current findings concerning the biological role of superoxide are connected with its signaling function [13]. Below are several examples from numerous up-to-date works. Superoxide enhanced the migration of monocytes across blood–brain barrier upon its exposition to cerebral endothelial cells [14]. Kulisz et al. [15] suggested that superoxide participated together with hydrogen peroxide in the activation of phosphorylation of p38 MAP kinase during hypoxia in cardiomyocytes. Zhang et al. [16] has found that superoxide activated myocardial mitochondrial ATP-sensitive potassium channels. Similarly, superoxide and hydrogen peroxide enhanced channel activity in rat and cat cerebral arteriols [17]. Unfortunately, the mechanisms of superoxide signaling are mainly unknown, and therefore, the detailed consideration of signaling functions of superoxide is outside the scope of this book.

21.2 HYDROXYL RADICAL

It has been thought for a long time that the major route from superoxide to reactive free radicals is the superoxide-dependent Fenton reaction (Reactions 1 and 2):

$$O_2^{\bullet -} + Fe^{3+} \Longrightarrow O_2 + Fe^{2+} \tag{1}$$

$$Fe^{2+} + H_2O_2 \Longrightarrow Fe^{3+} + HO^{\bullet} + HO^{-} \tag{2}$$

This mode of superoxide-dependent free radical-mediated damaging activity remains an important one although the nature of the generated reactive species (free hydroxyl radicals or perferryl, or ferryl ions) is still obscure. However, after the discovery of the fact that many cells produce nitric oxide in relatively large amounts (see below), it became clear that there is another and possibly a more potent mechanism of superoxide-induced free radical damage, namely, the formation of highly reactive peroxynitrite.

For a long time one question remained unanswered: the efficiency of the Fenton reaction as the in vivo producer of hydroxyl radicals due to the low rate of Reaction (2) (the rate constant is equal to $42.1 \, l \, mol^{-1} \, s^{-1}$ [18]). It is known that under in vitro conditions the rate of Fenton reaction can be sharply enhanced by chelators such as EDTA, but for a long time no effective in vivo chelators have been found. From this point of view new findings obtained by Chen and Schopfer [19] who found that peroxidases catalyze hydroxyl radical formation in plants deserve consideration. These authors showed that horseradish peroxidase (HRP) compound III is a catalyst of the Fenton reaction and that this compound is one to two orders of magnitude more active than Fe–EDTA.

Another possible pathway of accelerating the in vivo Fenton reaction has been proposed previously [20]. It was suggested that the level of catalytically active ferrous ions may be enhanced as a result of the interaction of superoxide with the [4Fe–4S] clusters of dehydratases such as aconitases. In accord with this mechanism, superoxide reacts with aconitase to oxidize ferrous ion inside of the [4Fe–4S] cluster. In the next step, the remaining ferrous ion is released from the cluster and is capable of participating in Reaction (2):

$$O_2^{\bullet -} + [2Fe^{2+}2Fe^{3+}-4S] + 2H^+ \Longrightarrow H_2O_2 + [Fe^{2+}3Fe^{3+}-4S] \tag{3}$$

$$[Fe^{2+}3Fe^{3+}-4S] \Longrightarrow [3Fe^{3+}-4S] + Fe^{2+} \tag{4}$$

The rate constant for Reaction (3) is in the range of 10^8 to $10^9 \, l \, mol^{-1} \, s^{-1}$ [20]. Therefore, Reactions (3) and (4) may significantly enhance the concentration of ferrous ions and make Fenton reaction a better competitor with the peroxynitrite-inducible damage [21]. The formation of hydroxyl radicals in the reaction of superoxide with mitochondrial aconitase has been confirmed by ESR spectroscopy [22]. Unfortunately, at present, it is very difficult to estimate the real effect of the interaction of superoxide with aconitase on hydroxyl radical production.

Formation of hydroxyl radicals has been suggested in many studies, which are considered in subsequent chapters in connection with the mechanisms of lipid peroxidation and protein and DNA destruction as well as the mechanisms of free radical pathologies. Furthermore, hydroxyl radical generation occurs under the conditions of iron overload and is considered below.

21.3 PERHYDROXYL RADICAL

Recently, perhydroxyl (hydroperoxyl) radical HOO$^\bullet$ has been wittily named by de Grey [23] as "the forgotten radical." Although the concentration of perhydroxyl radical should be many times lower than that of the superoxide (about 1000 times smaller at pH 7.8), this radical always presents in solution, in equilibrium with superoxide:

$$O_2^{\bullet-} + H^+ \Longleftrightarrow HOO^\bullet \quad pK_a(HOO^\bullet) = 4.8 \tag{5}$$

de Grey believes that despite a low concentration, perhydroxyl could be even more important radical than superoxide due to its greater reactivity in hydrogen abstraction reaction and therefore, it has been "forgotten" unfairly.

This conclusion is partly true because superoxide is unable to abstract hydrogen atom even from the most active bisallylic positions of unsaturated compounds, while perhydroxyl radical abstracts H atom from linoleic, linolenic, and arachidonic fatty acids with the rate constants of $1-3 \times 10^3 \, l \, mol^{-1} s^{-1}$ [24]. However, the superoxide damaging activity does not originate from hydrogen atom abstraction reactions but from one-electron reduction processes, leading to the formation of hydroxyl radicals, peroxynitrite, etc, and in these reactions perhydroxyl cannot compete with superoxide.

Another reason for neglecting perhydroxyl radical is a big difficulty to distinguish it from the much more abundant and more reactive peroxyl radicals. Nonetheless, in several works perhydroxyl radical was considered as a possible initiator of lipid peroxidation (see Chapter 25). It should be noted that at least two biological systems were described where the participation of perhydroxyl radicals seems to be possible. Thus, it has been shown [25,26] that perhydroxyl radical is able to abstract hydrogen atom from NADH (Reaction 6) and the glyceraldehyde-3-phosphate dehydrogenase–NADH (GAPDH–NADH) complex (Reaction 7).

$$HOO^\bullet + NADH \Longrightarrow H_2O_2 + NAD^\bullet, k_6 = 2 \times 10^5 \, l \, mol^{-1} s^{-1} \tag{6}$$

$$HOO^\bullet + GAPDH-NADH \Longrightarrow H_2O_2 + GAPDH-NAD^\bullet, k_7 = 2 \times 10^7 \, l \, mol^{-1} s^{-1} \tag{7}$$

As can be seen from Reaction (3), the free radical capable of oxidizing aconitase [2Fe^{2+}2Fe^{3+}–4S] center is HOO$^\bullet$ and not $O_2^{\bullet-}$.

21.4 NITRIC OXIDE

21.4.1 FORMATION AND LIFETIME OF NITRIC OXIDE

The discovery of nitric oxide in living organisms was a great event in the development of free radical studies in biology. NO is a gaseous neutral free radical with relatively long lifetime and at the same time is an active species capable of participating in many chemical reactions.

A great attention has been drawn to the discovery that NO is the endothelium-derived relaxing factor (EDRF) [27]. Now, it is known that nitric oxide is synthesized by many cells, including macrophages, endothelial cells, neutrophils, neurons, hepatocytes, and others. NO formation was experimentally shown in all tissues and organs, for example, the hypoxic lung [28] or the ischemic heart [29]. Nitric oxide has been shown to be involved in many physiological functions such as blood pressure regulation, inhibition of platelet aggregation, neurotransmission, etc. [30]. At the same time, enhanced NO concentrations exhibit cytotoxic and mutagenic effects.

Major producers of nitric oxide are constitutive and inducible isoforms of NO synthase (Chapter 22). However, it has been proposed that there are other enzymatic and nonenzymatic sources of NO generation. Godber et al. [31] found that xanthine oxidase reduced nitrite to nitric oxide under anaerobic conditions in the presence of NADH or xanthine. Nagase et al. [32] suggested that NO is formed in the reaction of hydrogen peroxide with D- and L-arginine. One of the important sources of NO production is S-nitrosothiols such as S-nitrosoglutathione and S-nitrosocysteine, which can participate in storage and transport of nitric oxide. It has been shown [33] that S-nitrosothiols are decomposed to form nitric oxide in the presence of transition metal ions and reductants. Trujillo et al. [34] has shown that the decomposition of S-nitrosothiols may be induced by superoxide generated by xanthine oxidase.

Lifetime of nitric oxide is an important parameter of its reactivity. Measurement of NO in intact tissue yielded a value in the order of 0.1 s [35] although preliminary estimates gave a much bigger lifetime. It has been accepted that the main reason for the rapid disappearance of NO in tissue is its reaction with dioxygen, which proceeds in aqueous solution with the following overall stoichiometry:

$$4NO + O_2 + 2H_2O \Longrightarrow 4NO_2^- + 4H^+ \quad (8)$$

However, this reaction is too slow with physiologically relevant NO concentrations, even though its rate in hydrophobic biological membranes can be about 300 times higher [36]. Therefore, it has been proposed [37] that the accelerated disappearance of nitric oxide in tissue is explained by its interaction with superoxide. The extravascular lifetime of nitric oxide is estimated to be in the range from 0.09 to >2 s, depending on the dioxygen concentration and distance from the vessel.

Brovkovych et al. [38] applied the electrochemical porphyrinic sensor technique for the direct measurement of NO concentrations in the single endothelial cell. It was found that NO concentration was the highest at the cell membrane (about $1\,\mu\mathrm{mol\,l}^{-1}$) and decreased exponentially with distance from the cell, becoming undetectable at the distance of 50 μm. Now we will consider the principal reactions of nitric oxide relevant to real biological systems.

21.4.2 Reaction with Dioxygen

NO reacts with dioxygen to form nitrite (Reaction 8) with the rate constant equal to $8-9 \times 10^6\,\mathrm{l}^2\,\mathrm{mol}^{-2}\,\mathrm{s}^{-1}$ [39,40]. Although this reaction is a rather slow one, Goldstein and Czapski [41] proposed that the first step (Reaction 9) could be fast, and the oxidation of NO by O_2 may be of importance in biological systems.

$$NO + O_2 \Longrightarrow O_2NO \quad (9)$$

21.4.3 Reaction with Superoxide and Inorganic Nitrogen Compounds

$$NO + O_2^{\bullet-} \Longrightarrow ONOO^- \quad (10)$$

Reaction of nitric oxide with superoxide is undoubtedly the most important reaction of nitric oxide, resulting in the formation of peroxynitrite, one of the main reactive species in free radical-mediated damaging processes. This reaction is a diffusion-controlled one, with the rate constant (which has been measured by many workers, see, for example, Ref. [41]), of about $2 \times 10^9 \, l \, mol^{-1} \, s^{-1}$. Goldstein and Czapski [41] also measured the rate constant for Reaction (11):

$$NO + HOO^\bullet \Longrightarrow ONOOH \qquad k_{11} = (3.2 \pm 0.3) \times 10^9 \, l \, mol^{-1} s^{-1} \qquad (11)$$

The chemistry of inorganic nitrogen compounds is very complicated, and therefore, it is difficult to prove which of these compounds is of a real importance in biological systems. In addition to NO and peroxynitrite, the formation of NO_2, N_2O_3, and NO^- might be of importance in biological systems. Some reactions of nitrogen oxide species are cited below.

$$4NO + O_2 \Longrightarrow 2N_2O_3 \qquad (12)$$
$$ONOOH \Longrightarrow HNO_3 \qquad (13)$$
$$ONOOH + NO \Longrightarrow NO_2 + HNO_2 \qquad (14)$$
$$NO_2 + NO \Longrightarrow N_2O_3 \qquad (15)$$
$$NO + ONOO^- \Longrightarrow NO_2 + NO_2^- \qquad (16)$$

It has been shown [42] that NO_2^- and NO_3^- are formed by stimulated macrophages simultaneously with nitric oxide supposedly via the interaction of NO with superoxide. Not all of the above mentioned reactions are rapid processes. For example, Reaction (16) is rather slow ($k_{16} < 1.3 \times 10^{-3} \, l \, mol^{-1} \, s^{-1}$ [43]).

21.4.4 REACTIONS WITH BIOMOLECULES

Nitric oxide is capable of reacting with some low-molecular-weight substrates and enzymes. (Reactions of NO with antioxidants are considered in Chapter 29). The interaction of NO with thiols is an important in vivo process due to the biological role of the nitrosothiols formed. It has been established that *S*-nitrosothiols play an important role in the storage and transport of nitric oxide. Two mechanisms of the reaction of NO with thiols have been proposed. Wink et al. [44] suggested that at the first step of the reaction, NO is oxidized by dioxygen to form the genuine nitrosation species N_2O_3, which reacts further with thiol:

$$4NO + O_2 \Longrightarrow 2N_2O_3 \qquad (12)$$
$$N_2O_3 + GSH \Longrightarrow GSNO + HNO_2 \qquad (17)$$

Gow et al. [45] proposed that NO may directly react with thiols because the formation of nitrosothiols is possible under anaerobic conditions in the presence of another dioxygen electron acceptor.

$$NO + RSH \Longrightarrow (RSNOH)^\bullet \qquad (18)$$
$$(RSNOH)^\bullet + O_2 \Longrightarrow RSNO + O_2^{\bullet -} \qquad (19)$$

In contrast to superoxide, which participates in one-electron transfer reactions as a reductant, nitric oxide is apparently able to oxidize various transition metal-containing proteins and enzymes. The study of NO reaction with hemoglobin has been started many years ago when numerous in vivo functions of nitric oxide were unknown. In 1976 and 1977 Maxwell and Caughey [46] and Hille et al. [47] showed by the use of ESR and optical spectroscopies that NO formed with hemoglobin the ferroheme–NO complex (nitrosylhemoglobin). There are two major pathways of the interaction of nitric oxide with hemoproteins: the reversible NO binding and NO-induced oxidation. (The reversible formation of the protein–NO complex in the reaction of NO with ferric hemoglobin, myoglobin, and ferric enzyme microperoxidase was demonstrated by Sharma et al. [48] in 1983.) Both reactions occur in two steps: binding of NO to the distal portion of the heme pockets and the rapid reaction of bound NO with iron atom to produce the Fe^{2+}(heme)NO complex or ferric heme and nitrate [49]:

$$NO + Fe^{2+}(heme) \Longleftrightarrow Fe^{2+}(heme)NO \qquad (20)$$

$$NO + Fe^{2+}(heme) \Longrightarrow Fe^{3+}(heme)H_2O + NO_3^- \qquad (21)$$

Both reactions are rapid, with the rate constants equal to $3-5 \times 10^7 \, l \, mol^{-1} \, s^{-1}$.

Herod et al. [50] has studied the kinetics and mechanism of oxyhemoglobin (HbO_2) and oxymyoglobin (MbO_2) oxidation by nitric oxide. At pH 7.0 the rate constants for these reactions were equal to $43.6 \pm 0.5 \times 10^6 \, l \, mol^{-1} \, s^{-1}$ for MbO_2 and $89 \pm 3 \times 10^6 \, l \, mol^{-1} \, s^{-1}$ for HbO_2. It has been suggested that these reactions proceed via the formation of intermediate peroxynitrito complexes, which were rapidly decomposed to the Met-form of proteins, for example:

$$NO + HbO_2 \Longleftrightarrow Hb(Fe^{3+})OONO \Longrightarrow MetHb + NO_3^-. \qquad (22)$$

MetMb is also able to bind reversibly NO, yielding a nitrosyl adduct [51]:

$$NO + metMb \Longleftrightarrow metMbNO \qquad (23)$$

Nitric oxide reacts not only with free hemoglobin but also with hemoglobin inside the erythrocyte. Although the reaction of NO with erythrocytes is rapid enough, its rate is about 650 times slower than that with free hemoglobin [52]. It has been suggested that the interaction of nitric oxide with hemoglobin may be limited by the diffusion of NO into the cell [52,53] or the resistance of the erythrocyte membrane to the NO uptake [54]. Although a major in vivo nitrating agent is probably peroxynitrite (see later), Gunther et al. [55] suggested that nitric oxide and not peroxynitrite is an intermediate in the nitration of tyrosine by prostaglandin H synthase.

At present, new developments challenge previous ideas concerning the role of nitric oxide in oxidative processes. The capacity of nitric oxide to oxidize substrates by a one-electron transfer mechanism was supported by the suggestion that its reduction potential is positive and relatively high. However, recent determinations based on the combination of quantum mechanical calculations, cyclic voltammetry, and chemical experiments suggest that E^0(NO/NO^-) $= -0.8 \pm 0.2$ V [56]. This new value of the NO reduction potential apparently denies the possibility for NO to react as a one-electron oxidant with biomolecules. However, it should be noted that such reactions are described in several studies. Thus, Sharpe and Cooper [57] showed that nitric oxide oxidized ferrocytochrome c to ferricytochrome c to form nitroxyl anion. These authors also proposed that the nitroxyl anion formed subsequently reacted with dioxygen, yielding peroxynitrite. If it is true, then Reactions (24) and (25) may represent a new pathway of peroxynitrite formation in mitochondria without the participation of superoxide.

$$NO + cyt.(Fe^{2+}) \Longrightarrow NO^- + cyt.(Fe^{3+}) \qquad (24)$$

$$NO^- + O_2 \Longrightarrow ONOO^- \qquad (25)$$

Furthermore, Laranjinha and Cadenas [58] have recently showed that nitric oxide oxidizes 3,4-dihydroxyphenylacetic acid (DOPAC) to form nitrosyl anion and the DOPAC semiquinone supposedly by one-electron transfer mechanism.

21.4.5 INTERACTION WITH ENZYMES

Nitric oxide is able to activate enzymes, be a physiological substrate, protect enzymes against free radical damage, or inhibit enzymatic activity. Probably, the most important enzymatic activity of nitric oxide is the interaction with soluble guanylyl cyclase. This enzyme is the main receptor for NO, and it mediates many physiological and pathophysiological functions such as vasodilation, platelet disaggregation, and neutral signaling through cGMP accumulation and protein kinase activation. For example, NO prevents oxidized LDL-stimulated p53 accumulation and apoptosis in macrophages via guanylyl cyclase stimulation [59].

Nitric oxide is a physiological substrate for mammalian peroxidases [myeloperoxide (MPO), eosinophil peroxide, and lactoperoxide), which catalytically consume NO in the presence of hydrogen peroxide [60]. On the other hand, NO does not affect the activity of xanthine oxidase while peroxynitrite inhibits it [61]. Nitric oxide suppresses the inactivation of CuZnSOD and NO synthase supposedly via the reaction with hydroxyl radicals [62,63]. On the other hand, SOD is able to modulate the nitrosation reactions of nitric oxide [64].

The inhibitory effect of nitric oxide on the enzymes aconitases (a family of dehydratases catalyzing the reversible isomerization of citrate and isocitrate) is probably an important physiological process. It has been found [65] that NO inactivated mitochondrial and cytosolic aconitases by the oxidation of the enzyme [4Fe–4S] cluster into the [3Fe–4S] cluster losing one Fe atom. In the case of cytosolic aconitase, the inactivated apo-form of enzyme is identical to iron-regulatory protein IRP-1, a RNA-binding protein, which is involved in iron and energy metabolism. The reaction of nitric oxide with aconitases proceeds with the intermediate formation of the iron–nitrosyl–thiol–aconitase complex with a $g \sim 2.04$ ESR signal [65,66]. In subsequent studies [67,68] the ability of nitric oxide to inactivate cytosolic aconitase has been confirmed although it was proposed that peroxynitrite may be a more important inactivated agent [69]. However, Bouton et al. [70] earlier concluded that only nitric oxide and not superoxide or peroxynitrite is able to convert aconitase into the iron responsible element-binding protein.

In addition to aconitases, nitric oxide is an inhibitor of many other enzymes such as ribonucleotide reductase [71], glutathione peroxidase [72,73], cytochrome c oxidase [74], NADPH oxidase [75], xanthine oxidase [76], and lipoxygenase [77] but not prostaglandin synthase [78]. (Mechanism of lipoxygenase inhibition by nitric oxide is considered in Chapter 26.) It is usually believed that NO inhibits enzymes by reacting with heme or nonheme iron or copper or via the S-nitrosilation or oxidation of sulfhydryl groups, although precise mechanisms are not always evident. By the use of ESR spectroscopy, Ichimori et al. [76] has showed that NO reacts with the sulfur atom coordinated to the xanthine oxidase molybdenum center, converting xanthine oxidase into a desulfo-type enzyme. Similarly, Sommer et al. [79] proposed that nitric oxide and superoxide inhibited calcineurin, one of the major serine and threonine phosphatases, by oxidation of metal ions or thiols.

21.5 OTHER REACTIVE NITROGEN OXIDE SPECIES

In contrast to nitric oxide, which is firmly identified in biological systems and for which numerous (but not all) functions are known, the participation of other nitrogen species in biological processes is still hypothetical. At present, the most interest is drawn to the very reactive nitroxyl anion NO^-. It has been shown that nitroxyl (or its conjugate acid, HNO) formed by the decomposition of Angeli's salt (Reaction 26) is a much more damaging species than NO itself [80].

$$N_2O_3^{2-} + H^+ \Longrightarrow HNO + NO_2^- \quad (26)$$

However, the in vivo sources of nitroxyl production remain uncertain. Some authors suggested that nitroxyl anion might be generated by NO synthases [81,82] or during the decomposition of nitrosothiols [83]. It has also been proposed [81] that the primary product of NO synthase is not nitric acid but nitroxyl anion, which is next oxidized by SOD to NO:

$$NO^- + Cu(II)SOD \Longleftrightarrow {}^\bullet NO + Cu(I)SOD \quad (27)$$

Another suggested mechanism of nitroxyl formation is the decomposition of peroxynitrite [84]. It is of interest that this proposal is connected with the old discussion of a possible role of singlet oxygen in biology. Singlet dioxygen $^1O_2$ is an extremely reactive species, but it forms only in highly exothermic reactions. Unfortunately, at present, many nominees for the sources of singlet oxygen production (for example, the reaction of superoxide with hydrogen peroxide) turn out to be the false ones, and therefore, the possibility of singlet oxygen formation in biological processes remains highly questionable. (Of course, it is not true for the processes initiated by light, for example, the carotene-sensibilized oxidation in the skin, which is mediated by the singlet oxygen.) Nonetheless, Khan et al. [84] recently suggested that the decomposition of peroxynitrite in acidic solution leads to the formation of nitroxyl radical (as peroxynitrous acid) and singlet oxygen:

$$ONOOH \Longrightarrow HNO + {}^1O_2 \quad (28)$$

However, it has been shown that this proposal is wrong and is explained by some analytical errors of HNO and $^1O_2$ detection [85,86].

It should be noted that a major difficulty in the detection of nitroxyl anion is explained by the impossibility to apply ESR spectroscopy because nitroxyl is not a free radical. Moreover, the use of spin traps such as iron N-methyl-D-glucamine dithiocarbamate (Fe-MGD) to distinguish NO and NO$^-$ production by NO synthase failed because both nitrogen species reacted with this spin trap [87].

Another free radical, which is supposedly formed in biological systems is the nitric dioxide $^\bullet NO_2$. This radical is much more reactive than nitric oxide; its rate constants with thiols, urate, and Trolox C are about $10^7-10^8 \, l\,mol^{-1}\,s^{-1}$ [88,89] (Table 21.1). It has been proposed [88] that thiols are dominant acceptors of NO$_2^\bullet$ in cells and tissues while urate is a major scavenger

TABLE 21.1
Bimolecular Rate Constants for the Reactions of $^\bullet NO_2$ Radical

	$k\,(\times 10^7)\,(l\,mol^{-1}\,s^{-1})$	Ref.
Glutathione	2	[88]
Cysteine	5	[88]
Urate	2	[88]
Trolox C	50	[89]

in plasma. Moreover, the rapid reaction of •NO$_2$ with urate makes doubtful the formation of N$_2$O$_3$ in cytoplasma from NO and •NO$_2$.

It has been suggested that •NO$_2$ might be formed by the oxidation of nitrite by numerous biological oxidants. Thus, Shibata et al. [90] reported that horse radish peroxidase (HRP) + hydrogen peroxide oxidized nitrite by the following mechanism:

$$\text{HRP} + \text{H}_2\text{O}_2 \Longrightarrow \text{Compound I} \tag{29}$$

$$\text{Compound I} + \text{NO}_2^- \Longrightarrow \text{Compound II} + {}^\bullet\text{NO}_2 \tag{30}$$

$$\text{Compound II} + \text{NO}_2^- \Longrightarrow \text{HRP} + {}^\bullet\text{NO}_2 \tag{31}$$

It has been proposed [91] that nitric dioxide radical formation during the oxidation of nitrite by HRP or lactoperoxidase (LPO) can contribute to tyrosine nitration and be involved in cell and tissue injuries. This proposal was supported in the later work [92] where it has been shown that •NO$_2$ formed in peroxide-catalyzed reactions is able to enter cells and induce tyrosyl nitration. Reszka et al. [93] demonstrated that •NO$_2$ mediated the oxidation of biological electron donors and antioxidants (NADH, NADPH, cysteine, glutathione, ascorbate, and Trolox C) catalyzed by lactoperoxidase in the presence of nitrite.

•NO$_2$ probably plays a more important role in nitrosating reactions in biological systems than proposed earlier. Thus Espey et al. [94] suggested that the oxidation of NO into N$_2$O$_3$ with the intermediate formation of •NO$_2$ could be more important in cells compared to aqueous solution. Furthermore, •NO$_2$ is a likely candidate in oxidative processes due to its ability to penetrate cells [95].

21.6 PEROXYNITRITE

21.6.1 Formation, Decomposition, and Reactions of Peroxynitrite

As noted earlier, peroxynitrite is formed with a diffusion-controlled rate from superoxide and nitric oxide (Reaction 10). As both these radicals are ubiquitous species, which present practically in all cells and tissues, peroxynitrite can be the most important species responsible for free radical-mediated damage in biological systems. Moreover, it is now known that NO synthases are capable of producing superoxide and nitric oxide simultaneously (see Chapter 22), greatly increasing the possible rate of peroxynitrite production. In addition, another enzyme xanthine dehydrogenase is also able to produce peroxynitrite in the presence of nitrite [96]. Peroxynitrite is a very reactive species capable of reacting with many biomolecules including the antioxidants ascorbic acid, vitamin E, and uric acid [97–99], thiols [100], DNA [101], phospholipids [102], etc (see later). Because of this, a great interest has been drawn to the mechanisms of peroxynitrite reactions.

In the last 10 to 15 years, many experimental and theoretical studies have been dedicated to the study of peroxynitrite reactions. Free radical and non-free radical mechanisms of peroxynitrite action have been proposed, which were discussed in numerous studies (see for example, Refs. [103–110]). In accord with non-radical mechanism an activated form of peroxynitrous acid is formed in the reaction of superoxide with nitric oxide, which is able to react with biomolecules without the decomposition to HO• and •NO$_2$ radicals.

$$\text{NO} + \text{O}_2^{\bullet-} \Longrightarrow \text{ONOO}^- \tag{10}$$

$$\text{ONOO}^- + \text{H}^+ \Longleftrightarrow \text{ONOOH} \tag{32}$$

$$\text{ONOOH} \Longleftrightarrow \text{ONOOH}^* \tag{33}$$

$$\text{ONOOH}^* + \text{S} \Longrightarrow \text{products} \tag{34}$$

On the other hand, in accord with the free radical mechanism peroxynitrite is dissociated into free radicals, which are supposed to be genuine reactive species. Although free radical mechanism was proposed as early as in 1970 [111], for some time it was not considered to be a reliable one because "a great confusion ensued during the next two decades because of misinterpretations of inconclusive experiments, sometimes stimulated by improper thermodynamic estimations" [85]. The latest experimental data supported its reliability [107–109]. Among them, the formation of dityrosine in the reaction with tyrosine and ^{15}N chemically induced dynamic nuclear polarization (CIDNP) in the NMR spectra of the products of peroxynitrite reactions are probably the most convincing evidences (see below).

It has been suggested that there are three major pathways for peroxynitrite to react with various substrates. With a pK value for peroxynitrite equal to 6.8 [103], peroxynitrite exists in the ionized or nonionized form depending on pH. Furthermore, the reactivity of peroxynitrite depends on its *cis*- and *trans*-conformations. Correspondingly, the following mechanisms for peroxynitrite decomposition have been proposed:

$$cis\text{-ONOO}^- + H^+ \Longleftrightarrow cis\text{-ONOOH} \tag{35}$$

$$cis\text{-ONOOH} \Longrightarrow HO^\bullet + {}^\bullet NO_2 \tag{36}$$

$$cis\text{-ONOOH} \Longleftrightarrow trans\text{-ONOOH} \tag{37}$$

$$trans\text{-ONOOH} \Longrightarrow NO_3^- + H^+ \tag{38}$$

It should be noted that different authors give slightly different versions of the above reactions, but there are two major points, which are always the same: (i) in basic solutions peroxynitrite exists only in the relatively stable *cis*-conformation, which prevents its rearrangement into nitrate and makes possible to dissociate into hydroxyl and nitrite radicals (Reaction 36). (ii) *Trans*-peroxynitrite exists principally in acidic solutions and rapidly rearranges into nitrate (Reaction 38).

Now, we will consider the major reactions of peroxynitrite with biomolecules. It was found that peroxynitrite reacts with many biomolecules belonging to various chemical classes, with the bimolecular rate constants from 10^{-3} to $10^8 \, l \, mol^{-1} \, s^{-1}$ (Table 21.2). Reactions of peroxynitrite with phenols were studied most thoroughly due to the important role of peroxynitrite in the in vivo nitration and oxidation of free tyrosine and tyrosine residues in proteins. In 1992, Beckman et al. [112] have showed that peroxynitrite efficiently nitrates 4-hydroxyphenylacetate at pH 7.5. van der Vliet et al. [113] found that the reactions of peroxynitrite with tyrosine and phenylalanine resulted in the formation of both hydroxylated and nitrated products. In authors' opinion the formation of these products was mediated by $^\bullet NO_2$ and $HO^\bullet$ radicals. Studying peroxynitrite reactions with phenol, tyrosine, and salicylate, Ramezanian et al. [114] showed that these reactions are of first-order in peroxynitrite and zero-order in phenolic compounds. These authors supposed that there should be two different intermediates responsible for the nitration and hydroxylation of phenols but rejected the most probable proposal that these intermediates should be $^\bullet NO_2$ and $HO^\bullet$.

Interestingly, that the reactions of peroxynitrite with phenols were accelerated in the presence of ferric and cupric ions [112,114]. Until now, there seems no explanation of transition metal effects in these reactions. We just wonder if it is possible that ferric and cupric ions are able to oxidize peroxynitrite:

$$Fe^{3+} \text{ or } Cu^{2+} + ONOOH \Longrightarrow Fe^{2+} \text{ or } Cu^+ + NO_2^+ + HO^\bullet \tag{39}$$

This could explain the appearance in reaction mixture a strong nitrating agent such as NO_2^+ [114].

TABLE 21.2
Bimolecular Rate Constants for the Reactions of Peroxynitrite with Biomolecules[a]

Substrate	k (l mol^{-1} s^{-1})
NO	$<1.3 \times 10^{-3}$[b]
DMPO	8.7
Uric acid	155[c]
Glutathione	183 ± 12[d]
Tryptophan	184[e]
Ascorbate	236
NADH	233 ± 27[d]
Ubiquinol Q_o	485 ± 54[d]
Acetaldehyde	680
Methionine	902
Cysteine	5.9×10^3
Bovine serum albumin (BSA)	5.0×10^3
Oxyhemoglobin	$(10.4 \pm 0.3) \times 10^{3}$[f]
Iodide	2.3×10^4
CO_2	3.0×10^4
Cytochrome c	2.3×10^5
Alcohol dehydrogenase	2.6–5.6×10^5
Horseradish peroxidase	3.2×10^6
Myeloperoxidase	2.0×10^7
N_2O_3	3.1×10^8[b]

[a]Ref. [104].
[b]Ref. [43].
[c]Ref. [110].
[d]Ref. [141].
[e]Ref. [121].
[f]Ref. [132].

As already mentioned, tyrosine nitration and oxidation is an important damaging process supposedly responsible for the stimulation of numerous pathologies associated with oxidative stress [115]. It is understandable that there are numerous nitrating agents (for example, NO, NO$^-$, N$_2$O$_3$, NO$_2$, NO$_2$Cl, etc.), which might participate in the in vivo nitration of tyrosine, but peroxynitrite is probably rightly considered to be the most important nitrating agent. Despite recent (obviously erroneous) suggestion that peroxynitrite is forming from superoxide and nitric oxide in the inactive *trans*-form and, therefore, is unimportant in in vivo nitration [116], recent studies [115,117] confirmed its prevailing role in in vivo processes.

Probably, the most convincing proof of free radical mechanism of peroxynitrite reactions is the formation of dityrosine [117,118]. It has been suggested [118] that the nitric dioxide radical is responsible for the formation of both 3-nitrotyrosine and dityrosine (Figure 21.1), however, hydroxyl radicals (which were identified in this system by ESR spectroscopy [119]) may also participate in this process. Pfeiffer et al. [118] proposed that dityrosine is predominantly formed at low fluxes of superoxide and nitric oxide, which corresponds to in vivo conditions, however, this observation was not confirmed by Sawa et al. [117].

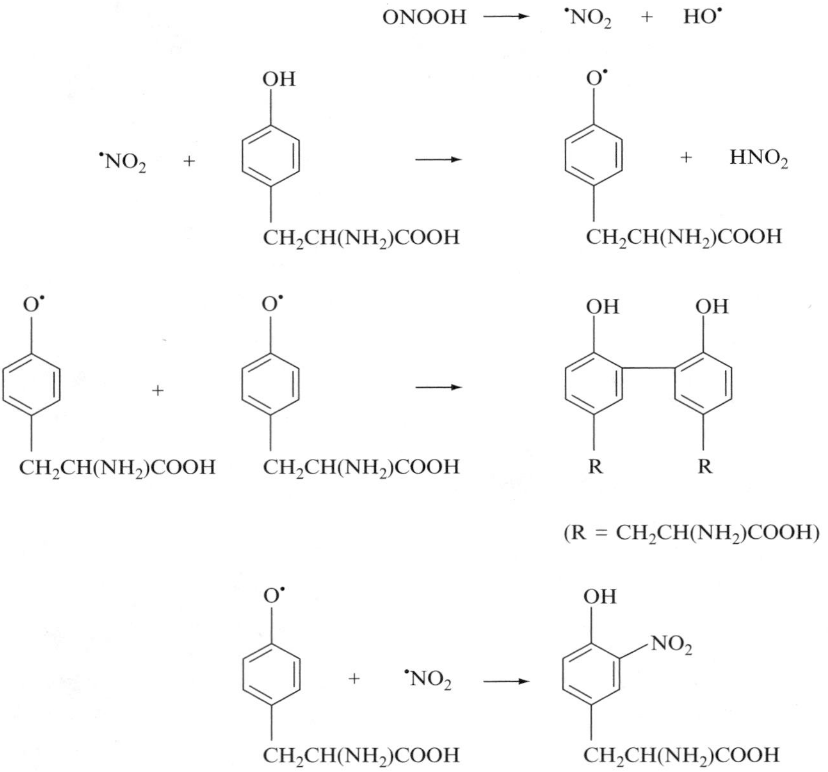

FIGURE 21.1 Scheme of the formation of 3-nitrotyrosine and dityrosine.

Peroxynitrite reacts with heme proteins such as prostacycline synthase (PGI_2), microperoxidase, and the heme–thiolate protein P450 to form a ferryl nitrogen dioxide complex as an intermediate [120]. Peroxynitrite also reacts with acetaldehyde with the rate constant of 680 $l\ mol^{-1}\ s^{-1}$ forming a hypothetical adduct, which is decomposed into acetate, formate, and methyl radicals [121]. The oxidation of NADH and NADPH by peroxynitrite most certainly occurs by free radical mechanism [122,123]. Kirsch and de Groot [122] concluded that peroxynitrite oxidized NADH by a one-electron transfer mechanism to form NAD˙ and superoxide:

$$ONOOH \iff [HO˙ + ˙NO_2] + NADH \Longrightarrow NAD˙ + HNO_2 \text{ or } H_2O \quad (40)$$
$$NAD˙ + O_2 \Longrightarrow NAD^+ + O_2˙^-. \quad (41)$$

However, Goldstein and Czapski [123] believe that in this reaction a genuine attacking species is hydroxyl radical, which reacts with NADH without superoxide generation.

Peroxynitrite easily oxidizes nonprotein and protein thiyl groups. In 1991, Radi et al. [102] have shown that peroxynitrite efficiently oxidizes cysteine to its disulfide form and bovine serum albumin (BSA) to some derivative of sulfenic acid supposedly via the decomposition to nitric dioxide and hydroxyl radicals. Pryor et al. [124] suggested that the oxidation of methionine and its analog 2-keto-4-thiomethylbutanic acid occurred by two competing mechanisms, namely, the second-order reaction of sulfide formation and the one-electron transfer reaction resulted in the ethylene formation supposedly without free radical formation. However, the formation of thiyl radicals in the reactions of peroxynitrite with thiolic compounds has been shown in the other studies [125–128].

The inactivation of enzymes containing the zinc–thiolate moieties by peroxynitrite may initiate an important pathophysiological process. In 1995, Crow et al. [129] showed that peroxynitrite disrupts the zinc–thiolate center of yeast alcohol dehydrogenase with the rate constant of $3.9 \pm 1.3 \times 10^5$ l mol^{-1} s^{-1}, yielding the zinc release and enzyme inactivation. Later on, it has been shown [130] that only one zinc atom from the two present in the alcohol dehydrogenase monomer is released in the reaction with peroxynitrite. Recently, Zou et al. [131] reported the same reaction of peroxynitrite with endothelial NO synthase, which is accompanied by the zinc release from the zinc–thiolate cluster and probably the formation of disulfide bonds between enzyme monomers. The destruction of zinc–thiolate cluster resulted in a decrease in NO synthesis and an increase in superoxide production. It has been proposed that such a process might be the mechanism of vascular disease development, which is enhanced by diabetes mellitus.

Peroxynitrite reacts with oxyhemoglobin in solution to yield methemoglobin with the rate constant of $10.4 \pm 0.3 \times 10^3$ l mol^{-1} s^{-1} [132]. The same reaction occurred in erythrocytes, indicating that peroxynitrite easily penetrates across the erythrocyte membrane. In addition to oxidation, peroxynitrite is apparently able to nitrate oxyhemoglobin. Minetti et al. [133] demonstrated the formation of tyrosyl radicals in this reaction by the use of ESR spectroscopy. Recently, Exner and Herold [134] have found that the reactions of peroxynitrite with oxyhemoglobin and oxymyoglobin are more complex than previously suggested. They showed that these reactions proceed in two steps: in the first step ferryl hemoglobin and ferryl myoglobin are formed, which are next oxidized by peroxynitrite to metHb and metMb. Peroxynitrite reacts with melatonin to form the melatoninyl radical cation as a primary product [135]. It was suggested that this radical was further converted into hydroxypyrrolo[2,3-b]indoles.

21.6.2 Reaction of Peroxynitrite with Carbon Dioxide

The reaction of peroxynitrite with the biologically ubiquitous CO_2 is of special interest due to the presence of both compounds in living organisms; therefore, we may be confident that this process takes place under in vivo conditions. After the discovery of this reaction in 1995 by Lymar [136], the interaction of peroxynitrite with carbon dioxide and the reactions of the formed adduct nitrosoperoxocarboxylate $ONOOCOO^-$ has been thoroughly studied. In 1996, Lymar et al. [137] have shown that this adduct is more reactive than peroxynitrite in the reaction with tyrosine, forming similar to peroxynitrite dityrosine and 3-nitrotyrosine. Experimental data were in quantitative agreement with free radical-mediated mechanism yielding tyrosyl and nitric dioxide radicals as intermediates and were inconsistent with electrophilic mechanism. The lifetime of $ONOOCOO^-$ was estimated as <3 ms, and the rate constant of Reaction (42) $k_{42} = 2 \times 10^3$ l mol^{-1} s^{-1}.

$$ONOO^- + CO_2 \Longrightarrow ONOOCOO^- \qquad (42)$$

$$ONOOCOO^- + \text{tyrosine} \Longrightarrow \text{dityrosine} + \text{3-nitrotyrosine} \qquad (43)$$

Subsequent studies suggested [138,139] that the decomposition of $ONOOCOO^-$ may proceed by both homolytic and heterolytic ways:

$$ONOOCOO^- \Longrightarrow {}^\bullet NO_2 + CO_3^{\bullet -} \qquad (44)$$

$$ONOOCOO^- \Longrightarrow NO_3^- + CO_2. \qquad (45)$$

Heterolytic mechanism is important in the absence of substrates and homolytic one occurs in the presence of oxidizable biomolecules. Bonini et al. [139] were able to identify $CO_3^{\bullet-}$ radical in the reaction of peroxynitrite with carbon dioxide by ESR spectroscopy.

Other very convincing evidences for free radical-mediated mechanism of decomposition and reactions of peroxynitrite and nitrosoperoxocarboxylate were demonstrated by Lehnig [140] with the use of CIDNP technique. This technique is based on the effects observed exclusively for the products of free radical reactions: their NMR spectra exhibit emission characterizing a radical pathway of their formation. Lehnig has found the enhanced emission in the ^{15}N NMR spectra of NO_3^- formed during the decomposition of both peroxynitrite and nitrosoperoxocarboxylate. This fact indicates that NO_3^- was formed from radical pairs [$^{\bullet}NO_2$, HO$^{\bullet}$] and [$^{\bullet}NO_2$, $CO_3^{\bullet-}$]. Emission was also observed in the reaction of both nitrogen compounds with tyrosine supposedly due to the formation of radical pair [$^{\bullet}NO_2$, tyrosyl$^{\bullet}$].

Similar to peroxynitrite, $ONOOCOO^-$ reacts with many biomolecules such as uric acid [110], oxyhemoglobin [133], melatonin [135], NADH, ubiquinol Q_o, and glutathione [141]. Reactions of $ONOOCOO^-$ with substrates in mitochondrial matrix is accompanied by protein nitration [141]. The reaction of $ONOOCOO^-$ with GSH was so rapid that glutathione inhibited tyrosine nitration by peroxynitrite in the presence of CO_2 [142]. The formation of $ONOOCOO^-$ increased the formation of 3-nitrotyrosine and decreased the formation of 3-hydroxytyrosine probably due to the enhanced selectivity of $CO_3^{\bullet-}$ compared to hydroxyl radicals [143].

21.6.3 EXAMPLES OF BIOLOGICAL FUNCTIONS OF NITRIC OXIDE AND PEROXYNITRITE

Ubiquitous nitric oxide and peroxynitrite participate in numerous physiological and pathophysiological functions; some of them are described in subsequent chapters. In this chapter we will consider several examples of processes mediated by these nitrogen species to illustrate the diversity of their biological activities. Peroxynitrite and NO influence the release of oxygen species and leukotriene B_4 (LTB$_4$) by FMLP-stimulated neutrophils in human blood [144]. Peroxynitrite decreased the antioxidant capacity of human plasma through the oxidation of ascorbic acid, uric acid, plasma SH groups, and ubihydroquinone [145]. Farias-Eisner et al. [146] have shown that nitric oxide + hydrogen peroxide exhibited cytotoxic effect against human ovarian cancer cell line. Surprisingly, superoxide and peroxynitrite were apparently not involved in the NO+H_2O_2-mediated cytotoxicity. It was proposed that nitric oxide enhanced the damaging effect of hydrogen peroxide via the inhibition of catalase activity.

It is widely recognized that peroxynitrite is an important factor of inflammatory processes. Zouki et al. [147] suggested that peroxynitrite might affect the inflammatory process via modulation of the surface expression of adhesion molecules on human neutrophils. As a result, peroxynitrite increases neutrophil adhesion to endothelial cells. Endothelial nitric oxide may function as an antioxidant by suppressing redox-sensitive gene expression in endothelial cells [148]. The enhancement of nitric oxide production in activated T cell leukemia cells led to the inhibition of cell respiration [149]. Peroxynitrite formed from NO and superoxide are believed to be a key modulator of nitric oxide-mediated effects in cartilage metabolism [150]. Peroxynitrite is capable of regulating soluble guanylyl cyclase, which is a key enzyme of nitric oxide–cGMP pathway [151]. Lipopolysaccharide (LPS)-induced renal oxidative injury depends at least partly on NO generation and subsequent peroxynitrite formation [152]. It is interesting that nitric oxide may exhibit renoprotective effect supposedly through the reaction with superoxide [153].

The above examples of biological activities of nitric oxide and peroxynitrite are of course just a small part of publications in this field. Although both nitrogen compounds can act as prooxidants in the in vitro and in vivo systems, peroxynitrite is rightly regarded to be the most toxic species. To fight against its toxicity, the inhibitors of nitric oxide or superoxide are usually applied to prevent the formation of peroxynitrite. Recently, another pathway has been proposed — to apply the catalysts of isomerization of $ONOO^-$ into NO_3^-, which diminish the decomposition of peroxynitrite to free radicals $HO^•$ and $^•NO_2$. Thus, Salvemini et al. [154] suggested that the Fe porphyrin 5,10,15,20-tetrakis(2,4,6-trimethyl-3,5-disulfonatophenyl)-porphyrinato iron (III) efficiently inhibited the damaging effect of peroxynitrite in cell cultures converting peroxynitrite into nitrate. This compound also inhibited inflammatory process in animals. It was concluded that the catalysts of peroxynitrite isomerization may be potential anti-inflammatory agents.

21.7 IRON

21.7.1 STRUCTURE AND ORIGINS OF "FREE" IRON

Iron is an extremely important element present in all living organisms; correspondingly, iron metabolism is well studied. Both iron deficiency and iron excess are origins of serious pathologies (iron-deficit anemias, hereditary hemochromatosis, thalassemia, etc.) associated with the overproduction of oxygen radicals. Free radical-mediated processes, characteristic of these pathologies, are considered in Chapter 31; here we will look at some mechanisms of toxic effects of iron.

As well known, most of the iron exists in an organism as the heme and nonheme complexes of biomolecules such as hemoglobin, myoglobin, cytochromes, numerous heme-containing enzymes, proteins of iron metabolism transferrin and ferritin, and so on. Under physiological conditions these compounds cannot, as a rule, catalyze free radical production, but under oxidative stress they are able to release "free" iron, which is believed to be an important initiator of free radical processes. However, it should be noted that some enzymes may probably catalyze free radical production under physiological conditions. Thus, Chen and Schopfer [19] showed that HRP compound III is a catalyst of the Fenton reaction and that this compound is one to two orders of magnitude more active than Fe–EDTA. It was also suggested [155] that purple acid phosphatases with binuclear iron centers are capable of catalyzing hydroxyl radical formation.

The structure of "free" iron, which is also called nontransferrin-bound iron (NTBI) [156], is unknown, although it probably consists of low-molecular iron complexes. It is possible that such complexes contain citrate or albumin ligands [157,158]. It has been suggested that NTBI is an effective initiator of free radical-mediated damaging processes, first of all, the Fenton reaction. In this case "free" iron must contain ferrous (Fe^{2+}) complexes although some authors suggest that NTBI contains ferric ions [158]. In this case NTBI can catalyze the formation of free radicals only in the presence of reductants.

The release of iron from ferritin can be induced by different factors. In 1984, Biemond et al. [159] have shown that stimulated leukocytes mobilize iron from human and horse ferritin. Release of iron was induced by superoxide because SOD inhibited this process. Similarly, the release of iron from ferritin can be induced by xanthine oxidase [160]; this process is believed to induce ischemia and inflammation. Under anaerobic conditions xanthine oxidase is also able to stimulate iron release from ferritin through superoxide-independent mechanism [161]. Another physiological free radical nitric oxide also stimulates iron release from ferritin [162].

In 1991, Brieland and Fantone [163] demonstrated that PMA-stimulated neutrophils stimulate ferrous ion mobilization from holosaturated transferrin but not from transferrin at physiological levels of iron saturation (about 32%) at pH 7.4. Decreasing pH drastically enhanced Fe^{2+} release from both holosaturated ferritin and from ferritin at physiological levels of iron saturation. Holosaturated transferrin potentiated oxygen radical-mediated endothelial cell injury supposedly through the release of ferrous ions [164].

Gardner et al. [165] have shown that the redox-cycling agent phenazine methosulfate (PMS), mitochondrial ubiquinol–cytochrome c oxidoreductase, or hypoxia inactivated aconitase in mammalian cells. It has been proposed that the inactivation of aconitase is mediated by superoxide produced by prooxidants because the overproduction of mitochondrial MnSOD protected aconitase from inactivation by the prooxidants mentioned above except hyperoxia. Later on, the reaction of superoxide with aconitases began to be considered as one of the most important ways to NTBI generation in vivo.

Ischemia–reperfusion is an accidental factor of iron toxic effects in cells and organs. For example, the hearts from iron-loaded rats showed an enhanced susceptibility to oxygen reperfusion damage [166]. The toxic effect of iron overloading was mediated by oxygen radical overproduction because it was inhibited by free radical scavenger (+)-cyanidanol. On the other hand, postischemic reperfusion of isolated rat hearts resulted in cytosolic iron release [167]. Hepatocytes and liver endothelial cells exhibited significant injury when incubated at low temperature (4°C). It was found that under these conditions a major origin of free radical-mediated damage was an increase in the cellular chelatable iron level [168], i.e., hypothermia can stimulate the release of "free" iron. Many iron-containing supplements used for the treatment of iron-deficient pathologies may also act as prooxidants. For example, it has been recently shown [169] that the administration of intravenous iron supplement ferric saccharate to healthy volunteers resulted in a more than fourfold increase in NTBI level and significant increase in superoxide production in the blood.

21.7.2 Iron-Stimulated Free Radical-Mediated Damaging Processes

There are numerous in vitro and in vivo studies, in which the damaging free radical-mediated effects of iron have been demonstrated. Many such examples are cited in the following chapters. However, recent studies [170,171] showed that not only iron excess but also iron deficiency may induce free radical-mediated damage. It has been shown that iron deficiency causes the uncoupling of mitochondria that can be the origin of an increase in mitochondria superoxide release. Furthermore, a decrease in iron apparently results in the reduction of the activity of iron-containing enzymes. Thus, any disturbance in iron metabolism may lead to the initiation of free radical overproduction.

The formation of hydroxyl or hydroxyl-like radicals in the reaction of ferrous ions with hydrogen peroxide (the Fenton reaction) is usually considered as a main mechanism of free radical damage. However, Qian and Buettner [172] have recently proposed that at high $[O_2]/[H_2O_2]$ ratios the formation of reactive oxygen species such as perferryl ion at the oxidation of ferrous ions by dioxygen (Reaction 46) may compete with the Fenton reaction (2):

$$Fe^{2+} + H_2O_2 \Longrightarrow Fe^{3+} + HO^{\bullet} + HO^{-} \qquad (2)$$

$$Fe^{2+} + O_2 \Longleftrightarrow [Fe^{2+}O_2] \Longleftrightarrow [Fe^{3+}O_2^{\bullet-}] \qquad (46)$$

There is no doubt that ferrous ions are very quickly oxidized under aerobic conditions and, therefore, this reaction most certainly might occur in vivo. However, there are still no evidences that the formed perferryl ions live long enough and reactive enough to react with biomolecules and not to dissociate forming superoxide:

$$[Fe^{2+}O_2] \Longleftrightarrow [Fe^{3+}O_2^{\bullet-}] \Longleftrightarrow Fe^{3+} + O_2^{\bullet-} \qquad (47)$$

(Of course, superoxide may reduce ferric to ferrous ions and by this again catalyze hydroxyl radical formation. Thus, the oxidation of ferrous ions could be just "a futile cycle," leading to the same Fenton reaction. However, the competition between the reduction of ferric ions by superoxide and the oxidation of ferrous ions by dioxygen depends on the one-electron reduction potential of the [Fe^{3+}/Fe^{2+}] pair, which varied from $+0.6$ to -0.4 V in biological systems [173] and which is difficult to predict.)

In addition to the well-known iron effects on peroxidative processes, there are also other mechanisms of iron-initiated free radical damage, one of them, the effect of iron ions on calcium metabolism. It has been shown that an increase in free cytosolic calcium may affect cellular redox balance. Stoyanovsky and Cederbaum [174] showed that in the presence of NADPH or ascorbic acid iron ions induced calcium release from liver microsomes. Calcium release occurred only under aerobic conditions and was inhibited by antioxidants Trolox C, glutathione, and ascorbate. It was suggested that the activation of calcium releasing channels by the redox cycling of iron ions may be an important factor in the stimulation of various hepatic disorders in humans with iron overload.

Iron overloading is widely used for studying the toxic effects of iron. Unfortunately, the experiments with cells are hampered by their relative impermeability to iron. Therefore, for iron overloading of cultured endothelial cells Balla et al. [175] used the complexes of iron with 8-hydroxyquinoline (HQ), which rapidly penetrated the plasma membrane. Such overloaded cells turn out to be extremely sensitive to oxygen radicals produced by PMA-stimulated granulocytes or generated intracellularly by menadione. Previously, it has been shown [176] that some regions of normal brain contain high concentrations of low-molecular-weight iron complexes, which may be responsible for the free radical-mediated damage in the brain. It was indeed found [177] that the iron loading of brain slices with the Fe^{3+}–HQ complex resulted in the sharp enhancement of lipid peroxidation. In the presence of ascorbate Fe^{3+}–HQ was reduced into catalytically active Fe^{2+}–HQ.

In many studies iron overloading of experimental animals has been used as a very informative model suitable for studying the toxic effects of iron. Dabbagh et al. [178] have shown that dietary iron overloading in rats resulted in a significant decrease in the antioxidants α-tocopherol and ascorbate in plasma and α-tocopherol, β-carotene, and ubiquinol in liver. Simultaneously, there was an increase in the levels of hepatic F_2-isoprostane and plasma lipoprotein cholesterol. Injection of iron–dextran complex significantly increased iron content in plasma, liver, kidney, and cellular cytosol in rats [179,180]. Iron overloading led to a decrease in α-tocopherol, β-carotene levels and, surprisingly, to decreasing the rate of superoxide, hydroxyl radical, and hydrogen peroxide production in isolated rat liver nuclei. In contrast, lipid peroxidation in kidney homogenates from iron-treated rats was significantly higher that in control animals.

Feeding rats with carbonyl iron caused an increase in iron and the level of 4-hydroxynonenal in the liver as well as changes in antioxidant enzymes [181]. Iron overload in rats resulted in the enhancement of microsomal lipid peroxidation and oxygen radical production by neutrophils and macrophages; both effects were inhibited by the administration of antioxidant flavonoid rutin [182]. Other antioxidants and free radical scavengers were also able to suppress free radical-mediated processes in iron-overloaded animals. Thus, Pietrangelo et al. [183] have shown that flavonoid silybin inhibited lipid peroxidation in mitochondria and hepatocytes from iron-overloaded rats. Galleano and Puntarulo [184] demonstrated that the administration of α-tocopherol to iron-overloading rats prevented the enhancement of lipid and protein oxidation. The most effective water-soluble antioxidant ascorbate is apparently able to suppress the in vivo damaging effects of "free" iron. Thus Berger et al. [185] have found that the main markers of oxidative stress in plasma from preterm infants with high levels of bleomycin-detectable iron (BDI) (such as F_2-isoprostanes and protein carbonyls) were not different from those from infants without BDI. It was concluded that the main reason for BDI inefficiency as an initiator of free radical formation was the presence of endogenous ascorbate because in the plasma devoid of ascorbate "free" iron was an effective initiator of free radical formation.

To study the effects of iron overloading on inflammatory cells, Muntane et al. [186] investigated the effect of iron–dextran administration on the acute and chronic phases of carrageenan-induced glanuloma. It was found that iron–dextran increased the iron content in plasma and stores, and enhanced lipid peroxidation and superoxide production by inflammatory cells. At the same time, iron–dextran had a beneficial effect on recovery from the anemia of inflammation. It has been suggested that iron overload may affect nitric oxide production in animals. For example, alveolar macrophages from iron-overloaded rats stimulated with LPS or interferon-γ diminished NO release compared to normal rats [187].

Iron-stimulated free radical-mediated processes are not limited to the promotion of peroxidative reactions. For example, Pratico et al. [188] demonstrated that erythrocytes are able to modulate platelet reactivity in response to collagen via the release of free iron, which supposedly catalyzes hydroxyl radical formation by the Fenton reaction. This process resulted in an irreversible blood aggregation and could be relevant to the stimulation by iron overload of atherosclerosis and coronary artery disease.

21.7.3 IRON-STIMULATED TOXIC EFFECTS OF PATHOGENIC FIBERS AND PARTICLES

It has been known for a long time that the inhalation of some mineral dusts and fibers leads to many pathophysiological disorders such as pulmonary fibrosis, lung cancer, mesothelioma, etc. Among them, exposure to asbestos fibers, natural fibrous mineral silicates, seems to be of particular danger. At present, a great deal of attention is drawn to toxic effects of asbestos due to its wide use in construction works. It has been proposed that free radical-mediated toxic effects of asbestos fibers depend on their iron contaminants. Recently, Korkina and co-workers [189–191] have shown that asbestos fibers are able to stimulate hydroxyl radical production in solutions and stimulate oxygen radical production by phagocytes. The important role of iron was proved by the fact that removal of iron ions from the fiber surface changed the structure of free radicals produced. Thus, it was found that after removal of iron ions from asbestos surface by the treatment with acids, a great enhancement of lucigenin-amplified chemiluminescence characterized superoxide production was observed, while untreated fibers produced luminol-amplified CL, which at least partly depended on hydroxyl radical production. On these grounds it was concluded that iron ions presented on the surface of asbestos fibers catalyzed superoxide-driven Fenton reaction.

Ghio et al. [192] also studied the effects of surface complexed iron ions on oxygen radical production by silicates. It has been shown that the ability of silica, crocidolite asbestos, kaolinite, and talc to catalyze the generation of oxygen radicals by ascorbate–hydrogen peroxide system, to stimulate respiratory burst and leukotriene B_4 release by alveolar macrophages, and to induce acute lung inflammation in rats increased with increasing the content of complexed ferric ions. Among three mineral particles studied (chrysotile, nemalite, and hematite), nemalite, which contained the highest level of ferrous ions on the surface, produced the largest amount of oxygen radicals and exhibited the strongest cytotoxic action on rabbit tracheal epithelial cells [193]. The addition of iron oxide (Fe_2O_3) promoted the crocidolite-induced development of mesothelioma [194]. Silica intratracheal instillation into rats induced an increase in the ionized ferric ions complexed on the surface of silica [195]. Correspondingly, iron concentration in bronchoalveolar fluid, lung tissue, and liver tissue as well as lipid peroxidation in the lung tissue increased.

Simeonova and Luster [196] studied the role of iron-catalyzed oxygen radical production in the asbestos-induced stimulation of α-TNF secretion by alveolar macrophages. It was found that the presence of iron on asbestos fibers was a critical factor of α-TNF secretion because the removal of iron by pretreatment with desferrioxamine or the addition of desferrioxamine to incubation mixture reduced the α-TNF response. α-TNF release was also inhibited by membrane-permeable hydroxyl radical scavengers tetramethylthiourea and dimethyl sulfoxide. Deshpande et al. [197] suggested that silica–aqueous solution interaction may lead to the generation of some factors responsible for the generation of free radicals and therefore, the damaging effect of silica on cells might not require direct particle–cell interaction. Porter et al. [198] reported that silica inhalation to rats resulted in the activation of alveolar macrophages and concomitant production of nitric oxide and oxygen radicals.

In addition to the aforementioned dusts and fibers, coal mine dusts may also stimulate oxygen radical production [199]. In this case hydroxyl radical production and lipid peroxidation also correlated well with the content of available surface iron. It has been proposed that free radical-mediated processes can be a casual cause of coal workers' pneumoconiosis due to exposure to coal dusts.

21.8 POSSIBLE EFFECTS OF MAGNETIC FIELD ON FREE RADICAL FORMATION

For long time magnetic fields (MFs) have been used in medical practice for the treatment of various pathologies on empirical grounds, although there are no proven mechanisms of their favorable or toxic effects. On the other hand during the last few years an increasing concern has been expressed about the possible health hazards due to exposure to electric and magnetic fields, which are impossible to avoid because of the extensive use of electricity. Epidemiological studies have been carried out to investigate the possible links of MF exposure to certain cancers, first of all, childhood leukemia, but no definite results were obtained [200]. A widespread use of cellular phones in the last decade added another potentially dangerous source of acute exposure of humans to radio frequency magnetic fields.

One of the important possible mechanisms of MF action on biological systems is the influence of free radical production. Chemical studies predict that MFs may affect free radical reactions through the radical pair mechanism [201]. A reaction between two free radicals can generate a free radical pair in the triplet state with parallel electron spins. In this state free radicals cannot recombine. However, if one of the electrons overturns its spin, then free radicals can react with one another to form a diamagnetic product. Such electron spin transition may be induced by an alternative MF.

These very simplified considerations suggest that MFs may affect free radical-mediated biological processes if free radical pairs are formed on certain steps as intermediates. At least in one study the effect of MF on enzymatic one-electron transfer reactions was explained on the basis of free radical pair mechanism. Taraban et al. [202] have shown that in the reactions catalyzed by HRP the rate of conversion of Compound I to Compound II decreased by 15% and the rate of conversion of Compound II to HRP decreased by 35% in the 750 G MF. It has been suggested that these effects of MF are explained by the formation of a triplet radical pair in the transfer of an electron from diamagnetic substrate to the low-spin ($S=1$) heme. Recently, another confirmation of possible formation of radical pairs in biological processes has been obtained. Lehnig [203] demonstrated that ^{15}N CIDNP was observed in ^{15}N NMR spectra of 3-nitro-N-acetyl-1-tyrosine and 1-nitrocyclohexa-2,5-dien-4-one formed during the nitration of N-acetyl-1-tyrosine with the $^{15}NO_2^-/H_2O_2$/HRP system. The occurrence of CINDP is a direct evidence of the formation of radical pairs in this process. It was suggested that the nitration products were formed by the recombination of ˙NO_2 and N-acetyl-1-tyrosinyl radicals.

The effects of MFs on free radical production by phagocytes were investigated in several studies. It was found [204] that oxygen radical production by PMA-stimulated rat peritoneal neutrophils increased by 12.4% in 0.1 mT (60 Hz) magnetic field. Khadir et al. [205] have shown that sinusoudal 60 Hz 22 mT MF did not affect nonstimulated superoxide production by resting PMNs. However, significant MF enhancement of superoxide production occurred in PMA-stimulated PMNs during the first minutes of MF exposure. MF (50 Hz) enhanced phagocytosis and superoxide production by 12-O-tetradecanophorbol-13-acetate (TPA)-stimulated murine macrophages [206].

In addition to oxygen radical production by phagocytes, MFs may affect other cellular processes. It is important that practically all such processes supposedly contain free radical-mediated stages. Thus, Katsir and Parola [207] have shown that low-frequency weak MFs enhanced the proliferation of chick embryo fibroblasts. Most importantly, the MF effect on proliferation was suppressed by antioxidants and antioxidant enzymes (vitamin E, catalase, and SOD), indicating the participation of free radicals at the stages influenced by MFs. Varani et al. [208] have shown that low frequency, low energy, pulsing MFs increased the capacity of A(2A) receptor agonists to stimulate cyclic AMP levels in human neutrophils. On the other hand, these agonists inhibited superoxide production by neutrophils; therefore, MFs might affect indirectly superoxide production through the adenosine receptor mechanism. MFs stimulated other free radical-mediated damaging processes. Thus Fiorani et al. [209] have shown that 0.5 mT MF increased the Fe(II)-ascorbate-induced decay in hexokinase activity and methemoglobin formation by rabbit erythrocytes. Low-level pulsed 2450 MHz radio frequency MF increased DNA single- and double-strand breaks in rat brain cells [210].

In addition to oxygen radicals, MFs can affect nitric oxide production. In 1993, Miura et al. [211] demonstrated that radio frequency burst-type 10 MHz MF significantly increased the production of cyclic GMP in rat cerebellum supernatant in the presence of L-arginine and NADPH. MF also increased nitric oxide production and the dilation of arteriols of the frog. All in vitro and in vivo effects of MF were inhibited by the inhibitors of NO synthase; on these grounds it has been proposed that MF activated NO synthase. Later on, the enhancement of NO production by 0.1 mT (60 Hz) MF has been shown in the in vivo experiments with mice injected with LPS [212]. After LPS administration, mice were given ferrous N-methyl-D-glucaminedithiocarbamate (Fe–MGD), a nitric oxide spin trap. After mice sacrifice, a strong ESR spectrum of the NO-spin adduct was observed in the livers of mice exposed to MF. It was concluded that MF did not induce NO production by itself but enhanced LPS-induced NO generation in vivo.

Thus, major MF effects registered in various in vitro systems are mainly associated with the enhancement of free radical-mediated damaging processes such as the stimulation of oxygen radical production by phagocytes, disruption of erythrocytes, DNA damage, etc. (It should be mentioned that one of the earliest biological effects of MFs was an increase in cytosolic free calcium in cells [213]. It was proposed that MFs induced functional disruption of the intramembranous portion of the calcium channel [214]; the role of free radicals in this process is unknown.) As there are still no other proven mechanisms of MF influence on free radical processes than the radical pair mechanism, we must assume that all examples of biological processes affected by MFs must proceed through the intermediate formation of free radical pairs.

In accord with in vitro studies we should expect that the in vivo effects of MFs must have been the damaging ones. There are only few experimental in vivo and ex vivo studies, which, nonetheless, support this suggestion. Recently, we have showed for the first time (personal communication by LG Korkina, IB Deeva, IB Afanas'ev) that MF effects may be much greater under pathophysiological than physiological conditions. We have studied the effects of a weak alternative magnetic field on leukocytes from Fanconi anemia (FA) patients, compared to

TABLE 21.3
Luminol-Amplified CL (mV) Produced by Leukocytes from FA Patients and their Relatives

	Spontaneous CL before MF Exposure	Spontaneous CL after MF Exposure	PMA-stimulated CL before MF Exposure	PMA-stimulated CL after MF Exposure
FA patients	3280 ± 1640	5090 ± 2140	4760 ± 2050	6300 ± 2750
FA relatives	350 ± 200	350 ± 200	1670 ± 300	1290 ± 400

MF effects on FA relatives and healthy donors. (Fanconi anemia is an autosomal recessive disease associated with the overproduction of free radicals, Chapter 31.) It has been shown earlier [215] that FA leukocytes produce the enhanced amount of hydroxyl or hydroxyl-like free radicals, which are probably formed by the Fenton reaction. It was suggested that MF would be able to accelerate hydroxyl radical production by FA leukocytes. Indeed, we found that MF significantly enhanced luminol-amplified CL produced by non-stimulated and PMA-stimulated FA leukocytes but did not affect at all oxygen radical production by leukocytes from FA relatives and healthy donors (Table 21.3). It is interesting that MF did not also affect the calcium ionophore A23187-stimulated CL by FA leukocytes, indicating the absence of the calcium-mediated mechanism of MF activity, at least for FA leukocytes.

Other examples of possible damaging effects of radio frequency MFs on humans are the MF effects on cellular phones. Moustafa et al. [216] suggested that acute exposure to the MFs of commercially available cellular phones for 1, 2, or 4 h significantly increased plasma lipid peroxidation and decreased the activities of SOD and glutathione peroxidase in erythrocytes.

REFERENCES

1. JM McCord, I Fridovich. *J Biol Chem* 243: 5753–5760, 1968.
2. M McCord, I Fridovich. *J Biol Chem* 244: 6056–6063, 1969.
3. JM McCord, I Fridovich. *J Biol Chem* 244: 6049–6055, 1969.
4. IB Afanas'ev. *Superoxide Ion: Chemistry and Biological Implications*, vol 1. Boca Raton, FL: CRC Press, 1989.
5. IB Afanas'ev. *Superoxide Ion: Chemistry and Biological Implications*, vol 2. Boca Raton, FL: CRC Press, 1990.
6. SJ Weiss. *J Biol Chem* 255: 9912–9917, 1980.
7. M Shibanuma, T Kuroki, K Nose. *J Cell Physiol* 136: 379–383, 1988.
8. SA Everett, MF Dennis, KB Patel, MR Stratford, P Wardman. *Biochem J* 317: 17–21, 1996.
9. E Ford, MN Hughes, P Wardman. *J Biol Chem* 277: 2430–2438, 2002.
10. P Biemond, HG Eijk, AJG Swaak, JF Koster. *J Clin Invest* 73: 1575–1579, 1984.
11. CE Thomas, SD Aust. *Free Radic Biol Med* 1: 293–300, 1985.
12. T Paul. *Arch Biochem Biophys* 382: 253–261, 2000.
13. T Finkel. *Curr Opin Cell Biol* 10: 248–253, 1998.
14. A van der Goes, D Wouters, SMA van der Pol, R Huizinga, E Ronken, P Adamson, J Greenwood, CD Dijkstra, HE de Vries. *FASEB J* 15, 1096/fj.00-0881fje, 2001.
15. A Kulisz, N Chen, NS Chandel, Z Shao, PT Schumacker. *Am J Physiol Lung Cell Mol Physiol* 282: L1324–L1329, 2002.
16. DX Zhang, Y-F Chen, WB Campbell, A-P Zou, GJ Gross, P-L Li. *Circ Res* 89: 1177–1183, 2001.
17. Y Liu, DD Gutterman. *Clin Exp Pharmacol Physiol* 29: 305–311, 2002.
18. TJ Hardwick. *Can J Chem* 35: 428–232, 1957.
19. S Chen, P Schopfer. *Eur J Biochem* 260: 726–735, 1999.
20. SI Liochev, I Fridovich. *Free Radic Biol Med* 16: 29–33, 1994.

21. SI Liochev, I Fridovich. *Free Radic Biol Med* 26: 777–778, 1999.
22. J Vasquez-Vivar, B Kalyanaraman, MC Kennedy. *J Biol Chem* 275: 14064–14069, 2000.
23. ADNJ de Grey. *DNA Cell Biol* 21: 251–257, 2002.
24. BHJ Bielski, RL Arudi, MW Sutherland. *J Biol Chem* 258: 4759–4761, 1983.
25. EJ Land, AJ Swallow. *Biochim Biophys Acta* 234: 34–42, 1971.
26. PC Chan, BHJ Bielski. *J Biol Chem* 255: 874–876, 1980.
27. RMJ Palmer, AG Ferrige, S Moncada. *Nature* 327: 524–526, 1987.
28. TD Le Cras, IF McMurtry. *Am J Physiol Lung Cell Mol Physiol* 280: L575–L582, 2001.
29. JL Zweier, P Wang, P Kuppusamy. *J Biol Chem* 270: 304–307, 1995.
30. S Moncada, PMJ Palmer, EA Higgs. *Pharmacol Rev* 43: 109–142, 1991.
31. BLJ Godber, JJ Doel, GP Sapkota, DR Blake, CR Stevens, R Eisenthal, R Harrison. *J Biol Chem* 275: 7757–7763, 2000.
32. S Nagase, K Takemura, A Ueda, A Hirayama, K Aoyagi, M Kondoh, A Koyama. *Biochem Biophys Res Commun* 233: 150–153, 1997.
33. RJ Singh, N Hogg, J Joseph, B Kalyanaraman. *J Biol Chem* 271: 18596–18603, 1996.
34. M Trujillo, MN Alvarez, G Peluffo, BA Freeman, R Radi. *J Biol Chem* 273: 7828–7834, 1998.
35. M Kelm, J Schrader. *Circ Res* 66: 1561–1575, 1990.
36. X Liu, MJ Miller, MS Joshi, DD Thomas, JR Lancaster, Jr. *Proc Natl Acad Sci USA* 95: 2175–2179, 1998.
37. DD Thomas, X Liu, SP Kantrow, JR Lancaster, Jr. *Proc Natl Acad Sci USA* 98: 355–360, 2001.
38. V Brovkovych, E Stolarczyk, J Oman, P Tomboulian, T Malinski. *J Pharm Biomed Anal* 19: 135–143, 1999.
39. HH Awad, DM Stanbury. *Int J Chem Kinet* 25: 375–381, 1993.
40. B Mayer, P Klatt, ER Werner, K Schmidt. *J Biol Chem* 270: 655–659, 1995.
41. S Goldstein, G Czapski. *Free Radic Biol Med* 19: 505–510, 1995.
42. RS Lewis, S Tamir, SR Tannenbaum, WM Deen. *J Biol Chem* 270: 29350–29355, 1995.
43. S Goldstein, G Czapski, J Lind, G Merenyi. *Chem Res Toxicol* 12: 132–136, 1999.
44. DA Wink, JA Cook, SY Kim, Y Vodovotz, R Pacelli, MC Krishna, A Russo, JB Mitchell, D Jourd'heuil, AM Miles, MB Grisham. *J Biol Chem* 272: 11147–11151, 1997.
45. AJ Gow, DG Buerk, H Ischiropoulos. *J Biol Chem* 272: 2841–2845, 1997.
46. JC Maxwell, WS Caughey. *Biochemistry* 15: 388–395, 1976.
47. R Hille, G Palmer, JS Olson. *J Biol Chem* 252: 403–405, 1977.
48. VS Sharma, RA Isaacson, ME John, MR Wareman, M Chevion. *Biochemistry* 22: 3897–3902, 1983.
49. RF Eich, T Li, DD Lemon, DH Doherty, SR Curry, JF Aitken, AJ Mathews, KA Johnson, RD Smith, GN Philips Jr, JS Olson. *Biochemistry* 35: 6976–6983, 1996.
50. S Herod, M Exner, T Nauser. *Biochemistry* 40: 3385–3395, 2001.
51. LE Laverman, A Wanat, J Oszajca, G Stochel, PC Ford, R van Eldik. *J Am Chem Soc* 123: 285–293, 2001.
52. X Liu, MJ Miller, MS Joshi, H Sadowska-Krowicka, DA Clark, JR Lancaster, Jr. *J Biol Chem* 273: 18709–18713, 1998.
53. X Liu, A Samouilov, JR Lancaster JR, JL Zweier. *J Biol Chem* 277: 26194–26199, 2002.
54. MW Vaughen, KT Huang, L Kuo, JC Liao. *J Biol Chem* 275: 2342–2348, 2000.
55. MR Gunther, LC Hsi, JF Curtis, JK Gierse, LJ Marnett, TE Eling, RP Mason. *J Biol Chem* 272: 17086–17090, 1997.
56. MD Bartberger, W Liu, E Ford, KM Miranda, C Switzer, JM Fukuto, PJ Farmer, DA Wink, KN Houk. *Proc Natl Acad Sci USA* 99: 10958–10963, 2002.
57. MA Sharpe, CE Cooper. *Biochem J* 332: 9–19, 1998.
58. J Laranjinha, E Cadenas. *J Neurochem* 81: 892–900, 2002.
59. A Heinloth, B Brune, B Fisher, J Galle. *Atherosclerosis* 162: 93–101, 2002.
60. HM Abu-Soud, SL Hazen. *J Biol Chem* 275: 37524–37532, 2000.
61. C Lee, X Liu, JL Zweier. *J Biol Chem* 275: 9369–9376, 2000.
62. YS Kim, S Han. *FEBS Lett* 479: 25–28, 2000.
63. KY Xu. *Biochim Biophys Acta* 1481: 156–166, 2000.

64. TM Hu, WL Hayton, MA Morse, SR Mallery. *Biochem Biophys Res Commun* 295: 1125–1134, 2002.
65. MC Kennedy, WE Antholine, H Beinert. *J Biol Chem* 272: 20340–20347, 1997.
66. AF Vanin. *Biokhimiya* 32: 228–232, 1967.
67. PR Gardner, G Costantino, C Szab, AL Salzman. *J Biol Chem* 272: 25071–25076, 1997.
68. LA Castro, RL Robalinho, A Cayota, R Meneghini, R Radi. *Arch Biochem Biophys* 359: 215–224, 1998.
69. G Cairo, R Ronchi, S Recalcati, A Campanella, G Minotti. *Biochemistry* 41: 7435–7442, 2002.
70. C Bouton, M Raveau, JC Drapier. *J Biol Chem* 271: 2300–2306, 1996.
71. M Lepoivre, JM Flaman, Y Henry. *J Biol Chem* 267: 22994–23000, 1992.
72. HG Holzhutter, R Wiesner, J Rathmann, R Stosser, H Kuhn. *Eur J Biochem* 245: 608–616, 1997.
73. M Asahi, J Fujii, K Suzuki, HG Seo, T Kuzuya, M Hori, M Tada, S Fujii, N Taniguchi. *J Biol Chem* 270: 21035–21039, 1995.
74. MW Cleeter, JM Cooper, VM Darley, S Mobcada, AH Schapira. *FEBS Lett* 345: 50–54, 1994.
75. RM Clancy, J Leszczynska-Piziak, SB Abramson. *J Clin Invest* 90: 1116–1121, 1992.
76. K Ichimori, M Fukahori, H Nakazawa, K Okamoto, T Nishino. *J Biol Chem* 274: 7763–7768, 1999.
77. VB O'Donnell, KB Taylor, S Parthasarathy, H Kuhn, D Koesling, A Friebe, A Bloodsworth, VM Darley-Usmar, BA Freeman. *J Biol Chem* 274: 20083–20091, 1999.
78. A Bloodsworth, VB O'Donnell, BA Freeman. *Arterioscler Thormb Vasc Biol* 20: 1707–1715, 2000.
79. D Sommer, S Coleman, SA Swanson, PM Stemmer. *Arch Biochem Biophys* 404: 271–278, 2002.
80. KM Miranda, MG Espey, K Yamada, M Krishna, N Ludwick, S-M Kim, D Jourd'heuil, MB Grisham, M Feelisch, JM Fukuto, DA Wink. *J Biol Chem* 276: 1720–1727, 2001.
81. HH Schmidt, H Hofmann, U Schindler, ZS Shutenko, DD Cunningham, M Feelisch. *Proc Natl Acad Sci USA* 93: 14492–14497, 1996.
82. KM Rusche, MM Spiering, MA Marletta. *Biochemistry* 37: 15503–15512, 1998.
83. DR Arnelle, JS Stamler. *Arch Biochem Biophys* 318: 279–285, 1995.
84. AU Khan, D Kovacic, A Kolbanovskiy, M Desai, K Frenkel, NE Geacintov. *Proc Natl Acad Sci USA* 97: 2984–2989, 2000.
85. G Merenyi, J Lind, G Czapski, S Goldstein. *Proc Natl Acad Sci USA* 97: 8216–8218, 2000.
86. GR Martinez, P Di Mascio, MG Bonini, O Augusto, K Briviba, H Sies, P Maurer, U Rothlisberger, S Herod, WH Koppenol. *Proc Natl Acad Sci USA* 97: 10307–10312, 2000.
87. AM Komarov, DA Wink, M Feelisch, HH Schmidt. *Free Radic Biol Med* 28: 739–742, 2000.
88. E Ford, MN Hughes, P Wardman. *Free Radic Biol Med* 32: 1314–1323, 2002.
89. MJ Davies, LG Forni, RL Willson. *Biochem J* 255: 513–522, 1988.
90. H Shibata, Y Kono, S Yamashita, Y Sawa, H Ochiai, K Tanaka. *Biochim Biophys Acta* 1230: 45–50, 1995.
91. A van der Vliet, JP Eiserich, B Halliwell, CE Cross. *J Biol Chem* 272: 7617–7625, 1997.
92. MG Espey, S Xavier, DD Thomas, KM Miranda, DA Wink. *Proc Natl Acad Sci USA* 99: 3481–3486, 2002.
93. KJ Reszka, Z Matuszak, CF. Chignell, J Dillon. *Free Radic Biol Med* 26: 669–678, 1999.
94. MG Espey, KM Miranda, DD Thomas, DA Wink. *J Biol Chem* 276: 30085–30091, 2001.
95. MG Espey, DD Thomas, KM Miranda, DA Wink. *Proc Natl Acad Sci USA* 99: 11127–11132, 2002.
96. BL Godber, JJ Doel, J Durgan, R Eisenthal, R Harrison. *FEBS Lett* 475: 93–96, 2000.
97. D Bartlett, DF Church, PL Bounds, VH Koppenol. *Free Radic Biol Med* 18: 85–92, 1995.
98. GL Squadrito, X Jin, WA Pryor. *Arch Biochem Biophys* 322: 53–59, 1995.
99. J Vasquez-Vivar, AM Santos, BC Junqueira, O Augusto. *Biochem J* 314: 869–876, 1996.
100. R Radi, JS Beckman, KM Bush, BA Freeman. *J Biol Chem* 266: 4244–4250, 1991.
101. PA King, VE Anderson, JO Edwards, G Gustafson, RC Plumb, JW Suggs. *J Am Chem Soc* 114: 5430–5432, 1992.
102. R Radi, JS Beckman, KM Bush, BA Freeman. *Arch Biochem Biophys* 288: 481–487, 1991.
103. WH Koppenol, JJ Moreno, WA Pryor, H Ischiropoulos, JS Beckman. *Chem Res Toxicol* 5: 834–842, 1992.

104. JP Crow, C Spruell, J Chen, C Gunn, H Ischiropoulos, M Tsai, CD Smith, R Radi, WH Koppenol, JS Beckman. *Free Radic Biol Med* 16: 331–338, 1994.
105. S Goldstein, GL Squadrito, WA Pryor, G Czapski. *Free Radic Biol Med* 21: 965–974, 1996.
106. S Pfeiffer, ACF Gorren, K Schmidt, ER Werner, B Hansert, BS Bohle, B Mayer. *J Biol Chem* 272: 3465–3470, 1997.
107. G Merenyi, J Lind, S Goldstein, G Czapski. *J Phys Chem* 103: 5685–5691, 1999.
108. GR Hodges, KU Ingold. *J Am Chem Soc* 121: 10695–10701, 1999.
109. OV Gerasimov, SV Lymar. *Inorg Chem* 38: 4317–4321, 1999.
110. GL Squadrito, R Cueto, AE Splenser, A Valavanidis, H Zhang, RM Uppu, WA Pryor. *Arch Biochem Biophys* 376: 333–337, 2000.
111. LR Mahoney. *J Am Chem Soc* 92: 4244–4245, 1970.
112. JS Beckman, H Ischiropoulos, L Zhu, M van der Woerd, CD Smith, J Chen, J Harrison, JC Martin, M Tsai. *Arch Biochem Biophys* 298: 438–445, 1992.
113. A van der Vliet, CA O'Neill, B Halliwell, CE Cross, H Kaur. *FEBS Lett* 339: 89–92, 1994.
114. M Ramezanian, S Padmaia, WH Koppenol. *Chem Res Toxicol* 9: 232–240, 1996.
115. CD Reiter, R.-J Teng, JS Beckman. *J Biol Chem* 275: 32460–32466, 2000.
116. S Pfeiffer, B Mayer. *J Biol Chem* 273: 27280–27285, 1998.
117. T Sawa, T Akaike, H Maeda. *J Biol Chem* 275: 32467–32474, 2000.
118. S Pfeiffer, K Schmidt, B Mayer. *J Biol Chem* 275: 6346–6352, 2000.
119. H Zhang, J Joseph, N Hogg, B Kalyanaraman. *Biochemistry* 40: 7675–7686, 2001.
120. M Mehl, A Daiber, S Herold, H Shoun, V Ullrich. *Nitric Oxide* 3: 142–152, 1999.
121. LS Nakao, D Ouchi, O Augusto. *Chem Res Toxicol* 12: 1010–1018, 1999.
122. M Kirsch, H de Groot. *J Biol Chem* 274: 24664–24670, 1999.
123. S Goldstein, G Czapski. *Chem Res Toxicol* 13: 736–741, 2000.
124. WA Pryor, X Jin, GL Squadrito. *Proc Natl Acad Sci USA* 91: 11175–11177, 1994.
125. O Augusto, RM Gatti, R Radi. *Arch Biochem Biophys* 310: 118–125, 1994.
126. J-N Lemercier, CL Squadrito, WA Pryor. *Arch Biochem Biophys* 321: 31–59, 1995.
127. X Shi, A Lenhart, Y Mao. *Biochem Biophys Res Commun* 203: 1515–1521, 1994.
128. RM Gatti, R Radi, O Augusto. *FEBS Lett* 348: 287–290, 1994.
129. JP Crow, JS Beckman, JM McCord. *Biochemistry* 34: 3544–3552, 1995.
130. A Daiber, D Frein, D Namgaladze, V Ullrich. *J Biol Chem* 277: 11882–11888, 2002.
131. M-H Zou, C Shi, RA Cohen. *J Clin Invest* 109: 817–826, 2002.
132. A Denicola, JM Souza, R Radi. *Proc Natl Acad Sci USA* 95: 3566–3571, 1998.
133. M Minetti, G Scorza, D Pietraforte. *Biochemistry* 38: 2078–2087, 1999.
134. M Exner, S Herold. *Chem Res Toxicol* 13: 287–293, 2000.
135. H Zhang, GL Squadrito, R Uppu, WA Pryor. *Chem Res Toxicol* 12: 526–534, 1999.
136. SV Lymar, JK Hurst. *J Am Chem Soc* 117: 8867–8868, 1995.
137. SV Lymar, Q Jiang, JK Hurst. *Biochemistry* 35: 7855–7861, 1996.
138. GL Squadrito, WA Pryor. *Free Radic Biol Med* 25: 392–403, 1998.
139. MG Bonini, R Radi, G Ferrer-Sueta, AM Da C Ferreira, O Augusto. *J Biol Chem* 274: 10802–10806, 1999.
140. M Lehnig. *Arch Biochem Biophys* 368: 303–318, 1999.
141. LB Valdez, S Alvarez, SL Arnaiz, F Shopfer, MC Carreras, JJ Poderoso, A Boveris. *Free Radic Biol Med* 29: 349–356, 2000.
142. M Kirsch, M Lehnig, HG Korth, R Sustmann, H de Groot. *Chemistry* 7: 3313–3320, 2001.
143. CX Santos, MG Bonini, O Augusto. *Arch Biochem Biophys* 377: 146–152, 2000.
144. MM Bednar, M Balazy, M Murphy, C Booth, SP Fuller, A Barton, J Bingham, L Golding, CE Gross. *J Leukoc Biol* 60: 619–624, 1996.
145. A van der Vliet, D Smith, CA O'Neill, H Kaur, V Darley-Usmar, CE Cross, B Halliwell. *Biochem J* 303: 295–301, 1994.
146. R Farias-Eisner, G Chaudhuri, E Aeberhard, FM Fukuno. *J Biol Chem* 271: 6144–6151, 1996.
147. C Zouki, S-L Zhang, JSD Chan, JG Filep. *FASEB J* 15: 25–27, 2001.
148. BS Wung, JJ Cheng, SK Shyue, DL Wang. *Arterioscler Thromb Vasc Biol* 21: 1941–1947, 2001.

149. B Beltran, M Quintero, E Garcia-Zaragoza, E O'Connor, JV Esplugues, S Moncada. *Proc Natl Acad Sci USA* 99: 8892–8897, 2002.
150. JY Jouzeau, S Pacquelet, C Boileau, E Nedelec, N Presle, P Netter, B Terlain. *Biorheology* 39: 201–214, 2002.
151. M Weber, N Lauer, A Mulsch, G Kojda. *Free Radic Biol Med* 31: 1360–1367, 2001.
152. C Zhang, LM Walker, PR Mayeux. *Biochem Pharmacol* 59: 203–209, 2000.
153. DS Majid, A Nishiyama. *Hypertension* 39: 293–297, 2002.
154. D Salvemini, ZQ Wang, MK Stern, MG Currie, TP Misko. *Proc Natl Acad Sci USA* 95: 2659–2663, 1998.
155. JC Sibille, K Doi, P Aisen. *J Biol Chem* 262: 59–62, 1987.
156. JJM Marx, RC Hider. *Eur J Clin Invest* 32 (Suppl 1): 1–2, 2002.
157. M Grootveld, JD Bell, H Halliwell, OI Aruoma, A Bomford, PJ Sadler. *J Biol Chem* 264: 4417–4423, 1989.
158. RC Hider. *Eur J Clin Invest* 32 (Suppl 1): 50–54, 2002.
159. P Biemond, HG van Eijk, AJ Swaak, JF Koster. *J Clin Invest* 73: 1576–1579, 1984.
160. P Biemond, AJ Swaak, CM Beindorff, JF Koster JF. *Biochem J* 239: 169–73, 1986.
161. DE Van Epps, S Greiwe, J Potter, J Goodwin. *Inflammation* 11: 59–72, 1987.
162. DW Reif, RD Simmons. *Arch Biochem Biophys* 283: 537–542, 1990.
163. JK Brieland, JC Fantone. *Arch Biochem Biophys* 284: 78–83, 1991.
164. JK Brieland, SJ Clarke, S Kurmiol, SH Phan, JC Fantone. *Arch Biochem Biophys* 94: 265–271, 1992.
165. PR Gardner, I Raineri, LB Epstein, CW White. *J Biol Chem* 270: 13399–13405, 1995.
166. AM van der Kraaij, LJ Mostert, HG van Eijk, JF Koster. *Circulation* 78: 442–449, 1988.
167. F Boucher, S Pucheu, C Coudray, A Favier, J Deleiris. *FEBS Lett* 302: 261–265, 1992.
168. U Rauen, F Petrat, T Li, H De Groot. *FASEB J* 14: 1953–1964, 2000.
169. TM Rooyakkers, ESG Stroes, MP Kooistra, EE van Faassen, RC Hider, TJ Rabelink, JJM Marx. *Eur J Clin Invest* 32 (Suppl 1): 9–16, 2002.
170. MD Knutson, PB Walter, BN Ames, FE Viteri. *J Nutr* 130: 621–628, 2000.
171. PB Walter, MD Knutson, A Paler-Martinez, S Lee, Y Xu, FE Viteri, BN Ames. *Proc Natl Acad Sci USA* 99: 2264–2269, 2002.
172. SY Qian, GR Buettner. *Free Radic Biol Med* 26: 144–1456, 1999.
173. P Geisser. *In Iron Therapy with Special Emphasis on Oxidative Stress*. Switzerland: Vifor Inc., 1998, p. 27.
174. DA Stoyanovsky, AI Cederbaum. *Free Radic Biol Med* 24: 745–753, 1998.
175. G Balla, GM Vercellotti, JW Eaton, HS Jacob. *J Lab Clin Med* 116: 546–555, 1990.
176. JMC Gutteridge, W Cao, M Chevion. *Free Rad Res Commun* 11: 317–320, 1991.
177. M Oubidar, M Boquillon, C Marie, C Bouvier, A Beley, J Bralet. *Free Radic Biol Med.* 21: 763–769, 1996.
178. AJ Dabbagh, T Mannion, SM Lynch, B Frei. *Biochem J* 300: 799–803, 1994.
179. M Galleano, S Puntarulo. *Toxicology* 93: 125–134, 1994.
180. M Galleano, SM Farre, JF Turrens, S Puntarulo. *Toxicology* 88: 141–149, 1994.
181. KE Brown, MT Kinter, TD Oberley, ML Freeman, HF Frierson, LA Ridnour, Y Tao, LW Oberley, DR Spitz. *Free Radic Biol Med* 24: 545–555, 1998.
182. IB Afanas'ev, EA Ostrachovitch, NE Abramova, LG Korkina. *Biochem Pharmacol* 50: 627–635, 1995.
183. A Pietrangelo, F Borella, G Casalgrandi, G Montosi, D Ceccarelli, D Gallesi, F Giovannini, A Gasparetto, A Masini. *Gastroenterology* 109: 1941–1949, 1995.
184. M Galleano, S Puntarulo. *Toxicology* 124: 73–81, 1997.
185. TM Berger, MC Polidori, A Dabbagh, PJ Evans, B Halliwell, JD Morrow, LJ Robers II, B Frei. *J Biol Chem* 272: 15656–15660, 1997.
186. J Muntane, P Puig-Parellada, MT Mitjavila. *J Lab Clin Med* 126: 435–443, 1995.
187. Y Zhang, RR Crichton, JR Boelaert, PG Jorens, AG Herman, RJ Ward, F Lallemand, P de Witte. *Biochem Pharmacol* 55: 21–25, 1998.

188. D Pratico, M Pasin, OP Barry, A Ghiselli, G Sabatino, L Iuliano, GA FitzGerald, F Violi. *Circulation* 99: 3118–3124, 1999.
189. LG Korkina, TB Suslova, ZP Cheremisina, BT Velichkovskii. *Studia Biophys* 126: 99–104, 1988.
190. LG Korkina, AD Durnev, TB Suslova, ZP Cheremisina, NO Daugel-Dauge, IB Afanas'ev. *Mutat Res* 265: 245–253, 1992.
191. TB Suslova, ZP Cheremisina, LG Korkina. *Environ Res* 66: 222–234, 1994.
192. AJ Ghio, TP Kennedy, AR Whorton, AL Crumbliss, GE Hatch, JR Hoidal. *Am J Physiol* 263: L511–L518, 1992.
193. C Guilianelli, A Baeza-Squiban, E Boisvieux-Ulrich, O Houcine, R Zalma, C Guennou, H Pezerat, F Marano. *Environ Health Perspect* 101: 436–442, 1993.
194. S Adachi, S Yoshida, K Kawamura, M Takahashi, H Uchida, Y Odagiri, K Takemoto. *Carcinogenesis* 15: 753–758, 1994.
195. AJ Ghio, RH Jaskot, GE Hatch. *Am J Physiol* 267: L686–L692, 1994.
196. PP Simeonova, MI Luster. *Am J Respir Cell Mol Biol* 12: 676–683, 1995.
197. A Deshpande, PK Narayanan, BE Lehnert. *Toxicol Sci* 67: 275–283, 2002.
198. DW Porter, L Millecchia, VA Robinson, A Hubbs, P Willard, D Pack, D Ramsey, J McLaurin, A Khan, D Landsittel, A Teass, V Castranova. *Am J Physiol Lung Cell Mol Physiol* 283: L485–L493, 2002.
199. NS Dalal, J Newman, D Pack, S Leonard, V Vallyathan. *Free Radic Biol Med* 18: 11–22, 1995.
200. A Lacy-Hulbert, JC Metcalfe, R Hesketh. *FASEB J* 12: 395–420, 1998.
201. B Brocklehurst, KA McLauchlan. *Int J Radiat Biol* 69: 3–24, 1996.
202. MB Taraban, TV Leshina, MA Anderson, CB Grissom. *J Am Chem Soc* 119: 5768–5769, 1997.
203. M Lehnig. *Arch Biochem Biophys* 393: 245–254, 2001.
204. S Roy, Y Noda, V Eckert, MG Traber, A Mori, R Liburdy, L Packer. *FEBS Lett* 376: 164–166, 1995.
205. R Khadir, JL Morgan, JJ Murray. *Biochim Biophys Acta* 1472: 359–367, 1999.
206. M Simko, S Droste, R Kriehuber, DG Weiss. *Eur J Cell Biol* 80: 562–566, 2001.
207. G Katsir, AH Parola. *Biochem Biophys Res Commun* 252: 753–756, 1998.
208. K Varani, S Gessi, S Merighi, V Iannotta, E Cattabriga, S Spisani, R Cadossi, PA Borea. *Br J Pharmacol* 136: 57–66, 2002.
209. M Fiorani, B Biagiarelli, F Vetrano, G Guidi, M Dacha, V Stocchi. *Bioelectromagnetics* 18: 125–131, 1997.
210. H Lai, NP Singh. *Int J Rad Biol* 69: 513–521, 1996.
211. M Miura, K Takayama, J Okada. *J Physiol* 461: 513–524, 1993.
212. T Yoshikawa, M Tanigawa, T Tanigawa, A Imai, H Hongo, M Kondo. *Pathophysiology* 7: 131–135, 2000.
213. JJ Carson, FS Prato, DJ Drost, LD Diesbourg, SL Dixon. *Am J Physiol* 259: C687–C692, 1990.
214. AD Rosen. *Biochim Biophys Acta* 1282: 149–155, 1996.
215. LG Korkina, EV Samochatova, AA Maschan, TB Suslova, ZP Cheremisina, IB Afanas'ev. *J Leukocyte Biol* 52: 357–362, 1992.
216. YM Moustafa, RM Moustafa, A Belacy, SH Abou-El-Ela, FM Ali. *J Pharm Biomed Anal* 26: 605–608, 2001.

22 Generation of Free Radicals by Prooxidant Enzymes

In this chapter the generation of free radicals, mainly superoxide and nitric oxide, catalyzed by "prooxidant" enzymes will be considered. Enzymes are apparently able to produce some other free radicals (for example, HO$^\bullet$ and $^\bullet$NO$_2$), although their formation is not always rigorously proved or verified. The reactions of such enzymes as lipoxygenase and cyclooxygenase also proceed by free radical mechanism, but the free radicals formed are consumed in their catalytic cycles and probably not to be released outside. Therefore, these enzymes are considered separately in Chapter 26 dedicated to enzymatic lipid peroxidation.

22.1 XANTHINE OXIDASE

22.1.1 MECHANISMS OF OXYGEN RADICAL PRODUCTION

The discovery of superoxide production by xanthine oxidase (XO) by McCord and Fridovich [1–3] was a pivotal point in the development of free radical studies in biology. These authors showed that XO catalyzed cytochrome c reduction, which was inhibited by SOD. (Actually, McCord and Fridovich used another enzyme, bovine erythrocyte carbonic anhydrase containing SOD as an admixture.) It should be mentioned that the possibility of free radical formation in the reactions catalyzed by XO has been suggested much earlier [4], but only in 1968–1978 superoxide was positively identified by using biological, physicochemical, and spectral analytical methods [5–7]. XO and xanthine dehydrogenase (XDH) are the two forms of the enzyme xanthine oxidoreductase. These forms of the enzyme are interconvertible and depend on the oxidation state of protein thiols. Xanthine oxidoreductase catalyzes the oxidation of xanthine to urate with concomitant reduction of dioxygen or NAD. The enzyme exists as a homodimer with subunits containing one FAD, one molibdopterin, and two 2F/2S clusters [8].

In 1974, Olson et al. [9] proposed a mechanism for the reactions catalyzed by XO. In accord with this mechanism six electrons are transferred from fully reduced enzyme through four redox centers during the oxidation of xanthine (Reaction (1)):

$$\text{XO(6)} \Rightarrow \text{XO(4)} \Rightarrow \text{XO(2)} \Rightarrow \text{XO(1)} \Rightarrow \text{XO(0)}$$
$$\Downarrow \qquad \Downarrow \qquad \Downarrow \qquad \Downarrow \qquad (1)$$
$$\text{H}_2\text{O}_2 \quad \text{H}_2\text{O}_2 \quad \text{O}_2^{\bullet-} \quad \text{O}_2^{\bullet-}$$

As seen from the above scheme, XO reduces dioxygen into hydrogen peroxide by two-electron reduction mechanism and into superoxide by one-electron reduction mechanism. The efficiency of superoxide production depends on the nature of the substrate (in addition to xanthine, XO oxidizes many other substrates such as purines, pteridines, and aliphatic and aromatic aldehydes), substrate concentration, low or high turnover of substrates, dioxygen concentration, and pH [10]. Percentage of superoxide production in a total electron flow (the percent univalent flux) is equal to 27% at low xanthine level and about 13% at its saturated level [10].

Porras et al. [11] and Nagano and Fridovich [12] proposed that the percent univalent flux from XO can be regulated by the concentrations of substrate and dioxygen. It was suggested that at the high concentrations of substrates and dioxygen the fully reduced XO steady state takes place and the divalent reduction of dioxygen predominates. In contrast, low substrate and dioxygen concentrations produce a less reduced steady state, which favors the univalent dioxygen reduction. This proposal was confirmed in the study of lumazine oxidation by XO, which proceeds with a low reaction rate [13]. It has been shown that with lumazine as the substrate of XO, the percent univalent flux may achieve 90%. Another way of increasing the percentage of superoxide production is the XO inhibition by its reaction product uric acid [14,15]. Inhibition of XO by uric acid apparently occurs under in vivo conditions. Thus, it has been shown [16] that physiological levels of uric acid inhibited XO activity in human plasma. Activity and superoxide production by XO are also inhibited by many metal ions such as Ag^+, Hg^{2+}, Zn^{2+}, Cu^{2+}, Cr^{6+}, etc. [17] or hyperoxia and active oxygen species [18].

At the beginning only XO and not XDH was considered as a superoxide producer. For example, in 1985 McCord [19] suggested that the conversion of XDH into XO is responsible for an increase in superoxide production in postischemic reperfusion injury. However, it has later been shown [20,21] that XDH itself is a producer of superoxide although not so effective as XO. Moreover, the efficiency of superoxide production differs for different types of the enzyme. Thus, 2.8 to 3.0 mol of superoxide were produced by chicken liver XDH, while superoxide production by bovine milk XDH was insignificant [21]. Sanders et al. [22] found that NADH oxidation by human milk and by bovine milk XDHs catalyzed superoxide production more rapidly than XO; this process was inhibited by NAD and diphenyleneiodonium but not by the established XO inhibitors allopurinol and oxypurinol.

It is widely accepted that superoxide is the only oxygen radical produced by XO. However, as early as in 1970, Beauchamp and Fridovich [23] suggested that XO is able to produce hydroxyl radicals in addition to superoxide through the reaction of superoxide with hydrogen peroxide (the Haber–Weiss reaction). As this reaction occurs only in the presence of iron ions (the Fenton reaction), XO may apparently produce hydroxyl radicals only indirectly in the presence of adventitious iron ions. Nonetheless, later on, Kuppusamy and Zweier [24] again proposed that XO directly generates hydroxyl radicals by the reduction of hydrogen peroxide. Their proposal was based on the identification of ESR spectra of DPMO—OH adducts. It was also shown that the DMPO—OH adduct was formed in the reaction of HO$^\bullet$ with DMPO and not during the decomposition of DMPO—OOH adduct and that desferrioxamine did not affect hydroxyl radical formation, excluding the possible effect of adventitial iron. However, despite these experimental data, conclusion about the ability of XO to generate hydroxyl radicals is apparently wrong. Lloyd and Mason [25] and Britigan et al. [26] have showed that that adventitious iron still presented in aforementioned experiments and was responsible for the hydroxyl radical generation by XO, because commercial XO samples and buffers contained iron. Recently, the indirect formation of hydroxyl radicals in the XO-catalyzed reaction has been shown in the presence of 1-methyl-3-nitro-1-nitrosoguanidine (MNNG), the inducer of gastric cancer [27]. It was suggested that MNNG reacted with the XO-produced hydrogen peroxide to form peroxynitrite, which decomposed to hydroxyl and nitric dioxide radicals. Thus, in this system hydroxyl radicals can probably be formed by a superoxide-independent way.

22.1.2 Reactions of XO with Organic and Inorganic Substrates

Superoxide generation is the most important way of catalytic XO activity, resulting in the free radical formation by the reduction of dioxygen (molecular oxygen). However, XO may reduce other inorganic and organic biomolecules under anaerobic and aerobic conditions. In the last case such molecules compete with dioxygen for electrons. One important example is the reaction with quinones. Winterbourn [28] has shown that many quinones are better acceptors of electrons from XO than dioxygen. Because of this, three main reactions occur when quinones are incubated with the substrate–XO system:

$$Q \xrightarrow{XO} Q^{\bullet -} \quad (2)$$

$$Q^{\bullet -} + O_2 \Longleftrightarrow Q + O_2^{\bullet -} \quad (3)$$

$$O_2 \xrightarrow{XO} O_2^{\bullet -} \quad (4)$$

Competition between dioxygen and quinones depends on the one-electron reduction potentials of quinones [29], and therefore, quinones may inhibit or stimulate superoxide production.

It is extremely important that the interaction of quinones with XO (Reaction (3)) is reversible that can lead to receiving erroneous results at the measurement of superoxide production by SOD-inhibitable cytochrome c reduction [28,29] (see also Chapter 27). Lusthof et al. [30] demonstrated that 2,5-bis(1-aziridinyl)-1,4-benzoquinones are directly reduced by XO. Interestingly at quinone concentrations greater than $25\,\mu\text{mol}\,\text{l}^{-1}$, quinones entirely suppressed one-electron reduction of dioxygen, and cytochrome c was completely reduced by the semiquinones formed. It is well known that cytochrome c and lucigenin are effective superoxide scavengers and due to that, these compounds are widely used in the quantitative assays of superoxide detection. Nonetheless, under certain experimental conditions they can be directly reduced by XO [31].

In Chapter 21 we already noted that in addition to superoxide, XO is able to produce another free radical, nitric oxide in the presence of nitrite. The relationship between XO and nitrogen oxide species is rather complicated because XO may both generate and react with NO and peroxynitrite. It has been known for a long time that XO catalyzes the reduction of nitrate to nitrite [32], but the further reduction of nitrite to nitric oxide by XO was shown just recently [33,34]. Furthermore, it was found that XDH catalyzed NO formation from nitrite in the presence of NADH about 50 times more efficiently than XO [35]. (This finding could be of importance because in vivo XDH is a predominant form, at least intracellularly.) Nitric oxide generation apparently occurs at the MoS group of the enzyme because its desulfuration and the conversion of MoS to the MoO group caused greater enzyme activity. Under aerobic conditions, XO generated both superoxide and nitric oxide. Thus, the reduction of nitrite by XO results in the simultaneous formation of NO and superoxide, which inevitably gives peroxynitrite, the most active biological damaging species.

Although the inhibition of XO by nitrogen oxide species has been shown by many authors, there is no mutual agreement on its mechanism and the relative role of NO and peroxynitrite. Houston et al. [36] concluded that only peroxynitrite is a direct inhibitor of XO and that the apparent NO inhibition of XO activity is a consequence of diminishing uric acid yield due to the reaction of uric acid with peroxynitrite. However, Ichimori et al. [37] demonstrated that nitric oxide inhibited XO and XDH under anaerobic conditions supposedly via desulfuration of the molybdenum MoS center. Another possible mechanism of XO inactivation is NO binding or destruction of the FeS centers [38]. The most convincing argument against the direct NO inhibitory effects is the rapid formation of peroxynitrite from superoxide and NO under aerobic conditions, which undoubtedly is a far more effective inhibitory agent of XO.

22.1.3 Free Radical-Mediated Biological Activity of Xanthine Oxidoreductase

Two forms of xanthine oxidoreductase namely XO and XDH are present in many human and animal cells and plasma, XDH and XO are the predominant species in cytoplasma and serum, respectively [39]. Damaging effects of XO-catalyzed superoxide production in postischemic tissues were demonstrated by many authors. For example, Chambers et al. [40] and Hearse et al. [41] have shown that the suppression of superoxide production by the administration of XO inhibitor allopurinol or SOD resulted in the reduction of infarct size in the dog and of the incidence of reperfusion-induced arrhythmia in the rat. Similarly, Charlat et al. [42] has also shown that allopurinol improved the recovery of the contractile function of reperfused myocardium in the dog. However, the use of allopurinol as the XO inhibitor has been questioned because this compound may affect oxygen radical formation not only as a XO inhibitor but as well as free radical scavenger [43]. Smith et al. [44] also showed that gastric mucosal injury depends on the oxygen radical production catalyzed by XO and iron.

Due to the important role of xanthine oxidoreductase in tissue injury, its functions in endothelial cells have now been thoroughly studied. XO presents in endothelial cells of many organs including the heart, bowel, liver, kidney, and brain where this enzyme is responsible for superoxide-mediated reperfusion injury [45]. XO is localized within the cytoplasm and on the endothelial cell surface. Such distribution suggests an important role of XO in cell–cell interactions involving the superoxide signaling [46,47]. It has been shown that superoxide production sharply increased after anoxia–reoxygenation in bovine pulmonary microvascular endothelial cells (BPMVE) [48], bovine pulmonary artery endothelial cells (BPAEC) [49], rat brain capillary endothelial cells (RBCEC) [50], and cerebral endothelial cells isolated from piglet cortex [51]. It is of importance that endothelial cells are able to release constitutively the XO activity into the extracellular medium [48]. On the other hand, circulating XO can bind to vascular cells, impairing cell functions by oxygen radical generation [52]. Earlier, Paler-Martinez et al. [53] proposed that the role of XO, as a generator of oxygen radicals, can be insignificant in human ECs due to the lack of activity towards xanthine. However, Zhang et al. [54] showed that both XO and XDH of human umbilical vein endothelial cells (HUVEC) efficiently catalyze superoxide production in the presence of NADH.

It is usually accepted that the augmentation of the XO activity in ischemic tissues undergoing reperfusion is a consequence of the formation of hypoxanthine from degradation of ATP in the presence of dioxygen. It has been confirmed by Xia and Zweier [55] who studied the mechanism of stimulation of the XO-catalyzed superoxide production in postischemic tissues. It was found that an increase in superoxide production in isolated rat hearts after reperfusion was triggered by the enhancement of hypoxanthine and xanthine levels due to the degradation of ATP during ischemia.

22.2 NADPH OXIDASES

If XO is an undoubted historical pioneer among free radical-producing enzymes, whose capacity to catalyze one-electron transfer reactions opened a new era in biological free radical studies, NADPH oxidase is undoubtedly the most important superoxide producer. This enzyme possesses numerous functions from the initiation of phagocytosis to cell signaling, and it is not surprising that its properties have been considered in many reviews during last 20 years [56–58].

There are two types of NADPH oxidases existing in phagocytic (neutrophils, monocytes, and macrophages) and nonphagocytic (endothelial cells, vascular smooth muscle cells [VSMCs], fibroblasts, etc.) cells. In accord with numerous studies, there is apparently no fundamental difference in the structures of enzymes of both types. However, while NADPH oxidase in phagocytic cells does not exist as a whole in "dormant" state and is synthesized only after stimulation with various organic and inorganic compounds and particles, NADPH oxidase in nonphagocytic cells exists as a complete enzyme. Correspondingly, modes of oxygen radical production in these cells significantly differ. The stimulation of NADPH oxidase in phagocytes results in "oxidative burst," i.e., the release of a relatively large amount of oxygen radicals, which is extinguished during a short time (5–30 min), while nonphagocytic cells produce oxygen radicals continuously but in smaller quantities.

22.2.1 Leukocyte NADPH Oxidase

Historically, leukocyte NADPH oxidase was the first discovered enzyme of this type; the existence of this enzyme in neutrophils was reported in 1962–1966 [59–61]. (As the evidence of superoxide production in biological systems has been obtained several years later, at that time the nature of oxygen species produced by leukocytes was of course unknown.) And only about 10 years later, Babior et al. [62] have shown that the activation of human neutrophils resulted in the production of superoxide. The structure of leukocyte NADPH has been widely discussed and well-established [57]. Superoxide production catalyzed by NADPH oxidase is described by Reaction (5):

$$2O_2 + NADPH \iff 2O_2^{\bullet-} + NADP^+ + H^+ \qquad (5)$$

In resting "dormant" state three of the five major components of the enzyme, $p40^{phox}$, $p47^{phox}$, and $p67^{phox}$ form a complex in the cytosol. The two other components, $p22^{phox}$ and $gp91^{phox}$, exist in secretory vesicles and special granules. (Secretory vesicles are small intracellular vesicles that fuse with plasma membrane and discharge their contents with response to the stimuli [57].) $p22^{phox}$ and $gp91^{phox}$ compose a heterodimeric flavohemoprotein cytochrome b_{558}. Upon exposure to stimuli cytosolic $p47^{phox}$ becomes phosphorylated and the whole cytosolic complex adds to cytochrome b_{558} in the membrane assembling of the active NADPH oxidase. A complete synthesis of the enzyme also requires two small guanine nucleotide-binding proteins cytosolic Rac2 and membranic Rap1A (Figure 22.1). It should be noted that Cross et al. [63] suggested that the activation of NADPH oxidase may not necessarily require the formation of a stable stoichiometric complex between the *phox* proteins. In this case $p67^{phox}$ or Rac2 are probably able to catalytically activate flavocytochrome.

Generation of superoxide occurs during the one-electron oxidation of cytochrome b_{558}, which contains one FAD and two functionally different heme groups with reduction potentials of -0.225 and -0.265 V [64,65]. It is usually accepted that the heme groups are responsible for the release of superoxide, however, Babior [57] pointed out that the rate of superoxide production by heme is too slow and that the flavin group of cytochrome b_{558} is a main electron carrier. The roles of the other NADPH oxidase components are now also well understood. $p47^{phox}$ is mainly responsible for the transfer of cytosolic complex from the cytosol to the membrane during NADPH oxidase activation. However, before the transfer of cytosolic components to the membrane, $p47^{phox}$ must be phosphorylated acquiring as many as nine phosphate residues [66]. Recently, Shiose and Sumimoto [67] demonstrated that instead of phosphorylation, $p47^{phox}$ can be activated by arachidonic acid. These authors have found that the high concentrations of arachidonic acid induce a direct interaction

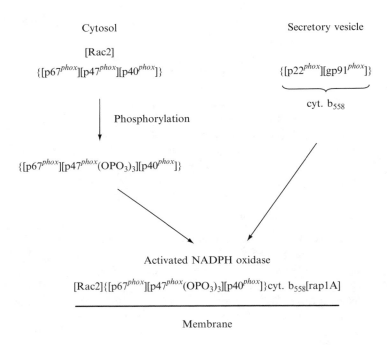

FIGURE 22.1 Synthesis of phagocytic NADPH oxidase upon activation.

between p47^{phox} and p22^{phox}, while at the low concentrations, arachidonic acid synergizes with p47^{phox} phosphorylation. The function of p67^{phox} is not so clear; probably, it participates in electron transfer to the flavin of cytochrome b_{558} [68]. The function of p40^{phox} is also uncertain; it seems to be able to inhibit or stimulate the enzyme activation [57].

The release of superoxide by the phagocytic cell is leaving behind the proton produced by Reaction (5). Correspondingly, the efflux of protons through proton channel must take place as the charge compensation pathway for electrogenic generation of superoxide by NADPH oxidase. Henderson [69] suggested that the gp91^{phox} subunit functions as a NADPH oxidase-associated proton channel and that the mechanism of proton transfer includes the protonation and deprotonation of some amino acid.

The mechanism of the activation of NADPH oxidase includes several enzymatic stages. Hazan et al. [70] has found that stimuli opsonized zymosan (OZ) and PMA activate cytosolic phospholipase A2 and NADPH oxidase. This activation is mediated by extracellular-signal-regulated kinases (ERKs), which in turn are activated through a protein kinase C (PKC)-dependent pathway in the case of PMA stimulation and through tyrosine kinase in the case of OZ stimulation. Another enzyme participating in NADPH oxidase activation is p38 mitogen-activated protein kinase (p38 MAPK). It has been shown [71] that the inhibition of p38 MAPK blocked the OZ-stimulated activation and translocation of Rac2 in bovine neutrophils.

The existence of nitric oxide synthase (NOS) in phagocytes (see below) provides a different kind of stimulation and the inhibition of NADPH oxidase. It has been found [72] that the low physiological concentrations of peroxynitrite formed from NO and superoxide stimulated superoxide production by PMA-activated human PMNs through the ERK MAPK pathway, while higher peroxynitrite concentrations inhibited it. Moreover, NADPH oxidase was inhibited by lidocaine, a sodium-blocker, in OZ-activated neutrophils through the suppression of p47^{phox} translocation [73].

As described earlier, major studies of phagocytic NADPH oxidase have been performed on human and animal neutrophils. However, NADPH oxidase presents in many other blood and tissue phagocytes such as monocytes, eosinophils, basophils, macrophages, etc. Many studies on superoxide production by these cells have already been considered earlier [74]. Recently, several studies have been carried out on the structure and activation mechanism of NADPH oxidase of eosinophils and monocytes. It is known that eosinophils produce more superoxide than neutrophils that could be a consequence of some structural changes in the enzyme. However, Someya et al. [75] has found that there is no structural difference in the NADPH oxidases of these leukocytes, but the contents of all NADPH oxidase components were 1.5 to 3.3-fold greater in eosinophils than neutrophils. Lynch et al. [76] demonstrated that leukotriene B_4-induced superoxide production by adherent and nonadherent eosinophils is mediated by p38 MAP kinase, MEK-1, PKC, and protein tyrosine kinases. Bankers-Fulbright et al. [77] found that stimuli IL-5, LTB_4, and PMA activated human eosinophils through a Zn-sensitive plasma membrane proton channel. Similar to neutrophils, the superoxide production by PMA and AA-stimulated eosinophils accompanied by the activation of proton and electron currents in these cells [78–80]. It has been recently shown [81] that monocyte NADPH oxidase is also activated by cytosolic phospholipase A(2) (cPLA(2)). This enzyme catalyzed the translocation and phosphorylation of $p47^{phox}$ and $p67^{phox}$ subunits via the production of arachidonic acid.

22.2.2 NADPH Oxidase of Nonphagocytic Cells

The major function of superoxide producing NADPH oxidase in phagocytes is the initiation of oxidative processes leading to the destruction of pathogens during phagocytosis. In addition, phagocytic NADPH oxidase catalyzes various free radical-mediated damaging processes considered in subsequent chapters. These functions are not the major ones for NADPH oxidase of nonphagocytic cells. Interesting hypothesis has been offered by Babior [57] who suggested that this enzyme might have arisen evolutionary through the mutation of an ancient NADPH oxidase of lower activity. Such phagocytic NADPH oxidase is present in many cells and fulfills numerous signaling functions.

The structure of nonphagocytic NADPH oxidase has been widely discussed. Although everybody agrees that both phagocytic and nonphagocytic enzymes represent the same enzymatic family, there are some differences depending on the type of cells [82]. For example, it has been demonstrated that endothelial cells [83,84], adventitial cells [85,86], cardiac myocytes [87], and the cells of rat ventilatory muscles [88] expressed all major subunits of phagocytic NADPH oxidase, while VSMCs and mesangial cells appear to express $p22^{phox}$, $p40^{phox}$, and Rac1, but not $gp91^{phox}$ [89]. Gorlach et al. [90] demonstrated that there is difference between the structures of endothelial NADPH oxidase and NADPH oxidase of VSMCs: the first one contains the $gp91^{phox}$ subunit while the second one does not. On the whole, there are five major differences in phagocytic and nonphagocytic NADPH oxidases: (a) the organization of enzyme in the cell, (b) the enzyme composition, (c) the nature of substrates, (d) the mode of stimulation, and (e) the mode and amount of superoxide production.

(a) As mentioned above, an entirely synthesized nonphagocytic NADPH oxidase is located in the vascular wall whereas phagocytic one is distributed between the cytosol and the membrane and is synthesized upon stimulation.

(b) Examples cited above already demonstrate a dependence of the enzyme structure on a cell type. There are numerous studies in which authors have studied the similarity and differences of phagocytic and nonphagocytic NADPH oxidases, and at present it is difficult to achieve a full agreement. As already noted, not all subunits of phagocytic NADPH enzyme are found in various nonphagocytic cells, although all subunits are expressed in porcine pulmonary artery endothelial cells (PAECs) [91] human, bovine and porcine endothelial cells [92], and coronary microvascular endothelial cells (CMECs) [93]. At the same time it has been suggested [94,95] that nonphagocytic enzymes contain not the gp91^{phox} subunit but its homologs Nox 1, Nox 3, Nox 4, and Nox 5. Souza et al. [96] confirmed that vascular NADPH oxidase does not possess the gp91^{phox} subunit identical to the phagocytic enzyme because there was no significant difference between superoxide production by aortas from wide-type and gp91^{phox}-deficient mice.

(c) In contrast to the phagocytic enzyme, vascular NADPH oxidase may catalyze superoxide production using both NADH and NADPH as substrates. Earlier studies suggested that NADH is the preferred substrate, however, it has been later proposed that this preference is due to the use of high concentrations of lucigenin in the CL detection of superoxide [82]. It was suggested [97] that lucigenin artificially overestimated superoxide production due to participating in the redox cycling and that this effect is especially high in the presence of NADH. In our opinion the criticism of lucigenin-amplified CL as an assay for superoxide detection is greatly exaggerated and could be completely wrong [98]. (This problem is discussed in detail in Chapter 32.) At present, the predominance of NADPH-stimulation of superoxide production by nonphagocytic enzyme is shown in human, bovine, and porcine endothelial cells [92], and VSMCs [99], although NADH and NADPH are equally effective in rat aortic homogenates and intact segments [100].

(d) Stimuli and the mechanism of stimulation of nonphagocytic NADPH oxidase are mainly different from those for phagocytic enzyme. Nonetheless, it has been shown that the nonphagocytic enzyme may be stimulated by some typical stimuli of leukocyte NADPH oxidase such as calcium, interleukin-1, and TNF-α in human fibroblasts [101,102], PMA in pulmonary artery endothelial cells [91], human endothelial cells [103], and murine vascular smooth cells [104], PMA in pig coronary arteries (supposedly through the activation of protein kinase C) [105], and TNF-α in rat aortic smooth muscle cells [106].

However, the most important stimulus of nonphagocytic NADPH oxidase is angiotensin II. It has been shown that neuropeptide angiotensin II, which exerts numerous effects on the cardiovascular system, stimulates superoxide production through the activation of NADPH oxidase in vascular cells [107–109]. For example, it has been shown that the NADPH oxidase activation and superoxide overproduction can be responsible for the angiotensin II-induced hypertension [110,111]. Pagano et al. [112,113] and Wang et al. [85] have studied the activation of NADPH oxidase by angiotensin II in aortic adventitial fibroblasts. They concluded that angiotensin II stimulates superoxide production by NADPH oxidase in these cells via p67^{phox} activation. The amount of superoxide generated in this system is apparently sufficient to create a barrier capable of inactivating nitric oxide by the formation of peroxynitrite. Angiotensin II stimulation of NADPH oxidase is also mediated by phospholipases. Thus Zafari et al. [114] have shown that angiotensin II stimulated phospholipase A2 in VSMC, releasing arachidonic acid, which in turn stimulated NADPH oxidase. Touyz and Schiffrin [115] reported that in human VSMC the angiotensin II stimulation of NADPH oxidase depended on phospholipase D.

There are various angiotensin II-dependent pathways of NADPH oxidase activation. Xie et al. [116] have found that angiotensin II induced the stimulation of osteopontin, an extracellular matrix protein, in cardiac microvascular endothelial cells, which is accompanied by the activation of p42/44 mitogen-activated protein kinase (p42/44 MAPK) and increased superoxide production. It has been suggested that p42/44 MAPK is a critical component of NADPH oxidase angiotensin II stimulation. Mollnau et al. [117] have studied the effect of angiotensin II stimulation of superoxide production under in vivo conditions. It was found that the angiotensin II treatment of rats resulted in the enhancement of superoxide production not only due to NADPH oxidase activation but also as a result of NO synthase uncoupling.

Li et al. [118] demonstrated that hydrogen peroxide stimulated superoxide production by NADPH oxidase from smooth muscle cells and fibroblasts. Although the mechanism of hydrogen peroxide stimulation is unclear, it has been supposed that hydrogen peroxide may induce arachidonic acid release, a known stimulus of NADPH oxidase, or activate the NADPH oxidase by phosphorylating the $p22^{phos}$ subunit. The discovery of hydrogen peroxide stimulation can be of importance because it shows the existence of a self-promoting cycle of oxygen radical production in nonphagocytic cells. The ubiquitous glycosphingolipid lactosylceramide (LacCer) is another stimulus of NADPH oxidase in human aortic smooth muscle cells. It has been shown [119] that LacCer stimulated NADPH-catalyzed superoxide production in these cells, resulting in the activation of the kinase cascade and the induction of cell proliferation.

Several other stimuli are also able to activate nonphagocytic NADPH oxidase. Thus high glucose levels and palmitate stimulated superoxide production by vascular aortic smooth muscle and endothelial cells supposedly through a protein kinase C-dependent activation of NADPH oxidase [120]. Lysophosphatidylcholine (LPC), an atherogenic lipid formed during the oxidation of low-density lipoproteins, stimulated superoxide production by NADPH oxidase in BAECs via a tyrosine kinase-dependent pathway [121]. Heinloth et al. [122] demonstrated that LDC and oxidized LDLs enhanced superoxide production by HUVEC and induced HUVEC proliferation through NADPH oxidase activation. Endothelin-1, a pro-atherosclerotic stimulus, induced $gp91^{phox}$ mRNA expression and augmented superoxide production in HUVECs [123]. Superoxide production was enhanced in cyclically stretched ventricular myocytes [124].

Holland et al. [125] have shown that the potent vascular smooth muscle cell mitogen and phospholipase A2 activator thrombin stimulated superoxide production in human endothelial cells, which was inhibited by the NADPH oxidase inhibitors. Similarly, thrombin enhanced the production of oxygen species and the expression of $p47^{phos}$ and Rac2 subunits of NADPH oxidase in VSMCs [126,127]. Greene et al. [128] demonstrated that the activator of NO synthase neuropeptide bradykinin is also able to stimulate NADPH oxidase in VSMCs. Similar to XO, NADPH oxidase enhanced superoxide production in pulmonary artery smooth muscle cells upon exposure to hypoxia [129].

(e) Although the capacity of NADPH oxidase to produce superoxide is now experimentally proven, there is disagreement in the estimation of superoxide concentration released by nonphagocytic cells. On the basis of lucigenin-amplified CL measurements, it has been usually accepted that vascular cells constitutively produce about 1% of superoxide generated by stimulated peritoneal macrophages [91]. However, as noted above, Zweier and coworkers [97,100] have recently concluded, that the use of lucigenin-amplified CL leads to the overestimation of superoxide generation due to the lucigenin redox cycling and that the real superoxide production can be several times lower. But, there are many contradictions in the interpretation of findings obtained by these and other authors. For example, the enhanced lucigenin-amplified

CL produced in the presence of exogenous NADPH or NADH, which has been studied in Refs. [97,100], may originate from other sources than superoxide generation [90]. In our opinion [98], the data obtained on the basis lucigenin CL measurement, especially with the use of small lucigenin concentrations provide the reliable estimate of superoxide concentration while the use of ESR spin technique underestimates it, particularly in vascular tissue and cells (see Chapter 32).

As noted earlier, a major role of nonphagocytic NADPH oxidase in cells is to produce active oxygen species accomplishing the role of second messengers. These signaling processes result in the development of vascular diseases, including hypertension and atherosclerosis. Below are cited several examples of NADPH oxidase catalyzed superoxide signaling functions. It has been shown that superoxide generated by NADPH oxidase of VSMCs regulates the protein kinase cascade by activating the mitogen-activated protein kinases p42/44 MAPKs [130]. Angiotensin II activation of JNK and p38 MAPK kinases was shown to be mediated by NADPH oxidase-catalyzed superoxide formation in VSMC [131]. In cardiac fibroblasts angiotensin II stimulated NADPH oxidase via an AT-1 receptor to produce oxygen radicals, which activated EPK1/2, p38 MAPK, and JNK kinases [132], while in VSMC kinase cascade included p38 MAPK but not ERK [133]. Xiao et al. [87] reported that NADPH oxidase is involved into α1-adrenoceptor-stimulated hypertrophic signaling through superoxide-mediated activation of ERK1/2 kinase in rat ventricular myocytes. Brar et al. [134] showed that NADPH oxidase of human airway smooth muscle cells regulated proliferation of these cells through the superoxide-mediated activation of redox-regulated transcription factor NF-κB.

Besides cell signaling, superoxide production by nonphagocytic cells may exhibit damaging activity through the interaction with nitric oxide to form peroxynitrite, toxic effects of which were considered in Chapter 21. On the other hand, a decrease in NO concentration may result in endothelial dysfunction due to reduction in endothelium-dependent vasorelaxations [135]. Furthermore, as already mentioned above, the interaction of superoxide with NO synthase in vascular tissue may lead to uncoupling of this enzyme and increasing superoxide production [117]. Therefore, the overproduction of superoxide may affect superoxide/nitric oxide by different ways, resulting in cell injury. It is interesting that some biomolecules may be useful for improving this balance. Thus, 17β-estradiol upregulated NO synthase expression and inhibited NADPH oxidase expression in human endothelial cells [136].

22.3 NITRIC OXIDE SYNTHASES

22.3.1 TYPES AND STRUCTURE OF NITRIC OXIDE SYNTHASES

There are three main distinct forms of NOSs expressed in animals, which are major enzymes responsible for NO generation: two constitutive enzymes neuronal NOS I (nNOS) and endothelial NOS III (eNOS), and one inducible enzyme NOS II (iNOS). These enzymes catalyze NO and L-citrulline formation from L-arginine (through the intermediate formation of N-hydroxyl-L-arginine) and are present in many cells including macrophages, Kupffer cells [137], hepatocytes [138], and neutrophils [139], but not eosinophils [140]. Recently, Bal-Price et al. [141] has found that significant amount of nitric oxide is produced by lipopolysaccharide/interferon-γ (LPS/IFN-γ)-activated rat microglia and astrocytes, with a steady state of NO equal to 0.5–0.7×10^{-6} mol l^{-1}. All NO synthases are homodimers containing a N-terminal oxygenase domain with binding sites for heme, 6R-tetrahydrobiopterin (H_4B), and L-arginine, and a C-terminal reductase domain with binding sites for FMN, FAD, and NADPH. The

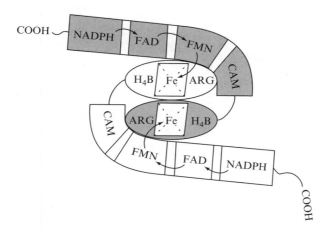

FIGURE 22.2 Structure of NO synthases and "domain swapping," [From U Siddhanta, A Presta, B Fa, D Wolan, DL Rousseau, DJ Stiehr. *J Biol Chem* 273: 18950–18958, 1998. With permission.]

enzymes also contain the calmodulin subunit located between oxygenase and reductase domains (Figure 22.2).

NO synthases perform numerous functions contributing to the regulation of systemic blood pressure, activating intracellular signaling pathways in VSMCs, resulting in vasorelaxation [142], participating in the toxic activity of macrophages against virus, bacteria, and tumor cells [143], etc. NO synthases play an important role under many pathophysiological conditions such as hypercholesterolemia and atherosclerosis where an impaired endothelial-dependent vasorelaxation takes place induced by oxidative vascular damage [144]. These biological functions of nitric oxide synthases under physiological and pathophysiological conditions are considered in the following chapters. Now, we will regard the structures of NO synthases.

Synthesis of nitric oxide occurs by accepting of electrons from NADPH by the flavins of a reductase domain and transferring them to the heme of an oxygenase domain [145]. In the case of iNOS it has been shown that the single reductase domain transfers electrons only to the adjacent one of the two oxygenase domains, supporting a normal rate of NO synthesis [145,146]. Siddhanta et al. [145] considered the effect of iNOS dimeric structure on the mechanism of electron transfer through iNOS. They proposed that there is possibly a barrier for electron transfer from the reductase to oxygenase domain located on the same subunit that results in the enzyme "domain swapping," which allows electrons to transfer between the reductase and oxygenase domains of adjacent subunits (Figure 22.2). Thus, dimerization helps to orient reductase and oxygenase domains properly for electron transfer.

22.3.2 Free Radical Production by Nitric Oxide Synthases

As mentioned earlier, extensive literature is dedicated to the study of functions of NO synthases under physiological and pathophysiological conditions. Much attention has been drawn to the capacity of these enzymes to generate free radicals. The mechanism of nitric oxide production by NO synthases was widely discussed and are presented in Figure 22.3 [147].

At the initial step electrons are transferred from NADPH to the oxidized FAD, reducing it to $FADH_2$. Disproportionation between flavins leads to the formation of two free radicals $FADH^\bullet$ and $FMNH^\bullet$. Electron transfer from $FMNH^\bullet$ to the heme results in the reduction of Fe^{3+} to Fe^{2+}, and the reduced heme becomes able to bind O_2 to form the intermediate

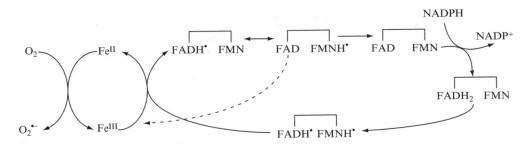

FIGURE 22.3 Mechanism of the catalytic cycle of NO synthase [From S Pou, L Keaton, W Surichamorn, GM Rosen. *J Biol Chem* 274: 9573–9580, 1999. With permission.]

$$\text{NADPH} \xrightarrow{e} \text{FADH}^\bullet/\text{FMNH}^\bullet \xrightarrow{e} \text{Heme Fe}^{2+} \xrightarrow{O_2} \text{Heme Fe}^{2+}O_2 \xrightarrow{e} \text{Heme Fe}^{2+}O_2^{\bullet-} \Rightarrow \text{Heme (Fe}^{4+}\text{O)}$$

$$\text{Heme (Fe}^{4+}\text{O)} + \underset{\underset{\text{NH}}{\|}}{\text{NH}_2\text{C}}\text{—NHCH}_2\text{CH}_2\text{CH}_2\text{CH(NH}_2)\text{COOH} \Rightarrow$$

$$\underset{\underset{\text{NH}}{\|}}{\text{HONHC}}\text{—NHCH}_2\text{CH}_2\text{CH}_2\text{CH(NH}_2)\text{COOH} \Rightarrow \text{NO} + \text{NH}_2\text{CONHCH}_2\text{CH}_2\text{CH}_2\text{CH(NH}_2)\text{COOH}$$

$\qquad\qquad$ *N*-Hydroxy-L-arginine $\qquad\qquad\qquad\qquad\qquad\qquad$ L-Citrulline

FIGURE 22.4 Mechanism of nitric oxide and L-citrulline production by NO synthases through the intermediate formation of *N*-hydroxyl-L-arginine.

Fe(II)O$_2$ oxygenated complex [148]. Binding L-arginine to the heme results in its oxidation and the formation of nitric oxide. It has been suggested [149] that similar to the reactions catalyzed by cytochromes P-450, the oxidation of L-arginine proceeds via the hydroxylation of L-arginine to *N*-hydroxyarginine (Figure 22.4). In accord with this mechanism ferric heme iron is converted into Fe(II)O$_2$ species (I) and further into the iron-peroxy intermediate (II), which after losing water forms the high valence oxo-iron species heme Fe(IV)O (III), capable of hydroxylating the substrate (Scheme 1).

$$\text{Heme Fe}^{III} \xrightarrow[O_2]{e^-} \text{heme Fe}^{II}O_2 \xrightarrow{e^-} \text{heme Fe}^{II}O_2^{\bullet-} \xrightarrow[-H_2O]{H^+} \text{heme (Fe}^{IV}\text{O)}$$
$\qquad\qquad\qquad\qquad\quad$ I $\qquad\qquad\qquad$ II $\qquad\qquad\quad$ III

SCHEME 1 (With permission from HM Abu-Soud, R Gachhui, FM Raushel, DJ Stuehr. *J Biol Chem* 272: 17349–17353, 1997.)

Because NO synthases belong to the same superfamily of enzymes as cytochrome P-450, they are able to produce not only nitric oxide (although it is undoubtedly their main function) but also other free radicals, first of all, superoxide. In 1992, Pou et al. [148] showed that brain nitric oxide synthase (NOS I) produced superoxide identified as a DMPO—OOH adduct in a calcium- or calmodulin-dependent manner. This finding was confirmed in numerous studies for all three isoforms of NO synthase. Although the structures of all the three NO oxidase isoforms are similar, there are differences in the mode of generation of nitric oxide and superoxide by these enzymes. Firstly it was proposed [148] that the absence of substrate L-arginine is an obligatory condition for superoxide production by NO synthase. However, subsequent studies demonstrated that all NO synthase isoforms can generate superoxide in addition to NO even at saturated substrate levels [147]. However, the sites of superoxide production are different in NO synthase isoforms. It has been proposed [149,150] that nNOS I and eNOS III isoforms produce superoxide by the oxidation of ferrous heme, while iNOS II generates superoxide at the flavin-binding sites of the reductase domain [151]. This difference is probably due to a weak coupling between $FMNH^{\bullet}$ and ferric heme in iNOS II (Figure 22.3) that makes possible for electrons to leak from the flavin domain to form superoxide even in the presence of substrate [147].

It has been shown that the activity of NO synthases is regulated by cofactors calcium binding protein calmodulin and tetrahydrobiopterin (H_4B). Abu-Soud et al. [149] have studied the effect of H_4B on the activity of neuronal nNOS I, using the isolated heme-containing oxygenase domain nNOSoxy. It was found that nNOSoxy rapidly formed an oxygenated complex in the reaction with dioxygen, which dissociated to produce superoxide (Reaction (6)):

$$nNOS(Fe^{2+}) + O_2 \Longleftrightarrow nNOS(Fe^{2+}O_2) \Longleftrightarrow nNOS(Fe^{3+}) + O_2^{\bullet-} \tag{6}$$

Surprisingly, H_4B accelerated the dissociation of the oxygenated complex even in the presence of bound L-arginine. However, there is another view on the role of H_4B in this system. Pou et al. [147] suggested that the low concentrations H_4B inhibited superoxide production by nNOS I, while at higher concentrations H_4B formed superoxide by autoxidation. Recently, the ability of H_4B to participate in one-electron transfer processes during the nNOS I and eNOS III catalysis was confirmed by the identification of the H_4B radical cation protonated at N5 atom [152,153]. The mechanism of action of another cofactor calmodulin has been studied by Daff et al. [154] who suggested that calmodulin controlled electron transfer via a large structural rearrangement.

It is possible that tetrahydrobiopterin is a critical subunit responsible for the production of NO but not other nitrogen oxides by NO synthases. In 1995, Pufahl et al. [155] found that NO synthase from murine macrophages catalyzed the oxidation of NG-hydroxyl-L-arginine in the presence of hydrogen peroxide to citrulline and nitrite or nitrate without NO formation. Later on, Schmidt et al. [156] suggested that NO synthases produce not nitric oxide but nitroxyl NO^-, which is oxidized by SOD into NO. Their conclusion was based on the fact that they cannot detect any NO in the absence of SOD but can detect other nitrogen oxides (N_2O and NH_2OH), which are probably not derived from nitric oxide. It was suggested that NO is formed by the oxidation of nitroxyl with SOD:

$$NO^- + Cu(II)SOD \Longrightarrow NO + Cu(I)SOD \tag{7}$$

Recent findings suggest a new role for H_4B in the mechanism of nitric oxide generation by NO synthases. It has been suggested [157] that a critical moment in performing hydroxylation of substrates by the enzyme is the transfer of second electron to the ferrous–dioxy complex heme ($Fe(II)O_2$) to form heme ($Fe(II)O_2^{\bullet-}$) (Scheme 1). H_4B is not an obligatory subunit, which is necessary for this process, because the NOS heme hydroxylates L-arginine even without receiving two electrons from NADPH. However, the absence of H_4B results in uncoupling between NADPH oxidation and product formation. Therefore, it is possible that H_4B may improve coupling between NADPH oxidation and product formation by the one-electron oxidation of H_4B [152,153], which can be a faster process than the transfer of an electron from the reductase domain. At the same time the formed H_4B cation radical can oxidize nitroxyl bound to the ferric heme into NO. Thus, it may be suggested that the essential function of H_4B is to convert the bound NO^- into nitric oxide [157].

Simultaneous generation of nitric oxide and superoxide by NO synthases results in the formation of peroxynitrite. As the reaction between these free radicals proceeds with a diffusion-controlled rate (Chapter 21), it is surprising that it is possible to detect experimentally both superoxide and NO during NO synthase catalysis. However, Pou et al. [147] pointed out that the reason is the fact that superoxide and nitric oxide are generated consecutively at the same heme iron site. Therefore, after superoxide production NO synthase must cycle twice before NO production. Correspondingly, there is enough time for superoxide to diffuse from the enzyme and react with other biomolecules.

In addition to nitric oxide, superoxide, and peroxynitrite, NO synthases are able to generate "secondary" free radicals because similar to cytochrome P-450 reductase, the reductase domain can transfer an electron from the heme to a xenobiotic. Thus it has been found [158,159] that neuronal NO synthase NOS I catalyzed the formation of $CH_3CH(OH)^{\bullet}$ radical from ethanol. It was suggested that the perferryl complex of NOS I is responsible for the formation of such secondary radicals. Miller [160] also demonstrated that 1,3-dinitrobenzene mediated the formation of superoxide by nNOS. It was proposed that the enhancement of superoxide production in the presence of 1,3-dinitrobenzene converted nNOS into peroxynitrite-produced synthase and may be a mechanism of neurotoxicity of certain nitro compounds.

Recent studies showed an importance of the association of NO synthases with heat shock protein 90 (hsp90). Song et al. [161] demonstrated that hsp90 enhanced nitric oxide production by neuronal NO synthase and inhibited superoxide production. This effect was not due to the direct reaction of hsp90 with superoxide because hsp90 did not affect superoxide production by XO. Pritchard et al. [162] suggested that normal low-density lipoproteins induce eNOS dysfunction by decreasing the association of hsp90 with the enzyme. On the other hand, chronic hypoxia increased the association of hsp90 with endothelial eNOS and enhanced nitric oxide production [163].

It should also be mentioned that superoxide and not nitric oxide production by eNOS may have implications for atherosclerosis and septic shock due to imbalance between NO and superoxide formation, for example due to an increase in TNF-α production [164]. These pathophysiological functions of NO synthases will be considered in detail in Chapter 31.

Completing the consideration of the mechanisms of superoxide and nitric oxide generation by NO synthases, it is necessary to discuss the studies by Xu [165–167] who, contrary to numerous findings cited above, proposed that superoxide generation by these enzymes is an artifact. This unexpected conclusion was based on the fact that this author registered superoxide production by inactivated nNOS as well as by its cofactors FAD, FMN, H_4B, and calmodulin. His findings were repudiated on the basis of uncorrected experimental data or their interpretation [168,169], but in our opinion Xu's data are simply irrelevant to the problem of superoxide production by active NO oxidases. The ability of many biomolecules to generate superoxide during autoxidation is well known and usually has no connection with their in vivo functions. On the other hand, the ability of NO synthases to produce superoxide during their catalytic cycles under in vitro and ex vivo conditions is decisively proven by the use of selective enzyme inhibitors.

22.3.3 MITOCHONDRIAL NITRIC OXIDE SYNTHASE

In contrast to well-described neuronal, endothelial, and inducible isoforms of NO synthase, mitochondrial NO synthase (mtNOS) is still not fully identified enzyme. As early as 1995, it has been reported that this NOS isoform is located to the inner mitochondrial membrane in brain, kidney, liver, and muscle [170,171]. Later on, the existence of mtNOS associated with the inner mitochondrial membrane and not with the matrix fraction was shown in rat liver mitochondria [172,173]. It has been shown that mtNOS is a constitutive enzyme, but there is no full agreement whether it is nNOS I, iNOS II, or a new NO synthase isoform. Lacza et al. [174] demonstrated that antibodies against eNOS III but not nNOS I or iNOS II showed positive immunoblotting and therefore mtNOS should be the eNOS-like isoform. In contrast, Kanai et al. [175] suggested that mtNOS is nNOS because NO production was absent in the mitochondria of knockout mice for nNOS I, but not iNOS II or eNOS III. Lopez-Figueroa et al. [176] demonstrated the in vivo production of nitric oxide within mitochondria.

The functions of mtNOS in mitochondria have been studied (see Chapter 23). Ghafourifar et al. [177] found that the calcium-induced stimulation of mtNOS caused the release of cytochrome c from mitochondria and induced apoptosis. On the other hand, the same group of authors [178] showed that the production of NO by mtNOS and superoxide in mitochondria resulted in the formation of peroxynitrite and stimulated calcium release, indicating the existence of a "feedback loop" which prevents calcium overload in mitochondria.

22.4 PEROXIDASES

22.4.1 Mechanism of Free Radical Production

Peroxidases are a group of enzymes that catalyze the oxidation of hydrogen peroxide and various hydroperoxides. There are two major groups of peroxidases: mammalian peroxidase family (superfamily II) and plant, fungal, and bacterial peroxidase family (superfamily I) [179]. Mammalian peroxidases, which include myeloperoxidase (MPO), lactoperoxidase (LPO), and thyroid peroxidase (TPO) greatly differ from peroxidases of superfamily I in structure in having a covalent binding between the heme group and protein matrix. The most important representative of superfamily II is undoubtedly MPO, which is a major neutrophil protein, constituting 2–5% of neutrophil dry weight. MPO are also present in monocytes and eosinophils (where is named eosinophil peroxidase (EPO)). MPO is a major component of the azurophilic cytoplasmic granules and is released during phagocytosis. It has been shown that the mechanism of catalytic activity of heme peroxidases is practically identical; therefore, the other enzymes, horseradish peroxidase (HRP) and LPO are widely used in numerous in vitro studies. It has been known for a long time that peroxidases are able to catalyze the one-electron oxidation of xenobiotics and biomolecules to free radicals. However, in contrast to the above considered XO, NADPH oxidase, and NO synthase, which produce superoxide as a major active oxygen species capable of initiating free radical processes, peroxidases are able to generate free radicals from substrates as a result of a direct interaction with the heme groups.

The mechanism of free radical formation in the reactions catalyzed by MPO and other heme peroxidase may be presented as follows [180]:

$$MPO + H_2O_2 \Longrightarrow Compound\ I + H_2O \qquad (8)$$

$$Compound\ I + RH \Longrightarrow Compound\ II + R^\bullet \qquad (9)$$

$$Compound\ II + RH \Longrightarrow MPO + R^\bullet + H_2O \qquad (10)$$

In accord with this mechanism, a single two-electron oxidation of the enzyme into Compound I by hydrogen peroxide (Reaction (8)) is followed by two one-electron steps: Reaction (9), in which substrate RH is oxidized to a radical $R^\bullet$ and Compound I is reduced to Compound II and Reaction (10), in which Compound II is reduced to native MPO, completing the catalytic cycle. At high hydrogen peroxide concentrations Compound II can be converted in Compound III [181]:

$$\text{Compound II} + H_2O_2 \Longrightarrow \text{Compound III} \tag{11}$$

Recovering of Compound II involved the reduction of Compound III to Fe(II)MPO and the reaction of Fe(II)MPO with hydrogen peroxide:

$$\text{Compound III} \Longrightarrow \text{Fe(II)MPO} \tag{12}$$

$$\text{Fe(II)MPO} + H_2O_2 \Longrightarrow \text{Compound IIs} \tag{13}$$

At low concentrations of hydrogen peroxide Compound II is converted back to MPO.

The following structures of peroxide intermediates have been proposed [180]: Compound I contains a porphyrin π cation radical and a low-spin oxyferryl $Fe^{IV}O$ center; Compound II is the one-electron reduction product having supposedly the structure of $Fe^{IV}OH$; Compound III, which can be obtained from native Fe(III) peroxidase with superoxide, from reduced Fe(II) peroxidase with dioxygen, and from Compound II with hydrogen peroxide [182], is a $Fe(II)O_2$ compound. The subsequent studies showed that Compound I may exist in two different forms as $Fe^{IV}O$ porphyrin π cation radical and $Fe^{IV}O$ protein radical [183–185]. It was proposed [185,186] that protein radical present in Compound I of MPO and LPO may be related to covalent binding of the heme group to the protein or participates in enzyme inactivation.

22.4.2 Oxidation of Phenolic Compounds

Several compounds can be oxidized by peroxidases by a free radical mechanism. Among various substrates of peroxidases, L-tyrosine attracts a great interest as an important phenolic compound containing at 100–200 μmol l^{-1} in plasma and cells, which can be involved in lipid and protein oxidation. In 1980, Ralston and Dunford [187] have shown that HRP Compound II oxidizes L-tyrosine and 3,5-diiodo-L-tyrosine with pH-dependent reaction rates. Ohtaki et al. [188] measured the rate constants for the reactions of hog thyroid peroxidase Compounds I and II with L-tyrosine (Table 22.1) and showed that Compound I was reduced directly to ferric enzyme. Thus, in this case the reaction of Compound I with L-tyrosine proceeds by two-electron mechanism. In subsequent work these authors have shown [189] that at physiological pH TPO catalyzed the two-electron oxidation not only L-tyrosine but also D-tyrosine, N-acetyltyrosinamide, and monoiodotyrosine, whereas diiodotyrosine was oxidized by a one-electron mechanism.

Similar to other phenols, L-tyrosine is oxidized by peroxidases to phenolic free radical [190]:

$$4\text{-}HOC_6H_4CH_2CH(NH_2)COOH \Longrightarrow {}^{\bullet}OC_6H_4CH_2CH(NH_2)COOH \tag{14}$$
$$\text{Tyrosine} \qquad \text{Tyr}^{\bullet}$$

Coupling of tyrosyl radical yields o,o'-dityrosine:

$$2\text{Tyr}^{\bullet} \Longrightarrow o,o'\text{-dityrosine} \tag{15}$$

Heinecke et al. [191] studied the oxidation of L-tyrosine by the H_2O_2–MPO system and showed that the main product of this reaction is dityrosine. They have also found that tyrosine successfully competed with chloride as a substrate for MPO that points out at the possibility of in vivo oxidation of tyrosine by MPO even in the presence of big physiological concentration (0.10–0.1 mol l^{-1}) of chloride in human blood. It was also suggested that the tyrosyl radical formed at the catalytic oxidation of tyrosine by peroxidases may interact with

TABLE 22.1
Rate Constants (l mol^{-1} s^{-1}) for the Reactions of Compounds I and II with Organic Compounds

Peroxidase	Substrate	Compound I	Compound II
Thyroid peroxidase	L-Tyrosine	7.5×10^{4a} [188]	430 [188]
Thyroid peroxidase	L-Tyrosine		90 [189]
HRP	L-Tyrosine	5.0×10^4 [187]	1.1×10^3 [194]
Lactoperoxidase	L-Tyrosine	$>1.1 \times 10^5$ [188]	1.03×10^4 [195]
MPO	L-Tyrosine	7.72×10^5 [193]	1.57×10^4 [193]
MPO	Dityrosine	$(1.12 \pm 0.01) \times 10^5$ [193]	$(7.5 \pm 0.3) \times 10^2$ [193]
Thyroid peroxidase	D-Tyrosine		260 [189]
Thyroid peroxidase	N-Acetyltyrosinamide		9.0×10^3 [189]
Thyroid peroxidase	Monoiodotyrosine		1.8×10^3 [189]
Thyroid peroxidase	Diiodotyrosine		1.5×10^4 [189]
HRP	Peroxynitrite		3×10^{6a} [212]
MPO	Peroxynitrite		2×10^{7a} [211]
Lactoperoxidase	Trolox C		7.7×10^3 [238]
HRP	Trolox C		2.1×10^4 [238]
HRP	2-Aminofluorene	$(0.5-1) \times 10^9$ [210]	$(1.7 \pm 0.2) \times 10^8$ [210]
HRP	Ascorbate		5×10^2 [241]
Myeloperoxidase	Ascorbate	$(1.1 \pm 0.1) \times 10^6$ [211]	$(1.1 \pm 0.2) \times 10^4$ [211]
Lactoperoxidase	Ascorbate		3.5×10^3 [211]
Myeloperoxidase	Clozapine	$(1.5 \pm 0.1) \times 10^6$ [211]	$(4.8 \pm 0.1) \times 10^4$ [211]
HRP	Luminol		3.6×10^4 [205]
Lactoperoxidase	Luminol		2.5×10^4 [205]

^{a}Two-electron transfer reaction.

protein radicals yielding tyrosylated proteins. On these grounds, Francis et al. [192] suggested that oxidative tyrosylation of high-density lipoproteins by peroxidases can enhance the removal of cholesterol from fibroblasts and macrophage foam cells.

Marquez and Dunford [193] have studied the kinetics of L-tyrosine oxidation by MPO. They measured the rate constants for the reactions of MPO compounds I and II with tyrosine and dityrosine and found out that, comparing with HRP, LPO, and TPO, MPO is the most effective catalyst of tyrosine oxidation at physiological pH (Table 22.1). Furthermore, the rate constant for Reaction (9) with tyrosine turns out to be comparable with that for Reaction (16), confirming the possibility for tyrosine to compete in blood plasma with chloride, which is considered to be the major MPO substrate and a potent oxidizing agent against invading bacteria and viruses.

$$\text{Compound I} + \text{HCl} \Longrightarrow \text{MPO} + \text{HOCl} + \text{H}_2\text{O} \tag{16}$$

Tien [196] has studied the oxidation of dipeptides, tripeptides, and polypeptides of tyrosine by MPO and found that the rates of their reactions with MPO Compounds I and II decreased with increasing substrate size. He suggested that such reactions might be responsible for the formation of atherosclerotic plaques. Interestingly that LPO but not MPO or EPO undergoes a radical dimerization in the reaction with hydrogen peroxide, in which two tyrosine residues couple to form a dityrosine cross-linking dimeric enzyme [185]. Using an ESR rapidly mixing flow system, McCormick et al. [197] was able to register the ESR spectrum of tyrosyl radical in the

HRP-catalyzed oxidation of tyrosine. ESR spectra of the spin-adducts of tyrosyl radical with 2-methyl-2-nitrosopropane were also obtained in the reactions catalyzed by MPO and LPO.

Various hydroxyl and amino derivatives of aromatic compounds are oxidized by peroxidases in the presence of hydrogen peroxide, yielding neutral or cation free radicals. Thus the phenacetin metabolites *p*-phenetidine (4-ethoxyaniline) and acetaminophen (*N*-acetyl-*p*-aminophenol) were oxidized by LPO or HRP into the 4-ethoxyaniline cation radical and neutral *N*-acetyl-4-aminophenoxyl radical, respectively [198,199]. In both cases free radicals were detected by using fast-flow ESR spectroscopy. Catechols, Dopa methyl ester (dihydroxyphenylalanine methyl ester), and 6-hydroxy-Dopa (trihydroxyphenylalanine) were oxidized by LPO mainly to *o*-semiquinone free radicals [200]. Another catechol derivative adrenaline (epinephrine) was oxidized into adrenochrome in the reaction catalyzed by HRP [201]. This reaction can proceed in the absence of hydrogen peroxide and accompanied by oxygen consumption. It was proposed that the oxidation of adrenaline was mediated by superoxide. HRP and LPO catalyzed the oxidation of Trolox C (an analog of α-tocopherol) into phenoxyl radical [202]. The formation of phenoxyl radicals was monitored by ESR spectroscopy, and the rate constants for the reaction of Compounds II with Trolox C were determined (Table 22.1).

Similarly, LPO oxidized the estrogen 17β-estradiol to phenoxyl radical [203]. It is important that the phenoxyl radical formed was reactive enough to abstract a hydrogen atom from glutathione, ascorbate, and NADH. The latter was oxidized by dioxygen yielding superoxide. The same processes were observed during the catalytic oxidation of fluorescent dye 2′,7′-dichlorofluorescein (DCF) by HRP [204]. As in the case of 17β-estradiol [203], the phenoxyl radical formed from 2′,7′-dichlorofluorescein oxidized glutathione and NADH, forming further superoxide and hydrogen peroxide. This enzymatic chain process makes doubtful the use of DCF for measuring superoxide and hydrogen peroxide formation in cells and tissues. It is known that peroxidases are the efficient catalysts of luminol oxidation. Nakamura and Nakamura [205] measured the rates of one-electron oxidation of luminol by HRP, *Arthromyces ramosus* peroxidase, and LPO (Table 22.1); they concluded that luminol is oxidized by Compound II and not Compound I.

22.4.3 Oxidation of Glutathione and Other Organic Substrates

Peroxidases easily oxidize glutathione by both one-electron (LPO, HRP) and two-electron (thyroid peroxide) pathways [207,208]. It was found that peroxidases cannot directly oxidize glutathione, and that this process is mediated by the interaction of tyrosyl radicals of enzymes with GSH [207]. Similarly, glutathione was oxidized by HRP during co-oxidation with flavonoids by flavonoid phenoxyl radicals [209]. HRP rapidly oxidized 2-aminofluorene with rate constants of $10^8 - 10^9 \, l \, mol^{-1} \, s^{-1}$ (Table 22.1) supposedly to form the substrate cation radical as an intermediate [210]. Hsuanyu and Dunford [211] measured the rate constants for the reactions of MPO Compounds I and II with clozapine and ascorbate (Table 22.1) and concluded that ascorbate was a free radical scavenger and a competitive inhibitor of the oxidation of this diazepine derivative. Morehouse et al. [212] demonstrated that some natural and synthetic water-soluble porphyrins were oxidized to corresponding cation radicals in the reactions catalyzed by HRP, MPO, and LPO. Huang et al. [213] studied the oxidation of hydroxyurea by HRP. It was found that nitroxide radical and a C-nitroso compound are the intermediates of this reaction, which yielded nitric oxide together with carbon dioxide, ammonia, nitrate, and nitrite as final products.

22.4.4 Reactions with Inorganic Compounds

It has already been noted that hypochlorous acid HOCl is a main physiological substrate of MPO, playing an important cytotoxic role against invading bacteria and viruses. In 1976,

TABLE 22.2
Rate Constants ($l\ mol^{-1}\ s^{-1}$) for the Reactions of Compounds I and II with Inorganic Compounds

Peroxidase	Substrate	Compound I	Compound II
MPO	Chloride	$(4.7 \pm 0.1) \times 10^6$ [206]	
HRP	Iodide	4.2×10^3 [205]	
Lactoperoxidase	Iodide	1.6×10^7 [205]	
HRP	NO	$(7.0 \pm 0.3) \times 10^5$ [253]	$(0.7–1.3) \times 10^6$ [253]
MPO	Nitrite	$(2.0 \pm 0.2) \times 10^6$ [244]	$(5.5 \pm 0.1) \times 10^2$ [244]
HRP	Peroxynitrite		3×10^{6a} [249]
MPO	Peroxynitrite		2×10^{7a} [248]

[a]Two-electron transfer reaction.

Harrison and Schultz [214] suggested that MPO is able to catalyze the oxidation of chloride to HOCl. Later on, it was found that the rate constant for Reaction (16) is relatively high (Table 22.2) that together with very high (0.10–0.1 $mol\ l^{-1}$) concentrations of chloride in human blood makes HOCl an effective damaging agent of biomolecules. It is interesting that only MPO is an effective oxidant of chloride ion while HRP and LPO react with chloride very slowly.

Another important example of catalytic oxidation of inorganic compounds by peroxidases is the catalysis of iodide oxidation by TPO. TPO is involved in the biosynthesis of thyroid hormone and catalyzes the reactions of iodination and coupling in the thyroid gland. Magnusson et al. [215] considered two possible pathways of iodination: the formation of enzyme-bound hypoiodite and the formation of free hypoiodide (Reactions (17) and (18)):

$$\text{Compound I} + I^- \Longrightarrow \text{Compound}(I^-) \quad (17)$$

$$\text{Compound}(I^-) + H_2O \Longrightarrow \text{TPO} + \text{HOI} + HO^- \quad (18)$$

Taurog et al. [216] showed that contrary to previous suggestions, both iodination and coupling are catalyzed by the oxoferryl porphyrin π-cation radical of TPO Compound I and not the oxoferryl protein radical. HRP catalyzed the oxidation of bisulfite to sulfate with the intermediate formation of sulfur trioxide radical anion $SO_3^{\bullet-}$ [217]; HPO, MPO, LPO, chloroperoxidase, NADH peroxidase, and methemoglobin oxidized cyanide to cyanyl radical [218].

22.4.5 Mechanism of the Interaction of Peroxidases with Hydroperoxides. Role of Superoxide and Hydroxyl Radicals in Reactions Catalyzed by Peroxidases

It has been pointed out earlier that peroxidases oxidize hydrogen peroxide by two-electron transfer mechanism to form Compound I. Thus for MPO, we have:

$$\text{MPO} + H_2O_2 \Longrightarrow \text{Compound I} + H_2O \quad (8)$$

However, in addition to two-electron oxidation by native peroxidase, Compound I can oxidize hydrogen peroxide by one-electron mechanism:

$$\text{Compound I} + H_2O_2 \Longrightarrow \text{Compound II} + HOO^{\bullet} \quad (19)$$

TABLE 22.3
Rate Constants (l mol^{-1} s^{-1}) for Two-Electron Reaction (8) and One-Electron Reaction (19) of MPO and MPO Compound I with Hydroperoxides

Hydroperoxide	k_8	k_{19}
Hydrogen peroxide	2.3×10^7 [219]	3.5×10^4 [222]
	1.8×10^7 [220]	8.2×10^4 [220]
	$(1.4 \pm 0.1) \times 10^7$ [221]	$(4.4 \pm 0.2) \times 10^4$ [221]
Peroxyacetic acid	$(2.7 \pm 0.1) \times 10^6$ [221]	$(3.1 \pm 0.4) \times 10^3$ [221]
Ethyl hydroperoxide	2.8×10^5 [219]	
	$(5.2 \pm 0.2) \times 10^5$ [221]	$(1.0 \pm 0.01) \times 10^4$ [221]
t-Butyl hydroperoxide	$(7.3 \pm 0.1) \times 10^3$ [221]	$(1.3 \pm 0.01) \times 10^2$ [221]
3-Chloroperoxybenzoic acid	$(1.8 \pm 0.1) \times 10^7$ [221]	$(6.6 \pm 0.3) \times 10^4$ [221]
Cumene hydroperoxide	$(1.2 \pm 0.1) \times 10^6$ [221]	$(1.6 \pm 0.2) \times 10^4$ [221]

The rate constants for two-electron reaction (8) (k_8) and one-electron reaction (19) (k_{19}) are cited in Table 22.3. As seen from Table 22.3, k_8 values are about 100–1000 times greater than k_{19} values; therefore, the production of superoxide (or hydroperoxyl radical) by peroxidases might play an insignificant role compared to the two-electron oxidation of hydroperoxides.

At the same time the interaction of superoxide with MPO may affect a total superoxide production by phagocytes. Thus, the superoxide adduct of MPO (Compound III) is probably quantitatively formed in PMA-stimulated human neutrophils [223]. Edwards and Swan [224] proposed that superoxide production regulate the respiratory burst of stimulated human neutrophils. It has also been suggested that the interaction of superoxide with HRP, MPO, and LPO resulted in the formation of Compound III by a two-step reaction [225]. Superoxide is able to react relatively rapidly with peroxidases and their catalytic intermediates. For example, the rate constant for reaction of superoxide with Fe(III)MPO is equal to $1.1–2.1 \times 10^6$ l mol^{-1} s^{-1} [226], and the rate constants for the reactions of $O_2^{\bullet-}$ and HOO$^{\bullet}$ with HRP Compound I are equal to 1.6×10^6 and 2.2×10^8 l mol^{-1} s^{-1}, respectively [227]. Thus, peroxidases may change their functions, from acting as prooxidant enzymes and the catalysts of free radical processes, and acquire antioxidant catalase properties as shown for HRP [228] and MPO [229]. In this case catalase activity depends on the two-electron oxidation of hydrogen peroxide by Compound I.

In spite of a relatively slow rate of superoxide release by peroxidases (Reaction (19), Table 22.3), these processes apparently cannot be wholly excluded from the consideration of oxidative reactions catalyzed by peroxidases. For example, the participation of superoxide in adrenaline, melatonin, and tryptophan oxidation has been suggested [201,230]. On the other hand, Kettle et al. [231] demonstrated that superoxide may accelerate the MPO-catalyzed oxidation of antiinflammatory drugs by reducing Compound II back to the active enzyme. At the same time, considerable interest presents the possibility of superoxide-mediated formation of hydroxyl radicals in these processes. In Chapter 21, we already discussed the mechanism of hydroxyl radical production by the Fenton reaction catalyzed by the perferryl Compound III of HRP proposed by Chen and Schopfer [232]. However, it should be noted here that the role of hydroxyl radicals in peroxidase oxidative reactions was suggested much earlier. In 1957, Mason et al. [233] suggested that Compound III takes part in hydroxylation of organic compounds. In contrast, Winterbourn [234] concluded that MPO actually inhibited hydroxyl production by the reaction with hydrogen peroxide, making the Fenton reaction insignificant.

Subsequent studies confirmed that there is no reliable evidence of the MPO-catalyzed hydroxyl generation by neutrophils [235,236]. Kettle and Winterbourn [237] demonstrated that the hydroxylation of salicylate by stimulated neutrophils or the purified MPO, which yielded 2,5-dihydroxybenzoate, was unaffected by hydroxyl radical scavengers mannitol or DMSO. These authors suggested that an active peroxidative agent in this system was the reduced Compound III, formed by the following reactions:

$$\text{Fe(III)MPO} + O_2^{\bullet-} \Longrightarrow \text{MPO Compound III}\{\text{Fe(II)}O_2 \Longleftrightarrow \text{Fe(III)}O_2^{\bullet-}\} \quad (20)$$
$$\text{MPO Compound III} + O_2^{\bullet-} \Longrightarrow \text{MPOFe(II)}O_2^{\bullet-} + O_2 \quad (21)$$

It should be mentioned that the MPOFe(II)$O_2^{\bullet-}$ intermediate is very similar to the analogous hydroxylating species of cytochrome P450. Thus, the participation of hydroxyl radicals in the reactions catalyzed by peroxidases remains questionable.

22.4.6 PEROXIDASE-CATALYZED OXIDATIVE PROCESSES

As follows from data discussed above, peroxidases initiate free radical-mediated oxidative processes by different ways: producing free radicals by the direct interaction of Compounds I and II with biomolecules, releasing superoxide, and generating relatively active secondary free radicals (for example, tyrosyl radical), capable of reacting with neutral substrates (NADH, glutathione) to form new free radicals. The last of these reactions have been discussed in detail earlier [238]. It should be noted that not only organic free radicals may participate in free radical-mediated processes catalyzed by peroxidases. Thus Marquez and Dunford [206] have shown that MPO catalyzes chlorination of taurine, one of the most abundant amino acid, to chlorotaurine, by forming the enzyme–Cl$^-$ chlorinating complex as intermediate:

$$\text{Complex I} + \text{Cl}^- \Longrightarrow \text{Complex I} - \text{Cl}^- \quad (22)$$
$$\text{Complex I} - \text{Cl}^- + \text{taurine} \Longrightarrow \text{MPO} + \text{chlorotaurine} \quad (23)$$

Peroxidases are effective initiators of lipid peroxidation and LDL oxidation. These processes are considered in detail in Chapter 25.

22.4.7 INTERACTION WITH NITROGEN OXIDES

For a long time a major function of MPO in mammalians has been well established — the oxidation of chloride ions to HOCl, a strong bactericidal agent produced by neutrophils [239]. Just recently, another important property of MPO has been demonstrated — the participation in the nitric oxide metabolism. It was found that MPO and other peroxidases are capable of reacting with nitrogen species to form nitric oxide, playing a role of NO synthases. It has been shown that HRP is able to oxidize nitrite ion similar to the reactions with chloride or iodide [240]. Due to a significant concentration of nitrite in biological fluids, (which is about $1-200 \times 10^{-6}$ mol l^{-1} under physiological conditions and may be much more during inflammatory processes), such process is probably possible in vivo. The oxidation of nitrite can result in the formation of nitrogen dioxide free radical [240]:

$$\text{Compound I} + \text{NO}_2^- \Longrightarrow \text{Compound II} + {}^{\bullet}\text{NO}_2 \quad (24)$$
$$\text{Compound II} + \text{NO}_2^- \Longrightarrow \text{HPR} + {}^{\bullet}\text{NO}_2 \quad (25)$$

van der Vliet et al. [241] accepted this mechanism for the oxidation of nitrite by MPO and LPO and concluded that ${}^{\bullet}\text{NO}_2$ is capable of nitrating phenolic compounds such as tyrosine.

Furthermore, it was suggested that nitrite is a physiological substrate for mammalian peroxidases because it may compete with other physiological substrates (Cl^-, Br^-, and SCN^-) in the reactions catalyzed by MPO, EPO, and LPO.

The ability of MPO to catalyze the nitration of tyrosine and tyrosyl residues in proteins has been shown in several studies [241–243]. However, nitrite is a relatively poor nitrating agent, as evident from kinetic studies. Burner et al. [244] measured the rate constants for Reactions (24) and (25) (Table 22.2) and found out that although the oxidation of nitrite by Compound I (Reaction (24)) is a relatively rapid process at physiological pH, the oxidation by Compound II is too slow. Nitrite is a poor substrate for MPO, at the same time, is an efficient inhibitor of its chlorination activity by reducing MPO to inactive Complex II [245]. However, the efficiency of MPO-catalyzing nitration sharply increases in the presence of free tyrosine. It has been suggested [245] that in this case the relatively slow Reaction (26) ($k_{26} = 3.2 \times 10^5$ 1 mol^{-1} s^{-1} [246]) is replaced by rapid reactions of Compounds I and II with tyrosine, which accompanied by the rapid recombination of tyrosyl and $NO_2^{\bullet}$ radicals with a k_{27} equal to 3×10^9 1 mol^{-1} s^{-1} [246].

$$NO_2^{\bullet} + tyrosineH \Longrightarrow NO_2^- + tyrosine^{\bullet} + H^+ \qquad (26)$$

$$NO_2^{\bullet} + tyrosine^{\bullet} \Longrightarrow tyrosine\text{-}NO_2 \qquad (27)$$

Recently, Pfeiffer et al. [247] suggested that the peroxidase oxidation of nitrite may be the main source of nitration of protein tyrosine in macrophages. It was found that the maximal formation of protein-bound 3-nitrotyrosine in cytokine-activated murine macrophages was observed only after a 20–24 h poststimulation while NO was released at 6–9 h and superoxide was at 1–5 h. It follows that peroxynitrite formed from superoxide and NO plays a minor role in this process. These findings contradict a widely accepted opinion about the exclusive role of peroxynitrite in tyrosine nitration (Chapter 21). However, it should be noted that macrophage heme peroxidases are still poorly characterized, and therefore, further studies are needed to clarify their role in nitration processes. The nitrogen dioxide free radical generated by Reactions (24) and (25) is also able to oxidize other biomolecules such as melanin [248], NADH, NADPH, cysteine, glutathione, ascorbate, and Trolox C [249].

Another physiological substrate for peroxidases is peroxynitrite. Floris et al. [250] have shown that peroxynitrite rapidly reacts with HRP and MPO, forming Compound II without the intermediate generation of Compound I. These authors proposed that the intermediate complex Compound I–NO_2^- immediately decomposed to Compound II and $NO_2^{\bullet}$ (Reaction (28)).

$$MPO + ONOOH \Longrightarrow [Compound\ I\text{-}NO_2^-] \Longrightarrow Compound\ II + NO_2^{\bullet} + H^+ \qquad (28)$$

Grace et al. [251] demonstrated that the oxidation of peroxynitrite by HRP sharply accelerated in the presence of chlorogenic acid.

The most important physiological nitrogen substrate of peroxidases is undoubtedly nitric oxide. In 1996, Ishiropoulos et al. [252] suggested that nitric oxide is able to interact with HRP Compounds I and II. Glover et al. [253] measured the rate constants for the reactions of NO with HRP Compounds I and II (Table 22.2) and proposed that these reactions may occur in in vivo inflammatory processes. The interaction of NO with peroxidases may proceed by two ways: through the NO one-electron oxidation or the formation of peroxidase–NO complexes. One-electron oxidation of nitric oxide will yield nitrosonium cation NO^+ [253,254], which is extremely unstable and rapidly hydrolyzed to nitrite. On the other hand, in the presence of high concentrations of nitric oxide and the competitor ligand Cl^-, the formation of peroxidase–NO complexes becomes more favorable. It has been shown [255] that NO forms stable six-coordinate Fe(III) and Fe(II) nitrosyl complexes of MPO. Abu-Soud and Hazen [254] suggested that peroxidases may be considered as a catalytic sink for nitric oxide, limiting its activity in vivo.

The present interest in the mechanism of the interaction of nitric oxide with MPO is explained by the critical role of MPO in host defenses and inflammatory tissue injury. MPO and inducible NO synthase present in primary granules of neutrophils [256] and during phagocytosis are secreted together into the phagolysosome and extracellular space. It has been found that NO successfully competes with physiological concentrations of chloride for MPO; therefore, both HOCl and active nitrogen species ($^{\bullet}NO_2$) must participate in destroying invading bacteria. It is important that physiological levels of tyrosine and ascorbate accelerates NO reaction with MPO [257] supposedly via the rapid reactions of tyrosyl and ascorbate radicals with nitric oxide, similar to nitration by nitrite in the presence of tyrosine [245]. These findings also indicate the ability of MPO to affect the NO-dependent processes during inflammation.

In conclusion, it should be noted that XO, NADPH oxidase, NO synthase, and peroxidases are major but not only free radical-generating enzymes. For example, superoxide participates in the oxidative ring cleavage of various indoleamine derivatives by indoleamine 2,3-dioxygenase [258]; the ascorbate-dependent plasma membrane NADPH oxidase of synaptosomes is a source of neuronal superoxide [259] and so on. Such examples can undoubtedly be broadened. Furthermore, oxygen radical-producing enzymes of mitochondria and microsomes are considered in the next chapters. It was already pointed out that only enzymes responsible for the initiation of free radical processes were discussed in this chapter, while the enzymatic free radical oxidative processes where free radicals participate in all stages of catalytic cycle are considered in Chapter 26 dedicated to lipid peroxidation catalyzed by cyclooxygenase and lipoxygenase.

REFERENCES

1. JM McCord, I Fridovich. *J Biol Chem* 243: 5753–5760, 1968.
2. LM McCord, I Fridovich. *J Biol Chem* 244: 6056–6063, 1969.
3. JM McCord, I Fridovich. *J Biol Chem* 244: 6049–6055, 1969.
4. I Fridovich, P Handler. *J Biol Chem* 236: 1836–1840, 1961.
5. PF Knowles, JF Gibson, FM Pick, RC Bray. *Biochem J* 111: 53–56, 1969.
6. RC Bray, FM Pick, D Samuel. *Eur J Biochem* 15: 352, 1970.
7. GR Buettner, LW Oberley. *Biochem Biophys Res Commun* 83: 69–73, 1978.
8. V Massey, PE Brumby, H Komai, G Palmer. *J Biol Chem* 244: 1682–1691, 1969.
9. JS Olson, DP Ballou, G Palmer, V Massey. *J Biol Chem* 249: 4350–4362, 1974.
10. I Fridovich. *J Biol Chem* 245: 4053–4057, 1970.
11. AG Porras, JS Olson, G Palmer. *J Biol Chem* 256: 9096–9100, 1981.
12. T Nagano, I Fridovich. *J Free Radic Biol Med* 1: 39–42, 1095.
13. MD Davis, JS Olson, G Palmer. *J Biol Chem* 259: 3526–3533, 1984.
14. H Rubbo, R Radi, E Prodanov. *Biochim Biophys Acta* 1074: 386–392, 1991.
15. R Radi, S Tan, E Prodanov, RA Evans, DA Park. *Biochim Biophys Acta* 1122: 178–182, 1992.
16. S Tan, R Radi, F Gaudier, RA Evans, A Rivera, KA Kirk, DA Parks. *Pediatr Res* 34: 303–307, 1993.
17. AM Ghe, C Stefanilli, P Tsintiki, G Veschi. *Talanta* 32: 359–362, 1985.
18. LS Terada, CJ Beehler, A Banerjee, JM Brown, MA Grosso, AH Harken, JM McCord, JE Repine. *J Appl Physiol* 65: 2349–2353, 1988.
19. JM McCord. *N Engl J Med* 312: 159–163, 1985.
20. T Nishino, T Nishino, LM Schopfer, V Massey. *J Biol Chem* 264: 2518–2527, 1989.
21. CM Harris, V Massey. *J Biol Chem* 272: 8370–8379, 1997.

22. SA Sanders, R Eisenthal, R Harrison. *Eur J Biochem* 245: 541–548, 1997.
23. C Beauchamp, I Fridovich. *J Biol Chem* 245: 4641–4646, 1970.
24. P Kuppusamy, JL Zweier. *J Biol Chem* 264: 9880–9884, 1989.
25. RV Lloyd, RP Mason. *J Biol Chem* 265: 16733–16736, 1990.
26. BE Britigan, S Pou, GM Rosen, DM Lilleg, GB Buettner. *J Biol Chem* 265: 17533–17538, 1990.
27. T Mikuni, M Tatsuta. *Free Radic Res* 36: 641–647, 2002.
28. CC Winterbourn. *Arch Biochem Biophys* 209: 159–167, 1981.
29. IB Afanas'ev, LG Korkina, TB Suslova, SK Soodaeva. *Arch Biochem Biophys* 281: 245–250, 1990.
30. KJ Lusthof, W Richter, NJ de Mol, LH Janssen, W Verboom, DN Reinhoudt. *Arch Biochem Biophys* 277: 137–142, 1990.
31. IB Afanas'ev, EA Ostrachovitch, LG Korkina. *Arch Biochem Biophys* 366: 267–274, 1999.
32. M Dixon, S Thurlow. *Biochem J* 18: 989–992, 1924.
33. TM Millar, CR Stevens, N Benjamin, R Eisenthal, R Harrison, DR Blake. *FEBS Lett* 427: 225–228, 1997.
34. Z Zang, D Naughton, PG Winyard, N Benjamin, DR Blake, MCR Symons. *Biochem Biophys Res Commun* 249: 767–772, 1998.
35. BLJ Godber, JJ Doel, GP Sapkota, DR Blake, CR Stevens, R Eisenthal, R Harrison. *J Biol Chem* 275: 7757–7763, 2000.
36. M Houston, P Chumley, R Radi, H Rubbo, BA Freeman. *Arch Biochem Biophys* 355: 1–8, 1998.
37. K Ichimori, M Fukahori, H Nakazawa, K Okamoto, T Nishino. *J Biol Chem* 274: 7763–7768, 1999.
38. C Lee, X Liu, JL Zweier. *J Biol Chem* 275: 9369–9376, 2000.
39. A Kooij, HJ Schiller, M Schijns, CJF van Noorden, WM Frederriks. *Hepatology* 19: 1488–1495, 1994.
40. DE Chambers, DA Parks, G Patterson, R Roy, JM McCord, S Yoshida, LF Parmley, JM Downey. *J Mol Cell Cardiol* 17: 145–152, 1985.
41. DJ Hearse, AS Manning, JM Downney, DM Yellow. *Acta Physiol Scand Suppl* 548: 65–78, 1986.
42. ML Charlat, PG O'Neill, JM Egan, DR Abernethy, LH Michael, ML Myers, R Roberts, R Bolli. *Am J Physiol* 252: H566–H577, 1987.
43. DK Das, RM Engelman, R Clement, H Otani, MR Prasad, PS Rao. *Biochem Biophys Res Commun* 148: 314–319, 1987.
44. SM Smith, MB Grisham, EA Manci, DN Granger, PR Kvietys, JM Russell. *Gastroenterology* 92: 950–956, 1987.
45. DA Park, GB Bulkley, DN Granger, SR Hamilton, JM McCord. *Gastroenterology* 82: 9–15, 1982.
46. M Rouquette, S Page, R Bryant, M Banboubetra, CR Stevens, DR Blake, WD Whish, R Harrison, D Tosh. *FEBS Lett* 426: 397–401, 1998.
47. S Vickers, HJ Schiller, JE Hildreth, GB Bulkley. *Surgery* 124: 551–560, 1998.
48. CA Partridge, FA Blumenstock, AB Malik. *Arch Biochem Biophys* 294: 184–187, 1992.
49. JJ Zulueta, R Sawhney, TS Yu, CC Cote, PM Hassoun. *Am J Physiol* 272: L897–L902, 1997.
50. S Wu, N Tamaki, T Nagashima, M Yamaguchi. *Neurosurgery* 43: 577–583, 1998.
51. JW Beetsch, TS Park, LL Dugan, AR Shah, JM Gidday. *Brain Res* 786: 89–95, 1998.
52. M Houston, A Estevez, P Chumley, M Aslan, S Marklund, DA Parks, BA Freeman. *J Biol Chem* 274: 4985–4994, 1999.
53. A Paler-Martinez, PC Panus, PH Chumley, U Ryan, MM Hardy, BA Freeman. *Arch Biochem Biophys* 311: 79–85, 1994.
54. Z Zhang, DR Blake, CR Stevens, JM Kanczler, PG Winyard, MC Symons, M Benboubetra, R Harrison. *Free Radic Res* 28: 151–164, 1998.
55. Y Xia, IL Zweier. *J Biol Chem* 270: 18797–18803, 1995.
56. F Rossi. *Biochim Biophys Acta* 853: 65–89, 1986.
57. BM Babior. *Blood* 93: 1464–1476, 1999.
58. KK Griendling, D Sorescu, M Ushio-Fukai. *Circ Res* 86: 494–501, 2000.
59. ML Karnovsky. *Physiol Rev* 42: 143–168, 1962.
60. RJ Selvaraj, AJ Sbarra. *Nature* 211: 1272–1276, 1966.
61. F Rossi, M Zatti. *Br J Exp Pathol* 45: 548–559, 1964.
62. BM Babior, RS Kipnes, JT Curnutte. *J Clin Invest* 52: 741–744, 1973.

63. AR Cross, RW Erickson, JT Curnutte. *Biochem J* 341: 251–255, 1999.
64. Y Nisimoto, H Otsuka-Murakami, DJ Lambeth. *J Biol Chem* 270: 16428–16433, 1995.
65. AR Cross, J Rae, JT Curnutte. *J Biol Chem* 270: 17075–17079, 1995.
66. O Inanami, JL Johnson, JK McAdara, JE Benna, LR Faust, PE Newburger, BM Babior. *J Biol Chem* 273: 9539–9543, 1998.
67. A Shiose, H Sumimoto. *J Biol Chem* 275: 13793–13801, 2000.
68. AR Cross, JT Curnutte. *J Biol Chem* 270: 6543–6549, 1995.
69. LM Henderson. *J Biol Chem* 273: 33216–33223, 1998.
70. I Hazan, R Dana, Y Granot, R Levy. *Biochem J* 326: 867–876, 1997.
71. T Yamamori, O Inanami, H Sumimoto, T Akasaki, H Nagahata, M Kuwabara. *Biochem Biophys Res Commun* 293: 1571–1578, 2001.
72. C Lee, K Miura, X Liu, JL Zweier. *J Biol Chem* 275: 38965–38972, 2000.
73. K Arakawa, H Takahashi, S Nakagawa, S Ogawa. *Aneth Analg* 93: 1501–1506, 2001.
74. IB Afanas'ev. In: *Superoxide Ion: Chemistry and Biological Implications*, vol 2. CRC Press, Boca Raton, FL, pp. 101–116, 1990.
75. A Someya, K Nishijima, H Nunoi, S Irie, I Nagaoka. *Arch Biochem Biophys* 345: 207–213, 1997.
76. OT Lynch, MA Giembycz, PJ Barnes, MA Lindsay. *Br J Pharmacol* 34: 797–806, 2001.
77. JL Bankers-Fulbright, H Kita, GJ Gleich, SM O'Grady. *J Cell Physiol* 189: 306–315, 2001.
78. J Schrenzel, L Serrander, B Banfi, O Nusse, R Fouyouzi, DP Lew, N Demaurex, KH Krause. *Nature* 392: 734–737, 1998.
79. V Cherny, LM Henderson, W Xu, L Thomas, TE DeCoursey. *J Physiol* 535: 783–794, 2001.
80. TE DeCoursey, V Cherny, AG DeCoursey, W Xu, L Thomas L. *J Physiol* 535: 767–781, 2001.
81. X Zhao, EA Bey, FB Wientjes, MK Cathcart. *J Biol Chem* 277: 25385–25392, 2002.
82. KK Griendling, D Sorescu, M Ushio-Fukai. *Circ Res* 86: 494–501, 2000.
83. SA Jones, VB O'Donnell, JP Broughton, EJ Hughes, OTG Jones. *Am L Physiol* 271: H1626–H1634, 1996.
84. JW Meyer, JA Holland, LM Ziegler, MM Chang, G Beebe, ME Schmitt. *Endothelium* 7(1): 11–22, 1999.
85. HD Wang, PJ Pagano, Y Du, AJ Cayatte, MT Quinn, P Brecher, RA Cohen. *Circ Res* 82: 810–818, 1998.
86. U Bayraktutan, N Draper, D Lang, AM Shah. *Cardiovasc Res* 38: 256–262, 1998.
87. L Xiao, DR Pimentel, J Wang, K Singh, WS Colucci, DB Sawyer. *Am J Physiol Cell Physiol* 282: C926–C934, 2002.
88. D Javesghani, SA Magder, E Barreiro, MT Quinn, SN Hussain. *Am J Respir Care Med* 165: 412–418, 2002.
89. M Ushio-Fukai, AM Zafari, T Fukui, N Ishizaka, KK Griendling. *J Biol Chem* 271: 23317–23321, 1996.
90. A Gorlach, RP Brandes, K Nguyen, M Amidi, F Dehghani, R Busse. *Circ Res* 87: 26–32, 2000.
91. B Hohler, B Holzapfel, W Kummer. *Histochem Cell Biol* 114: 29–37, 2000.
92. JM Li, AM Shah. *Cardiovasc Res* 52: 477–486, 2001.
93. LM Li, AM Mullen, S Yun, F Wientjes, GY Brouns, AJ Thrasher, AM Shah. *Circ Res* 90: 143–150, 2002.
94. Y Suh, RS Arnold, B Lassegue, J Shi, X Xu, D Sorescu, AB Chung, KK Griendling, JD Lamberth. *Nature* 401: 79–82, 1999.
95. B Lassegue, D Sorescu, K Szoes, Q Yin, M Akers, Y Zhang, SL Grant, JD Lambeth, KK Griendling. *Circ Res* 88: 858–860, 2001.
96. HP Souza, FRM Laurindo, RC Ziegelstein, CO Berlowitz, JL Zweier. *Am J Physiol Heart Circ Physiol* 280: H658–H667, 2001.
97. M Janiszewski, HP Souza, X Liu, MA Pedro, JL Zweier, FR Laurindo. *Free Radic Biol Med* 32: 446–453, 2002.
98. T Munzel, IB Afanas'ev, A Kleschev, DG Harrison. *Arterioscler Throm Vasc Biol* 22: 1761–1768, 2002.
99. D Sorescu, MJ Somers, B Lassegue, S Grant, DG Harrison, KK Griendling. *Free Radic Biol Med* 30: 603–612, 2001.

100. HP Souza, X Liu, A Samouilov, P Kuppusamy, FRM Laurindo, JL Zweier. *Am J Physiol Heart Circ Physiol* 282: H466–H474, 2002.
101. B Meyer. *Adv Exp Med Biol* 387: 113–116, 1996.
102. B Meier, HH Radeke, S Selle, M Younes, H Sies, K Resch, GG Habermehl. *Biochem J* 263: 539–545, 1989.
103. JA Holland, RW O'Donnell, MM Chang, DK Johnson, LM Ziegler. *Endothelium* 7: 109–119, 2000.
104. MC Lavigne, HL Malech, SM Holland, TL Leto. *Circulation* 104: 79–84, 2001.
105. RP Brandes, M Barton, KM Philippens, G Schweitzer, A Mugge. *J Physiol* 500: 331–342, 1997.
106. GW De Keulenaer, RW Alexander, M Ushio-Fukai, N Ishizaka, KK Griendling. *Biochem J* 329: 653–657, 1998.
107. K Griendling, JD Ollerenshaw, CA Minieri, RW Alexander. *Circ Res* 74: 1141–1148, 1994.
108. S Rajagopalan, S Kurz, T Munzel, M Tarpey, BA Freeman, KK Griendling, DG Harrison. *J Clin Invest* 97: 1916–1923, 1996.
109. H Zhang, A Schmeisser, CD Garlichs, K Plotze, U Damme, A Mugge, WG Daniel. *Cardiovasc Res* 44: 215–222, 1999.
110. T Fukui, N Ishizaka, S Rajagopalan, JB Laursen, Q Capers, WR Taylor, DG Harrison, H de Leon, JN Wilcox, KK Griendling. *Circ Res.* 80: 45–51, 1997.
111. G Zalba, FJ Beaumont, GS Jose, A Fortuno, MA Fortuno, JC Etayo, J Diez. *Hypertension* 35: 1055–1061, 2000.
112. PJ Pagano, JK Clark, ME Cifuentes-Pagano, SM Clark, GM Callis, MT Quinn. *Proc Natl Acad Sci USA* 94: 14483–14488, 1997.
113. PJ Pagano, SJ Chanock, DA Siwik, WS Colucci, JK Clark. *Hypertension* 32: 331–337, 1998.
114. AM Zafari, M Ushio-Fukai, CA Minieri, M Akers, B Lassegue, KK Griendling. *Antioxid Redox Signal* 1: 167–179, 1999.
115. RM Touyz, EL Schiffrin. *Hypertension* 34: 976–982, 1999.
116. Z Xie, DR Pimental, S Lohan, A Vasertriger, C Pligavko, WS Colucci, K Singh. *J Cell Physiol* 188: 132–138, 2001.
117. H Mollnau, M Wendt, K Szoes, B Lassegue, E Schulz, M Oelze, H Li, M Bodenschatz, M August, AL Kleschyov, N Tsilimingas, U Walter, U Forstermann, T Meinertz, K Griendling, T Munzel. *Circ Res* 90: E58–E65, 2002.
118. WG Li, FJ Miller, Jr, HJ Zhang, DR Spitz, LW Oberley, NJ Weintraub. *J Biol Chem* 276: 29251–29256, 2001.
119. AK Bhunia, H Han, A Snowden, S Chatterjee. *J Biol Chem* 272: 15642–15649, 1997.
120. T Inoguchi, P Li, F Umeda, HY Yu, M Kakimoto, M Imamura, T Aoki, T Etoh, T Hashimoto, M Naruse, H Sano, H Utsumi, H Nawata. *Diabetes* 49: 1939–1945, 2000.
121. S Takeshita, N Inoue, D Gao, Y Rikitake, S Kawashima, R Tawa, H Sakurai, Yokoyama. *J Atheroscler Thromb* 7: 238–246, 2000.
122. A Heinloth, K Heermeier, U Raff, C Wanner, J Galle. *J Am Soc Nephrol* 11: 1819–1825, 2000.
123. N Duerrschmidt, N Wippich, W Goettsch, HJ Broemme, H Morawietz. *Biochem Biophys Res Commun* 269: 713–717, 2000.
124. DR Pimentel, JK Amin, L Xiao, T Miller, J Viereck, J Oliver-Krasinski, R Baliga, J Wang, DA Siwik, K Singh, P Pagano, WS Colucci, DB Sawyer. *Circ Res* 89: 453–460, 2001.
125. JA Holland, JW Meyer, MM Chang, RW O'Donnell, DK Johnson, LM Ziegler. *Endothelium* 6: 113–121, 1998.
126. C Patterson, J Ruef, NR Madamanchi, P Barry-Lane, Z Hu, C Horaist, CA Ballinger, AR Brasier, C Bode, MS Runge. *J Biol Chem* 274: 19814–19822, 1999.
127. RP Brandes, C Viedt, K Nguyen, S Beer, J Kreuzer, R Busse, A Gorlach. *Thromb Haemost* 85: 1104–1110, 2001.
128. EL Greene, V Velarde, AA Jaffe. *Hypertension* 35: 942–947, 2000.
129. C Marshall, AJ Mamary, AJ Verhoeven, BE Marshall. *Am J Respir Cell Mol Biol* 15: 633–644, 1996.
130. AS Baas, BC Berk. *Circ Res* 77: 29–36, 1995.

131. C Viedt, HI Krieger-Brauer, J Fei, C Elsing, W Kubler, J Kreuzer. *Arterioscler Thromb Vasc Biol* 20: 940–949, 2000.
132. M Sano, K Fukuda, T Sato, H Kawaguchi, M Suematsu, S Matsuda, S Koyasu, H Matsui, K Yamauchi-Takihara, M Harada, Y Saito, S Ogawa. *Circ Res* 89: 661–669, 2001.
133. M Ushio-Fukai, RW Alexander, M Akers, KK Griendling. *J Biol Chem* 273: 15022–15029, 1998.
134. SS Brar, TP Kennedy, AB Sturrock, TP Hueckstead, MT Quinn, TM Murphy, P Chitano, JR Hoidal. *Am J Physiol Lung Cell Mol Physiol* 282: L782–795, 2002.
135. TJ Guzik, NE West, E Black, D McDonald, C Ratnatunga, R Pillai, KM Channon. *Circ Res* 86: E85–E90, 2000.
136. AH Wagner, MR Schroeter, M Hecker. *FASEB J* 15: 2121–2130, 2001.
137. TR Billiar, RD Curran, DJ Stuehr, MA West, BG Bentz, RL Simmons. *J Exp Med* 169: 1467–1472, 1989.
138. RD Curran, TR Billiar, DJ Stuehr, K Hofmann, RL Simmons. *J Exp Med* 170: 1769–1774, 1989.
139. Y Yui, R Hattori, K Kosuga, H Eizawa, K Hiki, S Ohkawa, K Ohnigh, S Terao, C Kawai. *J Biol Chem* 266: 3369–3371, 1991.
140. DM Zardini, EJ Tschirhart. *Inflamm Res* 50: 357–361, 2001.
141. A Bal-Price, A Matthias, GC Brown. *J Neurochem* 80: 73–80, 2002.
142. LJ Ignarro, GM Buga, KS Wood, RE Byrn, G Chaudhuri. *Proc Natl Acad Sci USA* 84: 9265–9269, 1987.
143. J MacMicking, G Xie, C Nathan. *Ann Rev Immunol* 15: 323–350, 1997.
144. XJ Girerd, AT Hirsch, JP Cooke, VJ Dzau, MA Creager. *Circ Res* 67: 1301–1308, 1990.
145. U Siddhanta, A Presta, B Fa, D Wolan, DL Rousseau, DJ Stiehr. *J Biol Chem* 273: 18950–18958, 1998.
146. U Siddhanta, C Wu, HM Abu-Soud, J Zhang, DK Ghosh, DJ Stuehr. *J Biol Chem* 271: 7309–7312, 1996.
147. S Pou, L Keaton, W Surichamorn, GM Rosen. *J Biol Chem* 274: 9573–9580, 1999.
148. S Pou, WS Pou, DS Bredt, SH Snyder, GM Rosen. *J Biol Chem* 267: 24173–24176, 1992.
149. HM Abu-Soud, R Gachhui, FM Raushel, DJ Stuehr. *J Biol Chem* 272: 17349–17353, 1997.
150. Y Xia, AL Tsai, V Berka, JL Zweier. *J Biol Chem* 273: 25804–25808, 1998.
151. Y Xia, LJ Roman, BSS Masters, JL Zweier. *J Biol Chem* 273: 22635–22639, 1998.
152. AR Hurshman, C Krebs, DE Edmondson, BH Huynh, MA Marletta. *Biochemistry* 38: 15689–15696, 1999.
153. PP Schmidt, R Lange, AC Gorren, ER Werner, B Mayer, KK Andersson. *J Biol Inorg Chem* 6: 151–158, 2001.
154. S Daff, MA Noble, DH Craig, SL Rivers, SK Chapman, AW Munro, S Fujiwara, E Rozhkova, I Sagami, T Shimizu. *Biochem Soc Trans* 29: 147–152, 2001.
155. RA Pufahl, JS Wishnok, MA Marletta. *Biochemistry* 34: 1930–1941, 1995.
156. HHHW Schmidt, H Hofmann, U Schindler, ZS Shutenko, DD Cunningham, M Feelisch. *Proc Natl Acad Sci USA* 93: 14492–14497, 1996.
157. S Adak, Q Wang, DJ Stuehr. *J Biol Chem* 275: 33554–33561, 2000.
158. S Porasuphatana, P Tsai, S Pou, GM Rosen. *Biochem Biophys Acta* 1526: 95–104, 2001.
159. S Porasuphatana, P Tsai, S Pou, GM Rosen. *Biochem Biophys Acta* 1569 (1–3): 111–116, 2002.
160. RT Miller. *Chem Res Toxicol* 15: 927–934, 2002.
161. Y Song, AJ Cardounel, JL Zweier, Y Xia. *Biochemistry* 41: 10616–10622, 2002.
162. KA Pritchard, AW Ackerman, J Ou, M Curtis, DM Smalley, JT Fontana, MB Stemerman, WC Sessa. *Free Radic Biol Med* 33: 52–62, 2002.
163. Y Shi, JE Baker, C Zhang, JS Tweddell, J Su, KA Pritchard KA Jr. *Circ Res* 91: 300–306, 2002.
164. W Wang, S Wang, L Yan, P Madara, AD Cintron, RA Wesley, RL Danner. *J Biol Chem* 275: 16899–16903, 2000.
165. KY Xu. *FEBS Lett* 474: 252–253, 2000.
166. KY Xu. *FEBS Lett* 481: 306–307, 2000.
167. KY Xu. *Biochem Biophys Acta* 1481: 156–166, 2000.
168. J Vasquez-Vivar, B Kalyanaraman. *FEBS Lett* 481: 305–306, 2000.

169. B Mayer. *FEBS Lett* 481: 304, 2000.
170. L Kobzik, B Stringer, JL Balligand, MB Reid, JS Stamler. *Biochem Biophys Res Commun* 211: 375–381, 1995.
171. TE Bates, A Loesch, G Burnstock, JB Clark. *Biochem Biophys Res Commun* 213: 896–900, 1995.
172. P Ghafourifar, C Richter. *FEBS Lett* 418: 291–296, 1997.
173. C Giulivi, JJ Poderoso, A Boveris. *J Biol Chem* 273: 11038–11043, 1998.
174. Z Lacza, M Puskar, JP Figueroa, J Zhang, N Rajapakse, DW Busija. *Free Radic Biol Med* 31: 1609–1615, 2001.
175. AJ Kanai, LL Pearce, PR Clemens, LA Birder, MM VanBibber, S-Y Choi, WC de Groat, J Peterson. *Proc Natl Acad Sci* 98: 14126–14131, 2001.
176. MO Lopez-Figueroa, C Caamano, MI Morano, LC Ronn, H Akil, SJ Watson. *Biochem Biophys Res Commun* 272: 129–133, 2000.
177. P Ghafourifar, U Schenk, SD Klein, C Richter. *J Biol Chem* 274: 31185–31188, 1999.
178. U Bringold, P Ghafourfar, C Richter. *Free Radic Biol Med* 29: 343–348, 2000.
179. JE Harrison, J Schultz. *J Biol Chem* 251: 1371–1374, 1976.
180. H Kohler, H Lenzer. *Free Radic Biol Med* 6: 323–339, 1989.
181. H Jenzer, W Jones, H Kohler. *J Biol Chem* 261: 15550–15556, 1986.
182. HB Dunford, JS Stillman. *Coord Chem Rev* 19: 187–251, 1976.
183. PR Ortiz de Montellano. *Annu Rev Pharmacol Toxicol* 32: 89–107, 1992.
184. A Taurog, ML Dorris, DR Doerge. *Arch Biochem Biophys* 330: 24–32, 1996.
185. OM Lardinois, KF Medzihradszky, PR Ortiz de Montellano. *J Biol Chem* 274: 35441–35448, 1999.
186. OM Lardinois, PR Ortiz de Montellano. *Biochem Biophys Res Commun* 270: 199–202, 2000.
187. IM Ralston, HB Dunford. *Can J Biochem* 58: 1270–1276, 1980.
188. S Ohtaki, H Nakagawa, M Nakamura, I Yamazaki. *J Biol Chem* 257: 761–766, 1982.
189. S Ohtaki, H Nakagawa, M Nakamura, I Yamazaki. *J Biol Chem* 257: 13398–13403, 1982.
190. RC Sealy, L Harman, PR West, RP Mason. *J Am Chem Soc* 107: 3401–3406, 1985.
191. JW Heinecke, W Li, HL Daehnke III, JA Goldstein. *J Biol Chem* 268: 4069, 1993.
192. GA Francis, AJ Mendez, EL Bierman, JW Heinecke. *Proc Natl Acad Sci USA* 90: 6631–6635, 1993.
193. LA Marquez, HB Dunford. *J Biol Chem* 270: 30434–30440, 1995.
194. IM Ralston, HB Dunford. *Can J Biochem* 56: 1115–1119, 1978.
195. GS Bayse, AW Michaels, M Morrison. *Biochim Biophys Acta* 284: 34–42, 1972.
196. M Tien. *Arch Biochem Biophys* 367: 61–66, 1999.
197. ML McCormick, JP Gaut, TS Lin, BE Britigan, GR Buettner, JW Heinecke. *J Biol Chem* 273: 32030–32037, 1998.
198. PR West, LS Harman, PD Josephy, RP Mason. *Biochem Pharmacol* 33: 2933–2936, 1984.
199. V Fischer, LS Harman, PR West, RP Mason. *Chem Biol Interact* 60: 115–127, 1986.
200. D Metodiewa, K Reszka, HB Dunford. *Arch Biochem Biophys* 274: 601–608, 1989.
201. S Adak, U Bandyopadhyay, D Bandyopadhyay, RK Banerjee. *Biochemistry* 37: 16922–16933, 1988.
202. M Nakamura. *J Biochem* 108: 245–249, 1990.
203. HJ Sipe Jr, SJ Jordan, PM Hanna, RP Mason. *Carcinigenesis* 15: 2637–2643, 1994.
204. C Rota, YC Fann, RP Mason. *J Biol Chem* 274: 28161–28168, 1999.
205. M Nakamura, S Nakamura. *Free Radic Biol Med* 24: 537–544, 1998.
206. LA Marquez, HB Dunford. *J Biol Chem* 269: 7950–7956, 1994.
207. M Nakamura, I Yamazaki, S Ohtaki, S Nakamura. *J Biol Chem* 261: 13923–13927, 1986.
208. LS Harmon, DK Carver, J Schreiber, RP Mason. *J Biol Chem* 261: 1642–1648, 1986.
209. G Galati, T Chan, B Wu, PJ O'Brien. *Chem Res Toxicol* 12: 521–525, 1999.
210. J Huang, HB Dunford. *Arch Biochem Biophys* 287: 257–262, 1991.
211. Y Hsuanyu, HB Dunford. *Arch Biochem Biophys* 368: 413–420, 1999.
212. KM Morehouse, HJ Sipe Jr, RP Mason. *Arch Biochem Biophys* 273: 158–164, 1989.
213. J Huang, EM Sommers, DB Kim-Shapiro, SB King. *J Am Chem Soc* 124: 3473–3480, 2002.
214. JE Harrison, J Schultz. *J Biol Chem* 251: 1371–1374, 1976.
215. RP Magnusson, A Taurog, ML Dorris. *J Biol Chem* 259: 13783–13790, 1984.

216. A Taurog, ML Dorris, DR Doerge. *Arch Biochem Biophys* 330: 24–32, 1996.
217. C Mottley, TB Trice, RP Mason. *Mol Pharmacol* 22: 732–737, 1982.
218. SN Moreno, K Stolze, EG Janzen, RP Mason. *Arch Biochem Biophys* 265: 267–271, 1988.
219. GJM Bolscher, R Wever. *Biochim Biophys Acta* 788: 1–10, 1984.
220. LA Marquez, JT Huang, HB Dunford. *Biochemistry* 33: 1447–1454, 1994.
221. PG Furtmuller, U Burner, W Jantschko, G Regelsberger, C Obinger. *FEBS Lett* 44: 139–143, 2000.
222. H Hoogland, HL Dekker, C van Riel, A Kuilenburg, AO Muijsers, R Wever. *Biochem Biophys Acta* 955: 337–345, 1988.
223. CC Winterbourn, RC Garcia, AW Segal. *Biochem J* 228: 583–592, 1985.
224. SW Edwards, TF Swan. *Biochem J* 237: 601–604, 1986.
225. D Metodiewa, HB Dunford. *Arch Biochem Biophys* 272: 245–253, 1989.
226. AJ Kettle, DF Sangster, JM Gebicki, CC Winterbourn. *Biochim Biophys Acta* 956: 58–62, 1988.
227. BNJ Bielski, JM Gebicki. *Biochim Biophys Acta* 364: 233–237, 1974.
228. CJ Baker, K Deahl, J Domek, EW Orlandi. *Arch Biochem Biophys* 382: 232–237, 2000.
229. AJ Kettle, CC Winterbourn. *Biochemistry* 40: 10204–10212, 2001.
230. VF Ximenes, LH Catalani, A Campa. *Biochem Biophys Res Commun* 287: 130–134, 2001.
231. AJ Kettle, GA Gedye, CC Winterbourn. *Biochem.Pharmacol* 45: 2003–2010, 1993.
232. S Chen, P Schopfer. *Eur J Biochem* 260: 726–735, 1999.
233. HS Mason, L Onopryenko, D Buhler. *Biochim Biophys Acta* 24: 225–226, 1957.
234. CC Winterbourn. *J Clin Invest* 78: 545–550, 1986.
235. BE Britigan, DJ Hassett, GM Rosen, DR Hamill, MS Cohen. *Biochem J* 264: 447–455, 1989.
236. MS Cohen, BE Britigan, DJ Hassett, GM Rosen. *Free Radic Biol Med* 5: 81–88, 1988.
237. AJ Kettle, CC Winterbourn. *J Biol Chem* 269: 17146–17151, 1994.
238. PJ O'Brien. *Free Radic Biol Med* 4: 169–183, 1988.
239. MB Hampton, AJ Kettle, CC Winterbourn. *Blood* 92: 3007–3017, 1998.
240. H Shibata, Y Kono, S Yamashita, Y Sawa, H Ochiai, K Tanaka. *Biochim Biophys Acta* 1230: 45–50, 1995.
241. A van der Vliet, JP Eiserich, B Halliwell, CE Cross. *J Biol Chem* 272: 7617–7625, 1997.
242. JP Eiserich, M Hristova, CE Cross, AD Jones, BA Freeman, B Halliwell, A van der Vliet. *Nature* 391: 393–397, 1998.
243. JB Sampson, YZ Ye, H Rosen, JS Beckman. *Arch Biochem Biophys* 356: 207–213, 1998.
244. U Burner, PG Furtmuller, AJ Kettle, WH Koppenol, C Obinger. *J Biol Chem* 275: 20597–20601, 2000.
245. CJ van Dalen, CC Winterbourn, R Senthilmohan, AJ Kettle. *J Biol Chem* 275: 11638–11644, 2000.
246. WA Prutz, H Monig, J Butler, EJ Land. *Arch Biochem Biophys* 243: 125–134, 1985.
247. S Pfeiffer, A Lass, K Schmidt, B Mayer. *J Biol Chem* 276: 34051–34058, 2001.
248. KJ Reszka, Z Matuszak, CF Chignell. *Free Radic Biol Med* 25: 208–219, 1998.
249. KJ Reszka, Z Matuszak, CF Chignell, J Dillon. *Free Radic Biol Med* 26: 669–678, 1999.
250. R Floris, SR Piersma, G Yang, P Jones, R Wever. *Eur J Biochem* 215: 767–775, 1993.
251. SC Grace, MG Salgo, WA Pryor. *FEBS Lett* 426: 24–28, 1998.
252. H Ishiropoulos, J Nelson, D Duran, A Al-Mehdi. *Free Radic Biol Med* 20: 373–381, 1996.
253. RE Glover, V Koshkin, HB Dunford, RP Mason. *Nitric Oxide* 3: 439–444, 1999.
254. HM Abu-Soud, SL Hazen. *J Biol Chem* 275: 37524–37532, 2000.
255. HM Abu-Soud, SL Hazen. *J Biol Chem* 275: 5425–5430, 2000.
256. TJ Evans, LDK Buttery, A Carpenter, DP Springall, JM Polak, J Cohen. *Proc Natl Acad Sci USA* 93: 9553–9558, 1996.
257. JP Eiserich, S Baldus, ML Brennan, W Ma, C Zhang, A Tousson, L Castro, AJ Lusis, WM Nauseef, CR White, BA Freeman. *Science* 296: 2391–2394, 2002.
258. T Taniguchi, M Sono, F Hirata, O Hayaishi, M Tamura, K Hayashi, T Iizuka, Y Ishimura. *J Biol Chem* 254: 3288–3294, 1979.
259. FJ Martin-Romero, Y Gutierrez-Martin, F Henao, C Gutierrez-Merino. *J Neurochem* 82: 604–614, 2002.

23 Production of Free Radicals by Mitochondria

Although the first manifestations of hydrogen peroxide formation by mitochondria [1,2] have been taken with caution, the following studies [3,4] confirmed these findings. It has been reasonable to suggest that hydrogen peroxide could be a product of superoxide dismutation and that superoxide is produced by mitochondria. Such proposal was indeed confirmed by Loschen et al. [5] and Boveris and Cadenas [6], who showed that hydrogen peroxide is a stoichiometric precursor of mitochondrial hydrogen peroxide.

The production of superoxide by mitochondria strongly depends on the mitochondrial respiratory chain. As is well known, respiratory chain (Figure 23.1) consists of a series of electron carriers (pyridine nucleotide, flavoproteins, iron–sulfur proteins, ubiquinone, and cytochromes), which are arranged on the mitochondrial inner membrane according to their redox potentials from -320 to $+380$ mV. It is usually accepted that up to 97% of dioxygen consumption by mitochondria is converted by cytochrome oxidase into water and only 1–3% is leaked to be reduced into superoxide. Analysis of the literature data on mitochondrial oxygen radical production shows that there are big differences between results obtained by different authors. One of the reasons for this disagreement is difficulty of the quantitative detection of oxygen species in mitochondria. Therefore, we will start with the consideration of the analytical techniques applied.

23.1 DETECTION OF ACTIVE OXYGEN SPECIES IN MITOCHONDRIA

Controversial results of oxygen radical detection in mitochondria have been described in the literature. Owing to experimental difficulties many authors were obliged to work with submitochondrial particles instead of the whole mitochondria. However, it is quite possible that oxygen radical production by submitochondrial particles may be artificially enhanced due to exposure to oxygen. On the other hand, some analytical methods of superoxide detection such as cytochrome c reduction cannot be used due to the direct reduction of cytochrome by mitochondrial components.

In earlier studies [5,6] superoxide detection in mitochondria was equated to hydrogen peroxide formation. However, while it is quite possible that superoxide is a stoichiometric precursor of mitochondrial hydrogen peroxide, it is understandable that the level of hydrogen peroxide may be decreased due to the reactions with various mitochondrial oxidants. Moreover, superoxide level can be underestimated due to the reaction with mitochondrial MnSOD. Several authors [7,8] assumed that mitochondrial superoxide production may be estimated through cyanide-resistant respiration, which supposedly characterizes univalent dioxygen reduction. This method was applied for the measurement of superoxide production under in vitro normoxic and hyperoxic conditions, in spite of the finding [7] that cyanide-resistant respiration reflects also the oxidation of various substrates (lipids, amino acids, and nucleotides). Earlier,

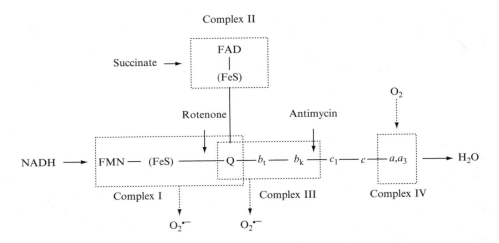

FIGURE 23.1 Mitochondrial respiratory chain. (Adapted from JF Turrens, BA Freeman, JG Levitt, JD Crapo. *Arch Biochem Biophys* 217: 401–410, 1982.)

it has been suggested that superoxide can be measured in mitochondria by the SOD-inhibitable oxidation of adrenaline to adrenochrome and the reduction of acetylated cytochrome c [8–10]. It is believed that the acetylation of cytochrome c prevents its reoxidation by mitochondrial cytochrome oxidase and direct reduction by NADH- or succinate-dependent cytochrome c reductase. A very interesting method has been proposed by Ksenzenko et al. [11], who used ESR spectroscopy for the measurement of superoxide production in mitochondria on the basis of Tiron (1,2-dihydroxybenzene-3,5-sulfonate) semiquinone formed in the reaction with superoxide.

As in the case of superoxide production by prooxidant enzymes described in Chapter 22, reliable data on mitochondrial superoxide production have been received by the use of lucigenin-amplified CL [12]. Lucigenin cation easily accumulates in mitochondria with the negative mitochondrial membrane potential and is able to penetrate the inner membrane of mitochondria. Therefore, this method permits to measure superoxide production by the whole mitochondria and mitochondrial superoxide production in cells. For example, Esterline and Trush [13] registered a significant lucigenin-amplified CL produced by the mitochondria of unstimulated rat alveolar macrophages that cannot be due to the activity of dormant NADPH oxidase.

23.2 RATES AND MECHANISM OF SUPEROXIDE PRODUCTION IN MITOCHONDRIA

Despite a long-time studying of superoxide production by mitochondria, an important question is still debated: does mitochondria produce superoxide under physiological conditions or superoxide release is always a characteristic of some pathophysiological disorders resulting in the damage of normal mitochondrial functions? Uncertainties in this question arise due to the different results obtained with the use of respiratory inhibitors and different analytical methods.

Usually, mitochondrial superoxide production is registered only after the incubation of submitochondrial particles with respiratory inhibitors, first of all, rotenone and antimycin. Under such conditions, superoxide production may achieve about $1\,\text{nmol}\,\text{l}^{-1}\,\text{min}^{-1}$ per mg of protein [14]. The rates of superoxide production depend on both tissue oxygen tension and the redox state of respiratory chain carriers. For example, hyperoxia increases superoxide production in rat lung mitochondria [7] and lung submitochondrial particles [8]. It has also been shown that hydrogen peroxide production by mitochondria depends on the mitochondrial metabolic state, and is higher in State 4, which is characterized by a slow respiratory rate and a high reduction level of electron carriers, and lower in State 3 where electron carriers are largely oxidized [15]. As superoxide is a precursor of hydrogen peroxide, the same may be said about the formation rates of hydrogen peroxide.

The most convincing proof of mitochondrial superoxide production under physiological conditions was presented by Li et al. [12] who has found that the mitochondria isolated from monocyte or macrophages generated significant lucigenin-amplified CL in the presence of substrates and without any respiratory inhibitors. It should be stressed that the efficiency of superoxide detection depends on the capacity of superoxide scavengers to achieve the sites of superoxide production in mitochondria. It has been shown [16,17] that the superoxide release takes place in the mitochondrial matrix or the intermembrane space. Therefore, lucigenin, which is able to enter the mitochondrial matrix, can quantitatively register superoxide. However, it should be noted again that superoxide concentration under physiological condition might be underestimated by mitochondrial MnSOD [18].

Knowledge of the sites of superoxide production is very important. Despite numerous studies carried out during the last 25 to 30 years (see Ref. [19]), the positions of specific sites of mitochondrial superoxide production are still controversial. It has been proposed [20] that the superoxide generator in mitochondrial membrane must be placed between the rotenone- and the antimycin-sensitive sites (Figure 23.1). Furthermore, there are probably at least two sites of mitochondrial superoxide production: the NADH ubiquinone reductase (Complex I) and the ubiquinol–cytochrome c reductase (Complex III). It is believed that ubiquinone can be one of the major producers of superoxide because the depletion of endogenous ubiquinone resulted in the formation of diminishing hydrogen peroxide [21]. Furthermore, Complexes I and III, which contain ubiquinone as a major component, are the effective producers of superoxide and hydrogen peroxide [20].

Now, we may consider in detail the mechanism of oxygen radical production by mitochondria. There are definite thermodynamic conditions, which regulate one-electron transfer from the electron carriers of mitochondrial respiratory chain to dioxygen: these components must have the one-electron reduction potentials more negative than that of dioxygen $E_0([O_2^{\cdot-}/O_2]) = -0.16\,\text{V}$. As the reduction potentials of components of respiratory chain are changed from -0.320 to $+0.380\,\text{V}$, it is obvious that various sources of superoxide production may exist in mitochondria. As already noted earlier, the two main sources of superoxide are present in Complexes I and III of the respiratory chain; in both of them, the role of ubiquinone seems to be dominant. Although superoxide may be formed by the one-electron oxidation of ubisemiquinone radical anion (Reaction (1)) [10,22] or even neutral semiquinone radical [9], the efficiency of these ways of superoxide formation in mitochondria is doubtful.

$$Q^{\cdot-} + O_2 \Longleftrightarrow Q + Q_2^{\cdot-} \tag{1}$$

Reaction (1) is a reversible process, and it can be a source of superoxide if only its equilibrium is shifted to the right. The estimation of the equilibrium constant for this reaction in aqueous solution is impossible because the reduction potential of water-insoluble ubiquinone in water is of course undetectable. However, Reaction (1) occurs in the mitochondrial membrane and therefore, the data for the aqueous solutions are irrelevant for the measurement of its equilibrium. Some time back we studied Reaction (1) in aprotic media and found out that K_1 is about 0.4 [23]. As the ubiquinone concentration in mitochondria is very high (it is about

100 times higher than the concentrations of other components of respiratory chain), the equilibrium of Reaction (1) is most probably shifted to the left; i.e., ubiquinone is a scavenger and not a producer of superoxide.

However, ubihydroquinone, a two-electron reduced form of ubiquinone, can produce superoxide on reaction with molecular oxygen:

$$QH_2 + O_2 \Longleftrightarrow QH^{\bullet} + O_2^{\bullet-} + H^+ \qquad (2)$$

Such a process is supposed to occur within the limits of Q-cycle mechanism (Figure 23.2). In accord with this scheme ubihydroquinone reduced dioxygen in Complex III, while superoxide producers in Complex I could be FMN or the FeS center [12]. Zhang et al. [24] also suggested that the Q-cycle mechanism is responsible for the superoxide production by the succinate–cytochrome c reductase in bovine heart mitochondria and that FAD of succinate dehydrogenase is another producer of superoxide. Young et al. [25] concluded that, in addition to Complex III, flavin-containing enzymes and FeS centers are also the sites of superoxide production in liver mitochondria.

As mentioned above, in an earlier work, Nohl et al. [9] suggested that neutral ubisemiquinone reduced dioxygen to superoxide (this suggestion was dropped in subsequent studies of these authors). Although the participation of neutral semiquinone in the reduction of dioxygen is impossible, the observation of these authors might be interpreted as the support of a role of ubihydroquinone in mitochondrial superoxide production. If neutral semiquinone is indeed formed in mitochondria via the protonation of semiquinone radical anion (Reaction (3)), then it might be easily reduced to ubihydroquinone anion (Reaction (4)), a genuine reductant of dioxygen (Reaction (5)).

$$Q^{\bullet-} + H^+ \Longleftrightarrow QH^{\bullet} \qquad (3)$$

$$QH^{\bullet} \xrightarrow{e} QH^- \qquad (4)$$

$$QH^- + O_2 \Longrightarrow QH^{\bullet} + O_2^{\bullet-} \qquad (5)$$

Casteilla et al. [26] suggested that mitochondrial superoxide production is modulated by uncoupling proteins. It has also been proposed [27] that the production of superoxide by

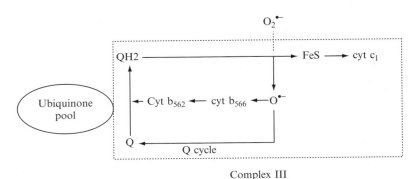

Complex III
(Ubiquinone-cytochrome c reductase)

FIGURE 23.2 Redox cycling of ubiquinone in mitochondria. (Q-cycling mechanism). (Adapted from Y Li, H Zhu, MA Trush. *Biochim Biophys Acta* 1428: 1–12, 1999.)

Complex I can be regulated by phosphorylation. Demin et al. [28] studied superoxide generation by Complex III using the kinetics model of electron transfer from succinate to cytochrome c.

23.3 FREE RADICAL-MEDIATED DAMAGE TO MITOCHONDRIA

Free radical-mediated damage to mitochondria originates from two sources: active oxygen species (superoxide, hydrogen peroxide, and secondary radicals) formed in mitochondria itself and free radicals, which attack mitochondria from outside. Early studies demonstrated that active oxygen species produced by xanthine oxidase impaired the ability of mitochondria to take and retain calcium [29], decreased the State 3 respiration [30], and reduced the pyruvate-induced mitochondrial respiration [31]. Other exogenous sources of free radical-mediated damage are xenobiotics and drugs, which can be reduced by mitochondria. Thus as early as 1970, Iyanagi and Yamazaki [32] have shown that mitochondrial NADH oxidase catalyzed one-electron reduction of quinones to semiquinones. Davies and Doroshow [33,34] demonstrated that the anthracycline antibiotics doxorubicin and daunorubicin are reduced to semiquinones by Complex I of cardiac mitochondria and enhance the production of superoxide, hydrogen peroxide, and probably hydroxyl radicals. Menadione also stimulated superoxide production and calcium release from rat liver mitochondria [35]. Kohda and Gemba [36] suggested that the cephaloridine-induced nephrotoxicity in rats might be a consequence of the enhanced superoxide production in kidney cortical mitochondria. Another stimulator of superoxide production in mitochondria is ethinyl estradiol, a strong promoter of hepatocarcinogenesis. It was demonstrated [37] that ethinyl estradiol increased superoxide production by rat liver mitochondria under in vitro and in vivo conditions.

Hennet et al. [38] demonstrated that TNF-α induced mitochondrial superoxide production in L929 cells. Superoxide generation was followed by a decrease in mitochondrial dehydrogenase activity and cellular ATP level. Corda et al. [39] have shown that TNF-α enhanced oxygen radical production by mitochondria in endothelial cells was mediated by ceramide-dependent protein kinase. Another endogenous mediator of oxygen radical production in mitochondria is the nuclear nick sensor enzyme, poly(ADP)–ribose synthase (PARS). Virag et al. [40] concluded that mitochondrial alterations induced by oxidative stress might be, to a significant degree, related to PARS activation rather than to the direct action of free radicals. The derivative of sphingolipid ceramide, N-acetylsphingosine, increased mitochondrial hydrogen peroxide production, which was prevented by blocking electron transport at Complexes I and II but enhanced by antimycin, an inhibitor of Complex III [41].

Calcium oxalate monohydrate responsible for the formation of most kidney stones significantly increased mitochondrial superoxide production in renal epithelial cells [42]. Recombinant human interleukin IL-β induced oxygen radical generation in alveolar epithelial cells, which was suppressed by mitochondrial inhibitors 4'-hydroxy-3'-methoxyacetophenone and diphenylene iodinium [43]. Espositio et al. [44] found that mitochondrial oxygen radical formation depended on the expression of adenine nucleotide translocator Ant1. Correspondingly, mitochondria from skeletal muscle, heart, and brain from the Ant1-deficient mice sharply increased the production of hydrogen peroxide.

As a rule, oxygen radical overproduction in mitochondria is accompanied by peroxidation of mitochondrial lipids, glutathione depletion, and an increase in other parameters of oxidative stress. Thus, the enhancement of superoxide production in bovine heart submitochondrial particles by antimycin resulted in a decrease in the activity of cytochrome c oxidase through the peroxidation of cardiolipin [45]. Iron overload also induced lipid peroxidation and a decrease in mitochondrial membrane potential in rat liver mitochondria [46]. Sensi et al. [47] demonstrated that zinc influx induced mitochondrial superoxide production in postsynaptic neurons.

It has been shown in Chapter 22 that mitochondria contains nitric oxide synthase (mtNOS) and produces nitric oxide. It is not surprising that the formation of highly reactive peroxynitrite by the interaction of superoxide with NO leads to various damaging effects in mitochondria. Ghafourifar et al. [48,49] suggested that calcium-induced mtNOS activation resulted in the formation of peroxynitrite, which caused the release of cytochrome c and enhanced mitochondrial lipid peroxidation. Riobo et al. [50] demonstrated that nitric oxide induced mitochondrial production of superoxide, hydrogen peroxide, and peroxynitrite and resulted in the inhibition of NADH-cytochrome c reductase activity due to the inhibition of Complex I.

Sarkela et al. [51] studied the effect of endogenously produced NO on the production of oxygen species by rat liver mitochondria. They found that the stimulation of mtNOS by L-arginine enhanced hydrogen peroxide production and diminished mitochondrial respiratory rate. It was suggested that NO increased the steady-state reduction of electron carriers, which in turn enhanced the one-electron reduction of dioxygen. Poderoso et al. [52] proposed that there are three major pathways of NO decay in mitochondria: the reversible inhibition of cytochrome c oxidase, the formation of peroxynitrite, and the oxidation of ubihydroquinone. It has been shown [53] that the reversible inhibition of cytochrome c oxidase by nitric oxide in mitochondrial membranes depends on the dioxygen concentration.

23.4 MECHANISMS OF MITOCHONDRIA PROTECTION FROM FREE RADICAL-MEDIATED DAMAGE

There are different pathways for the regulation and suppression of free radical production in mitochondria. Thus, Skulachev [54,55] proposed that the overproduction of free radicals in mitochondria may be suppressed by molecular processes such as the maintenance of low dioxygen levels, strong uncoupling, and the release of mitochondrial suicide proteins AIF protease, and cytochrome c. Endogenous antioxidants (ubihydroquinone, glutathione, and vitamin E) and the antioxidant enzymes (first of all, MnSOD) play an important role in preventing free radical-mediated damage. For example, Lass and Sohal [56] have shown that the oral administration of α-tocopherol to mice suppressed superoxide production by submitochondrial particles; the same results were obtained under in vitro conditions. In contrast, the administration of ubiquinone Q_{10} had no effect on mitochondrial superoxide production. Furthermore, defensive mechanisms in mitochondria against free radical damage depend on the relative activities of antioxidant enzymes. Mitochondrial glutathione peroxidase apparently plays a key role in destroying hydrogen peroxide due to the absence of catalase in mitochondria of most of the animal cells [57]. For regeneration of reduced glutathione (an important mitochondrial antioxidant) and maintaining the activities of glutathione reductase and peroxidase systems mitochondria requires NADPH. Jo et al. [58] have shown that $NADP^+$-dependent isocitrate dehydrogenase (ICDH) is a major NADPH producer in mitochondria and therefore, an important factor in fighting against free radical-induced damage. Correspondingly, the decreased expression of ICDH significantly enhanced free radical production, DNA fragmentation, and lipid peroxidation in mitochondria. It has been found [59] that nitric oxide inactivated ICDH supposedly through S-nitrosylation of its cysteine residues.

23.5 PRECONDITIONING AS A PROTECTION OF MITOCHONDRIA FROM OXYGEN RADICAL-MEDIATED DAMAGE

The phenomenon of ischemic preconditioning (PC) is already known for a long time. Thus in 1986, Murry et al. [60] showed that the short periods of ischemia protected against subsequent ischemic injury. A great interest in understanding the mechanism of ischemic preconditioning was stimulated by its cardioprotective effects against subsequent lethal ischemia. Numerous hypotheses of this paradoxical phenomenon have been proposed, most of which are related to the effects of reactive oxygen species formation. At least, four hypotheses of the free radical-mediated mechanisms of preconditioning have been proposed [61,62]: (a) the formation of hydrogen peroxide by microsomal NADH oxidoreductase [63]; (b) the hypoxia activation of sarcolemmal NADPH oxidase [64]; (c) the regulation of mitochondrial ATP-sensitive K^+ (K_{ATP}) channels by oxygen species [65]; and (d) the hypoxia-stimulated increase in mitochondrial oxygen radical production [66]. Two of these hypotheses (c) and (d) (which are now considered as the most convincing ones) suggest that mitochondria plays a key role in preconditioning. However, it remains questionable whether the formation of oxygen species increases or decreases during hypoxia [67,68].

There is now consensus that crucial role in PC stimulation belongs to opening mitochondrial (mito) K_{ATP} channels, but opinions are divided on the role of oxygen species. For example, Hoek et al. [69] suggested that PC protection is associated with the attenuation of oxygen radical formation during reperfusion. Ozcan et al. [70] also found that the opener-mediated mitochondrial protection was enhanced in the presence of antioxidant enzymes SOD and catalase. However, an increase in the production of oxygen species during PC has been demonstrated in other studies. It was found that the treatment of human cardiomyocytes [71] and THP-1 cells [72] with diazoxide, a selective opener of mito K_{ATP} channels resulted in a sharp increase in oxygen radical formation, which was supposedly an origin of the cardioprotective activity of diazoxide. These findings confirmed the previous data [73,74]. Minners et al. [75] also demonstrated that moderate oxygen radical production by mitochondria promoted PC-induced cytoprotection through uncoupling of mitochondrial oxidation from phosphorylation.

It has been suggested [76] that the activation of protein kinases (for example, protein kinase C or tyrosine kinase) may stimulate the phosphorylation of K_{ATP} channels, causing channel opening. However, Pain et al. [77] concluded that it is the K_{ATP} channel opening that increases the production of oxygen species, which triggers the entrance into a PC state. It is possible that the disagreement in the effects of hypoxia on oxygen radical production obtained by different authors could be related to the differences in preparations and analytical methods. Thus Sham [61] pointed out that oxygen radical production measured in isolated mitochondria must certainly differ from that in cells and tissue due to the presence of other than mitochondrial sources of oxygen radicals in cells.

It is important that mitochondrial oxygen radical production depends on the type of mitochondria. Recently, Michelakis et al. [78] demonstrated that hypoxia and the proximal inhibitors of electron transport chain (rotenone and antimycin) decreased mitochondrial oxygen radical production by pulmonary arteries and enhanced it in renal arteries. This difference is probably explained by a lower expression of the proximal components of electron transport chain and a greater expression of mitochondrial MnSOD in pulmonary arteries compared to renal arteries.

In 1992, Geng et al. [79] demonstrated that the interferon-γ- and TNF-α-induced nitric oxide production in vascular smooth muscle cells blocked mitochondrial respiration in these cells. Because it was showed [67] that the endogenous nitric oxide production was enhanced during ischemia, it is not surprising that NO can affect mitochondrial oxygen radical production under hypoxic conditions. Thus, Sasaki et al. [68] have shown that NO directly activated K_{ATP} channels and enhanced the ability of diazoxide to open them. Hypoxia-induced mitochondrial oxygen radical production also affects calcium level in pulmonary arterial myocytes. It was found [80] that oxygen radicals generated in the proximal region of mitochondrial electron transport chain during hypoxia act as second messengers to trigger the enhancement of calcium level in myocytes. Another important effect of increasing mitochondrial oxygen radical production induced by opening of K_{ATP} channels is the activation of extracellular signal-regulated kinases (ERKs) [72]. Increase in the hypoxia-induced mitochondrial generation of reactive oxygen species also results in the phosphorylation of p38 mitogen-activated protein kinase (MAPK) [81].

23.6 MITOCHONDRIAL OXYGEN RADICAL PRODUCTION AND APOPTOSIS

At present, the study of apoptosis, physiological cell death, is probably the most rapidly growing part of biological studies relevant to free radical research. Apoptosis is a two-stage process, in which mitochondria plays an important role [82]. The first stage is characterized by the attack of mitochondria by various physiological and pathophysiological stimuli, which increases mitochondrial membrane permeability. Mitochondrial permeability leads to the formation of a dynamic multiprotein complex at the site between the inner and outer mitochondrial membranes. Inhibition of the formation of this complex (for example, by mitochondrial expression of oncoprotein Bcl-2) prevents the development of apoptosis. At the second stage, mitochondrial dysfunction is accompanied by uncoupling of the respiratory chain and overproduction of free radicals.

23.6.1 REACTIVE OXYGEN SPECIES AS MEDIATORS OF APOPTOSIS

Free radicals have been implicated as important modulators of apoptosis in many studies, although the experiments carried out under hypoxic conditions suggested that apoptosis might occur in their absence. There are numerous apoptotic pathways induced by different stimuli, and reactive oxygen species may affect them in different ways. It has been proposed that p53 gene-induced apoptosis is characterized by the transcriptional induction of redox-related genes, the formation of reactive oxygen species, and the oxidative degradation of mitochondrial components, culminating in cell death [83]. Johnson et al. [84] have showed that p53 regulated the level of reactive oxygen species in vascular smooth muscle cells (VSMCs), which are generated concomitantly with an increase in p53 levels and the onset of apoptosis. The functional role of oxygen species in the development of p53-stimulated apoptosis was supported by the inhibitory effects of antioxidants (pyrrolidine dithiocarbamate (PDTC), N-acetylcysteine, and catalase) on both oxygen radical generation and apoptosis. Cai and Jones [85] also found that the overproduction of oxygen radicals occurred in staurosporine-treated HL60 cells, and that the release of cytochrome c, a key event in the development of apoptosis, was associated with inhibited mitochondrial respiration and stimulated superoxide production. However, they pointed out that another key factor of apoptosis, the activation of "death protease" caspase 3 was independent of oxygen radical formation.

Nonetheless, the stimulation of apoptosis by reactive oxygen species has been shown in many other studies with the use of different apoptotic stimuli. For example, it has been suggested that oxygen species mediate the transforming growth factor beta (TGF-β)-induced apoptosis in fetal hepatocytes [86,87]. In this case TGF-β-induced apoptosis was suppressed by free radical scavengers (ascorbate and PDTC). A decrease in superoxide production inhibited T cell apoptosis [88]. Hsieh et al. [89] have shown that oxidized LDL induced apoptosis in VSMC, which was mediated by reactive oxygen species. As the formation of oxygen species was inhibited by rotenone and nordihydroguaiaretic acid (the lipoxygenase inhibitor), mitochondrial and lipoxygenase pathways can be involved. Blatt et al. [90] have studied the effect of proapoptotic 1,4-benzodiazepine Bz-423 on the transformed Ramon B cells. It was found that Bz-423 induced superoxide formation, which functioned as an upstream signal that initiated cytochrome *c* release, mitochondrial depolarization, and caspase activation. Priault et al. [91] demonstrated that the expression of proapoptotic protein Bax induced the oxidation of mitochondrial lipids. It was concluded that Bax-induced lipid peroxidation, and not superoxide or hydrogen peroxide formation is relevant to Bax-stimulated apoptosis.

The *ras* protooncogene is a well-known modulator of apoptosis. For example, it has been demonstrated [92] that apoptosis occurred in the protein kinase C (PKC)-inhibited *ras*-transformed NIH/3T3 cells. The treatment with *N*-acetylcysteine and culturing the cells in low oxygen conditions suppressed the apoptotic response to PKC inhibition. It was concluded that oxygen species apparently mediated the *ras*-induced apoptosis, but they are not only obligatory mediators because continued cell progression is also necessary for the induction of apoptosis under these conditions. Another effective apoptotic stimulus is *Vibrio vulnificus* cytolysin (VVC), a water-soluble polypeptide isolated from a marine bacteria, which causes wound infections and septicemia [93]. It was proposed that VVC-stimulated apoptosis in human vascular endothelial cells is triggered by superoxide elevation, which is accompanied by the release of cytochrome *c*, the activation of caspase 3, the cleavage of poly(ADP-ribose) polymerase, and DNA fragmentation. Zinc induced apoptosis in mammary cancer cells, which was apparently mediated by the formation of oxygen species and the p53 and Fas/Fas induction [94]. The enhancement of superoxide production was demonstrated during apoptosis induced by 7-β-hydroxycholesterol and 7-ketocholesterol [95]. *Tert*-butylhydroperoxide also exhibited proapoptotic effect inducing the oxidation of pyridine nucleotides, the mitochondrial production of reactive oxygen species, and an increase in mitochondrial free calcium in cells [96]. Free radical damage stimulated the onset of mitochondrial permeability transition, a critical event in the development of apoptosis. Superoxide (as a precursor of hydrogen peroxide) and peroxynitrite supposedly induced selective and nonselective apoptosis in transformed and nontransformed fibroblasts [97].

Colquhoun and Schumacher [98] have shown that γ-linolenic acid and eicosapentaenoic acid, which inhibit Walker tumor growth in vivo, decreased proliferation and apoptotic index in these cells. Development of apoptosis was characterized by the enhancement of the formation of reactive oxygen species and products of lipid peroxidation and was accompanied by a decrease in the activities of mitochondrial complexes I, III, and IV, and the release of cytochrome *c* and caspase 3-like activation of DNA fragmentation. Earlier, a similar apoptotic mechanism of antitumor activity has been shown for the flavonoid quercetin [99]. Kamp et al. [100] suggested that the asbestos-induced apoptosis in alveolar epithelial cells was mediated by "iron-derived" oxygen species, although authors did not hypothesize about the nature of these species (hydroxyl radicals, hydrogen peroxide, or iron complexes?).

Recently, the apoptotic effect of glucose has been shown. Russell et al. [101] demonstrated that high glucose level stimulated the formation of reactive oxygen species by mitochondria and apoptosis in primary neurons. Kang et al. [102] also found that high ambient glucose concentrations induced apoptosis in murine and human mesangial cells by an oxidant-dependent mechanism. The study of cyanide-induced injury in rat primary cortical (CX) and mesencephalic (MC) neurons showed that cyanide can produce both apoptosis or necrosis [103]. Cyanide-stimulated generation of nitric oxide and superoxide in MC cells produced necrosis of these cells, while a lower level of oxidative stress in cyanide-treated CX cells led to cytochrome *c* release, caspase activation, and apoptosis.

23.6.2 Mechanisms of the Activation of Apoptosis by Reactive Oxygen Species

After considering experimental data relevant to the participation of mitochondrial reactive oxygen species in the development of apoptosis, we may now regard mechanisms of the oxygen species activation of apoptosis. Although major events and main participants of oxygen radical-mediated apoptotic pathways are well known, there is no full agreement in their roles and the succession of the events.

As described earlier, superoxide is a well-proven participant in apoptosis, and its role is tightly connected with the release of cytochrome c. It has been proposed that a switch from the normal four-electron reduction of dioxygen through mitochondrial respiratory chain to the one-electron reduction of dioxygen to superoxide can be an initial event in apoptosis development. This proposal was supported by experimental data. Thus, Petrosillo et al. [104] have shown that mitochondrial-produced oxygen radicals induced the dissociation of cytochrome c from bovine heart submitochondrial particles supposedly via cardiolipin peroxidation. Similarly, it has been found [105] that superoxide elicited rapid cytochrome c release in permeabilized HepG2 cells. In contrast, it was also suggested [106] that it is the release of cytochrome c that inhibits mitochondrial respiration and stimulates superoxide production.

Superoxide may activate other apoptotic pathways with or without stimulating the cytochrome c release. It is of interest that superoxide and hydrogen peroxide might induce apoptosis by different ways [106]. Thus, von Harsdorf et al. [107] demonstrated that superoxide-induced apoptosis in cardiomyocytes was not accompanied by cytochrome c release and was associated with an increase in apoptotic p53 protein level. In studying drug-induced apoptosis in tumor cells Hirpara et al. [108] showed that such drugs stimulated superoxide production which was followed by cytochrome c release. They proposed that apoptosis depended on acidification triggered by hydrogen peroxide and cytochrome c release, with both factors acting as signals for the activation of caspase cascade. Petit et al. [109] observed the enhancement of superoxide production during Fas- and ceramide-induced apoptosis in Jurkat cells, which is accompanied by NADH and NADPH depletion before the onset of apoptosis. It has been shown [110] that exposure of hippocampal neurons to the glutamate receptor agonist N-methyl-D-aspartate (NMDA) resulted in cytochrome c release, the delayed superoxide production, and apoptosis. Decrease in antioxidant enzymes may also contribute to the development of apoptosis [111]. Recently, Huang et al. [112] have shown that the inhibition of SOD in human leukemia cells by estrogen derivatives resulted in the superoxide-mediated damage of mitochondrial membranes, the release of mitochondrial cytochrome c, and apoptosis of cancer cells. Echtay et al. [113] demonstrated that superoxide exhibited uncoupling of mitochondria by activating the proton transport mechanism of uncoupling proteins at the matrix side of the mitochondrial inner membrane.

23.6.3 Protection Against Apoptosis Activated by Reactive Oxygen Species

The biggest difficulty in understanding the role of reactive oxygen species in apoptosis are the findings showing that the same oxygen radicals, which activate apoptosis may, under certain conditions, inhibit it. Correspondingly, antioxidants, which predictably must inhibit oxygen radical-stimulated apoptosis, are also able to activate apoptosis under certain conditions. For example, superoxide, an effective promoter of apoptosis, at the same time may suppress Fas-mediated apoptosis [114]. It has been suggested that reactive oxygen species are able to inhibit caspases [115], exhibiting antiapoptotic instead of proapoptotic affects. Furthermore, superoxide generation may activate the transcription factor NF-κB, which suppresses TNF-α-induced apoptosis [116].

Skulachev [117] proposed that the released cytochrome c oxidizes superoxide and, by this, exhibits an antioxidant function. This proposal was supported by recent experimental findings by Atlante et al. [118], who suggested that cytochrome c released from mitochondria by oxygen species protected mitochondria through a feedback-like process oxidizing superoxide. The most important physiological inhibitor of apoptosis is multifunctional protein Bcl-2, which is localized on the mitochondrial outer membrane. When overexpressed, Bcl-2 protects cells from free radical-mediated damage [119,120]. It has been shown that Bcl-2 prevents entering of cytochrome c into the cytosol probably by blocking the release of cytochrome c [85,121] or by direct reaction with cytochrome [122]. Despite the well-proven antiapoptotic effect of Bcl-2, the mechanism of its antioxidant activity is not fully understood. Interestingly, Esposti et al. [123] showed that Bcl-2 expression actually enhanced the level of hydrogen peroxide in lymphoma cells but at the same time protected these cells against ceramide and TNF-induced apoptosis. It has been suggested that the total Bcl-2 antiapoptotic effect be due to the enhanced formation of mitochondrial NADPH.

Many antioxidants and free radical scavengers are shown to be able to inhibit oxygen radical-mediated apoptosis; some examples are cited below. Mitochondrial MnSOD is one of the primary antioxidants, which might affect the development of apoptosis. Correspondingly, it has been shown [124] that the overexpression of MnSOD protected against apoptosis in murine fibrosarcoma cells. Recently, it has been shown [125] that the protein kinase B-regulated Forkhead transcription factor FOXO3 protects quiescent cells from apoptosis by increasing MnSOD messenger RNA and MnSOD protein. Another important microsomal antioxidant enzyme is phospholipid hydroperoxide glutathione peroxidase (PHGPx). Nomura et al. [126] have found that the overexpression of mitochondrial PHGPx prevented the release of cytochrome c, the activation of caspase 3, and apoptosis in rat basophile leukemia cells treated with 2-deoxyglucose. In subsequent work [127] these authors showed that the release of cytochrome c in these cells was triggered by the formation of cardiolipin hydroperoxide.

Classic antioxidants, vitamin E, vitamin C, and others can suppress the activation of apoptosis. For example, ascorbic acid prevented cytochrome c release and caspase activation in human leukemia cells exposed to hydrogen peroxide [128]. Pretreatment with N-acetylcysteine, ascorbate, and vitamin E decreased homocysteine thiolactone-induced apoptosis in human promyelocytic leukemia HL-60 cells [129]. Resveratrol protected rat brain mitochondria from anoxia–reoxygenation damage by the inhibition of cytochrome c release and the reduction of superoxide production [130]. However, it should be mentioned that the proapoptotic effect of ascorbate, gallic acid, or epigallocatechin gallate has been shown in the same human promyelocytic leukemia cells [131].

Guo et al. [132] demonstrated that vitamin E and selenium protected against superoxide-induced apoptosis of cultured fibroblasts. Vitamin E significantly decreased apoptosis induced by 7β-hydroxycholesterol and 7-ketocholesterol [133]. Dominguez-Rodriguez et al. [134] showed the in vivo antiapoptotic effect of vitamin E in murine peritoneal macrophages treated with adriamycin. Intratracheal application of liposomal α-tocopherol to anesthetized rats exposed to severe hypoxia increased antiapoptotic defense by the overexpression of genes encoding MnSOD and CuZnSOD, Bcl-2, and heat shock 70 proteins [135]. Thioredoxin, which contains two cysteine residues, is able to react with reactive free radicals and inhibit free radical-mediated damaging processes such as lipid peroxidation. Correspondingly, thioredoxin inhibited oxidative stress-induced apoptosis in neuronal cells, although in this case antiapoptotic effect of thioredoxin may also depend on the induction of MnSOD [136]. As mentioned above, zinc is an apoptosis stimulus in mammary cancer cells [94]. On the other hand, the depletion of intracellular zinc in airway epithelial cells increased lipid peroxidation and activated caspase 3 and apoptosis [137].

23.7 MITOCHONDRIAL NITROGEN OXIDE PRODUCTION AND APOPTOSIS

Similar to reactive oxygen species, nitric oxide, peroxynitrite, and other nitrogen oxide species produced by mitochondria are able to stimulate or inhibit apoptosis. Proapoptotic effect of nitric oxide was probably first shown by Albina et al. [138], who demonstrated NO-induced apoptosis in macrophages. Since then, NO-stimulated apoptosis was found in many primary cell types such as macrophages, pancreatic cells, thymocytes, and neurons [139]. NO-stimulated apoptosis in vascular smooth muscle cells apparently involves the activation of soluble guanylyl cyclase [140]. It has been proposed that apoptosis by nitric oxide is a consequence of DNA damage, which leads to the accumulation of proapoptotic tumor suppressor protein p53 [141]. Cheng et al. [142] found that nitric oxide induced apoptosis in neural progenitor cells, which was mediated by the activation of p38 MAP kinase, poly(ADP–ribose) polymerase, and caspase 3. Antiapoptotic protein Bcl-2 protected progenitor cells against NO-induced apoptosis and inhibited the activation of p38 MAP kinase. Nitric oxide also stimulated apoptosis in human and rat pulmonary artery smooth muscle cells by activating K(Ca) and K(v) channels in the plasma membrane [143].

The above examples point out at the direct stimulation of apoptosis by nitric oxide. At the same time, the exclusively rapid reaction of NO with superoxide always suggests the possibility of peroxynitrite participation in this process [141]; correspondingly, the role peroxynitrite in the stimulation of apoptosis has been considered. Bonfoco et al. [144] has found that the producers of low peroxynitrite concentrations during the exposure of cortical neurons to the low level of NMDA or the use of peroxynitrite donors resulted in an apoptosis in neurons, while the high concentrations of peroxynitrite induced necrotic cell damage. The formation of peroxynitrite is apparently responsible for NO-stimulated apoptosis in superoxide-generating transformed fibroblasts because nontransformed cells, which do not produce superoxide, were not affected by nitric oxide [145]. It is of interest that proapoptotic effect of peroxynitrite may depend on the cell type. Thus, the formation of peroxynitrite enhanced the NO-induced apoptosis in glomerular endothelial cells, while superoxide inhibited the formation of ceramide and apoptosis in these cells exposed to nitric oxide probably due to peroxynitrite formation [146]. The origin of this discrepancy is unknown. Cerielo et al. [147] studied the stimulation of apoptosis by acute hyperglycemia in working rat hearts. It was found that high glucose levels raised nitric oxide and superoxide production, which supposedly yielded peroxynitrite; the last by itself or through the formation of nitrotyrosine induced apoptosis in rat hearts.

Despite its well-characterized proapoptotic properties, nitric oxide turns out to be a two-faced Janus, showing in some cases antiapoptotic effect. In 1994, an important work by Mannick et al. [148] demonstrated that NO inhibited apoptosis in human B lymphocytes. Kim et al. [139] supposed that there are four major mechanisms of antiapoptotic activity of nitric oxide: (a) NO can oxidize intracellular GSH and by this induce heat shock proteins HSP32 and HSP70, which protect cells from TNF-α-stimulated apoptosis. (b) Activation of guanylyl cyclase by nitric oxide may generate cGMP with subsequent diminishing of cellular calcium concentration, one of the key signals of apoptosis. This mechanism may also involve the cGMP-dependent activation of protein kinase and the inhibition of caspase activation; (c) Nitric oxide may inhibit caspase activation through S-nitrosylation of its cysteine residue. (d) Nitric oxide may inhibit the cytochrome c release. Recently, it has also been proposed [149] that NO is able to suppress the expression of proapoptotic Bcl-2 binding protein BNIP3 and inhibit hepatocyte apoptosis. Heinloth et al. [150] showed that NO diminished cytochrome c release, p53 accumulation, and apoptosis in human macrophages induced by oxidized LDL supposedly via the activation of soluble guanylyl cyclase and the cGMP formation.

In conclusion, it should be stressed that the competition between pro- and antiapoptotic effects of nitric oxide must probably depends on its relevant levels [137]: the low physiological levels of NO principally suppress the apoptotic pathway by several mechanisms, whereas the higher rates of NO production may overcome cellar protective mechanisms and stimulate apoptosis. Furthermore, the simultaneous formation of nitric oxide and superoxide increases the possibility of apoptosis activation due to the formation of peroxynitrite.

REFERENCES

1. PK Jensen. *Biochim Biophys Acta* 157: 167–174, 1966.
2. PC Hinkle, RA Butow, E Racker, B Chance. *J Biol Chem* 242: 5169–5173, 1967.
3. G Loschen, L Flohe, B Chance. *FEBS Lett* 18: 261–263, 1971.
4. A Boveris, N Oshino, B Chance. *Biochem J* 134: 707–716, 1973.
5. G Loschen, A Azzi, C Richter, L Flohe. *FEBS Lett* 42: 68–72, 1974.
6. A Boveris, E Cadenas. *FEBS Lett* 54: 311–314, 1975.
7. BA Freeman, JD Crapo. *J Biol Chem* 256: 10986–10992, 1981.
8. JF Turrens, BA Freeman, JG Levitt, JD Crapo. *Arch Biochem Biophys* 217: 401–410, 1982.
9. H Nohl, L Gille, K Schonheit, Y Liu. *Free Radic Biol Med* 20: 207–213, 1996.
10. K Staniek L Gille, AV Kozlov, H Nohl. *Free Radic Res* 36: 381–387, 2002.
11. M Ksenzenko, AA Konstantinov, GB Khomutov, AN Tikhonov, EK Ruuge. *FEBS Lett* 155: 19–24, 1983.
12. Y Li, H Zhu, MA Trush. *Biochim Biophys Acta* 1428: 1–12, 1999.
13. RL Esterline, MA Trush. *Biochem Biophys Res Commun* 159: 584–591, 1989.
14. JF Turrens, A Boveris. *Biochem J* 191: 421–427, 1980.
15. A Boveris, E Cadenas. In: LW Oberley, (ed.) *Superoxide Dismutase*, vol II. Boca Raton, FL: CRC Press, 1982, pp. 16–30.
16. D Han, E Williams, E Cadenas. *Biochem J* 353: 411–416, 2001.
17. J St-Pierre, JA Buckingham, SJ Roebuck, MD Brand. *J Biol Chem* 277: 44784–44790, 2002.
18. S Raha, GE McEachern, AT Myint, BH Robinson. *Free Radic Biol Med* 29: 170–180, 2000.
19. IB Afanas'ev. In: *Superoxide Ion: Chemistry and Biological Implications*, vol 2, Boca Raton, FL: CRC Press, 1990, Chapter 2.
20. E Cadenas, A Boveris, CI Ragan, AOM Stoppani. *Arch Biochem Biophys* 180: 248–257, 1977.
21. A Boveris, E Cadenas, AOM Stoppani. *Biochem J* 156: 435–440, 1976.
22. H Nohl, AV Kozlov, K Staniek, L Gille. *Biorg Chem* 29: 1–13, 2001.
23. IB Afanas'ev. In: *Superoxide Ion: Chemistry and Biological Implications*, vol I, Boca Raton, FL: CRC Press, 1989, p. 161.
24. L Zhang, L Yu, CA Yu. *J Biol Chem* 273: 33972–33976, 1998.
25. T Young, C Cunningham, S Bailey. *Arch Biochem Biophys* 405: 65, 2002.
26. L Casteilla, M Rigoulet, L Penicaud. *IUBMB Life* 52: 181–188, 2001.
27. S Raha, AT Myint, L Johnstone, BH Robnson. *Free Radic Biol Med* 32: 421–430, 2002.
28. OV Demin, BN Kholodenko, VP Skulachev. *Mol Cell Biochem* 184: 21–33, 1998.
29. EJ Harris, R Booth, MB Cooper. *FEBS Lett* 146: 267–272, 1982.
30. JM Braughler, LA Duncan, T Goodman. *J Neurochem* 45: 1288–1293, 1985.
31. G Guarnieri, C Muscari, C Ceconi, F Flamigni, CM Caldarera. *J Mol Cell Cardiol* 15: 859–862, 1983.
32. T Iyanagi, I Yamazaki. *Biochim Biophys Acta* 216: 282–294, 1970.
33. KJA Davies, JH Doroshow. *J Biol Chem* 261: 3060–3067, 1986.
34. JH Doroshow, KJA Davies. *J Biol Chem* 261: 3068–3074, 1986.
35. GA Moore, PJ O'Brien, S Orrenius. *Xenobiotica* 16: 873–882, 1986.
36. Y Kohda, M Gemba. *Biochem Pharmacol* 64: 543–549, 2002.
37. J Chen, Y Li, JA Lavigne, MA Trush, JD Yager. *Toxicol Sci* 51: 224–235, 1999.
38. T Hennet, C Richter, E Peterhans. *Biochem J* 289: 587–592, 1993.
39. S Corda, C Laplace, E Vicaut, J Duranteau. *Am J Respir Cell Mol Biol* 24: 762–768, 2001.
40. L Virag, AL Salzman, C Szabo. *J Immunol* 161: 3753–3759, 1998.
41. C Garcia-Ruiz, A Colell, M Mari, A Morales, JC Fernandez-Checa. *J Biol Chem* 272: 11369–11377, 1997.
42. FD Khand, MP Gordge, WG Robertson, AA Noronha-Dutra, JS Hothersall. *Free Radic Biol Med* 32: 1339–1350, 2002.
43. JJ Haddad. *Eur Cytokine Netw* 13: 250–260, 2002.
44. LA Espositio, S Melov, A Panov, BA Cottrell, DC Wallace. *Proc Natl Acad Sci USA* 96: 4820–4825, 1999.

45. G Paradies, G Petrosillo, M Pistolese, FM Ruggiero. *FEBS Lett* 466: 323–326, 2000.
46. A Masoni, T Trenti, D Ceccarelli-Stanzani, E Ventura. *Biochim Biophys Acta* 810: 20–26, 1985.
47. SL Sensi, HZ Yin, SG Carriedo, SS Rao, JH Weiss. *Proc Natl Acad Sci USA* 96: 2414–2419, 1999.
48. P Ghafourifar, U Schenk, SD Klein, C Richter. *J Biol Chem* 274: 31185–31188, 1999.
49. U Bringold, P Ghafourfar, C Richter. *Free Radic Biol Med* 29: 343–348, 2000.
50. NA Riobo, E Clementi, M Melani, A Boveris, E Cadenas, S Moncada, JJ Poderoso. *Biochem J* 359: 139–145, 2001.
51. TM Sarkela, J Berthiaume, S Elfering, AA Gybina, C Giulivi. *J Biol Chem* 276: 6945–6949, 2001.
52. JJ Poderoso, C Lisdero, F Schopfer, N Riobo, MC Carreras, E Cadenas, A Boveris. *J Biol Chem* 274: 37709–37716, 1999.
53. S Shiva, PS Brookes, RP Patel, PG Anderson, VM Darley-Usmar. *Proc Natl Acad Sci USA* 98: 7212–7217, 2001.
54. VP Skulachev. *Quart Rev Biophys* 29: 169–202, 1996.
55. VP Skulachev. *Bioscience Rep* 17: 347–366, 1997.
56. A Lass, RS Sohal. *FASEB J* 14: 87–94, 2000.
57. RS Esworthy, YS Ho, FF Chu. *Arch Biochem Biophys* 340: 59–63, 1997.
58. S-H Jo, M-K Son, H-J Koh, S-M Lee, I-H Song, Y-O Kim, Y-S Lee, K-S Jeong, WB Kim, J-W Park, BJ Song, T-L Huhe. *J Biol Chem* 276:16168–16176, 2001.
59. E Yang, C Richter, J Chun, T Huh, S Kang, J Park. *Free Radic Biol Med* 33: 927, 2002.
60. CE Murry, RB Jennings, KA Reimer. *Circulation* 74: 1124–1136, 1986.
61. JSK Sham. *Circ Res* 91: 649–651, 2002.
62. PP Dzeja, EL Holmuhamedov, C Ozcan, D Pucar, A Jahangir, A Terzic. *Circ Res* 89: 744–746, 2001.
63. KM Mohazzab, MS Wolin. *Am J Physiol* 267: L823–L831, 1994.
64. C Marshall, AJ Mamary, AJ Verhoeven, BE Marshall. *Am J Respir Cell Mol Biol* 15: 633–644, 1996.
65. SL Archer, J Huang, T Henry, D Peterson, EK Weir. *Circ Res* 73: 1100–1112, 1993.
66. GB Waypa, NS Chandel, PT Schumacker. *Circ Res* 88: 1259–1266, 2001.
67. JL Zweier, P Wang, P Kuppusamy. *J Biol Chem* 270: 304–307, 1995.
68. M Sasaki, T Sato, A Ohler, B O'Rourke, E Marban. *Circulation* 101: 439–445, 2000.
69. TLV Hoek, LB Becker, Z-H Shao, C-Q Li, PT Schumacker. *Circ Res* 86: 541–548, 2000.
70. C Ozcan, M Bienengraeber, PP Dzeja, A Terzic. *Am J Physiol Heart Circ Physiol* 282: H531–H539, 2002.
71. RA Forbes, C Steenberg, E Murphy. *Circ Res* 88: 802–809, 2001.
72. L Samavati, MM Monick, S Sanlioglu, GR Buettner, LW Oberley, GW Hunninghake. *Am J Physiol Cell Physiol* 283: C273–C281, 2002.
73. CP Baines, M Goto, JM Downey. *J Mol Cell Cardiol* 29: 207–216, 1997.
74. I Tritto, D D'Andrea, N Eramo, A Scognaminglio, C De Simone, A Violante, A Esposito, M Chiariello, G Ambrosio. *Circ Res* 80: 743–748, 1997.
75. J Minners, L Lacerda, J McCarthy, JJ Meiring, DM Yello, MN Sack. *Circ Res* 89: 787–792, 2001.
76. Y Liu, B O'Rourke. *Circ Res* 88: 750–752, 2001.
77. T Pain, X-M Yang, SD Critz, Y Yue, A Nakano, GS Liu, G Heusch, MV Cohen, JM Downey. *Circ Res* 87: 460–466, 2000.
78. ED Michelakis, V Hampl, A Nsair, XC Wu, G Harry, A Haromy, R Gurtu, SL Archer. *Circ Res* 90: 1307–1315, 2002.
79. Y Geng, GK Hansson, E Holme. *Circ Res* 71: 1268–1276, 1992.
80. GB Waypa, JD Marks, MM Mack, C Boriboun, PT Mungai, PT Schumacker. *Circ Res* 91: 719–726, 2002.
81. A Kulisz, N Chen, NS Chandel, Z Shao, PT Schumacker. *Am J Physiol Lung Cell Mol Physiol* 282: L1324–L1329, 2002.
82. G Kroemer, B Dallaporta, M Resche-Rigon. *Ann Rev Physiol* 60: 619–642, 1998.
83. K Polyak, Y Xia, JL Zweier, KW Kinzler, B Vogfestein. *Nature* 389: 300–305, 1997.
84. TM Johnson, Z-X Yu, VJ Ferrans, RA Lowenstein, T Finkel. *Proc Natl Acad Sci USA* 93: 11848–11852, 1996.

85. J Cai, DP Jones. *J Biol Chem* 273: 11401–11404, 1998.
86. A Sanchez, AM Alvarez, M Benito, I Fabregat. *J Biol Chem* 271: 7416–7422, 1996.
87. B Herrera, AM Alvarez, A Sanchez, M Fernandez, C Roncero, M Benito, I Fabregat. *FASEB J* 15: 741–751, 2001.
88. DA Hildeman, T Mitchell, TK Teague, P Henson, NJ Day, J Kappler, PC Marrack. *Immunity* 10: 735–744, 1999.
89. C Hsieh, M Yen, C Yen, Y Lau. *Cardiovasc Res* 49: 135–145, 2001.
90. NB Blatt, JJ Bednarski, RE Warner, F Leonetti, KM Johnson, A Boitano, R Yung, BC Richardson, KJ Johnson, JA Ellman, AW Opipari Jr, GD Glick. *J Clin Invest* 110: 1123–1132, 2002.
91. M Priault, JJ Bessoule, A Grelaud-Coq, N Camougrand, S Manon. *Eur J Biochem* 269: 5440–5450, 2002.
92. JS Liou, C-Y Chen, JS Chen, DV Faller. *J Biol Chem* 275: 39001–39011, 2000.
93. K-B Kwon, J-Y Yang, D-G Ryu, H-W Rho, J-S Kim, J-W Park, H-R Kim, B-H Park. *J Biol Chem* 276: 47518–47523, 2001.
94. M Provinciali, A Donnini, K Argentati, G Di Stasio, B Bartozzi, G Bernardini. *Free Radic Biol Med* 32: 431–445, 2002.
95. C Miguet-Alfonsi, C Prunet, S Monier, G Bessede, S Lemaire-Ewing, A Berthier, F Menetrier, D Neel, P Gambert, G Lizard. *Biochem Pharmacol* 64: 527–541, 2002.
96. JJ Lemasters, AL Nieminen, T Qian, LC Trost, SP Elmore, Y Nishimura, RA Crowe, WE Cascio, CA Bradham, DA Brenner, B Herman. *Biochim Biophys Acta* 1366: 177–196, 1998.
97. B Ivanovas, A Zerweck, G Bauer. *Anticancer Res* 22: 841–856, 2002.
98. A Colquhoun, RI Schumacher. *Biochim Biophys Acta* 1533: 207–219, 2001.
99. Y Wei, X Zhao, Y Kariya, H Fukata, K Teshigawara, A Uchida. *Cancer Res* 54: 4952–4957, 1994.
100. DW Kamp, V Panduri, SA Weitzman, N Chandel. *Mol Cell Biochem* 234–235: 153–160, 2002.
101. JW Russell, D Golovoy, AM Vincent, P Mahendru, JA Olzmann, A Mentzer, EL Feldman. *FASEB J* 16: 1738–1748, 2002.
102. BP Kang, S Frencher, V Reddy, A Kessler, A Malhotra, LG Meggs. *Am J Physiol Renal Physiol.* 2002 Nov 5; [epub ahead of print]
103. K Prabhakaran, L Li, JL Borowitz, GE Isom. *J Pharmacol Exp Ther* 303: 510–519, 2002.
104. G Petrosillo, FM Ruggiero, M Pistolese, G Paradies. *FEBS Lett* 509: 435–438, 2001.
105. M Madesh, G Hajnoczky. *J Cell Biol* 156: 1003–1016, 2001.
106. PF Li, R Dietz, R von Harsdorf. *Circulation* 96: 3602–3609, 1997.
107. R von Harsdorf, PF Li, R Dietz. *Circulation* 99: 2934–2941, 1999.
108. JL Hirpara, M-V Clement, S Pervaiz. *J Biol Chem* 276: 514–521, 2001.
109. PX Petit, MC Gendron, N Schrantz, D Metivier, G Kroemer, Z Maciorowska, F Sureau, S Koester. *Biochem J* 353: 357–367, 2001.
110. CM Luetjens, NT Bui, B Sengpiel, G Munstermann, M Poppe, AJ Krohn, E Bauerbach, J Kriegistein, HM Prehn. *J Neurosci* 20: 5715–5723, 2000.
111. MM Briehl, IA Cotgreave, G Powis. *Death and Differentiation* 2: 41–46, 1995.
112. P Huang, L Feng, EA Oldham, MJ Keating, W Plunkett. *Nature* 407: 390–395, 2000.
113. KS Echtay, MP Murphy, RA Smith, DA Talbot, MD Brand. *J Biol Chem* 277: 47129–47135, 2002.
114. MV Clement, I Stamenkovic. *EMBO J.* 15: 216–225, 1996.
115. MB Hampton, S Orrenius. *FEBS Lett* 414: 552–556, 1997.
116. DJ van Antwerp, SJ Martin, T Kafri, DR Green, IM Verma. *Science* 274: 787–789, 1996.
117. VP Skulachev. *FEBS Lett* 423: 275–280, 1998.
118. A Atlante, P Calissano, A Bobba, A Azzariti, E Marra, S Passarella. *J Biol Chem* 275: 37159–37166, 2000.
119. DJ Kane, R Sarafian, H Anton, GJ Butler, VT Selverstone, DE Bredesen. *Science* 262: 1274–1277, 1993.
120. DM Hockenbery, ZN Oltvai, XM Yin, CL Milliman, SJ Korsmeyer. *Cell* 75: 241–251, 1993.
121. J Yang, X Liu, K Bhalla, CN Kim, AM Ibrado, J Cai, T-I Peng, DP Jones, X Wang. *Science* 275: 1129–1132, 1997.
122. S Kharbanda, P Pandey, I Schefield, S Israels, R Roncinske, K Bharti, Z-M Yuan, S Saxena, R Weichselbaum, C Nalin, D Kufe. *Proc Natl Acad Sci USA* 94: 6932–6942, 1997.

123. MD Esposti, I Hatzinisiriou, H McLennan, S Ralph. *J Biol Chem* 274: 29831–29837, 1999.
124. KK Kiningham, TD Oberley, S Lin, CA Mattingly, DK St Clair. *FASEB J* 13: 1601–1610, 1999.
125. GJ Kops, TB Dansen, PE Polderman, I Saarloos, KW Wirtz, PJ Coffer, TT Huang, JL Bos, RH Medema, BM Burgering. *Nature* 419: 316–321, 2002.
126. K Nomura, IH Koumura, M Arai, Y Nakagawa. *J Biol Chem* 274: 29294–29302, 1999.
127. K Nomura, H Imai, T Koumura, T Kobayashi, Y Nakagawa. *Biochem J* 351: 183–193, 2000.
128. T Gruss-Fischer, I Fabian. *Biochem Pharmacol* 63: 1325–1335, 2002.
129. RF Huang, SM Huang, BS Lin, CY Hung, HT Lu. *J Nutr* 132: 2151–2156, 2002.
130. R Zini, C Morin, A Bertelli, AA Bertelli, JP Tillement. *Life Sci* 71: 3091–3108, 2002.
131. W Zhang, K Hashimoto, GY Yu, H Sakagami. *Anticancer Res* 22: 219–224, 2002.
132. L Guo, AN Xue, SQ Wang, JY Chen, YD Wu, B Zhang. *Biomed Environ Sci* 14: 241–247, 2001.
133. C Miguet-Alfonsi, C Prunet, S Monier, G Bessede, S Lemaire-Ewing, A Berthier, F Menetrier, D Neel, P Gambert, G Lizard. *Biochem Pharmacol* 64: 527–541, 2002.
134. JR Dominguez-Rodriguez, PC Gomez-Contreras, G Hernandez-Flores, JM Lerma-Diaz, A Carranco, R Cervantes-Munguia, S Orbach-Arbouys, A Bravo-Cuella. *Anticancer Res* 21: 1869–1872, 2001.
135. T Minko, A Stefanov, V Pozharov. *J Appl Physiol* 93: 1550–1560, 2002.
136. T Andoh, PB Chock, CC Chiueh. *J Biol Chem* 277: 9655–9660, 2002.
137. J Carter, A Truong-Tran, D Grosser, L Ho, R Ruffin, P Zalewski. *Biochem Biophys Res Commun* 297: 1062–1066, 2002.
138. JE Albina, S Cui, RB Mateo, JS Reichner. *J Immunol* 150: 5080–5085, 1993.
139. Y-M Kim, C Bombeck, TR Billiar. *Circ Res* 84: 253–256, 1999.
140. IM Lincoln, TL Cornwell, P Komalavilas, N Boerth. *Meth Enzymol* 269: 149–166, 1996.
141. UK Messmer, M Ankarcrona, P Nicotera, B Brune. *FEBS Lett* 355: 23–26, 1994.
142. A Cheng, SL Chan, O Milhavet, S Wang, MP Mattson. *J Biol Chem* 276: 43320–43327, 2001.
143. S Krick, O Platoshyn, M Sweeney, SS McDaniel, S Zhang, LJ Rubin, JX Yuan. *Am J Physiol Heart Circ Physiol* 282: H184–H193, 2002.
144. E Bonfoco, D Krainc, M Ankarcrona, P Nicotera, SA Lipton. *Proc Natl Acad Sci USA* 92: 7162–7166, 1995.
145. S Heigold, C Sers, W Bechtel, B Ivanovas, R Schafer, G Bauer. *Carcinogenesis* 23: 929–941, 2002.
146. A Pautz, R Franzen, S Dorsch, B Boddinghaus, VA Briner, J Pfeilschifer, A Huwiler. *Kidney Int* 61: 790–796, 2002.
147. A Cerielo, L Quagliaro, M D'Amico, C Di Filippo, R Marfella, F Nappo, L Berrino, F Rossi, D Guigliano. *Diabetes* 51: 1076–1082, 2002.
148. JB Mannick, K Asano, K Izumi, E Kieff, JS Stamler. *Cell* 79: 1137–1146, 1994.
149. R Zamora, L Alarcon, Y Vodovotz, B Betten, PK Kim, KF Gibson, TR Billiar. *J Biol Chem* 276: 46887–46895, 2001.
150. A Heinloth, B Brune, B Fischer, J Galle. *Atherosclerosis* 162: 93–101, 2002.

24 Production of Free Radicals by Microsomes

The discovery of oxygen radical production by microsomes was made during the same "gold age" of free radical studies in biology (the late 1960s to the beginning of 1970s) as the discovery of enzymatic production of superoxide, mitochondrial production of reactive oxygen species, and production of oxygen radicals by phagocytes. Microsomes from animals, plants, and microorganisms contain the mixed function oxidase system, which consists of flavoprotein NADPH-cytochrome P-450 reductase and NADH cytochrome b_5 reductase. This system is able to oxidize numerous substrates including drugs, carcinogens, antioxidants, pesticides, alcohols, steroids, lipid hydroperoxides, etc., and therefore, it is not surprising that its reactions are mediated by free radicals.

24.1 MICROSOMAL NADPH–CYTOCHROME P-450 REDUCTASE AND NADH CYTOCHROME b_5 REDUCTASE

The primary function of flavoprotein NADPH–cytochrome P-450 reductase is the hydroxylation of various substrates, which occurs during electron transfer from NADPH to cytochrome P-450 [1]:

$$\text{NADPH} \Longrightarrow \text{FAD} \Longrightarrow \text{FMN} \Longrightarrow \text{cyt.P-450} \tag{1}$$

The role of every component of this process is well established. The interaction with a substrate takes place at cytochrome P-450 in accord with Reaction (2):

$$\text{RH} + \text{O}_2 + \text{NADPH} + \text{H}^+ \Longrightarrow \text{ROH} + \text{NADP}^+ + \text{H}_2\text{O} \tag{2}$$

It has been proposed [2] (Figure 24.1) that after binding to cytochrome, the substrates such as epoxides, N-oxides, nitro compounds, and lipid hydroperoxides accept two electrons and are reduced to the compounds RH(H)_2. In contrast, the oxidizable substrates react with the oxygenated P-450 complex $(\text{RH})\text{Fe}^{2+}\text{O}_2 \Leftrightarrow (\text{RH})\text{Fe}^{3+}\text{O}_2^{\bullet-}$. After transfer the second electron substrate RH is hydroxylated to ROH and cytochrome P-450 is oxidized to the starting Fe^{3+} state, completing the catalytic cycle. It is possible that hydroxylation proceeds through the formation of hydroxyl and carbon radicals [3], but a true role of free radicals at the final stages of hydroxylation is still obscure.

While cytochrome P-450 catalyzes the interaction with substrates, a final step of microsomal enzymatic system, flavoprotein NADPH-cytochrome P-450 reductase catalyzes the electron transfer from NADPH to cytochrome P-450. As is seen from Reaction (1), this enzyme contains one molecule of each of FMN and FAD. It has been suggested [4] that these flavins play different roles in catalysis: FAD reacts with NADPH while FMN mediates electron

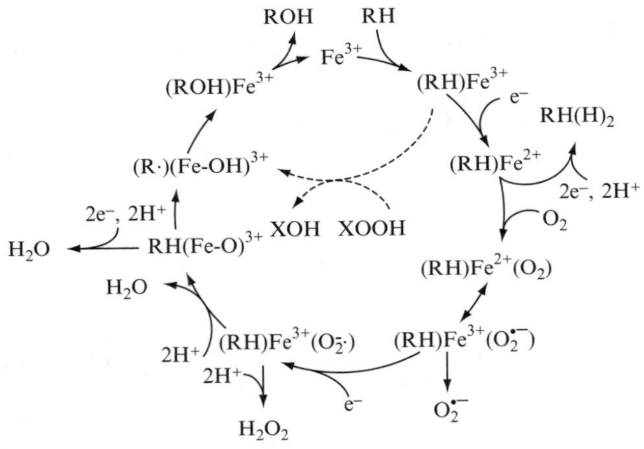

FIGURE 24.1 Mechanism of hydroxylation and reduction of substrates by cytochrome P-450. (From TD Porter, MJ Coon. *J Biol Chem* 266: 13469–13472, 1991. With permission.)

transfer to P-450. This proposal was supported by the measurement of midpoint reduction potentials of flavins, which were found to be equal to -190 and $-328\,\text{mV}$ for FMN and FAD, respectively [5]. It has also been shown [4] that the protein stabilizes neutral semiquinone free radicals formed as a result of reduction of flavins by NADPH and that the air-stable FMN semiquinone takes part in electron transfer as a $FMNH^\bullet - FAD$ complex. It should be mentioned that Minotti and Gennaro [6] proposed that in heart microsomes, which do not contain cytochrome P-450, NADPH oxidation can be mediated by electron transfer from reductase to nonheme iron.

It has been believed that P-450 reduction by NADPH cytochrome P-450 reductase is a biphasic process, but it was recently shown [7] that some P-450 cytochromes are reduced with single-exponential kinetics and that the presence of substrate is not an obligatory condition for the reduction of all P-450 forms. Thus, the kinetics of reduction of various ferric P-450 cytochromes possibly depends on many factors such as substrate, rate-limiting step, etc.

Another component of microsomal mixed function oxidase system is NADH cytochrome b_5 reductase. In 1971, Estabrook and his coworkers [8,9] proposed that cytochrome P-450 may be reduced by NADH through NADH cytochrome b_5 reductase and cytochrome b_5. It was suggested that cytochrome b_5 may supply the second electron to cytochrome P-450 for substrate oxidation. Iyanagi et al. [10] demonstrated that electron transfer from flavin to cytochrome b_5 in the cytochrome b_5 reductase proceeds in two successive one-electron steps. Bonfils et al. [11] suggested that to start electron transfer from cytochrome b_5 to cytochrome P-450, both cytochromes had to form a 1:1 complex and be incorporated into micelles. The kinetic study of rat liver microsomal electron transfer showed [12] that there are two independent pathways of electron transfer from NADH to cytochrome R-450: the first one is through cytochrome P-450 reductase and the second one is through cytochrome b_5 reductase and cytochrome b_5. Thus, four components (NADPH-cytochrome P-450 reductase, cytochrome P-450, NADH-cytochrome b_5 reductase, and cytochrome b_5) participate in electron transfer by microsomal mixed function oxidase system. The first electron is most probably supplied by NADPH-cytochrome P-450 reductase and the second electron comes from cytochrome b_5, which is apparently the rate-limiting step of the overall monooxygenase reaction [13].

24.2 PRODUCTION OF FREE RADICALS BY MICROSOMES

Although it is still unclear whether the formation of oxidized and hydroxylated products, which is the main pathway of catalytic activities of cytochrome-R-450 reductase, is mediated by free radicals, mitochondrial enzymes are certainly able to produce oxygen radicals as the side products of their reactions. It has been proposed in earlier studies [14,15] that superoxide and hydroxyl radicals (the last in the presence of iron complexes) are formed as a result of the oxidation of reduced NADPH–cytochrome-P-450 reductase:

$$\text{P-450 reductase (red)} + O_2 \Longrightarrow \text{P-450 reductase (oxid)} + O_2^{\bullet -} \qquad (3)$$

$$2 O_2^{\bullet -} + 2H^+ \Longrightarrow H_2O_2 + O_2 \qquad (4)$$

$$H_2O_2 + Fe^{2+}(\text{complex}) \Longrightarrow HO^{\bullet} + HO^- + Fe^{3+}(\text{complex}) \qquad (5)$$

Superoxide generation was detected via the NADPH-dependent SOD-inhibitable epinephrine oxidation and spin trapping [15,16]. Grover and Piette [17] proposed that superoxide is produced equally by both FAD and FMN of cytochrome P-450 reductase. However, from comparison of the reduction potentials of FAD (-328 mV) and FMN (-190 mV) one might expect FAD to be the most efficient superoxide producer. Recently, the importance of the microsomal cytochrome b_{558} reductase-catalyzed superoxide production has been shown in bovine cardiac myocytes [18].

Another superoxide producer in microsomes is the oxygenated complex of cytochrome P-450 (Figure 24.1). To study the superoxide production by microsomes, Kuthan et al. [19] used the reduction of partly succinoylated cytochrome c because native cytochrome c was directly reduced by cytochrome P-450 reductase and therefore cannot be applied for this purpose. It has been demonstrated [20] that carbon monoxide inhibited 75% of superoxide production by microsomal cytochrome P-450 system; therefore, the most of superoxide must be formed at the decomposition of oxygenated cytochrome R-450 complex. These authors also showed that the stoichiometry of superoxide and hydrogen peroxide production is close to 2:1, indicating that all hydrogen peroxide is formed through the dismutation of superoxide. Recently, Fleming et al. [21] showed that cytochrome P-450 is responsible for oxygen radical generation in coronary endothelial cells.

There is some difference in the ability of rat and human liver microsomes to produce oxygen radicals. It was found [22] that in the presence of NADPH or NADH human microsomes produced superoxide and hydrogen peroxide at rates of 20% to 30% of those observed for rat microsomes. A decrease in the production of oxygen species is probably explained by a threefold lower content of cytochrome P-450 in the human liver microsomes. Rasba-Step et al. [23] also showed that NADH similar to NADPH is able to stimulate superoxide formation in rat liver microsomes although in this case the rate of superoxide generation was about 20% to 30% of NADPH. NADH-dependent superoxide production by microsomal NADH-cytochrome b_{558} reductase was demonstrated in calf pulmonary artery smooth muscle [24]. Puntarulo and Cederbaum [25] have shown that not all forms of cytochrome P-450 were equally effective in oxygen radical production, with the CYP3A4 form of P-450 as the most active cytochrome P-450 inducer of superoxide production.

If the mechanism of superoxide production in microsomes by NADPH-cytochrome P-450 reductase, NADH-cytochrome b_5 reductase, and cytochrome P-450 is well documented, it cannot be said about microsomal hydroxyl radical production. There are numerous studies, which suggest the formation of hydroxyl radicals in various mitochondrial preparations and by isolated microsomal enzymes. It has been shown that the addition of iron complexes to microsomes stimulated the formation of hydroxyl radicals supposedly via the Fenton reaction [5]. (Such a system is widely used for the initiation of microsomal lipid peroxidation, Chapter 25.) Iron complexes accelerate the generation of oxygen radicals by microsomes, although the nature of reactive oxygen radical produced (hydroxyl or hydroxyl-like "crypto-hydroxyl radicals") is still uncertain. Morehouse et al. [26] showed that Fe^{3+}(DTPA) and Fe^{3+}(EDTA) complexes greatly enhanced superoxide production and NADPH oxidation by NADPH–cytochrome P-450 reductase. These authors also demonstrated the formation of hydroxyl radical in the presence of iron complexes.

Iron complexes or microsomal nonheme iron are undoubtedly obligatory components in the microsomal oxidation of many organic compounds mediated by hydroxyl radicals. In 1980, Cohen and Cederbaum [27] suggested that rat liver microsomes oxidized ethanol, methional, 2-keto-4-thiomethylbutyric acid, and dimethylsulfoxide via hydrogen atom abstraction by hydroxyl radicals. Then, Ingelman-Sundberg and Ekstrom [28] assumed that the hydroxylation of aniline by reconstituted microsomal cytochrome P-450 system is mediated by hydroxyl radicals formed in the superoxide-driven Fenton reaction. Similar conclusion has been made for the explanation of inhibitory effects of pyrazole and 4-methylpyrazole on the microsomal oxidation of ethanol and DMSO [29].

It is obvious that the sites of hydroxyl (or hydroxyl-like) and superoxide productions must coincide. Winston and Cederbaum [30] demonstrated that the oxidation of hydroxyl radical scavengers by purified NADPH-cytochrome P-450 reductase and cytochrome P-450 did not change after cytochrome P-450 addition that pointed out at cytochrome reductase as the site of hydroxyl radical production. Puntarulo and Cederbaum [31] confirmed the formation of hydroxyl radicals at the interaction of NADPH-cytochrome P-450 reductase with the ferric (EDTA) complex by the use of chemiluminescent method. In contrast, Ingelman-Sundberg and Johansson [32] concluded that the oxidation of ethanol occurred at the cytochrome P-450 site and not via the cytochrome P-450-specific mechanism but by hydroxyl radicals formed in the Fenton reaction with nonheme iron. Later on, Terelius and Ingelman-Sundberg [33] showed that small amounts of ferric (EDTA) react with hydrogen peroxide formed by cytochrome P-450, while higher ferric (EDTA) concentrations uncouple electron transport chain inducing the reductase-dependent formation of hydroxyl radicals.

Recent studies suggest that many factors may affect hydroxyl radical generation by microsomes. Reinke et al. [34] demonstrated that the hydroxyl radical-mediated oxidation of ethanol in rat liver microsomes depended on phosphate or Tris buffer. Cytochrome b_5 can also participate in the microsomal production of hydroxyl radicals catalyzed by NADH–cytochrome b_5 reductase [35,36]. Considering the numerous demonstrations of hydroxyl radical formation in microsomes, it becomes obvious that this is not a genuine enzymatic process because it depends on the presence or absence of "free" iron. Consequently, in vitro experiments in buffers containing iron ions can significantly differ from real biological systems.

24.3 MICROSOMAL FREE RADICAL-MEDIATED OXIDATIVE PROCESSES

There are various pathways for free radical-mediated processes in microsomes. Microsomes can stimulate free radical oxidation of various substrates through the formation of superoxide and hydroxyl radicals (the latter in the presence of iron) or by the direct interaction of chain electron carriers with these compounds. One-electron reduction of numerous electron acceptors has been extensively studied in connection with the conversion of quinone drugs and xenobiotics in microsomes into reactive semiquinones, capable of inducing damaging effects in humans. (In 1980s, the microsomal reduction of anticancer anthracycline antibiotics and related compounds were studied in detail due to possible mechanism of their cardiotoxic activity and was discussed by us earlier [37]. It has been shown that semiquinones of anthracycline are able to participate in redox cycling to produce superoxide. The possible mechanisms of anthracycline-induced lipid peroxidation are considered in Chapter 25.)

On the other hand, microsomes may also directly oxidize or reduce various substrates. As already mentioned, microsomal oxidation of carbon tetrachloride results in the formation of trichloromethyl free radical and the initiation of lipid peroxidation. The effect of carbon tetrachloride on microsomes has been widely studied in connection with its cytotoxic activity in humans and animals. It has been shown that CCl_4 is reduced by cytochrome P-450. For example, by the use of spin-trapping technique, Albani et al. [38] demonstrated the formation of the $CCl_3^{\bullet}$ radical in rat liver microsomal fractions and in vivo in rats. McCay et al. [39] found that carbon tetrachloride metabolism to $CCl_3^{\bullet}$ by rat liver accompanied by the formation of lipid dienyl and lipid peroxydienyl radicals. The incubation of carbon tetrachloride with liver cells resulted in the formation of the $CO_2^{\bullet -}$ free radical (identified as the $PBN-CO_2$ radical spin adduct) in addition to trichoromethyl radical [40]. It was found that glutathione rather than dioxygen is needed for the formation of this additional free radical. The formation of trichloromethyl radical caused the inactivation of hepatic microsomal calcium pump [41].

Microsomal oxidation of amines and phenols may proceed by different ways. For example, it has been shown [42] that phentermine (2-methyl-1-phenyl-2-propylamine) is hydroxylated to N-hydroxyphentermine by rat liver cytochrome P-450 system through a normal cytochrome P-450 way:

$$PhCH_2CMe_2NH_2 \xRightarrow{P-450} PhCH_2CMe_2NOH \qquad (6)$$

Then, N-hydroxyphentermine supposedly reacts with superoxide generated by NADPH–cytochrome P-450 reductase and forms the final product 2-methyl-2-nitro-1-phenylpropane:

$$PhCH_2CMe_2NHOH \xRightarrow{O_2^{\bullet -}} PhCH_2CMe_2NO \xRightarrow{O_2^{\bullet -}} PhCH_2CMe_2NO_2 \qquad (7)$$

(It should be noted that Reaction (7) had to be considered as a purely hypothetical one.)

Manno et al. [43] observed the formation of superoxide during the oxidation of arylamines by rat liver microsomes. Noda et al. [44] demonstrated that microsomes are able to oxidize hydrazine into a free radical. In contrast, hepatic cytochrome P-450 apparently oxidizes paracetamol (4′-hydroxyacetanilide) to N-acetyl-p-benzoquinone imine by a two-electron mechanism [45]. Younes [46] proposed that superoxide mediated the microsomal S-oxidation of thiobenzamide.

Nitrobenzyl chlorides are also reduced by microsomes through one-electron reduction mechanism. Moreno et al. [47] suggested that p- and o-nitrobenzyl chlorides are reduced by rat hepatic microsomes to unstable radical anions, which are decomposed to form benzyl radicals under anaerobic conditions. However, in the presence of dioxygen the radical anions of these compounds participate in "futile" redox cycling yielding superoxide (Figure 24.2). In contrast to p- and o-nitrobenzyl chlorides, m-nitrobenzyl chloride was reduced by microsomes to a relatively stable m-nitrobenzyl radical anion.

The above examples show the ability of microsome reductases to oxidize substrates in the processes where the first step is a one-electron reduction, which may or may not be accompanied by superoxide formation. However, cytochrome P-450 can directly oxidize some substrates including amino derivatives. For example, mitochondrial oxidation (dehydrogenation) of 1,4-dihydropyridines apparently proceeds by two mechanisms: via hydrogen atom abstraction or one-electron oxidation [48–50]. Guengerich and Bocker [49] have shown that

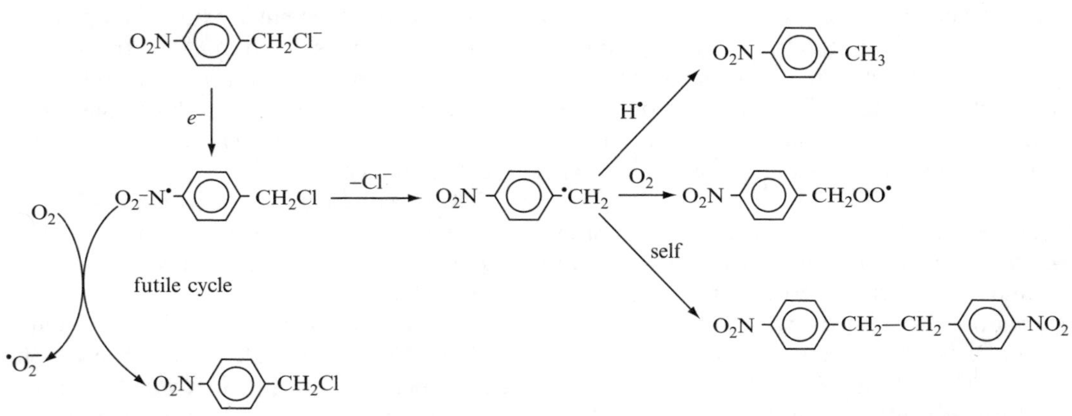

FIGURE 24.2 Mechanism of reduction of nitrobenzyl chlorides by microsomes. (From SNJ Moreno, J Schreiber, RP Mason. *J Biol Chem* 261: 7811–7815, 1991. With permission.)

the dehydrogenation of 1,4-dihydro-2,6-dimethyl-4-phenyl-3,5-pyridinedicarboxylic dimethyl ester or 1,4-dihydro-2,6-dimethyl-3,5-pyridinedicarboxylic acid diethyl ester (Hantzsch esters) by microsomes proceeded without significant kinetic hydrogen isotope effect. On these grounds it has been concluded that the dihydropyridines studied are oxidized by a P-450-mediated one-electron transfer mechanism (Figure 24.3). Similarly, the same mechanism was proposed for the dehydrogenation of nifedipine (1,4-dihydro-2,6-dimethyl-4-(2-nitrophenyl)-3,5-pyridinedicarboxylic acid dimethyl ester), which also proceeded with low kinetic hydrogen isotope effect [50].

Although the oxidation (hydroxylation) of hydrocarbons is usually believed to occur via hydrogen atom abstraction [51], the one-electron transfer mechanism of cytochrome P-450 catalyzed oxidation has also been proposed for the oxidation of *N,N*-dialkylanilines [52]. This mechanism (Figure 24.4) is generally preferred for the substrates with low reduction

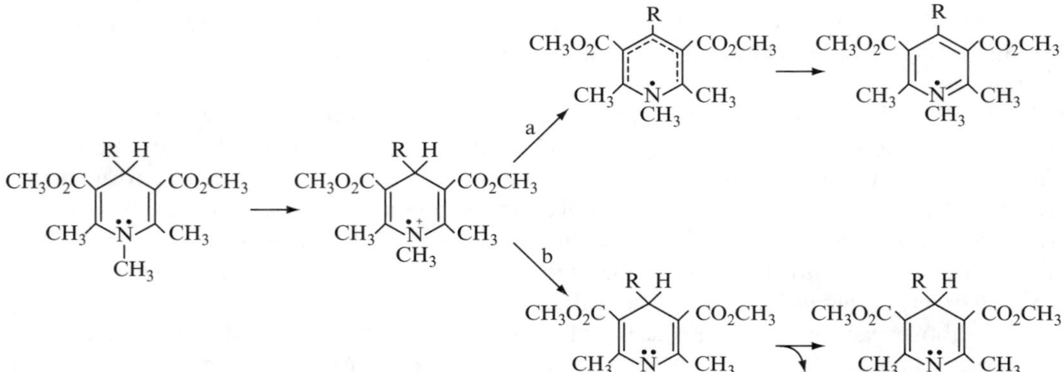

FIGURE 24.3 One-electron transfer mechanism of P-450-mediated oxidation of dihydropyridines. (From FP Guengerich, RH Bocker. *J Biol Chem* 263: 8168–8188, 1988. With permission.)

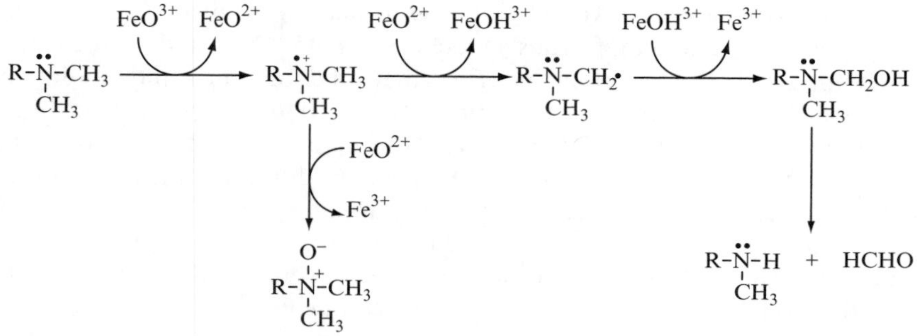

FIGURE 24.4 Postulated pathway for P450-catalyzed tertiary amine N-dealkylation (shown for N-demethylation of N,N-dimethylamine). (From FP Guengerich, CH Yun, TL Macdonald. *J Biol Chem* 271: 27321–27329, 1996. With permission.)

potentials. Sushkov et al. [53] have shown that microsomes demethylated 2-dimethylamino-3-chloro-1,4-naphthoquinone; this process was accompanied by the formation of corresponding semi- and hydroquinone.

In addition to a well-known NADPH-dependent hydroxylation mechanism (Reaction (2)), cytochrome P-450 is able to catalyze the oxidation of substrates by peroxygenase mechanism (Reaction (8)) where XOOH presents the peroxy compound acting as the oxygen donor.

$$RH + XOOH \xrightarrow{P\text{-}450} ROH + XOH \qquad (8)$$

Such reaction with aromatic hydroperoxides has been studied by Coon and coworkers [54]. Instead of hydroperoxides, iodosobenzenes, and iodobezene acetates may participate in Reaction (8) as the oxygen donors [55]. These authors proposed that the reaction of cytochrome P-450 with iodosobenzene proceeds to form an iron–oxo intermediate complex containing only one oxygen atom derived from the substrate.

Microsomes are capable of oxidizing not only organic substrates but also inorganic ones. An interesting example is the metabolism of bisulfite (aqueous sulfur dioxide) in microsomes. Although mitochondrial sulfite oxidase is responsible for the in vivo oxidation of bisulfite by a two-electron mechanism, cytochrome P-450 is also able to reduce bisulfite to the sulfur dioxide radical anion [56]:

$$SO_2 + \text{ferrous P-450} \Longrightarrow SO_2^{\bullet-} + \text{ferric P-450} \qquad (9)$$

24.4 FORMATION AND REACTIONS OF NITROGEN OXYGEN SPECIES IN MICROSOMES

In 1989, Servent et al. [57] demonstrated that microsomes denitrated glyceryl trinitrate in the presence of NADPH to form a mixture of glyceryl dinitrates and glyceryl mononitrates. They proposed that glyceryl trinitrate was oxidized by cytochrome P-450 and that this process was accompanied by the formation of nitric oxide. Following papers supported this proposal, showing the formation of nitrogen reactive species during the microsomal oxidation of various nitro compounds. For example, under anerobic conditions rat liver microsomes catalyzed the oxidation of *p*-hexyloxy-benzamidoxime to corresponding arylamide and nitrite anion NO_2^- in the presence of NADPH [58]. The formation of nitric oxide in this reaction was proven by the detection of ferrous cytochrome P-450–NO and ferrous cytochrome P-420–NO complexes. It has also been suggested that other amidines and amidoximes could be precursors of NO generation by microsomes. In the following paper [59] these authors demonstrated that cytochrome P-450 catalyzed the denitration of N-hydroxyl-L-arginine (NOHA) to form NO, NO_2, and citrulline. NO synthase did not contribute to this process because NO synthase inhibitors such as N-methylarginine and N-nitroarginine failed to inhibit this reaction. In contrast, the inhibitors of cytochrome P-450 (CO, miconazole, and others) suppressed the NOHA denitration. Similarly, microsomal P-450 generated nitric oxide during the oxidation of 18-nitro-oxyandrostenedione [60].

The constituent of paint, 2-nitropropane, exhibiting genotoxicity and hepatocarcinogenicity was oxidized by liver microsomes forming nitric oxide, which was identified as a ferrous–NO complex [61]. Clement et al. [62] concluded that superoxide may participate in the microsomal oxidation of N-hydroxyguanidines, which produced nitric oxide, urea, and the cyanamide derivative. Caro et al. [63] suggested that the oxidation of ketoxime acetoxime to nitric oxide by microsomes enriched with P-450 isoforms might be mediated by hydroxyl or hydroxyl-like radicals.

The formation of nitric oxide in microsomes results in the inhibition of microsomal reductase activity. It has been found that the inhibitory effect of nitric oxide mainly depend on the interaction with cytochrome P-450. NO reversibly reacts with P-450 isoforms to form the P-450–NO complex, but at the same time it irreversibly inactivates the cytochrome P-450 via the modification of its thiol residues [64]. Incubation of microsomes with nitric oxide causes the inhibition of 20-HETE formation from arachidonic acid [65], the generation of reactive oxygen species [66], and the release of catalytically active iron from ferritin [67].

REFERENCES

1. JL Vermilion, DP Ballou, V Massey, MJ Coon. *J Biol Chem* 256: 266–277, 1981.
2. TD Porter, MJ Coon. *J Biol Chem* 266: 13469–13472, 1991.
3. JT Groves, GA McClusky, RE White, MJ Coon. *Biochem Biophys Res Commun* 81: 154–160, 1978.
4. JL Vermilion, MJ Coon. *J Biol Chem* 253: 8812–8819, 1978.
5. T Iyanagi, N Makino, HS Mason. *Biochemistry* 13: 1701–1710, 1974.
6. G Minotti, M Di Gennaro. *Arch Biochem Biophys* 282: 270–274, 1990.
7. FP Guengerich, WW Johnson. *Biochemistry* 36: 14741–14750, 1997.
8. BS Cohen, RW Estabrook. *Arch Biochem Biophys* 143: 54–65, 1971.
9. A Hildebrandt, RW Estabrook. *Arch Biochem Biophys.* 143: 66–79, 1971.
10. T Iyanagi, S Watanabe, KF Anan. *Biochemistry* 23: 1418–1425, 1984.
11. C Bonfils, C Balny, P Maurel. *J Biol Chem* 256: 9457–9465, 1981.
12. GJ Fisher, JL Gaylor. *J Biol Chem* 257: 7449–7455, 1982.
13. H Taniguchi, Y Imai, R Sato. *Arch Biochem Biophys* 232: 585–596, 1994.
14. K-L Fong, PB McCay, JL Poyer, BB Keele, H Misra. *J Biol Chem* 248: 7792–7797, 1973.
15. CS Lai, TA Grover, LH Piette. *Arch Biochem Biophys* 193: 373–378, 1979.
16. RC Sealy, HM Swartz, PL Olive. *Biochem Biophys Res Commun* 82: 68–684, 1978.
17. TA Grover, LH Piette. *Arch Biochem Biophys* 212: 105–114, 1981.
18. KM Mohazzab-H, PM Kaminski, MS Wolin. *Circulation* 96: 614–620, 1997.
19. H Kuthan, V Ulrich, RW Estabrook. *Biochem J* 203: 551–558, 1982.
20. H Kuthan, V Ullrich. *Eur J Biochem* 126: 583–588, 1982.
21. I Fleming, UR Michaelis, D Bredenkotter, B Fisslthaler, F Dehghani, RP Brandes, R Busse. *Circ Res* 88: 44–51, 2001.
22. J Rasba-Step, AI Cederbaum. *Mol Pharmacol* 45: 150–157, 1994.
23. J Rasba-Step, NJ Turro, AI Cederbaum. *Arch Biochem Biophys* 300: 391–400, 1993.

24. KM Mohazzab, MS Wolin. *Am J Physiol* 267: L823–L831, 1994.
25. S Puntarulo, AI Cederbaum. *Free Radic Biol Med* 24: 1324–1330, 1998.
26. LA Morehouse, CE Thomas, SD Aust. *Arch Biochem Biophys* 232: 366–377, 1984.
27. G Cohen, AI Cederbaum. *Arch Biochem Biophys* 199: 438–447, 1980.
28. M Ingelman-Sundberg, G Ekstrom. *Biochem Biophys Res Commun* 106: 625–631, 1982.
29. AI Cederbaum, L Berl. *Arch Biochem Biophys* 216: 530–543, 1982.
30. GW Winston, AI Cederbaum. *J Biol Chem* 258: 1508–1513, 1983.
31. S Puntarulo, AI Cederbaum. *Arch Biochem Biophys* 258: 510–518, 1987.
32. M Ingelman-Sundberg, I Johansson. *J Biol Chem* 259: 6447–6458, 1984.
33. Y Terelius, M Ingelman-Sundberg. *Biochem Pharmacol* 37: 1383–1389, 1988.
34. LA Reinke, DR Moore, PB McCay. *Arch Biochem Biophys* 348: 9–14, 1997.
35. MX Yang, AI Cederbaum. *Arch Biochem Biophys* 324: 282–292, 1995.
36. DN Rao, MX Yang, JM Lasker, AI Cederbaum. *Mol Pharmacol* 49: 814–821, 1996.
37. IB Afanas'ev. In: *Superoxide Ion: Chemistry and Biological Implications*, vol 2. Boca Raton, FL: CRC Press, Chapter 3, 1990.
38. E Albani, KA Lott, TF Slater, A Stier, MC Symons, A Tomasi. *Biochem J.* 204: 593–603, 1982.
39. PB McCay, EK Lai, JL Poyer, CM DuBose, EG Janzen. *J Biol Chem* 259: 2135–2143, 1984.
40. HD Connor, LB Lacagnin, KT Knecht, RG Thurman, RP Mason. *Mol Pharmacol* 37: 443–451, 1990.
41. SP Srivastava, NQ Chen, LJ Holtzman. *J Biol Chem* 265: 8392–8399, 1990.
42. JD Duncan, EW Stefano, GT Miwa, AK Cho. *Biochemistry* 24: 4155–4161, 1985.
43. M Manno, C Ioannides, GG Gibson. *Toxicol Lett* 25: 121–130, 1985.
44. A Noda, H Noda, K Ohno, T Sendo, A Misaka, Y Kanazawa, R Isobe, M Hirata. *Biochem Biophys Res Commun* 133: 1086–1091, 1985.
45. R van der Straat, RM Vromans, P Bosman, J de Vries, NP Vermeulen. *Chem Biol Interact* 64: 267–280, 1988.
46. M Younes. *Experientia* 41: 479–481, 1985.
47. SNJ Moreno, J Schreiber, RP Mason. *J Biol Chem* 261: 7811–7815, 1991.
48. O Augusto, HS Beilan, PRO de Montellano. *J Biol Chem* 257: 11288–11295, 1982.
49. FP Guengerich, RH Bocker. *J Biol Chem* 263: 8168–8188, 1988.
50. FP Guengerich. *Chem Res Toxicol* 3: 21–26, 1990.
51. FP Guengerich, TL Macdonald. *FASEB J* 4: 2453–2459, 1990.
52. FP Guengerich, CH Yun, TL Macdonald. *J Biol Chem* 271: 27321–27329, 1996.
53. DG Sushkov, GV Rumyanzeva, LM Weiner. *Biokhimia* 52: 1898–11906, 1985 (in Russian).
54. RC Blake, MJ Coon. *J Biol Chem.* 255: 4100–4111, 1980.
55. RC Blake II, MJ Coon. *J Biol Chem* 264: 3694–3701, 1989.
56. C Mottley, LS Harman, RP Mason. *Biochem Pharmacol* 34: 3005–3008, 1985.
57. D Servent, M Delaforge, C Ducrocq, D Mansuy, M Lenfant. *Biochem Biophys Res Commun* 163: 1210–1216, 1989.
58. V Andronik-Lion, JL Boucher, M Delaforge, Y Henry, D Mansuy. *Biochem Biophys Res Commun* 185: 452–458, 1992.
59. JL Boucher, A Genet, S Vadon, M Delaforge, Y Henry, D Mansuy. *Biochem Biophys Res Commun* 187: 880–886, 1992.
60. M Delaforge, A Piffeteau, JL Boucher, A Viger. *J Pharmacol Exp Ther* 274: 634–640, 1995.
61. C Kohl, P Morgan, A Gescher. *Chem Biol Interact* 97: 175–184, 1995.
62. B Clement, JL Boucher, D Mansuy, A Harsdorf. *Biochem Pharmacol* 58: 439–445, 1999.
63. AA Caro, AI Cederbaum, DA Stoyanovsky. *Nitric Oxide* 5: 413–424, 2001.
64. Y Minamiyama, S Takemura, S Imaoka, Y Funae, Y Tanimoto, M Inoue. *J Pharmacol Exp Ther* 283: 1479–1485, 1997.
65. M Alonso-Galicia, HA Drummond, KK Reddy, JR Falck, RJ Roman. *Hypertension* 29: 320–325, 1997.
66. D Gergel, V Misik, P Riesz, AI Cederbaum. *Arch Biochem Biophys* 337: 239–250, 1997.
67. S Puntarulo, AI Cederbaum. *Arch Biochem Biophys* 340: 19–26, 1997.

25 Nonenzymatic Lipid Peroxidation

Lipid peroxidation is probably the most studied oxidative process in biological systems. At present, Medline cites about 30,000 publications on lipid peroxidation, but the total number of studies must be much more because Medline does not include publications before 1970. Most of the earlier studies are in vitro studies, in which lipid peroxidation is carried out in lipid suspensions, cellular organelles (mitochondria and microsomes), or cells and initiated by simple chemical free radical-produced systems (the Fenton reaction, ferrous ions + ascorbate, carbon tetrachloride, etc.). In these in vitro experiments reaction products (mainly, malondialdehyde (MDA), lipid hydroperoxides, and diene conjugates) were analyzed by physicochemical methods (optical spectroscopy and later on, HPLC and EPR spectroscopies). These studies gave the important information concerning the mechanism of lipid peroxidation, the structures of reaction products, etc.

Of course, the most important question is how these in vitro findings correspond to in vivo oxidative processes. It may be assumed that the data obtained in in vitro experiments can be used for understanding the mechanisms of real biological oxidative processes. However, the results obtained by studying much more important in vivo lipid peroxidation were for some time questioned. These doubts were based on the possibility of artificial formation of MDA and diene conjugates during the experiments. Fortunately, at present it became possible to study lipid peroxidation directly under ex vivo and in vivo conditions. (For example, Chamulitrat et al. [1] reported the in vivo EPR evidence of the formation of adducts of fatty acid-derived free radicals with DMPO in bile of rats dosed with this spin trap.) Comparison of in vitro and in vivo findings gives an opportunity to understand better the mechanism of lipid peroxidation.

25.1 INITIATION OF NONENZYMATIC LIPID PEROXIDATION

25.1.1 HO$^\bullet$ AND $O_2^{\bullet-}$ AS INITIATORS OF LIPID PEROXIDATION

Lipid peroxidation may proceed by both enzymatic and nonenzymatic pathways. Enzymatic peroxidation is catalyzed by enzymes such as lipoxygenases and cyclooxygenases (COXs) and is considered in Chapter 26.* One of the most important questions in the study of the mechanisms of lipid peroxidation is the characteristic of an initiation stage. Obvious

---

*It should be noted that the many free radical-producing enzymes (xanthine oxidase, peroxidases, NADPH oxidase of phagocytosing and nonphagocytosing cells, microsomal P-450 reductase, etc.) are the initiators of lipid peroxidation under both in vitro and in vivo conditions. However, these enzymes participate in lipid peroxidation only on the initiation stage and therefore, we are considering them as the initiators of nonenzymatic peroxidation. At the same time, in the enzyme-catalyzed lipid peroxidation lipoxygenases and cyclooxygenases catalyze all stages of oxidative process.

candidates for initiators are of course major physiological free radicals superoxide and nitric oxide. It is known that superoxide is not able to abstract a hydrogen atom even from the very reactive bisallylic methylene groups [2], although its conjugated acid HOO$^\bullet$ (perhydroxyl radical) is more active in abstraction reactions and probably capable of initiating lipid peroxidation [3] (see also Chapter 21). However, as a pK_a value for HOO$^\bullet$ is equal to 4.88 in aqueous solution, the equilibrium of Reaction (1) is practically completely shifted to the right at physiological pH.

$$HOO^\bullet \Longleftrightarrow H^+ + O_2^{\bullet-} \quad (1)$$

Aikens and Dix [4] have shown that the concentration of lipid hydroperoxides formed during the oxidation of linoleic acid by HOO$^\bullet$/O$_2^{\bullet-}$ in aqueous solution decreased by about 20 times at pH changing from 1.8 (100% HOO$^\bullet$) to 7.0 (about 1% HOO$^\bullet$). Although these authors believed that perhydroxyl-dependent lipid peroxidation may play a certain role under in vivo conditions, one should agree with the conclusion that its role is insignificant [3]. Recently, it has been pointed out [5] that the acidic pH may also enhance iron-mediated lipid peroxidation in cells by an increase in iron solubility. In addition to acidity, the initiation of lipid peroxidation by HOO$^\bullet$/O$_2^{\bullet-}$ depends also on the solution ionic strength [6]. However, it should be remembered that lipid peroxidation occurs in aprotic media (lipid membranes) where a pK_a value for perhydroxyl radical is much higher (about 8) and therefore, HOO$^\bullet$-initiated processes become more probable.

Bedard et al. [7] studied quantitatively the initiation of the peroxidation of human low-density lipoproteins (LDL) with HOO$^\bullet$/O$_2^{\bullet-}$. In accord with the above findings the initiation rate increased when pH decreased from 7.6 to 6.5. It was suggested that initiation occurred via hydrogen atom abstraction by perhydroxyl radical from endogenous α-tocopherol, which in this process exhibited prooxidant and not antioxidant properties. Neutral, positively, and negatively charged alkyl peroxyl free radicals were the more efficient initiators of LDL peroxidation compared to superoxide.

Thus, superoxide itself is obviously too inert to be a direct initiator of lipid peroxidation. However, it may be converted into some reactive species in superoxide-dependent oxidative processes. It has been suggested that superoxide can initiate lipid peroxidation by reducing ferric into ferrous iron, which is able to catalyze the formation of free hydroxyl radicals via the Fenton reaction. The possibility of hydroxyl-initiated lipid peroxidation was considered in earlier studies. For example, Lai and Piette [8] identified hydroxyl radicals in NADPH-dependent microsomal lipid peroxidation by EPR spectroscopy using the spin-trapping agents DMPO and phenyl-*tert*-butylnitrone. They proposed that hydroxyl radicals are generated by the Fenton reaction between ferrous ions and hydrogen peroxide formed by the dismutation of superoxide. Later on, the formation of hydroxyl radicals was shown in the oxidation of NADPH catalyzed by microsomal NADPH-cytochrome P-450 reductase [9,10].

However, subsequent studies demonstrated that the formation of hydroxyl radicals, even if it takes place during lipid peroxidation, is of no real importance. Beloqui and Cederbaum [11] have found that although the glutathione–glutathione peroxidase system suppressed hydroxyl radical generation during the oxidation of 4-methylmercapto-2-oxo-butyrate, it exhibited a much smaller effect on microsomal lipid peroxidation. Therefore, hydroxyl radical formation is apparently unimportant in this process. Other authors also pointed out at an unimportant role of hydroxyl radicals in the initiation of microsomal lipid peroxidation [12–14]. For example, it has been shown that Fe(EDTA), a most efficient catalyst of hydroxyl radical formation by the Fenton reaction, inhibited microsomal and liposomal lipid peroxidation, while the weak catalysts of this reaction Fe(ADP) and Fe(ATP) enhanced it [13].

The mechanism of iron-initiated superoxide-dependent lipid peroxidation has been extensively studied by Aust and his coworkers [15–18]. It was found that superoxide produced by xanthine oxidase initiated lipid peroxidation, but this reaction was not inhibited by hydroxyl radical scavengers and, therefore the formation of hydroxyl radicals was unimportant. Lipid peroxidation depended on the Fe^{3+}/Fe^{2+} ratio, with 50:50 as the optimal value [19]. Superoxide supposedly stimulated peroxidation both by reducing ferric ions and oxidizing ferrous ions. As superoxide is able to release iron from ferritin, superoxide-promoted lipid peroxidation can probably proceed under in vivo conditions [16,20].

In contrast to earlier data [21] obtained by the use of cytochrome c reduction assay for detecting superoxide, a high level of superoxide was detected in microsomal lipid peroxidation with the aid of lucigenin-amplified CL method [22]. It was suggested that superoxide reduced Fe^{3+}–ADP or (in the presence of antibiotic doxorubicin) Fe^{3+}–ADP–Dox complexes to generate catalytically active Fe^{2+}–ADP complex. Importance of superoxide as an initiator of microsomal peroxidation was confirmed by the study of inhibitory effects of two antioxidants, rutin and the copper–rutin complex $Cu(Rut)CL_2$ [23]. It was found that the inhibitory effect of the copper–rutin complex on lipid peroxidation was nine times higher than that of rutin and excellently correlated with its effect on lucigenin-amplified CL. As the enhanced inhibitory activity of copper–rutin complex apparently depended on its acquired additional SOD activity, these findings indicate an important role of superoxide in the initiation of microsomal lipid peroxidation.

Thus, despite the inability to initiate lipid peroxidation directly, superoxide might participate on the stage of initiation by indirect pathways. Many earlier studies during 1970 to 1990 have been dedicated to the study of superoxide-mediated lipid peroxidation. (These studies are reviewed in Ref. [24].) Unfortunately, the inability of superoxide to initiate lipid peroxidation directly was not recognized in these studies, and many authors suggested that superoxide reacted with unsaturated compounds by abstracting a hydrogen atom. For example, Thomas et al. [25] suggested that the acceleration of xanthine oxidase-mediated peroxidation of linoleic acid by hydroperoxyoctadecadienoic acid is due to the reaction of this peroxidation product with superoxide, forming a new peroxyl radical. However, it has later been shown that superoxide reacts with hydrogen peroxide and hydroxyperoxides only via a proton-abstracting mechanism [26,27].

In 1977, Kellogg and Fridovich [28] showed that superoxide produced by the XO–acetaldehyde system initiated the oxidation of liposomes and hemolysis of erythrocytes. Lipid peroxidation was inhibited by SOD and catalase but not the hydroxyl radical scavenger mannitol. Gutteridge et al. [29] showed that the superoxide-generating system (aldehyde–XO) oxidized lipid micelles and decomposed deoxyribose. Superoxide and iron ions are apparently involved in the NADPH-dependent lipid peroxidation in human placental mitochondria [30]. Ohyashiki and Nunomura [31] have found that the ferric ion-dependent lipid peroxidation of phospholipid liposomes was enhanced under acidic conditions (from pH 7.4 to 5.5). This reaction was inhibited by SOD, catalase, and hydroxyl radical scavengers. Ohyashiki and Nunomura suggested that superoxide, hydrogen peroxide, and hydroxyl radicals participate in the initiation of liposome oxidation. It has also been shown [32] that SOD inhibited the chain oxidation of methyl linoleate (but not methyl oleate) in phosphate buffer.

The regulation of superoxide formation by SOD can affect both in vivo and ex vivo lipid peroxidation. Thus, SOD inhibited lipid peroxidation in cats following regional intestinal ischemia and reperfusion [33]. Similarly, the treatment of rats with polyethylene glycol superoxide dismutase (PEG-SOD) prevented the development of lipid peroxidation in hepatic ischemia–reperfusion injury [34]. Interesting data have been reported by Bartoli et al. [35]. They showed that SOD depletion in the liver of rats feeding with a copper-deficient diet resulted in a decrease in polyunsaturated fatty acids, an increase in monosaturated acid contents, and a decrease in in vitro microsomal lipid peroxidation. They proposed that the loss of SOD led to the oxidation of polyunsaturated acids and the enhancement of the synthesis of more saturated fatty acids by cells. Such phenomenon probably occurs in fast-growing hepatomas, which exhibit a noticeable saturation of fatty acids and diminishing SOD activity, which can be an origin of hepatocarcinigenesis. Superoxide supposedly initiated acetaldehyde-mediated lipid peroxidation in the pathogenesis of alcohol-induced liver injury [36].

It is of interest that under certain conditions SOD can manifest both inhibitory and stimulatory effects on lipid peroxidation. Thus, Nelson et al. [37] showed that small SOD concentrations (up to 5 μg ml^{-1}) were protective in the reoxygenated isolated myocardium but very high doses (50 μg ml^{-1}) exacerbated the injury. These authors hypothesized that superoxide may not only initiate lipid peroxidation but also inhibited it, reacting with free radicals:

$$O_2^{\bullet-} + LO^{\bullet} \Longrightarrow O_2 + LO^- \tag{2}$$

$$O_2^{\bullet-} + LOO^{\bullet} \Longrightarrow O_2 + LOO^- \tag{3}$$

In such a case the inhibitory or stimulatory effect of SOD must indeed depend on its concentration.

25.1.2 NO, NO Metabolites, and HOCL

The role of another physiological free radical NO in lipid peroxidation is probably even more complicated than that of superoxide. Similar to superoxide, NO is incapable of abstracting a hydrogen atom from unsaturated substrates and similar to superoxide, NO may form various reactive species capable of initiating lipid peroxidation. During 1992 to 1993, it has been shown that the effect of nitric oxide on lipid peroxidation is mostly an inhibitory one. Thus, Jessup et al. [38] found that NO can oxidize LDL only together with superoxide, and rather protective in macrophage-mediated LDL oxidation. Nitric oxide inhibited LDL oxidation, the formation of TBAR products, and lipid hydroperoxides [39]. Yates et al. [40] studied the effect of mouse peritoneal macrophages on human LDL. Macrophages were stimulated with interferon-γ and TNF-α, resulting in an increase in NO production by tenfold. It was found that nitric oxide was inhibitory to macrophage-induced LDL oxidation. Laskey and Mathews [41] compared the effects of peroxynitrite and nitric oxide on the peroxidation of phosphatidylcholine liposomes. They found that peroxynitrite caused significant liposome peroxidation characterized by the increased formation of hydroperoxy- and hydroxyeicosatetraenoic acids (HETEs) and F$_2$-isoprostanes, while nitric oxide inhibited both iron- and peroxynitrite-initiated lipid peroxidation.

Thus the competition between stimulatory and inhibitory effects of NO depends on the competition between two mechanisms: the direct interaction of NO with free radicals formed in lipid peroxidation and the conversion of NO into peroxynitrite or other reactive NO metabolites. Based on this suggestion, Freeman and his coworkers [42–44] concluded that the prooxidant and antioxidant properties of nitric oxide depend on the relative concentrations of NO and oxygen. It was supposed that the prooxidant effect of nitric oxide originated from its reaction with dioxygen and superoxide:

$$2NO + O_2 \Longrightarrow 2^{\bullet}NO_2 \tag{4}$$

$$NO + O_2^{\bullet-} \Longrightarrow ONOO^- \tag{5}$$

(It should be noted that Reaction (4) is not a one-stage process.) Both free radical $^\bullet NO_2$ and highly reactive peroxynitrite are the initiators of lipid peroxidation although the elementary stages of initiation by these compounds are not fully understood. (Crow et al. [45] suggested that *trans*-ONOO$^-$ is protonated into *trans* peroxynitrous acid, which is isomerized into the unstable *cis* form. The latter is easily decomposed to form hydroxyl radical.) Another possible mechanism of prooxidant activity of nitric oxide is the modification of unsaturated fatty acids and lipids through the formation of active nitrated lipid derivatives.

Initiation of lipid peroxidation by nitric oxide and peroxynitrite has been studied in many publications. Darley-Usmar et al. [46] showed that simultaneous production of superoxide and nitric oxide by sydnonimine SIN-1 initiated the peroxynitrte-mediated peroxidation of LDL. Later on, it has been shown [47] that various nitrogen species formed during inflammation may react with unsaturated fatty acids, forming nitrated oxidation products. Peroxynitrite formed with linoleic acid nitrated lipids (their formation was inhibited by SOD, ferric-EDTA, and bicarbonate) and oxidized lipids (which were inhibited by SOD and bicarbonate, but not ferric-EDTA). The same nitrated lipid products were formed in the reactions of linoleic acid with nitric dioxide $^\bullet NO_2$. Peroxynitrite and its by-products reacted with the unsaturated lipid components and damaged surfactant proteins, forming conjugated dienes and MDA [48]. Shi et al. [49] studied the oxidation of phospholipids in rat brain synaptosomes by peroxynitrite generated from 3-morpholinosydnonimine SIN-1. This reaction resulted in the formation of phospholipid hydroperoxides, including phosphatidylcholine and phosphatidylethanolamine hydroperoxides. Endogenous α-tocopherol potently inhibited peroxidation and was very rapidly oxidized to α-tocopheryl quinone. Another effective antioxidant was uric acid.

As mentioned earlier, when NO concentration exceeds that of superoxide, nitric oxide mostly exhibits an inhibitory effect on lipid peroxidation, reacting with lipid peroxyl radicals. These reactions are now well studied [42–44]. The simplest suggestion could be the participation of NO in termination reaction with peroxyl radicals. However, it was found that NO reacts with at least two radicals during inhibition of lipid peroxidation [50]. On these grounds it was proposed that LOONO, a product of the NO recombination with peroxyl radical LOO$^\bullet$ is rapidly decomposed to LO$^\bullet$ and $^\bullet NO_2$ and the second NO reacts with LO$^\bullet$ to form nitroso ester of fatty acid (Reaction (7), Figure 25.1). Alkoxyl radical LO$^\bullet$ may be transformed into a nitro epoxy compound after rearrangement (Reaction (8)). In addition, LOONO may be hydrolyzed to form fatty acid hydroperoxide (Reaction (6)). Various nitrated lipids can also be formed in the reactions of peroxynitrite and other NO metabolites.

It is important that NO is a much more effective scavenger of peroxyl radicals than α-tocopherol because the rate constant for the reaction of LOO$^\bullet$ with NO is equal to 2×10^9 l mol^{-1} s^{-1} and that for the reaction with α-tocopherol is much smaller (about 2×10^5 l mol^{-1} s^{-1}). Therefore, the in vivo concentrations of nitric oxide (up to 2 μmol) may effectively compete with endogenous concentrations of α-tocopherol. Antioxidant function of nitric oxide in LDL is also determined by the NO capacity to traverse the LDL surface and penetrate the lipid core of the LDL particle [51]. D'Ischia et al. [52] found that nitric oxide suppressed lipid peroxidation in rat brain homogenates. It was suggested that NO inhibited the formation of oxidation products by decomposing primary lipid peroxide such as 15-HPETE. Competition between the stimulatory and inhibitory effects of nitric oxide may lead to different results in different organs. Thus, at high oxygen concentrations in lung lining fluid, the NO metabolite NO$_2$ may predominantly mediate lipid peroxidation. In contrast, within inflamed hypoxic organs nitration reactions may terminate free radical processes [50].

Another reactive species, which is also able to initiate lipid peroxidation, is hypochlorous acid HOCl generated by the myeloperoxidase (MPO)–hydrogen peroxide–chloride ion system. Although this system is mainly considered below in the section dedicated to LDL

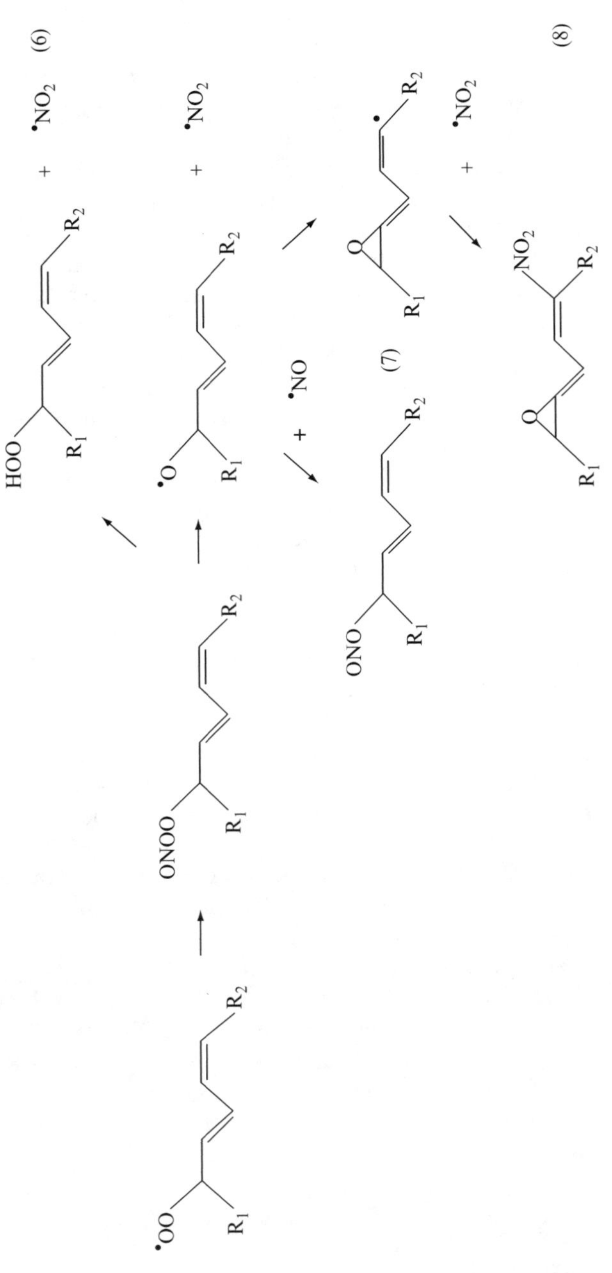

FIGURE 25.1 Mechanism of the inhibitory effect of nitric oxide on lipid peroxidation. (Adapted from VB O'Donnell, BA Freeman. *Circ Res* 88: 12–21, 2001.)

oxidation, here we will look at the MPO-catalyzed chlorination of some other lipid compounds. Thus, Thukkani et al. [53] studied the chlorination of plasmalogens, which are glycerophospholipids present in the plasma membranes of mammalian tissues. It was found that PMA-stimulated neutrophils chlorinated these compounds through the MPO-dependent mechanism to form α-chloro fatty aldehydes, 2-chlorohexadecanal, and 2-chlorooctadecanal, targeting 16 and 18 carbon vinyl ether-linked aliphatic groups of neutrophil plasmalogens. It should be noted that α-chloro fatty aldehydes play a physiological role participating in neutrophils recruitment.

25.1.3 Xenobiotics as Initiators of Lipid Peroxidation

Many organic and inorganic compounds including drugs, components of food, ozone, environment contaminants, etc. can initiate lipid peroxidation. There are two major mechanisms of xenobiotic-dependent lipid peroxidation: these compounds could be reductants, i.e., compounds which are able to reduce some biological molecules or they could be reduced by enzymes or biological substrates and initiate lipid peroxidation in their reduced forms. A well-known example of first mechanism is the ascorbate-dependent iron-initiated lipid peroxidation where ascorbic acid reacts as a prooxidant, reducing ferric ions or complexes into ferrous ones. It should be mentioned that the other classic antioxidants such as α-tocopherol, glutathione, and SOD might also initiate lipid peroxidation under certain conditions (Chapter 29).

The second mechanism is realized when organic or inorganic compounds are reduced by endogenous reductants (for example, by NADH or NADPH and the other components of mitochondrial or microsomal respiratory chains). The typical compounds are anthracycline antibiotics and carbon tetrachloride. CCl_4 is easily reduced by microsomes to the free radical $CCl_3^{\bullet}$, which is able to abstract a hydrogen atom from unsaturated lipids and initiate lipid peroxidation. Because of this, the CCl_4-initiated lipid peroxidation is a reliable and frequently applied model system for the study of in vitro iron-independent lipid peroxidation and the effects of antioxidants (see for example Ref. [54]).

Anthracycline antibiotics such as doxorubicin (andriamycin) are powerful anticancer drugs. These compounds contain an anthraquinone moiety, which makes them strong prooxidants (Figure 25.2). A great interest in the prooxidant activity of anthracyclines has been caused by a suggestion that this activity is the major reason of anthracycline cardiotoxicity, which limits a success of the treatment of cancer patients with these antibiotics. Owing to that, many studies have been dedicated to the investigation of anthracycline-stimulated lipid peroxidation. In 1977, Goodman and Hochstein [55] showed that anthracycline antibiotics

R is daunosamine

Adriamycin

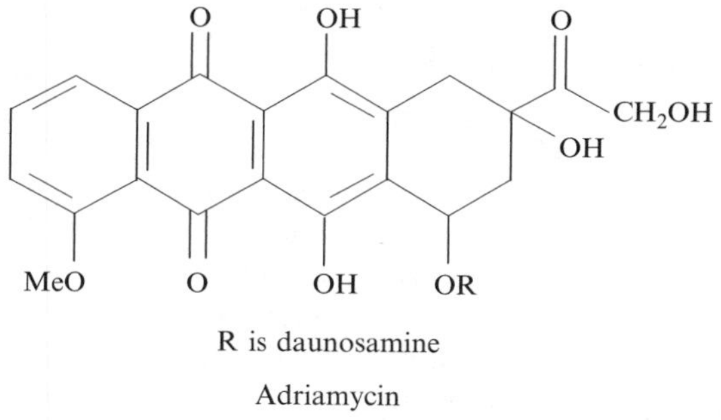

FIGURE 25.2 Adriamycin (Doxorubicin).

doxorubicin and daunorubicin stimulated microsomal NADPH–cytochrome P-450 reductase-mediated lipid peroxidation. This work was followed by numerous studies of the effects of anthracyclines on in vitro and in vivo lipid peroxidation [56]. It has been widely accepted that anthracyclines are reduced by mitochondrial or microsomal respiratory chain components to radical anions, which are able to reduce dioxygen to superoxide, a genuine initiator of lipid peroxidation:

$$\text{Anthr} + e \Longrightarrow \text{Anthr}^{\bullet -} \qquad (9)$$

$$\text{Anthr}^{\bullet -} + O_2 \Longrightarrow \text{Anthr} + O_2^{\bullet -} \qquad (10)$$

However, later on, this simple mechanism has been questioned. Although Reactions (9) and (10) can occur, anthracyclines are also capable of reacting irreversibly with superoxide [57,58]. This reaction apparently results in deglycosidation, making the redox cycling of anthracyclines (Reactions (9) and (10)) less probable. Moreover, the anthracycline stimulation of lipid peroxidation takes place probably only in the presence of iron ions. On these grounds, it has been proposed that genuine initiation species are the iron–anthracycline complexes [59,60]. It seems that not only the so-called "free iron" may react with anthracyclines forming iron–anthracycline complexes but also heme-containing proteins, for example myoglobin [61]. Interesting observations have been made by Vile and Winterbourn [62] who found out that adriamycin-stimulated microsomal lipid peroxidation was efficiently inhibited by α-tocopherol at high oxygen pressure, whereas α-tocopherol became relatively ineffective at low oxygen pressure.

Mimnaugh et al. [63] showed that endogenous antioxidant glutathione effectively inhibited adriamycin-dependent lipid peroxidation. Later on, the mechanism of inhibitory effect of glutathione was studied by Powell and McCay [64]. These authors demonstrated that glutathione inhibited adriamycin-stimulated lipid peroxidation if only microsomes contained α-tocopherol and did not affect it in the α-tocopherol-depleted microsomes. It was suggested that α-tocopherol is needed as a shuttle between aqueous phase containing glutathione and lipid membranes where lipid peroxidase takes place. Yen et al. [65] found that cardiac damage was suppressed by MnSOD overexpression, suggesting that superoxide and not hydroxyl or hydroxyl-like free radicals contributes to adriamycin toxicity. In recent work Konorev et al. [66] showed that the toxic effects of adriamycin were enhanced by bicarbonate, which increased the superoxide-mediated cardiomyocyte injury and enhanced intracellular loading of adriamycin.

At present, new data have been obtained concerning the role of lipid peroxidation and other free radical-mediated processes in adriamycin-induced cardiotoxicity [67]. An important objection to the radical-mediated mechanism of its cardiotoxicity is a well-known fact that while the chelator ICRF-187 (dexrazoxane) and antioxidants vitamin E and N-acetylcysteine suppressed cardiotoxicity in rats and mice and in in vitro model systems, they could not prevent or significantly reduce it in larger animals (dogs) and were ineffective in clinical trials. Therefore, the iron-independent mechanism of adriamycin-induced cardiotoxicity has been proposed. Furthermore, it has been found that adriamycin may actually decrease and not increase the myocardial release of hydroperoxides and conjugated dienes, reacting with free radicals [68]. (It should be mentioned that the irreversible reaction between superoxide and adriamycin has been shown much earlier [69].) Nonetheless, in the last work Minotti et al. [67] concluded that despite the criticism, the iron-initiated free radical-mediated mechanism plays an important role in adriamycin-stimulated cardiotoxicity, especially on its acute stage where the effects of chelators and antioxidants are prominent. These authors proposed to use free radical scavengers able to react with a wide range of free radicals, for example the spin-trapping agents nitroxides, for the suppression of adriamycin toxicity.

Many other substances possessing prooxidant properties are also able to initiate lipid peroxidation. Several examples are cited below. Nephrotoxicity of allopurinol, a drug widely used for the treatment of hyperuricacidemia, apparently depends on an increase in kidney lipid peroxidation [70]. Mineral particles and fibers can catalyze the production of oxygen radicals and initiate oxidative processes in biological systems [71]. Administration of the mixtures of dioxins containing 2,3,7,8-tetrachlorodibenzo-p-dioxin to rats resulted in an increase in superoxide production, lipid peroxidation, and DNA single-strand breaks in the hepatic and brain tissues [72]. The treatment of cultured human skin fibroblasts with thiram, a widely used dithiocarbamate fungicide, resulted in the depletion of reduced glutathione and a decrease in the glutathione reductase (GR) activity [73]. Decrease in the GSH content and GR activity were supposedly the main reason for thiram-induced lipid peroxidation in these cells.

In contrast to transition metals iron and copper, which are well-known initiators of in vitro and in vivo lipid peroxidation (numerous examples of their prooxidant activities are cited throughout this book), the ability of nontransition metals to catalyze free radical-mediated processes seems to be impossible. Nonetheless, such a possibility is suggested by some authors. For example, it has been suggested that aluminum toxicity in human skin fibroblasts is a consequence of the enhancement of lipid peroxidation [74]. In that work MDA formation was inhibited by SOD, catalase, and vitamins E and C. It is possible that in this case aluminum is an indirect prooxidant affecting some stages of free radical formation.

25.1.4 PHAGOCYTES AS INITIATORS OF LIPID PEROXIDATION

It is well known that neutrophils, monocytes, macrophages, and other phagocytes produce superoxide upon activation with various stimuli and therefore, are potential initiators of lipid peroxidation. In 1985, Carlin and Arfors [75,76] showed that leukocytes initiate the oxidation of unsaturated lipids. Surprisingly, the leukocyte-initiated peroxidation of linoleic acid was not inhibited by SOD and, therefore, apparently was not initiated by superoxide, while liposome peroxidation was mediated by superoxide. No convincing explanations were given.

Rodenas et al. [77] studied PMN-stimulated lipid peroxidation of arachidonic acid. As MDA formation was inhibited both with L-arginine (supposedly due to the formation of excess NO) and DTPA (an iron ion chelator), it was concluded that about 40% of peroxidation was initiated by hydroxyl radicals formed via the Fenton reaction and about 60% was mediated by peroxynitrite. However, it should be noted that the probability of hydroxyl radical-initiated lipid peroxidation is very small (see above). Phagocyte-mediated LDL oxidation is considered below.

25.2 LIPID PEROXIDATION OF UNSATURATED FATTY ACIDS

Unsaturated fatty acids are probably the most abundant oxidizable endogenous substrates. In the past it was erroneously believed that unsaturated fatty acids are just products of lipid peroxidation. Now, it has been shown that they have dietary origin. Family of unsaturated fatty acids includes linoleic (C_{18}), arachidonic (C_{20}), docosahexaenoic (C_{22}), and other fatty acids containing two, three, four, five, or six double bonds. Some acids can be in vivo converted into others; for example, linoleic acid can be metabolized to linolenic and eicosatrienoic acids [78].

All unsaturated fatty acids contain highly reactive allylic positions, which are easily attacked by hydroxyl and peroxyl free radicals. Bielski et al. [3] studied the reactivity of unsaturated acid with perhydroxyl radical by the stopped flow technique. These authors found that the HOO• radical reacts with compounds having the bisallylic hydrogen atoms (linoleic, linolenic, and arachidonic acids) with rate constants equal to $1-3 \times 10^3$ l mol^{-1} s^{-1}, while there was no reaction with compounds having one double bond (oleic acid) or two conjugated double bonds (9,11- and 10,12-octadecadienoic acids). Extreme importance of bisallylic positions for the oxidizability of cellular lipids has later been confirmed by Wagner et al. [79] who measured the formation of lipid free radicals in murine leukemia cells by ESR spin-trapping method. Thus, an initiation step of the oxidation of unsaturated compounds must be the formation of allylic free radicals, which is followed by the addition of dioxygen. The fate of the peroxyl radicals formed depends on many factors including the lifetime of a peroxyl radical, its reactivity, the structures of neighboring molecules, etc. Peroxyl radicals can abstract a hydrogen atom from neighbors to form hydroperoxides or rearrange into cyclic free radicals; the last pathway leads to very important prostanoic compounds (Figure 25.3). Hydroperoxides of unsaturated fatty acids are unstable compounds and decomposed (metabolized) into various products, the most important and abundant ones are MDA and 4-hydroxynonenal (4-HNE).

4-HNE was discovered in peroxidation of unsaturated fatty acids by Esterbauer and co-workers [80]. It has been suggested that 4-HNE was formed in NADPH-dependent microsomal peroxidation exclusively from arachidonic acid in polar phospholipids [81]. 4-HNE is a very toxic compound, which stimulates many damaging processes in a living organism. Despite numerous studies of its role in physiological and pathophysiological processes, the mechanism of in vivo 4-HNE formation is not fully understood. It is now recognized that arachidonic acid is not the only precursor of 4-HNE formation; linoleic acid is another candidate [82]. One of the possible routes of the conversion of linoleic acid to 4-HNE was proposed by Schneider et al. [82]. In accord with this mechanism, hydroperoxides of linoleic acid (9-hydroperoxy-10, 12-octadecadienoic acid (9-HPODE) and 13-hydroperoxy-9,11-octadecadienoic acid (13-HPODE)) are oxidized into 4-hydroperoxy-2-nonenal (4-HPNE), a precursor of 4-HNE. (This mechanism is shown in Figure 25.4 [82] for the oxidation of 13-HPODE.)

9-Hydroxy-10,12-octadecadienoic acid, which is formed by the reduction of 9-HPODE, was identified in the erythrocyte membrane phospholipid of diabetic patients [83]. It was suggested that this compound was formed as a result of glucose-induced oxidative stress in the reaction of hydroxyl radicals with linoleic acid.

25.3 CHOLESTEROL OXIDATION

The oxidation of cholesterol is one of the most important peroxidation processes, which takes place in plasma and LDL. Although cholesterol oxidation is undoubtedly a marker of the enhanced oxidative stress under some pathophysiological conditions, this process apparently occurs to some degree in healthy persons. In 1989, Yamamoto and Niki [84] demonstrated that about 3 nmol l^{-1} cholesteryl ester hydroperoxides (CE-OOH), mainly cholesteryl linoleate hydroperoxides (Ch18:2-OOH) was formed in blood plasma from healthy individuals. It is interesting that the ratio of CE-OOH/CE (CE is a cholesteryl ester) increases in the range: humans < Sprague–Dawley rats < Nagase analbuminemic rats, and this order possibly correlates with the lifespans of humans and rats [85].

The oxidation of cholesteryl linoleate (Ch18:2), a major cholesteryl ester in human blood plasma, results in the formation of main primary oxidation products: cholesteryl 13-hydroperoxy-9Z,11E-octadecadienoate (13ZE-Ch18:2-OOH), cholesteryl 13-hydroperoxy-9E, 11E-octadecadienoate (13EE-Ch18:2-OOH), cholesteryl 9-hydroperoxy-10E,12Z-octadecadienoate (9EZ-Ch18:2-OOH), and cholesteryl 9-hydroperoxy-10E,12E-octadecadienoate (9EE-Ch18:2-OOH) (Figure 25.5). As in the peroxidation of unsaturated fatty acids, the attack of peroxyl radicals is directed on the bisallylic methylene group of cholesterol

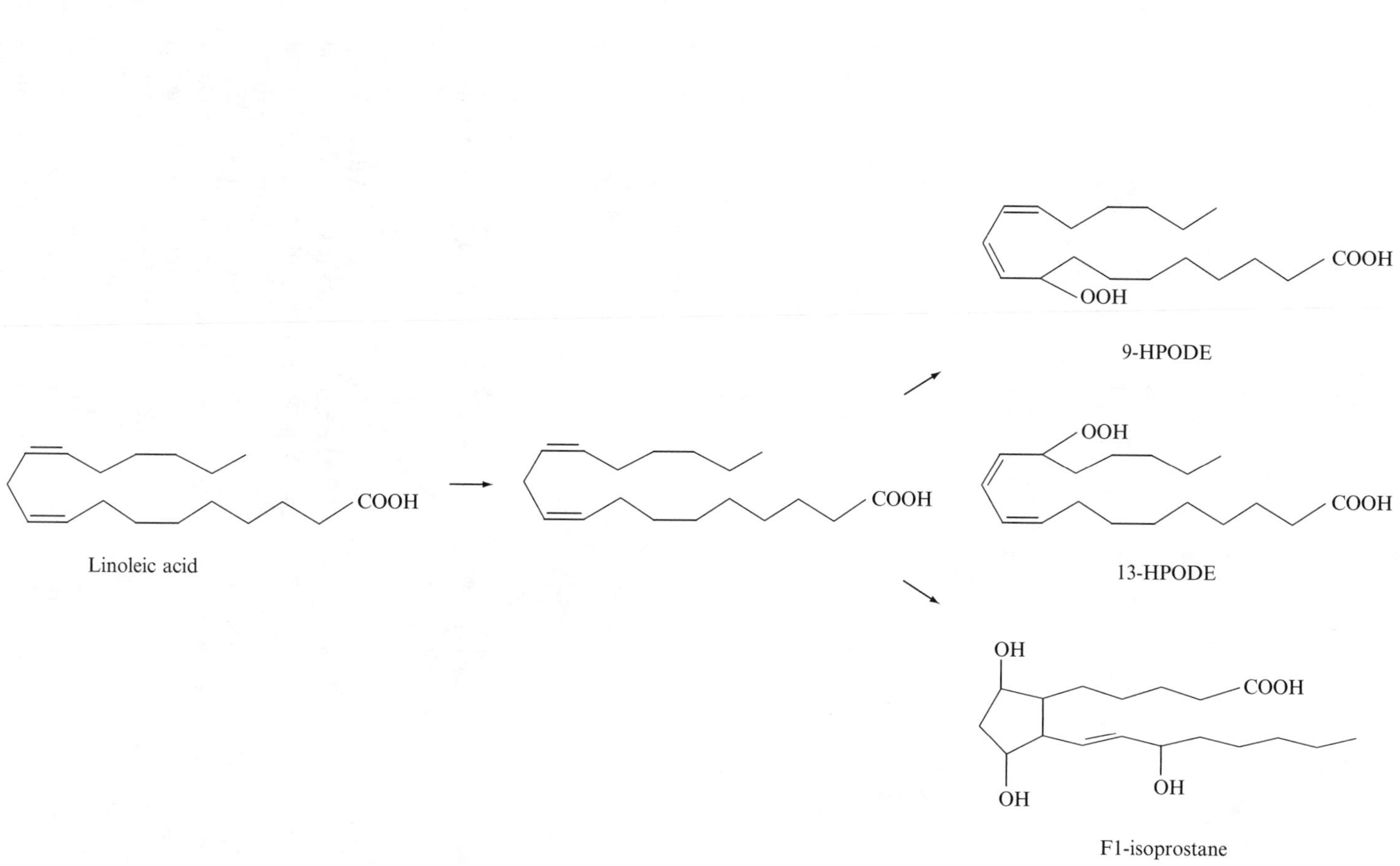

FIGURE 25.3 The formation of linear hydroperoxides and cyclic prostanoids during peroxidation of unsaturated acids with bisallylic positions.

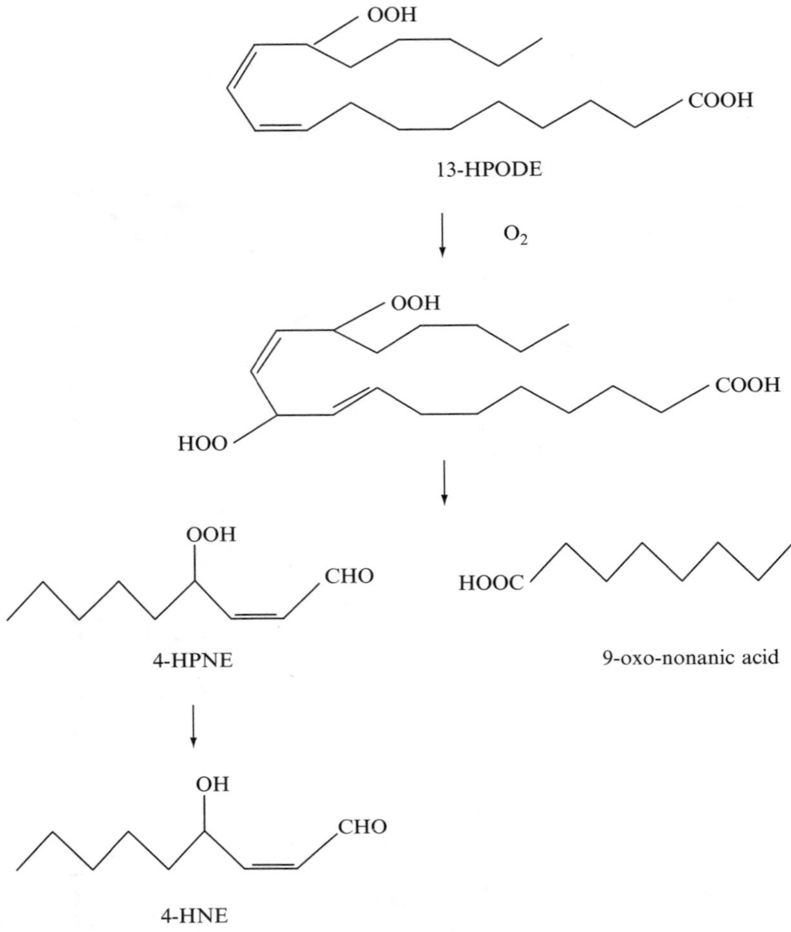

FIGURE 25.4 Mechanism of conversion of 13-hydroperoxy-9,11-octadecadienoic acid (13-HPODE) into 4-hydroxynonenal (4-HNE). (Adapted from C Schneider, KA Tallman, NA Porter, AR Brash. *J Biol Chem* 276: 20831–20838, 2001. With permission.)

molecule. The mechanism of the formation of isomeric cholesterol hydroperoxides is also similar to that for lipid peroxidation of fatty acids. Mashima et al. [86] suggested that the nonenzymatic oxidation of cholesteryl esters is a predominant pathway of lipid peroxidation in blood plasma from healthy humans because the major product of lipoxygenase-catalyzed oxidation 13ZE-Ch18:2-OOH (Chapter 26) was not a major product of nonenzymatic peroxidation. Furthermore, the occurrence of nonenzymatic lipid peroxidation in the blood of healthy humans is supported by the presence of isoprostanes, the nonenzymatic oxidation products of arachidonic acid (see below). However, it should be noted that the amount of isoprostanes in plasma from healthy humans is equal to 0.1–0.3 nmol l^{-1}, which is significantly lower than the plasma Ch18:2-OOH level equal to 13.6 nmol l^{-1} [86].

Although significance of cholesterol hydroperoxides formation under physiological conditions is still unknown, they are apparently very important factors in the development of many pathophysiological disorders. Thus it has been shown [87] that 13ZE-Ch18:2-OOH exists in vivo in atherosclerotic lesions and is the primary toxin of oxidized human LDL.

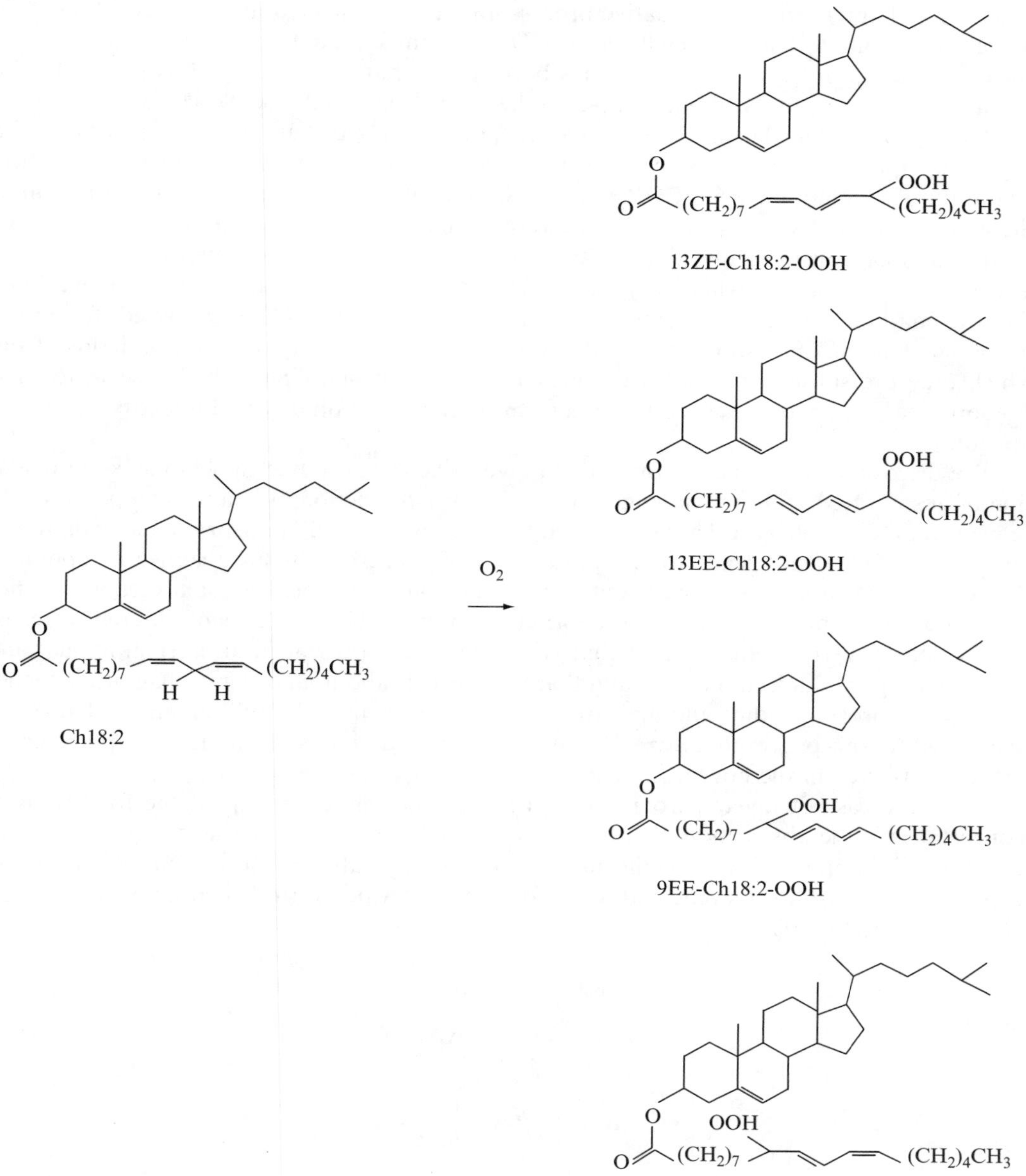

FIGURE 25.5 Oxidation of cholesteryl linoleate (Ch18:2).

25.4 THE FORMATION OF PROSTANOIDS

The unique characteristic of free peroxyl radicals formed from unsaturated fatty acids is their ability to transform into cyclic radicals. This reaction is of utmost importance because it leads to highly biologically active compounds. Enzymatic oxidation of arachidonic acid catalyzed by COX results in the formation of prostaglandins having various physiopathological functions (Chapter 26). This oxidative process might be considered as "civilized" biological oxidation because it is strictly regulated by COX to form a limited number of physiologically active products. However, in 1975, it has been found that total endogenous prostaglandin production in humans is about tenfold higher than that could be expected from COX-catalyzed process [88]. Then, Morrow et al. [89] demonstrated that a series of prostaglandin F_2-like compounds are formed in vivo in humans by the non-COX-catalyzed free radical oxidation of arachidonic acid. COX inhibitors, as a rule, did not suppress the formation of these compounds (F_2-isoprostanes). On the other hand, the administration of prooxidants (diquat and carbon tetrachloride) to rats resulted in up to 200-fold increase in circulating levels of F_2-isoprostanes, while butylated hydroxytoluene (BHT) nearly completely inhibited their formation. It has been found that in contrast to the COX-catalyzed formation of prostaglandins, F_2-isoprostanes are initially formed in situ on phospholipids, from which they are subsequently released, presumably by phospholipases [90]. Owing to that, F_2-isoprostanes may enhance oxidant injury by affecting the fluidity and integrity of cellular membranes.

The discovery of products formed by in vivo nonenzymatic free radical-mediated oxidation of unsaturated acids and having pathophysiological functions was a turning point in free radical studies in biology. It has been shown that the "uncivilized" nonenzymatic oxidation of unsaturated fatty acids leads to the production of numerous highly reactive compounds. Peroxidation of arachidonic (eicosatetraenoic), eicosapentaenoic, docosahexaenoic, α-linolenic, and γ-linolenic acids leads to the formation of probably hundreds of different isoprostane molecules [91]. It is important that isoprostanes are present in all normal animal and human biological fluids and tissues and that their level is about an order of magnitude higher than that of prostaglandins and increases dramatically in animals and humans under oxidative stress. All this proves that nonenzymatic lipid peroxidation is incompletely suppressed by antioxidants even in the normal state [92].

As in the case of linear peroxidation products, the initiation step of the formation of isoprostanes is the abstraction of a hydrogen atom from unsaturated acids by a radical of initiator. Initiation is followed by the addition of oxygen to allylic radicals and the cyclization of peroxyl radicals into bicyclic endoperoxide radicals, which form hydroperoxides reacting with hydrogen donors.

$$R_i^{\bullet} + AA \Longrightarrow R_iH + AA^{\bullet} \tag{11}$$

$$AA^{\bullet} + O_2 \Longrightarrow AAOO^{\bullet} \tag{12s}$$

$$AAOO^{\bullet} \Longleftrightarrow \underset{O}{\overset{O}{\diagdown}}AA^{\bullet} \tag{13}$$

$$\underset{O}{\overset{O}{\diagdown}}AA^{\bullet} + O_2 \Longrightarrow \underset{O}{\overset{O}{\diagdown}}AAOO^{\bullet} \Longrightarrow \underset{O}{\overset{O}{\diagdown}}AAOOH \tag{14}$$

Finally, hydroperoxides are reduced to trihydroxy compounds (Figure 25.6). As seen from Figure 25.6, the oxidation of AA resulted in the formation of four F_2-isoprostane regioisomers, each of which is a mixture of eight racemic diastereomers. It is important that the level of F_2-isoprostanes in normal human plasma and urine are one to two orders of magnitude higher than the level of COX-derived prostaglandins.

As follows from the above mechanism, to form stable trihydroxy compounds, bicyclic endoperoxides must be reduced. It was found that glutathione may be an efficient reductant

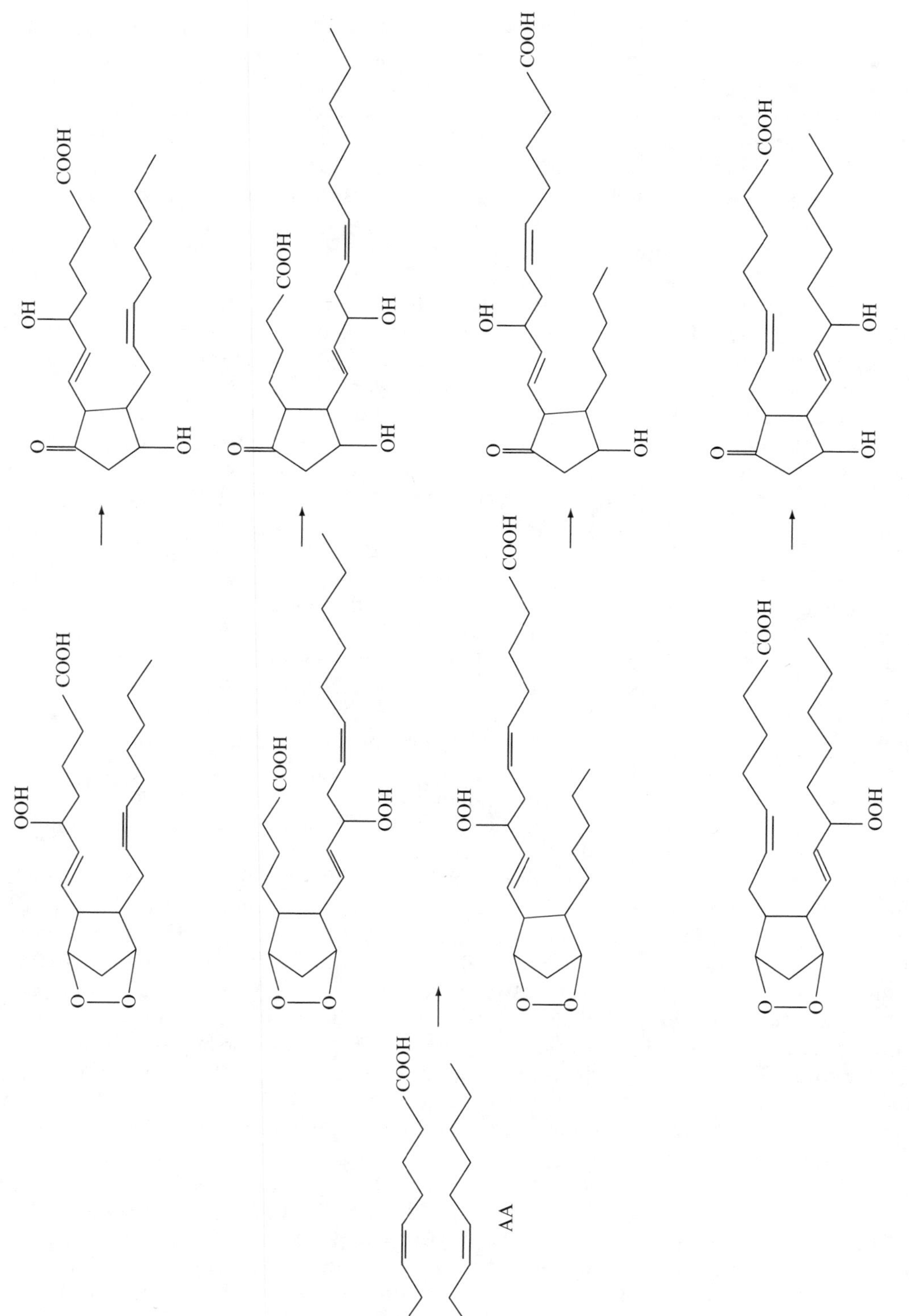

FIGURE 25.6 The formation of F_2-isoprostane by the oxidation of arachidonic acid.

[93]. But, endoperoxides are not very stable compounds, and therefore, there is always the competition between their reduction and rearrangement. Therefore, in addition to F_2-isoprostanes the compounds with a prostane D-ring and E-ring (D_2- and E_2-isoprostanes) and thromboxane-like compounds are also formed in vivo as esterified to phospholipids and in free forms by rearrangement of endoperoxides [91] (Figure 25.6 and Figure 25.7). However, the products of "uncivilized" free radical AA oxidation are not only a mixture of F_2-, D_2-, and E_2-isoprostanes. Unstable endoperoxides may also rearrange into acyclic γ-ketoaldehydes [94] termed E_2 and D_2 isolevuglandins (IsoLG) (Figure 25.7). It has been found that the in vitro oxidation of arachidonic acid catalyzed by iron/ADP/ascorbate resulted in approximately the same amount of IsoLG and F_2-isoprostanes, with D_2-, and E_2-isoprostanes as the major oxidation product [92]. This suggests that IsoLG formation could be of biological importance. Isolevuglandins are extremely reactive compounds and therefore in vivo they immediately form protein adducts [92]. Owing to their remarkable reactivity their formation might play an important role in settlings of oxidant injury.

Although Reactions (11)–(14) apparently describe correctly the in vivo mechanism of isoprostane formation, some questions remain unanswered. For example, the structure of an initiator still remains unknown. It was proposed that "oxygen-centered radicals such as peroxide and superoxide can react with unsaturated bonds of arachidonic acid, leading to the formation of as many as four different bicycloendoperoxide intermediates ... " [91]. It is difficult to agree with such a conclusion because the addition reactions are not typical for superoxide (2) and the radical of initiator must abstract a hydrogen atom from AA and not to add to double bond. (It should be also noted that peroxide is not a free radical.) It has also been suggested that peroxynitrite could be a good candidate as an initiator of AA oxidation [95]. Peroxynitrite is not also a free radical, but it can be a precursor of hydroxyl radicals formed during its decomposition. In addition to hydroxyl radicals, the most probable initiating free radicals should be peroxyl radicals. Lipid peroxyl radicals are formed by numerous pathways in biological systems, but maybe the most temping route is the formation of peroxyl radicals by the COX-catalyzed oxidation of arachidonic acid.

One should expect that arachidonic acid is not a unique unsaturated fatty acid able to be oxidized by "uncivilized" nonenzymatic pathway because any unsaturated fatty acid containing bisallylic methylene groups can probably be oxidized into isoprostanes. At present, the formation of isoprostanes was shown for the compounds with three, four, five, and six double bonds (γ-linolenic, arachidonic, eicosapentaenoic, and docosahexaenoic acids (DHAs), respectively) under both in vitro and in vivo conditions [91,96]. Nonenzymatic oxidation of DHA is of a special importance for oxidative injury in brain because DHA is highly enriched in brain gray matter. Similar to AA, DHA oxidation may lead to the formation of all classes of isoprostanes (named neuroprostanes [96]) (Figure 25.8) but their number is much greater. Furthermore, DHA is a highly oxidizable compound (having six double bonds against four in AA); due to that, the amount of neuroprostanes in brain exceeds the levels of isoprostanes formed from AA by 3.4 times. Neuroprostanes were identified in vivo in human and rat brain tissue [97]. It was found that the level of neuroprostanes is significantly enhanced in patients with Alzheimer's disease [96].

In the last decade numerous studies were dedicated to the study of biological role of nonenzymatic free radical oxidation of unsaturated fatty acids into isoprostanes. This task is exclusively difficult due to a huge number of these compounds (maybe many hundreds). Therefore, unfortunately, the study of several isoprostanes is not enough to make final conclusions even about their major functions. F_2-isoprostanes were formed in plasma and LDL after the treatment with peroxyl radicals [98]. It is interesting that their formation was observed only after endogenous ascorbate and ubiquinone-10 were exhausted, despite the presence of other antioxidants such as urate or α-tocopherol. LDL oxidation was followed by

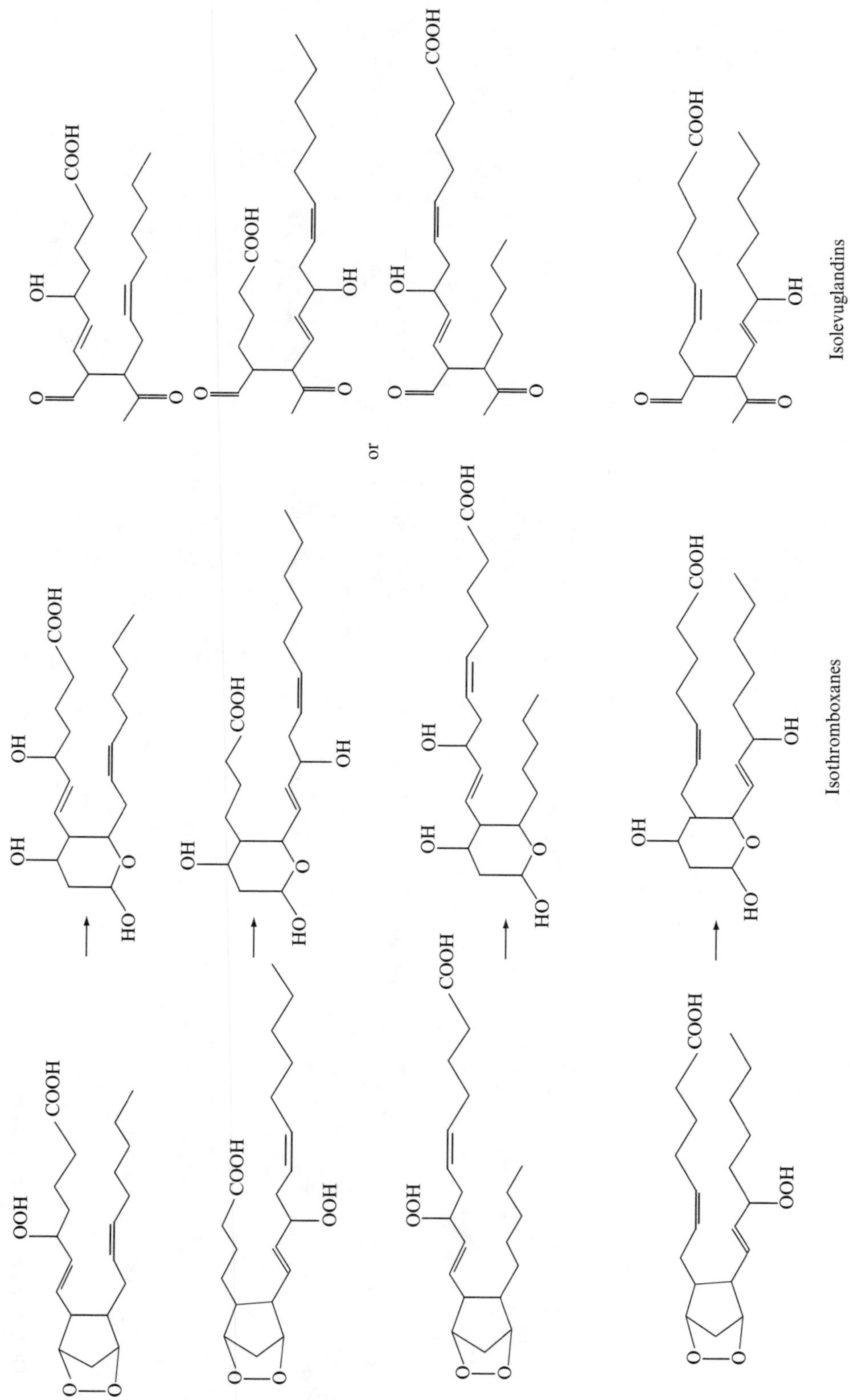

FIGURE 25.7 The formation of isothromboxanes and isolevuglandins.

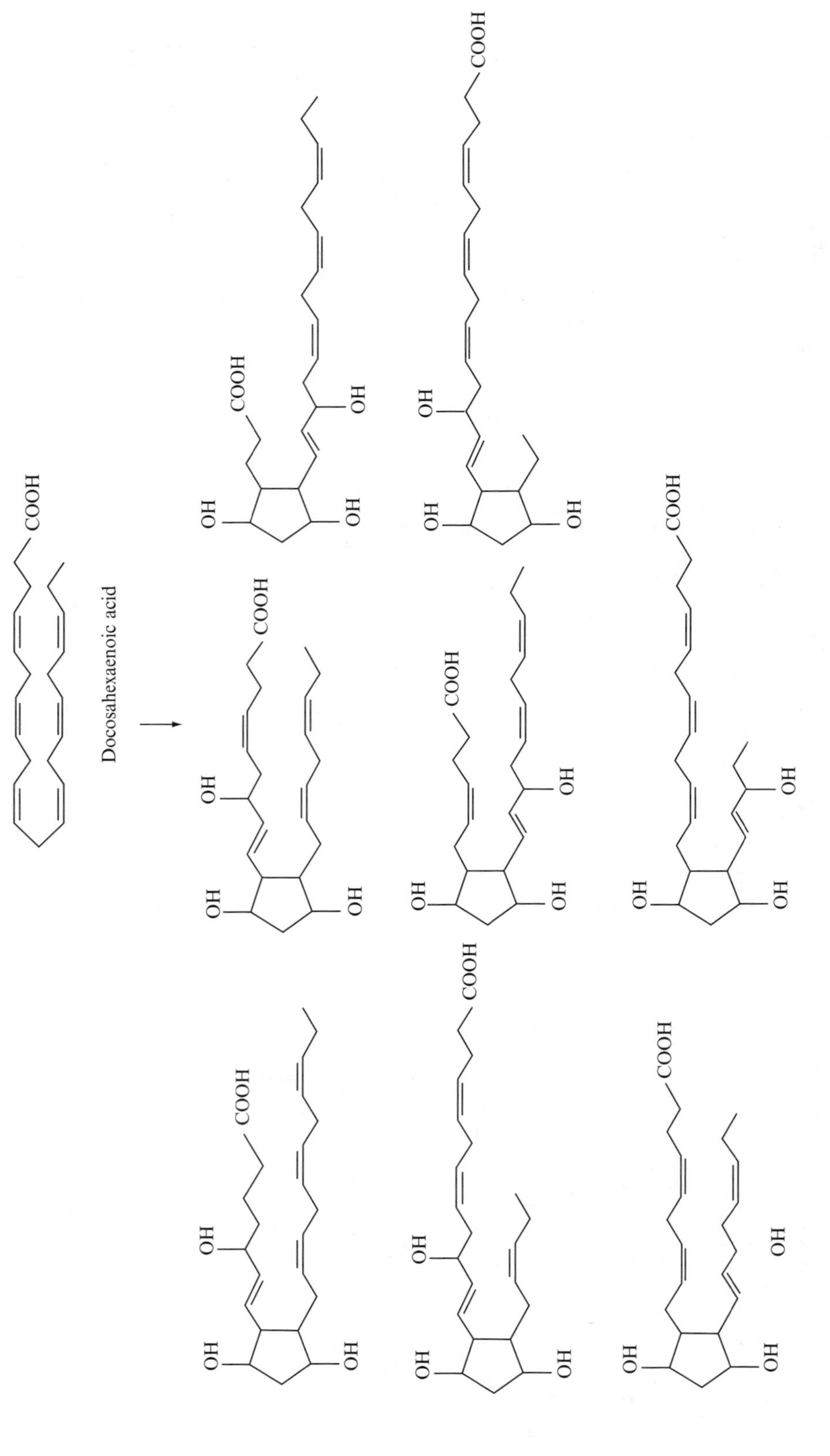

FIGURE 25.8 The formation of neuroprostanes.

rapid formation of lipid hydroperoxides and isoprostanes and an increase in LDL electronegativity characterized its atherogenic modification. As F_2-isoprostanes are potent vasoconstrictors and able to modulate platelet aggregation, their formation during LDL oxidation points out their potential role in cardiovascular disease. The formation of PGF_2-isoprostanes (with 8-epi-PGF_2 as a major component) was also registered during copper-catalyzed LDL oxidation [99].

Many findings show the important role of isoprostanes in oxidative stress-initiated pathologic disorders. Thus, even physiological concentrations of 8-epi-PGF_2 stimulated cell proliferation, DNA synthesis, and endothelium-1 mRNA and protein expression in bovine aortic endothelial cells [100], indicating pathophysiologic significance of F_2-isoprostanes during oxidant injury. The enhanced level of F_2-isoprostanes was observed in porcine vascular muscle cells under hyperglycemic conditions indicating their possible contribution to the complications of diabetes mellitus and cardiovascular disease [101]. In humans, significant increase in urinary concentrations of the F_2-isoprostane metabolite or of 8-epi-PGF_2 was observed in patients with scleroderma (various pathologies, which include limited disease with refractory digital ulceration or pulmonary hypertension and diffuse disease) [102], noninsulin-dependent diabetes mellitus [103], hepatorenal syndrome [104], or hypercholesterolemic patients [105]. In the last of the studies mentioned, it has been shown that vitamin E supplementation decreased urinary 8-epi-PGF_2 while COX inhibitors aspirin and indobufen had no effect. This points out the aspirin-insensitive free radical-mediated mechanism of isoprostane formation. Enhanced urinary levels of isoprostanes were also found in patients with pulmonary hypertension [106].

Isoprostanes apparently play an important role in different types of atherosclerotic processes [107]. It was found [108] that human atherosclerotic lesions contain the enhanced content of isoprostanes together with racemic hydroxy linoleate isomers, the major products of linoleic acid oxidation. Cyrus et al. [109] showed that developing extensive atherosclerosis LDL receptor-deficient mice on a high-fat diet had increased urinal levels of 8,12-isoprostane $F_{2\alpha}$-VI and 2,3-donor-thromboxane B_2. Atherosclerosis and these markers of in vivo lipid peroxidation and platelet activation were efficiently suppressed by vitamin E and platelet inhibitor indomethacin supplementation. Tangirala et al. [110] found that apolipoprotein E, a multifunctional protein synthesized by hepatocytes and macrophages reduced the progression of atherosclerosis in mice via antioxidant mechanism, suppressing the formation of 8,12-isoprostane $F_{2\alpha}$-VI. Enhanced levels of the protein adducts of reactive isolevuglandins were found in plasma from patients with atherosclerosis or end-stage renal disease, suggesting an abnormally high degree of oxidative injury associated with these pathologies [111].

Although a main pathway to isoprostanes is undoubtedly nonenzymatic peroxidation of unsaturated fatty acids, it has been shown that these products can be also formed in COX-catalyzed reactions. Thus, under certain conditions the in vitro and in vivo formation of 8-epi-PGF_2 can be inhibited by COX inhibitors. For example, indomethacin inhibited 8-epi-PGF_2 formation in platelets stimulated with calcium ionophore, AA, or thrombin [112]. Later on, the COX-catalyzed generation of isoprostanes was shown in isolated rat kidney glomeruli [113], which was inhibited by COX inhibitors and was not affected by free radical inhibitors (BHT and NDGA) or prooxidants (menadione or methylviologen). An increase in the 8-epi-PGF_2 concentration in human endothelial cells upon reoxygenation was also inhibited by indomethacin and aspirin [114].

Thus, the formation of isoprostanes is rightly considered as a unique noninvasive method for the estimation of in vivo lipid peroxidation. However, the use of 8-*iso*-$PGF_{2\alpha}$ isoprostane (now known as $iPF_{2\alpha}$-III) as an index of nonenzymatic in vivo peroxidation has some limitations due to the possibility of its formation by COX-1- and COX-2-catalyzed reactions. Pratico et al. [115] have shown that 8-*iso*-PGF$_{2\alpha}$ was formed by both free radical and COX-catalyzed pathways (although in the latter case as a minor product). However, these authors found that another isoprostane isomer IPF$_{2\alpha}$-1 (now known as iPF$_{2\alpha}$-VI) is also present in human urine and even more abundant than 8-*iso*-PGF$_{2\alpha}$. Furthermore, there is no evidence for COX-catalyzed formation of this isoprostane because the COX inhibitors, aspirin and indomethacin, did not suppress the ex vivo IPF$_{2\alpha}$-1 formation in volunteers.

Delani et al. [116] suggested that 8-*iso*-PGF$_{2\alpha}$ can be used as a quantitative marker of enhanced oxidative stress during coronary reperfusion. In subsequent paper [117] the determination of urinal IPF$_{2\alpha}$-1 in humans was successfully used as an index of in vivo lipid peroxidation in several free radical pathologies including alcohol-induced liver disease and cirrhosis. It was also found that alcohol enhanced the IPF$_{2\alpha}$-1 level in healthy volunteers indicating that oxidative stress precedes and initiates alcohol-induced liver disease.

As mentioned earlier, isoprostanes, which are formed by nonenzymatic lipid peroxidation, are the analogs of prostaglandins formed in the processes catalyzed by prostaglandin H synthase. Similarly, the lineal compounds isomeric to leukotrienes derived from 5-lipoxygenase-catalyzed arachidonate oxidation should be formed by the nonenzymatic oxidation of fatty acids. Indeed, Harrison and Murphy [118] isolated eicosanoid 5,12-dihydroxy-6,8,10,14-eicosatetraenoic acid from the oxidation of 1-hexadecanoyl-2-arachidonoyl-glycerophosphocholine catalyzed by the copper–hydrogen peroxide system. They termed such eicosanoids as B$_4$-isoleukotrienes and proposed that these compounds might mediate tissue damage through the activation of B$_4$-leukotriene receptors on target cells.

25.5 OXIDATION OF LOW-DENSITY LIPOPROTEINS

25.5.1 Mechanism of LDL Oxidation

Low-density lipoproteins in plasma and arterial wall are susceptible to oxidation to form oxidized LDL, which are thought to promote the development of atherosclerosis. LDL particles have a density of about 1.05, a molecular weight of about 2.5×10^6, and a diameter of about 20 nm [119]. LDL composition from different donors varies widely; an average LDL particle contains about 1200 molecules of unsaturated acids and antioxidants: about six molecules of α-tocopherol, about 0.53 molecule of γ-tocopherol, about 0.33 molecule of β-carotene, and about 0.18 molecule of lycopene [120]. Rapid oxidation of LDL is started only after the depletion of tocopherols and carotenoids [121].

Free radical oxidation of LDL has been thoroughly studied. Traditionally well-known chain mechanism of oxidation of organic compounds (Reactions (15)–(18)) is complicated in the case of LDL by the dual role of α-tocopherol.

$$R_i^\bullet + O_2 \Longrightarrow R_iOO^\bullet \tag{15}$$

$$R_iOO^\bullet + LDL \Longrightarrow R_iOOH + LDL^\bullet \tag{16}$$

$$LDL^\bullet + O_2 \Longrightarrow LDLOO^\bullet \tag{17}$$

$$LDLOO^\bullet + LDL \Longrightarrow LDLOOH + LDL^\bullet \tag{18}$$

(Here $R_i^\bullet$ is the radical of initiator.)

Similar to the other antioxidants present in LDL, α-tocopherol is able to react with peroxyl radicals:

$$LDLOO^\bullet + \alpha\text{-TocH} \Longrightarrow LDLOOH + \alpha\text{-Toc}^\bullet \tag{19}$$

However, it has been suggested that in contrast to traditional view about the inactivity of tocopheroxyl radical, α-Toc$^\bullet$ is capable of participating in chain propagation. This mechanism was discussed in detail [121–124]. It has been proposed that at low free radical fluxes and in the absence of ascorbate or ubihydroquinone (both antioxidants are supposedly able to regenerate α-tocopherol) tocopheroxyl radical abstracts a hydrogen atom from the bisallylic position of unsaturated compounds:

$$\alpha\text{-Toc}^\bullet + \text{LDL} \Longrightarrow \alpha\text{-TocH} + \text{LDL}^\bullet \qquad (20)$$

As mentioned earlier, ascorbate and ubihydroquinone regenerate α-tocopherol contained in a LDL particle and by this may enhance its antioxidant activity. Stocker and his coworkers [123] suggest that this role of ubihydroquinone is especially important. However, it is questionable because ubihydroquinone content in LDL is very small and only 50% to 60% of LDL particles contain a molecule of ubihydroquinone. Moreover, there is another apparently much more effective co-antioxidant of α-tocopherol in LDL particles, namely, nitric oxide [125]. It has been already mentioned that nitric oxide exhibits both antioxidant and prooxidant effects depending on the $O_2^{\bullet-}$/NO ratio [42]. It is important that NO concentrates up to 25-fold in lipid membranes and LDL compartments due to the high lipid partition coefficient, charge neutrality, and small molecular radius [126,127]. Because of this, the value of $O_2^{\bullet-}$/NO ratio should be very small, and the antioxidant effect of NO must exceed the prooxidant effect of peroxynitrite. As the rate constants for the recombination reaction of NO with peroxyl radicals are close to diffusion limit (about 10^9 l mol^{-1} s^{-1} [125]), NO will inhibit both Reactions (7) and (8) and by that spare α-tocopherol in LDL oxidation.

25.5.2 Initiation of LDL Oxidation

There are three major initiation pathways for LDL oxidation: the transition metal-catalyzed reactions, the enzymatic (lipoxygenase)-catalyzed oxidation (Chapter 26), and LDL oxidation by the cells, producers of oxygen radicals. The best-known in vitro model of initiation is copper-initiated LDL peroxidation [128,129]. Ziozenkova et al. [129] studied in detail the kinetics and mechanism of copper-initiated LDL oxidation. These authors demonstrated that LDL oxidation depends on copper concentration and that α-tocopherol exhibits both antioxidant and prooxidant effects on different stages of this process. It has been suggested that cupric ions are able to bind protein and that the formed Cu^{2+}–protein complex is reduced by lipid hydroperoxides to Cu^+–protein free radical, capable of initiating LDL oxidation. It is surprising that copper is a much better catalyst of LDL oxidation than iron, which was shown to catalyze LDL oxidation only in the presence of homocysteine [130]. Difference in the mechanisms of LDL oxidation by copper and iron ions was discussed by Lynch and Frei [131]. These authors have shown that LDL easily reduce Cu^{2+} ions but not Fe^{3+} ions. Correspondingly, Cu^{1+} is able to catalyze LDL oxidation without any additional reductants, while in order to reduce Fe^{3+} into catalytically active Fe^{2+} superoxide is needed as an additional reductant. Owing to that, iron-catalyzed LDL oxidation is inhibited by SOD while copper-catalyzed reaction is not [131]. It has been suggested that cupric ions may be reduced by α-tocopherol, although the rate constant for the reduction of Cu^{2+} to Cu^{1+} by α-tocopherol (2.9 l mol^{-1} s^{-1} [132]) is apparently too small.

Recently, Batthyany et al. [133] pointed out that the reduction of cupric ions bound to apolipoprotein B-100 by endogenous LDL components might be an initiation step in copper-mediated LDL oxidation. They suggested that this reaction proceeds to form cuprous ion and the protein-tryptophanyl free radical; the latter was identified on the basis of EPR spectrum with spin-trap 2-methyl-2-nitrosopropane.

Much attention has been focused on the study of LDL oxidation initiated by phagocytes (leukocytes and macrophages) and other superoxide-producing cells. It has been proposed that monocyte-derived macrophages may initiate in vivo LDL oxidation because they are present in arterial lesions and are able to generate oxygen radicals [134]. In 1985, Cathcart et al. [135] showed that the incubation of LDL with monocytes or PMNs resulted in LDL oxidation making them toxic to proliferating fibroblasts. The activation of monocytes enhanced LDL oxidation, while BHT, vitamin E, and glutathione inhibited it. Important results were obtained by Heinecke et al. [136], who demonstrated that arterial smooth muscle cells are able to oxidize LDL. These cells are not phagocytes and produce only about 10% superoxide in comparison with leukocytes. Nonetheless, they modify LDL in the presence of copper or iron ions. SOD but not hydroxyl radical scavengers inhibited LDL oxidation, indicating that superoxide initiated oxidation. It was suggested that the superoxide-dependent LDL oxidation mediated by smooth muscle cells may contribute to biological modification of LDL, leading to foam cell formation and atherogenesis. As already mentioned, human PMNs are also able to oxidize LDL by superoxide-dependent mechanism [137]. It was suggested that under the in vivo conditions of oxidative stress, PMNs contribute to foam cell formation by a scavenger receptor-dependent process at lesion sites. All stimuli, which activated superoxide generation by PMNs (PMA, FMLP, lipopolysaccharide, and opsonized zymosan), significantly increased LDL oxidation [138].

An enzyme responsible for superoxide-mediated LDL oxidation is most certainly macrophage NADPH oxidase because the oxidation was inhibited by NADPH oxidase inhibitors. Inhibitors of xanthine oxidase or NO synthase did not affect LDL oxidation [139]. It is interesting that LDL on their own stimulate superoxide release from macrophage-like cells and human monocyte-derived macrophages initiating LDL oxidation in the presence of copper ions [140]. Under these conditions, macrophage 15-lipoxygenase was also activated, as determined by the release of 15-HETE and 13-hydroxyoctadecadienoic acid (13-HODE). However, the contribution of 15-lipoxygenase into LDL oxidation is apparently insignificant because monocyte-derived macrophages from patients with chronic granulomatous disease (CGD) that lack NADPH oxidase but possess 15-lipoxygenase activity failed to oxidize LDL. Nonetheless, 15-lipoxigenase as well as MPO can probably participate in monocyte-macrophage-mediated LDL oxidation [134,141]. Moreover, Mabile et al. [142] suggested that mitochondria may participate directly or indirectly in the generation of superoxide by endothelial cells.

Recently, Bey and Cathcart [143] considered again the mechanism of monocyte-stimulated LDL oxidation. These authors used the antisense oligodeoxyribonucleotide (ODN) designed to target $p47_{phox}$ mRNA, the cytosolic component of phagocyte NADPH oxidase, for elucidation of the role of this enzyme in LDL oxidation. It was found that $p47_{phox}$ antisense ODN efficiently inhibited the formation of MDA and cholesteryl-HPODE oxidation products during LDL oxidation by the zymosan-stimulated human monocytes. LDL oxidation was completely dependent on the activation of the monocytes and was blocked by SOD. It was concluded [143] that different findings obtained earlier at the study of superoxide role in monocyte-mediated LDL oxidation depended on the presence or absence of free transition metal ions. Metal ion-dependent monocyte-mediated LDL oxidation was found to be independent of superoxide generation [144,145], while metal ion-independent one is not. Actually, these authors do not believe that the completely metal ion-independent monocyte-stimulated LDL oxidation is possible, and they think that the metal centers of enzymes and some metal complexes may participate in in vivo LDL oxidation. However, such trace amounts of metal ions are apparently unable to change superoxide-initiated mechanism of the process in contrast to significant amounts of free metal ions. In conclusion, it should be stressed that the presence of free metal ions in in vivo systems is doubtful; therefore, the significance of free metal ion-dependent LDL oxidation remains unclear.

Thus, superoxide is an indubitable precursor of phagocyte-mediated LDL oxidation. However, as we have seen above, it must be converted into the other more active species to become capable of initiating lipid peroxidation. Taking into account the ability of copper ions to initiate in vitro LDL oxidation, it has been suggested that transition metal ions (copper and iron) could participate in in vivo monocyte-mediated peroxidation [146]. As the presence of free copper or iron ions under in vivo conditions is unlikely, it was proposed that the copper-containing plasma protein ceruloplasmin could be the origin of needed copper ions. Indeed, Mukhopadhyay et al. [147] found that ceruloplasmin markedly enhanced LDL oxidation by phagocytes. In subsequent work [148] these authors showed that exogenous superoxide reduced a single copper atom in ceruloplasmin, the same copper atom that is needed for LDL oxidation. In accord with these findings, the superoxide-mediated copper-dependent LDL oxidation by smooth muscle and endothelial cells was inhibited by SOD. In addition to superoxide, other reactive species such as hydrogen peroxide may play a certain role in LDL oxidation by macrophages [149]. For example, macrophage-mediated LDL oxidation may also proceed by a thiol-dependent mechanism [150]. Garner et al. [151] demonstrated that metal (copper and iron)-dependent LDL oxidation by macrophages was augmented by thiol production. However, macrophages were able to accelerate metal-dependent LDL oxidation when cellular thiol production was insignificant in the absence of extracellular cysteine.

25.5.3 Role of Nitric Oxide and Peroxynitrite in the Initiation of LDL Oxidation

It has been already pointed out that nitric oxide exhibits antioxidant effect in LDL oxidation at the $NO/O_2^{\bullet-}$ ratio $\gg 1$. Under these conditions the antioxidant effect of NO prevails on the prooxidant effect of peroxynitrite. Although some earlier studies suggested the possibility of NO-mediated LDL oxidation [152,153], these findings were not confirmed [154]. On the other hand, at lower values of $NO/O_2^{\bullet-}$ ratio the formed peroxynitrite becomes an efficient initiator of LDL modification. Beckman et al. [155] suggested that peroxynitrite rapidly reacts with tyrosine residues to form 3-nitrotyrosine. Later on, Leeuwenburgh et al. [156] found that 3-nitrotyrosine was formed in the reaction of peroxynitrite with LDL. The level of 3-nitrotyrosine sharply differed for healthy subjects and patients with cardiovascular diseases: LDL isolated from the plasma of healthy subjects contained a very low level of 3-nitrotyrosine (9 ± 7 μmol/mol^{-1} of tyrosine), while LDL isolated from aortic atherosclerotic intima had a 90-fold higher level (840 ± 140 μmol/mol^{-1} of tyrosine). It has been proposed that peroxynitrite formed in the human artery wall is able to promote LDL oxidation in vivo.

Thus, despite its well-known antiatherogenic effect, nitric oxide may promote atherogenesis by virtue of its ability to form peroxynitrite. Recently, Trostchansky et al. [157] compared the effects of peroxynitrite and nitric oxide on LDL oxidation. Peroxynitrite initiated the formation of conjugated dienes and cholesteryl ester hydroperoxides during LDL oxidation. At the same time, physiologically relevant fluxes of nitric oxide strongly inhibited peroxynitrite-stimulated oxidative processes. The authors suggested that the balance between the rates of peroxynitrite and NO production in the vascular wall determine the extent of LDL oxidation. Important characteristic of oxidized LDL is their ability to interact with scavenger receptors. Guy et al. [158] have shown that the effect of peroxynitrite on receptor binding of LDL is comparable to that of LDL oxidized by cupric ions. It was suggested that peroxynitrite-mediated LDL oxidation may represent a physiologically relevant model.

25.5.4 Initiation of LDL Oxidation by MPO

Thus, physiological free radicals superoxide and nitric oxide produced by phagocytes and nonphagocytes are responsible for the two major pathways of LDL oxidation: transition metal-dependent and peroxynitrite-dependent mechanisms. However, there is another mode of phagocyte-mediated oxidative damage of LDL, which depends on MPO. This enzyme is secreted by activated phagocytes and oxidizes LDL, producing hydrogen peroxide and hypochlorous acid (HOCl) [154,159]. As already discussed (Chapter 22), MPO catalyzes free radical oxidation of organic compounds by a classical peroxidase cycle:

$$\text{MPO} + \text{H}_2\text{O}_2 \Longrightarrow \text{Compound I} + \text{H}_2\text{O} \tag{21}$$

$$\text{Compound I} + \text{Cl}^- + \text{H}^+ \Longrightarrow \text{MPO} + \text{HOCl} \tag{22}$$

$$\text{Compound I} + \text{RH} \Longrightarrow \text{Compound II} + \text{R}^\bullet + \text{H}^+ \tag{23}$$

$$\text{Compound II} + \text{RH} + \text{H}^+ \Longrightarrow \text{MPO} + \text{R}^\bullet + \text{H}_2\text{O} \tag{24}$$

The reaction of HOCl with LDL results in apoB modification with little oxidation of lipids by attacking amino groups of apoB lysine residues and the formation of N-chloramines, with the latter as the major products of HOCl reaction with LDL:

$$\text{HOCl} + \text{RNH}_2 \Longrightarrow \text{RNHCl} + \text{H}_2\text{O} \tag{25}$$

Chloramines change LDL charge characteristics, inducing uncontrolled uptake of modified LDL by macrophages and the formation of cholesterol-engorged foam cells [160]. The ability of HOCl to modify LDL was confirmed in a model system [161].

It follows from the above that the neutrophil-mediated LDL oxidation may occur by both NADPH oxidase- and MPO-dependent mechanisms. It was recently demonstrated [162] that the rates of formation of phosphatidylcholine and cholesteryl ester hydroperoxides during LDL oxidation by PMA-stimulated neutrophils of MPO-knockout mice were about 66% and 44% of those by wild-type neutrophils. In both cases LDL oxidation was inhibited by SOD. These findings suggest that superoxide mediates both NADPH oxidase- and MPO-dependent pathways of oxidation by stimulated neutrophils.

The formation of chloramines is an initial step of another mechanism of oxidative modification of LDL. It has been shown that the MPO–hydrogen peroxide–chloride system reacts with L-tyrosine to form p-hydroxyphenylacetaldehyde [163]. As activated neutrophils release both MPO and hydrogen peroxide, it was suggested that neutrophils can stimulate the formation of p-hydroxyphenylacetaldehyde by producing chloramines as intermediates during the oxidation of LDL [164].

$$\text{MPO} + \text{H}_2\text{O}_2 \Longrightarrow \text{Compound I} + \text{H}_2\text{O} \tag{21}$$

$$\text{Compound I} + \text{Cl}^- + \text{H}^+ \Longrightarrow \text{MPO} + \text{HOCl} \tag{22}$$

$$\text{HOC}_6\text{H}_4\text{CH}_2\text{CH}(\text{NH}_2)\text{COOH} + \text{HOCl} \Longrightarrow \text{HOC}_6\text{H}_4\text{CH}_2\text{CH}(\text{NHCl})\text{COOH} \tag{26}$$

$$\text{L-Tyrosine} \qquad\qquad\qquad \text{Monochloramine}$$

This process probably occurs in vivo because the adduct of ethanolamine and p-hydroxyphenylacetaldehyde is abundant in the phospholipids of LDL exposed to activated neutrophils and tyrosine.

Another mode of MPO-catalyzed LDL oxidation is the formation of 3-chlorotyrosine. Hazen and Heinecke [165] have shown that the interaction of MPO–hydrogen peroxide–Cl$^-$ system with LDL resulted in the chlorination of protein tyrosyl residues. HOCl is presumably an intermediate of this reaction because 3-chlorotyrosine formation was also observed in the reaction with reagent HOCl. Furthermore, 3-chlorotyrosine was not detected in LDL oxidized by hydroxyl radicals, copper, iron, peroxynitrite, and other oxidants. The level of 3-chlorotyrosine was much higher in atherosclerotic tissue and in LDL isolated from atherosclerotic intima. It was proposed that the formation of 3-chlorotyrosine indicates that the MPO system of phagocytes may catalyze the chlorination of LDL in human atherosclerotic lesions.

HOCl is probably not a single active MPO oxidant able to chlorinate LDL. Hazen et al. [166] have shown that such a powerful oxidant as molecular chlorine is formed under in vitro conditions during the reaction of MPO–hydrogen peroxide–chloride system of phagocytes with LDL. They pointed out that there is an equilibrium between HOCl and Cl_2, which is shifted to the right under acidic conditions:

$$HOCL + H^+ \Longleftrightarrow Cl_2 + H_2O \qquad (27)$$

It was found that the exposure of LDL to the $MPO-H_2O_2-Cl^-$ system resulted in the formation of chlorinated sterols.

$$Cholesterol \xrightarrow{MPO} HOCl \Longrightarrow Cl_2 \Longrightarrow cholesterol\ chlorohydrins + dichlorocholesterol \qquad (28)$$

The same products were formed in the reaction of LDL with molecular chlorine but not with HOCl. These authors suggested that the formation of molecular chlorine might occur in vivo because a pH value could be less than 4 during macrophage phagocytosis. In addition, atherosclerotic tissue may also be relatively acidic due to impaired oxygen diffusion and hypoxia.

As mentioned earlier, MPO–hydrogen peroxide–chloride system of phagocytes induces the formation of lipid peroxidation products in LDL but their amount is small [167–169]. It was proposed that HOCL can decompose the lipid hydroperoxides formed to yield alkoxyl radicals [170]. It was also suggested that chloramines formed in this process decompose to free radicals, which can initiate lipid peroxidation [171].

It follows from the above that MPO may catalyze the formation of chlorinated products in media containing chloride ions. Recently, Hazen et al. [172] have shown that the same enzyme catalyzes lipid peroxidation and protein nitration in media containing physiologically relevant levels of nitrite ions. It was found that the interaction of activated monocytes with LDL in the presence of nitrite ions resulted in the nitration of apolipoprotein B-100 tyrosine residues and the generation of lipid peroxidation products 9-hydroxy-10,12-octadecadienoate and 9-hydroxy-10,12-octadecadienoic acid. In this case there might be two mechanisms of MPO catalytic activity. At low rates of nitric oxide flux, the process was inhibited by catalase and MPO inhibitors but not SOD, suggesting the MPO initiation.

$$Monocyte \Longrightarrow O_2^{\cdot -} \Longrightarrow H_2O_2 \qquad (29)$$

$$H_2O_2 + NO_2^- \xrightarrow{MPO} {}^\cdot NO_2(?) \Longrightarrow LDL\ nitration\ and\ peroxidation \qquad (30)$$

or

$$HOCl + NO_2^- \Longrightarrow NO_2Cl \Longrightarrow LDL\ nitration\ and\ peroxidation \qquad (31)$$

However, at high rates of nitric oxide flux, the formation of nitrated and oxidized products became insensitive to the presence of catalase or MPO inhibitors but increasingly inhibited by SOD, suggesting the participation of peroxynitrite. (It is interesting that Reaction (30) might be a one-electron reduction of hydrogen peroxide by nitrite ion. If such a process really takes place, then it is a modified Fenton reaction, and we might expect the side formation of hydroxyl radicals.)

The initiation of LDL oxidation by reactive nitrogen species is probably an important in vivo process because LDL recovered from human atherosclerotic aorta is enriched in nitrotyrosine. Nitrogen species produced by the MPO–hydrogen peroxide–nitrate system of monocytes is able to form nitrated LDL–NO_2, which are easily degraded by macrophages, resulting in the massive cholesterol deposition and foam cell formation [173]. It was concluded that in this case LDL modification and conversion into a high-uptake form occurred at physiological chloride levels in the absence of transition metal ions and in the presence of nitrate ions.

There is probably one more mechanism of MPO-mediated lipid peroxidation. Kettle and Candaeis [174] have studied the oxidation of tryptophan by neutrophil MPO. They suggested that tryptophan, which is present in plasma at the similar concentration as tyrosine and has a similar one-electron reduction potential, can contribute to oxidative stress at inflammation sites. It was proposed that the formed tryptophan free radicals may stimulate oxidative stress during inflammation.

25.5.5 MAJOR PRODUCTS AND PATHOPHYSIOLOGICAL EFFECTS OF LDL OXIDATION

As mentioned earlier, oxidation of LDL is initiated by free radical attack at the diallylic positions of unsaturated fatty acids. For example, copper- or endothelial cell-initiated LDL oxidation resulted in a large formation of monohydroxy derivatives of linoleic and arachidonic acids at the early stage of the reaction [175]. During the reaction, the amount of these products is diminished, and monohydroxy derivatives of oleic acid appeared. Thus, monohydroxy derivatives of unsaturated acids are the major products of the oxidation of human LDL. Breuer et al. [176] measured cholesterol oxidation products (oxysterols) formed during copper- or soybean lipoxygenase-initiated LDL oxidation. They identified chlolest-5-ene-3β, 4α-diol, cholest-5-ene-3β, 4β-diol, and cholestane-3β, 5α, 6α-triol, which are present in human atherosclerotic plaques.

Oxidized LDL are considered to be one of the major factors associated with the development of atherosclerosis. The earliest event is the transport of LDL into the arterial wall where LDL, being trapped in subendothelial space, are oxidized by oxygen radicals produced by endothelial and arterial smooth muscle cells. The oxidation of LDL in the arterial wall is affected by various factors including hemodynamic forces such as shear stress and stretch force. Thus, it has been shown [177] that stress force imposed on vascular smooth muscle cells incubated with native LDL increased the MDA formation by about 150% concomitantly with the enhancement of superoxide production. It was suggested that oxidation was initiated by NADPH oxidase-produced superoxide and depended on the presence of metal ions.

Increased LDL oxidation takes place during infection and inflammation [178]. Interestingly, LDL oxidation occurs during the acute-phase response, a host reaction to infection and inflammation, which is protection mechanism for the suppression of systemic injury. Therefore, LDL oxidation has possibly a beneficial purpose, by scavenging reactive free radicals and preventing membrane damage. On the other hand, increasing LDL oxidation during infection and inflammation may stimulate atherogenesis and increase the incidence of coronary artery disease in patients with chronic infections and inflammatory disorders.

It has also been shown that LDL oxidation is increased in diabetes. In this connection, Mowri et al. [179] studied the effect of glucose on metal ion-dependent and -independent LDL oxidation. They found that pathophysiological glucose concentrations enhanced copper- and iron-induced LDL oxidation measured via the formation of conjugated dienes. In contrast, glucose had no effect on metal-independent free radical LDL oxidation. Correspondingly, metal ion-dependent but not free radical-mediated oxidation was completely inhibited by SOD. These findings suggest that prooxidant effect of glucose on metal ion-dependent LDL oxidation is due to the reduction of metal ions.

Contrary to LDL, high-density lipoproteins (HDL) prevent atherosclerosis, and therefore, their plasma levels inversely correlate with the risk of developing coronary artery disease. HDL antiatherogenic activity is apparently due to the removal of cholesterol from peripheral tissues and its transport to the liver for excretion. In addition, HDL acts as antioxidants, inhibiting copper- or endothelial cell-induced LDL oxidation [180]. It was found that HDL lipids are oxidized easier than LDL lipids by peroxyl radicals [181]. HDL also protects LDL by the reduction of cholesteryl ester hydroperoxides to corresponding hydroperoxides. During this process, HDL specific methionine residues in apolipoproteins AI and AII are oxidized [182].

REFERENCES

1. W Chamulitrat, SJ Jordan, RP Mason. *Arch Biochem Biophys* 299: 361–367, 1992.
2. IB Afanas'ev. In: *Superoxide Ion: Chemistry and Biological Implications*, vol 1. CRC Press, Boca Raton, FL, 1989, p. 241.
3. BHJ Bielski, RL Arudi, MW Sutherland. *J Biol Chem* 258: 4759–4761, 1983.
4. J Aikens, TA Dix. *J Biol Chem* 266: 15091–15099, 1991.
5. FQ Schafer, GR Buettner. *Free Radic Biol Med* 28: 1175–1181, 2000.
6. J Aikens, TA Dix. *Chem Res Toxicol* 5: 263–267, 1992.
7. Bedard L, Young MJ, Hall D, Paul T, Ingold KU. *J Am Chem Soc* 123: 12439–12448, 2001.
8. CS Lai, LH Piette. *Biochem Biophys Res Commun* 78: 51–59, 1977.
9. C Kai, TA Grover, LH Piette. *Arch Biochem Biophys* 193: 373–378, 1979.
10. M Ingelman-Sundberg, A-L Hagbjork. *Xenobiotica* 12: 673–686, 1982.
11. O Beloqui, AI Cederbaum. *Biochem Pharmacol* 35: 2663–2669, 1986.
12. JM Gutteridge. *FEBS Lett* 150: 454–458, 1982.
13. GF Vile, CC Winterbourn. *FEBS Lett* 215: 151–154, 1987.
14. A Bast, MHM Steeghs. *Experientia* 42: 555–556, 1986.
15. M Tien, BA Svingen, SD Aust. *Fed Proc* 40: 179–182, 1981.
16. CE Thomas, LA Morehouse, SD Aust. *J Biol Chem* 260: 3275–3280, 1985.
17. G Minotti, SD Aust. *J Biol Chem* 262: 1098–1104, 1987.
18. DM Miller, TA Grover, N Nayini, SD Aust. *Arch Biochem Biophys* 301: 1–7, 1993.
19. JM Braughler, LA Duncan, RL Chase. *J Biol Chem* 261: 10282–10289, 1989.
20. JF Koster, RG Slee. *FEBS Lett* 199: 85–88, 1986.
21. LA Morehouse, CE Thomas, SD Aust. *Arch Biochem Biophys* 232: 366–372, 1984.
22. IB Afanas'ev, AI Dorozhko, NI Polozova, NS Kuprianova, AV Brodskii, EA Ostrachovitch, LG Korkina. *Arch Biochem Biophys* 302, 200–205, 1993.
23. IB Afanas'ev, EA Ostrakhovitch, LG Korkina. *FEBS Lett* 425, 256–258, 1998.
24. IB Afanas'ev. In: *Superoxide Ion: Chemistry and Biological Implications*, vol 2. CRC Press, Boca Raton, FL, 144, 1990.
25. MJ Thomas, KS Mehl, WA Pryor. *J Biol Chem* 257: 8343–8347, 1982.
26. IB Afanas'ev, NS Kuprianova, AV Letuchaia. In: *Oxygen Radicals in Chemistry and Biology*, W Bors, M Saran, D Tait, (eds.) Walter de Gruyter, Berlin, 1984, p. 17.
27. IB Afanas'ev, NS Kuprianova. *J Chem Soc Perkin Trans 2* 1351–1355, 1985.
28. EW Kellogg III, I Fridovich. *J Biol Chem* 252: 6721–6728, 1977.
29. JM Gutteridge, AP Beard, GJ Quinlan. *Biochem Biophys Res Commun* 117: 901–907, 1983.
30. J Klimek. *Biochim Biophys Acta* 958: 31–39, 1988.
31. T Ohyashiki, M Nunomura. *Biochim Biophys Acta* 1484: 241–250, 2000.
32. V Roginsky, T Barsukova. *Chem Phys Lett* 111: 87–91, 2001.
33. M Younes, A Mohr, MH Schoenberg, FW Schildberg. *Res Exp Med* 187: 9–17, 1987.

34. WD Nguyen, DH Kim, HB Alam, HS Provido, JR Kirkpatrick. *Crit Care* 3: 127–130, 1999.
35. GM Bartoli, B Giannattasio, P Palozza, A Cittadini. *Biochim Biophys Acta* 966: 214–221, 1988.
36. S Shaw, E Jayatilleke. *Biochem Biophys Res Commun* 143: 984–990, 1987.
37. SK Nelson, SK Bose, JM McCord. *Free Radic Biol Med* 16: 195–201, 1994.
38. W Jessup, D Mohr, SP Gieseg, RT Dean, R Stocker. *Biochim Biophys Acta* 1180: 73–82, 1992.
39. N Hogg, B Kalyanaraman, J Joseph, A Struck, S Parthasarathy. *FEBS Lett* 334, 170–174, 1993.
40. MT Yates, LE Lambert, JP Whitten, I McDonald, M Mano, G Ku, SJ Mao. *FEBS Lett* 309, 135–138, 1992.
41. RE Laskey, WR Mathews. *Arch Biochem Biophys* 330: 193–198, 1996.
42. H Rubbo, R Radi, M Trujillo, R Telleri, B Kalyanaraman, S Barnes, M Kirk, BA Freeman. *J Biol Chem* 269: 26066–26075, 1994.
43. VB O'Donnell, PH Chumley, N Hogg, A Bloodsworth, VM Darley-Usmar, BA Freeman. *Biochemistry* 36: 15216–15223, 1997.
44. A Bloodsworth, VB O'Donnell, BA Freeman. *Arterioscler Thormb Vasc Biol* 20: 1707–1715, 2000.
45. JP Crow, C Spruell, J Chen, C Gunn, H Ischiropoulos, M Tsai, CD Smith, R Radi, WH Koppenol, JS Beckman. *Free Radic Biol Med* 16: 331–339, 1994.
46. VM Darley-Usmar, N Hogg, VJ O'Leary, MT Wilson, S Moncada. *Free Radic Res Commun* 17: 9–20, 1992.
47. VB O'Donnell, JP Eiserich, PH Chumley, MJ Jablonsky, NR Krishna, M Kirk, S Barnes, VM Darley-Usmar, BA Freeman. *Chem Res Toxicol* 12: 83–92, 1999.
48. IY Haddad, H Ischiropoulos, BA Holm, JS Beckman, JB Baker, S Matalon. *Am J Physiol* 265: L555–L564, 1993.
49. H Shi, N Noguchi, Y Xu, E Niki. *Biochem Biophys Res Commun* 257: 651–656, 1999.
50. VB O'Donnell, BA Freeman. *Circ Res* 88: 12–21, 2001.
51. A Denicola, C Batthyany, E Lissi, BA Freeman, H Rubbo, R Radi. *J Biol Chem* 277: 932–936, 2002.
52. M d'Ischia, A Palumbo, F Buzzo. *Nitric Oxide* 4: 4–14, 2000.
53. AK Thukkani, F-F Hsu, JR Crowley, RB Wysolmerski, CJ Albert, DA Ford. *J Biol Chem* 277: 3842–3849, 2002.
54. IB Afanas'ev, AI Dorozhko, AV Brodskii, VA Kostyuk, AI Potapovitch. *Biochem Pharmacol* 38: 1763–1769, 1989.
55. J Goodman, P Hochstein. *Biochem Biophys Res Commun* 77: 797–801, 1977.
56. IB Afanas'ev. In: *Superoxide Ion: Chemistry and Biological Implications*, vol 2. CRC Press, Boca Raton, FL, 1990, p. 27.
57. IB Afanas'ev, NI Polozova. *J Chem Soc Perkin Trans 2*, 835–838, 1987.
58. IB Afanas'ev, NI Polozova, NS Kuprianova, VI Gunar. *Free Radic Res Commun.* 3, 141–145, 1987.
59. K Sugioka, H Nakano, T Noguchi, J Tsuchiya, M Nakano. *Biochem Biophys Res Commun* 100: 1251–1255, 1981.
60. JMC Gutteridge. *Biochem Pharmacol* 32: 1949–1952, 1983.
61. LC Trost, KB Wallace. *Biochem Biophys Res Commun* 204: 23–29, 1994.
62. GF Vile, CC Winterbourn. *Biochem Pharmacol* 37: 2893–2897, 1988.
63. EG Mimnaugh, MA Trush, TE Gram. *Biochem Pharmacol* 30, 2797, 1981.
64. SR Powell, PB McCay. *Free Radic Biol Med* 18: 159–168, 1995.
65. HC Yen, TD Oberley, S Vichitbandha, YS Ho, DK St Clair. *J Clin Invest* 98: 1253–1260, 1997.
66. EA Konorev, H Zhang, J Joseph, MC Kennedy, B Kalyanaraman. *Am J Physiol Heart Circ Physiol* 279: H2424–H2430, 2000.
67. G Minotti, G Cairo, E Monti. *FASEB J* 13: 199–212, 1999.
68. G Minotti, C Mancuso, A Frustaci, A Mordente, SA Santini, AM Calafiore, G Liberi, N Gentiloni. *J Clin Invest* 98: 650–661, 1996.
69. IB Afanas'ev, NI Polozova, GI Samokhvalov. *Bioorgan Chem* 9: 434–439, 1980.
70. Y Suzuki, J Sudo. *Jpn J Pharmacol* 45: 271–279, 1987.
71. LG Korkina, TB Suslova, ZP Cheremisina, BT Velichkovski. *Stud Biophysica* 126: 99–104, 1988.

72. EA Hassoun, F Li, A Abushaban, SJ Stohs. *J Appl Toxicol* 21: 211–219, 2001.
73. C Cereser, S Boget, P Parvaz, A Revol. *Toxicology* 163: 153–162, 2001.
74. R Anane, EE Creppy. *Hum Exp Toxicol* 20: 477–481, 2001.
75. G Carlin. *J Free Radic Biol Med* 1: 255–261, 1985.
76. G Carlin, KE Arfors. *J Free Radic Biol Med* 1: 437–442, 1985.
77. J Rodenas, T Carbonell, MT Mitjavila. *Free Radic Biol Med* 28: 374–380, 2000.
78. L Lucchi, S Banni, MP Melis, E Angioni, G Carta, V Casu, R Rapana, A Ciuffreda, FP Corongiu, A Albertazzi. *Kidney Int* 58: 1695–1702, 2000.
79. BA Wagner, GR Buettner, CP Burns. *Biochemistry* 33: 4449–4454, 1994.
80. A Benedetti, M Comporti, H Esterbauer. *Biochim Biophys Acta* 620: 281–296, 1980.
81. H Esterbauer, A Benedetti, J Lang, R Fulceri, G Fauler, M Comporti. *Biochim Biophys Acta* 876: 154–166, 1986.
82. C Schneider, KA Tallman, NA Porter, AR Brash. *J Biol Chem* 276: 20831–20838, 2001.
83. M Inouye, T Mio, K Sumino. *Biochim Biophys Acta* 1438: 204–212, 1999.
84. Y Yamamoto, E Niki. *Biochem Biophys Res Commun* 165: 988–993, 1989.
85. Y Yamamoto, K Wakabayashi, M Nagano. *Biochem Biophys Res Commun* 189: 518–523, 1992.
86. R Mashima, K Onodera, Y Yamamoto. *J Lipid Res* 41: 109–115, 2000.
87. GM Chisolm, G Ma, KC Irwin, LL Martin, KG Gunderson, LF Linberg, DW Morel, PE DiCorleto. *Proc Natl Acad Sci USA* 91, 11452–11456, 1994.
88. DH Nugteren. *J Biol Chem* 250: 2808–2812, 1975.
89. JD Morrow, KE Hill, RF Burk, TN Nammour, KF Badr, LJ Roberts II. *Proc Natl Acad Sci USA* 87: 9383–9387, 1990.
90. JD Morrow, JA Awad, HJ Boss, IA Blair, LJ Roberts II. *Proc Natl Acad Sci USA* 89: 10721–10725, 1992.
91. LJ Janssen. *Am J Physiol Lung Cell Mol Physiol* 280: L1067–L1082, 2001.
92. LJ Roberts, RG Salomon, JD Morrow, CJ Brame. *FASEB J* 13: 1157–1168, 1999.
93. LD Morrow, LJ Roberts II, VC Daniel, O Mirotchnechenko, L Swift, RF Burk. *Arch Biochem Biophys* 353: 160–171, 1998.
94. RG Salomon, DB Miller, MG Zagorski, DJ Coughlin. *J Am Chem Soc* 106: 6049–6060, 1984.
95. RC Van der Veen, LJ Roberts. *J Neuroimmunol* 95: 1–4, 1999.
96. LJ Roberts II, TJ Montine, W.R Markesbery, AR Tapper, P Hardy, S Chemtob, WD Dettbarn, JD Morrow. *J Biol Chem* 273: 13605–13612, 1998.
97. EE Reich, WE Zackert, CJ Brame, Y Chen, LJ Roberts II, DL Hachey, TJ Montine, JD Morrow. *Biochemistry* 39: 2376–2383, 2000.
98. SM Lynch, JD Morrow, LJ Roberts, B Frei. *J Clin Invest* 93: 998–1004, 1994.
99. NK Gopaul, J Nourooz-Zadeh, AI Mallet, EE Anggard. *Biochem Biophys Res Commun* 200: 338–343, 1994.
100. T Yura, M Fukunaga, R Khan, GN Nassar, KF Badr, A Montero. *Kidney Int* 56: 471–478, 1999.
101. R Natarajan, L Lanting, N Gonzales, J Nadler. *Am J Physiol* 271: H159–H165, 1996.
102. CM Stein, SB Tanner, JA Awad, LJ Roberts II, JD Morrow. *Arthritis Rheum* 39: 1146–1150, 1996.
103. Y Murai, T Hishinuma, N Suzuki, J Satoh, T Toyota, M Mizugaki. *Prostaglandins Leukot Essent Fatty Acids* 62: 173–181, 2001.
104. JD Morrow, TA Minton, CR Mukundan, MD Campbell, WE Zackert, VC Daniel, KF Badr, IA Blair. *J Biol Chem* 269: 4317–4326, 1994.
105. G Davi, P Alessandrini, A Mezzetti, G Minotti, T Bucciarelli, F Costantini, F Cipollone, GB Bon, G Ciabattoni, C Patrono. *Arterioscler Throm Vasc Biol* 17: 3230–3235, 1997.
106. JL Cracowski, C Cracowski, G Bessard, JL Pepin, J Bessard, C Schwebel, F Stanke-Labesque, C Pison. *Am J Respir Crit Care Med* 164: 1038–1042, 2001.
107. A Oguogho, H Kritz, O Wagner, H Sinzinger. *Prostaglandins Leukot Essent Fatty Acids* 64: 167–171, 2001.
108. C Gniwotta, JD Morrow, LJ Roberts II, H Kuhn. *Arterioscler Throm Vasc Biol* 17: 3236–3241, 1997.
109. T Cyrus, LX Tang, J Rokach, GA FitzGerald, D Pratico. *Circulation* 104: 1940–1945, 2001.

110. RK Tangirala, D Pratico, GA FitzGerald, S Chun, K Tsukamoto, C Maugeais, DC Usher, E Pure, DJ Rader. *J Biol Chem* 276: 261–266, 2001.
111. RG Salomon, E Batyreva, K Kaur, DL Sprecher, MJ Schreiber, JW Crabb, MS Penn, AM DiCorletoe, SL Hazen, EA Pondez. *Biochim Biophys Acta* 1485: 225–235, 2000.
112. T Klein, F Reutter, H Schweer, HW Seyberth, RM Nusing. *J Pharmacol Exp Ther* 282: 1658–1665, 1997.
113. T Klein, K Neuhaus, F Reutter, RM Nusing. *Br J Pharmacol* 133: 643–650, 2001.
114. MT Watkins, GM Patton, HM Soler, H Albadawi, DE Humphries, JE Evans, H Kadowaki. *Biochem J* 344: 747–754, 1999.
115. D Pratico, OP Barry, JA Lawson, M Adiyaman, S-W Hwang, SP Khanapure, L Luliano, J Rokach, GA FitzGerald. *Proc Natl Acad Sci USA* 95, 3449–3454, 1998.
116. N Delani, MP Reilly, D Pratico, JA Lawson, JF McCarthy, AE Wood, ST Ohnishi, DJ Fitzgerald, GA FitzGerald. *Circulation* 95: 2492–2499, 1997.
117. EA Meagher, OP Barry, A Burke, MR Lucey, JA Lawson, J Rokach, GA FitzGerald. *J Clin Invest* 104: 805–813, 1999.
118. KA Harrison, RC Murphy. *J Biol Chem* 270: 17273–17278, 1995.
119. H Esterbauer, M Dieber-Rotheneder, G Waeg, G Striegl, G Jurgens. *Chem Res Toxicol* 3: 77–92, 1990.
120. H Esterbauer, M Dieber-Rotheneder, G Striegl, G Waeg. *Am J Clin Nutr* 53: 314S–321S, 1991.
121. KU Ingold, VW Bowry, R Stocker, C Walling. *Proc Natl Acad Sci USA* 90: 45–49, 1993.
122. JM Upston, J Neuzil, PK Witting, R Alleva, R Stocker. *J Biol Chem* 272: 30067–30074, 1997.
123. JM Upston, AC Terentis, R Stocker. *FASEB J* 13: 977–994, 1999.
124. L Bedard, MJ Young, D Hall, T Paul, KU Ingold. *J Am Chem Soc* 123: 12439–12448, 2001.
125. H Rubbo, R Radi, D Anselm, M Kirk, S Barnes, J Butler, JP Eiserich, BA Freeman. *J Biol Chem* 275: 10812–10818, 2000.
126. T Malinski, Z Taha, S Grunfeld, S Patton, M Kapturczak, P Tomboulian. *Biochem Biophys Res Commun* 193: 1076–1082, 1993.
127. X Liu, MJ Miller, M Joshi, D Thomas, JR Lancaster. *Proc Natl Acad Sci USA* 95: 2175–2179, 1998.
128. H Esterbauer, J Gebicke, H Puhl, G Jungen. *Free Radic Biol Med* 13: 341–390, 1992.
129. O Ziozenkova, A Sevanian, PM Abuja, P Ramos, H Esterbauer. *Free Radic Biol Med* 24: 607–633, 1998.
130. K Hirano, T Ogihara, M Miki, H Yasuda, H Tamai, N Kawamura, M Mino. *Free Radic Res* 21: 267–276, 1994.
131. SM Lynch, B Frei. *J Biol Chem* 270: 5158–5163, 1995.
132. Y Yoshida, J Tsuchiya, E Niki. *Biochim Biophys Acta* 1200: 85–92, 1994.
133. C Batthyany, CX Santos, H Botti, C Cervenansky, R Radi, O Augusto, H Rubbo. *Arch Biochem Biophys* 384: 335–340, 2000.
134. GM Chisolm, SL Hazen, PL Fox, MK Cathcart. *J Biol Chem* 274: 25959–25962, 1999.
135. MK Cathcart, DW Morel, GM Chisolm. *J Leukoc Biol* 38: 341–350, 1985.
136. JW Heinecke, L Baker, H Rosen, A Chait. *J Clin Invest* 77: 757–761, 1986.
137. E Wieland, A Brandes, VW Armstrong, M Oellerich. *Eur J Clin Chem Clin Biochem* 31: 725–731, 1993.
138. C Scaccini, I Jialal. *Free Radic Biol Med* 16: 49–55, 1994.
139. GM Wilkins, D.S.Leake. *Biochim Biophys Acta* 1211: 69–78, 1994.
140. M Aviram, M Rosenblat, A Etzioni, R Levy. *Metabolism* 45: 1069–1079, 1996.
141. VA Folcik, R Aamir, MK Cathcart. *Arteriosc Throm Vas Biol* 17: 1954–1961, 1997.
142. L Mabile, O Meilhac, I Escargueil-Blanc, M Troly, MT Pieraggi, R Salvayre, A Negre-Salvayre. *Arterioscler Thromb Vasc Biol* 17: 1575–1582, 1997.
143. EA Bey, MK Cathcart. *J Lipid Res* 41: 489–495, 2000.
144. B Garner, RT Dean, W Jessup. *Biochem J* 301: 421–428, 1994.
145. L Kritharides, W Jessup, RT Dean. *Arch Biochem Biophys* 323: 127–137, 1995.
146. X Xing, J Baffic, CP Sparrow. *J Lipid Res* 39: 2201–2208, 1998.

147. CK Mukhopadhyay, E Ehrenwald, PL Fox. *J Biol Chem* 271: 14773–14778, 1996.
148. CK Mukhopadhyay, PL Fox. *Biochemistry* 37: 14222–14229, 1998.
149. GM Wilkins, DS Leake. *Biochim Biophys Acta* 1215: 250–258, 1994.
150. A Graham, JL Wood, VJ O'Leary, D Stone. *Free Radic Res* 21: 295–308, 1994.
151. B Garner, D van Reyk, RT Dean, W Jessup. *J Biol Chem.* 272: 6927–6935, 1997.
152. JM Wang, SN Chow, JK Lin. *FEBS Lett* 342: 171–176, 1994.
153. GJ Chang, P Woo, HM Honda, LJ Ignarro, L Young, JA Berliner, LL Demer. *Arterioscler Thromb* 14: 1808–1814, 1994.
154. AC Carr, MR McCall, B Frei. *Arterioscler Thormb Vasc Biol* 20: 1716–1723, 2000.
155. JS Beckman, J Chen, H Ischiropoulos, JP Crow. *Meth Enzymol* 233, 229–240, 1994.
156. C Leeuwenburgh, MM Hardy, SL Hazen, P Wagner, S Oh-ishi, UP Steinbrecher, LW Heinecke. *J Biol Chem* 272: 1433–1436, 1997.
157. A Trostchansky, C Batthyany, H Botti, R Radi, A Denicola, H Rubbo. *Arch Biochem Biophys* 395: 225–232, 2001.
158. RA Guy, GF Maguire, I Crandall, PW Connelly, KC Kain. *Atherosclerosis* 155: 19–28, 2001.
159. E Malle, G Waeg, R Schreider, EF Grone, W Sattler, H-J Grone. *Eur J Biochem* 267: 4495–4503, 2000.
160. LJ Hazell, R Stocker. *Biochem J* 290: 165–172, 1993.
161. EA Gorbatenkova, GM Artmann, OM Panasenko. *Membr Cell Biol* 13: 537–546, 2000.
162. N Noguchi, K Nakano, Y Aratani, H Koyama, T Kodama, E Niki. *J Biochem (Tokyo)* 127: 971–976, 2000.
163. SL Hazen, FF Hsu, JW Heinecke. *J Biol Chem* 271: 1861–1867, 1996.
164. SL Hazen, A d'Avignon, MM Anderson, FF Hsu, JW Heinecke. *J Biol Chem* 273: 4997–5005, 1998.
165. SL Hazen, JW Heinecke. *J Clin Invest* 99: 2075–2081, 1997.
166. SL Hazen, FF Hsu, K Duffin, JW Heinecke. *J Biol Chem* 271: 23080–23088, 1996.
167. OM Panasenko, SA Evgina, RK Aidyraliev, VI Sergienko, YA Vladimirov. *Free Radic Biol Med* 16: 143–148, 1994.
168. LJ Hazell, JJ van den Berg, R Stocker. *Biochem J* 302: 297–304, 1994.
169. A Jereich, JS Fabjan, S Tschabuschnig, AV Smirnova, L Horakova, M Hayn, H Auer, H Guttenberger, HJ Leis, F Tatzber, G Waeg, RJ Schaur. *Free Radic Biol Med.* 24: 1139–1148, 1998.
170. OM Panasenko. *Biofactors* 6: 181–190, 1998.
171. LJ Hazell, MJ Davies, R Stocker. *Biochem J* 339: 489–495, 1999.
172. SL Hazen, R Zhang, Z Shen, W Wu, EA Podrez, JC MacPherson, D Schmitt, SN Mitra, C Mukhopadhyay, Y Chen, PA Cohen, HF Hoff, HM Abu-Soud. *Circ Res* 85: 950–958, 1999.
173. EA Podrez, D Schmitt, HF Hoff, SL Hazen. *J Clin Invest* 103: 1547–1560, 1999.
174. AL Kettle, LP Candaeis. *Redox Rep* 5: 179–184, 2000.
175. T Wang, WG Yu, WS Powell. *J Lipid Res* 33: 525–537, 1992.
176. O Breuer, S Dzeletovic, E Lund, U Diczfalusy. *Biochim Biophys Acta* 1302: 145–152, 1996.
177. N Inoue, S Kawashima, K Hirata, Y Rikitake, S Takeshita, W Yamochi, H Akita, M Yokoyama. *Am J Physiol* 274: H1928–H1932, 1998.
178. RA Memon, I Staprans, M Noor, WM Holleran, Y Uchida, AH Moser, KR Feingold, C Grunfeld. *Arterioscler Throm Vasc Biol* 20: 1536–1542, 2000.
179. H Mowri, B Frei, JF Keaney. *Free Radic Biol Med* 29: 814–824, 2000.
180. S Parthasaraty, J Barnett, LJ Fong. *Biochim Biophys Acta.* 1044: 275–283, 1990.
181. VW Bowry, KK Stanley, R Stocker. *Proc Natl Acad Sci USA* 89: 10316–10320, 1992.
182. B Garner, PK Witting, AR Waldeck, JK Christison, M Raftery, R Stocker. *J Biol Chem* 273: 6080–6087, 1998.

26 Enzymatic Lipid Peroxidation

As noted in Chapter 25, enzymatic lipid peroxidation is characterized by the involvement of enzymes at different stages of lipid peroxidation and not only at the initiation. Furthermore, the products (leukotrienes, prostaglandins, etc.), which are selectively formed in the reactions catalyzed by lipoxygenases (LOXs) and prostaglandin H synthases, display important physiological and pathophysiological properties, while the products of nonenzymatic lipid peroxidation are mostly the damaging agents. In this chapter, we will consider the mechanisms of lipid peroxidation and free radical production catalyzed by LOXs and prostaglandin H synthases.

26.1 LIPID PEROXIDATION AND FREE RADICAL PRODUCTION CATALYZED BY LOXs

26.1.1 MECHANISM OF LOX CATALYSIS

LOXs as well as prostaglandin H synthases (cyclooxygenases) are enzymes catalyzing the oxidation of unsaturated fatty acids containing *cis* double bonds. LOXs are widely distributed in plants, fungi, and animals. The LOX superfamily consists of several main enzymes, whose physiological and pathological functions depend on the position of a double bond of the substrate [1]. 5-LOX produces precursors for leukotrienes involved in inflammation and allergic responses, 12-LOX oxygenases arachidonic acid (AA) at carbon-12 and responsible for hypertrophy, proliferation, and hypertensive actions in vascular endothelium, smooth muscle cells, platelets, and leukocytes. Mammalian 15-LOX, which corresponds to plant 13-LOX, oxidizes AA, 20-carbon fatty acids, and mitochondrial phospholipids.

The most extraordinary aspect of LOX activity is a dependence of the mechanism of action on their functions. There are at least two major LOX functions: (i) the formation of signaling molecules and (ii) peroxidation reactions.

(i) When LOXs participate in signaling pathways by forming signaling molecules, a single hydroperoxide is synthesized. For example, the activation of 5-LOX of human leukocytes produces leukotrienes, which provoke together with dihydroxyeicosanoids bronchoconstriction and inflammation. In this case, the LOX product is an intermediate in the signaling pathway. However, the LOX product or its reduced hydroxy derivative may be also the end product of the synthesis of 12-hydroxyeicosatetraenoic acid (12-HETE) by the platelet 12-LOX.
(ii) When LOXs catalyze the peroxidation reactions of membrane lipids, a mixture of hydroperoxides is formed, which are responsible for structural changes of the membrane. In this case the structures of hydroperoxides is not important for the perturbation of membrane structure.

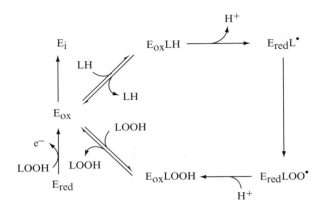

FIGURE 26.1 Dioxygenase cycle 15-LOX. (From VB O'Donnell, KB Taylor, S Parthasarathy, H Kuhn, D Koesling, A Friebe, A Bloodsworth, VM Darley-Usmar, BA Freeman. *J Biol Chem* 274: 20083–20091, 1999. With permission.)

LOXs are proteins containing a single atom of nonheme iron in catalytic center, with the ferric enzyme in an active form. The free radical-mediated mechanism of LOX-catalyzed process may be presented as follows (see also Figure 26.1):

$$\text{Enz}(Fe^{2+}) + LOO^\bullet \Longrightarrow \text{Enz}(Fe^{3+})LOO^- \quad (1)$$

$$\text{Enz}(Fe^{3+})LOO^- + H^+ \Longrightarrow \text{Enz}(Fe^{3+}) + LOOH \quad (2)$$

$$\text{Enz}(Fe^{3+}) + LH \Longrightarrow \text{Enz}(Fe^{2+})L^\bullet + H^+ \quad (3)$$

$$\text{Enz}(Fe^{2+})L^\bullet + O_2 \Longrightarrow \text{Enz}(Fe^{2+})LOO^\bullet \Longleftrightarrow \text{Enz}(Fe^{3+})LOO^- \quad (4)$$

$$\text{Enz}(Fe^{2+})LOO^\bullet \Longleftrightarrow \text{Enz}(Fe^{2+}) + LOO^\bullet \quad (5)$$

$$LOO^\bullet + LH \Longrightarrow LOOH + L^\bullet \quad (6)$$

$$L^\bullet + O_2 \Longrightarrow LOO^\bullet \quad (7)$$

In accord with this mechanism, free peroxyl radical of the reaction product hydroperoxide activates the inactive ferrous form of enzyme (Reaction (1)). Then, active ferric enzyme oxidizes substrate to form a bound substrate radical, which reacts with dioxygen (Reaction (4)). The bound peroxyl radical may again oxidize ferrous enzyme, completing redox cycling, or dissociate and abstract a hydrogen atom from substrate (Reaction (6)).

Free radical mechanism of LOX catalyzing reactions was first proposed by de Groot et al. [2] for the oxidation of polyunsaturated fatty acids containing a 1,4-*cis,cis*-pentadiene system by soybean LOX (15-LOX). These authors were able to spin-trap the carbon-centered free radical from linoleic acid. Later on, as an alternative to free radical mechanism, the organoiron-mediated pathway was proposed by Corey and Nagata [3]. However, free radical mechanism of LOX catalysis has been confirmed in subsequent studies. Thus Chamulitrat and Mason [4] demonstrated the formation of free radicals in the reactions catalyzed by soybean LOX under physiological conditions by the method of rapid mixing, continuous-flow EPR spectroscopy. Based on the measurement of kinetic deuterium isotope effects, Glickman and Klinman [5] concluded that during the catalytic oxidation of linoleate, dioxygen directly interacted with the LOX substrate radical to form hydroperoxyl radical. Hydrogen atom abstraction mechanism was further confirmed in the study of reactions of synthetic ferrous complex, a model of LOX [6].

Chamulitrat and Mason [4] also demonstrated that the ESR spectra of free radicals formed in LOX-catalyzed reactions did not exhibit any *g*-anisotropy, suggesting that radicals are free and are not enzyme-bound. It has been suggested that radicals that formed DMPO spin adducts are peroxyls. However, Dikalov and Mason [7] recently reinvestigated these data and concluded that attempted spin-trapping with DMPO of peroxyl radicals formed in LOX-catalyzed reactions results in the formation of alkoxyl radical adducts as a result of the rearrangement of peroxyl radicals. Actually, at present, it is impossible to judge with certainty what kind of free radicals, free or bound to the enzyme mediates the oxygenation of substrate. Nonetheless, peroxyl radicals are probably the only oxidants in this system able to oxidize ferrous form of the enzyme (Reaction (1)).

Inhibition and stimulation of LOX activity occurs as a rule by a free radical mechanism. Riendeau et al. [8] showed that hydroperoxide activation of 5-LOX is product-specific and can be stimulated by 5-HPETE and hydrogen peroxide. NADPH, FAD, Fe^{2+} ions, and Fe^{3+}(EDTA) complex markedly increased the formation of oxidized products while NADH and 5-HETE were inhibitory. Jones et al. [9] also demonstrated that another hydroperoxide 13(*S*)-hydroperoxy-9,11(*E,Z*)-octadecadienoic acid (13-HPOD) (formed by the oxidation of linoleic acid by soybean LOX) activated the inactive ferrous form of the enzyme. These authors suggested that 13-HPOD attached to LOX and affected its activation through the formation of a protein radical. Werz et al. [10] showed that reactive oxygen species produced by xanthine oxidase, granulocytes, or mitochondria activated 5-LOX in the Epstein–Barr virus-transformed B-lymphocytes.

Free radical scavengers inactivate LOXs by the interaction with their free radical intermediates or by reducing the enzyme active ferric form to inactive ferrous form. It has been shown that flavonoids, typical free radical scavengers are the effective inhibitors of neutrophil 5-LOX [11]. Van der Zee et al. [12] showed that the inhibition of soybean LOX by nordihydroguaiaretic acid, *p*-aminophenol, catechol, hydroquinone, and phenidone accompanied by the reduction of its ferric form and the formation of inhibitor free radicals. Analogs of natural unsaturated acids containing triple bonds irreversibly inactivate soybean LOX-1 supposedly via the abstraction of bisallylic hydrogen atoms and the formation of hydroxyl radicals [13]. Another group of LOX inhibitors is iron chelators. Abeysinghe et al. [14] demonstrated that lipophilic hydroxypyridinones inhibited 5-LOX probably by removing iron from an active center of the enzyme.

26.1.2 LOX-Catalyzed Oxidation of Unsaturated Acids

LOX-catalyzed oxidation of AA yields an unstable epoxy intermediate leukotriene A (LTA_4), which is enzymatically converted to leukotriene B (LTB_4) [15] (Figure 26.2). It has been proposed that these processes are mediated by free radicals. Rabbit reticulocyte lipoxygenase catalyzed the oxidation of AA to 15(*S*)-hydroperoxyeicosatetraenoic acid (15-HPETE) by the hydrogen abstraction from C-13 and to 12(*S*)-hydroperoxyeicosatetraenoic acid (12-HPETE) by hydrogen abstraction from C-10 [16]. The same LOX further catalyzes 15-HPETE into leukotriene B. MacMillan et al. [17] found that two mammalian lipoxygenases 5-LOX and 15-LOX are responsible for leukotriene formation by human eosinophils. 5-LOX begins the leukotriene biosynthetic pathway by the oxygenation of AA to 5(*S*)-hydroperoxyeicosatetraenoic acid (5-HPETE) with subsequent dehydration of this intermediate into LTA_4 (Figure 26.3). Next, 15-LOX oxidizes both 15-HPETE and 5-HPETE to LTA_4 (Figure 26.4). Rat peritoneal monocytes and human leukocytes are also able to oxidize AA into the dihydroxy derivative, 5(*S*),15(*S*)-dihydroxy-6,13-*trans*-8,11-*cis*-eicosatetraenoic acid (5,15-DiHETE). It has been shown [18] that 5,15-DiHETE was formed by a double oxidation of AA catalyzed by both 5-LOX and 15-LOX lipoxygenases (Figure 26.5). It is interesting that the

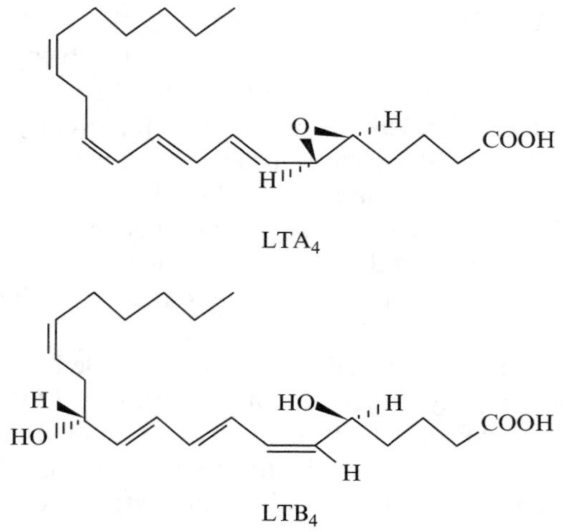

FIGURE 26.2 Leukotrienes LTA$_4$ and LTB$_4$.

stereochemistry of hydroperoxides formed by lipoxygenase can be pH-dependent. Thus, the formation of (9S)-hydroperoxide by the soybean LOX-1-catalyzed oxidation of linoleic acid depended on pH, becoming insignificant at pH < 6.

An important characteristic of mammalian 15-LOX is its capacity to oxidize the esters of unsaturated acid in biological membranes and plasma lipoproteins without their hydrolysis to free acids. Jung et al. [19] found that human leukocyte 15-LOX oxidized phosphatidylcholine at carbon-15 of the AA moiety. Soybean and rabbit reticulocyte 15-LOXs were also active while human leukocyte 5-LOX, rat basophilic leukemia cell 5-LOX, and rabbit platelet 12-LOX were inactive. It was suggested that the oxygenation of phospholipid is a unique property of 15-LOX. However, Murray and Brash [20] showed that rabbit reticulocyte

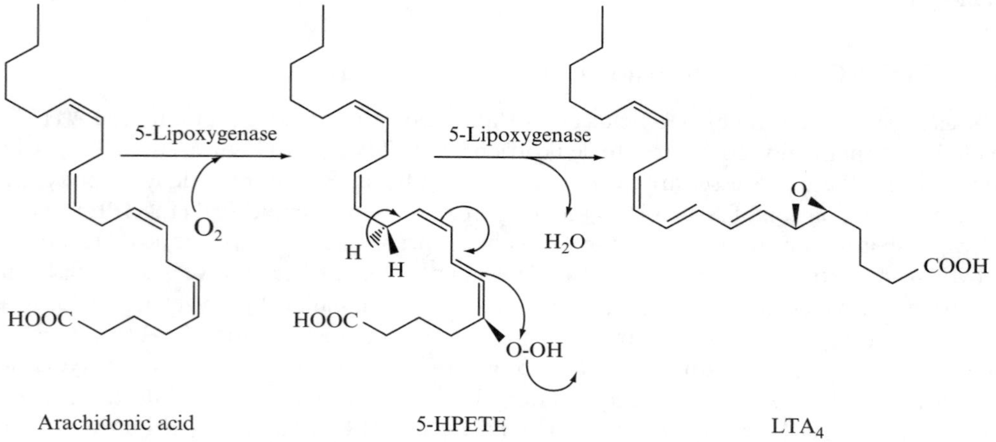

FIGURE 26.3 Formation of leukotriene LTA$_4$ by the LOX-5-catalyzed oxidation of arachidonic acid. (From DK MacMillan, E Hill, A Sala, E Sigal, T Shuman, PM Henson, RC Murphy. *J Biol Chem* 269: 26663–26668, 1994. With permission.)

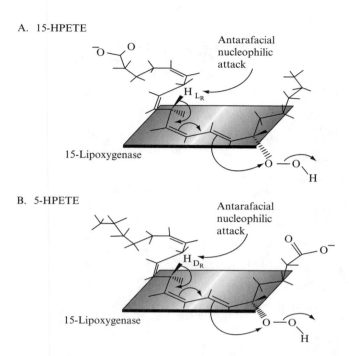

FIGURE 26.4 Oxidation of 5-HPETE and 15-HPETE by LOX-15. (From DK MacMillan, E Hill, A Sala, E Sigal, T Shuman, PM Henson, RC Murphy. *J Biol Chem* 269: 26663–26668, 1994. With permission.)

LOX catalyzed the oxidation of arachidonoylphosphatidylcholine at both carbon-12 and carbon-15. Later on, it has been found [21] that reticulocyte lipoxygenase oxidized rat liver mitochondrial membranes, beef heart submitochondrial particles, rat liver endoplasmic membranes, and erythrocyte plasma membranes without preliminary release of unsaturated acids by phospholipases.

Schnurr et al. [22] showed that rabbit 15-LOX oxidized beef heart submitochondrial particles to form phospholipid-bound hydroperoxy- and keto-polyenoic fatty acids and induced the oxidative modification of membrane proteins. It was also found that the total oxygen uptake significantly exceeded the formation of oxygenated polyenoic acids supposedly due to the formation of hydroxyl radicals by the reaction of ubiquinone with lipid 15-LOX-derived hydroperoxides. However, it is impossible to agree with this proposal because it is known for a long time [23] that quinones cannot catalyze the formation of hydroxyl radicals by the Fenton reaction. Oxidation of intracellular unsaturated acids (for example, linoleic and arachidonic acids) by lipoxygenases can be suppressed by fatty acid binding proteins [24].

26.1.3 LOX-Catalyzed Oxidation of Low-Density Lipoproteins

Besides the oxidation of unsaturated acids such as arachidonic and linoleic acids, LOXs are able to oxidize other substrates. One of the most important oxidative processes catalyzed by LOXs is the oxidation of low-density lipoproteins (LDL). (Nonenzymatic LDL oxidation has been discussed in detail in Chapter 25.) As already mentioned, the oxidation of LDL in the arterial intimal space is an important step in the development of atherogenesis. Now, we will consider the involvement of LOXs in this process. In 1989, Parthasarathy et al. [25] found that

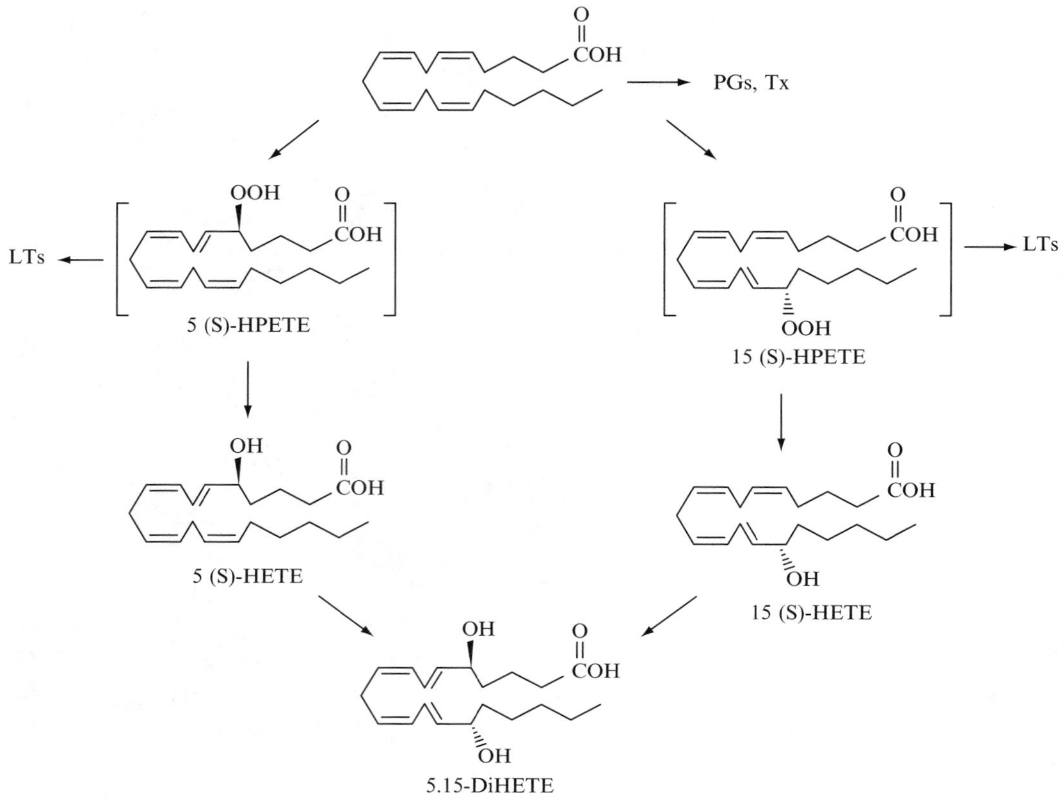

FIGURE 26.5 Formation of 5,15-DiHETE in the oxidation of arachidonic acid by LOX-5 and LOX-15. (From RL Maas, J Turk, JA Oates, AR Brash. *J Biol Chem* 257: 7056–7067, 1982. With permission.)

the endothelial cell-induced oxidation of LDL was inhibited by LOX inhibitors while cyclooxygenase inhibitors were inactive. This fact proves that in this case cellular LOX activity was involved in LDL oxidation.

LOX-catalyzed oxidation of LDL has been studied in subsequent studies [26,27]. Belkner et al. [27] showed that LOX-catalyzed LDL oxidation was not restricted to the oxidation of lipids but also resulted in the cooxidative modification of apoproteins. It is known that LOX-catalyzed LDL oxidation is regio- and enantio-specific as opposed to free radical-mediated lipid peroxidation. In accord with this proposal Yamashita et al. [28] showed that LDL oxidation by 15-LOX from rabbit reticulocytes formed hydroperoxides of phosphatidylcholine and cholesteryl esters regio-, stereo-, and enantio-specifically. Sigari et al. [29] demonstrated that fibroblasts with overexpressed 15-LOX produced bioactive "minimally modified" LDL, which is probably responsible for LDL atherogenic effect in vivo. Ezaki et al. [30] found that the incubation of LDL with 15-LOX-overexpressed fibroblasts resulted in a sharp increase in the cholesteryl ester hydroperoxide level and a lesser increase in free fatty acid hydroperoxides.

The mechanism of LOX-catalyzed LDL oxidation is still not clearly understood [31]. On one hand, it has been proposed that LDL oxidation may be initiated by oxygen radicals, which are released from the active site of the enzyme. On the other hand, the formation of lipid peroxide by direct oxygenation of unsaturated acids without the participation of free radicals has been suggested. It is important that the last mechanism of LDL peroxidation can be realized only in the presence of transition metals copper and iron because lipid peroxides are stable in their absence. O'Leary et al. [31] proposed that synergistic interaction may exist between lipid hydroperoxides formed in LDL by LOX and cupric ions or heme proteins.

Belkner et al. [32] demonstrated that 15-LOX oxidized preferably LDL cholesterol esters. Even in the presence of free linoleic acid, cholesteryl linoleate continued to be a major LOX substrate. It was also found that the depletion of LDL from α-tocopherol has not prevented the LDL oxidation. This is of a special interest in connection with the role of α-tocopherol in LDL oxidation. As the majority of cholesteryl esters is normally buried in the core of a lipoprotein particle and cannot be directly oxidized by LOX, it has been suggested that LDL oxidation might be initiated by α-tocopheryl radical formed during the oxidation of α-tocopherol [33,34]. Correspondingly, it was concluded that the oxidation of LDL by soybean and recombinant human 15-LOXs may occur by two pathways: (a) LDL-free fatty acids are oxidized enzymatically with the formation of α-tocopheryl radical, and (b) the α-tocopheryl-mediated oxidation of cholesteryl esters occurs via a nonenzymatic way. *Pro* and *con* proofs related to the prooxidant role of α-tocopherol were considered in Chapter 25 in connection with the study of nonenzymatic lipid oxidation and in Chapter 29 dedicated to antioxidants. It should be stressed that comparison of the possible effects of α-tocopherol and nitric oxide on LDL oxidation does not support importance of α-tocopherol prooxidant activity. It should be mentioned that the above data describing the activity of cholesteryl esters in LDL oxidation are in contradiction with some earlier results. Thus in 1988, Sparrow et al. [35] suggested that the 15-LOX-catalyzed oxidation of LDL is accelerated in the presence of phospholipase A2, i.e., the hydrolysis of cholesterol esters is an important step in LDL oxidation.

26.1.4 LOX-Catalyzed Cooxidation of Substrates

One of the most convincing proofs of free radical mechanism of the LOX catalysis is the ability of LOX systems to oxidize various organic molecules to their free radicals. Numerous substrates are cooxidized by LOXs to form free radicals as intermediates. Lund et al. [36] showed that cooxidation of cholesterol with linoleic acid by soybean LOX under aerobic conditions resulted in the formation of a mixture of 7-oxygenated products. It was suggested that this reaction proceeded by a nonenzymatic pathway, in which the carbon-centered radical of linoleic acid abstracted a hydrogen atom directly from cholesterol molecule. By similar mechanism, linoleic acid–mammalian 15-LOX system supposedly oxidizes NADH and NADPH [37]. It was suggested that NADPH is oxidized by a free radical formed during enzymatic cycle while NADH oxidation does not occur at the active site of the enzyme. LOX-catalyzed oxidation of Trolox C [38] and diethylstilbestol, a synthetic carcinogenic estrogen, [39] in the presence of hydrogen peroxide yielded corresponding phenoxyl radicals, semiquinones, and quinones.

One of numerous examples of LOX-catalyzed cooxidation reactions is the oxidation and demethylation of amino derivatives of aromatic compounds. Oxidation of such compounds as 4-aminobiphenyl, a component of tobacco smoke, phenothiazine tranquillizers, and others is supposed to be the origin of their damaging effects including reproductive toxicity. Thus, LOX-catalyzed cooxidation of phenothiazine derivatives with hydrogen peroxide resulted in the formation of cation radicals [40]. Soybean LOX and human term placenta LOX catalyzed the free radical-mediated cooxidation of 4-aminobiphenyl to toxic intermediates [41]. It has been suggested that demethylation of aminopyrine by soybean LOX is mediated by the cation radicals and neutral radicals [42]. Similarly, soybean and human term placenta LOXs catalyzed N-demethylation of phenothiazines [43] and derivatives of *N,N*-dimethylaniline [44] and the formation of glutathione conjugate from ethacrynic acid and *p*-aminophenol [45,46].

LOX–hydrogen peroxide system catalyzed the conversion of 5,6-dihydroxyindole and 5,6-dihydroxyindole-2-carboxylic acid, which are important intermediates of melanogenesis, into melanin pigments [47].

26.1.5 Oxygen Radical Formation by LOX-Catalyzed Reactions

In 1985, Fischer and Adams [48] found that LOX inhibitors such as nordihydroguaiaretic acid inhibited TPA-induced chemiluminescent response in mouse epidermal cells. These findings suggested that at least a part of TPA-induced superoxide production by these cells was due to LOX-catalyzed arachidonate oxidation. Kukreja et al. [49] showed that the oxidation of linoleic and arachidonic acids by LOX is accompanied by superoxide formation. As superoxide production was observed only in the presence of NADH and NADPH, it was suggested that superoxide was formed by the oxidation of NAD and NADP radicals. This proposal has been confirmed in a subsequent study [50]. Jahn and Hansch [51] showed that superoxide generation by the arachidonic-stimulated human platelets was inhibited by LOX inhibitors, suggesting the LOX pathway of superoxide formation. Similarly, Tanaka et al. [52] found that the formation of superoxide by arachidonate, leukotriene B4, or C5a-stimulated human leukocytes depended on LOX inhibition. Chamulitrat et al. [53] demonstrated superoxide formation during the oxidation of polyunsaturated acids by soybean LOX using spin-trapping with DMPO. Superoxide production was also observed during soybean LOX-mediated cooxidation of reduced glutathione with linoleic and arachidonic acids [54].

LOX-dependent superoxide production was also registered under ex vivo conditions [55]. It has been shown that the intravenous administration of lipopolysaccharide to rats stimulated superoxide production by alveolar and peritoneal macrophages. O'Donnell and Azzi [56] proposed that a relatively high rate of superoxide production by cultured human fibroblasts in the presence of NADH was relevant to 15-LOX-catalyzed oxidation of unsaturated acids and was independent of NADPH oxidase, prostaglandin H synthase, xanthine oxidase, and cytochrome P-450 activation or mitochondrial respiration. LOX might also be involved in the superoxide production by epidermal growth factor-stimulated pheochromocytoma cells [57].

Thus, LOX-catalyzed oxidative processes are apparently effective producers of superoxide in cell-free and cellular systems. (It has also been found that the arachidonate oxidation by soybean LOX induced a high level of lucigenin-amplified CL, which was completely inhibited by SOD; LG Korkina and TB Suslova, unpublished data.) It is obvious that superoxide formation by LOX systems cannot be described by the traditional mechanism (Reactions (1)–(7)). There are various possibilities of superoxide formation during the oxidation of unsaturated compounds; one of them is the decomposition of hydroperoxides to alkoxyl radicals. These radicals are able to rearrange into hydroxylalkyl radicals, which form unstable peroxyl radicals, capable of producing superoxide in the reaction with dioxygen.

$$ROOH + Fe^{2+} \Longrightarrow RO^{\bullet} + HO^{-} + Fe^{3+} \tag{8}$$

$$RO^{\bullet} \Longleftrightarrow HOR^{\bullet} \tag{9}$$

$$HOR^{\bullet} + O_2 \Longrightarrow HOROO^{\bullet} \Longrightarrow O_2^{\bullet -} \tag{10}$$

This mechanism must of course be considered purely hypothetical.

26.1.6 Effects of Nitrogen Oxides on LOX-Catalyzed Processes

At present, there is growing interest in the study of the effects of nitrogen oxides on LOX-catalyzed processes. It has been shown that nitric oxide is able to inhibit LOX-catalyzed lipid peroxidation [58,59]. Kanner et al. [58] proposed that the inhibitory effect of nitric oxide may be explained by: (a) the reduction of the active ferric form of the enzyme to the inactive ferrous form; (b) the competition for the iron side available for exogenous ligands; and (c) the reaction with enzymatic free radicals. Later on, the mechanism of NO interaction with LOXs has been discussed in detail. Rubbo et al. [60] proposed that the inhibition of LOX-catalyzed liposome and LDL oxidation by nitric oxide depended on the reaction with alkoxyl and peroxyl radicals formed during the enzymatic catalytic cycle. However, nitric oxide may both activate and inhibit LOX [61,62]. Holzhutter et al. [62] suggested that NO rapidly and reversibly binds to the inactive ferrous form of LOX with subsequent slow irreversible conversion of the Fe^{2+}–NO complex into the active ferric form.

Nonetheless, the stimulatory effect of nitric oxide is probably not very important at low physiological NO concentrations. O'Donnell et al. [63] proposed a scheme for the interaction of 15-LOX with low nitric oxide concentrations, which included three major pathways (Figure 26.6). (i) The interaction of NO with the enzyme during activation by hydroperoxide, which is accompanied by the formation of nitroxyl anion NO^-. (ii) Reaction of NO with $E_{red}LOO^{\bullet}$ radical. (iii) The formation of the E–Fe^{2+}–NO complex at higher NO concentrations, which decomposes forming the active form of enzyme E_{ox}. It was also pointed out that the reaction of nitric oxide with 15-LOX inhibited the NO-dependent activation of soluble guanylate cyclase and cGMP production [73].

Nitric oxide is not an only nitrogen compound, affecting LOX reactions. Thus it has been found that alkylhydroxylamines and N-hydroxyurea derivatives inhibited soybean LOX, reducing it to the inactive ferrous form [64]. It is interesting that peroxynitrite was unable

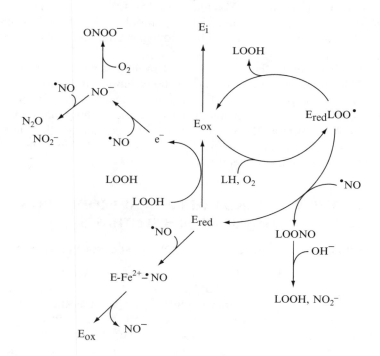

FIGURE 26.6 Possible mechanisms of NO interaction with LOX-15. (From VB O'Donnell, KB Taylor, S Parthasarathy, H Kuhn, D Koesling, A Friebe, A Bloodsworth, VM Darley-Usmar, BA Freeman. *J Biol Chem* 274: 20083–20091, 1999. With permission.)

to inhibit 12-LOX [52]. Correspondingly, superoxide reversed the NO inhibition of LOX, forming peroxynitrite.

26.1.7 Comments on Pathophysiological Activities of LOXs

The important role of LOXs in pathophysiological processes is well known. Here, we will consider just some data concerning the involvement of LOXs in the development of atherosclerosis. Damaging effects of LOX-catalyzed processes are also discussed in Chapter 31. It has been shown that LOX activity is strongly affected by oxidative stress in cells. For example, LOX activity in cardiac myocytes characterized by the HETE formation was insignificant under normoxic conditions but greatly increased after hypoxia–reoxygenation [66]. Cell injury was observed simultaneously with the HETE production. It was concluded that the arachidonate–LOX metabolic pathway plays an important role in reoxygenation-induced myocardial cell injury.

The expression of 15-LOX in atherosclerotic lesions is one of the major causes of LDL oxidative modification during atherosclerosis. To obtain the experimental evidence of a principal role of 15-LOX in atherosclerosis under in vivo conditions, Kuhn et al. [67] studied the structure of oxidized LDL isolated from the aorta of rabbits fed with a cholesterol-rich diet. It was found that specific LOX products were present in early atherosclerotic lesions. On the later stages of atherosclerosis the content of these products diminished while the amount of products originating from nonenzymatic lipid peroxidation increased. It was concluded that arachidonate 15-LOX is of pathophysiological importance at the early stages of atherosclerosis. Folcik et al. [68] demonstrated that 15-LOX contributed to the oxidation of LDL in human atherosclerotic plaques because they observed an increase in the stereospecificity of oxidation in oxidized products. Arachidonate 15-LOX is apparently more active in young human lesions and therefore, may be of pathophysiological importance for earlier atherosclerosis. In advanced human plaques nonenzymatic lipid peroxidation products prevailed [69].

O'Donnell et al. [70] found that LOX and not cyclooxygenase, cytochrome P-450, NO synthase, NADPH oxidase, xanthine oxidase, ribonucleotide reductase, or mitochondrial respiratory chain is responsible for TNF-α-mediated apoptosis of murine fibrosarcoma cells. 15-LOX activity was found to increase sharply in heart, lung, and vascular tissues of rabbits by hypercholesterolemia [71]. Schnurr et al. [72] demonstrated that there is an inverse regulation of 12/15-LOXs and phospholipid hydroperoxide glutathione peroxidases in cells, which balanced the intracellular concentration of oxidized lipids.

26.2 LIPID PEROXIDATION AND FREE RADICAL PRODUCTION CATALYZED BY PROSTAGLANDIN H SYNTHASES

26.2.1 Mechanism of Reactions Catalyzed by Prostaglandin H Synthases

Prostaglandin endoperoxide H synthases (PGHS) are enzymes, which catalyze the oxidation of AA by dioxygen into prostaglandin PGH_2, the first step of biosynthesis of prostanoids. PGH_2 is further transformed by other enzymes into prostaglandins and thromboxane (Figure 26.7). There are two isoforms of prostaglandin H synthase: the constitutive PGHS-1 and inducible PGHS-2, which despite sharing only 60% sequence identity, have very similar catalytic sites. Although the reason for the existence of two PGHS isoenzymes is unknown, both isoenzymes are frequently coexpressed in the same cell and independently channel prostanoids to the extracellular milieu and the nucleus. Both isoenzymes are heme-containing glycosylated proteins with two catalytic sites. PGHS-2 is undetectable in most mammalian

Enzymatic Lipid Peroxidation

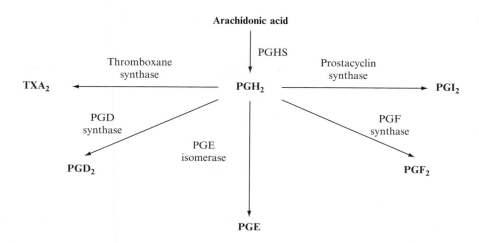

FIGURE 26.7 Products of PGHS-catalyzed oxidation of arachidonic acid.

tissues, but is rapidly expressed in fibroblasts, endothelial cells, and monocytes in response to growth factors, tumor promotors, hormones, bacterial endotoxin, and cytokines. In contrast to PGHS-2, PGHS-1 may be found in most tissues.

PGHS-1 and PGHS-2 exhibit two enzymatic activities: a cyclooxygenase activity, converting AA into prostaglandin PGG_2, and peroxidase activity transforming PGG_2 into prostaglandin PGH_2 and other prostaglandins. The PGHS-catalyzed oxidation of AA should be considered a special type of free radical lipid peroxidation. By the use of ESR spectroscopy, the formation of the carbon-centered AA free radical was shown in the reaction of AA with ram seminal vesicle microsomes possessing high PGHS activity [74]. It has been shown in subsequent studies [75,76] that the coupling between the peroxidase and cyclooxygenase activities of PGHS resulted in the formation of tyrosyl radicals. Rapid electronic spectroscopy study demonstrated that the interaction of PGHS with PGG_2 led to the formation of an intermediate (Complex I) within 2 ms with the rate constant of $1.4 \times 10^7 \, l \, mol^{-1} \, s^{-1}$. During the next 170 ms Complex I is converted into Complex II containing the tyrosyl cation radical (the rate constant is $65 \, s^{-1}$) [75]. Karthein et al. [76] suggested that tyrosyl radical is able to abstract a hydrogen atom from AA. This suggestion was questioned because the time course of the production of tyrosyl radical seems to be not correlated with the time course of metabolism of AA [77]. But later on, it has been demonstrated in anaerobic experiments that tyrosyl radical is indeed able to abstract hydrogen atom from AA [78,79]. Furthermore, the rapid kinetics study showed that the time courses of tyrosyl radical formation actually coincide with the heme redox state changes in both PGHS-1 and PGHS-2 [80]. Therefore, the mechanism of AA oxidation catalyzed by PGHS may be presented as follows (Figure 26.8) [79]. Tyrosyl radical Fe(IV)Tyr$^\bullet$ is formed by an internal electron transfer in the peroxidase Compound I:

$$PGHSFe(V)Tyr \Longleftrightarrow PGHSFe(IV)Tyr^\bullet \qquad (11)$$

Then, tyrosyl radical reacts with bound AA to form a bound AA free radical:

$$PGHSFe(IV)Tyr^\bullet + AA \Longrightarrow PGHSFe(IV)TyrAA^\bullet \qquad (12)$$

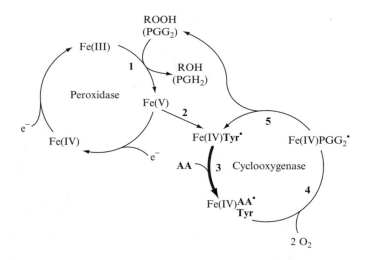

FIGURE 26.8 Mechanism of PGHS-catalyzed arachidonate oxidation. (From A Tsai, G Palmer, G Hiao, DC Swinney, RJ Kulmacz. *J Biol Chem* 273: 3888–3894, 1998. With permission.)

This radical reacts with dioxygen and rearranges into a PGG_2 free radical:

$$PGHS\ Fe(IV)TyrAA^{\bullet} + 2O_2 \Longrightarrow PGHSFe(IV)PGG_2^{\bullet} \tag{13}$$

In the last step of cyclooxygenase cycle tyrosyl radical is regenerated and PGG_2 is released:

$$PGHSFe(IV)PGG_2^{\bullet} \Longrightarrow PGHSFe(IV)Tyr^{\bullet} + PGG_2 \tag{14}$$

It has been suggested that only one tyrosine residue, tyrosine 385, is oxidized into tyrosyl free radical in cyclooxygenase cycle. This suggestion was confirmed by NO trapping of this tyrosyl radical generated by prostaglandin H synthase [81]. It was found that the stoichiometry of AA oxidation by prostaglandin H synthase AA/O_2 is equal to ca. 2 [82,83].

Interesting data have been obtained by the modification of PGHS-1. Tsai et al. [84] found that the replacement of the enzyme heme group by manganese protoporphyrin IX preserved the cyclooxygenase activity but sharply decreased peroxidase activity of the enzyme. The difference between native and modified enzymes is thought to originate from a very slow rate of the substitution of a water molecule at the Mn^{3+} center by hydroperoxide. This process becomes a rate-limiting step that retards the peroxidase reaction.

Thus, cyclooxygenase activity of prostaglandin H synthases is a physiological free radical enzymatic oxidation, despite the suggestion that cyclization step may also proceed by carbocation mechanism [85]. However, PGHS activity is regulated, stimulated, or inhibited by free radicals or free radical scavengers. For example, nitric oxide (a free radical) is apparently able to inhibit PGHS-1 activity [86], while peroxynitrite, the product of the reaction of NO with superoxide activates the cyclooxygenase activity of both isoenzymes [87].

26.2.2 Production and Interaction with Oxygen Radicals

Similar to LOXs, cyclooxygenases may catalyze superoxide production in the presence of NADH and NADPH [49]. It has been shown [88] that prostaglandin H synthase produced oxygen radicals and hydrogen peroxide during the transformation of 2(3)-*tert*-butyl-4-

hydroxyanisol into 2-*tert*-butylhydroquinone. Vasopressin stimulated cyclooxygenase-dependent superoxide production after brain injury [89]. Didion et al. [90] suggested that cyclooxygenase is responsible for an increase in superoxide levels in cerebral blood vessels because it was greatly augmented upon exposure to AA and bradykinin. In addition, superoxide production was enhanced by the inhibition of endogenous SOD with diethyldithiocarbamate. It was also demonstrated [91] that LPS stimulation of murine peritoneal macrophages resulted in the production of nitric oxide and prostaglandin PGE_2, which was accompanied by superoxide formation and the increased expression of COX-2. It has been suggested that superoxide production might stimulate the AA release and PGE_2 formation by COX-2. Minegishi et al. [92] proposed that $PGF_{2\alpha}$ induced the superoxide-mediated luteolysis (the loss of the capacity to synthesize and secrete progesterone and the loss of luteal cells) in rats.

Reactive oxygen species can also inhibit cyclooxygenase activity. Thus Adler et al. [93] showed that oxygen radicals and hydrogen peroxide generated by xanthine oxidase decreased PGE_2 and TxB_2 production by glomerular mesangial cells. Cyanide-induced production of active oxygen species inhibited COX-2 activity (as well as the activities of LOX and NO synthase) in cerebellar granule cells [94]. Superoxide is possibly involved in the regulation of PGE_2 production in mouse peritoneal macrophages [95].

26.2.3 Cyclooxygenase-Catalyzed Cooxidation of Substrates

Similar to peroxidases (Chapter 22) and LOXs (see above), cyclooxygenases are capable of catalyzing the oxidation of substrates during the reduction of PGG_2 to PGH_2. Potter and Hinson [96] proposed that prostaglandin H synthase catalyzed the oxidation of acetaminophen by both one-electron and two-electron mechanisms. Formosa et al. [91] showed that 3-methylindole, which causes a highly tissue- and species-selective lesion of the lung, is cooxidized with AA by PGHS. Similarly, PGHS (and LOX) oxidized presumably carcinogenic food antioxidant 2(3)-*tert*-butyl-4-hydroxyanisole (BHA), producing 2-*tert*-butylbenzoquinone [97]. Caffeic acid, which is a moderate stimulator of PGHS cyclooxygenase activity, reacted with PGHS Compound II with the rate constant of $1.25 \pm 0.1 \times 10^6 \, l \, mol^{-1} \, s^{-1}$ [97]. Samokyszyn et al. [98] suggested that PGHS oxidized all *trans* (E)-retinoic acid by a free radical mechanism. Using spin-trapping, these authors showed in subsequent work [99] that PGHS-catalyzed oxidation of retinoic acids resulted in the formation of C4 carbon-centered free radicals, which subsequently reacted with dioxygen to form retinoid-derived peroxyl radicals.

Parman et al. [100] studied the mechanism of PGHS-catalyzed oxidation of phenytoin (diphenylhydantoin), an anticonvulsant drug, which exhibits teratogenic effects in humans and animals. It was suggested that phenytoin and its analogs are oxidized into imidyl free radicals, which are converted to carbon-centered free radicals with an isocyanate substituent by ring opening. The latter can covalently bind to embryonic macromolecules with double bonds including DNA and proteins and initiate macromolecular damage (Figure 26.9). Moreover, the authors proposed another mechanism of free radical-mediated damage with the participation of oxygen radicals (Figure 26.9). The formation of free radicals of phenytoin in PGHS-catalyzed oxidation was supported by EPR spin-trapping experiments.

26.2.4 Effects of Reactive Nitrogen Species on Prostaglandin H-Catalyzed Processes

It has already been mentioned earlier that similar to LOXs, prostaglandin H synthases can be activated or inhibited by reactive nitrogen species. Nitric oxide may exhibit the inhibitory [58,65,86,101–104] or stimulatory effects [105–110] on PGHSs. Inhibitory effects depend on the ability of nitric oxide to reduce the ferric enzyme to the inactive ferrous form, competition

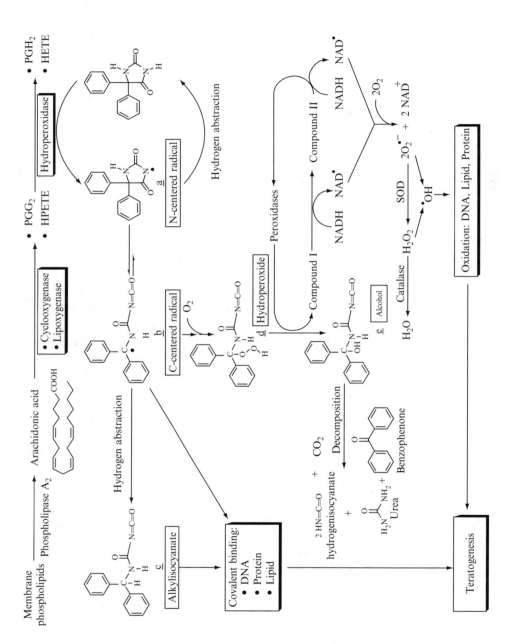

FIGURE 26.9 Hypothetical free radical mechanism of PGHS-catalyzed oxidation of phenytoin. (From T Parman, G Chen, PR Wells. *J Biol Chem* 273: 25079–25088, 1998. With permission.)

for the iron site available for exogenous ligands, the interaction of NO with peroxyl radicals [58], or regulation of the expression of cyclooxygenase-2 [103].

Among possible directions of the interaction of nitric oxide with PGSH-1, namely, the formation of nitrosyl complex with ferric heme [112], the reaction with tyrosyl radical [81], and the participation in catalytic cycle as a reducing peroxidase substrate [113]) the last of the mentioned mechanisms is the most probable in vivo pathway [111]. Indeed, the formation of a nitrosyl complex is unlikely due to the low in vivo NO concentration and the suppression of reaction of NO with tyrosyl radical by AA. It was also found that the consumption of nitric oxide by PGHS-1 during platelet aggregation inhibited cGMP generation.

Although many experimental data showed the stimulatory effect of nitric oxide on prostaglandin H synthase [105–110], a genuine stimulator of PGHS under the conditions studied was probably peroxynitrite and not NO [87]. Peroxynitrite, an inorganic hydroperoxide, is able to stimulate PGHS as a substrate for peroxidase activity and an activator of cyclooxygenase activity. It is important that prostaglandin H synthase and nitric oxide synthase can be simultaneously stimulated in cells. Thus, it is possible that there is "concerted" mechanism of superoxide- and peroxynitrite-mediated stimulation of prostaglandin production by three prooxidant enzymes (NADPH oxidase, NO synthase, and prostaglandin H synthase) (Figure 26.10). Recently, Beharka et al. [114] proposed that the inhibitory effect of vitamin E on increased PGHS-2-mediated prostaglandin production in macrophages from old mice is a consequence of the suppression of peroxynitrite-induced cyclooxygenase activity. Boulos et al. [115] found out that in the absence of AA peroxynitrite inhibits cyclooxygenase activity in human platelets by the nitration of tyrosine residues.

26.3 OXIDATION BY LINOLEATE DIOL SYNTHASE

LOXs and prostaglandin H synthases are the most important but not the only enzyme-catalyzed oxidative processes in biological systems. One of the similar enzymes is linoleate diol oxidase (LDS), which was isolated from the fungus *Gaeumannomyces graminis*. In contrast to prostaglandin H synthases, LDS exhibits hydroperoxide isomerase and not peroxidase activity (Figure 26.11). As follows from Figure 26.11, LDS catalyzes the oxidation of linoleic acid into two products: $8R$-hydroperoxyoctadecadienoic acid (8-HPODE) and $7S,8S$-dihydroxyoctadecadienoic acid (7,8-D-HODE). A hypothetical mechanism of LDS catalytic activity is presented in Figure 26.12 [116]. In accord with this mechanism the catalytic cycle is initiated by the reduction of 8-HPODE to 8-HODE, yielding ferryl oxygen complex containing porphyrin π-cation radical (Compound I). Compound I was rapidly reduced to Compound II, and tyrosyl free radical is formed. Tyrosyl radical abstracts hydrogen atom from linoleic acid (18:2) and the alkyl radical formed adds to dioxygen yielding peroxyl radical. The formed peroxyl radical is converted into 8-HPODE and regenerates tyrosyl radical. Linoleate diol synthase has important physiological function in funguses.

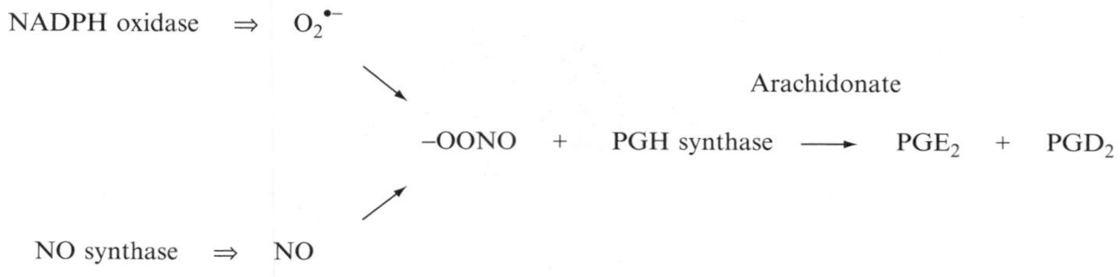

```
NADPH oxidase  ⇒  O₂•⁻
                     ↘
                              Arachidonate
                  −OONO  +  PGH synthase  ⟶  PGE₂  +  PGD₂
                     ↗
NO synthase  ⇒  NO
```

FIGURE 26.10 Superoxide- and NO-mediated stimulation of prostaglandin production.

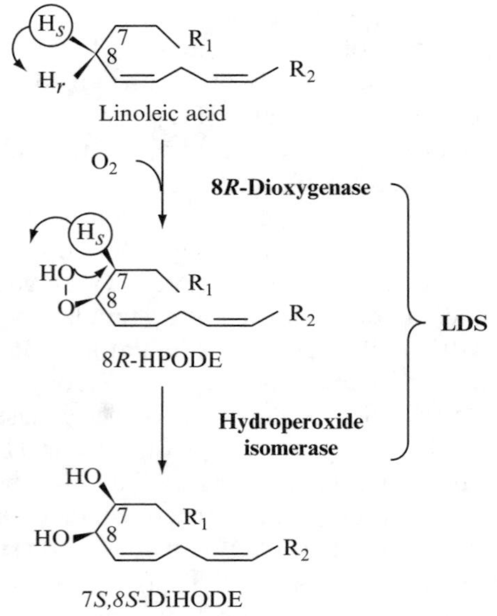

FIGURE 26.11 Oxygenation of linoleic acid by linoleate diol synthase. (From C Su, M Sahlin, EH Oliw. *J Biol Chem* 273: 20744–20751, 1998. With permission.)

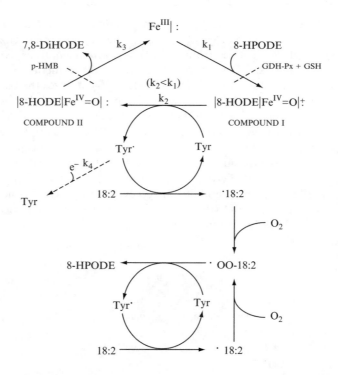

FIGURE 26.12 Hypothetical mechanism of catalytic activity of linoleate diol synthase. (From C Su, M Sahlin, EH Oliw. *J Biol Chem* 273: 20744–20751, 1998. With permission.)

REFERENCES

1. AR Brash. *J Biol Chem* 274: 23679–23682, 1999.
2. JJMC de Groot, GJ Garssen, JFG Vliegenthart, J Boldingh. *Biochim Biophys Acta* 326: 279–284, 1973.
3. EJ Corey, R Nagata. *J Am Chem Soc* 109: 8107–8108, 1987.
4. W Chamulitrat, RP Mason. *J Biol Chem* 264: 20968–20973, 1989.
5. MH Glickman, JP Klinman. *Biochemistry* 35: 12882–12892, 1996.
6. CR Goldsmith, RT Jonas, TD Stack. *J Am Chem Soc* 124: 83–96, 2002.
7. SI Dikalov, RP Mason. *Free Radic Biol Med* 30: 187–197, 2001.
8. D Riendeau, D Denis, LY Choo, DF Nathaniel. *Biochem J* 263: 565–572, 1989.
9. GD Jones, L Russell, VM Darley-Usmar, D Stone, MT Wilson. *Biochemistry* 35: 7197–7203, 1996.
10. O Werz, D Szellas, D Steinhilber. *Eur J Biochem* 267: 1263–1269, 2000.
11. MA Moroney, MJ Alcaraz, RA Forder, F Carey, JR Hoult. *J Pharm Pharmacol* 40: 787–792, 1988.
12. J Van der Zee, TE Eling, RP Mason. *Biochemistry* 28: 8363–8367, 1989.
13. WF Nieuwenhuizen, A Van der Kerk-Van Hoof, JH van Lenthe, RC Van Schaik, K Versluis, GA Veldink, JF Vliegenhart. *Biochemistry* 36: 4480–4488, 1997.
14. RD Abeysinghe, PJ Roberts, CE Cooper, KH MacLean, RC Hider, JB Porter. *J Biol Chem* 271: 7965–7972, 1996.
15. B Samuelsson. *Science* 220: 568–575, 1983.
16. RW Bryant, T Schewe, SM Rapoport, JM Bailey. *J Biol Chem* 260: 3548–3555, 1985.
17. DK MacMillan, E Hill, A Sala, E Sigal, T Shuman, PM Henson, RC Murphy. *J Biol Chem* 269: 26663–26668, 1994.
18. RL Maas, J Turk, JA Oates, AR Brash. *J Biol Chem* 257: 7056–7067, 1982.
19. G Jung, DC Yang, A Nakao. *Biochem Biophys Res Commun* 130: 559–566, 1985.
20. JJ Murray and AR Brash. *Arch Biochem Biophys* 265: 514–523, 1988.
21. H Kuhn, J Belkner, R Wiesner, AR Brash. *J Biol Chem* 265: 18351–18361, 1990.
22. K Schnurr, M Hellwing, B Seidemann, P Jungblut, H Kuhn, SM Rapoport, T Schewe. *Free Radic Biol Med* 20: 11–21, 1996.
23. IB Afanas'ev. *Superoxide Ion: Chemistry and Biological Implications*, vol 1, Chapter 9. Boca Raton, FL: CRC Press, 1989.
24. BA Ek, DP Cistola, JA Hamilton, TL Kaduce, AA Spector. *Biochim Biophys Acta* 1346: 75–85, 1997.
25. S Parthasarathy, E Wieland, D Steinberg. *Proc Natl Acad Sci USA* 86: 1046–1050, 1989.
26. MK Cathcart, AK McNally, GM Chisolm. *J Lipid Res* 32: 63–70, 1991.
27. J Belkner, R Wiesner, J Rathman, J Barnett, E Sigal, H Kuhn. *Eur J Biochem* 213: 251–261, 1993.
28. H Yamashita, A Nakamura, N Noguchi, E Niki, H Kuhn. *FEBS Lett* 445: 287–290, 1999.
29. F Sigari, C Lee, JL Witztum, PD Reaven. *Arterioscler Thromb Vasc Biol* 17: 3639–3645, 1997.
30. M Ezaki, JL Witztum, D Steinberg. *J Lipid Res* 36: 1996–2004, 1995.
31. VJ O'Leary, A Graham, D Stone, VM Darley-Usmar. *Free Radic Biol Med* 20: 525–532, 1996.
32. J Belkner, H Stender, H Kuhn. *J Biol Chem* 273: 23225–23232, 1998.
33. JM Upston, J Neuzil, PK Witting, R Alleva, R Stocker. *J Biol Chem* 272: 30067–30074, 1997.
34. J Neuzil, JM Upston, PK Witting, KF Scott, B Stocker. *Biochemistry* 37: 9203–9210, 1998.
35. CP Sparrow, S Parthasarathy, D Steinberg. *J Lipid Res* 29: 745–753, 1988.
36. E Lund, U Diszfalusy, I Bjoerkhem. *J Biol Chem* 267: 12462–12467, 1992.
37. VB O'Donnell, H Kuhn. *Biochem J* 327: 203–208, 1997.
38. E Nunez-Delicado, A Sanchez-Ferrer, F Garcia-Carmona. *Biochim Biophys Acta* 1335: 127–134, 1997.
39. E Nunez-Delicado, A Sanchez-Ferrer, F Garcia-Carmona. *Arch Biochem Biophys* 348: 411–414, 1997.
40. M Perez-Gilabert, A Sanchez-Ferrer, F Garcia-Carmona. *Biochem Pharmacol* 47: 2227–2232, 1994.
41. K Datta, PM Sherblom, AP Kulkarni. *Drug Metab Disposition* 25: 196–205, 1997.

42. M Perez-Gilabert, A Sanchez-Ferrer, F Garcia-Carmona. *Free Radic Biol Med* 23: 548–555, 1997.
43. AV Rajadhyaksha, V Reddy, CG Hover, AP Kulkarni. *Teratog Carcinog Mutagen* 19: 211–222, 1999.
44. CG Hover, AP Kulkarni. *Chem Biol Interact* 124: 191–203, 2000.
45. AP Kulkarni, M Sajan. *Arch Biochem Biophys* 371: 220–227, 1999.
46. X Yang, AP Kulkarni. *Toxicol Lett* 111: 253–261, 2000.
47. C Blarzino, L Mosca, C Foppoli, R Coccia, C De Marco, MA Rosen. *Free Radic Biol Med* 26: 446–453, 1999.
48. SM Fischer, LM Adams. *Cancer Res* 45: 3130–3136, 1985.
49. RC Kukreja, HA Kontos, ML Hess, EF Ellis. *Circ Res* 59: 612–619, 1986.
50. P Roy, SK Roy, A Mitra, AP Kulkarni. *Biochim Biophys Acta* 1214: 171–179, 1994.
51. B Jahn, GM Hansch. *Int Arch Allergy Appl Immunol* 93: 73–80, 1991.
52. K Tanaka, M Abe, N Shigematsu. *Int Arch Allergy Appl Immunol* 98: 361–369, 1992.
53. W Chamulitrat, MF Hughes, TE Eling, RP Mason. *Arch Biochem Biophys* 290: 153–159, 1991.
54. P Roy, MP Sajan, AP Kulkarni. *J Biochem Toxicol* 10: 111–120, 1995.
55. H Takahashi, M Abe, S Hashimoto, K Takayama, M Miyazaki. *Am J Respir Cell Mol Biol* 8: 291–298, 1993.
56. VB O'Donnell, A Azzi. *Biochem J* 318: 805–812, 1996.
57. EM Mills, K Takeda, ZX Yu, V Ferrans, Y Katagiri, H Jiang, MC Lavigne, TL Leto, G Guroff. *J Biol Chem* 273: 22165–22168, 1998.
58. H Kanner, S Harel, R Granit. *Lipids* 27: 46–49, 1992.
59. MT Yates, LE Lambert, JP Whitten, JP McDonald, M Mano, G Ku, SGT Mao. *FEBS Lett* 309: 135–138, 1992.
60. H Rubbo, S Parthasarathy, S Barnes, M Kirk, B Kalyanaraman, BA Freeman. *Arch Biochem Biophys* 324: 15–25, 1995.
61. R Wiesner, J Rathmann, HG Holzhutter, R Stosser, K Mader, H Nolting, H Kuhn. *FEBS Lett* 389: 229–232, 1996.
62. HG Holzhutter, R Wiesner, J Rathmann, R Stosser, H Kuhn. *Eur J Biochem* 245: 608–616, 1997.
63. VB O'Donnell, KB Taylor, S Parthasarathy, H Kuhn, D Koesling, A Friebe, A Bloodsworth, VM Darley-Usmar, BA Freeman. *J Biol Chem* 274: 20083–20091, 1999.
64. W Chamulitrat, RP Mason, D Riendeau. *J Biol Chem* 267: 9574–9579, 1992.
65. Y Fujimoto, S Tagano, K Ogawa, S Sakuma, T Fujita. *Prostaglandins Leukot Essent Fatty Acids* 59: 95–100, 1998.
66. T Kuzuya, S Hoshida, Y Kim, H Oe, M Hori, T Kawada, M Tada. *Cardiovasc Res* 27: 1056–1060, 1993.
67. H Kuhn, J Belkner, S Zaiss, T Fahrenklemper, S Wohlfeil. *J Exp Med* 179: 1903–1911, 1994.
68. VA Folcik, RA Nivar-Aristy, LP Krajewski, MK Cathcart. *J Clin Invest* 96: 504–510, 1995.
69. H Kuhn, D Heydeck, I Hugou, C Gniwotta. *J Clin Invest* 99: 888–893, 1997.
70. VB O'Donnell, S Spycher, A Azzi. *Biochem J* 310: 133–141, 1995.
71. LM Bailey, AN Makheja, TH Simon. *Atherosclerosis* 113: 247–258, 1995.
72. K Schnurr, A Borchert, H Kuhn. *FASEB J* 13: 143–154, 1999.
73. A Bloodsworth, VB O'Donnell, BA Freeman. *Arterioscler Thormb Vasc Biol* 20: 1707–1715, 2000.
74. RP Mason, B Kalyanaraman, BE Tainer, TE Eling. *J Biol Chem* 255: 5019–5022, 1980.
75. R Dietz, W Nastainczyk, HH Ruf. *Eur J Biochem* 171: 321–328, 1988.
76. R Karthein, R Dietz, W Nastainczyk, HH Ruf. *Eur J Biochem* 171: 313–320, 1988.
77. G Lassmann, R Odenwaller, JF Curtis, JA DeGray, RP Mason, LJ Marnett, TE Eling. *J Biol Chem* 266: 20045–20055, 1991.
78. A Tsai, RJ Kulmacz, G Palmer. *J Biol Chem* 270: 10503–10508, 1995.
79. A Tsai, G Palmer, G Hiao, DC Swinney, RJ Kulmacz. *J Biol Chem* 273: 3888–3894, 1998.
80. AL Tsai, G Wu, G Palmer, B Bambai, JA Koehn, PJ Marshall, RJ Kulmacz. *J Biol Chem* 274: 21695–21700, 1999.
81. DC Goodwin, MR Gunther, LC Hsi, BC Crews, TE Eling, RP Mason, LJ Marnett. *J Biol Chem* 273: 8903–8909, 1998.
82. M Bakovic, HB Dunford. *Prostaglandins Leukot Essent Fatty Acids* 53: 423–431, 1995.

83. AL Tsai, G Wu, RJ Kulmacz. *Biochemistry* 36: 13085–13094, 1997.
84. A Tsai, C Wei, HK Baek, RJ Kulmacz, HE Van Wart. *J Biol Chem* 272: 8885–8894, 1997.
85. AM Dean, FM Dean. *Protein Sci* 8: 1087–1098, 1999.
86. RK Upmacis, RS Deeb, DP Hajjar. *Biochemistry* 38: 12505–12513, 1999.
87. LM Landino, BC Crews, MD Timmons, JD Morrow, LJ Marnett. *Proc Natl Acad Sci USA* 93: 15069–15074, 1996.
88. PAEL Schilderman, JMS Van Maanen, EJ Smeets, F Ten Hoor, CS Kleinjans. *Carcinigenesis* 14: 347–353, 1993.
89. WM Armstead. *Brain Res* 910: 19–28, 2001.
90. SP Didion, CA Hathaway, FM Faraci. *Am J Physiol Heart Circ Physiol* 281: H1697–H1703, 2001.
91. PJ Formosa, TM Bray, S Kubow. *Can J Physiol Pharmacol* 66: 1524–1530, 1988.
92. K Minegishi, M Tanaka, O Nishimura, S Tanagaki, K Miyakoshi, H Ishimoto, Y Yoshimura. *Am J Physiol Endocrinol Metab* 283: E1308–E1315, 2002.
93. S Adler, RAK Stahl, PJ Baker, YP Chen, PM Pritzl, WG Couser. *Am J Physiol* F743–F749, 1987.
94. PG Gunasekar, JL Borowitz, GE Isom. *J Pharm Exp Ther* 285: 236–241, 1998.
95. J Martinez, T Sanchez, JJ Moreno. *Free Radic Res* 32: 303–311, 2000.
96. DW Potter, JA Hinson. *J Biol Chem* 262: 974–980, 1987.
97. M Bakovic, HB Dunford. *Prostaglandins Leukot Essent Fatty Acids* 51: 337–345, 1994.
98. VM Samokyszyn, T Chen, KR Maddipati, TJ Franz, PA Lehman, RV Lloyd. *Chem Res Toxicol* 8: 807–815, 1995.
99. MA Freyaldenhove, RV Lloyd, VM Samokyszyn. *Chem Res Toxicol* 9: 677–681, 1996.
100. T Parman, G Chen, PR Wells. *J Biol Chem* 273: 25079–25088, 1998.
101. J Stadler, BG Harbrecht, M Di Silvio, RD Curran, ML Jordan, RL Simmons, TR Billiar. *J Leukoc Biol* 53: 165–172, 1993.
102. L Minghetti, E Polazzi, A Nicolini, C Creminon, G Levi. *J Neurochem* 66: 1963–1970, 1996.
103. A Habib, C Bernard, M Lebret, C Creminon, B Esposito, A Tedgui, J Maclouf. *J Immunol* 158: 3845–3851, 1997.
104. O Kosonen, H Kankaanranta, U Malo-Ranta, A Ristimaki, E Moilanen. *Br J Pharmacol* 125: 247–254, 1998.
105. D Salvemini, K Seibert, JL Masferrer, TP Misko, MG Currie, P Needleman. *J Clin Invest* 93: 1940–1947, 1994.
106. D Salvemini D, MG Currie, V Mollace. *J Clin Invest* 97: 2562–2568, 1996.
107. AB Motta, ET Gonzalez, I Rudolph, MF Gimeno. *Prostaglandins Leukot Essential Fatty Acids* 60: 73–76, 1999.
108. ST Davidge, PN Baker, MK Laughlin, JM Roberts. *Circ Res* 77: 274–283, 1995.
109. L Manfield, D Jang, GA Murrell. *Inflamm Res* 45: 254–258, 1996.
110. FJ Hughes, LD Buttery, MV Hukkanen, A O'Donnell, J Maclouf, JM Polak. *J Biol Chem* 274: 1776–1782, 1999.
111. VB O'Donnell, B Coles, MJ Lewis, BC Crews, LJ Marnett, BA Freeman. *J Biol Chem* 275: 38239–38244, 2000.
112. AL Tsai, C Wei, RJ Kulmacz. *Arch Biochem Biophys* 313: 367–372, 1994.
113. JF Curtis, NG Reddy, RP Mason, B Kalyanaraman, TE Eling. *Arch Biochem Biophys* 335: 369–376, 1996.
114. AA Beharka, D Wu, M Serafini, SN Meydani. *Free Radic Biol Med* 32: 503–511, 2002.
115. C Boulos, H Jiang, M Balazy. *J Pharmacol Exp Ther* 293: 222–229, 2000.
116. C Su, M Sahlin, EH Oliw. *J Biol Chem* 273: 20744–20751, 1998.

27 Oxidation of Proteins

Similar to lipids the oxidation of proteins has already been studied for more than 20 years. Before discussing the data on protein oxidation, it should be mentioned that many associated questions were already considered in previous chapters. For example, the oxidation of lipoproteins, which is closely connected with the problems of nonenzymatic lipid peroxidation was discussed in Chapter 25. Many questions on the interaction of superoxide and nitric oxide with enzymes including the inhibition of enzymatic activities of prooxidant and antioxidant enzymes are considered in Chapters 22 and 30. Therefore, the findings reported in those chapters should be taken into account for considering the data presented in this chapter.

27.1 FREE RADICAL MECHANISMS OF PROTEIN OXIDATION

In earlier studies the in vitro transition metal-catalyzed oxidation of proteins and the interaction of proteins with free radicals have been studied. In 1983, Levine [1] showed that the oxidative inactivation of enzymes and the oxidative modification of proteins resulted in the formation of protein carbonyl derivatives. These derivatives easily react with dinitrophenyl-hydrazine (DNPH) to form protein hydrazones, which were used for the detection of protein carbonyl content. Using this method and spin-trapping with PBN, it has been demonstrated [2,3] that protein oxidation and inactivation of glutamine synthetase (a key enzyme in the regulation of amino acid metabolism and the brain L-glutamate and γ-aminobutyric acid levels) were sharply enhanced during ischemia- and reperfusion-induced injury in gerbil brain.

27.1.1 Free Radical Initiation of Protein Oxidation

Dean et al. [4] has studied the reaction of radiolytically generated hydroxyl radicals with bovine serum albumin (BSA) by chromatographic methods. It was found that in the presence of dioxygen this reaction led to the BSA fragmentation and formation of low molecular weight peptides. In the absence of dioxygen (under anaerobic conditions) hydroxyl radicals resulted in protein crosslinking. Other free radicals (superoxide, hydroperoxyl radical, and peroxyl radicals) did not react with BSA. However, similar to hydroxyl radical, peroxyl radicals induced a greater susceptibility of proteins to enzymatic hydrolysis. It has also been found that free radicals increased proteolysis in mitochondria and macrophages. Dean et al. [4] suggested that proteolysis might thus be considered as a component of cellular defense system. In 1993, Simpson et al. [5] found that hydroxyl radicals oxidized protein tyrosine residues to DOPA; it has been suggested that protein-bound DOPA is the main reducing moiety formed in oxidized proteins.

Davies and his co-workers [6,7] studied in detail the interaction of various reactive oxygen species with proteins. It was found that the hydroxyl radicals formed covalently bound protein aggregates with most proteins, but practically no fragmentation was observed. Drastic difference occurred between the effects of hydroxyl radical and superoxide alone and the simultaneous effect of $HO^\bullet + O_2^{\bullet-}$. Superoxide alone did not cause aggregation, fragmentation, or some other changes in protein structure. However, the $HO^\bullet + O_2^{\bullet-}$ combination induced a more extensive protein fragmentation even than hydroxyl radical alone [6]. Furthermore, this combination caused a multifold increase in proteolytic susceptibility of oxidatively damaged BSA [7]. These findings are really confusing because superoxide and hydroxyl radicals react with each other with a diffusion-controlled rate to form inactive hydroxyl anion and dioxygen:

$$O_2^{\bullet-} + HO^\bullet \Longrightarrow O_2 + HO^- \qquad (1)$$

Therefore, the enhancement of protein damage by the $HO^\bullet + O_2^{\bullet-}$ combination is only possible if both free radicals react with proteins independently.

The possibility of protein modification by active oxygen species has been studied with various proteins. Jones et al. [8] concluded that physiological concentrations of superoxide or hydroxyl radicals cannot modulate the functions of heart fatty acid-binding protein. In contrast, hydroxyl radicals affected myofibrillar calcium-ATPase activity and protein structure in rat heart [9]. Suzuki et al. [10] also found that active oxygen species produced by xanthine oxidase decreased calcium-stimulated ATPase activity and thiol group content in rat heart myofibrils. These effects of oxygen species were completely prevented by SOD + catalase. It has also been demonstrated that the xanthine–xanthine oxidase system inhibited creatine kinase activity in rat heart myofibrils that could be an origin of cardiac dysfunction [11]. Active oxygen species produced by xanthine oxidase oxidized the thiol groups of protein human plasma and of BSA [12]. Protein thiol groups can be considered as sacrificial antioxidants capable of preventing plasma lipid peroxidation. Neuzil et al. [13] suggested that protein oxidation by oxygen radicals may occur by chain mechanism.

Recently, Czapski et al. [14] studied the protein oxidation of human umbilical-vein endothelial cells (HUVEC) by PMA-stimulated human neutrophils and horseradish peroxidase–hydrogen peroxide system. Active oxygen species produced by neutrophils oxidized extracellular proteins but not the intracellular proteins of HUVEC, suggesting that the oxygen radical-mediated oxidation is restricted to proteins in the medium. Echtay et al. [15] demonstrated that superoxide increased mitochondrial proton conductance through the activation of uncoupling proteins. It has been suggested that the interaction of superoxide with uncoupling proteins could be a mechanism for the reduction of oxygen radical levels inside mitochondria. Besides superoxide and hydroxyl radicals, a very active $CO_2^{\bullet-}$ radical may participate in protein oxidation. For example, it has been shown that $CO_2^{\bullet-}$ rapidly reacts with galactose oxidase (rate constant $\geq 6.5 \times 10^8$ l mol^{-1} s^{-1}) [16]. Much earlier Favaudon et al. [17] showed that $CO_2^{\bullet-}$ induced cleavage of disulfide bonds in aponeocarzinostatin, an aporiboflavin-binding protein, and bovine immunoglobulin.

Many prooxidants initiate protein oxidation via the formation of oxygen radicals and other free radicals. In 1990, Fagian et al. [18] showed that calcium ions + diamide, a thiol oxidant, decreased submitochondrial membrane potential probably due to the formation of protein aggregates via the thiol-crosslinking. Free radical-mediated modification of glyceraldehyde-3-phosphate dehydrogenase (GAPDH) occurred by both nitric oxide- and superoxide-initiated pathways. It has been found that GAPDH was covalently modified by NAD in the presence of nitric oxide [19]. On the other hand, NADH-induced GAPDH modification was stimulated by thiols probably through superoxide formation [20]. Free thiyl radicals formed by FAD-dependent oxidation of the 2-oxo acid dehydrogenase complex inactivated dehydrogenase catalytic intermediates [21].

27.1.2 METAL ION-CATALYZED OXIDATION OF PROTEINS

Many prooxidant systems and free radicals (Figure 27.1) are able to initiate protein oxidation, but it has been proposed that the metal-catalyzed oxidation is a major pathway of protein modification under normal conditions [22]. During 1980 to 1990, Stadtman and his coworkers (their works reviewed in Ref. [23]) proposed the mechanism of metal ion-catalyzed protein oxidation. It has been shown [2] that the inactivation of glutamine synthetase by different metal catalyzed oxidation systems was mediated by hydrogen peroxide and ferrous ions. On these grounds, it was suggested that ferrous ions bind to a metal-binding site of the enzyme to form the ferrous ion–enzyme complex, which, reacting with hydrogen peroxide, generates hydroxyl or hydroxyl-like free radicals. These active oxygen radicals reacted preferentially with the side chains of amino acid residues, yielding carbonyl derivatives, inducing the loss of catalytic activity and increasing protein susceptibility to proteolytic degradation.

Metal-dependent initiation of protein oxidation may have various origins. Thus Davies et al. [24] demonstrated that protein hydroperoxides formed in the presence of free radical-producing systems decomposed by ferrous complexes, initiating novel processes of fragmentation and rearrangement of amino acid residues. For example, the degradation of a hydroperoxide derivative of glutamic acid resulted in its decarboxylation and the formation of $CO_2^{\bullet-}$ radical. Similarly, hydroperoxides of unsaturated fatty acids drastically enhanced the oxidative modification of proteins initiated by the ascorbate–iron system [25]. It was also shown that the ability of hydroperoxides to promote carbonyl formation depended on the degree of unsaturation and increased in the order of linoleate < linolenate < arachidonate. It should be noted that the most preferred targets for protein oxidation were lysine residues.

BSA was oxidized by the cupric–quercetin complex probably through binding to tryptophan residue [26]. Iwai et al. [27] showed that the iron regulatory protein 2 (IRP2) responsible

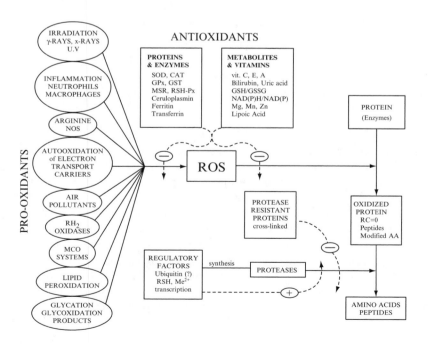

FIGURE 27.1 Prooxidants and antioxidants in protein oxidation and proteolysis of oxidized proteins. (From ER Stadtman, RL Levine. *Ann NY Acad Sci* 899: 191–208, 2000. With permission.)

for the regulation of expression of genes of iron metabolism is oxidized by the systems containing iron ions. As a result, significant iron-dependent carbonylation of IRP2 containing arginine, lysine, proline, and threonine residues was observed. Iron can stimulate protein oxidation in patients suffering from iron-deficient anemia after the treatment with iron-containing drugs. Thus, the administration of iron saccharide to hemodialysis patients induced an increase in protein oxidation in the blood [28].

Cupric ions catalyzed the oxidative deamination of the BSA lysine residue by various carbohydrates at physiological pH and temperature [29]. It was also found that the levels of α-aminoadipic-δ-semialdehyde residue, the oxidative deamination product of lysine, were much higher in streptozotocin-induced diabetic rats than in control animals. Efficiency of transition metal-stimulated protein oxidation may be decreased by ligands capable of forming inactive iron–ligand complexes. Thus Breccia et al. [30] demonstrated that polyethyleneimine protected muscle lactate dehydrogenase from oxidation by the cupric- and ferrous–hydrogen peroxide systems through the formation of inactive metal–polyethyleneimine complexes.

Although metal-catalyzed protein oxidation is undoubtedly a very effective oxidative process, the origin of free metal ions under in vivo conditions is still uncertain (see Chapter 21). However, protein oxidation can probably be initiated by metal-containing enzymes. Mukhopadhyay and Chatterjee [31] have shown that NADPH-stimulated oxidation of microsomal proteins was mediated by cytochrome P-450 and occurred in the absence of free metal ions. It is important that in contrast to metal ion-stimulated oxidation of proteins, ascorbate inhibited and not enhanced P-450-dependent protein oxidation reacting with the oxygenated P-450 complex. The following mechanism of P-450-dependent oxidation of the side chain protein amino acid residues has been proposed:

$$(P450)Fe^{2+} + O_2^{\bullet-}O_2 \Longleftrightarrow (P450)Fe^{2+}O_2^{\bullet-} \tag{2}$$

$$RCH_2NH_2 + (P450)Fe^{2+}O_2^{\bullet-} + H^+ \Longrightarrow RCH(\cdot)NH_2 + (P450)Fe^{3+} + H_2O_2 \tag{3}$$

$$RCH(\cdot)NH_2 \Longrightarrow RCH=NH \tag{4}$$

$$RCH=NH + H_2O \Longrightarrow RCHO + NH_3 \tag{5}$$

Choi et al. [32] showed that NADPH stimulated the modification of microsomal proteins, which was probably dependent on lipid peroxidation and inhibited by iron chelators.

27.1.3 Protein Oxidation Initiated by Peroxynitrite, Nitric Oxide, and Hypoclorite

Numerous studies demonstrate that peroxynitrite is a very reactive compound, which easily reacts with many biomolecules including proteins (see Chapter 21). Earlier studies [33,34] have already suggested that peroxynitrite might play an important role in nitration and oxidation of proteins. Later on, it was found that the summary effect of peroxynitrite on proteins turns out to be even more complicated. Berlett et al. [35] found that peroxynitrite reacts with glutamine synthetase by two ways: nitrating tyrosine residues and oxidizing methionine residues. Both reactions depended on carbon dioxide. In the absence of CO_2, peroxynitrite oxidized methionine residues, but at physiological CO_2 concentrations peroxynitrite nitrated tyrosine residues, while the oxidation of methionine residues was inhibited. It has been suggested that the effect CO_2 on the interaction of peroxynitrite with proteins is explained by the formation of nitrosoperoxocarboxylate $ONOOCOO^-$, which is an effective nitrating agent.

In contrast to the findings obtained in Ref. [34], it was concluded [36,37] that exposure of proteins to peroxynitrite leads to a very small increase in carbonyl content at physiological pH and CO_2 concentration. At the same time, carbonyl contents in glutamine synthetase and BSA increased in the absence of CO_2. These data show the importance of CO_2 in the regulation of prooxidant effect of peroxynitrite on proteins. Szabo et al. [38] suggested that endogenously produced peroxynitrite induced protein oxidation in mitochondria and nucleus of immunostimulated macrophages. Similarly, Di Stasi et al. [39] suggested that peroxynitrite induced tyrosine nitration and increased tyrosine phosphorylation of proteins from synaptosomes. However, Pfeiffer et al. [40] later concluded that peroxynitrite apparently is not an important factor of protein tyrosine nitration in cytokine-activated murine macrophages.

Recently, Gunther et al. [41] proposed that nitric oxide may directly react with enzymes without intermediate formation of peroxynitrite. It is known that the oxidation of arachidonic acid by prostaglandin H oxidase is mediated by the formation of enzyme tyrosyl radical (see Chapter 26). Correspondingly, it has been suggested that NO is able to react with this radical to form the tyrosine iminoxyl radical and then nitrotyrosine. Therefore, the NO-dependent nitration of protein tyrosine residue may occur without the formation of peroxynitrite or other nitrogen oxides.

Another initiator of protein oxidation of biological significance is hypochlorite (hypochlorous acid). Hypochlorite is produced in vivo by myeloperoxidase from hydrogen peroxide and chloride (Chapter 22). In 1995, Domigan et al. [42] showed that the myeloperoxidase–hydrogen peroxide–chloride system of stimulated human neutrophils chlorinated tyrosine residue in the peptide Gly-Gly-Tyr-Arg. It has been proposed that hypochlorite initially reacts with the peptide amino group yielding chloramine, which further reacts with tyrosine residue converting it into chlorotyrosine. Pattison and Davies [43] determined the rate constants for the reaction of hypochlorous acid with protein side chains (Table 27.1). It is seen that the rate constants for the reaction of HOCl with peptide bonds varied by four orders of magnitude.

HOCl-mediated protein oxidation accelerates under pathophysiological conditions. Thus, proteins from extracellular matrix obtained from advanced human atherosclerotic lesions contained the enhanced levels of oxidized amino acids (DOPA and dityrosine) compared to healthy arterial tissue [44]. It was also found that superoxide enhanced the prooxidant effect of hypochlorite in protein oxidation supposedly by the decomposition of chloramines and chloramides forming nitrogen-centered free radicals and increasing protein fragmentation [45]. In addition to chlorination, hypochlorite is able to oxidize proteins. The most readily oxidized amino acid residue of protein is methionine. Methionine is reversibly oxidized by many oxidants including hypochlorite to methionine sulfide and irreversibly to methionine sulfone [46]:

$$CH_3SCH_2CH_2CH(NH_2)COOH \Longleftrightarrow CH_3S(O)CH_2CH_2CH(NH_2)COOH \quad (6)$$
$$\text{Methionine} \qquad \text{Methionine sulfide}$$
$$CH_3S(O)CH_2CH_2CH(NH_2)COOH \Longrightarrow CH_3S(O_2)CH_2CH_2CH(NH_2)COOH \quad (7)$$
$$\text{Methionine sulfone}$$

TABLE 27.1
Rate Constants for the Reaction of Hypochlorous Acid with Proteins [43]

	Rate Constant ($\times 10^7$) (l mol^{-1} s^{-1})
Methionine residue	3.8
Cysteine residue	3.0
Cystine residue	0.016
Histidine residue	ca. 0.01

Chen et al. [47] demonstrated that the reaction of HOCl with cytochrome c increased cytochrome peroxidase activity by the oxidation of the methionine residue. Methionine oxidation also significantly decreased the efficiency of cytochrome c as a mitochondrial electron carrier. HOCl, HOBr, and HOI are also able to oxidize (Fe^{II})cytochrome c [48].

27.2 SOME EXAMPLES OF OXIDATIVE PROCESSES WITH THE PARTICIPATION OF PROTEINS

Oxidation of hemoproteins hemoglobin (Hb) and myoglobin (Mb) has been extensively studied and discussed in many reviews (see, for example, Refs. [49,50]). In contrast to other proteins, free radicals have an additional target in hemoproteins for attack, namely, metal atom. For example, in 1982, Weiss [51] suggested that the oxidation of Hb to metHb in erythrocytes by PMA-stimulated neutrophils was mediated by superoxide or hydrogen peroxide. It has been established that free radical attack on the heme group of hemoproteins involves one- or two-electron steps and leads to the denaturation of globin structure. Oxyhemoglobin (HbO$_2$) was oxidized to metHb via the oxidation of ferrous ion to ferric ion with further conversion to hemichrome [52]. Hb and Mb redox reactions may stimulate various pathophysiological processes in the living organism [50].

There is a group of enzymes, which are especially prone to oxidative damage initiated by active oxygen and nitrogen species. These proteins contain metal complexes susceptible to free radical attack. The inactivation or destruction of two such enzymes, aconitase containing the [4Fe–4S] cluster and alcohol dehydrogenase containing the zinc dithiolate center by superoxide and peroxynitrite was considered in Chapter 21. Another enzyme susceptible to oxidative damage is calcineurin. Calcineurin (protein phosphatase 2B) is the calmodulin-binding serine/threonine protein phosphatase, which plays a critical role in the coupling of calcium signals to cellular responses [53]. In 1996, Wang et al. [54] suggested that superoxide inactivated calcineurin reacting with its Fe^{2+}–Zn^{2+} catalytic center. These authors found that the activity of calcineurin increased in the presence of SOD apparently due to the dismutation of superoxide. Although it has been suggested that native calcineurin exists in a redox-insensitive Fe^{3+}–Zn^{2+} form [55], recent studies confirmed that native calcineurin is a catalytically active Fe^{2+}–Zn^{2+} protein [56,57]. It is interesting that nitric oxide suppressed the inhibitory effect of superoxide [58]. Therefore, the formed peroxynitrite is apparently unable to inactivate the enzyme. Another confirmation of free radical-mediated oxidative damage of calcineurin was obtained from the study of the effects of antioxidants. Thus, it has been shown [56] that the inhibition of the phosphatase activity of calcineurin by superoxide, hydrogen peroxide, or glutathione disulfide was reversed in the presence of ascorbate, α-lipoic acid, N-acetylcysteine, and glutathione.

Agbas et al. [59] showed that superoxide modified and inactivated neuronal receptor-like protein complex having glutamate/N-methyl-D-aspartate receptor characteristics. It was found that superoxide produced by the xanthine–xanthine oxidase system strongly inhibited L-[3H]glutamate binding, because superoxide is a more active inhibitor than hydrogen peroxide. The authors proposed that superoxide modified protein cysteine residues. Membrane-bound anion exchange protein 2 (AE2) was found to be able to transport superoxide to the extracellular matrix in addition to its major function to mediate chloride–bicarbonate exchange [60]. These authors have shown that active oxygen species stimulated the protein and AE2 mRNA expression.

27.3 COMPETITION BETWEEN PROTEIN AND LIPID OXIDATION

Free radical attack on cells can be directed on many biological molecules including lipids, proteins, and nucleic acids. Several studies suggested that proteins could be the most vulnerable targets. Caraceni et al. [61,62] concluded that during postanoxic reoxygenation protein oxidation apparently occurs before lipid peroxidation or nucleic acid oxidation in rat hepatocytes. It was also found [61] that after anoxia hepatocytes produced reactive oxygen species, and that reoxygenation injury was correlated with the production of oxygen species but not lipid peroxidation. In subsequent work these authors showed [62] that metal ion-catalyzed protein oxidation is an important factor in postanoxic reoxygenation-induced injury of hepatocytes while cellular DNA and RNA free radical-mediated damage is unimportant during the early stages of reoxygenation. Oxidative modification of proteins characterized by an increase in carbonyl groups was supposed to be an early event following exposure of endothelial cells to hydrogen peroxide [63]. Gieseg et al. [64] demonstrated that peroxyl radicals initiated protein oxidation in cells when no lipid peroxidation was still detected. It has been also suggested that proteins are among the primary and critical targets in free radical-mediated cytolysis [65].

In contrast to numerous literature data, which indicate that protein oxidation, as a rule, precedes lipid peroxidation, Parinandi et al. [66] found that the modification of proteins in rat myocardial membranes exposed to prooxidants (ferrous ion/ascorbate, cupric ion/*tert*-butylhydroperoxide, linoleic acid hydroperoxide, and soybean lipoxygenase) accompanied lipid peroxidation initiated by these prooxidant systems.

27.4 INHIBITION OF PROTEIN OXIDATION BY ANTIOXIDANTS AND FREE RADICAL SCAVENGERS

Similar to lipids, proteins can be protected from oxidation by antioxidants and free radical scavengers. For example, the formation of protein carbonyls in rat liver microsomes was inhibited by glutathione [67]. Vitamin E inhibited protein oxidation in rat skeletal muscle [68]. Forsmark-Andree et al. [69] found that Fe^{3+}ADP and ascorbate initiated both protein carbonylation and lipid peroxidation in beef heart submitochondrial particles, which was inhibited by endogenous ubiquinone but not vitamin E. Jain and Palmer [70] found that vitamin E inhibited the glycation of hemoglobin in erythrocytes supposedly by indirect way, namely through the suppression of the formation of lipid peroxidation products. Similarly, Onorato et al. [71] proposed that pyridoxamine, an inhibitor of advanced glycation reactions inhibited protein modification by reacting with active oxygen species formed during lipid peroxidation. Teng et al. [72] showed that proteins in heart homogenates were much more resistant to peroxynitrite-mediated nitration than proteins in brain homogenates. These authors concluded that the resistance of heart proteins to oxidation depended on the presence in heart homogenates of urate, which provided a significant antioxidant defense against peroxynitrite- and nitric oxide-mediated oxidation.

The effects of antioxidants on protein oxidation were also studied in animal experiments. Barja et al. [73] demonstrated that feeding guinea pigs with vitamin C decreased endogenous protein oxidative damage in the liver. Administration of the mixture of antioxidants containing Trolox C, ascorbic palmitate, acetylcysteine, β-carotene, ubiquinones 9 and 10, and (+)-catechin in addition to vitamin E and selenium to rats inhibited heme protein oxidation of kidney homogenates more efficiently than vitamin E + selenium [74].

27.5 REPAIRING AND PROTEOLYSIS OF OXIDIZED PROTEINS

In contrast to nucleic acids, which can be repaired after oxidative damage by excision and insertion mechanisms (see Chapter 28), the repair of oxidized proteins does not occur except the oxidized sulfur-containing amino acid residues [22]. Instead, oxidized proteins are degraded to amino acids by various endogenous proteases. It has been already shown that protein methionine residues (Met) are easily converted to methionine sulfoxide derivatives (MetO) (Reaction (6)). It is important that Reaction (6) can be reversed by the thioredoxin-dependent peptide methionine sulfoxide reductase (msrA) widely distributed in animal tissues and bacteria. It has been suggested [22] that an antioxidant cycle exists, which is able to repair oxidized methionine residues in proteins:

$$\text{MsrA MeO} + \text{thioredoxin(SH)}_2 \xrightarrow{\text{mrsA}} \text{Met} + \text{thioredoxin(S-S)} + H_2O \quad (8)$$

$$\text{thioredoxin(S-S)} + \text{NADPH} + H^+ \xrightarrow{\text{thioredoxin reductase}} \text{thioredoxin(SH)}_2 + \text{NADP}^+ \quad (9)$$

Another interesting mechanism of the repair of amino acid residues of LDL apolipoprotein B100 by flavonoids has been recently described [75]. The authors of this work suggested that LDL-bound quercetin (but not rutin) repaired the tyrosine free radical by intramolecular electron transfer.

Oxidation and free radical-induced degradation and modification of proteins may be a cause of proteolysis [4,76,77]. Of special importance was a finding that free radical modification of proteins (BSA) led to an increase in their susceptibility to enzymatic proteolysis [78]. It was proposed that this phenomenon might be of biological significance as a pathway for removing damaging proteins. In series of publications, Davies and his coworkers [6,7,79,80] thoroughly studied the relationship between protein oxidation and proteolysis. It was found [6] that many hydroxyl radical-modified proteins are proteolytically degraded up to 50 times faster than untreated proteins. Oxidation accelerated proteolysis of erythrocytes [79] and denatured proteins [7]. While various proteolytic enzymes are apparently involved in the degradation of oxidatively damaged proteins in bacteria, the multicatalytic proteinase complex proteasome is responsible for the proteolysis of oxidized proteins in rat liver cells and rabbit, human, and bovine erythrocytes and reticulocytes [80].

Cervera and Levine [81] studied the mechanism of oxidative modification of glutamine synthetase from *Escherichia coli*. It was found that active oxygen species initially caused inactivation of the enzyme and generated a more hydrophilic protein, which still was not a substrate for the protease. Continuous action of oxygen species resulted in the formation of oxidized protein subjected to the proteolytic attack of protease.

REFERENCES

1. RL Levine. *J Biol Chem* 258: 11823–11827, 1983.
2. RL Levine, CN Oliver, RM Funk, ER Stadtman. *Proc Natl Acad Sci USA* 78: 2120–2124, 1981.
3. CN Oliver, PE Starke-Reed, ER Stadtman, GJ Liu, JM Carney, RA Floyd. *Proc Natl Acad Sci USA* 87: 5144–5147, 1990.
4. RT Dean, SM Thomas, G Vince, SP Wolff. *Biomed Biochim Acta* 45: 1363–1373, 1986.
5. JA Simpson, SP Gieseg, RT Dean. *Biochim Biophys Acta* 1156: 190–196, 1993.
6. KJ Davies. *J Biol Chem* 262: 9895–9901, 1987.
7. KJ Davies, SW Lin, RE Pacifici. *J Biol Chem* 262: 9914–9920, 1987.
8. RM Jones, MR Prasad, DK Das. *Mol Cell Biochem* 98: 161–167, 1990.
9. V Robert, S Ayoub, G Berson. *Am J Physiol* 261: H1785–H1790, 1991.
10. S Suzuki, M Kaneko, DC Chapman, NS Dhalla. *Biochim Biophys Acta* 1074: 95–101, 1991.
11. M Kaneko, H Masuda, H Suzuki, Y Matsumoto, A Kobayashi, N Yamazaki. *Mol Cell Biochem* 125: 163–169, 1993.

12. R Radi, KM Bush, TP Cosgrove, BA Freeman. *Arch Biochem Biophys* 286: 117–126, 1991.
13. J Neuzil, JM Gebicki, R Stocker. *Biochem J* 293: 601–606, 1993.
14. GA Czapski, D Avram, DV Sakharov, KW Wirtz, JB Strosznajder, EH Pap. *Biochem J* 365: 897–902, 2002.
15. KS Echtay, D Roussel, J St-Pierre, MB Jekabsons, S Cadenas, LA Stuart, JA Harper, SJ Roebuck, A Morrison, S Pickering, JC Clapman, MD Brand. *Nature* 415: 96–99, 2002.
16. CD Borman, C Wright, MB Twitchett, GA Salmon, AG Sykes. *Inorg Chem* 41: 2158–2163, 2002.
17. V Favaudon, H Tourbez, C Houeelevin, JM Lhoste. *Biochemistry* 29: 10978–10989, 1990.
18. MM Fagian, L Pereira-da-Silva, IS Martins, AE Vercesi. *J Biol Chem* 265: 19955–19960, 1990.
19. LJ McDonald, J Moss. *Proc Natl Acad Sci USA* 90: 6238–6241, 1993.
20. J Rivera-Nieves, WC Thompson, RL Levine, J Moss. *J Biol Chem* 274: 19525–19531, 1999.
21. VI Bunik, C Sievers. *Eur J Biochem* 269: 5004–5015, 2002.
22. ER Stadtman, RL Levine. *Ann NY Acad Sci* 899: 191–208, 2000.
23. ER Stadtman. *Free Radic Biol Med* 9: 315–325, 1990.
24. MJ Davies, S Fu, RT Dean. *Biochem J* 305: 643–649, 1995.
25. HHF Refsgaard, L Tsai, ER Stadtman. *Proc Nat Acad Sci USA* 97: 611–616, 2000.
26. MS Ahmed, K Ainley, JH Parish, SM Hadi. *Carcinogenesis* 15: 1627–1630, 1994.
27. K Iwai, SK Drake, NB Wehr, AM Weissman, T LaVaute, N Minato, D Klausner, RL Levine, TA Rouault. *Proc Nat Acad Sci USA* 95: 4924–4928, 1998.
28. D Tovbin, D Mazor, M Vorobiov, C Chaimovitz, N Meyerstein. *Am J Kidney Dis* 40: 1005–1012, 2002.
29. M Akagawa, T Sasaki, K Suyama. *Eur J Biochem* 269: 5451–5458, 2002.
30. JD Breccia, MM Anderson, R Hatti-Kaul. *Biochim Biophys Acta* 1570: 165–173, 2002.
31. CK Mukhopadhyay, IB Chatterjee. *J Biol Chem* 269: 13390–13397, 1994.
32. D Choi, B Leininger-Muller, YC Kim, P Leroy, M Wellman. *Free Radic Res* 36: 893–903, 2002.
33. H Ischiropoulos, L Zhu, JS Beckman. *Arch Biochem Biophys* 298: 446–451, 1992.
34. RL Vinor, AFR Huhmer, DJ Biglow, C Schoneich. *Free Radic Res* 24: 243–259, 1996.
35. BS Berlett, RL Levine, ER Stadtman. *Proc Natl Acad Sci USA* 95: 2784–2789, 1998.
36. H Ischiropoulos, AB Al-Medi. *FEBS Lett* 364: 279–282, 1995.
37. M Tien, BS Berlett, RL Levine, PB Chock, ER Stadman. *Proc Natl Acad Sci USA* 96, 7809–7814, 1999.
38. C Szabo, M O'Connor, AL Salzman. *FEBS Lett* 409: 147–150, 1997.
39. AM Di Stasi, C Mallozzi, G Macchia, TC Petrucci, M Minetti. *J Neurochem* 73: 727–735, 1999.
40. S Pfeiffer, A Lass, K Schmidt, B Mayer. *J Biol Chem* 276: 34051–34058, 2001.
41. MR Gunther, BE Sturgeon, RP Mason. *Toxicology* 177: 1–9, 2002.
42. NM Domigan, TS Charlton, MW Duncan, CC Winterbourn, AJ Kettle. *J Biol Chem* 270: 16542–16548, 1995.
43. DI Pattison, MJ Davies. *Chem Res Toxicol* 14: 1453–1464, 2001.
44. AA Woods, SM Linton, MJ Davies. *Biochem J* 2002 Nov 28, Pt (epub ahead of print).
45. CL Hawkins, MD Rees, MJ Davies. *FEBS Lett* 510: 41–44, 2002.
46. W Vogt. *Free Radic Biol Med* 18: 93–105, 1995.
47. Y-R Chen, LJ Deterding, BE Sturgeon, KB Tomer, RP Mason. *J Biol Chem* 277: 29781–29791, 2002.
48. WA Prurz, R Kissinger, T Nauser, WH Koppenol. *Arch Biochem Biophys* 389: 110–122, 2001.
49. H Chen, AL Tappel, RC Boyle. *Free Radic Biol Med* 14: 509–517, 1993.
50. AI Alayash, RP Patel, RE Cashon. *Antioxid Redox Signal* 3: 13–27, 2001.
51. SJ Weiss. *J Biol Chem* 257: 2947–2953, 1982.
52. CC Winterbourn. In: L Packer, AN Glazer, (eds.) *Methods in Enzymology*. New York: Academic Press, 1990, vol 186, pp. 265–272.
53. CB Klee, H Ren, X Wang. *J Biol Chem* 273: 13367–13370, 1998.
54. X Wang, VC Culotta, CB Klee. *Nature* 383: 434–437, 1996.
55. L Yu, A Haddy, J Golbeck, J Yao, F Rusnak. *Biochemistry* 36: 10727–10734, 1997.
56. D Sommer, KL Fakata, SA Swanson, PM Stemmer. *Eur J Biochem* 267: 2312–2322, 2000.
57. D Namgaladze, HW Hofer, V Ullrich. *J Biol Chem* 277: 5962–5969, 2002.

58. V Ullrich, M Bachschmid. *Biochem Biophys Res Commun* 278: 1–8, 2000.
59. A Agbas, X Chen, O Hong, KN Kumar, EK Michaelis. *Free Radic Biol Med* 32: 512–524, 2002.
60. JL Turi, I Jaspers, LA Dailey, MC Madden, LE Brighton, JD Carter, E Nozik-Grayck, CA Piantadosi, AJ Ghio. *Am J Physiol Lung Cell Mol Physiol* 283: L791–L798, 2002.
61. P Caraceni, ER Rosenblum, DH Van Thiel, AB Borle. *Am J Physiol* 266: G799–G806, 1994.
62. P Caraceni, N Fe Maria, HS Ryu, A Colantoni, L Roberts, ML Maidt, Q Pye, M Bernardi, DH Van Thiel, RA Floyd. *Free Radic Biol Med* 23: 339–344, 1997.
63. HP Ciolino, RL Levine. *Free Radic Biol Med* 22: 1277–1282, 1997.
64. S Gieseg, S Duggan, JM Gebicki. *Biochem J* 350 (Pt 1): 215–218, 2000.
65. DM Richards, RT Dean, W Jessup. *Biochim Biophys Acta* 946: 281–948, 1988.
66. NL Parinandi, CW Zwizinski, HHO Schmid. *Arch Biochem Biophys* 289: 118–124, 1991.
67. JR Palamanda, JP Kehler. *Arch Biochem Biophys* 293: 103–109, 1992.
68. AZ Reznick, E Witt, M Matsumoto, L Packer. *Biochem Biophys Res Commun* 189: 801–806, 1992.
69. P Forsmark-Andree, G Dallner, L Ernster. *Free Radic Biol Med* 19: 749–757, 1995.
70. SK Jain, M Palmer. *Free Radic Biol Med* 22: 593–596, 1997.
71. JM Onorato, AJ Jenkins, SR Thorpe, JW Baynes. *J Biol Chem* 275: 21177–21184, 2000.
72. RJ Teng, YZ Ye, DA Parks, JS Beckman. *Free Radic Biol Med* 33: 1243–1249, 2002.
73. G Barja, M Lopez-Torres, R Perez-Campo, C Rojas, S Cadenas, J Prat, R Pamplona. *Free Radic Biol Med* 17: 105–115, 1994.
74. CA Knudsen, AL Tappel, JA North. *Free Radic Biol Med* 20: 165–173, 1996.
75. P Filipe, P Morliere, LK Patterson, GL Hug, JC Maziere, C Maziere, JP Fernandes, R Santus. *Biochemistry* 41: 11057–11064, 2002.
76. K Nakamura, ER Stadtman. *Proc Natl Acad Sci USA* 81: 2001–2015, 1984.
77. GS Vince, RT Dean. *FEBS Lett* 216: 253–256, 1987.
78. SP Wolff, RT Dean. *Biochem J* 234: 399–403, 1986.
79. KJ Davies, AL Goldberg. *J Biol Chem* 262: 8220–8226, 1987.
80. T Grune, T Reinheckel, KJA Davies. *J Biol Chem* 271: 15504–15509, 1996.
81. J Cervera, RL Levine. *FASEB J* 2: 2591–2595, 1988.

28 DNA Oxidative Damage

The amount of work dedicated to the study of DNA oxidative damage is large enough to successfully compete with the studies on lipid peroxidation (Chapters 25 and 26). DNA molecule is an easy target for the attack of many oxidants including free radicals, transition metal ions, and active oxygen and nitrogen species. On the other hand, specific effective repairing mechanisms protect nucleic acids from oxidative damage. This chapter is dedicated to the consideration of mechanisms of free radical damage and repairing DNA.

28.1 HYDROXYL RADICAL-MEDIATED DNA DAMAGE

28.1.1 Mechanism of Reactions of Hydroxyl Radicals with DNA

The reactions of hydroxyl radicals with DNA have been thoroughly studied in connection with the damaging effects of ionizing radiation on DNA. (The problem of the formation of hydroxyl radicals in biological systems has been discussed in Chapter 21.) Although it is doubtful that free hydroxyl radicals may be formed in cells and tissue without external irradiation, the results of radiobiological studies can be compared with the effects on DNA of the reactive oxygen species (ferryl and perferryl ions or hydroxyl radicals formed by site-specific mechanism) generated by the Fenton reaction and the other oxygen radical-produced systems. An important peculiarity of the free radical-mediated damage to a DNA molecule is the multiple ways of free radical attack. The most common directions are base modifications [1–3]. (Some of the base modifications are shown in Figure 28.1). Another important direction of DNA damage under oxidative stress is strand cleavage. Hydroxyl radicals easily abstract a hydrogen atom from the DNA deoxyribose sugar moiety, resulting in the formation of single DNA strands (SSBs) [4]. SSBs are not necessarily lethal and can be enzymatically repaired because DNA is held together by the other strand. Cell death originates from double-stranded breaks (DSBs), which occur near to each other on both strands. DSBs may be the result of a multiple hydroxyl radical attack [5].

There is a competition between the addition of hydroxyl radicals to DNA bases and hydrogen atom abstraction from DNA sugars. It has been found that addition to the double bonds of single-stranded polyribonucleotide is much more preferable to H-abstraction reaction [6]. However, hydrogen atom abstraction becomes more favorable for double-stranded nucleic acids [7]. Although hydrogen atom abstraction can occur from all the sugar carbon atoms, the abstraction from C-4' is probably the most important (Reaction (1), Figure 28.2) [4]. As seen from this figure, one of the possible ways of DNA disruption under aerobic condition is the formation of DNA-3'-phosphoglycolate (I), propenal base (II), and DNA-5'-phosphate (III).

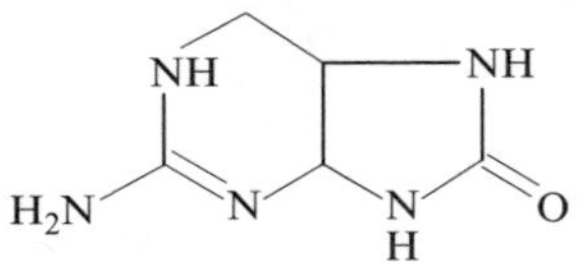

8-Oxoguanine

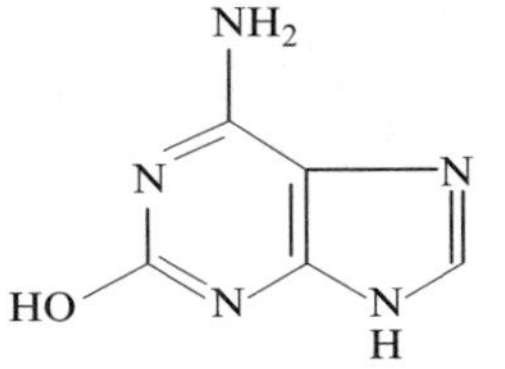

2-Hydroxyadenine

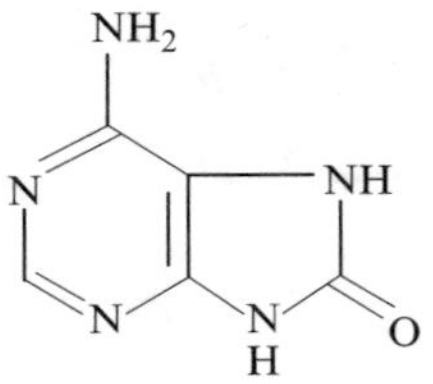

8-Oxyadenine

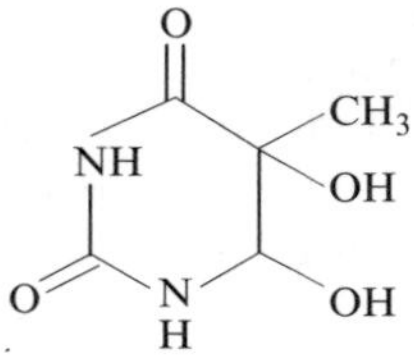

Thymine glycol

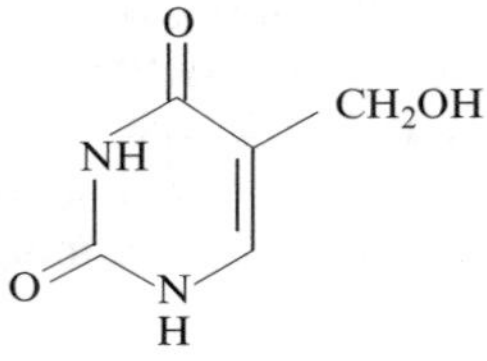

5-Hydroxymethylthyluracil

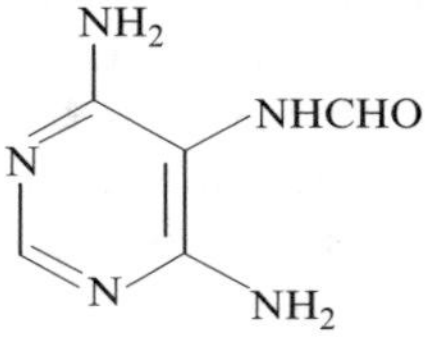

4,6-Diamino-5-formamidopyridine

FIGURE 28.1 Structures of base modifications.

28.1.2 Reactions of Hydroxyl and Hydroxyl-Like Radicals Produced by the Fenton Reaction with Nucleic Acids

Transition metal-catalyzed decomposition of hydrogen peroxide (the Fenton reaction) has been frequently used for studying hydroxyl-mediated DNA damage. As discussed earlier (Chapter 21), this system is not specific for the production only free hydroxyl radicals but also "hydroxyl-like" ferryl and perferryl complexes. Correspondingly, DNA free radical damage in this system may depend on all such active oxygen species. Therefore, although the effects of radiolytically produced hydroxyl radicals and those produced by the Fenton reaction are quite similar, there might be some differences, for example, due to the direct participation of iron ions in product formation [8].

In 1981, Floyd [9] demonstrated that ferrous ions form Fe^{2+}–DNA complexes, which are genuine catalysts of hydroxyl radical generation in the presence of hydrogen peroxide. Hydroxyl radicals measured as DMPO–OH complexes attack DNA to form aldehydes. In subsequent studies Park and Floyd [10,11] showed that hydroxyl radicals generated by the thiol/Fe^{3+}/dioxygen system induce DNA strand breaks and the formation of 8-hydroxy-2′-deoxyguanosine (8-OHdG). Ascorbate/iron mixture (the Udenfriend system), which produces hydroxyl radicals via the Fenton reaction, is also able to damage PBR322 plasmid DNA [12]. Aruoma et al. [13] suggested that damage to the DNA bases by the ferric ion chelate/hydrogen peroxide system was mediated by hydroxyl radicals, although the mechanism of their formation in this system is uncertain. Nunoshiba et al. [14] recently demonstrated the hydroxyl radical-mediated DNA damage under in vivo conditions. The augmentation of hydroxyl radical production and an increase in oxidative DNA lesions such as 7,8-dihydro-8-oxoguanine and 1,2-dihydro-2-oxoadenine have been detected in aerobically grown *Escherichia coli* cells lacking in both SOD and the repressor of iron uptake (Fur). It was concluded that the enhancement of iron and superoxide in this *E. coli* strain caused an increase in the ferrous ion level and hydroxyl radical formation via the Fenton reaction.

The data presented above are based on the assumption that hydroxyl-mediated DNA damage must depend on iron ion and hydrogen peroxide contents. Itoh et al. [15] showed that

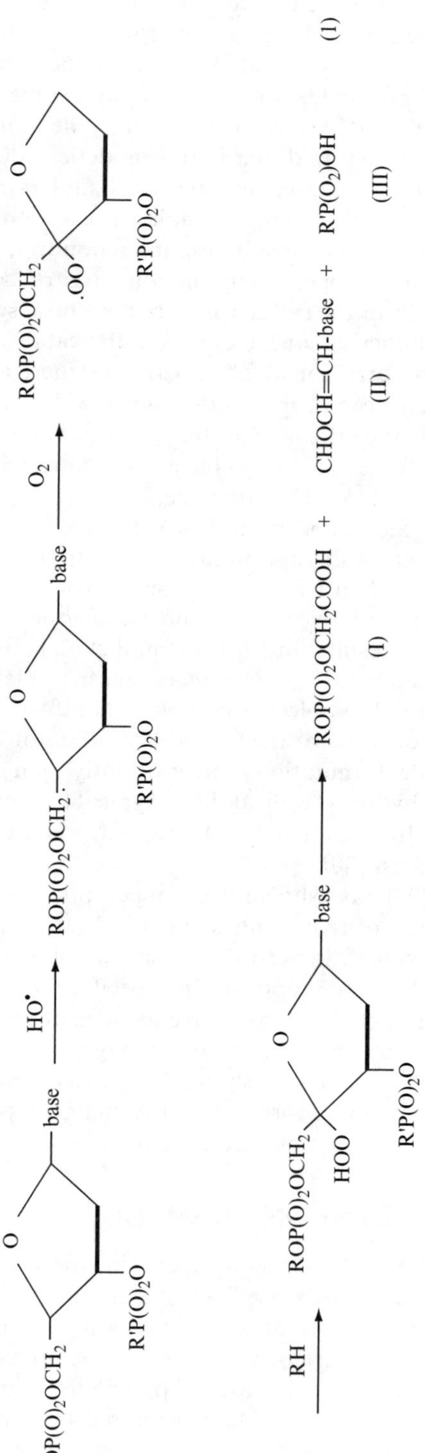

DNA Oxidative Damage

FIGURE 28.2 Reaction 1.

ferrous or ferric ions markedly enhanced the free radical damage of mitochondrial mtDNA in rat hepatoma cells. The substitution of zinc finger by iron caused DNA damage under in vitro and in vivo conditions [16]. Iron-catalyzed DNA oxidation increased with the augmentation of hydrogen peroxide level in liposomal system [17]. Important results concerning the in vivo effects of iron on mitochondria DNA have been recently demonstrated by Ames and his coworkers [18]. These authors have found that both iron deficiency and iron excess damaged mtDNA in rats. While the effect of iron overloading is well understandable, the interpretation of damaging effect of iron deficiency is not so simple. It has been suggested that this effect may depend on various factors such as uncoupling mitochondria, gastrointestinal upregulation of iron absorption, hepatic copper accumulation, loss of activity of iron-containing repair enzymes, and the changes in the cellular iron homeostasis system.

Despite long-standing hypothesis that ferrous ions, the catalysts in the Fenton reaction, are formed by the reduction of ferric ions with superoxide, there are still reservations in the efficiency of this reaction. It has been proposed that superoxide is able to release ferrous ions from storage proteins or enzymatic [4Fe–4S] clusters [19] (see Chapter 21). An increase in free iron released by superoxide from [4Fe–4S] clusters of damaged dehydrogenases greatly enhanced hydrogen peroxide-induced DNA damage.

One of the most surprising facts concerning DNA damage by hydroxyl radicals produced via the Fenton reaction is that DNA disruption catalyzed with the iron ion/hydrogen peroxide system is about 50 times slower than with copper ion/hydrogen peroxide [20]. This is really surprising because in any other oxidative systems the use of copper ions as the catalysts of the Fenton reaction gives very poor results. In 1981, Samuni et al. [21] proposed "a site-specific" mechanism for the copper-catalyzed Fenton reaction, in which the catalytically active copper compound is a cuprous–biomolecule complex. The formation of tertnary complex of CuL_2^+ with DNA is of importance for the Cu-catalyzed Fenton reaction because unbound CuL_2^+ may catalyze superoxide dismutation more efficiently than hydrogen peroxide decomposition and the formation of hydroxyl radicals [22]. Typical hydroxyl radical scavengers *tert*-butanol or formate do not affect Cu-catalyzed DNA degradation that supports the site-specific mechanism of the process [20].

Gutteridge and Halliwell [23] have shown that copper–phenanthroline complex catalyzed DNA disruption in the presence of reductants such as NADH, 2-mercapoethanol, or superoxide produced by xanthine oxidase. In accord with site-specific mechanism hydroxyl radical scavengers did not suppress DNA disruption. Site-specific mechanism has been proposed for the copper + hydrogen peroxide DNA oxidative damage at polyguanosines [24]. Aruoma et al. [25] demonstrated that major products of damage to the DNA bases induced by copper ion + hydrogen peroxide were cytosine glycol, thymine glycol, 8-hydroxyadenine, and 8-hydroxyguanine. Parsons and Morrison [26] found that the copper/Dopa/ascorbate system induced DNA breaks in human melanoma cells.

28.2 SUPEROXIDE-DEPENDENT DNA DAMAGE

Numerous studies suggest an important role of superoxide in the oxidation of DNA. Because it has long been shown that superoxide itself is virtually unreactive with DNA [27], it is a common knowledge that its role is to reduce ferric ions to ferrous ions, which are genuine catalysts of the Fenton reaction. Therefore, it is usually accepted that the superoxide initiation of DNA damage includes, as an obligatory step, the formation of hydroxyl or hydroxyl-like radicals. On the other hand, Dix et al. [28] suggested that the perhydroxyl radical can abstract 5′-hydrogen atom from the deoxyribose ring, and that the 5′-hydrogen abstraction mechanism may be specific one for the reaction of $HOO^\bullet$. However, owing to a low concentration of perhydroxyl radicals at physiological pH, the significance of their reactions in biological systems remains uncertain. Cunningham et al. [29] demonstrated that the incubation of rodent and human cells with KO_2 (which dissociates to $O_2^{\cdot-}$ and K^+ in aqueous solution) resulted in the single-strand DNA breaks in these cells, but authors do not exclude the formation of hydroxyl radicals.

28.3 DNA DAMAGE BY ENZYMATIC SUPEROXIDE PRODUCTION

Xanthine oxidase, a widely used source of superoxide, has been frequently applied for the study of the effects of superoxide on DNA oxidation. Rozenberg-Arska et al. [30] have shown that xanthine oxidase plus excess iron induced chromosomal and plasmid DNA injury, which was supposedly mediated by hydroxyl radicals. Ito et al. [31] compared the inactivation of *Bacillus subtilis* transforming DNA by potassium superoxide and the xanthine–xanthine oxidase system. It was found that xanthine oxidase but not KO_2 was a source of free radical mediated DNA inactivation apparently due to the conversion of superoxide to hydroxyl radicals in the presence of iron ions. Deno and Fridovich [32] also supposed that the single strand scission formation after exposure of DNA plasmid to xanthine oxidase was mediated by hydroxyl radical formation. Oxygen radicals produced by xanthine oxidase induced DNA strand breakage in promotable and nonpromotable JB6 mouse epidermal cells [33].

Grishko et al. [34] studied mitochondrial DNA damage in pulmonary vascular endothelial cells induced by external superoxide producer xanthine oxidase and intercellular producer menadione. These authors found that both xanthine oxidase- and menadione-generated oxygen radicals caused severe damage to the mitochondrial genome in lung endothelial cells with no significant effects on nuclear DNA. These results confirm previous data [35,36] on a greater sensitivity of mtDNA to oxidative stress compared with nuclear DNA due to the lack of protective histone proteins.

28.3.1 PHAGOCYTE-STIMULATED DNA DAMAGE

As phagocytes are effective producers of superoxide, one can expect that the stimulation of oxygen radical production by phagocytosing cells may result in DNA oxidation. This proposal was supported by numerous experimental data. One of the first studies discussing the role of superoxide in the induction of DNA strand scission was published by Lesko et al. [37] in 1980. Birnboim and his coworkers [38–40] demonstrated that PMA-stimulated human leukocytes induced DNA strand breakage. It has been suggested that superoxide- and hydrogen peroxide-induced DNA damage could be distinguished by the use of catalase and SOD [40]. Furthermore, it was found that the PMA-stimulated strand breaks caused by superoxide are not readily repaired in contrast to those induced by radiolytically formed hydroxyl radicals. Floyd et al. [41] has also showed that the stimulation of human granulocytes with tetradeconylphorbol acetate (TPA) resulted in the formation of high levels of 8-OHdG in the DNA of the treated cells.

Jackson et al. [42] studied the base damage in calf thymus DNA induced by PMA-stimulated neutrophils in the presence or absence of exogenous iron ions. Several oxidation products such as cytosine glycol, thymine glycol, 4,6-diamino-5-formamidopyrimidine, 8-hydroxyadenine, 2,6-diamino-4-hydroxy-5-formamidopyridine, and 8-hydroxyguanine were identified. It has been proposed that DNA damage was mediated by hydroxyl radicals because the yield of base modifications increased with increasing added iron ions and diminished in the presence of iron chelators. Moreover, the chemical structures of base modifications were the same as those formed by radiolytically generated hydroxyl radicals. Lewis et al. [43] concluded that oxidative DNA damage induced by activated macrophages depended not only on hydrogen peroxide but also on lipoxygenase products of arachidonic acid metabolism. Oxygen radicals released by human neutrophil-like dimethylsulfoxide-differentiated HL60 cells induced DNA damage characterized by the formation of 8-OHdG, which was supposedly mediated by hydroxyl radicals [44]. It is interesting that lipophilic chelator 1,10-phenantroline either inhibited or stimulated DNA strand breaks in human granulocytes [45]. It has been proposed that the effect of 1,10-phenantroline depended on the formation of its DNA-damaging copper complex or inhibiting iron complex.

Thus, all the above studies suggest the transformation of superoxide produced by phagocytes into hydroxyl radicals via an iron-dependent mechanism during oxygen radical mediated DNA damage. However, it has been recently proposed that the conversion of superoxide into hydroxyl radicals may occur without iron ions. Thus Shen and Hazen [46] found that exposure of DNA to stimulated neutrophils or eosinophils in the presence of iron chelators resulted in the formation of 8-OHdG, which was inhibited by peroxidase inhibitors, hypochlorite scavengers, catalase, and SOD. It has been suggested that the peroxidase/hydrogen peroxide/chloride system of leukocytes is responsible for hydroxyl radical-mediated DNA damage.

Another mechanism of myeloperoxidase-initiated DNA damage by neutrophils has been proposed by Henderson et al. [47]. These authors showed that the myeloperoxidase/hydrogen peroxide/chloride system catalyzed the chlorination of 2'-deoxycytidine into 5-chloro-2'-deoxycytidine. As HOCl is unable to react with 2'-deoxycytidine in the absence of Cl$^-$, it was concluded that in this case a genuine active intermediate was molecular chlorine formed by Reactions (2) and (3):

$$Cl^- + H_2O_2 + H^+ \Longrightarrow HOCl + H_2O \qquad (2)$$
$$HOCl + H^+ + Cl^- \Longleftrightarrow Cl_2 + H_2O \qquad (3)$$

Furthermore, it was found that stimulated human neutrophils are able to produce 5-chloro-2'-deoxycytidine and that the myeloperoxidase system generates just the same levels of 5-chlorocytosine in DNA and RNA in vitro (Reaction (4), Figure 28.3). It is possible that myeloperoxidase-generated chlorinated products may modify nuclear acids of pathogens and nuclear acids in host cells during inflammation. Hawkins et al. [48] suggested that DNA oxidation may be initiated by protein chloramines formed in the reaction of HOCl with histones in the nucleosome.

An interesting example of DNA damage by superoxide and hydrogen peroxide produced by microbes has been recently described by Huycke et al. [49]. These authors have showed that reactive oxygen species produced by *Enterococcus faecalis*, a microorganism of the human intestinal tract, oxidized DNA in Chinese hamster ovary and intestinal epithelial cells.

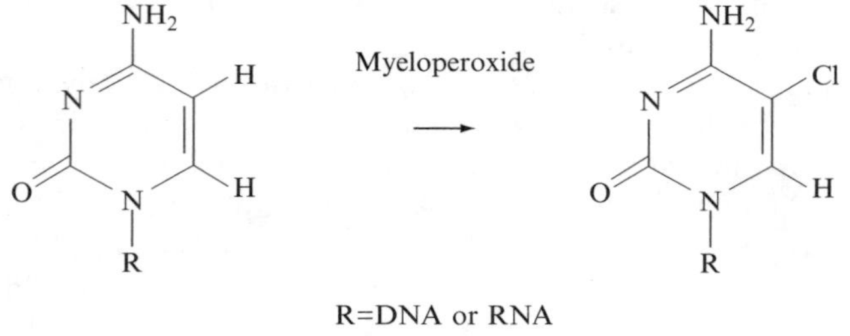

(4)

FIGURE 28.3 Myeloperoxidase-catalyzed chlorination of DNA and RNA (Reaction 4).

28.4 REACTIVE OXYGEN SPECIES AS MEDIATORS OF DRUG- AND XENOBIOTIC-INDUCED DNA DAMAGE

Many organic and inorganic compounds, fibers, and particles are capable of damaging nucleic acids by generating reactive oxygen species via the reduction of dioxygen. These stimuli include different classes of organic compounds, classic prooxidants (anticancer antibiotics, various quinones, asbestos fibers, and so on), and even antioxidants, which can be oxidized in the presence of transition metal ions.

28.4.1 Effects of Prooxidants

Anticancer antibiotic anthracyclines are efficient prooxidants capable of initiating many oxidative processes such as overproduction of superoxide, lipid peroxidation, and free radical-mediated DNA damage. It is believed that the prooxidant activity of anthracyclines and particularly the most important antibiotic doxorubicin (Adriamycin) is a cause of their adverse side effects, first of all cardiotoxicity. Although it is known that doxorubicin intercalates easily to DNA inhibiting both RNA transcription and DNA replication [50], a major damaging effect on DNA induced by anthracyclines is mediated by free radicals [51,52]. The reduction of doxorubucin to semiquinone by NADPH cytochrome P-450 reductase stimulated DNA damage, which was supposedly mediated by the formation of oxygen radicals [53]. Similar mechanism of oxygen radical-mediated DNA damage has been proposed by Rowley and Halliwell [54]. As the formation of hydroxyl radicals had to occur in the presence of iron ions, subsequent works on anthracycline-stimulated DNA damage were centered on the effects of iron–anthracycline complexes. Muindi et al. [55] have shown that the Fe^{3+}–doxorubicin complex and ternary DNA–Fe^{3+}–doxorubicin complexes acquired the ability to cleave DNA. Similarly, Eliot et al. [56] demonstrated the oxidative destruction of DNA by the 2:1 Fe^{3+}–doxorubicin complex.

Another anticancer antibiotic able to induce DNA damage is bleomycin. It has been shown that superoxide enhanced DNA breakage by bleomycin [57]. Sausville et al. [58] demonstrated the important role of ferrous ions and dioxygen in the degradation of DNA by bleomycin. Iron chelators and Cu^{2+}, Zn^{2+}, and Co^{2+} ions inhibited DNA degradation by bleomycin + Fe^{2+}. It has been suggested that DNA degradation depended on the oxidation of Fe^{2+}–bleomycin–DNA complex [59]. "Activated" bleomycin (bleomycin + Fe^{2+}) induced the oxygen-dependent DNA strand scission, resulting in the cleavage of the deoxyribose C3′—C4′ bond and the production of propenals, 5′-phosphate, and 3′-phosphoglycolate [60].

Many other antibiotics and drugs are able to induce free radical-mediated DNA damage. For example, the hydroxy derivative of anticancer drug VP-16 generated DNA nicking supposedly by iron-dependent free radical mechanism [61]. Protein-containing anticancer antibiotic neocarzinostatin induced DNA disruption probably via a hydrogen atom abstraction from the deoxyribose C-5′ [62]. The reaction, accompanied by superoxide formation, however, did not participate in DNA damage. Shoji et al. [63] showed that tumor necrosis factor-α (TNF-α) induced DNA damage in L929 cells in the presence of actinomycin D supposedly due to the increased mitochondrial formation of oxygen radicals. Ye and Bodell [64] have shown that microsomal activation of antiestrogen drug tamoxifen and its metabolite 4-hydroxytamoxifen resulted in DNA damage, demonstrated by an increase in 8-OHdG level, which was inhibited by antioxidant enzymes and free radical scavengers. Free radical-mediated DNA damage may be a cause of tamoxifen carcinogenic effect.

Many quinones (including anthracyclines considered above) are well-known prooxidants that makes them potential DNA damaging agents. It has been shown that menadione (2-methyl-1,4-naphthoquinone), a redox cycling quinone, induced single- and double-strand

DNA breaks in human MCF-7 cells [65]. Damaging effect of menadione was probably mediated by hydroxyl radicals as it was demonstrated by ESR spin-trapping method. The analogs of menadione 2-methylmethoxynaphthoquinone and 2-chloromethylnaphthoquinone also stimulated DNA damage through the formation of superoxide and other free radicals [66]. Similar effects have been shown for hydroquinone, catechol, benzoquinone, and benzenetriol [67,68].

Polyunsaturated fatty acids such as linoleic and arachidonic acids induced specific genotoxic damage through DNA disruption [69]. DNA damage appears to depend on the degree of fatty acid unsaturation because monounsaturated oleic acid was inactive. ESR spin-trapping demonstrated superoxide formation in this process. 5-Aminolevulinic acid, a heme precursor, produced oxidative DNA damage including strand breaks and 8-OHdG formation, which was mediated by superoxide and transition metals [70].

Lynn et al. [71] demonstrated the damaging effect of arsenite on DNA. It has been shown that arsenite at low concentrations increased DNA oxidative damage in vascular smooth muscle cells (VSMCs) that can be a cause of arsenite-induced atherosclerosis. Bruskov et al. [72] found that heat induced the formation of 8-oxoguanine in DNA solution at pH 6.8, which was supposedly mediated by oxygen radicals.

28.4.2 Damaging Effects of Antioxidants

Inhibition of free radical-mediated DNA damage by antioxidants has been shown in many studies (see below). However, many antioxidants and chelators, which can be oxidized by dioxygen in the presence of transition metals or form reactive metal ion complexes, are also able to stimulate DNA oxidation. Kanabus-Kaminka et al. [73] showed that cysteamine, a radioprotector and free radical scavenger, N-acetylcysteine (NAC), and ascorbate may induce DNA strand breaks under certain conditions by forming hydrogen peroxide during oxidation. Similar conclusion was later drawn by Oikawa et al. [74], who demonstrated the NAC induced metal-dependent hydrogen peroxide formation and, subsequently, the damage to cellular and isolated DNA.

However, particularly effective prooxidants turn out to be the transition metal complexes of flavonoids containing the easily oxidizable phenolic hydroxyl groups. One of the most effective DNA-damaging flavonoids is quercetin, which induces strand scission in DNA in the presence of cupric ions [75–77]. The mechanism of DNA destruction in the presence of quercetin is still uncertain, although the reduction of cupric into cuprous ions inside the Cu–quercetin or Cu–quercetin–DNA complexes with subsequent formation of oxygen radicals seems to be the most probable mechanism. Other flavonoids such as kaempferol [78], morin, and naringenin [79] are also able to induce nuclear DNA damage. In many cases, antioxidant enzymes and hydroxyl radical scavengers did not influence or even enhance DNA damage. It has been suggested that DNA damage was mediated by hydroxyl radicals via a site-specific mechanism; another possibility is the formation of peroxyl radicals during the oxidation of nuclear membrane lipids [80].

Glutathione may also initiate DNA damage. Thomas et al. [80] showed that glutathione-stimulated DNA strand breakage was inhibited by SOD and catalase, while DNA damage in human lymphocytes incubated with glutathione was not. This difference is not easily understandable. Giulivi and Cadenas [81] studied the damaging effect of mitochondrial glutathione on mitochondrial DNA, based on the 8-HOdG formation. The mechanism of DNA oxidation mediated by the reduction of Cu–DNA complexes by glutathione has been proposed.

The possible prooxidant effects of a major lipophilic antioxidant vitamin E (α-tocopherol) have already been discussed in Chapter 25. Yamashita et al. [82] showed that α-tocopherol induced extensive DNA damage including base modification and strand breakage in the presence of cupric ions. Catalase and bathocuproine inhibited DNA oxidation, and ESR spin-trapping confirmed the formation of hydroxyl radicals. It has been suggested that DNA damage was mediated by hydroxyl radicals through a site-specific mechanism.

Vitamin A (retinol) is not a classic antioxidant although it is frequently related to a group of "antioxidant vitamins" E, C, and A. Murata and Kawanishi [83] found that retinol and its derivative retinal induced the formation of 8-HOdG in HL-60 cells. This process was supposedly mediated by hydroxyl radicals formed from hydrogen peroxide (the product of superoxide dismutation) and endogenous transition metal ions.

DNA damage by iron chelators has been shown by Cragg et al. [84]. These authors showed that various iron chelators affected the hydrogen peroxide-mediated DNA damage in iron-loaded liver cells in different ways. Thus, desferrithiocin was protective and desferrioxamine had no effect, while 1,2-dimethyl-3-hydroxypyrid-4-one (L1) enhanced oxidative DNA damage. It has been suggested that the oxidative effect of L1 depended on stoichiometry of Fe—L1 complexes.

28.4.3 Damaging Effects of Mineral Dusts and Fibers

The stimulation of oxidative processes by mineral dusts and fibers is well known. Their DNA damaging effects may originate from direct interaction with nucleic acids or cells and from the stimulation of oxygen radical release by phagocytes. For example, Livingston et al. [85] showed that asbestos fibers induced sister chromatid exchanges in Chinese hamster ovarian fibroblast cells. On the other hand, mineral dusts and fibers are able to stimulate oxygen radical production by phagocytes [86]. Damaging effects of mineral dusts and fibers strongly depend on absorbed iron ions. It has been shown that removing iron ions from the particle surface sharply decreased luminol-amplified CL (characterized the formation of hydroxyl or hydroxyl-like radicals, which are responsible for oxidative damage) and enhanced lucigenin-amplified CL (characterized the formation of innocuous superoxide) by phagocytes [87].

Lund and Aust [88] showed that DNA SSBs induced by asbestos fibers are directly related to asbestos iron content. Correspondingly, the iron chelator desferrioxamine completely inhibited DNA damage. Crocidolite, one of the most carcinogenic asbestos fibers, induced the formation of 8-OHdG in cellular DNA of PMA-differential human promyelocytic leukemia cells HL60 [89]. Since 8-OHdG formation was not correlated with oxygen radical generation, it was proposed that crocidolite might convert innocuous superoxides into reactive hydroxyl radicals. Dong et al. [90] demonstrated that crocidolite and chrysotile asbestos fibers damaged DNA in rat pleural mesothelial cells. Fung et al. [91] confirmed the formation of 8-OHdG after the treatment of DNA from rat pleural mesothelial cells and human mesothelial cell line with crocidolite asbestos.

28.5 DNA DAMAGE BY REACTIVE NITROGEN SPECIES

Prooxidant activity of nitric oxide, peroxynitrite, and other reactive nitrogen oxides has been already discussed in previous the chapters (21, 25, and 27) in connection with their oxidative effects on lipids, proteins, and other biomolecules. Similarly, reactive nitrogen species are capable of inducing DNA oxidative damage. In 1993, Delaney et al. [92] suggested that the levels of endogenous nitric oxide generated by interleukin-1β-induced NO synthase are sufficiently high to cause DNA damage in rat islets of Langerhans and HIT-T15 cells. In subsequent work [93] these authors showed that SOD did not inhibit DNA damage; therefore, it was concluded that superoxide and peroxynitrite did not participate in oxidation. However, deRojas-Walker et al. [94] demonstrated that peroxynitrite formed from NO and superoxide is responsible for the oxidative damage in macrophage DNA. Similarly, Salgo et al. [95] showed that peroxynitrite caused SSBs in pBR322 supercoiled DNA, although hydroxyl radical scavengers did not inhibit DNA damage. In contrast, Inoue and Kawanishi [96] demonstrated that hydroxyl radical scavengers inhibited 8-OHdG formation in DNA oxidative damage by peroxynitrite. Epe et al. [97] concluded that DNA damage induced by peroxynitrite is different from that caused by hydroxyl radicals or singlet dioxygen.

Peroxynitrite is apparently able to damage DNA by various ways. Thus, Treyakova et al. [98] showed that the peroxynitrite treatment of pUC19 plasmid resulted in the formation of single-strand breaks from the direct sugar damage as well as nucleobase modifications. Carbon dioxide increased the damage of nucleobases and suppressed deoxyribose oxidation. When synthetic oligonucleotides were used in the reaction with peroxynitrite, 8-nitroguanine (8-nitro-G) was identified in the reaction mixture. In contrast to the reaction of peroxynitrite with DNA, where 8-nitro-G is rapidly formed by the decomposition of 8-nitroguanosine, the latter was identified in the reaction of peroxynitrite with cellular RNA [99]. Watanabe et al. [100] showed that activated microphages increased the 8-OHdG level and DNA single-strands in cocultured hepatocytes producing nitric oxide and peroxynitrite. Phoa and Epe [101] demonstrated that compared to peroxynitrite, DNA damage by nitric oxide is relatively small and that NO can even be protective against DNA single-strand break formation caused by hydrogen peroxide. However, at the same time NO is able to inhibit selectively the repair of oxidative DNA base modifications.

A very important consequence of DNA single-strand breakage is the activation of poly (ADP–ribose) synthetase (PARS). PARS is a protein-modifying and nucleotide-polymerizing enzyme, which causes the cleavage of NAD^+ into ADP–ribose and nicotinamide. Then, PARS polymerizes ADP–ribose into poly(ADP) ribose. Peroxynitrite is a very potent trigger of DNA strand breakage and an activator of PARS [102]. Szabo et al. [102] proposed that DNA damage and PARS activation play a central role in peroxynitrite-mediated cell injury. Peroxynitrite-induced PARS activation could be important in the development of hemorrhagic shock [103,104], in vitro fibroblast injury, and in vivo arthritis development [105].

Another possible DNA damaging nitrogen species is nitroxyl anion. As discussed in Chapter 21, NO^- is a very reactive species capable of reacting with many biomolecules, although the probability of its formation in biological systems is still questionable. Ohshima et al. [106] suggested that nitroxyl anion is responsible for DNA strand breakage and deoxyribose oxidation during the decomposition of Angeli's salt. The conversion of NO^- to NO by electron acceptors, ferricyanide and 4-HO-TEMPO, inhibited DNA damage by Angeli's salt, supporting the proposal that nitric oxide is not a mediator of this process [107].

28.6 REPAIR OF FREE RADICAL-MEDIATED DNA DAMAGE

There is an enzymatic mechanism of DNA damage repair, which is responsible for recognition, discharge, and replacement of damaging structures with normal ones. On the other hand, carbon radicals formed on the DNA backbone can be repaired by the interaction with antioxidants. Practically all traditional antioxidants and free radical scavengers have been described to inhibit free radical-mediated DNA damage.

As in the oxidation of many other biomolecules, ascorbic acid causes both inhibitory and stimulatory action on DNA oxidation by free radicals depending on its concentration, the presence of iron, etc. For example, the protective effect of ascorbic acid on hydrogen peroxide-stimulated DNA damage in human lymphocytes increased with its increasing concentration [108]. On the other hand, it was found that ascorbate significantly enhanced DNA damage in rat liver nuclei in the presence of iron ions in a concentration-dependent manner [109]. However, it is possible that ascorbate is mainly protective in vivo. Thus, Fraga et al. [110] showed that dietary ascorbic acid protects human sperm from endogenous oxidative DNA damage.

Salgo and Pryor [111] studied the effect of Trolox C (a water-soluble analog of vitamin E) on peroxynitrite-mediated DNA damage in rat thymocytes. They proposed that peroxynitrite mediated the formation of TBAR products, which caused the DNA–protein crosslinks. The latter were inhibited by the posttreatment of cells with Trolox. However, Trolox produced no effects on hydrogen peroxide- or bleomycin-induced DNA damage in human lymphocytes [108].

Flavonoids and flavones seem to be the most widely studied inhibitors of free radical DNA oxidation. In 1992, Korkina et al. [87] showed that flavonoid rutin was a more effective inhibitor of chromatid breaks in cultured lymphocytes than ascorbic acid. Wei et al. [112] studied the inhibition of 8-OHdG formation in calf thymus DNA by soybean isoflavone genistein when exposed to UV irradiation or the hydroxyl radical-producing Fenton reaction. Similarly, Yen and Lai [113] demonstrated the inhibition of peroxynitrite-induced DNA degradation by isoflavones genistein and daidzein. It has been proposed that the inhibition of DNA damage by genistein suggests its potential anticarcinogenic activity. Noroozi et al. [114] studied the inhibitory effects of flavonoids on the hydrogen peroxide-initiated oxidative DNA damage to human lymphocytes. The efficiency of flavonoids decreased in the range of luteolin > myricetin > quercetin > kaempferol > quercitrin > apigenin > quercetin-3-glucoside > rutin. Most of the flavonoids were more effective inhibitors than ascorbic acid.

As discussed above, flavonoids not only inhibit but also cause DNA oxidation. Ohshima et al. [115] compared the antioxidant and prooxidant effects of flavonoids on DNA damage induced by nitric oxide, peroxynitrite, and nitroxyl anion. Only flavonoids containing an *ortho*-trihydroxyl group in the B ring (delphinidine, epigallocatechin gallate, and myricetin) or in the A ring (for example, quercetagetin) were able to enhance single-strand breakage induced by a NO-producing system. It has been suggested that the prooxidant effect of these flavonoids was due to the formation of some reactive nitrogen species in their reaction with nitric oxide, although the mechanism of such reactions is unknown. Later on, Johnson and Loo [116] showed that epigallocatechin gallate and quercetin inhibited hydrogen peroxide- and peroxynitrite-induced DNA damage at low concentrations but enhanced it at high concentrations.

Abalea et al. [117] demonstrated a high efficiency of flavonoid myricetin at the inhibition of iron-induced DNA damage in primary rat hepatocyte cultures. It was found that myricetin not only efficiently suppressed the accumulation of DNA oxidation products but also activated DNA repair pathways. Quercetin and myricetin but not rutin and kaempferol suppressed hydrogen peroxide-mediated DNA strand breakage in human colonocyte Caco-2 cells [118]. However, Aherne and O'Brien [119] showed that all three flavonoids myricetin, quercetin, and rutin significantly protected Caco-2 and Hep G2 cells against hydrogen peroxide-mediated DNA damage. In subsequent studies these authors have studied the mechanism of protection by quercetin and rutin against *tert*-butylhydroperoxide- and menadione-induced DNA damage in Caco-2 cells [120]. It was found that the iron chelator desferrioxamine, quercetin, and rutin but not 2,6-di-*tert*-4-methylphenol (BHT) protected against *tert*-BOOH-induced DNA single strand break formation. It has been shown earlier [121] that the difference between the inhibitory effects on free radical-mediated processes between flavonoids quercetin and rutin and the classic antioxidant BHT is due to the ability of flavonoids to chelate iron ions. Thus, the effects of flavonoids and desferrioxamine indicate iron-initiated free radical-mediated mechanism of *tert*-BOOH-induced DNA damage. In the case of menadione BHT was capable of reducing DNA damage, suggesting the contribution of peroxyl radicals in this process.

Antioxidant enzymes also protect against free radical-mediated DNA degradation. For example, SOD and catalase as well as PARS inhibitors suppressed alloxan- and streptozotocin-induced islet DNA strand breaks [122]. Uric acid inhibited single-strand DNA breaks induced by the xanthine oxidase–acetaldehyde–ferrous ions and hematin–hydroperoxide systems [123]. Tomasetti et al. [124] proposed that ubihydroquinone-10 and ubiquinone-10 may diminish DNA damage in human lymphocytes. Hydroxytyrosol, a component of virgin olive oil, has been shown to be highly protective against DNA damage by peroxynitrite [125].

Metallothioneins (MTs), low molecular weight proteins, have been shown to exhibit antioxidant properties in various free radical-mediated processes. Thus, it has long ago been shown that metallothionein inhibited DNA radiation damage [126] and hydroxyl radical-mediated DNA degradation [127]. In contrast, Muller et al. [128] demonstrated the ability of Cd/Zn-MD to stimulate DNA strand breaks. Cai et al. [129] studied the effects of Zn-MD on DNA damage induced by copper and iron ions in the presence of hydrogen peroxide and ascorbate. These authors found that Zn-MD inhibited cupric ion-stimulated DNA double-strand breaks but was ineffective in protecting DNA from iron-induced damage. It has been suggested that the protective effect of Zn-MD on DNA degradation may be explained by the sequestering of copper to prevent copper-catalyzed formation of oxygen radicals and by free radical scavenging. MT inhibited plasmid DNA damage by peroxynitrite [130]. Another zinc-containing protein, a sensitive to apoptosis gene (SAG) protein, a metal chelator, and a reactive oxygen species scavenger, markedly protected single strand breaks in supercoiled plasmid DNA and 8-OHdG formation in calf thymus DNA induced by peroxynitrite [131].

REFERENCES

1. OI Aruoma, B Halliwell, E Gajewski, M Dizdaroglu. *J Biol Chem* 264: 20509–20512, 1989.
2. B Halliwell, OI Aruoma. *FEBS Lett* 281: 9–19, 1991.
3. R Meneghini. *Free Radic Biol Med* 23: 783–792, 1997.
4. AP Breen, JA Murphy. *Free Radic Biol Med* 18: 1033–1077, 1995.
5. JF Ward. *Radiat Res* 104: S103–S111, 1985.
6. DJ Deeble, D Schulz, C von Sonntag. *Int J Radiat Biol* 49: 915–926, 1986.
7. G Scholes, RL Willson, M Ebert. *J Chem Soc Chem Commun.* 17–18, 1969.
8. ES Henle, S Linn. *J Biol Chem* 272: 19095–19098, 1997.
9. RA Floyd. *Biochem Biophys Res Commun* 99: 1209–1215, 1981.
10. JW Park, RA Floyd. *Arch Biochem Biophys* 312: 289–291, 1994.
11. JW Park, RA Floyd. *Biochim Biophys Acta* 1336: 263–268, 1997.
12. JE Schneider, MM Browning, RA Floyd. *Free Radic Biol Med* 5: 287–295, 1988.
13. OI Aruoma, B Halliwell, E Gajewski, M Dizdaroglu. *J Biol Chem* 264: 20509–20512, 1989.
14. T Nunoshiba, F Obata, AC Boss, S Oikawa, T Mori, S Kawanishi, K Yamamoto. *J Biol Chem* 274: 34832–34837, 1999.
15. H Itoh, T Shioda, T Matsura, S Koyama, T Nakanishi, G Kajiyama, T Kawasaki. *Arch Biochem Biophys* 313: 120–125, 1994.
16. D Conte, S Narindrasorasak, B Sarkar. *J Biol Chem* 271: 5125–5130, 1996.
17. Z Djuric, DW Potter, BG Taffe, GM Strasburg. *J Biochem Mol Toxicol* 15: 114–119, 2001.
18. PB Walter, MD Knutson, A Paler-Martinez, S Lee, Y Xu, FE Viteri, BN Ames. *Proc Natl Acad Sci USA* 99: 2264–2269, 2002.
19. K Keyer, JA Imlay. *Proc Natl Acad Sci USA* 93: 13635–13640, 1996.
20. R Stoewe, WA Prutz. *Free Radic Biol Med* 3: 97–105, 1987.
21. A Samuni, M Chevion, G Czapski. *J Biol Chem* 256: 12632–12635, 1981.
22. S Goldstein, G Czapski. *J Am Chem Soc* 108: 2244–2250, 1986.
23. JM Gutteridge, B Halliwell. *Biochem Pharmacol* 31: 2801–2805, 1982.
24. JL Sagripant, KH Kraemer. *J Biol Chem* 264: 1729–1734, 1989.
25. OI Aruoma, B Halliwell, E Gajewski, M Dizdaroglu. *Biochem J* 273: 601–604, 1991.
26. PG Parsons, LE Morrison. *Cancer Res* 42: 3783–3788, 1982.
27. BHJ Bielski, DE Cabelli, RL Aradi, AB Ross. *J Phys Chem Ref Data* 14: 1041–1100, 1985.

28. TA Dix, KM Hess, MA Medina, RW Sullivan, SL Tilly, TL Webb. *Biochemistry* 35: 4578–4583, 1996.
29. ML Cunningham, JG Peak, MJ Peak. *Mutation Res* 184: 217–222, 1987.
30. M Rozenberg-Arska, BS van Asbeck, TFJ Martens, J Verhoff. *J Gen Microbiol* 131: 3325–3330, 1985.
31. A Ito, NI Krinski, ML Cunningham, MJ Peak. *Free Radic Biol Med* 3: 111–118, 1987.
32. R-Y Deno, I Fridovich. *Free Radic Biol Med* 6: 123–129, 1989.
33. D Muehlematter, R Larsson, P Cerutti. *Carcinogenesis* 9: 239–245, 1988.
34. V Grishko, M Solomon, GL Wilson, SP LeDoux, MN Gillespie. *Am J Physiol Lung Cell Mol Physiol* 280: L1300–L1308, 2001.
35. FM Yakes, B van Houten. *Proc Natl Acad Sci USA* 94: 514–519, 1997.
36. SW Ballinger, C Patterson, C-N Yan, R Doan, DL Burow, GG Young, FM Yakes, BV Houten, CA Ballinger, BA Freeman, MS Runge. *Circ Res* 86: 960–966, 2000.
37. SA Lesko, RJ Lorentzen, POP Ts'o. *Biochemistry* 19: 3023–3028, 1980.
38. HC Birnboim. *Science* 215: 1247–1249, 1982.
39. HC Birnboim, M Kaminska. *Proc Natl Acad Sci USA* 82: 6820–6824, 1985.
40. HC Birnboim. *Carcinogenesis* 7: 1511–1517, 1986.
41. RA Floyd, JJ Watson, J Harris, M West, PK Wong. *Biochem Biophys Res Commun* 137: 841–846, 1986.
42. JH Jackson, E Gajewski, IU Schraufstatter, PA Hyslop, AF Fuciarelli, CG Cochrane, M Dizdaroglu. *J Clin Invest* 84: 1644–1649, 1989.
43. JG Lewis, T Hamilton, DO Adams. *Carcinogenesis* 7: 813–818, 1986.
44. T Takeuchi, M Nakajima, K Morimoto. *Carcinogenesis* 17: 1543–1548, 1996.
45. HC Birnboim. *Arch Biochem Biophys* 294: 17–22, 1992.
46. Z Shen, SL Hazen. *Biochemistry* 39: 5474–5482, 2000.
47. JP Henderson, J Byun, JW Heinecke. *J Biol Chem* 274: 33440–33448, 1999.
48. CL Hawkins, DI Pattison, MJ Davies. *Biochem J* 365: 605–615, 2002.
49. MM Huycke, V Abrams, DR Moore. *Carcinogenesis* 23: 529–536, 2002.
50. MJ Waring. *Nature* 219: 1320–1325, 1968.
51. K Handa, S Sato. *Gann* 66: 43–47, 1975.
52. NR Bachur, SL Gordon, MV Gee. *Mol Pharmacol* 13: 901–910, 1977.
53. V Berlin, WA Haseltine. *J Biol Chem* 256: 4747–4756, 1981.
54. DA Rowley, B Halliwell. *Biochim Biophys Acta* 761: 86–93, 1983.
55. JR Muindi, BK Sinha, I Gianni, CE Myers. *FEBS Lett* 172: 226–230, 1984.
56. H Eliot, L Gianni, C Myers. *Biochemistry* 23: 928–936, 1984.
57. R Ishida, T Takahashi. *Biochem Biophys Res Commun* 66: 1432–1438, 1975.
58. EA Sausville, J Peisach, SB Horowitz. *Biochem Biophys Res Commun* 73: 814–822, 1976.
59. EA Sausville, J Peisach, SB Horwitz. *Biochemistry* 17: 2740–2746, 1978.
60. GH McGall, LE Rabow, J Stubbe. *J Am Chem Soc* 109: 2836–2837, 1987.
61. BK Sinha, HM Eliot, B Kalyanaraman. *FEBS Lett* 227: 240–244, 1988.
62. DH Chin, IH Goldberg. *Biochemistry* 25: 1009–1015, 1986.
63. Y Shoji, Y Uedono, H Ishikura, N Takeyama, T Tanaka. *Immunology* 84: 543–548, 1995.
64. Q Ye, WJ Bodell. *Carcinogenesis* 17: 1747–1750, 1996.
65. LM Nutter, EO Ngo, GR Fisher, PL Gutierrez. *J Biol Chem* 267: 2474–2480, 1992.
66. C Giulivi, E Cadenas. *Biochem J* 301: 21–30, 1994.
67. JG Lewis, W Stewart, DO Adams. *Cancer Res* 48: 4762–4765, 1988.
68. TW Yu, D Anderson. *Mutat Res* 379: 201–210, 1997.
69. TM de Kok, F Vaarwerk, I Zwingman, JM va Maanen, JC Kleinjans. *Carcinogenesis* 15: 1399–1404, 1994.
70. J Onuki, MHG Medeiros, EJH Bechara, P Di Mascio. *Biochim Biophys Acta* 1225: 259–263, 1994.
71. S Lynn, J-R Gurr, H-T Lai, K-Y Jan. *Circ Res* 86: 514–519, 2000.
72. VI Bruskov, LV Malakhova, ZK Masalimov, AV Chernikov. *Nucleic Acids Res* 30: 1354–1363, 2002.
73. JM Kanabus-Kaminka, M Feeley, HC Birnboim. *Free Radic Biol Med* 4: 141–145, 1988.

74. S Oikawa, K Yamada, N Yamashita, S Tada-Oikawa, S Kawanishi. *Carcinogenesis* 20: 1485–1490, 1999.
75. A Rahman, Shahabuddin, SM Hadi, JH Parish, K Ainley. *Carcinogenesis* 10: 1833–1839, 1989.
76. F Fazal, A Rahman, J Greensill, K Ainley, SM Hadi, JH Parish. *Carcinogenesis* 11: 2005–2009, 1990.
77. A Rahman, F Fazal, J Greensill, K Ainley, JH Parish, SM Hacli. *Mol Cell Biochem* 111: 3–11, 1992.
78. SC Sahu, GC Gray. *Cancer Lett* 85: 159–164, 1994.
79. SC Sahu, GC Gray. *Food Chem Toxicol* 35: 443–447, 1997.
80. S Thomas, JE Lowe, V Hadjivassiliou, RG Knowles, IC Green, MH Green. *Biochem Biophys Res Commun* 243: 241–245, 1998.
81. C Giulivi, E Cadenas. *Biochim Biophys Acta* 1366: 265–274, 1998.
82. N Yamashita, M Murata, S Inoue, MJ Burkitt, L Milne, S Kawanishi. *Chem Res Toxicol* 11: 855–862, 1998.
83. M Murata, S Kawanishi. *J Biol Chem* 275: 2003–2008, 2000.
84. L Cragg, RP Hebbel, W Miller, A Solovey, S Selby, H Enright. *Blood* 92: 632–638, 1998.
85. GK Livingston, WN Rom, MV Morris. *J Environ Pathol Toxicol* 4: 473–482, 1980.
86. BT Velichkovskyi, YA Vladimirov, LG Korkina, TB Suslova. *Vestn Akad Nauk SSSR* (1982), 45–48 (in Russian).
87. LG Korkina, AD Durnev, TB Suslova, ZP Cheremisina, NO Daugel-Dauge, IB Afanas'ev. *Mutat Res* 265: 245–253, 1992.
88. LG Lund, AE Aust. *Carcinogenesis* 13: 637–642, 1992.
89. T Takeuchi, K Morimoto. *Carcinogenesis* 15: 635–639, 1994.
90. H Dong, A Buard, A Renier, F Levy, L Saint-Etienne, MC Jaurand. *Carcinogenesis* 15: 1251–1255, 1994.
91. H Fung, YW Kow, B Van Houten, BT Mossman. *Carcinogenesis* 18: 825–832, 1997.
92. CA Delaney, MH Green, JE Lowe, IC Green. *FEBS Lett* 333: 291–295, 1993.
93. CA Delaney, IC Green, JE Lowe, JM Cunningham, AR Butler, L Renton, I D'Costa, MH Green. *Mutat Res* 375: 137–146, 1997.
94. T deRojas-Walker, S Tamir, H Ji, JS Wishnok, SR Tannenbaum. *Chem Res Toxicol* 8: 473–477, 1995.
95. MG Salgo, K Stone, GL Squadrito, JR Battista, WA Pryor. *Biochem Biophys Res Commun* 25: 1025–1030, 1995.
96. S Inoue, S Kawanishi. *FEBS Lett* 28: 86–88, 1995.
97. B Epe, D Ballmaier, I Roussyn, K Broviba, H Sies. *Nucleic Acids Res.* 24: 4105–4110, 1996.
98. NY Treyakova, S Burney, B Pamir, JS Wishnok, PC Dedon, GN Wogan, SR Tannenbaum. *Mutat Res* 447: 287–303, 2000.
99. M Masuda, H Nishino, H Ohshima. *Chem Biol Interact* 139: 187–197, 2002.
100. N Watanabe, S Miura, S Zeki, H Ishii. *Free Radic Biol Med* 30: 1019–1028, 2001.
101. N Phoa, B Epe. *Carcinogenesis* 23: 469–475, 2002.
102. C Szabo, B Zigngarelli, M O'Connor, AL Salzman. *Proc Natl Acad Sci USA* 93: 1753–1758, 1996.
103. C Szabo. *Shock* 9: 341–344, 1998.
104. S Cuzzocrea, B Zingarelli, AP Caputi. *Shock* 9: 336–340, 1998.
105. C Szabo, L Virag, S Cuzzocrea, GS Scott, P Hake, MP O'Connor, B Zingarelli, A Salzman, E Kun. *Proc Natl Acad Sci USA* 95: 3867–3872, 1998.
106. H Ohshima, I Gilibert, F Bianchini. *Free Radic Biol Med* 26: 1305–1313, 1999.
107. L Chazotte-Aubert, S Oikawa, I Gilibert, F Bianchini, S Kawanishi, H Ohshima. *J Biol Chem* 274: 20909–20915, 1999.
108. D Anderson, TW Yu, BJ Phillips, P Schmezer. *Mutat Res* 307: 261–273, 1994.
109. ML Hu, MK Shih. *Free Radic Res* 26: 585–592, 1997.
110. CG Fraga, PA Motchnik, MK Shigenaga, HJ Helbock, RA Jacob, BN Ames. *Proc Natl Acad Sci USA* 88: 11003–11007, 1991.
111. MG Salgo, WA Pryor. *Arch Biochem Biophys* 333: 482–488, 1996.
112. H Wei, Q Cai, RO Rahn. *Carcinogenesis* 17: 73–77, 1996.

113. GC Yen, HH Lai. *Food Chem Toxicol* 40: 1433–1440, 2002.
114. M Noroozi, WJ Angerson, MEJ Lean. *Am J Clin Nutr* 67: 1210–1218, 1998.
115. H Ohshima, Y Yoshie, S Auriol, I Gilbert. *Free Radic Biol Med* 25: 1057–1065, 1998.
116. MK Johnson, G Loo. *Mutat Res* 459: 211–218, 2000.
117. V Abalea, J Cillard, M-P Dubos, O Sergent, P Gillard, I Morel. *Free Radic Biol Med* 26: 1457–1466, 1999.
118. SJ Duthie, VL Dobson. *Z Ernahrungswiss* 38: 28–34, 1999.
119. SA Aherne, NM O'Brien. *Nutr Cancer* 34: 160–166, 1999.
120. SA Aherne, NM O'Brien. *Free Radic Biol Med* 29: 507–514, 2000.
121. IB Afanas'ev, AI Dorozhko, AV Brodskii, VA Kostyuk, AI Potapovitch. *Biochem Pharmacol* 38: 1763–1769, 1989.
122. Y Uchigata, H Yamamoto, A Kawamura, H Okamoto. *J Biol Chem* 257: 6084–6088, 1982.
123. AM Cohen, RE Aberdroth, P Hochstein. *FEBS Lett* 174: 147–150, 1984.
124. M Tomasetti, GP Littarru, R Stocker, R Alleva. *Free Radic Biol Med* 27: 1027–1032, 1999.
125. M Deiana, OI Aruoma, M de LP Bianchi, JPE Spencer, H Kaur, B Halliwell, R Aeschbach, S Banni, MA Dessi, FP Corongiu. *Free Radic Biol Med* 26: 762–769, 1999.
126. CL Greenstock, CP Jinot, RP Whitehouse, MD Sargent. *Free Rad Res Commun* 2: 233–239, 1987.
127. J Abel, N De Ruiter. *Toxicol Lett* 47: 1991–1996, 1989.
128. T Muller, R Schuckelt, L Jaenicke. *Arch Toxicol* 65: 20–26, 1991.
129. L Cai, J Koropatnick, MG Cherian. *Chem Biol Interact* 96: 143–155, 1995.
130. L Cai, JB Klein, YJ Kang. *J Biol Chem* 275: 38957–38960, 2000.
131. SY Kim, JH Lee, ES Yang, IS Kil, JW Park. *Biochem Biophys Res Commun* 301: 671–674, 2003.

29 Antioxidants

There are three major types of compounds, which can directly suppress free radical formation: antioxidants, free radical scavengers, and chelators. (Antioxidants and free radical scavengers are usually considered to be synonyms, although not always. For example, ethanol is a hydroxyl radical scavenger but was never regarded as an antioxidant. Antioxidant is the oldest term, which at the beginning had been applied for the description of inhibitors of oxidative processes, which are able to react with peroxyl radicals. Now, this term is frequently applied to all free radical inhibitors.) In addition to direct antioxidants, there are two other important groups of free radical inhibitors: antioxidant enzymes (Chapter 25) and the compounds possessing indirect antioxidative properties. Compounds having indirect antioxidative properties may affect the formation of free radicals by an indirect way, for example by inhibiting the activity of prooxidant enzymes.

Direct free radical inhibitors suppress free radical formation by reacting with free radicals to form new inactive radicals (Reactions (1) and (2)) or chelating catalytically active transition metals to form inactive complexes:

$$HO^{\bullet} + AntH \Longrightarrow H_2O + Ant^{\bullet} \tag{1}$$
$$ROO^{\bullet} + AntH \Longrightarrow ROOH + Ant^{\bullet} \tag{2}$$

Recently, Kirsch and De Groot [1] pointed out at another possible mode of antioxidant activity, namely, "repairing" of biomolecules by the interaction of the forming free radicals with antioxidants (Reactions (3) and (4)).

$$ROO^{\bullet} \text{ or } HO^{\bullet} + RH \Longrightarrow ROOH \text{ or } H_2O + R^{\bullet} \tag{3}$$
$$R^{\bullet} + AntH \Longrightarrow RH + Ant^{\bullet} \tag{4}$$

These authors supposed that the "repairing" function of antioxidants may be even more important compared to scavenging reactive free radicals. However, although Reactions (3) and (4) may occur in biological systems, it is very difficult to estimate their importance. First of all, there is always a competition between the "repairing" Reaction (4) and the reaction of the biomolecule free radical $R^{\bullet}$ with dioxygen (Reaction (5)):

$$R^{\bullet} + O_2 \Longrightarrow ROO^{\bullet} \tag{5}$$
$$ROO^{\bullet} + AntH \Longrightarrow ROOH + Ant^{\bullet} \tag{2}$$

As the rate constant of Reaction (5) is many orders of magnitude higher than that of Reaction (4), the probability of "repairing" function seems to be very small.

The best known and undoubtedly most important natural antioxidants are α-tocopherol (vitamin E) and ascorbic acid (vitamin C), which are major biological antioxidants in lipid and aqueous phases, respectively. There is a tremendous amount of publications and reviews (for example, Refs. [2–9]) dedicated to chemical and biological properties of these vitamins; the most important studies are considered below.

29.1 VITAMIN E

29.1.1 ANTIOXIDANT AND PROOXIDANT ACTIVITY OF VITAMIN E

α-Tocopherol is a lipid-soluble phenolic derivative (Figure 29.1), having a very active hydroxyl group, which is responsible for a great antioxidant capacity of this vitamin. Indeed, it has been shown that the rate constants for the reactions of α-tocopherol with $HOO^•$ and $ROO^•$ radicals (Reactions (6) and (7)) are sufficiently high (about $2 \times 10^5\,l\,mol^{-1}\,s^{-1}$) [10].

$$HOO^• + \alpha\text{-TocH} \Longrightarrow HOOH + \alpha\text{-Toc}^• \quad (6)$$

$$ROO^• + \alpha\text{-TocH} \Longrightarrow ROOH + \alpha\text{-Toc}^• \quad (7)$$

$$O_2^{•-} + \alpha\text{-TocH} \Longrightarrow HOO^- + \alpha\text{-Toc}^• \quad (8)$$

$$O_2^{•-} + \alpha\text{-TocH} \Longrightarrow HOO^• + \alpha\text{-Toc}^- \quad (9)$$

In contrast, α-tocopherol is not a scavenger of superoxide. Earlier, it has been suggested that superoxide is able to abstract a hydrogen atom from α-tocopherol with a rate constant of about 10^3 to $10^4\,l\,mol^{-1}\,s^{-1}$ (Reaction (8); the incorrectness of these data is discussed in Ref. [2]). Actually, superoxide is unable to abstract a hydrogen atom even from very active organic compounds [2], and can only deprotonize α-tocopherol with rate constants of $6\,l\,mol^{-1}\,s^{-1}$ in 85% ethanol [10] and $0.6\,l\,mol^{-1}\,s^{-1}$ in acetonitrile [11].

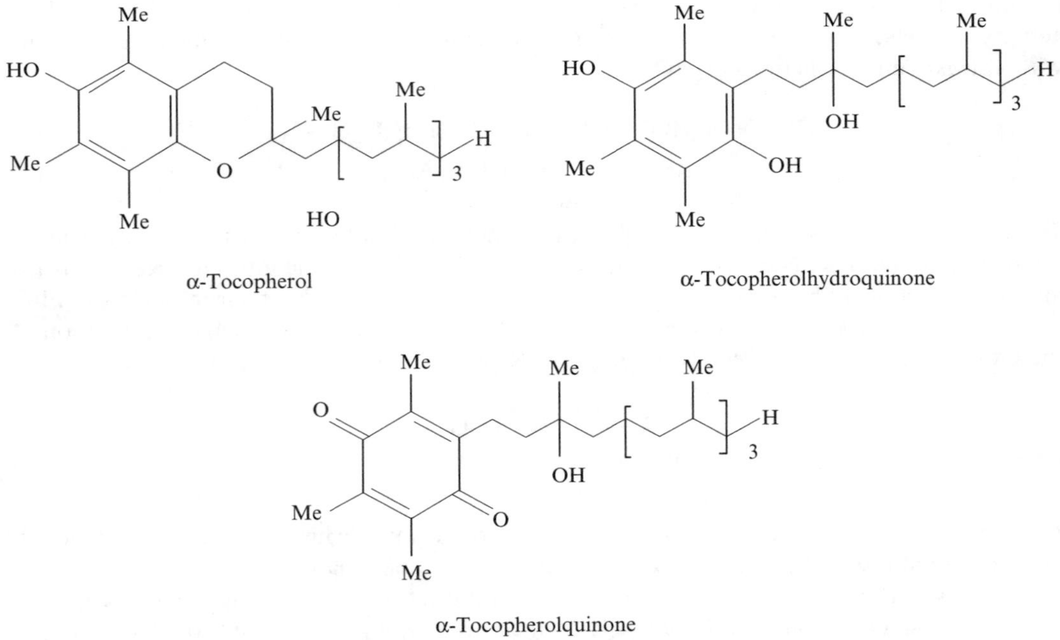

FIGURE 29.1 Members of vitamin E "family."

It is surprising, but the in vivo free radical scavenging activity of α-tocopherol has been questioned. (These disagreements are thoroughly discussed in earlier studies by McCay and coworkers [12,13].) Nonetheless, convincing chemical and biochemical findings for free radical scavenging activity of α-tocopherol were later obtained [14–16]. For example, it was shown that α-tocopheroxyl radical identified by its ESR spectrum was formed during autooxidation of lipids [15] and in the iron-initiated peroxidation of phosphatidylcholine liposomes [16]. Burton et al. [17] stated that vitamin E is the only lipid-soluble antioxidant in human blood plasma and erythrocytes.

Similar to many other antioxidants, α-tocopherol may manifest prooxidant effects under certain conditions. In 1986 Terao and Matsushita [18] have shown that α-tocopherol enhanced autoxidation of linoleic acid. Stocker and coworkers [7] studied in detail the prooxidant effect of α-tocopherol on LDL oxidation. It was suggested that α-tocopheroxyl radical is able to abstract a hydrogen atom from unsaturated fatty acids and initiate lipid peroxidation. Stimulatory effect of α-tocopherol was inhibited by many antioxidants and free radical scavengers [19]. Recently, it has been found that α-tocopherol enhanced in vivo lipid peroxidation in cigarette smokers consuming a high polyunsaturated fat diet [20]. Similar to α-tocopherol, both antioxidant and prooxidant activities have been found for its model compound Trolox C (Figure 29.2) [21]. This water-soluble analog of vitamin E stimulated or inhibited copper-initiated LDL oxidation depending on the time of Trolox addition, but its effect was always antioxidative when oxidation was initiated by peroxyl radicals.

The mechanism of prooxidant effect of α-tocopherol in aqueous lipid dispersions such as LDLs has been studied [22]. This so-called tocopherol-mediated peroxidation is considered in detail in Chapter 25, however, in this chapter we should like to return once more to the question of possible prooxidant activity of vitamin E. The antioxidant effect of α-tocopherol on lipid peroxidation including LDL oxidation is well established in both in vitro and in vivo systems (see, for example, Refs. [3,4] and many other references throughout this book). However, Ingold et al. [22] suggested that despite its undoubted high antioxidant efficiency in homogenous solution α-tocopherol can become a chain transfer agent in aqueous LDL

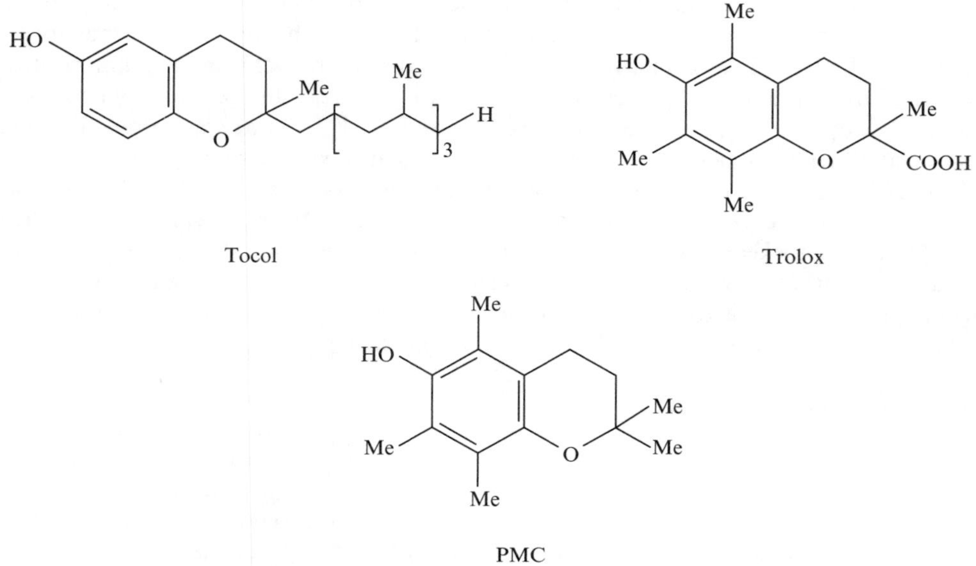

FIGURE 29.2 Synthetic analogs α-tocopherol.

suspensions. This proposal was based on experimental findings that LDL oxidation proceeds more effectively in the presence of α-tocopherol than in its absence. In contrast, available kinetic and thermodynamic data indicate that the abstraction reactions of α-tocopheroxyl radical are very slow.

$$\alpha\text{-Toc}^{\bullet} + \text{LH} \Longrightarrow \alpha\text{-TocH} + \text{L}^{\bullet} \qquad (10)$$

(Here, LH is a lipid-containing bisallylic methylene groups.)

The rate constant for the reaction of α-tocopheroxyl radical with unsaturated fatty acids is very small ($k_{10} = 0.02$–$0.08 \, \text{l mol}^{-1} \, \text{s}^{-1}$ [23,24]. The dissociation energy of the α-Toc–H bond is also relatively low (76 kcal mol^{-1} [25]). In contrast, the rate constants for the reactions of free radicals with α-tocopherol are about 10^6 to 10^7 times higher (for example, k_7 is about $2 \times 10^5 \, \text{l mol}^{-1} \, \text{s}^{-1}$ [10]). Therefore, the antioxidant activity of α-tocopherol must prevail on its activity as a chain transfer agent. However, Ingold et al. [22] pointed out that a mean lifetime of tocopheroxyl radical in the LDL particle could be as high as 12.5 s, and due to this Reaction (10) might be significant.

Although vitamin E and α-tocopherol are frequently considered to be synonyms, vitamin E is actually a name corresponding to a group of natural phenolic compounds comprising four tocopherols (α, β, δ, γ, distinguished by a number of methyl substituents) and four tocotrienols. In addition, it has been assumed that the by-products of α-tocopherol oxidation α-tocopherolquinone and α-tocopherolhydroquinone (Figure 29.1) can also be the very effective inhibitors of lipid peroxidation [26]. Shi et al. compared the antioxidant activities of α-tocopherol, α-tocopherolhydroquinone, and ubihydroquinone Q_{10} in several model systems [27]. It is interesting that although the relative reactivities of α-tocopherolhydroquinone and ubihydroquinone Q_{10} toward galvinoxyl (a stable phenoxyl radical) and peroxyl radicals were much greater than that of α-tocopherol, the latter was the most efficient antioxidant in the oxidation of methyl linoleate. Both α-tocopherolhydroquinone and ubihydroquinone Q_{10} reduced α-tocopheroxyl radical into α-tocopherol. Neuzil et al. [28] found that α-tocopherolhydroquinone effectively inhibited LDL oxidation. This compound associated with LDL, reduced the LDL's ubiquinone to ubihydroquinone, and suppressed the formation of α-tocopheroxyl radical. Appenroth et al. [29] showed that the metabolite of γ-tocopherol LLU-α [2,7,7-trimethyl-2-(carboxyethyl)-6-hydroxychroman] is a more effective inhibitor of iron-stimulated lipid peroxidation and luminol- or lucigenin-amplified CL than α- and γ-tocopherols. These authors also showed that LLU-α was protective against Tl-stimulated nephrotoxicity at least partly due to its antioxidant activity.

It should be noted that pharmacological vitamin E is not a free natural *RRR*-α-tocopherol or synthetic *All rac* α-tocopherol but its acetate ester. α-Tocopheryl acetate has the phenolic hydroxyl group blocked and therefore, is not a genuine antioxidant, but this compound is very rapidly hydrolyzed in vivo into α-tocopherol. It is interesting that the biological activity of α-tocopheryl acetate is the same as that of α-tocopherol in humans but significantly lower in rats [30]. ("A man is not a rat!" Professor KU Ingold.)

29.1.2 Biological Activity

As already mentioned, a great deal of work has been dedicated to the study of biological activity of vitamin E under physiological and pathophysiological conditions. The efficiency of vitamin E in suppressing free radical-mediated damage and the complete absence of toxicity together with its important vitamin activity makes this compound a potential important medicine in the treatment of many pathologies associated with the overproduction of free radicals.

α-Tocopherol is the most abundant and probably most effective in vivo antioxidant from all the forms of vitamin E. For a long time this fact was not fully understood because from the chemical point of view, missing one or two methyl substituents in the phenolic nucleus, should not strongly affect the free radical scavenging activity of the tocopherol molecule. Nonetheless, it was found that in humans α-tocopherol preferably appears in the plasma after passage of all components of vitamin E through the liver [31]. Later on, it has been shown that the plasma preference for α-tocopherol is a consequence of the selection by the hepatic α-tocopherol transfer protein [32]. It is now known that two α-tocopherol-binding proteins (TBP) are responsible for the regulation of α-tocopherol concentrations in plasma (30 kDa TBP) and for its intracellular distribution (15 kDa TBP) [33]. The 30-kDa TBP is unique to hepatocytes while 15 kDa presents in all major tissues.

Although α-tocopherol is considered to be the most effective antioxidant form of vitamin E group, the effects of other tocopherols continue to draw attention. Thus, Li et al. [34] studied the effects of α-, γ-, and δ-tocopherols on human platelets. It was found that all tocopherols are equally efficient in suppression of lipid peroxidation and platelet aggregation, but their combination in the concentrations found in nature was more potent than individual tocopherols. The activities of different tocopherols may significantly differ in the processes initiated by peroxynitrite and other nitrogen species. Thus, Christen et al. [35] found that γ-tocopherol was a more effective inhibitor of liposomal peroxidation (but not LDL oxidation) initiated by peroxynitrite than α-tocopherol. Furthermore, the mechanisms of inhibitory action of these tocopherols are different: α-tocopherol reacts as a free radical scavenger oxidizing into α-tocopherolquinone while γ-tocopherol is nitrated at 5-position by a nucleophilic mechanism forming o-quinone as the end-product. Christen et al. believe that nucleophilic trapping of lipid-soluble nitrogen species and other electrophilic mutagens may be still unknown nonantioxidant inhibitory activity of γ-tocopherol and some other classic antioxidants. It was suggested that the presence of both tocopherols is needed in vivo for optimal protection from free radical-mediated damage. The authors also argue that γ-tocopherol is more important than α-tocopherol for the prevention of cardiovascular disease.

It is important that there is equilibrium in the distribution of vitamin E between the plasma and erythrocytes in a living organism, with the content of vitamin in plasma about threefold higher [12]. The membrane structure is a critical factor for recognition of how much vitamin E the membrane may absorb. It means that notwithstanding how much vitamin was consumed, its content in the membrane is inherently limited.

Numerous studies demonstrate inhibitory effects of vitamin E on free radical-mediated processes. For example, Zhang et al. [36] showed that the pretreatment of isolated rat hepatocytes with α-tocopheryl succinate protected mitochondria from oxidative damage. In another study [37] these authors showed that α-tocopheryl succinate inhibited rotenone-induced mitochondrial lipid peroxidation. Cachia et al. [38] found that the normal content of α-tocopherol in LDL (the α-tocopherol/apoB molar ratio is 6 to 8) is important for a decrease in monocyte superoxide production, which is involved in LDL oxidation. Beharka et al. [39] showed that vitamin E inhibited cyclooxygenase activity in macrophages from old mice, which responsible for the production of proinflammatory prostaglandin PGE_2 through the interaction with peroxynitrite. Vitamin E supplementation diminished enzymatic and nonenzymatic lipid peroxidation in rats (measured by the levels of F_2-isoprostanes and $PGF_{2\alpha}$ metabolite) in blood, urine, and liver [40]. Intraperitoneal administration of α-tocopherol to rats suppressed ascorbate-initiated lipid peroxidation in mitochondria and microsomes isolated from rat liver [41]. Similarly, vitamin E effectively inhibited carbon tetrachloride-initiated lipid peroxidation in mice [42]. It was found that intravenous therapy with vitamin E-containing liposomes decreased mouse mortality by nearly 90% when a lethal dose of carbon tetrachloride was given. Endogenous vitamin E is apparently the most efficient antioxidant in rat brain homogenate because it is first depleted before membrane lipid and membrane-bound proteins showed some oxidative destruction [43]. Vitamin E was effective in suppression of oxidative damage to rat blood and tissue in in vivo experiments in diets supplemented with both fat-soluble and water-soluble antioxidants (selenium, Trolox C, ascorbate palmitate, acetylcysteine, ubiquinone Q_o, ubiquinone Q_{10}, β-carotene, canthaxanthin, and (+)-catechin) [44].

To estimate the importance of structural factors on the inhibitory activity of vitamin E in free radical-mediated damaging processes, Kaneko et al. [45] studied protection by α-tocopherol and its model analogs tocol, 2,2,5,7,8-pentamethylchroman-6-ol (PMC), and trolox C (Figure 29.2) against linoleic acid hydroperoxide-induced toxicity to cultured human umbilical vein cells. Preincubation of cells with antioxidants resulted in efficient protection by α-tocopherol and PMC but not with tocol and Trolox C. Although the loss of three methyl substituents in tocol may affect the reactivity of the phenolic hydroxyl in this compound, free radical scavenging activity of water-soluble Trolox C should be equal to that of α-tocopherol. Therefore, the inhibitory effects of phenolic compounds in cells depend not only on their reactivity as free radical scavengers but also on their incorporation rate into cells.

The efficiency of vitamin E in the suppression of free radical-mediated damage induced by iron overload has been studied in animals and humans. Galleano and Puntarulo [46] showed that iron overload increased lipid and protein peroxidation in rat liver. Vitamin E supplementation successfully suppressed these effects and led to an increase in α-tocopherol, ubiquinone-9, and ubiquinone-10 contents in liver. Important results were obtained by Roob et al. [47] who found that vitamin E supplementation attenuated lipid peroxidation (measured as plasma MDA and plasma lipid peroxides) in patients on hemodialysis after receiving iron hydroxide sucrose complex intravenously during hemodialysis session. These findings support the proposal that iron overload enhances free radical-mediated damage in humans.

29.1.3 Effects of Vitamin E Supplementation in Aging and Heart Diseases

Many studies are dedicated to the study of favorable effects of vitamin E on various pathological disorders in humans and animals. As mentioned, the effects of vitamin E supplementation during the treatment of various pathological processes are mostly considered in Chapter 29. In this chapter, I would like only to draw attention to some findings describing the effects of vitamin E in aging and heart diseases because these data characterize its in vivo antioxidant activity. Unfortunately, the data on the protective effects of vitamin E in humans and animals remain highly controversial. For example, it was found that healthy, very old people with the highest plasma vitamin E level have the lowest risk of cardiovascular events [48]. On the other hand, no significant effects of vitamin E on lipid peroxidation in healthy people measured via urinary 4-hydroxynonenal and isoprostane formation was found [49].

Recently, it has been found that the content of vitamin E in the aorta of old rats is extremely high (about 70 times greater) as compared with young animals [50]. An increase in vitamin E was paralleled by an increase in superoxide production. The authors assumed that vitamin E content increases with age in order to diminish oxidative damage to vascular tissue. These findings also suggest that high levels of vitamin E can be accumulated from a normal diet to suppress oxidative stress-associated vascular aging. Critical consideration of the effects of vitamin E, vitamin C, and β-carotene on oxidative damage in humans is given by McCall and Frei [51]. These authors concluded that there are scarce evidences of positive effects of these antioxidants on lipid damage in both smokers and nonsmokers. However, much more favorite conclusions were made by Pryor [52] based on clinical trials on the supplementation of vitamin E to patients with heart disease. He pointed out that there is "little doubt that vitamin E provides significant protection both to those (patients) without diagnosed heart conditions and to those with proven heart disease." Furthermore, it was concluded that the supplemental level of vitamin E from 100 to 800 IU a day is safe. It must be also noted that surprisingly good results have been obtained in Cambridge Heart Antioxidant Study (CHAOS) [6], where it was found that high doses (400–800 IU/day) of vitamin E produced a 77% reduction in the occurrence of myocardial infraction and a 50% reduction in all cardiovascular events.

29.1.4 SYNTHETIC ANALOGS OF VITAMIN E

For a long time attempts have been made to develop synthetic analogs of vitamin E with improved antioxidant activities or different physicochemical properties. The most known synthetic analog of α-tocopherol is probably Trolox (Trolox C, Figure 29.2), which was synthesized with the purpose to have a water-soluble antioxidant with the properties of vitamin E. Some findings obtained with the use of Trolox are cited above. Recently, the novel analog of vitamin E has been synthesized, which contains two active parts: the α-tocopheryl moiety and the NO synthase inhibitor pharmacophore [53] (Figure 29.3). This compound turns out to be a very efficient antioxidant (with I_{50} value equal to 0.29 μmol l^{-1} for the inhibition of iron-initiated lipid peroxidation of rat brain homogenate) and a neuroprotective agent.

α-Tocopherol may exhibit physiological functions distinct from its antioxidant effects, and acts as a signaling molecule in vascular smooth muscle cells (5,6), suppressing cell–cell adhesion [54], inactivating protein kinase c [55], etc., but similar topics are naturally out of the scope of this book.

29.2 VITAMIN C

29.2.1 ANTIOXIDANT AND PROOXIDANT ACTIVITY OF VITAMIN C

Vitamin C (ascorbic acid) is probably the most known vitamin in the world. Its legendary fame is based on the two events: its exceptionally important role in the treatment of scurvy and Linus Pauling's proposal to use the huge doses of ascorbic acid for the prevention of common cold. The latter proposal, based obviously on the antioxidant properties of ascorbic acid, generated numerous studies and was frequently disputed, but many people (me including) successfully apply ascorbic acid for the treatment of starting stage of common cold.

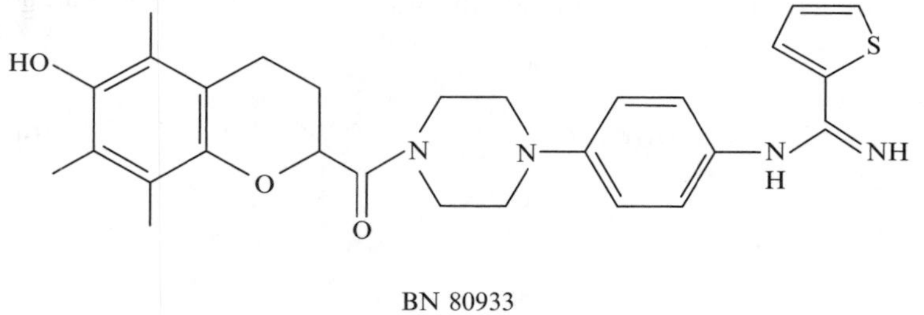

BN 80933

FIGURE 29.3 Analog of vitamin E-containing the α-tocopheryl moiety and the NO synthase inhibitor pharmacophore.

Ascorbic acid (which mostly present in biological systems as ascorbate anion ($pK_a = 4.25$) has a very active hydroxyl group and therefore, is a very efficient free radical scavenger (Figure 29.4). On the other hand, ascorbic acid is the very reactive reductant easily reducing ferric into ferrous ions, the catalysts of the Fenton reaction. Oxidation and reduction reactions of ascorbic acid with numerous oxidants and reductants are widely studied [2]. It is interesting that similar to α-tocopherol (see, above) ascorbic acid reacts with superoxide very slowly ($k = 0.32 \pm 0.08$ l mol^{-1} s^{-1}) and only by deprotonation [11] but relatively quickly with perhydroxyl radical ($k = 1.6 \times 10^4$ l mol^{-1} s^{-1}) [56]. Thus, ascorbic acid may be an antioxidant or a prooxidant depending on conditions. It is interesting that the efficacy of ascorbic acid in the treatment of scurvy is due to its "prooxidant" properties because ascorbic acid is needed to reduce the active center metal ions of hydroxylases and oxygenases involved in the biosynthesis of procollagen, carnitine, and neurotransmitters. Depletion in ascorbic acid decreases the activities of these enzymes and causes the development of scurvy [57].

In vitro antioxidant and prooxidant properties of ascorbic acid have been clearly demonstrated. It is understandable that the competition between antioxidant and prooxidant activities of ascorbic acid depends on the rates of Reactions (11) and (12).

$$AH_2 + R^\bullet \Longrightarrow AH^\bullet + RH \qquad (11)$$

$$AH_2 + Fe^{3+} \Longrightarrow AH^\bullet + Fe^{2+} + H^+ \qquad (12)$$

Therefore, the total effect of ascorbic acid will depend on many factors, first of all, its concentration and the concentrations of iron ions. For example, ascorbic acid in low concentrations enhanced lipid peroxidation in rat lung microsomes but inhibited it at higher (above 4 mM) concentrations [58]. Similarly, ascorbic acid stimulated iron-dependent liposomal lipid peroxidation at low ferrous ion concentrations and inhibited it at higher ferrous ion concentrations [59]. Opposite effects of ascorbic acid on free radical generation are typical not only for lipid peroxidation. For example, ascorbic acid induced sister-chromatid exchanges in cultured mammalian cells [60] but inhibited oxygen radical-mediated mutagenic effect of fibers and particles on human lymphocytes [61]. In vitro prooxidant effects of ascorbic acid have also been shown in hydroperoxide-initiated lipid peroxidation in rat liver microsomes [62] and human lung cells [63]. Recently, it has been shown [64] that ascorbate increased "free" iron content and in vitro lipid peroxidation in the serum from iron-loading guinea pig.

29.2.2 Biological Activity

There are contradictory data on the effects of dietary ascorbic acid on free radical-mediated damage in animals. Barja et al. [65] found that the administration of 660 mg/kg vitamin C to guinea pigs for 5 weeks significantly decreased the levels of protein carbonyls and lipid peroxidation products. On the other hand, the administration of 500 mg/kg vitamin C to

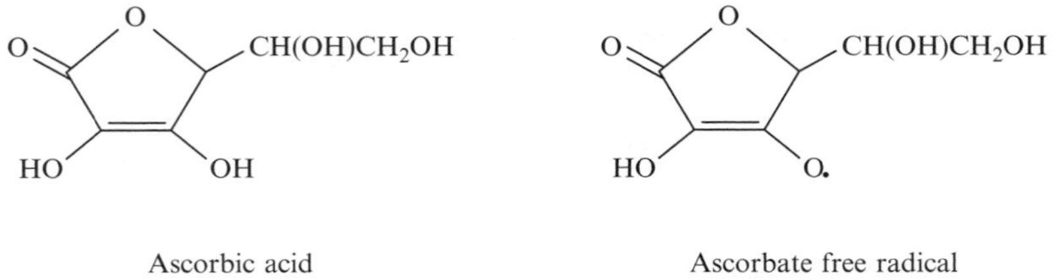

FIGURE 29.4 Ascorbic acid and ascorbate free radical.

rats for 4 days markedly induced hepatic cytochrome P-450-linked monooxygenases and stimulated the formation of large amounts of superoxide [66]. Origins of antioxidant or prooxidant activity of vitamin C in these animal experiments are still unknown.

Despite numerous examples of prooxidant effects of ascorbic acid under the in vitro conditions, there are still no irrefutable evidences of its prooxidant activity in humans. It might be expected that the most pronounced prooxidant effect of ascorbic acid will be observed under iron-overloading conditions. Thus, it has been proposed that ascorbate supplementation can be harmful for persons with plasma containing "free" iron (nontransferrin-bound bleomycin-chelatable iron). This form of iron was detected in patients with iron overload such as hemochromatosis [67] or rheumatoid arthritis patients [68]. However, recent study [69] has shown that despite the high levels of ascorbic acid and "free" iron in the plasma of preterm infants, there was no difference in the lipid peroxidation end products such as F_2-isoprostanes in the plasma of those infants and infants without iron overload. This work and other studies [8] showed that even under iron-overloading conditions ascorbic acid mainly exhibits antioxidant effect although in the pathologies associated with iron overload, for example in thalassemia, ascorbic acid supplementation could be dangerous [70] (see Chapter 29).

Recently, Carr and Frei reviewed studies on the antioxidant and prooxidant effects of ascorbic acid [8]. These authors pointed out that a "highly controversial" work by Podmore and coworkers [71] who found that the prooxidant effect of ascorbic acid supplementation to healthy volunteers is much questionable. These authors demonstrated that of the 44 in vivo studies, 38 showed the antioxidant effect of ascorbic acid, 14 showed no change, and only six showed the enhancement of oxidative damage after ascorbate supplementation. It was concluded that ascorbic acid is an antioxidant in biological fluids, animals, and humans, both with and without iron supplementation.

Ellis et al. [72] recently studied the effects of short- and long-term vitamin C therapy in the patients with chronic heart failure (CHF). It was found that oxygen radical production and TBAR product formation were higher in patients with CHF than in control subjects. Both short-term (intravenous) and long-term (oral) vitamin C therapy exhibited favorable effects on the parameters of oxidative stress in patients: the treatments decreased oxygen radical formation and the level of lipid peroxidation and improved flow-mediated dilation in brachial artery. However, there was no correlation between changes in endothelial function and oxidative stress.

29.2.3 Interaction Between Vitamins E and C

For a long time a great interest has been drawn to the study of in vitro (and later in vivo) interaction between lipid-soluble vitamin E and water-soluble vitamin C. As far back as 1941, Columbic and Mattill [73] showed that ascorbic acid enhanced antioxidant effects of tocopherols in lard and cottonseed oil. Later on, Tappel [74] suggested that the α-tocopheroxyl radical may react with ascorbate to regenerate vitamin E in biological systems. Eleven years later, Packer et al. [75] measured the rate constant for Reaction (13), which turns out to be high enough ($1.55 \times 10^6 \, l \, mol^{-1} \, s^{-1}$).

$$\alpha\text{-Toc}^\bullet + \text{Asc} \Longrightarrow \alpha\text{-Toc} + \text{Asc}^\bullet \tag{13}$$

Thus, vitamin C is able to replenish vitamin E, making the latter a much more efficient free radical inhibitor in lipid membranes. In addition, it has been suggested [9] that ascorbic acid can directly interact with the plasma membrane giving electrons to a trans-plasma membrane oxidoreductase activity. This ascorbate reducing capacity is apparently transmitted into and across the plasma membrane.

The interaction of vitamin C with vitamin E has been studied in numerous publications [12]. Leung et al. [76] found that the total inhibitory effect of relatively small concentrations of both vitamins on iron-initiated peroxidation of phospholipid liposomes was generally equal to the sum of their individual effects. However, the synergistic effect was observed at vitamin higher concentrations. Niki et al. [77] studied the reaction of α-tocopheroxyl radical with ascorbic acid and glutathione by ESR spectroscopy. Scarpa et al. [16] showed that the antioxidant effect of α-tocopherol on peroxidation of soybean phosphatidylcholine liposomes was maintained as long as ascorbate acid was present. The rate constant for Reaction (13) in this heterogeneous system was found to be equal to $5 \times 10^5 \, \text{l mol}^{-1} \, \text{s}^{-1}$. Negre-Salvayre et al. [78] showed that the triple combination of α-tocopherol, ascorbic acid, and bioflavonoid rutin is most effective in the suppression of superoxide formation and lipid peroxidation. Inhibitory effects of ascorbic acid on the peroxidation of dilinoleoylphosphatidylcholine lamellar liposomes initiated by azobis(2,4-dimethylvaleronitrile) were studied in combination with α-tocopherol and trolox [79]. Ascorbate was an effective inhibitor of peroxidation in aqueous phase but a very poor one in lipid phase. At the same time, ascorbate was an excellent synergist with α-tocopherol and trolox. Synergistic effect of ascorbic acid and α-tocopherol has also been shown in human blood plasma [80,81]. It should be mentioned that the enzymatic mechanisms of recycling of ascorbic acid may also take place in living systems. For example, the NADH-dependent reductase of ascorbate free radical is able to recycle ascorbic acid at the inner face of the plasma membrane [82].

In the recent review Carr et al. [54] considered potential antiatherogenic mechanisms of α-tocopherol and ascorbic acid. These authors concluded that these antioxidants are able to inhibit LDL oxidation, leukocyte adhesion to the endothelium, and vascular endothelial dysfunction. They also believe that ascorbic acid is more effective than α-tocopherol in the inhibition of these pathophysiological processes due to its capacity of reacting with a wide spectrum of oxygen and nitrogen free radicals and its ability to regenerate α-tocopherol.

Both vitamin E and vitamin C are able to react with peroxynitrite and suppress its toxic effects in biological systems. For example, it has been shown [83] that peroxynitrite efficiently oxidized both mitochondrial and synaptosomal α-tocopherol. Ascorbate protected against peroxynitrite-induced oxidation reactions by the interaction with free radicals formed in these reactions [84].

29.3 FLAVONOIDS

Flavonoids are a group of naturally occurring, low molecular weight polyphenols of plant origin, which are derivatives of benzo-γ-pyrone. These compounds are present in fruits and vegetables regularly consumed by humans. The main sources of flavonoids are apples, onions, berries, tea, beer, and wine. Most of these compounds belong to four main groups: flavones (I), flavonols (II), flavanone (III), and flavanols (IV) (Figure 29.5). Members of these groups differ by the number and the positions of hydroxyl substituents in rings A and B. However, there are flavonoids, which do not belong to these groups but are of biological and pharmacological importance, for example, catechin, (+)-cianidanol, and others.

Huge literature on biological functions of flavonoids and their antioxidant and free radical scavenging activities successfully competes with work on antioxidant effects of vitamins E and C. Flavonoids have been reported to exert multiple biological effects and exhibit antiinflammatory, antiallergic, antiviral, and anticancer activities [85–89]. However, considering flavonoids as the inhibitors of free radical-mediated processes, two types of their reactions should be discussed: flavonoids as free radical scavengers (antioxidants) and flavonoids as metal chelators.

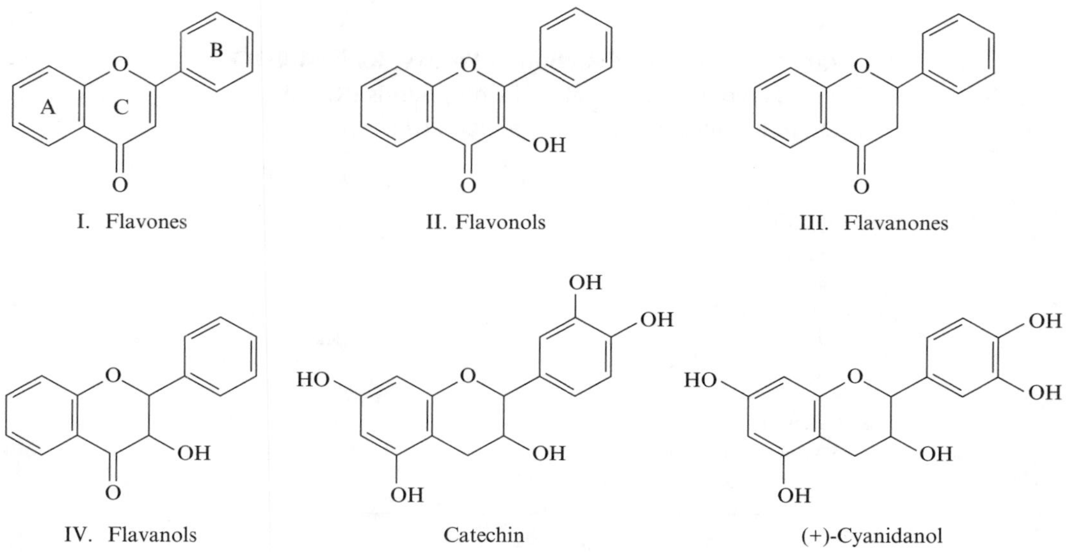

FIGURE 29.5 Major classes of flavonoids.

29.3.1 FREE RADICAL SCAVENGING ACTIVITY

Antioxidant activity of flavonoids has already been shown about 40 years ago [90,91]. (Early data on antioxidant flavonoid activity are cited in Ref. [92].) Flavonoids are polyphenols, and therefore, their antioxidant activity depends on the reactivity of hydroxyl substituents in hydrogen atom abstraction reactions. As in the case of vitamins E and C, the most studied (and most important) reactions are the reactions with peroxyl radicals [14], hydroxyl radicals [15], and superoxide [16].

$$ROO^{\bullet} + FlOH \Longrightarrow ROOH + FlO^{\bullet} \qquad (14)$$
$$HO^{\bullet} + FlOH \Longrightarrow H_2O + FlO^{\bullet} \qquad (15)$$

(The values of rate constants for these reactions are cited in Tables 29.1–29.3.) (As expected, hydroxyl radicals react with flavonoids with a diffusion-controlled reaction rate about 10^9 $l\, mol^{-1}\, s^{-1}$, while the rate constants for significantly less reactive peroxyl radicals are usually of $(0.1-1) \times 10^7\, l\, mol^{-1}\, s^{-1}$. As seen from Table 29.3, flavonoids react with superoxide with the rate constants of $(0.1-5) \times 10^4\, l\, mol^{-1}\, s^{-1}$. This reaction is of great importance because the capacity of flavonoids to scavenge superoxide makes these compounds useful pharmaceutical agents for the treatment of the diseases associated with free radical overproduction. However, superoxide cannot abstract a hydrogen atom even from the most active bisallylic methylene groups. Therefore, the most probable mechanism of this reaction is a concerted abstraction of a proton and a hydrogen atom by superoxide from o-hydroxyls of the flavonoid molecule (Reaction (16), Figure 29.6) [93].

As mentioned above, in contrast to classic antioxidant vitamins E and C, flavonoids are able to inhibit free radical formation as free radical scavengers and the chelators of transition metals. As far as chelators are concerned their inhibitory activity is a consequence of the formation of transition metal complexes incapable of catalyzing the formation of hydroxyl radicals by the Fenton reaction. In addition, as shown below, some of these complexes, for example, iron– and copper–rutin complexes, may acquire additional antioxidant activity.

TABLE 29.1
Rate Constants for Reaction of Peroxyl Radical ROO$^\cdot$ with Flavonoids and Related Compounds (Kinetic Chemiluminescence Experiments) [102]

Compound	$k_{15} \times 10^7$ (l mol^{-1} s^{-1})
Quercetin	2.1
Dihydroquercetin	1.9
Luteolin	2.2
Catechin	0.66
Fisetin	1.2
Naringenin	0.00034
Kaempferol	0.1
Caffeic acid	1.5
3,5-ditert-butylcatechol	1.9
Nordihydroguaiaretic acid	1.0

The ability of flavonoids (quercetin and rutin) to react with superoxide has been shown in both aqueous and aprotic media [59,94]. Then, the inhibitory activity of flavonoids in various enzymatic and nonenzymatic superoxide-producing systems has been studied. It was found that flavonoids may inhibit superoxide production by xanthine oxidase by both the scavenging of superoxide and the inhibition of enzyme activity, with the ratio of these two mechanisms depending on the structures of flavonoids (Table 29.4). As seen from Table 29.4, the data obtained by different authors may significantly differ. For example, in recent work [107] it was found that rutin was ineffective in the inhibition of xanthine oxidase that contradicts the previous results [108,109]. The origins of such big differences are unknown.

In addition to xanthine oxidase, flavonoids are able to inhibit the activity of a wide range of enzymes. These inhibitory effects of flavonoids may depend both on their free radical scavenging and chelating properties. Thus, it has been shown that flavonoids inhibit

TABLE 29.2
Rate Constants for the Reaction of Hydroxyl Radical with Flavonoids and Related Compounds (Pulse-Radiolytic Experiments)

Compound	$k_{14} \times 10^9$ l mol^{-1} s^{-1}	Ref.
Baicalin	77	[103][a]
(+)-Catechin	2.2	[104]
(−)-Epicatechin	1.0	[104]
Pycnogenol	1.8	[104]
(−)-Epigallocatechin	4.7	[104]
(−)-Epicatechin gallate	5.8	[104]
(−)-Epigallocatechin gallate	7.1	[104]
Propylgallate	3.1	[104]
β-Glucogallin	4.4–16	[104]
Pentagalloyl glucose	37–71	[104]

[a]Competition experiments with DMPO.

TABLE 29.3
Rate Constants for Reaction of Superoxide with Flavonoids and Related Compounds (Pulse-Radiolysis Experiments)

Compound	$k_{16} \times 10^4$ (l mol^{-1} s^{-1})	Ref.
Quercetin	4.7	[105]
	17	[106]
Rutin	5.1	[105]
	5.0	[106]
Hesperetin	0.59	[105]
Hesperidin	2.8	[105]
Kaempferol	0.24	[105]
Galangin	0.088	[105]
Fisetin	1.3	[105]
Catechin	1.8	[105]
(+)-Catechin	6.4	[104]
(−)-Epicatechin	6.8	[104]
Pycnogenol	43	[104]
(−)-Epigallocatechin	41	[104]
(−)-Epicatechin gallate	43	[104]
(−)-Epigallocatechin gallate	65	[104]
Propylgallate	26	[104]
β-Glucogallin	65	[104]
Pentagalloyl glucose	103	[104]
Biacalin	320	[103][a]
Dihydroquercetin	15	[106]

[a]Competition experiments with DMPO.

cyclooxygenase [95], lipoxygenase [96,97], microsomal monooxygenase [98], and glutathione S-transferase [99]. Beyeler et al. [98] proposed that the inhibition by flavonones and (+)-cyanidanol of microsomal monooxygenase activity was a consequence of the formation of a complex with cytochrome P450 via ligand binding. Hodnick et al. [100] showed that many flavonoids having a catechol moiety in the ring B inhibited mitochondrial succinoxidase and HADN oxidase activities. It was suggested that the primary inhibition site was complex I

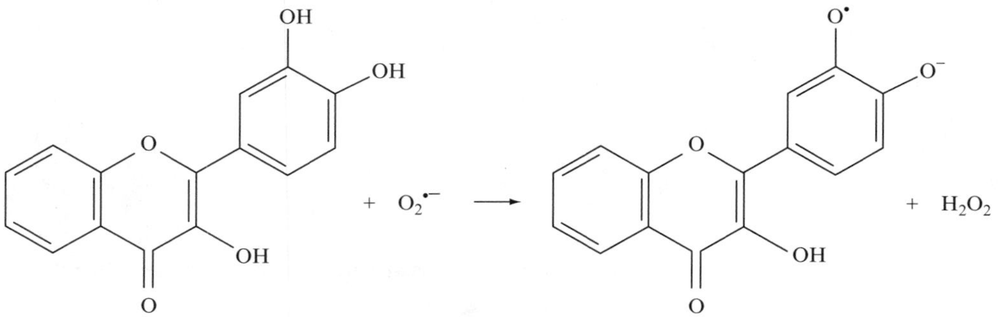

FIGURE 29.6 Mechanism of reaction of superoxide with flavonoids having two *o*-hydroxyl substituents.

TABLE 29.4
I_{50} Values (μmol l^{-1}) of Flavonoids for the Inhibition of Xanthine Oxidase and Scavenging Superoxide Ion

No	Flavonoid	I_{50} for Xanthine Oxidase Inhibition	I_{50} for Superoxide Scavenging	Ref.
1	($\pm$) Taxifolin	>100	1.73 $\pm$ 0.12	[108]
2	(+)-Catechin	>100	1.61 $\pm$ 0.04; 46.0 $\pm$ 5.0	[108,110]
3	(−)-Epicatechin	>100	1.59 $\pm$ 0.08; 11.8 $\pm$ 1.0	[108,110]
4	(−)-Epigallocatechin	>100	0.48 $\pm$ 0.02	[108]
5	4′-hydroxyflavanone	>30	>18	[108]
6	Naringenin	>50	>50	[108]
7	7-hydroxyflavanone	38 $\pm$ 7	>100	[108]
8	Chrysin	0.84 $\pm$ 0.13	1.87 $\pm$ 0.21	[108]
			73.2 $\pm$ 2.3	[110]
9	Apigenin	0.70 $\pm$ 0.23	1.33 $\pm$ 0.04	[108]
10	Luteolin	0.55 $\pm$ 0.04	1.13 $\pm$ 0.16	[108]
11	Baicalein	2.79 $\pm$ 0.01	2.72 $\pm$ 0.02	[108]
			21.2 $\pm$ 4.0	[110]
		3.12	370.33	[111]
12	3-Hydroxyflavone	>100	>100	[108]
			>300	[110]
13	Galangin	1.80 $\pm$ 0.07	6.74 $\pm$ 0.32	[108]
			151.8 $\pm$ 41.9	[110]
14	Kaempferol	1.06 $\pm$ 0.03	0.84 $\pm$ 0.04	[108]
			24.5 $\pm$ 4.6	[110]
15	Quercetin	2.62 $\pm$ 0.13	1.63 $\pm$ 0.02	[108]
		10.1 $\pm$ 1.1		
			42.3 $\pm$ 1.8[a)]	[109]
			12.5 $\pm$ 0.9 [b)]	[109]
			51.8 $\pm$ 2.8	[110]
16	Rutin	52.2 $\pm$ 0.6	10.6 $\pm$ 1.6	[108]
		37.8 $\pm$ 0.9		
			42.7 $\pm$ 1.4[a]	[109]
			15.4 $\pm$ 0.6[b]	[109]
			20.4 $\pm$ 1.5	[110]
17	Fisetin	4.33 $\pm$ 0.19	1.84 $\pm$ 0.07	[108]
			12.2 $\pm$ 3.3	[110]
18	Morin	10.1 $\pm$ 0.70	9.1 $\pm$ 0.08	[108]
			188.8 $\pm$ 27.8	[110]
19	Myricetin	2.38 $\pm$ 0.13	0.33 $\pm$ 0.03	[108]
		16.5 $\pm$ 1.3		
			20.1 $\pm$ 1.1[a]	[109]
			14.3 $\pm$ 0.8[b]	[109]
			12.5 $\pm$ 0.5	[110]
20	Baicalin	215.19	224.12	[111]
21	Wogonin	157.38	300.1	[111]

[a]Measured on the basis of NBT reduction by xanthine oxidase.
[b]Generation of superoxide by the NADH + phenazine methosulfate system was measured on the basis of NBT reduction.

(NADH–ubiquinone reductase) and the second one was complex II (succinate–ubiquinone reductase). Chiesi and Schwaller [101] found that quercetin and tannin inhibited neuronal constitutive endothelial NO synthase.

It should be noted that the number and positions of hydroxyl groups and the size of conjugated system are the important factors of flavonoid antioxidant and prooxidant activity. While hydroxyl radicals react with flavonoids indiscriminately, the most reactive compounds, capable of reacting with peroxyl and especially superoxide, are having two o-hydroxyl groups in the B-ring. (For superoxide it supports the proposal for a concerted mechanism.) On the other hand, the occurrence of three hydroxyl groups in B-ring leads to prooxidant activity in some flavonoids.

Can flavonoids react with nitric oxide, another physiological free radical? This is an important question due to the participation of NO in many pathophysiological processes. It has been suggested that nitric oxide reacts with phenols including α-tocopherol by abstracting hydrogen atom and producing phenoxyl radicals [112]. Acker et al. [113] studied the reaction of NO (gas) with several flavonoids including rutin, for which the rate constant was about $10 \, l \, mol^{-1} \, s^{-1}$. The mechanism of this reaction is unknown. Kim et al. [114] investigated the effects of flavonoids on NO production by LPS-stimulated macrophages. They have found that flavonoids apigenin, luteolin, tectorigenin, and quercetin inhibited nitric oxide production by these cells, but the most probable mechanism of flavonoid activity is the suppression of induced NO synthase and not the interaction with NO. Flavonoids catechin, epicatechin, and taxifolin also inhibited NO production by interferon-γ-stimulated macrophages [115]. In this work it has been suggested that the inhibitory effects of flavonoids might be mediated by the combination of NO scavenging and the inhibition of induced NO synthase activity or iNOS mPNA expression.

29.3.2 Protection Against Free Radical-Mediated Damage

The ability of flavonoids to scavenge superoxide is one of major mechanisms of their protective activity against cellular free radical-mediated damage. As far back as 1980, Berton et al. [116] found that quercetin suppressed the concanavalin A-induced activation of PMNs. Later on, the suppression of cellular damage by flavonoids has been studied in numerous papers. Thus it has been shown that quercetin, kaempferol, catechin, and taxifolin reduced cytotoxicity of superoxide and hydrogen peroxide in Chinese hamster lung fibroblast V79 cells [117]. Quercetin, luteolin, and 5,7,3′,4′-tetrahydroxy-3-methoxy flavone suppressed fMLP- and PMA-stimulated superoxide production by human neutrophils [118]. Ishige et al. [119] studied the effects of flavonoids on neuronal cells in model system (the mouse hippocampal cell line HT-22), in which exogenous glutamate was applied to inhibit cystine uptake and deplete intracellular glutathione that stimulated oxygen radical production and ultimately caused neuronal death. It was found that glutamate toxicity was suppressed only by flavonoids with the hydroxylated C-3 carbon atom and unsaturated C-ring.

Dietary flavonoids epicatechin and kaempferol protected neurons against the oxidized LDL-induced apoptosis involved c-Jun N-terminal kinase (JNK), c-Jun, and captase-3 [120]. Flavonoid silibinin (used for the treatment of liver disease) was an effective scavenger of hypochlorite ($I_{50} = 7 \, \mu mol \, l^{-1}$) but not superoxide ($I_{50} > 200 \, \mu mol \, l^{-1}$) [121]. Apparently through scavenging HOCl, silibinin inhibited the formation of leukotrienes LTB4 and LTC4/D4/F4 produced by human granulocytes (which are especially important in inflammatory reactions), PGE$_2$ formation by human monocytes, TXB$_2$ formation by human platelets, and 6-K-PGf1-α formation by human endothelial cells. It has been concluded that inflammation in humans can be suppressed by the administration of usual clinical doses of silibinin.

Flavonoid baicalein, which is believed to be one of the most important components of Japanese Kampo (traditional herbal) medicine, was found to be an effective scavenger of superoxide and hydroxyl radicals and the inhibitor of iron-induced in vivo lipid peroxidation in gerbils [122].

The effects of flavonoids on in vitro and in vivo lipid peroxidation have been thoroughly studied [123]. Torel et al. [124] found that the inhibitory effects of flavonoids on autoxidation of linoleic acid increased in the order fustin < catechin < quercetin < rutin = luteolin < kaempferol < morin. Robak and Gryglewski [109] determined I_{50} values for the inhibition of ascorbate-stimulated lipid peroxidation of boiled rat liver microsomes. All the flavonoids studied were very effective inhibitors of lipid peroxidation in model system, with I_{50} values changing from 1.4 μmol l^{-1} for myricetin to 71.9 μmol l^{-1} for rutin. However, as seen below, these I_{50} values differed significantly from those determined in other in vitro systems. Terao et al. [125] described the protective effect of epicatechin, epicatechin gallate, and quercetin on lipid peroxidation of phospholipid bilayers.

Numerous studies were dedicated to the effects of flavonoids on microsomal and mitochondrial lipid peroxidation. Kaempferol, quercetin, 7,8-dihydroxyflavone and D-catechin inhibited lipid peroxidation of light mitochondrial fraction from the rat liver initiated by the xanthine oxidase system [126]. Catechin, rutin, and naringin inhibited microsomal lipid peroxidation, xanthine oxidase activity, and DNA cleavage [127]. Myricetin inhibited ferric nitrilotriacetate-induced DNA oxidation and lipid peroxidation in primary rat hepatocyte cultures and activated DNA repair process [128].

29.3.3 COMPARISON OF FREE RADICAL SCAVENGING AND CHELATING ACTIVITIES

Free radical scavenging and chelating activities of quercetin and rutin have been studied [59], comparing the effects of these flavonoids on liposomal and microsomal lipid peroxidation with that of classic antioxidant 3,5-di-*tert*-butyl-4-hydroxytoluene (BHT), which has no chelating activity. Both flavonoids inhibited lipid peroxidation of lecithin liposomes and NADPH- and carbon tetrachloride-dependent peroxidation of rat liver microsomes. However, the efficiency of flavonoids (especially rutin) was much higher in NADPH-stimulate than in carbon tetrachloride-stimulated peroxidation, in contrast to the absence of any difference in the inhibitory effects of BHT (Table 29.5). It is known that microsomal NADPH-dependent lipid peroxidation is catalyzed by NADPH cytochrome P-450 reductase in the presence iron ions. On the other hand, carbon tetrachloride is oxidized to the $CCl_3\cdot$ radical by cytochrome P-450 in the absence of iron ions. Therefore, the findings cited in Table 29.5 suggest that rutin and quercetin inhibit the CCl_4-dependent microsomal lipid peroxidation by scavenging mechanism and NADPH-dependent peroxidation by chelating mechanism.

TABLE 29.5
I_{50} Values (μmol l^{-1}) of Flavonoids and BHT for the Inhibition of NADPH-Dependent- (I) and CCl_4-Dependent (II) Microsomal Lipid Peroxidation

	I_{50} for (I)	I_{50} for (II)
Quercetin	4.5	6.0
Rutin	16	116
BHT	1.25	1.3

Adapted from: IB Afanas'ev, AI Dorozhko, AV Brodskii, VA Kostyuk, AI Potapovitch. *Biochem Pharmacol* 38: 1763–1769, 1989.

Kozlov et al. [129] studied the mechanism of inhibitory action of several chelators including rutin, EDTA, phenanthrolone, and ADP on lipid peroxidation in rat brain homogenates. It was found that the inhibitory effects of these compounds well correlated with the rates of ferrous ion oxidation in solution and brain homogenates. These findings confirm the proposal that rutin inhibits iron-dependent lipid peroxidation by chelating mechanism forming Fe^{2+}–rutin complex and oxidizing an active ferrous ion into inactive ferric ion inside the complex. Chelating mechanism of the inhibitory activity of flavonoids was also suggested in the experiments with iron-loaded rat hepatocytes [130]. The ability of three flavonoids catechin, quercetin, and diosmetin to remove iron from iron-loaded hepatocytes correlated well with the inhibition of lipid peroxidation and intracellular enzyme release from hepatocytes. In the subsequent work [131] these authors showed that flavonoid myricetin inhibited lipid peroxidation in iron-treated rat hepatocytes forming, as an intermediate, relatively stable phenoxyl radical.

Ex vivo study of iron-overloaded rats [132] confirmed the suppression of iron-stimulated lipid peroxidation with rutin. Iron-overloaded rats were administrated interperitoneal injections of rutin solution during 10 days, and the levels of nonheme iron and MDA in liver microsomes were determined (Table 29.6). There are two interesting findings in these experiments. The first one is that rutin does not affect lipid peroxidation in normal animals (control group) but sharply diminished the MDA content in IOL rats. The second one is that despite the inhibition of lipid peroxidation in IOL rats, rutin did not change the level of nonheme iron. Thus, in these ex vivo animal experiments rutin inhibited lipid peroxidation mainly by chelating mechanism forming the inactive iron–rutin complexes and not by removing active iron from rats. It should be noted that Acker et al. [133] was unable to find difference in the inhibitory effects of flavonoids on iron-dependent and iron-independent lipid peroxidation.

The above findings are supported in the other studies of the inhibitory effects of flavonoids on iron-stimulated lipid peroxidation. Quercetin was found to be an inhibitor of iron-stimulated hepatic microsomal lipid peroxidation ($I_{50} = 200\, \mu mol\, l^{-1}$) [134]. Flavonoids eriodictyol, luteolin, quercetin, and taxifolin inhibited ascorbate and ferrous ion-stimulated MDA formation and oxidative stress (measured by fluorescence of 2′,7′-dichlorodihydrofluorescein) in cultured retinal cells [135]. It should be mentioned that in recent work Heijnen et al. [136] revised the structure–activity relationship for the protective effects of flavonoids against lipid peroxidation.

It must be noted that the inhibitory effects of flavonoids and other antioxidants in nonhomogenous biological systems can depend not only on their reactivities in reactions with free radicals (the chain-breaking activities) but also on the interaction with biomembranes. Thus, Saija et al. [137] compared the antioxidant effects and the interaction with biomembranes of four flavonoids quercetin, hesperetin, naringen, and rutin in iron-induced

TABLE 29.6
Effects of Rutin on the Nonheme Iron and MDA Levels in Liver Microsomes of Iron Overloaded (IOL) Rats [132]

Groups of Rats	Nonheme Iron (nmol mg protein^{-1})	MDA (nmol mg protein^{-1})
Control group	18.4 ± 5.1	3.0 ± 0.9
Control + rutin	19.6 ± 4.0	4.3 ± 2.7
IOL group	60.1 ± 10.4	4.9 ± 0.1
IOL + rutin	67.0 ± 1.5	1.2 ± 0.3

linoleate peroxidation and in the autoxidation of rat cerebral membranes [137]. These authors found that quercetin showed the deepest interaction with biomembranes supposedly due to planar conformation while rutin was unable to penetrate lipid membrane due to low liposolubility ($\log k' = -0.115$). Owing to this, rutin exhibited decreased antioxidant activity in the autoxidation of rat cerebral membranes (where free radicals were generated within membranes) in comparison with the peroxidation of linoleate in solution. However, the findings cited in Table 29.7 seem to contradict that conclusion because rutin remains a much more effective antioxidant than lipid-soluble hesperetin and naringenin, and therefore, free radical scavenging activity probably continues to be the most important factor of antioxidant activity of flavonoids.

29.3.4 Inhibition of Free Radical-Mediated Damage in Cells

Several studies discuss the inhibitory effect of flavonoids on free radical formation in erythrocytes. Maridonneau-Parini et al. [138] developed experimental model for free radical-mediated damage in erythrocytes based on the incubation of cells with phenazine methosulfate (PMS, an intracellular generator of oxygen radicals) and diethyldithiocarbamate (DDC, SOD inhibitor). In this system the enhanced free radical damage was characterized by an increase in lipid peroxidation and passive potassium permeability. Although all the flavonoids studied decreased free radical damage in erythrocytes (measured via the reduction of nitroblue tetrazolium (NBT)), with morin and rutin as the most efficient inhibitors, not all of them suppressed the oxygen radical-stimulated potassium permeability. Furthermore, flavonoids affected differently membrane lipid peroxidation: for example, kaempferol inhibited peroxidation, rutin had no effect, and myricetin exhibited prooxidant activity. (The activity of myricetin is not surprising because, as mentioned above, flavonoids with three hydroxyl groups in the B ring such as myricetin are easily oxidized to produce oxygen radicals.)

Quercetin and rutin suppressed photosensitized hemolysis of human erythrocytes with I_{50} values equal to 40 μmol l^{-1} and 150 μmol l^{-1}, respectively [139]. Suppression of photohemolysis was accompanied by inhibition of lipid peroxidation. Morin inhibited oxygen radical-mediated damage induced by superoxide or peroxyl radicals to the human cells in the cardiovascular system, erythrocytes, ventricular myocytes, and saphenous vein endothelial cells [140]. Rutin protected against hemoglobin oxidation inside erythrocytes stimulated by prooxidant primaquine [141].

It is known that the toxic effects of some solid particles and fibers (latex, zeolite, asbestos fibers, etc.) depend on the stimulation of free radical production. Therefore, the inhibitory effects of flavonoids on particle-mediated damaging processes might be expected. Thus, rutin

TABLE 29.7
I_{50} Values (μmol l^{-1}) of Flavonoids for Iron-Induced Linoleate Peroxidation (I) and Autoxidation of Rat Cerebral Membranes (II) and Log Capacity factor log k'

Flavonoids	I_{50} for (I)	I_{50} for (II)	log k'
Quercetin	28.61	3.09	0.510
Hesperetin	17.20	148.27	0.530
Naringenin	565	321.8	0.458
Rutin	6.77	27.50	−0.115

Adapted from: A Saija, M Scalese, M Lanza, D Marzullo, F Bonina, F Castelli. *Free Radic Biol Med* 19: 481–486, 1995.

(and iron–rutin complex, see below) inhibited particle-induced free radical-mediated cellular processes [142]. Korkina et al. [61] studied the effect of rutin on oxygen radical production by peritoneal macrophages stimulated with asbestos chrysotile fibers and zeolite particles. Rutin very efficiently inhibited luminol-amplified CL produced by these cells with I_{50} values equal to about 0.5 µmol l^{-1} compared to ascorbic acid with I_{50} equal to about 50 µmol l^{-1}. Rutin also suppressed cytogenetic effects of latex and zeolite particles and chrysotile fibers on cultured lymphocytes [61]. Rutin and quercetin inhibited superoxide production and free radical-mediated processes such as the formation of TBAR products and LDL release induced by asbestos-stimulated macrophages [143]. Later on, it has been suggested [106] that rutin and quercetin protect erythrocytes from asbestos-dependent free radical-mediated damage owing to their chelating properties, attaching to the asbestos surface on the sites responsible for free radical generation. Dihydroquercetin was inactive in this process. Asbestos-induced injury of peritoneal macrophages and erythrocytes was also effectively inhibited by green tea extract and its major flavonoid components (−)-epicatechin gallate and (−)-epigallocatechin gallate [144].

29.3.5 INHIBITION LDL OXIDATION AND ENZYMATIC LIPID PEROXIDATION

Flavonoids exhibit protective action against LDL oxidation. It has been shown [145] that the pretreatment of macrophages and endothelial cells with tea flavonoids such as theaflavin digallate diminished cell-mediated LDL oxidation probably due to the interaction with superoxide and the chelation of iron ions. Quercetin and epicatechin inhibited LDL oxidation catalyzed by mammalian 15-lipoxygenase, and are much more effective antioxidants than ascorbic acid and α-tocopherol [146]. Luteolin, rutin, quercetin, and catechin suppressed copper-stimulated LDL oxidation and protected endogenous urate from oxidative degradation [147]. Quercetin was also able to suppress peroxynitrite-induced oxidative modification of LDL [148].

As already noted, the effects of flavonoids on enzymatic lipoxygenase- and cyclooxygenase-catalyzed lipid peroxidation are quite complicated. Robak et al. [149] has shown that many flavonoids stimulated cyclooxygenase, although some of them are the inhibitors of soybean lipoxygenase. Ratty et al. [97] found that all flavonoids studied (quercetin, quercetrin, rutin, myricetin, morin, and others) inhibited to some degree soybean lipoxygenase activity. Quercetin was the most potent inhibitor of lipoxygenase activity in liposomal suspension (about 42%) while other flavonoids inhibited the enzyme by about 14–23%.

29.3.6 OTHER EXAMPLES OF PROTECTIVE ACTIVITY OF FLAVONOIDS AGAINST FREE RADICAL-MEDIATED DAMAGE IN BIOLOGICAL SYSTEMS

In addition to already considered inhibitory effects of flavonoids on free radical-mediated damage in cell-free and cellular systems, flavonoids showed protective activity against many other damaging processes in the biological systems under oxidative stress. Thus quercetin administrated intraperitoneally to rats before irradiation with UVA light decreased MDA level significantly and enhanced SOD and catalase activities in the liver [150]. *Ginkgo biloba* extract containing 33% flavone glycosides, mostly quercetin and kaempferol derivatives, significantly inhibited cutaneous blood flux (the indicator of skin inflammatory level) [151]. Another flavonoid, with good efficacy in the treatment against UV-radiation-induced oxidative stress in the skin, is the green tea constituent (−)-epigallocatechin-3-gallate [152]. It was found that the application of this flavonoid before UV exposure markedly decreased hydrogen peroxide and nitric oxide production in both epidermis and dermis. (−)-Epigallocatechin-3-gallate pretreatment also suppressed UV-induced infiltration of inflammatory leukocytes into the skin, which are the major producers of oxygen radicals, and inhibited UV-induced epidermal lipid peroxidation. Morin hydrate minimized free radical-mediated damage to cardiovascular cells by anticancer drugs [153].

Feeding rabbits with the citrus flavonoid naringin, a potent cholesterol-lowering agent, did not affect plasma and hepatic lipid peroxidation but significantly increased the level of plasma vitamin E, enhanced SOD and catalase activities, and upregulated the gene expressions of SOD, catalase, and GSH-peroxidase [154]. Pretreatment with quercetin reduced MDA and nitric oxide levels in the brain of rats subjected to endotoxin-induced shock [155]. Supplementation of a rat diet with 4(G)-α-glucopyranosylrutin (G-rutin) significantly decreased urinary excretion of thymine glycol and thymidine glycol, the indices of DNA base damage in the whole body, and the protein carbonyl contents, the characteristics of protein oxidation [156].

The application of flavonoids for the treatment of various diseases associated with free radical overproduction is considered in Chapter 29. However, it seems useful to discuss here some studies describing the activity of flavonoids under certain pathophysiological conditions. Oral pretreatment with rutin of rats, in which gastric lesions were induced by the administration of 100% ethanol, resulted in the reduction of the area of gastric lesions [157]. Rutin was found to be an effective inhibitor of TBAR products in the gastric mucosa induced by 50% ethanol [158]. Rutin and quercetin were active in the reduction of azoxymethanol-induced colonic neoplasma and focal area of dysplasia in the mice [159]. Chemopreventive effects of quercetin and rutin were also shown in normal and azoxymethane-treated mouse colon [160]. Flavonoids exhibited radioprotective effect on γ-ray irradiated mice [161], which was correlated with their antioxidative activity. Dietary flavones and flavonols protected against the toxicity of the environmental contaminant dioxin [162]. Rutin inhibited ovariectomy-induced osteopenia in rats [163].

29.3.7 Antioxidant Effect of Metal–Flavonoid Complexes

Although the above data show the efficiency of flavonoids as antioxidants, free radical scavengers, and chelators in suppression of free radical-mediated damage in biological systems, it is desirable to enhance their antioxidant potential without radical altering their structures and losing nontoxic properties of natural products. We suggested that this can be achieved by the use of transition metal–flavonoid complexes, in which the flavonoid molecule acquires an additional superoxide-dismuting center without the formation of new covalent bonds. It has been known for a long time that polyphenols form stable complexes with many two- and three-valence metal ions (see, for example, Ref. [164]). Although the structure of metal–flavonoid complexes is not well determined, it is usually accepted that they have the 2:1 or even 3:1 flavonoid–metal ion stoichiometry; however, in real diluted biological solutions a 1:1 complex might be of importance.

In 1989, we showed [142] that the $Fe^{2+}(rutin)_2$ complex is a more effective inhibitor than rutin of asbestos-induced erythrocyte hemolysis and asbestos-stimulated oxygen radical production by rat peritoneal macrophages. Later on, to evaluate the mechanisms of antioxidant activities of iron–rutin and copper–rutin complexes, we compared the effects of these complexes on iron-dependent liposomal and microsomal lipid peroxidation [165]. It was found that the iron–rutin complex was by two to three times a more efficient inhibitor of liposomal peroxidation than the copper–rutin complex, while the opposite tendency was observed in NADPH-dependent microsomal peroxidation. On the other hand, the copper–rutin complex was much more effective than the iron–rutin complex in the suppression of microsomal superoxide production, indicating that the copper–rutin complex indeed acquired additional SOD-dismuting activity because superoxide is an initiator of NADPH-dependent microsomal lipid peroxidation (Chapter 23). The participation of superoxide in liposomal peroxidation where there are no superoxide producers is virtually impossible; as a result, the iron–rutin complex exhibited a higher scavenging activity in this peroxidative process than its copper analog.

Superoxide-dismuting activity of copper–rutin complex was confirmed by comparison of the inhibitory effects of this complex and rutin on superoxide production by xanthine oxidase and microsomes (measured via cytochrome c reduction and by lucigenin-amplified CL, respectively) with their effects on microsomal lipid peroxidation [166]. An excellent correlation between the inhibitory effects of both compounds on superoxide production and the formation of TBAR products was found, but at the same time the effect of copper–rutin complex was five to nine times higher due to its additional superoxide dismuting capacity.

Enhanced antioxidant activity of metal–rutin complexes has been confirmed in subsequent in vitro and ex vivo studies [167]. In accord with previous findings iron–rutin and copper–rutin complexes were more effective inhibitors than rutin of superoxide production by xanthine oxidase, microsomal lipid peroxidation, and lucigenin-amplified CL produced by microsomes (Table 29.8). (Only exception was a weak effect of iron–rutin complex on microsomal peroxidation probably due to partial dissociation of the complex.) Iron–rutin and copper–rutin complexes also efficiently inhibited oxygen radical production by zymosan-stimulated rat peritoneal macrophages in the presence and absence of prooxidant antibiotic bleomycin, where the copper–rutin complex was always the most effective inhibitor. These in vitro data were compared with the ex vivo findings obtained in the experiments with bleomycin-treated rats. This anticancer antibiotic possesses strong inflammatory and fibrotic activity, which, at least in part, is derived from the stimulation of free radical production. It was found that the administration of copper–rutin complex to bleomycin-treated rats was the most effective in the suppression of inflammatory and fibrotic processes, sharply increasing the macrophage–neutrophil ratio, decreasing the wet ling weight, lowering the content of total protein, and decreasing hydroxyproline concentration. Similar to copper–rutin complex, iron–rutin complex was quite effective in suppression of lung edema [167]. Thus, both metal–rutin complexes studied turn out to be the effective suppressors of free radical-mediated pathophysiological processes that makes them promising potential pharmaceutical agents.

Iron-, copper-, and zinc complexes of rutin, dihydroquercetin, and green tea epicatechins were found to be much more efficient inhibitors than parent flavonoids of toxic effects of chrysotile asbestos fibers on peritoneal macrophages and erythrocytes [168]. It was proposed that in this case the enhanced activity of metal–flavonoid complexes was increased by the absorption on chrysotile fibers.

O'Brien and coworkers [169,170] found that iron complexes of flavonoids and catechols were much more effective than the noncomplexed parent compounds at preventing the

TABLE 29.8
I_{50} Values (μmol l^{-1}) for Inhibitory Effects of Metal–Rutin Complexes and Rutin on Cytochrome c Reduction by Xanthine Oxidase (I), Iron-Catalyzed Microsomal Lipid Peroxidation (II), and Lucigenin-Amplified Microsomal CL (III) [167]

	(I)	(II)	(III)
Rutin	35 ± 0.3	90 ± 10	20 ± 3
Fe(rut)Cl$_3$	22 ± 2	>500	23 ± 3
Cu(rut)Cl$_2$	2.5 ± 0.2	10 ± 3	3 ± 1

hypoxic injury of hepatocytes and suppressing hydrogen peroxide formation. These findings were also explained by the complexation of the compounds studied with iron ions that increased their SOD activity. These authors also suggested that both ferric–ferrous and hydroquinone–semiquinone redox cycling may be involved in SOD activity of iron–catechol complexes. In subsequent work [171] it has been shown that iron–catecholamine and iron–catechol complexes are more effective superoxide scavengers than noncomplexed catechols and catecholamines. Brown et al. [172] found that the flavonoids capable of complexing copper ions (quercetin, rutin, and luteolin) manifested enhanced activity in the inhibition of copper-induced LDL peroxidation. It was also found that the formed copper–flavonoid complexes had no prooxidant activities.

29.3.8 Comments on Prooxidant Activity of Flavonoids

As discussed above, many flavonoids are effective inhibitors of free radical-mediated processes due to their free radical scavenging and chelating properties. However, flavonoids are not a homogenous group of compounds with similar chemical properties, and as a result, some of them are able to exhibit prooxidant activity by reducing dioxygen to superoxide and other oxygen radicals. For example, Laughton et al. [173] found that hydroxyl radicals were formed during autoxidation of quercetin and myricetin. Canada et al. [174] confirmed the formation of hydroxyl radicals by flavonoid oxidation using DMPO as a spin-trapping agent.

The competition between antioxidant and prooxidant activity of flavonoids depends firstly on their chemical structure. If we suppose that the oxidation of flavonoids (Reaction (17)) takes place by one-electron transfer mechanism, then it must depend on the capacity of flavonoids to donate an electron, i.e., on their one-electron oxidation potentials.

$$\text{FlOH} + O_2 \Longrightarrow \text{FlO}^\bullet + O_2^{\bullet -} + H^+ \tag{17}$$

Bors et al. [175] determined the rate constants and equilibrium constants for the reactions of flavonoids with ascorbate (Reaction (18)) by a pulse-radiolysis method and on their basis calculated the one-electron oxidation potentials of flavonoids (Table 29.9).

$$\text{FlO}^- + \text{AH}^\bullet \Longleftrightarrow \text{FlO}^\bullet + \text{AH}^- \tag{18}$$

It is interesting that only dihydroquercetin has the redox potential much lower than that of ascorbate (282 mV) and kaempferol, luteolin, and fisetin have the values close to ascorbate.

TABLE 29.9
Rate Constants ($l\,mol^{-1}\,s^{-1}$) and Equilibrium Constants for the Reactions of Flavonoids with Ascorbate and One-Electron Oxidation Potentials (mV) of Flavonoids

Flavonoid	k_{18} (×10^5)	k_{-18} (×10^5)	$K_{eq} = k_{18}/k_{-18}$	E_1° (V)
Dihydroquercetin	120 ± 25	1.60 ± 0.40	74.4	83
Kaempferol	28.0 ± 14.0	52.0 ± 7.0	0.54	209
Luteolin	1.55 ± 0.35	99.0 ± 17.0	0.016	299
Fisetin	0.385 ± 0.045	0.865 ± 0.180	0.44	214
Quercetin	0.016 ± 0.003	47.5 ± 10.5	0.00033	398
Rutin	0.52 ± 0.015	12.5 ± 0.5	0.041	275

Adapted from: W Bors, C Michel, S Schikova. *Free Radic Biol Med* 19: 45–52, 1995.

Therefore, depending on their structures, flavonoids may reduce the ascorbate radical or oxidize ascorbate, exhibiting antioxidant or prooxidant properties in the systems containing flavonoids and ascorbate together.

Although no good quantitative correlation between redox potentials of flavonoids and their prooxidant activities still was not documented, a relationship between the prooxidant toxicity of flavonoids to HL-60 cells and redox potentials apparently takes place [176]. However, there is a simple characteristic of possible prooxidant activity of flavonoids, which increases with an increase in reactive hydroxyl groups in the B ring. From this point of view, the prooxidant activity of flavonoids should increase in the range: kaempferol < quercetin < myricetin (Figure 29.7). Thus, for many flavonoids the ratio of their antioxidant and prooxidant activities must depend on the competition between Reactions (14) and (15) and Reaction (17).

Possible prooxidant effect of catechins from green tea, (−)-epicatechin, (−)-epicatechin gallate, (−)-epigallocatechin, and (−)-epigallocatechin gallate may originate from their autoxidation, which is accompanied by the formation of superoxide and semiquinone free radicals [177]. Autoxidation of catechins was accelerated by cupric ions. Some studies suggest that transition metal ions are catalysts of flavonoid oxidation, but this process again strongly depends on the flavonoid structure. For example, flavonols myricetin, quercetin, fisetin, and kaempferol inhibit metal-induced lipid peroxidation, while flavones luteolin, apigenin, and chrysin were antioxidative at low iron concentrations but prooxidative at high iron concentrations [178]. Some other flavonoids such as naringenin also enhanced iron-dependent lipid peroxidation [179]. It is also possible that antioxidant and prooxidant effects of flavonoids depend on the nature of metal–flavonoid complexes. As shown above, "tight" iron– and copper–rutin complexes behave themselves as even more effective antioxidants than parent flavonoids. However, when metal–flavonoid complexes not so tightly bound, the flavonoid molecule is easily oxidized inside of the complex.

Another mode of prooxidant activity of flavonoids was recently observed by Galati et al. [180]. These authors have shown that flavone apigenin and flavanones naringin and naringenin are responsible for the extensive glutathione oxidation in the presence of peroxidase. It was proposed that the phenoxyl radicals formed during the oxidation of flavonoids react with GSH, producing superoxide and oxidized glutathione as final products:

$$FlO^{\bullet} + GSH \Longrightarrow FlOH + GS^{\bullet} \tag{19}$$

$$GS^{\bullet} + GS^{-} \Longrightarrow GSSG^{\bullet -} \tag{20}$$

$$GSSG^{\bullet -} + O_2 \Longrightarrow GSSG + O_2^{\bullet -} \tag{21}$$

Cytotoxic prooxidant effects of flavonoids can also be a consequence of their enzymatic oxidation. For example, it was found that quercetin was oxidized by lactate peroxide to form semiquinone and quinone [181].

29.4 PHENOLIC COMPOUNDS OTHER THAN FLAVONOIDS

There are numerous other polyphenolic compounds possessing in vitro and in vivo antioxidative activity. Several examples of these compounds are cited below. One of nonflavonoid polyphenols of particular interest is resveratrol (3,5,4'-trihydroxy-*trans*-stilbene, Figure 29.8), which has been identified as a potential cancer chemopreventive agent and an antimutagen [182]. It has been found that resveratrol is the efficient inhibitor of cyclooxygenase and the inhibitor of free radical-mediated cellular processes. For example, resveratrol is a better free radical scavenger than α-tocopherol or ascorbic acid but has nearly the same activity as

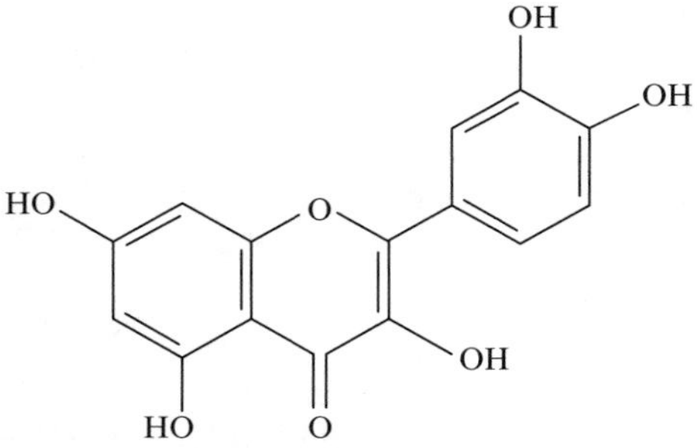

Quercetin

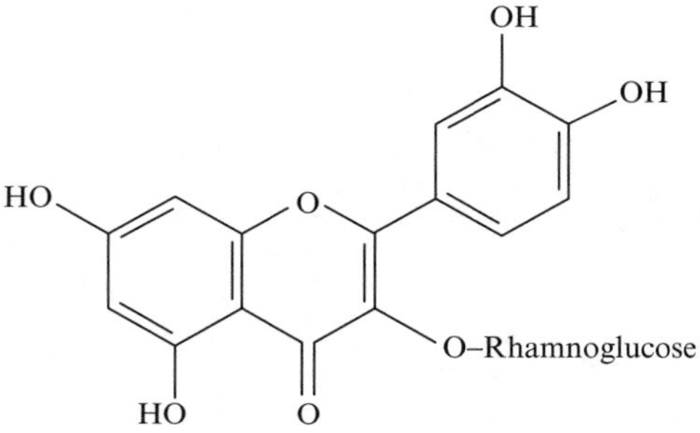

Rutin

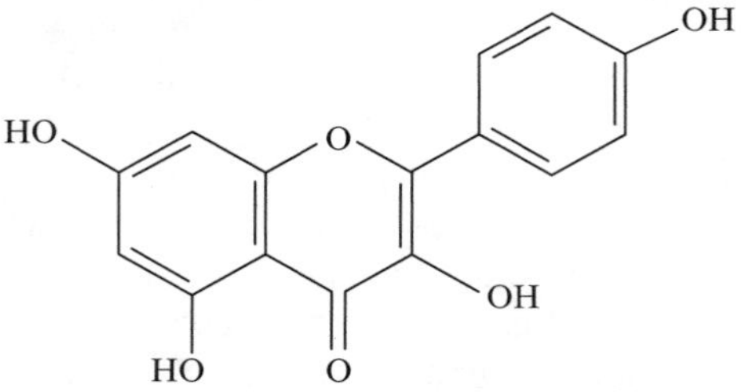

Kaempferol

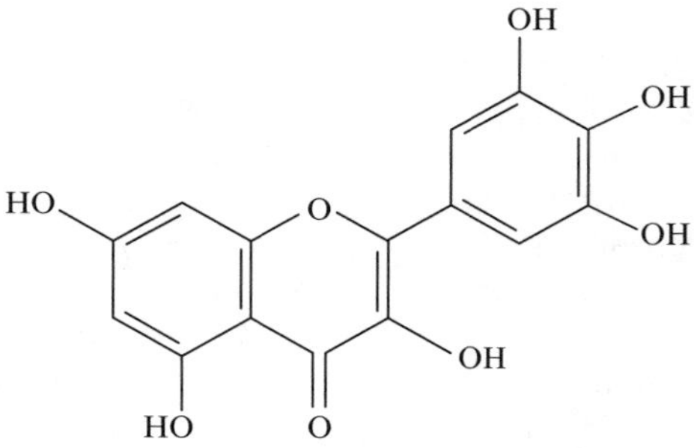

Myricetin

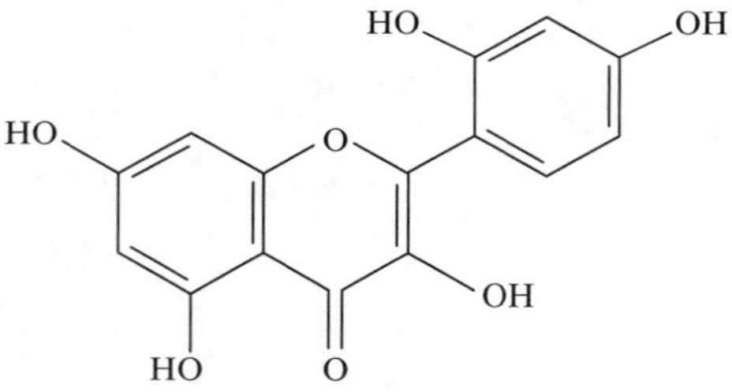

Morin

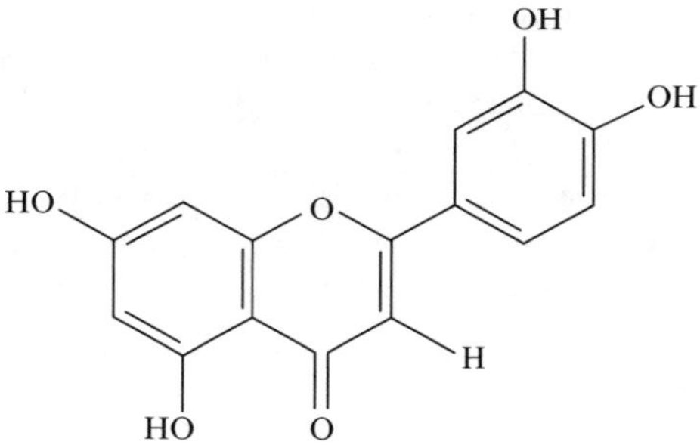

Luteolin

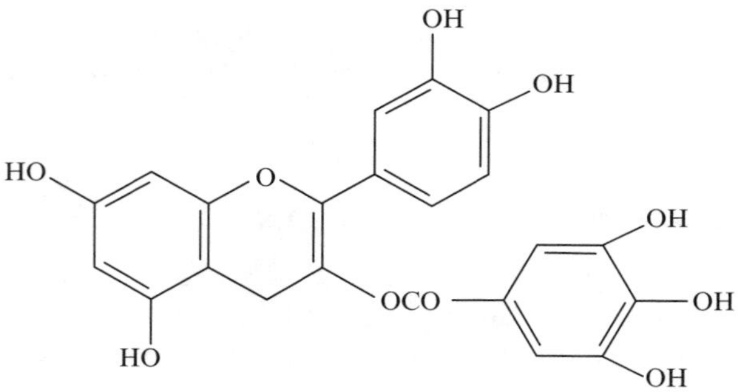

(−)-Epigallocatechin-3-gallate

FIGURE 29.7 Structures of flavonoids possessing different antioxidant and prooxidant properties.

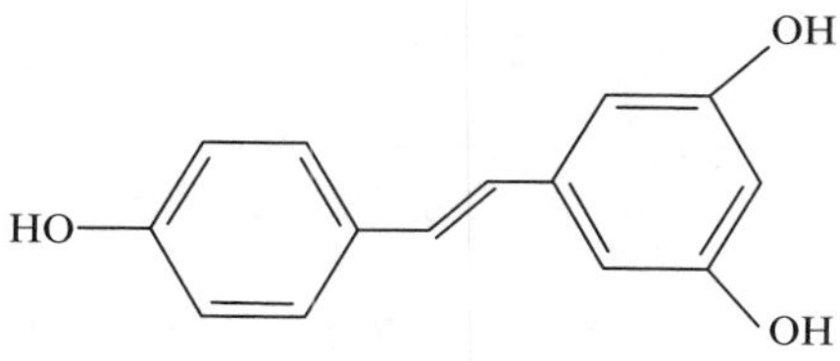

Resveratrol

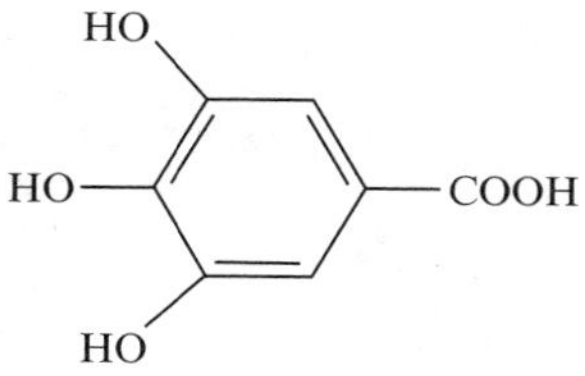

Gallic acid

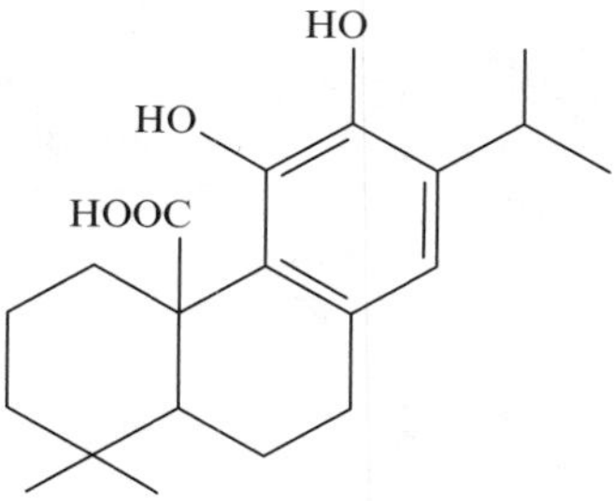

Carnosic acid

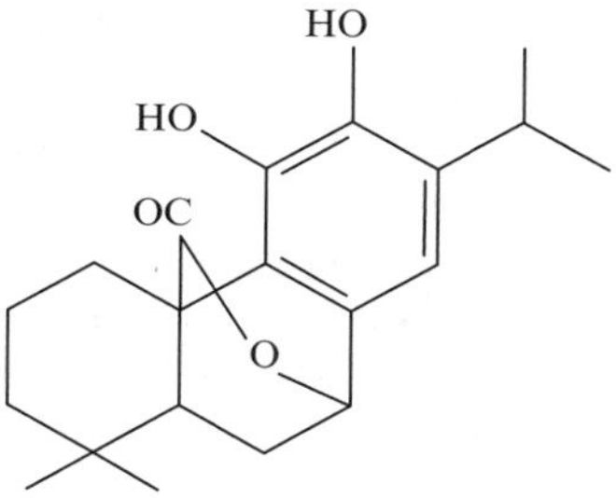

Carnosol

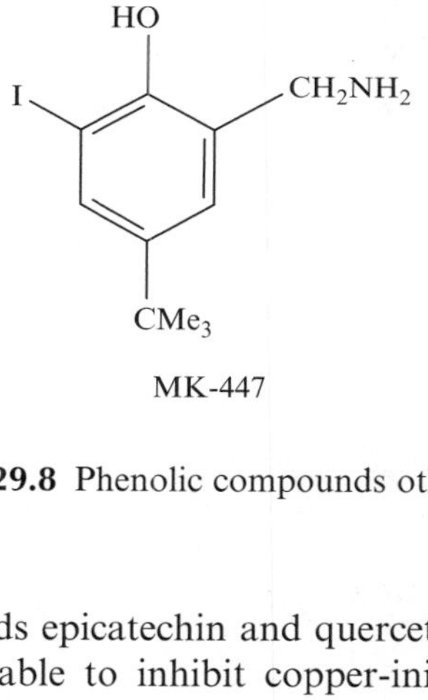

MK-447

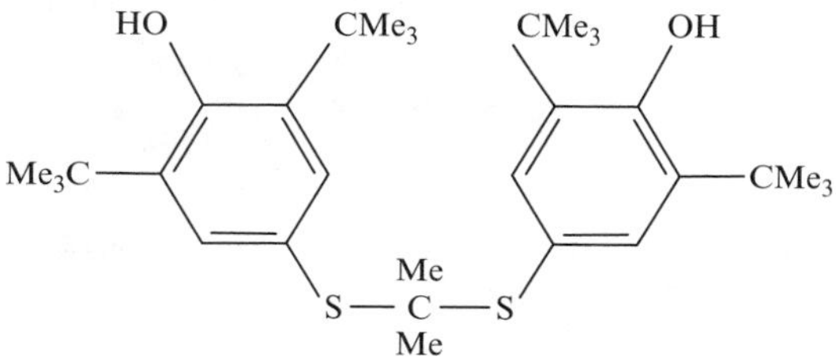

Probucol

FIGURE 29.8 Phenolic compounds other than flavonoids.

flavonoids epicatechin and quercetin [183]. In 1993, Frankel et al. [184] showed that resveratrol is able to inhibit copper-initiated LDL oxidation. Fremont et al. [185] found that resveratrol is even a more effective inhibitor of copper-induced LDL oxidation compared to flavonoids. Brito et al. [186] have shown that resveratrol inhibits peroxynitrite- and ferrylmyoglobin-stimulated LDL oxidation. In the latter case the reduction of ferrylmyoglobin into metmyoglobin was observed.

A comparison with its different derivatives shows that 4′-OH is not a sole reactive group responsible for the antioxidant activity of resveratrol, while the *trans*-conformation is absolutely necessary for the inhibition of cell proliferation [187]. However, similar to flavonoids resveratrol may exhibit prooxidant properties, for example to promote DNA fragmentation, although its prooxidant activity seems to be unimportant under physiological conditions [188].

There is the often-overlooked group of polyphenols containing the derivatives of hydroxycinnamic and hydroxybenzoic acids such as caffeic, chlorogenic, and gallic acids (Figure 29.8), which occur in food and exhibit certain antioxidant activity. For example, some caffeic acid esters isolated from propolis from honeybee hives showed antiinflammatory activity. Wu et al. [189] showed that propyl gallate, traditionally used as a preservative in food and fuel, is an effective hepatoprotector with antioxidative activity. Thus, propyl gallate protected hepatocytes and hepatic vascular endothelial cells against superoxide produced by xanthine oxidase or menadione and inhibited lipid peroxidation. Cos et al. [190] measured the antioxidant activities of many phenolic acid derivatives. It has been found that gallic, protocatechuic, caffeic, and chlorogenic acids are good inhibitors of microsomal lipid peroxidation with I_{50} values of 1.5–11.5 μmol l^{-1}. (Corresponding I_{50} values for quercetin and rutin in their experiments were 0.95 and 26.5 μmol l^{-1}.) These findings show that phenolic acids are good competitors of traditional flavonoids as natural antioxidants.

It has been shown [191] that flavonolignans (the derivatives of flavonoids) inhibited superoxide release by PMA-stimulated human leukocytes. Efficiency of flavonolignans such as silybin, isosilybin, and silychristin was comparable with that of α-tocopherol. Major polyphenolic components of rosemary extract carnosic acid and carnosol (Figure 29.8) rapidly reacted with trichloroperoxyl and hydroxyl radicals and inhibited liposomal and microsomal lipid peroxidation [192]. However, similar to some flavonoids these compounds exhibited prooxidant effect enhancing bleomycin-stimulated DNA damage. Gossypol, another polyphenolic plant pigment and male contraceptive efficiently inhibited myocardial phospholipid peroxidation at low concentrations [193].

Among many synthetic phenols with antioxidative activity, two compounds, namely, the nonsteroidal antiinflammatory drug 2-aminomethyl-4-*tert*-butyl-6-iodophenol (MK-447) and probucol are of certain interest (Figure 29.8). Cheesman and Forni [194] have shown that MK-447 inhibited rat liver microsomal lipid peroxidation, and therefore, its antiinflammatory function may be a consequence of its free radical scavenging activity. Probucol, a pharmacological agent now widely applied in heart therapy, has two phenolic hydroxyl groups and therefore, should be an effective antioxidant. Indeed, probucol is the effective inhibitor of LDL oxidation [195,196], although its inhibitory effect is independent of the reactions with superoxide and peroxyl radicals [197]. Probucol exhibits synergistic effect in combination with ascorbic acid (see below).

29.5 THIOLS

All reduced thiols contain very active SH groups and therefore, are potentially efficient antioxidants. There are numerous endogenous and synthetic thiols, which have already been studied and applied as antioxidative drugs and food supplements, but the most important antioxidant thiols are undoubtedly lipoic acid and glutathione.

29.5.1 LIPOIC ACID

Lipoic acid (the other names are α-lipoic acid or thioctic acid) (Figure 29.9) is a natural compound, which presents in most kinds of cells. Lipoic acid (LA) is contained in many food products, in particular in meat, but it is also synthesized in human organism from fatty acids. Earlier, it has been shown that in humans lipoic acid functions as a component of the pyruvate dehydrogenase complex. However, later on, attention has been drawn to the possible antioxidant activity of the reduced form of lipoic acid, dihydrolipoic acid (DHLA) (Figure 29.9).

Chemical, biological, and pharmacological properties of lipoic acid as well as its therapeutic effects in several diseases (diabetes mellitus, liver cirrhosis, polyneuritis, etc.) are reviewed [198,199]. It is evident from the chemical structures of LA and DHLA that only DHLA may be an efficient scavenger of all oxygen radicals, while LA should be active only in the reactions with highly reactive hydroxyl radicals. On the other hand, DHLA must be easily

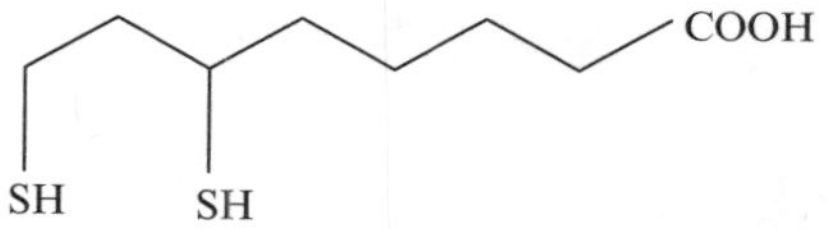

LA

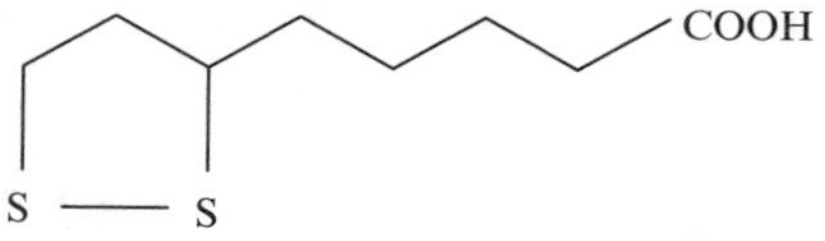

DHLA

γ-L-Glutamyl-L-cysteinyl-glycine
Glutathione (GSH)

$HSCH_2CH(NHAc)COOH$
N-Acetylcysteine (NAC)

FIGURE 29.9 Major thiolic antioxidants.

oxidized especially in the presence of transition metal ions. Antioxidant and prooxidant properties of LA and DHLA have been considered earlier [200].

In 1988 Bast and Haenen [201] reported that both LA and DHLA did not affect iron-stimulated microsomal lipid peroxidation. However, Scholich et al. [202] found that DHLA inhibited NADPH-stimulated microsomal lipid peroxidation in the presence of iron–ADP complex. Inhibitory effect was observed only in the presence of α-tocopherol, suggesting that some interaction takes place between these two antioxidants. Stimulatory and inhibitory effects of DHLA have also been shown in other transition metal-stimulated lipid peroxidation systems [203,204]. Later on, the ability of DHLA (but not LA) to react with water-soluble and lipid-soluble peroxyl radicals has been proven [205]. But it is possible that the double (stimulatory and inhibitory) effect of DHLA on lipid peroxidation originates from subsequent reactions of the DHLA free radical, capable of participating in new initiating processes.

Conflicting data were also received for the reactions of LA and DHLA with hydroxyl radicals and superoxide. Suzuki et al. [206] found that both LA and DHLA inhibited the formation of DMPO–OH adducts formed in the Fenton reaction. However, Scott et al. [207] concluded that only LA is a powerful scavenger of hydroxyl radicals while DHLA accelerated iron-catalyzed hydroxyl radical formation and lipid peroxidation.

Neither Suzuki et al. [206] nor Scott et al. [207] found any effect of LA on superoxide production by xanthine oxidase. Scott et al. also concluded that DHLA is incapable of reacting with superoxide. The last conclusion seems highly improbable. The ability of superoxide to react with thiols with the rate constants equal to 10^5 to $10^6\,l\,mol^{-1}\,s^{-1}$ has been shown in chemical studies [208]. Dikalov et al. [209] estimated the rate constant for the reaction of DHLA with superoxide as $(4.8 \pm 2) \times 10^5\,l\,mol^{-1}\,s^{-1}$ using the competition experiments with spin trap DMPO, which is very close to the previous value of $(7.3 \pm 0.24) \times 10^5\,l\,mol^{-1}\,s^{-1}$ reported for this reaction [210]. Negative results obtained by Scott et al. [207] are probably explained by the use of unreliable NBT assay for superoxide detection [211].

The recent work by Winterbourn and Metodiewa [211] demonstrated that the above values for the rate constant of reaction of DHLA with superoxide might be overestimated. These authors studied the reactions of superoxide with several thiols, glutathione, cysteine, cysteamine, penicillamine, *N*-acetylcysteine, dithiothreitol, and captopril and found that thiols reacted with superoxide by a chain mechanism with the regeneration of superoxide. They suggested that the rate constants for the reactions of thiols with superoxide could not be more than $10^3\,l\,mol^{-1}\,s^{-1}$.

These findings may or may not be applied to DHLA. The chain mechanism of reaction of thiols with superoxide suggests that the first step of this reaction is the oxidation of thiols:

$$O_2^{\bullet -} + GSH \Longrightarrow [GSO_2H]^{\bullet -} \Longrightarrow GSO^{\bullet} + HO^{\bullet -} \qquad (22)$$

(Here, thiol is glutathione.) On subsequent steps $O_2^{\bullet-}$ is regenerated by one-electron oxidation of GSSG:

$$GSO^{\bullet} + GS^- \Longrightarrow GSO^- + GS^{\bullet} \qquad (23)$$

$$GS^{\bullet} + GS^- \Longrightarrow GSSG^{\bullet-} \qquad (20)$$

$$GSSG^{\bullet-} + O_2 \Longrightarrow GSSG + O_2^{\bullet-} \qquad (21)$$

Regeneration of superoxide during the oxidation of thiols hints at the possible prooxidant effect of these antioxidants. This suggestion was recently confirmed by Mottley and Mason [212] who have showed that superoxide was formed in the oxidation of DHLA by horseradish peroxidase in the presence of phenol. However, DHLA is dithiolic compound and the other mechanisms such as the concerted mechanism, which has been proposed earlier for flavonoids may be realized (Figure 29.6).

LA and DHLA exhibited opposite effects on oxygen radical production by lipoxygenase, NADPH oxidase, and phagocytes [200]. It is interesting that the activity of lipoic acid was, as a rule, stimulatory in these oxygen radical-producing systems: LA stimulated SOD-inhibitable lucigenin-amplified CL produced during the peroxidation of linolenic acid catalyzed by soybean lipoxygenase and the cytochrome c reduction by leukocyte NADPH oxidase. Similarly, LA stimulated spontaneous superoxide production by leukocytes and macrophages. In contrast, LA inhibited oxygen radical production by latex- and PMA-stimulated leukocytes and macrophages probably due its incapability to compete with these stimuli for the activation of phagocytic NADPH oxidase. On the contrary, DHLA was always inhibitory in cell-free and cellular superoxide-generating systems due to direct scavenging of superoxide.

It should be mentioned that the inhibition of superoxide overproduction and lipid peroxidation by lipoic acid has been recently shown in animal models of diabetes mellitus. The administration of LA to streptozotocin-diabetic rats suppressed the formation of lipid peroxidation products [213]. In another study the supplementation of glucose-fed rats with lipoic acid suppressed aorta superoxide overproduction as well as an increase in blood pressure and insulin resistance [214].

Several studies suggest that LA and DHLA form complexes with metals (Mn^{2+}, Cu^{2+}, Zn^{2+}, Cd^{2+}, and Fe^{2+}/Fe^{3+}) [215–218]. However, in detailed study of the interaction of LA and DHLA with iron ions no formation of iron–LA complexes was found [217]. As vicinal dithiol, DHLA must undoubtedly form metal complexes. However, the high prooxidant activity of DHLA makes these complexes, especially with transition metals, highly unstable. Indeed, it was found that the Fe^{2+}–DHLA complex is formed only under anaerobic conditions and it is rapidly converted into Fe^{3+}–DHLA complex, which in turn decomposed into Fe^{2+} and LA [217]. Because of this, the Fe^{3+}/DHLA system may initiate the formation of hydroxyl radicals in the presence of hydrogen peroxide through the Fenton reaction. Lodge et al. [218] proposed that the formation of Cu^{2+}–DHLA complex suppressed LDL oxidation. However, these authors also found that this complex is unstable and may be prooxidative due to the intracomplex reduction of Cu^{2+} ion.

29.5.2 Glutathione

Another important thiolic antioxidant is glutathione. Glutathione is a tripeptide (γ-L-glutamyl-L-cysteinyl-glycine, GSH), which is the most abundant thiol presenting in mammalian cells with concentrations of 1 to 10 mmol l^{-1}. Glutathione is synthesized by two enzymes, γ-glutamylcysteine synthase and glutathione synthase. There are many functions of glutathione in an organism such as the participation in metabolism, transport, catalysis, maintenance of thiol moieties of proteins, etc. As a redox-cycling compound, glutathione may function in two ways: as a free radical scavenger and as an electron donor in the enzymatic redox cycle of glutathione peroxidase and glutathione reductase catalyzing the reduction of peroxides. (The last function of glutathione is considered in Chapter 30.)

Ascorbic acid and GSH are the most important in vivo water-soluble antioxidants, which may manifest synergistic effect (see below). The antioxidant activity of glutathione is especially important in the brain, which contains relatively low levels of SOD, catalase, and glutathione peroxidase. In this case the alteration of GSH metabolism may contribute to the pathogenesis of neurodegenerative diseases such as Parkinson's disease [219]. It is possible that one of the protective functions of GSH in brain is the interaction with nitric oxide. It has been shown that GSH reacts with NO to form S-nitrosoglutathione [220]. Canals et al. [221] found that the depletion of glutathione by the GSH synthesis inhibitor L-buthionine-(S,R)-sulfoximine in fetal midbrain cultures drastically enhanced the toxic effect of nitric oxide in these cells. Recently, Ford et al. [222] showed that glutathione and cysteine are the effective scavengers of another important biological nitrogen radical NO_2, reacting with radical with rate constants of about 2×10^7 and 5×10^7 $l\,mol^{-1}\,s^{-1}$, respectively.

Since glutathione is synthesized in cells in relatively huge amounts, it is seldom applied as pharmacological antioxidant. Furthermore, the mechanism of its antioxidant activity is not so simple as that of vitamins E and C. The major reason is that the GS$^\cdot$ radical formed during scavenging of free radicals by GSH does not disappear by dimerization but participates in the chain reaction, producing superoxide (Reactions (20)–(23)). Furthermore, it has recently been shown that contrary to previous findings the rate constant for the reaction of GSH with superoxide is relatively small (200–1000 $l\,mol^{-1}\,s^{-1}$) [211,223].

It has been shown that glutathionyl radicals may participate in various potentially damaging processes. Sampath and Caughey [224] showed that the addition of GSH to aerobic solution of hemoglobin resulted in the heme and protein oxidation. Similar to dihydrolipoic acid, glutathione is easily oxidized by horseradish peroxidase (HRP) to the glutathionyl radical [225]. On the other hand, glutathione may intervene into the redox cycling of oxidizable drugs reducing the formed radical cations (RC) [226].

$$RC^{\cdot +} + GSH \iff RC + GS^{\cdot} + H^+ \qquad (24)$$

It is interesting that the equilibrium of Reaction (24) is actually shifted to the left (K_{24} is of the order of 10^{-4}, but due to the rapid removing GS$^\cdot$ in following reactions (for example, Reaction (20)) from reaction mixture, glutathione is a good reductant of drug radical cations.

Very interesting mechanism of the enhanced glutathione antioxidant activity in the presence of SOD was offered by Winterbourn [227]. She pointed out that the interaction of all free radicals with glutathione resulted in the formation of superoxide as the only final active species "a free radical sink." Therefore, the mixture of glutathione together with SOD may be considered as a "universal" scavenger of free radicals of any structure.

Unfortunately, later findings questioned the reliability of free radical sink hypothesis. Using the known values of rate constants, Wardman [228] calculated that in well oxygenated tissue not more than 20% of GS$^\cdot$ radicals are able to react with GS$^-$ to form GSSG$^{\cdot -}$ (Reaction (20)) and generate superoxide by Reaction (21). Furthermore, in the presence of ascorbate this amount fall down to 3%. Similarly, Sturgeon et al. [229] showed that in the tyrosyl radical-generating system, containing ascorbate and glutathione, ascorbate successfully competed with glutathione to form the ascorbate radical and diminished GS$^\cdot$ formation. Thus, the GSH + SOD system cannot apparently function as effective "universal" scavenger of free radicals.

29.5.3 N-Acetylcysteine and Tetradecylthioacetic Acid

N-Acetylcysteine (NAC) is a thiolic compound, which is applied in clinical practice since the mid-1950s. As NAC has been used for the treatment of some inflammatory disorders connected, for example, with redox cycling of xenobiotics or exposure to environmental pollutants and cigarette smoke, it has been proposed that this compound is possessing antioxidant activity similar to other thiols. It has been shown that NAC reacts with free radicals to form thiyl radicals and NAC disulfide, the latter as a final product [230]. NAC is an excellent scavenger of hydroxyl radicals (the rate constant is equal to 1.36×10^{10} $1\,mol^{-1}\,s^{-1}$ [231]. There were different estimates of its rate constant with superoxide, but as in the case of other thiols, this reaction is a rather slow one with the rate constant from 10 to 10^3 $1\,mol^{-1}\,s^{-1}$ [211]. Nonetheless, NAC is apparently able to suppress superoxide-mediated toxic effects in cells [232]. Another possible mode of NAC antioxidant activity is the enhancement of cellular concentrations of cysteine and glutathione [230]. It has recently been shown [233] that NAC is able to suppress redox regulation of mitogen-activated protein kinase (MAPK(p38))-mediated proinflammatory cytokine production in the alveolar epithelium. NAC induced intracellular accumulation of GSH and reduced the concentration of GSSG.

There are other synthetic and natural thiolic compounds possessing antioxidant activity. One such compound is tetradecylthioacetic acid (TTA), which inhibited the iron–ascorbate-induced microsomal lipid peroxidation [234]. Its Se analog exhibited even a more profound antioxidative effect.

29.6 UBIQUINONES

Ubiquinones (coenzymes Q) Q_9 and Q_{10} are essential cofactors (electron carriers) in the mitochondrial electron transport chain. They play a key role shuttling electrons from NADH and succinate dehydrogenases to the cytochrome b–c_1 complex in the inner mitochondrial membrane. Ubiquinones are lipid-soluble compounds containing a redox active quinoid ring and a "tail" of 50 (Q_{10}) or 45 (Q_9) carbon atoms (Figure 29.10). The predominant ubiquinone in humans is Q_{10} while in rodents it is Q_9. Ubiquinones are especially abundant in the mitochondrial respiratory chain where their concentration is about 100 times higher than that of other electron carriers. Ubihydroquinone Q_{10} is also found in LDL where it supposedly exhibits the antioxidant activity (see Chapter 23).

Contemporary interest in ubiquinones is explained by their potential antioxidant activity and the possibility of using these nontoxic natural compounds as pharmaceutical agents. But it should be noted that ubiquinones are not vitamins and that they are synthesized in humans. Taking into account a high level of ubiquinones in mitochondria, the effective supplementation of ubiquinones to fight against free radical-mediated damage seems to be a hard task.

There are two kinds of redox interactions, in which ubiquinones can manifest their antioxidant activity: the reactions with quinone and hydroquinone forms. It is assumed that the ubiquinone–ubisemiquinone pair (Figure 29.10) is an electron carrier in mitochondrial respiratory chain. There are numerous studies [235] suggesting that superoxide is formed during the one-electron oxidation of ubisemiquinones (Reaction (25)). As this reaction is a reversible one, its direction depends on one-electron reduction potentials of semiquinone and dioxygen.

$$Q^{\bullet-} + O_2 \Longleftrightarrow Q + O_2^{\bullet-} \qquad (25)$$

However, ubiquinones are water nonsoluble compounds, and it is impossible to measure their reduction potentials in aqueous solution. Fortunately, it is possible to measure directly the

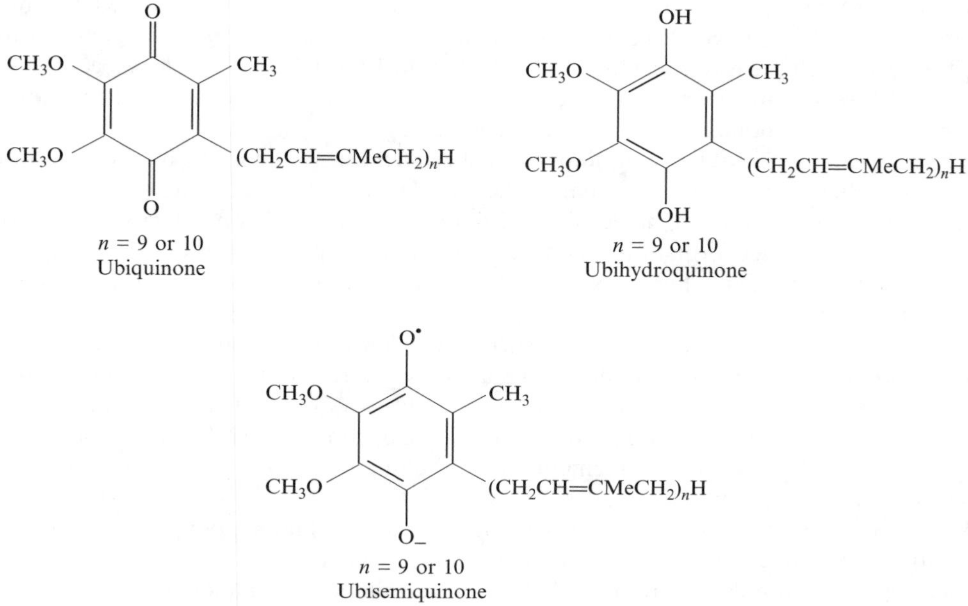

FIGURE 29.10 Ubiquinone, ubihydroquinone, and ubisemiquinone.

equilibrium constants K_{25} for Reaction (25) in aprotic media. It was found that $K_{25} = 2$–3 for ubiquinones in dimethylformamide [236]. Accepting these values as a rough estimate of the equilibrium constants in lipid membranes and taking into account that ubiquinone concentrations in mitochondria are much greater than that of dioxygen, one may suggest that the equilibrium of Reaction (25) in biological systems is sharply shifted to the left. As semiquinones are inactive radical-anions, ubiquinones may indeed exhibit antioxidant activity as superoxide scavengers under in vivo conditions.

As already mentioned, another mechanism of antioxidant activity of ubiquinones is scavenging of free radicals by ubihydroquinones (Reaction (26)):

$$QH + ROO^{\bullet} \Longrightarrow Q^{\bullet} + ROOH \qquad (26)$$

(This mechanism is now considered to be of importance for the protection of LDL against oxidation stress, Chapter 25.) The antioxidant effect of ubiquinones on lipid peroxidation was first shown in 1980 [237]. In 1987 Solaini et al. [238] showed that the depletion of beef heart mitochondria from ubiquinone enhanced the iron–adriamycin-initiated lipid peroxidation whereas the reincorporation of ubiquinone in mitochondria depressed lipid peroxidation. It was concluded that ubiquinone is able to protect mitochondria against the prooxidant effect of adriamycin. Inhibition of in vitro and in vivo liposomal, microsomal, and mitochondrial lipid peroxidation has also been shown in studies by Beyer [239] and Frei et al. [240]. Later on, it was suggested that ubihydroquinones inhibit lipid peroxidation only in cooperation with vitamin E [241]. However, simultaneous presence of ubihydroquinone and vitamin E apparently is not always necessary [242], although the synergistic interaction of these antioxidants may take place (see below). It has been shown that the enzymatic reduction of ubiquinones to ubihydroquinones is catalyzed by NADH-dependent plasma membrane reductase and NADPH-dependent cytosolic ubiquinone reductase [243,244].

Peroxyl radicals are not only ones, which are able to react with ubihydroquinones. Poderoso et al. [245] showed that the short-chain ubihydroquinones Q_o and Q_2 are oxidized by nitric oxide with the rate constants of 0.49×10^4 and 1.6×10^4 l mol^{-1} s^{-1}, respectively. The reaction apparently proceeded by one-electron transfer mechanism because the formation of intermediate semiquinone radicals has been registered.

The antioxidant activity of ubiquinones has been demonstrated in the in vitro and in vivo studies. Takahashi et al. [246] compared the prooxidant effects of hydrogen peroxide on hepatocytes isolated from rats injected with ubiquinone Q_{10} and control rats. It was found that the TBAR product formation was nearly completely suppressed in Q_{10} hepatocytes and that these cells exhibited a higher cell viability and lower release of lactate dehydrogenase than control hepatocytes. Recently, it has been found that ubihydroquinone Q_{10} improved platelet mitochondrial function and protected the cells from oxidative injury [247]. In contrast, ubiquinone was practically ineffective under similar conditions. Ubiquinone Q_{10} can apparently exert neuroprotective effect since it has been shown that its oral administration to rats attenuated striatal lesions induced by 3-nitropropinic acid [248]. Raitakari et al. [249] concluded that dietary supplementation of ubiquinone Q_{10} decreased ex vivo LDL oxidation in humans. However, the last analysis of their own and literature data performed by Kaikkonen et al. [250] showed that there are no reliable data supporting the existence of any antioxidant effect of Q_{10} under in vitro or ex vivo conditions at high radical flux. Inefficiency of Q_{10} supplementation probably depends on its very small plasma concentration at high radical flux conditions and the absence of regeneration mechanism for ubiquinone in plasma ex vivo or isolated LDL. On the other hand, it has been suggested that ubiquinone Q_{10} may exhibit antioxidant activity under low radical flux conditions mainly through the regeneration of vitamin E. Rosenfeldt et al. [251] concluded that the administration of Q_{10} to the elderly improved the efficiency of mitochondrial energy production and reduced free radical-mediated myocardial damage.

The administration of Q_{10} or quercetin to rats protected against endotoxin-induced shock in rat brain [252]. It was found that the pretreatment with these antioxidants diminished the shock-induced increase in brain MDA and nitric oxide levels. Interesting data have been obtained by Yamamura et al. [253] who showed that ubiquinone Q_{10} is able to play a double role in mitochondria. It was found that on the one hand, Q_{10} enhanced the release of hydrogen peroxide from antimycin A- or calcium-treated mitochondria, but on the other hand, it inhibited mitochondrial lipid peroxidation. It was proposed that Q_{10} acts as a prooxidant participating in redox signaling and as an antioxidant suppressing permeability transition and cytochrome c release.

From chemical point of view, efficient free radical scavengers must contain substituents with the very weak C—H, O—H, or S—H bonds, from which reactive free radicals are able to abstract a hydrogen atom. It can be seen that the antioxidants discussed above (ascorbic acid, α-tocopherol, ubihydroquinones, glutathione, etc) fall under this category. However, many other compounds manifest free radical scavenging activity in in vitro and in vivo systems.

29.7 QUINONES

In the case of ubiquinones we have already considered the ability of quinones to react with superoxide and other free radicals. Naphthoquinones, vitamin K and its derivatives, especially menadione, are the well known producers of superoxide through redox cycling with dioxygen. However, in 1985, Canfield et al. [254] have shown that vitamin K quinone reduced the oxidation of linoleic acid while vitamin K hydroquinone stimulated lipid peroxidation. Surprisingly, later on, conflicting results were reported by Vervoort et al. [255] who found that only hydroquinones of vitamin K and its analogs inhibited microsomal lipid peroxidation.

29.8 URIC ACID

Another physiologically important antioxidant is uric acid. Uric acid (2,6,8-purinetrione) contains two active hydroxyl groups in the purine heterocycle (Figure 29.11), which are responsible for its free radical scavenging activity. In 1981, Ames et al. [256] showed that the physiological plasma concentrations of uric acid protected erythrocytes against free radical-mediated damage. Antioxidant effects of uric acid have been shown in many other in vitro and in vivo systems. Free radical scavenging activity of uric acid was supported by the identification of urate free radicals [257]. Uric acid was found to be a major antioxidant in human airway mucosal surfaces [258].

In 1998, Schlotte et al. [259] showed that uric acid inhibited LDL oxidation. However, subsequent studies showed that in the case of copper-initiated LDL oxidation uric acid behaves itself as prooxidant [260,261]. It has been suggested that in this case uric acid enhances LDL oxidation by the reduction of cupric into cuprous ions and that the prooxidant effect of uric acid may be prevented by ascorbate. On the other hand, urate radicals formed during the interaction of uric acid with peroxyl radicals are able to react with other compounds, for example, flavonoids [262], and by that participate in the propagation of free radical damaging reactions. In addition to the inhibition of oxygen radical-mediated processes, uric acid is an effective scavenger of peroxynitrite [263].

29.9 STEROIDS

Some steroid molecules (estrone, estradiol, and estriol) have phenolic hydroxyl in the ring A (Figure 29.12) and therefore, are able to react as free radical scavengers. In 1987, Japanese authors [264,265] showed that all these compounds inhibited iron–adriamycin- or iron–ADP–ascorbate-dependent phospholipid and liposomal lipid peroxidation. Later on, most attention was drawn to the study of antioxidative properties of estradiol-17β (estrogen E_2); it has been proposed that E_2 antioxidant activity may contribute to cardioprotection observed after estrogen therapy in postmenopausal women. The necessity for the phenolic hydroxyl has been shown by studying the effects of several estrogens on LDL oxidation. It was found [266]

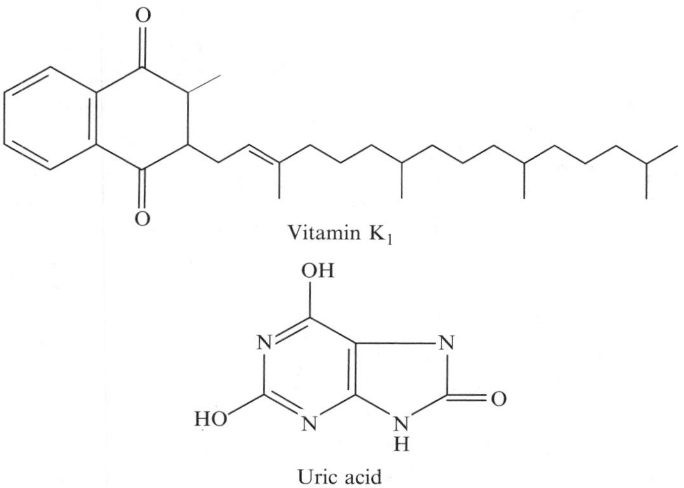

FIGURE 29.11 Vitamin K and uric acid.

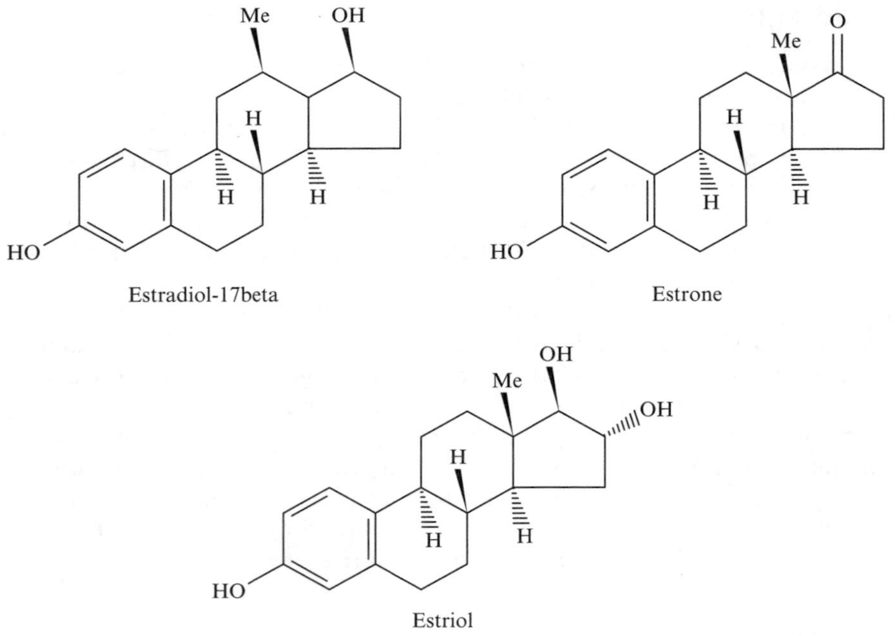

FIGURE 29.12 Antioxidant steroids.

that only estrdiol-17β but not testosterone (having no phenolic hydroxyl) or estrdiol-17β benzoate (with the blocked HO group) inhibited lipid peroxidation. The ability of estradiol-17β to inhibit LDL oxidation and DNA damage has been shown in several studies [267–270]. Estradiol-17β also inhibited lipid peroxidation in rat liver microsomes [271]. Ayres et al. [270] concluded that the inhibitory effect of estradiol-17β is due to its interaction with hydrogen peroxide, superoxide, and hydroperoxyl radical. Ascorbic acid drastically enhanced the antioxidant activity of estradiol-17β in human aortic endothelial cells [272].

Winterle et al. [273] determined the rate constant for the reaction of estradiol-17β with peroxyl radicals as $10^5 \, 1 \, mol^{-1} s^{-1}$ and the dissociation energy of the phenolic O—H bond as $85 \pm 2 \, kcal \, mol^{-1}$. However, it is most surprising that the rate constants of the phenoxyl radical $E_2O^\cdot$ for hydrogen atom abstraction from α-tocopherol and ascorbic acid are found to be unusually high (10^8–$10^9 \, 1 \, mol^{-1} \, s^{-1}$). It means that estradiol-17β is extremely rapidly regenerated in the presence of α-tocopherol and ascorbic acid, but at the same time, $E_2O^\cdot$ is able to initiate oxidative processes in the absence of these antioxidants. There are the other steroid analogs containing phenolic hydroxyl groups, which are also capable of exhibiting antioxidant activity. For example, LY-139478, a nonsteroidal estradiol receptor modulator, was about a 15-fold more effective free radical scavenger than estradiol-17β during the inhibition of copper-initiated LDL peroxidation [274].

Another group of steroids, whose antioxidative properties are widely discussed in literature, is 21-amino steroids (lazaroids). In 1987, Braughler et al. [275] reported the results of their investigation of antioxidative properties of two lazaroids, 21-[4-(2,6-di-1-pyrrolidinyl-4-pyrimidinyl)-1-piperazinyl]-16α-methylpregna-1,4,9(11)-tiene-3,20-dione monomethane sulfonate (U74006F) and 21-[4-(3,6-bis(diethylamino)-2-pyridinyl)-1-piperazinyl]-16α-methylpregna-1,4,9(11)-triene-3,20-dione hydrochloride (U74500A) (Figure 29.13). These authors concluded that 21-amino steroids are potent inhibitors of iron-dependent lipid peroxidation and that their efficacy may be compared with that of α-tocopherol. However, as seen from Figure 29.13, lazaroids have no active substituents capable of reacting with free radicals. In subsequent work [276] it has been shown that despite a great efficacy in iron-dependent lipid peroxidation, U74006F and U74500A are considerably less effective free radical scavengers compared to α-tocopherol.

It is obvious that lazaroids cannot be effective scavengers of free radicals excepting highly reactive hydroxyl radicals. Therefore, the "second generation" lazaroid U-78517F (Figure 29.13) was synthesized, which was supposed to have a much greater antioxidant activity. This

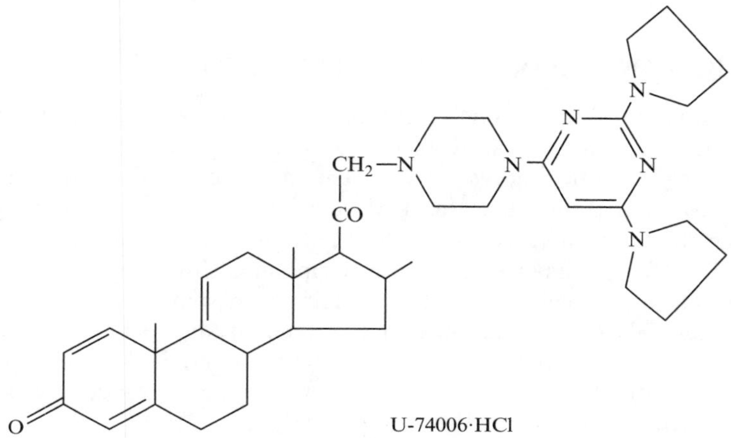

U-74006·HCl

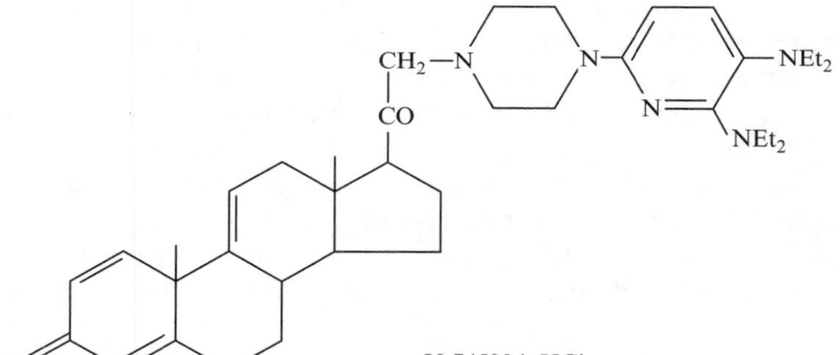

U-74500A·HCl

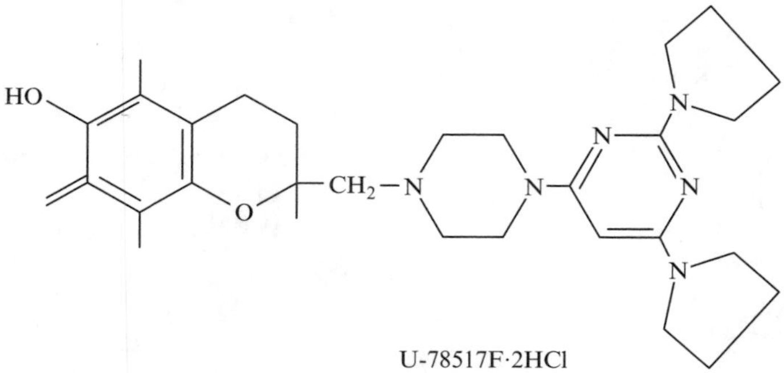

U-78517F·2HCl

FIGURE 29.13 Major lazaroids.

compound indeed turned out to be an effective in vitro and in vivo free radical scavenger [277,278]. U-83836E also effectively inhibited free radical-mediated damage in endothelial cells injured by hypoxia–reoxygenation [279]. Furthermore, it was found that lazaroids can inhibit free radical damaging processes by reacting with peroxynitrite. Thus, Fici et al. [280] found that lazaroids U-74006F and U-74500A suppressed peroxynitrite toxic effect in rat cerebellar granule cells with I_{50} values of about 100 μmol l^{-1}.

It seems worthy to comment on the name "second generation" lazaroid. Looking at the structure of U-83836E, one may wonder why this compound is associated with 12-amino steroids at all, whereas it is just another model of α-tocopherol and should be compared not with "first generation" lazaroids but with the synthetic analogs of vitamin E (see above).

29.10 CALCIUM ANTAGONISTS AND β-BLOCKERS

Calcium antagonists and β-adrenergic receptor blockers are well-known drugs widely used for the treatment of cardiovascular disorders. It has been suggested that some of these compounds also exhibit antioxidative activity. The mechanisms of antioxidative activity of calcium antagonists and β-blockers are, as a rule, uncertain because their structures are widely varied and free radical scavenging substituents such as phenolic hydroxyl groups are rarely present. Therefore, in many cases these compounds should be considered as indirect antioxidants, which may affect free radical production in some ways without direct interaction with free radicals or be metabolized into genuine radical scavengers. In addition, it seems that the antioxidant activity of calcium antagonists and β-blockers is independent of their therapeutic activities because both active and inactive stereoisomers of these compounds are the equally active antioxidants [281].

Mak and Weglicki [282] have found that calcium blockers (antagonists) nifedipine, verapamil, and some others (Figure 29.14) inhibited iron-dependent lipid peroxidation of sarcolemmal membranes. Furthermore, it has been found that calcium blockers are able to inhibit LDL oxidation. Thus, calcium antagonists of dihydropyridine-type (verapamil, nifedipine, amlodipine, isradipine, and lacidepine) inhibited copper-initiated LDL oxidation although their inhibitory effect was smaller comparing to vitamin E [283]. Rojstaczer and Triggle [284] measured the inhibitory effects of several calcium antagonists on copper- and monocyte-induced LDL oxidation and found out that the inhibitory activity of these compounds decreased in the range of felodipine > 2-chloro analog of nifedipine > nifedipine > amlodipine, verapamil, nitrendipine > diltiazem. Similarly, calcium blockers inhibited LDL

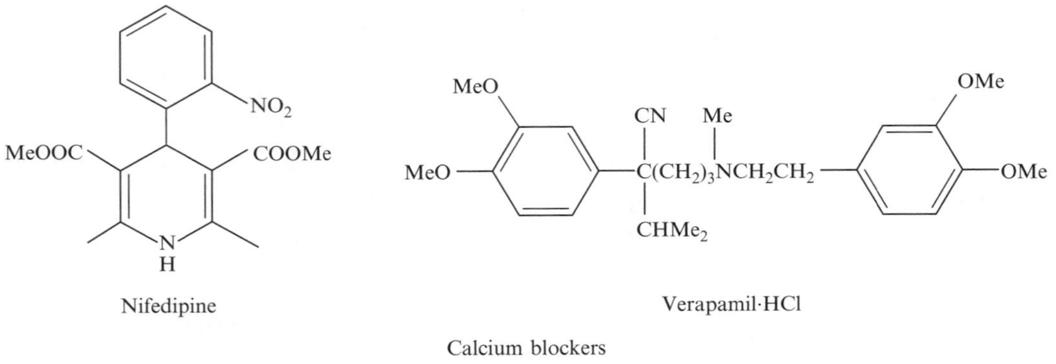

FIGURE 29.14 Specimens of calcium blockers.

degradation. It was suggested that the antioxidant effects of these compounds strongly depended on the 2-substitutent of the phenyl ring. Sobala et al. [285] found that calcium antagonists inhibited the oxidation of both native LDL and their advanced glycation end products but affect very slightly initial glycation reactions, such as Amadori product formation.

Napoli et al. [286] found that the nifedipine treatment of stroke-prone spontaneously hypertensive rats (SPSHR) suppressed the plasma and LDL oxidation and the formation of oxidation-specific epitopes and increased the survival of rats independently of blood pressure modification. Their results suggest that the protective effects of calcium blockers of dihydropyridine-type on cerebral ischemia and stroke may, at least in part, depend on their antioxidant activity. In vivo antioxidant effect of nilvadipine on LDL oxidation has been studied in hypertensive patients with high risk of atherosclerosis [287]. It was found that there was a significant decrease in the level of LDL cholesterol oxidation in patients after nilvadipine treatment.

Calcium antagonists are able to affect nitric oxide production and suppress the peroxynitrite-induced damage. Thus, nifedipine enhanced the bioavailability of endothelial NO in porcine endothelial cell cultures supposedly through an antioxidative mechanism [288]. Pretreatment with nisoldipine, a vascular-selective calcium blocker of dihydropyridine-type, of confluent bovine aortic endothelial cells suppressed the peroxynitrite-induced GSH loss and increased cell survival [289].

Similar to dihydropyridine calcium blockers, many β-adrenoreceptor antagonists exhibit antioxidant activity. Mak and Weglinski [290] showed that the pretreatment of canine myocytic sarcolemmal membranes with β-adrenoreceptor antagonists (propranolol, pindolol, metoprolol, atenolol, or sotalol) (Figure 29.15) inhibited superoxide-induced sarcolemmal

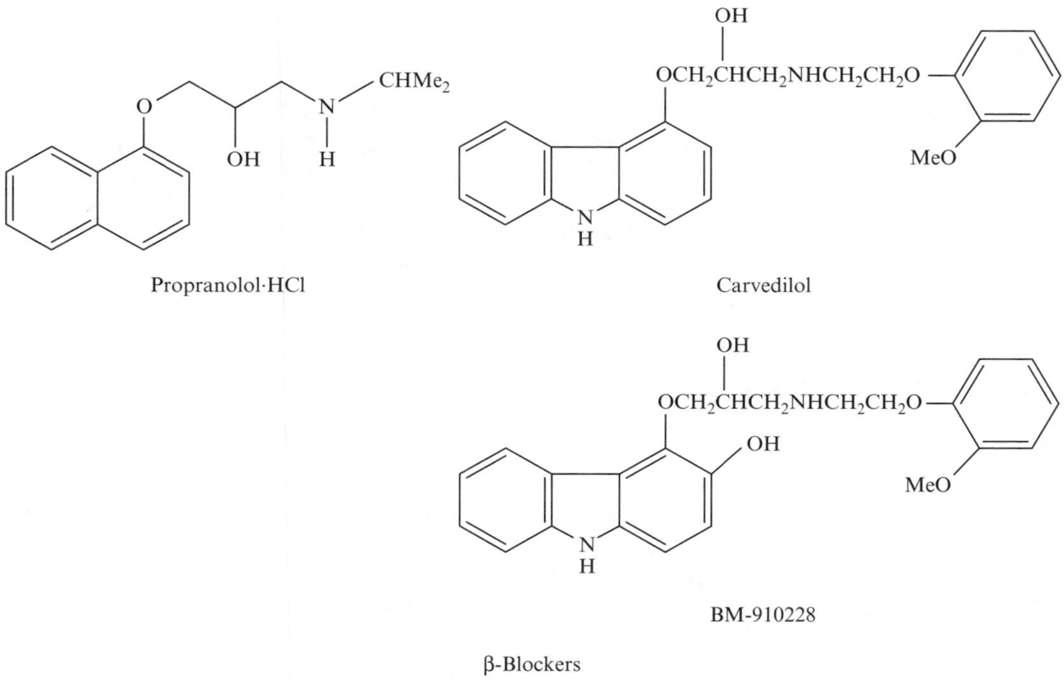

FIGURE 29.15 β-Blockers.

peroxidation. Jenkins et al. [291] showed that propranolol, dilevolol, labevolol, and metoprolol inhibited to some degree lipid peroxidation in the homogenates or liposomes of adult rat hearts, while atenolol was completely inactive. It has been supposed that the inhibitory efficiency of these compounds was related to their lipophilicity. Janero et al. [292] suggested that the antioxidant effect of propranolol is not a consequence of its free radical scavenging activity but is due the inhibition of xanthine oxidase, one of major producer of superoxide.

The absence of substituents with free radical scavenging properties in most of the β-blockers makes doubtful their efficacy as powerful antioxidants. Arouma et al. [293] tested the antioxidative properties of several β-blockers in reactions with superoxide, hydroxyl radicals, hydrogen peroxide, and hypochlorous acid. It was demonstrated that most of the compounds tested were inactive in these experiments. Nonetheless, propranolol, verapamil, and flunarizine effectively inhibited iron–ascorbate-stimulated microsomal lipid peroxidation and all drugs (excluding flunarizine) were effective scavengers of hydroxyl radicals. Contrary to Janero et al. [292], these authors did not find the inhibition of xanthine oxidase by propranolol. It was concluded that β-blockers are not the effective in vivo antioxidants.

Despite this criticism, many studies considering the antioxidant effects of β-adrenoreceptor antagonists continue to be published. Special attention is now drawn to β-blocker carvedilol and its hydroxylated derivative (BM 910228 or SB 211475) (Figure 29.15). These compounds seem to be the most efficient antioxidants among β-adrenoreceptor antagonists and calcium blockers and, in some systems, among all known antioxidants. Yue et al. [294] found that carvedilol inhibited macrophage- and copper-induced LDL oxidation with I_{50} values of 3.8 and 17 μmol l^{-1}, respectively. Under the same conditions, propranolol and atenolol were inactive. Carvedilol similarly inhibited iron-initiated lipid peroxidation in the rat brain homogenate (I_{50} value is equal to 8.1 μmol l^{-1}), while corresponding I_{50} values for propranolol, atenolol, labetalol, and some other β-blockers were over 1.0 mmol l^{-1} [295]. Carvedilol was again a much more effective inhibitor of lipid peroxidation and glutathione depletion in bovine endothelial cells compared to other β-blockers (propranolol, labetalol, pindolol, atenolol, and celiprolol) [296]. In addition, carvedilol protected against oxygen radical-mediated cell damage measured via lactate dehydrogenase release ($I_{50} = 4.1$ μmol l^{-1}).

High antioxidative activity carvedilol has been shown in isolated rat heart mitochondria [297] and in the protection against myocardial injury in postischemic rat hearts [281]. Carvedilol also preserved tissue GSL content and diminished peroxynitrite-induced tissue injury in hypercholesterolemic rabbits [298]. Habon et al. [299] showed that carvedilol significantly decreased the ischemia–reperfusion-stimulated free radical formation and lipid peroxidation in rat hearts. Very small I_{50} values have been obtained for the metabolite of carvedilol SB 211475 in the iron–ascorbate-initiated lipid peroxidation of brain homogenate (0.28 μmol l^{-1}), mouse macrophage-stimulated LDL oxidation (0.043 μmol l^{-1}), the hydroxyl-initiated lipid peroxidation of bovine pulmonary artery endothelial cells (0.15 μmol l^{-1}), the cell damage measured by LDL release (0.16 μmol l^{-1}), and the promotion of cell survival (0.13 μmol l^{-1}) [300]. SB 211475 also inhibited superoxide production by PMA-stimulated human neutrophils.

The mechanism of high antioxidant activity of carvedilol is widely discussed in literature. There are puzzling data on its reactivity in the reactions with free radicals of different structure. For example, Kramer and Wegliski [281] demonstrated that carvedilol and BM 910228 practically did not influence the hydroxyl radical production measured via the DMPO—OH adduct formation. At the same time, both compounds were the extremely efficient "chain-breaking" antioxidants in the reaction with peroxyl radicals. These findings of course question a high free radical scavenging activity of carvedilol. Nonetheless, Oettl et al. [301] suggested that carvedilol inhibited the peroxyl radical-initiated LDL oxidation through a radical-scavenging mechanism. Unfortunately, nobody was able to measure the rate constants for the reactions of carvedilol with free radicals. And furthermore, Noguchi et al. [302] showed that carvedilol had a poor reactivity toward phenoxyl, alkoxyl, and peroxyl radicals in acetontrile solution. Moreover, it did not inhibit the peroxyl radical-initiated oxidation of linoleate or phosphatidylcholine liposomal membranes or the oxidation of isolated LDL initiated by peroxyl radicals or copper ions. At the same time, carvedilol completely inhibited iron-stimulated peroxidation of methyl linoleate micelles.

As free radical-scavenging activity of carvedilol seems to be doubtful, two other possible mechanisms of its antioxidative effects have been discussed: iron chelating and membrane biophysical interaction. Oettl et al. [301] and Noguchi et al. [302] suggested that carvedilol is able to chelate iron ions with binding constant of about $10^5 \, l \, mol^{-1}$. However, the carvedilol structure is not typical of efficient chelators (see below). On the other hand, it has been found that carvedilol binds to liposomal membranes with the binding constant of the order of $10^4 \, l \, mol^{-1}$ [303]. (Similarly, the membrane-stabilizing mechanism of antioxidative activity of β-blockers has been proposed for propranolol [304] and the metabolite SB 211475 [305]. In the latter work it was suggested that the intercalation of SB 211475 into the glycerol phosphate–hydrocarbon interface makes the membrane more resistant to lipid peroxidation.)

Regarding the proposed mechanisms of carvedilol antioxidative activity, membrane stabilization through the biophysical interaction of carvedilol with the membrane seems to be the most reliable one. However, a higher antioxidant activity of the metabolite SB 211475 leads to another explanation. In contrast to the parent carvedilol, SB 211475 has the active free radical scavenging phenolic hydroxyl, which is apparently responsible for its enhanced antioxidant activity. Thus, we may suggest that the in vivo antioxidant activity of carvedilol is due to its converting into active metabolites, which, for example, may be formed in the reactions with "primary" free radicals such as hydroxyl radicals.

Completing the discussion on the antioxidant properties of β-blockers, we should like to comment on the study of in vivo effects of carvedilol in humans. Recently, it has been shown [306] that the administration of small doses of carvedilol (3.125 mg twice a day) significantly decreased oxygen radical production by blood leukocytes and monocytes and the levels of m- and o-tyrosine supposedly formed as a result of oxidative conversion of phenylalanine. Thus, regardless of the mechanisms, carvedilol does exhibit in vivo antioxidant activity.

29.11 PYRROLOPYRIMIDINES

Pyrrolopyrimidines (Figure 29.16) were synthesized and studied as effective in vitro and in vivo antioxidants possessing neuroprotective activity in brain injury and ischemia models due to the improved ability to penetrate blood–brain barrier and achieve the neural tissue compared to the other antioxidants, for example, lazaroids. Greater efficacy of pyrrolopyrimidines has been shown in permanent and temporary focal cerebral ischemia [307]. Pyrrolopyrimidines are also very effective inhibitors of lipid peroxidation: for example, IC_{50} values for pyrrolopyrimidines are $0.3-10.6 \, \mu mol \, l^{-1}$ compared to $IC_{50} = 26.4 \, \mu mol \, l^{-1}$ for lazaroid U-74006F [307]. It was demonstrated [308] that pyrrolopyrimidines are able to inhibit peroxynitrite-induced damaging processes such as tyrosine nitration in erythrocytes. Lauderback et al. [309] found the enhanced antioxidant activity of pyrrolopyrimidine U101033E in aqueous cell-free solution.

Similar to some other antioxidants, pyrrolopyrimidines do not contain active free radical scavenging groups such as phenolic or thiolic substituents. At present, at least two different mechanisms of their antioxidant activity have been proposed [307]. It was suggested that pyrrolopyrimidines, which are electron donating compounds, can be oxidized by hydroxyl or peroxyl radicals or hydroxylated by cytochrome P-450 forming phenolic metabolite

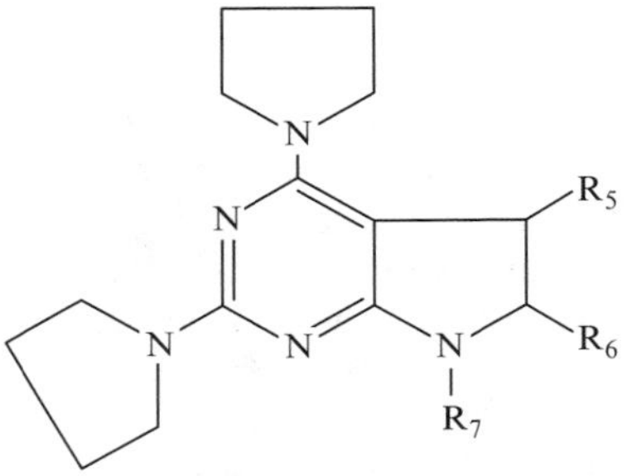

                              Pyrrolopyrimidines

U-87663			U-89843		
R_5 = H	R_6 = Ph	R_7 = Me	R_5 = H	R_6 = Me	R_7 = Me

FIGURE 29.16 Pyrrolopyrimidines.

(Figure 29.17). The first mechanism is untypical one for classical antioxidants and probably may be considered as a novel mode of antioxidant activity. The second one is typical for indirect antioxidants.

29.12 NADPH

NADPH might be considered as an indirect antioxidant due to its function as the reductant of various oxidized substrates, for example in the re-reduction of GSSG into GSH. However, it has recently been shown [1] that NADPH possesses the free radical scavenging activity reacting with such free radicals as $CO_3^{•-}$, $NO_2^•$, $ROO^•$ and $RO^•$. It should be noted that the efficiency of NADPH may increase under in vivo conditions due to its regeneration by numerous enzymes.

29.13 β-CAROTENE

β-Carotene (Figure 29.18) is a nutrient presented in fresh fruits and vegetables. For some time this unsaturated hydrocarbon and its derivatives carotenoids were considered to be the efficient antioxidants and free radical scavengers. Indeed, it is seen from its chemical structure that β-carotene should be very reactive in the reactions with free radicals. For example, it rapidly reacts with the $^•NO_2$ radical by one-electron transfer mechanism with the rate constant equal to 1.1×10^9 l mol^{-1} s^{-1} and with the glutathione free radical by addition mechanism with the rate constant equal to 2.2×10^8 l mol^{-1} s^{-1} [310]. Burton and Ingold [311] demonstrated the unusual high antioxidant and free radical scavenging activity of β-carotene. This work was followed by many in vivo studies of antioxidant effects of β-carotene [312]. Other modes of possible antioxidant action of β-carotene are the suppression of singlet oxygen formation [313] and the reaction with peroxynitrite [314].

In spite of a high reactivity of β-carotene in free radical reactions and marked antioxidant effects in in vitro systems, β-carotene did not show itself as an effective in vivo antioxidant. Furthermore, recent clinical trials suggested that the administration of β-carotene may be useless or even harmful to patients with heart and some other diseases, especially to smokers. One might suspect that one of the major reasons of toxic in vivo effects of β-carotene might be the formation of prooxidative compounds during β-carotene oxidation. In contrast to

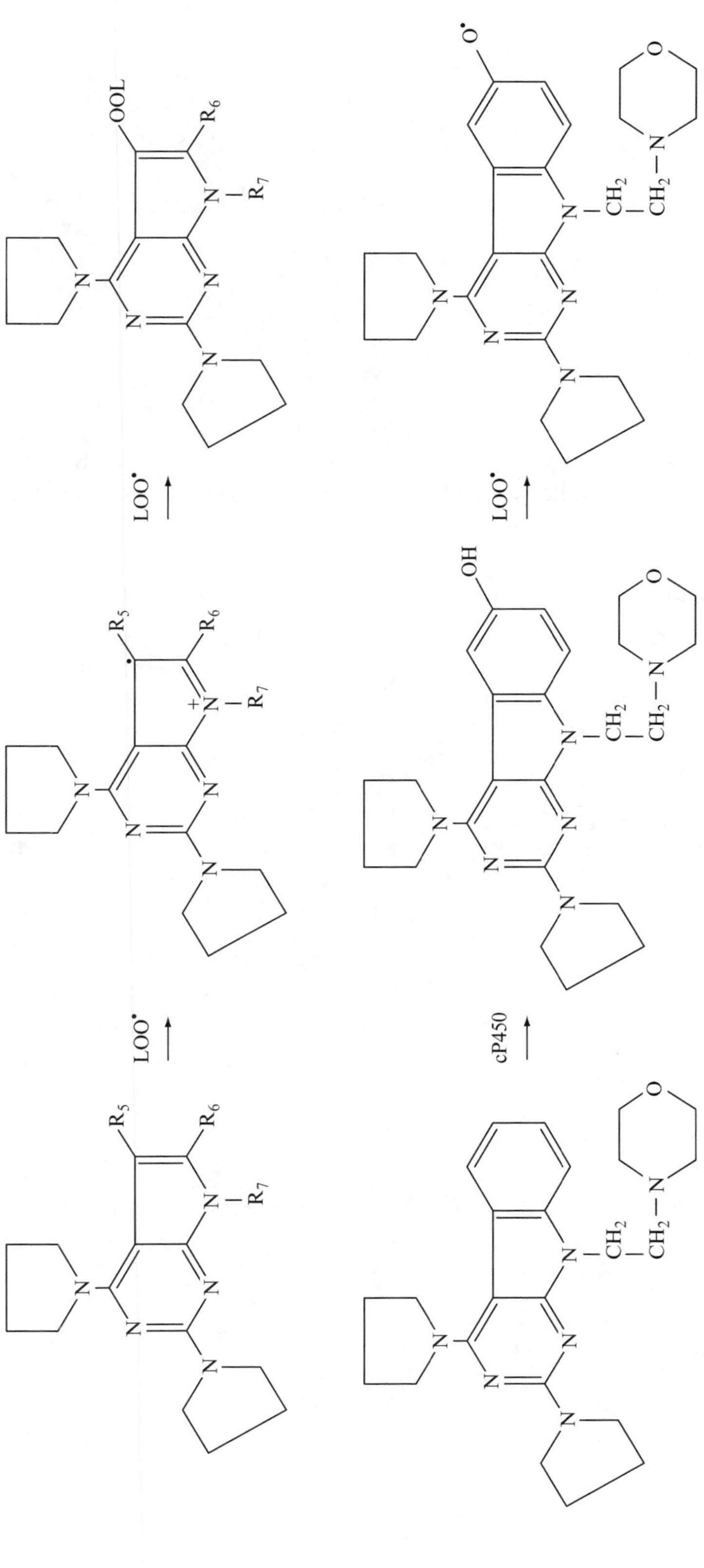

Antioxidants

FIGURE 29.17 Mechanism of hydroxylation of pyrrolopyrimidines by cytochrome P-450.

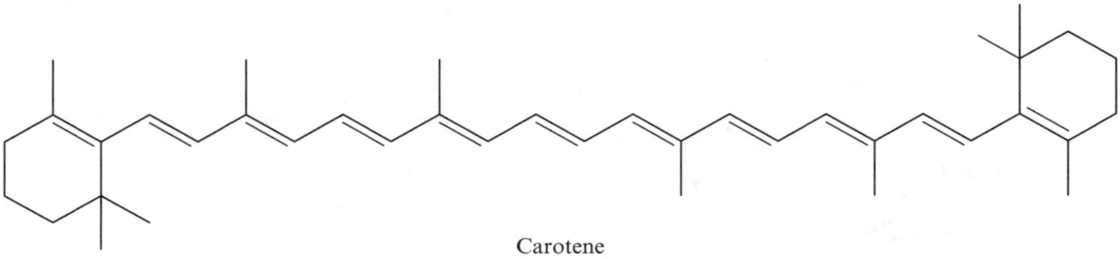

[Figure: Structure of Carotene]

Carotene

FIGURE 29.18 Structure of carotene.

α-tocopherol, which forms a relatively inactive α-tocopheroxyl radical in the reactions with free radicals, β-carotene free radical formed in oxidative processes is rapidly decomposed to form potentially reactive compounds. It is possibly the reason why β-carotene manifests prooxidant activity at 100% oxygen [311]. This proposal is supported by recent work [315] where it was found that oxidized β-carotene and the carotenoid lycopene induced oxidative DNA damage.

29.14 MELATONIN

Melatonin, N-acetyl-5-methoxytryptamine (Figure 29.19), is a pineal hormone, which is synthesized from tryptophan mainly in pineal gland. Last 10 years showed a big interest in the possible antioxidative activity of melatonin, especially under in vivo conditions [316,317]. It has been shown that melatonin is an effective scavenger of hydroxyl radicals, nitric oxide, and peroxynitrite [316]. Melatonin is a moderate inhibitor of the iron-initiated peroxidation of brain phospholipid liposomes ($IC_{50} = 210$ μmol l^{-1}) [318] and soybean phosphatidylcholine liposomes [319], although the early conclusion that melatonin may be twice as effective as trolox [320] is wrong. These findings also suggest that melatonin is able to inhibit homocystine-induced lipid peroxidation of brain homogenates (which suggests the possibility of using melatonin as therapeutic agent in reducing cardiovascular disease) [321]. Baydas et al. [322] also proposed that melatonin is a more efficient inhibitor of lipid peroxidation in rats with streptozocin-induced diabetes mellitus compared to vitamin E.

Despite a very optimistic estimate of antioxidant activity of melatonin [323], the thorough 1999 study [324] showed that melatonin possessed no free radical scavenging activity against peroxyl radicals in homogenous and heterogeneous systems but exhibited some inhibitory activity in iron-initiated lipid peroxidation. Summarizing the up-to-date findings on antioxidative activity of melatonin, the following might be said:

(1) Melatonin is a good scavenger of hydroxyl radicals reacting with them with the rate constant of about $10^{10}\, l^{-1}\, mol^{-1}$ [316]. However, it should be noted that practically all

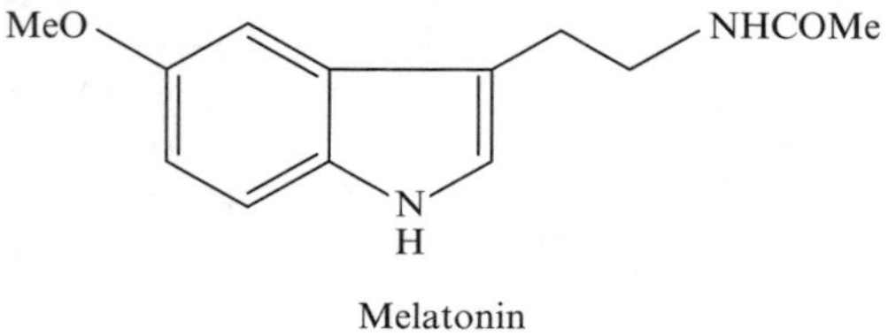

[Figure: Structure of Melatonin]

Melatonin

FIGURE 29.19 Structure of melatonin.

antioxidants and many biological substrates possess the same hydroxyl radical scavenging activity because the most rate constants of hydroxyl radical reactions are diffusion-controlled ones. Furthermore, the role of hydroxyl radicals as the initiators of peroxidative processes is doubtful (Chapter 25).

(2) Despite some literature data, melatonin is not a free radical scavenger because it does not contain the substituents with a weak C—H or O—H bonds. Suggestions that melatonin may reduce free radicals through a one-electron transfer mechanism are very interesting but they still were not confirmed by any kinetic measurements.

(3) Melatonin probably manifests some chelating activity although its structure is very different from the structures of effective chelators (see below).

(4) Melatonin might be an indirect in vivo antioxidant, getting metabolized into a genuine free radical scavenger in some enzymatic processes.

29.15 EBSELEN

Although selen is a very toxic element, some of its derivatives exhibit a relatively low toxicity and certain antioxidant activity. The most well known selen-containing antioxidant is ebselen (2-phenyl-1,2-benzoisoselenazol-3-(2H)-one, Figure 29.20), which was found to be an effective inhibitor of numerous free radical-mediated processes. The principal mode of the antioxidant activity of ebselen is apparently its activity as a glutathione peroxidase mimic (Chapter 30), although ebselen may probably also act as an antioxidant. There are early studies showing the ability of ebselen to inhibit microsomal lipid peroxidation [325,326]. Subsequent studies demonstrated that the antioxidative activity of ebselen is derived from the reduction of hydroperoxides and not from direct scavenging of free radicals [327,328]. This conclusion corresponds well to the ebselen chemical structure without efficient free radical scavenging substituents.

Thomas and Jackson [329] have shown that ebselen inhibited copper-initiated LDL oxidation in the presence of glutathione. Noguchi et al. [330] showed that the inhibitory effect of ebselen on copper-initiated LDL was also observed without glutathione, while in the case of radical-initiated LDL oxidation ebselen was inactive. However, it is possible that ebselen may inhibit both copper- and peroxyl radical-initiated LDL oxidation although in the latter case the inhibitory effect of ebselen depends on the size of peroxyl radical flux [331]. The inhibitory effect of ebselen on LDL oxidation also depends on its ability to reduce LDL hydroperoxides.

Another mode of antioxidant activity of ebselen is the suppression of peroxynitrite formation. Ebselen rapidly reacts with peroxynitrite to form ebselen Se-oxide [332]. Ebselen Se-oxide is reduced by glutathione [333]; therefore, ebselen can be regenerated under in vivo conditions by redox cycling (Figure 29.20).

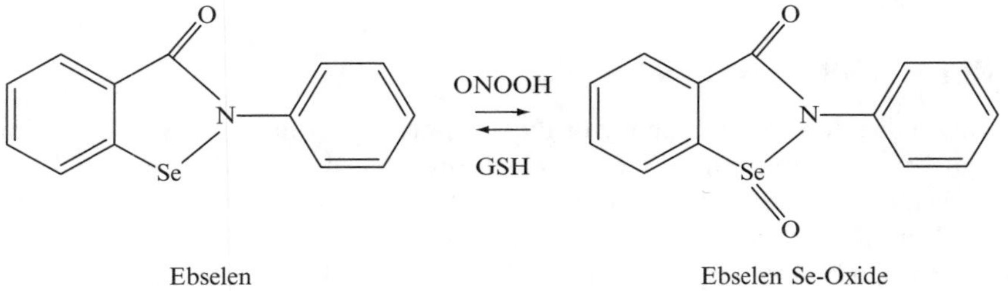

FIGURE 29.20 Reaction of ebselen with peroxynitrite.

29.16 METALLOTHIONEINS AND ZINC

Metallothioneins (MT) are unique 7-kDa proteins containing 20 cysteine molecules bounded to seven zinc atoms, which form two clusters with bridging or terminal cysteine thiolates. A main function of MT is to serve as a source for the distribution of zinc in cells, and this function is connected with the MT redox activity, which is responsible for the regulation of binding and release of zinc. It has been shown that the release of zinc is stimulated by MT oxidation in the reaction with glutathione disulfide or other biological disulfides [334]. MT redox properties led to a suggestion that MT may possesses antioxidant activity. The mechanism of MT antioxidant activity is of a special interest in connection with the possible antioxidant effects of zinc. (Zinc can be substituted in MT by some other metals such as copper or cadmium, but Ca–MT and Cu–MT exhibit manly prooxidant activity.)

Cai et al. [335] showed that MT inhibited copper-induced DNA damage and the reduction of cupric ions into cuprous ions in the presence of hydrogen peroxide and ascorbate. It is believed that the protective effect of MT against DNA damage could be about five times higher compared to glutathione [336]. MT seem to be efficient inhibitors of peroxynitrite–DNA damage and LDL oxidation [337]. Kumari et al. [338] found that hippocampal metallothionein isoforms are scavengers of hydroxyl radicals and superoxide.

Despite the conclusions in the cited literature about direct MT interaction with free radicals, the mechanism of MT antioxidant activity remains obscure. Markant and Pallauf [339] concluded that cysteine groups and not zinc are responsible for the inhibition of lipid peroxidation in hepatocytes. Maret and Vallee [340,341] also questioned the possibility of direct scavenging of free radicals by MT and suggested that zinc release is a major mechanism of antioxidant effects of metallothioneins.

Thus, the mechanism of MT antioxidant activity might be connected with the possible antioxidant effect of zinc. Zinc is a nontransition metal and therefore, its participation in redox processes is not really expected. The simplest mechanism of zinc antioxidant activity is the competition with transition metal ions capable of initiating free radical-mediated processes. For example, it has recently been shown [342] that zinc inhibited copper- and iron-initiated liposomal peroxidation but had no effect on peroxidative processes initiated by free radicals and peroxynitrite. These findings contradict the earlier results obtained by Coassin et al. [343] who found no inhibitory effects of zinc on microsomal lipid peroxidation in contrast to the inhibitory effects of manganese and cobalt. Yeomans et al. [344] showed that the zinc–histidine complex is able to inhibit copper-induced LDL oxidation, but the antioxidant effect of this complex obviously depended on histidine and not zinc because zinc sulfate was ineffective. We proposed another mode of possible antioxidant effect of zinc [345]. It has been found that Zn and Mg aspartates inhibited oxygen radical production by xanthine oxidase, NADPH oxidase, and human blood leukocytes. The antioxidant effect of these salts supposedly was a consequence of the acceleration of spontaneous superoxide dismutation due to increasing medium acidity.

29.17 METALLOPORPHYRINS

Recently, the high inhibitory efficiency of metalloporphyrins has been shown in lipid peroxidation of rat brain homogenates [346]. It was found that manganese and cobalt porphyrins were very effective inhibitors of lipid peroxidation while iron and especially zinc porphyrins had very weak inhibitory activity, if any. For example, I_{50} values were equal to 21, 29, 212, 946 μmol l^{-1} for CoTBAP, MnTBAP, FeTBAP, and ZnTBAP, respectively, where TBAP is 5,10,15,20-tetrakis [4-carboxyphenyl]porphyrin; similar values were obtained for other porphyrin derivatives.

The inhibition of lipid peroxidation by metalloporphyrins apparently depends on metal ions because only compounds with transition metals were efficient inhibitors. Therefore, the most probable mechanism of inhibitory effects of metalloporphyrins should be their dismuting activity. Manganese metalloporphyrins seem to be more effective inhibitors than Trolox ($I_{50} = 204\ \mu mol\ l^{-1}$) and rutin ($I_{50} = 112\ \mu mol\ l^{-1}$), and practically equal to SOD ($I_{50} = 15\ \mu mol\ l^{-1}$). The mechanism of inhibitory activity of manganese and zinc metalloporphyrins might be compared with that of copper– and iron–flavonoid complexes [167,168], which exhibited enhanced antiradical properties due to additional superoxide-dismuting activity.

29.18 LACTATE

Interest in the possible antioxidant activity of lactate appeared in connection with its production during exhaustive exercise accompanied by increasing free radical formation. Grousard et al. [347] proposed that under such conditions lactate might suppress oxidative stress by reacting with oxygen radicals. As seen from the lactate structure (Figure 29.21), this compound cannot be a scavenger of most free radicals excluding hydroxyl radicals. In accord with this suggestion, Grousard et al. determined high reactivity of lactate in reaction with hydroxyl radicals and its inability to inhibit lipid peroxidation. Surprisingly, they found that lactate scavenged superoxide by an unknown mechanism.

29.19 ASPIRIN

Aspirin (acetylsalicylic acid) (Figure 29.21) is a widely applied drug for reducing ischemic cardiovascular events in patients with coronary artery disease, hypertension, or at cardiovascular risk. It is believed that the main protective function of aspirin is the inhibition of cyclooxygenase; however, it has been recently proposed that aspirin may possess additional antioxidant activity [348]. It was found that long-term aspirin treatment of normotensive and hypertensive rats resulted in a decrease in vascular superoxide production by the inhibition of NADPH oxidase activity.

$CH_3CH(OH)COO^-$

Lactate

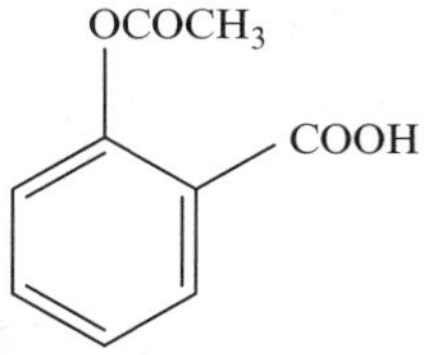

Aspirin

$NH_2C(=NH)$-$NHCH_2CH_2CH_2$-$C(H)(NH_2)$-COOH

L-Arginine

$NH_2CH_2CH_2SO_3H$

Taurine

FIGURE 29.21 Structures of lactate, aspirin, L-arginine, and taurine.

29.20 AMINO ACIDS

Similar to other indirect antioxidants, L-arginine (Figure 19.21), the substrate of NO synthase, has no free radical scavenging substituents. However, it has been proposed [349] that L-arginine inhibits myocardial contractility in buffer-perfused rat hearts by suppressing oxygen radical generation by xanthine oxidase. It was proposed that this inhibitory effect of L-arginine may be explained by the direct reaction with superoxide because D-arginine and the inhibitor of NO synthase N(G)-nitro-L-arginine manifested similar cardioprotective effects in this system.

Another amino acid taurine (Figure 29.21) is an effective scavenger of hypochlorous acid, which is known to participate in tissue damage associated with reperfusion injury mediated by neutrophils. Thus, taurine protected against the cytotoxic action of HOCl in neuronal cells [350]. Furthermore, taurine was found to reduce hydroxyl radical-induced damage to DNA [351]. Vohra and Hui [352] showed that the pretreatment of cultured neutrons with taurine suppressed lipid peroxidation and the loss of glutathione peroxidase activity induced in these cells by carbon tetrachloride.

29.21 VITAMIN B_6

Vitamin B6 (pyridoxine) and its derivative pyridoxamine are apparently able to inhibit superoxide production, reduce lipid peroxidation and glycosylation in high glucose-exposed erythrocytes [353]. It was suggested that the suppression of oxidative stress in erythrocytes may be a new mechanism by which these natural compounds inhibit the development of complication in diabetes mellitus.

29.22 TARGETED ANTIOXIDANTS

A new method to suppress oxygen radical production by mitochondrial respiratory chain has recently been proposed, namely, the use of antioxidants having a lipophilic cation [354–356]. For this purpose ubiquinone and tocopherol analogs (named MitoQ and MitoVitE, Figure 29.22) containing triphenylphosphonium cation have been synthesized. These so-called mitochondrial-targeted antioxidants are able to inhibit selectively mitochondrial oxidative damage and prevent free radical-mediated cell death. It has been shown that because of the large mitochondrial membrane potential, triphenylphosphonium cation accumulated within mitochondria with the antioxidant moiety inserted into the lipid bilayer. MitoVitE and MitoQ inhibited iron-stimulated mitochondrial lipid peroxidation and prevented the disruption of mitochondrial function. In addition, MitoQ protected mammalian cells from hydrogen peroxide-induced apoptosis.

29.23 NATURAL ANTIOXIDANT MIXTURES

At present, considerable interest is drawn to the use of natural mixtures of antioxidants isolated from various vegetable materials. Some authors claim that such mixtures manifest stronger antioxidant effects than individual components due to synergistic interactions. It is of course quite possible, but it should be noted that synergistic interactions are not a single mechanism of the interaction between components; for example, the simultaneous presence of the antioxidant and prooxidant flavonoids might diminish summary antioxidant effect of the mixture. Furthermore, natural mixtures contain, as a rule, some unknown compounds, which affect the summary effect by unknown manner.

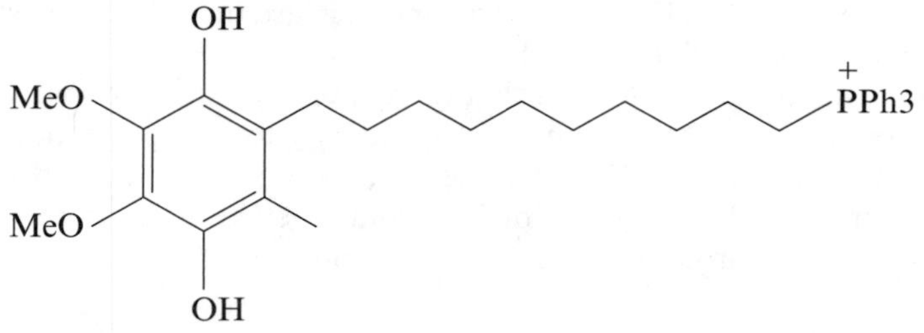

Antioxidants

[Chemical structure of Mito Q]

Mito Q

[Chemical structure of Mito Vit E]

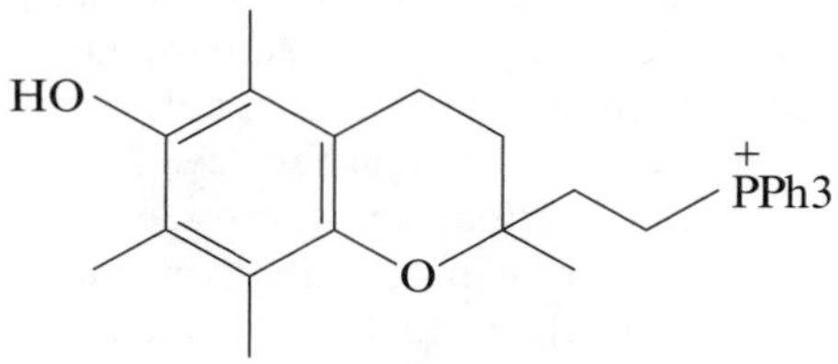

Mito Vit E

FIGURE 29.22 Structures of targeted antioxidants.

There are now probably thousands of supposedly antioxidant food supplements recommended for the treatment and prophylaxis of various pathophysiological conditions. The consideration of all these "antioxidant" compositions is of course outside the scope of this book. Now we will just comment on several products, in which antioxidant activity was shown in in vitro and in vivo studies. One of these natural antioxidant mixtures is pycnogenol. Pycnogenol obtained from the bark of the French maritime pine *Pinus maritima* is a mixture of many polyphenols, among them flavonoids (catechin, epicatechin, and taxifolin), condensed flavonoids (procyanidins), and phenolic acids (caffeic, ferulic, and p-hydroxybensoic acids). It was found that pycnogenol efficiently inhibited many free radical-mediated processes and protected vitamin E and glutathione from free radical attacks in cells [357–359]. Another well-known food supplement rich in flavonoids is *Ginkgo biloba* extract (GBE), a natural product, which also exhibits strong antioxidant activity [360]. Shi and Niki [361] compared free radical scavenging activity of GBE with that of individual antioxidants. It was found that the activity was diminished in the range: propyl gallate > α-tocopherol > quercetin > GBE ≅ kaempferol. It has also been shown that GBE efficiently protected human LDL against copper-initiated lipid peroxidation [362].

Prasad et al. [363] studied the use of garlic as an antioxidant for the prevention of hypercholesterolemic atherosclerosis in rats. It was found that supplementation with garlic decreased some parameters of oxidative stress such as aortic MDA and chemiluminescence in rabbits fed with cholesterol diet. NAO, a natural antioxidant isolated and purified from spinach containing flavonoids and coumaric acid derivatives, significantly improved the survival of rats subjected to lipopolysaccharide treatment [364]. NAO also suppressed MDA level in the heart, indicating the antioxidant mechanism of NAO protective activity. It has been suggested [365] that grape extract reduced ischemia–reperfusion injury in rats by functioning as in vivo antioxidant. Leonard et al. [366] also concluded that fruit and vegetable juices are efficient inhibitors of lipid peroxidation initiated by hydroxyl radicals. Olive oil was supposed to decrease the LDL cholesterol level and increase the HDL cholesterol level in the heart by diminishing LDL oxidation [367].

In conclusion, it should be mentioned that there are numerous proposals for the application of various food products having antioxidant activity for the protection against various free radical-mediated pathologies. For example, it has been suggested that nutritional interventions such as increasing dietary intake of fruits and vegetables can decrease the age-related declines in brain functions probably via the suppression of oxidative stress [368].

29.24 CHELATORS

Chelators of transition metals, mainly iron and copper, are usually considered as antioxidants because of their ability to inhibit free radical-mediated damaging processes. Actually, the so-called "chelating therapy" has been in the use probably even earlier than "antioxidant therapy" because it is an obvious pathway to treat the development of pathologies depending on metal overload (such as calcium overload in atherosclerosis or iron overload in thalassemia) with compounds capable of removing metals from an organism. Understanding of chelators as antioxidants came later when much attention was drawn to the possibility of in vivo hydroxyl radical formation via the Fenton reaction:

$$Fe^{2+} + H_2O_2 \Longrightarrow Fe^{3+} + HO^{\bullet} + HO^{-} \qquad (27)$$

It was found that various chelators are able to affect the hydroxyl radical formation in a different way, namely, some chelators (EDTA) can increase the rate of the Fenton reaction, while others (rutin) can inhibit it through the formation of inactive iron–chelator complexes.

At present, many natural and synthetic chelators have been studied as potential pharmacological agents for the treatment of iron or copper overload under various pathophysiological conditions. (The chelating activity of flavonoids was already discussed above and the application of chelators in thalassemia and some other pathologies is considered in Chapter 31.) There are specific thermodynamic demands to chelators to be efficient antioxidants [369]. We will consider just several chelators of potential therapeutic importance.

Chelators of iron, which are now widely applied for the treatment of patients with thalassemia and other pathologies associated with iron overload, are the intravenous chelator desferal (desferrioxamine) and oral chelator deferiprone (L1) (Figure 19.23, see also Chapter 31). Desferrioxamine (DFO) belongs to a class of natural compounds called siderophores produced by microorganisms. The antioxidant activity of DFO has been studied and compared with that of synthetic hydroxypyrid-4-nones (L1) and classic antioxidants (vitamin E). It is known that chronic iron overload in humans is associated with hepatocellular damage. Therefore, Morel et al. [370] studied the antioxidant effects of DFO, another siderophore pyoverdin, and hydroxypyrid-4-ones on lipid peroxidation in primary hepatocyte culture. These authors found that the efficacy of chelators to inhibit iron-stimulated lipid peroxidation in hepatocytes decreased in the range of DFO > hydroxypyrid-4-ones > pyoverdin. It seems that other siderophores are also less effective inhibitors of lipid peroxidation than DFO [371].

Ponka et al. [372] showed that pyridoxal isonicotinoyl hydrazone (PIH, Figure 19.23) is an iron chelating agent. Numerous studies showed the possibility of using this chelator for the treatment of iron overload disease [373]. In subsequent studies the antioxidant activity of PIN has been confirmed. For example, Hermes-Lima et al. [374,375] showed that PIN protected plasmid pUC-18 DNA and 2-deoxyribose against hydroxyl radical damage.

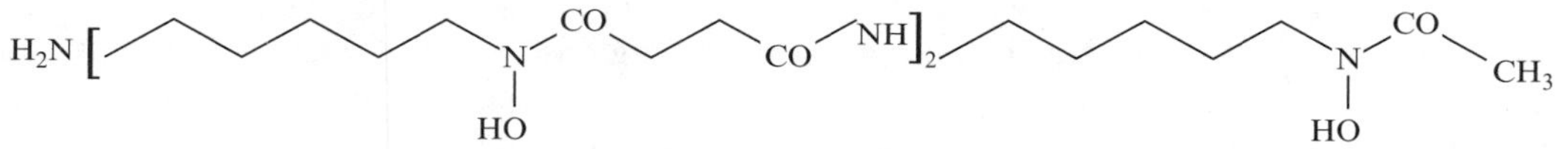

FIGURE 29.23 Chelators.

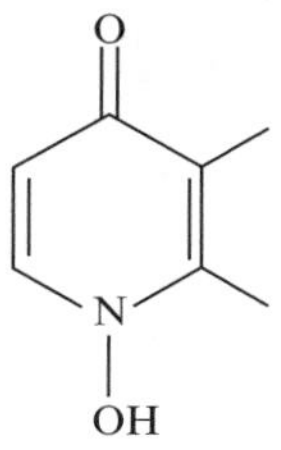

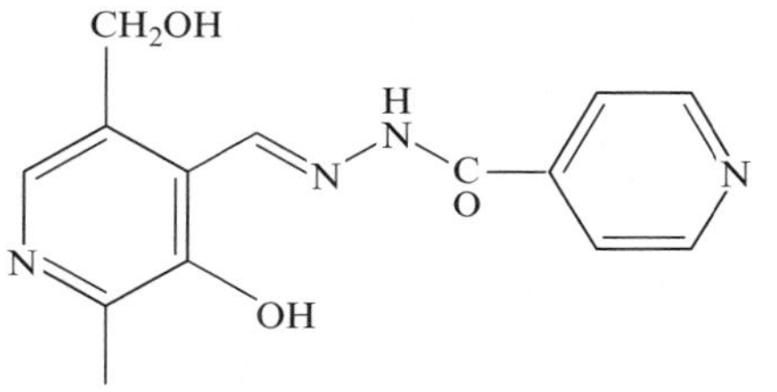

29.25 SYNERGISTIC INTERACTION OF ANTIOXIDANTS, FREE RADICAL SCAVENGERS, AND CHELATORS

As already mentioned above, the simultaneous presence of several free radical inhibitors may or may not exhibit synergistic effect. In 1991, Negre-Salvayre et al. [376] showed that the compositions of rutin and ascorbic acid or rutin, ascorbic acid, and α-tocopherol synergistically inhibited superoxide production and lipid peroxidation in linoleic acid ufasomes and human erythrocytes. In subsequent studies [377,378] these authors demonstrated the synergistic interaction of rutin, ascorbic acid, and α-tocopherol during the inhibition of copper- or UV radiation-stimulated LDL oxidation, glutathione oxidation, and ATP depletion in cultured endothelial cells. Kalyanaraman et al. [379] observed the synergistic effect of probucol + ascorbic acid combination on LDL oxidation. In this case the mechanism of synergistic effect was explained by the regeneration of probucol via the reaction of probucol phenoxyl radical with ascorbic acid similar to the above mentioned regeneration of α-tocopherol by ascorbate. It has also been shown [380] that flavonoids 7-monohydroxyethylrutoside, fisetin, and naringenin are able to substitute α-tocopherol during regeneration of glutathione and correspondingly enhance its protective effect against microsomal lipid peroxidation.

Recently, the possible synergistic interaction between flavonoids has been thoroughly discussed in connection with the cardioprotective effect of red wine and purple grape juice.

Although it has been shown earlier [381] that the red wine phenolic compound *trans*-resveratrol and flavonoid quercetin are protective against platelet aggregation and can suppress eicosanoid synthesis, they were not effective as components of the whole red wine or purple grape juice. This difference has been explained by the synergistic effects of flavonoids presented in these products. It was suggested that the antioxidant activity of flavonoids of red wine and purple grape juice is responsible for the suppression of platelet aggregation, an increase in platelet-derived NO release, a decrease in superoxide production [382], and the inhibition of collagen-induced hydrogen peroxide production [383].

REFERENCES

1. M Kirsch, H de Groot. *FASEB J* 15: 1569–1574, 2001.
2. IB Afanas'ev. In *Superoxide Ion: Chemistry and Biological Implications*. Vol 1. CRC Press, Boca Raton, FL, Chaper 10, 1989.
3. CK Chow. *Free Radic Biol Med* 11: 215–232, 1991.
4. WA Pryor. *Free Radic Biol Med* 28: 141–164, 2000.
5. R Brigelius-Flohe, M Traber. *FASEB J* 13: 1145–1155, 1999.
6. JF Keaney Jr, DI Simon, JE Freedman. *FASEB J* 13: 965–976, 1999.
7. JM Upston, AC Terentis, R Stocker. *FASEB J* 13: 977–994, 1999.
8. A Carr, B Frey. *FASEB J* 13:1007–1024, 1999.
9. LM May. *FASEB J* 13: 995–1006, 1999.
10. RL Arudi, MW Sutherland, BHJ Bielski. In *Oxy Radicals and Their Scavenger Systems*. Vol 1, G Cohen, RA Greenwald (Eds). Elsevier, Amsterdam, 1983.
11. IB Afanas'ev, VV Grabovetskii, NS Kuprianova. *J Chem Soc, Perkin Trans 2*. 281–284, 1987.
12. PB McCay, K-L Fong, EK Lai, MM King. In Tocopherol, *Oxygen and Biomembranes*. C de Duve, O Hayashi (Eds.), Elsevier, Amsterdam, 41–57, 1978.
13. PB McCay. *Ann Rev Nutr*. 5: 323–340, 1985.
14. T Doba, GW Burton, KU Ingold. *Biochim Biophys Acta*. 835: 298–303, 1985.
15. P Lambelet, J Loliger. *Chem Phys Lipids*. 35: 185–198, 1984.
16. M Scarpa, A Rigo, M Maiorino, F Ursini, C Gregolin. *Biochim Biophys Acta*. 801: 215–219, 1984.
17. GW Burton, A Joyce, KU Ingold. *Arch Biochem Biophys*. 221: 281–290, 1983.
18. J Terao, S Matsushota. *Lipids*. 21: 255–260, 1986.
19. VW Bowry, D Mohr, J Cleary, R Stocker. *J Biol Chem*. 270: 5756–5763, 1995.
20. RB Weinberg, BS VanderWerken, RA Anderson, JE Stegner, MJ Thomas. *Arterioscler Thromb Vasc Biol* 21: 1029–1033, 2001.
21. R Albertini, PM Abuja. *Free Radic Res*. 30: 181–188,1999.
22. KU Ingold, VW Bowry, R Stocker, C Walling. *Proc Natl Acad Sci USA*. 90: 45–49, 1993.
23. S Nagaoka, Y Okauchi, S Urano, U Nagashima, K Mukai. *J Am Chem Soc*. 112, 9821–8924, 1990.
24. AA Remorova, VA Roginskii. *Kinet Catal. (Engl Transl)* 32, 726–731, 1991.
25. MEJ Coronel, AJ Colussi. *Int J Chem Kinet*. 20, 749–752, 1988.
26. A Bindoli, M Valente, L Cavallini. *Biochem Int*. 10: 753–761, 1985.
27. H Shi, N Noguchi, E Niki. *Free Radic Biol Med*. 27: 334–346, 1999.
28. J Neuzil, PK Witting, R Stocker. *Proc Natl Acad Sci USA* 94 ; 7885–7890, 1997.
29. D Appenroth, E Karge, G Kiessling, WJ Wechter, K Winnefeld, C Fleck. *Toxicol Lett*. 122: 255–265, 2001.
30. GW Burton, KU Ingold, DO Foster, SC Cheng, A Webb, L Hughes, E Lusztyk. *Lipids*. 23: 834–840, 1988.
31. MG Traber, H Sies. *Annu Rev Nutr*. 16: 321–347, 1996.
32. A Hosomi, M Arita, Y Sato, C Kiyose, T Ueda, O Igarashi, H Arai, K Inoue. *FEBS Lett*. 409: 105–108, 1997.
33. AK Dutta-Roy. *Food Chem Toxicol*. 37: 967–971, 1999.
34. D Li, T Saldeen, F Romeo, JL Mehta. *J Cardiovasc Pharmacol Ther*. 6: 155–161, 2001.

35. S Christen, AA Woodall, MK Shigenaga, PT Southwell-Keely, MW Duncan, BN Ames. *Proc Natl Acad Sci USA* 94: 3217–3222, 1997.
36. LG Zhang, FA Nicholls-Grzemski, MA Tirmenstein, MW Fariss. *Chem Biol Interact*.138: 267–284, 2001.
37. LG Zhang, MA Tirmenstein, FA Nicholls-Grzemski, MW Fariss. *Arch Biochem Biophys*. 393: 87–96, 2001.
38. O Cachia, CL Leger, B Descomps. *Atherosclerosis*. 138: 263–269, 1998.
39. AA Beharka, D Wu, M Serafini, SN Meydani. *Free Radic Biol Med*. 32:503–511, 2002.
40. E Sodergren, J Cederberg, S Basu, B Vessby. *J Nutr* 130: 10–14, 2000.
41. M Gavazza, A Catala. *Mol Cell Biochem*. 225: 121–128, 2001.
42. T Yao, S Degli-Esposti, L Huang, R Arnon, A Spangenberger, MA Zern. *Am J Physiol*. 267: G476–G484, 1995.
43. KL Linseman, P Larson, JM Braughler, JM McCall. *Biochem Pharmacol*. 45: 1477–1482, 1993.
44. H Chen, AL Tappel. *Free Radic Biol Med*. 18: 949–953, 1995.
45. T Kaneko, K Kaji, M Matsuo. *Free Radic Biol Med*. 16: 405–409, 1994.
46. M Galleano, S Puntarulo. *Toxicology*.124:73–81, 1997.
47. JM Roob, G Khoschsorur, A Tiran, JH Horina, H Holzer, BM Winklhofer-Roob. *J Am Soc Nephrol*. 11: 539–649, 2000.
48. A Mezzetti, G Zuliani, F Romano, F Costantini, SD Pierdomenico, F Cuccurullo, R Fellin. *J Am Geriatr Soc* 49: 533–537, 2001.
49. EA Meagher, OP Barry, JA Lawson, J Rokach, GA FitzGerald. *JAMA* 285: 1178–1182, 2001.
50. B van der Loo, R Labugger, CP Aebischer, JN Skepper, M Bachschmid, V Spitzer, J Kilo, L Altwegg, V Ullrich, TF Luscher. *Circulation* 105: 1635–1638, 2002.
51. MR McCall, B Frei. *Free Radic Biol Med* 26: 1034–1053, 1999.
52. WA Pryor. *Free Radic Biol Med* 28: 141–164, 2000.
53. P-E Chabrier, M Auguet, B Spiinnewyn, et al. (15 authors). *Proc Natl Acad Sci USA* 96: 10824–10829, 1999.
54. AC Carr, BZ Zhu, B Frey. *Circ Res* 87: 349–354, 2000.
55. T Egger, A Hammer, A Wintersperger, D Goti, E Malle, W Satter. *J Neurochem* 79: 1169–1182, 2001.
56. DE Cabelli, BHJ Bielski. *J Phys Chem* 87: 1809–1812, 1983.
57. M Levine. *New Engl J Med* 314: 892–902, 1986.
58. M Gulumian, TA Kilroe-Smith. *Environ Res* 44: 254–259, 1987.
59. IB Afanas'ev, AI Dorozhko, AV Brodskii, VA Kostyuk, AI Potapovitch. *Biochem Pharmacol* 38: 1763–1769, 1989.
60. AB Weitberg. *Mutat Res* 191: 53–56, 1987.
61. LG Korkina, AD Durnev, TB Suslova, ZP Cheremisina, NO Daugel-Dauge, IB Afanas'ev. *Mutat Res* 265: 245–253, 1992.
62. DC Laudicina, LJ Marnett. *Arch Biochem Biophys* 278: 73–80, 1990.
63. P Zhang, ST Omaye. *Toxicol in vivo* 15: 13–24, 2001.
64. M Kapsokefalou, DD Miller. *Br J Nutr* 85: 681–687, 2001.
65. G Barja, M Lopez-Torres, R Perez-Campo, C Rojas, S Cadenas, J Prat, R Pamplona. *Free Radic Biol Med* 17: 105–117,1994.
66. M Paolini, L Pozzetti, GF Pedulli, E Marchesi, G Cantelli-Forti. *Life Sci* 64: PL 273–278, 1999.
67. JM Gutteridge, DA Rowley, E Griffiths, B Halliwell. *Clin Sci* (*Lond*). 68: 463–467, 1985.
68. JMC Gutteridge, DA Rowley, B Halliwell. *Biochem J* 206, 605–609, 1982.
69. TM Berger, MC Polidori, A Dabbagh, PJ Evans, B Halliwell, JD Morrow, LJ Robers II, B Frei. *J Biol Chem* 272: 15656–15660, 1997.
70. MA Livrea, MA Livrea, L Tesoriere, AD Pintaudi, A Calabrese, A Maggio, HJ Freisleben, D D'Arpa, R D'Anna, A Bongiorno. *Blood* 88: 3608–3614, 1996.
71. ID Podmore, HR Griffiths, KE Herbert, N Mistry, P Mistry, J Lunes. *Nature* 392: 559, 1998.
72. GR Ellis, RA Anderson, D Lang, DJ Blackman, RH Morris, J Morris-Thurgood, IF McDowell, SK Jackson, MJ Lewis, MP Frenneaux. *J Am Coll Cardiol* 36: 1474–1482, 2000.
73. C Colombic, HA Mattill. *J Am Chem Soc* 63: 1279–1280, 1941.

74. AI Tappel. *Geriatrics* 23: 97–105, 1968.
75. JE Packer, TF Slator, RF Willson. *Nature* 278: 737–739, 1979.
76. HW Leung, MJ Vang, RD Mavis. *Biochim Biophys Acta* 664: 266–272, 1981.
77. E Niki, J Tsuchiya, Rtanimura, Y Kamiya. *Chem Lett* 789–792, 1982.
78. A Negre-Salvayre, A Affany, C Hariton, R Salvayre. *Pharmacology* 42: 262–272, 1991.
79. T Doba, GW Burton, KU Ingold. *Biochim Biophys Acta* 835: 298–303, 1985.
80. B Frey, R Stocker, BN Ames. *Proc Natl Acad Sci USA* 85: 9748–9752, 1988.
81. MK Sharma, GR Buettner. *Free Radic Biol Med* 14: 649–653, 1993.
82. JM May, Z Qu, CE Cobb. *Free Radic Biol Med* 31: 117–124, 2001.
83. GT Vatassery, WE Smith, HT Quach. *J Nutr* 128:152–157, 1998.
84. M.Kirsch, H.de Groot. *J Biol Chem* 275: 16702–16708, 2000.
85. B Havsteen. *Biochem Pharmacol* 32: 1141–1148, 1983.
86. JW Critchfield, ST Butera, TM Folks. *AIDS Res Hum Retroviruses* 12: 39–46, 1996.
87. SC Chan, YS Chang, JP Wang, SC Chen, SC Kuo. *Planta Med* 64: 153–158, 1998.
88. H Cheong, SY Ryu, MH Oak, SH Cheon, GS Yoo, KM Kim. *Arch Pharm Res* 21: 478–480, 1998.
89. H Wei, X Zhang, J Zhao, Z Wang, D Bickers, M Lebwohl. *Free Radic Biol Med* 26: 1427–1435, 1999.
90. DE Pratt. *J Food Sci* 30: 737–741, 1965.
91. L Cavallini, A Bindoli, N Siliprandi. *Pharamacol Res Commun* 10: 133–137, 1978.
92. RA Larson. *Phytochemistry* 27: 969–978, 1988.
93. IB Afanas'ev. In: *Superoxide Ion: Chemistry and Biological Implications*. Vol 2. CRC Press, Boca Raton, FL, p. 244, 1990.
94. I Ueno, M Kohno, K Haraikawa, I Hirono. *J Pharm Din.* 7: 798–804, 1984.
95. RJ Gryglewski, R Korbut, J Robak, J Swies. *Biochem Pharmacol* 36: 317–322, 1987.
96. T Yoshimoto, M Furukawa, S Yamamoto, T Horie, S Watanabe-Kohno. *Biochem Biophys Res Commun* 116: 612–618, 1983.
97. AK Ratty, J Sunamoto, NP Das. *Biochem Pharmacol* 37: 989–997, 1988.
98. S Beyeler, B Testa, D Perrissoud. *Biochem Pharmacol* 37: 1971–1979, 1988.
99. M Merlos, RM Sanchez, J Camarasa, T Adzet. *Experiemtia* 47: 616–619, 1991.
100. WF Hodnick, CW Bohmont, C Capps, RS Pardini. *Biochem Pharmacol* 36: 2873–2874, 1987.
101. M Chiesi, P Schwaller. *Biochem Pharmacol* 49: 495–501, 1995.
102. VA Belyakov, VA Roginsky, W Bors. *J Chem Soc, Perkin Trans 2*, 2319–2326, 1995.
103. H Shi, B Zhao, W Xin. *Biochem Mol Int* 35: 981–994, 1995.
104. W Bors, C Michel. *Free Radic Biol Med* 27: 1413–1426, 1999.
105. S Jovanovic, S Steenken, M Tosic, B Majanovic, MGJ Simic. *J Am Chem Soc* 116: 4846–4851, 1994.
106. VA Kostyuk, AI Potapovich. *Arch Biochem Biophys* 355: 43–48, 1998.
107. L Selloum, S Reichl, M Muller, L Sebihi, J Arnhold. *Arch Biochem Biophys* 395: 49–56, 2001.
108. P Cos, L Ying, M Calomme, JP Hu, K Cimanga, B Van Poel, L Pieters, AJ Vlietinck, DV Berghe. *J Nat Prod* 61: 71–76, 1998.
109. J Robak, RJ Gryglewski. *Biochem Pharmacol* 37: 837–841, 1988.
110. Y Hanasaki, S Ogawa, S Fukui. *Free Radic Biol Med* 16: 845–850, 1994.
111. DE Shieh, LT Liu, CC Lin. *Anticancer Res* 20: 2861–2865, 2000.
112. Janzen, AL Wilcox, V Manoharan. *J Org Chem* 58: 3597–3598,1993.
113. SA van Acker, MN Tromp, GR Haenen, WJ van der Vijgh, A Bast. *Biochem Biophys Res Commun* 214: 755–759, 1995.
114. HK Kim, BS Cheon, YH Kim, SY Kim, HP Kim. *Biochem Pharmacol* 58: 759–765, 1999.
115. YC Park, G Rimbach, C Saliou, G Valacchi, L Packer. *FEBS Lett* 465: 93–97, 2000.
116. G Berton, C Schneider, D Romeo. *Biochim Biophys Acta* 595: 47–55, 1980.
117. T Nakayama, M Yamada, T Osawa, S Kawakishi. *Biochem Pharmacol* 45: 265–267,1993.
118. HW Lu, K Sugahara, Y Sagara, N Masuoka, Y Asaka, M Manabe, H Kodama. *Arch Biochem Biophys* 393: 73–77, 2001.
119. K Ishige, D Schubert, Y Sagara. *Free Radic Biol Med* 30: 433–446, 2001.
120. H Schroeter, JP Spenser, C Rice-Evans, RJ Williams. *Biochem J* 358: 547–557, 2001.
121. Dehmlow C, Murawski N, H de Groot. *Life Sci* 58: 1591–1600, 1996.

122. H Hamada, M Hiramatsu, R Edamatsu, A Mori. *Arch Biochem Biophys* 306: 261–266, 1993.
123. LG Korkina, IB Afanas'ev. In: *Advances in Pharmacology*. Academic Press, San Diego, Vol 38: 151–163, 1997.
124. J Torel, J Cillard. *Phytochemistry* 25: 383–387, 1986.
125. J Terao, M Piskula, Q Yao. *Arch Biochem Biophys* 308, 278–284, 1994.
126. T Decharneux, F Dubois, C Beauloye, S Wattiaux-De Coninck, R Wattiaux. *Biochem Pharmacol* 44: 1243–1248, 1992.
127. A Russo, R Acquaviva, A Campisi, V Sorrenti, C Di Giacomo, G Virgata, ML Barcellona, A Vanella. *Cell Biol Toxicol* 16: 91–98, 2000.
128. V Abalea, J Cillard, M-P. Dubos, O Sergent, P Cillard, I Morel. *Free Radic Biol Med* 26: 1457–1466, 1999.
129. AV Kozlov, EA Ostrachovitch, IB Afanas'ev. *Biochem Pharmacol* 47, 795–799, 1994.
130. I Morel, G Lescoat, P Cogrel, O Sergent, N Pasdeloup, P Brissot, P Cillard, J Cillard. *Biochem Pharmacol* 45: 13–19, 1993.
131. I Morel, V Abalea, O Sergent, P Cillard, J Cillard. *Biochem Pharmacol* 55: 1399–1404, 1998.
132. IB Afanas'ev, EA Ostrachovitch, NE Abramova, LG Korkina. *Biochem Pharmacol* 50: 627–635, 1995.
133. SA van Acker, GP van Balen, DJ van den Berg, A Bast, WJ van der Vijgh. *Biochem Pharmacol* 56: 935–943, 1998.
134. M Das, PK Ray. *Biochem Int* 17: 203–211, 1988.
135. FM Areias, AC Rego, CR Olivera, RM Seabra. *Biochem Pharmacol* 62: 111–118, 2001.
136. CGM Heijnen, GRMM Haenen, RM Oostveen, EM Stalpers, A Bast. *Free Radic Res* 36: 575–581, 2002.
137. A Saija, M Scalese, M Lanza, D Marzullo, F Bonina, F Castelli. *Free Radic Biol Med* 19: 481–486, 1995.
138. I Maridonneau-Parini, P Braquet, RP Garay. *Pharm Res Commun* 18: 61–73, 1986.
139. Y Sorata, U Takahama, M Kimura. *Biochim Biophys Acta* 799: 313–317, 1984.
140. TW Wu, LH Zeng, J Wu, KP Fung. *Biochem Pharmacol* 47: 1099–1103, 1994.
141. LN Grinberg, EA Rachmilewitz, H Newmark. *Biochem Pharmacol* 48: 643–649, 1994.
142. IB Afanas'ev, LG Korkina, KK Briviba, VI Gunar, BT Velichkovskii. In: Proceedings of 4th Biennial General Meeting of the Society for Free Radical Research "Medical and Chemical Aspects of Free Radicals," O. Hayashi et al. (Eds), Elsevier, Amsterdam, Vol 1, 515–518, 1989.
143. VA Kostyuk, AI Potapovich, SD Speransky, GT Maslova. *Free Radic Biol Med* 21: 487–493, 1996.
144. VA Kostyuk, AI Potapovich, EN Vladykovskaya, M Hiramatsu. *Planta Med* 66: 762–764, 2000.
145. H Yoshida, T Ishikawa, H Hosoai, M Suzukawa, M Ayaori, T Hisada, S Sawada, A Yonemura, K Higashi, T Ito, K Nakajima, T Yamashita, K Tomiyasu, M Nishiwaki, F Ohsuzu, H Nakamura. *Biochem Pharmacol* 58:1695–1703, 1999.
146. EL da Silva, DS Abdalla, J Terao. *IUBMB Life* 49: 289–295, 2000.
147. P Filipe, V Lanca, JN Silva, P Morliere, R Santus, A Fernandes. *Mol Cell Biochem* 221: 79–87, 2001.
148. J Terao, S Yamaguchi, M Shirai, M Miyoshi, JH Moon, S Oshima, T Inakuma, T Tsushida, Y Kato. *Free Radic Res* 35: 925–931, 2001.
149. J Robak, Z Diniec, H Rzadkowska-Bodalska, W Olechnowicz-Stepien, W Cisowski. *Pol J Pharmacol Pharm* 38: 483–491, 1986.
150. IM Erden, A Kahraman. *Toxicology* 154: 21–29, 2000.
151. J Hibatallah, C Carduner, MC Poelman. *J Pharm Pharmacol* 51:1435–1440, 1999.
152. SK Katiyar, F Afaq, A Perez, H Mukhtar. *Carcinogenesis* 22: 287–294, 2001.
153. LD Kok, YP Wong, TW Wu, HC Chan, TT Kwok, KP Fung. *Life Sci* 67: 91–99, 2000.
154. SM Jeon, SH Bok, MK Jang, MK Lee, KT Nam, YB Park, SJ Rhee, MS Choi. *Life Sci* 69: 2855–2866, 2001.
155. HM Abd El-Gawad, AE Khalifa. *Pharmacol Res* 43: 257–263, 2001.
156. R Funabiki, K Takeshita, Y Miura, M Shibasato, T Nagasawa. *J Agric Food Chem* 47:1078–1082, 1999.
157. C Perez Guerrero, MJ Martin, E Marhuenda. *Gen Pharmacol* 25: 575–580, 1994.

158. C La Casa, I Villegas, de la LC Alarson, V Motilva, CMJ Martin. *J Ethnopharmacol* 71: 45–53, 2000.
159. EE Deschner, J Ruperto, G Wong, HL Newmark. *Carcinogenesis* 12: 1193–1196, 1991.
160. K Yang, SA Lamprecht, Y Liu, H Shinozaki, K Fan, D Leung, H Newmark, VE Steele, GJ Kelloff, M Lipkin. *Carcinogenesis* 21:1655–1660, 2000.
161. K Shimoi, S Masuda, M Furugori, S Esaki, N Kinae. *Carcinogenesis* 15: 2669–2672, 1994.
162. H Ashida, I Fukuda, T Yamashita, K Kanazawa. *FEBS Lett* 476: 213–217, 2000.
163. MN Horcajada-Molteni, V Crespy, V Coxam, MJ Davicco, C Remesy, JP Barle. *J Bone Miner Res* 15: 2251–2258, 2000.
164. KN Raymond, SS Isied, LD Brown, FR Fronczek, JH Nibert. *J Am Chem Soc* 98: 1767–1774, 1976.
165. IB Afanas'ev, AI Dorozhko, NI Polozova, NS Kuprianova, AV Brodskii, EA Ostrachovitch, LG Korkina. *Arch Biochem Biophys* 302, 200–205, 1993.
166. IB Afanas'ev, EA Ostrakhovitch, LG Korkina. *FEBS Lett* 425, 256–258, 1998.
167. IB Afanas'ev, EA Ostrakhovitch, EV Mikhal'chik, GA Ibragimova, LG Korkina. *Biochem Pharmacol* 61: 677–684, 2001.
168. VA Kostyuk, AI Potapovich, EN Vladykovskaya, LG Korkina, IB Afanas'ev. *Arch Biochem Biophys* 385, 129–137, 2001.
169. ZS Zhao, S Khan, PJ O'Brien. *Biochem Pharmacol* 56: 825–830,1998.
170. AG Siraki, J Smythies, PJ O'Brien. *Neurosci Lett* 296: 37–40, 2000.
171. MY Moridani, PJ O'Brien. *Biochem Pharmacol* 62:1579–1585, 2001.
172. JE Brown, H Khodr, RC Hider, C Rice-Evans. *Biochem J* 330: 1173–1178,1998.
173. MJ Laughton, B Halliwell, PJ Evans, JRS Hoult. *Biochem Pharmacol* 38: 2859–2865, 1989.
174. AT Canada, E Giannella, TD Nguyen, RP Mason. *Free Radic Biol Med* 9: 441–449, 1990.
175. W Bors, C Michel, S Schikova. *Free Radic Biol Med* 19: 45–52, 1995.
176. E Sergediene, K Jonsson, H Szymusiak, B Tyrakowska, IMCM Rietjens, N Cenas. *FEBS Lett* 462: 392–396,1999.
177. M Mochizuki, S Yamazaki, K Kano, T Ikeda. *Biochim Biophys Acta* 1569: 35–44, 2002.
178. N Sugihara, T Arakawa, M Ohnishi, K Furuno. *Free Radic Biol Med* 27: 1313–1323, 1999.
179. RJ Rodriguez, CL Miranda, JF Stevens, ML Deinzer, DR Buhler. *Food Chem Toxicol* 39: 437–445, 2001.
180. G Galati, T Chan, B Wu, PJ O'Brien. *Chem Res Toxicol* 12: 521–525, 1999.
181. D Metodiewa, AK Jaiswal, N Cenas, E Dickancaite, J Segura-Aguilar. *Free Radic Biol Med* 26:107–116,1999.
182. M Jang, L Cai, GO Udeani, KV Slowing, cf. Thomas, CWW Beecher, HHS Fong, NR Farnsworth, AD Kinghorn, RG Mehta, RC Moon, JM Pezzuto. *Science* 275: 218–220, 1997.
183. S Stojanovic, H Sprinz, O Brede. *Arch Biochem Biophys* 391: 79–89, 2001.
184. EN Frankel, AL Waterhouse, JE Kinsella. *Lancet* 341: 1103–1104, 1993.
185. L Fremont, L Belguendouz, S Delpal. *Life Sci* 64: 2511–2521, 1999.
186. P Brito, LM Almeida, TCP Dinis. *Free Radic Res* 36: 621–631, 2002.
187. LA Stivala, M Savio, F Carafoli, P Perucca, L Bianchi, G Maga, L Forti, UM Pagnoni, A Albini, E Prosperi, V Vannini. *J Biol Chem* 276: 22586–22594, 2001.
188. MJ Burkitt, J Duncan. *Arch Biochem Biophys* 381: 253–263, 2000.
189. TW Wu, KP Fung, LH Zeng, J Wu, H Nakamura. *Biochem Pharmacol* 48: 419–422,1994.
190. P Cos, P Rajan, I Vedernikiva, M Calomme, L Pieters, AJ Vlietinck, K Augustus, A Haemers, DV Berghe. *Free Radic Res* 36: 711–716, 2002.
191. Z Varga, A Czompa, G Kakuk, S Antus. *Phytother Res* 15: 608–612, 2001.
192. OI Aruoma, B Halliwell, R Aeschbach, J Loligers. *Xenobiotica* 22: 257–268, 1992.
193. DR Janero, B Burghardt. *Biochem Pharmacol* 37: 3335–3342, 1988.
194. KH Cheeseman, LG Forni. *Biochem Pharmacol* 37: 4225–4233, 1988.
195. C Breugnot, C Maziere, S Salmon, M Auclair, R Santus, P Morliere, A Lenaers, JC Maziere. *Biochem Pharmacol* 40: 1975–1980, 1990.
196. M Kuzuya, M Naito, C Funaki, T Hayashi, K Asai, F Kuzuya. *J Lipid Res* 32: 197–205, 1991.

197. D Bonnefont-Rousselot, C Segaud, D Jore, J Delattre, M Gardes-Albert. *Radiat Res* 151: 343–353, 1999.
198. *Lipoic Acid in Health and Disease*, J Fuchs, L Packer, G Zimmer, (Eds). New York: Marcel Dekker, 1997.
199. L Packer, S Roy, CK Sen. In: H Sies (ed.) *Antioxidants in Disease, Mechanisms and Therapy*. San Diego: Academic Press, 1997, pp 79–96.
200. IB Afanas'ev, II Afanas'ev, LG Korkina. In: J Fuchs, L Packer, G Zimmer (Eds.), *Lipoic Acid in Health and Disease*. New York: Marcel Dekker, 1997, pp. 455–464.
201. A Bast, GRMM Haenen. *Biochim Biophys Acta* 963: 558–561, 1988.
202. H Scholich, ME Murphy, H Sies. *Biochim Biophys Acta* 1001: 256–261, 1989.
203. SK Jonas, PA Riley, RL Willson. *Biochem J* 264: 651–655, 1989.
204. T von Zglinicki, I Wiswedel, L Trumper, W Augustin. *Mech Ageing Dev* 57: 233–246, 1991.
205. VE Kagan, A Shvedova, E Serbinova, S Khan, C Swanson, R Powell, L Packer. *Biochem Pharmacol* 44: 1637–1649, 1992.
206. YL Suzuki, M Tsuchiya, L Packer. *Free Radic Res Commun* 15: 255–263, 1991.
207. BC Scott, OI Aruoma, PI Evans, C O'Neill, A Van der Vliet, CE Cross, H Tritschler, B Halliwell. *Free Radic Res* 20: 119–133, 1994.
208. IB Afanas'ev. *Superoxide Ion: Chemistry and Biological Implications*. Vol 2. Boca Raton, FL: CRC Press, 1990, pp 138–139.
209. S Dikalov, VV Khramtsov, G Zimmer. In: J Fuchs, L Packer, G Zimmer (eds), *Lipoic Acid in Health and Disease*. New York: Marcel Dekker, 1997, pp 47–66.
210. BHJ Bielski, DE Cabelli, RL Arudi, AB Ross. *J Phys Chem Ref Data* 4: 1041–1101, 1985.
211. CC Winterbourn, D Metodiewa. *Free Radic Biol Med* 27: 322–328, 1999.
212. C Mottley, RP Mason. *J Biol Chem* 276: 42677–42683, 2001.
213. IG Obrosova, AG Minchenko, V Marinescu, L Fathallah, A Kennedy, CM Stockert, RN Frank, MJ Stevens. *Diabetologia* 44: 1102–1110, 2001.
214. A El Midaoui, J de Champlain. *Hypertension* 39: 303–307, 2002.
215. H Sigel, B Prijs, DB McCornick, JCH Shih. *Arch Biochem Biophys* 187: 208–214, 1978.
216. L Muller, H Menzel. *Biochim Biophys Acta* 1052: 386–391, 1990.
217. II Afanas'ev, LG Korkina, AV Kozlov, IB Afanas'ev. *Phys Chem Biol Med* 3: 26–31, 1996.
218. JK Lodge, MG Traber, L Packer. *Free Radic Biol Med* 25: 287–297, 1998.
219. R Dringen, JM Gutterer, J Hirrlinger. *Eur J Biochem* 267: 4912–4916, 2000.
220. DA Wink, JA Cook, SY Kim, Y Vodovotz, R Pacelli, MC Krishna, A Russo, JB Mitchell, D Jourd'heuil, AM Miles, MB Grisham. *J Biol Chem* 272: 11147–11151, 1997.
221. S Canals, MJ Casarejos, S de Bernardo, E Rodriguez-Martin, MA Mena. *J Neurochem* 79: 1183–1195, 2001.
222. F Ford, MN Hughes, P Wardman. *Free Radic Biol Med* 32: 1314–1323, 2002.
223. CM Jones, A Lawrence, P Wardman, MJ Burkitt. *Free Radic Biol Med* 32: 982–990, 2002.
224. V Sampath, WS Caughey. *J Am Chem Soc* 107: 4076–4078, 1985.
225. D Ross, K Norbeck, P Moldeus. *J Biol Chem* 260: 15028–15032, 1985.
226. I Wilson, P Wardman, GM Cohen, MD Doherty. *Biochem Pharmacol* 35: 21–22, 1986.
227. CC Winterbourn. In: *The Oxygen Paradox* KJA Davies, F Ursini (eds), Cleup University Press, 1995, pp 23–32.
228. P Wardman. In: *Biothiols in Health and Disease*, L Parker, E Cadenas (eds), Marcel Dekker, New York, 1995, pp 1–19.
229. BE Sturgeon, HJ Sipe Jr, DP Barr, JT Corbett, JG Martinez, RP Mason. *J Biol Chem* 273: 30116–30121, 1998.
230. P Moldeus, IA Congreave, M Berggren. *Respiration* 50: s31–s42, 1986.
231. OI Aruoma, B Halliwell, BM Hoey, J Butler. *Free Radic Biol Med* 6: 593–597, 1989.
232. AF Junod, L Junod, G Grichting. *Agents Actions* 22: 177–183, 1987.
233. JJ Haddad. *Biochem Pharmacol* 63: 305–320, 2002.
234. ZA Muna, BJ Bolann, X Chen, J Songstad, RK Berge. *Free Radic Biol Med* 28: 1068–1078, 2000.

235. IB Afanas'ev. In: *Superoxide Ion: Chemistry and Biological Implications.* Vol 2. CRC Press, Boca Raton, FL, 1990, p. 12.
236. IB Afanas'ev. In *Superoxide Ion: Chemistry and Biological Implications.* Vol 1. CRC Press, Boca Raton, FL, 1989, Chapter 7.
237. K Takeshige, R Takayanagi, S Minakami. *Biochem J* 192: 861–866, 1980.
238. G Solaini, L Landi, P Pasquali, GA Rossi. *Biochem Biophys Res Commun* 147: 572–580, 1987.
239. RE Beyer. *Free Radic Biol Med* 5: 297–303, 1988.
240. B Frei, MC Kim, BN Ames. *Proc Natl Acad Sci USA* 87: 4878–4883, 1990.
241. V Kagan, E Serbinova, L Packer. *Biochem Biophys Res Commun* 169: 851–857, 1990.
242. P Forsmark, F Aberg, B Norling, K Nordenbrand, G Dallner, L Ernster. *FEBS Lett* 285: 39–43, 1991.
243. JM Villalba, F Navarro, F Cordoba, A Serrano, A Arroyo, FL Crane, P Navas. *Proc Natl Acad Sci USA* 92: 4887–4891, 1995.
244. T Takahashi, T Okamoto, T Kishi. *J Biochem* 119: 256–263, 1996.
245. JJ Poderoso, MC Carreras, F Schopfer, CL Lisdero, NA Riobo, C Giulivi, AD Boveris, A Boveris, E Cadenas. *Free Radic Biol Med* 26: 925–935, 1999.
246. T Takahashi, T Hohda, N Sugimoto, S Mizobuchi, T Okamoto, K Mori, T Kishi. *Biol Pharm Bull* 22: 1226–1233, 1999.
247. MM Pich, A Castagnoli, A Biondi, A Bernacchia, PL Tazzari, M D'Aurello, GP Castelli, G Formiggini, R Conte, C Bovina, G Lenaz. *Free Radic Res* 36: 429–436, 2002.
248. RT Matthews, L Yang, S Browne, M.Baik, M F Beal. *Proc Natl Acad Sci USA* 95: 8892–8897, 1998.
249. OT Raitakari, RJ McCredie, P Witting, KA Griffiths, J Letters, D Sullivan, R Stocker, DS Celermajer. *Free Radic Biol Med* 28: 1100–1105, 2000.
250. J Kaikkonen, T-P Tuomainen, K Nyyssonen, YT Salonen. *Free Radic Res* 36: 389–397, 2002.
251. FL Rosenfeldt, S Pepe, A Ninnane, P Nagley, M Rowland, R Qu, S Marasco, W Lyon, D Esmore. *Ann NY Acad Sci* 959: 355–359, 2002.
252. HM Abd El-Gawad, AE Khalifa. *Pharmacol Res* 43: 257–263, 2001.
253. T Yamamura, H Otani, Y Nakao, R Hattori, M Osako, H Imamura, DK Das. *Antioxid Redox Signal* 3:103–112, 2001.
254. LM Canfield, LA Davy, GL Thomas. *Biochem Biophys Res Commun* 128: 211–219, 1985.
255. LM Vervoort, JE Ronden, HH Thijssen. *Biochem Pharmacol* 54: 871–876, 1997.
256. BN Ames, R Cathart, E Schwiers, P Hochstein. *Proc Natl Acad Sci USA* 78: 6858–6862, 1981.
257. KR Maples, RP Mason. *J Biol Chem* 263: 1709–1712, 1988.
258. DB Peden, R Hohman, ME Brown, RT Maso, Cberkebile, HM Fales, MA Kaliner. *Proc Natl Acad Sci USA* 87: 7638–7642, 1990.
259. V Schlotte, A Sevanian, P Hochstein, KU Weithmann. *Free Radic Biol Med* 25: 839–847, 1998.
260. PM Abuja. *FEBS Lett* 446: 305–308, 1999.
261. M Bagnati, C Perugini, C Cau, R Bordone, E Albano, G Bellomo. *Biochem J* 340: 143–152, 1999.
262. R Santus, LK Patterson, P Filipe, P Morliere, GL Hug, A Fernandes, JC Maziere. *Free Radic Res* 35: 129–136, 2001.
263. DC Hooper, CS Scott, A Zborek, T Mikheeva, RB Kean, H Koprowski, SV Spitsin. *FASEB J* 14: 691–698, 2000.
264. K Sugioka, Y Shimosegawa, M Nakano. *FEBS Lett* 210: 37–39, 1987.
265. M Nakano, K Sugioka, I Naito, S Takekoshi, E Niki. *Biochem Biophys Res Commun* 142: 919–924, 1987.
266. LA Huber, E Scheffler, T Poll, R Ziegler, HA Dresel. *Free Radic Res Commun* 8: 167–173, 1990.
267. SA Ayres, M Tang, MTR Subbiah. *J Lab Clin Med* 128: 367–375, 1996.
268. M Tang, MTR Subbiah. *Biochim Biophys Acta* 1299: 155–159, 1996.
269. GT Shwaery, JA Vita, JR Keaney, Jr. *Atherosclerosis* 138: 255–262, 1998.
270. S Ayres, W Abplanalp, LH Liu, MTR Subbiah. *Am J Physiol* 274: E1002–E1008, 1998.
271. W Klinger, A Lupp, E Karge, H Baumbach, F Eichhorn, A Feix, F Fuldner, S Gernhardt, L Knels, B Kost, G Mertens, F Werner, M Oettel, W Romer, S Schwarz, W Elger, B Schneider. *Toxicol Lett* 128: 129–144, 2002.

272. J Hwang, H Peterson, HN Hodis, B Choi, A Sevanian. *Atherosclerosis* 150: 275–284, 2000.
273. JS Winterle, T Mill, T Harris, RA Goldbeck. *Arch Biochem Biophys* 392: 233–244, 2001.
274. AK Rattan, Y Arad. *Atherosclerosis* 136: 305–314, 1998.
275. JM Braughler, JF Pregenzer, RL Chase, LA Duncan, EJ Jacobsen, JM McCall. *J Biol Chem* 262: 10438–10440, 1987.
276. JM Braughler, JF Pregenzer. *Free Radic Biol Med* 7: 125–130, 1989.
277. W Zhao, JS Richardson, MJ Mombourquette, FA Weil. *Free Radic Biol Med* 19: 21–30, 1995.
278. GM Campo, F Squadrito, S Campo, D Altavilla, A Avenoso, M Ferlito, G Squadrito, AP Caputi. *Free Radic Res* 27: 577–590, 1997.
279. K Mertsch, T Grune, S Kunstmann, B Wiesner, AM Ladhoff, WG Siems, RF Haseloff, IE Blasig. *Biochem Pharmacol* 56: 945–954, 1998.
280. GJ Fici, JS Althaus, PF VonVoigtlander. *Free Radic Biol Med* 22: 223–228, 1997.
281. JH Kramer, WB Weglicki. *Free Radic Biol Med* 21: 813–825, 1996.
282. IT Mak, WB Weglicki. *Circ Res* 66:1449–1452, 1990.
283. E Lupo, R Locher, B Weisser, W Vetter. *Biochem Biophys Res Commun* 203: 1803–1808, 1994.
284. N Rojstaczer, DJ Triggle. *Biochem Pharmacol* 51: 141–150, 1996.
285. G Sobala, EJ Menzel, H Sinzinger. *Biochem Pharmacol* 61: 373–379, 2001.
286. C Napoli, S Salomone, T Godfraind, W Palinski, DM Capuzzi, G Palumbo, FP D'Armiento, R Donzelli, F de Nigris, RL Capizzi, M Mancini, JS Gonnella, A Bianchi. *Stroke* 30: 1907–1915, 1999.
287. M Inouye, T Mio, K Sumino. *Eur J Clin Pharamacol* 56: 35–41, 2000.
288. R Berkels, G Egink, TA Marsen, H Bartels, R Roesen, W Klaus. *Hypertension* 37: 240–245, 2001.
289. IT Mak, J Zhang, WB Weglinski. *Pharmacol Res* 45: 27–33, 2002.
290. IT Mak, WB Weglinski. *Circ Res* 63: 262–266, 1988.
291. RR Jenkins, CM Del Signore, P Sauer, S Skelly. *Lipids* 27: 539–542, 1992.
292. DR Janero, R Lopez, J Pittman, B Burghardt. *Life Sci* 44: 1579–1588, 1989.
293. OI Aruoma, C Smith, R Cecchini, PJ Evans, B Halliwell. *Biochem Pharmacol* 42: 735–743, 1991.
294. TL Yue, PJ McKenna, PG Lysko, RR Ruffolo Jr, GZ Feuerstein. *Atherosclerosis* 97: 209–216, 1992.
295. TL Yue, HY Cheng, PG Lysko, PJ McKenna, R Feuerstein, JL Gu, KA Lysko, LL Davis, G Feuerstein. *J Pharmacol Exp Ther* 263: 92–98, 1992.
296. TL Yue, PJ Mckenna, JL Gu, HY Cheng, RR Ruffolo Jr, GZ Feuerstein. *Hypertension* 22: 922–928, 1993.
297. DJ Santos, AJ Moreno. *Biochem Pharmacol* 61: 155–164, 2001.
298. XL Max, BL Lopez, GL Liu, TA Christopher, F Gao, Y Guo, CZ Feuerstein, RR Ruffolo Jr, FC Barone, TL Yue. *Circ Res* 80: 894–901, 1997.
299. T Habon, E Szabados, G Kesmarky, R Halmosi, T Past, B Sumegi, K Toth. *Cardiovasc Res* 52: 153–160, 2001.
300. TL Yue, PJ Mckenna, PG Lysko, JL Gu, KA Lysko, RR Ruffolo Jr, GZ Feuerstein. *Eur J Pharmacol* 251: 237–243, 1994.
301. K Oettl, J Greilberger, K Zangger, E Haslinger, G Reibnegger, G Jurgens. *Biochem Pharmacol* 62: 241–248, 2001.
302. N Noguchi, K Nishino, E Niki. *Biochem Pharmacol* 59: 1069–1076.
303. HY Cheng, CS Randall, WW Holl, PP Constantinides, TL Yue, GZ Feuerstein. *Biochim Biophys Acta* 1284: 20–28, 1996.
304. R Anderson, G Ramafi, AJ Theron. *Biochem Pharmacol* 52: 341–349, 1996.
305. PG Lysko, KA Lysko, CL Webb, G Feuerstein, PE Mason, MF Walter, RP Mason. *Biochem Pharmacol* 56: 1645–1656, 1998.
306. P Dandona, R Karne, H Ghanim, W Hamouda, A Aljada, CH Magsino, Jr. *Circulation* 101: 122–124, 2000.
307. ED Hall, PK Andrus, SL Smith, TJ Fleck, HM Scherch, BS Lutzke, GA Sawada, JS Althaus, PF Vonvoigtlander, GE Padbury, PG Larson, JR Palmer, GL Bundy. *J Pharmacol Exp Ther* 281: 895–904, 1997.
308. TT Rohn, MT Quinn. *Eur J Pharamacol* 353: 329–336, 1998.

309. CM Lauderback, AM Breier, J Hackett, S Varadarajan, J Goodlett-Mercer, DA Butterfield. *Biochim Biophys Acta* 1501, 149–161, 2000.
310. SA Everett, MF Dennis, KB Patel, S Maddix, SC Kundu, RL Willson. *J Biol Chem* 271, 3988–3994, 1996.
311. GW Burton, KU Ingold. *Science* 224: 569–573, 1984.
312. NI Krinsky. *Ann NY Acad Sci* 854: 443–447, 1998.
313. CS Foote, RW Denny. *J Am Chem Soc* 90: 6233–6235,1968.
314. OM Panasenko, VS Sharov, K Briviba, H Sies. *Arch Biochem Biophys* 373: 302–305, 2000.
315. SL Yeh, ML Hu. *Free Radic Res* 35: 203–213, 2001.
316. RJ Reiter, DX Tan, D Acuna-Castroviejo, S Burkhardt, M Karbownik. *Curr Top Biophys* 24: 171–183, 2000.
317. RJ Reiter, DX Tan, LC Manchester, W Qi. *Cell Biochem Biophys* 34: 237–256, 2001.
318. KA Marshall, RJ Reiter, B Poeggeler, OI Arouma, B Halliwell. *Free Radic Biol Med* 21: 307–315, 1996.
319. MA Livrea, L Tesoriere, D D'Arpa, M Morreale. *Free Radic Biol Med* 23, 706–711, 1997.
320. C Pieri, M Marra, F Moroni, R Recchioni, F Marcheselli. *Life Sci* 55: 271–276, 1994.
321. C Osuna, RJ Reiter, JJ Garcia, M Karbownik, DX Tan, JR Calvo, LC Mancheter. *Pharmacol Toxicol* 90: 32–37, 2002.
322. G Baydas, H Canatan, A Turkoglu. *J Pineal Res* 32: 225–230, 2002.
323. DX Tan, RJ Reiter, LC Manchester, MT Yan, M El-Sawi, RM Sainz, JC Mayo, R Kohen, M Allegra, R Hardeland. *Curr Top Med Chem* 2: 181–197, 2002.
324. F Antunes, LRC Barclay, KU Ingold, M King, FQ Norris, JC Scaiano, F Xi. *Free Radic Biol Med* 26: 117–128, 1999.
325. A Muller, E Caddenas, P Graf, H Sies. *Biochem Pharmacol* 33: 3235–3239, 1984.
326. M Hayashi, TF Slater. *Free Rad Res Commun* 2: 179–185, 1986.
327. M Maiorino, A Roveri, F Ursini. *Arch Biochem Biophys* 295: 404–409, 1992.
328. N Noguchi, Y Yoshida, H Kaneda, Y Yamamoto, E Niki. *Biochem Pharmacol* 44: 39–44, 1992.
329. CE Thomas, RL Jackson. *J Pharmacol Exp Ther* 256: 1182–1188, 1991.
330. N Noguchi, N Gotoh, E Niki. *Biochim Biophys Acta* 1213: 176–182, 1994.
331. A Lass, P Witting, R Stocker, H Esterbauer. *Biochim Biophys Acta* 1303: 111–118, 1996.
332. H Masumoto, H Sies. *Chem Res Toxicol* 9: 262–267, 1996.
333. RS Glass, R Farooqui, M Sabahi, KW Ehler. *J Org Chem* 54: 1092–1097, 1989.
334. W Maret. *Proc Natl Acad Sci USA*. 91: 237–241, 1994.
335. L Cai, J Koropatrick, MG Cherian. *Chem Biol Interact* 96: 143–155, 1995.
336. L Cai, G Tsiapalis, MG Cherian. *Chem Biol Interact* 115: 141–151, 1998.
337. L Cai, JB Klein, YJ Kang. *J Biol Chem* 275: 38957–38960, 2000.
338. MV Kumari, M Hiramatsu, M Ebadi. *Cell Mol Biol* (Noisy-le-grand). 46: 627–636, 2000.
339. A Markant, J Pallauf. *J Trace Elem Med Biol* 10: 88–95, 1996.
340. W Maret, BL Vallee. *Proc Natl Acad Sci USA* 95: 3478–3482, 1998.
341. C Jacob, W Maret, BL Valee. *Proc Natl Acad Sci USA* 96: 1910–1914, 1999.
342. MP Zago, PI Oteiza. *Free Radic Biol Med* 31: 266–274, 2001.
343. M Coassin, F.Ursini, A Bindoli. *Arch Biochem Biophys* 299: 330–334, 1992.
344. VC Yeomans, AR Rechner, CA Rice-Evans. *Free Radic Res* 36: 717–718, 2002.
345. IB Afanas'ev, TB Suslova, ZP Cheremisina, NE Abramova, LG Korkina. *Analyst* 120: 859–862, 1995.
346. BJ Day, I Batinic-Haberle, JD Crapo. *Free Radic Biol Med* 26: 730–736, 1999.
347. C Groussard, I Morel, M Chevanne, M Monnier, J Cillard, AJ Delamarche. *J Appl Physiol* 89: 169–175, 2000.
348. R Wu, D Lamontagne, J de Champlain. *Circulation* 105: 387–392, 2002.
349. A Lass, A Suessenbacher, G Wolkart, B Mayer, F Brunner. *Mol Pharmacol* 61: 1081–1088, 2002.
350. S Kearns, R Dawson, Jr. *Adv Exp Med Biol* 483: 563–570, 2000.
351. SA Messina, R Dawson, Jr. *Adv Exp Med Biol* 483: 355–367, 2000.
352. BP Vohra, X Hui. *Arch Physiol Biochem* 109: 90–94, 2001.
353. SK Jain, Lim G. *Free Radic Biol Med* 30: 232–237, 2001.

354. RAJ Smith, CM Porteous, CV Coulter, MP Murphy. *Eur J Biochem* 263: 709–716, 1999.
355. GF Kelso, CM Porteous, CV Coulter,,G Hughes, VK Porteous, EC Ledgerwood, RAJ Smith, MP Murphy. *J Biol Chem* 276: 4588–4596, 2001.
356. GF Kelso, CM Porteous, G Hughes, EC Ledgerwood, AM Gane, RAJ Smith, MP Murphy. *Ann NY Acad Sci* 959: 263–274, 2002.
357. L Packer, G Rimbach, F Virgili. *Free Radic Biol Med* 27:704–724, 1999.
358. K-J Cho, C-H Yun, L Packer, A-S Chung. *Ann NY Acad Sci* 928: 141–156, 2001.
359. J Kim, J Chehade, JL Pinnas, AD Mooradian. *Nutrition* 16: 1079–1081, 2000.
360. L Marcocci, L Packer, MT Droy-Lefaix, A Sekaki, M Gardes-Albert. *Methods Enzymol* 234: 462–475, 1994.
361. H Shi, E Niki. *Lipids* 33: 365–370, 1998.
362. LJ Yan, MT Droy-Lefaix, L Packer. *Biochem Biophys Res Commun* 212: 360–366, 1995.
363. K Prasad, SV Mantha, J Kalra, P Lee. *J Cardiovasc Pharmacol Ther* 2: 309–320, 1997.
364. V Ben-Shaul, L Lomnitski, A Nyska, Y Zurovsky, M Bergman, S Grossman. *Toxicol Lett* 123: 1–10, 2001.
365. J Cui, GA Cordis, A Tosaki, N Maulik, DK Das. *Ann NY Acad Sci* 957: 302–307, 2002.
366. SS Leonard, D Cutler, M Ding, V Vallyathan, V Castranova, X Shi. *Ann Clin Lab Sci* 32: 193–200, 2002.
367. C de la Lastra Alarcon, MD Barranco, V Motilva, HM Herrerias. *Curr Pharm Des* 7: 933–950, 2001.
368. RL Galli, B Shukitt-Hale, KA Youdim, JA Joseph. *Ann NY Acad Sci* 959: 128–132, 2002.
369. J-B Galey. In: *Antioxidants in Disease. Mechanisms and Therapy*. H Sies (ed.), Academic Press, San Diego, 1997, pp 167–203.
370. I Morel, J Cillard, G Lescoat, O Sergent, N Pasdeloup, AZ Ocaktan, MA Abdallah, P Brissot, P Cillard. *Free Radic Biol Med* 13: 499–508, 1992.
371. S Rachidi C Coudray, P Baret, G Gelon, J-L Pierre, A Favier. *Biol Trace Elem Res* 41: 77–87, 1994.
372. P Ponka, J Borova, J Neuwirt, O Fuchs. *FEBS Lett* 97: 317–321, 1979.
373. DR Richardson, P Ponka. *J Lab Clin Med* 131: 306–315, 1998.
374. M Hermes-Lima, E Nagy P Ponka, HM Schulman. *Free Radic Biol Med* 25: 875–880, 1998.
375. M Hermes-Lima, P Ponka, HM Schulman. *Biochim Biophys Acta* 1523: 154–160, 2000.
376. A Negre-Salvayre, A Affany, C Hariton, R Salvayre. *Pharmacology* 42: 262–272, 1991.
377. A Negre-Salvayre, L Mabile, J Delchambre, R Salvayre. *Biol Trace Elem Res* 47: 81–91, 1995.
378. A Schmitt, R Salvayre, J Delchambre, A Negre-Salvayre. *Br J Pharmacol* 116: 1985–1990, 1995.
379. B Kalyanaraman, VM Darley-Usmar, J Wood, J Joseph, S Parthasarathy. *J Biol Chem* 267: 6789–6795, 1992.
380. FAA van Acker, O Schouten, GRMM Haenen, WJF.van der Vijgh, A Bast. *FEBS Lett* 473: 145–148, 2000.
381. CR Pace-Asciak, S Hahn, EP Diamandis, G Soleas, DM Goldberg. *Clin Chim Acta* 235: 207, 219, 1995.
382. JE Freedman, C Parker III, L Li, JA Perlman, B Frei, V Ivanov, LR Deak, JD Iafrati, JD Folts. *Circulation* 103: 2792–2798, 2001.
383. P Pignatelli, FM Pulcinelli, A Celestini, L Lenti, A Ghiselli, PP Gazzaniga, F Violi. *Am J Clin Nutr* 72: 1150–1155, 2000.

30 Antioxidant Enzymes

30.1 SUPEROXIDE DISMUTASE: THE LATEST DEVELOPMENTS

Superoxide dismutase (SOD) is an antioxidant enzyme of great importance for the regulation of free radical-mediated processes in biological systems. The discovery of SOD or more exactly, the identification of erythrocuprein as a superoxide-dismuting enzyme by McCord and Fridovich [1] in 1969 was a pivotal step in the development of free radical studies in biology. Literature on SOD is overwhelming. At present about 24,000 references are cited in Medline, three books entitled *Superoxide Dismutase* (edited by L.W. Oberley) have been published by CRC Press from 1982 to 1985, and five International Conferences on Superoxide and Superoxide Dismutase have been held until 1989. Now, we will consider just the latest findings relevant to the mechanisms of antioxidant and prooxidant activities of SODs and their biological role.

30.1.1 MECHANISM OF SUPEROXIDE-DISMUTING ACTIVITY OF SODs

The mechanisms of superoxide-dismuting activity of SODs are well established. Dismutation of superoxide occurs at copper, manganese, or iron centers of SOD isoenzymes CuZnSOD, MnSOD, or FeSOD. These isoenzymes were isolated from a variety of sources, including humans, animals, microbes, etc. In the case of CuZnSOD, dismutation process consists of two stages: the one-electron transfer oxidation of superoxide by cupric form (Reaction (1)) and the one-electron reduction of superoxide by cuprous form (Reaction (2)).

$$O_2^{\bullet-} + Cu(II)ZnSOD \Longrightarrow O_2 + Cu(I)ZnSOD \qquad (1)$$

$$O_2^{\bullet-} + Cu(I)ZnSOD + 2H^+ \Longrightarrow H_2O_2 + Cu(II)ZnSOD \qquad (2)$$

Similar reactions are catalyzed by Mn and Fe centers of MnSOD and FeSOD. It is obvious that before participation in Reaction (2), superoxide must be protonized to form hyperoxyl radical $HOO^{\bullet}$ by an outer-sphere or an intra-sphere mechanisms. All stages of dismuting mechanism, including the measurement of elementary rate constants, have been thoroughly studied earlier (see, for example, Ref. [2]).

It is well known that most of the antioxidant enzymes and substrates can exhibit prooxidant activity under certain conditions, mainly because many stages of the reactions catalyzed by such enzymes are reversible. The question of possible prooxidant effects of SODs and the ability of SODs to react with the other substrates than superoxide have been studied for a long time. It is known that CuZnSOD is inactivated by the hydrogen peroxide formed. Hodgson and Fridovich [3] proposed that this inactivation depends on the reaction of hydrogen peroxide with the oxidized form Cu(II)ZnSOD yielding the "bound" hydroxyl radicals.

$$Cu(II)ZnSOD + H_2O_2 \Longleftrightarrow Cu(I)ZnSOD + O_2^{\bullet-} + 2H^+ \qquad (3)$$

$$Cu(I)ZnSOD + H_2O_2 \Longrightarrow (HO^{\bullet})\text{-}Cu(II)ZnSOD + HO^- \qquad (4)$$

The formed "bound" HO• is supposedly a reactive species capable of reacting with a neighboring histidine at the active site and inactivates the enzyme. This prooxidant action was referred to by the authors as the peroxidase activity of SOD.

However, it has been shown that the incubation of SOD with hydrogen peroxide in the presence of spin-trap DMPO resulted in the appearance of the ESR spectrum of DMPO–OH [4,5], apparently indicating the formation of free hydroxyl radicals during the inactivation of SOD. Nonetheless, Sankarapandi and Zweier [6] disputed this proposal, pointing out that the effects of hydroxyl radical scavengers ethanol, formate, and azide on the DMPO–OH spin adduct formation did not correspond to those, which should be in the case of free hydroxyl radicals. They suggested that in accordance with the Hodgson and Fridovich [3] proposal hydroxyl radicals formed by Reaction (4) remain bound to cupric ion and may participate in the substrate oxidation and the formation of DMPO–OH. Importantly, this reaction proceeded only in the presence of bicarbonate, which supposedly facilitated connection of the hydrogen peroxide molecule to the active site.

However, the existence of an extremely reactive "bound" hydroxyl radical is questionable because it is difficult to understand why it does not immediately react with adjacent molecules (most of the reactions of hydroxyl radicals proceed with the rates close to a diffusion limit). Therefore, the mechanism proposed by Zhang et al. [7,8] seems to be much more convincing. They suggested that the genuine oxidizing free radical formed during SOD inactivation is the bicarbonate radical anion $CO_3^{•-}$, which is formed as a result of the oxidation of bicarbonate. It has also been suggested that DMPO–OH is formed by the addition of water to an intermediate of the reaction of DMPO with $CO_3^{•-}$ via a nucleophilic or electron transfer mechanism.

Another mode of SOD prooxidant activity has been proposed by Offer et al. [9]. In 1973, Rotilio et al. [10] showed that SOD can readily oxidize ferrocyanide. Offer et al. [9] found that low SOD concentrations inhibited superoxide-induced oxidation of ferrocyanide, but SOD becomes prooxidative at higher concentrations. As this reaction did not require hydrogen peroxide, it was suggested that the prooxidant effect of enhanced SOD concentrations might be explained by decreasing the steady state of superoxide and the direct oxidation of ferrocyanide by SOD.

SOD is able to accelerate the oxidation of different substrates such as hydroquinone [11] or tetracyclic catechol [12]. den Hartog et al. [13] showed that high concentrations of CuZnSOD catalyzed the hydroxylation of coumarin 3-carboxylic acid. Winterbourn et al. [14] demonstrated that CuZnSOD (but not MnSOD) stimulated the autoxidation of low molecular weight thiols such as amino thiols, cysteine, cysteamine, and, to a less degree, glutathione. These authors suggested that thiol oxidation occurs by a two-electron transfer mechanism rather than through the mediation with superoxide; however, the observed acceleration by adventitious metal ions remained unexplained. The oxidation of substrates by CuZnSOD may be accompanied by the inactivation and the destruction of enzyme and the release of copper [11]. Hydrogen peroxide inactivation of CuZnSOD can be suppressed by nitric oxide [15]. Auchere and Capeillere-Blandin [16] demonstrated that CuZnSOD is easily oxidized by MPO–hydrogen peroxide–chloride system.

30.1.2 SOD Mimics and SOD Modification

One of the major drawbacks of SOD as putative pharmaceutical agent is its small plasma half-life. It has been suggested that this shortcoming might be corrected by the chemical modification of SOD, for example, by binding to water-soluble polymers [17] or the entrapment in liposomes [18]. However, it seems that the membrane-permeable, low molecular weight compounds to be more promising for the use as SOD mimics. Two major types of

SOD mimics have been developed: transition metal complexes and organic nitroxides. Among metal complexes, salen–manganese complexes [19], the salicylate iron complex [20], and Mn(III) meso-tetrakis(N-ethylpyridinium-2-yl)porphyrin [21] are thought to be effective SOD mimics. SOD dismuting activity has also been suggested for copper–rutin complex [22]. However, it has been suggested that the metal-containing SOD mimics might dissociate in the presence of proteins, losing SOD activity and releasing toxic metal ions [23].

Another type of SOD mimics is organic nitroxides 2,2,6,6-tetramethylpiperidinoxyl (TPO or TEMPO) and 3-carbamoyl-2,2,5,5-tetramethylpyrrolidinoxyl (3-CP). In 1988, Samuni et al. [24] suggested that these compounds may catalyze superoxide dismutation. Before discussing the mechanism of superoxide dismutation by SOD mimics, it is necessary to consider the difference between scavenging and dismuting modes of superoxide inhibition. Weiss et al. [25] stressed the well-known fact that the compound could be a SOD mimic if only it catalytically destroys superoxide, while superoxide scavengers react with superoxide stoichiometrically. Using stopped-flow kinetic method, these authors concluded that nitroxides TEMPO and TEMPOL (4-hydroxy-2,2,6,6-tetramethylpiperidinoxyl) have no SOD-dismuting activity. However, this conclusion has been contradicted in subsequent studies. Krishna et al. [26] showed that the limitations of stopped-flow method used in Ref. [25] did not permit to distinguish the catalytic activity of nitroxides and spontaneous superoxide dismutation.

The mechanism of catalytic superoxide dismutation by nitroxides can be presented as follows [27]:

$$RNO^{\bullet} + HOO^{\bullet} \Longrightarrow RNO^{+} + HOO^{-} \quad (5)$$

$$RNO^{+} + O_2^{\bullet-} \Longrightarrow RNO^{\bullet} + O_2 \quad (6)$$

In this mechanism, $RNO^{\bullet}$ and RNO^{+} are nitroxide radical and its oxoammonium cation. The rate constants for Reactions (5) and (6) determined by pulse radiolysis are equal to: $k_5 = 1.2 \pm 0.1 \times 10^8$ for TPO and $1.3 \pm 0.1 \times 10^6 \, l \, mol^{-1} \, s^{-1}$ for 3-CP and $k_6 = 3.4 \pm 0.2 \times 10^9$ for TPO and $5.0 \pm 0.2 \times 10^9 \, l \, mol^{-1} \, s^{-1}$ for 3-CP.

High values of reaction rates for the two dismutation steps confirm the ability of both nitroxides TPO and 3-CP to be SOD mimics. However, as follows from the above mechanism, hydroperoxyl radical and not superoxide must participate in the first dismutation step (Reaction (5)). (As expected, a rate constant for the reaction of nitroxides with superoxide is very low $<10^3 \, l \, mol^{-1} \, s^{-1}$ [27].) Therefore, superoxide had to be protonated before participating in Reaction (5), which will diminish the total catalytic process at physiological pH and increase it at lower pH values.

30.1.3 Biological Effects of SOD

Biological effects of SODs are widely studied in many in vitro, ex vivo, and in vivo experiments; many of these studies have already been considered throughout this book. Examples of SOD application for the treatment of "free radical" pathologies are discussed in Chapter 31. Some studies related to the effects of SODs on free radical processes in cells and tissues are discussed below.

The role of CuZnSOD on oxygen radical production in cerebral vessels has been studied. Didion et al. [28] demonstrated that endogenous CuZnSOD diminished superoxide levels in rabbit cerebral blood vessels and affected nitric oxide- and cyclooxygenase-mediated responses in cerebral microcirculation. A subsequent study by the same group [29] showed increased superoxide production and vascular dysfunction in CuZnSOD-deficient mice. Chang et al. [30] suggested that superoxide induced cytokines, which activated microglial cells and that CuZnSOD may exhibit neuroprotective function by suppression of microglial activation. Marikovsky et al. [31] showed that overexpressing CuZnSOD in transgenic mice induced a significant increase in the release of TNF-α and metalloproteinases from activated peritoneal elicited macrophages that points out an important role of SOD in inflammation.

At present, much attention is focused on the role of extracellular SOD (EC-SOD) [32], which is produced and secreted to the extracellular space by smooth muscle cells. Stralin and Marklund [33] pointed out that various factors including histamine, vasopressin, angiotensin II, and others may affect EC-SOD expression and change the susceptibility of the vascular wall to superoxide. Oury et al. [34] has showed that hyperoxia caused depletion of pulmonary EC-SOD: exposure to 100% dioxygen resulted in a significant decrease in EC-SOD levels in the lungs and bronchoalveolar lavage fluid of mice.

The relationship between EC-SOD expression and NO production in cells is of a great physiological and pathophysiological significance. In 1992, Oury et al. [35] demonstrated that EC-SOD increased oxygen toxicity in central nervous system (CNS) by the inhibition of superoxide-mediated inactivation of nitric oxide. This conclusion is obviously erroneous one because, as it is well known that the interaction of superoxide and nitric oxide results in the formation of a very toxic peroxynitrite. Indeed, the same authors recently showed that EC-SOD promoted nitric oxide vasodilation by dismuting superoxide [36]. On the other hand, it has been found that nitric oxide can downregulate the synthesis of EC-SOD by smooth muscle cells [37].

The important role of MnSOD in suppression of mitochondrial overproduction of oxygen radicals has already been discussed in Chapter 23 and other chapters of this book. Placed within the mitochondrial matrix, MnSOD catalyzes the dismutation of superoxide including cytosolic superoxide, which can be scavenged by the mitochondria via a polarized inner membrane. It has been shown that overexpression of MnSOD prevents apoptosis induced by TNF-α [38] or peroxynitrite [39]. Kiningham et al. [40] showed that MnSOD overexpression protected murine fibrosarcoma cells against cell death mediated by rotenone, adriamycin, or poly(ADP–ribose) polymerase. MnSOD overexpression has also been found to protect lung epithelial cells against hyperoxic injury [41]. MnSOD overexpression can be stimulated in endothelial cells by cytokines IL-1 and TNF-α, LPS [42], and vascular endothelial growth factor [43].

30.2 SUPEROXIDE REDUCTASES

Recent discovery of superoxide reductases (SORs), the enzymes capable of reducing superoxide to hydrogen peroxide, added a second type of enzymes, which are able to catalytically destroy superoxide in living organisms. These enzymes have been found in anerobic microorganisms and are supposedly needed for their protection against reactive oxygen species. SORs apparently substitute SODs, although the advantages of superoxide reduction before superoxide dismutation in anerobic organisms are still unclear.

Two types of SORs have been firstly described by Lombard et al. [44] and Jenney et al. [45]. The first one is a small protein called desulfoferrodoxin (Dfx) found in anerobic sulfate-reducing bacteria *Desulfoarculus baarsii* containing two protein domains: iron center I and iron center II [44]. Iron center II is supposed to be responsible for the superoxide reducing activity. Another SOR has been isolated from anerobic archaea, *Pyrococcus furiosus*, which has a unique mononuclear iron center [45]. Lombard et al. [46] and Jovanovic et al. [47] also demonstrated that the *Treponema pallidum* protein of *T. pallidum* belongs to a new class of SORs.

The kinetics and the mechanism of superoxide reduction by SORs have been studied by several researchers. It was suggested that SORs react with superoxide via an inner-sphere mechanism, binding superoxide at ferrous center to form a ferric–hydroperoxo intermediate [46,48–50]. The rate constant for this reaction is equal to 10^8–10^9 $l\ mol^{-1}\ s^{-1}$ [46,49]. This reaction is the first stage of the process, which is accompanied by the slow decomposition of intermediate with a rate of 500–5000 s^{-1} [49].

There is still a possibility that SOR activity of the enzymes studied could be adventitious [50], but on the whole, the experimental results are very convincing [51,52]. However, the questions remained related to catalytic mechanism of SORs. In contrast to SODs, these enzymes are unable to complete the catalytic cycle by themselves because they need reductants for reducing the oxidized form of an enzyme. Therefore, it is possible that the mechanism of superoxide reduction by SORs may change from catalytic to stoichiometric one depending on the presence of additional reductants.

30.3 CATALASE

For many years heme-containing enzyme catalase, which decomposes hydrogen peroxide to water and dioxygen, has been considered to be one of the major antioxidant enzymes. Similar to peroxidases, catalase forms intermediates: CatFe(III)H$_2$O$_2$ complex (Compound I), Cat-Fe(II)H$_2$O$_2$ (Compound II), and inactive superoxo complex of native enzyme CatFe(III)O$_2^{\cdot-}$ (Compound III) during the enzymatic cycle (see, for example, Ref. [53]). It is usually accepted that the damaging effect of hydrogen peroxide depends on the formation of hydroxyl or hydroxyl-like free radicals by the Fenton reaction. Therefore, the first target of hydroxyl radical attack should be the catalase itself, and the enzyme must be protected against the damaging effect of hydrogen peroxide.

Indeed, it has been shown that catalase is protected by NADPH. It was found that one molecule of NADPH tightly bound to each of the four subunits of catalase in human erythrocytes [54]. Kirkman et al. [55] demonstrated that physiologically realistic concentrations of NADPH effectively protected catalase against the damaging effect of hydrogen peroxide. It was proposed that NADPH protects catalase by suppression of the formation of inactive enzyme form Compound II. Various mechanisms of NADPH protective activity, including the formation of the intermediate state [56] and the one-electron reduction of Compound II through electron tunneling [57], as well as other pathways besides the prevention of Compound II formation [58] have been proposed.

Catalase (alone or together with SOD) has been applied in many experimental investigations; some examples are given below. In 1991, Koerner et al. [59] showed that catalase + SOD reduced postischemic myocardial dysfunction in rabbits. Mao et al. [60] found that the high levels of SOD catalyzed the formation of hydroxyl radicals in the presence of iron ions. The use of SOD–catalase conjugates resulting in 80% normalization of normal mechanical function of reperfused rat hearts, in which free SOD was toxic or failed to give any protection. Rojanasakul et al. [61] proposed that hydroxyl radicals are formed in hydrogen peroxide-treated cells. To fight iron-mediated hydroxyl radical formation, these authors applied transferrin–catalase conjugate, which exhibited much improved protective effects on cells under oxidative stress compared to free catalase. It has been shown that catalase and vitamins E and C protected rat hepatocytes against bromobenzene-induced toxicity mediated by hydrogen peroxide [62]. Catalase and SOD protected renal epithelial cells against free radical injury induced by oxalate or calcium oxalate monohydrate crystals [63].

The treatment of rats with acute lung injury, induced by the intra-alveolar formation of IgG immune complexes of BSA, with catalase caused substantial protection or failed to prevent lung injury depending on the time of catalase administration [64]. Catalase in cytosol and mitochondria of the cells transfected with human catalase cDNA is capable of protecting cells against oxidative stress [65]. Han et al. [66] showed that catalase inhibited hydrogen peroxide upregulation of iNOS expression through NFκB activation in peritoneal macrophages.

30.4 GLUTATHIONE REDOX CYCLE

Glutathione redox cycle consists of NADPH, reduced glutathione GSH, glutathione reductase (GR), and glutathione peroxidase (GP) (Reactions (7) and (8)):

$$\text{NADPH} + \text{GSSR} + \text{H}^+ \xRightarrow{\text{Glutathione reductase}} \text{NADP}^+ + 2\text{GSH} \quad (7)$$

$$2\text{GSH} + \text{H}_2\text{O}_2 \xRightarrow{\text{Glutathione peroxidase}} \text{GSSG} + 2\text{H}_2\text{O} \quad (8)$$

Glutathione peroxidase is a selenium-dependent enzyme, which rapidly detoxifies hydrogen peroxide and various hydroperoxides. Suttorp et al. [67] showed that the impairment of glutathione cycle resulted in an increase in the injury of pulmonary artery endothelial cells. Glutathione cycle protected against endothelial cell injury induced by 15-HPETE, an arachidonate metabolite produced by 15-lipoxygenase-catalyzed oxidation [68].

Glutathione peroxidase plays an especially important role in mitochondria where catalase is absent. There are two kinds of glutathione peroxidases, namely classic glutathione peroxidase (cGP) and phospholipid hydroperoxide glutathione peroxidase (PHGP). The latter is the only known intracellular antioxidant enzyme capable of directly reducing phospholipid peroxides and cholesterol in the membrane. Thus, it has been shown that PHGP expression protects cells from lipid hydroperoxide-mediated injury [69,70]. Arai et al. [71] demonstrated that PHGP expression suppressed cell death caused by mitochondrial injury due to potassium cyanide or rotenone.

Glutathione cycle plays an important protective role against the reactive drug intermediates. For example, the GP mimetic ebselen synergistically enhanced the cardioprotection by SOD mimetic MnTBAP in doxorubicin-treated rat cardiomyocytes [72]. Glutathione peroxidase exhibits strong protective effect against ischemic reperfusion injury. Yoshida et al. [73] showed that the glutathione peroxidase knockout mice is more susceptible to myocardial ischemia and reperfusion damage than normal mice. On the other hand, the overexpression of glutathione peroxidase decreased necrotic and apoptotic cell death in transgenic mice [74]. In subsequent work Ishibashi et al. [75] demonstrated that the overexpression of glutathione peroxidase modulated inflammatory response in transgenic mice. Peng and Li [76] showed that the induction of glutathione and the enzymes of glutathione cycle by the chemoprotective agent 3H-1,2-dithiole-3-thione protected against oxidative cell injury in rat cardiomyocytes.

30.5 THIOREDOXIN REDUCTASE

Mammalian thioredoxin reductases are a family of selenium-containing pyridine nucleotide-disulfide oxidoreductases. These enzymes catalyze NADPH-dependent reduction of the redox protein thioredoxin (Trx), which contains a redox-active disulfide and dithiol group and by itself may function as an efficient cytosolic antioxidant [77]. One of the functions of Trx/thioredoxin reductase system is the NADPH-catalyzed reduction of protein disulfide [78]:

$$\text{NADPH} + (\text{thioredoxin reductase})\text{S}_2 + \text{H}^+ \Longrightarrow \text{NADP}^+ + (\text{thioredoxin reductase})(\text{SH})_2 \quad (9)$$

$$(\text{thioredoxin reductase})(\text{SH})_2 + \text{TrxS}_2 \Longrightarrow (\text{thioredoxin reductase})\text{S}_2 + \text{Trx}(\text{SH})_2 \quad (10)$$

$$\text{Trx}(\text{SH})_2 + (\text{protein})\text{S}_2 \Longrightarrow \text{TrxS}_2 + (\text{protein})(\text{SH})_2 \quad (11)$$

Mammalian thioredoxin reductase is able to reduce many substances in addition to thioredoxin such as insulin, vitamin K, alloxan, and others, while *Escherichia coli* enzyme is a highly specific for thioredoxins [79]. Nikitovic and Holmgren [80] have also shown that Trx/thioredoxin reductase system cleaves S-nitrosoglutathione to form glutathione and nitric oxide.

In 1986, the antioxidant effects of thioredoxin reductase were studied by Schallreuter et al. [81]. It has been shown that thioredoxin reductase was contained in the plasma membrane surface of human keratinocytes where it provided skin protection against free radical mediated damage. Later on, the reductive activity of Trx/thioredoxin reductase system has been shown for the reduction of ascorbyl radical to ascorbate [82], the redox regulation of NFκB factor [83], and in the regulation of nitric oxide–nitric oxide synthase activities [84,85].

30.6 PHASE 2 ANTIOXIDANT ENZYMES AND ALKENAL/ONE OXIDOREDUCTASE

Phase 2 enzymes (glutathione transferase, glucuronosyltransferases, NADPH:quinone reductase, heme oxygenase 1, and some others) are usually considered as protective enzymes against electrophilic carcinogens. However, it is now recognized that many Phase 2 enzymes exhibit indirect antioxidant properties. Gao et al. [86] found that the sulforaphane induction of Phase 2 enzymes protected human adult retinal pigment epithelial cells from prooxidant effects of menadione, *tert*-butyl hydroperoxide, 4-hydroxynonenal, and peroxynitrite. Similar protective effects were demonstrated for human keratinocytes and murine leukemia cells. Yang et al. [87] showed that glutathione S-transferases protect cells against lipid peroxidation by catalyzing GSH-dependent reduction of hydrogen peroxide and phospholipid hydroperoxides.

Dick et al. [88] studied the antioxidant effects of NADPH-dependent alkenal/one oxidoreductase (AO), also known as leukotriene B_4 12-hydroxydehydrogenase, 15-oxoprostaglandin 13-reductase, and dithiolethione-inducible gene-1. It was found that AO catalyzed the hydrogenation of many α,β-unsaturated aldehydes and ketones and protected against cytotoxic action of 4-hydroxy-2-nonenal.

REFERENCES

1. JM McCord, I Fridovich. *J Biol Chem* 244: 6049–6053, 1969.
2. IB Afanas'ev. In: *Superoxide Ion: Chemistry and Biological Implications*, vol 2. CRC Press, Boca Raton, FL, 1990, pp. 157–161.
3. EK Hodgson, I Fridovich. *Biochemistry* 14: 5294–5303, 1975.
4. MB Yim, PB Chock, ER Stadtman. *Proc Natl Acad Sci USA* 87: 5006–5010, 1990.
5. K Sato, T Akaike, M Kohno, M Ando, H Maeda. *J Biol Chem* 267: 25371–25377, 1992.
6. S Sankarapandi, JL Zweier. *J Biol Chem* 274: 34576–34583, 1999.
7. H Zhang, J Joseph, C Felix, B Kalyanaraman. *J Biol Chem* 275: 14038–14045, 2000.
8. H Zhang, J Joseph, M Gurney, D Becker, B Kalyanaraman. *J Biol Chem* 277: 1012–1020, 2002.
9. T Offer, A Russo, A Samuni. *FASEB J* 14: 1215–1223, 2000.
10. G Rotilio, L Morpurgo, L Calabrese, B Mondovi. *Biochim Biophys Acta* 302: 229–235, 1973.
11. Y Li, P Kuppusamy, JL Zweir, MA Trush. *Mol Pharmacol* 49: 412–421, 1996.
12. C Nebot, M Moutet, P Huet, J-Z Xu, J-C Yadan, J Chaudiere. *Anal Biochem* 214: 442–451, 1993.
13. GJ den Hartog, GR Haenen, E Vegt, WJ van der Vijgh, A Bast. *Chem Biol Interact* 145: 33–39, 2003.
14. CC Winterbourn, AV Peskin, HN Parsons-Mair. *J Biol Chem* 277: 1906–1911, 2002.
15. YS Kim, S Han. *FEBS Lett* 479: 25–28, 2000.
16. F Auchere, C Capeillere-Blandin. *Free Radic Res* 36: 1185–1198, 2002.
17. T Ogino, M Inoue, Y Ando, M Awai, H Maeda, Y Morino. *Int J Peptide Res* 32: 153–159, 1988.
18. T Yusa, JD Crapo, BA Freeman. *J Appl Physiol* 57: 1674–1681, 1984.

19. SR Doctrow, K Huffman, CB Marcus, W Musleh, A Bruce, M Baudry, B Malfroy. *Adv Pharmacol* 38: 247–269, 1997.
20. D Jay, EG Jay, MA Medina. *Arch Med Res* 30: 93–96, 1999.
21. I Spasojevic, I Batinic-Haberle, JS Reboucas, YM Idemori, I Fridovich. *J Biol Chem* 278: 6831–6837, 2003.
22. IB Afanas'ev, EA Ostrakhovitch, LG Korkina. *FEBS Lett* 425, 256–258, 1998.
23. MC Krishna, A Russo, JB Mitchell, S Goldstein, H Dafni, A Samuni. *J Biol Chem* 271: 26026–26031, 1996.
24. A Samuni, CM Krishna, P Riesz, E Finkelstein, A Russo. *J Biol Chem* 263: 17921–17924, 1988.
25. RH Weiss, AG Flickinger, WJ Rivers, MM Hardy, KW Aston, US Ryan, DP Riley. *J Biol Chem* 268: 23049–23054, 1993.
26. MC Krishna, A Russo, JB Mitchell, S Goldstein, H Dafni, A Samuni. *J Biol Chem* 271: 26026–26031, 1996.
27. S Goldstein, G Merenyi, A Russo, A Samuni. *J Am Chem Soc* 125: 789–795, 2003.
28. SP Didion, CA Hathaway, FM Faraci. *Am J Physiol Heart Circ Physiol* 281: H1697–H1703, 2001.
29. SP Didion, MJ Ryan, LA Didion, PE Fegan, CD Sigmund, FM Faraci. *Circ Res* 91: 938–944, 2002.
30. SC Chang, MC Kao, MT Fu, CT Lin. *Free Radic Biol Med* 31: 1084–1089, 2001.
31. M Marikovsky, V Ziv, N Nero, C Harris-Cerruti, O Mahler. *J Immunol* 170: 2993–3001, 2003.
32. T Fukai, RJ Folz, U Landmesser, DG Harrison. *Cardiovasc Res* 55: 239–249, 2002.
33. P Stralin, SL Marklund. *Am J Physiol Heart Circ Physiol* 281: H1621–H1629, 2001.
34. TD Oury, LM Schaefer, CL Fattman, A Choi, KE Weck, SC Watkins. *Am J Physiol Lung Cell Mol Physiol* 283: L777–L784, 2002.
35. TD Oury, YS Ho, CA Piantadosi, JD Crapo. *Proc Natl Acad Sci USA* 89: 9715–9719, 1992.
36. IT Demchenko, TD Oury, JD Crapo, CA Piantadosi. *Circ Res* 91: 1031–1037, 2002.
37. P Stralin, H Jacobsson, SL Marklund. *Biochim Biophys Acta* 1619: 1–8, 2003.
38. SK Manna, HJ Zhang, T Yan, LW Oberley, BB Aggarwal. *J Biol Chem* 273: 13245–13254, 1998.
39. JN Keller, MS Kindy, FW Holtsberg, DK St Clair, HC Yen, A Germeyer, SM Steiner, AJ Bruce-Keller, JB Hutchins, MP Mattson. *J Neurosci* 18: 687–697, 1998.
40. KK Kiningham, TD Oberley, S-M Lin, CA Mattingly, DK St Clair. *FASEB J* 13: 1601–1610, 1999.
41. AM Ilizarov, HC Koo, JA Kazzaz, LL Mantell, Y Li, R Bhapat, S Pollack, S Horowitz, JM Davis. *Am J Respir Cell Mol Biol* 24: 436–441, 2001.
42. GA Visner, SE Chesrown, J Monnier, US Ryan, HS Nick. *Biochem Biophys Res Commun* 188: 453–462, 1992.
43. MR Abid, JC Tsai, KC Spokes, SS Deshpande, K Irani, WC Aird. *FASEB J* 10.1096/fj.01-0338fje.
44. M Lombard, M Fontecave, D Touati, V Niviere. *J Biol Chem* 275: 115–121, 2000.
45. FE Jenney Jr, MF Verhagen, X Cui, MW Adams. *Science* 286: 306–309, 1999.
46. M Lombard, D Touati, M Fontecave, V Niviere. *J Biol Chem* 275: 27021–27026, 2000.
47. T Jovanovic, C Ascenso, KR Hazlett, R Sikkink, C Krebs, R Litwiller, LM Benson, I Moura, JJ Moura, JD Radolf, BH Huynh, S Navlor, F Rusnak. *J Biol Chem* 275: 28439–28448, 2000.
48. MD Clay, FE Jenney Jr, PL Hagedoorn, GN George, MW Adams, MK Johnson. *J Am Chem Soc* 124: 788–805, 2002.
49. V Niviere, M Lombard, M Fontecave, C Houee-Levin. *FEBS Lett* 497: 171–173, 2001.
50. JP Emerson, ED Coulter, DE Cabelli, RS Phillips, DM Kurtz Jr. *Biochemistry* 41: 4348–4357, 2002.
51. JA Imlay. *J Biol Inorg Chem* 7: 659–663, 2002.
52. MW Adams, FE Jenney Jr, MD Clay, MK Johnson. *J Biol Inorg Chem* 7: 647–652, 2002.
53. IB Afanas'ev. In: *Superoxide Ion: Chemistry and Biological Implications*, vol 2., CRC Press, Boca Raton, FL, 1990, pp. 154–156.
54. HN Kirkman, GF Gaetani. *Proc Natl Acad Sci USA* 81: 4343–4347, 1984.
55. HN Kirkman, S Galiano, GF Gaetani. *J Biol Chem* 262: 660–666, 1987.
56. A Hillar, P Nicholls, J Switals, PC Loewen. *Biochem J* 300: 531–539, 1994.
57. O Almarsson, A Sinha, E Gopinath, TC Bruce. *J Am Chem Soc* 115: 7093–7102, 1993.
58. HN Kirkman, M Rolfo, AM Ferraris, GF Gaetani. *J Biol Chem* 274: 13908–13914, 1999.
59. JE Koerner, BA Anderson, RC Dage. *J Cardiovasc Pharmacol* 17: 185–191, 1991.
60. GD Mao, PD Thomas, GD Lopaschuk, MJ Poznansky. *J Biol Chem* 268: 416–420, 1993.

61. Y Rojanasakul, X Shi, D Deshpande, WW Liang, LY Wang. *Biochim Biophys Acta* 1315: 21–28, 1996.
62. J Wu, K Karlsson, A Danielsson. *J Hepatol* 26: 669–677, 1997.
63. S Thamilselvan, KJ Byer, RL Hackett, SR Khan. *J Urol* 164: 224–229, 2000.
64. RL Warner, NM Bless, CS Lewis, E Younkin, L Beltran, R Cuo, KJ Johnson, J Varani. *Free Radic Biol Med* 29: 8–16, 2000.
65. J Bai, AM Rodriguez, JA Melendez, AI Cederbaum. *J Biol Chem* 274: 26217–26224, 1999.
66. YJ Han, YG Kwon, HT Chung, SK Lee, RL Simmons TR Billiar, YM Kim. *Nitric Oxide* 5: 504–513, 2001.
67. N Suttorp, W Toepfer, L Roka. *Am J Physiol* 251: C671–C680, 1986.
68. H Ochi, I Morita, S Murota. *Arch Biochem Biophys* 294: 407–411, 1992.
69. G Sun, H Kojima, S Komura, N Ohishi, K Yagi. *Biochem Mol Biol Int* 42: 957–963, 1997.
70. K Yagi, Y Shidoji, S Komura, H Kojima, N Ohishi. *Biochem Biophys Res Commun* 45: 528–533, 1998.
71. M Arai, H Imai, T Koumura, M Yoshida, K Emoto, M Umeda, N Chiba, Y Nakagawa. *J Biol Chem* 274: 4924–4933, 1999.
72. EA Konorev, MC Kennedy, B Kalyanaraman. *Arch Biochem Biophys* 368: 421–428, 1999.
73. T Yoshida, N Maulik, RM Engelman, YS Ho, JL Magnenat, JA Rousou, JE Flack III, D Deaton, DK Das. *Circulation* 96: II-216–II-220, 1997.
74. N Ishibashi, O Prokopenko, M Weisbrot-Lefkowitz, KR Reuhl, O Mirochnitchenko. *Brain Res Mol Brain Res* 109: 34–44, 2002.
75. N Ishibashi, O Prokopenko, KR Reuhl, O Miroshnitchenko. *J Immunol* 168: 1926–1933, 2002.
76. X Peng, Y Li. *Pharamacol Res* 45: 491–497, 2002.
77. R Goldman, DA Stoyanovsky, BW Day, VE Kagan. *Biochemistry* 34: 4765–4772, 1995.
78. ESJ Arner, A Holmgren. *Eur J Biochem* 267: 6102–6109, 2000.
79. A Holmgren, M Bjornstedt. *Meth Enzymol* 252: 199–208, 1995.
80. D Nikitovic, A Holmgren. *J Biol Chem* 271: 19180–19185, 1996.
81. KU Schallreuter, JM Wood. *Biochem Biophys Res Commun* 136: 630–637, 1986.
82. JM May, CE Cobb, S Mendiratta, KE Hill, RF Burk. *J Biol Chem* 273: 23039–23045, 1998.
83. T Hayashi, Y Ueno, T Okamoto. *J Biol Chem* 268: 11380–11388, 1993.
84. PJ Ferret, E Soum, O Negre, EE Wollman, D Fradelizi. *Biochem J* 346: 759–765, 2000.
85. LE Shao, T Tanaka, R Gribi, J Yu. *Ann NY Acad Sci* 962: 140–150, 2002.
86. X Gao, AT Dinkova-Kostova, P Talaly. *Proc Natl Acad Sci USA* 98: 15221–15226, 2001.
87. Y Yang, JZ Cheng, SS Singhal, M Saini, U Pandya, S Awasthi, YC Awasthi. *J Biol Chem* 276: 19220–19230, 2001.
88. RA Dick, M-K Kwak, TR Sutter, TW Kensler. *J Biol Chem* 276: 40803–40810, 2001.

31 Free Radicals and Oxidative Stress in Pathophysiological Processes

For a long time the formation of free radicals and the development of oxidative stress in living organisms have been considered as exclusively damaging factors, leading to various pathophysiological events. This is not surprising because numerous studies showed that the development of many diseases is associated with the overproduction of free radicals. Although this fact cannot of course be an undeniable proof of the predominant role of free radicals or oxidative stress in these pathologies, an excess in the formation of free radicals and other reactive oxygen and nitrogen species pointed out at least a casual intervention of all these reactive compounds in disease development. However, we now know that not only the overproduction but also the insufficient formation of free radicals could also be a cause of pathological disorders. Therefore, the main approach to the treatment and prevention of pathologies associated or accompanied by free radical formation should be the regulation and not simple suppression of their formation.

At present, numerous free radical studies related to many pathologies have been carried out. The amount of these studies is really enormous and many of them are too far from the scope of this book. The main topics of this chapter will be confined to the mechanism of free radical formation and oxidative processes under pathophysiological conditions. We will consider the possible role of free radicals in cardiovascular disorders, cancer, anemias, inflammation, diabetes mellitus, rheumatoid arthritis, and some other diseases. Furthermore, the possibilities of antioxidant and chelating therapies will be discussed.

31.1 CARDIOVASCULAR DISEASES

31.1.1 ISCHEMIA–REPERFUSION

Hypoxia (lack of oxygen relative to metabolic needs) and reoxygenation (reintroduction of oxygen in hypoxic tissue) are two important origins of cellular injury [1]. It is believed that reoxygenation injury is a cause of circulatory shock, myocardial ischemia, and stroke. (It is well known that myocardial infraction and stroke, diseases due to ischemia, are common cases of morbidity and mortality.) The enhancement of oxygen radical production during reoxygenation has long been recognized as a cause of reoxygenation injury. The formation of oxygen radicals under these pathophysiological conditions has been suggested based on the inhibitory effects of antioxidant enzymes and hydroxyl radical scavengers [2–5] and later on with the aid of ESR spin-trapping method [6].

Cellular hypoxia may lead to different degrees of cell injury. Sublethal hypoxia can be reversed without apparent consequences and can even be followed by enhanced resistance to reoxygenation injury (named conditioning). Severe hypoxia–reoxygenation results in cell death, necrosis, and apoptosis. It is of utmost importance to identify the sources responsible for free radical production during ischemia–reperfusion. In 1985, McCord [7] proposed a mechanism of superoxide formation in reperfusion. It is known that during ischemia xanthine dehydrogenase is converted to xanthine oxidase and ATP is transformed to hypoxanthine and xanthine. Thus, ischemia induces the formation of xanthine–xanthine oxidase system, a producer of superoxide. McCord's work has been followed by many studies with emphasis on the importance of the xanthine oxide-catalyzed oxygen radical formation in ischemia–reperfusion. As superoxide is a rather inactive free radical, it has been proposed that it be converted to hydroxyl radicals, stimulating, for example, the release of ferrous ions from ferritin [8].

Later on, the importance of xanthine oxidase as the producer of reoxygenation injury was questioned at least in the cells with low or no xanthine oxidase activity. Thus, it has been shown that human and rabbit hearts, which possess extremely low xanthine oxidase activity, nonetheless, develop myocardial infractions and ischemia–reperfusion injury [9]. However, recent studies supported the importance of the xanthine oxidase-catalyzed oxygen radical generation. It has been showed that xanthine oxidase is partly responsible for reoxygenation injury in bovine pulmonary artery endothelial cells [10], human umbilical vein and lymphoblastic leukemia cells [11], and cerebral endothelial cells [12]. Zwang et al. [11] concluded that xanthine dehydrogenase may catalyze superoxide formation without conversion to xanthine oxidase using NADH instead of xanthine as a substrate.

Xanthine oxidase is not the only source of reactive species in ischemia–reoxygenation injury. Another source of oxygen radicals is NADPH oxidase. For example, it has been shown that endothelial NADPH oxidase produced reactive oxygen species in lungs exposed to ischemia [13]. (The role of NADPH oxidase as a producer of oxygen radicals in tissue is considered below.)

Mitochondrial production of oxygen and nitrogen reactive species is also increased during hypoxia–reoxygenation injury. Anoxia (i.e., the absence of dioxygen) causes the reduction of respiratory electron carriers and consequently, increases superoxide production by mitochondria. Hypoxia also decreases MnSOD activity and protein or heme expression that causes an increase in oxygen radical production. On the other hand, reoxygenation of hypoxic mitochondria induces mitochondrial complex I dysfunction and also augments superoxide production [14]. Kim et al. [15] recently showed that cerebral ischemia–reperfusion in mice caused the enhancement of mitochondrial superoxide production in neurons, which was increased in mutant mice deficient in MnSOD.

It was earlier thought that activated neutrophils do not play an important role in reoxygenation injury [1]. However, Duilio et al. [16] pointed out that this conclusion was drawn from the experiments with brief episodes of ischemia resulted in myocardial stunning, while neutrophil-mediated damage is expected after prolonged ischemia associated with myocardial infraction. These authors demonstrated that neutrophils were a major source of oxygen radicals in hearts reperfused under in vivo conditions after prolonged ischemia.

There are numerous examples of a critical role of reactive oxygen and nitrogen species in ischemic injury. Thus, Manevich et al. [17] showed that the abrupt cessation of flow of flow-adapted endothelial cells (simulated ischemia) caused an oxidative burst accompanied by the production of oxygen radicals. It has long been known that oxidative processes occur in acute renal ischemia–reperfusion [18]. Recently, Noiri et al. [19] concluded that peroxynitrite rather than superoxide is responsible for lipid peroxidation and DNA damage during renal damage in ischemic acute renal failure (ARF).

31.1.2 ATHEROSCLEROSIS

Oxidative stress is involved in the pathogenesis of many cardiovascular pathologies, including hypercholesterolemia, atherosclerosis, hypertension, diabetes mellitus, and heart failure. It is a common knowledge that excess formation of reactive oxygen and nitrogen species is linked with vascular lesion formation and functional defects. An important factor of atherosclerosis development is the dysfunction of vascular endothelium induced by the reaction of superoxide with nitric oxide. It has been shown long ago that superoxide inactivated the endothelium-derived relaxing factor (EDRF) [20]. Now EDRF is identified as nitric oxide; therefore, it is understandable that EDRF inactivation is a result of the interaction of superoxide with nitric oxide, which leads to the formation of peroxynitrite, a genuine reactive species responsible for endothelial dysfunction. Thus the inhibition of nitric oxide production by this mechanism leads to endothelial dysfunction, one of the earliest steps in the atherosclerotic process.

The disturbance of balance between superoxide and nitric oxide occurs in a variety of common disease states. For example, altered endothelium-dependent vascular relaxation due to a decrease in NO formation has been shown in animal models of hypertension, diabetes, cigarette smoking, and heart failure [21]. Miller et al. [22] suggested that a chronic animal model atherosclerosis closely resembles the severity of atherosclerosis in patients. On the whole, the results obtained in humans, for example, in hypertensive patients [23] correspond well to animal experiments. It is important that endothelium-dependent vascular relaxation in patients may be improved by ascorbic acid probably through the reaction with superoxide.

At present, a great attention is given to the role of angiotensin II (Ang II), a principal effector of renin angiotensin system, in the development of atherosclerosis and other cardiovascular diseases [24]. Ang II induces oxidative stress through AT_1 receptor and NADPH oxidase activation [25,26]; the blockade of Ang II type 1 receptor prevents aortic superoxide production in rats in vivo [27]. Rueckschloss et al. [28] showed that Ang II induced proatherosclerotic oxidative stress in human endothelial cells by the expression of NADPH oxidase. The blockade of AT_1 receptor reduced oxidative stress and could be useful in the treatment of patients with coronary disease. Zhang et al. [29] demonstrated that AT_2 receptor mediated the vasodilation of coronary arterioles. Ushio-Fukai et al. [30] showed that Ang II stimulated intracellular oxygen radical production, leading to vascular smooth muscle hypertrophy.

The role of NADPH oxidase in the stimulation of oxidative stress during atherosclerosis has been widely discussed. Recently, Barry-Lane et al. [31] demonstrated that mice lacking the p47phox gene had lower aortic levels of superoxide production as compared to wild-type mice. Although Hsich et al. [32] and Kirk et al. [33] did not find any change in superoxide production under similar conditions, Griendling and Harrison [34] explained these contradictions by difference in experimental conditions. Subsequent studies have shown that the suppression of NADPH oxidase or SOD activities resulted in the inhibition or enhancement of superoxide production, respectively. Thus, Bendal et al. [35] demonstrated the principal role of a gp91phox-containing NADPH oxidase in angiotensin II-induced cardiac hypertrophy in mice. Wang et al. [36] showed that the formation of superoxide and 3-nitrotyrosine increased in aortic endothelium and adventitia of Ang II-treated wild-type mice but not in the mice expressing SOD. However, the in vivo effects of Ang II were supposedly mediated by the other reactive oxygen species than superoxide or peroxynitrite. Nonetheless, Landmesser et al. [37] demonstrated that a decrease in the in vivo Ang II hypertensive effect in mice depended on NADPH oxidase-produced superoxide because it was markedly diminished in mice deficient in $p47^{phox}$. Hathaway et al. [38] also found that the development of moderately severe atherosclerosis and regression of atherosclerosis in monkeys depended on the level of superoxide production and NADPH oxidase activity. Chatterjee [39] proposed that glycosphingolopid lactosylceramide contributes to atherosclerosis through the activation of NADPH oxidase. Zimmerman et al. [40] demonstrated the mediation by superoxide of Ang II effects on central nervous system (CNS). It was found that the expression of SOD in the brain of mice injected with Ang II abolished the changes in blood pressure and heart rate.

It is well known that atherogenesis is characterized by LDL oxidative modification. Kuhn et al. [41] studied the mechanism of LDL oxidation on the different stages of atherogenesis. Their data suggest that 15-lipoxygenase is involved in the oxidation of LDL in the early stages of atherogenesis, while nonenzymatic arachidonate oxidation becomes more important in the later stages. An increase in the products of linoleic acid oxidation (hydroxy fatty acids) has been observed in the LDL of atherosclerotic patients [42].

31.1.3 MITOCHONDRIA AND NADPH OXIDASE AS INITIATORS OF OXYGEN RADICAL OVERPRODUCTION IN HEART DISEASES

The above data suggest an important role of reactive oxygen species in the development of heart diseases. This suggestion has been supported by many studies, which also demonstrated a potential efficacy of antioxidants, free scavengers, and chelators in the treatment of these diseases. Mitochondrial oxygen radical overproduction can probably be one of the critical causes.

For example, Ide et al. [43] demonstrated the involvement of oxygen radicals produced by mitochondria in heart failure myocytes from adult mongrel dogs, which could be responsible for contractile dysfunction and structural damage to the myocardium. In a subsequent study [44], using the same animal model, these authors suggested that hydroxyl radical formation contributed to left ventricular failure. This work has been commented as an important evidence of significance of mitochondrial oxygen radical overproduction in heart injury [45]. However, it should be noted that the formation of hydroxyl radicals in the failing myocardium is still questionable. Unfortunately, Ide et al. [44] failed to demonstrate definitely that the hydroxyl spin-trapping adduct was formed in the reaction of hydroxyl radicals with a spin trap and not during the decomposition of the hydroperoxyl spin adduct originating from superoxide, for example, by conversion of hydroxyl radicals into ethanol radicals. Ballinger et al. [46] also proposed that reactive oxygen and nitrogen species contribute to mitochondrial oxidant production and DNA damage and stimulate the progression of atherosclerotic lesions in humans.

MacCarthy et al. [47] showed that the treatment of isolated ejecting hearts from aortic-banded guinea pigs with ascorbic acid and desferrioxamine restored left verticular relaxant responses to NO agonists. On these grounds, it was proposed that an increase in oxygen radical production is responsible for the impaired endothelial regulation of left verticular relaxation. Hamilton et al. [48] showed that the inhibition of NADPH oxidase by apocynin improved endothelial function in rat and human blood vessels by increasing nitric oxide bioavailability, assuming that NADPH oxidase may be a novel target for drug intervention in cardiovascular disease. Azumi et al. [49] demonstrated an importance of NADPH oxidase-stimulated superoxide production in directional coronary atherectomy specimens from patients with stable and unstable angina pectoris. They suggested that oxygen radicals initiate LDL oxidation and probably play important role in pathogenesis of coronary artery disease.

31.1.4 ANTIOXIDANTS AND HEART DISEASE

From earlier times when it has first been established that ischemia and hypoxia are the potential causes of coronary heart disease and myocardial infraction [50,51], antioxidants have been considered as the possible candidates for the treatment of these diseases. Therefore, it is not surprising that the well-known antioxidants vitamin E and SOD + catalase have been applied in the first place in the prevention and therapy studies [50,52]. Recent studies of animal models supported early proposals about the beneficial effects of vitamin E administration for he prevention of endothelial dysfunction in heart failure [53]. Some data on protective effects of vitamin E and C in heart diseases have already been considered in Chapter 29. Finnish researchers also demonstrated that the supplementation of diet with vitamins E, C, carotene, and selenium increased the resistance of atherogenic lipoproteins to oxidative stress in human plasma [54] and that high serum vitamin E and selenium levels in local diet correlated with low mortality from coronary heart disease in people from northern Finland [55].

As noted in Chapter 29, Pryor's estimate [56] of results of clinical trials on the treatment of heart diseases with vitamin E is mainly positive. However, the final conclusions are not so optimistic. Shihabi et al. [57] suggested that one of main reasons of the failure of large clinical trials could be inefficiency of vitamin E to suppress free radical damage initiated by superoxide. Indeed, the present studies show that the most important directions of free radical attack on tissue and cells are mediated by peroxynitrite and hypochlorous acid. As peroxynitrite is formed by the reaction of superoxide with nitric oxide, the enhanced superoxide production in atherosclerosis might be a critical factor of free radical injury. However, as it is rightly noted by Shihabi et al. [57], vitamin E is not an effective scavenger of superoxide. Actually, vitamin E (α-tocopherol) cannot react with superoxide at all by free radical abstraction mechanism because superoxide is able only to deprotonate α-tocopherol to form even a more active perhydroxyl radical $HOO^\bullet$ (Chapter 29). Of course, α-tocopherol is the most effective scavenger of hydroxyl, peroxyl, alkoxyl, and alkyl radicals and, therefore, can be useful for suppressing free radical damage to lipids and other biomolecules but is ineffective on the superoxide-mediated initiation stages of free radical damage. The latest data on the 6-year supplementation of vitamin E + vitamin C to healthy hypercholesterolemic persons showed attenuation of the progression of carotid atherosclerosis especially in male patients [58]. Nonetheless, the benefits of vitamin E supplementation for fighting against cardiovascular disease are still not clear [59].

It should be stressed that among numerous natural and synthetic antioxidants, not many are effective superoxide scavengers. These are quinones, some polyphenolic compounds, and probably, thiols (Chapter 29). Therefore, it is not surprising that the use of traditional antioxidants such as vitamin E in clinical trials still did not give successful results. From these considerations, one may expect that flavonoids can exhibit better cardioprotector activity. For example, Wu et al. [60] showed that the flavonoid purpurogallin protected rabbit against myocardial ischemia–reperfusion injury. Shao et al. [61] demonstrated that another flavonoid baicalein protected cardiomyocytes against oxidative stress by scavenging superoxide formed during ischemia–reperfusion. Recently, cardioprotective activity of ubiquinones has been confirmed in ischemia–reperfusion experiments with rats, however, an increase in rat myocardium resistance to reperfusion damage was found only for water-soluble ubiquinone derivatives [62]. Another possible antioxidant capable of protecting heart proteins against peroxynitrite-mediated protein oxidation is urate [63].

It has been found that the 3-hydroxy-3-methylglutaryl-CoA (HMG CoA) inhibitors statins (atorvastatin, pravastatin, and cerivastatin), widely prescribed cholesterol-lowering agents, are able to inhibit phorbol ester-stimulated superoxide formation in endothelial-intact segments of the rat aorta [64] and suppress angiotensin II-mediated free radical production [65]. Delbose et al. [66] found that statins inhibited NADPH oxidase-catalyzed PMA-induced superoxide production by monocytes. It was suggested that statins can prevent or limit the involvement of superoxide in the development of atherosclerosis. It is important that statin therapy may be useful in prevention of cardiac hypertrophy, a major cause of mortality worldwide. Takemoto et al. [67] suggested that statins decreased superoxide production in rats treated with Ang II infusion by the inhibition of Rho proteins. Cerivastatin protected against the hypertension-based stroke in stroke-prone spontaneously hypertensive rats and inhibited superoxide production and inflammation in brain [68]. Treatment of hypercholesterolemic patients with atorvastatin reduced both LDL cholesterol and superoxide levels [69]. Similarly, Lubrano et al. [70] found that the atorvastatin treatment of patients with familial hypercholesterolemia reduced total plasma and LDL-cholesterol, NO_2/NO_3, and malondialdzhyde (MDA) levels. Khan et al. [71] studied the effect of irbesartan, an Ang type 1 receptor inhibitor, on vascular oxidative stress in patients with coronary artery disease. The treatment with irbesartan significantly decreased peroxidation and superoxide production in patients. Recently, it has been found that the iron chelator desferrioxamine improves NO-mediated, endothelium-dependent vasodilation in patients with coronary artery disease [72]. Thus, it is possible that iron availability is responsible for impaired NO activity in atherosclerosis.

31.1.5 Hypertension

The possible involvement of free radicals in the development of hypertension has been suspected for a long time. In 1988, Salonen et al. [73] demonstrated the marked elevation of blood pressure for persons with the lowest levels of plasma ascorbic acid and serum selenium concentrations. In subsequent studies these authors confirmed their first observations and showed that the supplementation with antioxidant combination of ascorbic acid, selenium, vitamin E, and carotene resulted in a significant decrease in diastonic blood pressure [74] and enhanced the resistance of atherogenic lipoproteins in human plasma to oxidative stress [75]. Kristal et al. [76] demonstrated that hypertention is accompanied by priming of PMNs although the enhancement of superoxide release was not correlated with the levels of blood pressure. Russo et al. [77] showed that essential hypertension patients are characterized by higher MDA levels and decreased SOD activities.

The mechanism of free radical-stimulated hypertension has been studied in animal experiments. Kerr et al. [78] demonstrated the enhancement of superoxide production and eNOS mRNA expression in spontaneously hypertensive stroke-prone rats. It was suggested that a major producer of superoxide was NO synthase. In contrast, it has been shown that NADPH oxidase is mainly responsible for the enhanced superoxide production in mineralocorticoid (deoxycorticosrerone acetate, DOCA) hypertensive rats characterized by low Ang II level [79,80]. Beswick et al. [80] found that xanthine oxidase and uncoupled endothelial NO synthase had no effect on superoxide production in the DOCA-salt rats, while the inhibition of NADPH oxidase significantly decreased the formation of superoxide in aortic rings. Chen et al. [81] suggested that the reduction of oxidative stress in stroke-prone spontaneously hypertensive rats after supplementation with vitamin E and C was associated with decreasing NADPH oxidase activity and increasing SOD activity. Ulker et al. [82] also showed that endothelial NADPH oxidase is a major source of superoxide production in spontaneously hypertensive rats and that vitamins E and C normalize endothelial dysfunction by regulation of endothelial NO synthase and NADPH oxidase activities. Increase in vascular cell adhesion molecule-1 (VCAM-1) expression, which contributes to vascular dysfunction in hypertension, can be suppressed in DOCA-salt hypertension rats by gene transfer of MnSOD or eNOS [83]. Enhanced superoxide production and arterial pressure in spontaneous hypertensive rats can be reduced by the treatment with the stable free radical Tempol [84].

The accumulation of inflammatory cells (leukocytes and macrophages) in the arterial wall is another inducer of hypertension. For example, it has been shown that NO blockade in the aortas in L-Nitro arginine methyl ester(NAME)-treated rats resulted in the accumulation of macrophages in arteries accompanied by an increase in superoxide production and stimulated the expression of specific adhesion molecules [85].

31.1.5.1 Preeclampsia

Preeclampsia is a hypertensive disorder of pregnancy leading to maternal hypertension. This disease is one of the most significant health disorders of human pregnancy. Contemporary studies demonstrated the importance of oxidative stress in preeclampsia, including the enhancement of lipid peroxidation, formation of isoprostane 8-iso-$PGF_{2\alpha}$, an increase in nitrotyrosine and nitrotyrosine residues in placenta, and an increase in xanthine oxidase activity [86]. Recently, Walsh et al. [87] and Barden et al. [88] confirmed the enhanced formation of isoprostanes in preeclamptic placentas, which was correlated with the elevated MDA formation. Sikkema et al. [89] showed that superoxide formation was significantly increased in the placental tissue of preeclamptic women while SOD activity was apparently reduced. Furthermore, it has been shown that N-formyl-Met-Leu-Phe-(FMLP) and PMA-stimulated neutrophils isolated from preeclamptic women during the third trimester produced much more oxygen radicals than normal cells [90]. Vaughan and Walsh [91] proposed that placental oxidative stress in preeclampsia can be a consequence of maternal hyperlipidemia and increased iron levels.

31.1.6 HYPERGLYCEMIA AND DIABETES MELLITUS

31.1.6.1 Hyperglycemia-Induced Oxidative Stress in Diabetes

It has been shown that vascular disease accounts for most of the clinical complications of diabetes mellitus. As hyperglycemia is a starting event of the development of diabetes mellitus, the mechanisms of free radical formation stimulated by glucose have been thoroughly studied. Glycation of proteins is an important process taking place in diabetic atherosclerosis. Hunt [92] considered the possible involvement of oxygen radicals in protein glycation. This process consists of the formation of Schiff bases in the reaction of glucose with proteins, which are transformed into so-called Amadori products; the latter again react with proteins to form fluorescent adducts called Maillard products or advanced glycation endproducts (AGE) (Reaction (1) and (2)):

$$\text{Glucose} + \text{protein} \Longrightarrow \underset{\text{This is Schiff base}}{\text{RCH(OH)CH(OH)CH}=\text{N}-\text{Pr}}$$

$$\Longleftrightarrow \underset{\text{Schiff base}}{\text{RCH(OH)C(OH)}=\text{CHNH}-\text{Pr}} \Longrightarrow$$

$$\Longrightarrow \underset{\text{Amadori product}}{\text{RCH(OH)C(O)CH2NH}-\text{Pr}} \Longleftrightarrow \underset{\text{Protein enediol}}{\text{RC(OH)}=\text{C(OH)CH2NH}-\text{Pr}} \quad (1)$$

$$\text{Amadori product} + \text{protein} \Longrightarrow \text{AGE (Maillard products)} \quad (2)$$

It has been proposed [92] that oxygen radicals may be formed in the stage of glycoxidation during the transition metal oxidation of protein enediol.

At present, the importance of the involvement of oxygen radicals in earlier stages of hyperglycemia has been demonstrated. Three major pathways of hyperglycemia-stimulated diabetic complications have been proposed: the activation of protein kinase C, the increased formation of glucose-derived AGE and the increased flux of glucose by the aldose reductase pathway. However, Brownlee and coworkers [93] have recently suggested that all these pathways occur due to the same reason: hyperglycemia-induced mitochondrial superoxide overproduction. It has been also shown that hyperglycemia-induced mitochondrial superoxide production activates hexosamine pathway that may affect gene expression and protein function, contributing to diabetic complications [94].

Furthermore, hyperglycemia increases oxygen radical production by mitochondrial electron transport chain in human platelets that may be an important factor of platelet dysfunction in patients with diabetes mellitus [95]. Mitochondrially produced oxygen radicals are also responsible for the activation of the cytokine-sensitive transcription factor NF-κB in insulin producing cells, and the overexpression of MnSOD is beneficial for the survival of these cells [96]. Recently, Brodsky et al. [97] demonstrated that glucose reduced NO production and increased superoxide production by endothelial cells; both events can be restored to control levels by cell-penetrated SOD mimic. It was also proposed that uncoupled mtNOS is an important source of superoxide in endothelial cells. Gupta et al. [98] also showed that the hyperglycemia-induced augmentation of endothelial superoxide production inhibited the stimulatory effect of nitric oxide on vascular Na^+–K^+–ATPase activity in smooth muscle cells.

The above findings suggest an important role of mitochondria-derived reactive oxygen species in hyperglycemia and diabetes mellitus. However, other oxygen radical-producing enzymes can also be of importance in the development of diabetes mellitus. Christ et al. [99] has showed that exposure of porcine coronary segments to high glucose increased the expression of $p22^{phox}$, indicating an importance of NADPH oxidase in chronic hyperglycemia-induced endothelial dysfunction. HMG CoA inhibitor atorvastatin inhibited glucose-induced superoxide production. This fact suggests that the beneficial effects of statins in diabetic patients could be at least partly explained by their activities as superoxide scavengers. Guzik et al. [100] studied vascular superoxide production in human blood vessels from diabetic patients with coronary artery disease compared to nondiabetic patients. These authors demonstrated that vessels from diabetic patients generated the enhanced amount of superoxide from two major sources: vascular NADPH oxidase and unblocked endothelial NO synthase.

NADH oxidase and upregulated expression of $p22^{phox}$ mRNA are supposed to be the sources of superoxide overproduction in type 2 diabetic rats [101]. Inoguchi et al. [102] showed that high glucose + palmitate stimulated oxygen radical production via protein kinase-dependent activation of NADPH oxidase in cultured vascular aortic smooth muscle and endothelial cells. Hink et al. [103] suggested that NO synthase and NADPH oxidase are responsible for the overproduction of superoxide in streptozotocin-induced diabetic rats. Venugopal et al. [104] found that under hyperglycemic conditions superoxide release from human monocytes depends on NADPH oxidase activity and not the mitochondrial superoxide production. These authors proposed that hyperglycemia activated protein kinase C-α (PKC-α) and that α-tocopherol inhibited superoxide production by the suppression of PKC-α activity.

An increase in hypoglycemia-induced oxygen radical production must affect NO-dependent phenomena such as vasodilation, angiogenesis, and vascular maintenance through the interaction of superoxide with nitric oxide. Damaging effect of oxygen radicals on NO-dependent processes has been shown, but its mechanism seems not be limited only to the direct reaction of superoxide with nitric oxide. Cosentino et al. [105] found that prolonged exposure of human aortic endothelial cells to high glucose increased eNOS gene expression and NO release but simultaneously significantly augmented superoxide generation. Trachtman et al. [106] suggested that high glucose caused the inhibition of nitric oxide production in cultured rat mesangial cells by depletion of arginine, the substrate of NO synthase. In bovine aortic endothelial cells hyperglycemia inhibited the activity of endothelial NO synthase by 67% supposedly via mitochondria superoxide overproduction [107]. Similarly, Noyman et al. [108] showed that the pretreatment of bovine aortic endothelial cells with high glucose resulted in a decrease in eNOS protein expression.

Napoli et al. [109] found that glycosylated LDL are more susceptible than normal LDL to lipid peroxidation, which is mediated by superoxide. Correspondingly, glycosylated LDL produced the enhanced amount of superoxide on the treatment of aortic rings compared to normal LDL [110]. Stronger impairment of endothelium-dependent dilation by oxidized glycosylated LDL can be due to the interaction of superoxide with nitric oxide.

It is important that acute hyperglycemia may be a cause of oxidative damage not only in diabetic patients but also in healthy subjects. Marfella et al. [111] showed that blood pressure and nitrotyrosine content significantly increased in healthy subjects undergoing a hyperglycemic glucose clamp test. These authors suggested that nitrotyrosine, a marker of oxidative stress, was formed via a peroxynitrite-dependent mechanism, although the peroxynitrite-independent mechanism of tyrosine nitration also cannot be excluded [112].

31.1.6.2 Free Radical-Mediated Processes in Diabetes Mellitus

Earlier studies on oxygen radical-mediated processes in diabetes mellitus were concentrated on the study of overproduction of oxygen radicals by leukocytes. Thus, it has been shown that PMNs from diabetic patients exhibited significantly increased unstimulated superoxide production [113,114]. Wierusz-Wysocka et al. [114] found that the PMN-enhanced superoxide generation correlated well with the level of glycosylated hemoglobin HbA_1. These results have been later confirmed for type 2 diabetic patients [115]. It was also suggested that some intracellular factors might be responsible for PMN priming and that the priming PMNs contribute to chronic inflammation in diabetes. The relationship between enhanced superoxide production, endothelial dysfunction, and HbA_1 level has been also shown in an animal model of diabetic rats [116]. Josefsen et al. [117] demonstrated that the circulating monocytes in newly diagnosed type 1 diabetes patients were activated, which could play a pathogenic role in β-cell destruction.

Enhanced superoxide production by monocytes of patients with hypertriglyceridemia and diabetes has also been shown [118]. Orie et al. [119] found that the activation of mononuclear leukocytes with FMLP or phytohemagglutinin significantly increased oxygen radical production in type 2 diabetic patients compared to controls. It is interesting that the stimulation of diabetic platelets from type 1 diabetic children with platelet activating factor (PAF) resulted in a decrease in oxygen radical production in diabetic group with respect to controls [120]. All the above data point out at the involvement of oxygen radicals produced by unstimulated and stimulated leukocytes in diabetes development.

Lipid peroxidation is another free radical-mediated process enhanced in diabetes mellitus. It should be noted that some data obtained in animal models of diabetes could be misleading and not related to real diabetic state. For example, the enhanced intracellular generation of hydroxyl radicals has been shown in widely applied streptozotocin-induced model of diabetes in rats [121]. However, Lubec et al. [122] later showed that streptozotocin itself and not the diabetic state is responsible for the formation of hydroxyl radicals in this model.

Nonetheless, lipid and protein oxidation certainly occurs in diabetic patients. Dominguez et al. [123] found that plasma MDA and protein carbonyl group levels were enhanced in type 1 diabetic children and adolescents, while the α-tocopherol:total lipid ratio decreased. Furthermore, systemic oxidative stress presenting upon early diabetes onset was apparently augmented in early adulthood. MDA content was significantly enhanced in erythrocytes from noninsulin-dependent diabetes (NIDDM) patients with and without nephropathy whereas SOD and catalase activities decreased [124]. The level of oxidative stress was greater in NIDDM patients without nephropathy. Kesavulu et al. [125] has also found enhanced lipid peroxidation and diminished SOD activity in NIDDH patients. Inouye et al. [126] proposed that 9-hydroxy linoleic acid is a marker of phospholipid peroxidation in erythrocytes of diabetic patients. It has been suggested that this product is formed via hydroxylation of linoleic acid with hydroxyl radicals. Altavilla et al. [127] showed that the inhibition of lipid peroxidation restored impaired vascular endothelial growth factor expression and improved wound healing diabetic mice. DNA damage is also a consequence of oxidative stress in diabetes mellitus. Thus, Collins et al. [128] found the enhanced levels of strand breaks and oxidized pyrimidines in diabetic patients.

It has already been noted above that hyperglycemia may affect nitric oxide production in different ways. Mohan and Das [129] demonstrated that NO prevented β-cell damage in the model of alloxan-induced diabetes in rats. However, far more important data on the role of nitric oxide in diabetes development were obtained in the study of diabetic patients. Thus, Kedziora-Kornatowska [130] showed that there is difference in superoxide and NO production by the granulocytes from NIDDM patients: NO production was increased in diabetic patients with and without diabetic nephropathy, while superoxide production was increased or decreased in the same patients, respectively.

31.1.6.3 Antioxidants and Diabetes Mellitus

As in the case of other cardiovascular diseases, the possibility of antioxidant treatment of diabetes mellitus has been studied in both animal models and diabetic patients. The treatment of streptozotocin-induced diabetic rats with α-lipoic acid reduced superoxide production by aorta and superoxide and peroxynitrite formation by arterioles providing circulation to the region of the sciatic nerve, suppressed lipid peroxidation in serum, and improved lens glutathione level [131]. In contrast, hydroxyethyl starch desferrioxamine had no effect on the markers of oxidative stress in diabetic rats. Lipoic acid also suppressed hyperglycemia and mitochondrial superoxide generation in hearts of glucose-treated rats [132].

Sanders et al. [133] found that although quercetin treatment of streptozotocin diabetic rats diminished oxidized glutathione in brain and hepatic glutathione peroxidase activity, this flavonoid enhanced hepatic lipid peroxidation, decreased hepatic glutathione level, and increased renal and cardiac glutathione peroxidase activity. In authors' opinion the partial prooxidant effect of quercetin questions the efficacy of quercetin therapy in diabetic patients. (Antioxidant and prooxidant activities of flavonoids are discussed in Chapter 29.) Administration of endothelin antagonist J-104132 to streptozotocin-induced diabetic rats inhibited the enhanced endothelin-1-stimulated superoxide production [134]. Interleukin-10 preserved endothelium-dependent vasorelaxation in streptozotocin-induced diabetic mice probably by reducing superoxide production by xanthine oxidase [135].

At present, antioxidants are extensively studied as supplements for the treatment diabetic patients. Several clinical trials have been carried out with vitamin E. In 1991, Ceriello et al. [136] showed that supplementation of vitamin E to insulin-requiring diabetic patients reduced protein glycosylation without changing plasma glucose, probably due to the inhibition of the Maillard reaction. Then, Paolisso et al. [137] found that vitamin E decreased glucose level and improved insulin action in noninsulin-dependent diabetic patients. Recently, Jain et al. [138] showed that vitamin E supplementation increased glutathione level and diminished lipid peroxidation and HbA_1 level in erythrocytes of type 1 diabetic children. Similarly, Skyrme-Jones et al. [139] demonstrated that vitamin E supplementation improved endothelial vasodilator function in type 1 diabetic children supposedly due to the suppression of LDL oxidation. Devaraj et al. [140] used the urinary F2-isoprostane test for the estimate of LDL oxidation in type 2 diabetics. They also found that LDL oxidation decreased after vitamin E supplementation to patients.

Keenoy et al. [141] treated type 1 diabetic patients with Daflon 500, a mixture of flavonoids diosmin (90%) and hesperdin (10%). It was found that flavonoid therapy resulted in a decrease in the levels of the HbA_{1c} hemoglobin and the in vitro oxidability of non-HDL lipoproteins. Lipoic acid was found to improve microcirculation in patients with diabetic polyneuropathy [142]. De Mattia et al. [143] showed that the treatment of type 2 diabetic patients with glicazide, a derivative of sulfonylurea possessing antioxidant properties, improved antioxidant status and nitric oxide-mediated vasodilation in patients. Gargiulo et al. [144] demonstrated that antidiabetic agent metformin diminished superoxide overproduction in type 2 diabetic patients.

31.2 CANCER, CARCINOGENESIS, FREE RADICALS, AND ANTIOXIDANTS

In 1965–1967 a great interest has been attached to the possible role of free radicals in cancer after studies by Emanuel and his coworkers who reported the excessive production of free radicals in tumor cells (see, for example, Ref. [145]). On these grounds the authors suggested to apply antioxidant therapy for the treatment of cancer patients. Unfortunately, experimental proofs of overproduction of free radicals in cancer tissue turn out to be erroneous [146]. A new interest in the role of free radicals in cancer development emerged after the discovery of superoxide and superoxide dismutases.

Numerous excellent reviews on the possible role of oxygen radicals in cancer and carcinogenesis have been published 10–20 years ago [147–153]. Earliest studies have been much concerned with the role of SOD in tumor cells. Despite some contradictory results, it is general conclusion that tumor cells are usually characterized by lowered CuZnSOD activity and always by lowered MnSOD activity [147]. The origin of SOD declining in cancer cells is unknown. It has been suggested that MnSOD is not induced in cancer immortalized cells in response to oxidative stress, but the reason of this is uncertain [154].

31.2.1 MECHANISMS OF FREE RADICAL REACTIONS IN TUMOR CELLS

There is a major difference between the role of free radicals in cancer and other pathologies such as cardiovascular diseases, hypertension, diabetes mellitus, etc. In contrast to the latter diseases where the sources of free radical overproduction are well established (vascular cells and macrophages in cardiovascular diseases and leukocytes in inflammation), the origin and the levels of free radical production in tumor cells are still uncertain.

Most researchers agreed that oxygen radicals participate in both initiation and promotion of cancer. At the initiation stage oxygen radicals together with various carcinogens may change normal cellular genetic material to that of neoplastic genetic composition. At promotion stage the participation of free radicals was firstly suggested on the grounds of the effects of organic peroxides, which promoted cancer development [150]. As discussed in Chapter 28, the reaction of oxygen radicals with DNA results in the formation of specific adducts such as 8-OHdG. This is an important step of carcinogenesis because 8-OHdG is carcinogenic and its formation correlates in many cases with cancer development [155]. Thus, Wei and Frenkel [156] suggested that the 8-OHdG formation in mice is related to tumor promoting activity of phorbol ester-type tumor promoters. Ogawa et al. [157] proposed that oxygen radicals mediated the formation of 8-OHdG during the synthetic estrogen induction of hepatocellular carcinomas in rats.

It is of importance that oxygen radical generation by leukemia cells can be enhanced or diminished. In 1986, Mazzone et al. [158] was unable to receive stable data on superoxide production by leukemic blast cells from patients with acute nonlymphocytic leukemia (ALL) and AML chronic myeloid leukemia. However, later on, Korkina et al. [159] demonstrated that nonstimulated and latex-stimulated blood leukocytes from acute nonlymphocytic leukemia (ALL) and chronic myeloid leukemia (AML) children produced surprisingly low levels of oxygen radicals but had high CuZnSOD and MnSOD activities. At the same time, chemotherapy and following irradiation exposure sharply increased oxygen radical production measured via luminol-amplified CL (indicating possibly the enhanced generation of hydroxyl or hydroxyl-like radicals) but drastically decreased SOD activities. It has also been shown that the overproduction of oxygen radicals after chemotherapy and irradiation in the blood of ALL and AML children can be diminished by supplementation of food supplement Bio-Normalizer prepared by the fermentation of *Carica papaya*.

Enhanced level of typical hydroxyl radical-induced base lesions and the diminished content of antioxidant enzymes were also observed in lymphocytes of ALL children [160]. Similarly, the levels of MDA and 8-OHdG were elevated while SOD and catalase activities diminished in chronic lymphocytic leukemia (CLL) cells [161]. Babbs [162] proposed that the enhanced hydroxyl radical generation in feces points out at intracolonic oxygen radical formation as one of the factors for the high incidence of cancer in the colon and rectum. It has been suggested that the 8-OHdG lesions induced by oxygen radicals and generated by inflammatory leukocytes or carcinogens participate in the mutation of cancer-related genes [163]. Another possible role of superoxide in cancer development has been suggested by Pervaiz et al. [164]. These authors proposed that intracellular superoxide may regulate tumor cell response to drug-induced cell death by affecting the caspase activation pathway.

The importance of inflammatory phagocytes in cancer promotion has been suggested for a long time. Thus, Chong et al. [165] demonstrated that unstimulated and stimulated peritoneal macrophages induced tumor cell DNA strand breaks supposedly through the generation of oxygen radicals and arachidonate metabolites. Trulson et al. [166] showed that blood monocytes from patients with renal cancer produced the enhanced amount of superoxide. It has also been found that the superoxide mediated 12-*O*-tetradecanoylphorbol-13-acetate TPA-induced promotion of neoplastic transformation in mouse epidermal JB6 cells [167]. Kim et al. [168] studied the mutagenicity of oxygen and nitrogen free radicals produced by stimulated leukocytes. They showed that TPA-stimulated human promyelocytic leukemia cells (HL-60) and LPS/IFN-γ-stimulated murine macrophages induced 8-OHdG formation in transgenic Chinese hamster ovary cells mediated by oxygen and nitrogen radicals.

Thus, oxygen radical production by leukocytes can be responsible for cancer development. However, the levels of leukocyte oxygen radical generation depend on the type of cancer. For example, PMNs and monocytes from peripheral blood of patients with lung cancer produced a diminished amount of superoxide [169]. Timoshenko et al. [170] observed the reduction of superoxide production in bronchial carcinoma patients after the incubation of neutrophils with concanavalin A or human lectin, while neutrophils from breast cancer patients exhibited no change in their activity. Chemotherapy of lung and colorectal carcinoma patients also reduced neutrophil superoxide production. Human ALL and AML cells produced, as a rule, the diminished amounts of superoxide in response to PMA or FMLP [171]. On the other hand total SOD activity was enhanced in AML cells but diminished in ALL cells, while MnSOD in AML cells was very low. It has been proposed that decreased superoxide production may be responsible for susceptibility to infections in cancer patients.

One should expect that nitrogen reactive species could also affect cancer development. Similar to oxygen radicals, the effects of nitrogen species depend on many factors, cell types, cancer types, etc. Thus, it has been shown that both nitric oxide and peroxynitrite formation increased in erythrocytes from breast cancer patients with a subsequent increase in the membrane rigidity under oxidative stress [172]. It was supposed that these mechanical changes may cause shortening of the lifespan of erythrocytes and stimulate toxic anemia in cancer patients. On the other hand, nitric oxide inhibited tumor cell growth in IFN-γ-activated rat neutrophils [173]. Haklar et al. [174] measured the content of reactive oxygen and nitrogen species in cancerous tissues from colon and breast carcinoma cases. They found that all reactive species including nitric oxide and peroxynitrite increased in cancerous colon tissues with hypochlorite making a major contribution, while only superoxide significantly increased in breast carcinoma. Thomas et al. [175] found that the TNF-α inhibition of proliferation of pancreatic cancer cells was mediated by superoxide formation while endogenously generated nitric oxide suppressed this effect.

Numerous studies demonstrated that lipid peroxidation significantly decreased in cancer cells and tissues (Ref. [176] and references therein). It has been proposed that this can be a consequence of a decrease in the content of highly unsaturated fatty acids, the concentration of cytochrome p-450, and the contents of NADPH, SOD, and catalase in tumors. Cheeseman et al. [176] suggested that the reduction of lipid peroxidation in tumors may depend on both the expression of malignant transformation and cell division. It should be mentioned that Boyd and McGuire [177] demonstrated that there is a correlation between lipid peroxidation and breast cancer risk in premenopausal women.

In agreement with the above consideration of the role of oxidative stress in cancer development, it was found that tumor cells (thymocytes) are more sensitive to oxidative stress than normal thymocytes [178]. There are apparently the other free radical-mediated damaging processes, which can be more intensive in tumors. For example, it has been found that metHb formation was significantly elevated in cancer patients [179].

31.2.2 The Treatment of Cancer with Prooxidants

The ambiguity of cancer therapy directed on the regulation of free radical status of tumors is because both prooxidants and antioxidants are able to act in two ways, suppressing or stimulating cancer growth. One of the most efficient ways for fighting cancer is the use of prooxidants, capable of destroying tumor cells. As stated above, superoxide produced by inflammatory leukocytes participates in cancer promotion; nonetheless, it has been described that the superoxide-producing systems can be applied for cancer treatment.

As early as 1958, Haddow et al. [180] demonstrated that xanthine oxidase exhibited anticancer activity in mice. Later on, Yoshikawa et al. [181] showed that oxygen radicals produced by the xanthine–xanthine oxidase system suppressed carcinomas growth in experimental rabbit model. The antitumor effect of this system was inhibited by SOD and catalase. However, therapeutic activity of native xanthine oxidase was compromised by its high binding affinity to blood vessels that may cause systemic vascular damage. Because of this, Sawa et al. [182] proposed to apply poly(ethylene glycol)-conjugated xanthine oxidase (PEG-XO) for tumor-targeting chemotherapy. It was found that PEG-XO exhibited highly tumoritropic accumulation in tumor-bearing mice and that the PEG-XO treatment resulted in significant suppression of tumor growth. It is of utmost importance that macromolecules such as PEG-XO preferentially accumulate in solid tumor and not in healthy tissues.

Yamaguichi et al. [183] proposed that intracellular SOD is probably an important factor in protecting leukemic and cancer cells against superoxide and irradiation. Therefore, the damaging effect of superoxide might be enhanced by suppressing SOD activity in cancer cells. Thus, Huang et al. [184] showed that some estrogen derivatives selectively killed human leukemia cells but not normal lymphocytes through SOD inhibition and apoptosis induction. The authors proposed that targeting SOD may be a promising approach to the selective killing of cancer cells. It has also been suggested that malignant cells are more dependent on SOD because they have higher superoxide and lower SOD levels [185].

The importance of superoxide-mediated damage to cancer cells was also demonstrated in the experiments with overexpressed mitochondrial MnSOD. Hirose et al. [186] showed that the overexpression of mitochondrial MnSOD enhanced the survival of human melanoma cells exposed to cytokines IL-1 and TNF-α, anticancer antibiotics doxorubicin and mitomycin C, and γ-irradiation. Similarly, Motoori et al. [187] found that overexpression of MnSOD reduced the levels of reactive oxygen species in mitochondria, the intracellular production of 4-hydroxy-2-nonenal, and prevented radiation-induced cell death in human hepatocellular carcinoma cells. Interestingly, MnSOD expression was not influenced by NO production stimulated by irradiation. MnSOD deficiency also enhanced TPA-induced oxidative stress in a skin cancer model [188].

It is possible that the anticancer activity of the most effective anticancer antibiotics such as adriamycin (doxorubicin) depends on their ability to generate oxygen radicals. Mechanisms of free radical-mediated anticancer activity of anthracycline antibiotics and their analogs have been thoroughly studied and reviewed (see, for example, Refs. [189,190]). (The production of oxygen radicals by anthracyclines is also considered in Chapters 21 and 25.) Both prooxidant and antioxidant activities have been shown for flavonoids in normal and tumor cells (see below and Chapter 21). We have already considered earlier the cytotoxic effects of flavonoids against tumor cells [191]. Many flavonoids exhibit antiproliferative and antileukemic effects and suppress cancer growth. It is possible that prooxidant activity of flavonoids might be responsible for their anticancer effects, but such a suggestion still has no experimental support.

Similar to flavonoids, other antioxidants and cancer chemopreventive agents are able to exhibit prooxidant action. Thus, the antioxidant *N*-acetylcysteine was found to increase the formation of 8-OHdG in human leukemia cell line HL-60 [192]. It was suggested that NAC-initiated DNA damage is mediated by reactive oxygen species formed by NAC + copper ions. Spallholz et al. [193] suggested that the anticarcinogenic activity of L-selenomethionine, L-Se-methylselenocysteine, and some other methylated selenium compounds depended on the reduction of selenium compounds in the presence of glutathione to methylselenol and the reduction by the last compound of dioxygen to superoxide.

It has been suggested that tamoxifen, one of the most effective therapeutic and chemopreventive agent for breast cancer, modulates protein kinase C through oxidative stress in breast cancer cells [194]. Unfortunately, most breast cancers initially responsive to tamoxifen treatment later become resistant. Schiff et al. [195] suggested that the conversion of breast tumors to a tamoxifen-resistant phenotype is associated with oxidative stress and depends on significantly enhanced SOD activity in tumors.

31.2.3 THE TREATMENT OF CANCER WITH ANTIOXIDANTS

As the involvement of free radicals in both the initiation and promotion stages of cancer has been demonstrated (see above), it is reasonable to suggest that antioxidants may be applied for cancer treatment and prevention. Therefore, already for a long time, the effects of many antioxidants and antioxidant enzymes on cancer development have been studied under in vitro and in vivo conditions. Naturally from the beginning, much attention has been drawn to the study of possible SOD effects. Unfortunately, natural SOD has a short circulation lifetime and unable to penetrate into cells. Therefore, various SOD mimics were studied as potential anticancer drugs.

In 1983, Kensler et al. [196] demonstrated that copper complex copper(II)(3,5-diisopropylsalicylic acid)$_2$ (CuDIPS) exhibited SOD-mimetic activity, suppressing TPA-promoted skin tumors in mice initiated with 7,12-dimethylbenz[a]anthracene. Although Oberley et al. [197] questioned the SOD-mimetic mechanism of CuDIPS anticancer activity, the other studies also supported the SOD mimetic activity of CuDIPS at both initiation and promotion stages of carcinogenesis [198,199]. It should also be noted that in spite of the inefficiency of natural SOD to penetrate the cellular membrane, Armato et al. [200] showed that exogenous SOD is able to suppress the stimulation of neonatal rat hepatocyte growth by tumor promotors. Much earlier, Kennedy [201] also concluded that SOD as well as SOD mimics can decrease the induction of malignant transformation, at least in vitro. Burdon [202] suggested that SOD and catalase are able to reduce the proliferation of hamster and rat fibroblasts (nontransformed or oncogene transformed). Glaves [203] showed that SOD inhibited the release of lung carcinoma cells in rats. Cullen et al. [204] suggested that the overexpression of MnSOD may be effective in growth suppression of pancreatic cancer.

In earlier years the effects of low-molecular-weight antioxidants (vitamins E, C, and polyphenols) have been also extensively studied. Prasad and Rama [205] reviewed their studies concerning the possible role of vitamin E in cancer treatment. It is of interest that among various forms of vitamin E, only vitamin E succinate was effective in the inhibition of differentiation and growth of mouse melanoma cells. It was also pointed out that vitamin E is able to enhance radiation damage in tumor cells under in vitro and in vivo conditions. London et al. [206] proposed that supplementation with vitamin E may reduce the risk of developing breast cancer in women. Possible anticarcinogenic effect of vitamin C was suggested based on its ability to block the formation nitrosamines, the inductors of various tumors in animals [207]. However, the relevance of these animal studies in the prevention of cancer development in humans is uncertain.

A more widely accepted practice was to supplement animals or cancer patients with a group of so-called "antioxidant vitamins" A, C, and E plus carotene or some other carotenoids such as licopene. (Actually, vitamin A and carotenoids are not genuine antioxidants and their addition to "antioxidant group" may or may not improve total therapeutic or prophylactic effects but not due to their free radical scavenging properties.) In 1984, Kahl [208] considered the role of various antioxidants including vitamins E and C in cancer prevention but did not find real benefits of their supplementation. Rotstein and Slaga [209] studied the effects of antioxidants such as reduced glutathione, 2(3)-*tert*-butyl-4-hydroxyanisole (BHA), vitamin E, CuDIPS, and NAC on tumor progression in the murine skin multistage carcinogenesis model. Among these antioxidants only glutathione inhibited tumor progression to a significant degree. Synthetic phenolic antioxidants 3,5-di-*tert*-butyl-4-hydroxytoluene (BHT) and BHA have also been studied as potential cancer chemopreventive agents [210]. These compounds are efficient free radical scavengers, but their well known toxicity and ability to damage different tissues or even to promote cancer makes doubtful their usefulness for cancer treatment.

Thus, the above consideration of earlier studies dedicated to the study of antioxidants as potential drugs for cancer treatment did not show any significant favorable effects of antioxidant treatment. It could be the consequence of a double role of prooxidants and antioxidants because both of them are capable of inhibiting and promoting cancer development. However, the studies of antioxidants for cancer treatment have not been stopped, and to the present day about 10,000 references are cited by Medline on this subject. Of course, it is impossible to discuss all these studies in this book; therefore, we will consider here only the several latest ones.

Radiation is one of the most important known environmental stimuli of cancer development. This environmental factor becomes especially dangerous for humans living in the areas affected by irradiation from nuclear accidents. Earlier we found that the administration of a mixture of vitamin E and α-lipoic acid to children living in the area of Chernobyl nuclear accident significantly and synergistically suppressed leukocyte oxygen radical overproduction [211]. Thus α-lipoic acid and α-lipoic acid + vitamin E supplements may be of interest as antioxidant preventive agents for the treatment of radiation-induced cancer development.

Several contemporary studies seem to suggest the inhibitory effect of vitamin E on human prostate cancer and some other tumor growth. Huang et al. [212] demonstrated a strong inverse correlation between γ-tocopherol and the risk of developing prostate cancer. Ni et al. [213] showed that vitamin E succinate inhibited human prostate cancer cell growth by the regulation of multiple molecules of the cell cycle. Malafa et al. [214] showed that vitamin E succinate increased melanoma dormancy and inhibited melanoma angiogenesis via the inhibition of vascular endothelial growth factor. Pace et al. [215] concluded that the supplementation of cancer patients receiving cisplatin chemotherapy with vitamin E decreased the incidence and severity of peripheral neurotoxicity.

In some cases the protective effects of vitamin C on cancer development and oxidative stress in tumor cells has been observed. Thus, it was found that experimental gastric tumors apparently produced more oxygen radicals than normal gastric tissue and that the administration of ascorbic acid reduced oxidative stress and gastric tumor incidence in rats [216]. Supplementation of various antioxidants (including vitamin C, α-lipoic acid, NAC, reduced glutathione, and vitamin E) caused a decrease in reactive oxygen species production in cancer patients [217]. However, Jacobs et al. [218] showed that only long-duration supplementation of vitamin E but not vitamin C may reduce the risk of bladder cancer mortality.

In addition to their possible prooxidant activity (see above) polyphenols and flavonoids may influence cancer cells via their antioxidant properties. Recently, Jang et al. [219] studied cancer chemopreventive activity of resveratrol, a natural polyphenolic compound derived from grapes (Chapter 29). These authors showed that resveratrol inhibited the development of preneoplastic lesions in carcinogen-treated mouse mammary glands in culture and inhibited tumorigenesis in a mouse skin cancer model. Flavonoids silymarin and silibinin also exhibited antitumor-promoting effects at the stage I tumor promotion in mouse skin [220] and manifested antiproliferative effects in rat prostate cancer cells [221].

31.3 INFLAMMATION

31.3.1 Mechanisms of Free Radical-Mediated Inflammatory Processes

Since the work by Babior et al. [222], who demonstrated the extracellular release of superoxide and hydrogen peroxide by neutrophils during phagocytosis, a great deal of attention has been drawn to the role of reactive oxygen species in inflammation. It became clear that reactive oxygen species produced by neutrophils and macrophages not only kill microbes but are also responsible for the tissue injury in acute inflammation. The mechanisms of oxygen radical production by phagocytes are now well studied (see, for example, earlier reviews [223,224]). The mechanisms of production of reactive oxygen and nitrogen species by NADPH oxidase and other prooxidant enzymes are considered in Chapter 22.

As early as in 1976, Oyanagui [225] suggested that macrophage-produced superoxide initiated inflammation in rat carrageenan food–edema model. Subsequent studies demonstrated different ways of free radical participation in inflammatory processes. It has been shown that superoxide (but not hydrogen peroxide) is able to stimulate the production of interleukin 1-like factors from human peripheral blood monocytes and PMNs [226]. These findings suggest that there is a feedback mechanism of inflammation, in which reactive oxygen species stimulate the enhanced LI 1-like factor production that, in turn, increases the formation of oxygen radicals.

Critical stage of inflammation is the starting of lipid peroxidation and the formation of bioactive eicosanoids. It is now known that lipoxygenases, cyclooxygenases, cytochrome P450 monooxygenases, and peroxidases are enzymatic catalysts of these processes (Chapter 26). Recently, using a peritonitis model of inflammation with MPO knockout mice, Zhang et al. [227] showed that the peritinitis-stimulated formation of F_2-isoprostanes, hydroxy- and hydroperoxy-eicosatetraenoic acids, hydroxy- and hydroperoxy-octadecadienoic acids, and their precursors, arachidonic and linoleic acids was suppressed in this model. It was concluded that myeloperoxidase could be a major enzymatic catalyst of lipid peroxidation at inflammatory sites.

Another pathway of the initiation of lipid peroxidation is the formation of peroxynitrite from superoxide and nitric oxide. Kausalya and Nath [228] found that the FMLP-stimulated injury and killing of endothelial cells in PMN–endothelial cell coculture was inhibited by a NO donor. However, other authors demonstrated the stimulatory effect of nitric oxide in inflammation. Thus, Farivar et al. [229] suggested that NO produced by cardiac fibroblasts might participate in inflammatory cardiac diseases. Zhang et al. [227] pointed out that MPO-dependent formation of reactive nitrogen species might be a preferred pathway for the initiation of lipid peroxidation at the inflammatory sites. Liu et al. [230] suggested that excessive NO production as a consequence of NO synthase induction by glial cells is responsible for brain inflammation and neurodegeneration.

Inflammatory processes play an important role in development of many pathologies. The participation of inflammatory cells in the development of cardiovascular disease and cancer has already been discussed above. Below, we will consider the other pathologies, in which inflammation plays important and even deciding role.

31.3.2 RHEUMATOID ARTHRITIS

In rheumatoid arthritis (RA) the chronic inflammatory state exists in synovium where it is characterized by the presence of monocytes in the pannus and PMNs in pannus and the synovial fluid. It has been suggested that the oxygen radical production by PMNs and lipid peroxidation products are responsible for the destruction of synovial tissues in rheumatoid arthritis [231]. However, although the role of oxygen radicals in the initiation of free radical damage in this disease is well established, there are different data on the levels of oxygen radical production by RA neutrophils. Thus, in contrast to studies demonstrating an increase in phagocytic superoxide production in rheumatic patients [232–235], no significant difference has been found in other studies [236–240]. Davis et al. [241] even found depressed superoxide production by neutrophils from patients with rheumatoid arthritis and neuropenia.

The observed differences in the levels of production of oxygen radicals by neutrophils and monocytes depend on the nature of stimuli and some other reasons. However, in the case of RA phagocytes the most important factor might be the prior priming of cells by TNF-α [233,238] or IL-8 [235] in synovial fluid. For example, we have shown that spontaneous oxygen radical production by RA neutrophils and monocytes is considerably enhanced in comparison with healthy donors, whereas there was no significant difference between oxygen radical production by PMA-stimulated RA and normal neutrophils [234]. NADPH oxidase is a major enzyme responsible for oxygen radical production in RA and normal neutrophils and monocytes. However, it was found [234] that superoxide production by RA neutrophils was reduced in the presence of NO synthase inhibitor NMMA and enhanced by L-arginine, while superoxide production by monocytes decreased in the presence of mitochondrial inhibitors rotenone and antimycin A. Thus, NO synthase in neutrophils and mitochondria in monocytes could be the additional producers of oxygen radicals in RA phagocytes.

Reactive nitrogen species are another factor of free radical damage in rheumatoid arthritis, although their role is less studied than that of oxygen radicals. Stichtenoth and Frolich [242] pointed out that the inhibition of nitric oxide synthesis had beneficial effects in humans. Mazzetti et al. [243] found that IL-1β stimulated NO production in RA chondrocytes. We demonstrated that NO synthase of RA neutrophils generated the enhanced amount of peroxynitrite [234]. Nitric oxide and oxygen radicals are also important inducers of death of human osteoarthritic synoviocytes [244].

It is of special interest that rheumatoid arthritis is one of the first examples of the extensive antioxidant treatment of human patients. In previous years the most recommended pharmaceutical antioxidant agent has been SOD. In 1986, Wilsman [245] reviewed the results of 10 years of presumably successful clinical experience with CuZnSOD treatment of inflammatory disorders including RA. Niwa et al. [246] recommended the application of liposomal

CuZnSOD, which showed no toxicity and had various advantages compared to native SOD. However, later on, Flohe [247] concluded that the systematic treatment of rheumatoid arthritis by SOD yielded disappointing results.

In subsequent years the study of CuZnSOD and MnSOD have been prolonged in animal inflammatory models. Dowling et al. [248] studied the effects of human recombinant MnSOD and CuZnSOD in adjuvant mediated paw edema and carrageenan-induced synovitis models. They demonstrated significant advantage of MnSOD in the suppression of inflammation in these models although MnSOD exhibited both antiinflammatory and inflammatory action depending on its concentration. Corvo et al. [249] demonstrated that the major limitation of native SOD — its rapid elimination from the circulation, can be overcome by the use of small-sized SOD PEG-liposomes. Iyama et al. [250] suggested that the transfer of extracellular SOD gene may be an effective form of RA therapy. SOD low molecular mimetic M40403 is apparently able to attenuate chronic inflammation in rat collagen-induced arthritis [251].

Treatment of RA patients with various drugs was shown to affect free radical production by phagocytic cells. It was found that the therapy with gold compounds [252] and piroxicam [237] diminished superoxide production in RA patients. Surprisingly, Hurst et al. [253] reported that successful therapy with penicillamine or sodium aurothiomalate is accompanied by an increase in superoxide production and serum thiol levels. The mechanism of this phenomenon is unknown. Mur et al. [254] demonstrated that antirheumatic medication of RA patients caused reducing cytokine priming of superoxide generation.

Antioxidant vitamins and chelators have been studied as the inhibitors of free radical production in RA. In contrast to cardiovascular diseases, the effects of vitamin E and C on RA development were not widely investigated. However, Harper et al. [255] recently demonstrated that the treatment of vasculitis patients with vitamins E and C reduced superoxide generation by neutrophils and may have an important role as adjuvant therapy. Antioxidative flavonoids are apparently the efficient inhibitors of oxygen radical production by RA phagocytes. Wittenborg et al. [256] reported the results of epidemiological study in RA patients treated with oral enzyme-combination product Phlogenzym containing trypsin, bromelain, and flavonoid rutin. These authors concluded that the treatment of patients with this medicine might be even more successful that with nonsteroidal antiinflammatory drugs. We compared the effects of SOD, catalase, rutin, mannitol, and desferrioxamine on oxygen radical production by neutrophils from RA patients and normal subjects [234]. It is interesting that only SOD and rutin exhibited strong inhibitory effects while catalase, mannitol (hydroxyl radical scavenger), and desferrioxamine were ineffective. These findings show that RA bloodstream leukocytes mainly produce superoxide and that the iron-catalyzed formation of hydroxyl or hydroxyl-like radicals is of little importance in RA. It is interesting to compare these data with the effects of rutin and mannitol on neutrophils from Fanconi anemia patients where iron-catalyzed hydroxyl radical formation is of importance (see below).

It is known that peroxynitrite is able to induce DNA strand breakage, which activates nuclear enzyme poly(ADP–ribose) synthase (PARS). Szabo et al. [257] showed that the inhibition of PARS by oral treatment with lipophilic inhibitor 5-iodo-6-amino-1,2-benzopyrone delayed the onset of arthritis in rats. It is possible that infrared pulse laser therapy can be useful for the treatment of RA patients [258].

31.3.3 Lung Diseases

Overproduction of oxygen radicals by inflammatory leukocytes in asthmatic patients has been shown for a long time [259,260]. Vargas et al. [261] suggested that peripheral granulocytes of patients with allergic bronchial asthma produced oxygen radicals in a hyperreactive state compared to control. Overproduction of superoxide and supposedly hydroxyl radicals in blood cells from asthmatic children was correlated with the severity of asthma [262]. Plaza et al. [263] pointed out at importance of the platelet lipoxygenase pathway of free radical production by platelets from asthmatic patients. These authors also supposed that there is difference in oxygen radical production in aspirin-tolerant and aspirin-intolerant patients. Enhanced lipid peroxidation in mild asthmatics has also been demonstrated by the measurement of plasma isoprostane level [264] and the elevation of exhaled ethane concentration [265].

The latest studies show that reactive nitrogen species play even more important role in asthma development. It was found that exhaled nitrogen oxide, an indicator of eosinophilic airway inflammation, is drastically enhanced in asthmatic patients. Correspondingly, it has been shown that lung damage is characterized by the augmentation of nitrotyrosine and iNOS expression in neutrophils, eosinophils, and macrophages in the airways of asthmatic patients [266].

Lung tissue injury is also mediated by reactive oxygen and nitrogen species in another inflammatory lung disease, acute respiratory distress syndrome (ARDS) [267]. Lamb et al. [268] showed that bronchoalveolar lavage fluid from ARDS patients contained the enhanced levels of dityrosine, chlorotyrosine, and nitrotyrosine, the products of neutrophil-mediated hydroxylation, nitration, and chlorination. These findings indicate that enhanced oxidative stress in the lungs of ARDS patients causes increased oxidative protein damage. Sittipunt et al. [269] demonstrated the expression of iNOS at high levels in alveolar macrophages and the accumulation of nitric oxide end products in the lungs of ARDS patients. It is possible that the suppression of oxidative stress and the protection of the lungs of ARDS patients might be achieved by the application of antioxidants. Thus, Ortolani et al. [270] showed that the administration of rutin and NAC to patients with early ARDS reduced the amount of expired ethane, decreased MDA concentration, and increased GSH level in the epithelial lining fluid.

Reactive oxygen and nitrogen species are also involved in the pathogenesis of asbestos-mediated pulmonary diseases. Kamp et al. [271] demonstrated that oxygen radicals mediate asbestos-induced toxicity to pulmonary cells including alveolar macrophages, epithelial cells, mesothelial cells, and endothelial cells. Quinlan et al. [272] showed that the inhalation of chrysotile asbestos by rats led to the appearance of inflammation and proliferation markers in the lungs. Dorger et al. [273] recently showed the dual role of iNOS in acute asbestos-induced lung injury. These authors found that iNOS deficiency enhanced inflammation in mice but diminished free radical-mediated lung damage.

Cystic fibrosis is the most common lethal autosomal-recessive disease, in which oxidative stress takes place at the airway surface [274]. This disease is characterized by chronic infection and inflammation. Enhanced free radical formation in cystic fibrosis has been shown as early as 1989 [275] and was confirmed in many following studies (see references in Ref. [274]). Contemporary studies also confirm the importance of oxidative stress in the development of cystic fibrosis. Ciabattoni et al. [276] demonstrated the enhanced in vivo lipid peroxidation and platelet activation in this disease. These authors found that urinary excretion of the products of nonenzymatic lipid peroxidation $PGF_{2\alpha}$ and TXB_2 was significantly higher in cystic fibrotic patients than in control subjects. It is of importance that vitamin E supplementation resulted in the reduction of the levels of these products of peroxidation. Exhaled ethane, a noninvasive marker of oxidative stress, has also been shown to increase in cystic fibrosis patients [277].

It was found that nitric oxide plays a very important role in cystic fibrosis. In contrast to most inflammatory airway diseases (for example, asthma), exhaled NO decreased in cystic fibrosis, and nitric oxide deficiency may contribute to the bronchial obstruction [278]. Bebok et al. [279] demonstrated that NO levels in the vicinity of airway cells during inflammation are sufficient to nitrate cystic fibrosis transmembrane conductance regulator (CFTR), causing its degradation and decreased function. It has been suggested that CFTR degradation may account for cystic fibrosis development occurring in chronic inflammatory lung diseases. Morrissey et al. [280] concluded that cystic fibrosis is characterized by disordered NO airway metabolism and protein nitration.

It has been shown that lung macrophages from patients with systemic sclerosis (SS) produced the elevated levels of nitric oxide, superoxide, and peroxynitrite and expressed the enhanced level of iNOS [281]. NAC administration reduced peroxynitrite production and might be possibly recommended for the treatment SS patients. Solans et al. [282] found the significant enhancement of lipid peroxidation in erythrocytes from SS patients. Cracowski et al. [283] showed that in vivo lipid peroxidation was enhanced in scleroderma spectrum disorders including SS and undifferentiated connective tissue disease.

Antioxidant therapy might be promising medication for the treatment of some lung disorders. For example, lecithinized phosphatidylcholine–CuZnSOD suppressed the development of bleomycin-induced pulmonary fibrosis in mice [284]; these findings could be of relevance for the treatment of bleomycin-stimulated pulmonary fibrosis in humans. Davis et al. [285] recently demonstrated that the treatment of premature infants with recombinant human CuZnSOD may reduce early pulmonary injury.

31.3.4 Skin Inflammation

The formation of reactive oxygen and nitrogen species is of significant importance in the pathogenesis of skin inflammation. In 1987, Niwa et al. [286] showed that that the level of lipid peroxides and SOD activity considerably increased in the skin lesions of patients with severe skin diseases. Polla et al. [287] found that peripheral blood monocytes but not peripheral blood neutrophils from patients with atopic dermatitis were primed for superoxide production. They suggested that the in vivo monocyte priming and enhanced superoxide production may be responsible for the pathogenesis of this skin disease. On the other hand, Greenacre et al. [288] demonstrated that the induction of iNOS in neutrophils and protein nitration are important stimuli of cutaneous inflammation in rats.

One of the common inducers of skin inflammation is skin exposure to UV irradiation. It has been suggested that exposure to ultraviolet B (UVB) radiation may be a cause of cNOS and xanthine oxidase activation of human keratinocytes and that such photo-induced response may be involved in the pathogenesis of sunburn erythema and inflammation [289]. Antioxidant (−)-epigallocatechin-3-gallate from green tea decreased UVB exposure-induced skin injury and might be possibly useful for protection against oxygen radical-mediated UVB-stimulated inflammatory dermatoses, photoaging, and photocarcinogenesis [290]. Another indirect antioxidant ebselen has been found to inhibit TPA-induced TBAR formation in mouse skin and probably should be studied as a potential drug for the treatment of inflammation-associated carcinogenesis [291].

Shingu et al. [292] found that sera from patients with systemic lupus erythematosus and psoriasis vulgaris were resistant to hydrogen peroxide-mediated complement activation. These authors suggested that reactive oxygen species produced by cytokine-activated fibroblasts have an important role in inflammation and subsequent tissue damage at skin lesions. Er-rali et al. [293] demonstrated the enhanced superoxide production by dermal fibroblasts of psoriatic patients. Although psoriasis is characterized by the overproduction of reactive oxygen species, peculiarity of this disease is that the most common efficient therapeutic agent applied for psoriasis treatment is prooxidant anthralin. The mechanism of skin inflammatory response and antipsoriatic efficiency of anthralin is uncertain, but it might be suggested that its effects are due to the formation of anthralin and oxygen free radicals

[294]. It has been proposed that anthralin-associated toxicity might be suppressed by systematic antioxidant administration [295]. The treatment of psoriatic patients with another drug dithranol resulted in a decrease in in vivo oxygen radical generation [296].

31.3.5 Brain Inflammatory Diseases

Contemporary data suggest that inflammation in brain is closely associated with the pathogenesis of degenerative neurologic pathologies such as Alzheimer's disease, Parkinson's disease, multiple sclerosis, and amyotrophic lateral sclerosis. Brain inflammation is mediated by the activation of glial cells, which produce a variety of proinflammatory and neurotoxic factors and reactive oxygen and nitrogen species [230]. These brain inflammatory disorders are considered below.

31.3.5.1 Multiple Sclerosis

The release of reactive oxygen species by monocytes or macrophages is supposed to be an important factor for the destruction of CNS white matter in multiple sclerosis (MS) pathogenesis [297,298]. Glabinski et al. [299] demonstrated an increase in PMA-stimulated superoxide production in the blood of MS patients. Subsequent studies showed an importance of excessive NO formation by macrophages, microglia, and astrocytes in demyelinating lesions in multiple sclerosis [300] probably through the upregulation of inducible NO synthase in MS lesions [301]. As NO is a relatively unreactive reagent, it has been suggested that nitrotyrosine residues in brain tissues from MS patients are probably formed in reactions with peroxynitrite [302]. Thus, Hooper et al. [303] demonstrated that the administration of uric acid, a natural scavenger of peroxynitrite, to mice with experimental allergic encephalomyelitis (an animal model of MS) suppressed the invasion of inflammatory cells into CNS and prevented development of disease. Mijkovic et al. [304] and Nazliel et al. [305] have found that the level of nitric oxide metabolites was enhanced in cerebrospinal fluid of MS patients.

31.3.5.2 Amyotrophic Lateral Sclerosis

Amyotrophic lateral sclerosis (ALS) is a fatal motor neuron degenerative disease, which characterized by a single-site mutation in the CuZnSOD gene. This fact suggests a possible damaging role of free radicals in ALS. In 1999, Liu et al. [306] obtained an in vivo evidence for the enhanced levels of hydroxyl radicals and hydrogen peroxide and the reduced level of superoxide in ALS mutant mice. These findings indicate the transformation of innocuous superoxide into highly reactive hydroxyl radicals in ALS in the absence of SOD. (Compare the similar phenomenon in Fanconi anemia patients discussed below.) An early event of ALS is mitochondrial dysfunction; therefore, the overexpression of mitochondrial antioxidant genes MnSOD and GPX4 and preincubation with spin trap 5,5-dimethyl-1-pyrroline-N-oxide (DMPO) prevented mutant SOD1-mediated motor neuron cell death and increased ALS-like transgenic mouse survival [307].

31.3.5.3 Alzheimer's Disease

Alzheimer's disease (AD), a major dementing disorder of the elderly, draws an exclusive interest of researchers working in the field of degenerative neurologic pathologies. The study of oxidative stress in Alzheimer's disease development is attaching a special interest; correspondingly, numerous reviews on this subject have been published only during the last few years [308–312]. Reactive oxygen and nitrogen species can be produced in AD by different sources. It has been established that neurotic plagues in the brain of AD patients contain aluminosilicate deposits. Evans et al. [313,314] suggested that alumosilicates may stimulate the production of reactive oxygen species by phagocytes in Alzheimer's disease. Iron and copper ions might also be important initiators of oxygen radical production because these metals are present in significant concentrations in AD neutrophils and chelators are able to suppress oxidative stress in Alzheimer's disease [309]. Mitochondrial anomalies can be also the source of free radical overproduction in AD brain [308].

It has been established that amyloid β-peptide (Aβ) formed by the proteolytic cleavage of the transmembrane amyloid precursor protein and, particularly its Aβ(1–42) form, play a central role in the AD pathogenesis. Aβ(1–42) is able to induce lipid peroxidation and initiate the formation of free radicals and reactive aldehydes [311]. The formation of isoprostanes, another marker of oxidative stress, is also increasing in the presence of Aβ [310,311]. Montine et al. [315] demonstrated that lateral ventricular fluid from AD patients contains significantly enhanced levels of F_2-isoprostanes. In contrast to the above findings, Pratico et al. [316] showed that an increase in lipid peroxidation and the formation of isoprostanes preceded Aβ formation in an animal model of Alzheimer's disease. Tuppo et al. [317] found that the concentrations of $iPF_{2\alpha}$-III isoprostane (the product of nonenzymatic lipid peroxidation) and TXB_2 thromboxane (the product of enzymatic peroxidation of arachidonic acid) were essentially elevated in urine of patients with probable Alzheimer's disease. It was also suggested that peroxynitrite is a nitration agent of proteins, causing the enhancement of nitrotyrosine and dityrosine formation in AD brain [310]. Xie et al. [318] demonstrated that peroxynitrite is a mediator of the toxicity of Aβ(1–42)-activated microglia in Alzheimer's disease. Aβ is also able to activate NADPH oxidase in microglia, resulting in the production of superoxide [319].

The above data demonstrate an important, possibly even critical role of oxidative stress in AD pathogenesis. Therefore, it is reasonable to suggest that antioxidant administration could be useful for the treatment of AD patients. Grundman [320] recently summarized the results of clinical trial, in which vitamin E was administrated to AD patients with moderately severe disease. It has been concluded that the treatment with vitamin E may delay and slow disease progress in these patients.

31.3.5.4 Parkinson's Disease

The hallmark of Parkinson's disease (PD) is the destruction of dopaminergic neurons in the substantia nigra. It has been suggested that the metabolism of dopamine may be a cause of free radical generation [321]. However, there are other sources of free radical overproduction in PD such as an increase in lipid peroxidation, a decrease in the level of reduced glutathione, an increase in iron, and the inhibition of mitochondrial Complex I activity [322]. For example, Berman and Hasting [323] found that dopamine quinone is able to stimulate mitochondrial dysfunction. A recent study suggested that the loss of dopaminergic neurones in substantia nigra with PD may be the consequence of the inflammation-induced proliferation of microglia and macrophages expressing iNOS [324].

There were attempts to treat PD patients with antioxidants [325], but it seems that no encouraging results have been achieved. As the changes of glutathione metabolism plays an important role in the pathogenesis of neurodegenerative diseases, it was suggested that the supplementation with glutathione could be useful for PD treatment. However, GSH penetrates the blood–brain barrier poorly; therefore, the treatment with glutathione precursors or analogs could be more promising [321].

Oxidative stress induced by reactive oxygen species is described for many other brain disorders of inflammatory and noninflammatory nature. Some examples of such disorders are given below. Quick and Dugan [326] demonstrated superoxide-mediated damage of neurons in a model of ataxia–telangiectasia, an autosomal recessive disorder in children. The product of nonenzymatic lipid peroxidation isoprostane 8-iso-PGF$_{2\alpha}$ induced oxidant stress-induced cerebral microvascular injury and brain damage [327]. Svenungsson et al. [328] showed that the levels of nitric metabolites and proinflammatory cytokines increased in neuropsychiatric lupus erythematosus. Fluid percussion brain injury was accompanied by superoxide production through COX-2 activation [329]. An increase in the levels of isoprostane are also observed in spinal cord injury [330] and traumatic brain injury [331].

31.3.6 Kidney Diseases

In 1986, Diamond et al. [332] demonstrated an important role of oxygen radicals in aminonucleoside nephrosis and showed protective effects of SOD and allopurinol. Similarly, Adachi et al. [333] showed that SOD inhibited lipid peroxidation in kidneys of rats with glomerular nephritis. Recent studies confirm the participation of free radicals in various kidney disorders. Zhou et al. [334] showed an increase in plasma NO levels and lipid peroxidation in plasma and erythrocytes in chronic glomerulonephritis patients. At the same time, SOD, catalase, and glutathione peroxidase activities in erythrocytes of these patients were significantly decreased. In contrast, total and renal NOS activities were found to be reduced in glomerulonephritis rat model of chronic renal disease supposedly due to an increase in circulating endogenous NOS inhibitors [335]. Walpen et al. [336] pointed out that nitric oxide is a critical mediator of several forms glomerulonephritis. These authors showed that the inhibition of NO synthesis in a rat model of glomerulonephritis resulted in the reduction of macrophage inflammatory protein 2 mRNA expression and attenuated neutrophil infiltration in the glomerulus. Increased formation of nitric oxide and oxygen radicals has been shown in acute renal failure (ARF) during sepsis [337]. It has been suggested that the superoxide level increased in kidney during ARF due to a decrease in extracellular SOD that in turn decreased vascular NO and caused renal vasoconstriction. Correspondingly, protective effect of the SOD mimic metalloporphyrin on renal function has been found.

Lucchi et al. [338] found that the concentrations of conjugated linolenic acid (CLA) and conjugated dienes were substantially reduced in the erythrocytes of patients with chronic renal failure (CRF); at the same time CLA significantly increased in the plasma and adipose tissue of the end-stage patients. Increased CLA levels in the end-stage CRF patients are supposedly a consequence of reduced metabolism of CLA to conjugated dienes. Chen et al. [339] showed that IgA from patients with IgA nephropathy may induce oxidative injury in glomerular mesangial cells of patients with IgA nephropathy. Thamilselvan et al. [340] suggested that the oxalate-induced free radical-mediated damage in renal epithelial cells promoted the calcium oxalate stone formation.

31.3.7 Liver Diseases

A number of early in vitro studies demonstrated a considerable role of free radicals in liver injury (see, for example, *Proceedings of International Meeting on Free Radicals in Liver Injury* [341]). Later on, it was shown that chronic inflammation in the liver-induced oxidative DNA damage stimulated chronic active hepatitis and increased the risk of hepatocarcinogenesis [342,343]. Farinati et al. [344] showed that 8-OHdG content increased in circulating leukocytes of patients with chronic hepatitis C virus (HCV) infection. DNA oxidative damage is supposedly an early event of HCV-related hepatitis. The formation of isoprostanes in the liver of carbon tetrachloride-treated rats can be suppressed by the administration of vitamin E [345].

An important factor in pathogenesis of chronic hepatitis C is iron overload. Casaril et al. [346] found that even a mild increase in iron content caused additional free radical-mediated damage due to viral infection. It has long been suggested that oxygen radicals may directly mediate cell damage in cirrhosis liver [347]. Kono et al. [348] proposed that oxygen radicals produced by NADPH oxidase of hepatic Kupffer cells have a predominant role in the pathogenesis of early alcohol-induced hepatitis. Yamamoto et al. [349] suggested that the copper-catalyzed formation of hydroxyl radicals in the liver might be responsible for the pathogenesis of acute hepatitis in Long-Evans Cinnamon rats, which similarly to patients with Wilson's disease, accumulate excess of copper.

31.3.8 Septic Shock, Pancreatitis, and Inflammatory Bowel Disease

The formation of reactive oxygen and nitrogen species has been reported in many other inflammatory disorders. We quote the most interesting (in our opinion) examples below. Similar to the other inflammatory diseases, the overproduction of superoxide and nitric oxide has been shown under septic conditions [350,351]. Fukuyama et al. [352] demonstrated the formation of peroxynitrite in chronic renal failure patients with septic shock, which has been suggested to be responsible for nitration of tyrosine residues. Macarthur et al. [353] proposed that the inactivation of catecholamines by superoxide plays an important role in the pathogenesis of septic shock. It has been suggested that antioxidants ascorbic acid and NAC may improve the function of lymphocytes, which are important targets of endotoxins contributing to oxygen radical overproduction by septic shock [354].

Free radicals are supposed to have a significant role in the progression of acute pancreatitis. The involvement of free radicals was firstly demonstrated in many animal models [355,356]. Later on, it has been shown that the levels of superoxide and lipid peroxides increased in the blood from patients with acute pancreatitis [357]. Rahman et al. [358] found enhanced urinary nitrite excretion in patients with severe acute pacreatitis. It was suggested that this fact is not simply a reflection of systemic inflammation but probably a consequence of the endotoxin-mediated upregulation of inducible NO synthase.

Gionchetti et al. [359] found that superoxide production by circulating PMNs in untreated patients with ulcerative colitis and Crohn's disease in remission was significantly lower compared to controls. Interestingly, later on, Oldenburg et al. [360] showed that oxygen radical production by the whole blood in patients with the same pathologies has been enhanced. This difference could depend on the use different methods of oxygen radical detection, cytochrome c reduction in Ref. [359] and luminol-amplified CL in Ref. [360]. Miller et al. [361] showed that superoxide production, lipid peroxidation, and tyrosine nitration were considerably enhanced in aortas from patients with abdominal aortic aneurysm undergoing surgical repair. Increase in superoxide production was due to the expression of NADPH oxidase.

31.4 IRON-CATALYZED PATHOPHYSIOLOGICAL DISORDERS

The catalytic effects of iron on free radical processes in biological systems have been considered in Chapter 21. It has been recognized for a long time that the appearance of "free" iron in blood and tissues may initiate cell injury and cause various pathophysiological disorders. One of the first examples of the enhancement of oxygen radical production catalyzed by high levels of nonprotein-bound iron in humans has been reported by Gutteridge et al. [362] for the patients with neuronal ceroid lipofuscinoses. In subsequent work by these authors [363] an increase in oxygen radical production catalyzed by "free" iron has been demonstrated for patients with idiopathic hemochromatosis. Aust and White [364] proposed that ischemia–reperfusion tissue damage is stimulated by iron release during reperfusion and can be ameliorated by the use of chelators. It has also been suggested that iron-catalyzed oxygen radical formation takes place after reperfusion of the ischemic bowel [365].

Iron-catalyzed free radical-mediated damage is supposed to be an important factor of many diseases including those considered above. Sayre et al. [366] reviewed the role of iron-mediated processes in Alzheimer's disease, Parkinson's disease, and some other neurodegenerative diseases. Hepatic iron deposition was found in chronic hepatitis C. Thus, Kageyama et al. [367] identified enhanced hepatic iron accumulation and lipid peroxidation in patients with hepatitis C that were decreased during interferon therapy. Jung et al. [368] demonstrated in a model of sepsis that hepatocellular iron reduced the formation of nitrite and nitrate and S-nitrosothiols and increased the production of reactive oxygen species. Carmine et al. [369] found children with acute lymphoblastic leukemia undergoing chemotherapy characterized by the enhanced plasma levels of iron that can be related to toxic side effects. Ogihara et al. [370] also found nonprotein-bound iron in cerebrospinal fluid of newborn infants with hypoxic ischemic encephalopathy as well as higher levels of dityrosine and ascorbic acid. These authors suggested that iron-catalyzed formation of hydroxyl radicals occurs in the CNS of children during asphyxiation. Below, we will consider the other pathologies associated with iron-stimulated free radical damage.

31.4.1 Hemochromatosis and Hemodialysis of Patients

Hereditary hemochromatosis is a pathophysiological disorder characterized by disordered iron metabolism and the accumulation of excessive iron in body organs. Hemochromatosis is the origin of cirrhosis and liver failure; moreover, the risk of hepatocellular carcinoma is increased by 200 times in hemochromatosis patients [371]. Stimulation of free radical formation in hemochromatosis patients has been shown earlier [363]. Increased lipid peroxidation has also been shown in the kidneys of IOL animals [372,373]. Young et al. [374] demonstrated that enhanced lipid peroxidation is accompanied by reduced antioxidant status in hemochromatosis patients. Using model of chronic experimental hemosiderosis in rats, Zhou et al. [375] confirmed that chronic iron overload increased oxygen radical production and renal injury. Hussain et al. [376] showed that the formation of reactive oxygen and nitrogen species by iron overload in hemochromatosis resulted in mutations in the p53 tumor suppressor gene, the most prominent genetic alteration in the development of human cancer.

Oxidative stress and inflammation are elevated in hemodialysis patients, which, at least partly, might be initiated by intravenous iron administration. Thus, Tovbin et al. [377] showed that the administration of iron saccharide to hemodialysis patients increased the blood level of oxidized proteins. Similarly, an increase in the levels of free ("nontransferrin-bound" or "labile") iron has been shown in the plasma of hemodialysis patients [378] including patients after intravenous iron saccharate infusion [379].

31.4.2 Thalassemia

Thalassemia (Tl) is a genetic disorder with wide variety of clinical phenotypes associated with impaired synthesis of hemoglobin. These phenotypes are varied from clinically silent heterozygous β-thalassemia to severe transfusion-dependent thalassemia major. The development of this pathology leads to chronic anemia and increased dietary absorption, which results in iron overload. Regular transfusion in patients increases the iron release from hemoglobin. The enhancement of free radical production and suppression of antioxidant level have been shown in patients with different forms of thalassemia. It has been suggested that the unstable hemoglobin, excess of α-Hb subunits, and the high levels of cytosolic and membrane-bound iron are major sources of free radical overproduction by thalassemic erythrocytes. The formation of highly reactive hydroxyl radicals was observed in β-thalassemic erythrocytes in the presence of ascorbate [380]. The formation of hydroxyl radicals is apparently typical for some other hemoglobinopathic erythrocytes such as sickle erythrocytes [381].

As hydroxyl or hydroxyl-like radicals are produced by the superoxide-driven Fenton reaction, superoxide overproduction must also occur in thalassemic cells. First, it has been shown by Grinberg et al. [382], who demonstrated that thalassemic erythrocytes produced the enhanced amount of superoxide in comparison with normal cells in the presence of prooxidant antimalarial drug primaquine. Later on, it has been found that the production of superoxide and free radical-mediated damage (measured through the MetHb/Hb ratio) was much higher in thalassemic erythrocytes even in the absence of prooxidants, although quinones (menadione, 1,4-naphthoquinone-2-methyl-3-sulfonate) and primaquine further increased oxidative stress [383]. Overproduction of superoxide was also observed in thalassemic leukocytes [384].

Overproduction of free radicals by erythrocytes and leukocytes and iron overload result in a sharp increase in free radical damage in Tl patients. Thus, Livrea et al. [385] found a twofold increase in the levels of conjugated dienes, MDA, and protein carbonyls with respect to control in serum from 42 β-thalassemic patients. Simultaneously, there was a decrease in the content of antioxidant vitamins C (44%) and E (42%). It was suggested that the iron-induced liver damage in thalassemia may play a major role in the depletion of antioxidant vitamins. Plasma thiobarbituric acid-reactive substances (TBARS) and conjugated dienes were elevated in β-thalassemic children compared to controls together with compensatory increase in SOD activity [386]. The development of lipid peroxidation in thalassemic erythrocytes probably depends on a decrease in reduced glutathione level and decreased catalase activity [387].

Another important characteristic of oxidative stress in thalassemia is LDL oxidative modification. Livrea et al. [388] showed that the concentration of hydroperoxides in LDL of thalassemia patients was equal to 22.60 ± 12.84 nmol/mg LDL protein compared to 6.25 ± 3.04 nmol/mg in control LDL. These authors proposed that the enhanced LDL oxidation in thalassemia was connected with the depletion of vitamin E in LDL. Interestingly, these findings contradict the suggestion about the prooxidant role of vitamin E (α-tocopherol) in LDL oxidation (Chapter 25). It was proposed that LDL oxidation could be the origin of atherogenetic risk in thalassemic patients.

The mechanism of the initiation of free radical-mediated damage in thalassemia is still uncertain. Some data indicate an important role of excess α-hemoglobin chains in this process. Thus, Scott et al. [389] showed that "model" beta-thalassemic cells with unpaired α-hemoglobin chains entrapped within normal erythrocytes generated significantly greater amounts of methemoglobin and intracellular hydrogen peroxide than did control cells. α-Chains are also the most efficient oxidants of LDL protein ApoB [390]. Another possible mechanism of free radical formation in thalassemic erythrocytes is ferric ion-mediated Hb oxidation [391]. It has been proposed that the release of small amounts of free ions from unpaired α-hemoglobin chains in β-thalassemic erythrocytes can initiate redox processes catalyzed by the oxidation of reduced glutathione.

The most successful up-to-date treatment of thalassemic patients is chelating therapy, which is based on patient's lifetime application of iron chelators. Removal of excess iron is supposed to be effective route for suppressing free radical-mediated damage. There is a great number of studies showing successful treatment of thalassemic patients with intravenous chelator desferal (desferrioxamine) and oral chelator deferiprone (L1). Biochemical studies show the efficacy of both chelators in removal of excess iron. For example, the incubation of thalassemic erythrocytes with 0.5 mmol l^{-1} L1 during 6 h resulted in 96% removal of membrane-free iron [392]. It was demonstrated that L1 is able to remove pathologic deposits of chelatable iron from thalassemic and sickle erythrocyte membranes. Animal study confirmed removing of pathologic free iron deposits from murine thalassemic erythrocytes and improvement of erythrocyte survival after L1 therapy [393].

Unfortunately, the iron complexes of both chelators desferal and L1 are able to catalyze the formation of oxygen radicals [394,395]. Cragg et al. [395] also showed that L1 exposure markedly enhanced free radical-mediated DNA damage in iron-loaded liver cells. It has been suggested that the prooxidant:antioxidant ratio of L1 activity depends on the composition of complexes formed: a 1:3 Fe/L1 is supposed to be inactive in the production of free radicals while the generation of radicals is possible at lower Fe/L1 ratios [395]. But it should be noted that in real biological systems there is always equilibrium between iron–chelator complexes of different composition.

Taking into account the possible prooxidant effect of chelators, the simultaneous application of antioxidants and desferal or L1 might be useful in the treatment of thalassemic patients. There are two major free radical-mediated damaging processes during chelating therapy, which could be stopped by the application of antioxidants: the formation of free radicals by the ferrous ion–chelator complexes and the peroxidation of lipids. Tesoriere et al. [396] studied the effect of oral supplementation of vitamin E on lipid peroxidation in plasma and the oxidative damage to LDL and erythrocytes in 15 β-thalassemia intermedia patients. Low level of vitamin E and high level of MDA in plasma were normalized after 3 months, while the low level of vitamin E in LDL became normal only after 9 months; however, the level of conjugated dienes remained twice higher than control. Thus, in authors' opinion oral treatment with vitamin E improves the antioxidant:oxidant balance in plasma, LDL, and erythrocytes in β-thalassemia intermedia patients.

Bioflavonoid rutin (vitamin P) was proposed to be useful antioxidant for the suppression of free radical production in thalassemia [383]. Grinberg et al. [397] showed that rutin protected hemoglobin inside the erythrocyte from primaquine-induced free radical-mediated attack. We compared the protective effects of rutin and L1 on menadione-stimulated superoxide production by normal and thalassemic erythrocytes [383]. The most interesting finding is the difference found between the inhibitory effects of these compounds on normal and pathologic erythrocytes. Rutin was equally efficient in the suppression superoxide overproduction by normal and thalassemic erythrocytes while L1 diminished superoxide production by thalassemic cells and did not affect it in the case of normal cells (Figure 31.1). This fact is a strong support of a traditional point of view that iron overload is a main source of free radical damage in thalassemic erythrocytes.

31.4.3 SICKLE CELL DISEASE

It is known that erythrocytes from patients with sickle cell anemia contain various types of abnormal iron deposits [398], which could be the origin of the overproduction of oxygen radicals in these cells. Indeed, Hebbel et al. [399] has showed that sickle erythrocytes spontaneously generate approximately twice as much superoxide as normal erythrocytes. Later on, it has been shown that these cells are also able to generate hydroxyl radicals catalyzed by three types of "iron," preexisting free iron, free iron released during oxidative stress, and iron that cannot be chelated with desferrioxamine [400].

It has been proposed that a major source of oxygen radicals in sickle erythrocytes is mutant hemoglobin HbS. However, although HbS showed an accelerated autoxidation rate under in vitro conditions, its in vivo oxidative activity was not determined. Sheng et al. [401] suggested that the observed oxidation rate of HbS is exaggerated by adventitious iron. Dias-Da-Motta et al. [402] proposed that another source of enhanced superoxide production in sickle cells are monocytes; in contrast, there is no difference in superoxide release by sickle

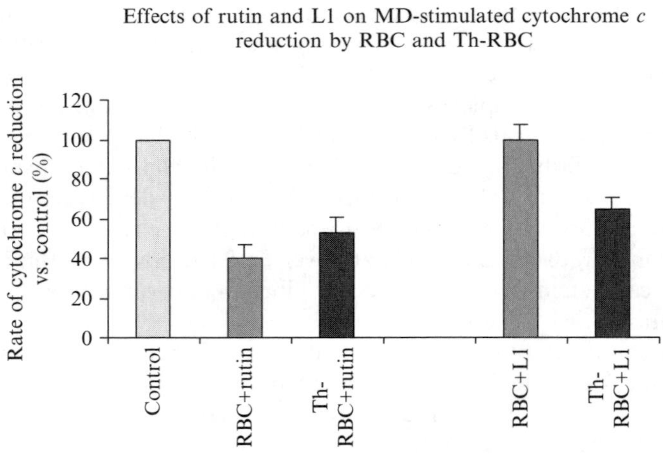

Effects of rutin and L1 on MD-stimulated cytochrome c reduction by RBC and Th-RBC

FIGURE 31.1 Effects of rutin and L1 on menadione-stimulated cytochrome c reduction normal and Tl erythrocytes. (From IB Afanas'ev et al.*Transfusion Sci* 23: 237–238, 2000.)

and normal neutrophils. These authors suggested that enhanced monocyte superoxide generation caused the inactivation of nitric oxide (via the formation of peroxynitrite) and additional tissue damage. Nath et al. [403] demonstrated an increase in lipid peroxidation in the kidney of transgenic sickle mouse.

However, the latest findings completely changed the conception of the mechanism of free radical damage in sickle cell disease. Aslan et al. [404] showed that in contrast to previous reports [399,400] there is no significant difference in the rates of superoxide and hydrogen peroxide production as well as in the levels of lipid peroxides between HbA and HbS cells. Instead, it has been concluded that increased reactive oxygen species production in tissue is possibly a major factor of impaired NO signaling and corresponding tissue damage. It has also been suggested that hypoxia–reperfusion associated with sickle cell disease resulted in the release of xanthine oxidase into circulation from hepatic cells. Superoxide produced by xanthine oxidase reacts with NO and induced vascular dysfunction. A new mechanism appears to allow considering sickle cell disease as a chronic inflammatory disease [405], in which iron-catalyzed oxygen radical production is not a major damaging factor. Recently Aslan et al. [406] demonstrated the increased rates of tissue reactive oxygen species production, liver and kidney NOS2 expression, and tissue nitrotyrosine formation in humans with sickle cell disease and murine sickle model.

31.4.4 Fanconi Anemia

Fanconi anemia (FA) is a rare usually fatal autosomal recessive disease with a life expectancy of about 16 years. Although no differences in the activity of antioxidant enzymes (SOD, catalase, and glutathione peroxidase) and the levels of reduced glutathione in FA were found in comparison with controls, enhanced oxygen radical production have been shown in FA cells. Thus, Nagasawa and Little [407] suggested that the suppression of cytotoxic effect of mitomycin C on FA fibroblasts by SOD pointed out at superoxide-mediated cell damage. Scarpa et al. [408] showed that superoxide production by FA erythrocytes was 2.3 times higher than in controls while SOD activities were very similar. Later on, Malorni et al. [409] found a decrease in the SOD and catalase activities and probable increase in superoxide production by FA erythrocytes.

However, the major producers of oxygen radicals in Fanconi anemia are leukocytes and not erythrocytes. It has been found that blood and bone marrow FA leukocytes produced much more oxygen radicals than controls without stimulation or stimulated with concanavalin A, SiO_2, latex, and opsonized zymosan [410]. Another important finding was that FA leukocytes not only released enhanced amount of oxygen radicals but that the concentration of hydroxyl or hydroxyl-like radicals in these cells was also increased. (This conclusion was made comparing the levels of lucigenin- and luminol-amplified CL by FA and normal leukocytes.) As hydroxyl radicals are to be formed by the iron-catalyzed decomposition of hydrogen peroxide (the Fenton reaction), these findings point out at the importance of iron-mediated free radical damage in FA.

The importance of iron in the induction of oxidative stress in FA has already been suggested. In 1987, Joenje et al. [411] supposed that reactive oxygen species may be involved in the induction of chromosomal damage in FA. In 1989, Porfirio et al. [412] showed that the iron chelator desferrioxamine reduced spontaneous chromosomal breakage of FA cells. The involvement of iron-stimulated hydroxyl radical formation is also seen from the different effects of free radical scavengers and iron chelators on FA and normal leukocytes. Thus, hydroxyl radical scavengers rutin and mannitol were much more effective inhibitors of oxygen radical release by FA leukocytes compared to normal cells, while SOD was about five to ten times more effective in normal leukocytes [410]. It is important that the administration of nontoxic flavonoid rutin (vitamin P) to FA children resulted in the significant reduction of oxygen radical overproduction by blood leukocytes, a decrease in the amount of chromosomal aberration, and the improvement of hematological characteristics and patients' health [410,413]. On these grounds rutin has been recommended for the treatment of FA patients. It should be noted that earlier hypersensitivity toward iron has been also shown for FA lymphoblastoid cells [414].

While "free" iron is a catalyst of hydroxyl or hydroxyl-like radical overproduction by FA leukocytes, the other stimuli might also exist, which are responsible for the enhancement of the formation of superoxide, a precursor of hydroxyl radicals in the superoxide-dependent Fenton reaction. Thus, Schultz and Shahidi [415] showed that such a stimulus could be TNF-α, which was detected in the plasma of FA patients but not in healthy donors. Another factor of enhanced oxidative stress in FA might be a low thioredoxin level, which may cause an increasing DNA damage [416].

It has already been noted above that oxygen radical-mediated DNA damage is an important element of oxidative stress in FA patients. Takeuchi and Morimoto [417] demonstrated the increased formation of 8-OHdG, a product of DNA oxidation, in FA lymphoblasts. Clarke et al. [418] showed that increased sensitivity of FA cells to mitomycin C-induced apoptosis is due to oxygen radical overproduction and not DNA crosslinking. In conclusion, the work by Pagano et al. [419] should be mentioned, which stressed the significance of oxygen radical-mediated DNA damage in FA cells.

31.5 BLOOM'S SYNDROME AND DOWN SYNDROME

Bloom's syndrome is a rare autosomal recessive disease characterized by a high level of spontaneous chromosomal aberrations and sister chromatid exchange (SCE). Bloom's syndrome involves exhibiting numerous clinical features including predisposition to cancer. The importance of oxidative stress in Bloom's syndrome follows from the overproduction of superoxide, an increase in free radical-mediated damaging processes, and SOD induction [420]. The capacity to produce elevated levels of oxygen radicals probably induces the spontaneous chromosomal instability of Bloom's cells and is responsible for the high incidence of neoplasia in Bloom' patients [421]. It should be noted that Emerit and Cerutti [422] applied exogenous SOD for the suppression of oxidative stress in Down's syndrome, but found no change in SCE rates.

Down's syndrome caused by trisomy 21 is associated with premature aging, congenital anomalies, and neurodevelopmental impairment. This neurodegenerative disorder is also characterized by oxygen radical overproduction and enhanced SOD activity. Thus Colton et al. [423] demonstrated the enhanced superoxide production by microglia from trisomy 16 mice, an animal model of Down's syndrome. Busciglio and Yankner [424] showed that Down's neurons produced the three- to fourfold increase in the amount of reactive oxygen species and the elevated level of lipid peroxidation that preceded neuronal death. It has been suggested that the defective repair of oxidative damage in mitochondrial DNA is responsible for the defective mitochondrial electron transport and the overproduction of superoxide in Down's syndrome [425]. Schuchmann and Heinemann [426] studied superoxide generation in cultured hippocampal neurons from trisomy 16 mice (Ts16). They suggested that elevated superoxide production by Ts16 neurons is probably caused by a deficient Complex I of mitochondrial electron transport chain, which finally leads to neuronal cell death. Recently, Capone et al. [427], using lucigenin-amplified CL, confirmed that mitochondrial superoxide production increased in Down's syndrome.

31.6 OTHER EXAMPLES OF FREE RADICAL FORMATION IN PATHOPHYSIOLOGICAL DISORDERS

Some other examples of free radical formation in various pathologies are discussed below. (Of course, they are only few examples among many others, which can be found in literature.) Mitochondrial diseases are associated with superoxide overproduction [428] and cytochrome c release [429]. For example, mitochondrial superoxide production apparently contributes to hippocampal pathology produced by kainate [430]. It has been found that erythrocytes from iron deficiency anemia are more susceptible to oxidative stress than normal cells but have a good capacity for recovery [431]. The beneficial effects of treatment of iron deficiency anemia with iron dextran and iron polymaltose complexes have been shown [432,433].

Shin et al. [434] found that rutin and harmaline (1-methyl-7-methoxy-3,4-dihydro-β-carboline) exhibited beneficial protective effects against the development of the surgically induced reflux oesophagitis. Enhanced levels of urinary NO metabolites and TBAR products were found in migraine sufferers [435]. A significant decrease in NO synthesis and MDA increase have been demonstrated in PMNs of schizophrenia patients [436]. Selvam [437] suggested that oxalate-induced membrane injury in calcium oxalate stone disease is mediated by superoxide and hydroxyl and peroxyl radicals. Yang et al. [438] demonstrated that the production of reactive oxygen species significantly increased in fatty liver mitochondria of obese mice. Dandona et al. [439] also showed an increase in reactive oxygen species produced by leukocytes of obese subjects and demonstrated that dietary restriction and weight loss reduced both reactive oxygen species generation by leukocytes and oxidative damage to lipids, proteins, and amino acids.

31.7 FREE RADICALS IN AGING

31.7.1 THE LATEST DEVELOPMENTS

Since Harman's famous work [440] on free radical theory of aging in 1956, numerous studies have been dedicated to the development of his theory, which were reviewed by many authors (see, for example, early reviews, Refs. [441–444]). There is of course no necessity and possibility to discuss these reviews, which have been perfectly considered by their authors.

The connection between free radical generation and aging has been demonstrated in early and contemporary studies. Thus, the difference in oxygen radical production in cells from young and old organisms has been shown [445,446]. Martins et al. [447] found a significant increase in the formation of reactive oxygen species by granulocytes of healthy persons from 40 to 69 years of age in comparison with younger men. On the contrary, Phillips et al. [448] showed that enhanced oxygen radical production occurs both in older and younger humans, but in contrast to damaging effects in older age, it could be a normal physiological response in youth when oxygen radicals act as signal transducers. Ide et al. [449] showed that the plasma TBARS and urinary 8-iso-PGF$_{2\alpha}$ levels were higher in healthy young men compared to premenopausal women. These authors concluded that enhanced oxidative stress might be one of the important factors of atherosclerotic diseases in men. Positive correlation probably exists between mammalian life span and cellular resistance to oxidative stress [450]. In contrast to previous proposal, no difference was found between the effects of ubiquinone homologies on superoxide production and longevity in different mammals [451].

However, it should be noted that aging not always results in an increase in the formation of reactive oxygen species especially if other pathophysiological dysfunctions occur. For example, it has been shown that trauma and age-related decline in neutrophil function cause a decrease in superoxide production and the immune response to bacteria in the elderly [452]. Csiszar et al. [453] demonstrated that aging drastically changed the prooxidant:antioxidant balance in young and aged rats. It was found that coronary arteriols of aged rats were characterized by the enhanced superoxide and 3-nitrotyrosine formation, increased iNOS mRNA expression, and decreased eNOS and COX-1 expressions. It has been concluded that aging contributes to the development of oxidative stress through impairing of NO-mediated dilations. Drew and Leeuwenburgh [454] recently discussed the role of reactive nitrogen species and nitration in aging.

Despite inconclusive data at present, the study of antioxidants in protection against age-related disorders remains promising. Diet supplementation with antioxidants resulted in a significant increase in the mean life span of laboratory animals and reduced the risk of arteriosclerosis [455]. Van der Loo et al. [456] demonstrated the accumulation of vitamin E in plasma and major organs of 3-year-old rats not susceptible to atherosclerosis. It was concluded that the accumulation of vitamin E in aged rats could be a compensatory mechanism of self-regulatory protective adaptation against cardiovascular aging. Cherubuni et al. [457] also suggested that an appropriate level of vitamin E and a low level of LDL oxidation in the elderly might be important for achieving advanced age without development of atherosclerosis.

31.7.2 THE MITOCHONDRIAL THEORY OF AGING

Harman's theory of aging predicts that the cells continuously exposed to reactive oxygen species are progressively damaged. It has been suggested that oxidative damage to mitochondria, which is a power source of oxygen radicals, can be a major determinant of the rate of aging [458]. The theory postulates that oxygen radicals induce mutations in the mitochondrial DNA (mtDNA) and that the accumulation of these mutations leads to errors in the mtDNA-encoded polypeptides, defective electron transport, and the suppression of oxidative phosphorylation [459,460]. These disturbances may stimulate increased production of reactive oxygen species and result in a "vicious" circle [461]. The best confirmation of mitochondrial theory of aging is findings that mitochondrial superoxide production correlates with rates of aging of various species [462,463]. Lenaz et al. [460] suggested that Complex I is a primary target of superoxide attack in aging. Harman [464] also proposed that defective mitochondrial Complex I in aging and Alzheimer's disease is responsible for a decrease in endothelium calcium and a following increase in mitochondrial calcium pool, which amplifies superoxide formation.

Recent development of mitochondrial theory of aging is so-called reductive hotspot hypothesis. De Grey [465] proposed that the cells with suppressed oxidative phosphorylation survive by reducing dioxygen at the plasma membrane rather than at the mitochondrial inner membrane. Plasma membrane redox system is apparently an origin of the conversion of superoxide into hydroxyl and peroxyl radicals and LDL oxidation. Morre et al. [466] suggested that plasma membrane oxidoreductase links the accumulation of lesions in mitochondrial DNA to the formation of reactive oxygen species on the cell surface.

REFERENCES

1. C Li, RM Jackson. *Am J Physiol Cell Physiol* 282: C227–C241, 2002.
2. SW Werns, MJ Shea, EM Driscoll, C Cohen, GD Abrams, B Pitt, BR Lucchesi. *Circ Res* 56: 895–898, 1985.
3. ML Hess, NH Manson, E Okabe. *Can J Physiol Pharmacol* 60: 1382–1389, 1982.
4. DK Das, RM Engelman, JA Rousou, RH Breyer, H Otani, S Lemeshow. *Basic Res Cardiol* 81: 155–166, 1986.
5. N Watanabe, M Inoue, Y Morino. *Biochem Pharmacol* 38: 3477–3483, 1989.
6. JL Zweier, P Kuppusamy, S Thompson Gorman, D Klunk, GA Lutty. *Am J Physiol* 266: C700–C709, 1994.
7. JM McCord. *N Engl J Med* 312: 159–163, 1985.
8. SW Werns, MJ Shea, BR Lucchesi. *J Free Radic Biol Med* 1: 103–110, 1985.
9. M Muxfelt, W Schaper. *Basic Res Cardiol* 82: 486–492, 1987.
10. JJ Zulueta, R Sawhney, TS Yu, CC Cote, PM Hassoun. *Am J Physiol* 272: L897–L902, 1997.
11. Z Zhang, DR Blake, CR Stevens, JM Kanczler, PG Winyard, MC Symons, M Benboubetra, R Harrison. *Free Radic Res* 28: 151–164, 1998.
12. JW Beetsch, TS Park, LL Dugan, AR Shah, JM Gidday. *Brain Res* 786: 89–95, 1998.
13. AB Al-Mehdi, G Zhao, C Dodia, K Tozawa, K Costa, V Muzykantov, C Ross, F Blecha, M Dinauer, AB Fisher. *Circ Res* 83: 730–737, 1998.
14. AJ Kowaltowski, AE Vercesi. *Free Radic Biol Med* 26: 463–471, 1999.
15. GW Kim, T Kondo, N Noshita, PH Chan. *Stroke* 33: 809–815, 2002.
16. C Duilio, G Ambrosio, P Kuppusamy, A Dipaula, LC Becker, JL Zweier. *Am J Physiol Heart Circ Physiol* 280: H2649–H2657, 2001.
17. Y Manevich, A Al-Mechdi, V Muzykantov, AB Fisher. *Am J Physiol Heart Circ Physiol* 280: H2126–H2135, 2001.
18. MS Paller, JR Hoidal, TF Ferris. *J Clin Invest* 74: 1156–1164, 1984.
19. E Noiri, A Nakao, K Uchida, H Tsukahara, M Ohno, T Fujita, S Brodsky MS Goligorsky. *Am J Physiol Renal Physiol* 281: F948–F957, 2001.
20. GM Rubanyi, PM Vanhoutte. *Am J Physiol* 250: H822–H827, 1986.
21. H Cai, DG Harrison. *Circ Res* 87: 840–844, 2000.
22. FJ Miller Jr, DD Gutterman, CD Rios, DD Heistad, BL Davidson. *Circ Res* 82: 1298–1305, 1998.
23. U Solzbach, B Hornig, M Jeserich, H Just. *Circulation* 96: 1513–1519, 1997.
24. G Nickenig, DG Harrison. *Circulation* 105: 393–396, 2002.
25. KK Griendling, CA Minieri, JD Ollerenshaw, RW Alexander. *Circ Res* 74: 1141–1148, 1994.
26. JB Laursen, S Rajagopalan, Z Galis, M Tarpey, BA Freeman, DG Harrison. *Circulation* 95: 588–593, 1997.
27. M Usui, K Egashra, H Tomita, M Koyanagi, M Katoh, H Shimokawa, M Takeya, T Yoshimura, K Matsushima, A Takeshita. *Circulation* 101: 305–310, 2000.
28. U Rueckschloss, MT Quinn, J Holtz, H Morawietz. *Arterioscler Thromb Vasc Biol* 22: 1845–1851, 2002.
29. C Zhang, TW Hein, W Wang, L Kuo. *Circ Res* 92: 322–329, 2003.

30. M Ushio-Fukai, AM Zafari, T Fukui, N Ishizaka, KK Griendling. *J Biol Chem* 271: 23317–23321, 1996.
31. PA Barry-Lane, C Patterson, M van Der Merwe, Z Hu, SM Holland, ET Yeh, MS Runge. *J Clin Invest* 108: 1513–1522, 2001.
32. E Hsich, HB Segal, PJ Pagano, FE Rey, B Paigen, J Deleonardis, RF Hoyt, SM Holland, T Finkel. *Circulation* 101: 1234–1236, 2000.
33. EA Kirk, MC Dinauer, H Rosen, A Chait, JW Heinecke, RC LeBoeuf. *Arterioscler Thromb Vasc Biol* 20: 1529–1535, 2000.
34. KK Griendling, DG Harrison. *J Clin Invest* 108: 1423–1424, 2001.
35. JK Bendal, AC Cave, C Heymes, N Gall, AM Shah. *Circulation* 105: 293–296, 2002.
36. HD Wang, DG Johns, S Xu, RA Cohen. *Am J Physiol Heart Circ Physiol* 282: H1697–H1702, 2002.
37. U Landmesser, H Cai, S Dikalov, L McCann, J Hwang, H Jo, SM Holland, DG Harrison. *Hypertension* 40: 511–515, 2002.
38. CA Hathaway, DD Heistad, DJ Piegors, FJ Miller Jr. *Circ Res* 90: 277–283, 2002.
39. S Chatterjee. *Arterioscler Thromb Vasc Biol* 18: 1523–1533, 1998.
40. MC Zimmerman, E Lazartigues, JA Lang, P Sinnayah, IM Ahmad, DR Spitz, RL Davisson. *Circ Res* 91: 1038–1045, 2002.
41. H Kuhn, J Belkner, S Zaiss, T Fahrenklemper, S Wohlfeil. *J Exp Med* 179: 1903–1911, 1994.
42. W Jira, G Spiteller, W Carson, A Schramm. *Chem Phys Lipids* 91: 1–11, 1998.
43. T Ide, H Tsutsui, S Kinugawa, H Utsumi, D Kang, N Hattori, K Uchida, K Arimura, K Egashira, A Takeshita. *Circ Res* 85: 357–363, 1999.
44. T Ide, H Tsutsui, S Kinugawa, N Suematsu, S Hayashidani, S Ichikawa, H Utsumi, Y Machida, K Egashira, A Takeshita. *Circ Res* 86: 152–157, 2000.
45. DB Sawyer, WS Colucci. *Circ Res* 86: 119–120, 2000.
46. SW Ballinger, C Patterson, CA Knight-Lozano, DL Burow, CA Conklin, Z Hu, J Reuf, C Horaist, R Lebovitz, GC Hunter, K McIntyre, MS Runge. *Circulation* 106: 544–549, 2002.
47. PA MacCarthy, DJ Grieve, J-M Li, C Dunster, FJ Kelly, AM Shah. *Circulation* 104: 2967–2974, 2001.
48. CA Hamilton, MJ Brosnan, S Al-Benna, G Berg, AF Dominiczak. *Hypertension* 40: 755–762, 2002.
49. H Azumi, N Inoue, Y Ohashi, MTerashima, T Mori, H Fujita, K Awano, K Kobayashi, K MaedA, K Hata, T Shinke, S Kobayashi, K Hirata, S Kawashima, H Itabe, Y Hayashi, S Imajoh-Ohmi, H Itoh, M Yokoyama. *Arterioscler Thromb Vasc Biol* 22: 1838–1844, 2002.
50. C Guarnieri, F Flamigni, CM Caldarera. *J Mol Cell Cardiol* 12: 797–808, 1980.
51. Y Gauduel, MA Duvelleroy. *J Mol Cell Cardiol* 16: 459–470, 1984.
52. M Shlafer, E Chazov, V Saks, G Rona, PF Kane, MM Kirsh. *J Thorac Cardiovasc Surg* 83: 830–839, 1982.
53. J Bauersachs, I Fleming, D Fraccarollo, R Busse, G Ertl. *Cardiovasc Res* 51: 344–350, 2001.
54. K Nyyssonen, E Porkkala, R Salonen, H Korpela, JT Salonen. *Eur J Clin Nutr* 48: 633–642, 1994.
55. PV Luoma, S Nayha, K Sikkila, J Hassi. *J Intern Med* 237: 49–54, 1995.
56. WA Pryor. *Free Radic Biol Med* 28: 141–164, 2000.
57. A Shihabi, W-G Li, FJ Miller, JR, NL Weintraub. *Am J Physiol Heart Circ Physiol* 282: H797–H802, 2002.
58. RM Salonen, K Nyyssönen, J Kaikkonen, E Porkkala-Sarataho, S Voutilainen, TH Rissanen, T-P Tuomainen, V-P Valkonen, U Ristonmaa, H-M Lakka, M Vanharanta, JT Salonen, HE. Poulsen. *Circulation* 107: 947–953, 2003.
59. I Jialal, S Devaraj. *Circulation* 107: 926–928, 2003.
60. TW Wu, J Wu, L-H Zeng, J-X Au, D Carey, KP Fung. *Life Sci* 54: PL23–PL28, 1994.
61. Z-H Shao, TLV Hoek, Y Qin, LB Becker, PT Schumacker, C-Q Li, L Dey, E Barth, H Halpern, GM Rosen, C-S Yuan. *Am J Physiol Heart Circ Physiol* 282: H999–H1006, 2002.
62. VL Lakomkin, OV Korkina, VG Tsyplenkova, AA Timoshin, EK Ruuge, VI Kapel'ko. *Kardiologiia* 42: 51–55, 2002 (in Russian).
63. RJ Teng, YZ Ye, DA Parks, JS Beckman. *Free Radic Biol Med* 33: 1243–1249, 2002.

64. AH Wagner, T Kohler, U Ruckschloss, I Just, M Hecker. *Arterioscler Thromb Vasc Biol* 20: 61–69, 2000.
65. S Wassmann, U Laufs, AT Baumer, K Muller, C Konkol, H Sauer, M Bohm, G Nickenig. *Mol Pharmacol* 59: 646–654, 2001.
66. S Delbosc, M Morena, F Djouad, C Ledoucen, B Descomps, JP Cristol. *J Cardiovasc Pharmacol* 40: 611–617, 2002.
67. M Takemoto, K Node, H Nakagami, Y Liao, M Grimm, Y Takemoto, M Kitakaze, JK Liao. *J Clin Invest* 108: 1429–1437, 2001.
68. S Kawashima, T Yamashita, Y Miwa, M Ozaki, M Namiki, T Hirase, N Inoue, KI Hirata, M Yokoyama. *Stroke* 34: 157–163, 2003.
69. V Sanguigni, P Pignatelli, D Caccese, FM Pulcinelli, L Lenti, R Magnaterra, F Martini. *Thromb Haemost* 87: 796–801, 2002.
70. V Lubrano, C Vassalle, C Blandizzi, M Del Tacca, C Palombo, A L'Abbate, S Baldi, A Natali. *Eur J Clin Invest* 33: 117–125, 2003.
71. BV Khan, S Navalkar, QA Khan, ST Rahman, S Parthasarathy. *J Am Coll Cardiol* 38: 1662–1667, 2001.
72. SJ Duffy, ES Biegelsen, M Holbrook, JD Russell, N Gokce, JF Keaney Jr, JA Vita. *Circulation* 103: 2799–2804, 2001.
73. JT Salonen, R Salonen, M Ihanainen, M Parviainen, R Seppanen, M Kantola, K Sepparien, R Rauramaa. *Am J Clin Nutr* 48: 1226–1232, 1988.
74. R Salonen, H Korpela, K Nyyssonen, E Porkkala, JT Salonen. *Life Chem Rep.* 12: 65–68, 1994.
75. K Nyyssonen, E Porkkala, R Salonen, H Korpela, JT Salonen. *Eur J Clin Nutr* 48: 633–642, 1994.
76. B Kristal, R Shurtz-Swirski, J Chezar, J Manaster, R Levy, G Shapiro, I Weissman, SM Shasha, S Sela. *Am J Hypertens* 11: 921–928, 1998.
77. C Russo, O Olivieri, D Girelli, G Faccini, ML Zenari, S Lombardi, R Corrocher. *J Hypertens* 16: 1267–1271, 1998.
78. S Kerr, MJ Brosnan, M McIntyre, JL Reid, AF Dominiczak, CA Hamilton. *Hypertension* 33: 1353–1358, 1999.
79. MJ Somers, K Mavromatis, ZS Galis, DG Harrison. *Circulation* 101: 1722–1728, 2000.
80. 61. **11217.** RA Beswick, AM Dorrance, R Leite, RC Webb. *Hypertension* 38: 1107–1111, 2001.
81. X Chen, RM Touyz, JB Park, EL Schiffrin. *Hypertension* 38: 606–611, 2001.
82. S Ulker, PP McKeown, U Bayraktutan. *Hypertension* 41: 534–539, 2003.
83. L Crockett, DH Wang, JJ Galligan, GD Fink, AF Chen. *Arterioscler Thromb Vasc Biol* 22: 249–255, 2002.
84. T Shokoji, A Nishiyama, Y Fujisawa, H Hitomi, H Kiyomoto, N Takahashi, S Kimura, M Kohno, Y Abe. *Hypertension* 41: 266–273, 2003.
85. W Gonzales, V Fontaine, ME Pueyo, N Laquay, D Messika-Zeitoun, M Philippe, J-F Arnal, M-P Jacob, J-B Michel. *Hypertension* 36: 103–109, 2000.
86. CA Hubel. *Proc Soc Exp Biol Med* 2222: 222–235, 1999.
87. SW Walsh, JE Vaughan, Y Wang, LJ Roberts II. *FASEB J* 14: 1289–1296, 2000.
88. A Barden, J Ritchie, B Walters, C Micharl, J Rivera, T Mori, K Croft, L Beilin. *Hypertension* 38: 803–808, 2001.
89. M Sikkema, BB van Rijn, A Franx, HW Bruinse, R de Roos, ES G.Stroes, EE van Faassen. *Placenta* 22: 304–308, 2001.
90. VM Lee, PA Quinn, SC Jennings, LL Ng. *J Hypertens* 21: 395–402, 2003.
91. JE Vaughan, SW Walsh. *Hypertens Pregnancy* 21: 205–223, 2002.
92. JV Hunt. In: *Free Radicals in the Environment, Medicine and Toxicology*, H Nohl, H Esterbauer, C Rice-Evans (eds), Richelieu Press, London, 1994, pp 137–162.
93. T Nishikawa, D Edelstein, XL Du, S Yamagishi, T Matsumura, Y Kaneda, MA Yorek, D Beebe, PJ Oates, HP Hammes, I Giardino, M Brownlee. *Nature* 404: 787–790, 2000.
94. XL Du, D Edelstein, L Rossetti, IG Fantus, H Goldberg, F Ziyadeh, J Wu, M Brownlee. *Proc Natl Acad Sci USA*. 97: 12222–12226, 2000.
95. SI Yamagishii, D Edelstein, Xl Du, M Brownlee. *Diabetes* 50: 1491–1494, 2001.

96. AK Azevedo-Martins, S Lortz, S Lenzen, R Curi, DL Eizirik, M Tiedge. *Diabetes* 52: 93–101, 2003.
97. SV Brodsky, S Gao, H Li, MS Goligorsky. *Am J Physiol Heart Circ Physiol* 283: H2130–H2139, 2002.
98. S Gupta, E Chough, J Daley, P Oates, K Tornheim, NB Ruderman, JF Keaney, Jr. *Am J Physiol Cell Physiol* 282: C560–C566, 2002.
99. M Christ, J Bauersachs, C Liebetrau, M Heck, A Gunther, M Wehling. *Diabetes* 51: 2648–2652, 2002.
100. TJ Guzik, S Mussa, D Gastaldi, J Sadowski, C Ratnatunga, R Pillai, KM Channon. *Circulation* 105: 1656–1662, 2002.
101. YK Kim, MS Lee, SM Son, IJ Kim, WS Lee, BY Rhim, KW Hong, CD Kim. *Diabetes* 51: 522–527, 2002.
102. T Inoguchi, P Li, F Umeda, HY Yu, M Kakimoto, M Imamura, T Aoki, T Etoh, T Hashimoto, M Naruse, H Sano, H Utsumi, H Nawata. *Diabetes* 49: 1939–1945, 2000.
103. U Hink, H Li, H Mollnau, M Oelze, E Matheis, M Hartmann, M Skatchkov, F Thaiss, RA Stahl, A Warnholtz, T Meinertz, K Griendling, DG Harrison, U Forstermann, T Munzel. *Circ Res* 88: E14–E22, 2001.
104. SK Venugopal, S Devaraj, T Yang, I Jialal. *Diabetes* 51: 3049–3054, 2002.
105. F Cosentino, K Hishikawa, ZS Katusic, TF Luscher. *Circulation* 96: 25–28, 1997.
106. H Trachtman, S Futterweit, DL Crimmins. *J Am Soc Nephrol* 8: 1276–1282, 1997.
107. XL Du, D Edelstein, S Dimmeler, Q Ju, C Sui, M Brownlee. *J Clin Invest* 108: 1341–1348, 2001.
108. I Noyman, M Marikovsky, S Sasson, A Stark, K Bernath, R Seger, Z Madar. *Nitric Oxide* 7: 187, 2002.
109. C Napoli, M Triggiani, G Palumbo, M Condorelli, M Chiariello, G Ambrosio. *Basic Res Cardiol* 92: 96–105, 1997.
110. J Galle, R Schneider, B Winner, C Lehmann-Bodem, R Schinzel, G Munch, E Conzelmann, C Wanner. *Atherosclerosis* 138: 65–77, 1998.
111. R Marfella, L Guagliaro, F Nappo, A Ceriello, D Giugliano. *J Clin Invest* 108: 635–636, 2001.
112. S Pennathur, JW Heinecke. *J Clin Invest* 108: 636, 2001.
113. B Wierusz-Wysocka, H Wysocki, H Siekierka, A Wykretowicz, A Szczepanik, R Klimas. *J Leukocyte Biol* 42: 519–523, 1987.
114. B Wierusz-Wysocka, A Wykretowicz, H Byks, K Sadurska, H Wysocki. *Diabetes Res Clin Pract* 21: 109–114, 1993.
115. R Shurtz-Swirski, S Sela, AT Herskovits, SM Shasha, G Shapiro, L Nasser, B Kristal. *Diabetes Care* 24: 104–110, 2001.
116. L Rodrigues-Manas, J Angulo, C Peiro, JL Llergo, A Sanchez-Ferrer, P Lopez-Doriga, CF Sanchez-Ferrez. *Br J Pharmacol* 123: 1495–1502, 1998.
117. K Josefsen, H Nielsen, S Lorentzen, P Damsbo, K Buschard. *Clin Exp Immunol* 98: 489–493, 1994.
118. K Hiramatsu, S Arimori. *Diabetes* 37: 832–838, 1988.
119. NN Orie, W Zide, M Tepel. *Exp Clin Endocrinol Diabetes* 108: 175–180, 2000.
120. L Tiano, A Kantar, G Falcioni, GP Littarru, V Cherubini, R Fiorini. *Prostaglandins Other Lipid Mediat* 62: 351–366, 2000.
121. GM Pieper, P Langenstroer, W Siebeneich. *Cardiovasc Res* 34: 145–156, 1997.
122. B Lubec, M Hermon, H Hoeger, G Lubec. *FASEB J* 12: 1581–1587, 1998.
123. C Dominguez, E Ruiz, M Gussinye, A Carrascosa. *Diabetes Care* 21: 1736–1742, 1998.
124. KZ Kedziora-Kornatowska, M Luciak, J Blaszczyk, W Pawlak. *Nephrol Dial Transplant* 13: 2829–2832, 1998.
125. MM Kesavulu, R Giri, RB Kameswara, C Apparao. *Diabetes Metab* 26: 387–392, 2000.
126. M Inouye, T Mio, K Sumino. *Biochim Biophys Acta* 1438: 204–212, 1999.
127. D Altavilla, A Saitta, D Cucinotta, M Galeano, B Deodato, M Colonna, V Torre, G Russo, A Sardella, G Urna, GM Campo, V Cavallari, G Squadrito, F Squadrito. *Diabetes* 50: 667–674, 2001.
128. AR Collins, K Raslova, M Somorovska, H Petrovska, A Ondrusova, B Vohnout, R Fabry, M Dusinska. *Free Radic Biol Med* 25: 373–377, 1998.

129. IK Mohan, UN Das. *Free Radic Biol Med* 25: 757–765, 1998.
130. KZ Kedziora-Kornatowska. *IUBMB Life* 48: 359–362, 1999.
131. LJ Coppey, JS Gellett, EP Davidson, JA Dunlap, DD Lund, MA Yorek. *Diabetes* 50: 1927–1937, 2001.
132. AE Midaoui, A Elimadi, L Wu, PS Haddad, J de Champlain. *Am J Hypertens* 16: 173–179, 2003.
133. RA Sanders, FM Rauscher, JB Watkins JB III. *J Biochem Mol Toxicol* 15: 143–149, 2001.
134. N Kanie, K Kamata. *Br J Pharmacol* 135: 1935–1942, 2002.
135. CA Gunnett, DD Heistad, FM Faraci. *Diabetes* 51: 1931–1937, 2002.
136. A Ceriello, D Giugliano, A Quatraro, C Donzella, G Dipalo, PJ Lefebvre. *Diabetes Care* 14: 68–72, 1991.
137. G Paolisso, A D'Amore, D Giugliano, A Ceriello, M Varricchio, F D'Onofrio. *Am J Clin Nutr* 57: 650–656, 1993.
138. SK Jain, R McVie, T Smith. *Diabetes Care* 23: 1389–1394, 2000.
139. RA Skyrme-Jones, RC O'Brien, KL Berry, IT Meredith. *J Am Coll Cardiol* 36: 94–102, 2000.
140. S Devaraj, SV Hirany, RF Burk, I Jialal. *Clin Chem* 47: 1974–1979, 2001.
141. BMY Keenoy, J Vertommen, I De Leeuw. *Diabetes Nutr Metab* 12: 256–263, 1999.
142. E Haak, KH Usadel, K Kusterer, P Amini, R Frommeyer, HJ Tritschler, T Haak. *Exp Clin Endocrinol Diabetes* 108: 168–174, 2000.
143. G De Mattia, O Laurenti, D Fava. *J Diabetes Complications* 17(2S): 30–35, 2003.
144. P Gargiulo, D Caccese, P Pignatelli, C Brufani, F De Vito, R Marino, R Lauro, F Violi, U Di Mario, V Sanguigni. *Diabetes Metab Res Rev* 18: 156–159, 2002.
145. A Saprin, E Klochko, K Kruglikova, V Chibrikin, N Emanuel. *Dokl Akad Nauk SSSR* 167: 222–226, 1965.
146. HM Swartz, PL Gutierrez. *Science* 198: 936–938, 1997.
147. L Oberley. In: *Superoxide Dismutase*, LW Oberley (ed.), Vol 2. CRC Press, Boca Raton, FL, 1982, pp 127–165.
148. TW Kensler, MA Trush. In: *Superoxide Dismutase*. LW Oberley (ed.), Vol 3. CRC Press, Boca Raton, FL, 1985, pp 191–236.
149. TW Kensler, BG Taffe. *Adv Free Radic Biol Med* 2: 347–387, 1986.
150. RA Floyd. *FASEB J* 4: 2587–2597, 1990.
151. LW Oberley, TD Oberley. *Mol Cell Biochem* 84: 147–153, 1988.
152. PA Riley. *Phil Trans R Soc Lond* B311: 679–689, 1985.
153. PA Cerutti. *Lancet* 344: 862–863, 1994.
154. LW Oberley, TD Oberley. *Mol Cell Biochem* 84: 147–153, 1988.
155. RA Floyd. *Carcinogenesis* 11: 1447–1450, 1990.
156. H Wei, K Frenkel. *Carcinogenesis* 14: 1195–1201, 1993.
157. T Ogawa, S Higashi, Y Kawarada, R Mizumoto. *Carcinogenesis* 16: 831–836, 1995.
158. A Mazzone, G Ricevuti, SC Rizzo, S Sachi. *Int J Tissue React* 8: 493–496, 1986.
159. L Korkina, JA Osato, I Chivileva, E Samochatova, Z Cheremisina, I Afanas'ev. *Nutrition* 11: 555–558, 1995.
160. S Senturker, B Karahalil, M Inal, H Yilmaz, H Muslumanoglu, G Gedikoglu, M Dizdaroglu. *FEBS Lett* 416: 286–290, 1997.
161. AM Oltra, F Carbonell, C Tormos, A Iradi, GT Saez. *Free Radic Biol Med* 30: 1286–1292, 2001.
162. CF Babbs. *Free Radic Biol Med* 8: 191–200, 1990.
163. P Cerutti. *Lancet* 344: 862–863, 1994.
164. S Pervaiz, JK Ramalingam, JL Hirpara, M Clement. *FEBS Lett* 459: 343–348, 1999.
165. YC Chong, GH Heppner, LA Paul, AM Fulton. *Cancer Res* 49: 6652–6657, 1985.
166. A Trulson, S Nilsson, E Brekkan, P Venge. *Inflammation* 18: 99–105, 1994.
167. Y Nakamura, TD Gindhart, D Winterstein, I Tomita, JL Seed, NH Colburn. *Carcinogenesis* 9: 203–207, 1988.
168. HW Kim, A Murakami, MV Williams, H Ohigashi. *Carcinogenesis* 24: 235–241, 2003.
169. N Hara, Y Ichinose, H Asoh, T Yano, M Kawasaki, M Ohta. *Cancer* 69: 1682–1687, 1992.
170. AV Timoshenko, K Kayser, P Drings, G Kolb, K Havemann, HJ Gabius. *Anticancer Res* 13: 1789–1792, 1993.

171. M Kato, H Minakami, M Kuroiwa, Y Kobayashi, S Oshima, K Kozawa, A Morikawa, H Kimura. *Hematol Oncol* 21: 11–16, 2003.
172. G Deliconstantinos, V Villiotou, JC Stavrides, N Salemes, J Gogas. *Anticancer Res* 15: 1435–1446, 1995.
173. T Yamashita, T Uchida, A Araki, F Sendo. *Int J Cancer* 71: 223–230, 1997.
174. G Haklar, E Sayin-Ozveri, M Yuksel, AO Aktan, AS Yalcin. *Cancer Lett* 165: 219–224, 2001.
175. WJ Thomas, DL Thomas, JA Knezetic, TE Adrian. *Pancreas* 24: 161–168, 2002.
176. KH Cheeseman, M Collins, K Proudfoot, TF Slater, GW Burton, AC Webb, KU Ingold. *Biochem J* 235: 507–514, 1986.
177. NF Boyd, V McGuire. *Free Radic Biol Med* 10: 185–190, 1991.
178. P Palozza, G Agostara, E Piccioni, GM Bartoli. *Arch Biochem Biophys* 312: 88–94, 1994.
179. F Della Rovere, A Granata, M Broccio, A Zirilli, G Broccio. *Anticancer Res* 15: 2089–2095, 1995.
180. A Haddow, G De Lamirande, F Bergel, RC Bray, DA Gilbert. *Nature* 182: 1144–1146, 1958.
181. T Yoshikawa, S Kokura, K Tainaka, Y Naito, M Kondo. *Cancer Res* 55: 1617–1620, 1995.
182. T Sawa, J Wu, T Akaike, H Maeda. *Cancer Res* 60: 666–671, 2000.
183. S Yamaguichi, S Sakurada, M Nagumo. *Free Radic Biol Med* 17: 389–395, 1994.
184. P Huang, L Feng, EA Oldham, MJ Keating, W Plunkett. *Nature* 407: 390–395, 2000.
185. EA Hileman, G Achanta, P Huang. *Expert Opin Ther Targets* 5: 697–710, 2001.
186. K Hirose, DL Longo, JJ Openheim, K Matsushima. *FASEB J* 7: 361–368, 1993.
187. S Motoori, HJ Majima, M Ebara, H Kato, F Hirai, S Kakinuma, C Yamaguchi, T Ozawa, T Nagano, H Tsujii, H Saisho. *Cancer Res* 61: 5382–5388, 2001.
188. Y Zhao, TD Oberley, L Chaiswing, SM Lin, CJ Epstein, TT Huang, D St Clair. *Oncogene* 21: 3836–3846, 2002.
189. IB Afanas'ev. In: *Superoxide Ion: Chemistry and Biological Implications*. Vol 2. CRC Press, Boca Raton, FL, 1990, Chapter 3.
190. BK Sinha, EG Mimnaugh. *Free Radic Biol Med* 8: 567–581, 1990.
191. LG Korkina, IB Afanas'ev. In: *Antioxidants in Disease. Mechanisms and Therapy*, H Sies (ed.), Academic Press, San Diego, CA, 1997, pp 151–163.
192. S Oikawa, K Yamada, N Yamashita, S Tada-Oikawa, S Kawanishi. *Carcinogenesis* 20: 1485–1490, 1999.
193. JE Spallholz, BJ Shriver, TW Reid. *Nutr Cancer* 40: 34–41, 2001.
194. U Gundimeda, ZH Chen, R Gopalakrishna. *J Biol Chem* 271: 13504–13514, 1996.
195. R Schiff, P Reddy, M Ahotupa, E Coronado-Heinsohn, M Grim, SG Hilsenbeck, R Lawrence, S Deneke, R Herrera, GC Chamness, SAW Fuqua, PH Brown, CK Osborne. *J Natl Cancer Inst* 92: 1926–1934, 2000.
196. TW Kensler, DM Bush, WJ Kozumbo. *Science* 221: 75–77, 1983.
197. LW Oberley, SW Leuthauser, RF Pasternack, TD Oberley, L Schutt, JR Sorenson. *Agents Actions* 15: 535–538, 1984.
198. PA Egner, TW Kensler. *Carcinogenesis* 6: 1167–1172, 1985.
199. V Solanski, L Yotti, MK Logani, TJ Slaga. *Carcinogenesis* 5: 129–131, 1984.
200. U Armato, PG Andreis, F Romano. *Carcinogenesis* 5: 1547–1555, 1984.
201. AR Kennedy. In: *Vitamin and Cancer — Human Cancer Prevention by Vitamins and Micronutrients*, FL Meyskens, KM Prasad (eds), Humana Press, Clifton, NJ, 1985, pp 51–64.
202. RH Burdon. *Proc Roy Soc Edinburgh*. 99B: 169–176, 1992.
203. D Glaves. *Inv Metastasis* 6: 101–111, 1986.
204. JJ Cullen, C Weydert, MM Hinkhouse, J Ritchie, FE Domann, D Spitz, LW Oberley. *Cancer Res* 63: 1297–1303, 2003.
205. KN Prasad, BN Rama. In: *Vitamins, Nutrition, and Cancer*, KN Prasad (ed.), Karger, Basel, 1984, pp 76–104.
206. RS London, L Murphy, KE Kitlowski. *J Am Coll Nutr* 4: 559–564, 1985.
207. E Bright-See. *Semin Oncol* 10: 294–298, 1983.
208. R Kahl. *J Environ Sci Health* C4: 47–92, 1984.
209. JB Rotstein, TJ Slaga. *Carcinogenesis* 9: 1547–1551, 1998.
210. G Hocman. *Int J Biochem* 20: 639–651, 1988.

211. LG Korkina, IB Afanas'ev, AT Diplock. *Biochem Soc Trans* 21: 314S, 1993.
212. HY Huang, AJ Alberg, EP Norkus, SC Hoffman, GW Comstock, KJ Helzlsouer. *Am J Epidemol* 157: 335–344, 2003.
213. J Ni, M Chen, Y Zhang, R Li, J Huang, S Yeh. *Biochem Biophys Res Commun* 300: 357–363, 2003.
214. MP Malafa, FD Fokum, L Dmith, A Louis. *Ann Surg Oncol* 9: 1023–1032, 2002.
215. A Pace, A Savarese, M Picardo, V Maresca, U Pacetti, G Del Monte, A Biroccio, C Leonetti, B Jandolo, F Cognetti, L Bove. *J Clin Oncol* 21: 927–931, 2003.
216. CP Oliveira, P Kassab, FP Lopasso, HP Souza, M Janiszewski, FR Laurindo, K Iriya, AA Laudanna. *World J Gastroendoterol* 9: 446–448, 2003.
217. G Mantovani, A Maccio, C Madeddu, L Mura, E Massa, G Gramignano, MR Lusso, V Murgia, P Camboni, L Ferreli. *J Cell Mol Med* 6: 570–582, 2002.
218. EJ Jacobs, AK Henion, PJ Briggs, CJ Connell, ML McCullough, CR Jonas, C Rodriguez, EE Calle, MJ Thun. *Am J Epidemiol* 156: 1002–1010, 2002.
219. M Jang, L Cai, GO Udeani, KV Slowing, cf. Thomas, CWW Beecher, HHS Fong, NR Farnsworth, AD Kinghorn, RG Mehta, RC Moon, JM Pezzuto. *Science* 275: 218–220, 1997.
220. J Zhao, Y Sharma, R Agarwal. *Mol Carcinog* 26: 321–333, 1999.
221. A Tyagi, N Bhatia, MS Condon, MC Bosland, C Agarwal, R Agarwal. *Prostate* 53: 211–217, 2002.
222. BM Babior, RS Kipnes, JT Curnitte. *J Clin Invest* 52: 741–745, 1973.
223. KJ Johnson, PA Ward. In: *Superoxide Dismutase*, LW Oberley (ed.), CRC Press, Boca Raton, FL, 1982, pp 130–142.
224. IB Afanas'ev. In: *Superoxide Ion: Chemistry and Biological Implications*. Vol 2. CRC Press, Boca Raton, FL, 1990, Chapter 4.
225. Y Oyanagui. *Biochem Pharmacol* 25: 1465–1472, 1976.
226. T Kasama, K Kobayashi, T Fukushima, M Tabata, I Ohno, M Negishi, H Ide, T Takahashi, Y Niwa. *Clin Immunol Immunopathol* 53: 439–448, 1989.
227. R Zhang, M-L Brennan, Z Shen, JC MacPherson, D Schmitt, CE Molenda, SL Hazen. *J Biol Chem* 277: 46116–46122, 2002.
228. S Kausalya, J Nath. *J Leukocyte Biol* 64: 185–191, 1998.
229. RS Farivar, AV Chobanian, P Brecher. *Circ Res* 78: 759–768, 1996.
230. B Liu B, HM Gao, JY Wang, GH Jeohn, CL Cooper, JS Hong. *Ann NY Acad Sci* 962: 318–331, 2002.
231. J Lunec, SP Halloran, AG Write, T Dormandz. *J Rheumatol* 8: 233–245, 1981.
232. P Eggleton, L Wang, J Penhallow, N Crawford, KA Brown. *Ann Rheum Dis* 54: 916–923, 1995.
233. R Miesel, R Hartung, H Kroeger. *Inflammation* 20: 427–438, 1996.
234. EA Ostrakhovitch, IB Afanas'ev. *Biochem Pharmacol* 62: 743–746, 2001.
235. J El Benna, G Hayem, PM Dang, M Fay, S Chollet-Martin, C Elbim, O Meyer, MA Gougerot-Pocidalo. *Inflammation* 26: 273–278, 2002.
236. Y Ozaki, T Ohashi, Y Niwa. *Inflammation* 10: 119–130, 1986.
237. DE Van Epps, S Greiwe, J Potter, J Goodwin. *Inflammation* 11: 59–72, 1987.
238. IC Kowanko, A Ferrante, G Clemente, PP Youssef, M Smith. *J Clin Immunol* 16: 216–221, 1996.
239. A Den Broeder, GJ Wanten, WJ Oyen, T Naber, P Van Riel, P Barrera. *J Rheumatol* 30: 232–237, 2003.
240. A Den Broeder, GJ Wanten, WJ Oyen, T Naber, P Van Riel, P Barrera. *J Rheumatol* 30: 232–237, 2003.
241. P Davis, C Johnston, J Bertouch, G Starkebaum. *Ann Rheum Dis* 46: 51–54, 1987.
242. DO Stichtenoth, JC Frolich. *Br J Rheum* 37: 246–257, 1998.
243. I Mazzetti, B Grigolo, L Pulsatelli, P Dolzani, T Silvestri, L Roseti, R Meliconi, A Facchini. *Clin Sci (Lond)* 101: 593–599, 2001.
244. DV Jovanovic, F Mineau, K Notoya, P Reboul, J Martel-Pelletier, JP Pelletier. *J Rheumatol* 29: 2165–2175, 2002.
245. KM Wilsmann. In: *Superoxide and Superoxide Dismutase in Chemistry, Biology and Medicine*, G Rotilio (ed.), Elsevier, Amsterdam, 1985, pp 500–507.
246. Y Niwa, K Somiya, AM Michelson, K Puget. *Free Radic Res Commun* 1: 137–153, 1985.
247. L Flohe. *Mol Cell Biochem* 84: 123–131, 1988.

248. EJ Dowling, CL Chandler, AW Claxson, C Lillie, DR Blake. *Free Rad Res Commun* 18: 291–298, 1993.
249. ML Corvo, OC Boerman, WJ Oyen, L Van Bloois, ME Cruz, DJ Crommelin, G Storm. *Biochim Biophys Acta* 1419: 325–334, 1999.
250. S Iyama, T Okamoto, T Sato, N Yamauchi, Y Sato, K Sasaki, M Takahashi, M Tanaka, T Adachi, K Kogawa, J Kato, S Sakamaki, Y Niitsu. *Arthritis Rheum* 44: 2160–2167, 2001.
251. D Salvemini, E Mazzon, L Dugo, I Serraino, A De Sarro, AP Caputi, S Cuzzocrea. *Arthritis Rheum* 44: 2909, 2001.
252. P Davis, C Johnston. *Inflammation* 10: 311–320, 1986.
253. NP Hurst, AL Bell, G Nuki. *Ann Rheum Dis* 45: 37–43, 1986.
254. E Mur, A Zabernigg, W Hilbe, W Eisterer, W Halder, J Thaler. *Clin Exp Rheumatol* 15: 233–237, 1997.
255. L Harper, SL Nuttall, U Martin, CO Savage. *Rheumatology* 41: 274–278, 2002.
256. A Wittenborg, PR Bock, J Hanisch, R Saller, B Schneider. *Arzneimittelforschung* 50: 728–738, 2000.
257. C Szabo, L Virag, S Cuzzocrea, GS Scott, P Hake, MP O'Connor, B Zingarelli, A Salzman, E Kun. *Proc Natl Acad Sci USA* 95: 3867–3872, 1998.
258. EA Ostrakhovitch, O Ilich-Stoianovich, IB Afanas'ev. *Vestn Ross Akad Med Nauk* 5: 23–27, 2001 (in Russian).
259. BZ Joseph, JM Routes, L Borish. *Inflammation* 17: 361–370, 1993.
260. NN Jarjour, WJ Calhoun. *J Lab Clin Med* 123: 131–136, 1994.
261. L Vargas, PJ Patino, F Montoya, AC Vanegas, A Echavarria, OD de Garcia. *Inflammation* 22: 45–54, 1998.
262. KR Shanmugasundaram, SS Kumar, S Rajajee. *Clin Chim Acta* 305: 107–114, 2001.
263. V Plaza, J Prat, J Rosello, E Ballester, I Ramis, J Mullol, E Gelpi, JL Vives-Corrons, C Picado. *Thorax* 50: 490–496, 1995.
264. LG Wood, Fitzgerald, PG Gibson, DM Cooper, ML Carg. *Lipids* 35: 967–974, 2000.
265. P Paredi, SA Kharitonov, PJ Barnes. *Am J Respir Crit Care Med* 162: 1450–1454, 2000.
266. D Saleh, P Ernst, S Lim, Pj Barnes, A Giaid. *FASEB J* 12: 929–937, 1998.
267. CG Cochrane, R Spragg, SD Revak. *J Clin Invest* 71: 754–761, 1983.
268. NJ Lamb, JM Gutteridge, C Baker, TW Evans, GJ Quinlan. *Crit Care Med* 27: 1738–1744, 1999.
269. C Sittipunt, KP Steinberg, JT Ruzinski, C Myles, S Zhu, RB Goodman, LD Hudson, S Matalon, TR Martin. *Am J Respir Crit Care Med* 163: 503–510, 2001.
270. O Ortolani, A Conti, AR De Gaudio, M Masoni, G Novelli. *Shock* 13: 14–18, 2000.
271. DW Kamp, P Graceffa, WA Pryor, SA Weitzman. *Free Radic Biol Med* 12: 293–315, 1992.
272. TR Quinlan, KA BeruBe, JP Marsh, YM Janssen, P Taishi, KO Leslie, D Hemenway, PT O'Shaughnessy, P Vacek, BT Mossman. *Am J Pathol* 147: 728–739, 1995.
273. M Dorger, AM Allmeling, R Kiefmann, A Schropp, F Krombach. *Free Radic Biol Med* 33: 491–501, 2002.
274. A van der Vliet, JP Eiserich, GP Marelich, B Halliwell, CE Cross. In: *Antioxidants in Disease. Mechanisms and Therapy*, H Sies (ed.), Academic Press, San Diego, 1997, pp. 491–513.
275. B Salh, K Webbs, PM Guyan, JP Day, D Wickens, J Griffin, JM Braganza, TL Dormandy. *Clin Chim Acta* 181: 65–74, 1989.
276. G Ciabattoni, G Davi, M Collura, L Iapichino, F Pardo, A Ganci, R Romagnoli, J Maclouf, C Patrono. *Am J Respir Crit Care Med* 162: 1195–1201, 2000.
277. P Paredi, SA Kharitonov, D Leak, PL Shah, D Cramer, ME Hodson, PJ Barnes. *Am J Respir Crit Care Med* 161: 1247–1251, 2000.
278. MJ Mhanna, T Ferkol, RJ Martin, IA Dreshaj, AM van Heeckeren, TJ Kelley, MA Haxhiu. *Am J Respir Cell Mol Biol* 24: 621–626, 2001.
279. Z Bebok, K Varga, JK Hicks, CJ Venglarik, T Kovacs, L Chen, KM Hardiman, JF Collawn, EJ Sorscher, S Matalon. *J Biol Chem* 277: 43041–43049, 2002.
280. BM Morrissey, K Schilling, JV Weil, PE Silkoff, DM Rodman. *Arch Biochem Biophys* 406: 33–39, 2002.

281. P Failli, L Palmieri, C D'Alfonso, L Giovannelli, S Generini, AD Rosso, A Pignone, N Stanflin, S Orsi, L Zilletti, M Matucci-Cerinic. *Nitric Oxide* 7: 277–282, 2002.
282. R Solan, C Motta, R Sola, AE La Ville, J Lima, P Simeon, N Montella, L Armadans-Gil, V Fonollosa, M Vilardell. *Arthritis Rheum* 43: 894–900, 2000.
283. JL Cracowski, C Marpeau, PH Carpentier, B Imbert, M Hunt, F Stanke-Labesque, G Bessard. *Arthritis Rheum* 44: 1143–1148, 2001.
284. K Tamagawa, Y Taooka, A Maeda, K Hiyama, S Ishioka, M Yamakido. *Am J Respir Crit Care Med* 161: 1279–1284, 2000.
285. JM Davis, RB Parad, T Michele, E Allred, A Price, W Rosenfeld. *Pediatrics* 111: 469–476, 2003.
286. Y Niwa, T Kanoh, T Sakane, H Soh, S Kawai, Y Miyachi. *Life Sci* 40: 921–927, 1987.
287. BS Polla, RA Ezekowitz, DY Leung. *J Allergy Clin Immunol* 89: 545–551, 1992.
288. SA Greenacre, FA Rocha, A Rawlingson, S Meinerikandathevan, RN Poston, E Ruiz, B Halliwell, SD Brain. *Br J Pharmacol* 136: 985–994, 2002.
289. G Deliconstantinos, V Villiotou, JC Stavrides. *Biochem Pharmacol* 51: 1727–1738, 1996.
290. SK Katiyar, MS Matsui, CA Elmets, H Mukhtar. *Photochem Photobiol* 69: 148–153, 1999.
291. Y Nakamura, Q Feng, T Kumagai, K Torikai, H Ohigashi, T Osawa, N Noguchi, E Niki, K Uchida. *J Biol Chem* 277: 2687–2694, 2002.
292. M Shingu, S Nonaka, M Nobunaga, N Ahamadzadeh. *Dermatologica* 179: 107–112, 1989.
293. A Er-rali, M Charveron, JL Bonafe. *Skin Pharmacol* 6: 253–258, 1993.
294. K Muller. *Biochem Pharmacol* 53: 1215–1221, 1997.
295. RW Lange, DR Germolec, JF Foley, MI Luster. *J Leukoc Biol* 64: 170–176, 1998.
296. GM Kavanagh, JL Burton, VO Donnell. *Br J Dermatol* 134: 234–237, 1996.
297. KPW Hammann, MP Dierich, HC Hoff. *Acta Neurol Scand* 68: 151–156, 1983.
298. M Fisher, PH Levine, BH Weiner, CH Vaudreuil, A Natale, MH Johnson, JJ Hoogasian. *Inflammation* 12: 123–131, 1988.
299. A Glabinski, NS Tawsek, G Bartosz. *Acta Neurol Scand* 88: 174–177, 1993.
300. P Sarchielli, A Orlacchio, F Vininanza, GP Pelliccioli, M Tognoloni, C Saccardi, V Gallai. *J Neuroimmunol* 80: 76–86, 1997.
301. L Bo, TM Dawson, S Wesselingh, S Mork, S Choi, PA Kong, D Hanley, BD Trapp. *Ann Neurol* 36: 778–786, 1994.
302. AH Cross, PT Maning, RM Keeling, RE Schmidt, TP Misko. *J Neuroimmunol* 88: 45–56, 1998.
303. DC Hooper, CS Scott, A Zborek, T Mikheeva, RB Kean, H Koprowski, SV Spitsin. *FASEB J* 14: 691–998, 2000.
304. DJ Miljkovic, J Drulovic, V Trajkovic, S Mesaros, I Dujmovic, D Maksimivic, T Samardzic, N Stojsavljevic, Z Levic, MM Stojkovic. *Eur J Neurol* 9: 413–418, 2002.
305. B Nazliel, D Taskiran, C Irkec, FZ Kutay, S Pogun. *J Clin Neurosci* 9: 530–532, 2002.
306. D Liu, J Wen, J Liu, L Li. *FASEB J* 13: 2318–2328, 1999.
307. R Liu, SW Flanagan, LW Oberley, D Gozal, M Qiu. *J Nuerochem* 80: 488–500, 2002.
308. Y Christen. *Am J Clin Nutr* 71(suppl): 621S–629S, 2000.
309. MA.Smith, CA Rottkamp, A Nunomura, AK Raina, G Perry. *Biochim Biophys Acta* 1502: 139–144, 2000.
310. DA Butterfield, CM Lauderback. *Free Radic Biol Med* 32: 1050–1060, 2002.
311. DA Butterfield, A Castegna, CM Lauderback, J Drake. *Neurobiol Aging* 23: 655–6664, 2002.
312. D Harman. *Ann NY Acad Sci* 959: 384–395, 2002.
313. PH Evans. *Neurobiol Aging* 9: 225–226, 1988.
314. PH Evans, E Peterhaus, T Burge, J Klinowski. *Dementia* 3: 1–6, 1992.
315. TJ Montine, WR Markesbery, W Zackert, SC Sanchez, LJ Robers II, JD Morrow. *Am J Pathol* 155: 863–868, 1999.
316. D Pratico, K Uryu, S Leight, JQ Trojanoswki, VM Lee. *J Neurosci* 21: 4183–4187, 2001.
317. EE Tuppo, LJ Forman, BW Spur, RE Chan-Ting, A Chopra, TA Cavalieri. *Brain Res Bull* 54: 565–568, 2001.
318. Z Xie, M Wei, TE Morgan, P Fabrizio, D Han, CE Finch, VD Longo. *J Neurosci* 22: 3484–3492, 2002.

319. S Shimohama, H Tanino, N Kawakami, N Okamura, H Kodama, T Yamaguchi, T Hayakawa, A Nunomura, S Chiba, G Perry, MA Smith, S Fujimoto. *Biochem Biophys Res Commun* 273: 5–9, 2000.
320. M Grundman. *Am J Clin Nutr* 71(suppl): 630S–636S, 2000.
321. JB Schulz, J Lindenau, J Seyfried, J Dichgans. *Eur J Biochem* 267: 4904–4911, 2000.
322. DPR Muller. In: *Antioxidants in Disease. Mechanisms and Therapy*, H Sies (ed.), Academic Press, San Diego, CA, 1997, pp 557–580.
323. SB Berman, TG Hastings. *J Neurochem* 73: 1127–1137, 1999.
324. MM Iravani, K Kashefi, P Mander, S Rose, P Jenner. *Neuroscience* 110: 49–58, 2002.
325. JD Grimes, MN Hassan, J Thakar. *Can J Neurol Sci* 14: 483–487, 1987.
326. KL Quick, LL Dugan. *Ann Neurol* 49: 627–635, 2001.
327. S Brault, AK Martinez-Bermudez, AM Marrache, F Gobeil Jr, X Hou, M Beauchamp, C Quiniou, G Almazan, C Lachance, J Roberts II, DR Varma, S Chemtob. *Stroke* 34: 776–782, 2003.
328. E Svenungsson, M Andersson, L Brundin, R van Vollenhoven, M Khademi. *Ann Rheum* 60: 372–379, 2001.
329. M Kulkarni, WM Armstead. *J Neurotrauma* 19: 965–973, 2002.
330. D Liu, L Li, L Augustus. *J Neurochem* 77: 1036–1047, 2001.
331. H Bayir, VE Kagan, YY Tyurina, V Tyurin, RA Ruppel, PD Adelson, SH Graham, K Janesko, RS Clark, PM Kochanek. *Pediatr Res* 51: 571–578, 2002.
332. JR Diamond, JV Bonventre, MJ Karnovsky. *Kidney Int* 29: 478–483, 1986.
333. T Adachi, M Fukuta, Y Ito, K Hirano, M Sugiura, K Sugiura. *Biochem Pharmacol* 35: 341–345, 1986.
334. JF Zhou, JX Chen, HC Shen, D Cai. *Biomed Environ Sci* 15: 233–244, 2002.
335. L Wagner, A Riggleman, A Erdely, W Couser, C Baylis. *Kidney Int* 62: 532–536, 2002.
336. S Walpen, KF Beck, L Schaefer, I Raslik, W Eberhardt, Rm Schaefer, J Pfeilschifter. *FASEB J* 15: 571–573, 2001.
337. W Wang, S Jittikanont, SA Falk, P Li, L Feng, PE Gengaro, BD Poole, RP Bowler, BJ Day, JD Crapo, RW Schrier. *Am J Physiol Renal Physiol* 284: F532–F537, 2003.
338. L Lucchi, S Banni, MP Melis, E Angioni, G Carta, V Casu, R Rapana, A Ciuffreda, FP Corongiu, A Albertazzi. *Kidney Int* 58: 1695–1702, 2000.
339. HC Chen, JY Guh, JM Chang, YH Lai. *Nephron* 88: 211–217, 2001.
340. S Thamilselvan, SR Khan, M Menon. *Urol Res* 31: 3–9, 2003.
341. *Free Radicals in Liver Injury*, G Poli, KH Cheeseman, MU Dianzani, TF Slater (eds), IRL Press, Oxford, 1985.
342. TM Hagen, S Huang, J Curnutte, P Fowler, V Martinez, CM Wehr, BN Ames, FV Chisari. *Proc Natl Acad Sci USA* 91: 12808–12812, 1994.
343. R Shimoda, M Nagashima, M Sakamoto, N Yamaguchi, S Hirohashi, J Yokota, H Kasai. *Cancer Res* 54: 3171–3172, 1994.
344. F Farinati, R Cardin, P Degan, N De Maria, RA Floyd, DH Van Thiel, R Naccarato. *Free Radic Biol Med* 27: 1284–1291, 1999.
345. E Sodergren, J Cederberg, B Vessby, S Basu. *Eur J Nutr* 40: 10–16, 2001.
346. M Casaril, AM Stanzial, P Tognella, M Pantalena, F Capra, R Colombari, R Corrocher. *Hepatogastroenterology* 47: 220–223, 2000.
347. GD Nadrarni, NB D'Souza. *Biochem Med Metab Biol* 40: 42–45, 1988.
348. H Kono, I Rusyn, M Yin, E Gabele, S Yamashina, A Dikalova, MB Kadiiska, HD Connor, RP Mason, BH Segal, BU Bradford, SM Holland, RG Thurman. *J Clin Invest* 106: 867–872, 2000.
349. H Yamamoto, T Watanabe, H Mizuno, K Endo, T Hosokawa, A Kazusaka, R Gooneratne, S Fujita. *Free Radic Biol Med* 30: 547–554, 2001.
350. AA Vlessis, RK Goldman, DD Trunkey. *Br J Surg* 82: 870–876, 1995.
351. JB Ochoa, AO Udekwu, TR Billiar, RD Curran, FB Cerra, RL Simmons, AB Peitzman. *Ann Surg* 214: 621–626, 1991.
352. N Fukuyama, Y Takebayashi, M Hida, H Ishida, K Ichimori, H Nakazawa. *Free Radic Biol Med* 22: 771–774, 1997.

353. H Macarthur, TC Westfall, DP Riley, TP Misko, D Salvemini. *Proc Natl Acad Sci USA.* 97: 9753–9758, 2000.
354. M Fuente, VM Victor. *Free Radic Res* 35: 73–84, 2001.
355. PM Guyan, S Uden, JM Braganza. *Free Radic Biol Med* 8: 347–354, 1990.
356. D Closa, G Hotter, J Rosello-Catafau, O Bulbena, L Fernandez-Cruz, E Gelpi. *Dig Dis Sci* 39: 1537–1543, 1994.
357. K Tsai, S-S Wang, T-S Chen, C-W Kong, F-Y Chang, S-D Lee, F-J Lu, *Gut* 42: 850–855, 1998.
358. SH Rahman, BJ Ammori, M Larvin, MJ McMahon. *Gut* 52: 270–274, 2003.
359. P Gionchetti, M Campieri, C Guarnieri, A Belluzzi, C Brignola, E Bertinelli, M Ferretti, M Miglioli, L Barbara. *Dig Dis Sci* 39: 550–554, 1994.
360. B Oldenburg, H van Kats-Renaud, JC Koningsberger, GP van Berge Henegouwen, BS van Asbeck. *Clin Chim Acta* 310: 151–156, 2001.
361. FJ Miller Jr, WJ Sharp WJ, X Fang, LW Oberley, TD Oberley, NL Weintraub. *Arterioscler Thromb Vasc Biol* 22: 560–565, 2002.
362. JMC Gutteridge, DA Rowley, B Halliwell, T Westermarck. *Lancet* 2(8296): 459–460, 1982.
363. JMC Gutteridge, DA Rowley, E Griffiths, B Halliwell. *Clin Sci* 68: 463–467, 1985.
364. SD Aust, BC White. *Adv Free Radic Biol Med* 1: 1–17, 1985.
365. LA Hernandez, MB Grisham. *Am J Physiol* 253: G49–G53, 1987.
366. LM Sayre, G Perry, MA Smith. *Curr Opin Cell Biol* 3: 220–225, 1999.
367. F Kageyama, Y Kobayashi, T Kawasaki, S Toyokuni, K Uchida, H Nakamura. *Am J Gastroenterol* 95: 1041–1050, 2000.
368. M Jung, JC Drapier, H Weidenbach, L Renia, L Oliveira, A Wang, HG Beger, AK Nusser. *J Hepatol* 33: 387–394, 2000.
369. TC Carmine, P Evans, G Bruchelt, R Evans, R Handgretinger, D Niethammer, B Halliwell. *Cancer Lett* 94: 219–226, 1995.
370. T Ogihara, K Hirano, H Ogihara, K Misaki, M Hiroi, T Morinobu, HS Kim, S Ogawa, R Ban, M Hasegawa, H Tamai. *Pediatr Res* 2003 Jan 15 [epubication ahead of print].
371. C Niederau, R Fischer, A Sonnenberg, W Stremmel, HJ Trampisch, G Strohmeyer. *N Engl J Med* 313: 1256–1262, 1985.
372. L Goldberg, LE Martin, A Batchelor. *Biochem J* 83: 291–298, 1962.
373. NE Preece, PF Evans, JA Howarth, LJ King, DV Parker. *Toxicol Appl Pharmacol* 93: 89–100, 1988.
374. IS Young, TG Trouton, JJ Torney, D McMaster, ME Callender, ER Trimble. *Free Radic Biol Med* 16: 393–397, 1994.
375. XJ Zhou, Z Laszik, XQ Wang, FG Silva, ND Vaziri. *Lab Invest* 80: 1905–1914, 2000.
376. SP Hussain, K Raja, PA Amstad, M Sawyer, LJ Trudel, GN Wogan, LJ Hofseth, PG Shields, TR Billar, C Trautwein, T Hohler, PR Galle, DH Phillips, R Markin, AJ Marrogi, CC Harris. *Proc Natl Acad Sci USA* 97: 12770–12775, 2000.
377. D Tovbin, D Mazor, M Vorobiov, C Chaimovitz, N Meyerstein. *Am J Kidney Dis* 40: 1005–1012, 2002.
378. BP Esposito, W Breuer, I Slotki, ZI Cabantchik. *Eur J Clin Invest* 32: 42–49, 2002.
379. MP Kooistra, S Kersting, I Gosriwatana, S Lu, J Nijhoff-Schutte, RC Hider, JJM Marx. *Eur J Clin Invest* 32: 36–41, 2002.
380. LN Grinberg, EA Rachmilewitz, N Kitrossky, M Chevion. *Free Radic Biol Med* 18: 611–615, 1995.
381. T Repka, RP Hebbel. *Blood* 78: 2753–2758, 1991.
382. LN Grinberg, O Shalev, A Goldfarb, EA Rachmilewtz. *Biochim Biophys Acta* 1139, 248–250, 1992.
383. IB Afanas'ev, II Afanas'ev, IB Deeva, LG Korkina. *Transfusion Sci* 23: 237–238, 2000.
384. L Korkina, C De Luca, I Deeva, S Perrotta, B Nobili, S Passi, P Puddu. *Transfusion Sci* 23: 253–254, 2000.
385. MA Livrea, L Tesoriere, AD Pintaudi, A Calabrese, A Maggio, HJ Freisleben, D D'Arpa, R D'Anna, A Bongiorno. *Blood* 88: 3608–3614, 1996.
386. A Meral, P Tuncel, E Surmen-Gur, R Ozbek, E Ozturk, U Gunay. *Pediatr Hematol Oncol* 17: 687–693, 2000.
387. D Chakraborty, M Bhattacharyya. *Clin Chim Acta* 305: 123–129, 2001.

388. MA Livrea, L Tesorieri, A Maggio, D D'Arpa, AM Pintaudi, E Pedone. *Blood* 92: 3936–3942, 1998.
389. MD Scott, JJM Van den Berg, T Repka, P Rouyer-Feessard, RP Hebbel, Y Beuzard, B Lubin. *J Clin Invest* 91: 1706–1712, 1993.
390. SM Altamentova, N Shaklai. *Biofactors* 8: 169–172, 1998.
391. MD Scott, JW Eaton. *Br J Haematol* 91: 811–819, 1995.
392. O Shalev, T Repka, A Goldfarb, L Grinberg, A Abrahamov, NF Olivieri, EA Rachmilewitz, RP Hebbel. *Blood* 86: 2008–2013, 1995.
393. PV Browne, O Shalev, FA Kuypers, C Brugnara, A Solovey, N Mohandas, SL Schrier, RP Hebbel. *J Clin Invest* 100: 1459–1464, 1997.
394. AV Kozlov, EA Ostrachovitch, IB Afanas'ev. *Biochem Pharmacol* 47: 795–799, 1994.
395. L Cragg, RP Hebbel, W Miller, A Solovey, S Selby, H Enright. *Blood* 92: 632–638, 1998.
396. L Tesoriere, D D'Arpa, D Butera, M Allergra, D Renda, A Maggio, A Bongiorno, MA Livrea. *Free Radic Res* 34: 529–540, 2001.
397. LN Grinberg, EA Rachmilewitz, H Newmark. *Biochem Pharmacol* 48: 643–649, 1994.
398. P Browne, O Shalev, RP Hebbel. *Free Radic Biol Med* 24: 1040–1048, 1998.
399. RP Hebbel, JW Eaton, M Balasingam, MH Steinberg. *J Clin Invest* 70: 1253–1259, 1982.
400. T Repka, RP Hebbel. *Blood* 78: 2753–2758, 1991.
401. K Sheng, M Shariff, RP Hebbel. *Blood* 91: 3467–3470, 1998.
402. P Dias-Da-Motta, VR Arruda, MN Muscara, ST Saad, G De Nucci, FF Costa, A Condino-Neto. *Br J Haematol* 93: 333–340, 1996.
403. KA Nath, JP Grande, JJ Haggard, AJ Croatt, ZS Katusic, A Solovey, RP Hebbel. *Am J Pathol* 158: 893–903, 2001.
404. M Aslan, TM Ryan, B Adler, TM Townes, DA Parks, A Thompson, A Tousson, MT Gladwin, RP Patel, MM Tarpey, I Batinic-Haberle, CR White, BA Freeman. *Proc Natl Acad Sci USA* 98: 15215–15220, 2001.
405. JR Lancaster, Jr. *Proc Natl Acad Sci USA* 99: 552–553, 2002.
406. M Aslan, TM Ryan, TM Townes, L Coward, MC Kirk, S Barnes, CB Alexander, SS Rosenfeld, BA Freeman. *J Biol Chem* 278: 4194–4204, 2003.
407. H Nagasawa, JB Little. *Carcinogenesis* 4: 795–799, 1983.
408. M Scarpa, A Rigo, F Momo, G Isacchi, G Novelli, B Dallapiccola. *Biochem Biophys Res Commun* 130: 127–132, 1985.
409. W Malorni, E Straface, G Pagano, D Monti, A Zatterale, D Del Principe, IB Deeva, C Franceschi, R Masella, LG Korkina. *FEBS Lett* 468: 125–128, 2000.
410. LG Korkina, EV Samochatova, AA Maschan, TB Suslova, ZP Cheremisina, IB Afanas'ev. *J Leukocyte Biol* 52: 357–362, 1992.
411. H Joenje, AW Nieuwint, AB Oostra, F Arwert, H de Koning, KJ Roozendaal. *Cancer Genet Cytogenet* 25: 37–45, 1987.
412. B Porfirio, G Ambroso, G Giannella, G Isacchi, B Dallapiccola. *Hum Genet* 83: 49–51, 1989.
413. LG Korkina, EV Samochatova, AA Maschan, T Suslova, Z Cheremisina, IB Afanas'ev. *Drugs Today* 28: 165–169, 1992.
414. M Poot, O Gross, B Epe, M Pflaum, H Hoehn. *Exp Cell Res* 222: 262–268, 1996.
415. JC Schultz, NT Shahidi. *Am J Hematol* 42: 196–201, 1993.
416. W Ruppitsch, C Meisslitzer, M Hirsch-Kauffmann, M Schweiger. *FEBS Lett* 422: 99–102, 1998.
417. T Takeuchi, K Morimoto. *Carcinogenesis* 14: 1115–1120, 1993.
418. AA Clarke, NJ Philpott, EC Gordon-Smith, TR Rutherford. *Br J Haematology* 96: 240–247, 1997.
419. G Pagano, A Zatterale, LG Korkina. *Blood* 93: 1116–1118, 1999.
420. TM Nicotera, J Notaro, S Notaro, J Schumer, AA Sandberg. *Cancer Res* 49: 5239–5243, 1989.
421. Nicotera T, Thusu K, Dandona P. *Cancer Res* 53(21): 5104–5107, 1993.
422. I Emerit, P Cerutti. *Proc Natl Acad Sci USA* 78: 1868–1872, 1981.
423. CA Colton, JB Yao, D Gilbert, ML Oster-Granite. *Brain Res* 519: 236–242, 1990.
424. J Busciglio, BA Yankner. *Nature* 378: 776–779, 1995.
425. N Druzhyna, RG Nair, SP LeDoux, GL Wilson. *Mutat Res* 409: 81–89, 1998.
426. S Schuchmann, U Heinemann. *Free Radic Biol Med* 28: 235–250, 2000.

427. G Capone, P Kim, S Jovanovich, L Payne, L Freund, K Welch, E Miller, M Trush. *Life Sci* 70: 2885–2895, 2002.
428. S Pitkanen, BH Robinson. *J Clin Invest* 98: 345–351, 1996.
429. Y Umaki, T Mitsui, I Endo, M Akaike, T Matsumoto. *Acta Neuropathol* 103: 163–170, 2002.
430. LP Liang, YS Ho, M Patel. *Neuroscience* 101: 563–570, 2000.
431. M Bartal, D Mazor, A Dvilansky, N Meyerstein. *Acta Haematol* 90: 94–98, 1993.
432. J Muntane, P Puig-Parellada, MT Mitjavila. *J Lab Clin Med* 126: 435–443, 1995.
433. P Jacobs, L Wood, AR Bird. *J Hematol* 5: 77–83, 2000.
434. YK Shin, UD Sohn, MS Choi, C Kum, SS Sim, MY Lee. *Auton Autacoid Pharmacol* 22: 47–55, 2002.
435. I Ciancarelli, M Tozzi-Ciancarelli, C Di Massimo, C Marini, A Carolei. *Cephalalgia* 23: 39–42, 2003.
436. N Srivastava, MK Barthwal, PK Dalal, AK Agarwal, D Nag, RC Srimal, PK Seth, M Dikshit. *Psychopharmacology (Berl)* 158: 140–145, 2001.
437. R Selvam. *Urol Res* 30: 35–47, 2002.
438. S Yang, H Zhu, Y Li, H Lin, K Gabrielson, MA Trush, AM Diehl. *Arch Biochem Biophys* 378: 259–268, 2000.
439. P Dandona, P Mohanty, H Ghanim, A Aljada, R Browne, W Hamouda, A Prabhala, A Afzal, R Garg. *J Clin Endocrinol Metab* 86: 355–362, 2001.
440. D Harman. *J Gerontol* 11: 298–300, 1956.
441. RJ Mehlhorn, G Cole. *Adv Free Radic Biol Med* 1: 165–223, 1985.
442. RS Sohal, RG Allen. In: *Molecular Biology of Aging*. AD Woodhead, AD Blackett, A Hollsender (eds), Plenum Publishing, 1983, pp 75–104.
443. D Harman. In: *Free Radicals, Aging, and Degenerative Diseases*, JE Johnson, Jr., R Walford, D Harman, J Miquel (eds), 1986, pp 3–49. (Liss, New York)
444. J Miquel, J Fleming. In: *Free Radicals, Aging, and Degenerative Diseases*, JE Johnson Jr, R Walford, D Harman, J Miquel (eds), 1986, pp 51–74. (Liss, New York)
445. RG Allen, BP Keogh, M Tresini, GS Gerhard, C Volker, RJ Pignolo, J Horton, VJ Cristofalo. *J Biol Chem* 272: 24805–24812, 1997.
446. H Chen, D Cangello, S Benson, J Folmer, H Zhu, MA Trush, BR Zirkin. *Exp Gerontol* 36: 1361–1373, 2001.
447. CM Martins, RAL Prates, RA Pereira, NCGerzstein, JA Nogueira-Machado. *Gerontology* 48: 354–359, 2002.
448. M Phillips, RN Cataneo, J Greenberg, R Gunawardena, F Rahbari-Oskoui. *Clin Chim Acta* 328: 83–86, 2003.
449. T Ide, H Tsutsui, N Ohashi, S Hayashidani, N Suematsu, M Tsuchihashi, H Tamai, A Takeshita. *Arterioscler Thromb Vasc Biol* 22: 438–442, 2002.
450. P Kapahi, ME Boulton, TBL Kirkwood. *Free Radic Biol Med* 26: 495–500, 1999.
451. A Lass, S Agarwal, RS Sohal. *J Biol Chem* 272: 19199–19204, 1997.
452. SK Butcher, V Killampalli, H Chahal, E Kaya Alpar, JM Lord. *Biochem Soc Trans* 31: 449–451, 2003.
453. A Csiszar, Z Ungvari, JG Edwards, P Kaminski, MS Wolin, A Koller, G Kaley. *Circ Res* 90: 1159–1166, 2002.
454. B Drew, C Leeuwenburgh. *Ann NY Acad Sci* 959: 66–81, 2002.
455. J Miquel. *Ann NY Acad Sci* 959: 508–516, 2002.
456. B van der Loo, R Labugger, CP Aebischer, JN Skepper, M Bachschmid, V Spitzer, J Kilo, L Altwegg, V Ullrich, TF Luscher. *Circulation* 105: 1635–1638, 2002.
457. A Cherubini G Zuliani, F Costantini, SD Pierdomenico, S Volpato, A Mezzetti, P Mecocci, S Pezzuto, M Bregnocchi, R Fellin, U Senin. *J Am Geriatr Soc* 49: 651–654, 2001.
458. J Miquel, AC Economos, J Fleming, JE Johnson Jr. *Exp Gerontol* 15: 575–591, 1980.
459. AW Linnane, S Marzuki, T Ozawa, M Tanaka. *Lancet* 1(8639): 642–645, 1989.
460. G Lenaz, C Bovina, M D'Aurelio, R Fato, G Formiggini, ML Genova, G Giuliano, MM Pich, U Paolucci, GP Castelli, B Ventura. *Ann NY Acad Sci* 959: 199–213, 2002.
461. T Ozawa. *Physiol Rev* 77: 425–464, 1997.

462. G Barja, S Cadenas, C Rojas, R Pers-Campo, M Lopez-Torres. *Free Radic Res* 21: 317–327, 1994.
463. HH Ku, RS Sohal. *Mech Ageing Dev* 72: 67–76, 1993.
464. D Harman. *Ann NY Acad Sci* 959: 384–395, 2002.
465. ADNJ de Grey. *Eur J Biochem* 269: 2003–2009, 2002.
466. DM Morre, G Lenaz, DJ Morre. *J Exp Biol* 203: 1513–1521, 2000.

32 Comments on Contemporary Methods of Oxygen and Nitrogen Free Radical Detection

Detailed analysis of the experimental methods of reactive oxygen and nitrogen species detection is outside the scope of this book. However, the consideration of the most important contemporary analytical assays is necessary because the reliability of the data already considered strongly depends on the reliability of the methods applied.

32.1 DETECTION OF SUPEROXIDE IN BIOLOGICAL SYSTEMS

Early methods of superoxide detection are well known and described in many books and reviews. They include cytochrome c reduction, nitroblue tetrazolium reduction, spin trapping, etc. (see, for example, Ref. [1]). The most efficient assays are based on the ability of superoxide to reduce some compounds by one-electron transfer mechanism because such processes (Reaction (1)) proceed with high rates [2]:

$$O_2^{\bullet-} + A \Longrightarrow O_2 + A^{\bullet-} \tag{1}$$

However, to be a quantitative assay of superoxide detection, Reaction (1) had to be an exothermic reaction, i.e., the difference between the one-electron reduction potentials of reagents $\Delta E° = E°[O_2^{\bullet-}/O_2] - E°[A^{\bullet-}/A]$ must be <0. In this case the rate constants of Reaction (1) will be sufficiently high ($10^8 - 10^9$ l mol^{-1} s^{-1}). Among traditionally applied assays, three compounds satisfy this condition: cytochrome c, lucigenin, and tetranitromethane (Table 32.1).

32.1.1 CYTOCHROME c REDUCTION

SOD-inhibitable one-electron reduction of ferric cytochrome c is probably the most frequently used method of superoxide detection.

$$O_2^{\bullet-} + \text{cyt. } c(\text{III}) \Longrightarrow O_2 + \text{cyt. } c(\text{II}) \tag{2}$$

Reaction (2) is an outer-sphere exothermic process (ΔE^0 is about -0.4 V) and therefore, the equilibrium of this reaction is completely shifted to the right, i.e., the reoxidation of reduced cytochrome c by dioxygen is impossible. However, the rate constant for Reaction (2) ($2.6 \pm 0.1 \times 10^5$ l mol^{-1} s^{-1}) is unexpectedly low for the exothermic one-electron transfer

TABLE 32.1
$\Delta E°$ and Rate Constants for Reaction (1) with Cytochrome c, Lucigenin, and Tetranitromethane

	ΔE_o (V)	k_1 (l mol^{-1} s^{-1})	Ref.
Cytochrome c	About -0.4	$2.6 \pm 0.1 \times 10^5$	[3]
Lucigenin	-0.35	10^8	[4]
Tetranitromethane		2×10^9	[5]

reactions (usually 10^8–10^9 l mol^{-1} s^{-1}) that limits the ability of cytochrome c to compete with other electron acceptors such as quinones.

Possible errors due to the competition of cytochrome c reduction with the reversible reduction of quinones by superoxide are frequently neglected. For example, it has been found that quinones (Q), benzoquinone (BQ), and menadione (MD) enhanced the SOD-inhibitable cytochrome c reduction by xanthine oxidase [6]. This seems to be a mystery because only menadione may enhance superoxide production by redox cycling ($E°$[MD$^{•-}$]/[MD] = -0.20 V against $E°$[O$_2^{•-}$]/[O$_2$] = -0.16 V) via Reactions (3) and (4), whereas for benzoquinone ($E°$[BQ$^{•-}$]/[BQ] = $+0.099$ V) Reaction (4) completely shifted to the left.

$$Q \xrightarrow{\text{Enzyme}} Q^{•-} \qquad (3)$$

$$Q^{•-} + O_2 \Longleftrightarrow Q + O_2^{•-} \qquad (4)$$

This proposal was supported by the measurement of superoxide production by lucigenin-amplified CL (see below); it was found that only menadione enhanced CL in this system while benzoquinone inhibited CL [6].

Increase in cytochrome c reduction in the presence of benzoquinone could be due to Reaction (5).

$$BQ^{•-} + \text{cyt. } c(\text{III}) \Longrightarrow BQ + \text{cyt. } c(\text{II}) \qquad (5)$$

However, a more discouraged fact is that benzoquinone accelerated SOD-inhibitable part of cytochrome c reduction, which is usually considered as a reliable proof of superoxide formation. Such a phenomenon has been first shown by Winterbourn [7], who suggested that SOD may shift the equilibrium of Reaction (4) to the right even for nonredox cycling quinones. The artificial enhancement of superoxide production by SOD in the presence of quinones was demonstrated in the experiments with lucigenin-amplified CL, in which benzoquinone was inhibitory [6].

The efficiency of superoxide assays strongly depend on the nature of superoxide producers. Significant difficulties arise in the detection of superoxide in cells and tissue. Cytochrome c is unable to penetrate cell membranes and therefore, can be used only for the measurement of extracellular superoxide. Furthermore, SOD-inhibitable cytochrome c reduction is difficult to apply in nonphagocytic cells and tissue due to the complications of measuring low rates of superoxide release, direct reduction of cytochrome c by cellular enzymes, the reoxidation of reduced cytochrome by hydrogen peroxide, etc. [8]. Moreover, in nonphagocytic cells superoxide is formed exclusively inside the cells and is not released outside as in phagocytes. These circumstances severely limit the number of analytical methods, which can be used for superoxide detection in vasculature.

It has earlier been suggested to make cytochrome *c* a more specific reagent for superoxide detection by its acetylation or succinoylation [9–11]. It was proposed that acetylation and succinoylation must cause a greater decrease in the reaction of cytochrome *c* with NADPH cytochrome P-450 reductase than with superoxide due to a decrease in the electrostatic charge of native cytochrome *c* [12]. However, the rate constant for the most selective succinoylated cytochrome *c* became about 10% of native cytochrome [13], making this assay even less sensitive.

32.1.2 Spin-Trapping

In 1974, Harbour et al. [14] demonstrated that superoxide reacts with 5,5-dimethyl-1-pyrroline-*N*-oxide (DMPO), forming a spin-adduct DMPO–OOH, which is easily identified by its ESR spectrum (Figure 32.1) Unfortunately, this spin-adduct is rather unstable (the half-life of DMPO–OOH is changed from 27 s at pH 9 to 91 s at pH 5 [15]). Another disadvantage of using DMPO is a low rate of reaction with superoxide (see below). Nonetheless, the application of spin trapping is of a great worth — it is the only direct method of superoxide identification as free radical in biological systems.

Spin trapping has been widely used for superoxide detection in various in vitro systems [16]: this method was applied for the study of microsomal reduction of nitro compounds [17], microsomal lipid peroxidation [18], xanthine–xanthine oxidase system [19], etc. As DMPO–OOH adduct quickly decomposes yielding DMPO–OH, the latter is frequently used for the measurement of superoxide formation. (Discrimination between spin trapping of superoxide and hydroxyl radicals by DMPO can be performed by the application of hydroxyl radical scavengers, see below.) For example, Mansbach et al. [20] showed that the incubation of cultured enterocytes with menadione or nitrazepam in the presence of DMPO resulted in the formation of DMPO–OH signal, which supposedly originated from the reduction of DMPO–OOH adduct by glutathione peroxidase.

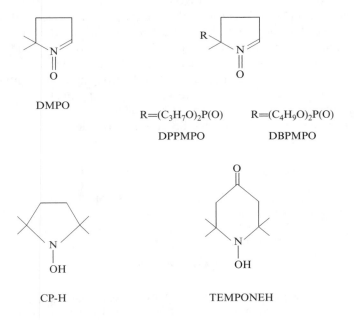

FIGURE 32.1 Spin traps.

Several methods were developed for diminishing the shortcomings of DMPO application. Thus, Souza et al. [21] developed procedure, which permits to take into account DMPO–OH decay during ESR measurement in rat aorta rings. Although this method may reduce the inaccuracy of measuring ESR signal, it cannot rule out the most important shortcoming — a very low rate of DMPO reaction with superoxide. It has also recently been proposed to use β-cyclodextrins to augment the stability of DMPO–OOH complexes [22]. It was found that the half-life of DMPO–OOH was sevenfold enhanced in the presence of methylated β-cyclodextrin.

Another approach to this problem is a search for the other more effective spin traps. Frejaville et al. [23] demonstrated that the half-life of spin-adduct of superoxide with 5-(diethoxyphosphoryl)-5-methyl-1-pyrroline-N-oxide (DEMPO) is about tenfold longer than that of DMPO–OOH. Despite a much more efficiency of this spin trap, its hydrophilic properties limit its use for superoxide detection in lipid membranes. Stolze et al. [24] studied the efficiency of some lipophilic derivatives of DEMPO in the reaction with superoxide. These authors demonstrated a higher stability of superoxide spin-adducts with 5-(di-n-propoxyphosphoryl)-5-methyl-1-pyrroline-N-oxide (DPPMPO) and 5-(di-n-butoxyphosphoryl)-5-methyl-1-pyrroline-N-oxide (DBPMPO) (lifetimes are equal to 7 and 8 min in phosphate buffer and 28 and 16 min in toluene and ethanol).

Another type of spin traps, which have been recommended for the detection of superoxide, are the derivatives of hydroxylamine. In 1982, Rosen et al. [25] showed that superoxide is able to oxidize the hydroxylamine derivative 2-ethyl-1-hydroxy-2,5,5-trimethyl-3-oxazolidine (OXANOH) to corresponding free radical 2-ethyl-1-hydroxy-2,5,5-trimethyl-3-oxazolidinoxyl (OXANO). Although this radical is very stable and easily identified by its ESR spectrum, it is also easily reduced by ascorbic acid and other reductants. Furthermore, OXANOH and other hydroxylamines are oxidized by dioxygen in the presence of transition metal ions to form superoxide, and therefore, superoxide detection must be carried out in the presence of chelators.

Later on, other hydroxylamine derivatives such as 1-hydroxy-2,2,6,6-tetramethyl-4-oxo-piperidine (TEMPONEH) and 1-hydroxy-3-carboxy-pyrrolidine (CP-3) have been used for superoxide detection [26]. It was found that these spin traps react with both superoxide and peroxynitrite and that they might be applied for quantification of these reactive species [27]. The CP-3 radical is less predisposed to reduction by ascorbic acid and therefore is probably more suitable for superoxide detection in biological systems.

Unfortunately, due to the above shortcomings of hydroxylamine derivatives as spin traps, the uncertainties of the mechanism of their reactions with superoxide are added. Although it is supposed that nitroxide radicals are formed by oxidation with superoxide (Reaction (6)), this reaction cannot be an elemental stage because superoxide cannot abstract a hydrogen atom.

$$R_2NOH + O_2^{\bullet-} + H^+ \Longrightarrow R_2NO^\bullet + H_2O_2 \qquad (6)$$

It has earlier been proposed (Ref. [2], p. 70) that this reaction may proceed through the deprotonation–oxidation mechanism:

$$R_2NOH + O_2^{\bullet-} \Longrightarrow R_2NO^- + HOO^\bullet \qquad (7)$$
$$R_2NO^- + O_2 \Longrightarrow R_2NO^\bullet + O_2^{\bullet-} \qquad (8)$$

But in this case spin trapping with hydroxylamine may artificially enhance superoxide production through redox cycling. On the other hand, hydroxylamines are probably able to react directly with perhydroxyl $HOO^\bullet$.

$$R_2NOH + HOO^{\bullet} \Longrightarrow R_2NO^{\bullet} + H_2O_2 \qquad (9)$$

However, the concentration of perhydroxyl is about 1000 times less than superoxide at physiological pH (Chapter 21), and therefore Reaction (9) is apparently of no importance.

There is also a big uncertainty in published rate constants for Reaction (6). In Refs. [25–27] these rate constants are found to be of $10^3 - 10^4 \, l \, mol^{-1} \, s^{-1}$. However, these values are apparently overestimated because Bielski et al. [28] earlier showed that the rate constant for the reaction of superoxide with hydroxylamine does not exceed $30 \, l \, mol^{-1} \, s^{-1}$. Thus, the use of hydroxylamines as spin traps for superoxide detection has several disadvantages.

32.1.3 Lucigenin-Amplified CL as a Sensitive and Specific Assay of Superoxide Detection

Lucigenin (bis-N-methylacridinium)-amplified CL, which is produced by Reactions (10), (11), or Reactions (12), (11), is probably the most specific assay of superoxide detection.

$$O_2^{\bullet -} + Luc^{2+} \Longrightarrow O_2 + Luc^+ \qquad (10)$$

$$O_2^{\bullet -} + Luc^+ \Longrightarrow LucO_2 \text{ (decomposition)} \Longrightarrow h\nu \qquad (11)$$

$$\text{reduced enzyme} + Luc^{2+} \Longrightarrow \text{oxidized enzyme} + Luc^+ \qquad (12)$$

$$Luc^+ + O_2 \Longrightarrow \diagup\!\!\!\!\diagdown Luc^{2+} + O_2^{\bullet -} \qquad (13)$$

CL is produced during the decomposition of excited dioxetane intermediate LucO$_2$. Recently, Okajima and Ohsaka [29] confirmed that simultaneously electrogenerated $O_2^{\bullet -}$ and Luc^+ produced CL by the decomposition of LucO$_2$.

The efficiency and specificity of this method depends on the irreversibility of the whole process due to a high rate constant and favorable thermodynamics of Reaction (10) [4] and a high rate of subsequent Reaction (11) (which is the recombination of a free radical anion and a free radical cation with the diffusion rate constant of about $10^9 \, l \, mol^{-1} \, s^{-1}$).

Since 1960, when Totter et al. [30] proposed that lucigenin-amplified CL produced in the reactions catalyzed by xanthine oxidase is mediated by oxygen radicals, this method has been widely applied for superoxide detection in cell-free and cellular systems and later on in tissue. Weimann et al. [31] observed the superoxide-mediated lucigenin-amplified CL during NADPH oxidation in rat liver microsomes. Gyllenhammar [32] showed that lucigenin-amplified CL can be used for the study of kinetics of superoxide generation by stimulated neutrophils. Kahl et al. [33] applied lucigenin-amplified CL for the determination of antioxidant efficacy by the suppression of microsomal superoxide production. It has also been shown that the intensity of lucigenin-amplified CL produced by microsomes can be significantly enhanced in the presence of detergents [34,35].

Lucigenin-amplified CL has also been widely used for the detection and measurement of superoxide production by phagocytic and nonphagocytic cells and tissue. It turned out to be an especially effective assay for measuring weak intracellular superoxide production by nonphagocytic cells and tissue where other methods such as cytochrome c reduction are ineffective [8]. Below, we are quoting just several studies among numerous examples of successful application of lucigenin-amplified CL for the detection of superoxide production. Thus, Omar et al. [36] applied this method for the measurement of superoxide production in vascular tissue. With the aid of lucigenin-amplified CL the enhanced superoxide production has been registered during many pathophysiological disorders such as hypercholesterolemia [37], atherosclerosis [38], diabetes mellitus [39], angiotensin II-mediated hypertension [40], etc. Miller et al. [41] showed that superoxide levels (measured by lucigenin-amplified CL) were

2.5 times higher in the abdominal aortic aneurysm segments compared to the adjacent nonaneurysmal aorta segments from patients undergoing surgical repair.

Important studies were performed by Trush and coworkers [42], who showed the advantages of applying lucigenin-amplified CL for the measurement of superoxide production by mitochondria in unstimulated monocytes and macrophages as well as by isolated mitochondria [43,44]. Later on, these authors have shown that mitochondrial superoxide production measured by lucigenin-amplified CL increased in the liver of rats treated with the promoter of hepatocarcinogenesis ethinyl estradiol [45], in liver from obese mice [46], and in children with Down syndrome [47].

However, the use of lucigenin-amplified CL has been recently criticized due to its possible overestimation of superoxide production via redox cycling. A principal reason for this proposal was the fact that in some cases the level of superoxide measured by CL was higher than that determined by other methods such as cytochrome c reduction or spin trapping. This criticism has been started by Liochev and Fridovich [48], who found that lucigenin enhanced the reduction of cytochrome c in the reactions catalyzed by xanthine oxidase and some other enzymes. These authors concluded that lucigenin enhanced superoxide production via redox cycling. Artificial enhancement of superoxide production due to the redox cycling of lucigenin has also been proposed by other authors [49–53].

The reliability of lucigenin-amplified CL as a highly selective and sensitive assay of superoxide detection has been recently reviewed [54,55]. There are three major facts, which supposedly point out at the redox cycling of lucigenin: the augmentation of cytochrome c reduction, an increase in CL with increasing lucigenin concentration, and the enhancement of dioxygen consumption in the presence of lucigenin. Redox cycling of lucigenin consists of two stages (Reactions (12) and (13)). There is no doubt that lucigenin can be reduced by some enzymatic systems to lucigenin semiquinone (Reaction (12)); for example, it has been shown that lucigenin is directly reduced by xanthine oxidase with NADH as a substrate but not with xanthine [4]. However, thermodynamic consideration shows that the second stage of redox cycling Reaction (13) is unfavorable.

In 1969, Legg and Hercules [56] measured the difference between the one-electron reduction potentials of lucigenin and dioxygen in DMF, $\Delta E° = E°[\text{Luc}^{•-}/\text{Luc}] - E°[O_2^{•-}/O_2] = 0.6\,\text{V}$. Estimate of this difference in aqueous solution yields $\Delta E° = 0.35\,\text{V}$ [4]. It means that the equilibrium of Reaction (10) is completely shifted to the right, i.e., the back reaction (Reaction (13)) is virtually impossible. (As already mentioned, the following Reaction (11) must have a diffusion rate constant of about $10^9\,\text{l mol}^{-1}\,\text{s}^{-1}$ and will shift even more the equilibrium of Reaction (10) to the right.) Thus, the acceleration of cytochrome c reduction by lucigenin in enzymatic reactions observed in Ref. [48] could not be a consequence of lucigenin redox cycling. Correspondingly, it has been suggested that in this system lucigenin is acting as a mediator of electron transfer from superoxide to cytochrome [4].

It should be mentioned that Spasojevic et al. [57] recently determined the two-electron reduction potential of lucigenin in water as $-0.14\,\text{V}$. As this value is close to the one-electron reduction potential of dioxygen $E°[O_2^{•-}/O_2] = -0.16\,\text{V}$, these authors regarded their finding as a support for lucigenin redox cycling. However, it has been demonstrated long ago that two-electron reduction potentials cannot be used for the calculation of equilibrium for one-electron transfer processes [58].

Another factor, which is considered in some studies [49–53] to be a proof of lucigenin redox cycling, is an increase in CL intensity with increasing lucigenin concentration. Actually, there are many reasons for a dependence of the CL response on lucigenin concentrations except redox cycling. For example, the amount of lucigenin could be insufficient to scavenge all superoxide produced by Reaction (10) or the rate of penetration of the cellular membrane by lucigenin is too low. Surprisingly, only redox cycling has been considered in literature as an origin of the dependence of lucigenin-amplified CL on lucigenin concentration.

Analysis of literature data shows that in many cases the intensity of lucigenin-amplified CL does not depend on lucigenin concentration at all (Figure 4B of Ref. [4]; CL produced by xanthine oxidase was independent of 50–200 μM lucigenin. Figures 1A and 2A of Ref. [51], CL produced by vascular homogenates and artery rings without the addition of NADH or NADPH was independent of 5–250 μM lucigenin.) There are also examples of level off or even a decrease in CL with increasing lucigenin concentration [32], which of course disagree with the possibility of redox cycling. In recent work Li and Shah [59] also showed that lucigenin-amplified CL produced by NADPH-stimulated endothelial cells very slowly increased or leveled off in the 10–400 μM concentration range of lucigenin. It is interesting that the NADH-stimulated CL sharply increased under the same conditions, pointing out a completely different mechanism of lucigenin-amplified CL. Nonredox cycling mechanism of lucigenin-amplified CL in these cells was supported by findings that there was the same dependence of superoxide formation measured by SOD-inhibitable cytochrome c reduction on cytochrome concentration. (We believe that nobody still suggested the possibility of cytochrome c redox cycling.)

Since redox cycling of lucigenin must stimulate dioxygen consumption by Reaction (13), it has been proposed that comparison of superoxide formation and dioxygen consumption might be useful for the verification of conditions where redox cycling can be neglected. Trush and coworkers [55,60] demonstrated that in most systems dioxygen consumption is negligible at lucigenin concentrations lower than 100 μM and increases in higher concentrations. But it does not mean that an increase in dioxygen consumption always originates from the redox cycling of lucigenin. Thus, Souza et al. [21] found that superoxide production by rat aorta homogenates in the presence of NADPH or NADH was about three times lower than dioxygen consumption and that superoxide production in aorta rings was five to six times lower than dioxygen consumption. These authors pointed out that dioxygen consumption could not be reliable measure of redox cycling.

The most convincing proof of the reliability of lucigenin-amplified CL as the method of superoxide detection is its comparison with other methods (Table 32.2). It is seen that the excellent correlations between various assays have been obtained by different authors. A comparison of these correlations shows one typical error frequently made at the interpretation of lucigenin CL data. We have shown [61] that the proportionality coefficients (slopes in Table 32.2) (Figure 32.2) for the correlations between lucigenin-amplified CL and cytochrome c reduction and epinephrine oxidation differ in different systems. Therefore, it is impossible to use the calibration data received for one system in calculations for other systems. For example, Sohn et al. [53] demonstrated an excellent correlation between lucigenin-amplified CL and cytochrome c reduction measurements of superoxide generation by xanthine–xanthine oxidase. Then, they used this correlation for the construction of the "expected" values for CL in the NADH/cell lysate system and found out that experimental values were much higher than the "expected" ones. However, they did not take into account that proportionality coefficients for these two systems are different and therefore, the calculated "expected" values are wrong ones.

32.1.4 Other Chemiluminescent Methods of Superoxide Detection

Another well-known CL amplifier, which is also frequently used for superoxide detection in biological systems, is luminol (5-amino-2,3-dihydro-1,4-phthalazinedione). It has been proposed that luminol semiquinone reacts with superoxide to form the peroxide intermediate, whose decomposition is accompanied by chemiluminescence [62].

TABLE 32.2
Correlations between Lucigenin-Amplified CL and SOD-Inhibited Cytochrome c Reduction or Epinephrine Oxidation

Superoxide Producer	Correlation Coefficient	Slope	Ref.
Xanthine–xanthine oxidase	0.979	41.5	[61]
Xanthine–xanthine oxidase	0.99		[59]
Xanthine–xanthine oxidase	0.997		[48]
NADH-xanthine oxidase	0.978	0.47	[61]
MCLA-amplified CL[a])	0.998		[53]
PMA-stimulated neutrophils	0.978	92	[61]
PMA-stimulated monocytes	0.992	105	[61]
TPA-stimulated monocyte/macrophages	0.98		[59]
PMA-stimulated WBC	0.930	10.8	[61]
Zymosan-stimulated WBC	0.994	5500	[61]
NADH-stimulated SMP[b])	0.97 (epinephrine)		[43]

[a]Calculated from data presented in Ref. [53] (see text).
[b]SMP are submitochondrial particles.

$$\text{Lum} \xRightarrow{\text{oxidant}} \text{Lum}^\bullet \quad (14)$$

$$\text{Lum}^\bullet + O_2^{\bullet -} + H^+ \Longrightarrow \text{LumO}_2 \quad (15)$$

$$\text{LumO}_2 \xRightarrow{\text{decomposition}} h\nu \quad (16)$$

$$\text{Lum}^\bullet + O_2 \Longrightarrow \text{Lum}^+ + O_2^{\bullet -} \quad (17)$$

Although the use of luminol-amplified CL for superoxide detection in cell-free and cellular systems yielded many important data, which were confirmed by the other analytical methods, these data are more questionable than in the case of lucigenin-amplified CL. At the first stage (Reaction (14)) luminol must be oxidized to luminol semiquinone, and therefore, the whole

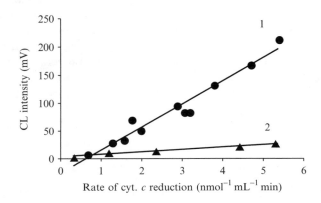

FIGURE 32.2 Correlations between maximal chemiluminescence intensity (CL) and the rates of cytochrome c reduction by the xanthine–XO and NADH–XO systems [61]. Line 1, xanthine–XO and line 2, NADH–XO (in this case CL values were multiplied by 10.)

process is dependent on a suitable oxidant. Very good oxidants of luminol are peroxides (HRP and myeloperoxidase), the latter is probably responsible for luminol-amplified CL produced by neutrophils. Such CL response is inhibited by SOD, confirming the participation of superoxide (Reaction (15)) [63,64].

There are also other drawbacks of luminol CL assay, for example, the possibility of superoxide generation by Reaction (17) and a dependence of the decomposition rate of Reaction (16) and the quantum yield of luminescence from pH [62]. Nonetheless, very interesting findings might be obtained from comparison of the measurement of lucigenin- and luminol-amplified CL. It has been suggested [63] that the oxidants responsible for the reduction of luminol by xanthine oxidase could be hydroxyl radicals or ferryl or perferryl complexes formed by the Fenton reaction. In this case superoxide participates in the stimulation of luminol-amplified CL in two stages, reducing ferric ions to ferrous ions and participating in Reaction (15). We suggested [65] that comparison of lucigenin and luminol CL produced by neutrophils from healthy people and patients with iron-overloading pathologies can be used for the characterization of iron-induced damage in these pathologies (Chapter 31). Recently, Rose and Waite [66] confirmed that ferrous ions indeed induced luminol-amplified CL in the presence of dioxygen.

In addition to superoxide and hydroxyl radicals, luminol produces CL in the reaction with peroxynitrite [67]. To discriminate between superoxide- and peroxynitrite-induced CL, the use of lucigenin-amplified CL has been recommended [68] because peroxynitrite does not interfere in this assay. Another way is to apply the inhibitors of peroxynitrite to distinguish between superoxide- and peroxynitrite-induced luminol CL.

At present, other CL amplifiers are recommended for the detection of superoxide in cells and tissue such as coelenterazine (2-(4-hydroxybenzyl)-6-(4-hydroxyphenyl)-8-benzyl-3,7-dihydroimidazo[1,2-α]pyrazin-3-one]) and its analogs CLA (2-methyl-6-phenyl-3,7-dihydroimidazo[1,2-α]pyrazin-3-one]) and MCLA [2-methyl-6-(4-methoxyphenyl)-3,7-dihydroimidazo[1,2-α]pyrazin-3-one]). It has been suggested that the origin of CL produced by these compounds is the oxidation of the acetamidopyrazine moiety [69,70]. Unfortunately, to our knowledge, there are still no reliable thermodynamic and kinetic data to validate the application of the above CL amplifiers for superoxide detection. Reichl et al. [71] proposed to use the photoprotein pholasin for the detection of superoxide and myeloperoxide activity in stimulated neutrophils.

32.1.5 Adrenochrome (Epinephrine) Oxidation

One of the oldest methods of superoxide detection is the oxidation of epinephrine [72]. This method has the typical disadvantages of oxidizable compounds due to the possibility of the nonsuperoxide-mediated oxidation of epinephrine. Still, SOD-inhibitable epinephrine oxidation might be used as a superoxide assay [72].

32.1.6 Nitroblue Tetrazolium Reduction

Nitroblue tetrazolium (NBT, 3,3′-(3,3′-dimethoxy-1,1′-biphenyl-4,4′-diyl)bis-2-(4-nitrophenyl)-5-phenyl-2H-tetrazolium dichloride) is reduced by superoxide to formazan as a final product, which can be measured spectrophotometrically [73]. Although the rate constant for NBT reduction by superoxide is moderately high $5.88 \pm 0.12 \times 10^4$ l mol^{-1} s^{-1} [74], the formation of formazan is not a simple one-electron transfer process, and the final product is formed as a result of disproportionation of intermediate free radicals. Similar to cytochrome c, NBT is easily reduced by the other reductants; that confines its application for superoxide detection. Moreover, similar to epinephrine, NBT free radical is apparently capable of reducing dioxygen to superoxide [75]. WST-1 (4-[3-(4-iodophenyl)-2-(4-nitrophenyl)-2H-5-tetrazoliol]-1,3-benzene disulfonate sodium salt), a water-soluble NBT analog, is also recommended for the application in spectrophotometric assay of superoxide [76,77]. Another sulfonated tetrazolium XTT ((2,3-bis(2-methoxy-4-nitro-5-sulfophenyl)-2-tetrazolium 5-carboxanilide) also forms a water-soluble formazan and is not directly reduced by xanthine or glucose oxidases. However, XTT might be reduced by NADPH oxidases that makes doubtful its application for the detection of superoxide in cells [78].

32.1.7 Fluorescent Methods

Two fluorescent probes dihydroethidium (DHE) and dichlorodihydrofluorescein (DCFH) are used for superoxide detection in biological systems [54]. Both assays are subjected to some drawbacks such as the oxidation of cytochrome c, the dismutation of superoxide in the presence of DHE [79], and the interaction of DCFH with hydrogen peroxide, hydroperoxides, or peroxynitrite. Particularly big enhancement of DCF formation due to the oxidation of cytochrome c is observed when cytochrome is released from mitochondria during cell death [80]. Rota et al. [81,82] concluded that DCF fluorescence could not be a reliable assay of superoxide detection in cells because superoxide is formed during the DCFH oxidation by peroxidases. It has also been shown that DCFH is oxidized by heme, hemoglobin, myoglobin, and cytochrome c [83]. However, recent work by Caldefie-Chezet et al. [84] showed that the measurements of superoxide production by PMNs with the use of 2'-7'-dichlorofluorescin diacetate flow cytometry correlated with the data obtained by lucigenin- and luminol-amplified CL assays.

32.1.8 Interaction with Aconitase

Gardner and Fridovich [85] proposed that the inactivation of aconitase might be used as an assay of superoxide formation in cells. The mechanism of the interaction of superoxide with aconitases has been considered in Chapter 21. As follows from data presented in that chapter, peroxynitrite is also able to inactivate aconitases rapidly; therefore, this method cannot be a specific assay of superoxide detection.

32.2 DETECTION OF HYDROXYL RADICALS

Several methods have been recommended for the detection of hydroxyl radicals in biological systems. A main uncertainty of hydroxyl radical detection is that in contrast to superoxide, we are not always sure that the reactive oxygen radicals under investigation are free hydroxyl radicals and not "hydroxyl-like" species such as ferryl or perferryl ions. Well-known methods based on the ability of hydroxyl radicals to hydroxylate aromatic compounds, degrade deoxyribose [86], decarboxylate ^{14}C-benzoic acid [87], and produce ethylene from methionine [88,89] have been proposed, but they cannot unquestionably differentiate between free hydroxyl and "hydroxyl-like" radicals. The development of ESR spin-trapping methods [90,91] allows answering positively on the question concerning the formation of free hydroxyl radicals, at least, under the in vitro conditions. In contrast to trapping of superoxide, the rate constant for trapping of hydroxyl radicals by DMPO is very high (3.4×10^9 l mol^{-1} s^{-1} [16]), and therefore, DMPO spin trapping is a very effective direct method of hydroxyl radical detection. The possible confusion with DMPO–OH formed at DMPO–OOH decomposition is easily resolved by the use of ethanol; in its presence the formation of hydroxyl radicals is confirmed by their conversion into radicals CH$_3$CH(OH)$^{\bullet}$. However, there are other complications, which can compromise the DMPO spin trapping of hydroxyl radicals. For example, it has been demonstrated that the Fe^{3+}(DETAPAC) complex is able to react directly with DMPO to form DMPO–OH adduct [92].

Oosthuizen and Greyling [93] recently investigated the possibility of using chemiluminescent methods for hydroxyl radical detection. These authors concluded that the lifetime of hydroxyl radicals (10^{-9} s) is too short to produce a meaningful level of CL. However, in the presence of carbonate the significant levels of luminol- and MCLA-amplified CL were observed supposedly due to Reaction (18), in which the formed much more stable radical $CO_3^{\bullet -}$ is capable of interacting with luminol or MCLA.

$$HO^\bullet + HCO_3^- \Longrightarrow H_2O + CO_3^{\bullet -} \tag{18}$$

32.3 DETECTION OF NITRIC OXIDE

At present, several studies are dedicated to the development of methods for nitric oxide and peroxynitrite detection (see for example Ref. [94] for review). Now, in addition to previous analytical methods such as the reaction of NO with oxyhemoglobin, yielding nitrate and metHb, and Griess reaction (the reaction of nitrite with sulfanilamide and N-(1-naphthyl)-ethylenediamine), there are some more new specific assays. One of them is based on the formation of paramagnetic complex with iron-diethyldithiocarbamate (Fe-DETC). For example, Vanin et al. [95] showed that the incubation of LPS-stimulated macrophages with DETC resulted in the appearance of an ESR spectrum of NOFe(DETC)$_2$. In 1997, Xia and Zweier [96] demonstrated that nitric oxide produced by NO synthase formed NOFe(DETC)$_2$ while NO$^-$, another active nitrogen species, which can supposedly be produced by NO synthase, was undetectable. In contrast, Komarov et al. [97] found that iron N-methyl-D-glucamine dithiocarbamate (Fe-MGD) formed paramagnetic complex with NO$^-$ and NO$^+$ under aerobic conditions. Correspondingly, these authors questioned the possibility to discriminate the formation of nitric oxide and other nitrogen species by the use of iron–dithiocarbamate complexes. However, subsequent researchers [98,99] concluded that the accurate use of Fe^{2+}–MGD allows to register only nitric oxide formation because no ESR signal was observed when nitroxyl ion was generated from Angeli's salt. Nonetheless, it should be taken into account that the redox chemistry of iron–dithiocarbamate complexes is very complicated.

Another important ESR method of NO detection is based on the interaction of NO with the stable radical phenyl-4,4,5,5-tetramethylimidazoline-1-oxyl (PTIO) and its derivatives carboxy-PTIO or trimethylammonio-PTIO [100]. It is interesting that in this assay NO reduces stable free radical PTIO to another stable free radical PTI and the NO$_2$ radical, and the reaction can be monitored by both a decrease in the ESR PTIO spectrum and an increase in the ESR PTI spectrum [101].

$$PTIO(NO^\bullet)(\equiv NO) + {}^\bullet NO \Longrightarrow PTIO(NO^\bullet)(\equiv N) + {}^\bullet NO_2 \tag{19}$$

Kelm et al. [102] proposed a simple method of simultaneous spectroscopic detection of nitric oxide and superoxide based on NO-induced oxidation of oxyhemoglobin to methemoglobin and superoxide-mediated reduction of cytochrome c.

32.4 DETECTION OF PEROXYNITRITE

Dihydrorhodamine 123 (DHR) and DCFH are widely used for the detection of peroxynitrite. Peroxynitrite oxidizes DHR and DCFH with efficiency equal to 38% and 44%, respectively

[103]. Possel et al. [104] also concluded that DCFH is more sensitive to peroxynitrite oxidation than DHR. Unfortunately, both compounds are oxidized by other reactive species (for example, DCFH by superoxide and DHR by HOCl), and therefore, their use for peroxynitrite detection must be confirmed by the other methods.

As mentioned above, the formation of peroxynitrite might be measured by luminol-amplified CL [67]. CL response was greatly augmented by bicarbonate and inhibited by SOD. Radi et al. [67] proposed the following mechanism of carbonate-enhanced luminol CL stimulated by peroxynitrite:

$$ONOO^- + HCO_3^- + H^+ \Longrightarrow ONOOC(O)O^- + H_2O \qquad (20)$$
$$ONOOC(O)O^- + LH^- \Longrightarrow L^{\bullet-} + O_2^{\bullet-} + NO^- + CO_2 + H^+ \qquad (21)$$

Luminol semiquinone is further oxidized to luminol endoperoxide, which elicited CL at decomposition. It should be added that in our experiments peroxynitrite-stimulated luminol CL in cells was enhanced in the presence of the NO synthase substrate L-arginine and sharply diminished in the presence of NO synthase inhibitors. Thus, the application of substrates and inhibitors of NO synthases may discriminate luminol-amplified CL stimulated by superoxide and peroxynitrite.

The formation of peroxynitrite in cells and tissue is frequently characterized by the formation of nitrotyrosine. The formation of nitrotyrosine is not a very specific assay of peroxynitrite detection because the other nitrogen oxide may also take part in this process, but peroxynitrite is undoubtedly the most efficient nitrating agent. (Mechanism of tyrosine nitration by peroxynitrite and other reactive nitrogen compounds has been considered in Chapters 21 and 22.)

REFERENCES

1. *Handbook of Methods for Oxygen Radical Research*. RA Greenwald (ed.) Boca Raton, FL: CRC Press, 1985.
2. IB Afanas'ev. In: *Superoxide Ion: Chemistry and Biological Implications*, vol I, Boca Raton, FL: CRC Press, 1989.
3. IB Afanas'ev. In: *Superoxide Ion: Chemistry and Biological Implications*, vol 2, Boca Raton, FL: CRC Press, 1990, p. 153.
4. IB Afanas'ev, EA Ostrachovitch, LG Korkina. *Arch Biochem Biophys* 366: 267–274, 1999.
5. GR Hodges, MJ Young, T Paul, KU Ingold. *Free Radic Biol Med* 29: 434–441, 2000.
6. IB Afanas'ev, LG Korkina, TB Suslova, SK Soodaeva. *Arch Biochem Biophys* 281: 245–250, 1990.
7. CC Winterbourn. *Arch Biochem Biophys* 209: 159–167, 1981.
8. MM Tarpey, CR White, E Suarez, G Richardson, R Radi, BA Freeman. *Circ Res* 84: 1203–1211, 1999.
9. A Azzi, C Montecucco, C Richter. *Biochem Biohys Res Commun* 65: 597–603, 1975.
10. H Kuthan, H Tsuji, H Graft, V Ullrich, J Werringloer, RW Estabrook. *FEBS Lett* 91: 343–345, 1978.
11. JL Daval, JF Ghersi,-Egea, J Oillet, V Koziel. *J Cereb Blood Flow Metab* 15: 71–77, 1995.
12. E Finkelstein, GM Rosen, SE Patton, MS Cohen, EJ Rauckman. *Biochem Biophys Res Commun* 102: 1008–1015, 1981.
13. H Kuthan, V Ullrich, RW Estabrook. *Biochem J* 203: 551–558, 1982.
14. JR Harbour, V Chow, JR Bolton. *Can J Chem* 52: 3549–3553, 1974.
15. GR Buettner, LW Oberlay. *Biochem Biophys Res Commun* 83: 69–74, 1978.
16. E Finkelstein, GM Rosen, EJ Rauckman. *Arch Biochem Biophys* 200: 1–16, 1980.
17. RC Sealy, HM Swartz, PL Olive. *Biochem Biophys Res Commun* 82: 680–684, 1978.
18. GM Rosen, EJ Rauckman. *Proc Natl Acad Sci USA* 78: 7346–7349, 1981.

19. I Ueno, M Kohno, K Yoshihira, I Hirono. *J Pharm Dyn* 7: 563–569, 1984.
20. CM Mansbach II, GM Rosen, CA Rahn, KE Strauss. *Biochim Biophys Acta* 888: 1–9, 1986.
21. HP Souza, X Liu, A Samouilov, P Kuppusamy, FRM Laurindo, JL Zweier. *Am J Physiol Heart Circ Physiol* 282: H466–H474, 2002.
22. H Karoui, A Rockenbauer, S Pietri, P Tordo. *Chem Commun* 3030–3031, 2002.
23. C Frejaville, H Karoui, B Tuccio, F Le Moigne, M Culcasi, S Pietri, R Lauricella, P Tordo. *J Med Chem* 38: 258–265, 1995.
24. K Stolze, N Udilova, H Nohl. *Free Radic Biol Med* 29: 1005–1014, 2000.
25. GM Rosen, E Finkelstein, EJ Rauckman. *Arch Biochem Biophys* 215: 367–378, 1982.
26. S Dikalov, M Skatchkov, E Bassenge. *Biochem Biophys Res Commun* 231: 701–704, 1997.
27. S Dikalov, M Skatchkov, B Fink, E Bassenge. *Nitric oxide* 1: 423–431, 1997.
28. BHJ Bielski, RL Arudi, DE Cabelli, W Bors. *Anal Biochem* 142: 207–211, 1984.
29. T Okajima, T Ohsaka. *Luminescence* 18: 49–57, 2003.
30. JR Totter, E Castro de Dugros, C Rivero. *J Biol Chem* 235: 1839–1842, 1960.
31. A Weimann, AG Hildebrandt, R Kahl. *Biochem Biophyd Res Commun* 125: 1033–1038, 1984.
32. H Gyllenhammar. *J Immunol Methods* 97: 209–211, 1987.
33. R Kahl, A Weimann, S Weinke, AG Hildebrandt. *Arch Toxicol* 60: 158–162, 1987.
34. J Storch, E Ferber. *Anal Biochem* 169: 262–267, 1988.
35. H Ischiropoulos, T Kumae, Y Kikkawa. *Biochem Biophys Res Commun* 161: 1042–1048, 1989.
36. HA Omar, PD Cherry, MP Mortelliti, T Burke-Wolin, MS Wolin. *Circ Res* 69: 601–608, 1991.
37. Y Ohara, TE Peterson, DG Harrison. *J Clin Invest* 91: 2546–2551, 1993.
38. FJ Miller Jr, DD Gutterman, CD Rios, DD Heistad, BL Davidson. *Circ Res* 82: 1298–1305, 1998.
39. PS Tsao, J Niebauer, R Buitrago, PS Lin, BY Wang, JP Cooke, YD Chen, GM Reaven. *Arterioscler Thromb Vasc Biol* 18: 847–953, 1998.
40. S Raiagopalan, S Kurz, T Munzel, M Tarpey, BA Freeman, KK Griendling, DG Harrison. *J Clin Invest* 97: 1916–1923, 1996.
41. FJ Miller Jr, WJ Sharp, X Fang, LW Oberley, TD Oberley, NL Weintraub. *Arterioscler Thromb Vasc Biol* 22: 560–565, 2002.
42. Y Li, MA Trush. *Biochem Biophys Res Commun* 253: 295–299, 1998.
43. Y Li, KH Stansbury, H Zhu, MA Trush. *Biochem Biophys Res Commun* 262: 80–87, 1999.
44. Y Li, H Zhu, MA Trush. *Biochim Biophys Acta* 1428: 1–12, 1999.
45. J Chen, Y Li, JA Lavigne, MA Trush, JD Yager. *Toxicol Sci* 51: 224–235, 1999.
46. S Yang, H Zhu, Y Li, H Lin, K Gabrielson, MA Trush, AM Diehl. *Arch Biochem Biophys* 378: 259–268, 2000.
47. G Capone, P Kim, S Jovanovich, L Payne, L Freund, K Welch, E Miller, M Trush. *Life Sci* 70: 2885–2895, 2002.
48. SI Liochev, I Fridovich. *Arch Biochem Biophys* 337: 115–120, 1997.
49. MM Tarpey, CR White, E Suarez, G Richardson, R Radi, BA Freeman. *Circ Res* 84: 1203–1211, 1999.
50. M Barbacanne, J Souchard, B Darblade, J Iliou, F Nepveu, B Pipy, F Bayard, J-F Arnal. *Free Radic Biol Med* 29: 388–396, 2000.
51. M Janiszewski, HP Souza, X Liu, MA Pedro, JL Zweier, FR Laurindo. *Free Radic Biol Med* 32: 446–453, 2002.
52. HP Souza, X Liu, A Samouilov, P Kuppusamy, FRM Laurindo, JL Zweier. *Am J Physiol Heart Circ Physiol* 282: H466–H474, 2002.
53. HY Sohn, M Keller, T Gloe, P Crause, U Pohl. *Free Radic Res* 32: 265–272, 2000.
54. T Munzel, IB Afanas'ev, AL Kleschyov, DG Harrison. *Arterioscler Thromb Vasc Biol* 22: 1761–1768, 2002.
55. MA Trush, Y Li. In: *Luminescence Biotechnology*, K van Dyke, C van Dyke, K Woodfork (eds.), Boca Raton, FL: CRC Press, 2002, pp. 287–303.
56. KD Legg, DM Hercules. *J Am Chem Soc* 91: 1902–1907, 1969.
57. I Spasojevic, SI Liochev, I Fridovich. *Arch Biochem Biophys* 373: 447–450, 2000.
58. YA Ilan, G Czapski, D Meisel. *Biochim Biophys Acta* 430: 209–224, 1976.
59. J-M Li, AM Shah. *Cardiovasc Res* 52: 477–486, 2001.

60. Y Li, H Zhu, P Kuppusamy, V Roubaud, JL Zweier, MA Trush. *J Biol Chem* 273: 2015–2023, 1998.
61. IB Afanas'ev EA Ostrakhovitch, EV Mikhal'chik, LG Korkina. *Luminescence* 16: 305–307, 2001.
62. G Merenyi, S Lind. *J Am Chem Soc* 102: 5830–5835, 1980.
63. CD Allred, J Margetts, HR Hill. *Biochim Biophys Acta* 631: 380–385, 1980.
64. R Lock, A Johansson, K Orselius, C Dahlgren. *Anal Biochem* 173: 450–455, 1988.
65. LG Korkina, EV Samochatova, AA Maschan, TB Suslova, ZP Cheremisina, IB Afanas'ev. *J Leukocyte Biol* 52: 357–362, 1992.
66. AL Rose, TD Waite. *Anal Chem* 73: 5909–5920, 2001.
67. R Radi, TP Cosgrove, JC Beckman, BA Freeman. *Biochem J* 290: 51–57, 1993.
68. L Castro, MN Alvarez, R Radi. *Arch Biochem Biophys* 333: 179–188, 1996.
69. T Goto, S Inoue, S Sugiura, K Nishikawa, M Isobe, Y Abe. *Tetrahedron Lett* 37: 4035–4038, 1968.
70. K Teranishi, O Shimomura. *Anal Biochem* 249: 37–43, 1997.
71. S Reichl, A Vocks, M Petkovic, J Schiller, J Arnhold. *Free Radic Res* 35: 723–733, 2001.
72. HP Misra, I Fridovich. *J Biol Chem* 247: 3170–3175, 1972.
73. C Auclair, E Voisin. In: *Handbook of Methods for Oxygen Radical Research*, RA Greenwald, (ed.). Boca Raton, FL: CRC Press, 1985, pp. 123–132.
74. BHJ Bielski, GG Shiue, S Bajuk. *J Phys Chem* 84: 830–834, 1980.
75. C Auclair, M Torres, J Hakim. *FEBS Lett* 89: 26–30, 1978.
76. AS Tan, MV Berridge. *J Immunol Methods* 238: 59–68, 2000.
77. H Ukeda, T Shimamura, M Tsubouchi, Y Harada, Y Nakai, M Sawamura. *Anal Sci* 18: 1151–1154, 2002.
78. L Benov, I Fridovich. *Anal Biochem* 310: 186–190, 2002.
79. L Benov, L Sztejnberg, I Fridovich. *Free Radic Biol Med* 25: 826–831, 1998.
80. MJ Burkitt, P Wardman. *Biochem Biophys Res Commun* 282, 329–333, 2001.
81. C Rota, CF Chignell, RP Mason. *Free Radic Biol Med* 27: 873–881, 1999.
82. C Rota, YC Fann, RP Mason. *J Biol Chem* 274: 28161–28168, 1999.
83. T Ohashi, A Mizutani, A Murakami, S Kojo, T Ishii, S Taketani. *FEBS Lett* 511: 21–27, 2002.
84. F Caldefie-Chezet, S Walrand, C Moinard, A Tridon, J Chassagne, MP Vasson. *Clin Chim Acta* 319: 9–17, 2002.
85. PR Gardner, I Fridovich. *J Biol Chem* 267: 8757–8763, 1992.
86. B Halliwell, JMC Gutteridge. In: *Handbook of Methods for Oxygen Radical Research*, RA Greenwald, (ed.) Boca Raton, FL: CRC Press, 1985, pp. 177–180.
87. GW Winston, AI Cederbaum. In: *Handbook of Methods for Oxygen Radical Research*, RA Greenwald, (ed.) Boca Raton, FL: CRC Press, 1985, pp. 169–175.
88. RJ Youngman, EF Elstner. In: *Handbook of Methods for Oxygen Radical Research*, RA Greenwald, (ed.) Boca Raton, FL: CRC Press, 1985, pp. 165–168.
89. GD Lawrence. In: *Handbook of Methods for Oxygen Radical Research*, RA Greenwald, (ed.) Boca Raton, FL: CRC Press, 1985, pp. 157–163.
90. I Yamazaki, LH Piette. *J Biol Chem* 265: 13589–13594, 1990.
91. GR Buettner. In: *Handbook of Methods for Oxygen Radical Research*, RA Greenwald, (ed.) Boca Raton, FL: CRC Press, 1985, pp. 151–155.
92. MJ Burkitt. *Free Radic Res Commun* 18: 43–57, 1993.
93. MMJ Oosthuizen, D Greyling. *Redox Rep* 6: 105–116, 2001.
94. MM Tarpey, I Fridovich. *Circ Res* 89: 224–236, 2001.
95. AF Vanin, PI Mordvintcev, S Hauschildt, A Mulsch. *Biochim Biophys Acta* 1177: 37–42, 1993.
96. Y Xia, JL Zweier. *Proc Natl Acad Sci USA* 94: 12705–12710, 1997.
97. AM Komarov, DA Wink, M Feelisch, HH Schmidt. *Free Radic Biol Med* 28: 739–742, 2000.
98. Y Xia, AJ Cardounel, AF Vanin, JL Zweier. *Free Radic Biol Med* 29: 793–797, 2000.
99. AF Vanin, X Liu, A Samouilov, RA Stukan, JL Zweier. *Biochim Biophys Acta* 1474: 365–377, 2000.
100. T Akaike, M Yoshida, Y Miyamoto, K Sato, M Kohno, K Sasamoto, K Miyazaki, S Ueda, H Maeda. *Biochemistry* 32: 827–832, 1993.

101. YM Janssen, R Soultanakis, K Steece, E Heerdt, RJ Singh, J Joseph, B Kalyanaraman. *Am J Physiol* 275: L1100–L1109, 1998.
102. M Kelm, R Dahmann, D Wink, M Feelisch. *J Biol Chem* 272: 9922–9932, 1997.
103. JP Crow. *Nitric oxide* 1: 145–157, 1997.
104. H Possel, H Noack, W Augustin, G Keilhoff, G Wolf. *FEBS Lett* 416: 175–178, 1997.

Index

A

Acceptor
  alkyl radicals, 649–651
  alkylsulfonyl radical, 422–424
  dioxygen, 670–671
  peroxyl radicals, 27, 491–511
Acetaminophen, 718, 801
N-acetylcysteine (NAC), 738, 739, 741, 764, 814, 826, 864, 918, 923, 928
Acid
  C–H bond dissociation energy, 321
  decarboxylation, 322–324
  enthalpy of formation, 17
  oxidation, 320–323
  product of oxidation, 16–17
Acid catalysis, 279, 389–390, 561–562, 581
Aconitase, 676, 681, 690, 814, 960
Activation energy, calculation, 216–218
Activation energy of reaction
  alkyl radical + RH, 222
  decomposition of hydroperoxide
    bimolecular, 155, 164
    in polymer, 445, 446, 448
    unimolecular, 153
  decomposition of
    alcoxyl radical, 74
    alkylsulfonyl peracid, 425
    alkylsulfonyl radical, 421
    bis(1,1-dimethylethylperoxide), 91
    bis(1-methyl-1-phenylperoxide), 92
    diacetylperoxides, 93
    dialkylperoxides, 92
    dibenzoylperoxide, 94
    N-alkoxyamines, 655–656
    peroxyalkyl radical, 55
    tetroxide, 57
  dioxygen with transition metals, 381–382
  disproportionation
    peroxyl macroradicals, 434
    peroxyl radicals, 59–65, 346
  disulfides with ROOH, 582
  epoxidation of olefins by
    oxidation, 55
    sulfonic peracid, 428
  factors influencing, 219–232, 242–250
  free radical abstraction

$Am^\bullet$ + cumene, 526
$Am^\bullet$ + ArOH, 604
$AmO^\bullet$ + AmH, 615–616, 631
$AmO^\bullet$ + ArOH, 613–615, 628–629, 638–640
$ArO^\bullet$ + cumene, 523–524
$ArO^\bullet$ + ionol, 609–610
$ArO^\bullet$ + ROOH, 647
p-Benzoquinone + AmH, 623–624
p-Benzoquinone + ArOH, 621–623
Peroxyl macroradical + RH, 437
$R^\bullet$ + ozone, 110
$R^\bullet$ + R_1H, 222
Radicals + ROOH, 174
$RO^\bullet$ + RH, 221
$RO_2^\bullet$ + acid, 322, 324
$RO_2^\bullet$ + alcohol, 273, 275–276
$RO_2^\bullet$ + amines, 337
$RO_2^\bullet$ + AmH, 505–507
$RO_2^\bullet$ + ArOH, 495, 617–618, 628, 630, 645–647
$RO_2^\bullet$ + ether, 294–295
$RO_2^\bullet$ + ester, 345, 351–353
$RO_2^\bullet$ + ketone, 315
$RO_2^\bullet$ + ozone, 110
$RO_2^\bullet$ + RH, 46–48, 220, 223, 233
free radical addition reaction, 241–251
  alcoxyl radicals to olefins, 245
  alkyl radicals to dioxygen, 34–39
  peroxyl radical to olefin, 49–55, 243–244
free radical decomposition
free radical reduction of ROOH, 256–257
free radical substitution $R^\bullet$ + ROOH, 253, 257
generation free radicals by reaction
  halogens with hydrocarbons, 114–121
  InH + dioxygen, 529–532
  nitrogen dioxide with hydrocarbons, 111–113
  ozone with hydrocarbons, 104–108
  RH + dioxygen, 139–144, 347
  2 RH + dioxygen, 142, 144, 278, 289
  ROOH + RH, 156–157, 165–168, 289
  ROOH + ArOH, 533–534
  ROOH + AmH, 535
intramolecular H transfer, 75, 238–240
phosphates + ROOH, 575–576
phosphites + $RO_2^\bullet$, 579–580, 592
sulfides + ROOH, 582, 584–585

Activation energy of reaction (*Continued*)
   sulfides + $RO_2^\bullet$, 586
   thiocarbamates + $RO_2^\bullet$, 591
   thiols + ROOH, 582
   thiophosphates + $RO_2^\bullet$, 591
Activation of dioxygen on surface, 398–399
Activation volume, 637
Acute nonlymphocytic leukemia (ALL), 915, 916
Acute pancreatitis, 928
Acute renal failure (ARF), 906, 927
Acute respiratory distress syndrome (ARDS), 923
Adrenaline, 718, 720
β-Adrenergic receptor blocker, 871, 872
After-effect, 27
Aging, 934, 935
Alcohol dehydrogenase, 687, 814
Alcohols
   bond dissociation energies, 262–263, 266
   co-oxidation, 270–272
   dipole momentum, 261
   formation enthalpy, 15
   free radical generation, 278–280
   mechanism of oxidation, 261–270
   product of ROOH decay, 14–15
Aldehydes
   bond dissociation energies, 300
   co-oxidation, 303, 305
   dipole momentum, 299
   free radical generation, 303, 304
   mechanism of oxidation, 299–303
   reactivity, 306–308
Aliphatic amine
   bond dissociation energies, 331–330
   mechanism of oxidation, 329–335
   reactivity, 333
Alkenal/one oxidoreductase, 901
Alkoxyl radicals, 7, 69, 72–75
Alkylphosphites, 573–579
Alkyl radical
   addition to dioxygen, 34–39
   intramolecular hydrogen transfer, 238–240
   reaction with ROOH, 168–172, 341, 442, 451
   reaction with ozone, 110
Allopurinol, 704, 765
Alzheimer's disease, 925
Amadori product, 911
Amides
   bond dissociation energies, 336
   dipole momentum, 338
   free radical generation
   mechanism of oxidation, 338–341
Aminyl radical, 511, 516–518, 525
Angeli's salt, 682, 828

Angiotensin II, 708, 710, 907
Antagonism of antioxidant action, 71
Anthracycline, 735, 750, 763, 825
Antimycin, 733, 735, 737
Antioxidants
   accepting alkyl radicals, 649–651
   accepting peroxyl radicals, 491–528
   capacity, 468–480
   classification, 466–468
   cyclic chain termination, 541–568
   decomposing hydroperoxides, 573–595
   efficiency, 468, 475–477
   kinetics inhibited autoxidation, 469–475, 477–480, 486, 487
   mechanism action, 469–475, 480–486
   metal complexes with phosphates, 591–595
   phosphorus-containing, 573–580
   sulfur-containing, 580–587
   strength, 468
   thiocarbamates, 588–591
   thiophosphates, 588–591
Apolipoprotein, 775, 783
Apolipoprotein B-100, 777, 816
Apoptosis, 738–742, 798, 849, 917
Arachidonic acid, 677, 705, 754, 765, 766, 769, 770, 772, 776, 782, 789, 793, 796, 799, 823, 826
Arginine, 675, 678, 710, 713, 765, 880
Arimoto, 7
Aromatic amines, 501–510
Arthritis, 843, 921
Arylphosphites, 573–579
Asbestos, 692, 739, 827, 852, 853, 855, 923
Ascorbate, 691, 763, 772, 777, 811, 814, 815, 826, 828, 839, 844, 856, 901
Ascorbic acid, 842, 843, 868, 907, 928
Aspirin, 775, 776, 879, 923
Asthma, 922
Ataxia-telangiectasia, 927
Atherogenesis, 778, 908
Atherosclerosis, 711, 714, 775, 776, 782, 783, 798, 826, 907, 909
Autoxidation
   hydrocarbons, 5, 10, 11, 173–179
   polymers, 443–444
Azo-compounds, 90, 95–97

B

Bach, 3, 5, 465
Backstrom, 6, 7, 465
Baicalein, 850, 909
Bartlett, 56, 251
Basophil, 707

Bateman, 7, 154
Bawn, 7
Baxendale, 7
Bayer, 5
β-Blocker, 872, 873
Bcl-2, 738, 740, 741, 742
Belousov, 386
Benson, 61
Benton, 5
p-Benzoquinone
  reaction with alkyl radicals
  reaction with amines
  reaction with hydroxylamines
  reaction with phenols
  reaction with, sterically hindered phenols
Bleomycin, 825, 855, 860, 924
Bleomycin-detectable iron (BDI), 691
Bloom's syndrome, 933
Bodenstein, 6, 7
Bolland, 6, 7, 137, 465
Bond dissociation energy (see Dissociation energy)
π-Bond effect, 12–14, 230, 243–244, 248
Boozer, 466
Bovine serum albumin (BSA), 686, 809, 810, 811, 816
Bradykinin, 709, 801

C

Cage effect in
  liquid, 94–101
  polymer, 431–433, 390–394, 627–641
Calcineurin, 814
Calcium blocker, 870, 871
Calmodulin, 711, 713
Cancer, 915, 916
3-Carbamoyl-2,2,5,5-tetramethylpyrrolidinoxyl (3-CP), 897
Carbon tetrachloride, 751, 757, 763
Carcinogenesis, 915
Cardiolipin, 735, 740, 741
Cardiomyocyte, 676, 737, 740
Cardiotoxicity, 764, 825
Cardiovascular disease, 905, 907
β-Carotene, 776, 840, 874, 876, 909
Carvedilol, 872
Caspase, 738–742, 849, 916
Catalase, 738, 759, 823, 854, 899, 913
Catalysis
  H_2O_2 decomposition by ferrous ions, 8, 360–362
  hydroperoxide decomposition, 363–369, 667
  initiation by acetylacetonate cobalt(II), 379
  of oxidation by

basis, 401–404
Co-Br catalyst, 382–386
heterogeneous, 396–400
nitroxyl radicals, 207–209
olefin epoxidation, 390–394
olefin oxidation to aldehydes by palladium salts, 394–396
phenol oxidation by ROOH by pyridine, 536
transition metals, 7, 359, 363–369, 382–383
Catechin, 844, 849
Cellular phone, 695
Ceruloplasmin, 779
Chain
  branching reaction, 6, 7
  decomposition of ROOH, 168–173, 341, 449–450
  initiation by
    halogens, 113–117
    initiators, 85–93
    ionizing radiation, 128–131
    nitrogen dioxide
    ozone, 101–109
    photoinitiation, 118–127
  mechanism of oxidation of
    alcohols, 261–272
    aldehydes, 300–303
    aliphatic amines, 331–333
    amides, 338–341
    esters, 343–349
    ethers, 283–287
    hydrocarbons, 6, 7, 27–32
    polymers, 431–457
    Sn(II) oxidation, 378–379
    sulfoxidation, 418–423
    trialkylphosphites, 577–578
Charge transfer complex, 142–144, 155–156
Chemically induced dynamic nuclear polarization (CIDNP), 684, 688, 693
Chemiluminescence (CL), 8, 27, 62–67, 101, 457, 692, 955, 956
Chernobyl, 919
Chloramine, 780, 781, 813
Chloride, 719, 723, 781
Cholesterol, 717, 766, 782, 795
Cholesterol hydroperoxide, 766, 768
Cholesteryl linoleate, 766
Christiansen, 465
Chronic myeloid leukemia (AML), 915, 916
Chronic renal failure (CRF), 927, 928
Ciamician, 5
(+)-Cianidanol, 844
Cirrhosis, 928
Citrulline, 675, 710, 754
Colitis, 928

Complex O$_2$ with metal chelates, 380
Concerted fragmentation, 86
Condit, 580
Conservation of orbital symmetry, 26
Cooper, 7, 137
Co-oxidation
   alcohols with aldehydes, 303, 305
   cyclohexanol with acids, 324
   hydrocarbons, 185–191
   hydrocarbons with
      alcohols, 191–196, 270–271
      aldehydes, 303, 305
      amines, 333
      esters, 350
      ethers, 287, 290
      intermediates, 204, 206
      ketones, 314
      monomers, 192–193
      products, 204–207
Copper-rutin complex, 759
Criegee, 4, 5
Critical concentration of antioxidant, 479–480
Critical fenomena in
   heterogeneous catalysis, 399–400
   inhibited hydrocarbon oxidation, 479–480
   inhibited polymer oxidation, 648
Crohn's disease, 928
Cross-disproportionation of peroxyl radicals, 196–199
Crossover experiment, 96–97
Cross-propagation in co-oxidation, 199–203
CuZnSOD, 681, 682, 713, 895, 896, 915, 922
Cyclic GMP, 694
Cyclic mechanisms of chain termination by
   aminyl radicals, 541–552
   aromatic amines, 541–552
   basis, 569
   nitroxyl radicals, 555–561, 652–658
   nitroxyl radicals with acid, 561–563
   quinones, 552–554
   quinones with acid, 563
   transition metal ions, 564–567
Cyclic peroxide, 12
Cyclicity of radical conversion, 23–24
Cyclooxygenase (COX), 770, 775, 776, 799, 800, 801, 803, 839, 847, 853, 879, 897
Cysteamine, 826
Cysteine, 686, 779, 864
Cystic fibrosis, 923
Cytochrome b$_{558}$, 705
Cytochrome b_5, 748, 750
Cytochrome c, 680, 703, 715, 732, 736, 738, 739, 740, 741, 742, 749, 814, 855, 862, 951

Cytochrome c oxidase, 735, 736
Cytochrome P-450, 675, 712, 747, 749, 750, 751, 752, 753, 812, 847, 850, 873, 917

D

Daunorubicin, 735, 764
Decarboxylation of carbonic acids, 320–324
Decker, 457
Decomposition of
   alkylsulfonylperacid, 424
   azo-compounds, 90
   hydrogen peroxide, 278
   hydroperoxide
      amides, 338, 340, 341
      bimolecular, 163–164
      catalytic, 363–367, 389, 397
      epoxidation, 390–394
      heterolytic, 368–370
      homolytic, 368–370
      unimolecular, 149, 150, 153
   hydroperoxide groups in polymers, 444–449
   initiator, 85–94
      concerted fragmentation, 86–87
      model of interacting oscillators, 87
      to molecular products, 88
      unimolecular, 85–86
   peroxides, 90–94
      anchimerically assisted, 87–88
      chain, 89, 168–173
      heterolytic, 88
      homolytic, 88
      reduction by ketyl radicals, 89
   peroxyl radicals, 35
   phenoxyl radical, 473
   phosphoranyl radicals, 577–578
   quinolide peroxides, 473
Degenerate chain branching, 6, 7
Degradation of polymer, 451–455, 660–662
Denison, 580
Denisov, 8, 141, 155, 315, 456
Density functional theory (DFT), 233–234
Desferrioxamine (desferal), 827, 829, 882, 910, 931
DHA (docosahexaenoic acid), 772
Di-(4-carboxybenzyl)hyponitrite, 675
Diabetes mellitus, 775, 782, 862, 911, 912, 913, 914
2′,7′-Dichlorofluorescein (DCF), 718, 960
5-(Diethoxyphosphoryl)-5-methyl-1-pyrroline-N-oxide (DEMPO), 954
Diethyldithiocarbamate (DDC), 801
Diffusion
   coefficient of O$_2$ in polymers, 438

Index

dioxygen in hydrocarbons, 32–34
dioxygen in polymers, 436–441
free radicals in polymer, 432
rotational in polymer, 643–644
Dihydroethidium (DHE), 960
Dihydrolipoic acid, 860
Dihydroperoxide, 6, 7, 11, 45–48, 447–448
Dihydropyridines, 751
Dihydrorhodamine 123 (DHR), 961
3,4-Dihydroxyphenylacetic acid, 681
1,2-Dimethyl-3-hydroxypyrid-4-one, deferiprone (L1), 827, 882, 930, 931
5,5-Dimethyl-1-pyrroline-N-oxide (DMPO), 702, 712, 757, 758, 791, 796, 820, 861, 896, 953, 960
5(S),15(S)-Dihydroxy-6,13-*trans*-8,11-*cis*-eicosatetraenoic acid (5,15-DiHETE), 791
Dinitrophenylhydrazine (DNPH), 809
Dioxygen
  addition to alkyl radical, 34–39
  diffusion in polymers, 436–441
  discovery, 5
  reaction with
    antioxidants, 472–473, 528, 530–532
    hydrocarbons, 137–143
  solubility, 33
Dipole momentum of
  alcohols, 261
  aldehydes, 299
  esters, 342
  ethers, 281
  hydroperoxides, 147
  peroxyl radical, 40
Disproportionation
  aminyl and hydroxyperoxyl radicals, 551
  nitroxyl and hydroxyperoxyl radical, 557–560
  peroxyl with aminyl radicals, 511
  peroxyl with phenoxyl radicals, 508–510
  peroxyl radicals, 7, 27–32, 57–65, 196–199, 267, 228, 302, 313, 332, 339, 346
Dissociation energies
  C–H bonds of acids, 321
  C–H bonds of alcohols, 262–263
  C–H bonds of aldehydes, 300
  C–H bonds of amines, 329–330
  C–H bonds of amides, 336
  C–H bonds of esters, 342
  C–H bonds of ethers, 282–283
  C–H bonds of ketones, 312
  C–O bond of peroxyl radical, 35, 42
  C–O bond of tetroxides, 57
  N–H bonds of aliphatic amines, 330

N–H bonds of aromatic amines, 502–504, 531–532
O–H bonds of hydroperoxides, 41, 147, 266, 291–292, 350
O–H bonds of hydroxylamines, 207
O–H bonds of phenols, 492, 495, 530–531
O–H bonds of semiquinone radicals, 619
O–O bond of hydroperoxide, 147
O–O bond of peroxide, 85
O–O bond of tetroxide, 57
Disulfides, 582–586
2,6-Di-tert-4-methylphenol (BHT), 770, 850, 919
Dityrosine, 685, 716
Docosahexaenoic acid, 765
Dopa, 718, 809
Dopamine, 926
Down syndrome, 933, 934
Doxorubicin, 735, 759, 763, 825, 917, 918
DSBs (double stranded breaks), 819
Dufresse, 465
Dusts, 693

E

Ebselen, 877, 900, 924
EDTA, 676, 750, 761, 851, 882
Efficiency of inhibition, 468–469, 476
Eicosanoid, 776, 920
Eicosatrienoic acid, 776
Einstein, 94
Electron affinity of atoms, 226–227, 229, 507–508, 525, 527
Emanuel, 8, 101
Endothelial cell, 690, 704, 707, 708, 710, 742, 778, 810, 815, 849, 868, 906
Endothelin-1, 709
Endothelium-derived relaxing factor, EDRF, 678, 907
Engler, 3–5
Enthalpy of equilibrium
  addition of $RO_2^\bullet$ to ketones, 315
  addition of $RO_2^\bullet$ to olefins, 241
  dimerization of NO_2, 111
  dioxygen with transition metal ions, 377
  H_2O_2 with transition metal ions, 360
  ROOH hydrogen bonding, 151–153, 340
  tetroxide formation, 56
Enthalpy of formation
  alcohols, 15
  alkoxyl radical, 26
  alkyl radicals, 26
  hydrocarbons, 12, 17, 26
  hydroperoxides, 9, 12, 26, 42–43, 148–149
  ketones, 17

Enthalpy of formation (*Continued*)
  peroxyl radical, 26
  tetroxide, 57
Enthalpy of reaction
  free radical abstraction
    Am$^\bullet$ + cumene, 526
    AmO$^\bullet$ + AmH, 615–616, 631
    AmO$^\bullet$ + ArOH, 613–615, 628–629, 638–640
    ArO$^\bullet$ + cumene, 523–524
    ArO$^\bullet$ + ionol, 609–610
    p-benzoquinone + AmH, 623–624
    p-benzoquinone + ArOH, 621–623
    RO$^\bullet$ + RH, 221
    RO$_2^\bullet$ + alcohol, 275–276
    RO$_2^\bullet$ + aldehyde, 309
    RO$_2^\bullet$ + amines, 337
    RO$_2^\bullet$ + AmH, 505–507
    RO$_2^\bullet$ + AmOH, 617
    RO$_2^\bullet$ + ArOH, 493, 495, 617–618
    RO$_2^\bullet$ + ether, 294–295
    RO$_2^\bullet$ + ketone, 312, 315
    RO$_2^\bullet$ + RH, 220, 223, 233, 309, 315
  free radical addition
    alkoxyl radical to olefin, 245
    peroxyl radical to olefin, 243–244
    peroxyl radical to ketone, 315
  free radical generation
    halogens with hydrocarbons, 116–121
    InH + dioxygen, 530–532
    nitrogen oxide with hydrocarbons, 112–113
    ozone with hydrocarbons, 107–108
    RH + dioxygen, 138, 140
    ROOH + ArOH, 533–534
    ROOH + AmH, 535
    ROOH + ROOH, 164
  free radical reduction of ROOH, 254
  free radical substitution of ROOH, 253, 257
  influence on activation energy, 219
  intramolecular H transfer, 238–240
  isomerisation, 75
Enthalpy of ROOH evaporation, 148–149
Entropy of reaction
  dioxygen with transition metal ions, 377
  dimerization of NO$_2$, 111
  hydrogen bonding of ROOH, 151–153, 340
  hydrogen peroxide with transition metal ions, 360
Entropy of
  ROOH formation, 148–149
  tetroxide formation, 56
Eosinophil, 707
Eosinophil peroxidase, 681, 715
Epigallocatechin gallate, 829, 924

Epinephrine, 749, 959
Epoxide formation, 13, 55, 306, 390–393
EPR spectroscopy, 8, 27, 101, 431
EPK1/2 kinase, 710
Equilibrium constant
  reaction of H$_2$O$_2$ with ketone, 280
  reaction of ROOH with ketone, 169–171, 319
  ROOH dimerization, 152
  ROOH hydrogen bonding, 153
  tetroxide formation, 56
Erythrocyte, 675, 680, 694, 759, 814, 852, 855, 867, 899, 930, 933
Esters
  C−H bond dissociation energy, 342
  dipole momentum, 342
  mechanism oxidation, 343–349
  multydipole interaction, 350–355
  reactivity, 349–355
Estrogen 17 β-estradiol, 718, 868
Ether
  C−H bond dissociation energy, 282–283
  Co-oxidation, 287–291
  dipole momentum, 281
  mechanism oxidation, 281–289
  polar effect, 294–296
  reactivity, 291–296
Ethinyl estradiol, 735
2-Ethyl-1-hydroxy-2,5,5-trimethyl-3-oxazolidine (OXANOH), 954
Evance, 7
Exchange reaction RO2 + R1OOH, 8, 187, 194–195
Extracellular-signal-regulated kinase (ERK)

F

Fanconi anemia, 694, 932, 933
Farmer, 4
Fenton reaction, 676, 692, 695, 702, 720, 750, 757, 758, 765, 782, 793, 819, 820, 822, 829, 842, 882
Ferritin, 675, 689, 754, 759
Ferryl, 676, 819
FeSOD, 895
Fibroblast, 694, 709, 741, 765, 794, 796
Flavanone, 844
Flavone, 844
Flavonoid, 829, 844, 846, 849, 850, 851, 852, 853, 854, 857, 909, 918
Flavonol, 844
Flavonolignans, 860
Force constant of the bond, 215, 224–225
Frank, 94
Franck-Condon principle, 121

Free valence migration in polymers, 432
Free valence persistence, 23

G

Gallic acid, 860
Garlic, 881
Gee, 6, 7, 137
Generation of free radicals, 6, 7, 24–25
Generation of free radicals on reaction
  dioxygen with
    alcohols, 278
    aldehides, 304
    antioxidants, 528, 530–532
    esters, 284
    ethers, 284, 289
    hydrocarbons, 7, 137–144
    ketones, 312–313
    polymer, 443–444
  halogens with hydrocarbons, 113–121
  hydrogen peroxide, 278–279
  hydroperoxide decomposition
    bimolecular, 7, 154–157
    unimolecular, 149, 150, 153
  hydroperoxide with
    acids, 389
    alcohols, 8, 157, 389
    antioxidants, 532–537
    carbonyl compounds, 8, 167–171, 279–280
    olefins, 8, 154–156
    phosphites, 573–576
    products of oxidation, 177–178
    sulfenic acid, 581
    sulfides, 580–582
  ionizing radiation, 129–131
  nitrogen dioxide with hydrocarbons, 110–113
  ozone with hydrocarbons, 101–109
Geometry of transition state in reaction
  $Am^\bullet$ + cumene, 526
  $AmO^\bullet$ + ArOH, 538–540
  $ArO^\bullet$ + cumene, 523–524
  Free radical abstraction, 233–235
  $RO_2^\bullet$ + alcohol, 274–276
  $RO_2^\bullet$ + aldehyde, 310
  $RO_2^\bullet$ + AmH, 504–507
  $RO_2^\bullet$ + ArOH, 494–495
  $RO_2^\bullet$ + ether, 295
  $RO_2^\bullet$ + ketone, 315
  $RO_2^\bullet$ + RH, 235
George, 457
Gibbs potential
  dioxygen with transition metal ions, 377
  hydrogen bonding of phenols, 499
  hydrogen peroxide with transition metal ions, 360
  $RO_2^\bullet$ + RH, 354–355
Ginkgo biloba, 881
Gladyshev, 670
Glomerulonephritis, 927
Glucose, 709, 739, 782, 911
Glutathione, 718, 736, 758, 764, 765, 770, 778, 795, 796, 814, 815, 826, 862, 863, 874, 900, 901, 914, 926
Glutamine synthetase, 809, 811, 812, 816
Glutathione peroxidase, 736, 741, 758, 863, 877, 900
Glutathione reductase, 736, 765, 863, 900
Glutathione transferase, 847, 901
Glyceraldehyde-3-phosphate dehydrogenase (GAPDH), 677, 810
Gomberg, 6, 7
Grotthus-Draper low, 120
Guanylyl cyclase, 681, 688, 742, 797
Guaraldi, 56

H

Haber, 3, 5, 7
Halogens, 113–121
Hammet, 213
Hantzsch esters, 752
Hartman, 5
ten Have, 465
Heart disease, 908, 909
Heat shock protein, 714, 742
Hemochromatosis, 843, 929
Hemoglobin, 679, 814, 815, 863, 929
Henry coefficient of dioxygen solubility in liquids, 33
Hepatitis, 927, 928
Hepatitis C virus, 927
Hepatocyte, 690, 775, 815
Heterogeneous inhibition, 665–671
Heterogeneous polymer oxidation, 456–458
High density lipoproteins (HDL), 783
Hinshelwood, 6, 7
Hock, 4, 5, 397
Horseradish peroxidase (HRP), 676, 683, 689, 693, 715, 718, 721, 810, 862
Howard, 8, 187, 466
Hughes, 7
Hydrazine, 751
Hydrocarbon
  enthalpy of formation, 12, 15, 17
  oxidation, 27–38, 72–75, 137–143, 173–180, 185–195, 363–367, 396–403

Hydrogen atom transfer
  alkoxyl radical, 236–239
  alkyl radical, 238–240
  peroxyl radical, 45–48, 236–238, 284, 447–448
Hydrogen bond, 147–148, 309, 340, 497–501, 509
Hydrogen peroxide, 3, 5, 7, 278–279, 360–362
Hydrogen peroxide, 719
Hydroperoxide, 144–181
  analysis, 145–146
  chain decomposition, 168–172
  decomposers, 573–594
  decomposition, 7, 13–14, 149–157, 280, 363–369
  dipole momentum, 147
  enthalpy of formation, 148–149
  hydrogen bond, 147, 150, 151
  products of decomposition, 13–18
  product of oxidation, 9–11
  reaction with antioxidants, 532–537
  structure, 146
  thermochemistry, 147–149
  yield, 9–11
12-Hydroxyeicosatetraenoic acid (12-HETE), 760, 778, 789
12(S)-hydroperoxyeicosatetraenoic acid (12-HPETE), 791
13-Hydroperoxy-9,11-octadecadienoic acid (13-HPODE), 766, 791
15(S)-hydroperoxyeicosatetraenoic acid (15-HPETE), 761, 791, 900
4-Hydroxy-2,2,6,6-tetramethylpiperidinoxyl (TEMPOL), 897, 910
4-Hydroxynonenal (4-HNE), 691, 766, 840, 901, 917
5-Hydroxyeicosatetraenoic acid (5-HETE), 791
5(S)-hydroperoxyeicosatetraenoic acid (5-HPETE), 791
8-Hydroxy-2′-deoxyguanosine (8-OHdG)
8-Hydroxyquinoline, 691
9-Hydroperoxy-10,12-octadecadienoic acid (9-HPODE), 766
N-Hydroxyl-L-arginine, 710
p-Hydroxyphenylacetaldehyde, 780
Hydroxyl radical, 7
Hydroxyl radical, 676, 686, 688, 702, 719, 720, 721, 749, 750, 754, 758, 759, 761, 765, 782, 791, 793, 809, 811, 816, 819, 823, 824, 825, 826, 827, 829, 845, 856, 861, 864, 869, 872, 873, 876, 877, 880, 895, 896, 899, 908, 913, 916, 922, 928, 931, 933, 960
Hypercholesterolemia, 711, 798, 910
Hyperglycemia, 911, 912, 913
Hypertension, 775, 910
Hypochlorite, 812, 813, 849
Hypochlorous acid, 718, 761, 780, 813, 880, 909
Hypoxia, 905, 906

I

ICRF-187, 764
Indictor, 390
Indomethacin, 775, 776
Induced oxidation, 5
Induction period
  oxidation of hydrocarbons, 470–475, 478–480, 486–487, 611, 618
  oxidation of polymers, 648–649
Infarct, 704
Infarction, 906, 908
Inflammation, 782, 789, 920, 924
Ingold, 8, 466
Inhibition coefficient
  nonstoichiometric, 545–549, 553, 557–560, 637–638
  stoichiometric, 588, 592, 658
Initiation (see Generation of free radicals)
Initiator, 27, 85–94
Insulin, 912, 914
Interferon-γ, 737, 760
Interleukin, 735
Intersecting parabolas method (IPM), 138, 140, 158–167, 213–236, 241–251, 273, 295, 494, 501, 504, 522–524, 528–529, 608, 612
Intramolecular isomerization of
  alkoxyl radical, 236–239
  alkyl radical, 238–240
  peroxyl radical, 6, 45, 48–49, 236–238, 284, 441–443, 447–448
Iodosobenzene, 753
Iron N-methyl-D-glucamine dithiocarbamate (Fe-MGD), 694, 961
Iron regulatory protein, 681, 811
Iron saccharate, 812, 929
Iron, 689, 690, 691, 692, 750, 763, 764, 765, 811, 822, 840, 843, 851, 882, 927, 928, 929
Iron-dextran, 692
Iron-diethyldithiocarbamate (Fe-DETC), 961
Iron-rutin complex, 851, 854, 855
Ischemia, 737, 873, 905, 906, 908
Ischemic preconditioning, 736
Isocitrate dehydrogenase (ICDH), 736
Isolevuglandin, 772
Isoprostane, 760, 768, 770, 775, 776, 840, 843, 911, 914, 920, 923, 926, 927
Ivanov, 5

Index

J

JNK kinase, 710, 849
Jorissen, 4

K

Kaempferol, 826, 849
Karpukhin, 8
Karpukhina, 603
K_{ATP} channel, 737, 738
Kennerly, 600
Ketone
    C−H bond dissociation energy, 312
    co-oxidation, 314
    enthalpy of formation, 17
    mechanism oxidation, 311
    product of oxidation, 16
    reaction with ROOH, 317–320
    reactivity, 314–317
Ketyl radical, 89, 263, 266
Kharash, 7
Khudyakov, 515
Kinetic isotope effect, 8, 43–44, 102, 138, 208, 322
Kinetics of
    alcohol oxidation, 263–270
    aldehyde oxidation, 300–303
    amide oxidation, 338
    amines oxidation, 331–333
    autoxidation, 173–178
    catalytic oxidation, 363–368
    co-oxidation, 185–187
    ester oxidation, 343–345
    ether oxidation, 283–287
    hydrocarbon oxidation, 6, 27–32
    ketone oxidation, 311–314
    macromolecules degradation, 451–456
    polymer oxidation, 433–436, 439–441, 447–449
Komissarov, 101, 418, 420
Kovalev, 665
Kucher, 411
Kupffer cell, 928

L

LacGer, 709
Lactate dehydrogenase, 812
Lactate, 879
Lactoperoxidase, 681, 683, 715, 718
Lang, 5
Latex, 853
Lavoisier, 5
Lazaroid, 868
Lebedev, 8

Leukemia, 675, 740, 741, 766, 792, 915, 929
Leukocyte, 791, 792, 796, 823, 873, 930, 933
Leukotriene A (LTA_4), 791
Leukotriene B (LTB_4), 707, 791, 796, 849
Liebich, 5
Linoleate diol oxidase (LDS), 803
Linoleic acid, 677, 761, 765, 766, 782, 790, 792, 793, 795, 796, 803, 826, 837, 850, 883
γ-Linolenic acid, 677, 739, 765, 766, 770, 862
Lipid peroxidation, 677, 739, 741, 750, 757, 758, 759, 760, 761, 763, 764, 765, 768, 770, 779, 781, 782, 789, 798, 799, 814, 815, 839, 842, 844, 850, 851, 854, 855, 860, 861, 865, 867, 858, 869, 872, 873, 878, 879, 882, 906, 913, 914, 917, 920, 921, 923, 929
Lipoic acid, 814, 860, 861, 862, 914, 919
Lipopolysaccharide, 796
Lipoxygenase (LOX), 681, 777, 778, 789, 791, 792, 793, 795, 796, 847, 853, 908, 923
Low-density lipoproteins (LDL), 758, 760, 761, 766, 768, 772, 775, 776, 777, 778, 779, 780, 781, 782, 793, 795, 797, 798, 837, 853, 859, 867, 868, 870, 872, 877, 908, 910, 912, 930
Lucigenin, 703, 709, 759, 796, 827, 855, 862, 933, 955, 956, 957
Lumazine, 702
Luminol, 692, 695, 718, 827, 853, 915, 957, 959, 962
Lupus erythematosus, 924
Lycopene, 776
Lymphocyte, 742, 842
Lysophosphatidylcholine (LPC), 709

M

Macrophage, 694, 707, 722, 742, 760, 775, 778, 779, 782, 796, 801, 813, 849, 855, 916
Magnetic field, 693, 694, 695
Mahoney, 607
Maillard products, 911, 914
Maizus, 8
Malondialdehyde (MDA), 757, 761, 765, 766, 782, 851, 853, 854, 866, 881, 910, 911, 913, 916, 930
Mayo, 6, 7, 141, 455
Mechanisms
    antioxidant action, 469–475
    cyclic chain termination, 541–568
    decomposition of hydroperoxides, 149–157, 338–340, 444–450
    decomposition of peroxides, 85–90
    hydrocarbon autoxidation, 175–178
    hydrocarbon oxidation, 27–32, 75–76

Mechanisms (*Continued*)
  phosphite oxidation, 577–578
  polymer oxidation, 433–436, 439–441, 447–449
  sulfoxidation, 418–424
Melatonin, 687, 720, 876, 877
Melville, 7, 137
Menadione, 735, 775, 823, 825, 860, 930, 952
Metal-desactivating antioxidants, 467
Metalloporphyrins, 878
Metallothioneins (MTs), 830, 878
Methionine, 686, 812, 813, 816
Microperoxidase, 679
Migration of a free valence in polymer, 432–433
Miller, 6, 7, 141
Mitomycin C, 917
MitoQ, 880
MitoVitE, 880
Mn(III) meso-tetrakis(N-ethylpyridinium-2-yl)porphyrin, 897
MnSOD, 732, 733, 736, 737, 741, 764, 895, 898, 906, 910, 912, 915, 917, 918, 922
MnTBAP, 900
Model of interacting oscillators, 87
Moiseev, 394
Molecular reaction, 25
Moloxide, 4
Monocyte, 676, 707, 778, 781, 791, 849, 873, 916
Morin, 826, 854
Moris, 7
3-Morpholinosydnonimine (SIN-1)
Moureu, 465
mtNOS, 714, 736
Multidipole interaction, 138, 142, 156, 250, 276–277, 350–355
Myeloperoxidase (MPO), 681, 715, 716, 719, 720, 721, 761, 780, 781, 782, 824, 920
Myocytes, 707, 709, 749, 798, 908
Myoglobin (Mb), 679, 764, 814
Myricetin, 829, 856

N

N,N-dialkylanilines, 752
NADH, 677, 686
NADH cytochrome b_5 reductase, 747, 750
NADPH, 686
NADPH-cytochrome P-450 reductase, 747, 749, 750, 758, 850
NADPH oxidase, 704, 705, 707, 709, 710, 778, 780, 782, 803, 879, 906, 907, 908, 910, 912, 921, 928
N-alkoxyamines, decay, 655–656

Nangia, 61
Nasse, 3
Necrosis, 739
Nephritis, 927
Nephrosis, 927
Neuron, 739, 880, 906, 926
Neuroprostanes, 772
Neutrophil, 675, 688, 689, 694, 707, 720, 780, 791, 810, 824, 849, 906
Nifedipine, 752, 870, 870
Nitrite, 683, 703, 721, 722
Nitrobenzyl chloride, 751
Nitroblue tetrazolium (NBT), 959, 852, 861
Nitrocompounds as alkyl radical acceptors, 650–651
Nitrogen dioxide NO_2, 682, 721, 722, 761, 863, 874
2-Nitropropane, 754
Nitrosoperoxocarboxylate, 812
Nitrosothiol, 679
Nitrotyrosine, 779, 911, 913, 923
Nitroxyl radical, reaction with
  alkyl radicals, 649–651
  aromatic amines, 615–616, 631
  cyclic chain termination, 555–563, 652–658
  phenols, 613–615, 638–640, 627–642
Nitroxyl, 680, 681, 713, 797, 828, 829
NO (nitric oxide), 675, 677–681, 683, 685, 688, 703, 706, 709–715, 722, 723, 736, 737, 741, 742, 754, 758, 760, 761, 777, 779, 796, 797, 798, 800, 801, 803, 810, 812, 813, 814, 815, 827, 828, 829, 849, 854, 871, 897, 898, 901, 907, 908, 910, 912, 914, 916, 921, 923, 925, 927, 932, 961
NO synthase I (NOS I), 710, 713, 714
NO synthase II (NOS II), 710, 713
NO synthase III (NOS III), 710, 713
NO synthase, 675, 687, 710, 803, 827, 849, 901, 910, 912, 921, 928
Nonstationary kinetics of autoxidation, 478
Nordihydroguaiaretic acid, 738, 791, 796
NTBI, 689
Nucleic acids, 819, 820
Nudenberg, 7
N^w-hydroxy-L-arginine (NHA), 675

O

8-OHdG, 820, 823, 824, 825, 827, 828, 829, 915, 916, 918
Olefin oxidation, 5, 49–55, 308–310, 390–395
Opeida, 203
Optimal inhibitor, 486–487
Oscillating oxidation, 386–389
Oscillation theory of decay, 86

Oxidation in microheterogeneous systems, 411–415
Oxidation of
  alcohols, 261–280
  aldehydes, 299–303
  by alkylsulfonic peracid, 425–429
  amides, 336–341
  amines, 329–336
  disulfide by hydroperoxide
  disulfide by $RO_2^{\bullet}$
  esters, 341–449
  ethers, 281–291
  hydrocarbons, 5, 7, 27–32, 75–76, 175–178
  in microheterogeneous system, 411–414
  by peracid, 308–310
  phosphine by hydroperoxide, 573–576
  phosphine by $RO_2^{\bullet}$, 577–579
  phosphite by hydroperoxide, 573–576, 591–594
  phosphite by $RO_2^{\bullet}$, 577–579
  polymers, 433–436, 456–458
  sulfide by hydroperoxide, 580–582
  sulfide by $RO_2^{\bullet}$, 583–587
Oxidative polymer degradation, 451–456, 659–662
Oxidized LDL, 709, 742, 779
Oxyhemoglobin, 687, 814
Ozone, 101–110
  chain decomposition, 108–110
  reaction with free radicals, 110
  reaction with hydrocarbon, 104–108

P

p38 mitogen-activated protein kinase (p38 MAPK), 676, 706, 709, 710, 738, 864
p53 gene, 738, 740
Paracetamol, 751
Park, 7
Parkinson's disease, 863, 926
Patterson, 600
Pchelintsev, 456
Peracid, 5, 308–311
Perferryl, 676, 690, 720, 819
Perhydroxyl radical, 677, 758, 765, 822, 954
Peroxide, 5, 12, 85–90
Peroxide theory, 4
Peroxyl radical
  addition to olefins, 7, 49–54
  cross-disproportionation with phenoxyl radicals, 508–510
  cross-disproportionation with aminyl radicals, 511
  cross-disproportionation with $RO_2^{\bullet}$, 196–197
  decomposition, 263
  disproportionation, 7, 27–32, 57–62, 196–199, 267, 228, 302, 313, 332, 339, 346
  EPR spectrum, 8, 27, 101
  isomerisation, 6, 45–48, 219–232, 236, 441–443
  reaction with
    acid, 322, 324
    addition, 241–251
    alcohols, 273, 275–276
    amines, 337
    antioxidants, 43–44, 187–203, 491–507
    aromatic amines, 505–507
    hydrocarbons, 46–48, 187–203, 220, 223, 233
    ketones, 315
    phenols, 495, 617–618, 628, 630, 645–647
    polymer, 433–435
  structure, 39–41
Peroxyl, 677, 759, 761, 766, 768, 772, 783, 790, 791, 796, 797, 803, 809, 815, 826, 845, 866, 868, 872, 873
Peroxynitrite, 676, 679, 680, 682, 683, 684, 686, 687, 703, 706, 713, 714, 722, 736, 739, 741, 742, 760, 761, 765, 779, 797, 798, 800, 803, 812, 813, 815, 827, 828, 829, 844, 859, 870, 871, 873, 877, 906, 909, 916, 920, 922, 961
Phagocytosis, 704, 707, 723, 920
Phenacetin, 718
Phenol
  critical concentration, 648
  hydrogen bonding, 497–501
  O—H bond dissociation energy, 492
  reaction with dioxygen, 528–531
  reaction with nitroxyl, 611–616, 628–640, 642
  reaction with peroxyl radical, 491–501, 642–643–647
  reaction with ROOH, 532–534
  reaction with phenoxyl, 608–610
Phenoxyl, reaction with
  decomposition, 473–474
  dioxygen, 527–532
  disproportionation, 512–518
  hydrocarbons, 519, 521–525
  hydroperoxide, 532–534, 628, 630, 642, 647
  peroxyl, 508–510
  phenol, 608–611
Phentermine, 751
Phenyl-4,4,5,5-tetramethylimiazoline-1-oxyl (PTIO), 961
Phenytoin, 801
Phosphatidylcholine, 761, 780, 792, 794
Phosphatidylethanolamine, 761
Phosphines, reaction with
  hydroperoxides, 573–577
  peroxyl radicals, 577–580

Phosphites, reaction with
    hydroperoxides, 573–577
    peroxyl radicals, 577–580
Phospholipase, 706
Phosphoranyl radical, decay, 577–578
Photoinitiation, 118–128
Photoinitiator, 124–127
Photooxidation, 6, 10–11
Photosensitizer, 123–124
PIH, 882
pK_a of ROOH, 149
Platelets, 692, 792, 849
PMNs, 778, 913, 916, 920, 921
Pobedimskii, 591
Polany-Semenov equation, 222
Polar effect, 231–232, 247, 273–274, 294–295, 310
Polarization of hydroperoxide, 147
Poly(ADP)-ribose synthase (PARS), 735, 739, 742, 828, 922
Polymeric peroxide, 13
Porphyrin, 689, 718
Preeclampsia, 911
Priestley, 5
Prilezhaev, 390
Priority of chain propagation, 24
Privalova, 8
Probucol, 860, 883
Propyl gallate, 860
Prostacycline synthase, 686
Prostaglandin endoperoxide H synthases (PGHS), 798, 800, 801, 803
Prostaglandin, 770, 776, 798, 799, 801
Prostaglandin H synthase, 789
Prostanoid, 768
Protein kinase C, 709, 737, 841, 911, 912, 918
Psoriasis, 924
Pycnogenol, 881
Pyridoxine, 880
Pyrrolopyrimidines, 873

Q

Quantum yield
    benzaldehyde photooxidation, 7
    hydrocarbon photooxidation, 27
    isomerization, 124
    photodissociation, 94–95, 126, 127
Quasi-stationary concentration of ROOH, 478, 602
Quercetin, 739, 811, 816, 826, 846, 849, 850, 853, 854, 856, 884, 914
Quinolide peroxide, 473
Quinones, 703, 952

Quinones
    alkyl radical acceptors, 649–653, 656–657
    cyclic chain termination, 552–555
    synergism with amines and phenols, 616–620

R

Rabinowitch, 94
Racemization, 97
Radiation yield, 130–131
Radical pair, 95–98
Radii of atoms, 228–229, 246–247
Radiolytic oxidation, 129
Rate constant
Rate constant of reaction
    chain propagation in oxidation
        alcohols, 267
        aldehydes, 302
        aliphatic amines, 332
        amides, 339
        esters, 345
        ethers, 288
        hydrocarbons, 27–32
        ketones, 313
        polymers, 436–437
    chain termination in oxidation
        alcohols, 267
        aldehydes, 303
        aliphatic amines, 332
        amides, 339
        esters, 343–349
        ethers, 288
        hydrocarbons, 27–32
        ketones, 313
        polymers, 436–437
    cross-disproportionation of peroxyl radicals, 196–199
    cross-propagation in co-oxidation, 201–202
    cyclic chain termination, 565–567
    decomposition of hydroperoxide
        bimolecular, 155, 164
        chain, 172–174, 341, 349–351
        in polymer, 445, 446, 448
        unimolecular, 153, 349
    decomposition of
        alcoxyl radicals, 74
        alkylhydroxyperoxyl, 268
        alkylsulfonyl peracid, 425
        alkylsulfonyl radical, 421
        bis(1,1-dimethylethylperoxide), 91
        bis(1-methyl-1-phenylperoxide), 92
        diacetylperoxides, 93
        dialkylperoxides, 92
        dibenzoylperoxide, 94

Index

N-alkoxyamines, 655–656
peroxyalkyl radicals, 55
phosphoranyl radicals, 577–578
sulfoxides, 587
tetroxides, 57
dioxygen with transition metals, 381–382
disproportionation of
aminyl and hydroxyperoxyl radicals, 551
nitroxyl and hydroxyperoxyl radical, 557–560
peroxyl macroradicals, 434
peroxyl with aminyl radicals, 511
peroxyl with phenoxyl radicas, 508–510
peroxyl radicals, 7, 27–32, 57–65, 196–199, 267, 228, 302, 313, 332, 339, 346
phenoxyl radicals, 610
epoxidation of olefins by sulfonic peracid, 428
free radical abstraction
Am$^\bullet$ + cumene, 526
Am$^\bullet$ + ArOH, 605–606
AmO$^\bullet$ + AmH, 615–616, 631
AmO$^\bullet$ + ArOH, 613–615, 628–629, 638–640
ArO$^\bullet$ + cumene, 523–524
ArO$^\bullet$ + ionol, 609–610
ArO$^\bullet$ + ROOH, 647
p-benzoquinone + AmH, 623–624
p-benzoquinone + ArOH, 621–623
ketyl radical + O$_2$, 266
peroxyl macroradical + RH, 437
radical with ROOH, 174
R + + ozone, 110
RO$^\bullet$ + RH, 73–74, 221
RO$_2^\bullet$ + acid, 322, 324
RO$_2^\bullet$ + alcohol, 273, 275–276
RO$_2^\bullet$ + aldehyde, 307
RO$_2^\bullet$ + amines, 334–335, 337
RO$_2^\bullet$ + AmH, 505–507
RO$_2^\bullet$ + ArOH, 495, 617–618, 628, 630, 645–647
RO$_2^\bullet$ + ethers, 292–295
RO$_2^\bullet$ + esters, 345, 351–354
RO$_2^\bullet$ + ketone, 315
RO$_2^\bullet$ + ozone, 110
RO$_2^\bullet$ + RH, 46–48, 205–206, 220, 233, 436, 437
RO$_2^\bullet$ + R$_1$OOH, 194–195
free radical addition reaction
alkyl radicals to dioxygen, 36–38
peroxyl radical to olefin, 50–54, 243–244, 308
phenoxyl + O$_2$, 527
sulfonyl radical + O$_2$, 420
free radical generation by reaction
halogens with hydrocarbons, 114–121
H$_2$O$_2$ with transition metals, 362
InH + dioxygen, 529–532

nitrogen dioxide with hydrocarbons, 111–113
ozone with hydrocarbons, 104–108
RH + dioxygen, 139–144, 304, 313, 347–348
2 RH + dioxygen, 142, 144, 304
ROOH + ketone, 169–171
ROOH + RH, 156–157, 165–168, 349
ROOH + ArOH, 533–534
ROOH + AmH, 535
ROOH + transition metals, 365–366
free radical recombination
aminyl radicals, 521
H$^\bullet$ + H$^\bullet$, 129
phenoxyl radicals, 512
R$^\bullet$ + R$^\bullet$, 70–71
R$^\bullet$ + RO$_2^\bullet$, 68
RO$_2^\bullet$ + Am$^\bullet$, 511–512
RO$_2^\bullet$ + ArO$^\bullet$, 508–510
free radical reduction of ROOH, 254–257
free radical substitution R$^\bullet$ + ROOH, 252–253, 257
hydrogen atom, 129
hydroperoxide decomposition by
phosphates, 575–576
phosphines, 575–576
disulfides, 582
sulfides, 582, 584–585
sulfoxides, 582
hydroxyl radical, 129, 371
intramolecular H transfer, 75, 238–240
phosphites + RO$_2^\bullet$, 579–560, 592–594
solvated electrons, 179
sulfides + RO$_2^\bullet$, 586
thiocarbamates + RO$_2^\bullet$, 591
thiophosphates + RO$_2^\bullet$, 591
transition metals with free radicals, 372–376
alkylsulfonyl radical, 422–424
Ratio $k_p/(2k_t)^{1/2}$ of chain oxidation
alcohols, 264
aldehydes, 301
aliphatic amines, 332
amides, 339
esters, 344
ethers, 285–287
hydrocarbons, 28–31
ketones, 313
polymers, 435
Reactivity of free radicals, 25–26
Recombination of radical pair, 94–101
Redox-regulated transcription factor (NF-κB), 710, 899, 901, 912
Reduction potential, 680
Reperfusion, 905, 906, 929
Resveratrol, 741, 857, 884, 920

Retinoic acid, 801
Rieche, 5
Rigid cage of polymer matrix, 627–641
Ring strain energy, 240
Rotenone, 733, 737
Rudakov, 213
Russell, 7
Rust, 6, 7, 11
Rutin, 759, 829, 844, 846, 850, 851, 852, 853, 854, 883, 922, 923, 931

S

Safiullin, 425
Salen-manganese complex, 897
Salicylate iron complex, 897
Sato, 225
Scheele, 5
Schrader, 4
Schwalm, 8
Sclerosis, 925
Scrambling of isotopes, 95–96
Seiberth, 5
Selenium, 741, 815, 909
Self-inhibition of oxidation, 179–181
Semenov, 6, 7, 213
Septic shock, 714, 928
Shilov, 5
Shlyapintokh, 8
Shonbein, 3, 5
Sickle cell disease, 931, 932
Silber, 5
Silybin, 691
SIN-1, 761
Singlet dioxygen, 682
Skibida, 8
Smooth muscle cell, 709, 738, 778
S-nitrosocysteine, 678
S-nitrosoglutathione, 675, 678
SOD, 740, 759, 760, 777, 778, 779, 783, 814, 823, 854, 856, 863, 895, 898, 907, 911, 913, 915, 917, 918, 921, 924, 952
SOD mimics, 896, 912, 918, 927
Solubility of dioxygen in solvents, 33
Solvating effect, 232–233, 277, 316, 318
SOTS-1 (di-(4-carboxybenzyl) hyponitrite
Spin multiplicity effect, 99
SSBs (single DNA breaks), 819, 827, 828
Stark-Einstein law, 120
Statins, 909, 912
Staudinger, 5
Stephence, 5
Steric hindrance, 250–251, 496–497, 524–525
Stern-Volmer equation, 66

Steroids, 867
Streptozotocin, 812, 829, 912, 913, 914
Substitution, 160, 172, 251–257
Sulfenic acid
Sulfides, reaction with
    hydroperoxide, 580–585
    peroxyl radical, 583, 586
Sulfoxidation, 417–423
Sulfoxide, decomposition, 587
Sulfur dioxide, 581
Superoxide dismutase (SOD), 740, 759, 760, 777, 778, 779, 783, 814, 823, 854, 856, 863, 895, 898, 907, 911, 913, 915, 917, 918, 921, 924, 952
Superoxide dismutase mimics, 896, 912, 918, 927
Superoxide, 675, 676, 677, 683, 689, 690, 701, 702, 703, 704, 705, 709, 712, 713, 714, 718–721, 731, 732, 733, 734, 736, 738, 739, 740, 742, 749, 750, 754, 758, 759, 760, 765, 778, 779, 796, 798, 800, 801, 809, 813, 814, 822, 823, 824, 826, 836, 839, 840, 842, 844, 845, 849, 855, 861, 863, 864, 871, 880, 883, 896, 898, 906–913, 915–923, 925, 930, 931, 951, 954
Superoxide reductase (SOR), 898
Susemihl, 5
Sutton, 4
Symonds, 8
Synergism of antioxidants action, 599–625
Systemic sclerosis, 924

T

Taft, 213
Tamoxifen, 825, 918
Taurine, 721, 880
TBAR products, 760, 829, 853, 854, 855, 924, 930
Tempol, 897, 910
2(3)-Tert-butyl-4-hydroxyanisole (BHA), 800, 801, 919
Tetradecylthioacetic acid (TTA)
Tetrahydrobiopterin, 710, 713
2,2,6,6-Tetramethylpiperidinoxyl (TEMPO), 897
Tetroxide, 55–57
Thalassemia, 929, 930, 931
Thenard, 5
Thiocarbamates, 588–590
Thioredoxin, 741, 816
Thioredoxin reductase, 900
Thiosulfinate, 588–590
Thrombin, 709
Thromboxane, 772, 775, 798
Thyroid peroxidase, 715, 716, 719
Tiron, 732

Tobolsky, 455
α-Tocopherol, 691, 736, 741, 758, 761, 764, 772, 776, 795, 826, 836, 837, 844, 868, 870, 912
α-Tocopherolhydroquinone, 838
Tocopheroxyl radical, 777, 837, 838
α-Tocopherol transfer protein (TBP), 839
α-Tocopherolquinone, 838, 839
β-Tocopherol, 839
δ-Tocopherol, 839
γ-Tocopherol, 776, 839
Topology of oxidation, 75–76, 480–486
Transferrin, 690
Transition metal catalysts, 7, 27, 360–385, 390–393
Transition state, 26, 44–45, 233–235, 508–511
Traube, 3 – 5
Trimolecular reaction, 8, 141–144, 278, 289, 304
Triplet energy, 125
Triplet repulsion, 225–226, 229, 242–243, 494, 496, 521
Trolox C, 691, 718, 795, 829, 837, 840, 841
Tryptophan, 720
Tsepalov, 8
Tumor, 915
Tumor necrosis factor-α (TNF-α), 760, 798, 825
Tyrosine, 680, 684, 687, 716, 717, 721, 722, 780, 800, 813, 816
Tyrosine kinase, 709, 737

U

Ubihydroquinone, 734, 736, 777, 830, 865, 866
Ubiquinone, 733, 734, 736, 772, 793, 815, 830, 838, 864, 865, 866, 909
Udenfriend system, 820
Urate, 683, 909
Uric acid, 701, 761, 829, 867

V

Van Tilborg, 582
Varlamov, 527
Vasil'ev, 8
Vasopressin, 801
Verapamil, 870

Villiger, 5
Viscosity, 98
Vitamin C, 741, 815, 835, 840, 841, 909, 910, 919, 920, 921
Vitamin E, 736, 741, 764, 775, 778, 803, 815, 835, 838, 840, 909, 910, 914, 919, 921, 923, 926, 931, 935
Vitamin E succinate, 919
Vitamin K, 866
Volume of transition state, 637

W

Weiss, 7
Williamson, 6, 7, 185
Willstatter, 7
Wirth, 5

X

Xanthine, 701, 704
Xanthine dehydrogenase, 701, 906
Xanthine oxidase, 675, 678, 681, 701, 759, 801, 810, 823, 830, 846, 855, 860, 872, 906, 914, 917, 952
Xanthine oxidoreductase, 701

Y

Yield
 epoxide, 427
 free radicals, 95, 98, 103, 369, 424, 431–432, 446
 hydroperoxide, 10–11, 49

Z

Zakharov, 384
Zeolite, 853
Zhabotinsky, 386
Ziegler, 7
Ziegler-Natta catalyst, 457
Zinc, 878
Zinc thiolate, 687
Zymosan, 706, 778